immuno
biology 5

THE IMMUNE SYSTEM IN HEALTH AND DISEASE

immuno biology 5

THE IMMUNE SYSTEM IN HEALTH AND DISEASE

Charles A. Janeway, Jr.

Yale University School of Medicine

Paul Travers

Anthony Nolan Research Institute, London

Mark Walport

Imperial College School of Medicine, London

Mark J. Shlomchik

Yale University School of Medicine

GARLAND PUBLISHING

ALERE FLAMMAM

·Taylor & Francis Group·

Vice President:	Denise Schanck
Text Editors:	Penelope Austin, Eleanor Lawrence
Managing Editor:	Sarah Gibbs
Editorial Assistant:	Mark Ditzel
Managing Production Editor:	Emma Hunt
Production Assistant:	Angela Bennett
New Media Editor:	Michael Morales
Copyeditor:	Len Cegielka
Indexer:	Liza Furnival
Illustration and Layout:	Blink Studio, London
Manufacturing:	Marion Morrow, Rory MacDonald

Distributors:
Inside North America: Garland Publishing, 29 West 35th Street,
New York, NY 10001-2299.
Inside Japan: Nankodo Co. Ltd., 42-6, Hongo 3-Chrome, Bunkyo-ku,
Tokyo, 113-8410, Japan.
Outside North America and Japan: Churchill Livingstone, Robert Stevenson House,
1–3 Baxter's Place, Leith Walk, Edinburgh, EH1 3AF.

ISBN 0 8153 3642 X (paperback) Garland
ISBN 0 4430 7098 9 (paperback) Churchill Livingstone
ISBN 0 4430 7099 7 (paperback) International Student Edition

Library of Congress Cataloging-in-Publication Data
Immunobiology : the immune system in health and disease / Charles A. Janeway, Jr. ...
[et al.].-- 5th ed.
       p. cm.
   Includes bibliographical references and index.
   ISBN 0-8153-3642-X (pbk.)
   1. Immunology. 2. Immunity. I. Janeway, Charles. II. Title.

QR181 .I454 2001
616.07'9--dc21

2001016039

This book was produced using QuarkXpress 4.11 and Adobe Illustrator 9.0

Published by Garland Publishing, a member of the Taylor & Francis Group,
29 West 35th Street, New York, NY 10001-2299.

Printed in the United States of America.
15 14 13 12 11 10 9 8 7 6 5 4 3 2 1

Preface to the fifth edition

This book is intended as an introductory text for use in immunology courses for medical students, advanced undergraduate biology students, graduate students, and scientists in other fields who want to know more about the immune system. It attempts to present the field of immunology from a consistent viewpoint, that of the host's interaction with an environment containing many species of potentially harmful microbes. The justification for this approach is that the absence of one or more components of the immune system is virtually always made clear by an increased susceptibility to one or more specific infections. Thus, first and foremost, the immune system exists to protect the host from infection, and its evolutionary history must have been shaped largely by this challenge. Other aspects of immunology, such as allergy, autoimmunity, graft rejection, and immunity to tumors, are treated as variations on this basic protective function. In these cases the nature of the antigen is the major variable.

In this, the fifth edition of this book, we have undertaken a major reorganization of the text. We did this for several reasons. The first was an attempt to include the mechanisms of innate immunity as a necessary precursor to the adaptive immune response. Innate immunity is now treated in a separate chapter early in the book. Readers of earlier versions of this text will find that the material on immunological methods that was formerly in Chapter 2 of previous editions is now more appropriately gathered in Appendix I, called the *Immunologists' Toolbox*. After discussing innate immunity the emphasis shifts to adaptive immunity, because we know far more about this subject thanks to the efforts of the vast majority of immunologists. The central theme of the subsequent text is clonal selection of lymphocytes.

Another new feature is that for many topics, we consider the two lymphoid lineages, B lymphocytes and T lymphocytes, together rather than separately. The reason for this major organizational shift was that it seemed to us that common mechanisms work in both cell types. Treating them together enables easier comparison and eliminates redundancies. The first instance of common mechanisms is lymphocyte recognition of antigen by means of receptors that result from RAG-dependent rearrangement of gene segments. The second shared theme is the similar course of development of B cells and T cells from a common precursor cell, B cells in the bone marrow, and T cells in the thymus, both of which are dependent on similar stromal cells. The third parallel is the promotion of the survival of T cells by self peptide:self MHC ligands and of B cell ligands that include secreted immunoglobulins, as originally proposed by Niels Jerne. We also treat activation and signaling through B-cell and T-cell receptors together, as their similarities again outweigh their differences.

But we have kept discussions of T-cell and B-cell activation and their effector functions separate, as completely different mechanisms are involved, B cells acting by secreting their antigen-specific receptors in the form of antibodies, and T cells by direct interaction with infected cells and by releasing cytokines that have profound effects on other cells of the immune system. Another distinction between activated B and T cells is the somatic genetic changes—somatic hypermutation and isotype switching—that B cells alone undergo.

Finally, we attempt to synthesize all of this information in Chapter 10, in which we treat the immune system as a whole, integrated system. In this chapter we ask, "How does the immune system in healthy individuals protect its host from an infection," speaking in general and specific terms. This chapter pulls together much of the information learned in the individual chapters, and synthesizes it into our view of the whole. In addition, we have placed much greater emphasis on the mucosal immune system, which now has a separate new section in Chapter 10. This material belongs in Chapter 10 because infection usually starts by breaching a mucosal barrier and, in most normal circumstances, the activation of the immune system begins there.

Case Studies in Immunology

An additional feature of this edition is the cross-referencing of diseases and immunological deficiencies to the third edition of the clinical companion text *Case Studies in Immunology* by Fred Rosen and Raif Geha. This text presents major topics in immunology as background to a selection of real clinical cases that serve to reinforce and extend the basic science. A marginal icon in *Immunobiology* provides the reader with a link to the relevant Case Study, where the science is applied in a clinical setting. There are five new cases in this third edition, bringing the total number to 35.

This edition of *Immunobiology* includes a CD-ROM containing 27 original immunological animations. We hope this exciting new dimension in presentation will help the reader to understand some of the complex pathways of the immune system. The choice of which topics to animate is a difficult one, but we hope we have chosen a sample where the process of animation helps to explain the subject better than a static image. As another new feature, the Garland Science website (http://www.garlandscience.com) now provides instructors who adopt *Immunobiology* with a link to Garland Science Classwire, where the textbook art can be found in a downloadable, web-ready format, as well as in Power Point-ready format.

The reorganization of the text was in large part facilitated by one of the original authors, Charles Janeway, and one new author, Mark Shlomchik. The later chapters which focus on various clinical situations are, in our opinion, much improved thanks to the efforts of one author who has been with us since the second edition, Mark Walport, who also contributed new material on mucosal immunity. The continued vitality of the book is largely due to the efforts of new authors, as well as the continuing efforts of the original authors, Paul Travers and Charles Janeway. Writing the fifth edition of this book has been very challenging, especially as one of the authors, Charles Janeway, has been ill for much of the past two years. He would like to thank his co-authors for their patience, and for carrying more than their fair share of the work, especially as they have other jobs that are even more important than writing.

This book would not exist without the able efforts of the 'Garland Crew.' Angela Bennett, Mark Ditzel, Emma Hunt, Sarah Gibbs, Denise Schanck and Richard Woof have been extremely helpful to all of us, and kept the book running. Many others have contributed to getting the book out. Eleanor Lawrence and Penny Austin are our invaluable editors and our substitute consciences, always wanting to get the book not just to read well, but to read correctly. The illustrations, although sketched mainly by the authors, are largely the work of Matthew McClements. The CD-ROM storyboards were created by Paul Travers and animated by Matthew McClements and Tom McElderry. Finally, we would like to acknowledge the efforts of Adam Sendroff, who markets the book in North America. All of these individuals have helped to make the book work, and we are truly grateful to them.

Finally, the authors would like to acknowledge certain individuals who are important to them personally. Charles Janeway would like to thank his long-suffering wife and family, Kim Bottomly, Katherine Anne Janeway, Hannah Heath Janeway, who deserves special thanks for accompanying him to England so we could finish the book, and Megan Gwynedd Janeway, his youngest daughter. He would like especially to acknowledge Kim, who has been a steadfast and faithful friend and companion for the last twenty-four years. Paul Travers would like to thank Rose Zamoyska for her infinite patience and support over the many editions of the text. Mark Walport would like to thank his wife, Julia, and children, Louise, Robert, Emily and Fiona for their patience and support while he has been occupied by matters immunological. Mark Shlomchik would like to thank his wife Ann Haberman for her constant support, patience, understanding and immunological advice and his daughter Mariel for all of the same (save the immunological advice!).

The authors would like to give great thanks to all those people who read parts or all of the chapters of the fourth edition, and advised us on how to make them better. They are listed by chapter on page vii. We would also like to absolve them of any responsibility for errors that are still in the text; the authors take full responsibility for all of them.

Charles A. Janeway, Jr   Paul Travers   Mark Walport   Mark J. Shlomchik

Acknowledgements

Text

We would like to thank the following experts who read parts or the whole of the fourth edition chapters indicated and provided us with invaluable advice in developing this fifth edition.

Chapter 2: Ivan Lefkovits, Basel Institute for Immunology, Switzerland; Anthony T. Vella, Oregon State University.

Chapter 3: Sherie Morrison, University of California, Los Angeles; Michael S. Neuberger, MRC Laboratory of Molecular Biology, Cambridge.

Chapter 4: Ian A. Wilson, The Scripps Research Institute, La Jolla; Peter Cresswell, Yale University School of Medicine; Mark M. Davis, Stanford University School of Medicine; Paul M. Allen, Washington University School of Medicine, St. Louis; John Trowsdale, Cambridge University.

Chapter 5: John C. Cambier, National Jewish Medical and Research Center, Denver; Dan R. Littman, Skirball Institute of Biomolecular Medicine, New York; Arthur Weiss, The University of California, San Francisco.

Chapter 6: Richard R. Hardy, Fox Chase Cancer Center, Philadelphia; John G. Monroe, University of Pennsylvania Medical Center; Max D. Cooper, Comprehensive Cancer Center, University of Alabama; David Nemazee, The Scripps Research Institute, La Jolla; Michel C. Nussenzweig, Rockefeller University, New York.

Chapter 7: Alexander Y. Rudensky, University of Washington School of Medicine; Johnathan Sprent, The Scripps Research Institute, La Jolla; Leslie J. Berg, University of Massachusetts Medical School; Adrian C. Hayday, Guy's King's St Thomas' Medical School, University of London; Mike Owen, Imperial Cancer Research Fund, London; Robert H. Swanborg, Washington State University; Steve C. Jameson, University of Minnesota.

Chapter 8: Donna Paulnock, University of Wisconsin; Tim Springer, Center for Blood Research, Harvard Medical School; Marc K. Jenkins, University of Minnesota; Jürg Tschopp, University of Lausanne; Ralph Steinman, The Rockefeller University, New York.

Chapter 9: Michael C. Carroll, The Center for Blood Research, Harvard Medical School; E. Sally Ward, University of Texas; Jeffrey Ravetch, Rockefeller University, New York; Garnett Kelsoe, Duke University Medical Center, Durham; Douglas Fearon, University of Cambridge.

Chapter 10: Alan Ezekowitz, Massachusetts General Hospital, Harvard Medical School; Eric Pamer, Yale University School of Medicine; Adrian C. Hayday, Guy's King's St Thomas' Medical School, University of London.

Chapter 11: Fred Rosen, Center for Blood Research, Harvard Medical School; Robin A. Weiss, Royal Free and University College Medical School, London.

Chapter 12: Raif S. Geha, Children's Hospital, Harvard Medical School; Hugh A. Sampson, Mount Sinai Medical Center, New York; Philip W. Askenase, Yale University School of Medicine; Jeffrey Ravetch, The Rockefeller University, New York.

Chapter 13: Diane Mathis, Harvard Medical School; Christopher C. Goodnow, John Curtin School of Medical Research, Canberra; Jeffrey Ravetch, The Rockefeller University, New York; Kathryn Wood, University of Oxford; Hugh Auchincloss, Massachusetts General Hospital, Harvard Medical School; Joseph E. Craft, Yale University School of Medicine; Jan Erikson, The Wistar Institute, University of Pennsylvania; Keith Elkon, Cornell University, New York; Fiona Powrie, University of Oxford.

Chapter 14: Thierry Boon, Ludwig Institute for Cancer Research, Brussels; Gerry Crabtree, Stanford University School of Medicine; Jeffrey A. Bluestone, University of Chicago.

Appendix II: Joost J. Oppenheim, National Cancer Institute–Frederick Cancer Research and Development Center, Maryland.

Appendix III: Jason Cyster, University of California, San Francisco; Craig Gerard, Children's Hospital, Harvard Medical School.

Immunobiology Animations

We would like to thank Hung-Sia Teh of the University of British Columbia and David A. Lawlor of the Rochester Institute of Technology, for reviewing these animations.

Photographs

The following photographs have been reproduced with the kind permission of the journal in which they originally appeared.

Chapter 1

Fig. 1.1 courtesy of Yale University Harvey Cushing/John Hay Whitney Medical Library.
Fig. 1.9 photo from *The Journal of Experimental Medicine* 1972, **135**:200-214. © 1972 The Rockefeller University Press.

Chapter 2

Fig. 2.10 photo from *FEBS Letters* 1989, **250**:78-84. © 1989 Elsevier Science.
Fig. 2.13 photo from *The Journal of Immunology* 1990, **144**:2287-2294. © 1990 The American Association of Immunologists.
Fig. 2.24 photos from *Blut* 1990, **60**:309-318. © 1990 Springer-Verlag.
Fig. 2.39 photo from *Nature* 1994, **367**:338-345. © 1994 Macmillan Magazines Limited.

Chapter 3

Fig. 3.1 photo from *Nature* 1992, **360**:369-372. © 1992 Macmillan Magazines Limited.
Fig. 3.4 photo from *Advances in Immunology* 1969, **11**:1-30. © 1969 Academic Press.
Fig. 3.8 panel a from *Science* 1990, **248**:712-719. © 1990 American Association for the Advancement of Science; panel b from *Structure* 1993, **1**:83-93 © 1993 Current Biology.
Fig. 3.10 from *Science* 1986, **233**:747-753. © 1986 American Association for the Advancement of Science.
Fig. 3.13 photos from *Science* 1996, **274**:209-219. © 1996 American Association for the Advancement of Science.
Fig. 3.14 panel a from *Journal of Biological Chemistry* 1998, **263**:10541-10544. © 1998 American Society for Biochemistry and Molecular Biology.
Fig. 3.18 from *Nature* 1997, **387**:630-634. © 1997 Macmillan Magazines Limited.
Fig. 3.27 from *Science* 1996, **274**:209-219. © 1996 American Association for the Advancement of Science.
Fig 3.28 from *Science* 1999, **286**:1913-1921. © 1999 American Association for the Advancement of Science.

Chapter 4

Fig. 4.23 top photo from the *European Journal of Immunology* 1988, **18**:1001-1008. © 1988 Wiley-VCH.

Chapter 5

Fig. 5.4 from *Science* 1995, **268**:533-539. © 1995 American Association for the Advancement of Science.
Fig. 5.7 model structure from *Cell* 1996, **84**:505-507. © 1996 Cell Press.
Fig. 5.18 photo from *Nature* 1996, **384**:188-192. © 1996 Macmillan Magazines Limited.

Chapter 7

Fig. 7.3 panel b from the *European Journal of Immunology* 1987, **17**:1473-1484. © 1987 VCH Verlagsgesellschaft mbH.
Fig. 7.10 photos from *Nature* 1994, **372**:100-103. © 1994 Macmillan Magazines Limited.
Fig. 7.32 photos from *International Immunology* 1996, **8**:1537-1548. © 1996 Oxford University Press.

Chapter 8

Fig. 8.2 bottom panel from *Nature* 1997, **388**:787-792. © 1997 Macmillan Magazines Limited.
Fig. 8.29 panel c from *Second International Workshop on Cell Mediated Cytoxicity*. Eds. P.A. Henkart, and E. Martz. © 1985 Plenum Press.
Fig. 8.37 panels a and b from *Second International Workshop on Cell Mediated Cytoxicity* . Eds. P.A. Henkart, and E. Martz. © 1985 Plenum Press; panel c from *Immunology Today* 1985, **6**:21-27. © 1985 Elsevier Science.

Chapter 9

Fig. 9.15 left panel from *The Journal of Immunolgy* 1989, **134**:1349-1359. © 1989 The American Association of Immunologists. Middle and right panels from *Annual Reviews of Immunolgy* 1989, **7**:91-109. © 1989 Annual Reviews.
Fig. 9.21 from *Nature* 1994, **372**:336-343. © 1994 Macmillan Magazines Limited.
Fig. 9.27 planar conformation from the *European Journal of Immunology* 1988, **18**:1001-1008. © 1988 Wiley-VCH.

Chapter 11

Fig. 11.6 top panels from *International Reviews of Experimental Pathology* 1986, **28**:45-78, edited by M.A. Epstein and G.W. Richter. © 1986, Academic Press.
Fig. 11.26 from *Cell* 1998, **93**:665-671. © 1998 Cell Press Limited.
Fig. 11.27 from the *Nature* 1995, **373**:117-122. © 1995 Macmillan Magazines Limited.

Chapter 13

Fig. 13.20 photo from *Cell* 1989, **59**:247-255. © Cell Press.
Fig. 13.34 photos from *The Journal of Experimental Medicine* 1992, **176**:1355-1364. © 1992 The Rockefeller University Press.

Chapter 14

Fig. 14.16 photos from *Mechanisms of Cytoxicity by Natural Killer Cells*, edited by R.B. Herberman and D.M. Callewaert © 1985 Academic Press.

Appendix I

Fig. A.39 from *Nature* 2000, **403**:503-511. © 2000 Macmillan Magazines Limited.

CONTENTS

List of Headings

Part IV   THE ADAPTIVE IMMUNE RESPONSE

Part V   THE IMMUNE SYSTEM IN HEALTH AND DISEASE

Chapter 12   Allergy and Hypersensitivity

The production of IgE.

Effector mechanisms in allergic reactions.

Hypersensitivity diseases.

Chapter 13 Autoimmunity and Transplantation

Autoimmune responses are directed against self antigens.

Responses to alloantigens and transplant rejection.

PART I

AN INTRODUCTION TO IMMUNOBIOLOGY AND INNATE IMMUNITY

Basic Concepts in Immunology

Immunology is a relatively new science. Its origin is usually attributed to **Edward Jenner** (Fig. 1.1), who discovered in 1796 that cowpox, or vaccinia, induced protection against human smallpox, an often fatal disease. Jenner called his procedure vaccination, and this term is still used to describe the inoculation of healthy individuals with weakened or attenuated strains of disease-causing agents to provide protection from disease. Although Jenner's bold experiment was successful, it took almost two centuries for smallpox vaccination to become universal, an advance that enabled the World Health Organization to announce in 1979 that smallpox had been eradicated (Fig. 1.2), arguably the greatest triumph of modern medicine.

When Jenner introduced vaccination he knew nothing of the infectious agents that cause disease: it was not until late in the 19th century that **Robert Koch** proved that infectious diseases are caused by **microorganisms**, each one responsible for a particular disease, or **pathology**. We now recognize four broad categories of disease-causing microorganisms, or **pathogens**: these are **viruses**, **bacteria**, pathogenic **fungi**, and other relatively large and complex eukaryotic organisms collectively termed **parasites**.

The discoveries of Koch and other great 19th century microbiologists stimulated the extension of Jenner's strategy of vaccination to other diseases. In the 1880s, **Louis Pasteur** devised a vaccine against cholera in chickens, and developed a rabies vaccine that proved a spectacular success upon its first trial in a boy bitten by a rabid dog. These practical triumphs led to a search for the mechanism of protection and to the development of the science of immunology. In 1890, **Emil von Behring** and **Shibasaburo Kitasato** discovered that the serum of vaccinated individuals contained substances—which they called **antibodies**—that specifically bound to the relevant pathogen.

A specific **immune response**, such as the production of antibodies against a particular pathogen, is known as an **adaptive immune response**, because it occurs during the lifetime of an individual as an adaptation to infection with that pathogen. In many cases, an adaptive immune response confers lifelong **protective immunity** to reinfection with the same pathogen. This

Fig. 1.1 Edward Jenner. Portrait by John Raphael Smith. Reproduced courtesy of Yale University, Harvey Cushing/John Hay Whitney Medical Library.

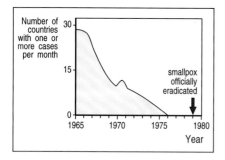

Fig. 1.2 The eradication of smallpox by vaccination. After a period of 3 years in which no cases of smallpox were recorded, the World Health Organization was able to announce in 1979 that smallpox had been eradicated.

distinguishes such responses from **innate immunity**, which, at the time that von Behring and Kitasato discovered antibodies, was known chiefly through the work of the great Russian immunologist **Elie Metchnikoff**. Metchnikoff discovered that many microorganisms could be engulfed and digested by **phagocytic cells**, which he called **macrophages**. These cells are immediately available to combat a wide range of pathogens without requiring prior exposure and are a key component of the innate immune system. Antibodies, by contrast, are produced only after infection, and are specific for the infecting pathogen. The antibodies present in a given person therefore directly reflect the infections to which he or she has been exposed.

Indeed, it quickly became clear that specific antibodies can be induced against a vast range of substances. Such substances are known as **antigens** because they can stimulate the *gen*eration of *anti*bodies. We shall see, however, that not all adaptive immune responses entail the production of antibodies, and the term antigen is now used in a broader sense to describe any substance that can be recognized by the adaptive immune system.

Both innate immunity and adaptive immune responses depend upon the activities of white blood cells, or **leukocytes**. Innate immunity largely involves **granulocytes** and macrophages. Granulocytes, also called polymorphonuclear leukocytes, are a diverse collection of white blood cells whose prominent granules give them their characteristic staining patterns; they include the **neutrophils**, which are phagocytic. The macrophages of humans and other vertebrates are presumed to be the direct evolutionary descendants of the phagocytic cells present in simpler animals, such as those that Metchnikoff observed in sea stars. Adaptive immune responses depend upon **lymphocytes**, which provide the lifelong immunity that can follow exposure to disease or vaccination. The innate and adaptive immune systems together provide a remarkably effective defense system. It ensures that although we spend our lives surrounded by potentially pathogenic microorganisms, we become ill only relatively rarely. Many infections are handled successfully by the innate immune system and cause no disease; others that cannot be resolved by innate immunity trigger adaptive immunity and are then overcome successfully, followed by lasting immunological memory.

The main focus of this book will be on the diverse mechanisms of adaptive immunity, whereby specialized classes of lymphocytes recognize and target pathogenic microorganisms or the cells infected with them. We shall see, however, that all the cells involved in innate immune responses also participate in adaptive immune responses. Indeed, most of the effector actions that the adaptive immune system uses to destroy invading microorganisms depend upon linking antigen-specific recognition to the activation of effector mechanisms that are also used in innate host defense.

In this chapter, we first introduce the cells of the immune system, and the tissues in which they develop and through which they circulate or migrate. In later sections, we outline the specialized functions of the different types of cells and the mechanisms by which they eliminate infection.

The components of the immune system.

The cells of the immune system originate in the **bone marrow**, where many of them also mature. They then migrate to guard the peripheral tissues, circulating in the blood and in a specialized system of vessels called the **lymphatic system**.

1-1   The white blood cells of the immune system derive from precursors in the bone marrow.

All the cellular elements of blood, including the red blood cells that transport oxygen, the platelets that trigger blood clotting in damaged tissues, and the white blood cells of the immune system, derive ultimately from the same **progenitor** or precursor cells—the **hematopoietic stem cells** in the bone marrow. As these stem cells can give rise to all of the different types of blood cells, they are often known as pluripotent hematopoietic stem cells. Initially, they give rise to stem cells of more limited potential, which are the immediate progenitors of red blood cells, platelets, and the two main categories of white blood cells. The different types of blood cell and their lineage relationships are summarized in Fig. 1.3. We shall be concerned here with all the cells derived from the common lymphoid progenitor and the myeloid progenitor, apart from the megakaryocytes and red blood cells.

The **myeloid progenitor** is the precursor of the granulocytes, macrophages, dendritic cells, and mast cells of the immune system. Macrophages are one of the three types of phagocyte in the immune system and are distributed widely in the body tissues, where they play a critical part in innate immunity. They are the mature form of **monocytes**, which circulate in the blood and differentiate continuously into macrophages upon migration into the tissues. **Dendritic cells** are specialized to take up antigen and display it for recognition by lymphocytes. Immature dendritic cells migrate from the blood to reside in the tissues and are both phagocytic and macropinocytic, ingesting large amounts of the surrounding extracellular fluid. Upon encountering a pathogen, they rapidly mature and migrate to lymph nodes.

Mast cells, whose blood-borne precursors are not well defined, also differentiate in the tissues. They mainly reside near small blood vessels and, when activated, release substances that affect vascular permeability. Although best known for their role in orchestrating allergic responses, they are believed to play a part in protecting mucosal surfaces against pathogens.

The **granulocytes** are so called because they have densely staining granules in their cytoplasm; they are also sometimes called **polymorphonuclear leukocytes** because of their oddly shaped nuclei. There are three types of granulocyte, all of which are relatively short lived and are produced in increased numbers during immune responses, when they leave the blood to migrate to sites of infection or inflammation. **Neutrophils**, which are the third phagocytic cell of the immune system, are the most numerous and most important cellular component of the innate immune response: hereditary deficiencies in neutrophil function lead to overwhelming bacterial infection, which is fatal if untreated. **Eosinophils** are thought to be important chiefly in defense against parasitic infections, because their numbers increase during a parasitic infection. The function of **basophils** is probably similar and complementary to that of eosinophils and mast cells; we shall discuss the functions of these cells in Chapter 9 and their role in allergic inflammation in Chapter 12. The cells of the myeloid lineage are shown in Fig. 1.4.

The **common lymphoid progenitor** gives rise to the lymphocytes, with which most of this book will be concerned. There are two major types of lymphocyte: **B lymphocytes** or **B cells**, which when activated differentiate into **plasma cells** that secrete antibodies; and **T lymphocytes** or **T cells**, of which there are two main classes. One class differentiates on activation into **cytotoxic T cells**, which kill cells infected with viruses, whereas the second class of T cells differentiates into cells that activate other cells such as B cells and macrophages.

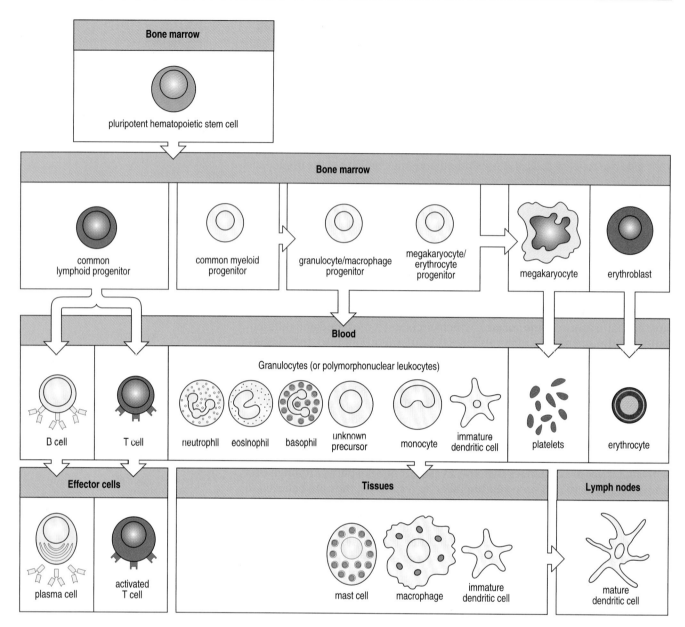

Fig. 1.3 All the cellular elements of blood, including the lymphocytes of the adaptive immune system, arise from hematopoietic stem cells in the bone marrow. These pluripotent cells divide to produce two more specialized types of stem cells, a common lymphoid progenitor that gives rise to the T and B lymphocytes responsible for adaptive immunity, and a common myeloid progenitor that gives rise to different types of leukocytes (white blood cells), erythrocytes (red blood cells that carry oxygen), and the megakaryocytes that produce platelets that are important in blood clotting. The existence of a common lymphoid progenitor for T and B lymphocytes is strongly supported by current data. T and B lymphocytes are distinguished by their sites of differentiation—T cells in the thymus and B cells in the bone marrow—and by their antigen receptors. Mature T and B lymphocytes circulate between the blood and peripheral lymphoid tissues. After encounter with antigen, B cells differentiate into antibody-secreting plasma cells, whereas T cells differentiate into effector T cells with a variety of functions. A third lineage of lymphoid-like cells, the natural killer cells, derive from the same progenitor cell but lack the antigen-specificity that is the hallmark of the adaptive immune response (not shown). The leukocytes that derive from the myeloid stem cell are the monocytes, the dendritic cells, and the basophils, eosinophils, and neutrophils. The latter three are collectively termed either granulocytes, because of the cytoplasmic granules whose characteristic staining gives them a distinctive appearance in blood smears, or polymorpho-nuclear leukocytes, because of their irregularly shaped nuclei. They circulate in the blood and enter the tissues only when recruited to sites of infection or inflammation where neutrophils are recruited to phagocytose bacteria. Eosinophils and basophils are recruited to sites of allergic inflammation, and appear to be involved in defending against parasites. Immature dendritic cells travel via the blood to enter peripheral tissues, where they ingest antigens. When they encounter a pathogen, they mature and migrate to lymphoid tissues, where they activate antigen-specific T lymphocytes. Monocytes enter tissues, where they differentiate into macrophages; these are the main tissue-resident phagocytic cells of the innate immune system. Mast cells arise from precursors in bone marrow but complete their maturation in tissues; they are important in allergic responses.

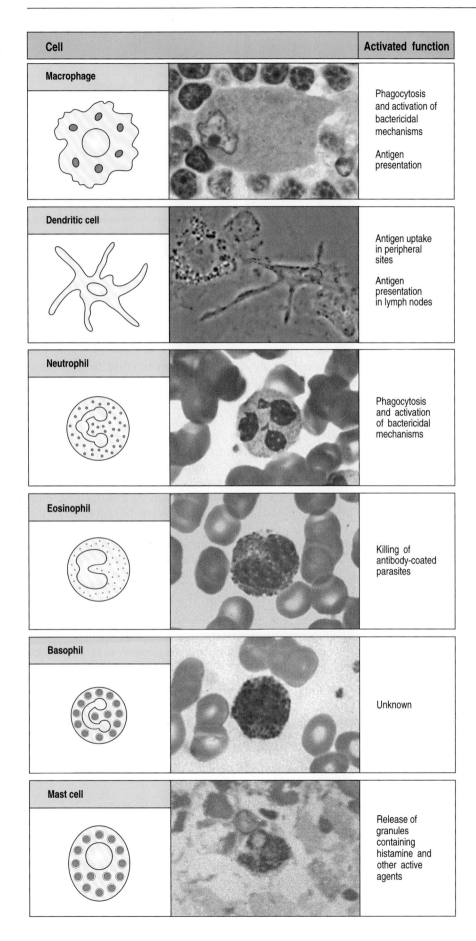

Cell		Activated function
Macrophage		Phagocytosis and activation of bactericidal mechanisms Antigen presentation
Dendritic cell		Antigen uptake in peripheral sites Antigen presentation in lymph nodes
Neutrophil		Phagocytosis and activation of bactericidal mechanisms
Eosinophil		Killing of antibody-coated parasites
Basophil		Unknown
Mast cell		Release of granules containing histamine and other active agents

Fig. 1.4 Myeloid cells in innate and adaptive immunity. Cells of the myeloid lineage perform various important functions in the immune response. The cells are shown schematically in the left column in the form in which they will be represented throughout the rest of the book. A photomicrograph of each cell type is shown in the center column. Macrophages and neutrophils are primarily phagocytic cells that engulf pathogens and destroy them in intracellular vesicles, a function they perform in both innate and adaptive immune responses. Dendritic cells are phagocytic when they are immature and take up pathogens; after maturing they act as antigen-presenting cells to T cells, initiating adaptive immune responses. Macrophages can also present antigens to T cells and can activate them. The other myeloid cells are primarily secretory cells that release the contents of their prominent granules upon activation via antibody during an adaptive immune response. Eosinophils are thought to be involved in attacking large antibody-coated parasites such as worms, whereas the function of basophils is less clear. Mast cells are tissue cells that trigger a local inflammatory response to antigen by releasing substances that act on local blood vessels. Photographs courtesy of N. Rooney and B. Smith.

Fig. 1.5 Lymphocytes are mostly small and inactive cells. The left panel shows a light micrograph of a small lymphocyte surrounded by red blood cells. Note the condensed chromatin of the nucleus, indicating little transcriptional activity, the relative absence of cytoplasm, and the small size. The right panel shows a transmission electron micrograph of a small lymphocyte. Note the condensed chromatin, the scanty cytoplasm and the absence of rough endoplasmic reticulum and other evidence of functional activity. Photographs courtesy of N. Rooney.

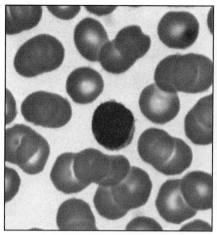
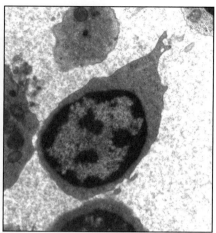

Most lymphocytes are small, featureless cells with few cytoplasmic organelles and much of the nuclear chromatin inactive, as shown by its condensed state (Fig. 1.5). This appearance is typical of inactive cells and it is not surprising that, as recently as the early 1960s, textbooks could describe these cells, now the central focus of immunology, as having no known function. Indeed, these small lymphocytes have no functional activity until they encounter antigen, which is necessary to trigger their proliferation and the differentiation of their specialized functional characteristics.

Lymphocytes are remarkable in being able to mount a specific immune response against virtually any foreign antigen. This is possible because each individual lymphocyte matures bearing a unique variant of a prototype antigen receptor, so that the population of T and B lymphocytes collectively bear a huge repertoire of receptors that are highly diverse in their antigen-binding sites. The **B-cell antigen receptor** (**BCR**) is a membrane-bound form of the antibody that the B cell will secrete after activation and differentiation to plasma cells. Antibody molecules as a class are known as **immunoglobulins**, usually shortened to **Ig**, and the antigen receptor of B lymphocytes is therefore also known as **membrane immunoglobulin** (**mIg**). The **T-cell antigen receptor** (**TCR**) is related to immunoglobulin but is quite distinct from it, as it is specially adapted to detect antigens derived from foreign proteins or pathogens that have entered into host cells. We shall describe the structures of these lymphocyte antigen receptors in detail in Chapters 3, 4, and 5, and the way in which their diversity of binding sites is created as lymphocytes develop in Chapter 7.

A third lineage of lymphoid cells, called **natural killer cells**, lack antigen-specific receptors and are part of the innate immune system. These cells circulate in the blood as large lymphocytes with distinctive cytotoxic granules (Fig. 1.6). They are able to recognize and kill some abnormal cells, for example some tumor cells and virus-infected cells, and are thought to be important in the innate immune defense against intracellular pathogens.

Natural killer (NK) cell

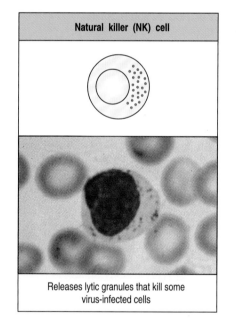

Releases lytic granules that kill some virus-infected cells

Fig. 1.6 Natural killer (NK) cells. These are large granular lymphocyte-like cells with important functions in innate immunity. Although lacking antigen-specific receptors, they can detect and attack certain virus-infected cells. Photograph courtesy of N. Rooney and B. Smith.

1-2   Lymphocytes mature in the bone marrow or the thymus.

The **lymphoid organs** are organized tissues containing large numbers of lymphocytes in a framework of nonlymphoid cells. In these organs, the interactions lymphocytes make with nonlymphoid cells are important either to lymphocyte development, to the initiation of adaptive immune responses, or to the sustenance of lymphocytes. Lymphoid organs can be divided broadly into **central** or **primary lymphoid organs**, where lymphocytes are generated,

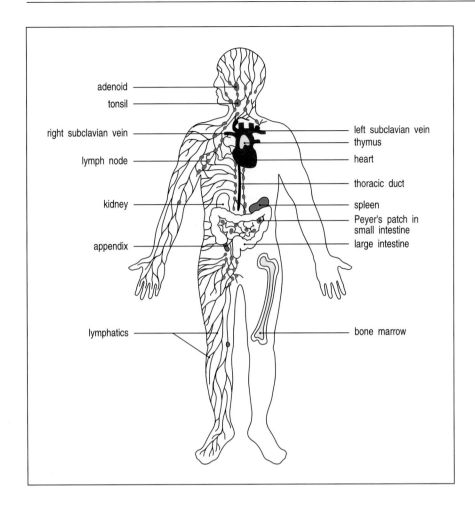

Fig. 1.7 The distribution of lymphoid tissues in the body. Lymphocytes arise from stem cells in bone marrow, and differentiate in the central lymphoid organs (yellow), B cells in bone marrow and T cells in the thymus. They migrate from these tissues and are carried in the bloodstream to the peripheral or secondary lymphoid organs (blue), the lymph nodes, the spleen, and lymphoid tissues associated with mucosa, like the gut-associated tonsils, Peyer's patches, and appendix. The peripheral lymphoid organs are the sites of lymphocyte activation by antigen, and lymphocytes recirculate between the blood and these organs until they encounter antigen. Lymphatics drain extracellular fluid from the peripheral tissues, through the lymph nodes and into the thoracic duct, which empties into the left subclavian vein. This fluid, known as lymph, carries antigen to the lymph nodes and recirculating lymphocytes from the lymph nodes back into the blood. Lymphoid tissue is also associated with other mucosa such as the bronchial linings (not shown).

and **peripheral** or **secondary lymphoid organs**, where adaptive immune responses are initiated and where lymphocytes are maintained. The central lymphoid organs are the bone marrow and the **thymus**, a large organ in the upper chest; the location of the thymus, and of the other lymphoid organs, is shown schematically in Fig. 1.7.

Both B and T lymphocytes originate in the bone marrow but only B lymphocytes mature there; T lymphocytes migrate to the thymus to undergo their maturation. Thus B lymphocytes are so-called because they are bone marrow derived, and T lymphocytes because they are thymus derived. Once they have completed their maturation, both types of lymphocyte enter the bloodstream, from which they migrate to the peripheral lymphoid organs.

1-3 The peripheral lymphoid organs are specialized to trap antigen, to allow the initiation of adaptive immune responses, and to provide signals that sustain recirculating lymphocytes.

Pathogens can enter the body by many routes and set up an infection anywhere, but antigen and lymphocytes will eventually encounter each other in the peripheral lymphoid organs—the lymph nodes, the spleen, and the mucosal lymphoid tissues (see Fig. 1.7). Lymphocytes are continually recirculating through these tissues, to which antigen is also carried from sites of infection, primarily within macrophages and dendritic cells. Within the lymphoid organs, specialized cells such as mature dendritic cells display the antigen to lymphocytes.

The **lymph nodes** are highly organized lymphoid structures located at the points of convergence of vessels of the lymphatic system, an extensive system of vessels that collects extracellular fluid from the tissues and returns it to the blood. This extracellular fluid is produced continuously by filtration from the blood, and is called **lymph**. The vessels are **lymphatic vessels** or **lymphatics** (see Fig. 1.7). **Afferent lymphatic vessels** drain fluid from the tissues and also carry antigen-bearing cells and antigens from infected tissues to the lymph nodes, where they are trapped. In the lymph nodes, B lymphocytes are localized in **follicles**, with T cells more diffusely distributed in surrounding **paracortical areas**, also referred to as **T-cell zones**. Some of the B-cell follicles include **germinal centers**, where B cells are undergoing intense proliferation after encountering their specific antigen and their cooperating T cells (Fig. 1.8). B and T lymphocytes are segregated in a similar fashion in the other peripheral lymphoid tissues, and we shall see that this organization promotes the crucial interactions that occur between B and T cells upon encountering antigen.

The **spleen** is a fist-sized organ just behind the stomach (see Fig. 1.7) that collects antigen from the blood. It also collects and disposes of senescent red blood cells. Its organization is shown schematically in Fig. 1.9. The bulk of the spleen is composed of **red pulp**, which is the site of red blood cell disposal. The lymphocytes surround the arterioles entering the organ, forming areas of **white pulp**, the inner region of which is divided into a **periarteriolar lymphoid sheath** (PALS), containing mainly T cells, and a flanking **B-cell corona**.

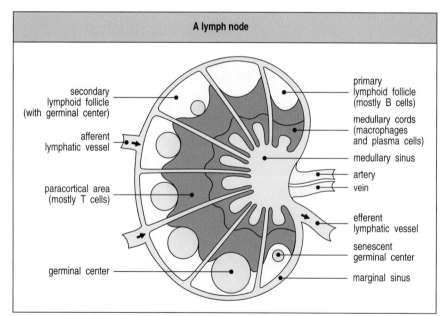
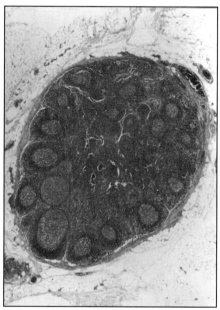

A lymph node

secondary lymphoid follicle (with germinal center)

afferent lymphatic vessel

paracortical area (mostly T cells)

germinal center

primary lymphoid follicle (mostly B cells)

medullary cords (macrophages and plasma cells)

medullary sinus

artery

vein

efferent lymphatic vessel

senescent germinal center

marginal sinus

Fig. 1.8 Organization of a lymph node. As shown in the diagram on the left, a lymph node consists of an outermost cortex and an inner medulla. The cortex is composed of an outer cortex of B cells organized into lymphoid follicles, and deep, or paracortical, areas made up mainly of T cells and dendritic cells. When an immune response is underway, some of the follicles contain central areas of intense B-cell proliferation called germinal centers and are known as secondary lymphoid follicles. These reactions are very dramatic, but eventually die out as senescent germinal centers. Lymph draining from the extracellular spaces of the body carries antigens in phagocytic dendritic cells and macrophages from the tissues to the lymph node via the afferent lymphatics. Lymph leaves by the efferent lymphatic in the medulla. The medulla consists of strings of macrophages and antibody-secreting plasma cells known as the medullary cords. Naive lymphocytes enter the node from the bloodstream through specialized postcapillary venules (not shown) and leave with the lymph through the efferent lymphatic. The light micrograph shows a section through a lymph node, with prominent follicles containing germinal centers. Magnification × 7. Photograph courtesy of N. Rooney.

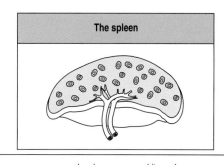

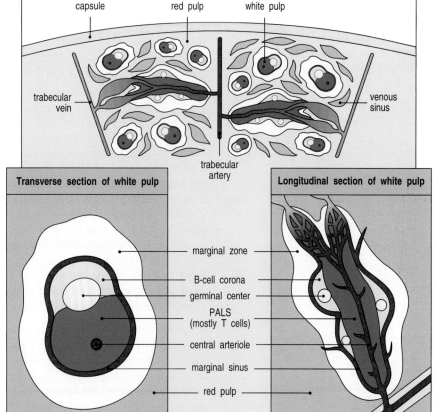

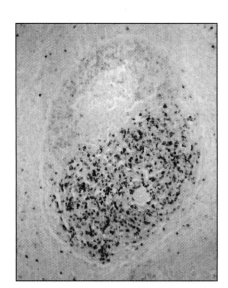

Fig. 1.9 Organization of the lymphoid tissues of the spleen. The schematic at top right shows that the spleen consists of red pulp (pink areas in the top panel), which is a site of red blood cell destruction, interspersed with lymphoid white pulp. An enlargement of a small section of the spleen (center) shows the arrangement of discrete areas of white pulp (yellow and blue) around central arterioles. Lymphocytes and antigen-loaded dendritic cells come together in the periarteriolar lymphoid sheath. Most of the white pulp is shown in transverse section, with two portions in longitudinal section. The bottom two schematics show enlargements of a transverse section (lower left) and longitudinal section (lower right) of white pulp. In each area of white pulp, blood carrying lymphocytes and antigen flows from a trabecular artery into a central arteriole. Cells and antigen then pass into a marginal sinus and drain into a trabecular vein. The marginal sinus is surrounded by a marginal zone of lymphocytes. Within the marginal sinus and surrounding the central arteriole is the periarteriolar lymphoid sheath (PALS), made up of T cells. The follicles consist mainly of B cells; in secondary follicles a germinal center is surrounded by a B-cell corona. The light micrograph at bottom left shows a transverse section of white pulp stained with hematoxylin and eosin. The T cells of the PALS stain darkly, while the B-cell corona is lightly stained. The unstained cells lying between the B- and T-cell areas represent a germinal center. Although the organization of the spleen is similar to that of a lymph node, antigen enters the spleen from the blood rather than from the lymph. Photograph courtesy of J.C. Howard.

The **gut-associated lymphoid tissues** (**GALT**), which include the **tonsils**, **adenoids**, and **appendix**, and specialized structures called **Peyer's patches** in the small intestine, collect antigen from the epithelial surfaces of the gastrointestinal tract. In Peyer's patches, which are the most important and

Fig. 1.10 Organization of typical gut-associated lymphoid tissue. As the diagram on the left shows, the bulk of the tissue is B cells, organized in a large and highly active domed follicle. T cells occupy the areas between follicles. The antigen enters across a specialized epithelium made up of so-called M cells. Although this tissue looks very different from other lymphoid organs, the basic divisions are maintained. The light micrograph shows a section through the gut wall. The dome of gut-associated lymphoid tissue can be seen lying beneath the epithelial tissues. Magnification × 16. Photograph courtesy of N. Rooney.

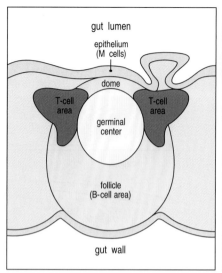
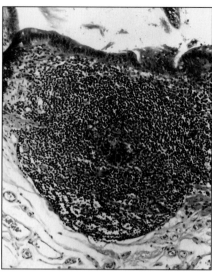

highly organized of these tissues, the antigen is collected by specialized epithelial cells called **multi-fenestrated** or **M cells**. The lymphocytes form a follicle consisting of a large central dome of B lymphocytes surrounded by smaller numbers of T lymphocytes (Fig. 1.10). Similar but more diffuse aggregates of lymphocytes protect the respiratory epithelium, where they are known as **bronchial-associated lymphoid tissue** (**BALT**), and other mucosa, where they are known simply as **mucosal-associated lymphoid tissue** (**MALT**). Collectively, the mucosal immune system is estimated to contain as many lymphocytes as all the rest of the body, and they form a specialized set of cells obeying somewhat different rules.

Although very different in appearance, the lymph nodes, spleen, and mucosal-associated lymphoid tissues all share the same basic architecture. Each of these tissues operates on the same principle, trapping antigen from sites of infection and presenting it to migratory small lymphocytes, thus inducing adaptive immune responses. The peripheral lymphoid tissues also provide sustaining signals to the lymphocytes that do not encounter their specific antigen, so that they continue to survive and recirculate until they encounter their specific antigen. This is important in maintaining the correct numbers of circulating T and B lymphocytes, and ensures that only those lymphocytes with the potential to respond to foreign antigen are sustained.

1-4   Lymphocytes circulate between blood and lymph.

Small B and T lymphocytes that have matured in the bone marrow and thymus but have not yet encountered antigen are referred to as **naive lymphocytes**. These cells circulate continually from the blood into the peripheral lymphoid tissues, which they enter by squeezing between the cells of capillary walls. They are then returned to the blood via the lymphatic vessels (Fig. 1.11) or, in the case of the spleen, return directly to the blood. In the event of an infection, lymphocytes that recognize the infectious agent are arrested in the lymphoid tissue, where they proliferate and differentiate into **effector cells** capable of combating the infection.

When an infection occurs in the periphery, for example, large amounts of antigen are taken up by dendritic cells which then travel from the site of infection through the afferent lymphatic vessels into the **draining lymph nodes** (see Fig. 1.11). In the lymph nodes, these cells display the antigen to

Fig. 1.11 Circulating lymphocytes encounter antigen in peripheral lymphoid organs. Naive lymphocytes recirculate constantly through peripheral lymphoid tissue, here illustrated as a lymph node behind the knee, a popliteal lymph node. Here, they may encounter their specific antigen, draining from an infected site in the foot. These are called draining lymph nodes, and are the site at which lymphocytes may become activated by encountering their specific ligand.

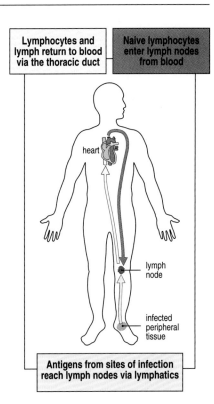

Lymphocytes and lymph return to blood via the thoracic duct

Naive lymphocytes enter lymph nodes from blood

heart

lymph node

infected peripheral tissue

Antigens from sites of infection reach lymph nodes via lymphatics

recirculating T lymphocytes, which they also help to activate. B cells that encounter antigen as they migrate through the lymph node are also arrested and activated, with the help of some of the activated T cells. Once the antigen-specific lymphocytes have undergone a period of proliferation and differentiation, they leave the lymph nodes as effector cells through the **efferent lymphatic vessel** (see Fig. 1.8).

Because they are involved in initiating adaptive immune responses, the peripheral lymphoid tissues are not static structures but vary quite dramatically depending upon whether or not infection is present. The diffuse mucosal lymphoid tissues may appear in response to infection and then disappear, whereas the architecture of the organized tissues changes in a more defined way during an infection. For example, the B-cell follicles of the lymph nodes expand as B lymphocytes proliferate to form germinal centers (see Fig. 1.8), and the entire lymph node enlarges, a phenomenon familiarly known as swollen glands.

Summary.

Immune responses are mediated by leukocytes, which derive from precursors in the bone marrow. A pluripotent hematopoietic stem cell gives rise to the lymphocytes responsible for adaptive immunity, and also to myeloid lineages that participate in both innate and adaptive immunity. Neutrophils, eosinophils, and basophils are collectively known as granulocytes; they circulate in the blood unless recruited to act as effector cells at sites of infection and inflammation. Macrophages and mast cells complete their differentiation in the tissues where they act as effector cells in the front line of host defense and initiate inflammation. Macrophages phagocytose bacteria, and recruit other phagocytic cells, the neutrophils, from the blood. Mast cells are exocytic and are thought to orchestrate the defense against parasites as well as triggering allergic inflammation; they recruit eosinophils and basophils, which are also exocytic. Dendritic cells enter the tissues as immature phagocytes where they specialize in ingesting antigens. These antigen-presenting cells subsequently migrate into lymphoid tissue. There are two major types of lymphocyte: B lymphocytes, which mature in the bone marrow; and T lymphocytes, which mature in the thymus. The bone marrow and thymus are thus known as the central or primary lymphoid organs. Mature lymphocytes recirculate continually from the bloodstream through the peripheral or secondary lymphoid organs, returning to the bloodstream through the lymphatic vessels. Most adaptive immune responses are triggered when a recirculating T cell recognizes its specific antigen on the surface of a dendritic cell. The three major types of peripheral lymphoid tissue are the spleen, which collects antigens from the blood; the lymph nodes, which collect antigen from sites of infection in the tissues; and the mucosal-associated lymphoid tissues (MALT), which collect antigens from the epithelial surfaces of the body. Adaptive immune responses are initiated in these peripheral lymphoid tissues: T cells that encounter antigen proliferate and differentiate into antigen-specific effector cells, while B cells proliferate and differentiate into antibody-secreting cells.

Principles of innate and adaptive immunity.

The macrophages and neutrophils of the innate immune system provide a first line of defense against many common microorganisms and are essential for the control of common bacterial infections. However, they cannot always eliminate infectious organisms, and there are some pathogens that they cannot recognize. The lymphocytes of the adaptive immune system have evolved to provide a more versatile means of defense which, in addition, provides increased protection against subsequent reinfection with the same pathogen. The cells of the innate immune system, however, play a crucial part in the initiation and subsequent direction of adaptive immune responses, as well as participating in the removal of pathogens that have been targeted by an adaptive immune response. Moreover, because there is a delay of 4–7 days before the initial adaptive immune response takes effect, the innate immune response has a critical role in controlling infections during this period.

1-5 Most infectious agents induce inflammatory responses by activating innate immunity.

Microorganisms such as bacteria that penetrate the epithelial surfaces of the body for the first time are met immediately by cells and molecules that can mount an innate immune response. Phagocytic macrophages conduct the defense against bacteria by means of surface receptors that are able to recognize and bind common constituents of many bacterial surfaces. Bacterial molecules binding to these receptors trigger the macrophage to engulf the bacterium and also induce the secretion of biologically active molecules. Activated macrophages secrete **cytokines**, which are defined as proteins released by cells that affect the behavior of other cells that bear receptors for them. They also release proteins known as **chemokines** that attract cells with chemokine receptors such as neutrophils and monocytes from the bloodstream (Fig. 1.12). The cytokines and chemokines released by macrophages in response to bacterial constituents initiate the process known as **inflammation**. Local inflammation and the phagocytosis of invading bacteria may also be triggered as a result of the activation of **complement** on the bacterial cell surface. Complement is a system of plasma proteins that activates a cascade of proteolytic reactions on microbial surfaces but not on host cells, coating these surfaces with fragments that are recognized and bound by phagocytic receptors on macrophages. The cascade of reactions also releases small peptides that contribute to inflammation.

Fig. 1.12 Bacterial infection triggers an inflammatory response.
Macrophages encountering bacteria in the tissues are triggered to release cytokines that increase the permeability of blood vessels, allowing fluid and proteins to pass into the tissues. They also produce chemokines that direct the migration of neutrophils to the site of infection. The stickiness of the endothelial cells of the blood vessels is also changed, so that cells adhere to the blood vessel wall and are able to crawl through it; first neutrophils and then monocytes are shown entering the tissue from a blood vessel. The accumulation of fluid and cells at the site of infection causes the redness, swelling, heat, and pain, known collectively as inflammation. Neutrophils and macrophages are the principal inflammatory cells. Later in an immune response, activated lymphocytes may also contribute to inflammation.

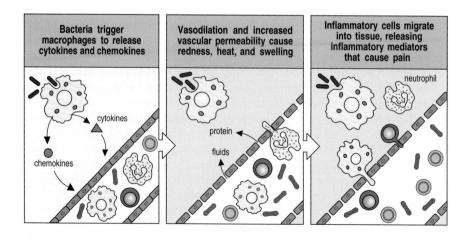

Inflammation is traditionally defined by the four Latin words *calor, dolor, rubor,* and *tumor,* meaning heat, pain, redness, and swelling, all of which reflect the effects of cytokines and other inflammatory mediators on the local blood vessels. Dilation and increased permeability of the blood vessels during inflammation lead to increased local blood flow and the leakage of fluid, and account for the heat, redness, and swelling. Cytokines and complement fragments also have important effects on the adhesive properties of the endothelium, causing circulating leukocytes to stick to the endothelial cells of the blood vessel wall and migrate between them to the site of infection, to which they are attracted by chemokines. The migration of cells into the tissue and their local actions account for the pain. The main cell types seen in an inflammatory response in its initial phases are neutrophils, which are recruited into the inflamed, infected tissue in large numbers. Like macrophages, they have surface receptors for common bacterial constituents and complement, and they are the principal cells that engulf and destroy the invading micro-organisms. The influx of neutrophils is followed a short time later by mono-cytes that rapidly differentiate into macrophages. Macrophages and neutrophils are thus also known as **inflammatory cells**. Inflammatory responses later in an infection also involve lymphocytes, which have mean-while been activated by antigen that has drained from the site of infection via the afferent lymphatics.

The innate immune response makes a crucial contribution to the activation of adaptive immunity. The inflammatory response increases the flow of lymph containing antigen and antigen-bearing cells into lymphoid tissue, while complement fragments on microbial surfaces and induced changes in cells that have taken up microorganisms provide signals that synergize in activating lymphocytes whose receptors bind to specific microbial antigens. Macrophages that have phagocytosed bacteria and become activated can also activate T lymphocytes. However, the cells that specialize in presenting antigen to T lymphocytes and initiating adaptive immunity are the dendritic cells.

1-6 Activation of specialized antigen-presenting cells is a necessary first step for induction of adaptive immunity.

The induction of an adaptive immune response begins when a pathogen is ingested by an **immature dendritic cell** in the infected tissue. These special-ized phagocytic cells are resident in most tissues and are relatively long-lived, turning over at a slow rate. They derive from the same bone marrow precursor as macrophages, and migrate from the bone marrow to their peripheral stations, where their role is to survey the local environment for pathogens. Eventually, all tissue-resident dendritic cells migrate through the lymph to the regional lymph nodes where they interact with recirculating naive lymphocytes. If the dendritic cells fail to be activated, they induce tolerance to the antigens of self that they bear.

The immature dendritic cell carries receptors on its surface that recognize common features of many pathogens, such as bacterial cell wall proteo-glycans. As with macrophages and neutrophils, binding of a bacterium to these receptors stimulates the dendritic cell to engulf the pathogen and degrade it intracellularly. Immature dendritic cells are also continually taking up extracellular material, including any virus particles or bacteria that may be present, by the receptor-independent mechanism of **macropinocytosis**. The function of dendritic cells, however, is not primarily to destroy pathogens but to carry pathogen antigens to peripheral lymphoid organs and there present them to T lymphocytes. When a dendritic cell takes up a pathogen in infected tissue, it becomes activated, and travels to a nearby lymph node. On activation,

Fig. 1.13 Dendritic cells initiate adaptive immune responses. Immature dendritic cells resident in infected tissues take up pathogens and their antigens by macropinocytosis and receptor-mediated phagocytosis. They are stimulated by recognition of the presence of pathogens to migrate via the lymphatics to regional lymph nodes, where they arrive as fully mature nonphagocytic dendritic cells. Here the mature dendritic cell encounters and activates antigen-specific naive T lymphocytes, which enter lymph nodes from the blood via a specialized vessel known from its cuboidal endothelial cells as a high endothelial venule (HEV).

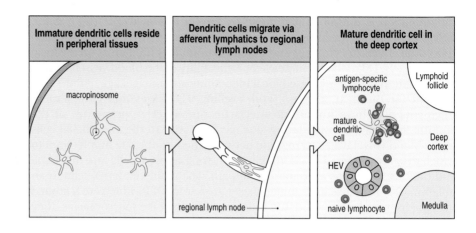

the dendritic cell matures into a highly effective **antigen-presenting cell** (**APC**) and undergoes changes that enable it to activate pathogen-specific lymphocytes that it encounters in the lymph node (Fig. 1.13). Activated dendritic cells secrete cytokines that influence both innate and adaptive immune responses, making these cells essential gatekeepers that determine whether and how the immune system responds to the presence of infectious agents. We shall consider the maturation of dendritic cells and their central role in presenting antigens to T lymphocytes in Chapter 8.

1-7 Lymphocytes activated by antigen give rise to clones of antigen-specific cells that mediate adaptive immunity.

The defense systems of innate immunity are effective in combating many pathogens. They are constrained, however, by relying on germline-encoded receptors to recognize microorganisms that can evolve more rapidly than the hosts they infect. This explains why they can only recognize microorganisms bearing surface molecules that are common to many pathogens and that have been conserved over the course of evolution. Not surprisingly, many pathogenic bacteria have evolved a protective capsule that enables them to conceal these molecules and thereby avoid being recognized and phagocytosed. Viruses carry no invariant molecules similar to those of bacteria and are rarely recognized directly by macrophages. Viruses and encapsulated bacteria can, however, still be taken up by dendritic cells through the nonreceptor-dependent process of macropinocytosis. Molecules that reveal their infectious nature may then be unmasked, and the dendritic cell activated to present their antigens to lympho-cytes. The recognition mechanism used by the lymphocytes of the adaptive immune response has evolved to overcome the constraints faced by the innate immune system, and enables recognition of an almost infinite diversity of antigens, so that each different pathogen can be targeted specifically.

Instead of bearing several different receptors, each recognizing a different surface feature shared by many pathogens, each naive lymphocyte entering the bloodstream bears antigen receptors of a single specificity. The specificity of these receptors is determined by a unique genetic mechanism that oper-ates during lymphocyte development in the bone marrow and thymus to generate millions of different variants of the genes encoding the receptor molecules. Thus, although an individual lymphocyte carries receptors of only one specificity, the specificity of each lymphocyte is different. This ensures that the millions of lymphocytes in the body collectively carry millions of

Fig. 1.14 Clonal selection. Each lymphocyte progenitor gives rise to many lymphocytes, each bearing a distinct antigen receptor. Lymphocytes with receptors that bind ubiquitous self antigens are eliminated before they become fully mature, ensuring tolerance to such self antigens. When antigen interacts with the receptor on a mature naive lymphocyte, that cell is activated and starts to divide. It gives rise to a clone of identical progeny, all of whose receptors bind the same antigen. Antigen specificity is thus maintained as the progeny proliferate and differentiate into effector cells. Once antigen has been eliminated by these effector cells, the immune response ceases.

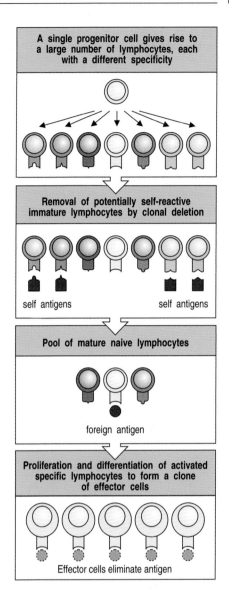

different antigen receptor specificities—the **lymphocyte receptor repertoire** of the individual. During a person's lifetime these lymphocytes undergo a process akin to natural selection; only those lymphocytes that encounter an antigen to which their receptor binds will be activated to proliferate and differentiate into effector cells.

This selective mechanism was first proposed in the 1950s by **Macfarlane Burnet** to explain why antibodies, which can be induced in response to virtually any antigen, are produced in each individual only to those antigens to which he or she is exposed. He postulated the preexistence in the body of many different potential antibody-producing cells, each having the ability to make antibody of a different specificity and displaying on its surface a membrane-bound version of the antibody that served as a receptor for antigen. On binding antigen, the cell is activated to divide and produce many identical progeny, known as a **clone**; these cells can now secrete **clonotypic** antibodies with a specificity identical to that of the surface receptor that first triggered activation and clonal expansion (Fig. 1.14). **Burnet** called this the **clonal selection theory**.

1-8 Clonal selection of lymphocytes is the central principle of adaptive immunity.

Remarkably, at the time that **Burnet** formulated his theory, nothing was known of the antigen receptors of lymphocytes; indeed the function of lymphocytes themselves was still obscure. Lymphocytes did not take center stage until the early 1960s, when **James Gowans** discovered that removal of the small lymphocytes from rats resulted in the loss of all known adaptive immune responses. These immune responses were restored when the small lymphocytes were replaced. This led to the realization that lymphocytes must be the units of clonal selection, and their biology became the focus of the new field of **cellular immunology**.

Clonal selection of lymphocytes with diverse receptors elegantly explained adaptive immunity but it raised one significant intellectual problem. If the antigen receptors of lymphocytes are generated randomly during the lifetime of an individual, how are lymphocytes prevented from recognizing antigens on the tissues of the body and attacking them? **Ray Owen** had shown in the late 1940s that genetically different twin calves with a common placenta were immunologically **tolerant** of one another's tissues, that is, they did not make an immune response against each other. **Sir Peter Medawar** then showed in 1953 that if exposed to foreign tissues during embryonic development, mice become immunologically tolerant to these tissues. **Burnet** proposed that developing lymphocytes that are potentially self-reactive are removed before they can mature, a process known as **clonal deletion**. He has since been proved right in this too, although the mechanisms of tolerance are still being worked out, as we shall see when we discuss the development of lymphocytes in Chapter 7.

Fig. 1.15 The four basic principles of clonal selection.

Postulates of the clonal selection hypothesis
Each lymphocyte bears a single type of receptor with a unique specificity
Interaction between a foreign molecule and a lymphocyte receptor capable of binding that molecule with high affinity leads to lymphocyte activation
The differentiated effector cells derived from an activated lymphocyte will bear receptors of identical specificity to those of the parental cell from which that lymphocyte was derived
Lymphocytes bearing receptors specific for ubiquitous self molecules are deleted at an early stage in lymphoid cell development and are therefore absent from the repertoire of mature lymphocytes

Clonal selection of lymphocytes is the single most important principle in adaptive immunity. Its four basic postulates are listed in Fig. 1.15. The last of the problems posed by the clonal selection theory—how the diversity of lymphocyte antigen receptors is generated—was solved in the 1970s when advances in molecular biology made it possible to clone the genes encoding antibody molecules.

1-9 The structure of the antibody molecule illustrates the central puzzle of adaptive immunity.

Antibodies, as discussed above, are the secreted form of the B-cell antigen receptor or BCR. Because they are produced in very large quantities in response to antigen, they can be studied by traditional biochemical techniques; indeed, their structure was understood long before recombinant DNA technology made it possible to study the membrane-bound antigen receptors of lymphocytes. The startling feature that emerged from the biochemical studies was that an antibody molecule is composed of two distinct regions. One is a **constant region** that can take one of only four or five biochemically distinguishable forms; the other is a **variable region** that can take an apparently infinite variety of subtly different forms that allow it to bind specifically to an equally vast variety of different antigens.

This division is illustrated in the simple schematic diagram in Fig. 1.16, where the antibody is depicted as a Y-shaped molecule, with the constant region shown in blue and the variable region in red. The two variable regions, which are identical in any one antibody molecule, determine the antigen-binding specificity of the antibody; the constant region determines how the antibody disposes of the pathogen once it is bound.

Each antibody molecule has a twofold axis of symmetry and is composed of two identical heavy chains and two identical light chains (Fig. 1.17). Heavy and light chains both have variable and constant regions; the variable regions of a heavy and a light chain combine to form an antigen-binding site, so that both chains contribute to the antigen-binding specificity of the antibody molecule. The structure of antibody molecules will be described in detail in Chapter 3, and the functional properties of antibodies conferred by their constant regions will be considered in Chapters 4 and 9. For the time being we are concerned only with the properties of immunoglobulin molecules as antigen receptors, and how the diversity of the variable regions is generated.

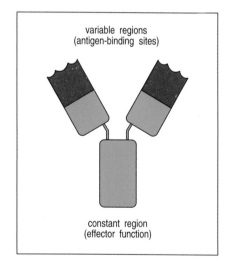

variable regions
(antigen-binding sites)

constant region
(effector function)

Fig. 1.16 Schematic structure of an antibody molecule. The two arms of the Y-shaped antibody molecule contain the variable regions that form the two identical antigen-binding sites. The stem can take one of only a limited number of forms and is known as the constant region. It is the region that engages the effector mechanisms that antibodies activate to eliminate pathogens.

1-10   Each developing lymphocyte generates a unique antigen receptor by rearranging its receptor genes.

How are antigen receptors with an almost infinite range of specificities encoded by a finite number of genes? This question was answered in 1976, when **Susumu Tonegawa** discovered that the genes for immunoglobulin variable regions are inherited as sets of **gene segments**, each encoding a part of the variable region of one of the immunoglobulin polypeptide chains (Fig. 1.18). During B-cell development in the bone marrow, these gene segments are irreversibly joined by DNA recombination to form a stretch of DNA encoding a complete variable region. Because there are many different gene segments in each set, and different gene segments are joined together in different cells, each cell generates unique genes for the variable regions of the heavy and light chains of the immunoglobulin molecule. Once these recombination events have succeeded in producing a functional receptor, further rearrangement is prohibited. Thus each lymphocyte expresses only one receptor specificity.

This mechanism has three important consequences. First, it enables a limited number of gene segments to generate a vast number of different proteins. Second, because each cell assembles a different set of gene segments, each cell expresses a unique receptor specificity. Third, because gene rearrangement involves an irreversible change in a cell's DNA, all the progeny of that cell will inherit genes encoding the same receptor specificity. This general scheme was later also confirmed for the genes encoding the antigen receptor on T lymphocytes. The main distinctions between B- and T-lymphocyte receptors are that the immunoglobulin that serves as the B-cell antigen receptor has two identical antigen-recognition sites and can also be secreted, whereas the T-cell antigen receptor has a single antigen-recognition site and is always a cell-surface molecule. We shall see later that these receptors also recognize antigen in very different ways.

The potential diversity of lymphocyte receptors generated in this way is enormous. Just a few hundred different gene segments can combine in different ways to generate thousands of different receptor chains. The diversity of lymphocyte receptors is further amplified by junctional diversity, created by adding or subtracting nucleotides in the process of joining the gene segments, and by the fact that each receptor is made by pairing two different variable chains, each encoded in distinct sets of gene segments. A thousand different chains of each type could thus generate 10^6 distinct antigen receptors through this **combinatorial diversity**. Thus a small amount of genetic material can encode a truly staggering diversity of receptors. Only a subset of these randomly generated receptor specificities survive the selective processes that shape the peripheral lymphocyte repertoire; nevertheless, there are lymphocytes of at least 10^8 different specificities in an individual at any one time. These provide the raw material on which clonal selection acts.

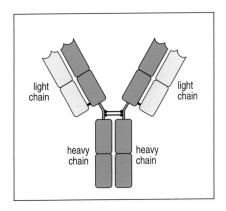

Fig. 1.17 Antibodies are made up of four protein chains. There are two types of chain in an antibody molecule: a larger chain called the heavy chain (green), and a smaller one called the light chain (yellow). Each chain has both a variable and a constant region, and there are two identical light chains and two identical heavy chains in each antibody molecule.

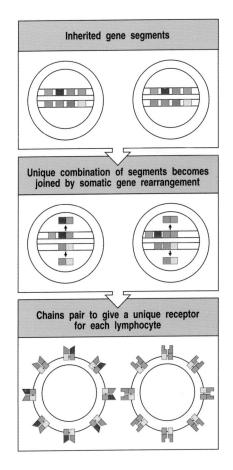

Fig. 1.18 The diversity of lymphocyte antigen receptors is generated by somatic gene rearrangements. Different parts of the variable regions of antigen receptors are encoded by sets of gene segments. During a lymphocyte's development, one member of each set of gene segments is joined randomly to the others by an irreversible process of DNA recombination. The juxtaposed gene segments make up a complete gene that encodes the variable part of one chain of the receptor, and is unique to that cell. This random rearrangement is repeated for the set of gene segments encoding the other chain. The rearranged genes are expressed to produce the two types of polypeptide chain. These come together to form a unique antigen receptor on the lymphocyte surface. Each lymphocyte bears many copies of its unique receptor.

1-11 Lymphocyte development and survival are determined by signals received through their antigen receptors.

Equally amazing as the generation of millions of specificities of lymphocyte antigen receptors is the shaping of this repertoire during lymphocyte development and the homeostatic maintenance of such an extensive repertoire in the periphery. How are the most useful receptor specificities selected, and how are the numbers of peripheral lymphocytes, and the percentages of B cells and T cells kept relatively constant? The answer seems to be that lymphocyte maturation and survival are regulated by signals received through their antigen receptors. Strong signals received through the antigen receptor by an immature lymphocyte cause it to die or undergo further receptor rearrangement, and in this way self-reactive receptor specificities are deleted from the repertoire. However, a complete absence of signals from the antigen receptor can also lead to cell death. It seems that in order to survive, lymphocytes must periodically receive certain signals from their environment via their antigen receptors. In this way, the body can ensure that each receptor is functional and regulate the number and type of lymphocytes in the population at any given time. These survival signals appear to be delivered by other cells in the lymphoid organs and must derive, at least in part, from the body's own molecules, the **self antigens**, as altering the self environment alters the lifespan of lymphocytes in that environment. Developing B cells in the bone marrow interact with stromal cells, while their final maturation and continued recirculation appears to depend on survival signals received from the B-cell follicles of peripheral lymphoid tissue. T lymphocytes receive survival signals from self molecules on specialized epithelial cells in the thymus during development, and from the same molecules expressed by dendritic cells in the lymphoid tissues in the periphery. The self ligands that interact with the T-cell receptor to deliver these signals are partially defined, being composed of known cell-surface molecules complexed with undefined peptides from other self proteins in the cell. These same cell-surface molecules function to present foreign intracellular antigens to T cells, as we shall explain in Section 1-16, and in Chapter 5. They select only a subset of T-cell receptors for survival, but these are the receptors most likely to be useful in responding to foreign antigens, as we shall see in Chapter 7.

Lymphocytes that fail to receive survival signals, and those that are clonally deleted because they are self-reactive, undergo a form of cell suicide called **apoptosis** or **programmed cell death**. Apoptosis, derived from a Greek word meaning the falling of leaves from the trees, occurs in all tissues, at a relatively constant rate in each tissue, and is a means of regulating the number of cells in the body. It is responsible, for example, for the death and shedding of skin cells, the turnover of liver cells, and the death of the oldest intestinal epithelial cells that are constantly replaced by new cells. Thus, it should come as no surprise that immune system cells are regulated through the same mechanism. Each day the bone marrow produces many millions of new neutrophils, monocytes, red blood cells, and lymphocytes, and this production must be balanced by an equal loss of these cells. Regulated loss of all these blood cells occurs by apoptosis, and the dying cells are finally phagocytosed by specialized macrophages in the liver and spleen. Lymphocytes are a special case, because the loss of an individual naive lymphocyte means the loss of a receptor specificity from the repertoire, while each newly matured cell that survives will contribute a different specificity. The survival signals received through the antigen receptors appear to regulate this process by inhibiting the apoptosis of individual lymphocytes, thus regulating the maintenance and composition of the lymphocyte repertoire. We shall return to the question of which ligands deliver these signals, and how they contribute to shaping and maintaining the receptor repertoire, in Chapter 7.

1-12 Lymphocytes proliferate in response to antigen in peripheral lymphoid organs, generating effector cells and immunological memory.

The large diversity of lymphocyte receptors means that there will usually be at least a few that can bind to any given foreign antigen. However, because each lymphocyte has a different receptor, the numbers of lymphocytes that can bind and respond to any given antigen is very small. To generate sufficient antigen-specific effector lymphocytes to fight an infection, a lymphocyte with an appropriate receptor specificity must be activated to proliferate before its progeny finally differentiate into effector cells. This **clonal expansion** is a feature common to all adaptive immune responses.

As we have seen, lymphocyte activation and proliferation is initiated in the draining lymphoid tissues, where naive lymphocytes and activated antigen-presenting cells can come together. Antigens are thus presented to the naive recirculating lymphocytes as they migrate through the lymphoid tissue before returning to the bloodstream via the efferent lymph. On recognizing its specific antigen, a small lymphocyte stops migrating and enlarges. The chromatin in its nucleus becomes less dense, nucleoli appear, the volume of both the nucleus and the cytoplasm increases, and new RNAs and proteins are synthesized. Within a few hours, the cell looks completely different and is known as a **lymphoblast** (Fig. 1.19).

The lymphoblasts now begin to divide, normally duplicating themselves two to four times every 24 hours for 3 to 5 days, so that one naive lymphocyte gives rise to a clone of around 1000 daughter cells of identical specificity. These then differentiate into effector cells (see Fig. 1.19). In the case of B cells, the differentiated effector cells, the plasma cells, secrete antibody; in the case of T cells, the effector cells are able to destroy infected cells or activate other cells of the immune system. These changes also affect the recirculation of antigen-specific lymphocytes. Changes in the cell-adhesion molecules they express on their surface allow effector lymphocytes to migrate into sites of infection or stay in the lymphoid organs to activate B cells.

After a naive lymphocyte has been activated, it takes 4 to 5 days before clonal expansion is complete and the lymphocytes have differentiated into effector cells. That is why adaptive immune responses occur only after a delay of several days. Effector cells have only a limited life-span and, once antigen is removed, most of the antigen-specific cells generated by the clonal expansion of small lymphocytes undergo apoptosis. However, some persist after the antigen has been eliminated. These cells are known as **memory cells** and form the basis of **immunological memory**, which ensures a more rapid and effective response on a second encounter with a pathogen and thereby provides lasting protective immunity.

The characteristics of immunological memory are readily observed by comparing the antibody response of an individual to a first or **primary immunization** with the response elicited in the same individual by a **secondary** or **booster immunization** with the same antigen. As shown in Fig. 1.20, the **secondary antibody response** occurs after a shorter lag phase, achieves a markedly higher level, and produces antibodies of higher **affinity**, or strength of binding, for the antigen. We shall describe the mechanisms of these remarkable changes in Chapters 9 and 10. The cellular basis of immunological memory is the clonal expansion and clonal differentiation of cells specific for the eliciting antigen, and it is therefore entirely antigen specific.

It is immunological memory that enables successful vaccination and prevents reinfection with pathogens that have been repelled successfully by an adaptive immune response. Immunological memory is the most important biological consequence of the development of adaptive immunity, although its cellular and molecular basis is still not fully understood, as we shall see in Chapter 10.

Fig. 1.19 Transmission electron micrographs of lymphocytes at various stages of activation to effector function. Small resting lymphocytes (top panel) have not yet encountered antigen. Note the scanty cytoplasm, the absence of rough endoplasmic reticulum, and the condensed chromatin, all indicative of an inactive cell. This could be either a T cell or a B cell. Small circulating lymphocytes are trapped in lymph nodes when their receptors encounter antigen on antigen-presenting cells. Stimulation by antigen induces the lymphocyte to become an active lymphoblast (center panel). Note the large size, the nucleoli, the enlarged nucleus with diffuse chromatin, and the active cytoplasm; again, T and B lymphoblasts are similar in appearance. This cell undergoes repeated division, which is followed by differentiation to effector function. The bottom panels show effector T and B lymphocytes. Note the large amount of cytoplasm, the nucleus with prominent nucleoli, abundant mitochondria, and the presence of rough endoplasmic reticulum, all hallmarks of active cells. The rough endoplasmic reticulum is especially prominent in plasma cells (effector B cells), which are synthesizing and secreting very large amounts of protein in the form of antibody. Photographs courtesy of N. Rooney.

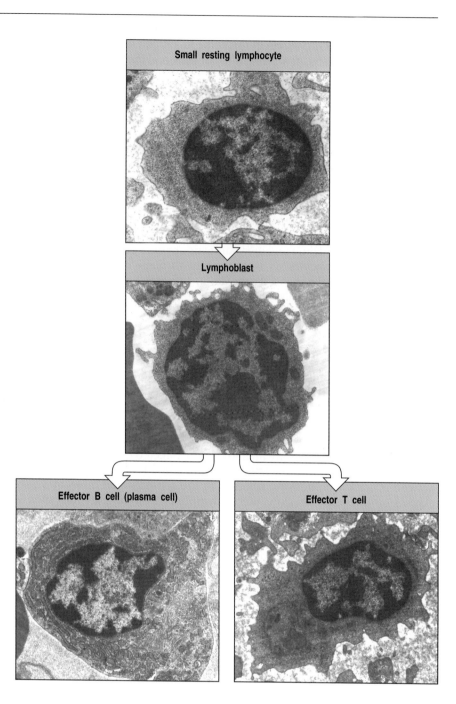

1-13   Interaction with other cells as well as with antigen is necessary for lymphocyte activation.

Peripheral lymphoid tissues are specialized not only to trap phagocytic cells that have ingested antigen (see Sections 1-3 and 1-6) but also to promote their interactions with lymphocytes that are needed to initiate an adaptive immune response. The spleen and lymph nodes in particular are highly organized for the latter function.

All lymphocyte responses to antigen require not only the signal that results from antigen binding to their receptors, but also a second signal, which is delivered by another cell. Naive T cells are generally activated by activated dendritic cells (Fig. 1.21, left panel) but for B cells (Fig. 1.21, right panel), the

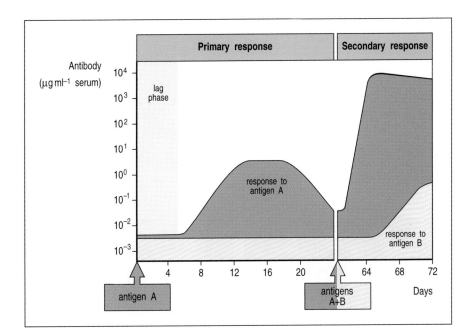

Fig. 1.20 The course of a typical antibody response. First encounter with an antigen produces a primary response. Antigen A introduced at time zero encounters little specific antibody in the serum. After a lag phase, antibody against antigen A (blue) appears; its concentration rises to a plateau, and then declines. When the serum is tested for antibody against another antigen, B (yellow), there is none present, demonstrating the specificity of the antibody response. When the animal is later challenged with a mixture of antigens A and B, a very rapid and intense secondary response to A occurs. This illustrates immunological memory, the ability of the immune system to make a second response to the same antigen more efficiently and effectively, providing the host with a specific defense against infection. This is the main reason for giving booster injections after an initial vaccination. Note that the response to B resembles the initial or primary response to A, as this is the first encounter of the animal with antigen B.

second signal is delivered by an armed effector T cell. Because of their ability to deliver activating signals, these three cell types are known as **professional antigen-presenting cells**, or often just **antigen-presenting cells**. They are illustrated in Fig. 1.22. Dendritic cells are the most important antigen-presenting cell of the three, with a central role in the initiation of adaptive immune responses (see Section 1-6). Macrophages can also mediate innate immune responses directly and make a crucial contribution to the effector phase of the adaptive immune response. B cells contribute to adaptive immunity by presenting peptides from antigens they have ingested and by secreting antibody.

Thus, the final postulate of adaptive immunity is that it occurs on a cell that also presents the antigen. This appears to be an absolute rule *in vivo*, although exceptions have been observed in *in vitro* systems. Nevertheless, what we are attempting to define is what does happen, not what can happen.

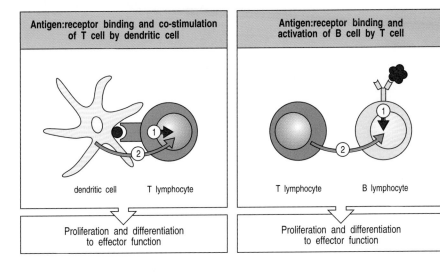

Fig. 1.21 Two signals are required for lymphocyte activation. As well as receiving a signal through their antigen receptor, mature naive lymphocytes must also receive a second signal to become activated. For T cells (left panel) it is delivered by a professional antigen-presenting cell such as the dendritic cell shown here. For B cells (right panel), the second signal is usually delivered by an activated T cell.

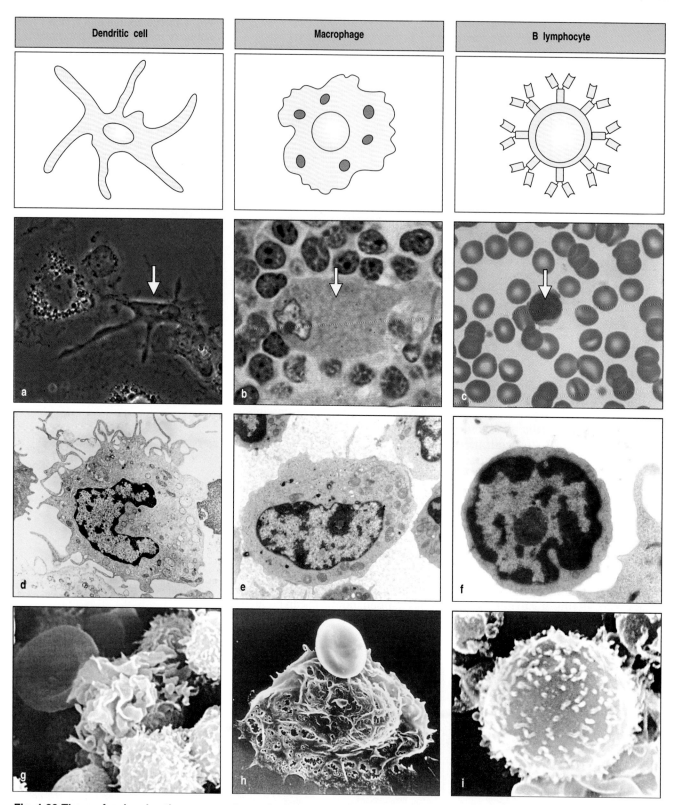

Fig. 1.22 The professional antigen-presenting cells. The three types of professional antigen-presenting cell are shown in the form in which they will be depicted throughout this book (top row), as they appear in the light microscope (second row; the relevant cell is indicated by an arrow), by transmission electron microscopy (third row) and by scanning electron microscopy (bottom row). Mature dendritic cells are found in lymphoid tissues and are derived from immature tissue dendritic cells that interact with many distinct types of pathogen. Macrophages are specialized to internalize extracellular pathogens, especially after they have been coated with antibody, and to present their antigens. B cells have antigen-specific receptors that enable them to internalize large amounts of specific antigen, process it, and present it. Photographs courtesy of R.M. Steinman (a); N. Rooney (b, c, e, f); S. Knight (d, g); P.F. Heap (h, i).

Summary.

The early innate systems of defense, which depend on invariant receptors recognizing common features of pathogens, are crucially important, but they are evaded or overcome by many pathogens and do not lead to immunological memory. The abilities to recognize all pathogens specifically and to provide enhanced protection against reinfection are the unique features of adaptive immunity, which is based on clonal selection of lymphocytes bearing antigen-specific receptors. The clonal selection of lymphocytes provides a theoretical framework for understanding all the key features of adaptive immunity. Each lymphocyte carries cell-surface receptors of a single specificity, generated by the random recombination of variable receptor gene segments and the pairing of different variable chains. This produces lymphocytes, each bearing a distinct receptor, so that the total repertoire of receptors can recognize virtually any antigen. If the receptor on a lymphocyte is specific for a ubiquitous self antigen, the cell is eliminated by encountering the antigen early in its development, while survival signals received through the antigen receptor select and maintain a functional lymphocyte repertoire. Adaptive immunity is initiated when an innate immune response fails to eliminate a new infection, and antigen and activated antigen-presenting cells are delivered to the draining lymphoid tissues. When a recirculating lymphocyte encounters its specific foreign antigen in peripheral lymphoid tissues, it is induced to proliferate and its progeny then differentiate into effector cells that can eliminate the infectious agent. A subset of these proliferating lymphocytes differentiate into memory cells, ready to respond rapidly to the same pathogen if it is encountered again. The details of these processes of recognition, development, and differentiation form the main material of the middle three parts of this book.

The recognition and effector mechanisms of adaptive immunity.

Clonal selection describes the basic operating principle of the adaptive immune response but not how it defends the body against infection. In the last part of this chapter, we outline the mechanisms by which pathogens are detected by lymphocytes and are eventually destroyed in a successful adaptive immune response. The distinct lifestyles of different pathogens require different response mechanisms, not only to ensure their destruction but also for their detection and recognition (Fig. 1.23). We have already seen that there are two different kinds of antigen receptor: the surface immunoglobulin of B cells, and the smaller antigen receptor of T cells. These surface receptors are adapted to recognize antigen in two different ways: B cells recognize antigen that is present outside the cells of the body, where, for example, most bacteria are found; T cells, by contrast, can detect antigens generated inside infected cells, for example those due to viruses.

The **effector mechanisms** that operate to eliminate pathogens in an adaptive immune response are essentially identical to those of innate immunity. Indeed, it seems likely that specific recognition by clonally distributed receptors evolved as a late addition to existing innate effector mechanisms to produce the present-day adaptive immune response. We begin by outlining the effector actions of antibodies, which depend almost entirely on recruiting cells and molecules of the innate immune system.

Fig. 1.23 The major pathogen types confronting the immune system and some of the diseases that they cause.

The immune system protects against four classes of pathogen		
Type of pathogen	Examples	Diseases
Extracellular bacteria, parasites, fungi	*Streptococcus pneumoniae* *Clostridium tetani* *Trypanosoma brucei* *Pneumocystis carinii*	Pneumonia Tetanus Sleeping sickness PC Pneumonia
Intracellular bacteria, parasites	*Mycobacterium leprae* *Leishmania donovani* *Plasmodium falciparum*	Leprosy Leishmaniasis Malaria
Viruses (intracellular)	Variola Influenza Varicella	Smallpox Flu Chickenpox
Parasitic worms (extracellular)	*Ascaris* *Schistosoma*	Ascariasis Schistosomiasis

1-14 Antibodies deal with extracellular forms of pathogens and their toxic products.

Antibodies, which were the first specific product of the adaptive immune response to be identified, are found in the fluid component of blood, or **plasma**, and in extracellular fluids. Because body fluids were once known as humors, immunity mediated by antibodies is known as **humoral immunity**.

As we have seen in Fig. 1.16, antibodies are Y-shaped molecules whose arms form two identical antigen-binding sites. These are highly variable from one molecule to another, providing the diversity required for specific antigen recognition. The stem of the Y, which defines the **class** of the antibody and determines its functional properties, takes one of only five major forms, or **isotypes**. Each of the five antibody classes engages a distinct set of effector mechanisms for disposing of antigen once it is recognized. We shall describe the isotypes and their actions in detail in Chapters 4 and 9.

The simplest and most direct way in which antibodies can protect from pathogens or their toxic products is by binding to them and thereby blocking their access to cells that they might infect or destroy (Fig. 1.24, left panels). This is known as **neutralization** and is important for protection against bacterial toxins and against pathogens such as viruses, which can thus be prevented from entering cells and replicating.

Binding by antibodies, however, is not sufficient on its own to arrest the replication of bacteria that multiply outside cells. In this case, one role of antibody is to enable a phagocytic cell to ingest and destroy the bacterium. This is important for the many bacteria that mare resistant to direct recognition by phagocytes; instead, the phagocytes recognize the constant region of the antibodies coating the bacterium (see Fig. 1.24, middle panels). The coating of pathogens and foreign particles in this way is known as **opsonization**.

The third function of antibodies is to activate a system of plasma proteins known as **complement**. The complement system, which we shall discuss in detail in Chapter 2, can also be activated without the help of antibodies on many microbial surfaces, and therefore contributes to innate as well as adaptive immunity. The pores formed by activated complement components directly destroy bacteria, and this is important in a few bacterial infections (see Fig. 1.24, right panels). However, the main function of complement, like that of the antibodies themselves, is to coat the surface of pathogens and enable phagocytes to engulf and destroy bacteria that they would otherwise not recognize. Complement also enhances the bactericidal actions of phagocytes; indeed it is so-called because it 'complements' the activities of antibodies.

Antibodies of different isotypes are found in different compartments of the body and differ in the effector mechanisms that they recruit, but all pathogens and particles bound by antibody are eventually delivered to phagocytes for ingestion, degradation, and removal from the body (see Fig. 1.24, bottom panels). The complement system and the phagocytes that antibodies recruit are not themselves antigen-specific; they depend upon antibody molecules to mark the particles as foreign. Antibodies are the sole contribution of B cells to the adaptive immune response. T cells, by contrast, have a variety of effector actions.

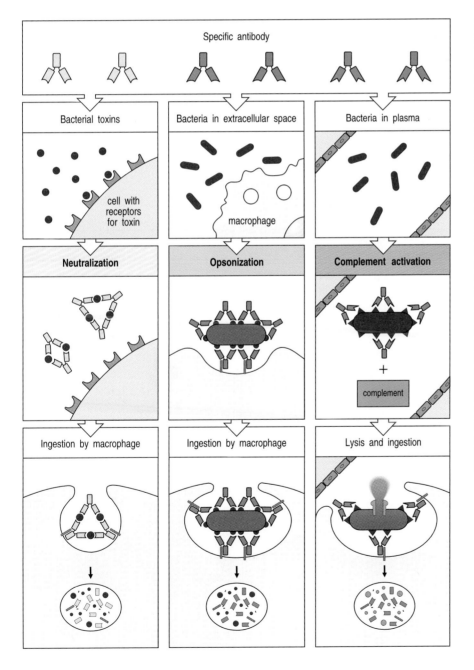

Fig. 1.24 Antibodies can participate in host defense in three main ways. The left panels show antibodies binding to and neutralizing a bacterial toxin, thus preventing it from interacting with host cells and causing pathology. Unbound toxin can react with receptors on the host cell, whereas the toxin:antibody complex cannot. Antibodies also neutralize complete virus particles and bacterial cells by binding to them and inactivating them. The antigen:antibody complex is eventually scavenged and degraded by macrophages. Antibodies coating an antigen render it recognizable as foreign by phagocytes (macrophages and neutrophils), which then ingest and destroy it; this is called opsonization. The middle panels show opsonization and phagocytosis of a bacterial cell. The right panels show activation of the complement system by antibodies coating a bacterial cell. Bound antibodies form a receptor for the first protein of the complement system, which eventually forms a protein complex on the surface of the bacterium that, in some cases, can kill the bacterium directly. More generally, complement coating favors the uptake and destruction of the bacterium by phagocytes. Thus, antibodies target pathogens and their toxic products for disposal by phagocytes.

1-15   T cells are needed to control intracellular pathogens and to activate B-cell responses to most antigens.

Pathogens are accessible to antibodies only in the blood and the extracellular spaces. However, some bacterial pathogens and parasites, and all viruses, replicate inside cells where they cannot be detected by antibodies. The destruction of these invaders is the function of the T lymphocytes, or T cells, which are responsible for the **cell-mediated immune responses** of adaptive immunity.

Cell-mediated reactions depend on direct interactions between T lymphocytes and cells bearing the antigen that the T cells recognize. The actions of cytotoxic T cells are the most direct. These recognize any of the body's cells that are infected with viruses, which replicate inside cells, using the biosynthetic machinery of the cell itself. The replicating virus eventually kills the cell, releasing new virus particles. Antigens derived from the replicating virus are, however, displayed on the surface of infected cells, where they are recognized by cytotoxic T cells. These cells can then control the infection by killing the infected cell before viral replication is complete (Fig. 1.25). Cytotoxic T cells typically express the molecule CD8 on their cell surfaces.

Other T lymphocytes that activate the cells they recognize are marked by the expression of the cell-surface molecule CD4 instead of CD8. Such T cells are often generically called helper T, or T_H cells, but this is a term that we will use for a specific subset of CD4 T cells. CD4 T lymphocytes can be divided into

Fig. 1.25 Mechanism of host defense against intracellular infection by viruses. Cells infected by viruses are recognized by specialized T cells called cytotoxic T cells, which kill the infected cells directly. The killing mechanism involves the activation of enzymes known as caspases, which cleave after aspartic acid. These in turn activate a cytostolic nuclease in the infected cell, which cleaves host and viral DNA. Panel a is a transmission electron micrograph showing the plasma membrane of a cultured CHO cell (the Chinese hamster ovary cell line) infected with influenza virus. Many virus particles can be seen budding from the cell surface. Some of these have been labeled with a monoclonal antibody that is specific for a viral protein and is coupled to gold particles, which appear as the solid black dots in the micrograph. Panel b is a transmission electron micrograph of a virus-infected cell (V) surrounded by cytotoxic T lymphocytes. Note the close apposition of the membranes of the virus-infected cell and the T cell (T) in the upper left corner of the micrograph, and the clustering of the cytoplasmic organelles in the T cell between the nucleus and the point of contact with the infected cell. Panel a courtesy of M. Bui and A. Helenius; panel b courtesy of N. Rooney.

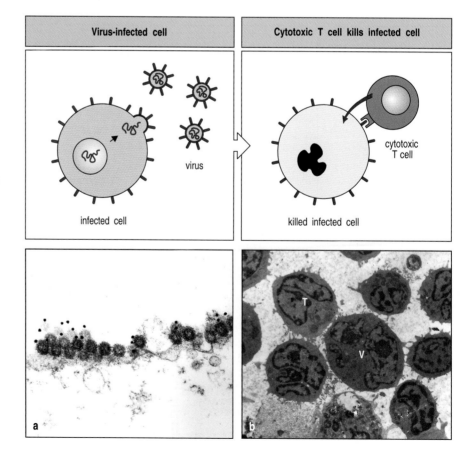

Infected macrophage	Activated infected macrophage

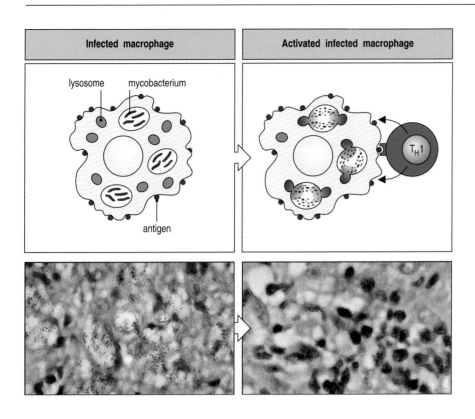

Fig. 1.26 Mechanism of host defense against intracellular infection by mycobacteria. Mycobacteria are engulfed by macrophages but resist being destroyed by preventing the fusion of the intracellular vesicles in which they reside with the lysosomes containing bactericidal agents; instead they persist and replicate in these vesicles. However, when a specific T$_H$1 cell recognizes an infected macrophage, it releases cytokines that activate the macrophage and induce lysosomal fusion and macrophage bactericidal activity. The light micrographs (bottom row) show resting (left) and activated (right) macrophages infected with mycobacteria. The cells have been stained with an acid-fast red dye to reveal mycobacteria. These are prominent as red-staining rods in the resting macrophages but have been eliminated from the activated macrophages. Photographs courtesy of G. Kaplan.

two subsets, which carry out different functions in defending the body, in particular from bacterial infections. The first subset of CD4 T lymphocytes is important in the control of intracellular bacterial infections. Some bacteria grow only in the intracellular membrane-bounded vesicles of macrophages; important examples are *Mycobacterium tuberculosis* and *M. leprae*, the pathogens that cause tuberculosis and leprosy. Bacteria phagocytosed by macrophages are usually destroyed in the lysosomes, which contain a variety of enzymes and antimicrobial substances. Intracellular bacteria survive because the vesicles they occupy do not fuse with the lysosomes. These infections can be controlled by a subset of CD4 T cells, known as a **T$_H$1 cells**, which activate macrophages, inducing the fusion of their lysosomes with the vesicles containing the bacteria and at the same time stimulating other antibacterial mechanisms of the phagocyte (Fig. 1.26). T$_H$1 cells also release cytokines and chemokines that attract macrophages to the site of infection.

T cells destroy intracellular pathogens by killing infected cells and by activating macrophages but they also have a central role in the destruction of extra-cellular pathogens by activating B cells. This is the specialized role of the second subset of CD4 T cells, called **T$_H$2 cells**. We shall see in Chapter 9, when we discuss the humoral immune response in detail, that only a few antigens with special properties can activate naive B lymphocytes on their own. Most antigens require an accompanying signal from helper T cells before they can stimulate B cells to proliferate and differentiate into cells secreting antibody (see Fig. 1.21). The ability of T cells to activate B cells was discovered long before it was recognized that a functionally distinct class of T cells activates macrophages, and the term helper T cell was originally coined to describe T cells that activate B cells. Although the designation 'helper' was later extended to T cells that activate macrophages (hence the H in T$_H$1), we consider this usage confusing and we will, in the remainder of this book, reserve the term helper T cells for all T cells that activate B cells.

1-16   T cells are specialized to recognize foreign antigens as peptide fragments bound to proteins of the major histocompatibility complex.

All the effects of T lymphocytes depend upon interactions with target cells containing foreign proteins. Cytotoxic T cells and T_H1 cells interact with antigens produced by pathogens that have have infected the target cell or that have been ingested by it. Helper T cells, in contrast, recognize and interact with B cells that have bound and internalized foreign antigen by means of their surface immunoglobulin. In all cases, T cells recognize their targets by detecting peptide fragments derived from the foreign proteins, after these peptides have been captured by specialized molecules in the host cell and displayed by them at the cell surface. The molecules that display peptide antigen to T cells are membrane glycoproteins encoded in a cluster of genes bearing the cumbersome name **major histocompatibility complex**, abbreviated to **MHC**.

The human **MHC molecules** were first discovered as the result of attempts to use skin grafts from donors to repair badly burned pilots and bomb victims during World War II. The patients rejected the grafts, which were recognized as being 'foreign.' It was soon appreciated from studies in mice that rejection was due to an immune response, and eventually genetic experiments using inbred strains of mice showed that rapid rejection of skin grafts is caused by differences in a single genetic region. Because they control the compatibility of tissue grafts, these genes became known as 'histocompatibility genes.' Later, it was found that several closely linked, and highly **polymorphic** genes specify histocompatibility, which led to the term major histocompatibility complex. The central role of the MHC in antigen recognition by T cells, which we shall discuss in Chapter 5, was discovered later still, revealing the true physiological function of the proteins encoded by the MHC. This, in turn, led to an explanation for the major effect on tissue compatibility for which they were named. We shall discuss these diverse functions of MHC molecules in Chapters 4, 5, 7, 8, and 13.

1-17   Two major types of T cell recognize peptides bound to proteins of two different classes of MHC molecule.

There are two types of MHC molecule, called MHC class I and MHC class II. These differ in several subtle ways but share most of their major structural features. The most important of these is formed by the two outer extracellular domains of the molecule, which combine to create a long cleft in which a single peptide fragment is trapped during the synthesis and assembly of the MHC molecule inside the cell. The MHC molecule bearing its cargo of peptide is then transported to the cell surface, where it displays the peptide to T cells (Fig. 1.27). The antigen receptors of T lymphocytes are specialized to recognize a foreign antigenic peptide fragment bound to an MHC molecule. A T cell with a receptor specific for the complex formed between that particular foreign peptide and MHC molecule can then recognize and respond to the antigen-presenting cell.

The most important differences between the two classes of MHC molecule lie not in their structure but in the source of the peptides that they trap and carry to the cell surface. **MHC class I molecules** collect peptides derived from proteins synthesized in the cytosol, and are thus able to display fragments of viral proteins on the cell surface (Fig. 1.28). **MHC class II molecules** bind peptides derived from proteins in intracellular vesicles, and thus display peptides derived from pathogens living in macrophage vesicles or internalized by phagocytic cells and B cells (Fig. 1.29). We shall see in Chapter 5 exactly how peptides from these different sources are made accessible to the two types of MHC molecule.

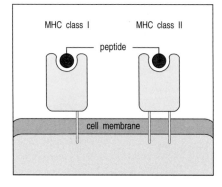

Fig. 1.27 MHC molecules on the cell surface display peptide fragments of antigens. MHC molecules are membrane proteins whose outer extracellular domains form a cleft in which a peptide fragment is bound. These fragments, which are derived from proteins degraded inside the cell, including foreign protein antigens, are bound by the newly synthesized MHC molecule before it reaches the surface. There are two kinds of MHC molecule—MHC class I and MHC class II—with related but distinct structures and functions.

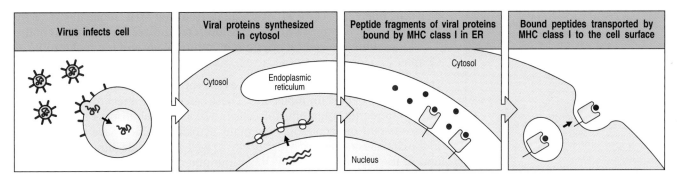

Fig. 1.28 MHC class I molecules present antigen derived from proteins in the cytosol. In cells infected with viruses, viral proteins are synthesized in the cytosol. Peptide fragments of viral proteins are transported into the endoplasmic reticulum (ER) where they are bound by MHC class I molecules, which then deliver the peptides to the cell surface.

Having reached the cell surface with their peptide cargo, the two classes of MHC molecule are recognized by different functional classes of T cell. MHC class I molecules bearing viral peptides are recognized by cytotoxic T cells, which then kill the infected cell (Fig. 1.30); MHC class II molecules bearing peptides derived from pathogens taken up into vesicles are recognized by T_H1 or T_H2 cells (Fig. 1.31).

The antigen-specific activation of these effector T cells is aided by co-receptors that distinguish between the two classes of MHC molecule; cytotoxic cells express the CD8 co-receptor that binds MHC class I molecules, whereas T_H1 and T_H2 cells express the CD4 co-receptor with specificity for MHC class II molecules. However, even before T cells have encountered the specific foreign antigen that activates them to differentiate into effector cells, they express the appropriate co-receptor to match their receptor specificity. The maturation of T cells into either CD8 or CD4 T cells reflects the testing of T-cell receptor specificity that occurs during development, and the selection

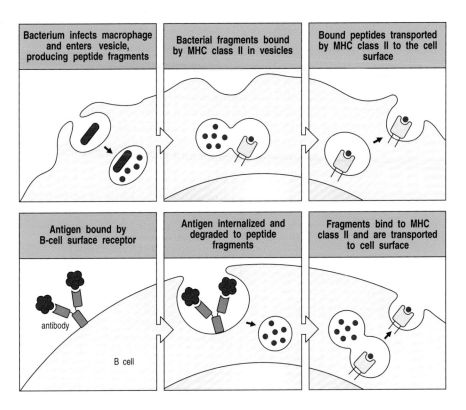

Fig. 1.29 MHC class II molecules present antigen originating in intracellular vesicles. Some bacteria infect cells and grow in intracellular vesicles. Peptides derived from such bacteria are bound by MHC class II molecules and transported to the cell surface (top row). MHC class II molecules also bind and transport peptides derived from antigen that has been bound and internalized by B-cell antigen receptor-mediated uptake into intracellular vesicles (bottom row).

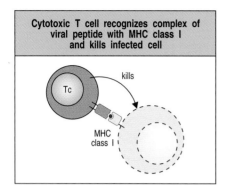

Fig. 1.30 Cytotoxic T cells recognize antigen presented by MHC class I molecules and kill the cell. The peptide:MHC class I complex on virus-infected cells is detected by antigen-specific cytotoxic T cells. Cytotoxic T cells are preprogrammed to kill the cells they recognize.

of T cells that can receive survival signals from self MHC molecules. Exactly how this selective process works, and how it maximizes the usefulness of the T cell repertoire is a central question in immunology and is a major topic of Chapter 7.

On recognizing their targets, the three types of T cell are stimulated to release different sets of effector molecules. These can directly affect their target cells or help to recruit other effector cells in ways we shall discuss in Chapter 8. These effector molecules include many cytokines, which have a crucial role in the clonal expansion of lymphocytes as well as in innate immune responses and in the effector actions of most immune cells; thus, understanding the actions of cytokines is central to understanding the various behaviors of the immune system.

1-18 Defects in the immune system result in increased susceptibility to infection.

We tend to take for granted the ability of our immune systems to free our bodies of infection and prevent its recurrence. In some people, however, parts of the immune system fail. In the most severe of these **immuno-deficiency diseases**, adaptive immunity is completely absent, and death occurs in infancy from overwhelming infection unless heroic measures are taken. Other less catastrophic failures lead to recurrent infections with particular types of pathogen, depending on the particular deficiency. Much has been learned about the functions of the different components of the human immune system through the study of these immunodeficiencies.

More than twenty-five years ago, a devastating form of immunodeficiency appeared, the **acquired immune deficiency syndrome**, or **AIDS**, which is itself caused by an infectious agent. This disease destroys the T cells that activate macrophages, leading to infections caused by intracellular bacteria and other pathogens normally controlled by these T cells. Such infections are the major cause of death from this increasingly prevalent immuno-deficiency disease, which is discussed fully in Chapter 11 together with inherited immunodeficiencies.

AIDS is caused by a virus, the **human immunodeficiency virus**, or **HIV**, that has evolved several strategies by which it not only evades but also subverts the protective mechanisms of the adaptive immune response. Such strategies are typical of many successful pathogens and we shall examine a variety of them in Chapter 11. The conquest of many of the world's leading diseases, including malaria and diarrheal diseases (the leading killers of children) as well as the more recent threat from AIDS, depends upon a better under-standing of the pathogens that cause them and their interactions with the cells of the immune system.

Fig. 1.31 T$_H$1 and helper T cells recognize antigen presented by MHC class II molecules. On recognition of their specific antigen on infected macrophages, T$_H$1 cells activate the macrophage, leading to the destruction of the intracellular bacteria (left panel). When helper T cells recognize antigen on B cells, they activate these cells to proliferate and differentiate into antibody-producing plasma cells (right panel).

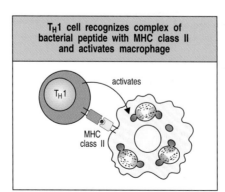

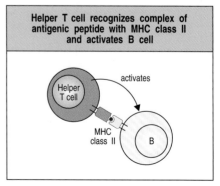

1-19 Understanding adaptive immune responses is important for the control of allergies, autoimmune disease, and organ graft rejection.

Many medically important diseases are associated with a normal immune response directed against an inappropriate antigen, often in the absence of infectious disease. Immune responses directed at noninfectious antigens occur in **allergy**, where the antigen is an innocuous foreign substance, in **autoimmune disease**, where the response is to a self antigen, and in **graft rejection**, where the antigen is borne by a transplanted foreign cell. What we call a successful immune response or a failure, and whether the response is considered harmful or beneficial to the host, depends not on the response itself but rather on the nature of the antigen (Fig. 1.32).

Allergic diseases, which include asthma, are an increasingly common cause of disability in the developed world, and many other important diseases are now recognized as autoimmune. An autoimmune response directed against pancreatic β cells is the leading cause of diabetes in the young. In allergies and autoimmune diseases, the powerful protective mechanisms of the adaptive immune response cause serious damage to the patient.

Immune responses to harmless antigens, to body tissues, or to organ grafts are, like all other immune responses, highly specific. At present, the usual way to treat these responses is with **immunosuppressive drugs**, which inhibit all immune responses, desirable or undesirable. If it were possible to suppress only those lymphocyte clones responsible for the unwanted response, the disease could be cured or the grafted organ protected without impeding protective immune responses. There is hope that this dream of antigen-specific **immunoregulation** to control unwanted immune responses could become a reality, since antigen-specific suppression of immune responses can be induced experimentally, although the molecular basis of this suppression is not fully understood. We shall see in Chapter 10 how the mechanisms of immune regulation are beginning to emerge from a better understanding of the functional subsets of lymphocytes and the cytokines that control them, and we shall discuss the present state of understanding of allergies, autoimmune disease, graft rejection, and immunosuppressive drugs in Chapters 12, 13, and 14.

Antigen	Effect of response to antigen	
	Normal response	Deficient response
Infectious agent	Protective immunity	Recurrent infection
Innocuous substance	Allergy	No response
Grafted organ	Rejection	Acceptance
Self organ	Autoimmunity	Self tolerance
Tumor	Tumor immunity	Cancer

Fig. 1.32 Immune responses can be beneficial or harmful depending on the nature of the antigen. Beneficial responses are shown in white, harmful responses in shaded boxes. Where the response is beneficial, its absence is harmful.

1-20   Vaccination is the most effective means of controlling infectious diseases.

Although the specific suppression of immune responses must await advances in basic research on immune regulation and its application, the deliberate stimulation of an immune response by immunization, or vaccination, has achieved many successes in the two centuries since Jenner's pioneering experiment.

Mass immunization programs have led to the virtual eradication of several diseases that used to be associated with significant morbidity (illness) and mortality (Fig. 1.33). Immunization is considered so safe and so important that most states in the United States require children to be immunized against up to seven common childhood diseases. Impressive as these accomplishments are, there are still many diseases for which we lack effective vaccines. And even where vaccines for diseases such as measles or polio can be used effectively in developed countries, technical and economic problems can prevent their widespread use in developing countries, where mortality from these diseases is still high. The tools of modern immunology and molecular biology are being applied to develop new vaccines and improve old

Fig. 1.33 Successful vaccination campaigns. Diphtheria, polio, and measles and its consequences have been virtually eliminated in the United States, as shown in these three graphs. SSPE stands for subacute sclerosing panencephalitis, a brain disease that is a late consequence of measles infection in a few patients. When measles was prevented, SSPE disappeared 15–20 years later. However, as these diseases have not been eradicated worldwide, immunization must be maintained in a very high percentage of the population to prevent their reappearance.

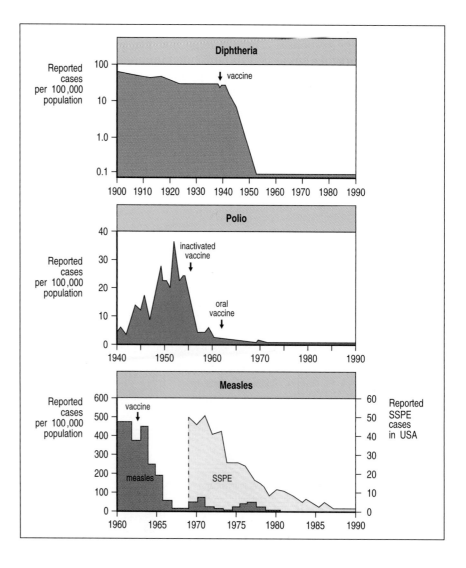

ones, and we shall discuss these advances in Chapter 14. The prospect of controlling these important diseases is tremendously exciting. The guarantee of good health is a critical step toward population control and economic development. At a cost of pennies per person, great hardship and suffering can be alleviated.

Summary.

Lymphocytes have two distinct recognition systems specialized for detection of extracellular and intracellular pathogens. B cells have cell-surface immunoglobulin molecules as receptors for antigen and, upon activation, secrete the immunoglobulin as soluble antibody that provides defense against pathogens in the extracellular spaces of the body. T cells have receptors that recognize peptide fragments of intracellular pathogens transported to the cell surface by the glycoproteins of the major histocompatibility complex (MHC). Two classes of MHC molecule transport peptides from different intracellular compartments to present them to distinct types of effector T cell: cytotoxic T cells that kill infected target cells, and T_H1 cells and helper T cells that mainly activate macrophages and B cells, respectively. Thus, T cells are crucially important for both the humoral and cell-mediated responses of adaptive immunity. The adaptive immune response seems to have engrafted specific antigen recognition by highly diversified receptors onto innate defense systems, which have a central role in the effector actions of both B and T lymphocytes. The vital role of adaptive immunity in fighting infection is illustrated by the immunodeficiency diseases and the problems caused by pathogens that succeed in evading or subverting an adaptive immune response. The antigen-specific suppression of adaptive immune responses is the goal of treatment for important human diseases involving inappropriate activation of lymphocytes, whereas the specific stimulation of an adaptive immune response is the basis of successful vaccination.

Summary to Chapter 1.

The immune system defends the host against infection. Innate immunity serves as a first line of defense but lacks the ability to recognize certain pathogens and to provide the specific protective immunity that prevents reinfection. Adaptive immunity is based on clonal selection from a repertoire of lymphocytes bearing highly diverse antigen-specific receptors that enable the immune system to recognize any foreign antigen. In the adaptive immune response, antigen-specific lymphocytes proliferate and differentiate into effector cells that eliminate pathogens. Host defense requires different recognition systems and a wide variety of effector mechanisms to seek out and destroy the wide variety of pathogens in their various habitats within the body and at its surface. Not only can the adaptive immune response eliminate a pathogen but, in the process, it also generates increased numbers of differentiated memory lymphocytes through clonal selection, and this allows a more rapid and effective response upon reinfection. The regulation of immune responses, whether to suppress them when unwanted or to stimulate them in the prevention of infectious disease, is the major medical goal of research in immunology.

General references.

Historical background

Burnet, F.M.: *The Clonal Selection Theory of Acquired Immunity.* London, Cambridge University Press, 1959.

Gowans, J.L.: **The Lymphocyte—a disgraceful gap in medical knowledge.** *Immunol. Today* 1996, **17**:288-291.

Landsteiner, K.: *The Specificity of Serological Reactions,* 3rd edn. Boston, Harvard University Press, 1964.

Metchnikoff, E.: *Immunity in the Infectious Diseases,* 1st edn. New York, Macmillan Press, 1905.

Silverstein, A.M.: *History of Immunology,* 1st edn. London, Academic Press, 1989.

Biological background

Alberts, B., Bray, D., Lewis, J., Raff, M., Roberts, K., and Watson, J.D.: *Molecular Biology of the Cell,* 3rd edn. New York, Garland Publishing, 1994.

Lodish, H., Berk, A., Zipursky, S.L., Matsudaira, P., Baltimore, D., and Darnell, J.: *Molecular Cell Biology,* 4th edn. New York, W.H. Freeman and Company, 2000.

Ryan, K.J., (ed): *Medical Microbiology,* 3rd edn. East Norwalk, CT, Appleton-Lange, 1994.

Stryer, L.: *Biochemistry,* 4th edn. New York, Freeman, 1995.

Primary journals devoted solely or primarily to immunology

Autoimmunity
Clinical and Experimental Immunology
Comparative and Developmental Immunology
European Journal of Immunology
Immunity
Immunogenetics
Immunology
Infection and Immunity
International Immunology
Journal of Autoimmunity
Journal of Experimental Medicine
Journal of Immunology
Nature Immunology
Regional Immunology
Thymus

Primary journals with frequent papers in immunology

Cell
Current Biology
EMBO Journal
Journal of Biological Chemistry
Journal of Cell Biology
Journal of Clinical Investigation
Molecular Cell Biology
Nature
Nature Cell Biology
Proceedings of the National Academy of Sciences, USA
Science

Review journals in immunology

Annual Reviews in Immunology
Contemporary Topics in Microbiology and Immunology
Current Opinion in Immunology
Immunogenetics Reviews
Immunological Reviews
Immunology Today
Proceedings of the International Congress of Immunology, **1-10**, *1971-1992, 1998.*
Research in Immunology
Seminars in Immunology
The Immunologist
The Immunologist, **3**: *Proceedings Issue, 9th International Congress of Immunology*

Advanced textbooks in immunology, compendia, etc.

Lachmann, P.J., Peters, D.K., Rosen, F.S., Walport, M.J. (eds): *Clinical Aspects of Immunology,* 5th edn. Oxford, Blackwell Scientific Publications, 1993.

Mak, T.W. and Simard, J.J.L.: *Handbook of Immune Response Genes.* New York, Plenum Press, 1998.

Paul, W.E. (ed): *Fundamental Immunology,* 4th edn. New York, Raven Press, 1998.

Roitt, I.M., and Delves, P.J. (eds): *Encyclopedia of Immunology,* 3rd edn. London/San Diego, Academic Press, 1992.

Rosen, F.S., and Geha, R.S.: *Case Studies in Immunology: A Clinical Companion,* 3rd edn. New York, Garland Publishing 2001.

Innate Immunity

Throughout this book we will examine the individual mechanisms by which the adaptive immune response acts to protect the host from pathogenic infectious agents. In this chapter, however, we will examine the role of those innate, nonadaptive defenses that form early barriers to infectious disease. The microorganisms that are encountered daily in the life of a normal healthy individual only occasionally cause perceptible disease. Most are detected and destroyed within minutes or hours by defense mechanisms that do not require a prolonged period of induction because they do not rely on the clonal expansion of antigen-specific lymphocytes: these are the mechanisms of **innate immunity**.

The time course and different phases of an encounter with a new pathogen are summarized in Fig. 2.1. The innate immune mechanisms act immediately, and are followed by **early induced responses**, which can be activated by infection but do not generate lasting protective immunity. Only if an infectious organism can breach these early lines of defense will an **adaptive immune response** ensue, with the generation of antigen-specific effector cells that specifically target the pathogen, and memory cells that can prevent reinfection with the same microorganism. The power of adaptive immune responses is due to their antigen specificity, which we will be studying in the following chapters. However, they harness, and also depend upon, many of the effector mechanisms used by the innate immune system, which we will describe in this chapter.

Whereas the adaptive immune system uses a large repertoire of receptors encoded by rearranging genes to recognize a huge variety of antigens (see Section 1-10), innate immunity depends upon germline-encoded receptors to recognize features that are common to many pathogens. In fact, as we will see, the mechanisms of innate immunity discriminate very effectively between host cells and pathogen surfaces, and this ability to discriminate between self and nonself, and to recognize broad classes of pathogens, contributes to the induction of an appropriate adaptive immune response.

In the first part of the chapter we will consider the fixed defenses of the body: the epithelia that line the internal and external surfaces of the body, and the phagocytes that can engulf and digest invading microorganisms. As well as killing microorganisms, the activities of some of these phagocytes induce the next phase of the early response, and ultimately, if the infection is not cleared, the adaptive immune response. The second part of the chapter is devoted to a system of plasma proteins known as the complement system. This important element of innate immunity interacts with microorganisms to promote their removal by phagocytic cells. Next, we take a closer look at the receptors used by the immune system to recognize pathogens, and the last part of the chapter describes how the activation of phagocytic cells at the beginning of the innate immune response to infection leads to the induced or adaptive immune response.

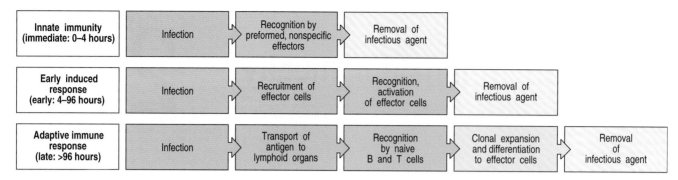

Fig. 2.1 The response to an initial infection occurs in three phases. The effector mechanisms that remove the infectious agent (for example, phagocytes and complement) are similar or identical in each phase, but the first two phases rely on recognition of pathogens by germline-encoded receptors of the innate immune system, whereas adaptive immunity uses variable antigen-specific receptors that are produced as a result of gene rearrangements. Adaptive immunity occurs late, because the rare B and T cells specific for the invading pathogen must undergo clonal expansion before they differentiate into effector cells that can clear the infection.

The front line of host defense.

Microorganisms that cause pathology in humans and animals enter the body at different sites and produce disease by a variety of mechanisms. Many different infectious agents can cause pathology, and those that do are referred to as **pathogenic microorganisms** or **pathogens**. Invasions by microorganisms are initially countered, in all vertebrates, by innate defense mechanisms that preexist in all individuals and act within minutes of infection. Only when the innate host defenses are bypassed, evaded, or overwhelmed is an induced or adaptive immune response required. In the first part of this chapter we will describe briefly the infectious strategies of microorganisms before examining the innate host defenses that, in most cases, prevent infection from becoming established. Thus we will look at the defense functions of the epithelial surfaces of the body, the role of antimicrobial peptides and proteins, and the defense of body tissues by macrophages and neutrophils, which bind and ingest invading microorganisms in a process known as **phagocytosis**.

2-1   Infectious agents must overcome innate host defenses to establish a focus of infection.

Our bodies are constantly exposed to microorganisms present in the environment, including infectious agents that have been shed from infected individuals. Contact with these microorganisms may occur through external or internal epithelial surfaces: the respiratory tract mucosa provides a route of entry for airborne microorganisms, the gastrointestinal mucosa for microorganisms in food and water; insect bites and wounds allow micro-organisms to penetrate the skin; and direct contact between individuals offers opportunities for infection of the skin and reproductive mucosa (Fig. 2.2).

In spite of this exposure, infectious disease is fortunately quite rare. The epithelial surfaces of the body serve as an effective barrier against most microorganisms, and are rapidly repaired if wounded. Furthermore, most of the microorganisms that do succeed in crossing the epithelial surfaces are efficiently removed by innate immune mechanisms that function in the

Routes of infection for pathogens			
Route of entry	**Mode of transmission**	**Pathogen**	**Disease**
Mucosal surfaces			
Airway	Inhaled droplets	Influenza virus *Neisseria meningitidis*	Influenza Meningococcal meningitis
Gastrointestinal tract	Contaminated water or food	*Salmonella typhi* Rotavirus	Typhoid fever Diarrhea
Reproductive tract	Physical contact	*Treponema pallidum*	Syphilis
External epithelia			
External surface	Physical contact	*Tinea pedis*	Athlete's foot
Wounds and abrasions	Minor skin abrasions Puncture wounds Handling infected animals	*Bacillus anthracis* *Clostridium tetani* *Pasteurella tularensis*	Anthrax Tetanus Tularemia
Insect bites	Mosquito bites (*Aedes aegypti*) Deer tick bites Mosquito bites (*Anopheles*)	Flavivirus *Borrelia burgdorferi* *Plasmodium* spp	Yellow fever Lyme disease Malaria

Fig. 2.2 Pathogens infect the body through a variety of routes.

underlying tissues. Thus in most cases these defenses, which we will examine in more detail in subsequent sections, prevent a site of infection from being established. It is difficult to know how many infections are repelled in this way, because there are no symptoms of disease. It is clear, however, that the microorganisms that a normal human being inhales or ingests, or that enter through minor wounds, are mostly held at bay or eliminated, since they seldom cause disease.

Infectious disease occurs when a microorganism succeeds in evading or overwhelming innate host defenses to establish a local site of infection and replication that allows its further transmission. In some cases, such as athlete's foot, the initial infection remains local and does not cause significant pathology. In other cases the infectious agent causes significant pathology as it spreads through the lymphatics or the bloodstream, or as a result of secreting toxins.

Pathogen spread is often countered by an inflammatory response that recruits more effector molecules and cells of the innate immune system from local blood vessels (Fig. 2.3), while inducing clotting farther downstream so that pathogens cannot spread through the blood. The induced responses of innate immunity act over several days while an adaptive immune response gets underway in response to pathogen antigens delivered to local lymphoid tissue. Such a response can target specific features of the pathogen and will usually clear the infection and protect the host against reinfection with the same pathogen.

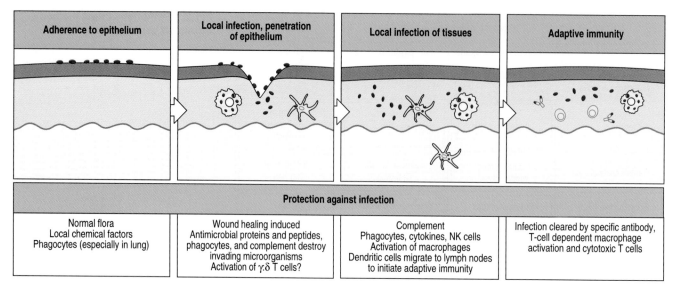

Adherence to epithelium	Local infection, penetration of epithelium	Local infection of tissues	Adaptive immunity
	Protection against infection		
Normal flora Local chemical factors Phagocytes (especially in lung)	Wound healing induced Antimicrobial proteins and peptides, phagocytes, and complement destroy invading microorganisms Activation of γ:δ T cells?	Complement Phagocytes, cytokines, NK cells Activation of macrophages Dendritic cells migrate to lymph nodes to initiate adaptive immunity	Infection cleared by specific antibody, T-cell dependent macrophage activation and cytotoxic T cells

Fig. 2.3 An infection and the response to it can be divided into a series of stages. These are illustrated here for an infectious microorganism entering through a wound in the skin. The infectious agent must first adhere to the epithelial cells and then cross the epithelium. A local innate immune response may prevent an infection from being established. If not, it helps to contain the infection and also delivers the infectious agent, carried in lymph and inside dendritic cells, to local lymph nodes. This initiates the adaptive immune response and eventual clearance of the infection. The role of γ:δ T cells is uncertain, as we will see in Section 2-28, and this is indicated by the question mark.

2-2 The epithelial surfaces of the body are the first defenses against infection.

Our body surfaces are defended by epithelia, which provide a physical barrier between the internal milieu and the external world that contains pathogens. Epithelial cells are held together by tight junctions, which effectively form a seal against the external environment. Epithelia comprise the skin and the linings of the body's tubular structures—the gastrointestinal, respiratory, and urinogenital tracts. Infections occur only when the pathogen can colonize or cross through these barriers, and since the dry, protective layers of the skin present a more formidable barrier, pathogen entry most often occurs through the internal epithelial surfaces. The importance of epithelia in protection against infection is obvious when the barrier is breached, as in wounds and burns, when infection is a major cause of mortality and morbidity. In the absence of wounding or disruption, pathogens normally cross epithelial barriers by binding to molecules on internal epithelial surfaces, or establish an infection by adhering to and colonizing these surfaces. This specific attachment allows the pathogen to infect the epithelial cell, or to damage it so that the epithelium can be crossed, or, in the case of colonizing pathogens, to avoid being dislodged by the flow of air or fluid across the epithelial surface. The internal epithelia are known as **mucosal epithelia** because they secrete a viscous fluid called mucus, which contains many glycoproteins called mucins. Microorganisms coated in mucus may be prevented from adhering to the epithelium, and in mucosal epithelia such as that of the respiratory tract, microorganisms can be expelled in the flow of mucus driven by the beating of epithelial cilia. The efficacy of mucus flow in clearing infection is illustrated by people with defective mucus secretion or inhibition of ciliary movement; they frequently develop lung infections caused by bacteria that colonize the epithelial surface. In the gut, peristalsis is an important mechanism for keeping both food and infectious agents moving through. Failure of peristalsis is typically accompanied by overgrowth of bacteria within the intestinal lumen.

Intrinsic epithelial barriers to infection	
Mechanical	Epithelial cells joined by tight junctions Longitudinal flow of air or fluid across epithelium Movement of mucus by cilia
Chemical	Fatty acids (skin) Enzymes: lysozyme (saliva, sweat, tears), pepsin (gut) Low pH (stomach) Antibacterial peptides; defensins (skin, gut), cryptidins (intestine)
Microbiological	Normal flora compete for nutrients and attachment to epithelium and can produce antibacterial substances

Fig. 2.4 Surface epithelia provide mechanical, chemical, and microbiological barriers to infection.

Our surface epithelia are more than mere physical barriers to infection; they also produce chemical substances that are microbicidal or inhibit microbial growth (Fig. 2.4). For example, the antibacterial enzyme lysozyme is secreted in tears and saliva. The acid pH of the stomach and the digestive enzymes of the upper gastrointestinal tract create a substantial chemical barrier to infection. Further down the intestinal tract, antibacterial and antifungal peptides called cryptidins or α-defensins are made by Paneth cells, which are resident in the base of the crypts in the small intestine beneath the epithelial stem cells. Related antimicrobial peptides, the β-defensins, are made by other epithelia, primarily in the skin and respiratory tract. Such antimicrobial peptides play a role in the immune defense of many organisms, including insects. They are cationic peptides that are thought to kill bacteria by damaging the bacterial cell membrane. Another type of antimicrobial protein is secreted into the fluid that bathes the epithelial surfaces of the lung. This fluid contains two proteins—surfactant proteins A and D—that bind to and coat the surfaces of pathogens so that they are more easily phagocytosed by macrophages that have left the subepithelial tissues to enter the alveoli of the lung. Coating of a particle with proteins that facilitate its phagocytosis is known as **opsonization** and we will meet several examples of this defense strategy in this chapter.

In addition to these defenses, most epithelial surfaces are associated with a normal flora of nonpathogenic bacteria that compete with pathogenic microorganisms for nutrients and for attachment sites on cells. The normal flora can also produce antimicrobial substances, such as the colicins (antibacterial proteins made by *Escherichia coli*) that prevent colonization by other bacteria. When the nonpathogenic bacteria are killed by antibiotic treatment, pathogenic microorganisms frequently replace them and cause disease.

2-3 After entering tissues, many pathogens are recognized, ingested, and killed by phagocytes.

If a microorganism crosses an epithelial barrier and begins to replicate in the tissues of the host, it is, in most cases, immediately recognized by the mononuclear phagocytes, or **macrophages**, that reside in tissues. Macrophages mature continuously from circulating monocytes that leave the circulation to migrate into tissues throughout the body (see Fig. 1.3). They are found in especially large numbers in connective tissue, in association with the gastrointestinal tract, in the lung (where they are found in both the interstitium and the alveoli), along certain blood vessels in the liver (where they are known as Kupffer cells), and throughout the spleen, where they remove senescent blood cells. The second major family of phagocytes—the **neutrophils**, or polymorphonuclear neutrophilic leukocytes (PMNs or polys)—are short-lived cells that are abundant in the blood but are not present

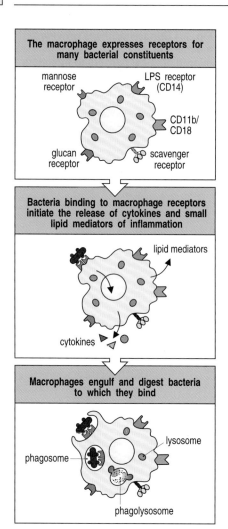

The macrophage expresses receptors for many bacterial constituents

mannose receptor

LPS receptor (CD14)

CD11b/ CD18

glucan receptor

scavenger receptor

Bacteria binding to macrophage receptors initiate the release of cytokines and small lipid mediators of inflammation

lipid mediators

cytokines

Macrophages engulf and digest bacteria to which they bind

lysosome

phagosome

phagolysosome

Fig. 2.5 Phagocytes bear several different receptors that recognize microbial components and induce phagocytosis. The figure illustrates five such receptors on macrophages—CD14, CD11b/CD18 (CR3), the macrophage mannose receptor, the scavenger receptor, and the glucan receptor, all of which bind bacterial carbohydrates. CD14 and CR3 are specific for bacterial lipopolysaccharide (LPS). The designation CD stands for 'cluster of differentiation,' a historical term that was coined to define cell-surface molecules that are recognized by a given set of monoclonal antibodies. This 'cluster of differentiation' then received a number, for example CD4, which stands only for the order of discovery. In general, each CD is associated with one or more functions, which were discovered through the effects on cell or tissue function of the antibodies that defined it. See Appendix II for a full listing of CDs.

in normal, healthy tissues. Both these phagocytic cells have a key role in innate immunity because they can recognize, ingest, and destroy many pathogens without the aid of an adaptive immune response. Macrophages are the first to encounter pathogens in the tissues but they are soon reinforced by the recruitment of large numbers of neutrophils to sites of infection.

Macrophages and neutrophils recognize pathogens by means of cell-surface receptors that can discriminate between the surface molecules displayed by pathogens and those of the host. These receptors, which we will examine in more detail later, include the macrophage mannose receptor, which is found on macrophages but not on monocytes or neutrophils, scavenger receptors, which bind many charged ligands, and CD14, a receptor for bacterial lipopolysaccharide (LPS) found predominantly on monocytes and macrophages (Fig. 2.5). Pathogens can also interact with macrophages and neutrophils through receptors for complement borne on these cells. As we will see in the second part of the chapter, the complement system is activated rapidly in response to many types of infection, producing complement proteins that opsonize the surface of pathogens as they enter the tissues.

Ligation of many of the cell-surface receptors that recognize pathogens leads to phagocytosis of the pathogen, followed by its death inside the phagocyte. Phagocytosis is an active process, in which the bound pathogen is first surrounded by the phagocyte membrane and then internalized in a membrane-bounded vesicle known as a **phagosome**, which becomes acidified. In addition to being phagocytic, macrophages and neutrophils have granules, called **lysosomes**, that contain enzymes, proteins, and peptides that can mediate an intracellular antimicrobial response. The phagosome fuses with one or more lysosomes to generate a **phagolysosome** in which the lysosomal contents are released to destroy the pathogen (see Fig. 2.5).

Upon phagocytosis, macrophages and neutrophils also produce a variety of other toxic products that help kill the engulfed microorganism (Fig. 2.6). The most important of these are hydrogen peroxide (H_2O_2), the superoxide anion (O_2^-), and nitric oxide (NO), which are directly toxic to bacteria. They are generated by lysosomal NADPH oxidases and other enzymes in a process known as the **respiratory burst**, as it is accompanied by a transient increase in oxygen consumption. Neutrophils are short-lived cells, dying soon after they have accomplished a round of phagocytosis. Dead and dying neutrophils are a major component of the **pus** that forms in some infections; bacteria that give rise to such infections are thus known as **pyogenic bacteria**. Macrophages, on the other hand, are long-lived and continue to generate new lysosomes. Patients with chronic granulomatous disease have a genetic deficiency of NADPH oxidase, which means that their phagocytes do not produce toxic oxygen derivatives and are less able to kill ingested microorganisms and clear the infection. People with this defect are unusually susceptible to bacterial and fungal infections, especially in infancy.

Macrophages can make this response immediately on encountering an infecting microorganism and this can be sufficient to prevent an infection from becoming established. The great cellular immunologist **Elie Metchnikoff** believed that the innate response of macrophages encompassed all host defense and, indeed, it is now clear that invertebrates, such as the sea star that he was studying, rely entirely on innate immunity for their defense against infection. Although this is not the case in humans and other vertebrates, the innate response of macrophages still provides an important front line of host defense that must be overcome if a microorganism is to establish an infection that can be passed on to a new host.

A key feature that distinguishes pathogenic from nonpathogenic microorganisms is their ability to overcome innate immune defenses. Pathogens have developed a variety of strategies to avoid being immediately destroyed

Class of mechanism	Specific products
Acidification	pH=~3.5 – 4.0, bacteriostatic or bactericidal
Toxic oxygen-derived products	Superoxide O_2^-, hydrogen peroxide H_2O_2, singlet oxygen $^1O_2^{\cdot}$, hydroxyl radical $OH^{\cdot}$, hypohalite OCl
Toxic nitrogen oxides	Nitric oxide NO
Antimicrobial peptides	Defensins and cationic proteins
Enzymes	Lysozyme—dissolves cell walls of some gram-positive bacteria. Acid hydrolases—further digest bacteria
Competitors	Lactoferrin (binds Fe) and vitamin B_{12}-binding protein

Fig. 2.6 Bactericidal agents produced or released by phagocytes on the ingestion of microorganisms. Most of these agents are made by both macrophages and neutrophils. Some of them are toxic; others, such as lactoferrin, work by binding essential nutrients and preventing their uptake by the bacteria. The same substances can be released by phagocytes interacting with large antibody-coated surfaces such as parasitic worms or host tissues. As these agents are also toxic to host cells, phagocyte activation can cause extensive tissue damage during an infection.

by macrophages. Many extracellular pathogenic bacteria coat themselves with a thick polysaccharide capsule that is not recognized by any phagocyte receptor. Other pathogens, for example mycobacteria, have evolved ways to grow inside macrophage phagosomes by inhibiting fusion with a lysosome. Without such devices, a microorganism must enter the body in sufficient numbers to simply overwhelm the immediate innate host defenses and establish a focus of infection.

The second important effect of the interaction between pathogens and tissue macrophages is activation of macrophages to release cytokines and other mediators that set up a state of inflammation in the tissue and bring neutrophils and plasma proteins to the site of infection. It is thought that the pathogen induces cytokine secretion by signals delivered through some of the receptors to which it binds, and we will see later how this occurs in response to LPS. Receptors that signal the presence of pathogens and induce cytokines also have another important role. This is to induce the expression of so-called co-stimulatory molecules on both macrophages and **dendritic cells**, another type of phagocytic cell present in tissues, thus enabling these cells to initiate an adaptive immune response (see Section 1-6).

The cytokines released by macrophages make an important contribution both to local inflammation and to other induced but nonadaptive responses that occur in the first few days of a new infection. We will be describing the role of individual cytokines in these induced responses in the last part of this chapter. However, since an inflammatory response is usually initiated within minutes of infection or wounding, we will outline here how it occurs and how it contributes to host defense.

2-4 Pathogen recognition and tissue damage initiate an inflammatory response.

Inflammation plays three essential roles in combating infection. The first is to deliver additional effector molecules and cells to sites of infection to augment the killing of invading microorganisms by the front-line macrophages. The second is to provide a physical barrier preventing the spread of infection, and the third is to promote the repair of injured tissue, a nonimmunological role that we will not discuss further. Inflammation at the site of infection is initiated by the response of macrophages to pathogens.

Inflammatory responses are operationally characterized by pain, redness, heat, and swelling at the site of an infection, reflecting three types of change in the local blood vessels. The first is an increase in vascular diameter, leading to increased local blood flow—hence the heat and redness—and a reduction in the velocity of blood flow, especially along the surfaces of small blood vessels. The second change is that the endothelial cells lining the blood vessel are activated to express adhesion molecules that promote the binding of circulating leukocytes. The combination of slowed blood flow and induced adhesion molecules allows leukocytes to attach to the endothelium and migrate into the tissues, a process known as extravasation, which we will describe in detail later. All these changes are initiated by the cytokines produced by activated macrophages. Once inflammation has begun, the first cells attracted to the site of infection are generally neutrophils. They are followed by monocytes, which differentiate into more tissue macrophages. In the later stages of inflammation, other leukocytes such as eosinophils and lymphocytes also enter the infected site. The third major change in the local blood vessels is an increase in vascular permeability. Instead of being tightly joined together, the endothelial cells lining the blood vessel walls become separated, leading to exit of fluid and proteins from the blood and their local accumulation in the tissue. This accounts for the swelling, or **edema**, and pain as well as the accumulation of plasma proteins that aid in host defense.

These changes are induced by a variety of inflammatory mediators released as a consequence of the recognition of pathogens. These include the lipid mediators of inflammation—**prostaglandins, leukotrienes**, and **platelet-activating factor** (PAF)—which are rapidly produced by macrophages through enzymatic pathways that degrade membrane phospholipids. Their actions are followed by those of the cytokines and chemokines (chemoattractant cytokines) that are synthesized and secreted by macrophages in response to pathogens. The cytokine **tumor necrosis factor-α** (TNF-α), for example, is a potent activator of endothelial cells.

As we will see in the next part of the chapter, another way in which pathogen recognition rapidly triggers an inflammatory response is through activation of the complement cascade. One of the cleavage products of the complement reaction is a peptide called C5a. C5a is a potent mediator of inflammation, with several different activities. In addition to increasing vascular permeability and inducing the expression of some adhesion molecules, it acts as a powerful chemoattractant for neutrophils and monocytes, and activates phagocytes and local **mast cells**, which are in turn stimulated to release granules containing the inflammatory molecule histamine and TNF-α.

If wounding has occurred, the injury to blood vessels immediately triggers two other protective enzyme cascades. The **kinin system** is an enzymatic cascade of plasma proteins that is triggered by tissue damage to produce several inflammatory mediators, including the vasoactive peptide **bradykinin**. This causes an increase in vascular permeability that promotes the influx of plasma proteins to the site of tissue injury. It also causes pain, which, although unpleasant to the victim, draws attention to the problem and leads to immobilization of the affected part of the body, which helps to limit the spread of any infectious agents. The **coagulation system** is another enzymatic cascade of plasma enzymes that is triggered following damage to blood vessels. This leads to the formation of a clot, which prevents any microorganisms from entering the bloodstream. Both these cascades have an important role in the inflammatory response to pathogens even if wounding or gross tissue injury has not occurred, as they are also triggered by endothelial cell activation. Thus, within minutes of the penetration of tissues by a pathogen, the inflammatory response causes an influx of proteins and cells that will control the infection. It also forms a physical barrier to limit the spread of infection and makes the host fully aware of what is going on.

Summary.

The mammalian body is susceptible to infection by many pathogens, which must first make contact with the host and then establish a focus of infection in order to cause disease. These pathogens differ greatly in their lifestyles, the structures of their surfaces, and means of pathogenesis, which therefore requires an equally diverse set of defensive responses from the host immune system. The first phase of host defense consists of those mechanisms that are present and ready to resist an invader at any time. The epithelial surfaces of the body keep pathogens out, and protect against colonization and against viruses and bacteria that enter through specialized cell-surface interactions, by preventing pathogen adherence and by secreting antimicrobial enzymes and peptides. Bacteria, viruses, and parasites that overcome this barrier are faced immediately by tissue macrophages equipped with surface receptors that can bind and phagocytose many different types of pathogen. This, in turn, leads to an inflammatory response, which causes the accumulation of plasma proteins, including the complement components that provide circulating or humoral innate immunity, as will be described in the next part of the chapter, and phagocytic neutrophils at the site of infection. Innate immunity provides a front line of host defense through effector mechanisms that engage the pathogen directly, act immediately on contact with it, and are unaltered in their ability to resist a subsequent challenge with either the same or a different pathogen. These mechanisms often succeed in preventing an infection from becoming established. If not, they are reinforced through the recruitment and increased production of further effector molecules and cells in a series of induced responses that we will consider later in this chapter. These induced innate responses often fail to clear the infection. In that case, macrophages and other cells activated in the early innate response help to initiate the development of an adaptive immune response.

The complement system and innate immunity.

Complement was discovered many years ago as a heat-labile component of normal plasma that augments the opsonization of bacteria by antibodies and allows antibodies to kill some bacteria. This activity was said to 'complement' the antibacterial activity of antibody, hence the name. Although first discovered as an effector arm of the antibody response, complement can also be activated early in infection in the absence of antibodies. Indeed, it now seems clear that complement first evolved as part of the innate immune system, where it still plays an important role.

The **complement system** is made up of a large number of distinct plasma proteins that react with one another to opsonize pathogens and induce a series of inflammatory responses that help to fight infection. A number of complement proteins are proteases that are themselves activated by proteolytic cleavage. Such enzymes are called zymogens and were first found in the gut. The digestive enzyme pepsin, for example, is stored inside cells and secreted as an inactive precursor enzyme, pepsinogen, which is only cleaved to pepsin in the acid environment of the stomach. The advantage to the host of not being autodigested is obvious.

In the case of the complement system, the precursor zymogens are widely distributed throughout body fluids and tissues without adverse effect. At sites of infection, however, they are activated locally and trigger a series of potent inflammatory events. The complement system activates through a triggered-enzyme cascade. In such a cascade, an active complement enzyme generated

Fig. 2.7 Schematic overview of the complement cascade. There are three pathways of complement activation: the classical pathway, which is triggered by antibody or by direct binding of complement component C1q to the pathogen surface; the MB-lectin pathway, which is triggered by mannan-binding lectin, a normal serum constituent that binds some encapsulated bacteria; and the alternative pathway, which is triggered directly on pathogen surfaces. All of these pathways generate a crucial enzymatic activity that, in turn, generates the effector molecules of complement. The three main conse-quences of complement activation are opsonization of pathogens, the recruitment of inflammatory cells, and direct killing of pathogens.

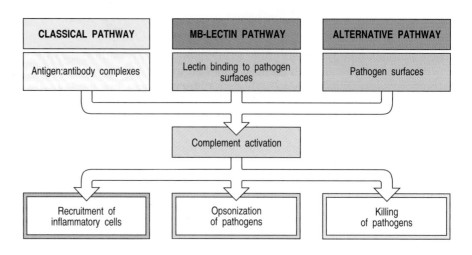

by cleavage of its zymogen precursor then cleaves its substrate, another complement zymogen, to its active enzymatic form. This in turn cleaves and activates the next zymogen in the complement pathway. In this way, the activation of a small number of complement proteins at the start of the pathway is hugely amplified by each successive enzymatic reaction, resulting in the rapid generation of a disproportionately large complement response. As might be expected, there are many regulatory mechanisms to prevent uncontrolled complement activation. The blood coagulation system is another example of a triggered-enzyme cascade. In this case, a small injury to a blood vessel wall can lead to the development of a large thrombus.

There are three distinct pathways through which complement can be activated on pathogen surfaces. These pathways depend on different molecules for their initiation, but they converge to generate the same set of effector molecules (Fig. 2.7). There are three ways in which the complement system protects against infection. First, it generates large numbers of activated complement proteins that bind covalently to pathogens, opsonizing them for engulfment by phagocytes bearing receptors for complement. Second, the small fragments of some complement proteins act as chemoattractants to recruit more phagocytes to the site of complement activation, and also to activate these phagocytes. Third, the terminal complement components damage certain bacteria by creating pores in the bacterial membrane.

2-5 Complement is a system of plasma proteins that interacts with pathogens to mark them for destruction by phagocytes.

In the early phases of an infection, the complement cascade can be activated on the surface of a pathogen through any one, or more, of the three pathways shown in Fig. 2.8. The **classical pathway** can be initiated by the binding of C1q, the first protein in the complement cascade, directly to the pathogen surface. It can also be activated during an adaptive immune response by the binding of C1q to antibody:antigen complexes, and is thus a key link between the effector mechanisms of innate and adaptive immunity. The **mannan-binding lectin pathway** (**MB-lectin pathway**) is initiated by binding of the mannan-binding lectin, a serum protein, to mannose-containing carbo-hydrates on bacteria or viruses. Finally, the **alternative pathway** can be initiated when a spontaneously activated complement component binds to the surface of a pathogen. Each pathway follows a sequence of reactions to generate a protease called a **C3 convertase**. These reactions are known as the 'early' events of complement activation, and consist of triggered-enzyme cascades

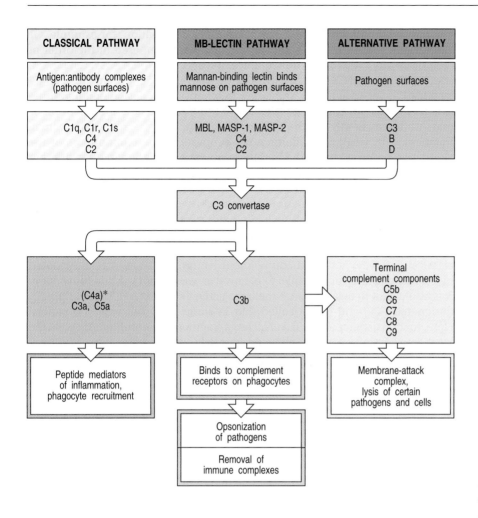

CLASSICAL PATHWAY	MB-LECTIN PATHWAY	ALTERNATIVE PATHWAY
Antigen:antibody complexes (pathogen surfaces)	Mannan-binding lectin binds mannose on pathogen surfaces	Pathogen surfaces
C1q, C1r, C1s C4 C2	MBL, MASP-1, MASP-2 C4 C2	C3 B D

C3 convertase

(C4a)* C3a, C5a	C3b	Terminal complement components C5b C6 C7 C8 C9
Peptide mediators of inflammation, phagocyte recruitment	Binds to complement receptors on phagocytes	Membrane-attack complex, lysis of certain pathogens and cells
	Opsonization of pathogens	
	Removal of immune complexes	

Fig. 2.8 Overview of the main components and effector actions of complement. The early events of all three pathways of complement activation involve a series of cleavage reactions that culminate in the formation of an enzymatic activity called a C3 convertase, which cleaves complement component C3 into C3b and C3a. The production of the C3 convertase is the point at which the three pathways converge and the main effector functions of complement are generated. C3b binds covalently to the bacterial cell membrane and opsonizes the bacteria, enabling phagocytes to internalize them. C3a is a peptide mediator of local inflammation. C5a and C5b are generated by cleavage of C5b by a C5 convertase formed by C3b bound to the C3 convertase (not shown in this simplified diagram). C5a is also a powerful peptide mediator of inflammation. C5b triggers the late events in which the terminal components of complement assemble into a membrane-attack complex that can damage the membrane of certain pathogens. C4a is generated by the cleavage of C4 during the early events of the classical pathway, and not by the action of C3 convertase, hence the *; it is also a peptide mediator of inflammation but its effects are relatively weak. Similarly, C4b, the large cleavage fragment of C4 (not shown), is a weak opsonin. Although the classical complement activation pathway was first discovered as an antibody-triggered pathway, it is now known that C1q can activate this pathway by binding directly to pathogen surfaces, as well as paralleling the MB-lectin activation pathway by binding to antibody that is itself bound to the pathogen surface. In the MB-lectin pathway, MASP stands for mannan-binding lectin-associated serine protease.

in which inactive complement zymogens are successively cleaved to yield two fragments, the larger of which is an active serine protease. The active protease is retained at the pathogen surface, and this ensures that the next complement zymogen in the pathway is also cleaved and activated at the pathogen surface. By contrast, the small peptide fragment is released from the site of the reaction and can act as a soluble mediator.

The C3 convertases formed by these early events of complement activation are bound covalently to the pathogen surface. Here they cleave C3 to generate large amounts of **C3b**, the main effector molecule of the complement system, and C3a, a peptide mediator of inflammation. The C3b molecules act as opsonins; they bind covalently to the pathogen and thereby target it for destruction by phagocytes equipped with receptors for C3b. C3b also binds the C3 convertase to form a **C5 convertase** that produces the most important small peptide mediator of inflammation, **C5a**, as well as a large active fragment, C5b, that initiates the 'late' events of complement activation. These comprise a sequence of polymerization reactions in which the terminal complement components interact to form a **membrane-attack complex**, which creates a pore in the cell membranes of some pathogens that can lead to their death.

The nomenclature of complement proteins is often a significant obstacle to understanding this system, and before discussing the complement cascade in more detail, we will explain the conventions, and the nomenclature used in this book. All components of the classical complement pathway and the membrane-attack complex are designated by the letter C followed by a number.

Functional protein classes in the complement system	
Binding to antigen:antibody complexes and pathogen surfaces	C1q
Binding to mannose on bacteria	MBL
Activating enzymes	C1r C1s C2b Bb D MASP-1 MASP-2
Membrane-binding proteins and opsonins	C4b C3b
Peptide mediators of inflammation	C5a C3a C4a
Membrane-attack proteins	C5b C6 C7 C8 C9
Complement receptors	CR1 CR2 CR3 CR4 C1qR
Complement-regulatory proteins	C1INH C4bp CR1 MCP DAF H I P CD59

Fig. 2.9 Functional protein classes in the complement system.

The native components have a simple number designation, for example, C1 and C2, but unfortunately, the components were numbered in the order of their discovery rather than the sequence of reactions, which is C1, C4, C2, C3, C5, C6, C7, C8, and C9. The products of the cleavage reactions are designated by added lower-case letters, the larger fragment being designated b and the smaller a; thus, for example, C4 is cleaved to C4b, the large fragment of C4 that binds covalently to the surface of the pathogen, and C4a, a small fragment with weak pro-inflammatory properties. The components of the alternative pathway, instead of being numbered, are designated by different capital letters, for example factor B and factor D. As with the classical pathway, their cleavage products are designated by the addition of lower-case a and b: thus, the large fragment of B is called Bb and the small fragment Ba. Finally, in the mannose-binding lectin pathway, the first enzymes to be activated are known as the **m**annan-binding lectin-**a**ssociated **s**erine **p**roteases MASP-1 and MASP-2, after which the pathway is essentially the same as the classical pathway. Activated complement components are often designated by a horizontal line, for example, $\overline{C2b}$; however, we will not use this convention. It is also useful to be aware that the large active fragment of C2 was originally designated C2a, and is still called that in some texts and research papers. Here, for consistency, we will call all large fragments of complement b, so the large active fragment of C2 will be designated C2b.

The formation of C3 convertase activity is pivotal in complement activation, leading to the production of the principal effector molecules, and initiating the late events. In the classical and MB-lectin pathways, the C3 convertase is formed from membrane-bound C4b complexed with C2b. In the alternative pathway, a homologous C3 convertase is formed from membrane-bound C3b complexed with Bb. The alternative pathway can act as an amplification loop for all three pathways, as it is initiated by the binding of C3b.

It is clear that a pathway leading to such potent inflammatory and destructive effects, and which, moreover, has a series of built-in amplification steps, is potentially dangerous and must be subject to tight regulation. One important safeguard is that key activated complement components are rapidly inactivated unless they bind to the pathogen surface on which their activation was initiated. There are also several points in the pathway at which regulatory proteins act on complement components to prevent the inadvertent activation of complement on host cell surfaces, hence protecting them from accidental damage. We will return to these regulatory mechanisms later.

We have now introduced all the relevant components of complement and are ready for a more detailed account of their functions. To help distinguish the different components according to their functions, we will use a color code in the figures in this part of the chapter. This is introduced in Fig. 2.9, where all the components of complement are grouped by function.

2-6 The classical pathway is initiated by activation of the C1 complex.

The classical pathway plays a role in both innate and adaptive immunity. As we will see in Chapter 9, the first component of this pathway, C1q, links the adaptive humoral immune response to the complement system by binding to antibodies complexed with antigens. C1q can, however, also bind directly to the surface of certain pathogens and thus trigger complement activation in the absence of antibody. C1q is part of the C1 complex, which comprises a single C1q molecule bound to two molecules each of the zymogens C1r and C1s. C1q is a calcium-dependent sugar-binding protein, a lectin, belonging to the **collectin** family of proteins, which contains both collagen-like and lectin domains hence the name collectin. It has six globular heads, linked together by a collagen-like tail, which surround the $(C1r:C1s)_2$ complex (Fig. 2.10).

Fig. 2.10 The first protein in the classical pathway of complement activation is C1, which is a complex of C1q, C1r, and C1s. C1q is composed of six identical subunits with globular heads and long collagen-like tails. The tails combine to bind to two molecules each of C1r and C1s, forming the C1 complex C1q:C1r$_2$:C1s$_2$. The heads can bind to the constant regions of immunoglobulin molecules or directly to the pathogen surface, causing a conformational change in C1r, which then cleaves and activates the C1s zymogen. Photograph ($\times$ 500,000) courtesy of K.B.M Reid.

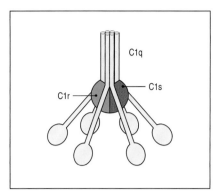

Binding of more than one of the C1q heads to a pathogen surface causes a conformational change in the (C1r:C1s)$_2$ complex, which leads to activation of an autocatalytic enzymatic activity in C1r; the active form of C1r then cleaves its associated C1s to generate an active serine protease.

Once activated, the C1s enzyme acts on the next two components of the classical pathway, cleaving C4 and then C2 to generate two large fragments, C4b and C2b, which together form the C3 convertase of the classical pathway. In the first step, C1s cleaves C4 to produce C4b, which binds covalently to the surface of the pathogen. The covalently attached C4b then binds one molecule of C2, making it susceptible, in turn, to cleavage by C1s. C1s cleaves C2 to produce the large fragment C2b, which is itself a serine protease. The complex of C4b with the active serine protease C2b remains on the surface of the pathogen as the C3 convertase of the classical pathway. Its most important activity is to cleave large numbers of C3 molecules to produce C3b molecules that coat the pathogen surface. At the same time, the other cleavage product, C3a, initiates a local inflammatory response. These reactions, which comprise the classical pathway of complement activation, are shown in schematic form in Fig. 2.11; the proteins involved, and their active forms, are listed in Fig. 2.12.

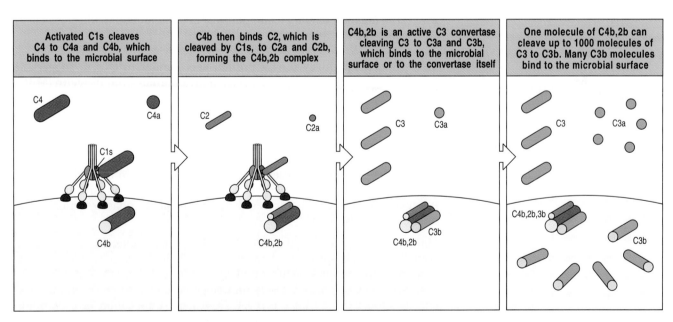

Fig. 2.11 The classical pathway of complement activation generates a C3 convertase that deposits large numbers of C3b molecules on the pathogen surface. The steps in the reaction are outlined here and detailed in the text. The cleavage of C4 by C1s exposes a reactive group on C4b that allows it to bind covalently to the pathogen surface. C4b then binds C2, making it susceptible to cleavage by C1s. The larger C2b fragment is the active protease component of the C3 convertase, which cleaves many molecules of C3 to produce C3b, which binds to the pathogen surface, and C3a, an inflammatory mediator.

Fig. 2.12 The proteins of the classical pathway of complement activation.

Proteins of the classical pathway of complement activation		
Native component	**Active form**	**Function of the active form**
C1 (C1q: C1r$_2$:C1s$_2$)	C1q	Binds directly to pathogen surfaces or indirectly to antibody bound to pathogens, thus allowing autoactivation of C1r
	C1r	Cleaves C1s to active protease
	C1s	Cleaves C4 and C2
C4	C4b	Covalently binds to pathogen and opsonizes it. Binds C2 for cleavage by C1s
	C4a	Peptide mediator of inflammation (weak activity)
C2	C2b	Active enzyme of classical pathway C3/C5 convertase: cleaves C3 and C5
	C2a	Precursor of vasoactive C2 kinin
C3	C3b	Many molecules of C3b bind to pathogen surface and act as opsonins. Initiates amplification via the alternative pathway. Binds C5 for cleavage by C2b
	C3a	Peptide mediator of inflammation (intermediate activity)

2-7   The mannan-binding lectin pathway is homologous to the classical pathway.

The MB-lectin pathway uses a protein very similar to C1q to trigger the complement cascade. This protein, called the **mannan-binding lectin** (**MBL**), is a collectin, like C1q. Mannan-binding lectin binds specifically to mannose residues, and to certain other sugars, which are accessible and arranged in a pattern that allows binding on many pathogens. On vertebrate cells, however, these are covered by other sugar groups, especially sialic acid. Thus, mannan-binding lectin is able to initiate complement activation by binding to pathogen surfaces. It is present at low concentrations in normal plasma of most individuals, and, as we will see in the last part of this chapter, its production by the liver is increased during the acute-phase reaction of the innate immune response.

Mannan-binding lectin, like C1q, is a six-headed molecule that forms a complex with two protease zymogens, which in the case of the mannan-binding lectin complex (MBL complex) are **MASP-1** and **MASP-2** (Fig. 2.13). MASP-1 and MASP-2 are closely homologous to C1r and C1s, and all four enzymes are likely to have evolved from gene duplication of a common precursor. When the MBL complex binds to a pathogen surface, MASP-1 and MASP-2 are activated to cleave C4 and C2. Thus the MB-lectin pathway initiates complement activation in the same way as the classical pathway, forming a C3 convertase from C2b bound to C4b. People deficient in mannan-binding lectin experience a substantial increase in infections during early childhood, indicating the importance of the MB-lectin pathway for host

Fig. 2.13 Mannan-binding lectin forms a complex with serine proteases that resembles the complement C1 complex. MBL forms clusters of two to six carbohydrate-binding heads around a central collagen-like stalk. This structure, easily discernible under the electron microscope (lower panels) has been described as looking like 'a bunch of tulips.' Associated with this complex are two serine proteases, MBL-associated serine protease (MASP)-1 and -2. The structural disposition of MASP proteins in the complex is not yet determined. On binding of MBL to bacterial surfaces, these serine proteases become activated and can then activate the complement system by cleaving and activating C4 and C2. Photograph courtesy of K.B.M. Reid.

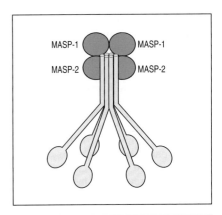

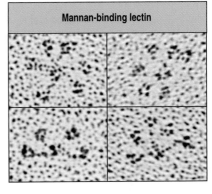

Mannan-binding lectin

defense. The age window of susceptibility to infections associated with mannan-binding lectin deficiency illustrates the particular importance of innate host defense mechanisms in childhood, before the child's adaptive immune responses are fully matured and after maternal antibodies transferred across the placenta and in colostrum have been lost.

2-8 Complement activation is largely confined to the surface on which it is initiated.

We have seen that the classical and MB-lectin pathways of complement activation are initiated by proteins that bind to pathogen surfaces. During the triggered-enzyme cascade that follows, it is important that activating events are confined to this same site, so that C3 activation also occurs on the surface of the pathogen, and not in the plasma or on host cell surfaces. This is achieved principally by the covalent binding of C4b to the pathogen surface. Cleavage of C4 exposes a highly reactive thioester bond on the C4b molecule that allows it to bind covalently to molecules in the immediate vicinity of its site of activation. In innate immunity, C4 cleavage is catalyzed by a C1 or MBL complex bound to the pathogen surface, and C4b can bind adjacent proteins or carbohydrates on the pathogen surface. If C4b does not rapidly form this bond, the thioester bond is cleaved by reaction with water and this hydrolysis reaction irreversibly inactivates C4b (Fig. 2.14). This helps to prevent C4b from diffusing from its site of activation on the microbial surface and becoming coupled to host cells.

C2 becomes susceptible to cleavage by C1s only when it is bound by C4b, and the C2b serine protease is thereby also confined to the pathogen surface, where it remains associated with C4b, forming a C3 convertase. The activation of C3 molecules thus also occurs at the surface of the pathogen. Furthermore, the C3b cleavage product is also rapidly inactivated unless it binds covalently by the same mechanism as C4b, and it therefore opsonizes only the surface on which complement activation has taken place.

2-9 Hydrolysis of C3 causes initiation of the alternative pathway of complement.

The third pathway of complement activation is called the alternative pathway because it was discovered as a second, or 'alternative,' pathway for complement activation after the classical pathway had been defined. This pathway can proceed on many microbial surfaces in the absence of specific antibody, and it leads to the generation of a distinct C3 convertase designated C3b,Bb. In contrast to the classical and MB-lectin pathways of complement activation, the alternative pathway does not depend on a pathogen-binding protein for its initiation; instead it is initiated through the spontaneous hydrolysis of C3,

Fig. 2.14 Cleavage of C4 exposes an active thioester bond that causes the large fragment, C4b, to bind covalently to nearby molecules on the bacterial cell surface. Intact C4 consists of an α, a β, and a γ chain with a shielded thioester bond on the α chain. This is exposed when the α chain is cleaved by C1s to produce C4b. The thioester bond (marked by an arrow in the third panel) is rapidly hydrolyzed (that is, cleaved by water), inactivating C4b unless it reacts with hydroxyl or amino groups to form a covalent linkage with molecules on the pathogen surface. The homologous protein C3 has an identical reactive thioester bond that is also exposed on the C3b fragment when C3 is cleaved by C2b. The covalent attachment of C3b and C4b enables these molecules to act as opsonins and is important in confining complement activation to the pathogen surface.

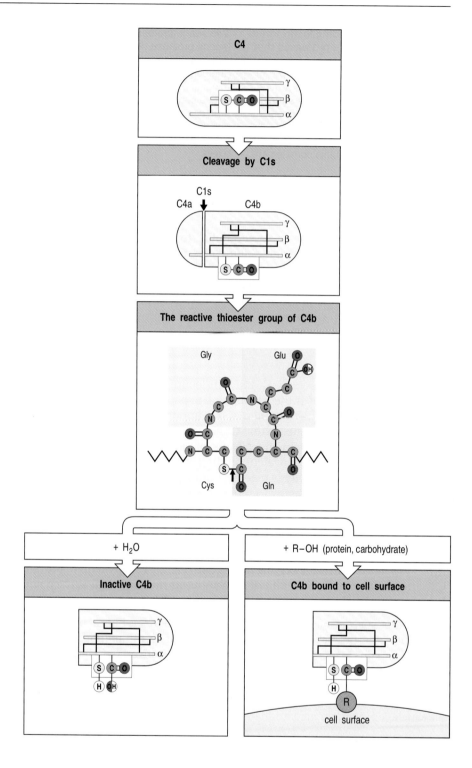

as shown in the top three panels of Fig. 2.15. The distinctive components of the pathway are listed in Fig. 2.16. A number of mechanisms ensure that the activation pathway will only proceed on the surface of a pathogen.

C3 is abundant in plasma, and C3b is produced at a significant rate by spontaneous cleavage (also known as 'tickover'). This occurs through the spontaneous hydrolysis of the thioester bond in C3 to form $C3(H_2O)$ which

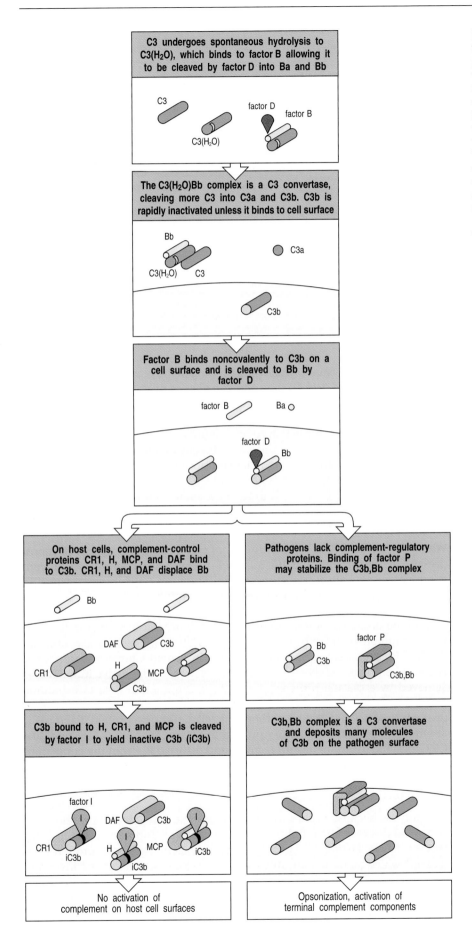

Fig. 2.15 Complement activated by the alternative pathway attacks pathogens while sparing host cells, which are protected by complement regulatory proteins. The complement component C3 is cleaved spontaneously in plasma to give C3(H₂O), which binds factor B and enables the bound factor B to be cleaved by factor D (top panel). The resulting soluble C3 convertase cleaves C3 to give C3a and C3b, which can attach to host cells or pathogen surfaces (second panel). Covalently bound C3b binds factor B, which in turn is rapidly cleaved by factor D to Bb, which remains bound to C3b to form a C3 convertase, and Ba, which is released (third panel). If C3b,Bb forms on the surface of host cells (bottom left panels), it is rapidly inactivated by complement-regulatory proteins expressed by the host cell: complement receptor 1 (CR1), decay-accelerating factor (DAF), and membrane cofactor of proteolysis (MCP). Host cell surfaces also favor binding of factor H from plasma. CR1, DAF, and factor H displace Bb from C3b, and CR1, MCP, and factor H catalyze the cleavage of bound C3b by the plasma protease factor I to produce inactive C3b (known as iC3b). Bacterial surfaces (bottom right panels) do not express complement-regulatory proteins and favor binding of factor P (properdin), which stabilizes the C3b,Bb convertase activity. This convertase is the equivalent of C4b,2b of the classical pathway (see Fig. 2.11).

Fig. 2.16 The proteins of the alternative pathway of complement activation.

Proteins of the alternative pathway of complement activation		
Native component	Active fragments	Function
C3	C3b	Binds to pathogen surface, binds B for cleavage by D, C3b,Bb is C3 convertase and C3b$_2$,Bb is C5 convertase
Factor B (B)	Ba	Small fragment of B, unknown function
	Bb	Bb is active enzyme of the C3 convertase C3b,Bb and C5 convertase C3b$_2$,Bb
Factor D (D)	D	Plasma serine protease, cleaves B when it is bound to C3b to Ba and Bb
Factor P (properdin)	P	Plasma protein with affinity for the C3b,Bb convertase on bacterial cells

has an altered conformation, allowing binding of the plasma protein **factor B**. The binding of B by C3(H_2O) then allows a plasma protease called **factor D** to cleave factor B to Ba and Bb, the latter remaining associated with C3(H_2O) to form the C3(H_2O)Bb complex. This complex is a fluid-phase C3 convertase, and although it is only formed in small amounts it can cleave many molecules of C3 to C3a and C3b. Much of this C3b is inactivated by hydrolysis, but some attaches covalently, through its reactive thioester group, to the surfaces of host cells or to pathogens. C3b bound in this way is able to bind factor B, allowing its cleavage by factor D to yield the small fragment Ba and the active protease Bb. This results in formation of the alternative pathway C3 convertase, C3b,Bb (see Fig. 2.15).

When C3b binds to host cells, a number of complement-regulatory proteins, present in the plasma and on host cell membranes combine to prevent complement activation from proceeding. These proteins interact with C3b and either prevent the convertase from forming, or promote its rapid dissociation (see Fig. 2.15). Thus, the complement receptor 1 (CR1) and a membrane-attached protein known as **decay-accelerating factor** (DAF or **CD55**) compete with factor B for binding to C3b on the cell surface, and can displace Bb from a convertase that has already formed. Convertase formation can also be prevented by cleaving C3b to its inactive derivative iC3b. This is achieved by a plasma protease, **factor I**, in conjunction with C3b-binding proteins that can act as cofactors, such as CR1 and **membrane cofactor of proteolysis** (MCP or **CD46**), another host cell membrane protein. **Factor H** is another complement-regulatory protein in plasma that binds C3b and, like CR1, it is able to compete with factor B and displace Bb from the convertase in addition to acting as a cofactor for factor I. Factor H binds preferentially to C3b bound to vertebrate cells as it has an affinity for the sialic acid residues present on these cells.

By contrast, because pathogen surfaces lack these regulatory proteins and sialic acid residues, the C3b,Bb convertase can form and persist. Indeed, this process may be favored by a positive regulatory factor, known as **properdin** or **factor P**, which binds to many microbial surfaces and stabilizes the convertase. Deficiencies in factor P are associated with a heightened susceptibility to infection with *Neisseria* species. Once formed, the C3b,Bb convertase rapidly cleaves yet more C3 to C3b, which can bind to the pathogen and either act as

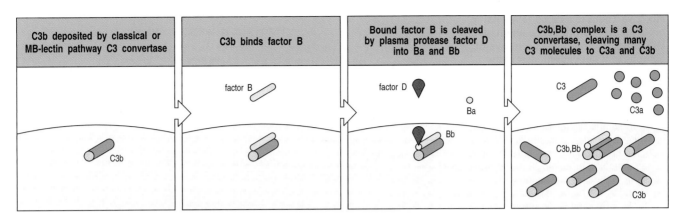

| C3b deposited by classical or MB-lectin pathway C3 convertase | C3b binds factor B | Bound factor B is cleaved by plasma protease factor D into Ba and Bb | C3b,Bb complex is a C3 convertase, cleaving many C3 molecules to C3a and C3b |

Fig. 2.17 The alternative pathway of complement activation can amplify the classical or the MB-lectin pathway by forming an alternative C3 convertase and depositing more C3b molecules on the pathogen. C3b deposited by the classical or MB-lectin pathways can bind factor B, making it susceptible to cleavage by factor D. The C3b,Bb complex is the C3 convertase of the alternative pathway of complement activation and its action, like that of C4b,2b, results in the deposition of many molecules of C3b on the pathogen surface.

an opsonin or reinitiate the pathway to form another molecule of C3b,Bb convertase. Thus, the alternative pathway activates through an amplification loop that can proceed on the surface of a pathogen, but not on a host cell. This same amplification loop enables the alternative pathway to contribute to complement activation initially triggered through the classical or MB-lectin pathways (Fig. 2.17).

The C3 convertases resulting from activation of the classical and MB-lectin pathways (C4b,2b) and from the alternative pathway (C3b,Bb) are apparently distinct. However, understanding of the complement system is simplified somewhat by recognition of the close evolutionary relationships between the different complement proteins. Thus the complement zymogens, factor B and C2, are closely related proteins encoded by homologous genes located in tandem in the major histocompatibility complex (MHC) on human chromosome 6. Furthermore, their respective binding partners, C3 and C4, both contain thioester bonds that provide the means of covalently attaching the C3 convertases to a pathogen surface. Only one component of the alternative pathway appears entirely unrelated to its functional equivalents in the classical and MB-lectin pathways; this is the initiating serine protease, factor D. Factor D can also be singled out as the only activating protease of the complement system to circulate as an active enzyme rather than a zymogen. This is both necessary for the initiation of the alternative pathway through spontaneous C3 cleavage, and safe for the host because factor D has no other substrate than factor B when bound to C3b. This means that factor D only finds its substrate at a very low level in plasma, and at pathogen surfaces where the alternative pathway of complement activation is allowed to proceed.

Comparison of the different pathways of complement activation illustrates the general principle that most of the immune effector mechanisms that can be activated in a nonclonal fashion as part of the early nonadaptive host response against infection have been harnessed during evolution to be used as effector mechanisms of adaptive immunity. It is almost certain that the adaptive response evolved by adding specific recognition to the original non-adaptive system. This is illustrated particularly clearly in the complement system, because here the components are defined, and the functional homo-logues can be seen to be evolutionarily related (Fig. 2.18).

Fig. 2.18 There is a close relationship between the factors of the alternative, MB-lectin, and classical pathways of complement activation. Most of the factors are either identical or the products of genes that have duplicated and then diverged in sequence. The proteins C4 and C3 are homologous and contain the unstable thioester bond by which their large fragments, C4b and C3b, bind covalently to membranes. The genes encoding proteins C2 and B are adjacent in the class III region of the MHC and arose by gene duplication. Factor H, CR1, and C4bp regulatory proteins share a repeat sequence common to many complement-regulatory proteins. The greatest divergence between the pathways is in their initiation: in the classical pathway the C1 complex binds either to certain pathogens or to bound antibody and in the latter serves to convert antibody binding into enzyme activity on a specific surface; in the MB-lectin pathway, mannan-binding lectin (MBL) associates with a serine protease, activating MBL-associated serine protease (MASP), to serve the same function as C1r:C1s; whereas in the alternative pathway this enzyme activity is provided by factor D.

Step in pathway	Protein serving function in pathway			Relationship
	Alternative (innate)	MB-lectin	Classical	
Initiating serine protease	D	MASP	C1s	Homologous (C1s and MASP)
Covalent binding to cell surface	C3b	C4b		Homologous
C3/C5 convertase	Bb	C2b		Homologous
Control of activation	CR1 H	CR1 C4bp		Identical Homologous
Opsonization	C3b			Identical
Initiation of effector pathway	C5b			Identical
Local inflammation	C5a, C3a			Identical
Stabilization	P	None		Unique

2-10 Surface-bound C3 convertase deposits large numbers of C3b fragments on pathogen surfaces and generates C5 convertase activity.

The formation of C3 convertases is the point at which the three pathways of complement activation converge, because both the classical pathway and MB-lectin pathway convertases C4b,2b, and the alternative pathway convertase C3b,Bb have the same activity, and they initiate the same subsequent events. They both cleave C3 to C3b and C3a. C3b binds covalently through its thioester bond to adjacent molecules on the pathogen surface; otherwise it is inactivated by hydrolysis. C3 is the most abundant complement protein in plasma, occurring at a concentration of 1.2 mg ml^{-1}, and up to 1000 molecules of C3b can bind in the vicinity of a single active C3 convertase (see Fig. 2.11). Thus, the main effect of complement activation is to deposit large quantities of C3b on the surface of the infecting pathogen, where it forms a covalently bonded coat that, as we will see, can signal the ultimate destruction of the pathogen by phagocytes.

The next step in the cascade is the generation of the C5 convertases. In the classical and the MB-lectin pathways, a C5 convertase is formed by the binding of C3b to C4b,2b to yield C4b,2b,3b. By the same token, the C5 convertase of the alternative pathway is formed by the binding of C3b to the C3 convertase

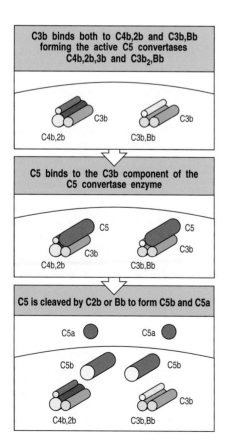

Fig. 2.19 Complement component C5 is cleaved when captured by a C3b molecule that is part of a C5 convertase complex. As shown in the top panel, C5 convertases are formed when C3b binds either the classical or MB-lectin pathway C3 convertase C4b,2b to form C4b,2b,3b, or the alternative pathway C3 convertase C3b,Bb to form C3b$_2$,Bb. C5 binds to the C3b in these complexes (center panel). The bottom panel shows how C5 is cleaved by the active enzyme C2b or Bb to form C5b and the inflammatory mediator C5a. Unlike C3b and C4b, C5b is not covalently bound to the cell surface. The production of C5b initiates the assembly of the terminal complement components.

to form C3b$_2$,Bb. C5 is captured by these C5 convertase complexes through binding to an acceptor site on C3b, and is then rendered susceptible to cleavage by the serine protease activity of C2b or Bb. This reaction, which generates C5b and C5a, is much more limited than cleavage of C3, as C5 can be cleaved only when it binds to C3b that is part of the C5 convertase complex. Thus, complement activation by both the alternative, MB-lectin and classical pathways leads to the binding of large numbers of C3b molecules on the surface of the pathogen, the generation of a more limited number of C5b molecules, and the release of C3a and C5a (Fig. 2.19).

2-11   Phagocyte ingestion of complement-tagged pathogens is mediated by receptors for the bound complement proteins.

The most important action of complement is to facilitate the uptake and destruction of pathogens by phagocytic cells. This occurs by the specific recognition of bound complement components by **complement receptors (CRs)** on phagocytes. These complement receptors bind pathogens opsonized with complement components: opsonization of pathogens is a major function of C3b and its proteolytic derivatives. C4b also acts as an opsonin but has a relatively minor role, largely because so much more C3b than C4b is generated.

The five known types of receptor for bound complement components are listed, with their functions and distributions, in Fig. 2.20. The best-characterized is the C3b receptor **CR1** (**CD35**), which is expressed on both macrophages and polymorphonuclear leukocytes. Binding of C3b to CR1 cannot by itself stimulate phagocytosis, but it can lead to phagocytosis in the presence of

Receptor	Specificity	Functions	Cell types
CR1 (CD35)	C3b, C4b iC3b	Promotes C3b and C4b decay Stimulates phagocytosis Erythrocyte transport of immune complexes	Erythrocytes, macrophages, monocytes, polymorphonuclear leukocytes, B cells, FDC
CR2 (CD21)	C3d, iC3b, C3dg Epstein–Barr virus	Part of B-cell co-receptor Epstein–Barr virus receptor	B cells, FDC
CR3 (CD11b/ CD18)	iC3b	Stimulates phagocytosis	Macrophages, monocytes, polymorphonuclear leukocytes, FDC
CR4 (gp150, 95) (CD11c/ CD18)	iC3b	Stimulates phagocytosis	Macrophages, monocytes, polymorphonuclear leukocytes, dendritic cells
C5a receptor	C5a	Binding of C5a activates G protein	Endothelial cells, mast cells, phagocytes
C3a receptor	C3a	Binding of C3a activates G protein	Endothelial cells, mast cells, phagocytes

Fig. 2.20 Distribution and function of receptors for complement proteins on the surfaces of cells. There are several different receptors specific for different bound complement components and their fragments. CR1 and CR3 are especially important in inducing phagocytosis of bacteria with complement components on their surface. CR2 is found mainly on B cells, where it is also part of the B-cell co-receptor complex and the receptor by which the Epstein–Barr virus selectively infects B cells, causing infectious mononucleosis. CR1 and CR2 share structural features with the complement-regulatory proteins that bind C3b and C4b. CR3 and CR4 are integrins; CR3 is known to be important for leukocyte adhesion and migration, while CR4 is only known to function in phagocytic responses. The C5a and C3a receptors are seven-span G protein-coupled receptors. FDC, follicular dendritic cells; these are not involved in innate immunity and are discussed in later chapters.

Fig. 2.21 The anaphylotoxin C5a can enhance phagocytosis of opsonized microorganisms. Activation of complement, either by the alternative or the MB-lectin pathway, leads to the deposition of C3b on the surface of the microorganism (left panel). C3b can be bound by the complement receptor CR1 on the surface of phagocytes, but this on its own is insufficient to induce phagocytosis (center panel). Phagocytes also express receptors for the anaphylotoxin C5a, and binding of C5a will now activate the cell to phagocytose microorganisms bound through CR1 (right panel).

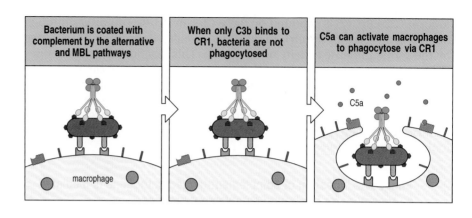

| Bacterium is coated with complement by the alternative and MBL pathways | When only C3b binds to CR1, bacteria are not phagocytosed | C5a can activate macrophages to phagocytose via CR1 |

other immune mediators that activate macrophages. For example, the small complement fragment C5a can activate macrophages to ingest bacteria bound to their CR1 receptors (Fig. 2.21). C5a binds to another receptor expressed by macrophages, the **C5a receptor**, which has seven membrane-spanning domains. Receptors of this type couple with intracellular guanine-nucleotide-binding proteins called G proteins, and the C5a receptor signals in this way. Proteins associated with the extracellular matrix, such as fibronectin, can also contribute to phagocyte activation; these are encountered when phagocytes are recruited to connective tissue and activated there. C3a, which has inflammatory activities similar to those of C5a, although it is a less potent chemoattractant, binds to its own specific receptor, the C3a receptor, which is homologous in structure to the C5a receptor.

Three other complement receptors—**CR2** (also known as **CD21**), **CR3** (**CD11b:CD18**), and **CR4** (**CD11c:CD18**)—bind to inactivated forms of C3b that remain attached to the pathogen surface. Like several other key components of complement, C3b is subject to regulatory mechanisms and can be cleaved into derivatives that cannot form an active convertase. One of the inactive derivatives of C3b, known as **iC3b** (see Section 2-9) acts as an opsonin in its own right when bound by the complement receptors CR2 or CR3. Unlike the binding of iC3b to CR1, the binding of iC3b to CR3 is sufficient on its own to stimulate phagocytosis. A second breakdown product of C3b, called **C3dg**, binds only to CR2. CR2 is found on B cells as part of a co-receptor complex that can augment the signal received through the antigen-specific immunoglobulin receptor. Thus a B cell whose antigen receptor is specific for a given pathogen will receive a strongly augmented signal on binding this pathogen if it is also coated with C3dg. The activation of complement can therefore contribute to producing a strong antibody response (see Chapters 6 and 9). This example of how an innate humoral immune response can contribute to activating adaptive humoral immunity parallels the contribution made by the innate cellular response of macrophages and dendritic cells to the initiation of a T-cell response, which we will discuss later in this chapter.

The central role of opsonization by C3b and its inactive fragments in the destruction of extracellular pathogens can be seen in the effects of various complement deficiency diseases. Whereas individuals deficient in any of the late components of complement are relatively unaffected, individuals deficient in C3 or in molecules that catalyze C3b deposition show increased susceptibility to infection by a wide range of extracellular bacteria, as we will see in Chapter 11.

2-12 Small fragments of some complement proteins can initiate a local inflammatory response.

The small complement fragments C3a, C4a, and C5a act on specific receptors (see Fig. 2.20) to produce local inflammatory responses. When produced in large amounts or injected systemically, they induce a generalized circulatory collapse, producing a shocklike syndrome similar to that seen in a systemic allergic reaction involving IgE antibodies (see Chapter 12). Such a reaction is termed **anaphylactic shock** and these small fragments of complement are therefore often referred to as **anaphylotoxins**. Of the three, C5a is the most stable and has the highest specific biological activity. All three induce smooth muscle contraction and increase vascular permeability, but C5a and C3a also act on the endothelial cells lining blood vessels to induce adhesion molecules. In addition, C3a and C5a can activate the mast cells that populate submucosal tissues to release mediators such as histamine and TNF-α that cause similar effects. The changes induced by C5a and C3a recruit antibody, complement, and phagocytic cells to the site of an infection (Fig. 2.22), and the increased fluid in the tissues hastens the movement of pathogen-bearing antigen-presenting cells to the local lymph nodes, contributing to the prompt initiation of the adaptive immune response.

C5a also acts directly on neutrophils and monocytes to increase their adherence to vessel walls, their migration toward sites of antigen deposition, and their ability to ingest particles, as well as increasing the expression of CR1 and CR3

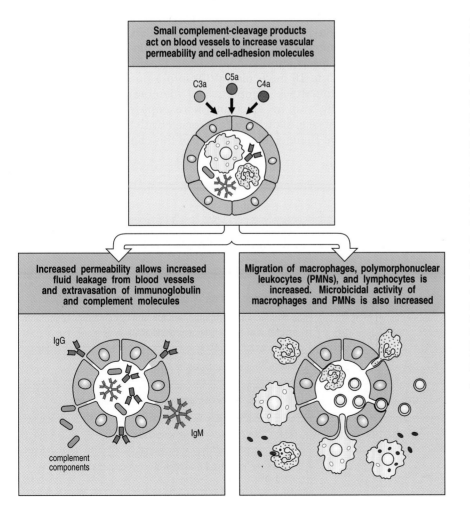

Fig. 2.22 Local inflammatory responses can be induced by small complement fragments, especially C5a. The small complement fragments are differentially active: C5a is more active than C3a, which is more active than C4a. They cause local inflammatory responses by acting directly on local blood vessels, stimulating an increase in blood flow, increased vascular permeability, and increased binding of phagocytes to endothelial cells. C5a also activates mast cells (not shown) to release mediators such as histamine and TNF-α that contribute to the inflammatory response. The increase in vessel diameter and permeability leads to the accumulation of fluid and protein. Fluid accumulation increases lymphatic drainage, bringing pathogens and their antigenic components to nearby lymph nodes. The antibodies, complement, and cells thus recruited participate in pathogen clearance by enhancing phagocytosis. The small complement fragments can also directly increase the activity of the phagocytes.

on the surfaces of these cells. In this way C5a and, to a smaller extent, C3a and C4a, act in concert with other complement components to hasten the destruction of pathogens by phagocytes. C5a and C3a signal through transmembrane receptors that activate G proteins; thus the action of C5a in attracting neutrophils and monocytes is analogous to that of chemokines, which also act via G proteins to control cell migration.

2-13   The terminal complement proteins polymerize to form pores in membranes that can kill certain pathogens.

One of the important effects of complement activation is the assembly of the terminal components of complement (Fig. 2.23) to form a membrane-attack complex. The reactions leading to the formation of this complex are shown schematically in Fig. 2.24. The end result is a pore in the lipid bilayer membrane that destroys membrane integrity. This is thought to kill the pathogen by destroying the proton gradient across the pathogen cell membrane.

The first step in the formation of the membrane-attack complex is the cleavage of C5 by a C5 convertase to release C5b (see Fig. 2.19). In the next stages, shown in Fig. 2.24, C5b initiates the assembly of the later complement components and their insertion into the cell membrane. First, one molecule of C5b binds one molecule of C6, and the C5b,6 complex then binds one molecule of C7. This reaction leads to a conformational change in the constituent molecules, with the exposure of a hydrophobic site on C7, which inserts into the lipid bilayer. Similar hydrophobic sites are exposed on the later components C8 and C9 when they are bound to the complex, allowing these proteins also to insert into the lipid bilayer. C8 is a complex of two proteins, C8β and C8α-γ. The C8β protein binds to C5b, and the binding of C8β to the membrane-associated C5b,6,7 complex allows the hydrophobic domain of C8α-γ to insert into the lipid bilayer. Finally, C8α-γ induces the polymerization of 10 to 16 molecules of C9 into a pore-forming structure called the membrane-attack complex. The membrane-attack complex, shown schematically and by electron microscopy in Fig. 2.24, has a hydrophobic external face, allowing it to associate with the lipid bilayer, but a hydrophilic internal channel. The diameter of this channel is about 100 Å,

Fig. 2.23 The terminal complement components assemble to form the membrane-attack complex.

The terminal complement components that form the membrane-attack complex		
Native protein	**Active component**	**Function**
C5	C5a	Small peptide mediator of inflammation (high activity)
	C5b	Initiates assembly of the membrane-attack system
C6	C6	Binds C5b, forms acceptor for C7
C7	C7	Binds C5b,6, amphiphilic complex inserts in lipid bilayer
C8	C8	Binds C5b,6,7, initiates C9 polymerization
C9	C9$_n$	Polymerizes to C5b,6,7,8 to form a membrane-spanning channel, lysing cell

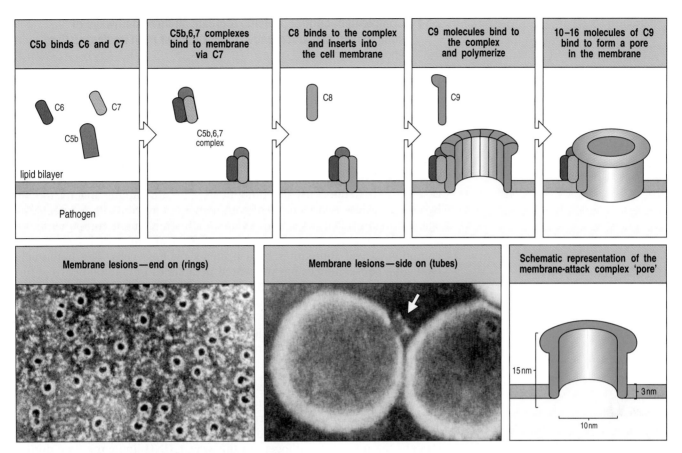

Fig. 2.24 Assembly of the membrane-attack complex generates a pore in the lipid bilayer membrane. The sequence of steps and their approximate appearance are shown here in schematic form. C5b triggers the assembly of a complex of one molecule each of C6, C7, and C8, in that order. C7 and C8 undergo conformational changes that expose hydrophobic domains that insert into the membrane. This complex causes moderate membrane damage in its own right, and also serves to induce the polymerization of C9, again with the exposure of a hydrophobic site. Up to 16 molecules of C9 are then added to the assembly to generate a channel of 100 Å diameter in the membrane. This channel disrupts the bacterial cell membrane, killing the bacterium. The electron micrographs show erythrocyte membranes with membrane-attack complexes in two orientations, end on and side on. Photographs courtesy of S. Bhakdi and J. Tranum-Jensen.

allowing the free passage of solutes and water across the lipid bilayer. The disruption of the lipid bilayer leads to the loss of cellular homeostasis, the disruption of the proton gradient across the membrane, the penetration of enzymes such as lysozyme into the cell, and the eventual destruction of the pathogen.

Although the effect of the membrane-attack complex is very dramatic, particularly in experimental demonstrations in which antibodies against red blood cell membranes are used to trigger the complement cascade, the significance of these components in host defense seems to be quite limited. To date, deficiencies in complement components C5–C9 have been associated with susceptibility only to *Neisseria* species, the bacteria that cause the sexually transmitted disease gonorrhea and a common form of bacterial meningitis. Thus, the opsonizing and inflammatory actions of the earlier components of the complement cascade are clearly most important for host defense against infection. Formation of the membrane-attack complex seems to be important only for the killing of a few pathogens, although, as we will see in Chapter 13, it might have a major role in immunopathology.

2-14 Complement control proteins regulate all three pathways of complement activation and protect the host from its destructive effects.

Given the destructive effects of complement, and the way in which its activation is rapidly amplified through a triggered-enzyme cascade, it is not surprising that there are several mechanisms to prevent its uncontrolled activation. As we have seen, the effector molecules of complement are generated through the sequential activation of zymogens, which are present in plasma in an inactive form. The activation of these zymogens usually occurs on a pathogen surface, and the activated complement fragments produced in the ensuing cascade of reactions usually bind nearby or are rapidly inactivated by hydrolysis. These two features of complement activation act as safeguards against uncontrolled activation. Even so, all complement components are activated spontaneously at a low rate in plasma, and activated complement components will sometimes bind proteins on host cells. The potentially damaging consequences are prevented by a series of complement control proteins, summarized in Fig. 2.25, which regulate the complement cascade at different points. As we saw in discussing the alternative pathway of complement activation (see Section 2-9) many of these control proteins specifically protect host cells while allowing complement activation to proceed on pathogen surfaces. The complement control proteins therefore allow complement to distinguish self from nonself.

The reactions that regulate the complement cascade are shown in Fig. 2.26. The top two panels show how the activation of C1 is controlled by a plasma serine proteinase inhibitor or **scrpin**, the C1 inhibitor (C1INH). C1INH binds the active enzyme C1r:C1s, and causes it to dissociate from C1q, which remains bound to the pathogen. In this way, C1INH limits the time during which active C1s is able to cleave C4 and C2. By the same means, C1INH limits the spontaneous activation of C1 in plasma. Its importance can be seen in the C1INH deficiency disease, **hereditary angioneurotic edema**, in which chronic spontaneous complement activation leads to the production of excess cleaved

Hereditary
Angioneurotic
Edema

Fig. 2.25 The proteins that regulate the activity of complement.

Control proteins of the classical and alternative pathways	
Name (symbol)	**Role in the regulation of complement activation**
C1 inhibitor (C1INH)	Binds to activated C1r, C1s, removing it from C1q
C4-binding protein (C4BP)	Binds C4b, displacing C2b; cofactor for C4b cleavage by I
Complement receptor 1 (CR1)	Binds C4b, displacing C2b, or C3b displacing Bb; cofactor for I
Factor H (H)	Binds C3b, displacing Bb; cofactor for I
Factor I (I)	Serine protease that cleaves C3b and C4b; aided by H, MCP, C4BP, or CR1
Decay-accelerating factor (DAF)	Membrane protein that displaces Bb from C3b and C2b from C4b
Membrane cofactor protein (MCP)	Membrane protein that promotes C3b and C4b inactivation by I
CD59 (protectin)	Prevents formation of membrane-attack complex on autologous or allogenic cells. Widely expressed on membranes

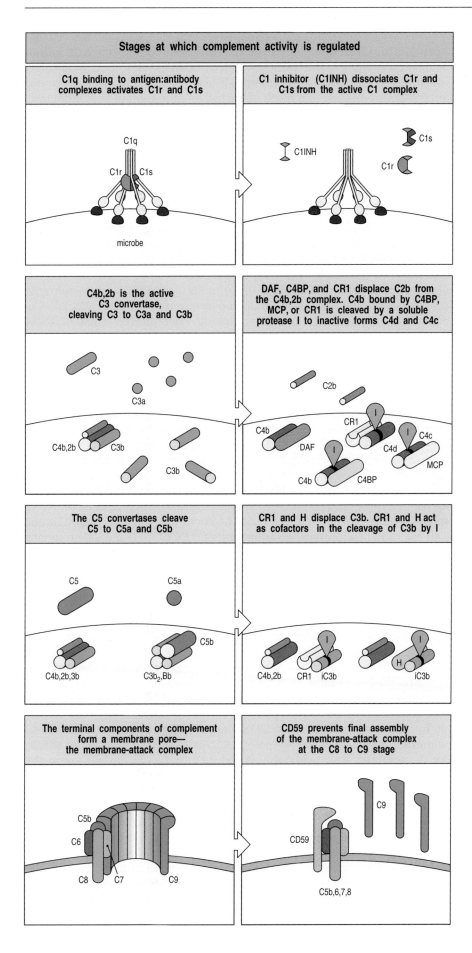

Fig. 2.26 Complement activation is regulated by a series of proteins that serve to protect host cells from accidental damage. These act on different stages of the complement cascade, dissociating complexes or catalyzing the enzymatic degradation of covalently bound complement proteins. Stages in the complement cascade are shown schematically down the left side of the figure, with the control reactions on the right. The alternative pathway C3 convertase is similarly regulated by DAF, CR1, MCP, and factor H, as shown in Fig. 2.15.

fragments of C4 and C2. The small fragment of C2, C2a, is further cleaved into a peptide, the C2 kinin, which causes extensive swelling—the most dangerous being local swelling in the trachea, which can lead to suffocation. Bradykinin, which has similar actions to C2 kinin, is also produced in an uncontrolled fashion in this disease, as a result of the lack of inhibition of another plasma protease, kallikrein, which is activated by tissue damage and also regulated by C1INH. This disease is fully corrected by replacing C1INH. The large activated fragments of C4 and C2, which normally combine to form the C3 convertase, do not damage host cells in such patients because C4b is rapidly inactivated in plasma (see Fig. 2.14) and the convertase does not form. Furthermore, any convertase that accidentally forms on a host cell is inactivated by the mechanisms described below.

The thioester bond of activated C3 and C4 is extremely reactive and has no mechanism for distinguishing an acceptor hydroxyl or amine group on a host cell from a similar group on the surface of a pathogen. A series of protective mechanisms, mediated by other proteins, has evolved to ensure that the binding of a small number of C3 or C4 molecules to host cell membranes results in minimal formation of C3 convertase and little amplification of complement activation. We have already encountered most of these mechanisms in the description of the alternative pathway (see Fig. 2.15), but we will consider them again here, as they are important regulators of the classical pathway convertase as well (see Fig. 2.26, second and third rows). The mechanisms can be divided into three categories. The first catalyze the cleavage of any C3b or C4b that does bind to host cells into inactive products. The complement-regulatory enzyme responsible is the plasma serine protease factor I; it circulates in active form but can only cleave C3b and C4b when they are bound to a cofactor protein. In these circumstances, factor I cleaves C3b, first into iC3b and then further to C3dg, thus permanently inactivating it. C4b is similarly inactivated by cleavage into C4c and C4d. There are two cell-membrane proteins that bind C3b and C4b and possess cofactor activity for factor I; these are CR1 and MCP (see Section 2-9). Microbial cell walls lack these protective proteins and cannot promote the breakdown of C3b and C4b. Instead, these proteins act as binding sites for factor B and C2, promoting complement activation. The importance of factor I can be seen in people with genetic factor I deficiency. Because of uncontrolled complement activation, complement proteins rapidly become depleted and such people suffer repeated bacterial infections, especially with ubiquitous pyogenic bacteria.

Factor I
Deficiency

There are also plasma proteins with cofactor activity for factor I. C4b is bound by a cofactor known as the **C4b-binding protein** (**C4BP**), which mainly acts as a regulator of the classical pathway in the fluid phase. C3b is bound in both the fluid phase and at cell membranes by a cofactor protein called factor H (see Section 2-9). Factor H is an important complement regulator at cell membranes, and at first sight it is not obvious how factor H can distinguish C3b bound to host cells or to a pathogen. However, the carbohydrate content of the cell membranes of bacterial pathogens differs from that of their hosts and this is the basis for the protective effect of factor H. Factor H has affinity for the terminal sialic acids of host cell membrane glycoproteins and this increases the binding of factor H to any C3b deposited on host cells. In contrast, factor H has a much lower affinity for C3b deposited on the cell walls of many bacteria, and factor B binds in preference, resulting in amplification of complement activation on bacterial cell surfaces. In effect, factor H and factor B compete for binding to C3b bound to cells. If factor B 'wins,' as is typically the case on a pathogen surface, then more C3b,Bb C3 convertase forms and complement activation is amplified. If factor H 'wins,' as is the case on cells of the host, then the bound C3b is catabolized by factor I to iC3b and C3dg and complement activation is inhibited.

The competition between factor H and factor B for binding to surface-bound C3b is an example of the second mechanism for inhibiting complement activation on host cell membranes. A number of proteins competitively inhibit the binding of C2 to cell-bound C4b and of factor B to cell-bound C3b, thereby inhibiting convertase formation. These proteins bind to C3b and C4b on the cell surface, and also mediate protection against complement through the third mechanism, which is to augment the dissociation of C4b,2b and C3b,Bb convertases that have already formed. Host cell membrane molecules that regulate complement through both these mechanisms include DAF (see Section 2-9) and CR1, which promotes dissociation of convertase in addition to its cofactor activity. All the proteins that bind the homologous C4b and C3b molecules share one or more copies of a structural element called the short consensus repeat (SCR), complement control protein (CCP) repeat, or (especially in Japan) the sushi domain.

In addition to the mechanisms for preventing C3 convertase formation and C4 and C3 deposition on cell membranes, there are also inhibitory mechanisms that prevent the inappropriate insertion of the membrane-attack complex into membranes. We saw in Section 2-13 that the membrane-attack complex polymerizes onto C5b molecules released from C5 convertase. This complex mainly inserts into cell membranes adjacent to the site of the C5 convertase, that is, close to the site of complement activation on a pathogen. However, some newly formed membrane-attack complexes may diffuse from the site of complement activation and insert into adjacent host cell membranes. Several plasma proteins bind to the C5b,6,7 complex and thereby inhibit its random insertion into cell membranes. The most important is probably C8β itself, when it binds to C5b,6,7 in the fluid phase. Host cell membranes also contain an intrinsic protein, **CD59** or **protectin**, which inhibits the binding of C9 to the C5b,6,7,8 complex (see Fig. 2.26, bottom row). CD59 and DAF are both linked to the cell surface by a phosphoinositol glycolipid (PIG) tail, like many other membrane proteins. One of the enzymes involved in the synthesis of PIG tails is encoded on chromosome X. In people with a somatic mutation in this gene in a clone of hematopoietic cells, both CD59 and DAF fail to function. This causes the disease **paroxysmal nocturnal hemoglobinuria**, which is characterized by episodes of intravascular red blood cell lysis by complement. Red blood cells that lack CD59 only are also susceptible to destruction as a result of spontaneous activation of the complement cascade.

Summary.

The complement system is one of the major mechanisms by which pathogen recognition is converted into an effective host defense against initial infection. Complement is a system of plasma proteins that can be activated directly by pathogens or indirectly by pathogen-bound antibody, leading to a cascade of reactions that occurs on the surface of pathogens and generates active components with various effector functions. There are three pathways of complement activation: the classical pathway, which is triggered directly by pathogen or indirectly by antibody binding to the pathogen surface; the MB-lectin pathway; and the alternative pathway, which also provides an amplification loop for the other two pathways. All three pathways can be initiated independently of antibody as part of innate immunity. The early events in all pathways consist of a sequence of cleavage reactions in which the larger cleavage product binds covalently to the pathogen surface and contributes to the activation of the next component. The pathways converge with the formation of a C3 convertase enzyme, which cleaves C3 to produce the active complement component C3b. The binding of large numbers of C3b molecules to the pathogen is the central event in complement activation. Bound complement components, especially bound C3b and its inactive fragments, are recognized by specific complement receptors on phagocytic cells, which engulf pathogens opsonized by C3b and its inactive fragments. The small cleavage fragments of C3, C4, and especially C5, recruit phagocytes to sites of infection and activate them by binding to specific trimeric G protein-coupled receptors. Together, these activities promote the uptake and destruction of pathogens by phagocytes. The molecules of C3b that bind the C3 convertase itself initiate the late events, binding C5 to make it susceptible to cleavage by C2b or Bb. The larger C5b fragment triggers the assembly of a membrane-attack complex, which can result in the lysis of certain pathogens. The activity of complement components is modulated by a system of regulatory proteins that prevent tissue damage as a result of inadvertent binding of activated complement components to host cells or spontaneous activation of complement components in plasma.

Receptors of the innate immune system.

Although the innate immune system lacks the specificity of adaptive immunity, it can distinguish nonself from self. We have already seen, in outline, how this is achieved in the complement system and in the response of macrophages to pathogens. In this part of the chapter we will look more closely at the receptors that activate the innate immune response, both those that recognize pathogens directly and those that signal for a cellular response. Proteins that recognize features common to many pathogens occur as secreted molecules and as receptors on cells of the innate immune system. Their general characteristics are contrasted with the antigen-specific receptors of adaptive immunity in Fig. 2.27. Unlike the receptors that mediate adaptive immunity, the receptors of the innate immune system are typically not clonally distributed; a given set of receptors will be present on all the cells of the same cell type. The

Fig. 2.27 The characteristics of receptors of the innate and adaptive immune systems are compared. The innate immune system uses receptors that are encoded by intact genes inherited through the germline, whereas the adaptive immune system uses antigen receptors encoded by genes that are assembled from individual gene segments during lymphocyte development, a process that leads to each individual cell expressing a receptor of unique specificity. As a result, receptors of the innate immune system are deployed nonclonally, whereas the antigen receptors of the adaptive immune system are clonally distributed on individual lymphocytes.

Receptor characteristic	Innate immunity	Adaptive immunity
Specificity inherited in the genome	Yes	No
Expressed by all cells of a particular type (e.g., macrophages)	Yes	No
Trigger immediate response	Yes	No
Recognize broad classes of pathogen	Yes	No
Encoded in multiple gene segments	No	Yes
Require gene rearrangement	No	Yes
Clonal distribution	No	Yes
Able to recognize a wide variety of molecular structures	No	Yes

binding of pathogens by these receptors gives rise to very rapid responses, which are put into effect without the delay imposed by the clonal expansion of cells needed in the adaptive immune response.

Receptors of the innate immune system mediate a number of different functions. Many are phagocytic receptors that stimulate ingestion of the pathogens they recognize. Some are chemotactic receptors, such as the f-Met-Leu-Phe receptor, which binds the *N*-formylated peptides produced by bacteria and guides neutrophils to sites of infection. A third function, which may be mediated by some of the phagocytic receptors as well as by specialized signaling receptors, is to induce effector molecules that contribute to the induced responses of innate immunity and molecules that influence the initiation and nature of any subsequent adaptive immune response. In this part of the chapter, we will first examine the recognition properties of the receptors that bind pathogens directly. We will then focus on an evolutionarily primitive recognition and signaling system, originally discovered in the fruit-fly *Drosophila melanogaster* on account of its role in embryonic development, but now known to play a key role in defense against infection in plants, insects, and vertebrates, including mammals. The receptor mediating these functions in *Drosophila* is known as Toll. The homologous proteins in mammals have been named the Toll-like receptors and activate phagocytes and tissue dendritic cells to respond to pathogens.

2-15   Receptors with specificity for pathogen surfaces recognize patterns of repeating structural motifs.

The surfaces of microorganisms typically bear repeating patterns of molecular structure. The innate immune system recognizes such pathogens by means of receptors that bind features of these regular patterns; these receptors are sometimes known as **pattern-recognition molecules**. The mannan-binding lectin that initiates the MB-lectin pathway of complement activation (see Section 2-5) is one such receptor, as shown by structural studies of its binding. As illustrated in Fig. 2.28, pathogen recognition and discrimination from self is due to recognition of a particular orientation of certain sugar residues, as well as their spacing, which is found only on pathogenic microbes and not on host cells. Other members of the collectin family also bind pathogens directly and function in innate immunity. As we saw in Section 2-6, the collectin C1q is able to bind directly to pathogen surfaces and initiate complement activation through the classical pathway. In addition, other collectins are made in the liver as part of the acute-phase response, which will be described in the last part of the chapter. The exact structures recognized by these other collectins have not yet been defined, but all collectins have multiple carbohydrate-recognition domains attached to a collagen helix and are thought to bind pathogen surfaces in a similar way to mannan-binding lectin.

The interaction of these soluble receptors with pathogens leads in turn to binding of the receptor:pathogen complex by phagocytes, either through direct interaction with the pathogen-binding receptor, or through receptors for complement, thus promoting phagocytosis and killing of the bound pathogen (see Section 2-3) and the induction of other cellular responses.

Phagocytes are also equipped with several cell-surface receptors that recognize pathogen surfaces directly. Among these are the **macrophage mannose receptor** (see Section 2-3). This receptor is a cell-bound C-type lectin that binds certain sugar molecules found on the surface of many bacteria and some viruses, including the human immunodeficiency virus (HIV). Its recognition properties are very similar to those of mannan-binding lectin (see Fig. 2.28) and, like mannan-binding lectin, it is a multipronged molecule with

Fig. 2.28 Mannan-binding lectin (MBL) binds to patterns of carbohydrate groups in the correct spatial orientation. MBL is a member of the collectin family of proteins, composed of between two to six clusters of carbohydrate-binding lectin domains that interact with each other via a collagen-like domain. Within each cluster are three separate binding sites that have a fixed orientation relative to each other; all three sites can therefore only bind when their ligands—mannose and fucose residues in bacterial cell-wall polysaccharides—have the appropriate spacing.

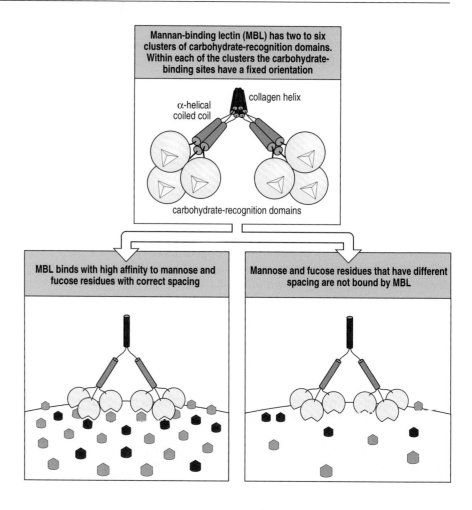

several carbohydrate-recognition domains. Because it is a transmembrane cell-surface receptor, however, it can function directly as a phagocytic receptor. A second set of phagocytic receptors, called **scavenger receptors**, recognize certain anionic polymers and also acetylated low-density lipoproteins. These receptors are a structurally heterogeneous set of molecules, existing in at least six distinct molecular forms. Some scavenger receptors recognize structures that are shielded by sialic acid on normal host cells. These receptors are involved in the removal of old red blood cells that have lost sialic acid, as well as in the recognition and removal of pathogens. There are other recognition targets, many of which still need to be characterized.

2-16 Receptors on phagocytes can signal the presence of pathogens.

In addition to triggering phagocytosis, binding of pathogens by macrophages can also trigger the induced responses of innate immunity, and responses that eventually lead to the induction of adaptive immunity. Not all the receptors on phagocytes appear to transmit such signals. The best-defined activation pathway of this type is triggered through a family of evolutionarily conserved transmembrane receptors that appear to function exclusively as signaling receptors. These receptors, known as the Toll receptors, were first described in the fruit-fly. They appear not to recognize and bind pathogens directly, but clearly are involved in signaling the appropriate response to different classes of pathogen. In the fruit-fly, the Toll receptor itself triggers the production of antifungal peptides in response to fungal infections, while a different member of the Toll family is involved in activating an antibacterial response. In mammals, a Toll-family protein, called **Toll-like receptor 4**, or **TLR-4**, signals the presence of LPS by associating with CD14, the macrophage receptor for LPS. TLR-4 is also involved in the immune response to at least one virus, respiratory syncytial virus, although in this case the nature of the stimulating ligand is not known. Another mammalian Toll-like receptor, TLR-2, signals the presence of a different set of microbial constituents, which include the proteoglycans of gram-positive bacteria, although how it recognizes these is not known. TLR-4 and TLR-2 induce similar but distinct signals, as shown by the distinct responses resulting from LPS signaling through TLR-4 and proteoglycan signaling through TLR-2. There are at least nine distinct proteins in this newly discovered family in mammals, and further functions of Toll-like receptors may soon be revealed as mice lacking one or other of these proteins are produced and analyzed.

2-17   The effects of bacterial lipopolysaccharide on macrophages are mediated by CD14 binding to Toll-like receptor 4.

Bacterial LPS is a cell-wall component of gram-negative bacteria that has long been known for its ability to induce a dramatic systemic reaction known as septic shock. Perhaps because of this, the best-characterized proteins in innate immunity are the plasma protein **LPS-binding protein** (**LBP**), and the receptor protein CD14 (see Section 2-3), which binds LBP-bound LPS. Both LBP and CD14 have leucine-rich repeat motifs. Although the structural details of the binding are not yet characterized, the LPS:LBP complex binds to CD14, which is either free in the plasma or bound to the cell surface by a phosphoinositol glycolipid tail (Fig. 2.29, top two panels). This binding triggers a cell response, but until recently the mechanism by which the signals were transduced across the cell membrane was unknown.

Mice were discovered that were genetically unresponsive to LPS and did not suffer from septic shock, but had no defects in LBP and CD14. The gene responsible was tracked down by positional cloning and turned out to be the gene for TLR-4, which had suffered inactivating mutations in these mice. The response to LPS could be restored by inserting a transgene encoding TLR-4 into the mouse germline, proving that the defect in TLR-4 was entirely responsible for the loss of responsiveness to LPS. It was subsequently found that TLR-4 binds to the CD14:LBP:LPS complex through a leucine-rich region in CD14's extracellular domain.

LPS responsiveness is most commonly assessed experimentally by the ability to induce LPS-mediated **septic shock**. This syndrome is the result of overwhelming secretion of cytokines, particularly of TNF-α, often as a result of an uncontrolled systemic bacterial infection. We will discuss the pathogenesis of septic shock later in this chapter, and will see that it is an undesirable consequence of the same effector actions of TNF-α that are important in containing local infections. The benefits of TLR-4 signaling are clearly illustrated by the mutant mice that lack TLR-4 function; although resistant to septic shock, they are highly sensitive to LPS-bearing pathogens such as *Salmonella typhimurium*, a natural pathogen of mice. Some people with gram-negative sepsis, an uncontrolled infection of the bloodstream with gram-negative bacteria, which indicates failure to contain an infection locally, have mutations in the TLR-4 gene, showing that TLR-4 is important in protecting against gram-negative sepsis in humans. Only a small fraction of patients have such mutations, however, so there may be other specific defects that contribute to a failure to respond adequately to gram-negative bacteria.

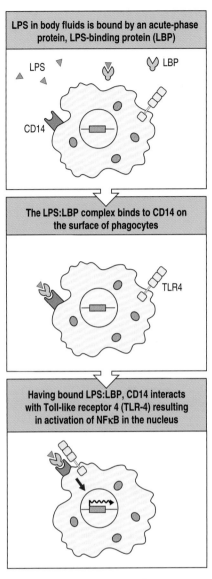

Fig. 2.29 Bacterial lipopolysaccharide signals through the Toll-like receptor TLR-4 to activate the transcription factor NFκB. TLR-4 is activated by the binding of bacterial lipopolysaccharide (LPS) through two other proteins. LPS is bound by the soluble LPS-binding protein, which then loads its bound LPS onto the phosphoinositol glycolipid-linked peripheral membrane protein CD14. This then triggers the membrane protein TLR-4 to signal to the nucleus, activating the transcription factor NFκB, which in turn activates genes involved in defense against infection.

When TLR-4 binds to CD14 complexed with its LBP:LPS ligand, TLR-4 sends a signal to the nucleus that activates the transcription factor NFκB (Fig. 2.29, bottom panel). We will describe the signaling pathway used by TLR-4 in detail in Chapter 6. This signaling pathway was first discovered as the pathway used by the Toll receptor to determine dorsoventral body pattern during embryogenesis in the fruit-fly *Drosophila*, and we will therefore call it the **Toll pathway**. The pathway was recently shown to participate in defense against infection in adult flies, and a similar pathway is used by plants in their defense against viruses. Thus the Toll pathway is an ancient signaling pathway that appears to be used in innate immune defense in most or all multicellular organisms.

2-18 Activation of Toll-like receptors triggers the production of pro-inflammatory cytokines and chemokines, and the expression of co-stimulatory molecules.

The Toll signaling pathway in adult flies induces the production of several antimicrobial peptides that contribute to the fly's defense against infection. In humans and all other vertebrates examined, activation of NFκB by the Toll pathway leads to the production of several important mediators of innate immunity, such as cytokines and chemokines. At the same time, Toll signaling induces molecules that are essential for the induction of adaptive immune responses. These are known as **co-stimulatory molecules**, and they will be considered in detail in Chapter 8. The co-stimulatory molecules, called B7.1 (CD80) and B7.2 (CD86), are cell-surface proteins that are expressed by both macrophages and tissue dendritic cells in response to LPS signaling through TLR-4. It is the presence of these molecules along with the microbial antigens presented by macrophages and dendritic cells (see Section 1-6) that activates the CD4 T cells required to initiate most adaptive immune responses. To encounter a CD4 T cell, the antigen-presenting dendritic cell must migrate to a nearby lymph node through which circulating T cells pass, and this migration is stimulated by cytokines such as TNF-α, which are also induced through signaling by TLR-4. Thus, the activation of adaptive immunity depends on molecules induced as a consequence of innate immune recognition and signaling (Fig. 2.30).

Substances such as LPS that induce co-stimulatory activity have been used for years in mixtures that are co-injected with protein antigens to enhance their immunogenicity. These substances are known as **adjuvants** (see Appendix I, Section A-4), and it was found empirically that the best adjuvants contained microbial components. Pathogen components that can induce macrophages and tissue dendritic cells to express co-stimulatory molecules

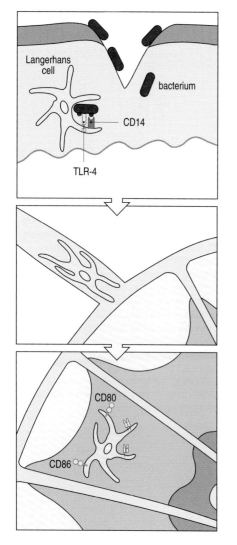

Fig. 2.30 Bacterial LPS induces changes in Langerhans' cells, stimulating them to migrate and initiate adaptive immunity to infection by activating CD4 T cells. Langerhans' cells are immature dendritic cells resident in the skin. In the case of a bacterial infection, they are activated by LPS via the Toll signaling pathway. This induces two types of change in the Langerhans' cells. The first is a change in behavior and location. From being resting cells in the skin they become activated migrating cells in the afferent lymphatics, and eventually fully mature dendritic cells in the regional lymph nodes. The second is a drastic alteration in their cell-surface molecules. Resting Langerhans' cells in the skin are highly phagocytic and macropinocytic, but lack the ability to activate T lymphocytes. Mature dendritic cells in the lymph nodes have lost the ability to take up antigen, but have gained the ability to stimulate T cells, by increasing the level of MHC molecules and by expressing the appropriate co-stimulatory molecules CD80 (B7.1) and CD86 (B7.2).

and cytokines include glycans, mannans, and the mycobacterial extract muramyl dipeptide. The receptor TLR-2 has recently been shown to be essential for the response to proteoglycans and it is thought that all these pathogen components are recognized by pattern-recognition molecules that then signal, often through Toll-like receptors, to induce co-stimulatory molecules and cytokines. The exact profile of cytokines produced varies according to the receptors involved and, as we will see in Chapters 8 and 10, this will in turn influence the functional character of the adaptive immune response that develops in their presence. In this way, the ability of the innate immune system to discriminate between different types of pathogen is used to ensure an appropriate type of adaptive immune response.

Summary.

The innate immune system uses a diversity of receptors to recognize and respond to pathogens. Those that recognize pathogen surfaces directly often bind to repeating patterns, for example, of carbohydrate or lipid moieties, that are characteristic of microbial surfaces but are not found on host cells. Some of these receptors, for example, the macrophage mannose receptor, directly stimulate phagocytosis, while others are produced as secreted molecules that promote the phagocytosis of pathogens by opsonization or by the activation of complement. The innate immune system receptors that recognize pathogens also have an important role in signaling for the induced responses responsible for local inflammation, the recruitment of new effector cells, the containment of local infection, and the initiation of an adaptive immune response. Such signals can be transmitted through a family of signaling receptors, known as the Toll-like receptors, that have been highly conserved across species and that serve to activate host defense through a signaling pathway that operates in most multicellular organisms. In mammals, Toll-like receptors also play a key role in enabling the initiation of adaptive immunity. TLR-4 detects the presence of gram-negative bacteria through its association with the peripheral membrane protein CD14, which is a receptor for bacterial LPS bound to its binding protein, LBP. TLR-2 signals in response to microbial proteoglycans, although how it recognizes them is not yet known. Ligation of TLR-2 or TLR-4 activates an evolutionarily ancient signaling pathway that leads to the activation of the transcription factor NFκB, and the induction of a variety of genes, including genes for cytokines, chemokines, and co-stimulatory molecules that play essential roles in directing the course of the adaptive immune response later in infection.

Induced innate responses to infection.

In this final part of this chapter we will look at the induced responses of innate immunity. These depend upon the cytokines and chemokines that are produced in response to pathogen recognition. We will therefore start with a brief overview of these proteins, followed by a description of how the macrophage-derived cytokines promote the phagocytic response through recruitment and production of fresh phagocytes and opsonizing molecules, while containing the spread of infection to the bloodstream through the activation of clotting mechanisms. We will also look at the role of the cytokines known as interferons, which are induced by viral infection, and at a class of lymphoid cells, known as natural killer (NK) cells, that are activated by interferons to contribute to innate host defense against viruses and other intracellular pathogens.

The induced innate responses either succeed in clearing the infection or contain it while an adaptive response develops. Adaptive immunity harnesses many of the same effector mechanisms that are used in the innate immune system, but is able to target them with greater precision. Thus antigen-specific T cells activate the microbicidal and cytokine-secreting properties of macrophages harboring pathogens, while antibodies activate complement, act as direct opsonins for phagocytes, and stimulate NK cells to kill infected cells. In addition, the adaptive immune response uses cytokines and chemokines, in a manner similar to that of innate immunity, to induce inflammatory responses that promote the influx of antibodies and effector lymphocytes to sites of infection. The effector mechanisms described here therefore serve as a primer for later chapters on adaptive immunity.

2-19 Activated macrophages secrete a range of cytokines that have a variety of local and distant effects.

Cytokines are small proteins (~25 kDa) that are released by various cells in the body, usually in response to an activating stimulus, and induce responses through binding to specific receptors. They can act in an autocrine manner, affecting the behavior of the cell that releases the cytokine, or in a paracrine manner, affecting the behavior of adjacent cells. Some cytokines can act in an endocrine manner, affecting the behavior of distant cells, although this depends on their ability to enter the circulation and on their half-life. **Chemokines** are a class of cytokines that have chemoattractant properties, inducing cells with the appropriate receptors to migrate toward the source of the chemokine. The cytokines secreted by macrophages in response to pathogens are a structurally diverse group of molecules and include interleukin-1 (IL-1), interleukin-6 (IL-6), interleukin-12 (IL-12), TNF-α, and the chemokine interleukin-8 (IL-8). The name **interleukin** (IL) followed by a number (for example IL-1, IL-2, and so on) was coined in an attempt to develop a standardized nomenclature for molecules secreted by, and acting on, leukocytes. However, this became confusing when an ever-increasing number of cytokines with diverse origins, structures, and effects were discovered, and although the IL designation is still used, it is hoped that eventually a nomenclature based on cytokine structure will be developed. The cytokines and their receptors are grouped according to their structure in the appendices at the end of this book (cytokines are listed in Appendix III and chemokines in Appendix IV). There are three major structural families: the hematopoietin family, which includes growth hormones as well as many interleukins with roles in both adaptive and innate immunity; the TNF family, which functions in both innate and adaptive immunity and includes some membrane-bound members; and the chemokine family, which we discuss below. Of the macrophage-derived interleukins shown in Fig. 2.31, IL-6 belongs to the large family of hematopoietins, TNF-α is obviously part of the TNF family, while IL-1 and IL-12 are structurally distinct. All have important local and systemic effects that contribute to both innate and adaptive immunity, and these are summarized in Fig. 2.31.

The recognition of different classes of pathogen may involve signaling through distinct receptors and result in some variation in the cytokines induced. The study of this is still in its infancy, but it is thought to be a way in which appropriate responses can be selectively activated as the released cytokines orchestrate the next phase of host defense. We will see how TNF-α, which is elicited by LPS-bearing pathogens, is particularly important in containing infection by these pathogens, and, how the release of different chemokines can recruit and activate different types of effector cells.

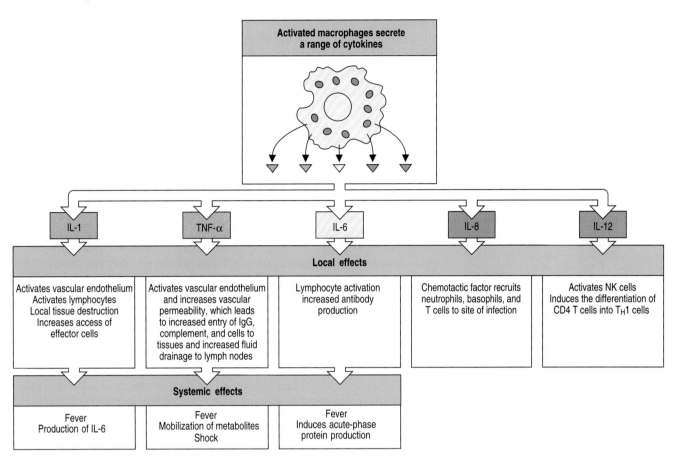

Fig. 2.31 Important cytokines secreted by macrophages in response to bacterial products include IL-1, IL-6, IL-8, IL-12, and TNF-α. TNF-α is an inducer of a local inflammatory response that helps to contain infections; it also has systemic effects, many of which are harmful (see Section 2-23). IL-8 is also involved in the local inflammatory response, helping to attract neutrophils to the site of infection. IL-1, IL-6, and TNF-α have a critical role in inducing the acute-phase response in the liver (see Section 2-24) and induce fever, which favors effective host defense in several ways. IL-12 activates natural killer (NK) cells and favors the differentiation of CD4 T cells into the T$_H$1 subset during adaptive immunity.

2-20   Chemokines released by phagocytes recruit cells to sites of infection.

Among the cytokines released in infected tissue in the earliest phases of infection are members of a family of chemoattractant cytokines known as chemokines. These molecules induce directed chemotaxis in nearby responsive cells and were discovered relatively recently. Because they were first detected in cytokine assays, they were initially named as interleukins: interleukin-8 (IL-8) was the first chemokine to be cloned and characterized, and it remains typical of this family (Fig. 2.32, top). All the chemokines are related in amino acid sequence and their receptors are all integral membrane proteins containing seven membrane-spanning helices. This structure is characteristic of receptors such as rhodopsin (Fig. 2.32, bottom) and the muscarinic acetyl-choline receptor, which are coupled to G proteins; the chemokine receptors also signal through coupled G proteins. Chemokines function mainly as chemoattractants for leukocytes, recruiting monocytes, neutrophils, and other effector cells from the blood to sites of infection. They can be released by many different types of cell and serve to guide cells involved in innate immunity and also the lymphocytes in adaptive immunity, as we will learn in Chapters 8–10. Some chemokines also function in lymphocyte development,

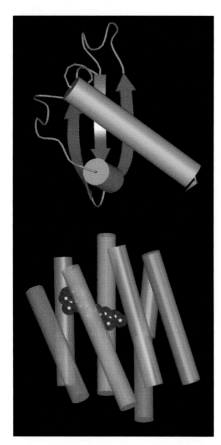

Fig. 2.32 Chemokines are a family of proteins of similar structure that bind to chemokine receptors, themselves part of a large family of G protein-coupled receptors. The chemokines are a large family of small proteins represented here by IL-8 (upper molecule). Each chemokine is thought to have a similar structure. The receptors for the chemokines are members of the large family of seven-span receptors, which also includes the photoreceptor protein rhodopsin and many other receptors. They have seven trans-membrane helices, and all members of this receptor family interact with G proteins. The only solved structure of a seven-span membrane protein is of the bacterial protein bacteriorhodopsin; it is depicted in the lower structure, showing the orientation of the seven trans-membrane helices (blue) with the bound ligand (in this case retinal) in red. Essentially all of this structure would be embedded within the cell membrane. Cylinders represent α helices and arrows β strands.

migration, and angiogenesis (the growth of new blood vessels). The properties of a variety of chemokines are listed in Fig. 2.33; quite why there are so many chemokines is not yet known, and neither is the exact role of each in defense against infection.

Members of the chemokine family fall mostly into two broad groups—CC chemokines with two adjacent cysteines near the amino terminus, and CXC chemokines, in which the equivalent two cysteine residues are separated by another amino acid. The two groups of chemokines act on different sets of receptors. CC chemokines bind to CC chemokine receptors, of which there are nine so far, designated CCR1–9. CXC chemokines bind to CXC receptors; there are five of these, CXCR1–5. These receptors are expressed on different cell types; in general, CXC chemokines with an Glu-Leu-Arg (ELR) tripeptide motif immediately before the first cysteine promote the migration of neutrophils. IL-8 is an example of this type of chemokine. Other CXC chemokines that lack this motif, such as the B-lymphocyte chemokine (BLC), guide lymphocytes to their proper destination. The CC chemokines promote the migration of monocytes or other cell types. An example is macrophage chemoattractant protein-1 (MCP-1). IL-8 and MCP-1 have similar, although complementary, functions: IL-8 induces neutrophils to leave the blood-stream and migrate into the surrounding tissues; MCP-1, in contrast, acts on monocytes, inducing their migration from the bloodstream to become tissue macrophages. Other CC chemokines such as RANTES may promote the infiltration into tissues of a range of leukocytes including effector T cells (see Section 10-8), with individual chemokines acting on different subsets of cells. The only known C chemokine (with only one cysteine) is called lymphotactin and is thought to attract T-cell precursors to the thymus. A newly discovered molecule called fractalkine is unusual in several ways: it has three amino acid residues between the two cysteines, making it a CX_3C chemokine; it is multi-modular; and it is tethered to the membrane of the cells that express it, where it serves both as a chemoattractant and as an adhesion protein. We will return to the discussion of chemokines in Chapter 10.

The role of chemokines such as IL-8 and MCP-1 in cell recruitment is twofold. First, they act on the leukocyte as it rolls along endothelial cells at sites of inflammation, converting this rolling into stable binding by triggering a change of conformation in the adhesion molecules known as leukocyte integrins. This allows the leukocyte to cross the blood vessel wall by squeezing between the endothelial cells, as we will see when we describe the process of extravasation. Second, the chemokines direct the migration of the leukocyte along a gradient of the chemokine that increases in concentration toward the site of infection. This is achieved by the binding of the small, soluble chemokines to proteoglycan molecules in the extracellular matrix and on endothelial cell surfaces, thus displaying the chemokines on a solid substrate along which the leukocytes can migrate.

Chemokines can be produced by a wide variety of cell types in response to bacterial products, viruses, and agents that cause physical damage, such as silica or the urate crystals that occur in gout. Thus, infection or physical damage to tissues sets in motion the production of chemokine gradients that can direct phagocytes to sites where they are needed. In addition, peptides that act as chemoattractants for neutrophils are made by bacteria themselves. All bacteria produce proteins with an amino-terminal N-formylated methionine, and, as discovered many years ago, the f-Met-Leu-Phe (fMLP) peptide is a potent chemotactic factor for inflammatory cells, especially neutrophils. The fMLP receptor is also a G protein-coupled receptor like the receptors for chemokines and for the complement fragments C5a, C3a, and C4a. Thus, there is a common mechanism for attracting neutrophils, whether by complement, chemokines, or bacterial peptides. Neutrophils are the first to arrive in large numbers at a site of infection, with monocytes and immature

Class	Chemokine	Produced by	Receptors	Cells attracted	Major effects
CXC	IL-8	Monocytes Macrophages Fibroblasts Keratinocytes Endothelial cells	CXCR1 CXCR2	Neutrophils Naive T cells	Mobilizes, activates and degranulates neutrophils Angiogenesis
	PBP β-TG NAP-2	Platelets	CXCR2	Neutrophils	Activates neutrophils Clot resorption Angiogenesis
	GROα, β, γ	Monocytes Fibroblasts Endothelium	CXCR2	Neutrophils Naive T cells Fibroblasts	Activates neutrophils Fibroplasia Angiogenesis
	IP-10	Keratinocytes Monocytes T cells Fibroblasts Endothelium	CXCR3	Resting T cells NK cells Monocytes	Immunostimulant Antiangiogenic Promotes T_H1 immunity
	SDF-1	Stromal cells	CXCR4	Naive T cells Progenitor (CD34$^+$) B cells	B-cell development Lymphocyte homing Competes with HIV-1
	BLC	Stromal cells	CXCR5	B cells	Lymphocyte homing
CC	MIP-1α	Monocytes T cells Mast cells Fibroblasts	CCR1, 3, 5	Monocytes NK and T cells Basophils Dendritic cells	Competes with HIV-1 Antiviral defense Promotes T_H1 immunity
	MIP-1β	Monocytes Macrophages Neutrophils Endothelium	CCR1, 3, 5	Monocytes NK and T cells Dendritic cells	Competes with HIV-1
	MCP-1	Monocytes Macrophages Fibroblasts Keratinocytes	CCR2B	Monocytes NK and T cells Basophils Dendritic cells	Activates macrophages Basophil histamine release Promotes T_H2 immunity
	RANTES	T cells Endothelium Platelets	CCR1, 3, 5	Monocytes NK and T cells Basophils Eosinophils Dendritic cells	Degranulates basophils Activates T cells Chronic inflammation
	Eotaxin	Endothelium Monocytes Epithelium T cells	CCR3	Eosinophils Monocytes T cells	Role in allergy
	DC-CK	Dendritic cells	?	Naive T cells	Role in activating naive T cells
C	Lymphotactin	CD8>CD4 T cells	?	Thymocytes Dendritic cells NK cells	Lymphocyte trafficking and development
CXXXC (CX₃C)	Fractalkine	Monocytes Endothelium Microglial cells	CX₃CR1	Monocytes T cells	Leukocyte–endothelial adhesion Brain inflammation

Fig. 2.33 Properties of selected chemokines. Chemokines fall mainly into two related but distinct groups: the CC chemokines, which in humans are mostly encoded in one region of chromosome 4, have two adjacent cysteine residues in their amino-terminal region; CXC chemokines, the genes for which are mainly found in a cluster on chromosome 17, have an amino acid residue between the equivalent two cysteines. These chemokines can be divided further by the presence or absence of an amino acid triplet (ELR; glutamic acid–leucine–arginine) preceding the first of these invariant cysteines. All the chemokines that attract neutrophils have this motif, while most of the other CXC chemokines, including the chemokines reacting with CXCR3, 4, and 5 lack it. A C chemokine with only one cysteine at this location, and fractalkine, a CX₃C chemokine, are encoded elsewhere in the genome. Each chemokine interacts with one or more receptors, and affects one or more types of cell. A comprehensive list of chemokines and their receptors is given in Appendix III.

dendritic cells being recruited later. The complement peptide C5a, and the chemokines IL-8 and MCP-1 also activate their respective target cells, so that not only are neutrophils and macrophages brought to potential sites of infection but, in the process, they are armed to deal with any pathogens they may encounter. In particular, neutrophils exposed to IL-8 and the cytokine TNF-α (see Fig. 2.37, and Section 2-23) are activated to produce the respiratory burst that generates oxygen radicals and nitric oxide, and to release their stored lysosomal contents, thus contributing both to host defense and to the tissue destruction and pus formation seen in local sites of infection with pyogenic bacteria.

Chemokines do not act alone in cell recruitment, which also requires the action of vasoactive mediators to bring leukocytes close to the blood vessel endothelium (see Section 2-4) and cytokines such as TNF-α to induce the necessary adhesion molecules on the endothelial cells. We will now turn to the molecules that mediate leukocyte–endothelium adhesion, and then describe the process of leukocyte extravasation step by step, as it is known to occur for neutrophils and monocytes.

2-21 Cell-adhesion molecules control interactions between leukocytes and endothelial cells during an inflammatory response.

The recruitment of activated phagocytes to sites of infection is one of the most important functions of innate immunity. Recruitment occurs as part of the inflammatory response and is mediated by cell-adhesion molecules that are induced on the surface of the local blood vessel endothelium. Before we consider the process of inflammatory cell recruitment we will first describe some of the cell-adhesion molecules involved.

A significant barrier to understanding cell-adhesion molecules is their nomenclature. Most cell-adhesion molecules, especially those on leukocytes, which are relatively easy to analyze functionally, were named after the effects of specific monoclonal antibodies against them, and were only later characterized by gene cloning. Their names therefore bear no relation to their structure; for instance, the **leukocyte functional antigens**, **LFA-1**, **LFA-2**, and **LFA-3**, are actually members of two different protein families. In Fig. 2.34, the adhesion molecules are grouped according to their molecular structure, which is shown in schematic form, alongside their different names, sites of expression, and ligands. Three families of adhesion molecules are important for leukocyte recruitment. The **selectins** are membrane glycoproteins with a distal lectinlike domain that binds specific carbohydrate groups. Members of this family are induced on activated endothelium and initiate endothelial–leukocyte interactions by binding to fucosylated oligosaccharide ligands on passing leukocytes (see Fig. 8.5). The next step in leukocyte recruitment depends on tighter adhesion, which is due to **intercellular adhesion molecules** (**ICAMs**) on the endothelium binding to heterodimeric proteins of the **integrin** family on leukocytes. We have already encountered two of the leukocyte integrins that function as complement receptors (CR3 and CR4). The leukocyte integrins important for extravasation are LFA-1 (α_L:β_2) and **Mac-1** (α_M:β_2; another name for CR3) and they bind to both **ICAM-1** and **ICAM-2** (Fig. 2.35). Strong adhesion between leukocytes and endothelial cells is promoted by the induction of ICAM-1 on inflamed endothelium and the activation of a conformational change in LFA-1 and Mac-1 in response to chemokines. The importance of the leukocyte integrins in inflammatory cell recruitment is illustrated by the disease **leukocyte adhesion deficiency**, which stems from a defect in the β_2 chain common to both LFA-1 and Mac-1. People with this disease suffer from recurrent bacterial infections and impaired healing of wounds.

Leukocyte Adhesion Deficiency

The activation of endothelium is driven by interactions with macrophage cytokines, particularly TNF-α, which induces rapid externalization of granules in the endothelial cells called **Weibel–Palade bodies**. These granules contain preformed **P-selectin**, which is thus expressed within minutes on the surface

	Name	Tissue distribution	Ligand
Selectins Bind carbohydrates. Initiate leukocyte–endothelial interaction P-selectin	P-selectin (PADGEM, CD62P)	Activated endothelium and platelets	PSGL-1, sialyl-Lewis^x
	E-selectin (ELAM-1, CD62E)	Activated endothelium	Sialyl-Lewis^x
Integrins Bind to cell-adhesion molecules and extracellular matrix. Strong adhesion LFA-1	$\alpha_L{:}\beta_2$ (LFA-1, CD11a/CD18)	Monocytes, T cells, macrophages, neutrophils, dendritic cells	ICAMs
	$\alpha_M{:}\beta_2$ (Mac-1, CR3, CD11b/CD18)	Neutrophils, monocytes, macrophages	ICAM-1, iC3b, fibrinogen
	$\alpha_X{:}\beta_2$ (CR4, p150.95, CD11c/CD18)	Dendritic cells, macrophages, neutrophils	iC3b
	$\alpha_5{:}\beta_1$ (VLA-5, CD49d/CD29)	Monocytes, macrophages	Fibronectin
Immunoglobulin superfamily Various roles in cell adhesion. Ligand for integrins ICAM-1	ICAM-1 (CD54)	Activated endothelium	LFA-1, Mac1
	ICAM-2 (CD102)	Resting endothelium, dendritic cells	LFA-1
	VCAM-1 (CD106)	Activated endothelium	VLA-4
	PECAM (CD31)	Activated leukocytes, endothelial cell–cell junctions	CD31

Fig. 2.34 Adhesion molecules in leukocyte interactions. Several structural families of adhesion molecules play a part in leukocyte migration, homing, and cell–cell interactions: the selectins, the integrins, and proteins of the immunoglobulin superfamily. The figure shows schematic representations of an example from each family, a list of other family members that participate in leukocyte interactions, their cellular distribution, and their ligand in adhesive interactions. The family members shown here are limited to those that participate in inflammation and other innate immune mechanisms. The same molecules and others participate in adaptive immunity and will be considered in Chapters 8 and 10. The nomenclature of the different molecules in these families is confusing because it often reflects the way in which the molecules were first identified rather than their related structural characteristics. Alternative names for each of the adhesion molecules are given in parentheses. Sulfated sialyl-Lewis^x, which is recognized by P- and E-selectin, is an oligosaccharide present on the cell-surface glycoproteins of circulating leukocytes. Sulfation can occur at either the sixth carbon atom of the galactose or the N-acetyl-glucosamine, but not both.

of local endothelial cells following production of TNF-α by macrophages. The same effect can be produced directly by exposing cultured human umbilical vein epithelial cells (HUVEC) to LPS, demonstrating that HUVEC can directly sense the presence of infection. Shortly after the appearance of P-selectin on the cell surface, mRNA encoding **E-selectin** is synthesized, and within 2 hours, the endothelial cells are mainly expressing E-selectin. Both these proteins interact with sulfated-sialyl-Lewis^x, which is present on the surface of neutrophils.

Resting endothelium carries low levels of ICAM-2, apparently in all vascular beds. This may be used by circulating monocytes to navigate out of the vessels and into their tissue sites, which happens continuously and essentially ubiquitously. However, upon exposure to TNF-α, local expression of ICAM-1 is strongly induced on the endothelium of small vessels within the infectious focus. This, in turn, binds to LFA-1 on circulating monocytes and polymorpho-nuclear leukocytes, in particular neutrophils, as shown in Fig. 2.35.

Cell-adhesion molecules have many other roles in the body, directing many aspects of tissue and organ development. In this brief description, we have considered only those that participate in the recruitment of inflammatory cells in the hours to days after the establishment of infection.

Fig. 2.35 Phagocyte adhesion to vascular endothelium is mediated by integrins. Vascular endothelium, when it is activated by inflammatory mediators, expresses two adhesion molecules—

ICAM-1 and ICAM-2. These are ligands for integrins expressed by phagocytes—$\alpha_L{:}\beta_2$ (also called LFA-1 or CD11a: CD18) and $\alpha_M{:}\beta_2$ (also called Mac-1, CR3, or CD11b:CD18).

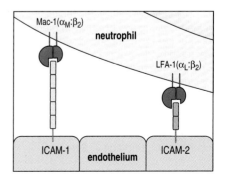

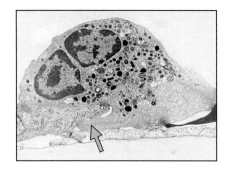

Fig. 2.36 Neutrophils leave the blood and migrate to sites of infection in a multistep process mediated through adhesive interactions that are regulated by macrophage-derived cytokines and chemokines. The first step (top panel) involves the reversible binding of leukocytes to vascular endothelium through interactions between selectins induced on the endothelium and their carbohydrate ligands on the leukocyte, shown here for E-selectin and its ligand the sialyl-Lewis^x moiety (s-Le^x). This interaction cannot anchor the cells against the shearing force of the flow of blood, and instead they roll along the endothelium, continually making and breaking contact. The binding does, however, allow stronger interactions, which occur as a result of the induction of ICAM-1 on the endothelium and the activation of its receptors LFA-1 and Mac-1 (not shown) on the leukocyte by contact with a chemokine like IL-8. Tight binding between these molecules arrests the rolling and allows the leukocyte to squeeze between the endothelial cells forming the wall of the blood vessel (to extravasate). The leukocyte integrins LFA-1 and Mac-1 are required for extravasation, and for migration toward chemoattractants. Adhesion between molecules of CD31, expressed on both the leukocyte and the junction of the endothelial cells, is also thought to contribute to extravasation. The leukocyte also needs to traverse the basement membrane; it penetrates this with the aid of a matrix metallo-proteinase enzyme that it expresses at the cell surface. Finally, the leukocyte migrates along a concentration gradient of chemokines (here shown as IL-8) secreted by cells at the site of infection. The electron micrograph shows a neutrophil extravasating between endothelial cells. The blue arrow indicates the pseudopod that the neutrophil is inserting between the endothelial cells. Photograph ($\times$ 5500) courtesy of I. Bird and J. Spragg.

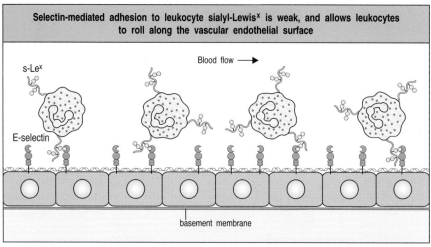

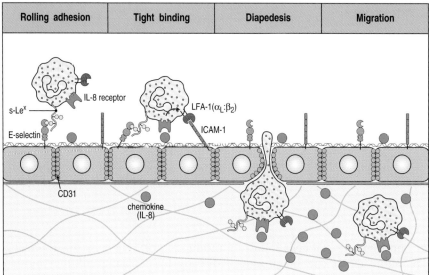

2-22   Neutrophils make up the first wave of cells that cross the blood vessel wall to enter inflammatory sites.

The physical changes that accompany the initiation of the inflammatory response have been described in Section 2-4; here we give a step-by-step account of how the required effector cells are recruited into sites of infection. Under normal conditions, leukocytes are restricted to the center of small blood vessels, where the flow is fastest. In inflammatory sites, where the vessels are dilated, the slower blood flow allows the leukocytes to move out of the center of the blood vessel and interact with the vascular endothelium. Even in the absence of infection, monocytes migrate continuously into the tissues, where they differentiate into macrophages; during an inflammatory response, the induction of adhesion molecules on the endothelial cells, as well as induced changes in the adhesion molecules expressed on leukocytes, recruit large numbers of circulating leukocytes, initially neutrophils and later monocytes, into the site of an infection. The migration of leukocytes out of blood vessels, a process known as **extravasation**, is thought to occur in four steps. We will describe this process as it is known to occur for monocytes and neutrophils (Fig. 2.36).

The first step involves selectins. P-selectin, which is carried inside endothelial cells in Weibel–Palade bodies, appears on endothelial cell surfaces within a few minutes of exposure to leukotriene B4, the complement fragment C5a, or histamine, which is released from mast cells in response to C5a. The appearance of P-selectin can also be induced by exposure to TNF-α or LPS, and both of these have the additional effect of inducing the synthesis of a second selectin, E-selectin, which appears on the endothelial cell surface a few hours later. These selectins recognize the sulfated-sialyl-Lewis^x moiety of certain leukocyte glycoproteins that are exposed on the tips of microvilli. The interaction of P-selectin and E-selectin with these glycoproteins allows monocytes and neutrophils to adhere reversibly to the vessel wall, so that circulating leukocytes can be seen to 'roll' along endothelium that has been treated with inflammatory cytokines (see Fig. 2.36, top panel). This adhesive interaction permits the stronger interactions of the next step in leukocyte migration.

This second step depends upon interactions between the leukocyte integrins LFA-1 and Mac-1 with molecules on endothelium such as ICAM-1, which is also induced on endothelial cells by TNF-α (see Fig. 2.36, bottom panel). LFA-1 and Mac-1 normally adhere only weakly, but IL-8 or other chemokines, bound to proteoglycans on the surface of endothelial cells, trigger a conformational change in LFA-1 and Mac-1 on the rolling leukocyte, which greatly increases its adhesive properties. In consequence, the leukocyte attaches firmly to the endothelium and rolling is arrested.

In the third step, the leukocyte extravasates, or crosses the endothelial wall. This step also involves LFA-1 and Mac-1, as well as a further adhesive interaction involving an immunoglobulin-related molecule called **PECAM** or **CD31**, which is expressed both on the leukocyte and at the intercellular junctions of endothelial cells. These interactions enable the phagocyte to squeeze between the endothelial cells. It then penetrates the basement membrane (an extracellular matrix structure) with the aid of proteolytic enzymes that break down the proteins of the basement membrane. The movement through the vessel wall is known as **diapedesis**, and enables phagocytes to enter the subepithelial tissues.

The fourth and final step in extravasation is the migration of leukocytes through the tissues under the influence of chemokines. As discussed in Section 2-20, chemokines such as IL-8 are produced at the site of infection and bind to proteoglycans in the extracellular matrix. They form a matrix-associated concentration gradient along which the leukocyte can migrate to the focus of infection. IL-8 is released by the macrophages that first encounter pathogens and recruits neutrophils, which enter the infected tissue in large numbers in the early part of the induced response. Their influx usually peaks within the first six hours of an inflammatory response, while monocytes can be recruited later, through the action of chemokines such as MCP-1. Once in an inflammatory site, neutrophils are able to eliminate many pathogens by phagocytosis. They act as phagocytic effectors in an innate immune response through receptors for complement and other opsonizing proteins of the innate immune system as well as by recognizing pathogens directly. In addition, as we will see in Chapter 9, they act as phagocytic effectors in humoral adaptive immunity. The importance of neutrophils is dramatically illustrated by diseases or treatments that severely reduce neutrophil numbers. Such patients are said to have **neutropenia**, and are very susceptible to infection with numerous pathogens. Restoring neutrophil levels in such patients by transfusion of neutrophil-rich blood fractions or by stimulating their production with specific growth factors largely corrects this susceptibility.

2-23   Tumor necrosis factor-α is an important cytokine that triggers local containment of infection, but induces shock when released systemically.

Inflammatory mediators also stimulate endothelial cells to express proteins that trigger blood clotting in the local small vessels, occluding them and cutting off blood flow. This can be important in preventing the pathogen from entering the bloodstream and spreading through the blood to organs all over the body. Instead, the fluid that has leaked into the tissue in the early phases of carries the pathogen enclosed in phagocytic cells, especially dendritic cells, via the lymph to the regional lymph nodes, where an adaptive immune response can be initiated. The importance of TNF-α in the containment of local infection is illustrated by experiments in which rabbits are infected locally with a bacterium. Normally, the infection will be contained at the site of the inoculation; if, however, an injection of anti-TNF-α antibody is also given to block the action of TNF-α, the infection spreads via the blood to other organs.

Once an infection spreads to the bloodstream, however, the same mechanisms by which TNF-α so effectively contains local infection instead become catastrophic (Fig. 2.37). The presence of infection in the bloodstream, known as **sepsis**, is accompanied by the release of TNF-α by macrophages in the liver, spleen, and other sites. The systemic release of TNF-α causes vasodilation and loss of plasma volume owing to increased vascular permeability, leading to shock. In **septic shock**, disseminated intravascular coagulation (blood clotting) is also triggered by TNF-α, leading to the generation of clots in many small vessels and the massive consumption of clotting proteins, thus causing the patient's ability to clot blood appropriately to be lost. This condition frequently leads to the failure of vital organs such as the kidneys, liver, heart, and lungs, which are quickly compromised by the failure of normal perfusion; consequently, septic shock has a high mortality rate.

Mice with a mutant TNF-α receptor gene are resistant to septic shock; however, such mice are also unable to control local infection. Although the features of TNF-α that make it so valuable in containing local infection are precisely those that give it a central role in the pathogenesis of septic shock, it is clear from the evolutionary conservation of TNF-α that its benefits in the former area far outweigh the devastating consequences of its systemic release.

2-24   Cytokines released by phagocytes activate the acute-phase response.

As well as their important local effects, the cytokines produced by macrophages have long-range effects that contribute to host defense. One of these is the elevation of body temperature, which is mainly caused by TNF-α, IL-1, and IL-6. These are termed **endogenous pyrogens** because they cause fever and derive from an endogenous source rather than from bacterial components. Fever is generally beneficial to host defense; most pathogens grow better at lower temperatures and adaptive immune responses are more intense at elevated temperatures. Host cells are also protected from the deleterious effects of TNF-α at raised temperatures.

The effects of TNF-α, IL-1, and IL-6 are summarized in Fig. 2.38. One of the most important of these is the initiation of a response known as the **acute-phase response** (Fig. 2.39). This involves a shift in the proteins secreted by the liver into the blood plasma and results from the action of IL-1, IL-6, and TNF-α on hepatocytes. In the acute-phase response, levels of some plasma proteins go down, while levels of others increase markedly. The proteins whose synthesis is induced by TNF-α, IL-1, and IL-6 are called **acute-phase**

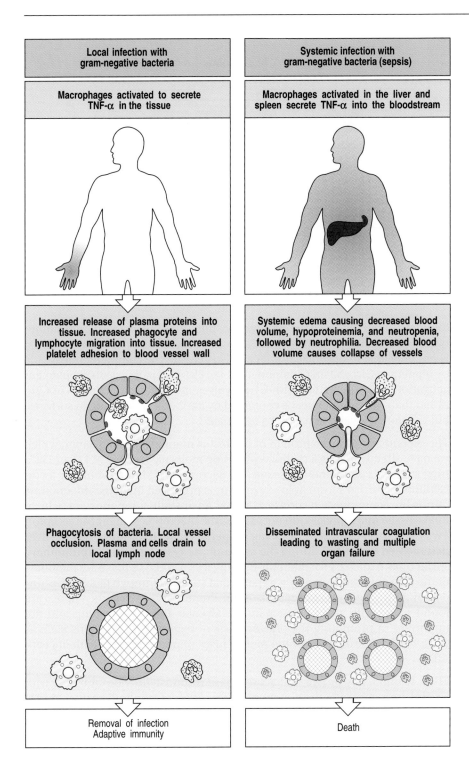

Local infection with gram-negative bacteria	Systemic infection with gram-negative bacteria (sepsis)
Macrophages activated to secrete TNF-α in the tissue	Macrophages activated in the liver and spleen secrete TNF-α into the bloodstream
Increased release of plasma proteins into tissue. Increased phagocyte and lymphocyte migration into tissue. Increased platelet adhesion to blood vessel wall	Systemic edema causing decreased blood volume, hypoproteinemia, and neutropenia, followed by neutrophilia. Decreased blood volume causes collapse of vessels
Phagocytosis of bacteria. Local vessel occlusion. Plasma and cells drain to local lymph node	Disseminated intravascular coagulation leading to wasting and multiple organ failure
Removal of infection Adaptive immunity	Death

Fig. 2.37 The release of TNF-α by macrophages induces local protective effects, but TNF-α can have damaging effects when released systemically. The panels on the left show the causes and consequences of local release of TNF-α, while the panels on the right show the causes and consequences of systemic release. Both left and right central panels illustrate the common effects of TNF-α, which acts on blood vessels, especially venules, to increase blood flow, to increase vascular permeability to fluid, proteins, and cells, and to increase endothelial adhesiveness for leukocytes and platelets. Local release thus allows an influx into the infected tissue of fluid, cells, and proteins that participate in host defense. Later, blood clots form in the small vessels, preventing spread of the infection via the blood, and the accumulated fluid and cells drain to regional lymph nodes where the adaptive immune response is initiated. When there is a systemic infection, or sepsis, with bacteria that elicit TNF-α production, then TNF-α is released into the blood by macrophages in the liver and spleen and acts in a similar way on all small blood vessels. The result is shock, disseminated intravascular coagulation with depletion of clotting factors and consequent bleeding, multiple organ failure, and frequently death. These effects require the presence of the TLR-4 protein on macrophages, which provides the initial signal in response to LPS.

proteins. Several of these are of particular interest because they mimic the action of antibodies, but, unlike antibodies, these proteins have broad specificity for pathogen-associated molecular patterns and depend only on the presence of cytokines for their production.

One of these proteins, **C-reactive protein**, is a member of the **pentraxin** protein family, so called because they are formed from five identical subunits. C-reactive protein is another example of a multipronged pathogen-recognition molecule, and binds to the phosphorylcholine portion of certain bacterial and fungal cell-wall lipopolysaccharides. Phosphorylcholine is also

Fig. 2.38 The cytokines TNF-α, IL-1, and IL-6 have a wide spectrum of biological activities that help to coordinate the body's responses to infection. IL-1, IL-6, and TNF-α activate hepatocytes to synthesize acute-phase proteins, and bone marrow endothelium to release neutrophils. The acute-phase proteins act as opsonins, while the disposal of opsonized pathogens is augmented by enhanced recruitment of neutrophils from the bone marrow. IL-1, IL-6, and TNF-α are also endogenous pyrogens, raising body temperature, which is believed to help eliminate infections. A major effect of these cytokines is to act on the hypothalamus, altering the body's temperature regulation, and on muscle and fat cells, altering energy mobilization to increase the body temperature. At elevated temperatures, bacterial and viral replication are decreased, while the adaptive immune response operates more efficiently.

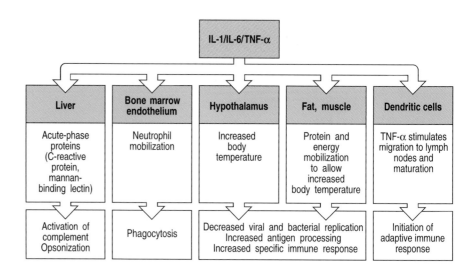

found in mammalian cell membrane phospholipids but in a form that cannot bind to C-reactive protein. When C-reactive protein binds to a bacterium, it is not only able to opsonize it but can also activate the complement cascade by binding to C1q, the first component of the classical pathway of complement activation. The interaction with C1q involves the collagen-like parts of C1q, rather than the globular heads contacted by antibody and pathogen surfaces, but the same cascade of reactions is initiated.

The second acute-phase protein of interest is mannan-binding lectin, which we have already met as a pathogen-binding molecule (see Fig. 2.13) and as a trigger for the complement cascade (see Section 2-7). Mannan-binding lectin is found in normal serum at low levels but is produced in increased amounts during the acute-phase response. It acts as an opsonin for monocytes, which, unlike tissue macrophages, do not express the macrophage mannose receptor. Two other proteins with opsonizing properties that are produced by the liver in increased quantities during the acute-phase response are the **pulmonary surfactants A** and **D**. Like mannan-binding lectin and C1q these are members of the collectin family, binding to pathogen surfaces through globular lectin-domains attached to a collagen-like stalk. Pulmonary surfactants A and D are found along with macrophages in the alveolar fluid of the lung and are important in promoting the phagocytosis of respiratory pathogens such as *Pneumocystis carinii*, one of the main causes of pneumonia in patients with AIDS.

Thus, within a day or two, the acute-phase response provides the host with several proteins with the functional properties of antibodies that bind a broad range of pathogens. However, unlike antibodies, they have no structural diversity and are made in response to any stimulus that triggers the release of TNF-α, IL-1, and IL-6, so their synthesis is not specifically induced and targeted.

A final distant effect of the cytokines produced by phagocytes is to induce a **leukocytosis**, an increase in circulating neutrophils. The neutrophils come from two sources: the bone marrow, from which mature leukocytes are released in increased numbers; and sites in blood vessels where they are attached loosely to endothelial cells. Thus, the effects of these cytokines contribute to the control of infection while the adaptive immune response is being developed. As shown in Fig. 2.30, TNF-α also has a role in this, as it stimulates the migration of dendritic cells from their sites in peripheral tissues to the lymph node, and their maturation into nonphagocytic but highly co-stimulatory antigen-presenting cells.

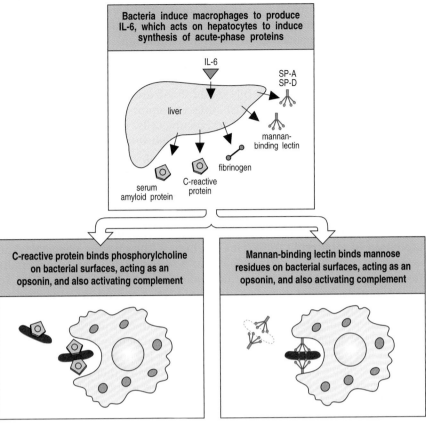

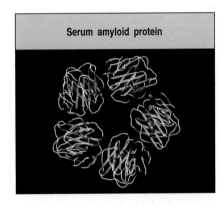

Fig. 2.39 The acute-phase response produces molecules that bind pathogens but not host cells. Acute-phase proteins are produced by liver cells in response to cytokines released by macrophages in the presence of bacteria. They include serum amyloid protein (SAP) (in mice but not humans), C-reactive protein (CRP), fibrinogen, and mannan-binding lectin (MBL). SAP and CRP are homologous in structure; both are pentraxins, forming five-membered discs, as shown for SAP (photograph on the right). CRP binds phosphorylcholine on certain bacterial and fungal surfaces but does not recognize it in the form in which it is found in host cell membranes. It both acts as an opsonin in its own right and activates the classical complement pathway by binding C1q to augment opsonization. MBL is a member of the collectin family, which includes C1q, which it resembles in its structure. We have already seen how MBL activates complement (see Section 2-7) and how it binds to pathogen surfaces (see Fig. 2.28). Like CRP, MBL can act as an opsonin in its own right, in addition to activating complement. SP-A and SP-D are surfactants A and D, both of which are collectins that coat bacterial surfaces, facilitating their phagocytosis. Photograph courtesy of J. Emsley.

2-25 Interferons induced by viral infection make several contributions to host defense.

Infection of cells with viruses induces the production of proteins that are known as **interferons** because they were found to interfere with viral replication in previously uninfected tissue culture cells. They are believed to have a similar role *in vivo*, blocking the spread of viruses to uninfected cells. These antiviral effector molecules, called **interferon-α (IFN-α)** and **interferon-β (IFN-β)**, are quite distinct from **interferon-γ (IFN-γ)**. This is not directly induced by viral infection, although it is produced later and does have an important role in the induced response to intracellular pathogens, as we will see below. IFN-α, actually a family of several closely related proteins, and IFN-β, the product of a single gene, are synthesized by many cell types following their infection by diverse viruses. Interferon synthesis is thought to occur in response to the presence of double-stranded RNA, as synthetic double-stranded RNA is a potent inducer of interferon. Double-stranded RNA, which is not found in

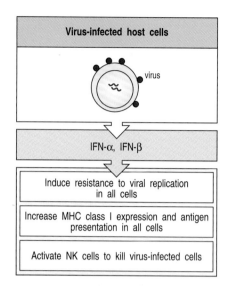

Virus-infected host cells

virus

IFN-α, IFN-β

Induce resistance to viral replication in all cells

Increase MHC class I expression and antigen presentation in all cells

Activate NK cells to kill virus-infected cells

Fig. 2.40 Interferons are antiviral proteins produced by cells in response to viral infection. The interferons (IFN)-α and -β have three major functions. First, they induce resistance to viral replication in uninfected cells by activating genes that cause the destruction of mRNA and inhibit the translation of viral and some host proteins. Second, they can induce MHC class I expression in most cell types in the body, thus enhancing their resistance to NK cells; they may also induce increased synthesis of MHC class I molecules in cells that are newly infected by virus, thus making them more susceptible to killing by CD8 cytotoxic T cells (see Chapter 8). Third, they activate NK cells, which then kill virus-infected cells selectively.

mammalian cells, forms the genome of some viruses and might be made as part of the infectious cycle of all viruses. Therefore, double-stranded RNA might be the common element in interferon induction.

Interferons make several contributions to defense against viral infection (Fig. 2.40). An obvious and important effect is the induction of a state of resistance to viral replication in all cells. IFN-α and IFN-β are secreted by the infected cell and then bind to a common cell-surface receptor, known as the interferon receptor, on both the infected cell and nearby cells. The interferon receptor, like many other cytokine receptors, is coupled to a Janus-family tyrosine kinase, through which it signals. This signaling pathway, which we will describe in detail in Chapter 6, rapidly induces new gene transcription as the Janus-family kinases directly phosphorylate signal-transducing activators of transcription known as STATs, which translocate to the nucleus where they activate the transcription of several different genes. In this way interferon induces the synthesis of several host cell proteins that contribute to the inhibition of viral replication. One of these is the enzyme oligoadenylate synthetase, which polymerizes ATP into a series of 2′–5′ linked oligomers (nucleotides in nucleic acids are normally linked 3′–5′). These activate an endoribonuclease that then degrades viral RNA. A second protein activated by IFN-α and IFN-β is a serine–threonine kinase called P1 kinase. This enzyme phosphorylates the eukaryotic protein synthesis initiation factor eIF-2, inhibiting translation and thus contributing to the inhibition of viral replication. Another interferon-inducible protein called Mx is known to be required for cellular resistance to influenza virus replication. Mice that lack the gene for Mx are highly susceptible to infection with the influenza virus, whereas genetically normal mice are not.

Another way interferons protect the host against viruses is by upregulating the cellular immune response to these pathogens. The adaptive immune response to viruses depends on their effective presentation to T cells as peptide fragments complexed with MHC class I molecules at the cell surface, and interferons promote this by inducing increased expression of these molecules. Interferons also activate natural killer (NK) cells to kill virus-infected cells and release cytokines. Although of lymphoid origin, NK cells do not have antigen-specific receptors and are therefore part of the innate immune system. It is not entirely clear what allows them to discriminate between infected and noninfected cells, but they possess both activating and inhibitory receptors. The latter inhibit killing when bound to normal MHC class I molecules and this means that the higher the expression of MHC class I on a cell surface, the more protected it is against destruction by NK cells. Therefore, interferons protect uninfected host cells from NK cells by upregulating class I MHC expression, while activating the NK cells to kill infected cells. Interferons also promote the release of effector cytokines by NK cells, as we will see in the next section.

2-26   Natural killer cells are activated by interferons and macrophage-derived cytokines to serve as an early defense against certain intracellular infections.

Natural killer cells (NK cells) develop in the bone marrow from the common lymphoid progenitor cell and circulate in the blood. They are larger than T and B lymphocytes, have distinctive cytoplasmic granules, and are functionally identified by their ability to kill certain lymphoid tumor cell lines *in vitro* without the need for prior immunization or activation. The mechanism of NK cell killing is the same as that used by the cytotoxic T cells generated in an adaptive immune response; cytotoxic granules are released onto the surface of the bound target cell, and the effector proteins they contain penetrate the

Fig. 2.41 Natural killer cells (NK cells) are an early component of the host response to virus infection. Experiments in mice have shown that IFN-α, IFN-β, and the cytokines TNF-α and IL-12 appear first, followed by a wave of NK cells, which together control virus replication but do not eliminate the virus. Virus elimination is accomplished when virus-specific CD8 T cells are produced. Without NK cells, the levels of some viruses are much higher in the early days of the infection, and can be lethal unless treated vigorously with antiviral compounds.

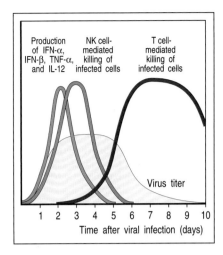

cell membrane and induce programmed cell death. However, NK cell killing is triggered by invariant receptors, and their known function in host defense is in the early phases of infection with several intracellular pathogens, particularly herpes viruses, the protozoan parasite *Leishmania*, and the bacterium *Listeria monocytogenes*.

NK cells are activated in response to interferons or macrophage-derived cytokines. Although NK cells that can kill sensitive targets can be isolated from uninfected individuals, this activity is increased by between twentyfold and one hundredfold when NK cells are exposed to IFN-α and IFN-β or to the NK cell-activating factor IL-12, which is one of the cytokines produced early in many infections. Activated NK cells serve to contain virus infections while the adaptive immune response generates antigen-specific cytotoxic T cells that can clear the infection (Fig. 2.41). At present the only clue to the physiological function of NK cells in humans comes from a rare patient deficient in NK cells who proved highly susceptible to early phases of herpes virus infection.

IL-12, in synergy with TNF-α, can also elicit the production of large amounts of IFN-γ by NK cells, and this secreted IFN-γ is crucial in controlling some infections before T cells have been activated to produce this cytokine. One example is the response to *Listeria monocytogenes*. Mice that lack T and B lymphocytes are initially quite resistant to this pathogen; however, antibody-mediated depletion of NK cells or neutralization of TNF-α or IFN-γ or their receptors renders these mice highly susceptible, so that they die a few days after infection before an adaptive immune response can be mounted.

2-27 NK cells possess receptors for self molecules that inhibit their activation against uninfected host cells.

If NK cells are to mediate host defense against infection with viruses and other pathogens, they must have some mechanism for distinguishing infected from uninfected cells. Exactly how this is achieved has not yet been worked out, but recognition of 'altered self' is thought to be involved. NK cells have two types of surface receptor that control their cytotoxic activity. One type is an 'activating receptor:' it triggers killing by the NK cell. Several types of receptor provide this activation signal, including calcium-binding C-type lectins that recognize a wide variety of carbohydrate ligands present on many cells. A second set of receptors inhibit activation, and prevent NK cells from killing normal host cells. These 'inhibitory receptors' are specific for MHC class I alleles, which helps to explain why NK cells selectively kill target cells bearing low levels of MHC class I molecules. Altered expression of MHC class I molecules may be a common feature of cells infected by intracellular pathogens, as many of these have developed strategies to interfere with the ability of MHC class I molecules to capture and display peptides to T cells. Thus, one possible mechanism by which NK cells distinguish infected from uninfected cells is by recognizing alterations in MHC class I expression (Fig. 2.42). Another is that they recognize changes in cell-surface glycoproteins induced by viral or bacterial infection.

Fig. 2.42 Possible mechanisms by which NK cells distinguish infected from uninfected cells. A proposed mechanism of recognition is shown. NK cells can use several different receptors that signal them to kill, including lectinlike activating receptors, or 'killer receptors,' that recognize carbohydrate on self cells. However, another set of receptors, called Ly49 in the mouse and killer inhibitory receptors (KIRs) in the human, recognize MHC class I molecules and inhibit killing by NK cells by overruling the actions of the killer receptors. This inhibitory signal is lost when cells do not express MHC class I and perhaps also in cells infected with virus, which might inhibit MHC class I expression or alter its conformation. Another possibility is that normal uninfected cells respond to IFN-α and IFN-β by increasing expression of MHC class I molecules, making them resistant to killing by activated NK cells. In contrast, infected cells can fail to increase MHC class I expression, making them targets for activated NK cells. Ly49 and KIR belong to different protein families—the C-type lectins in the case of Ly49 and the immunoglobulin superfamily for KIRs. The KIRs are made in two forms, p58 and p70, which differ by the presence of one immunoglobulin domain.

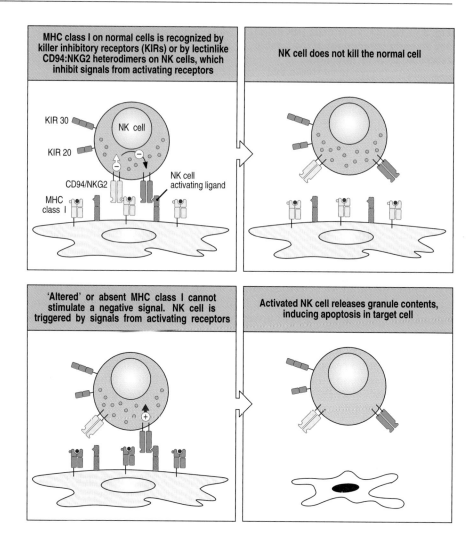

In mice, inhibitory receptors on NK cells are encoded by a multigene family of C-type lectins called Ly49. Different Ly49 receptors recognize different MHC class I alleles and are differentially expressed on different subsets of NK cells. Some NK cells express Ly49 receptors specific for nonself MHC alleles, but each cell expresses at least one receptor that can recognize an MHC class I allele expressed by the host. In humans, there are inhibitory receptors that recognize distinct HLA-B and HLA-C alleles (these are MHC class I alleles encoded by the B and C loci of the human MHC or Human Leukocyte Antigen gene complex). Although the MHC class I molecules of humans and mice are very similar, these human NK receptors are structurally different from those of the mouse, being members of the immunoglobulin gene superfamily; they are usually called p58 and p70, or killer inhibitory receptors (KIRs). In addition, human NK cells express a heterodimer of two C-type lectins, called CD94 and NKG2. The CD94:NKG2 receptor is also found in mice, and interacts with nonpolymorphic MHC-like molecules, HLA-E in man and Qa-1 in mice, that bind the leader peptides of other MHC class I molecules; thus this receptor may be sensitive to the presence of several different MHC class I alleles. Other inhibitory NK receptors specific for the products of the MHC class I loci are rapidly being defined, and all are members of either the immunoglobulin-like KIR family or the Ly49-like C-type lectins.

Signaling by the inhibitory NK receptors suppresses the killing activity of NK cells. This means that NK cells will not kill healthy genetically identical cells with normal expression of MHC class I molecules, such as the other cells of the body. Virus-infected cells, however, can become susceptible to killing by NK cells by a variety of mechanisms. First, some viruses inhibit all protein synthesis in their host cells, so synthesis of MHC class I proteins would be blocked in infected cells, even while being augmented by interferon in uninfected cells. The reduced level of MHC class I expression in infected cells would make them correspondingly less able to inhibit NK cells through their MHC-specific receptors, and therefore more susceptible to killing. Second, some viruses can selectively prevent the export of MHC class I molecules, which might allow the infected cell to evade recognition by the cytotoxic T cells of the adaptive immune response but would make it susceptible to killing by NK cells. There is also evidence that NK cells can detect the changes in MHC class I molecules that occur when they form complexes with peptides from proteins synthesized as a result of infection, instead of the self peptides from the proteins normally made by the cell. It is not known whether these peptides are recognized directly or whether they alter MHC conformation. Finally, virus infection alters the glycosylation of cellular proteins, perhaps allowing recognition by activating receptors to dominate or removing the normal ligand for the inhibitory receptors. Either of these last two mechanisms could allow infected cells to be detected even when the level of MHC class I expression had not been altered.

Clearly much remains to be learned about this innate mechanism of cytotoxic attack and its physiological relevance. The role of MHC molecules in allowing NK cells to detect intracellular infections is of particular interest as these same molecules govern the response of T cells to intracellular pathogens. It is possible that NK cells, which use a diverse set of nonclonotypic receptors to detect altered MHC, represent the modern remnants of the evolutionary forebears of T cells, which evolved rearranging genes that encode a vast repertoire of antigen-specific receptors geared to recognizing altered MHC.

2-28   Several lymphocyte subpopulations and 'natural antibodies' behave like intermediates between adaptive and innate immunity.

Receptor gene rearrangements are a defining characteristic of the lymphocytes of the adaptive immune system, and allow the generation of an infinite variety of receptors, each expressed by a different individual T or B cell (see Section 1-10). However, there are several minor lymphocyte subsets that express only a very limited diversity of receptors, encoded by a few common rearrangements. These lymphocytes do not need to undergo clonal expansion before responding effectively to the antigens they recognize, and therefore behave like intermediates between adaptive and innate immunity.

One such group of cells is the intraepithelial subset of γ:δ T cells. The γ:δ T cells are themselves a minor subset of T cells that express receptors that are distinct from the α:β receptors found on the majority of T cells involved in adaptive immunity. They were discovered as a consequence of having immunoglobulin-like receptors encoded by rearranged genes and their function remains obscure. One of their most striking features is their division into two highly distinct sets of cells. One set of γ:δ T cells is found in the lymphoid tissue of all vertebrates and, like B cells and α:β T cells, they display highly diversified receptors. By contrast, intraepithelial γ:δ T cells occur variably in different vertebrates, and commonly display receptors of very limited diversity, particularly in the skin and the female reproductive tract of mice, where the γ:δ T cells are essentially homogeneous in any one site. On the basis of this limited diversity of epithelial γ:δ T-cell receptors and their limited recirculatory behavior, it has been proposed that intraepithelial γ:δ T cells

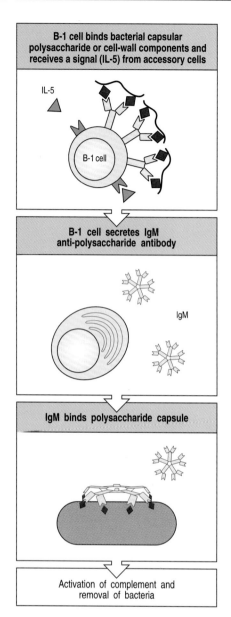

Fig. 2.43 CD5 B cells might be important in the response to carbohydrate antigens such as bacterial polysaccharides. These T-cell independent responses occur rapidly, with antibody appearing in 48 hours after infection, presumably because there is a high frequency of precursors of the responding lymphocytes so that little clonal expansion is required. In the absence of antigen-specific T-cell help, only IgM is made and, in mice, these responses therefore work mainly through the activation of complement, which is most efficient when the antibody is of the IgM isotype.

may recognize ligands that are derived from the epithelium in which they reside, but which are expressed only when a cell has become infected. Candidate ligands are heat-shock proteins, MHC class IB molecules, and unorthodox nucleotides and phospholipids, for all of which there is evidence of recognition by γ:δ T cells. Unlike α:β T cells, γ:δ T cells do not generally recognize antigen as peptides presented by MHC molecules; instead they seem to recognize their target antigens directly, and could potentially recognize and respond rapidly to molecules expressed by many different cell types. Recognition of molecules expressed as a consequence of infection, rather than of pathogen-specific antigens themselves, would distinguish γ:δ T cells from other lymphocytes and arguably place them at the intersection between innate and adaptive immunity. However, several recent studies of mice deficient in γ:δ T cells have revealed exaggerated responses to various pathogens and even to self tissues, rather than deficiencies in pathogen control and rejection. This has led to the suggestion that at least some γ:δ T cells have a regulatory role in modulating immune responses, a function that would be consistent with their demonstrated ability to secrete regulatory cytokines when activated. Which aspects of the phenotype of γ:δ-deficient mice are attributable to which subset of γ:δ T cells remains to be clarified.

Another subset of lymphocytes that express a limited diversity of receptors is the B-1 subset of B cells. B cells of this lineage are distinguished by the cell-surface protein CD5 and have properties quite distinct from those of conventional B cells that mediate adaptive humoral immunity. These so-called CD5 B cells, or B-1 cells, are in many ways analogous to epithelial γ:δ T cells: they arise early in ontogeny, they use a distinctive and limited set of gene rearrangements to make their receptors, they are self-renewing in the periphery, and they are the predominant lymphocyte in a distinctive microenvironment, the peritoneal cavity.

B-1 cells seem to make antibody responses mainly to polysaccharide antigens and can produce antibodies of the IgM class without needing T-cell help (Fig. 2.43). Although these responses can be augmented by T cells, they appear within 48 hours of the exposure to antigen, and the T cells involved are therefore not part of an antigen-specific adaptive immune response. The lack of an antigen-specific interaction with helper T cells might explain why immunological memory is not generated: repeated exposures to the same antigen elicit similar or decreased responses with each exposure. Thus, these responses, although generated by lymphocytes with rearranging receptors, resemble innate rather than adaptive immune responses.

As with γ:δ T cells, the precise role of B-1 cells in host defense is uncertain. Mice that are deficient in B-1 cells are more susceptible to infection with *Streptococcus pneumoniae* because they fail to produce an antibody against the phospholipid headgroup phosphorylcholine that effectively protects against this organism. A significant fraction of the B-1 cells can make antibodies of this specificity, and because no antigen-specific T-cell help is required, a potent response can be produced early in infection with this pathogen. Whether human B-1 cells have the same role is uncertain.

In terms of evolution, it is interesting to note that γ:δ T-cells seem to defend the body surfaces, whereas B-1 cells defend the body cavity. Both cell types are relatively limited in their range of specificities and in the efficiency of their responses. It is possible that these two cell types represent a transitional phase in the evolution of the adaptive immune response, guarding the two main compartments of primitive organisms—the epithelial surfaces and the body cavity. It is not yet clear whether they are still critical to host defense or whether they represent an evolutionary relic. Nevertheless, as each cell type is prominent in certain sites in the body and contributes to certain responses, they must be incorporated into our thinking about host defense.

Finally, there is a collection of antibodies known as 'natural antibody.' This 'natural IgM' is encoded by rearranged antibody genes that have not undergone somatic mutation. It makes up a considerable amount of the IgM circulating in humans and does not appear to be a result of an antigen-specific adaptive immune response to infection. It has a low affinity for many microbial pathogens, and is very highly cross-reactive, even binding to some self molecules. It is unknown whether this natural antibody has any role in host defense or which type of B cells produce it. Furthermore, it is not known whether it is produced in response to the normal flora of the epithelial surfaces of the body or in response to self. However, it might play a role in host defense by binding to the earliest infecting pathogens and clearing them before they become dangerous.

Summary.

Innate immunity can use a variety of induced effector mechanisms to clear an infection or, failing that, to hold it in check until the pathogen can be recognized by the adaptive immune system. These effector mechanisms are all regulated by germline-encoded receptor systems that are able to discriminate between noninfected self and infectious nonself ligands. Thus the phagocytes' ability to discriminate between self and pathogen controls its release of pro-inflammatory chemokines and cytokines that act together to recruit more phagocytic cells, especially neutrophils, which can also recognize pathogens, to the site of infection. Furthermore, cytokines released by tissue phagocytic cells induce fever, the production of acute-phase response proteins including the pathogen-binding mannan-binding lectin and the C-reactive proteins, and the mobilization of antigen-presenting cells that induce the adaptive immune response. Viral pathogens are recognized by the cells in which they replicate, leading to the production of interferons that serve to inhibit viral replication and to activate NK cells, which in turn can distinguish infected from noninfected cells. As we will see later in this book, cytokines, chemokines, phagocytic cells, and NK cells are all effector mechanisms that also are employed in an adaptive immune response that uses clonotypic receptors to target specific pathogen antigens.

Summary to Chapter 2.

The innate system of host defense against infection is made up of several distinct components. The first of these are barrier functions of the body's epithelia, which can prevent infection from becoming established altogether. Next, there are cells and molecules available to control or destroy the pathogen once it has breached the epithelial defenses. The most important of these are the tissue macrophages mediating cellular defense of the borders, and the complement system of proteins mediating humoral innate immunity of the tissue spaces and the blood. Understanding how the innate immune system recognizes pathogens is in its infancy, but structural studies, such as those of mannan-binding lectin, have begun to reveal in detail how innate immune receptors can distinguish pathogen surfaces from host cells. Furthermore, with the identification of the LPS receptor and its link to the human Toll-like receptor-4, it has now been determined how gram-negative pathogens as a class are recognized and responded to. Another member of the family of Toll-like proteins, the Toll-like receptor-2, responds to gram-positive pathogens by recognizing their proteoglycans. More information about recognition of many other classes of pathogen is likely to follow soon. Recognition by the innate immune system leads to elimination of invading pathogens through various effector mechanisms. Most of these have been known about for a long time; indeed, the elimination of microorganisms by phagocytosis was the first immune response to be observed. However, more is being learned all the time; the chemokines, for example, have only been known about for about 10 years, and over 50 chemokine proteins have now been discovered. The induction of powerful effector mechanisms on the basis of immune recognition through germline-encoded receptors clearly has some dangers. Indeed the double-edged sword embodied by the effects of the TNF-α protein—beneficial when it is released locally but disastrous when it is released systemically—illustrates the evolutionary knife-edge down which all innate mechanisms of host defense travel. The innate immune system can be viewed as a defense system that mainly frustrates the establishment of an infectious focus; however, when it is inadequate to this function, it can set the scene for the adaptive immune response, which forms an essential part of host defense in humans. Thus, having introduced the study of immunology with a consideration of innate immune function, we will now turn our attention to the adaptive immune response. This has been the focus of nearly all studies in immunology, because it is much easier to follow, and experiment with, reagents and responses that are specific for defined antigens.

General references.

Ezekowitz, R.A.B., and Hoffman, J.: **Innate immunity.** *Curr. Opin. Immunol.* 1998, **10**:9-53.

Fearon, D.T., and Locksley, R.M.: **The instructive role of innate immunity in the acquired immune response.** *Science* 1996, **272**:50-53.

Gallin, J.I., Goldstein, I.M., and Snyderman, R. (eds): *Inflammation—Basic Principles and Clinical Correlates*, 3rd edn. New York, Raven Press, 1999.

Mandell, G.L., Bennett, J.E., and Dolin, R. (eds): *Principles and Practice of Infectious Diseases*, 4th edn. New York, Churchill Livingstone, 1995.

Picker, L.J., and Butcher, E.C.: **Physiological and molecular mechanisms of lymphocyte homing.** *Annu. Rev. Immunol.* 1993, **10**:561-591.

Salyers, A.A., and Whitt, D.D.: *Bacterial Pathogenesis, A Molecular Approach.* Washington, DC, ASM Press, 1994.

Section references.

2-1   **Infectious agents must overcome innate host defenses to establish a focus of infection.**

Gibbons, R.J.: How microorganisms cause disease, in Gorbach, S.L., Bartlett, J.G., and Blacklow, N.R. (eds): *Infectious Diseases.* Philadelphia, W.B. Saunders Co., 1992.

2-2   **The epithelial surfaces of the body are the first defenses against infection.**

Krisanaprakornkit, S., Kimball, J.R., Weinberg, A., Darveau, R.P., Bainbridge, B.W., and Dale, B.A.: **Inducible expression of human beta-defensin 2 by** *Fusobacterium nucleatum* **in oral epithelial cells: multiple signaling pathways and role of commensal bacteria in innate immunity and the epithelial barrier.** *Infect. Immun.* 2000, **68**:2907-2915.

Ouellette, A.J.: **IV. Paneth cell antimicrobial peptides and the biology of the mucosal barrier.** *Am. J. Physiol.* 1999, **277**:G257-G261.

Podolsky, D.K.: **Mucosal immunity and inflammation. V. Innate mechanisms of mucosal defense and repair: the best offense is a good defense.** *Am. J. Physiol.* 1999, **277**:G495-G499.

2-3   **After entering tissues, many pathogens are recognized, ingested, and killed by phagocytes.**

Aderem, A., and Underhill, D.M.: **Mechanisms of phagocytosis in macrophages.** *Annu. Rev. Immunol.* 1999, **17**:593-623.

Zhang, P., Summer, W.R., Bagby, G.J., and Nelson, S.: **Innate immunity and pulmonary host defense.** *Immunol. Rev.* 2000, **173**:39-51.

2-4   **Pathogen recognition and tissue damage initiate an inflammatory response.**

Svanborg, C., Godaly, G., and Hedlund, M.: **Cytokine responses during mucosal infections: role in disease pathogenesis and host defence.** *Curr. Opin. Microbiol.* 1999, **2**:99-105.

2-5   **Complement is a system of plasma proteins that interacts with pathogens to mark them for destruction by phagocytes.**

Frank, M.M., and Fries, L.F.: **The role of complement in inflammation and phagocytosis.** *Immunol. Today* 1991, **12**:322-326.

Tomlinson, S.: **Complement defense mechanisms.** *Curr. Opin. Immunol.* 1993, **5**:83-89.

2-6   **The classical pathway is initiated by activation of the C1 complex.**

Cooper, N.R.: **The classical complement pathway. Activation and regulation of the first complement component.** *Adv. Immunol.* 1985, **37**:151-216.

Perkins, S.J., and Nealis, A.S.: **The quaternary structure in solution of human complement subcomponent $C1r_2Cls_2$.** *Biochem. J.* 1989, **263**:463-469.

Prodeus, A.P., Zhou, X., Maurer, M., Galli, S.J., and Carroll, M.C.: **Impaired mast cell-dependent natural immunity in complement C3-deficient mice.** *Nature* 1997, **390**:172-175.

2-7   The mannan-binding lectin pathway is homologous to the classical pathway.

Lu, J.: **Collectins: collectors of microorganisms for the innate immune system.** *Bioessays* 1997, **19**:509-518.

Vasta, G.R., Quesenberry, M., Ahmed, H., and O'Leary, N.: **C-type lectins and galectins mediate innate and adaptive immune functions: their roles in the complement activation pathway.** *Dev. Comp. Immunol.* 1999, **23**:401-420.

2-8   Complement activation is largely confined to the surface on which it is initiated.

Cicardi, M., Bergamaschini, L., Cugno, M., Beretta, A., Zingale, L.C., Colombo, M., and Agostoni, A.: **Pathogenetic and clinical aspects of C1 inhibitor deficiency.** *Immunobiology* 1998, **199**:366-376.

2-9   Hydrolysis of C3 causes initiation of the alternative pathway of complement.

Kolb, W.P., Morrow, P.R., and Tamerius, J.D.: **Ba and Bb fragments of Factor B activation: fragment production, biological activities, neoepitope expression and quantitation in clinical samples.** *Complement Inflamm.* 1989, **6**:175-204.

2-10   Surface-bound C3 convertase deposits large numbers of C3b fragments on pathogen surfaces and generates C5 convertase activity.

deBruijn, M.H.L., and Fey, G.M.: **Human complement component C3: cDNA coding sequence and derived primary structure.** *Proc. Natl. Acad. Sci. USA* 1985, **82**:708-712.

Volanakis, J.E.: **Participation of C3 and its ligand in complement activation.** *Curr. Top. Microbiol. Immunol.* 1989, **153**:1-21.

2-11   Phagocyte ingestion of complement-tagged pathogens is mediated by receptors for the bound complement proteins.

Ehlers, M.R.: **CR3: a general purpose adhesion-recognition receptor essential for innate immunity.** *Microbes. Infect.* 2000, **2**:289-294.

Fijen, C.A., Bredius, R.G., Kuijper, E.J., Out, T.A., De Haas, M., De Wit, A.P., Daha, M.R., and De Winkel, J.G.: **The role of Fcγ receptor polymorphisms and C3 in the immune defence against *Neisseria meningitidis* in complement-deficient individuals.** *Clin. Exp. Immunol.* 2000, **120**:338-345.

Linehan, S.A., Martinez-Pomares, L., and Gordon, S.: **Macrophage lectins in host defence.** *Microbes. Infect.* 2000, **2**:279-288.

Ross, G.D.: **Regulation of the adhesion versus cytotoxic functions of the Mac-1/CR3/αMβ2-integrin glycoprotein.** *Crit. Rev. Immunol.* 2000, **20**:197-222.

2-12   Small fragments of some complement proteins can initiate a local inflammatory response.

DiScipio, R.G., Daffern, P.J., Jagels, M.A., Broide, D.H., and Sriramarao, P.: **A comparison of C3a and C5a-mediated stable adhesion of rolling eosinophils in postcapillary venules and transendothelial migration *in vitro* and *in vivo*.** *J. Immunol.* 1999, **162**:1127-1136.

Foreman, K.E., Glovsky, M.M., Warner, R.L., Horvath, S.J., and Ward, P.A.: **Comparative effect of C3a and C5a on adhesion molecule expression on neutrophils and endothelial cells.** *Inflammation* 1996, **20**:1-9.

Gerard, C., Gerard, N.P.: **C5a anaphylatoxin and its seven transmembrane-segment receptor.** *Annu. Rev. Immunol.* 1994, **12**:775-808.

Hartmann, K., Henz, B.M., Kruger-Krasagakes, S., Kohl, J., Burger, R., Guhl, S., Haase, I., Lippert, U., and Zuberbier, T.: **C3a and C5a stimulate chemotaxis of human mast cells.** *Blood* 1997, **89**:2863-2870.

Hopken, U.E., Lu, B., Gerard, N.P., and Gerard, C.: **The C5a chemoattractant receptor mediates mucosal defence to infection.** *Nature* 1996, **383**:86-89.

2-13   The terminal complement proteins polymerize to form pores in membranes that can kill certain pathogens.

Bhakdi, S., and Tranum-Jensen, J.: **Complement lysis: a hole is a hole.** *Immunol. Today* 1991, **12**:318-320.

Esser, A.F.: **Big MAC attack: complement proteins cause leaky patches.** *Immunol. Today* 1991, **12**:316-318.

2-14   Complement control proteins regulate all three pathways of complement activation and protect the host from its destructive effects.

Kirschfink, M.: **Controlling the complement system in inflammation.** *Immunopharmacology* 1997, **38**:51-62.

Liszewski, M.K., Farries, T.C., Lublin, D.M., Rooney, I.A., and Atkinson, J.P.: **Control of the complement system.** *Adv. Immunol.* 1996, **61**:201-283.

Pangburn, M.K.: **Host recognition and target differentiation by factor H, a regulator of the alternative pathway of complement.** *Immunopharmacology* 2000, **49**:149-157.

Suankratay, C., Mold, C., Zhang, Y., Lint, T.F., and Gewurz, H.: **Mechanism of complement-dependent haemolysis via the lectin pathway: role of the complement regulatory proteins.** *Clin. Exp. Immunol.* 1999, **117**:442-448.

Suankratay, C., Mold, C., Zhang, Y., Potempa, L.A., Lint, T.F., and Gewurz, H.: **Complement regulation in innate immunity and the acute-phase response: inhibition of mannan-binding lectin-initiated complement cytolysis by C-reactive protein (CRP).** *Clin. Exp. Immunol.* 1998, **113**:353-359.

Zipfel, P.F., Jokiranta, T.S., Hellwage, J., Koistinen, V., and Meri, S.: **The factor H protein family.** *Immunopharmacology* 1999, **42**:53-60.

2-15   Receptors with specificity for pathogen surfaces recognize patterns of repeating structural motifs.

Fraser, I.P., Koziel, H., and Ezekowitz, R.A.: **The serum mannose-binding protein and the macrophage mannose receptor are pattern recognition molecules that link innate and adaptive immunity.** *Semin. Immunol.* 1998, **10**:363-372.

Gough, P.J., and Gordon, S.: **The role of scavenger receptors in the innate immune system.** *Microbes Infect.* 2000, **2**:305-311.

Kaisho, T., and Akira, S.: **Critical roles of toll-like receptors in host defense.** *Crit. Rev. Immunol.* 2000, **20**:393-405.

Lien, E., Sellati, T.J., Yoshimura, A., Flo, T.H., Rawadi, G., Finberg, R.W., Carroll, J.D., Espevik, T., Ingalls, R.R., Radolf, J.D., and Golenbock, D.T.: **Toll-like receptor 2 functions as a pattern recognition receptor for diverse bacterial products.** *J. Biol. Chem.* 1999, **274**:33419-33425.

Linehan, S.A., Martinez-Pomares, L., and Gordon, S.: **Macrophage lectins in host defence.** *Microbes Infect.* 2000, **2**:279-288.

2-16   Receptors on phagocytes can signal the presence of pathogens.

Kopp, E.B., and Medzhitov, R.: **The Toll-receptor family and control of innate immunity.** *Curr. Opin. Immunol.* 1999, **11**:13-18.

Medzhitov, R., and Janeway, C.A., Jr.: **How does the immune system distinguish self from nonself?** *Semin. Immunol.* 2000, **12**:185-188; discussion 257-344.

Medzhitov, R., and Janeway, C.A., Jr.: **The toll receptor family and microbial recognition.** *Trends Microbiol.* 2000, **8**:452-456.

2-17   The effects of bacterial lipopolysaccharide on macrophages are mediated by CD14 binding to Toll-like receptor 4.

Beutler, B.: **Endotoxin, toll-like receptor 4, and the afferent limb of innate immunity.** *Curr. Opin. Microbiol.* 2000, **3**:23-28.

Rhee, S.H., and Hwang, D.: **Murine TOLL-like receptor 4 confers lipopolysaccharide responsiveness as determined by activation of NFκB and expression of the inducible cyclooxygenase.** *J. Biol. Chem.* 2000, **275**:34035-34040.

Takeuchi, O., Hoshino, K., Kawai, T., Sanjo, H., Takada, H., Ogawa, T., Takeda,

K., and Akira, S.: **Differential roles of TLR2 and TLR4 in recognition of gram-negative and gram-positive bacterial cell wall components.** *Immunity* 1999, **11**:443-451.

Tapping, R.I., Akashi, S., Miyake, K., Godowski, P.J., and Tobias, P.S.: **Toll-like receptor 4, but not toll-like receptor 2, is a signaling receptor for escherichia and salmonella lipopolysaccharides.** *J. Immunol.* 2000, **165**:5780-5787.

2-18   Activation of Toll-like receptors triggers the production of pro-inflammatory cytokines and chemokines, and the expression of co-stimulatory molecules.

Bowie, A., and O'Neill, L.A.: **The interleukin-1 receptor/Toll-like receptor superfamily: signal generators for pro-inflammatory interleukins and microbial products.** *J. Leukoc. Biol.* 2000, **67**:508-514.

Brightbill, H.D., Libraty, D.H., Krutzik, S.R., Yang, R.B., Belisle, J.T., Bleharski, J.R., Maitland, M., Norgard, M.V., Plevy, S.E., Smale, S.T., Brennan, P.J., Bloom, B.R., Godowski, P.J., and Modlin, R.L.: **Host defense mechanisms triggered by microbial lipoproteins through toll-like receptors.** *Science* 1999, **285**:732-736.

2-19   Activated macrophages secrete a range of cytokines that have a variety of local and distant effects.

Larsson, B.M., Larsson, K., Malmberg, P., and Palmberg, L.: **Gram positive bacteria induce IL-6 and IL-8 production in human alveolar macrophages and epithelial cells.** *Inflammation* 1999, **23**:217-230.

Svanborg, C., Godaly, G., and Hedlund, M.: **Cytokine responses during mucosal infections: role in disease pathogenesis and host defence.** *Curr. Opin. Microbiol.* 1999, **2**:99-105.

2-20   Chemokines released by phagocytes recruit cells to sites of infection.

DeVries, M.E., Ran, L., and Kelvin, D.J.: **On the edge: the physiological and pathophysiological role of chemokines during inflammatory and immunological responses.** *Semin. Immunol.* 1999, **11**:95-104.

Scapini, P., Lapinet-Vera, J.A., Gasperini, S., Calzetti, F., Bazzoni, F., and Cassatella, M.A.: **The neutrophil as a cellular source of chemokines.** *Immunol. Rev.* 2000, **177**:195-203.

2-21   Cell-adhesion molecules control interactions between leukocytes and endothelial cells during an inflammatory response.

Jagels, M.A., Daffern, P.J., and Hugli, T.E.: **C3a and C5a enhance granulocyte adhesion to endothelial and epithelial cell monolayers: epithelial and endothelial priming is required for C3a-induced eosinophil adhesion.** *Immunopharmacology* 2000, **46**:209-222.

Muller, W.A., and Randolph, G.J.: **Migration of leukocytes across endothelium and beyond: molecules involved in the transmigration and fate of monocytes.** *J. Leukoc. Biol.* 1999, **66**:698-704.

Thompson, R.D., Wakelin, M.W., Larbi, K.Y., Dewar, A., Asimakopoulos, G., Horton, M.A., Nakada, M.T., and Nourshargh, S.: **Divergent effects of platelet-endothelial cell adhesion molecule-1 and beta 3 integrin blockade on leukocyte transmigration** *in vivo. J. Immunol.* 2000, **165**:426-434.

2-22   Neutrophils make up the first wave of cells that cross the blood vessel wall to enter inflammatory sites.

Frenette, P.S., and Wagner, D.D.: **Insights into selectin function from knock-out mice.** *Thromb. Haemost.* 1997, **78**:60-64.

Knott, P.G., Gater, P.R., Dunford, P.J., Fuentes, M.E., and Bertrand, C.P.: **Rapid up-regulation of CXC chemokines in the airways after Ag-specific CD4(⁺) T cell activation.** *J. Immunol.* 2001, **166**:1233-1240.

Lee, S.C., Brummet, M.E., Shahabuddin, S., Woodworth, T.G., Georas, S.N., Leiferman, K.M., Gilman, S.C., Stellato, C., Gladue, R.P., Schleimer, R.P., and Beck, L.A.: **Cutaneous injection of human subjects with macrophage inflammatory protein-1 alpha induces significant recruitment of neutrophils and monocytes.** *J. Immunol.* 2000, **164**:3392-3401.

Muruve, D.A., Barnes, M.J., Stillman, I.E., and Libermann, T.A.: **Adenoviral gene therapy leads to rapid induction of multiple chemokines and acute neutrophil-dependent hepatic injury** *in vivo. Hum. Gene. Ther.* 1999, **10**:965-976.

2-23   Tumor necrosis factor-α is an important cytokine that triggers local containment of infection, but induces shock when released systemically.

Hultgren, O., Eugster, H.P., Sedgwick, J.D., Korner, H., and Tarkowski, A.: **TNF/lymphotoxin-alpha double-mutant mice resist septic arthritis but display increased mortality in response to** *Staphylococcus aureus. J. Immunol.* 1998, **161**:5937-5942.

Krishnaswamy, G., Kelley, J., Yerra, L., Smith, J.K., and Chi, D.S.: **Human endothelium as a source of multifunctional cytokines: molecular regulation and possible role in human disease.** *J. Interferon Cytokine Res.* 1999, **19**:91-104.

Rigato, O., Ujvari, S., Castelo, A., and Salomao, R.: **Tumor necrosis factor α (TNF-α) and sepsis: evidence for a role in host defense.** *Infection* 1996, **24**:314-318.

Sriskandan, S., and Cohen, J.: **Gram-positive sepsis. Mechanisms and differences from gram-negative sepsis.** *Infect. Dis. Clin. North. Am.* 1999, **13**:397-412.

2-24   Cytokines released by phagocytes activate the acute-phase response.

Bopst, M., Haas, C., Car, B., and Eugster, H.P.: **The combined inactivation of tumor necrosis factor and interleukin-6 prevents induction of the major acute phase proteins by endotoxin.** *Eur. J. Immunol.* 1998, **28**:4130-4137.

Horn, F., Henze, C., and Heidrich, K.: **Interleukin-6 signal transduction and lymphocyte function.** *Immunobiology* 2000, **202**:151-167.

Ramadori, G., and Christ, B.: **Cytokines and the hepatic acute-phase response.** *Semin. Liver Dis.* 1999, **19**:141-155.

Uhlar, C.M., and Whitehead, A.S.: **Serum amyloid A, the major vertebrate acute-phase reactant.** *Eur. J. Biochem.* 1999, **265**:501-523.

2-25   Interferons induced by viral infection make several contributions to host defense.

Balachandran, S., Roberts, P.C., Brown, L.E., Truong, H., Pattnaik, A.K., Archer, D.R., and Barber, G.N.: **Essential role for the dsRNA-dependent protein kinase PKR in innate immunity to viral infection.** *Immunity* 2000, **13**:129-141.

Bogdan, C.: **The function of type I interferons in antimicrobial immunity.** *Curr. Opin. Immunol.* 2000, **12**:419-424.

Durbin, J.E., Fernandez-Sesma, A., Lee, C.K., Rao, T.D., Frey, A.B., Moran, T.M., Vukmanovic, S., Garcia-Sastre, A., and Levy, D.E.: **Type I IFN modulates innate and specific antiviral immunity.** *J. Immunol.* 2000, **164**:4220-4228.

Hayward, A.R., Chmura, K., and Cosyns, M.: **Interferon-gamma is required for innate immunity to** *cryptosporidium parvum* **in mice.** *J. Infect. Dis.* 2000, **182**:1001-1004.

Kadowaki, N., Antonenko, S., Lau, J.Y., and Liu, Y.J.: **Natural interferon alpha/beta-producing cells link innate and adaptive immunity.** *J. Exp. Med.* 2000, **192**:219-226.

2-26   Natural killer cells are activated by interferons and macrophage-derived cytokines to serve as an early defense against certain intracellular infections.

Biron, C.A., Nguyen, K.B., Pien, G.C., Cousens, L.P., and Salazar-Mather, T.P.: **Natural killer cells in antiviral defense: function and regulation by innate cytokines.** *Annu. Rev. Immunol.* 1999, **17**:189-220.

Carnaud, C., Lee, D., Donnars, O., Park, S.H., Beavis, A., Koezuka, Y., and Bendelac, A.: **Cutting edge: Cross-talk between cells of the innate immune system: NKT cells rapidly activate NK cells.** *J. Immunol.* 1999, **163**:4647-4650.

Salazar-Mather, T.P., Hamilton, T.A., and Biron, C.A.: **A chemokine-to-cytokine-to-chemokine cascade critical in antiviral defense.** *J. Clin. Invest.* 2000, **105**:985-993.

2-27 NK cells possess receptors for self molecules that inhibit their activation against uninfected host cells.

Lanier, L.L.: **NK cell receptors.** *Annu. Rev. Immunol.* 1998, **16**:359-393.

Renard, V., Cambiaggi, A., Vely, F., Blery, M., Olcese, L., Olivero, S., Bouchet, M., and Vivier, E.: **Transduction of cytotoxic signals in natural killer cells: a general model of fine tuning between activatory and inhibitory pathways in lymphocytes.** *Immunol. Rev.* 1997, **155**:205-221. (Plus other papers in this issue of Immunological Reviews, which is dedicated to NK receptors.)

Vales-Gomez, M., Reyburn, H., and Strominger, J.: **Molecular analyses of the interactions between human NK receptors and their HLA ligands.** *Hum. Immunol.* 2000, **61**:28-38.

2-28 Several lymphocyte subpopulations and 'natural antibodies' behave like intermediates between adaptive and innate immunity.

Carroll, M.C., and Prodeus, A.P.: **Linkages of innate and adaptive immunity.** *Curr. Opin. Immunol.* 1998, **10**:36-40.

Fagarasan, S., Watanabe, N., and Honjo, T.: **Generation, expansion, migration and activation of mouse B1 cells.** *Immunol. Rev.* 2000, **176**:205-215.

Godfrey, D.I., Hammond, K.J., Poulton, L.D., Smyth, M.J., and Baxter, A.G.: **NKT cells: facts, functions and fallacies.** *Immunol. Today* 2000, **21**:573-583.

Macpherson, A.J., Gatto, D., Sainsbury, E., Harriman, G.R., Hengartner, H., and Cramer, D.V.: **Natural antibodies and the host immune responses to xenografts.** *Xenotransplantation* 2000, **7**:83-92.

Seaman, W.E.: **Natural killer cells and natural killer T cells.** *Arthritis Rheum.* 2000, **43**:1204-1217.

Zinkernagel, R.M.: **A primitive T cell-independent mechanism of intestinal mucosal IgA responses to commensal bacteria.** *Science* 2000, **288**:2222-2226.

PART II

THE RECOGNITION OF ANTIGEN

Antigen Recognition by B-cell and T-cell Receptors

3

We have learned in Chapter 2 that the body is defended by innate immune responses, but these will only work to control pathogens that have certain molecular patterns or that induce interferons and other secreted yet non-specific defenses. Most crucially, they do not allow memory to form as they operate by receptors that are coded in the genome. Thus, innate immunity is good for preventing pathogens from growing freely in the body, but it does not lead to the most important feature of adaptive immunity, which is long-lasting memory of specific pathogen.

To recognize and fight the wide range of pathogens an individual will encounter, the lymphocytes of the adaptive immune system have evolved to recognize a great variety of different antigens from bacteria, viruses, and other disease-causing organisms. The antigen-recognition molecules of B cells are the **immunoglobulins**, or **Ig**. These proteins are produced by B cells in a vast range of antigen specificities, each B cell producing immunoglobulin of a single specificity (see Sections 1-8 to 1-10). Membrane-bound immunoglobulin on the B-cell surface serves as the cell's receptor for anti-gen, and is known as the **B-cell receptor** (**BCR**). Immunoglobulin of the same antigen specificity is secreted as **antibody** by terminally differentiated B cells—the plasma cells. The secretion of antibodies, which bind pathogens or their toxic products in the extracellular spaces of the body, is the main effector function of B cells in adaptive immunity.

Antibodies were the first molecules involved in specific immune recognition to be characterized and are still the best understood. The antibody molecule has two separate functions: one is to bind specifically to molecules from the pathogen that elicited the immune response; the other is to recruit other cells and molecules to destroy the pathogen once the antibody is bound to it. For example, binding by antibody neutralizes viruses and marks pathogens for destruction by phagocytes and complement, as described in Section 1-14. These functions are structurally separated in the antibody molecule, one part of which specifically recognizes and binds to the pathogen or antigen whereas the other engages different effector mechanisms. The antigen-binding region varies extensively between antibody molecules and is thus known as the **variable region** or **V region**. The variability of antibody molecules allows each antibody to bind a different specific antigen, and the total repertoire of anti-bodies made by a single individual is large enough to ensure that virtually any structure can be recognized. The region of the antibody molecule that engages the effector functions of the immune system does not vary in the same way and is thus known as the **constant region** or **C region**. It comes in five main forms, which are each specialized for activating different effector mechanisms. The membrane-bound B-cell receptor does not have these effector functions, as the C region remains inserted in the membrane of the B cell. Its function is as a receptor that recognizes and binds antigen by the V regions exposed on the surface of the cell, thus transmitting a signal that causes B-cell activation leading to clonal expansion and specific antibody production.

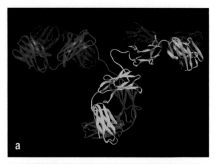

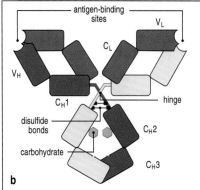

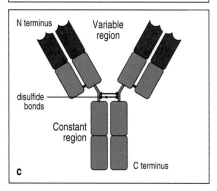

Fig. 3.1 Structure of an antibody molecule. Panel a illustrates a ribbon diagram based on the X-ray crystallographic structure of an IgG antibody, showing the course of the backbones of the polypeptide chains. Three globular regions form a Y. The two antigen-binding sites are at the tips of the arms, which are tethered to the trunk of the Y by a flexible hinge region. A schematic representation of the structure in a is given in panel b, illustrating the four-chain composition and the separate domains comprising each chain. Panel c shows a simplified schematic representation of an antibody molecule that will be used throughout this book. Photograph courtesy of A. McPherson and L. Harris.

The antigen-recognition molecules of T cells are made solely as membrane-bound proteins and only function to signal T cells for activation. These **T-cell receptors** (**TCRs**) are related to immunoglobulins both in their protein structure—having both V and C regions—and in the genetic mechanism that produces their great variability (see Section 1-10 and Chapter 4). However, the T-cell receptor differs from the B-cell receptor in an important way: it does not recognize and bind antigen directly, but instead recognizes short peptide fragments of pathogen protein antigens, which are bound to **MHC molecules** on the surfaces of other cells.

The MHC molecules are glycoproteins encoded in the large cluster of genes known as the **major histocompatibility complex** (**MHC**) (see Sections 1-16 and 1-17). Their most striking structural feature is a cleft running across their outermost surface, in which a variety of peptides can be bound. As we shall discuss further in Chapter 5, MHC molecules show great genetic variation in the population, and each individual carries up to 12 of the possible variants, which increases the range of pathogen-derived peptides that can be bound. T-cell receptors recognize features both of the peptide antigen and of the MHC molecule to which it is bound. This introduces an extra dimension to antigen recognition by T cells, known as **MHC restriction**, because any given T-cell receptor is specific not simply for a foreign peptide antigen, but for a unique combination of a peptide and a particular MHC molecule. The ability of T-cell receptors to recognize MHC molecules, and their selection during T-cell development for the ability to recognize the particular MHC molecules expressed by an individual, are topics we shall return to in Chapters 5 and 7.

In this chapter we focus on the structure and antigen-binding properties of immunoglobulins and T-cell receptors. Although B cells and T cells recognize foreign molecules in two distinct fashions, the receptor molecules they use for this task are very similar in structure. We will see how this basic structure can accommodate great variability in antigen specificity, and how it enables immunoglobulins and T-cell receptors to carry out their functions as the antigen-recognition molecules of the adaptive immune response.

The structure of a typical antibody molecule.

Antibodies are the secreted form of the B-cell receptor. An antibody is identical to the B-cell receptor of the cell that secretes it except for a small portion of the C-terminus of the heavy-chain constant region. In the case of the B-cell receptor the C-terminus is a hydrophobic membrane-anchoring sequence, and in the case of antibody it is a hydrophilic sequence that allows secretion. Since they are soluble, and secreted in large quantities, antibodies are easily obtainable and easily studied. For this reason, most of what we know about the B-cell receptor comes from the study of antibodies.

Antibody molecules are roughly Y-shaped molecules consisting of three equal-sized portions, loosely connected by a flexible tether. Three schematic representations of antibody structure, which has been determined by X-ray crystallography, are shown in Fig. 3.1. The aim of this part of the chapter is to explain how this structure is formed and how it allows antibody molecules to carry out their dual tasks—binding on the one hand to a wide variety of antigens, and on the other hand to a limited number of effector molecules and cells. As we will see, each of these tasks is carried out by separable parts of the molecule. The two arms of the Y end in regions that vary between different antibody molecules, the V regions. These are involved in antigen binding, whereas the stem of the Y, or the C region, is far less variable and is the part that interacts with effector cells and molecules.

All antibodies are constructed in the same way from paired heavy and light polypeptide chains, and the generic term immunoglobulin is used for all such proteins. Within this general category, however, five different **classes** of immunoglobulins—IgM, IgD, IgG, IgA, and IgE—can be distinguished by their C regions, which will be described more fully in Chapter 4. More subtle differences confined to the V region account for the specificity of antigen binding. We will use the IgG antibody molecule as an example to describe the general structural features of immunoglobulins.

3-1   IgG antibodies consist of four polypeptide chains.

IgG antibodies are large molecules, having a molecular weight of approximately 150 kDa, composed of two different kinds of polypeptide chain. One, of approximately 50 kDa, is termed the **heavy** or **H chain**, and the other, of 25 kDa, is termed the **light** or **L chain** (Fig. 3.2). Each IgG molecule consists of two heavy chains and two light chains. The two heavy chains are linked to each other by disulfide bonds and each heavy chain is linked to a light chain by a disulfide bond. In any given immunoglobulin molecule, the two heavy chains and the two light chains are identical, giving an antibody molecule two identical antigen-binding sites (see Fig. 3.1), and thus the ability to bind simultaneously to two identical structures.

Two types of light chain, termed **lambda** (λ) and **kappa** (κ), are found in antibodies. A given immunoglobulin either has κ chains or λ chains, never one of each. No functional difference has been found between antibodies having λ or κ light chains, and either type of light chain may be found in antibodies of any of the five major classes. The ratio of the two types of light chain varies from species to species. In mice, the average κ to λ ratio is 20:1, whereas in humans it is 2:1 and in cattle it is 1:20. The reason for this variation is unknown. Distortions of this ratio can sometimes be used to detect the abnormal proliferation of a clone of B cells. These would all express the identical light chain, and thus an excess of λ light chains in a person might indicate the presence of a B-cell tumor producing λ chains.

By contrast, the class, and thus the effector function, of an antibody, is defined by the structure of its heavy chain. There are five main **heavy-chain classes** or **isotypes**, some of which have several subtypes, and these determine the functional activity of an antibody molecule. The five major classes of immunoglobulin are **immunoglobulin M (IgM)**, **immunoglobulin D (IgD)**, **immunoglobulin G (IgG)**, **immunoglobulin A (IgA)**, and **immunoglobulin E (IgE)**. Their heavy chains are denoted by the corresponding lower-case Greek letter (μ, δ, γ, α, and ε, respectively). IgG is by far the most abundant immunoglobulin and has several subclasses (IgG1, 2, 3, and 4 in humans). Their distinctive functional properties are conferred by the carboxy-terminal part of the heavy chain, where it is not associated with the light chain. We will describe the structure and functions of the different heavy-chain isotypes in Chapter 4. The general structural features of all the isotypes are similar and we will consider IgG, the most abundant isotype in plasma, as a typical antibody molecule.

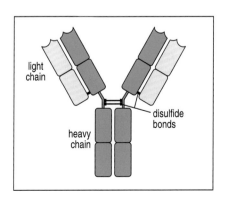

Fig. 3.2 Immunoglobulin molecules are composed of two types of protein chain: heavy chains and light chains. Each immunoglobulin molecule is made up of two heavy chains (green) and two light chains (yellow) joined by disulfide bonds so that each heavy chain is linked to a light chain and the two heavy chains are linked together.

3-2   Immunoglobulin heavy and light chains are composed of constant and variable regions.

The amino acid sequences of many immunoglobulin heavy and light chains have been determined and reveal two important features of antibody molecules. First, each chain consists of a series of similar, although not identical, sequences, each about 110 amino acids long. Each of these repeats corresponds to a discrete, compactly folded region of protein structure known as a protein domain. The light chain is made up of two such **immunoglobulin domains**, whereas the heavy chain of the IgG antibody contains four (see Fig. 3.1a). This suggests that the immunoglobulin chains have evolved by repeated duplication of an ancestral gene corresponding to a single domain.

The second important feature revealed by comparisons of amino acid sequences is that the amino-terminal sequences of both the heavy and light chains vary greatly between different antibodies. The variability in sequence is limited to approximately the first 110 amino acids, corresponding to the first domain, whereas the remaining domains are constant between immunoglobulin chains of the same isotype. The amino-terminal variable or **V domains** of the heavy and light chains (V_H and V_L, respectively) together make up the V region of the antibody and confer on it the ability to bind specific antigen, while the constant domains (**C domains**) of the heavy and light chains (C_H and C_L, respectively) make up the C region (see Fig. 3.1b, c). The multiple heavy-chain C domains are numbered from the amino-terminal end to the carboxy terminus, for example C_H1, C_H2, and so on.

3-3   The antibody molecule can readily be cleaved into functionally distinct fragments.

The protein domains described above associate to form larger globular domains. Thus, when fully folded and assembled, an antibody molecule comprises three equal-sized globular portions joined by a flexible stretch of polypeptide chain known as the **hinge region** (see Fig. 3.1b). Each arm of this Y-shaped structure is formed by the association of a light chain with the amino-terminal half of a heavy chain, whereas the trunk of the Y is formed by the pairing of the carboxy-terminal halves of the two heavy chains. The association of the heavy and light chains is such that the V_H and V_L domains are paired, as are the C_H1 and C_L domains. The C_H3 domains pair with each other but the C_H2 domains do not interact; carbohydrate side chains attached to the C_H2 domains lie between the two heavy chains. The two antigen-binding sites are formed by the paired V_H and V_L domains at the ends of the two arms of the Y (see Fig. 3.1b).

Proteolytic enzymes (proteases) that cleave polypeptide sequences have been used to dissect the structure of antibody molecules and to determine which parts of the molecule are responsible for its various functions. Limited digestion with the protease papain cleaves antibody molecules into three fragments (Fig. 3.3). Two fragments are identical and contain the antigen-binding activity. These are termed the **Fab fragments**, for Fragment antigen binding. The Fab fragments correspond to the two identical arms of the antibody molecule, which contain the complete light chains paired with the V_H and C_H1 domains of the heavy chains. The other fragment contains no antigen-binding activity but was originally observed to crystallize readily, and for this reason was named the **Fc fragment**, for Fragment crystallizable. This fragment corresponds to the paired C_H2 and C_H3 domains and is the part of the antibody molecule that interacts with effector molecules and cells. The functional differences between heavy-chain isotypes lie mainly in the Fc fragment.

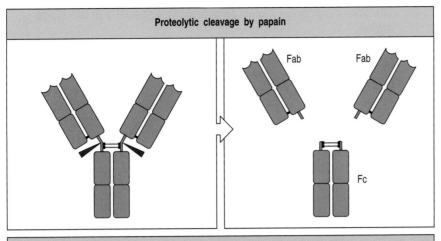

Proteolytic cleavage by papain

Fab   Fab

Fc

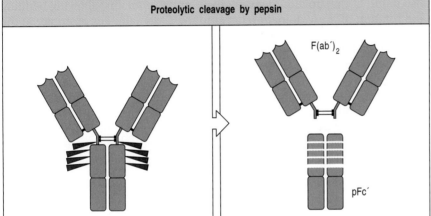

Proteolytic cleavage by pepsin

F(ab′)₂

pFc′

Fig. 3.3 The Y-shaped immunoglobulin molecule can be dissected by partial digestion with proteases. Papain cleaves the immunoglobulin molecule into three pieces, two Fab fragments and one Fc fragment (upper panels). The Fab fragment contains the V regions and binds antigen. The Fc fragment is crystallizable and contains C regions. Pepsin cleaves immunoglobulin to yield one F(ab′)₂ fragment and many small pieces of the Fc fragment, the largest of which is called the pFc′ fragment (lower panels). F(ab′)₂ is written with a prime because it contains a few more amino acids than Fab, including the cysteines that form the disulfide bonds.

The protein fragments obtained after proteolysis are determined by where the protease cuts the antibody molecule in relation to the disulfide bonds that link the two heavy chains. These lie in the hinge region between the C_H1 and C_H2 domains and, as illustrated in Fig. 3.3, papain cleaves the antibody molecule on the amino-terminal side of the disulfide bonds. This releases the two arms of the antibody as separate Fab fragments, whereas in the Fc fragment the carboxy-terminal halves of the heavy chains remain linked.

Another protease, pepsin, cuts in the same general region of the antibody molecule as papain but on the carboxy-terminal side of the disulfide bonds (see Fig. 3.3). This produces a fragment, the **F(ab′)₂ fragment**, in which the two antigen-binding arms of the antibody molecule remain linked. In this case the remaining part of the heavy chain is cut into several small fragments. The F(ab′)₂ fragment has exactly the same antigen-binding characteristics as the original antibody but is unable to interact with any effector molecule. It is thus of potential value in therapeutic applications of antibodies as well as in research into the functional role of the Fc portion.

Genetic engineering techniques also now permit the construction of many different antibody-related molecules. One important type is a truncated Fab comprising only the V domain of a heavy chain linked by a stretch of synthetic peptide to a V domain of a light chain. This is called **single-chain Fv**, named from Fragment variable. Fv molecules may become valuable therapeutic agents because of their small size, which allows them to penetrate tissues readily. They can be coupled to protein toxins to yield immunotoxins with potential application, for example, in tumor therapy in the case of a Fv specific for a tumor antigen (see Chapter 14).

3-4   The immunoglobulin molecule is flexible, especially at the hinge region.

The hinge region that links the Fc and Fab portions of the antibody molecule is in reality a flexible tether, allowing independent movement of the two Fab arms, rather than a rigid hinge. This has been demonstrated by electron microscopy of antibodies bound to **haptens**. These are small molecules of various sorts, typically about the size of a tyrosine side chain. They can be recognized by antibody but are only able to stimulate production of anti-hapten antibodies when linked to a larger protein **carrier** (see Appendix I, Section A-1). An antigen made of two identical hapten molecules joined by a short flexible region can link two or more anti-hapten antibodies, forming dimers, trimers, tetramers, and so on, which can be seen by electron microscopy (Fig. 3.4). The shapes formed by these complexes demonstrate that antibody molecules are flexible at the hinge region. Some flexibility is also found at the junction between the V and C domains, allowing bending and rotation of the V domain relative to the C domain. For example, in the antibody molecule shown in Fig. 3.1a, not only are the two hinge regions clearly bent differently, but the angle between the V and C domains in each of the two Fab arms is also different. This range of motion has led to the junction between the V and C domains being referred to as a 'molecular ball-and-socket joint.' Flexibility at both the hinge and V–C junction enables the binding of both arms of an antibody molecule to sites that are various distances apart, for example, sites on bacterial cell-wall polysaccharides. Flexibility at the hinge also enables the antibodies to interact with the antibody-binding proteins that mediate immune effector mechanisms.

Fig. 3.4 Antibody arms are joined by a flexible hinge. An antigen consisting of two hapten molecules (red balls in diagrams) that can cross-link two antigen-binding sites is used to create antigen:antibody complexes, which can be seen in the electron micrograph. Linear, triangular, and square forms are seen, with short projections or spikes. Limited pepsin digestion removes these spikes (not shown in the figure), which therefore correspond to the Fc portion of the antibody; the F(ab′)$_2$ pieces remain cross-linked by antigen. The interpretation of the complexes is shown in the diagrams. The angle between the arms of the antibody molecules varies, from 0° in the antibody dimers, through 60° in the triangular forms, to 90° in the square forms, showing that the connections between the arms are flexible. Photograph (× 300,000) courtesy of N.M. Green.

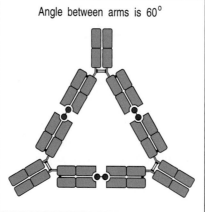
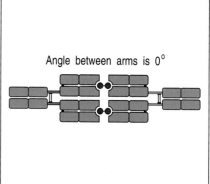

Angle between arms is 0°

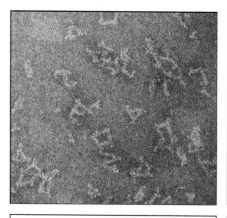

Angle between arms is 60°

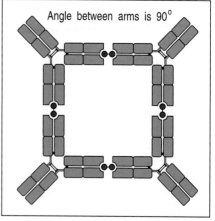

Angle between arms is 90°

3-5 The domains of an immunoglobulin molecule have similar structures.

As we saw in Section 3-2, immunoglobulin heavy and light chains are composed of a series of discrete protein domains. These protein domains all have a similar folded structure. Within this basic three-dimensional structure, there are distinct differences between V and C domains. The structural similarities and differences can be seen in the diagram of a light chain in Fig. 3.5. Each domain is constructed from two β sheets, which are elements of protein structure made up of strands of the polypeptide chain (β strands) packed together; the sheets are linked by a disulfide bridge and together form a roughly barrel-shaped structure, known as a β barrel. The distinctive folded structure of the immunoglobulin protein domain is known as the **immunoglobulin fold**.

Both the essential similarity of V and C domains and the critical difference between them are most clearly seen in the bottom panels of Fig. 3.5, where the cylindrical domains are opened out to reveal how the polypeptide chain folds to create each of the β sheets and how it forms flexible loops as it changes direction. The main difference between the V and C domains is that the V domain is larger, with an extra loop. We will see in Section 3-6 that the flexible loops of the V domains form the antigen-binding site of the immunoglobulin molecule.

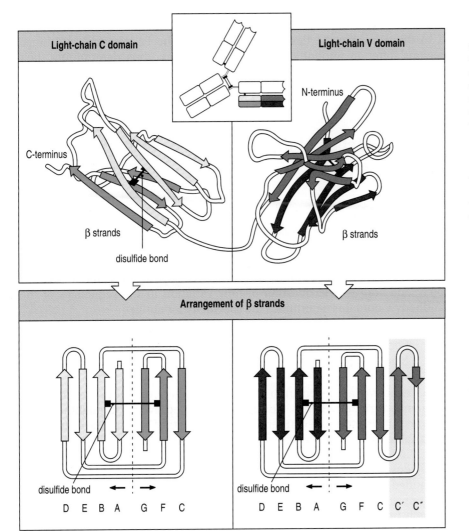

Fig. 3.5 The structure of immunoglobulin constant and variable domains. The upper panels show schematically the folding pattern of the constant (C) and variable (V) domains of an immunoglobulin light chain. Each domain is a barrel-shaped structure in which strands of polypeptide chain (β strands) running in opposite directions (antiparallel) pack together to form two β sheets (shown in yellow and green in the diagram of the C domain), which are held together by a disulfide bond. The way the polypeptide chain folds to give the final structure can be seen more clearly when the sheets are opened out, as shown in the lower panels. The β strands are lettered sequentially with respect to the order of their occurrence in the amino acid sequence of the domains; the order in each β sheet is characteristic of immunoglobulin domains. The β strands C′ and C″ that are found in the V domains but not in the C domains are indicated by a blue shaded background. The characteristic four-strand plus three-strand (C-region type domain) or four-strand plus five-strand (V-region type domain) arrangements are typical immuno-globulin superfamily domain building blocks, found in a whole range of other proteins as well as antibodies and T-cell receptors.

Many of the amino acids that are common to C and V domains of immunoglobulin chains lie in the core of the immunoglobulin fold and are critical to its stability. For that reason, other proteins having sequences similar to those of immunoglobulins are believed to form domains of similar structure, and in many cases this has been demonstrated by crystallography. These **immunoglobulin-like domains** are present in many other proteins of the immune system, and in proteins involved in cell–cell recognition in the nervous system and other tissues. Together with the immunoglobulins and the T-cell receptors, they make up the extensive **immunoglobulin superfamily**.

Summary.

The IgG antibody molecule is made up of four polypeptide chains, comprising two identical light chains and two identical heavy chains, and can be thought of as forming a flexible Y-shaped structure. Each of the four chains has a variable (V) region at its amino terminus, which contributes to the antigen-binding site, and a constant (C) region, which determines the isotype. The isotype of the heavy chain determines the functional properties of the antibody. The light chains are bound to the heavy chains by many non-covalent interactions and by disulfide bonds, and the V regions of the heavy and light chains pair in each arm of the Y to generate two identical antigen-binding sites, which lie at the tips of the arms of the Y. The possession of two antigen-binding sites allows antibody molecules to cross-link antigens and to bind them much more stably. The trunk of the Y, or Fc fragment, is composed of the carboxy-terminal domains of the heavy chains. Joining the arms of the Y to the trunk are the flexible hinge regions. The Fc fragment and hinge regions differ in antibodies of different isotypes, thus determining their functional properties. However, the overall organization of the domains is similar in all isotypes.

The interaction of the antibody molecule with specific antigen.

We have described the structure of the antibody molecule and how the V regions of the heavy and light chains fold and pair to form the antigen-binding site. In this part of the chapter we will look at the antigen-binding site in more detail. We will discuss the different ways in which antigens can bind to antibody and address the question of how variation in the sequences of the antibody V domains determines the specificity for antigen.

3-6   Localized regions of hypervariable sequence form the antigen-binding site.

The V regions of any given antibody molecule differ from those of every other. Sequence variability is not, however, distributed evenly throughout the V regions but is concentrated in certain segments of the V region. The distribution of variable amino acids can be seen clearly in what is termed a **variability plot** (Fig. 3.6), in which the amino acid sequences of many different antibody V regions are compared. Three segments of particular variability can be identified in both the V_H and V_L domains. They are designated **hypervariable regions** and are denoted HV1, HV2, and HV3. In the light chains these are roughly from residues 28 to 35, from 49 to 59, and from 92 to 103, respectively.

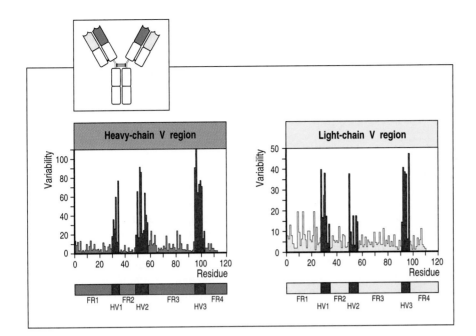

Fig. 3.6 There are discrete regions of hypervariability in V domains. A variability plot derived from comparison of the amino acid sequences of several dozen heavy-chain and light-chain V domains. At each amino acid position the degree of variability is the ratio of the number of different amino acids seen in all of the sequences together to the frequency of the most common amino acid. Three hypervariable regions (HV1, HV2, and HV3) are indicated in red and are also known as the complementarity-determining regions, CDR1, CDR2, and CDR3. They are flanked by less variable framework regions (FR1, FR2, FR3, and FR4, shown in blue or yellow).

The most variable part of the domain is in the HV3 region. The regions between the hypervariable regions, which comprise the rest of the V domain, show less variability and are termed the **framework regions**. There are four such regions in each V domain, designated FR1, FR2, FR3, and FR4.

The framework regions form the β sheets that provide the structural framework of the domain, whereas the hypervariable sequences correspond to three loops at the outer edge of the β barrel, which are juxtaposed in the folded domain (Fig. 3.7). Thus, not only is sequence diversity concentrated in particular parts of the V domain but it is localized to a particular region on the surface of the molecule. When the V_H and V_L domains are paired in the antibody molecule, the hypervariable loops from each domain are brought together, creating a single hypervariable site at the tip of each arm of the molecule. This is the binding site for antigen, the **antigen-binding site** or **antibody combining site**. The three hypervariable loops determine antigen specificity by forming a surface complementary to the antigen, and are more commonly termed the **complementarity-determining regions**, or **CDRs** (**CDR1, CDR2,** and **CDR3**). Because CDRs from both V_H and V_L domains contribute to the antigen-binding site, it is the combination of the heavy and the light chain, and not either alone, that determines the final antigen specificity. Thus, one way in which the immune system is able to generate antibodies of different specificities is by generating different combinations of heavy- and light-chain V regions. This means of producing variability is known as **combinatorial diversity**; we will encounter a second form of combinatorial diversity when we consider in Chapter 4 how the genes encoding the heavy- and light-chain V regions are created from smaller segments of DNA.

3-7   Antibodies bind antigens via contacts with amino acids in CDRs, but the details of binding depend upon the size and shape of the antigen.

In early investigations of antigen binding to antibodies, the only available sources of large quantities of a single type of antibody molecule were tumors of antibody-secreting cells. The antigen specificities of the tumor-derived antibodies were unknown, so many compounds had to be screened to identify ligands that could be used to study antigen binding. In general, the substances

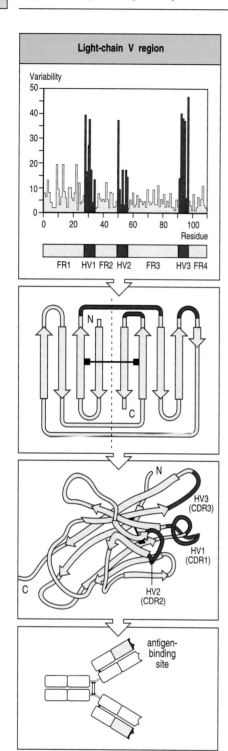

Light-chain V region

Variability

FR1   HV1 FR2 HV2   FR3   HV3 FR4

N

C

N

HV3
(CDR3)

HV1
(CDR1)

C

HV2
(CDR2)

antigen-
binding
site

Fig. 3.7 The hypervariable regions lie in discrete loops of the folded structure. When the hypervariable regions (CDRs) are positioned on the structure of a V domain it can be seen that they lie in loops that are brought together in the folded structure. In the antibody molecule, the pairing of a heavy and a light chain brings together the hypervariable loops from each chain to create a single hypervariable surface, which forms the antigen-binding site at the tip of each arm.

found to bind to these antibodies were haptens (see Section 3-4) such as phosphorylcholine or vitamin K_1. Structural analysis of complexes of antibodies with their hapten ligands provided the first direct evidence that the hypervariable regions form the antigen-binding site, and demonstrated the structural basis of specificity for the hapten. Subsequently, with the discovery of methods of generating monoclonal antibodies (see Appendix I, Section A-12), it became possible to make large amounts of pure antibodies specific for many different antigens. This has provided a more general picture of how antibodies interact with their antigens, confirming and extending the view of antibody–antigen interactions derived from the study of haptens.

The surface of the antibody molecule formed by the juxtaposition of the CDRs of the heavy and light chains creates the site to which an antigen binds. Clearly, as the amino acid sequences of the CDRs are different in different antibodies, so are the shapes of the surfaces created by these CDRs. As a general principle, antibodies bind ligands whose surfaces are complementary to that of the antibody. A small antigen, such as a hapten or a short peptide, generally binds in a pocket or groove lying between the heavy- and light-chain V domains (Fig. 3.8, left and center panels). Other antigens, such as a protein molecule, can be of the same size as, or larger than, the antibody molecule itself, and cannot fit into a groove or pocket. In these cases, the interface between the two molecules is often an extended surface that involves all of the CDRs and, in some cases, part of the framework region of the antibody (Fig. 3.8, right panel). This surface need not be concave but can be flat, undulating, or even convex.

3-8   Antibodies bind to conformational shapes on the surfaces of antigens.

The biological function of antibodies is to bind to pathogens and their products, and to facilitate their removal from the body. An antibody generally recognizes only a small region on the surface of a large molecule such as a polysaccharide or protein. The structure recognized by an antibody is called an **antigenic determinant** or **epitope**. Some of the most important pathogens have polysaccharide coats, and antibodies that recognize epitopes formed by the sugar subunits of these molecules are essential in providing immune protection from such pathogens. In many cases, however, the antigens that provoke an immune response are proteins. For example, protective antibodies against viruses recognize viral coat proteins. In such cases, the structures recognized by the antibody are located on the surface of the protein. Such sites are likely to be composed of amino acids from different parts of the polypeptide chain that have been brought together by protein folding. Antigenic determinants of this kind are known as **conformational** or **discontinuous epitopes** because the structure recognized is composed of segments of the protein that are discontinuous in the amino acid sequence of the antigen but are brought together in the three-dimensional structure. In contrast, an epitope composed of a single segment of polypeptide chain is termed a **continuous** or **linear epitope**. Although most antibodies raised

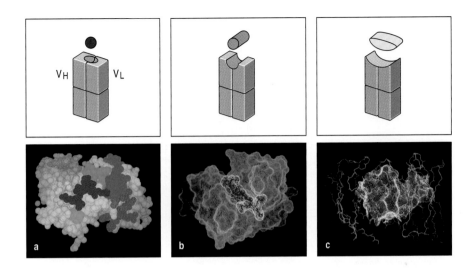

Fig. 3.8 Antigens can bind in pockets or grooves, or on extended surfaces in the binding sites of antibodies. The panels in the top row show schematic representations of the different types of binding site in a Fab fragment of an antibody: left, pocket; center, groove; right, extended surface. Below are examples of each type. Panel a: space-filling representation of the interaction of a small peptide antigen with the complementarity-determining regions (CDRs) of a Fab fragment as viewed looking into the antigen-binding site. Seven amino acid residues of the antigen, shown in red, are bound in the antigen-binding pocket. Five of the six CDRs (H1, H2, H3, L1, and L3) interact with the peptide, whereas L2 does not. The CDR loops are colored as follows: L2, magenta; L3, green; H1, blue; H2, pale purple; H3, yellow. Panel b: in a complex of an antibody with a peptide from the human immunodeficiency virus, the peptide (orange) binds along a groove formed between the heavy- and light-chain V domains (green). Panel c: complex between hen egg-white lysozyme and the Fab fragment of its corresponding antibody (HyHel5). Two extended surfaces come into contact, as can be seen from this computer-generated image, where the surface contour of the lysozyme molecule (yellow dots) is superimposed on the antigen-binding site. Residues in the antibody that make contact with the lysozyme are shown in full (red); for the rest of the Fab fragment only the peptide backbone is shown (blue). All six CDRs of the antibody are involved in the binding. Photographs a and b courtesy of I.A. Wilson and R.L. Stanfield. Photograph c courtesy of S. Sheriff.

against intact, fully folded proteins recognize discontinuous epitopes, some will bind peptide fragments of the protein. Conversely, antibodies raised against peptides of a protein or against synthetic peptides corresponding to part of its sequence are occasionally found to bind to the natural folded protein. This makes it possible, in some cases, to use synthetic peptides in vaccines that aim at raising antibodies against a pathogen protein.

3-9 Antigen–antibody interactions involve a variety of forces.

The interaction between an antibody and its antigen can be disrupted by high salt concentrations, extremes of pH, detergents, and sometimes by competition with high concentrations of the pure epitope itself. The binding is therefore a reversible noncovalent interaction. The forces, or bonds, involved in these noncovalent interactions are outlined in Fig. 3.9.

Electrostatic interactions occur between charged amino acid side chains, as in salt bridges. Interactions also occur between electric dipoles, as in hydrogen bonds, or can involve short-range van der Waals forces. High salt concentrations and extremes of pH disrupt antigen–antibody binding by weakening electrostatic interactions and/or hydrogen bonds. This principle is employed in the purification of antigens using affinity columns of immobilized antibodies, and vice versa for antibody purification (see Appendix I, Section A-5). Hydrophobic interactions occur when two hydrophobic surfaces come together to exclude water. The strength of a hydrophobic interaction is proportional to the surface area that is hidden from water. For some antigens, hydrophobic interactions probably account for most of the binding energy. In some cases, water molecules are trapped in pockets in the interface between antigen and antibody. These trapped water molecules may also contribute to binding, especially between polar amino acid residues.

The contribution of each of these forces to the overall interaction depends on the particular antibody and antigen involved. A striking difference between antibody interactions with protein antigens and most other natural protein–protein interactions is that antibodies possess many aromatic amino acids in their antigen-binding sites. These amino acids participate mainly in van der Waals and hydrophobic interactions, and sometimes in hydrogen bonds. In general, the hydrophobic and van der Waals forces operate over very

Fig. 3.9 The noncovalent forces that hold together the antigen:antibody complex. Partial charges found in electric dipoles are shown as δ^+ or δ^-. Electrostatic forces diminish as the inverse square of the distance separating the charges, whereas van der Waals forces, which are more numerous in most antigen–antibody contacts, fall off as the sixth power of the separation and therefore operate only over very short ranges. Covalent bonds never occur between antigens and naturally produced antibodies.

Noncovalent forces	Origin	
Electrostatic forces	Attraction between opposite charges	$\overset{\oplus}{-NH_3} \quad \overset{\ominus}{OOC-}$
Hydrogen bonds	Hydrogen shared between electronegative atoms (N,O)	$\underset{\delta^-}{>N} - \underset{\delta^+}{H} -- \underset{\delta^-}{O} = C<$
Van der Waals forces	Fluctuations in electron clouds around molecules oppositely polarize neighboring atoms	$\delta^+ \rightleftharpoons \delta^-$ / $\delta^- \rightleftharpoons \delta^+$
Hydrophobic forces	Hydrophobic groups interact unfavorably with water and tend to pack together to exclude water molecules. The attraction also involves van der Waals forces	

short ranges and serve to pull together two surfaces that are complementary in shape: hills on one surface must fit into valleys on the other for good binding to occur. In contrast, electrostatic interactions between charged side chains, and hydrogen bonds bridging oxygen and/or nitrogen atoms, accommodate specific features or reactive groups while strengthening the interaction overall.

For example, in the complex of hen egg-white lysozyme with the antibody D1.3 (Fig. 3.10), strong hydrogen bonds are formed between the antibody and a particular glutamine in the lysozyme molecule that protrudes between the V_H and V_L domains. Lysozymes from partridge and turkey have another amino acid in place of the glutamine and do not bind to the antibody. In the high-affinity complex of hen egg-white lysozyme with another antibody, HyHel5 (see Fig. 3.8c), two salt bridges between two basic arginines on the surface of the lysozyme interact with two glutamic acids, one each from the V_H CDR1 and CDR2 loops. Again, lysozymes that lack one of the two arginine residues show a 1000-fold decrease in affinity. Although overall surface complementarity must play an important part in antigen–antibody interactions, specific electrostatic and hydrogen-bonding interactions appear to determine antibody affinity. In most antibodies that have been studied at this level of detail, only a few residues make a major contribution to the binding energy. Genetic engineering by site-directed mutagenesis can further tailor an antibody's binding to its complementary epitope.

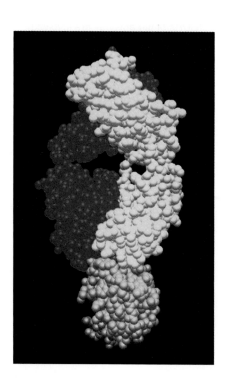

Fig. 3.10 The complex of lysozyme with the antibody D1.3. The interaction of the Fab fragment of D1.3 with hen egg-white lysozyme is shown, with the lysozyme in blue, the heavy chain in purple and the light chain in yellow. A glutamine residue of lysozyme, shown in red, protrudes between the two V domains of the antigen-binding site and makes hydrogen bonds important to the antigen–antibody binding. Original photograph courtesy of R.J. Poljak.

Summary.

X-ray crystallographic analysis of antigen:antibody complexes has demon-strated that the hypervariable loops (complementarity-determining regions) of immunoglobulin V regions determine the specificity of antibodies. With protein antigens, the antibody molecule contacts the antigen over a broad area of its surface that is complementary to the surface recognized on the antigen. Electrostatic interactions, hydrogen bonds, van der Waals forces, and hydrophobic interactions can all contribute to binding. Amino acid side chains in most or all of the hypervariable loops make contact with antigen and determine both the specificity and the affinity of the interaction. Other parts of the V region play little part in the direct contact with the antigen but provide a stable structural framework for the hypervariable loops and help determine their position and conformation. Antibodies raised against intact proteins usually bind to the surface of the protein and make contact with residues that are discontinuous in the primary structure of the molecule; they may, however, occasionally bind peptide fragments of the protein, and antibodies raised against peptides derived from a protein can sometimes be used to detect the native protein molecule. Peptides binding to antibodies usually bind in the cleft between the V regions of the heavy and light chains, where they make specific contact with some, but not necessarily all, of the hypervariable loops. This is also the usual mode of binding for carbohydrate antigens and small molecules such as haptens.

Antigen recognition by T cells.

In contrast to the immunoglobulins, which interact with pathogens and their toxic products in the extracellular spaces of the body, T cells only recognize foreign antigens that are displayed on the surfaces of the body's own cells. These antigens can derive from pathogens such as viruses or intracellular bacteria, which replicate within cells, or from pathogens or their products that cells have internalized by endocytosis from the extracellular fluid.

T cells can detect the presence of an intracellular pathogen because infected cells display on their surface peptide fragments derived from the pathogen's proteins. These foreign peptides are delivered to the cell surface by specialized host-cell glycoproteins. These are encoded in a large cluster of genes that were first identified by their powerful effects on the immune response to transplanted tissues. For that reason, the gene complex was called the **major histocompatibility complex (MHC)**, and the peptide-binding glycoproteins are still known as MHC molecules. The recognition of antigen as a small peptide fragment bound to an MHC molecule and displayed at the cell surface is one of the most distinctive features of T cells, and will be the focus of this part of the chapter. How peptide fragments of antigen become complexed with MHC molecules will be considered in Chapter 5.

In this part of the chapter we will describe the structure and properties of the T-cell antigen receptor, T-cell receptor, or TCR for short. As might be expected from their function as highly variable antigen-recognition structures, T-cell receptors are closely related to antibody molecules in the structure of their genes. There are, however, important differences between T-cell receptors and immunoglobulins that reflect the special features of antigen recognition by the T-cell receptor, and its lack of effector functions.

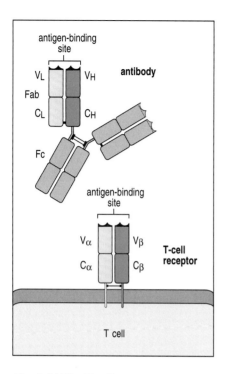

Fig. 3.11 The T-cell receptor resembles a membrane-bound Fab fragment. The Fab fragment of antibody molecules is a disulfide-linked heterodimer, each chain of which contains one immunoglobulin C domain and one V domain; the juxtaposition of the V domains forms the antigen-binding site (see Section 3-6). The T-cell receptor is also a disulfide-linked heterodimer, with each chain containing an immunoglobulin C-like domain and an immunoglobulin V-like domain. As in the Fab fragment, the juxtaposition of the V domains forms the site for antigen recognition.

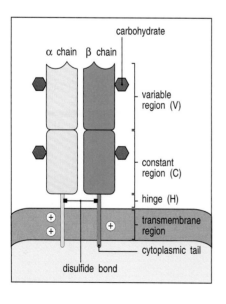

3-10   The antigen receptor on T cells is very similar to a Fab fragment of immunoglobulin.

T-cell receptors were first identified using monoclonal antibodies that bound only one cloned T-cell line but not others and that could specifically inhibit antigen recognition by that clone of T cells, or specifically activate them (see Appendix I, Section A-19). These **clonotypic** antibodies were then used to show that each T cell bears about 30,000 antigen-receptor molecules on its surface, each receptor consisting of two different polypeptide chains, termed the **T-cell receptor α (TCRα)** and **β (TCRβ) chains**, linked by a disulfide bond. These α:β heterodimers are very similar in structure to the Fab fragment of an immunoglobulin molecule (Fig. 3.11), and they account for antigen recognition by most T cells. A minority of T cells bear an alternative, but structurally similar, receptor made up of a different pair of polypeptide chains designated γ and δ. **γ:δ T-cell receptors** appear to have different antigen-recognition properties from the **α:β T-cell receptors**, and the function of γ:δ T cells in immune responses is not yet entirely clear. In the rest of this chapter, we shall use the term T-cell receptor to mean the α:β receptor, except where specified otherwise. Both types of T-cell receptor differ from the membrane-bound immunoglobulin that serves as the B-cell receptor: a T-cell receptor has only one antigen-binding site, whereas a B-cell receptor has two, and T-cell receptors are never secreted, whereas immunoglobulin can be secreted as antibody.

Our initial insights into the structure and function of the α:β T-cell receptor came from studies of cloned cDNA encoding the receptor chains. The amino acid sequences predicted from T-cell receptor cDNAs show clearly that both chains of the T-cell receptor have an amino-terminal variable (V) region with homology to an immunoglobulin V domain, a constant (C) region with homology to an immunoglobulin C domain, and a short hinge region containing a cysteine residue that forms the interchain disulfide bond (Fig. 3.12). Each chain spans the lipid bilayer by a hydrophobic transmembrane domain, and ends in a short cytoplasmic tail. These close similarities of T-cell receptor chains to the heavy and light immunoglobulin chains first enabled prediction of the structural resemblance of the T-cell receptor heterodimer to a Fab fragment of immunoglobulin.

Recently, the three-dimensional structure of the T-cell receptor has been determined. The structure is indeed similar to that of an antibody Fab fragment, as was suspected from earlier studies on the genes that encoded it. The T-cell receptor chains fold in much the same way as those of a Fab fragment (Fig. 3.13a), although the final structure appears a little shorter and wider. There are, however, some distinct differences between T-cell receptors and Fab fragments. The most striking difference is in the C_α domain, where the fold is unlike that of any other immunoglobulin-like domain. The half of the domain that is juxtaposed with the C_β domain forms a β sheet similar to that

Fig. 3.12 Structure of the T-cell receptor. The T-cell receptor heterodimer is composed of two transmembrane glycoprotein chains, α and β. The extracellular portion of each chain consists of two domains, resembling immunoglobulin V and C domains, respectively. Both chains have carbohydrate side chains attached to each domain. A short segment, analogous to an immunoglobulin hinge region, connects the immunoglobulin-like domains to the membrane and contains the cysteine residue that forms the interchain disulfide bond. The transmembrane helices of both chains are unusual in containing positively charged (basic) residues within the hydrophobic transmembrane segment. The α chains carry two such residues; the β chains have one.

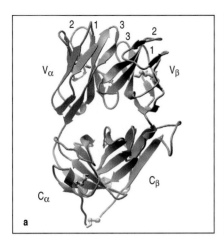
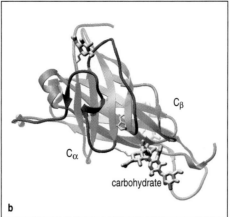
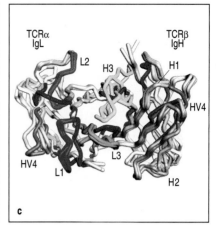

found in other immunoglobulin-like domains, but the other half of the domain is formed of loosely packed strands and a short segment of α helix (Fig. 3.13b). The intramolecular disulfide bond, which in immunoglobulin-like domains normally joins two β strands, in a C_α domain joins a β strand to this segment of α helix.

There are also differences in the way in which the domains interact. The interface between the V and C domains of both T-cell receptor chains is more extensive than in antibodies, which may make the hinge joint between the domains less flexible. And the interaction between the C_α and C_β domains is distinctive in being assisted by carbohydrate, with a sugar group from the C_α domain making a number of hydrogen bonds to the C_β domain (see Fig. 3.13b). Finally, a comparison of the variable binding sites shows that, although the complementarity-determining region (CDR) loops align fairly closely with those of antibody molecules, there is some displacement relative to those of the antibody molecule (Fig. 3.13c). This displacement is particularly marked in the V_α CDR2 loop, which is oriented at roughly right angles to the equivalent loop in antibody V domains, as a result of a shift in the β strand that anchors one end of the loop from one face of the domain to the other. A strand displacement also causes a change in the orientation of the V_β CDR2 loop in two of the seven V_β domains whose structures are known. As yet, the crystallographic structures of only seven T-cell receptors have been solved to this level of resolution, so it remains to be seen to what degree all T-cell receptors share these features, and whether there is more variability to be discovered.

3-11   A T-cell receptor recognizes antigen in the form of a complex of a foreign peptide bound to an MHC molecule.

Antigen recognition by T-cell receptors clearly differs from recognition by B-cell receptors and antibodies. Antigen recognition by B cells involves direct binding of immunoglobulin to the intact antigen and, as discussed in Section 3-8, antibodies typically bind to the surface of protein antigens, contacting amino acids that are discontinuous in the primary structure but are brought together in the folded protein. T cells, on the other hand, were found to respond to short contiguous amino acid sequences in proteins. These sequences were often buried within the native structure of the protein and thus could not be recognized directly by T-cell receptors unless some unfolding of the protein antigen and its 'processing' into peptide fragments had occurred (Fig. 3.14).

Fig. 3.13 The crystal structure of an α:β T-cell receptor resolved at 2.5 Å. In panels a and b the α chain is shown in pink and the β chain in blue. Disulfide bonds are shown in green. In panel a, the T-cell receptor is viewed from the side as it would sit on a cell surface, with the CDR loops that form the antigen-binding site (labeled 1, 2, and 3) arrayed across its relatively flat top. In panel b, the C_α and C_β domains are shown. The C_α domain does not fold into a typical immunoglobulin-like domain; the face of the domain away from the C_β domain is mainly composed of irregular strands of polypeptide rather than β sheet. The intramolecular disulfide bond joins a β strand to this segment of α helix. The interaction between the C_α and C_β domains is assisted by carbohydrate (colored grey and labeled on the figure), with a sugar group from the C_α domain making hydrogen bonds to the C_β domain. In panel c, the T-cell receptor is shown aligned with the antigen-binding sites from three different antibodies. This view is looking down into the binding site. The V_α domain of the T-cell receptor is aligned with the V_L domains of the antigen-binding sites of the antibodies, and the V_β domain is aligned with the V_H domains. The CDRs of the T-cell receptor and immunoglobulin molecules are colored, with CDRs 1, 2, and 3 of the TCR shown in red and the HV4 loop in orange. For the immunoglobulin V domains, the CDR1 loops of the heavy chain (H1) and light chain (L1) are shown in light and dark blue, respectively, and the CDR2 loops (H2, L2) in light and dark purple, respectively. The heavy-chain CDR3 loops (H3) are in yellow; the light-chain CDR3s (L3) are in bright green. The HV4 loops of the TCR (orange) have no hypervariable counterparts in immunoglobulins. Photographs courtesy of I.A. Wilson.

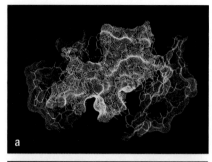

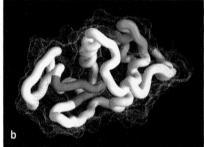

Fig. 3.14 Differences in the recognition of hen egg-white lysozyme by immunoglobulins and T-cell receptors. Antibodies can be shown by X-ray crystallography to bind epitopes on the surface of proteins, as shown in panel a, where the epitopes for three antibodies are shown on the surface of hen egg lysozyme (see also Fig. 3.10). In contrast, the epitopes recognized by T-cell receptors need not lie on the surface of the molecule, as the T-cell receptor recognizes not the antigenic protein itself but a peptide fragment of the protein. The peptides corresponding to two T-cell epitopes of lysozyme are shown in panel b, one epitope, shown in blue, lies on the surface of the protein but a second, shown in red, lies mostly within the core and is inaccessible in the folded protein. For this residue to be accessible to the T-cell receptor, the protein must be unfolded and processed. Panel a courtesy of S. Sheriff.

The nature of the antigen recognized by T cells became clear with the realization that the peptides that stimulate T cells are recognized only when bound to an MHC molecule. These cell-surface glycoproteins are encoded by genes within the major histocompatibility complex (MHC). The ligand recognized by the T cell is thus a complex of peptide and MHC molecule. The evidence for involvement of the MHC in T-cell recognition of antigen was at first indirect, but it has recently been proved conclusively by stimulating T cells with purified peptide:MHC complexes. The T-cell receptor interacts with this ligand by making contacts with both the MHC molecule and the antigen peptide.

3-12   T cells with different functions are distinguished by CD4 and CD8 cell-surface proteins and recognize peptides bound to different classes of MHC molecule.

T cells fall into two major classes that have different effector functions. The two classes are distinguished by the expression of the cell-surface proteins **CD4** and **CD8**. These two types of T cell differ in the class of MHC molecule they recognize. There are two classes of MHC molecule—**MHC class I** and **MHC class II**—which differ in their structure and expression pattern on

Fig. 3.15 The outline structures of the CD4 and CD8 co-receptor molecules. The CD4 molecule contains four immunoglobulin-like domains, as shown in diagrammatic form in panel a, and as a ribbon diagram of the structure in panel b. The amino-terminal domain, D_1, is similar in structure to an immunoglobulin V domain. The second domain, D_2, although clearly related to an immunoglobulin domain, is different from both V and C domains and has been termed a C2 domain. The first two domains of CD4 form a rigid rodlike structure that is linked to the two carboxy-terminal domains by a flexible link. The binding site for MHC class II molecules is thought to involve both the D_1 and D_2 domains. The CD8 molecule (panels a and c) is a heterodimer of an α and a β chain covalently linked by a disulfide bond; an alternative form of CD8 exists as a homodimer of α chains. The heterodimer is depicted in panel a, while the ribbon diagram in panel c is of the homodimer. CD8α and CD8β chains have very similar structures, each having a single domain resembling an immunoglobulin V domain and a stretch of polypeptide chain, believed to be in a relatively extended conformation, that anchors the V-like domain to the cell membrane.

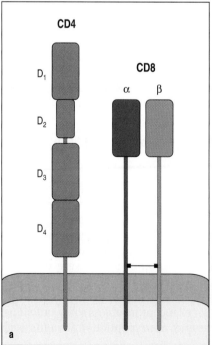

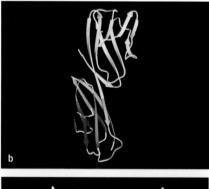

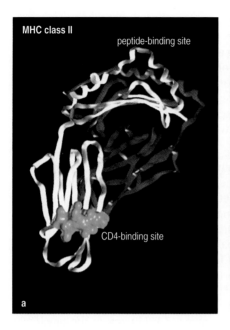

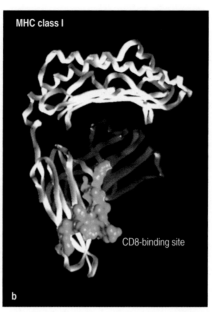

Fig. 3.16 The binding sites for CD4 and CD8 on MHC class II and class I molecules lie in the immunoglobulin-like domains. The binding sites for CD4 and CD8 on the MHC class II and class I molecules, respectively, lie in the immunoglobulin-like domains nearest to the membrane and distant from the peptide-binding cleft. In panel a, the binding site for CD4 is shown as a bright green surface. It lies at the base of the β_2 domain of an MHC class II molecule and is distant from the peptide-binding site at the top of the molecule. The α chain of the MHC class II molecule is purple, and the β chain is white. In panel b, the binding site for CD8 is also shown as a green surface, at the base of the α_3 domain of the MHC class I molecule. The α chain of the class I molecule is white, and the β_2-microglobulin is purple. Photographs courtesy of C. Thorpe.

tissues of the body (see Section 3-13). CD4 and CD8 were known as markers for different functional sets of T cells for some time before it became clear that they play an important part in the direct recognition of MHC class II and MHC class I molecules, respectively. CD4 binds to the MHC class II molecule and CD8 to the MHC class I molecule. During antigen recognition, depending on the type of T-cell, CD4 or CD8 molecules associate on the T-cell surface with the T-cell receptor and bind to invariant sites on the MHC portion of the composite MHC:peptide ligand. This binding is required for the T cell to make an effective response, and so CD4 and CD8 are called **co-receptors**.

CD4 is a single-chain molecule composed of four immunoglobulin-like domains (Fig. 3.15). The first two domains (D_1 and D_2) of the CD4 molecule are packed tightly together to form a rigid rod some 60 Å long, which is joined by a flexible hinge to a similar rod formed by the third and fourth domains (D_3 and D_4). CD4 binds MHC class II molecules through a region that is located mainly on a lateral face of the first domain, D_1. Because CD4 binds to a site on the β_2 domain of the MHC class II molecule that is well away from the site where the T-cell receptor binds (Fig. 3.16a), the CD4 molecule and the T-cell receptor can bind the same peptide:MHC class II complex. CD4 interacts strongly with a cytoplasmic tyrosine kinase called **Lck**, and can deliver this tyrosine kinase into close proximity with the signaling components of the T-cell receptor complex. This results in enhancement of the signal that is generated when the T-cell receptor binds its peptide:MHC class II ligand, as we will discuss further in Chapter 6. When CD4 and the T-cell receptor can simultaneously bind to the same MHC class II:peptide complex, the sensitivity of a T cell to antigen presented by MHC class II molecules is markedly increased; the T-cell in this case requires 100-fold less antigen for activation.

CD4 binding to an MHC class II molecule on its own is weak, and it is not clear whether such binding would be able to transmit a signal to the interior of the T cell. As shown in Fig. 3.17, CD4 can form homodimers through a site in the D_4 domain, which leaves the MHC-binding site free to interact with an MHC class II molecule. Thus, the CD4 dimer could cross-link two MHC class II molecules and thus the two T-cell receptors bound to them. Whether the dimerization of CD4 is important in its co-receptor function is not known at present.

Fig. 3.17 CD4 is capable of forming dimers. The structure of the extra-cellular domains of the CD4 molecule has been determined by X-ray crystallography. Two molecules of CD4 can interact with each other through their D_4 domains, forming homodimers. The site that binds MHC class II molecules remains available in such dimers.

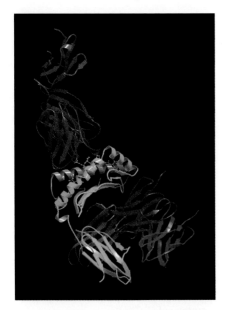

Fig. 3.18 CD8 binds to a site on MHC class I molecules distant from that to which the T-cell receptor binds. The relative positions of the T-cell receptor and CD8 molecules bound to the same MHC class I molecule can be seen in this hypothetical reconstruction of the interaction of an MHC class I molecule (the α chain is shown in green; β2-microglobulin (dull yellow) can be seen faintly in the background) with a T-cell receptor and CD8. The α and β chains of the T-cell receptor are shown in pink and purple, respectively. The CD8 structure is that of a CD8α homodimer, but is colored to represent the likely orientation of the subunits in the hetero-dimer, with the CD8α subunit in red and the CD8β subunit in blue. Photograph courtesy of G. Gao.

Although CD4 and CD8 both function as co-receptors, their structures are quite distinct. The CD8 molecule is a disulfide-linked heterodimer consisting of an α and a β chain, each containing a single immunoglobulin-like domain linked to the membrane by a segment of extended polypeptide chain (see Fig. 3.15). This segment is extensively glycosylated, which is thought to be impor-tant in maintaining this polypeptide in an extended conformation and protecting it from cleavage by proteases. CD8α chains can also form homo-dimers, although these are not found when the CD8β chains are present.

CD8 binds weakly to an invariant site in the α_3 domain of an MHC class I molecule (Fig. 3.16b), which is equivalent to the site in MHC class II molecules to which CD4 binds. Although only the interaction of the CD8α homodimer with MHC class I is so far known in detail, it is clear from this that the MHC class I binding site of the CD8 α:β heterodimer will be formed by the interaction of the CD8α and β chains. In addition, CD8 (most probably through the α chain) interacts with residues in the base of the α_2 domain of the MHC class I molecule. Binding in this way, CD8 leaves the upper surface of the MHC class I molecule exposed and free to interact simultaneously with a T-cell receptor, as shown in Fig. 3.18. Like CD4, CD8 also binds Lck through the cytoplasmic tail of the α chain and brings it into close proximity with the T-cell receptor. And as with CD4, the presence of CD8 increases the sensitivity of T cells to antigen presented by MHC class I molecules by about 100-fold. Thus, CD4 and CD8 have similar functions and bind to the same approximate location in MHC class I and MHC class II molecules even though the structures of the two co-receptor proteins are only distantly related.

3-13 The two classes of MHC molecule are expressed differentially on cells.

MHC class I and MHC class II molecules have a distinct distribution among cells that reflects the different effector functions of the T cells that recognize them (Fig. 3.19). MHC class I molecules present peptides from pathogens, commonly viruses, to CD8 cytotoxic T cells, which are specialized to kill any cell that they specifically recognize. As viruses can infect any nucleated cell, almost all such cells express MHC class I molecules, although the level of constitutive expression varies from one cell type to the next. For example, cells of the immune system express abundant MHC class I on their surface, whereas liver cells (hepatocytes) express relatively low levels (see Fig. 3.19). Nonnucleated cells, such as mammalian red blood cells, express little or no MHC class I, and thus the interior of red blood cells is a site in which an infec-tion can go undetected by cytotoxic T cells. As red blood cells cannot support viral replication, this is of no great consequence for viral infection, but it may be the absence of MHC class I that allows the *Plasmodium* species that cause malaria to live in this privileged site.

In contrast, the main function of the CD4 T cells that recognize MHC class II molecules is to activate other effector cells of the immune system. Thus MHC class II molecules are normally found on B lymphocytes, dendritic cells, and macrophages—cells that participate in immune responses—but not on other tissue cells (see Fig. 3.19). When CD4 T cells recognize peptides bound to MHC class II molecules on B cells, they stimulate the B cells to produce antibody Likewise, CD4 T cells recognizing peptides bound to MHC class II molecules on macrophages activate these cells to destroy the pathogens in their vesicles We shall see in Chapter 8 that MHC class II molecules are also expressed on specialized antigen-presenting cells in lymphoid tissues where naive T cells encounter antigen and are first activated. The expression of both MHC class I and MHC class II molecules is regulated by cytokines, in particular interfer-ons, released in the course of immune responses. Interferon-γ (IFN-γ), for

Tissue	MHC class I	MHC class II
Lymphoid tissues		
T cells	+++	+*
B cells	+++	+++
Macrophages	+++	++
Other antigen-presenting cells (eg Langerhans' cells)	+++	+++
Epithelial cells of the thymus	+	+++
Other nucleated cells		
Neutrophils	+++	−
Hepatocytes	+	−
Kidney	+	−
Brain	+	−†
Non-nucleated cells		
Red blood cells	−	−

Fig. 3.19 The expression of MHC molecules differs between tissues. MHC class I molecules are expressed on all nucleated cells, although they are most highly expressed in hematopoietic cells. MHC class II molecules are normally expressed only by a subset of hematopoietic cells and by thymic stromal cells, although they may be expressed by other cell types on exposure to the inflammatory cytokine interferon-γ. *In humans, activated T cells express MHC class II molecules, whereas in mice, all T cells are MHC class II-negative. † In the brain, most cell types are MHC class II-negative but microglia, which are related to macrophages, are MHC class II-positive.

example, increases the expression of MHC class I and MHC class II molecules, and can induce the expression of MHC class II molecules on certain cell types that do not normally express them. Interferons also enhance the antigen-presenting function of MHC class I molecules by inducing the expression of key components of the intracellular machinery that enables peptides to be loaded onto the MHC molecules.

3-14 The two classes of MHC molecule have distinct subunit structures but similar three-dimensional structures.

The two classes of MHC molecule differ from each other in their structure and also have different distributions on the cells of the body. Their different structures enable the two classes of MHC molecules to serve distinct functions in antigen presentation, binding peptides from different intracellular sites and activating different subsets of T cells, as we will see in Chapter 5. Despite their differences in subunit structure, however, MHC class I and class II molecules are closely related in overall structure. In both classes, the two paired protein domains nearest the membrane resemble immunoglobulin domains, whereas the two domains distal to the membrane fold together to create a long cleft, or groove, which is the site at which a peptide binds. Purified peptide:MHC class I and peptide:MHC class II complexes have been characterized structurally, allowing us to describe in detail both the MHC molecules themselves and the way in which they bind peptides.

Fig. 3.20 The structure of an MHC class I molecule determined by X-ray crystallography. Panel a shows a computer graphic representation of a human MHC class I molecule, HLA-A2, which has been cleaved from the cell surface by the enzyme papain. The surface of the molecule is shown, colored according to the domains shown in panels b–d and described below. Panels b and c show a ribbon diagram of that structure. Shown schematically in panel d, the MHC class I molecule is a hetero-dimer of a membrane-spanning α chain (molecular weight 43 kDa) bound noncovalently to β$_2$-microglobulin (12 kDa), which does not span the membrane. The α chain folds into three domains: α$_1$, α$_2$, and α$_3$. The α$_3$ domain and β$_2$-microglobulin show similarities in amino acid sequence to immunoglobulin C domains and have similar folded structures, whereas the α$_1$ and α$_2$ domains fold together into a single structure consisting of two segmented α helices lying on a sheet of eight antiparallel β strands. The folding of the α$_1$ and α$_2$ domains creates a long cleft or groove, which is the site at which peptide antigens bind to the MHC molecules. The transmembrane region and the short stretch of peptide that connects the external domains to the cell surface are not seen in panels a and b as they have been removed by the papain digestion. As can be seen in panel c, looking down on the molecule from above, the sides of the cleft are formed from the inner faces of the two α helices; the β-pleated sheet formed by the pairing of the α$_1$ and α$_2$ domains creates the floor of the cleft. We shall use the schematic representation in panel d throughout this text.

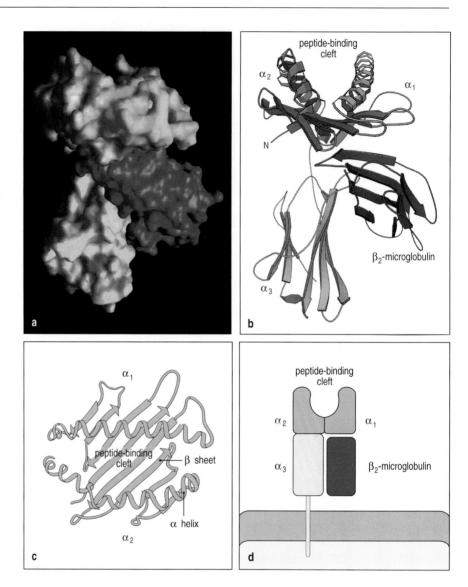

MHC class I structure is outlined in Fig. 3.20. MHC class I molecules consist of two polypeptide chains, a larger α chain encoded in the MHC genetic locus, and a smaller noncovalently associated chain, **β$_2$-microglobulin**, which is not polymorphic and is not encoded in the MHC locus. Only the class I α chain spans the membrane. The complete molecule has four domains, three formed from the MHC-encoded α chain, and one contributed by β$_2$-microglobulin. The α$_3$ domain and β$_2$-microglobulin have a folded structure that closely resembles that of an immunoglobulin domain. The most remarkable feature of MHC class I molecules is the structure of the folded α$_1$ and α$_2$ domains. These two domains form the walls of a cleft on the surface of the molecule; this is the site of peptide binding. They also are sites of polymorphisms that determine T-cell antigen recognition (see Chapter 5).

An MHC class II molecule consists of a noncovalent complex of two chains, α and β, both of which span the membrane (Fig. 3.21). The MHC class II α and β chains are both encoded within the MHC. The crystallographic structure of the MHC class II molecule shows that it is folded very much like the MHC class I molecule. The major differences lie at the ends of the peptide-binding cleft, which are more open in MHC class II molecules compared with MHC

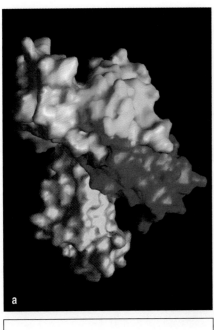

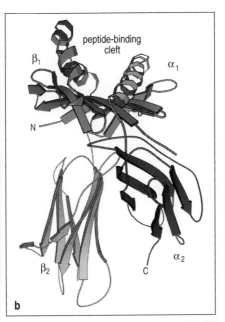

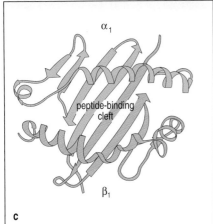

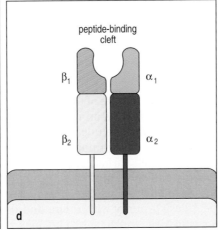

Fig. 3.21 MHC class II molecules resemble MHC class I molecules in overall structure. The MHC class II molecule is composed of two trans-membrane glycoprotein chains, α (34 kDa) and β (29 kDa), as shown schematically in panel d. Each chain has two domains, and the two chains together form a compact four-domain structure similar to that of the MHC class I molecule (compare with panel d of Fig. 3.20). Panel a shows a computer graphic representation of the surface of the MHC class II molecule, in this case the human protein HLA-DR1, and panel b shows the equivalent ribbon diagram. The α_2 and β_2 domains, like the α_3 and β_2-microglobulin domains of the MHC class I molecule, have amino acid sequence and structural similarities to immunoglobulin C domains; in the MHC class II molecule, the two domains forming the peptide-binding cleft are contributed by different chains and are therefore not joined by a covalent bond (see panels c and d). Another important difference, not apparent in this diagram, is that the peptide-binding groove of the MHC class II molecule is open at both ends.

class I molecules. The main consequence of this is that the ends of a peptide bound to an MHC class I molecule are substantially buried within the molecule, whereas the ends of peptides bound to MHC class II molecules are not. Again, the sites of major polymorphism are located in the peptide-binding cleft, which in the case of an MHC class II molecule are formed by the α_1, and β_1 domains (see Chapter 5).

In both MHC class I and class II molecules, bound peptides are sandwiched between the two α-helical segments of the MHC molecule (Fig. 3.22). The T-cell receptor interacts with this compound ligand, making contacts with both the MHC molecule and with the peptide fragment of antigen.

3-15 Peptides are stably bound to MHC molecules, and also serve to stabilize the MHC molecule on the cell surface.

An individual can be infected by a wide variety of different pathogens the proteins of which will not generally have peptide sequences in common. If T cells are to be alerted to all possible infections, then the MHC molecules on each cell (both class I and class II) must be able to bind stably to many different peptides. This behavior is quite distinct from that of other peptide-binding receptors, such as those for peptide hormones, which usually bind

Fig. 3.22 MHC molecules bind peptides tightly within the cleft. When MHC molecules are crystallized with a single synthetic peptide antigen, the details of peptide binding are revealed. In MHC class I molecules (panels a and c) the peptide is bound in an elongated conformation with both ends tightly bound at either end of the cleft. In the case of MHC class II molecules (panels b and d), the peptide is also bound in an elongated conformation but the ends of the peptide are not tightly bound and the peptide extends beyond the cleft. The upper surface of the peptide:MHC complex is recognized by T cells, and is composed of residues of the MHC molecule and the peptide. In representations c and d, the electrostatic potential of the MHC molecule surface is shown, with blue areas indicating a positive potential and red a negative potential.

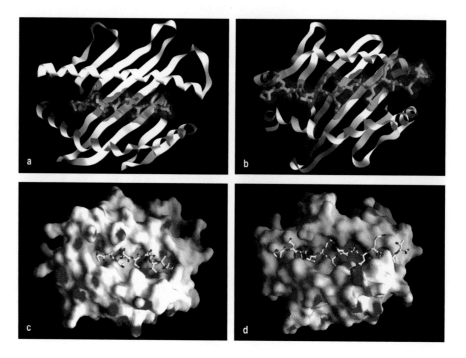

only a single type of peptide. The crystal structures of peptide:MHC complexes have helped to show how a single binding site can bind peptides with high affinity while retaining the ability to bind a wide variety of different peptides.

An important feature of the binding of peptides to MHC molecules is that the peptide is bound as an integral part of the MHC molecule's structure, and MHC molecules are unstable when peptides are not bound. The stability of peptide binding is important, because otherwise, peptide exchanges occurring at the cell surface would prevent peptide:MHC complexes from being reliable indicators of infection or of uptake of specific antigen. As a result of this stability, when MHC molecules are purified from cells, their bound peptides co-purify with them, and this has enabled the peptides bound by specific MHC molecules to be analyzed. The peptides are released from the MHC molecules by denaturing the complex in acid, and can then be purified and sequenced. Pure synthetic peptides can also be incorporated into previously empty MHC molecules and the structure of the complex determined, revealing details of the contacts between the MHC molecule and the peptide. From the sequences of peptides bound to specific MHC molecules, combined with structural analysis of the peptide:MHC complex, a detailed picture of the binding interactions has been built up. We will first discuss the peptide-binding properties of MHC class I molecules.

3-16   MHC class I molecules bind short peptides of 8–10 amino acids by both ends.

The binding of a peptide in the peptide-binding cleft of an MHC class I molecule is stabilized at both ends by contacts between atoms in the free amino and carboxy termini of the peptide and invariant sites that are found at each end of the cleft of all MHC class I molecules (Fig. 3.23). These contacts are thought to be the main stabilizing contacts for peptide:MHC class I complexes because synthetic peptide analogues lacking terminal amino and carboxyl groups fail to bind stably to MHC class I molecules. Other residues in the peptide serve as additional anchors. Peptides that bind to MHC class I molecules are usually 8–10 amino acids long. The peptide lies in an elongated conformation along the groove; variations in peptide length appear to be

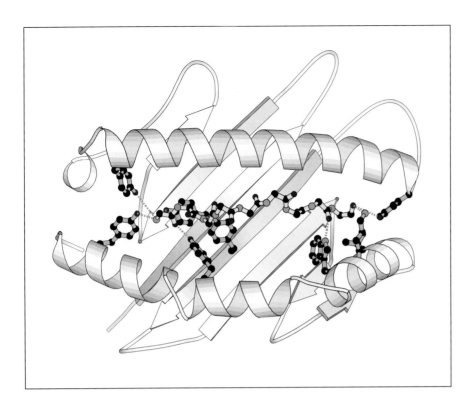

Fig. 3.23 Peptides are bound to MHC class I molecules by their ends. MHC class I molecules interact with the backbone of a bound peptide (shown in yellow) through a series of hydrogen bonds and ionic interactions (shown as dotted blue lines) at each end of the peptide. The amino terminus of the peptide is to the left; the carboxy terminus to the right. Black circles are carbon atoms; red are oxygen; blue are nitrogen. The amino acid residues in the MHC molecule that form these bonds are common to all MHC class I molecules and their side chains are shown in full (in gray) upon a ribbon diagram of the MHC class I groove. A cluster of tyrosine residues common to all MHC class I molecules forms hydrogen bonds to the amino terminus of the bound peptide, while a second cluster of residues forms hydrogen bonds and ionic interactions with the peptide backbone at the carboxy terminus and with the carboxy terminus itself.

accommodated, in most cases, by a kinking in the peptide backbone. However, two examples of MHC class I molecules where the peptide is able to extend out of the groove at the carboxy terminus suggest that some length variation may also be accommodated in this way.

These interactions give all MHC class I molecules their broad peptide-binding specificity. In addition, MHC molecules are highly polymorphic. There are hundreds of different versions, or alleles, of the MHC class I genes in the human population as a whole, and each individual carries only a small selection of them. The main differences between the allelic MHC variants are found at certain sites in the peptide-binding cleft, resulting in different amino acids in key peptide interaction sites in the different MHC variants. The consequence of this is that the different MHC variants preferentially bind different peptides. The peptides that can bind to a given MHC variant have the same or very similar amino acid residues at two or three particular positions along the peptide sequence. The amino acid side chains at these positions insert into pockets in the MHC molecule that are lined by the polymorphic amino acids. Because the binding of these side chains anchors the peptide to the MHC molecule, the peptide residues involved have been called **anchor residues**. Both the position and identity of these anchor residues can vary, depending on the particular MHC class I variant that is binding the peptide. However, most peptides that bind to MHC class I molecules have a hydrophobic (or sometimes basic) anchor residue at the carboxy terminus (Fig. 3.24). Changing an anchor residue can prevent the peptide from binding and, conversely, most synthetic peptides of suitable length that contain these anchor residues will bind the appropriate MHC class I molecule, in most cases irrespective of the amino acids at other positions in the peptide. These features of peptide binding enable an individual MHC class I molecule to bind a wide variety of different peptides, yet allow different MHC class I allelic variants to bind different sets of peptides.

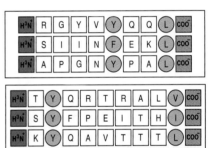

Fig. 3.24 Peptides bind to MHC molecules through structurally related anchor residues. Peptides eluted from two different MHC class I molecules are shown. The anchor residues (green) differ for peptides that bind different alleles of MHC class I molecules but are similar for all peptides that bind to the same MHC molecule. The upper and lower panels show peptides that bind to two different alleles of MHC class I molecules. The anchor residues that bind a particular MHC molecule need not be identical, but are always related (for example, phenyl-alanine (F) and tyrosine (Y) are both aromatic amino acids, whereas valine (V), leucine (L), and isoleucine (I) are all large hydro-phobic amino acids). Peptides also bind to MHC class I molecules through their amino (blue) and carboxy (red) termini.

3-17   The length of the peptides bound by MHC class II molecules is not constrained.

Peptide binding to MHC class II molecules has also been analyzed by elution of bound peptides and by X-ray crystallography, and differs in several ways from peptide binding to MHC class I molecules. Peptides that bind to MHC class II molecules are at least 13 amino acids long and can be much longer. The clusters of conserved residues that bind the two ends of a peptide in MHC class I molecules are not found in MHC class II molecules, and the ends of the peptide are not bound. Instead, the peptide lies in an extended conformation along the MHC class II peptide-binding groove. It is held in this groove both by peptide side chains that protrude into shallow and deep pockets lined by polymorphic residues, and by interactions between the peptide backbone and side chains of conserved amino acids that line the peptide-binding cleft in all MHC class II molecules (Fig. 3.25). Although there are fewer crystal structures of MHC class II-bound peptides than of MHC class I, the available data show that amino acid side chains at residues 1, 4, 6, and 9 of an MHC class II-bound peptide can be held in these binding pockets.

The binding pockets of MHC class II molecules are more permissive in their accommodation of different amino acid side chains than are those of the MHC class I molecule, making it more difficult to define anchor residues and predict which peptides will be able to bind particular MHC class II molecules (Fig. 3.26). Nevertheless, by comparing the sequences of known binding peptides, it is usually possible to detect a pattern of permissive amino acids for each of the different alleles of MHC class II molecules, and to model how the amino acids of this peptide sequence motif will interact with the amino acids that make up the peptide-binding cleft in the MHC class II molecule. Because the peptide is bound by its backbone and allowed to emerge from both ends of the binding groove there is, in principle, no upper limit to the length of peptides that could bind to MHC class II molecules. However, it appears that longer peptides bound to MHC class II molecules are trimmed

Fig. 3.25 Peptides bind to MHC class II molecules by interactions along the length of the binding groove. A peptide (yellow; shown as the peptide backbone only, with the amino terminus to the left and the carboxy terminus to the right), is bound by an MHC class II molecule through a series of hydrogen bonds (dotted blue lines) that are distributed along the length of the peptide. The hydrogen bonds toward the amino terminus of the peptide are made with the backbone of the MHC class II polypeptide chain, whereas throughout the peptide's length bonds are made with residues that are highly conserved in MHC class II molecules. The side chains of these residues are shown in gray upon the ribbon diagram of the MHC class II groove.

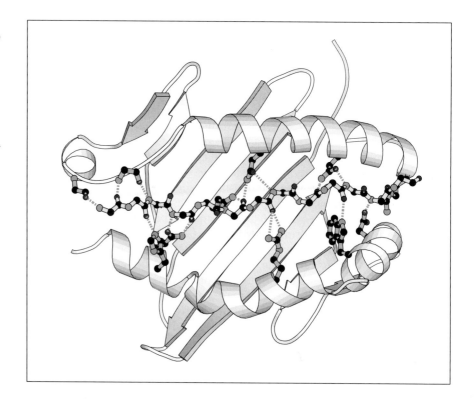

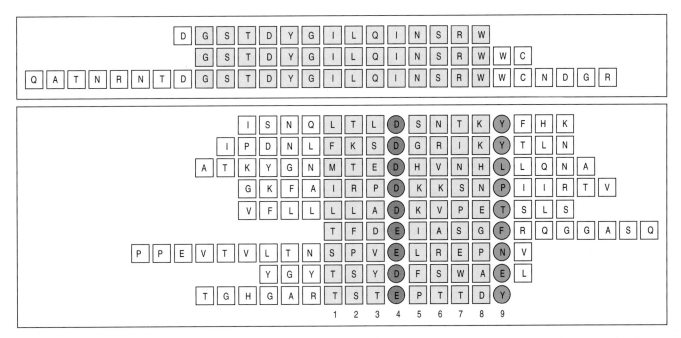

Fig. 3.26 Peptides that bind MHC class II molecules are variable in length and their anchor residues lie at various distances from the ends of the peptide. The sequences of a set of peptides that bind to the mouse MHC class II A^k allele are shown in the upper panel. All contain the same core sequence but differ in length. In the lower panel, different peptides binding to the human MHC class II allele HLA-DR3 are shown. The lengths of these peptides can vary, and so by convention the first anchor residue is denoted as residue 1. Note that all of the peptides share a negatively charged residue (aspartic acid (D) or glutamic acid (E)) in the P4 position (blue) and tend to have a hydrophobic residue (for example, tyrosine (Y), leucine (L), proline (P), phenylalanine (F)) in the P9 position (green).

by peptidases to a length of 13–17 amino acids in most cases. Like MHC class I molecules, MHC class II molecules that lack bound peptide are unstable, but the critical stabilizing interactions that the peptide makes with the MHC class II molecule are not yet known.

3-18 The crystal structures of several MHC:peptide:T-cell receptor complexes all show the same T-cell receptor orientation over the MHC:peptide complex.

At the time that the first X-ray crystallographic structure of a T-cell receptor was published, a structure of the same T-cell receptor bound to a peptide:MHC class I ligand was also produced. This structure (Fig. 3.27), which had been forecast by site-directed mutagenesis of the MHC class I molecule, showed the T-cell receptor aligned diagonally over the peptide and the peptide-binding groove, with the T-cell receptor α chain lying over the α_2 domain and the amino-terminal end of the bound peptide, the T-cell receptor β chain lying over the α_1 domain and the carboxy-terminal end of the peptide, with the CDR3 loops of both T-cell receptor α and T-cell receptor β meeting over the central amino acids of the peptide. The T-cell receptor is threaded through a valley between the two high points on the two surrounding α helices that form the walls of the peptide-binding cleft.

Analysis of other MHC class I:peptide:T-cell receptor complexes and of the single example so far of an MHC class II:peptide:T-cell receptor complex (Fig. 3.28) shows that all of them have a very similar orientation, particularly for the V_α domain, although some variability does occur in the location and orientation of the V_β domain. In this orientation, the V_α domain makes contact primarily with the amino terminus of the bound peptide, whereas the V_β domain contacts primarily the carboxy terminus of the bound peptide. Both

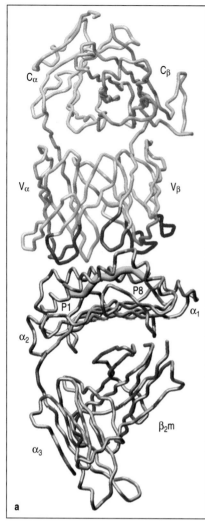

Fig. 3.27 The T-cell receptor binds to the MHC:peptide complex. Panel a: the T-cell receptor binds to the top of the MHC:peptide complex, straddling, in the case of the class I molecule shown here, both the α1 and α2 domain helices. The CDRs of the T-cell receptor are indicated in color; the CDR1 and CDR2 loops of the β chain in light and dark blue, respectively; and the CDR1 and CDR2 loops of the α chain in light and dark purple, respectively. The α chain CDR3 loop is in yellow while the β chain CDR3 loop is in green. The β chain HV4 loop is orange. Panel b: the outline of the T-cell receptor antigen-binding site (thick black line) is superimposed upon the top surface of the MHC:peptide complex (the peptide is shaded dull yellow). The T-cell receptor lies diagonally across the MHC:peptide complex, with the α and β CDR3 loops of the T-cell receptor (3α, 3β, yellow and green, respectively) contacting the center of the peptide. The α chain CDR1 and CDR2 loops (1α, 2α, light and dark purple, respectively) contact the MHC helices at the amino terminus of the bound peptide, whereas the β chain CDR1 and CDR2 loops (1β, 2β, light and dark blue, respectively) make contact with the helices at the carboxy terminus of the bound peptide. Courtesy of I.A. Wilson.

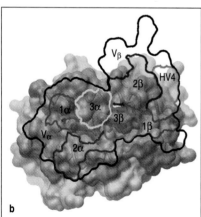

chains also interact with the α helices of the MHC class I molecule (see Fig. 3.27). The T-cell receptor contacts are not symmetrically distributed over the MHC molecule, so whereas the Vα CDR1 and CDR2 loops are in close contact with the helices of the MHC:peptide complex around the amino terminus of the bound peptide, the β-chain CDR1 and CDR2 loops, which interact with the complex at the carboxy terminus of the bound peptide, have variable contributions to the binding. This suggests that the Vα contacts are responsible for the conserved orientation of the T-cell receptor on the MHC:peptide complex.

Comparison of the three-dimensional structure of the T-cell receptor to that of the same T-cell receptor complexed to its MHC-peptide ligand could address the question of whether the T-cell receptor, like some other receptors, undergoes a conformational change, or 'induced fit,' in its three-dimensional structure when it binds its specific ligand. To date, there is no certain answer, owing to the limitations of the available structures. For one thing, all the T-cell receptors analyzed to date have either been bound to ligands that do not produce activation, or are bound to ligands that can activate them, but the comparable unliganded receptor structures are not available. Also, the crystals of the T-cell receptor are all formed at 0 °C or below, which locks the receptor into a single conformation. However, subtly different peptides can have strikingly different effects when the same T cell recognizes either of the two peptides complexed with MHC. This could be due to differences in how T-cell receptor conformation is altered by binding the two related yet different ligands. Recent evidence also suggests that the temperature at which the T-cell receptor binds to a particular peptide:MHC complex makes a large difference in the extent of T-cell receptor aggregation; protein conformation is affected by temperature, and so these differences may well result from a conformational change.

From an examination of these structures it is hard to predict whether the main binding energy is contributed by T-cell receptor contacts with the bound peptide, or by T-cell receptor contacts with the MHC molecule. It is known that alterations as simple as changing a leucine to isoleucine in the peptide are sufficient to alter the T-cell response from strong killing to absolutely no response at all. Studies show that mutations of single residues in the presenting MHC molecules can have the same effect. Thus, the specificity of T-cell recognition involves both the peptide and its presenting MHC molecule. This dual specificity underlies the MHC restriction of T-cell responses, a phenomenom that was observed long before the peptide-binding properties of MHC molecules were known. We will recount the story of how MHC restriction was discovered when we return to the issue of how MHC polymorphism affects antigen recognition by T cells in Chapter 5.

Another consequence of this dual specificity is a need for T-cell receptors to be able to interact appropriately with the antigen-presenting surface of MHC molecules. It appears that there is some inherent specificity for MHC molecules encoded in the T-cell receptor genes, as well as selection during T-cell development for a repertoire of receptors able to interact appropriately with the particular MHC molecules present in that individual. We will be discussing the evidence for this in Chapter 7.

3-19   A distinct subset of T cells bears an alternative receptor made up of γ and δ chains.

During the search for the gene for the T-cell receptor α chain, another T-cell receptorlike gene was unexpectedly discovered. This gene was named T-cell receptor γ, and its discovery led to a search for further T-cell receptor genes. Another receptor chain was identified using antibody to the predicted sequence of the γ chain and was called the δ chain. It was soon discovered that a minority population of T cells bore a distinct type of T-cell receptor made up of γ:δ heterodimers rather than α:β heterodimers. The development of these cells is described in Sections 7-13 and 7-14.

To date, there is no crystallographic structure of a γ:δ T-cell receptor, although it is expected to be similar in shape to α:β T-cell receptors. γ:δ T-cell receptors may be specialized to bind certain kinds of ligands, including heat-shock proteins and nonpeptide ligands such as mycobacterial lipid antigens. It seems likely that γ:δ T-cell receptors are not restricted by the 'classical' MHC class I and class II molecules. They may bind the free antigen, much as immunoglobulins do, and/or they may bind to peptides or other antigens presented by nonclassical MHC-like molecules. These are proteins that resemble MHC class I molecules but are relatively nonpolymorphic. We still know little about how γ:δ T-cell receptors bind antigen and thus how these cells function, and what their role is in immune responses. The structure and rearrangement of the genes for γ:δ T-cell receptors is covered in Sections 4-13 and 7-13 and the functions of γ:δ T cells are considered in Chapter 8.

Summary.

The receptor for antigen on most T cells, the α:β T-cell receptor, is composed of two protein chains, T-cell receptor α and T-cell receptor β, and resembles in many respects a single Fab fragment of immunoglobulin. T-cell receptors are always membrane-bound. α:β T-cell receptors do not recognize antigen in its native state, as do the immunoglobulin receptors of B cells, but recognize a composite ligand of a peptide antigen bound to an MHC molecule. MHC molecules are highly polymorphic glycoproteins encoded by genes in the major histocompatibility complex (MHC). Each MHC molecule binds a wide variety of different peptides, but the different variants each preferentially recognize sets of peptides with particular sequence and physical features. The peptide antigen is generated intracellularly, and bound stably in a peptide-binding cleft on the surface of the MHC molecule. There are two classes of MHC molecules and these are bound in their nonpolymorphic domains by CD8 and CD4 molecules that distinguish two different functional classes of α:β T cells. CD8 binds MHC class I molecules and can bind simultaneously to the same class I MHC:peptide complex being recognized by a T-cell receptor, thus acting as a co-receptor and enhancing the T-cell response; CD4 binds MHC class II molecules and acts as a co-receptor for T-cell receptors that recognise class II MHC:peptide ligands. T-cell receptors interact directly with both the antigenic peptide and polymorphic features of the MHC molecule that displays it, and this dual specificity underlies the

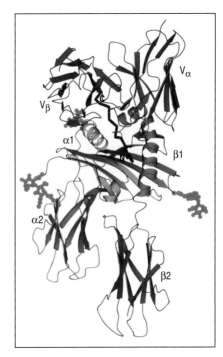

Fig. 3.28 The T-cell receptor interacts with MHC class I and MHC class II molecules in a similar fashion. The structure of a T-cell receptor binding to an MHC class II molecule has been determined, and shows the T-cell receptor binding to an equivalent site, and in an equivalent orientation, to the way that TCRs bind to MHC class I molecules (see Fig. 3.27). The structure of the molecules is shown in a cartoon form, with the MHC class II α and β chains shown in light green and orange respectively. Only the V_α and V_β domains of the T-cell receptor are shown, colored in blue. The peptide is colored red, while carbohydrate residues are indicated in gray. The TCR sits in a shallow saddle formed between the MHC class II α and β chain α-helical regions, at roughly 90° to the long axis of the MHC class II molecule and the bound peptide. Courtesy of E.L. Reinherz.

MHC restriction of T-cell responses. A second type of T-cell receptor, composed of a γ and a δ chain, is structurally similar to the α:β T-cell receptor but appears to bind different ligands, including nonpeptide ligands. It is not thought to be MHC restricted. It is found on a minority population of T cells, the γ:δ T cells, whose biological function is still not clear.

Summary to Chapter 3.

B cells and T cells use different, but structurally similar, molecules to recognize antigen. The antigen-recognition molecules of B cells are immunoglobulins, and are made both as a membrane-bound receptor for antigen, the B-cell receptor, and as secreted antibodies that bind antigens and elicit humoral effector functions. The antigen-recognition molecules of T cells, on the other hand, are made only as cell-surface receptors. Immunoglobulins and T-cell receptors are highly variable molecules, with the variability concentrated in that part of the molecule, the variable (V) region, that binds to antigen Immunoglobulins bind a wide variety of chemically different antigens, whereas the major α:β form of T-cell receptors will recognize only peptide fragments of foreign proteins bound to the MHC molecules that are ubiquitous on cell surfaces.

Binding of antigen by immunoglobulins has chiefly been studied using antibodies. The binding of antibody to its corresponding antigen is highly specific, and this specificity is determined by the shape and physicochemical properties of the antigen-binding site. The part of the antibody that elicits effector functions, once the variable part has bound an antigen, is located at the other end of the molecule from the antigen-binding sites, and is termed the constant region. There are five main functional classes of antibody, each encoded by a different type of constant region. As we will see in Chapter 9, these in turn interact with different components of the immune system to incite an inflammatory response and eliminate the antigen.

The T-cell receptor differs in several respects from the B-cell immunoglobulins. Among the most important of these differences is the absence of a secreted form of the receptor. This reflects the functional differences between T cells and B cells. B cells deal with pathogens and their protein products circulating within the body; secretion of a soluble antigen-recognition molecule by the activated B cell after antigen has been encountered enables them to mop up antigen effectively throughout the extracellular spaces of the body. T cells, on the other hand, are specialized for cell–cell interactions. They either kill cells that are infected with intracellular pathogens and that bear foreign antigenic peptides on their surface, or interact with cells of the immune system that have taken up foreign antigen and are displaying it on the cell surface. They thus have no requirement for a soluble, secreted receptor.

An additional distinctive feature of the T-cell receptor compared with immunoglobulins is that it recognizes a composite ligand made up of the foreign peptide bound to a self MHC molecule. This forces T cells to interact with infected body cells to become activated. Each T-cell receptor is specific for a particular combination of peptide and a self MHC molecule.

MHC molecules are encoded by a family of highly polymorphic genes and, although each individual expresses several, this represents only a small selection of all possible variants. During T-cell development, the T-cell receptor repertoire is selected so that the T cells of each individual recognize antigen only in conjunction with their own MHC molecules. Expression of multiple variant MHC molecules each with a different peptide-binding repertoire helps ensure that T cells from an individual will be able to recognize at least some peptides generated from nearly every pathogen.

General references.

Ager, A., Callard, R., Ezine, S., Gerard, C., and Lopez-Botet, M.: **Immune receptor supplement.** *Immunol. Today* 1996, **17**.

Davies, D.R., and Chacko, S.: **Antibody structure.** *Acc. Chem. Res.* 1993, **26**:421-427.

Frazer, K., and Capra, J.D.: Immunoglobulins: structure and function, in Paul W.E. (ed): *Fundamental Immunology*, 4th edn. New York, Raven Press, 1998.

Garcia, K.C., Teyton, L., and Wilson, I.A.: **Structural basis of T cell recognition.** *Annu. Rev. Immunol.* 1999, **17**:369-397.

Germain, R.N.: **MHC-dependent antigen processing and peptide presentation: providing ligands for T lymphocyte activation.** *Cell* 1994, **76**:287-299.

Honjo, T., and Alt, F.W. (eds): *Immunoglobulin Genes*, 2nd edn. London, Academic Press, 1996.

Moller, G. (ed): **Origin of major histocompatibility complex diversity.** *Immunol. Rev.* 1995, **143**:5-292.

Poljak, R.J.: **Structure of antibodies and their complexes with antigens.** *Mol. Immunol.* 1991, **28**:1341-1345.

Section references.

3-1 IgG antibodies consist of four polypeptide chains.

Edelman, G.M.: **Antibody structure and molecular immunology.** *Scand. J. Immunol.* 1991, **34**:4-22.

Faber, C., Shan, L., Fan, Z., Guddat, L.W., Furebring, C., Ohlin, M., Borrebaeck, C.A.K., and Edmundson, A.B.: **Three-dimensional structure of a human Fab with high affinity for tetanus toxoid.** *Immunotech.* 1998, **3**:253-270.

Harris, L.J., Larson, S.B., Hasel, K.W., Day, J., Greenwood, A., and McPherson, A.: **The 3-dimensional structure of an intact monoclonal antibody for canine lymphoma.** *Nature* 1992, **360**:369-372.

3-2 Immunoglobulin heavy and light chains are composed of constant and variable regions.

Han, W.H., Mou, J.X., Sheng, J., Yang, J., and Shao, Z.F.: **Cryo-atomic force microscopy—a new approach for biological imaging at high resolution.** *Biochem.* 1995, **34**:8215-8220.

3-3 The antibody molecule can readily be cleaved into functionally distinct fragments.

Porter, R.R.: **Structural studies of immunoglobulins.** *Scand. J. Immunol.* 1991, **34**:382-389.

Yamaguchi, Y., Kim, H., Kato, K., Masuda, K., Shimada, I., and Arata, Y.: **Proteolytic fragmentation with high specificity of mouse IgG—mapping of proteolytic cleavage sites in the hinge region.** *J. Immunol. Meth.* 1995, **181**:259-267.

3-4 The immunoglobulin molecule is flexible, especially at the hinge region.

Gerstein, M., Lesk, A.M., and Chothia, C.: **Structural mechanisms for domain movements in proteins.** *Biochem.* 1994, **33**:6739-6749.

Kim, J.K., Tsen, M.F., Ghetie, V., and Ward, E.S.: **Evidence that the hinge region plays a role in maintaining serum levels of the murine IgG1 molecule.** *Mol. Immunol.* 1995, **32**:467-475.

3-5 The domains of an immunoglobulin molecule have similar structures.

Barclay, A.N., Brown, M.H., Law, S.K., McKnight, A.J., Tomlinson, M.G., and van der Merwe, P.A. (eds): *The Leukocyte Antigen Factsbook*, 2nd edn. London, Academic Press, 1997.

Hsu, E., and Steiner, L.A.: **Primary structure of immunoglobulin through evolution.** *Curr. Opin. Struct. Biol.* 1992, **2**:422-430.

3-6 Localized regions of hypervariable sequence form the antigen-binding site.

Chitarra, V., Alzari, P.M., Bentley, G.A., Bhat, T.N., Eisele, J.L., Houdusse, A., Lescar, J., Souchon, H., and Poljak, R.J.: **3-dimensional structure of a heteroclitic antigen–antibody cross reaction complex.** *Proc. Nat. Acad. Sci. USA* 1993, **90**:7711-7715.

Gilliland, L.K., Norris, N.A., Marquardt, H., Tsu, T.T., Hayden, M.S., Neubauer, M.G., Yelton, D.E., Mittler, R.S., and Ledbetter, J.A.: **Rapid and reliable cloning of antibody variable regions and generation of recombinant single-chain antibody fragments.** *Tissue Antigens* 1996, **47**:1-20.

Johnson, G., and Wu, T.T.: **Kabat Database and its applications: 30 years after the first variability plot.** *Nucleic Acids Res.* 2000, **28**: 214-218.

Wu, T.T., and Kabat, E.A.: **An analysis of the sequences of the variable regions of Bence Jones proteins and myeloma light chains and their implications for antibody complementarity.** *J. Exp. Med.* 1970, **132**:211-250.

3-7 Antibodies bind antigens via contacts with amino acids in CDRs, but the details of binding depend upon the size and shape of the antigen.

&

3-8 Antibodies bind to conformational shapes on the surfaces of antigens.

Davies, D.R., and Cohen, G.H.: **Interactions of protein antigens with antibodies.** *Proc. Natl. Acad. Sci. USA* 1996, **93**:7-12.

Padlan, E.A.: **Anatomy of the antibody molecule.** *Mol. Immunol.* 1994, **31**:169-217.

Stanfield, R.L., and Wilson, I.A.: **Protein–peptide interactions.** *Curr. Opin. Struct. Biol.* 1995, **5**:103-113.

Wilson, I.A., and Stanfield, R.L.: **Antibody-antigen interactions: new structures and new conformational changes.** *Curr. Opin. Struct. Biol.* 1994, **4**:857-867.

3-9 Antigen–antibody interactions involve a variety of forces.

Braden, B.C., and Poljak, R.J.: **Structural features of the reactions between antibodies and protein antigens.** *FASEB J.* 1995, **9**:9-16.

Braden, B.C., Goldman, E.R., Mariuzza, R.A., and Poljak, R.J.: **Anatomy of an antibody molecule: structure, kinetics, thermodynamics and mutational studies of the antilysozyme antibody D1.3.** *Immunol. Rev.* 1998, **163**:45-57.

Ros, R., Schwesinger, F., Anselmetti, D., Kubon, M., Schäfer, R., Plückthun, A., and Tiefenauer, L.: **Antigen binding forces of individually addressed single-chain Fv antibody molecules.** *Proc. Natl. Acad. Sci. USA* 1998, **95**:7402-7405.

3-10 The antigen receptor on T cells is very similar to a Fab fragment of immunoglobulin.

Garcia, K.C., Degano, M., Stanfield, R.L., Brunmark, A., Jackson, M.R., Peterson, P.A., Teyton, L., and Wilson, I.A.: **An αβ T cell receptor structure at 2.5 Å and its orientation in the TCR-MHC complex.** *Science* 1996, **274**:209-219.

Housset, D., Mazza, G., Gregoire, C., Piras, C., Malissen, B., and Fontecilla-Camps, J.C.: **The three-dimensional structure of a T cell antigen receptor VαVβ domain reveals a novel arrangement of the Vβ domain.** *EMBO J.* 1997, **16**:4205-4216.

3-11 A T-cell receptor recognizes antigen in the form of a complex of a foreign peptide bound to an MHC molecule.

Davis, M.M., Boniface, J.J., Reich, Z., Lyons, D., Hampl, J., Arden, B., and Chien, Y.: **Ligand recognition by alpha beta T cell receptors.** *Annu. Rev. Immunol.* 1998, **16**:523-544.

Garboczi, D.N., Ghosh, P., Utz, U., Fan, Q.R., Biddison, W.E., and Wiley, D.C.: **Structure of the complex between human T-cell receptor, viral peptide and HLA-A2.** *Nature* 1996, **384**:134-141.

3-12 T cells with different functions are distinguished by CD4 and CD8 cell-surface proteins and recognize peptides bound to different classes of MHC molecule.

Gao, G.F., Tormo, J., Gerth, U.C., Wyer, J.R., McMichael, A.J., Stuart, D.I., Bell, J.I., Jones, E.Y., and Jakobsen, B.Y.: **Crystal structure of the complex between human CD8αα and HLA-A2.** *Nature* 1997, **387**:630-634.

Morrison, L.A., Lukacher, A.E., Braciale, V.L., Fan, D.P., and Braciale, T.J.: **Differences in antigen presentation to MHC class I- and class II-restricted influenza virus-specific cytolytic T-lymphocyte clones.** *J. Exp. Med.* 1986, **163**:903.

Wu, H., Kwong, P.D., and Hendrickson, W.A.: **Dimeric association and segmental variability in the structure of human CD4.** *Nature* 1997, **387**:427-530.

Zamoyska, R.: **CD4 and CD8: modulators of T cell receptor recognition of antigen and of immune responses?** *Curr. Opin. Immunol.* 1998, **10**:82-86.

3-13 The two classes of MHC molecule are expressed differentially on cells.

Steimle, V., Siegrist, C.A., Mottet, A., Lisowska-Grospierre, B., and Mach, B.: **Regulation of MHC class II expression by interferon-γ mediated by the trans-activator gene CIITA.** *Science* 1994, **265**:106-109.

3-14 The two classes of MHC molecule have distinct subunit structures but similar three-dimensional structures.

&

3-15 Peptides are stably bound to MHC molecules, and also serve to stabilize the MHC molecule on the cell surface.

Dessen, A., Lawrence, C.M., Cupo, S., Zaller, D.M., and Wiley, D.C.: **X-ray crystal structure of HLA-DR4 (DRA*0101, DRB1*0401) complexed with a peptide from human collagen II.** *Immunity* 1997, **7**:473-481.

Fremont, D.H., Hendrickson, W.A., Marrack, P., and Kappler, J.: **Structures of an MHC class II molecule with covalently bound single peptides.** *Science* 1996, **272**:1001-1004.

Fremont, D.H., Matsumura, M., Stura, E.A., Peterson, P.A. and Wilson, I.A.: **Crystal structures of two viral peptides in complex with murine MHC class 1 H-2Kb.** *Science* 1992, **257**:919-927.

Fremont, D.H., Monnaie, D., Nelson, C.A., Hendrickson, W.A., and Unanue, E.R.: **Crystal structure of I-Ak in complex with a dominant epitope of lysozyme.** *Immunity* 1998, **8**:305-317.

Madden, D.R., Gorga, J.C., Strominger, J.L. and Wiley, D.C.: **The three- dimensional structure of HLA-B27 at 2.1Å resolution suggests a general mechanism for tight peptide binding to MHC.** *Cell* 1992, **70**:1035-1048.

Murthy, V.L., and Stern, L.J.: **The class II MHC protein HLA-DR1 in complex with an endogenous peptide: implications for the structural basis of the specificity of peptide binding.** *Structure* 1997, **5**:1385-1396.

Stern, L.J. and Wiley, D.C.: **Antigenic peptide binding by class I and class II histocompatibility proteins.** *Structure* 1994, **2**:245-251.

3-16 MHC class I molecules bind short peptides of 8–10 amino acids by both ends.

Bouvier, M., and Wiley, D.C.: **Importance of peptide amino and carboxyl termini to the stability of MHC class I molecules.** *Science* 1994, **265**:398-402.

Weiss, G.A., Collins, E.J., Garboczi, D.N., Wiley, D.C., and Schreiber, S.L.: **A tricyclic ring system replaces the variable regions of peptides presented by three alleles of human MHC class I molecules.** *Chem. Biol.* 1995, **2**:401-407.

3-17 The length of the peptides bound by MHC class II molecules is not constrained.

Rammensee, H.G.: **Chemistry of peptides associated with MHC class I and class II molecules.** *Curr. Opin. Immunol.* 1995, **7**:85-96.

Rudensky, A.Y., Preston-Hurlburt, P., Hong, S.C., Barlow, A., and Janeway, C.A., Jr.: **Sequence analysis of peptides bound to MHC class II molecules.** *Nature* 1991, **353**:622.

3-18 The crystal structures of several MHC:peptide:T-cell receptor complexes all show the same T-cell receptor orientation over the MHC:peptide complex.

Ding, Y.H., Smith, K.J., Garboczi, D.N., Utz, U., Biddison, W.E., and Wiley, D.C.: **Two human T cell receptors bind in a similar diagonal mode to the HLA-A2/Tax peptide complex using different TCR amino acids.** *Immunity* 1998, **8**:403-411.

Garcia, K.C., Degano, M., Pease, L.R., Huang, M., Peterson, P.A., Leyton, L., and Wilson, I.A.: **Structural basis of plasticity in T cell receptor recognition of a self peptide-MHC antigen.** *Science* 1998, **279**:1166-1172.

Sant'Angelo, D.B., Waterbury, G., Preston-Hurlburt, P., Yoon, S.T., Medzhitov, R., Hong, S.C., and Janeway, C.A., Jr.: **The specificity and orientation of a TCR to its peptide-MHC class II ligands.** *Immunity* 1996, **4**:367-376.

Teng., M.K., Smolyar, A., Tse, A.G.D., Liu, J.H., Liu, J., Hussey, R.E., Nathenson, S.G., Chang, H.C., Reinherz, E.L., and Wang, J.H.: **Identification of a common docking topology with substantial variation among different TCR-MHC-peptide complexes.** *Curr. Biol.* 1998, **8**:409-412.

3-19 A distinct subset of T cells bears an alternative receptor made up of γ and δ chains.

Bukowski, J.F., Morita, C.T., and Brenner, M.B.: **Human gamma delta T cells recognize alkylamines derived from microbes, edible plants, and tea: implications for innate immunity.** *Immunity* 1999, **11**:57-65.

Chien, Y.H., Jores, R., and Crowley, M.P.: **Recognition by gamma/delta T cells.** *Annu. Rev. Immunol.* 1996, **14**:511-532.

Li, H., Lebedeva, M.I., Llera, A.S., Fields, B.A., Brenner, M.B., and Mariuzza, M.B.: **Structure of the Vδ domain of a human γ·δ T-cell antigen receptor.** *Nature* 1998, **391**:502-506.

Spada, F.M., Grant, E.P., Peters, P.J., Sugita, M., Melian, A., Leslie, D.S., Lee, H.K., van Donselaar, E., Hanson, D.A., Krensky, A.M., Majdic, O., Porcelli, S.A., Morita, C.T., and Brenner, M.B.: **Self-recognition of CD1 by gamma/delta T cells: implications for innate immunity.** *J. Exp. Med.* 2000, **191**:937-948.

The Generation of Lymphocyte Antigen Receptors

4

Lymphocyte antigen receptors, in the form of immunoglobulins on B cells and T-cell receptors on T cells, are the means by which lymphocytes sense the presence of antigens in their environment. The receptors produced by each lymphocyte have a unique antigen specificity, which is determined by the structure of their antigen-binding site, as described in Chapter 3. Because each person possesses billions of lymphocytes, these cells collectively provide the individual with the ability to respond to a great variety of antigens. The wide range of antigen specificities in the antigen receptor repertoire is due to variation in the amino acid sequence at the antigen-binding site, which is made up from the variable (V) regions of the receptor protein chains. In each chain the V region is linked to an invariant constant (C) region, which provides effector or signaling functions.

Given the importance of a diverse repertoire of lymphocyte receptors in the defense against infection, it is not surprising that a complex and elegant genetic mechanism has evolved for generating these highly variable proteins. Each receptor chain variant cannot be encoded in full in the genome, as this would require more genes for antigen receptors than there are genes in the entire genome. Instead, we will see that the V regions of the receptor chains are encoded in several pieces—so-called gene segments. These are assembled in the developing lymphocyte by somatic DNA recombination to form a complete V-region sequence, a mechanism known generally as **gene rearrangement**. Each type of gene segment is present in multiple copies in the germline genome. The selection of a gene segment of each type during gene rearrangement occurs at random, and the large number of possible different combinations accounts for much of the diversity of the receptor repertoire.

In the first two parts of this chapter we will describe the gene rearrangement mechanism that generates the V regions of immunoglobulin and T-cell receptor genes. The basic mechanism is common to both B cells and T cells, and involves many if not all of the same enzymes. We will describe the details of the enzymology of this recombination process, the evolution of which was probably critical to the evolution of the vertebrate adaptive immune system.

In B cells, but not T cells, the rearranged V region undergoes additional modification, known as **somatic hypermutation**. This does not occur until after B cells encounter and become activated by antigen. In these cells, the V regions of the assembled immunoglobulin genes undergo a high rate of point mutation that creates additional diversity within the expanding clone of B cells responding to antigen.

In the third part of the chapter we consider the limited, but functionally important, diversity of immunoglobulin C regions. The C regions of T-cell receptors do not show such diversity as they function only as part of a membrane-bound antigen receptor. Their role is to anchor and support the

V regions at the cell surface as well as linking the binding of antigen by the V regions to the receptor-associated intracellular signaling complex. The C regions of immunoglobulins also serve these functions but in addition the C regions of the heavy chain are responsible for the effector functions of the secreted immunoglobulins, or antibodies, made by activated B cells. These C_H regions come in several different versions, or isotypes, each of which has a different effector function. In B cells that have become activated by antigen, the heavy-chain V region can become associated with a different C_H region by a further somatic recombination event, in the process known as **isotype switching**. This enables the different heavy-chain C regions, each with a different function, to be represented among antibodies of the same antigen specificity.

The generation of diversity in immunoglobulins.

Virtually any substance can elicit an antibody response. Furthermore, the response even to a simple antigen bearing a single antigenic determinant is diverse, comprising many different antibody molecules each with a unique affinity, or binding strength, for the antigen and a subtly different specificity. The total number of antibody specificities available to an individual is known as the **antibody repertoire**, or **immunoglobulin repertoire**, and in humans is at least 10^{11}, perhaps many more. The number of antibody specificities present at any one time is, however, limited by the total number of B cells in an individual, as well as by each individual's encounters with antigens.

Before it was possible to examine the immunoglobulin genes directly, there were two main hypotheses for the origin of this diversity. The **germline theory** held that there is a separate gene for each different immunoglobulin chain and that the antibody repertoire is largely inherited. By contrast, **somatic diversification theories** proposed that the observed repertoire is generated from a limited number of inherited V-region sequences that undergo alteration within B cells during the individual's lifetime. Cloning of the immunoglobulin genes revealed that the antibody repertoire is, in fact, generated by DNA rearrangements during B-cell development. As we will see in this part of the chapter, a DNA sequence encoding a V region is assembled at each locus by selection from a relatively small group of inherited gene segments. Diversity is further enhanced by the process of somatic hypermutation in mature activated B cells. Thus the somatic diversification theory was essentially correct, although the concept of multiple germline genes embodied in the germline theory also proved true.

4-1   Immunoglobulin genes are rearranged in antibody-producing cells.

In nonlymphoid cells, the gene segments encoding the greater part of the V region of an immunoglobulin chain are some considerable distance away from the sequence encoding the C region. In mature B lymphocytes, however, the assembled V-region sequence lies much nearer the C region, as a consequence of gene rearrangement. Rearrangement within the immunoglobulin genes was originally discovered 25 years ago, when it first became possible to study the organization of the immunoglobulin genes in both B cells and nonlymphoid cells using restriction enzyme analysis and Southern blotting. In this procedure, chromosomal DNA is first cut with a restriction enzyme, and the DNA fragments containing particular V- and C-region sequences are identified by hybridization with radiolabeled DNA probes specific for the relevant DNA sequences. In germline DNA, from nonlymphoid cells, the V- and C-region

Fig. 4.1 Immunoglobulin genes are rearranged in B cells. The two photographs on the left (germline DNA) show a Southern blot of a restriction enzyme digest of DNA from nonlymphoid cells from a normal person. The locations of immunoglobulin DNA sequences are identified by hybridization with V- and C-region probes. The V and C regions are found in distinct DNA fragments in the nonlymphoid DNA. The two photographs on the right (B-cell DNA) are of the same restriction digest of DNA from peripheral blood lymphocytes from a patient with chronic lymphocytic leukemia (see Chapter 7), in which a single clone of B cells is greatly expanded. The malignant B cells express the V region from which the V-region probe was obtained and, owing to their predominance in the cell population, this unique rearrangement can be detected. In this DNA, the V and C regions are found in the same fragment, which is a different size from either the C- or the V-region germline fragments. Although not shown in this figure, a population of normal B lymphocytes has many different rearranged genes, so they yield a smear of DNA fragment sizes, which are not visible as a crisp band. Photograph courtesy of S. Wagner and L. Luzzatto.

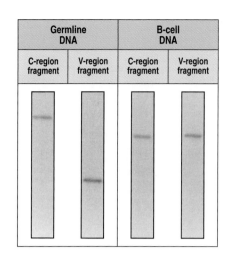

Germline DNA		B-cell DNA	
C-region fragment	V-region fragment	C-region fragment	V-region fragment

sequences identified by the probes are on separate DNA fragments. However, in DNA from an antibody-producing B cell these V- and C-region sequences are on the same DNA fragment, showing that a rearrangement of the DNA has occurred. A typical experiment using human DNA is shown in Fig. 4.1.

This simple experiment showed that segments of genomic DNA within the immunoglobulin genes are rearranged in cells of the B-lymphocyte lineage, but not in other cells. This process of rearrangement is known as **somatic recombination**, to distinguish it from the meiotic recombination that takes place during the production of gametes.

4-2   The DNA sequence encoding a complete V region is generated by the somatic recombination of separate gene segments.

The V region, or V domain, of an immunoglobulin heavy or light chain is encoded by more than one gene segment. For the light chain, the V domain is encoded by two separate DNA segments. The first segment encodes the first 95–101 amino acids of the light chain and is termed a **V gene segment** because it encodes most of the V domain. The second segment encodes the remainder of the V domain (up to 13 amino acids) and is termed a **joining** or **J gene segment**.

The rearrangements that lead to the production of a complete immunoglobulin light-chain gene are shown in Fig. 4.2 (center panel). The joining of a V and a J gene segment creates a continuous exon that encodes the whole of the light-chain V region. In the unrearranged DNA, the V gene segments are located relatively far away from the C region. The J gene segments are located close to the C region, however, and joining of a V segment to a J gene segment also brings the V gene close to a C-region sequence. The J gene segment of the rearranged V region is separated from a C-region sequence only by an intron. In the experiment shown in Fig. 4.1, the germline DNA fragment identified by the 'V-region probe' contains the V gene segment, and that identified by the 'C-region probe' actually contains both the J gene segment and the C-region sequence. To make a complete immunoglobulin light-chain messenger RNA, the V-region exon is joined to the C-region sequence by RNA splicing after transcription (see Fig. 4.2).

A heavy-chain V region is encoded in three gene segments. In addition to the V and J gene segments (denoted V_H and J_H to distinguish them from the light-chain V_L and J_L), there is a third gene segment called the **diversity** or D_H **gene segment**, which lies between the V_H and J_H gene segments. The process of recombination that generates a complete heavy-chain V region is shown in

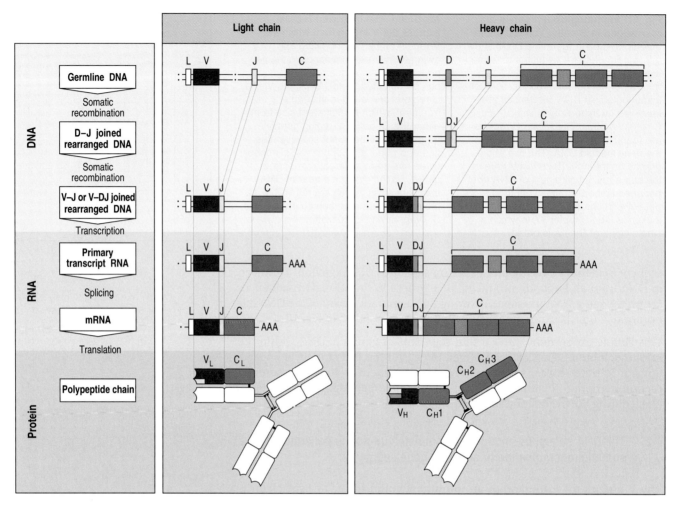

Fig. 4.2 V-region genes are constructed from gene segments. Light-chain V-region genes are constructed from two segments (center panel). A variable (V) and a joining (J) gene segment in the genomic DNA are joined to form a complete light-chain V-region exon. Immunoglobulin chains are extracellular proteins and the V gene segment is preceded by an exon encoding a leader peptide (L), which directs the protein into the cell's secretory pathways and is then cleaved. The light-chain C region is encoded in a separate exon and is joined to the V-region exon by splicing of the light-chain RNA to remove the L-to-V and the J-to-C introns. Heavy-chain V regions are constructed from three gene segments (right panel). First, the diversity (D) and J gene segments join, then the V gene segment joins to the combined DJ sequence, forming a complete V_H exon. A heavy-chain C-region gene is encoded by several exons. The C-region exons, together with the leader sequence, are spliced to the V-domain sequence during processing of the heavy-chain RNA transcript. The leader sequence is removed after translation and the disulfide bonds that link the polypeptide chains are formed. The hinge region is shown in purple.

Fig. 4.2 (right panel), and occurs in two separate stages. In the first, a D_H gene segment is joined to a J_H gene segment; then a V_H gene segment rearranges to DJ_H to make a complete V_H-region exon. As with the light-chain genes, RNA splicing joins the assembled V-region sequence to the neighboring C-region gene.

4-3   There are multiple different V-region gene segments.

For simplicity, we have so far discussed the formation of a complete immunoglobulin V-region sequence as though there were only a single copy of each gene segment. In fact, there are multiple copies of all of the gene segments in germline DNA. It is the random selection of just one gene segment of each type to assemble a V region that makes possible the great diversity of V regions among immunoglobulins. The numbers of functional gene segments of each type in the human genome, as determined by gene cloning and sequencing, are shown in Fig. 4.3. Not all the gene segments discovered are functional, as a proportion have accumulated mutations that prevent them from encoding a functional protein. These are termed 'pseudogenes.' Because there are many V, D, and J gene segments in germline DNA, no single one is essential. This reduces the evolutionary pressure on each gene segment to remain intact, and has resulted in a relatively large number of pseudogenes. Since some of these pseudogenes can undergo rearrangement just like a normal functional gene segment, a significant proportion of rearrangements will incorporate a pseudogene and thus be nonfunctional.

The immunoglobulin gene segments are organized into three clusters or genetic loci—the κ, λ, and heavy-chain loci. These are on different chromosomes and each is organized slightly differently, as shown in Fig. 4.4 for humans. At the λ light-chain locus, located on chromosome 22, a cluster of V_λ gene segments is followed by four sets of J_λ gene segments each linked to a single C_λ gene. In the κ light-chain locus, on chromosome 2, the cluster of V_κ gene segments is followed by a cluster of J_κ gene segments, and then by a single C_κ gene. The organization of the heavy-chain locus, on chromosome 14, resembles that of the κ locus, with separate clusters of V_H, D_H, and J_H gene segments and of C_H genes. The heavy-chain locus differs in one important way: instead of a single C-region, it contains a series of C regions arrayed one after the other, each of which corresponds to a different isotype. Generally, a cell expresses only one at a time, beginning with IgM. The expression of other isotypes, such as IgG, can occur through isotype switching, as will be described in Section 4-16.

The human V gene segments can be grouped into families in which each member shares at least 80% DNA sequence identity with all others in the family. Both the heavy-chain and κ-chain V gene segments can be subdivided into seven such families, whereas there are eight families of V_λ gene segments. The families can be grouped into clans, made up of families that are

Number of functional gene segments in human immunoglobulin loci			
Segment	Light chains		Heavy chain
	κ	λ	H
Variable (V)	40	30	65
Diversity (D)	0	0	27
Joining (J)	5	4	6

Fig. 4.3 The numbers of functional gene segments for the V regions of human heavy and light chains. These numbers are derived from exhaustive cloning and sequencing of DNA from one individual and exclude all pseudogenes (mutated and nonfunctional versions of a gene sequence). Owing to genetic polymorphism, the numbers will not be the same for all people.

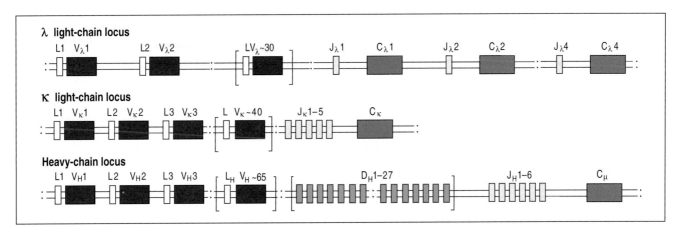

Fig. 4.4 The germline organization of the immunoglobulin heavy- and light-chain loci in the human genome. The genetic locus for the λ light chain (chromosome 22) has about 30 functional V_λ gene segments and four pairs of functional J_λ gene segments and C_λ genes. The κ locus (chromosome 2) is organized in a similar way, with about 40 functional V_κ gene segments accompanied by a cluster of five J_κ gene segments but with a single C_κ gene. In approximately 50% of individuals, the entire cluster of κ V gene segments has undergone an increase by duplication (not shown for simplicity). The heavy-chain locus (chromosome 14) has about 65 functional V_H gene segments and a cluster of around 27 D segments lying between these V_H gene segments and six J_H gene segments. The heavy-chain locus also contains a large cluster of C_H genes that are described in Fig. 4.18. For simplicity we have shown only a single C_H gene in this diagram without illustrating its separate exons, have omitted pseudogenes, and have shown all V gene segments in the same orientation. L, leader sequence. This diagram is not to scale: the total length of the heavy-chain locus is over 2 megabases (2 million bases), whereas some of the D segments are only six bases long.

more similar to each other than to families in other clans. Human V_H gene segments fall into three such clans. All of the V_H gene segments identified from amphibians, reptiles, and mammals also fall into the same three clans, suggesting that these clans existed in a common ancestor of these modern animal groups.

4-4   Rearrangement of V, D, and J gene segments is guided by flanking DNA sequences.

A system is required to ensure that DNA rearrangements take place at the correct locations relative to the V, D, or J gene segment coding regions. In addition, joins must be regulated such that a V gene segment joins to a D or J and not to another V. DNA rearrangements are in fact guided by conserved noncoding DNA sequences that are found adjacent to the points at which recombination takes place. These sequences consist of a conserved block of seven nucleotides—the **heptamer** 5′CACAGTG3′—which is always contiguous with the coding sequence, followed by a nonconserved region known as the **spacer**, which is either 12 or 23 nucleotides long. This is followed by a second conserved block of nine nucleotides—the **nonamer** 5′ACAAAAACC3′ (Fig. 4.5). The spacer varies in sequence but its conserved length corresponds to one or two turns of the DNA double helix. This brings the heptamer and non-amer sequences to the same side of the DNA helix, where they can be bound by the complex of proteins that catalyzes recombination. The heptamer–spacer–nonamer is called a **recombination signal sequence** (RSS).

Recombination only occurs between gene segments located on the same chromosome. It generally follows the rule that only a gene segment flanked by a RSS with a 12-base pair (bp) spacer can be joined to one flanked by a 23 bp spacer RSS. This is known as the **12/23 rule**. Thus, for the heavy chain, a D_H gene segment can be joined to a J_H gene segment and a V_H gene segment to a D_H gene segment, but V_H gene segments cannot be joined to J_H gene segments directly, as both V_H and J_H gene segments are flanked by 23 bp spacers and the D_H gene segments have 12 bp spacers on both sides (see Fig. 4.5).

It is now apparent, however, that, even though it violates the 12/23 rule, direct joining of one D gene segment to another can occur in most species. In humans, D–D fusion is found in approximately 5% of antibodies and is the major mechanism accounting for the unusually long CDR3 loops found in some heavy chains. By creating extra-long CDR3s and unusual amino acid combinations, these D–D fusions add further to the diversity of the antibody repertoire.

The mechanism of DNA rearrangement is similar for the heavy- and light-chain loci, although only one joining event is needed to generate a light-chain gene whereas two are needed to generate a complete heavy-chain gene. The commonest mode of rearrangement (Fig. 4.6, left panels) involves the

Fig. 4.5 Conserved heptamer and nonamer sequences flank the gene segments encoding the V regions of heavy (H) and light (λ and κ) chains. The spacer (white) between the heptamer (orange) and nonamer (purple) sequences is always either approximately 12 bp or approximately 23 bp, and joining almost always involves a 12 bp and a 23 bp recombination signal sequence.

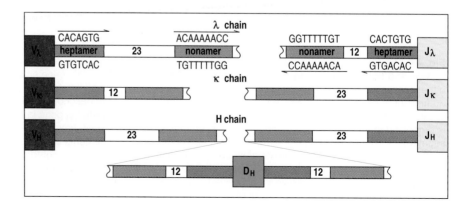

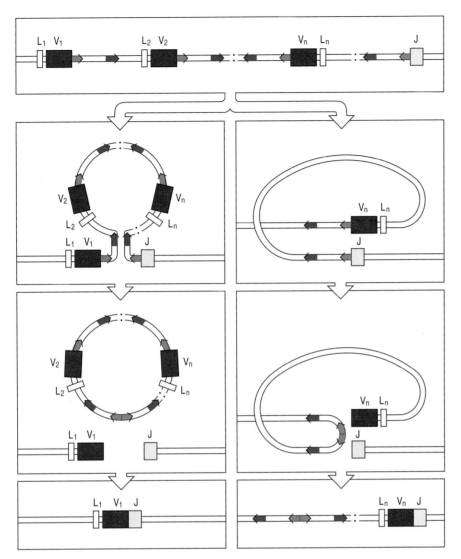

Fig. 4.6 V-region gene segments are joined by recombination. In every V-region recombination event, the signals flanking the gene segments are brought together to allow recombination to take place. For simplicity, the recombination of a light-chain gene is illustrated; for the heavy-chain gene, two separate recombination events are required to generate a functional V region. In some cases, as shown in the left panels, the V and J gene segments have the same transcriptional orientation. Juxtaposition of the recombination signal sequences results in the looping out of the intervening DNA. Heptamers are shown in orange, nonamers in purple, and the arrows represent the directions of the heptamer and nonamer recombination signals (see Fig. 4.5). Recombination occurs at the ends of the heptamer sequences, creating a signal joint and releasing the intervening DNA in the form of a closed circle. Subsequently, the joining of the V and J gene segments creates the coding joint. In other cases, illustrated in the right panels, the V and J gene segments are initially oriented in opposite transcriptional directions. Bringing together the signal sequences in this case requires a more complex looping of the DNA. Joining the ends of the two heptamer sequences now results in the inversion and integration of the intervening DNA. Again, the joining of the V and J segments creates a functional V-region exon.

looping-out and deletion of the DNA between two gene segments. This occurs when the coding sequences of the two gene segments are in the same orientation in the DNA. A second mode of recombination can occur between two gene segments that have opposite transcriptional orientations. This mode of recombination is less common, although such rearrangements account for about half of all V_κ to J_κ joins; the transcriptional orientation of half of the human V_κ gene segments is opposite to that of the J_κ gene segments. The mechanism of recombination is essentially the same, but the DNA that lies between the two gene segments meets a different fate (Fig. 4.6, right panels). When the RSSs in such cases are brought together and recombination takes place, the intervening DNA is not lost from the chromosome but is retained in an inverted orientation.

4-5   The reaction that recombines V, D, and J gene segments involves both lymphocyte-specific and ubiquitous DNA-modifying enzymes.

The molecular mechanism of V-region DNA rearrangement, or **V(D)J recombination**, is illustrated in Fig. 4.7. The 12 bp spaced and 23 bp spaced RSSs are brought together by interactions between proteins that specifically recognize the length of spacer and thus enforce the 12/23 rule for recombination. The DNA molecule is then broken in two places and rejoined in a different configuration. The ends of the heptamer sequences are joined precisely in a

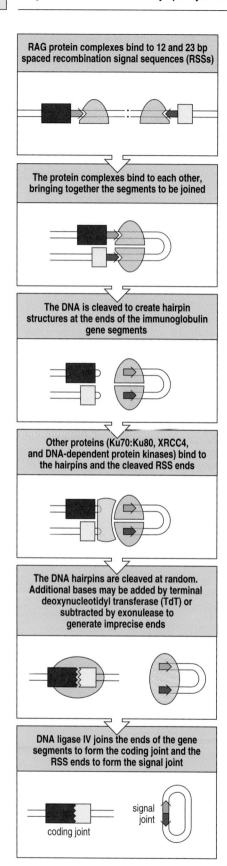

RAG protein complexes bind to 12 and 23 bp spaced recombination signal sequences (RSSs)

The protein complexes bind to each other, bringing together the segments to be joined

The DNA is cleaved to create hairpin structures at the ends of the immunoglobulin gene segments

Other proteins (Ku70:Ku80, XRCC4, and DNA-dependent protein kinases) bind to the hairpins and the cleaved RSS ends

The DNA hairpins are cleaved at random. Additional bases may be added by terminal deoxynucleotidyl transferase (TdT) or subtracted by exonuclease to generate imprecise ends

DNA ligase IV joins the ends of the gene segments to form the coding joint and the RSS ends to form the signal joint

coding joint

signal joint

Fig. 4.7 Enzymatic steps in the rearrangement of immunoglobulin gene segments. Rearrangement begins with the binding of RAG-1, RAG-2, and high mobility group (HMG) proteins (not shown). These RAG-1:RAG-2 complexes (domes, colored green or purple for clarity although they are identical at each recombination site) recognize the recombination signal sequences (arrows) flanking the coding sequences to be joined (red and yellow rectangles). These are then brought together (second panel), following which the RAG complex is activated to cut one strand of the double-stranded DNA precisely at the end of the heptamer sequences (third panel). The 5′ cut end of this DNA strand then reacts with the complementary uncut strand, breaking it to leave a double-stranded break at the end of the heptamer sequence, and forming a hairpin by joining to the cut end of its complementary strand on the other side of the break. Subsequently, through the action of additional essential proteins such as Ku70:Ku80 (indicated in blue) that join the complex (fourth panel) along with the RAG proteins, the DNA hairpin is cleaved at a random site to yield a single-stranded DNA end. This end is then modified by the action of TdT and exonuclease (indicated in pink, fifth panel), which randomly creates diverse, imprecise ends. Finally (sixth panel) the two heptamer sequences, which are not modified, are ligated to form the precise signal joint, while the coding joint is also ligated, both by the action of DNA ligase IV.

head-to-head fashion to form a **signal joint** in a circular piece of extra-chromosomal DNA, which is lost from the genome when the cell divides. The V and J gene segments, which remain on the chromosome, join to form what is called the **coding joint**. This junction is imprecise, and consequently generates much additional variability in the V-region sequence.

The complex of enzymes that act in concert to effect somatic V(D)J recombination is termed the **V(D)J recombinase**. The products of the two genes *RAG-1* and *RAG-2* (**recombination-activating genes**) comprise the lymphoid-specific components of the recombinase. This pair of genes is only expressed in developing lymphocytes while they are engaged in assembling their antigen receptors, as is described in more detail in Chapter 7. They are essential for V(D)J recombination. Indeed, these genes, when expressed together, are sufficient to confer on nonlymphoid cells such as fibroblasts the capacity to rearrange exogenous segments of DNA that contain appropriate RSSs; this is how *RAG-1* and *RAG-2* were initially discovered.

Although the RAG proteins are required for V(D)J recombination, they are not the only enzymes in the recombinase. The remaining enzymes are ubiquitously expressed DNA-modifying proteins that are involved in double-stranded DNA repair, DNA bending, or the modification of the ends of the broken DNA strands. They include the enzyme DNA ligase IV, the enzyme DNA-dependent protein kinase (DNA-PK), and Ku, a well-known autoantigen, which is a heterodimer (Ku 70:Ku 80) that associates tightly with DNA-PK.

V(D)J recombination is a multistep enzymatic process in which the first reaction is an endonucleolytic cleavage requiring the coordinated activity of both RAG proteins. Initially, two RAG protein complexes, each containing RAG-1, RAG-2, and high-mobility group proteins, recognize and align the two RSSs that are guiding the join (see Fig. 4.7). RAG-1 is thought to specifically recognize the nonamer of the RSS. At this stage, the 12/23 rule is established through mechanisms that are still poorly understood. The endonuclease activity of the RAG protein complexes then makes two single-strand DNA breaks at sites just 5′ of each bound RSS, leaving a free 3′-OH group at the end of each coding segment. This 3′-OH group then hydrolyzes the phosphodiester bond on the other strand, sealing the end of the double-stranded DNA to create a DNA 'hairpin' out of the gene segment coding region. This process simultaneously creates a flush double-stranded break at the ends of the two heptamer signal sequences. The DNA ends do not float apart, however, but are held tightly in a complex by the RAG proteins and other associated DNA repair enzymes until the join is completed. The two RSSs are precisely joined to form the signal joint. Coding joint formation is more complex. First, the DNA hairpin is nicked open by a single-stranded break, again by the RAG proteins. The nicking can happen at various points along the hairpin, which leads to sequence variability in the eventual joint. The DNA repair enzymes in the complex then modify the opened hairpins by removing nucleotides (by exonuclease activity) and by randomly adding nucleotides (by terminal deoxynucleotidyl transferase, TdT). It is not known if addition and deletion of nucleotides at the ends of coding regions occurs simultaneously or in a defined order. Finally, ligases such as DNA ligase IV join the processed ends together to generate a continuous double-stranded DNA, thus reconstituting a chromosome that includes the rearranged gene. This enzymatic process seems to create diversity in the joint between gene segments, while ensuring that the RSS ends are ligated without modification, and that unintended genetic damage such as a chromosomal break is avoided.

The recombination mechanism controlled by the RAG proteins shares many interesting features with the mechanism by which retroviral integrases catalyze the insertion of retroviral DNA into the genome, and also with the transposition mechanism used by transposons (mobile genetic elements that encode their own transposase, allowing them to excise and reinsert themselves in the genome). Even the structure of the *RAG* genes themselves, which lie close together in the chromosome and lack the usual mammalian introns, is reminiscent of a transposon. Indeed, it has recently been shown that the RAG complex can act as a transposase *in vitro*. These features have provoked speculation that the RAG complex originated as a transposase whose function was adapted by vertebrates to allow V gene segment recombination, thus leading to the advent of the vertebrate adaptive immune system. Consistent with this idea, no genes homologous to the *RAG* genes have been found in nonvertebrates.

The *in vivo* roles of the enzymes involved in V(D)J recombination have been established through natural or artificially induced mutations. Mice in which either of the *RAG* genes is knocked out suffer a complete block in lymphocyte development at the gene rearrangement stage. Mice lacking TdT do not add extra nucleotides to the joints between gene segments. A mutation that was discovered some time ago results in mice that make only trivial amounts of immunoglobulins or T-cell receptors. Such mice suffer from a **s**evere **c**ombined **i**mmune **d**eficiency—hence the name *scid* for this mutation. These mice have subsequently been found to have a mutation in the enzyme DNA-PK that prevents the efficient rejoining of DNA at gene segment junctions. Mutations of other proteins that are involved in DNA joining also give the *scid* phenotype.

Omenn Syndrome

4-6 The diversity of the immunoglobulin repertoire is generated by four main processes.

Antibody diversity is generated in four main ways. Two of these are consequences of the recombination process just discussed (see Sections 4-4 and 4-5) which creates complete immunoglobulin V-region exons during early B-cell development. The third is due to the different possible combinations of a heavy and a light chain in the complete immunoglobulin molecule. The fourth is a mutational process that occurs in mature B cells, acting only on rearranged DNA encoding the V regions.

The gene rearrangement that combines two or three gene segments to form a complete V-region exon generates diversity in two ways. First, there are multiple different copies of each type of gene segment, and different combinations of gene segments can be used in different rearrangement events. This

combinatorial diversity is responsible for a substantial part of the diversity of the heavy- and light-chain V regions. Second, **junctional diversity** is introduced at the joints between the different gene segments as a result of addition and subtraction of nucleotides by the recombination process. A third source of diversity is also combinatorial, arising from the many possible different combinations of heavy- and light-chain V regions that pair to form the antigen-binding site in the immunoglobulin molecule. The two means of generating combinatorial diversity alone could give rise, in theory, to approximately 3.5×10^6 different antibody molecules (see Section 4-7). Coupled with junctional diversity, it is estimated that as many as 10^{11} different receptors could make up the repertoire of receptors expressed by naive B cells. Finally, somatic hypermutation introduces point mutations into the rearranged V-region genes of activated B cells, creating further diversity that can be selected for enhanced binding to antigen. We will discuss these mechanisms at greater length in the following sections.

4-7 The multiple inherited gene segments are used in different combinations.

There are multiple copies of the V, D, and J gene segments, each of which is capable of contributing to an immunoglobulin V region. Many different V regions can therefore be made by selecting different combinations of these segments. For human κ light chains, there are approximately 40 functional V_κ gene segments and five J_κ gene segments, and thus potentially 200 different V_κ regions. For λ light chains there are approximately 30 functional V_λ gene segments and four J_λ gene segments, yielding 120 possible V_λ regions. So, in all, 320 different light chains can be made as a result of combining different light-chain gene segments. For the heavy chains of humans, there are 65 functional V_H gene segments, approximately 27 D_H gene segments, and 6 J_H gene segments, and thus around 11,000 different possible V_H regions ($65 \times 27 \times 6 \approx 11,000$). During B-cell development, rearrangement at the heavy-chain gene locus to produce any one of the possible heavy chains is followed by several rounds of cell division before light-chain gene rearrangement takes place. The particular combination of gene segments used to produce a heavy chain does not appear to restrict the choice of gene segments that can be recombined to assemble a light-chain variable region. Thus, in theory any one possible heavy chain can be produced together with any one possible light chain in a single B cell. As both the heavy- and the light-chain V regions contribute to antibody specificity, each of the 320 different light chains could be combined with each of the approximately 11,000 heavy chains to give around 3.5×10^6 different antibody specificities. This theoretical estimate of combinatorial diversity is based on the number of germline V gene segments contributing to functional antibodies (see Fig. 4.3); the total number of V gene segments is larger, but the additional gene segments are pseudogenes and do not appear in expressed immunoglobulin molecules.

In practice, combinatorial diversity is likely to be less than one might expect from the theoretical calculations above. One reason for this is that not all V gene segments are used at the same frequency; some are common in antibodies, while others are found only rarely. It is also clear that not every heavy chain can pair with every light chain; certain combinations of V_H and V_L regions result in failure to assemble a stable immunoglobulin molecule. Cells that have heavy and light chains that cannot pair may continue to undergo light-chain gene rearrangement until a suitable light chain is produced, or may be eliminated, but in both cases a heavy- and light-chain combination that does not pair is lost from the repertoire. Nevertheless, it is thought that most heavy and light chains can pair with each other, and that this type of combinatorial diversity has a major role in the formation of an immunoglobulin repertoire with a wide range of specificities. In addition, two further processes add greatly to repertoire diversity—imprecise joining of V, D, and J gene segments and somatic hypermutation.

4-8   Variable addition and subtraction of nucleotides at the junctions between gene segments contributes to diversity in the third hypervariable region.

Of the three hypervariable loops in the protein chains of immunoglobulins, two are encoded within the V gene segment DNA. The third (HV3 or CDR3, see Fig. 3.6) falls at the joint between the V gene segment and the J gene segment, and in the heavy chain is partially encoded by the D gene segment. In both heavy and light chains, the diversity of CDR3 is significantly increased by the addition and deletion of nucleotides at two steps in the formation of the junctions between gene segments. The added nucleotides are known as **P-nucleotides** and **N-nucleotides** and their addition is illustrated in Fig. 4.8.

P-nucleotides are so called because they make up palindromic sequences added to the ends of the gene segments. After the formation of the DNA hairpins as described in Section 4-5, the RAG protein complex catalyzes a single-stranded cleavage at a random point within the coding sequence but near the original point at which the hairpin was first formed. When this cleavage occurs at a different point from the initial break, a single-stranded tail is formed from a few nucleotides of the coding sequence plus the complementary nucleotides from the other DNA strand (see Fig. 4.8). In most light-chain gene rearrangements, DNA repair enzymes then fill in complementary nucleotides on the single-stranded tails which would leave short palindromic sequences at the joint, if the ends are rejoined without any further exonuclease activity (see below). In heavy-chain gene rearrangements and in some human light-chain genes, however, N-nucleotides are first added by a quite different mechanism.

N-nucleotides are so called because they are nontemplate-encoded. They are added by the enzyme terminal deoxynucleotidyl transferase (TdT) to single-stranded ends of the coding DNA after hairpin cleavage. After the addition of up to 20 nucleotides by this enzyme, the two single-stranded stretches at the ends of the gene segments form base pairs over a short region. Repair enzymes then trim off any nonmatching bases, synthesize

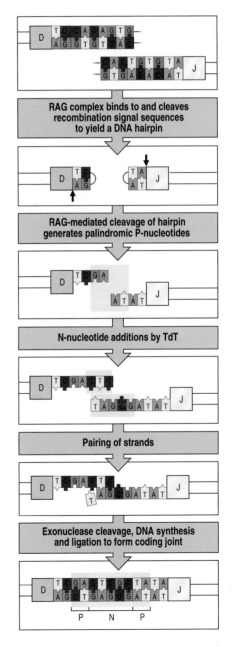

Fig. 4.8 The introduction of P- and N-nucleotides at the joints between gene segments during immuno-globulin gene rearrangement. The process is illustrated for a D_H to J_H rearrangement; however, the same steps occur in V_H to D_H and in V_L to J_L rearrangements. After formation of the DNA hairpins (see Fig. 4.7), the two heptamer sequences, as indicated by the outline, are ligated to form the signal joint (not shown here), while RAG proteins cleave the DNA hairpin at a random site to yield a single-stranded DNA end. Depending on the site of cleavage, this single-stranded DNA may contain nucleotides that were originally complementary in the double-stranded DNA and which therefore form short DNA palindromes, as indicated by the shaded box in the third panel. Such stretches of nucleotides that originate from the complementary strand are known as P-nucleotides. For example, the sequence GA at the end of the D segment shown is complementary to the preceding sequence TC. Where the enzyme terminal deoxynucleotidyl transferase (TdT) is present, nucleotides are added at random to the ends of the single-stranded segments (fourth panel), indicated by the shaded box surrounding these nontemplated, or N, nucleotides. The two single-stranded ends then pair (fifth panel). Exonuclease trimming of unpaired nucleotides and repair of the coding joint by DNA synthesis and ligation leaves both the P- and N-nucleotides present in the final coding joint (indicated by shading in the bottom panel). The randomness of insertion of P- and N-nucleotides makes an individual P-N region a valuable marker for following an individual B-cell clone as it develops, for instance in studies of somatic hypermutation (see Fig. 4.9).

complementary bases to fill in the remaining single-stranded DNA, and ligate it to the P-nucleotides (see Fig. 4.8). N-nucleotides are found especially in the V–D and D–J junctions of the assembled heavy-chain gene; they are less common in light-chain genes because TdT is expressed for only a short period in B-cell development, during the assembly of the heavy-chain gene, which occurs before that of the light-chain gene.

Nucleotides can also be deleted at gene segment junctions. This is accomplished by as yet unidentified exonucleases. Thus, the length of heavy-chain CDR3 can be even shorter than the smallest D segment. In some instances it is difficult, if not impossible, to recognize the D segment that contributed to CDR3 formation because of the excision of most of its nucleotides. Deletions may also erase the traces of P-nucleotide palindromes introduced at the time of hairpin opening. For this reason, many completed V(D)J joins do not show obvious evidence of P-nucleotides.

As the total number of nucleotides added by these processes is random, the added nucleotides often disrupt the reading frame of the coding sequence beyond the joint. Such frameshifts will lead to a nonfunctional protein, and DNA rearrangements leading to such disruptions are known as **nonproductive rearrangements**. As roughly two in every three rearrangements will be nonproductive, many B cells never succeed in producing functional immunoglobulin molecules, and junctional diversity is therefore achieved only at the expense of considerable wastage. We will discuss this further in Chapter 7.

4-9 Rearranged V genes are further diversified by somatic hypermutation.

The mechanisms for generating diversity described so far all take place during the rearrangement of gene segments in the initial development of B cells in the central lymphoid organs. There is an additional mechanism that generates diversity throughout the V region and that operates on B cells in peripheral lymphoid organs after functional immunoglobulin genes have been assembled. This process, known as **somatic hypermutation**, introduces point mutations into the V regions of the rearranged heavy- and light-chain genes at a very high rate, giving rise to mutant B-cell receptors on the surface of the B cells (Fig. 4.9). Some of the mutant immunoglobulin molecules bind antigen better than the original B-cell receptors, and B cells expressing them are preferentially selected to mature into antibody-secreting cells. This gives rise to a phenomenon called **affinity maturation** of the antibody population, which we will discuss in more detail in Chapters 9 and 10.

Somatic hypermutation occurs when B cells respond to antigen along with signals from activated T cells. The immunoglobulin C-region gene, and other genes expressed in the B cell, are not affected, whereas the rearranged V_H and V_L genes are mutated even if they are nonproductive rearrangements and are not expressed. The pattern of nucleotide base changes in nonproductive V-region genes illustrates the result of somatic hypermutation without selection for enhanced binding to antigen. The base changes are distributed throughout the V region, but not completely randomly: there are certain 'hotspots' of mutation that indicate a preference for characteristic short motifs of four to five nucleotides, and perhaps also certain ill-defined secondary structural features. The pattern of base changes in the V regions of expressed immunoglobulin genes is different. Mutations that alter amino acid sequences in the conserved framework regions will tend to disrupt basic antibody structure and are selected against. In contrast, the result of selection for enhanced binding to antigen is that base changes that alter amino acid sequences, and thus protein structure, tend to be clustered in the CDRs, whereas silent mutations that preserve amino acid sequence and do not alter protein structure are scattered throughout the V region.

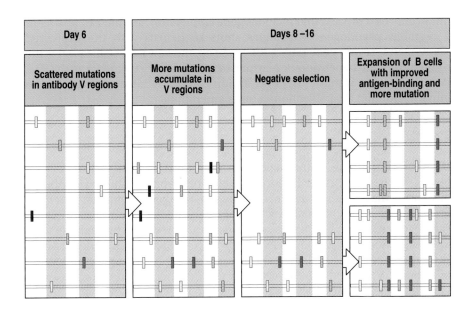

Day 6	Days 8–16		
Scattered mutations in antibody V regions	More mutations accumulate in V regions	Negative selection	Expansion of B cells with improved antigen-binding and more mutation

Fig. 4.9 Somatic hypermutation introduces variation into the rearranged immunoglobulin variable region that is subject to negative and positive selection to yield improved antigen binding. In some circumstances it is possible to follow the process of somatic hypermutation by sequencing immunoglobulin variable regions at different time points after immunization. The result of one such experiment is depicted here. Within a few days of immunization, it is found that the variable regions within a particular clone of responding B cells have begun to acquire mutations (first panel). Each variable region is represented by a horizontal line, on which the positions of the mutations are represented by vertical bars. These may be silent (yellow bars), neutral (pink bars), deleterious (red bars), or positive (blue bars). Over the course of the next week, more mutations accumulate (second panel). Those B cells whose variable regions have accumulated deleterious mutations and can no longer bind antigen die, a process of negative selection (third panel). Those B cells whose variable regions have acquired mutations that result in improved antigen binding are able to compete effectively for binding to the antigen, and receive signals that drive their proliferation and expansion, along with continued mutation (fourth panel). This process of mutation and selection can actually go through multiple cycles (not shown for simplicity) during the second and third weeks of the germinal center reaction. In this way, over time, the antigen-binding efficiency of the antibody response is improved.

The mechanism of somatic hypermutation is poorly defined, but there have been several new discoveries that shed some light. It is known that mutation requires the presence of **enhancers**, DNA sequences that enhance the transcription of immunoglobulin genes in B cells, as well as a transcriptional promoter. The promoter, and the sequences that are the target of mutation need not derive from immunoglobulin V genes, however. The generation of new mutations in V regions in mutating B cells has recently been shown to be accompanied by double-stranded breaks in the DNA which are thought to then be repaired in an error-prone way. In addition, it has recently been discovered that deficiency in an RNA editing enzyme called Activation Induced Cytidine Deaminase, blocks the accumulation of somatic hypermutations. The mechanism by which this enzyme contributes to hypermutation is unknown. Interestingly, deficiency of this enzyme also abrogates the rearrangement of C-region genes that underlies the immunoglobulin class switching seen in activated B cells (see Section 4-16).

4-10   In some species most immunoglobulin gene diversification occurs after gene rearrangement.

As we have seen in the preceding sections, a proportion of the immunoglobulin diversity in an adult human derives from the existence of a variety of germline gene segments, and a proportion from somatic alterations acquired during the lifetime of the individual. This particular combination of heritable and acquired components of diversity operates in several mammalian immune systems, including those of humans and mice. Other species achieve a mix of inherited and acquired diversity by different means. Overall, it would appear that there is strong selective pressure to generate sufficient diversity in the immune system to protect the organism from common pathogens, and several different mechanisms have evolved toward this end.

In birds, rabbits, cows, pigs, sheep, and horses there is little or no germline diversity in the V, D, and J gene segments that are rearranged to form the genes for the initial B-cell receptors, and the rearranged V-region sequences are identical or similar in most immature B cells. These B cells then migrate to specialized microenvironments, the best known of which is the bursa of Fabricius in chickens. Here, B cells proliferate rapidly, and their rearranged immunoglobulin genes undergo further diversification. In birds and rabbits this occurs by a process that includes **gene conversion**, in which an upstream V segment pseudogene exchanges short sequences with the expressed rearranged V-region gene (Fig. 4.10). In sheep and cows, diversification is the result of somatic hypermutation, which occurs in an organ known as the ileal Peyer's patch. Somatic hypermutation probably also contributes to immunoglobulin diversification in birds and rabbits.

Summary.

Diversity within the immunoglobulin repertoire is achieved by several means. Perhaps the most important factor that enables this extraordinary diversity is that V regions are encoded by separate gene segments, which are brought together by somatic recombination to make a complete V-region gene. Many different V-region gene segments are present in the genome of an individual, and thus provide a heritable source of diversity. Additional diversity, termed combinatorial diversity, results from the random recombination of separate V, D, and J gene segments to form a complete V-region exon. Variability at the joints between segments is increased by the insertion of random numbers of P- and N-nucleotides and by variable deletion of nucleotides at the ends of some coding sequences. The association of different light- and heavy-chain V regions to form the antigen-binding site of an immunoglobulin molecule contributes further diversity. Finally, after an

Fig. 4.10 The diversification of chicken immunoglobulins occurs through gene conversion. In chickens, all B cells express the same surface immunoglobulin (sIg) initially; there is only one active V, D, and J gene segment for the chicken heavy-chain gene and one active V and J gene segment for each of the light-chain genes (top left panel). Gene rearrangement can thus produce only a single receptor specificity. Immature B cells expressing this receptor migrate to the bursa of Fabricius, where the expression of sIg induces cell proliferation (center panels). Gene conversion events introduce sequences from adjacent V pseudogenes into the expressed gene, creating diversity in the receptors (bottom panels). Some of these gene conversions will inactivate the previously expressed gene (not shown). If a B cell can no longer express sIg after such a gene conversion, it is eliminated.

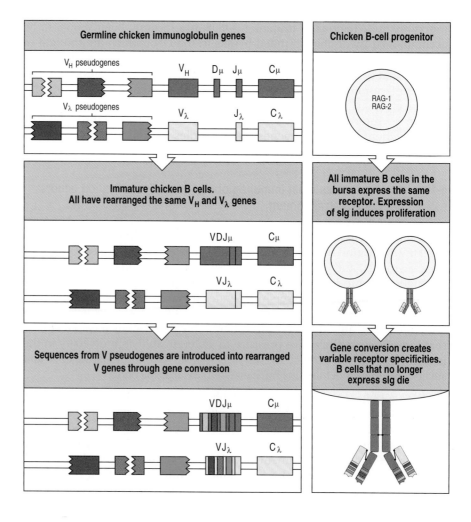

immunoglobulin has been expressed, the coding sequences for its V regions are modified by somatic hypermutation upon stimulation of the B cell by antigen. The combination of all these sources of diversity generates a vast repertoire of antibody specificities from a relatively limited number of genes.

T-cell receptor gene rearrangement.

The mechanism by which B-cell antigen receptors are generated is such a powerful means of creating diversity that it is not surprising that the antigen receptors of T cells bear structural resemblances to immunoglobulins and are generated by the same mechanism. In this part of the chapter we describe the organization of the T-cell receptor loci and the generation of the genes for the individual T-cell receptor chains.

4-11   The T-cell receptor loci comprise sets of gene segments and are rearranged by the same enzymes as the immunoglobulin loci.

Like immunoglobulin heavy and light chains, T-cell receptor α and β chains each consist of a variable (V) amino-terminal region and a constant (C) region (see Section 3-10). The organization of the TCRα and TCRβ loci is shown in Fig. 4.11. The organization of the gene segments is broadly homologous to that of the immunoglobulin gene segments (see Sections 4-2 and 4-3). The TCRα locus, like those for the immunoglobulin light chains, contains V and J gene segments (V_α and J_α). The TCRβ locus, like that for the immunoglobulin heavy-chain, contains D gene segments in addition to V_β and J_β gene segments.

The T-cell receptor gene segments rearrange during T-cell development to form complete V-domain exons (Fig. 4.12). T-cell receptor gene rearrangement takes place in the thymus; the order and regulation of the rearrangements will be dealt with in detail in Chapter 7. Essentially, however, the mechanics of gene rearrangement are similar for B and T cells. The T-cell receptor gene segments are flanked by heptamer and nonamer recombination signal sequences (RSSs) that are homologous to those flanking immunoglobulin gene segments (see Section 4-4 and Fig. 4.5) and are recognized by the same enzymes. All known defects in genes that control V(D)J recombination affect T cells and B cells equally, and animals with these genetic defects lack functional lymphocytes altogether (see Section 4-5).

Fig. 4.11 The germline organization of the human T-cell receptor α and β loci. The arrangement of the gene segments resembles that at the immunoglobulin loci, with separate variable (V), diversity (D), joining (J) gene segments, and constant (C) genes. The TCRα locus (chromosome 14) consists of 70–80 V_α gene segments, each preceded by an exon encoding the leader sequence (L). How many of these V_α gene segments are functional is not known exactly. A cluster of 61 J_α gene segments is located a considerable distance from the V_α gene segments. The J_α gene segments are followed by a single C gene, which contains separate exons for the constant and hinge domains and a single exon encoding the transmembrane and cytoplasmic regions (not shown). The TCRβ locus (chromosome 7) has a different organization, with a cluster of 52 functional V_β gene segments located distantly from two separate clusters each containing a single D gene segment, together with six or seven J gene segments and a single C gene. Each TCRβ C gene has separate exons encoding the constant domain, the hinge, the transmembrane region, and the cytoplasmic region (not shown). The TCRα locus is interrupted between the J and V gene segments by another T-cell receptor locus—the TCRδ locus (not shown here; see Fig. 4.15).

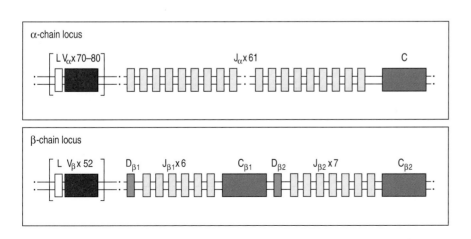

Fig. 4.12 T-cell receptor α- and β-chain gene rearrangement and expression. The TCRα- and β-chain genes are composed of discrete segments that are joined by somatic recombination during development of the T cell. Functional α- and β-chain genes are generated in the same way that complete immunoglobulin genes are created. For the α chain (upper part of figure), a V_α gene segment rearranges to a J_α gene segment to create a functional V-region exon. Transcription and splicing of the VJ_α exon to C_α generates the mRNA that is translated to yield the T-cell receptor α-chain protein. For the β chain (lower part of figure), like the immunoglobulin heavy chain, the variable domain is encoded in three gene segments, V_β, D_β, and J_β. Rearrangement of these gene segments generates a functional VDJ_β V-region exon that is transcribed and spliced to join to C_β; the resulting mRNA is translated to yield the T-cell receptor β chain. The α and β chains pair soon after their biosynthesis to yield the α:β T-cell receptor heterodimer. Not all J gene segments are shown, and the leader sequences preceding each V gene segment are omitted for simplicity.

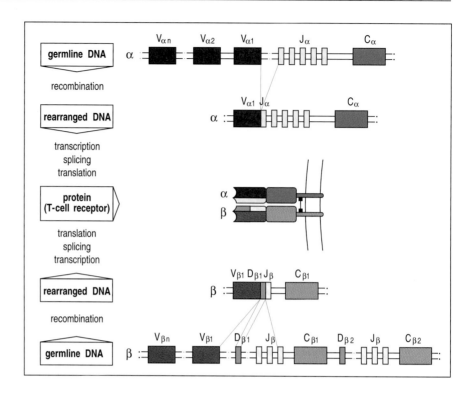

A further shared feature of immunoglobulin and T-cell receptor gene rearrangement is the presence of P- and N-nucleotides in the junctions between the V, D, and J gene segments of the rearranged TCRβ gene. In T cells, P- and N-nucleotides are also added between the V and J gene segments of all rearranged TCRα genes, whereas only about half the V–J joints in immunoglobulin light-chain genes are modified by N-nucleotide addition and these are often left without any P-nucleotides as well (see Section 4-8 and Fig. 4.13).

Fig. 4.13 The numbers of human T-cell receptor gene segments and the sources of T-cell receptor diversity compared with those of immunoglobulins. Note that only about half of human κ chains contain N-nucleotides. Somatic hypermutation as a source of diversity in immunoglobulins is not included in this figure.

Element	Immunoglobulin		α:β receptors	
	H	κ+λ	β	α
Variable segments (V)	65	70	52	~70
Diversity segments (D)	27	0	2	0
D segments read in 3 frames	rarely	–	often	–
Joining segments (J)	6	5(κ) 4(λ)	13	61
Joints with N- and P-nucleotides	2	50% of joints	2	1
Number of V gene pairs	3.4×10^6		5.8×10^6	
Junctional diversity	$\sim 3 \times 10^7$		$\sim 2 \times 10^{11}$	
Total diversity	$\sim 10^{14}$		$\sim 10^{18}$	

The main differences between the immunoglobulin genes and those encoding T-cell receptors reflect the fact that all the effector functions of B cells depend upon secreted antibodies whose different heavy-chain C-region isotypes trigger distinct effector mechanisms. The effector functions of T cells, in contrast, depend upon cell–cell contact and are not mediated directly by the T-cell receptor, which serves only for antigen recognition. Thus, the C regions of the TCRα and TCRβ loci are much simpler than those of the immuno-globulin heavy-chain locus. There is only one C_α gene and, although there are two C_β genes, they are very closely homologous and there is no functional distinction between their products. The T-cell receptor C-region genes encode only transmembrane polypeptides.

4-12   T-cell receptors concentrate diversity in the third hypervariable region.

The extent and pattern of variability in T-cell receptors and immunoglobulins reflect the distinct nature of their ligands. Whereas the antigen-binding sites of immunoglobulins must conform to the surfaces of an almost infinite variety of different antigens, and thus come in a wide variety of shapes and chemical properties, the ligand for the T-cell receptor is always a peptide bound to an MHC molecule. The antigen-recognition sites of T-cell receptors would therefore be predicted to have a less variable shape, with most of the variability focused on the bound antigenic peptide occupying the center of the surface in contact with the receptor.

In spite of differences in the sites of variability, the three-dimensional structure of the antigen-recognition site of a T-cell receptor looks much like that of an antibody molecule (see Sections 3-11 and 3-7, respectively). In an antibody, the center of the antigen-binding site is formed by the CDR3s of the heavy and light chains. The structurally equivalent third hypervariable loops (CDR3s) of the T-cell receptor α and β chains, to which the D and J gene segments contribute, also form the center of the antigen-binding site of a T-cell receptor; the periphery of the site consists of the equivalent of the CDR1 and CDR2 loops, which are encoded within the germline V gene segments for the α and β chains.

T-cell receptor loci have roughly the same number of V gene segments as do the immunoglobulin loci, but only B cells diversify rearranged V-region genes by somatic hypermutation. Thus, diversity in the CDR1 and CDR2 loops that comprise the periphery of the antigen-binding site will be far greater among antibody molecules than among T-cell receptors. This is in keeping with the fact that the CDR1 and CDR2 loops of a T-cell receptor will mainly contact the relatively less variable MHC component of the ligand rather than the highly variable peptide component (Fig. 4.14).

The structural diversity of T-cell receptors is mainly attributable to combinatorial and junctional diversity generated during the process of gene rearrangement. It can be seen from Fig. 4.13 that the variability in T-cell receptor chains is focused on the junctional region encoded by V, D, and J gene segments and modified by P- and N-nucleotides. The TCRα locus contains many more J gene segments than either of the immunoglobulin light-chain loci: in humans, 61 J_α gene segments are distributed over about 80 kb of DNA, whereas immunoglobulin light-chain loci have only five J gene segments at most (see Fig. 4.13). Because the TCRα locus has so many J gene segments, the variability generated in this region is even greater for T-cell receptors than for immunoglobulins. This region encodes the CDR3 loops in immunoglobulins and T-cell receptors that form the center of the antigen-binding site. Thus, the center of the T-cell receptor will be highly variable, whereas the periphery will be subject to relatively little variation.

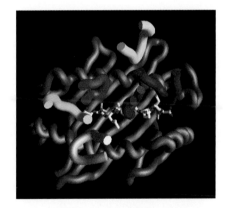

Fig. 4.14 The most variable parts of the T-cell receptor interact with the peptide bound to an MHC molecule. The positions of the CDR loops of a T-cell receptor are shown as the colored tubes in this figure superimposing onto the MHC:peptide complex. The CDR1 loops of the α and β chains of the TCR are shown in pale and dark blue respectively, the CDR2 loops shown in fawn and yellow respectively, and the CDR3 loops shown in pale and dark red. The CDR3 loops lie in the center of the interface between the TCR and the MHC:peptide complex, and make direct contacts with the antigenic peptide.

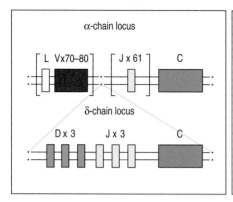

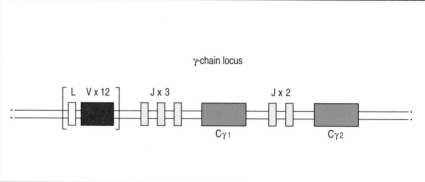

Fig. 4.15 The organization of the T-cell receptor γ- and δ-chain loci in humans. The TCRγ and TCRδ loci, like the TCRα and TCRβ loci, have discrete V, D, and J gene segments, and C genes. Uniquely, the locus encoding the δ chain is located entirely within the α-chain locus. The three D_δ gene segments, three J_δ gene segments, and the single δ C gene lie between the cluster of V_α gene segments and the cluster of J_α gene segments, whereas the V_δ gene segments are interspersed among the V_α gene segments; it is not known exactly how many V_δ gene segments there are, but there are at least four. The human TCRγ locus resembles the TCRβ locus in having two C genes each with its own set of J gene segments. The mouse γ locus (not shown) has a more complex organization and there are three functional clusters of γ gene segments, each containing V and J gene segments and a C gene. Rearrangement at the γ and δ loci proceeds as for the other T-cell receptor loci, with the exception that during TCRδ rearrangement both D segments can be used in the same gene. The use of two D segments greatly increases the variability of the δ chain, mainly because extra N-region nucleotides can be added at the junction between the two D gene segments as well as at the V–D and D–J junctions.

4-13   γ:δ T-cell receptors are also generated by gene rearrangement.

A minority of T cells bear T-cell receptors composed of γ and δ chains (see Section 3-19). The organization of the TCRγ and TCRδ loci (Fig. 4.15) resembles that of the TCRα and TCRβ loci, although there are important differences. The cluster of gene segments encoding the δ chain is found entirely within the TCRα locus, between the V_α and the J_α gene segments. Because all V_α gene segments are oriented such that rearrangement will delete the intervening DNA, any rearrangement at the α locus results in the loss of the δ locus. There are substantially fewer V gene segments at the TCRγ and TCRδ loci than at either the TCRα or TCRβ loci or at any of the immunoglobulin loci. Increased junctional variability in the δ chains may compensate for the small number of V gene segments and has the effect of focusing almost all of the variability in the γ:δ receptor in the junctional region. As we have seen, the amino acids encoded by the junctional regions lie at the center of the T-cell receptor binding site.

T cells bearing γ:δ receptors are a distinct lineage of T cells whose functions are at present unknown. The ligands for these receptors are also largely unknown (see Section 3-19). Some γ:δ T-cell receptors appear to be able to recognize antigen directly, much as antibodies do, without the requirement for presentation by an MHC molecule or processing of the antigen. Detailed analysis of the rearranged V regions of γ:δ T-cell receptors shows that they resemble the V regions of antibody molecules more than they resemble the V regions of α:β T-cell receptors.

4-14   Somatic hypermutation does not generate diversity in T-cell receptors.

When we discussed the generation of antibody diversity in Section 4-9, we saw that somatic hypermutation increases the diversity of all three complementarity-determining regions of both immunoglobulin chains.

Somatic hypermutation does not occur in T-cell receptor genes, so that variability of the CDR1 and CDR2 regions is limited to that of the germline V gene segments. All the diversity in T-cell receptors is generated during rearrangement and is consequently focused on the CDR3 regions.

Why T-cell and B-cell receptors differ in their abilities to undergo somatic hypermutation is not clear, but several explanations can be suggested on the basis of the functional differences between T and B cells. Because the central role of T cells is to stimulate both humoral and cellular immune responses, it is crucially important that T cells do not react with self proteins. T cells that recognize self antigens are rigorously purged during development (see Chapter 7) and the absence of somatic hypermutation helps to ensure that somatic mutants recognizing self proteins do not arise later in the course of immune responses. This constraint does not apply with the same force to B-cell receptors, as B cells usually require T-cell help to secrete antibodies. A B cell whose receptor mutates to become self reactive would, under normal circumstances, fail to make antibody for lack of self-reactive T cells to provide this help (see Chapter 9).

A further argument is that T cells already interact with a self component, namely the MHC molecule that makes up the major part of the ligand for the receptor, and thus might be unusually prone to developing self-recognition capability through somatic hypermutation. In this case, the converse argument can also be made: because T-cell receptors must be able to recognize self MHC molecules as part of their ligand, it is important to avoid somatic mutation that might result in the loss of recognition and the consequent loss of any ability to respond. However, the strongest argument for this difference between immunoglobulins and T-cell receptors is the simple one that somatic hypermutation is an adaptive specialization for B cells alone, because they must make very high-affinity antibodies to capture toxin molecules in the extracellular fluids. We will see in Chapter 10 that they do this through somatic hypermutation followed by selection for antigen binding.

Summary.

T-cell receptors are structurally similar to immunoglobulins and are encoded by homologous genes. T-cell receptor genes are assembled by somatic recombination from sets of gene segments in the same way as are the immunoglobulin genes. Diversity is distributed differently in immunoglobulins and T-cell receptors; the T-cell receptor loci have roughly the same number of V gene segments but more J gene segments, and there is greater diversification of the junctions between gene segments during gene rearrangement. Moreover, functional T-cell receptors are not known to diversify their V genes after rearrangement through somatic hypermutation. This leads to a T-cell receptor in which the highest diversity is in the central part of the receptor, which contacts the bound peptide fragment of the ligand.

Structural variation in immunoglobulin constant regions.

So far we have focused on the structural variation inherent in the assembly of the V regions of the antibody molecule and T-cell receptor. We have seen how this variation creates a diverse repertoire of antigen-specificities, and we have also considered how these variable regions are attached to constant regions in the monovalent heterodimeric T-cell receptor, and the Y-shaped four-chain structure of the divalent immunoglobulin molecule. However, we have discussed only the general structural features of the immunoglobulin C region as illustrated by IgG, the most abundant type of antibody in plasma (see Section 3-1). Immunoglobulins can be made in several different forms, or isotypes, and we now consider how this structural variation is generated by linking different heavy-chain constant regions to the same variable region. The C_H regions, which determine the class or isotype of the antibody and thus its effector functions, are encoded in separate genes located downstream of the V genes at the heavy-chain locus. Initially only the first of these genes, the C_μ gene, is expressed in conjunction with an assembled V gene. However, during the course of an antibody response activated B cells can switch to the expression of a different down-stream C_H gene by a process of somatic recombination known as isotype switching. In this part of the chapter we consider the structural features that distinguish the C_H regions of antibodies of the five major isotypes and confer on them their specialized functional properties as well as the mechanism of isotype switching. We also look at how alternative mRNA splicing allows the production of both membrane-bound and secreted forms of each isotype, and the simultaneous production of surface IgM and IgD in mature but naive B cells.

The use of isotype switching and alternative mRNA splicing to generate structural and functional variation is unique to the immunoglobulin heavy-chain locus and does not occur in T-cell receptor genes. This reflects the fact that immunoglobulins act as soluble molecules that must both bind antigen and recruit a variety of other effector cells and molecules to deal with it appropriately, whereas the T-cell receptor functions only as a membrane-bound receptor to activate an appropriate cellular immune response.

4-15   The immunoglobulin heavy-chain isotypes are distinguished by the structure of their constant regions.

The five main isotypes of immunoglobulin are IgM, IgD, IgG, IgE, and IgA. In humans, IgG antibodies can be further subdivided into four subclasses (IgG1, IgG2, IgG3, and IgG4), whereas IgA antibodies are found as two subclasses (IgA1 and IgA2). The IgG isotypes in humans are named in order of their abundance in serum, with IgG1 being the most abundant. The heavy chains that define these isotypes are designated by the lower-case Greek letters μ, δ, γ, ε, and α, as shown in Fig. 4.16, which also lists the major physical properties of the different human isotypes. IgM forms pentamers in serum, which accounts for its high molecular weight. Secreted IgA can occur as either a monomer or as a dimer.

Sequence differences between immunoglobulin heavy chains cause the various isotypes to differ in several characteristic respects. These include the number and location of interchain disulfide bonds, the number of attached oligosaccharide moieties, the number of C domains, and the length of the hinge region (Fig. 4.17). IgM and IgE heavy chains contain an extra C domain that replaces the hinge region found in γ, δ, and α chains. The absence of the hinge region does not imply that IgM and IgE molecules lack flexibility; electron micrographs of IgM molecules binding to ligands show that the Fab arms can bend relative to the Fc portion. However, such a difference in structure may have functional consequences that are not yet characterized. Different isotypes and subtypes also differ in their ability to engage various effector functions, as will be described in Section 4-18.

	Immunoglobulin								
	IgG1	**IgG2**	**IgG3**	**IgG4**	**IgM**	**IgA1**	**IgA2**	**IgD**	**IgE**
Heavy chain	γ_1	γ_2	γ_3	γ_4	μ	α_1	α_2	δ	ε
Molecular weight (kDa)	146	146	165	146	970	160	160	184	188
Serum level (mean adult mg ml^{-1})	9	3	1	0.5	1.5	3.0	0.5	0.03	5×10^{-5}
Half-life in serum (days)	21	20	7	21	10	6	6	3	2
Classical pathway of complement activation	++	+	+++	–	+++	–	–	–	–
Alternative pathway of complement activation	–	–	–	–	–	+	–	–	–
Placental transfer	+++	+	++	–/+	–	–	–	–	–
Binding to macrophages and other phagocytes	+	–	+	–/+	–	+	+	–	+
High-affinity binding to mast cells and basophils	–	–	–	–	–	–	–	–	+++
Reactivity with staphylococcal Protein A	+	+	–/+	+	–	–	–	–	–

Fig. 4.16 The properties of the human immunoglobulin isotypes. IgM is so called because of its size: although monomeric IgM is only 190 kDa, it normally forms pentamers, known as macroglobulin (hence the M), of very large molecular weight (see Fig. 4.23). IgA dimerizes to give a molecular weight of around 390 kDa in secretions. IgE antibody is associated with immediate-type hypersensitivity. When fixed to tissue mast cells, IgE has a much longer half-life than its half-life in plasma shown here. The activation of the alternative pathway of complement by IgA1 is caused not by its Fc portion but by its Fab portion. A subset of human V_H regions (clan 3, see Section 4-3) also binds staphylococcal Protein A through the Fab portion.

4-16   The same V_H exon can associate with different C_H genes in the course of an immune response.

The V-region exons expressed by any given B cell are determined during its early differentiation in the bone marrow and, although they may subsequently be modified by somatic hypermutation, no further V(D)J recombination occurs. All the progeny of that B cell will therefore express the same assembled V genes. By contrast, several different C-region genes can be expressed in the B cell's progeny as the cells mature and proliferate in the course of an immune response. Every B cell begins by expressing IgM as its B-cell receptor, and the first antibody produced in an immune response is always IgM. Later in the immune response, however, the same assembled V region may be expressed in IgG, IgA, or IgE antibodies. This change is known as **isotype switching**. It is stimulated in the course of an immune response by external

Fig. 4.17 The structural organization of the main human immunoglobulin isotype monomers. Both IgM and IgE lack a hinge region but each contains an extra heavy-chain domain. Note the differences in the numbers and locations of the disulfide bonds (black lines) linking the chains. The isotypes also differ in the distribution of N-linked carbohydrate groups, shown as turquoise hexagons.

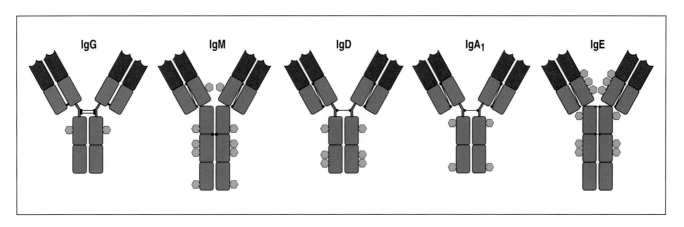

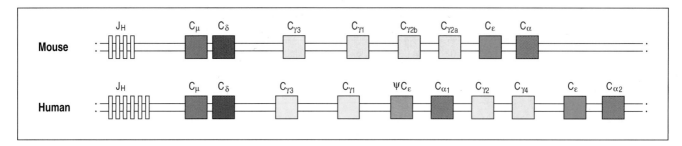

Fig. 4.18 The organization of the immunoglobulin heavy-chain C-region genes in mice and humans (not to scale). In humans, the cluster shows evidence of evolutionary duplication of a unit consisting of two γ genes, an ε gene and an α gene. One of the ε genes has become inactivated and is now a pseudogene (ψ); hence only one subtype of IgE is expressed. For simplicity, other pseudogenes are not illustrated, and the exon details within each C gene are not shown. The classes of immunoglobulins found in mice are called IgM, IgD, IgG1, IgG2a, IgG2b, IgG3, IgA, and IgE.

signals such as cytokines released by T cells or mitogenic signals delivered by pathogens, as we will discuss further in Chapter 9. Here we are concerned with the molecular basis of the isotype switch.

The immunoglobulin C_H genes form a large cluster spanning about 200 kb to the 3′ side of the J_H gene segments (Fig. 4.18). Each C_H gene is split into several exons (not shown in the figure), each corresponding to an individual immunoglobulin domain in the folded C region. The gene encoding the μ C region lies closest to the J_H gene segments, and therefore closest to the assembled V-region exon after DNA rearrangement. A complete μ heavy-chain transcript is produced from the newly rearranged gene. Any J_H gene segments remaining between the assembled V gene and the C_μ gene are removed during RNA processing to generate the mature mRNA. μ heavy chains are therefore the first to be expressed and IgM is the first immunoglobulin isotype to be expressed during B-cell development.

Immediately 3′ to the μ gene lies the δ gene, which encodes the C region of the IgD heavy chain. IgD is coexpressed with IgM on the surface of almost all mature B cells, although this isotype is secreted in only small amounts and its function is unknown. Indeed, mice lacking the C_δ exons seem to have essentially normal immune systems. B cells expressing IgM and IgD have not undergone isotype switching, which, as we will see, entails an irreversible change in the DNA. Instead, these cells produce a long primary transcript that is differentially cleaved and spliced to yield one of two distinct mRNA molecules. In one of these, the VDJ exon is linked to the C_μ exons to encode a μ heavy chain, and in the other the VDJ exon is linked to the C_δ exons to encode a δ heavy chain (Fig. 4.19). The differential processing of the long mRNA transcript is developmentally regulated, with immature B cells making mostly the μ transcript and mature B cells making mostly the δ form along with some of the μ transcript.

Switching to other isotypes occurs only after B cells have been stimulated by antigen. It occurs through a specialized nonhomologous DNA recombination mechanism guided by stretches of repetitive DNA known as **switch regions**. Switch regions lie in the intron between the J_H gene segments and the C_μ gene, and at equivalent sites upstream of the genes for each of the other heavy-chain isotypes, with the exception of the δ gene (Fig. 4.20, top panel). The μ switch region (S_μ) consists of about 150 repeats of the sequence $[(GAGCT)_n (GGGGGT)]$, where n is usually 3 but can be as many as 7. The sequences of the other switch regions (S_γ, S_α, and S_ϵ) differ in detail but all contain repeats of the GAGCT and GGGGGT sequences. Exactly how these repetitive sequences promote switch recombination is unclear because the enzyme(s) that promote switch recombination have not been identified; however, it is thought that the repetitive sequences might promote short stretches of homologous alignment that in turn promote DNA recombination.

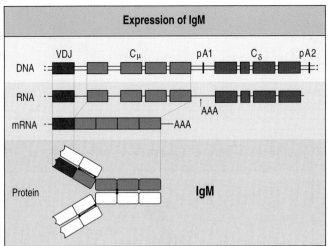

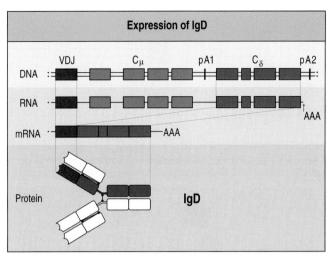

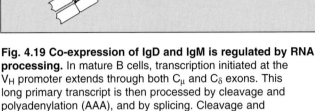

Fig. 4.19 Co-expression of IgD and IgM is regulated by RNA processing. In mature B cells, transcription initiated at the V_H promoter extends through both C_μ and C_δ exons. This long primary transcript is then processed by cleavage and polyadenylation (AAA), and by splicing. Cleavage and polyadenylation at the μ site (pA1) and splicing between C_μ exons yields an mRNA encoding the μ heavy chain (left panel). Cleavage and polyadenylation at the δ site (pA2) and a different pattern of splicing that removes the C_μ exons yields mRNA encoding the δ heavy chain (right panel). For simplicity we have not shown all the individual C-region exons.

When a B cell switches from coexpression of IgM and IgD to expression of an IgG subtype, DNA recombination occurs between S_μ and the S_γ of that IgG subtype. The C_μ and C_δ coding regions are deleted, and γ heavy-chain transcripts are made from the recombined gene. Figure 4.20 (left panels) illustrates switching to γ3 in the mouse. Some of the progeny of this IgG-producing cell may subsequently undergo a further switching event to produce a different isotype, for example IgA, as shown in the bottom panel of Fig. 4.20. Alternatively, as shown in the right panels of Fig. 4.20, the switch recombination may occur between S_μ and one of the switch regions downstream of the C_γ genes so that the cell switches from IgM to IgA or IgE (illustrated for IgA only). All switch recombination events produce genes that can encode a functional protein because the switch sequences lie in introns and therefore cannot cause frame shift mutations.

The enzymes that carry out isotype switching have not been clearly defined. However, we do know that DNA repair enzymes are involved since switching is markedly reduced in Ku protein knockouts; Ku proteins are also essential for the rejoining of DNA during V(D)J joining (see Section 4-5). Recently, it was discovered that deficiency in Activation Induced Cytidine Deaminase completely blocks isotype switching. As mentioned in Section 4-9, this deficiency also blocks somatic hypermutation. Activation Induced Cytidine Deaminise is thought to be an RNA editing enzyme and how it works to enable both hypermutation and switching is unknown at present. A deficiency in this enzyme in humans has now been associated with a form of immunodeficiency known as Hyper IgM type 2 syndrome, which is characterized by an absence of immunoglobulins other than IgM. A failure of T cells to activate isotype switching leads to a similar syndrome now classified as Hyper IgM type 1 (see Section 11-9).

Hyper IgM Immunodeficiency

Isotype switch recombination is unlike V(D)J recombination in several ways. First, all isotype switch recombination is productive; second, it uses different recombination signal sequences and enzymes; third, it happens after antigen stimulation and not during B-cell development in the bone marrow; and fourth, the switching process is not random but is regulated by external signals such as those provided by T cells, as will be discussed in Chapter 9.

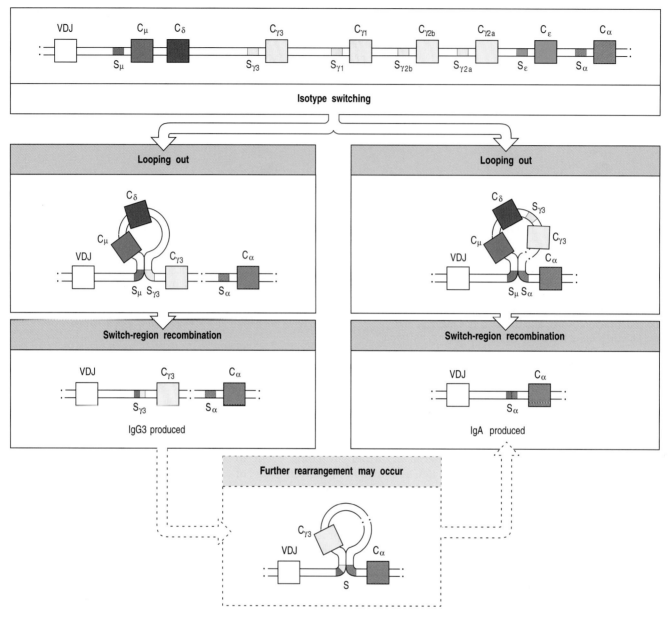

Fig. 4.20 Isotype switching involves recombination between specific switch signals. Repetitive DNA sequences that guide isotype switching are found upstream of each of the immunoglobulin C-region genes, with the exception of the δ gene. The figure illustrates switching in the mouse heavy-chain genes. Switching occurs by recombination between these repetitive sequences, or switch signals, with deletion of the intervening DNA. The initial switching event takes place from the μ switch region; switching to other isotypes can take place subsequently from the recombinant switch region formed after μ switching. S, switch region.

4-17   Transmembrane and secreted forms of immunoglobulin are generated from alternative heavy-chain transcripts.

Immunoglobulins of all heavy-chain isotypes can be produced either in secreted form or as a membrane-bound receptor. All B cells initially express the transmembrane form of IgM; after antigen stimulation, some of their progeny differentiate into plasma cells producing the secreted form of IgM, whereas others undergo isotype switching to express transmembrane immunoglobulins of a different isotype before switching to the production of secreted antibody of the new isotype.

The membrane forms of all isotypes are monomers comprised of two heavy and two light chains: IgM and IgA polymerize only when they are secreted. In its membrane-bound form the immunoglobulin heavy chain has a hydrophobic transmembrane domain of about 25 amino acid residues at the carboxy terminus, which anchors it to the surface of the B lymphocyte. This transmembrane domain is absent from the secreted form, whose carboxy terminus is a hydrophilic secretory tail. The two different carboxy termini of the transmembrane and secreted forms of immunoglobulin heavy chains are encoded in separate exons and production of the two forms is achieved by alternative RNA processing (Fig. 4.21). The last two exons of each C_H gene contain the sequences encoding the secreted and the transmembrane regions respectively; if the primary transcript is cleaved and polyadenylated at a site downstream of these exons, the sequence encoding the carboxy terminus of the secreted form is removed by splicing and the cell-surface form of immunoglobulin is produced. Alternatively, if the primary transcript is cleaved at the polyadenylation site located before the last two exons, only the secreted molecule can be produced. This differential RNA processing is illustrated for C_μ in Fig. 4.21, but occurs in the same way for all isotypes.

Although the production of membrane-bound and secreted versions of the heavy chain is achieved by similar mechanisms to those that allow the co-expression of surface IgM and IgD (see Fig. 4.19), these two instances of alternative RNA processing act at different stages in the life of the B cell, and on different primary transcripts. B cells make a long heavy-chain transcript that can be processed to give either transmembrane IgM or IgD before they are stimulated by antigen. A B cell that is activated ceases to coexpress IgD with IgM, either because μ and δ sequences have been removed as a consequence of an isotype switch or, in IgM-secreting plasma cells, because transcription from the V_H promoter no longer extends through the C_δ exons. In activated B cells that differentiate to become antibody-secreting plasma cells, much of the transcript is spliced to the secreted rather than transmembrane form of whichever isotype the B cell happens to express.

4-18 Antibody C regions confer functional specialization.

The secreted antibodies protect the body in a variety of ways, as we briefly outline here and discuss further in Chapter 9. In some cases it is enough for the antibody simply to bind antigen. For instance, by binding tightly to a toxin or virus, an antibody can prevent it from recognizing its receptor on a host cell. The V regions on their own are sufficient for this. The C region is essential, however, for recruiting the help of other cells and molecules to destroy and dispose of pathogens to which the antibody has bound, and it confers functionally distinct properties on each of the various isotypes.

The C regions of antibodies have three main effector functions. First, the Fc portions of different isotypes are recognized by specialized receptors expressed by immune effector cells. The Fc portions of IgG1 and IgG3 anti-bodies are recognized by **Fc receptors** present on the surface of phagocytic cells such as macrophages and neutrophils, which can thereby bind and engulf pathogens coated with antibodies of these isotypes. The Fc portion of IgE binds to a high-affinity Fcε receptor on mast cells, basophils, and activated eosinophils, enabling these cells to respond to the binding of specific antigen by releasing inflammatory mediators. Second, the Fc portions of antigen: antibody complexes can bind to complement (see Fig. 1.24) and initiate the complement cascade, which helps to recruit and activate phagocytes, can aid the engulfment of microbes by phagocytes, and can also directly destroy pathogens. Third, the Fc portion can deliver antibodies to places they would not reach without active transport. These include the mucus secretions, tears,

Fig. 4.21 Transmembrane and secreted forms of immunoglobulins are derived from the same heavy-chain sequence by alternative RNA processing. Each heavy-chain C gene has two exons (membrane-coding (MC) yellow) that encode the transmembrane region and cytoplasmic tail of the transmembrane form, and a secretion-coding (SC) sequence (orange) that encodes the carboxy terminus of the secreted form. In the case of IgD, the SC sequence is present on a separate exon, but for the other isotypes, including IgM as shown here, the SC sequence is contiguous with the last C-domain exon. The events that dictate whether a heavy-chain RNA will result in a secreted or transmembrane immunoglobulin occur during processing of the initial transcript. Each heavy-chain C gene has two potential polyadenylation sites (shown as pA_S and pA_m). In the upper panel, the transcript is cleaved and polyadenylated (AAA) at the second site (pA_m). Splicing between a site located between the $C_\mu4$ exon and the SC sequence, and a second site at the 5′ end of the MC exons, results in removal of the SC sequence and joining of the MC exons to the $C_\mu4$ exon. This generates the transmembrane form of the heavy chain. In the lower panel, the primary transcript is cleaved and polyadenylated at the first site (pA_S), eliminating the MC exons and giving rise to the secreted form of the heavy chain.

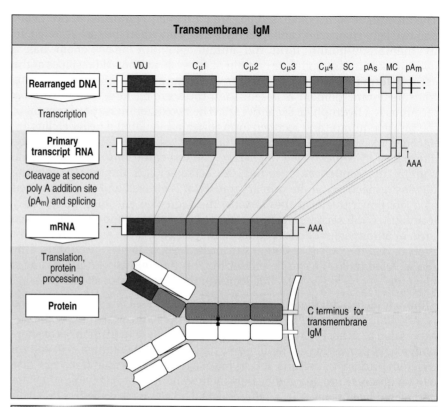

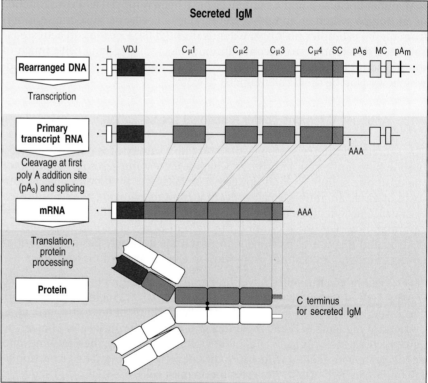

and milk (IgA), and the fetal blood circulation by transfer from the pregnant mother (IgG). In both cases, the Fc portion engages a specific receptor that leads to the active transport of the immunoglobulin through cells to reach different body compartments.

The role of the Fc portion in these effector functions can be demonstrated by studying enzymatically treated immunoglobulins that have had one or other domain of the Fc cleaved off (see Section 3-3) or, more recently, by genetic engineering, which permits detailed mapping of the exact amino acid residues within the Fc that are needed for particular functions. Many kinds of microorganism seem to have responded to the destructive potential of the Fc portion by manufacturing proteins that either bind to it or proteolytically cleave it, and so prevent the Fc region from working. Examples of these are Protein A and Protein G made by *Staphylococcus* species (Fig. 4.22), and Protein D of *Haemophilus* species. Researchers can exploit these proteins to help to map the Fc and as immunological reagents (see Appendix I, Section A-10). Not all immunoglobulin isotypes or subtypes have the same capacity to engage each of the effector functions. The differential capabilities of each isotype are summarized in Fig. 4.16. For example, IgG1 and IgG3 have higher affinity for the most common type of Fc receptor.

4-19   IgM and IgA can form polymers.

Although all immunoglobulin molecules are constructed from a basic unit of two heavy and two light chains, both IgM and IgA can form multimers (Fig. 4.23). IgM and IgA C regions contain a 'tailpiece' of 18 amino acids that contains a cysteine residue essential for polymerization. An additional separate 15 kDa polypeptide chain called the J chain promotes polymerization by linking to the cysteines of the tailpiece, which is found only in the secreted forms of the μ and α chains. (This J chain should not be confused with the J gene segment; see Section 4-2.) In the case of IgA, polymerization is required for transport through epithelia, as we discuss in Chapter 9. IgM molecules are found as pentamers, and occasionally hexamers (without J chain), in plasma, whereas IgA in mucous secretions, but not in plasma, is mainly found as a dimer (see Fig. 4.23).

The polymerization of immunoglobulin molecules is thought to be important in the binding of antibody to repetitive epitopes. The dissociation rate of an individual epitope from an individual antibody binding site influences the

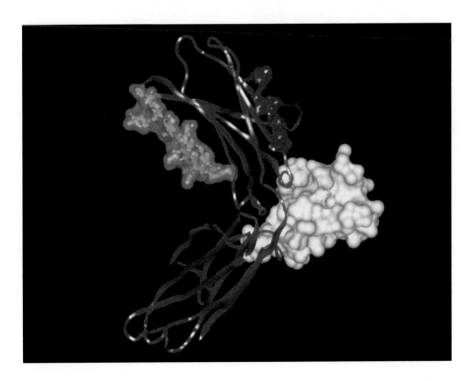

Fig. 4.22 Protein A of *Staphylococcus aureus* bound to a fragment of the Fc region of IgG. A fragment of the Fc portion of a single IgG heavy chain is complexed with a fragment of the immunoglobulin-binding Protein A from *Staphylococcus aureus*. The Fc fragment has two domains, C_H2 and C_H3, shown in magenta. A carbohydrate chain is attached to an asparagine residue in the C_H2 domain: all the atoms are shown and the surface is outlined in green. The fragment of Protein A (white) is bound between the two domains of the Fc fragment. The amino acids that bind to the complement component C1q (red) lie in the C_H2 domain. Photograph courtesy of C. Thorpe.

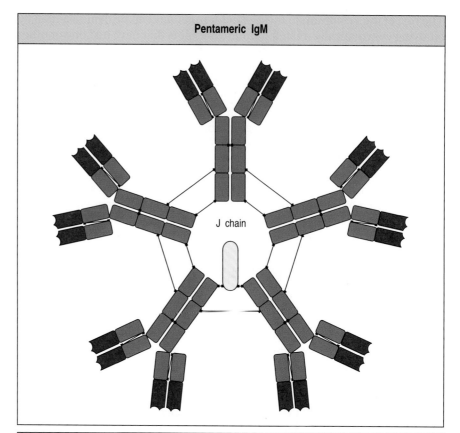

Pentameric IgM

J chain

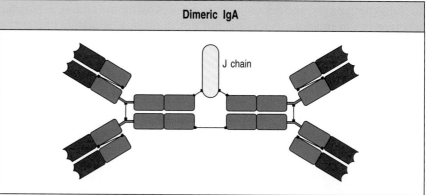

Dimeric IgA

J chain

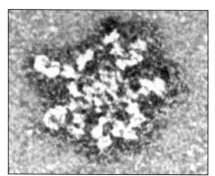

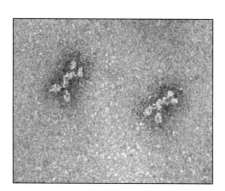

Fig. 4.23 The IgM and IgA molecules can form multimers.
IgM and IgA are usually synthesized as multimers in association with an additional polypeptide chain, the J chain. In pentameric IgM, the monomers are cross-linked by disulfide bonds to each other and to the J chain. The top left panel shows an electron micrograph of an IgM pentamer, showing the arrangement of the monomers in a flat disc. IgM can also form hexamers that lack a J chain but are more efficient in complement activation. In dimeric IgA, the monomers have disulfide bonds to the J chain as well as to each other. The bottom left panel shows an electron micrograph of dimeric IgA. Photographs (× 900,000) courtesy of K.H. Roux and J.M. Schiff.

strength of binding, or affinity, of that site: the lower the dissociation rate, the higher the affinity (see Appendix I, Section A-9). An antibody molecule has two or more identical antigen-binding sites, and if it attaches to two or more repeating epitopes on a single target antigen, it will only dissociate when all sites dissociate. The dissociation rate of the whole antibody from the whole antigen will therefore be much slower than the rate for the individual binding sites, giving a greater effective total binding strength, or **avidity**. This consideration is particularly relevant for pentameric IgM, which has ten antigen-binding sites. IgM antibodies frequently recognize repetitive epitopes such as those expressed by bacterial cell-wall polysaccharides, but the binding of individual sites is often of low affinity because IgM is made early in immune responses, before somatic hypermutation and affinity maturation. Multisite binding makes up for this, dramatically improving the overall functional binding strength.

4-20   Various differences between immunoglobulins can be detected by antibodies.

When an immunoglobulin is used as an antigen, it will be treated like any other foreign protein and will elicit an antibody response. Anti-immunoglobulin antibodies can be made that recognize the amino acids that characterize the isotype of the injected antibody. Such anti-isotypic antibodies recognize all immunoglobulins of the same isotype in all members of the species from which the injected antibody came.

It is also possible to raise antibodies that recognize differences in immuno-globulins from members of the same species that are due to the presence of multiple alleles of the individual C genes in the population (genetic polymorphism). Such allelic variants are called **allotypes**. In contrast to anti-isotypic antibodies, anti-allotypic antibodies will recognize immuno-globulin of a particular isotype only in some members of a species. Finally, as individual antibodies differ in their V regions, one can raise antibodies against unique sequence variants, which are called **idiotypes**.

A schematic picture of the differences between idiotypes, allotypes, and isotypes is given in Fig. 4.24. Historically, the main features of immuno-globulins were defined by using isotypic and allotypic genetic markers identified by antisera raised in different species or in genetically distinct members of the same species (see Appendix I, Section A-10). The independent segregation of allotypic and isotypic markers revealed the existence of separate heavy-chain, κ, and λ genes.

Summary.

The isotypes of immunoglobulins are defined by their heavy-chain C regions, each isotype being encoded by a separate C-region gene. The heavy-chain C-region genes lie in a cluster 3′ to the V-region gene segments. A productively rearranged V-region exon is initially expressed in association with μ and δ C$_H$ regions, but the same V-region exon can subsequently be associated with any one of the other isotypes by the process of isotype switching, in which the DNA is rearranged to place the V region 5′ to a different C-region gene. Unlike V(D)J recombination, isotype switching is always productive and occurs only in B cells activated by antigen. The immunological functions of the various isotypes differ; thus, isotype switching varies the response to the same antigen at different times or under different conditions. Immunoglobulin RNA can be processed in two different ways to produce either membrane-bound immunoglobulin, which acts as the B-cell receptor for antigen, or secreted antibody. In this way, the B-cell antigen receptor has the same specificity as the antibody that the B cell secretes upon activation.

Summary to Chapter 4.

Lymphocyte receptors are remarkably diverse, and developing B cells and T cells use the same basic mechanism to achieve this diversity. In each cell, functional genes for the immunoglobulin and T-cell receptor chains are assembled by somatic recombination from sets of separate gene segments that together encode the V region. The substrates for the joining process,

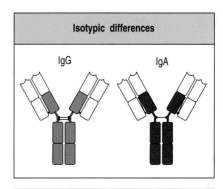

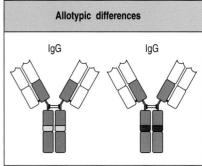

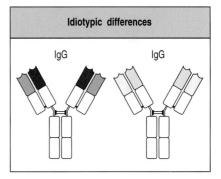

Fig. 4.24 Different types of variation between immunoglobulins. Differences between constant regions due to usage of different C-region genes are called isotypes; differences due to different alleles of the same C gene are called allotypes; differences due to particular rearranged V$_H$ and V$_L$ genes are called idiotypes.

arrays of V, D, and J gene segments, are similar among all the receptor loci, though there are some important differences in the details of their arrangement. The lymphoid-specific proteins, RAG-1 and RAG-2, direct the V(D)J recombination process in both T and B cells. These genes function in concert with other ubiquitous DNA-modifying enzymes and at least one other lymphoid-specific enzyme, TdT, to complete the joining process. As each type of gene segment is present in multiple, slightly different, versions, the random selection of gene segments for assembly is the source of enormous potential diversity. During the process of assembly, further diversity is introduced at the gene segment junctions through imprecise joining mechanisms. This diversity is concentrated in the DNA encoding the CDR3 loops of the receptors, which lie at the center of the antigen-binding site. The independent association of the two chains of immunoglobulins or T-cell receptors multiplies the overall diversity of the complete antigen receptor. In addition, mature B cells that are activated by antigen initiate a process of somatic point mutation of the V-region DNA, which creates numerous variants of the original assembled V region.

An important difference between immunoglobulins and T-cell receptors is that immunoglobulins exist in both membrane-bound (B-cell receptors) and secreted forms (antibodies). The ability to express both a secreted and membrane-bound form of the same molecule is due to differential splicing of mRNA to include exons that encode either the hydrophobic membrane anchor or the secreted tailpiece. Heavy-chain C regions contain three or four immunoglobulin domains, whereas the T-cell receptor chains have only one. Antibodies also have a variety of effector functions that are mediated by their C regions. Moreover, there are several alternative heavy-chain C regions for immunoglobulins, each with different effector functions. The same V region can be expressed along with different C regions through a process known as isotype switching. In this way, the progeny of a single B cell can express multiple different isotypes, thus maximizing the possible effector functions of any given antigen-specific antibody. The changes in immunoglobulin and T-cell receptor genes that occur during B-cell and T-cell development are summarized in Fig. 4.25.

Fig. 4.25 Changes in immunoglobulin and T-cell receptor genes that occur during B-cell and T-cell development and differentiation. Those changes that establish immunological diversity are all irreversible, as they involve changes in B-cell or T-cell DNA. Certain changes in the organization of DNA or in its transcription are unique to B cells. Somatic hypermutation has not been observed in functional T-cell receptors. The B-cell-specific processes, such as switch recombination, allow the same variable (V) region to be attached to several functionally distinct heavy-chain C regions, and thereby create functional diversity in an irreversible manner. By contrast, the expression of IgM versus IgD, and of membrane-bound versus secreted forms of all immunoglobulin types, can in principle be reversibly regulated.

Event	Process	Nature of change	Process occurs in: B cells	Process occurs in: T cells
V-region assembly	Somatic recombination of DNA	Irreversible	Yes	Yes
Junctional diversity	Imprecise joining, N-sequence insertion in DNA	Irreversible	Yes	Yes
Transcriptional activation	Activation of promoter by proximity to the enhancer	Irreversible but regulated	Yes	Yes
Switch recombination	Somatic recombination of DNA	Irreversible	Yes	No
Somatic hypermutation	DNA point mutation	Irreversible	Yes	No
IgM, IgD expression on surface	Differential splicing of RNA	Reversible, regulated	Yes	No
Membrane vs secreted form	Differential splicing of RNA	Reversible, regulated	Yes	No

General references.

Casali, P., and Silberstein, L.E.S. (eds): **Immunoglobulin gene expression in development and disease.** *Ann. N.Y. Acad. Sci.* 1995, **764**.

Fugmann, S.D., Lee, A.I., Shockett, P.E., Villey, I.J., and Schatz, D.G.: **The RAG proteins and V(D)J recombination: complexes, ends, and transposition.** *Annu. Rev. Immunol.* 2000, **18**:495-527.

Wagner, S.D., and Neuberger, M.S.: **Somatic hypermutation of immunoglobulin genes.** *Annu. Rev. Immunol.* 1996, **14**:441-457.

Section references.

4-1   Immunoglobulin genes are rearranged in antibody-producing cells.

Hozumi, N., and Tonegawa, S.: **Evidence for somatic rearrangement of immunoglobulin genes coding for variable and constant regions.** *Proc. Natl. Acad. Sci. USA* 1976, **73**:3628-3632.

Tonegawa, S., Brack, C., Hozumi, N., and Pirrotta, V.: **Organization of immunoglobulin genes.** *Cold Spring Harbor Symp. Quant. Biol.* 1978, **42**:921-931.

Waldmann, T.A.: **The arrangement of immunoglobulin and T-cell receptor genes in human lymphoproliferative disorders.** *Adv. Immunol.* 1987, **40**:247-321.

4-2   The DNA sequence encoding a complete V region is generated by the somatic recombination of separate gene segments.

Early, P., Huang, H., Davis, M., Calame, K., and Hood, L.: **An immunoglobulin heavy chain variable region gene is generated from three segments of DNA: VH, D and JH.** *Cell* 1980, **19**:981-992.

Tonegawa, S., Maxam, A.M., Tizard, R., Bernard, O., and Gilbert, W.: **Sequence of a mouse germ-line gene for a variable region of an immunoglobulin light chain.** *Proc. Natl. Acad. Sci. USA* 1978, **75**:1485-1489.

4-3   There are multiple different V-region gene segments.

Cook, G.P., and Tomlinson, I.M.: **The human immunoglobulin V-H repertoire.** *Immunol. Today* 1995, **16**:237-242.

Joho, R., Weissman, I.L., Early, P., Cole, J., and Hood, L.: **Organization of kappa light chain genes in germ-line and somatic tissue.** *Proc. Natl. Acad. Sci. USA* 1980, **77**:1106-1110.

Kofler, R., Geley, S., Kofler, H., and Helmberg, A.: **Mouse variable-region gene families—complexity, polymorphism, and use in nonautoimmune responses.** *Immunol. Rev.* 1992, **128**:5-21.

Maki, R., Traunecker, A., Sakano, H., Roeder, W., and Tonegawa, S.: **Exon shuffling generates an immunoglobulin heavy chain gene.** *Proc. Natl. Acad. Sci. USA* 1980, **77**:2138-2142.

Matsuda, F., and Honjo, T.: **Organization of the human immunoglobulin heavy-chain locus.** *Adv. Immunol.* 1996, **62**:1-29.

4-4   Rearrangement of V, D, and J gene segments is guided by flanking DNA sequences.

Grawunder, U., West, R.B., and Lieber, M.R.: **Antigen receptor gene rearrangement.** *Curr. Opin. Immunol.* 1998, **10**:172-180.

Max, E.E., Seidman, J.G., and Leder, P.: **Sequences of five potential recombination sites encoded close to an immunoglobulin kappa constant region gene.** *Proc. Natl. Acad. Sci. USA* 1979, **76**:3450-3454.

Muramatsu, M., Kinoshita, K., Fagarasan, S., Yamada, S., Shinkai, Y., and Honjo, T.: **Class switch recombination and hypermutation require activation-induced cytidine deaminase (AID), a potential RNA editing enzyme.** *Cell* 2000, **102**:553-563.

Sakano, H., Huppi, K., Heinrich, G., and Tonegawa, S.: **Sequences at the somatic recombination sites of immunoglobulin light-chain genes.** *Nature* 1979, **280**:288-294.

4-5   The reaction that recombines V, D, and J gene segments involves both lymphocyte-specific and ubiquitous DNA-modifying enzymes.

Agrawal, A., and Schatz, D.G.: **RAG1 and RAG2 form a stable postcleavage synaptic complex with DNA containing signal ends in V(D)J recombination.** *Cell* 1997, **89**:43-53.

Blunt, T., Finnie, N.J., Taccioli, G.E., Smith, G.C.M., Demengeot, J., Gottlieb, T.M., Mizuta, R., Varghese, A.J., Alt, F.W., Jeggo, P.A., and Jackson, S.P.: **Defective DNA-dependent protein kinase activity is linked to V(D)J recombination and DNA-repair defects associated with the murine–scid mutation.** *Cell* 1995, **80**:813-823.

Gu, Z., Jin, S., Gao, Y., Weaver, D.T., and Alt, F.W.: **Ku70-deficient embryonic stem cells have increased ionizing radiosensitivity, defective DNA end-binding activity, and inability to support V(D)J recombination.** *Proc. Natl. Acad. Sci. USA* 1997, **94**:8076-8081.

Li, Z.Y., Otevrel, T., Gao, Y.J., Cheng, H.L., Seed, B., Stamato, T.D., Taccioli, G.E., and Alt, F.W.: **The XRCC4 gene encodes a novel protein involved in DNA double-strand break repair and V(D)J recombination.** *Cell* 1995, **83**:1079-1089.

Oettinger, M.A., Schatz, D.G., Gorka, C., and Baltimore, D.: **RAG-1 and RAG-2, adjacent genes that synergistically activate V(D)J recombination.** *Science* 1990, **248**:1517-1523.

Shockett, P.E., and Schatz, D.G.: **DNA hairpin opening mediated by the RAG1 and RAG2 proteins.** *Mol. Cell. Biol.* 1999, **19**:4159-4166.

4-6   The diversity of the immunoglobulin repertoire is generated by four main processes.

Fanning, L.J., Connor, A.M., and Wu, G.E.: **Development of the immunoglobulin repertoire.** *Clin. Immunol. Immunopathol.* 1996, **79**:1-14.

Weigert, M., Perry, R., Kelley, D., Hunkapiller, T., Schilling, J., and Hood, L.: **The joining of V and J gene segments creates antibody diversity.** *Nature* 1980, **283**:497-499.

4-7   The multiple inherited gene segments are used in different combinations.

Lee, A., Desravines, S., and Hsu, E.: **IgH diversity in an individual with only one million B lymphocytes.** *Develop. Immunol.* 1993, **3**:211-222.

4-8   Variable addition and subtraction of nucleotides at the junctions between gene segments contributes to diversity in the third hypervariable region.

Gauss, G.H., and Lieber, M.R.: **Mechanistic constraints on diversity in human V(D)J recombination.** *Mol. Cell. Biol.* 1996, **16**:258-269.

Komori, T., Okada, A., Stewart, V., and Alt, F.W.: **Lack of N regions in antigen receptor variable region genes of TdT-deficient lymphocytes [published erratum appears in *Science* 1993, **262**:1957].** *Science* 1993, **261**:1171-1175.

Weigert, M., Gatmaitan, L., Loh, E., Schilling, J., and Hood, L.: **Rearrangement of genetic information may produce immunoglobulin diversity.** *Nature* 1978, **276**:785-790.

4-9   Rearranged V genes are further diversified by somatic hypermutation.

Betz, A.G., Rada, C., Pannell, R., Milstein, C., and Neuberger, M.S.: **Passenger transgenes reveal intrinsic specificity of the antibody hypermutation mechanism: clustering, polarity, and specific hot spots.** *Proc. Natl. Acad. Sci. USA* 1993, **90**:2385-2388.

McKean, D., Huppi, K., Bell, M., Straudt, L., Gerhard, W., and Weigert, M.: **Generation of antibody diversity in the immune response of BALB/c mice to influenza virus hemagglutinin.** *Proc. Natl. Acad. Sci. USA* 1984, **81**:3180-3184.

Neuberger, M.S., Ehrenstein, M.R., Klix, N., Jolly, C.J., Yelamos, J., Rada, C., and Milstein, C.: **Monitoring and interpreting the intrinsic features of somatic hypermutation.** *Immunol. Rev.* 1998, **162**:107-116.

Papavasiliou, F.N., and Schatz, D.G.: **Cell cycle regulated DNA double strand breaks in somatic hypermutation of immunoglobulin genes.** *Nature*, in press.

Sale, J.E., and Neuberger, M.S.: **TdT-accessible breaks are scattered over the immunoglobulin V domain in a constitutively hypermutating B cell line.** *Immunity* 1998, **9**:859-869.

Tomlinson, I.M., Walter, G., Jones, P.T., Dear, P.H., Sonnhammer, E.L.L., and Winter, G.: **The imprint of somatic hypermutation on the repertoire of human germline V genes.** *J. Mol. Biol.* 1996, **256**:813-817.

Weigert, M.G., Cesari, I.M., Yonkovich, S.J., and Cohn, M.: **Variability in the lambda light chain sequences of mouse antibody.** *Nature* 1970, **228**:1045-1047.

4-10 In some species most immunoglobulin gene diversification occurs after gene rearrangement.

Knight, K.L., and Crane, M.A.: **Generating the antibody repertoire in rabbit.** *Adv. Immunol.* 1994, **56**:179-218.

Reynaud, C.A., Bertocci, B., Dahan, A., and Weill, J.C.: **Formation of the chicken B-cell repertoire—ontogeny, regulation of Ig gene rearrangement, and diversification by gene conversion.** *Adv. Immunol.* 1994, **57**:353-378.

Reynaud, C.A., Garcia, C., Hein, W.R., and Weill, J.C.: **Hypermutation generating the sheep immunoglobulin repertoire is an antigen independent process.** *Cell* 1995, **80**:115-125.

Vajdy, M., Sethupathi, P., and Knight, K.L.: **Dependence of antibody somatic diversification on gut-associated lymphoid tissue in rabbits.** *J. Immunol.* 1998, **160**:2725-2729.

4-11 The T-cell receptor loci comprise sets of gene segments and are rearranged by the same enzymes as the immunoglobulin loci.

Rowen, L., Koop, B.F., and Hood, L.: **The complete 685-kilobase DNA sequence of the human β T cell receptor locus.** *Science* 1996, **272**:1755-1762.

Shinkai, Y., Rathbun, G., Lam, K.P., Oltz, E.M., Stewart, V., Mendelsohn, M., Charron, J., Datta, M., Young, F., Stall, A.M., and Alt, F.W.: **RAG-2 deficient mice lack mature lymphocytes owing to inability to initiate V(D)J rearrangement.** *Cell* 1992, **68**:855-867.

4-12 T-cell receptors concentrate diversity in the third hypervariable region.

Davis, M.M., and Bjorkman, P.J.: **T-cell antigen receptor genes and T-cell recognition** [published erratum appears in *Nature* 1988, 335:744]. *Nature* 1988, **334**:395-402.

Jorgensen, J.L., Esser, U., Fazekas de St. Groth, B., Reay, P.A., and Davis, M.M.: **Mapping T-cell receptor-peptide contacts by variant peptide immunization of single-chain transgenics.** *Nature* 1992, **355**:224-230.

4-13 γ:δ T-cell receptors are also generated by gene rearrangement.

Chien, Y.H., Iwashima, M., Kaplan, K.B., Elliott, J.F., and Davis, M.M.: **A new T-cell receptor gene located within the alpha locus and expressed early in T-cell differentiation.** *Nature* 1987, **327**:677-682.

Hayday, A.C., Saito, H., Gillies, S.D., Kranz, D.M., Tanigawa, G., Eisen, H.N., and Tonegawa, S.: **Structure, organization, and somatic rearrangement of T cell gamma genes.** *Cell* 1985, **40**:259-269.

Lafaille, J.J., DeCloux, A., Bonneville, M., Takagaki, Y., and Tonegawa, S.: **Junctional sequences of T cell receptor gamma delta genes: implications for gamma delta T cell lineages and for a novel intermediate of V-(D)-J joining.** *Cell* 1989, **59**:859-870.

Tonegawa, S., Berns, A., Bonneville, M., Farr, A.G., Ishida, I., Ito, K., Itohara, S., Janeway, C.A., Jr., Kanagawa, O., Kubo, R., et al.: **Diversity, development, ligands, and probable functions of gamma delta T cells.** *Adv. Exp. Med. Biol.* 1991, **292**:53-61.

4-14 Somatic hypermutation does not generate diversity in T-cell receptors.

Zheng, B., Xue, W., and Kelsoe, G.: **Locus-specific somatic hypermutation in germinal centre T cells.** *Nature* 1994, **372**:556-559.

4-15 The immunoglobulin heavy-chain isotypes are distinguished by the structure of their constant regions.

Davies, D.R., and Metzger, H.: **Structural basis of antibody function.** *Annu. Rev. Immunol.* 1983, **1**:87-117.

4-16 The same V_H exon can associate with different C_H genes in the course of an immune response.

Revy, P., Muto, T., Levy, Y., Geissmann, F., Plebani, A., Sanal, O., Catalan, N., Forveille, M., Dufourcq-Lagelouse, R., Gennery, A., Tezcan, I., Ersoy, F., Kayserili, H., Ugazio, A.G., Brousse, N., Muramatsu, M., Notarangelo, L.D., Kinoshita, K., Honjo, T., Fischer, A., and Durandy, A.: **Activation-induced cytidine deaminase (AID) deficiency causes the autosomal recessive form of the hyper-IgM syndrome (HIGM2).** *Cell* 2000, **102**:565-575.

Sakano, H., Maki, R., Kurosawa, Y., Roeder, W., and Tonegawa, S.: **Two types of somatic recombination are necessary for the generation of complete immunoglobulin heavy-chain genes.** *Nature* 1980, **286**:676-683.

Stavnezer, J.: **Immunoglobulin class switching.** *Curr. Opin. Immunol.* 1996, **8**:199-205.

4-17 Transmembrane and secreted forms of immunoglobulin are generated from alternative heavy-chain transcripts.

Early, P., Rogers, J., Davis, M., Calame, K., Bond, M., Wall, R., and Hood, L.: **Two mRNAs can be produced from a single immunoglobulin mu gene by alternative RNA processing pathways.** *Cell* 1980, **20**:313-319.

Peterson, M.L., Gimmi, E.R., and Perry, R.P.: **The developmentally regulated shift from membrane to secreted mu mRNA production is accompanied by an increase in cleavage-polyadenylation efficiency but no measurable change in splicing efficiency.** *Mol. Cell. Biol.* 1991, **11**:2324-2327.

Rogers, J., Early, P., Carter, C., Calame, K., Bond, M., Hood, L., and Wall, R.: **Two mRNAs with different 3′ ends encode membrane-bound and secreted forms of immunoglobulin mu chain.** *Cell* 1980, **20**:303-312.

4-18 Antibody C regions confer functional specialization.

Helm, B.A., Sayers, I., Higginbottom, A., Machado, D.C., Ling, Y., Ahmad, K., Padlan, E.A., and Wilson, A.P.M.: **Identification of the high affinity receptor binding region in human IgE.** *J. Biol. Chem.* 1996, **271**:7494-7500.

Jefferis, R., Lund, J., and Goodall, M.: **Recognition sites on human IgG for Fcg receptors—the role of glycosylation.** *Immunol. Lett.* 1995, **44**:111-117.

Sensel, M.G., Kane, L.M., and Morrison, S.L.: **Amino acid differences in the N-terminus of CH2 influence the relative abilities of IgG2 and IgG3 to activate complement.** *Mol. Immunol.* **34**:1019-1029.

4-19 IgM and IgA can form polymers.

Hendrickson, B.A., Conner, D.A., Ladd, D.J., Kendall, D., Casanova, J.E., Corthesy, B., Max, E.E., Neutra, M.R., Seidman, C.E., and Seidman, J.G.: **Altered hepatic transport of IgA in mice lacking the J chain.** *J. Exp. Med.* 1995, **182**:1905-1911.

Niles, M.J., Matsuuchi, L., and Koshland, M.E.: **Polymer IgM assembly and secretion in lymphoid and nonlymphoid cell-lines—evidence that J chain is required for pentamer IgM synthesis.** *Proc. Nat. Acad. Sci. USA* 1995, **92**:2884-2888.

4-20 Various differences between immunoglobulins can be detected by antibodies.

Capra, J.D., and Kehoe, J.M.: **Structure of antibodies with shared idiotypy: the complete sequence of the heavy chain variable regions of two immunoglobulin M anti-gamma globulins.** *Proc. Natl. Acad. Sci. USA* 1974, **71**:4032-4036.

Fields, B.A., Goldbaum, F.A., Ysern, X., Poljak, R.J., and Mariuzza, R.A.: **Molecular basis of antigen mimicry by an anti idiotope.** *Nature* 1995, **374**:739-742.

Wu, T.T., Kabat, E.A., and Bilofsky, H.: **Similarities among hypervariable segments of immunoglobulin chains.** *Proc. Natl. Acad. Sci. USA* 1975, **72**:5107-5110.

Antigen Presentation to T Lymphocytes

5

In an adaptive immune response, antigen is recognized by two distinct sets of highly variable receptor molecules—the immunoglobulins that serve as antigen receptors on B cells and the antigen-specific receptors of T cells. As we saw in Chapter 3, T cells recognize only antigens that are displayed on cell surfaces. These antigens may derive from pathogens that replicate within cells, such as viruses or intracellular bacteria, or from pathogens or their products that cells internalize by endocytosis from the extracellular fluid. T cells can detect the presence of intracellular pathogens because infected cells display on their surface peptide fragments derived from the pathogens' proteins. These foreign peptides are delivered to the cell surface by specialized host-cell glycoproteins, the **MHC molecules**, which are also described in Chapter 3. The MHC glycoproteins are encoded in a large cluster of genes that were first identified by their potent effects on the immune response to transplanted tissues. For that reason, the gene complex was termed the **major histocompatibility complex (MHC)**. We now know that within this region of the genome, in addition to those genes encoding the MHC molecules themselves, are many genes whose products are involved in the production of the MHC:peptide complexes.

We will begin by discussing the mechanisms of antigen processing and presentation, whereby protein antigens are degraded into peptides inside cells and the peptides are then carried to the cell surface bound to MHC molecules. We will see that the two different classes of MHC molecule, known as MHC class I and MHC class II, deliver peptides from different cellular compartments to the surface of the infected cell. Peptides from the cytosol are bound to MHC class I molecules and are recognized by CD8 T cells, whereas peptides generated in vesicles are bound to MHC class II molecules and recognized by CD4 T cells. The two functional subsets of T cells are thereby activated to initiate the destruction of pathogens resident in these two different cellular compartments. Some CD4 T cells activate naive B cells that have internalized specific antigen, and thus also stimulate the production of antibodies to extracellular pathogens and their products.

In the second part of this chapter we will see that there are several genes for each class of MHC molecule: that is, the MHC is polygenic. Each of these genes has many variants: that is, the MHC is also highly polymorphic. Indeed, the most remarkable feature of the MHC class I and II genes is their genetic variability. MHC polymorphism has a profound effect on antigen recognition by T cells, and the combination of polygeny and polymorphism greatly extends the range of peptides that can be presented to T cells by each individual and each population at risk from an infectious pathogen.

The generation of T-cell receptor ligands.

The protective function of T cells depends on their ability to recognize cells that are harboring pathogens or that have internalized pathogens or their products. T cells do this by recognizing peptide fragments of pathogen-derived proteins in the form of complexes of peptides and MHC molecules on the cell surface. Because the generation of peptides from an intact antigen involves modification of the native protein, it is commonly referred to as **antigen processing**, whereas the display of the peptide at the cell surface by the MHC molecule is referred to as **antigen presentation**. We have already described the structure of MHC molecules and seen how they bind peptide antigens in a cleft on their outer surface (see Sections 3-15 to 3-18). In this chapter we will look at how peptides are generated from pathogens present in the cytosol or in the vesicular compartment of cells and loaded onto MHC class I and MHC class II molecules, respectively, at different sites inside the cell. Both classes of MHC molecule must combine with a peptide before they can be stably expressed at the cell surface. Peptide-binding completes the folding and assembly of newly synthesized MHC class I molecules in the endoplasmic reticulum, while MHC class II molecules are prevented from binding peptides in the endoplasmic reticulum and are instead escorted to an endosomal compartment where loading with vesicular peptides occurs.

5-1 The MHC class I and class II molecules deliver peptides to the cell surface from two distinct intracellular compartments.

Infectious agents can replicate in either of two distinct intracellular compartments (Fig. 5.1). Viruses and certain bacteria replicate in the cytosol or in the contiguous nuclear compartment (Fig. 5.2, left panel), whereas many pathogenic bacteria and some eukaryotic parasites replicate in the endosomes and lysosomes that form part of the vesicular system (Fig. 5.2, center panel). The immune system has different strategies for eliminating pathogens from these two sites. Cells infected with viruses or with bacteria that live in the cytosol are eliminated by cytotoxic T cells; as mentioned in Chapter 3, these T cells are distinguished by the co-receptor molecule CD8. The function of **CD8 T cells** is to kill infected cells; this is an important means of eliminating sources of new viral particles and obligate cytosolic bacteria, thus freeing the host of infection.

Pathogens and their products in the vesicular compartments of cells are detected by a different class of T cell, distinguished by the co-receptor molecule CD4 (see Chapter 3). **CD4 T cells** are specialized to activate other cells and fall into two functional classes: T_H1 cells (sometimes known as inflammatory T cells), whose main function is to activate macrophages to kill the intravesicular pathogens they harbor; and T_H2 cells, or helper T cells, which activate B cells to make antibody. Microbial antigens may enter the vesicular compartment in either of two ways. Some bacteria, including the mycobacteria that cause tuberculosis and leprosy, invade macrophages and flourish in intracellular vesicles. Other bacteria proliferate outside cells, where they cause pathology by secreting toxins and other proteins. These bacteria and their toxic products can be internalized by phagocytosis, endocytosis, or macropinocytosis into the intracellular vesicles of cells that then present antigen to T cells. These include the dendritic cells that specialize in initiating T cell responses (see Section 1-6), macrophages that specialize in taking up particulate material (see Section 2-3), and B cells that efficiently internalize specific antigen by receptor-mediated endocytosis of the antigen bound to their surface immunoglobulin (Fig. 5.2, right panel).

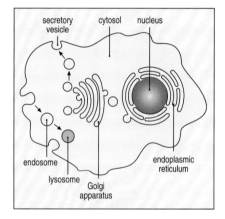

Fig. 5.1 There are two major intracellular compartments, separated by membranes. The first is the cytosol, which also communicates with the nucleus via the nuclear pores in the nuclear membrane. The second is the vesicular system, which comprises the endoplasmic reticulum, Golgi apparatus, endosomes, lysosomes, and other intracellular vesicles. The vesicular system can be thought of as continuous with the extracellular fluid. Secretory vesicles bud off from the endoplasmic reticulum and are transported via fusion with Golgi membranes to move vesicular contents out of the cell, whereas extracellular material is taken up by endocytosis into endosomes, which move it eventually into lysosomes, where it is degraded.

	Cytosolic pathogens	Intravesicular pathogens	Extracellular pathogens and toxins
	any cell	macrophage	B cell
Degraded in	Cytosol	Endocytic vesicles (low pH)	Endocytic vesicles (low pH)
Peptides bind to	MHC class I	MHC class II	MHC class II
Presented to	CD8 T cells	CD4 T cells	CD4 T cells
Effect on presenting cell	Cell death	Activation to kill intravesicular bacteria and parasites	Activation of B cells to secrete Ig to eliminate extracellular bacteria/toxins

Fig. 5.2 Pathogens and their products can be found in either the cytosolic or the vesicular compartment of cells. Left panel: all viruses and some bacteria replicate in the cytosolic compartment. Their antigens are presented by MHC class I molecules to CD8 T cells. Center panel: other bacteria and some parasites are taken up into endosomes, usually by specialized phagocytic cells such as macrophages. Here they are killed and degraded, or in some cases are able to survive and proliferate within the vesicle. Their antigens are presented by MHC class II molecules to CD4 T cells. Right panel: proteins derived from extracellular pathogens may enter the vesicular system of cells by binding to surface molecules followed by endocytosis. This is illustrated for proteins bound by the surface immunoglobulin of B cells (the endoplasmic reticulum and Golgi apparatus have been omitted for simplicity). The B cells present these antigens to CD4 helper T cells, which can then stimulate the B cells to produce antibody. Other types of cell can also internalize antigens in this way and are able to activate T cells.

To produce an appropriate response to infectious microorganisms, T cells need to be able to detect the presence of intracellular pathogens and to distinguish between foreign material coming from the cytosolic and vesicular compartments. This is achieved through the use of the different classes of MHC molecule. MHC class I molecules deliver peptides originating in the cytosol to the cell surface, where they are recognized by CD8 T cells. MHC class II molecules deliver peptides originating in the vesicular system to the cell surface, where they are recognized by CD4 T cells. As we saw in Section 3-12, CD8 and CD4 bind MHC class I and MHC class II molecules, respectively, and so help to ensure that the appropriate type of T cell is activated in response to a given pathogen.

5-2   Peptides that bind to MHC class I molecules are actively transported from the cytosol to the endoplasmic reticulum.

The antigen fragments that bind to MHC class I molecules are typically derived from viruses that take over the cell's biosynthetic mechanisms to make their own proteins. All proteins are made in the cytosol. The polypeptide chains of proteins destined for the cell surface, which include both classes of MHC molecule, are translocated during synthesis into the lumen of the endoplasmic reticulum. Here the chains must fold correctly, and assemble with each other if necessary, before the complete protein can be transported to the cell surface. Thus the peptide-binding site of the MHC class I molecule is formed in the lumen of the endoplasmic reticulum and is never exposed to the cytosol. This raised the question—how are peptides derived from viral proteins in the cytosol able to bind to MHC class I molecules for delivery to the cell surface?

The first clues came from mutant cells with a defect in antigen presentation by MHC class I molecules. Although both chains of MHC class I molecules are synthesized normally in these cells, the MHC class I proteins are present at abnormally low levels on the cell surface. This defect can be corrected by the addition of synthetic peptides to the medium bathing the cells, suggesting both that the mutation affects the supply of peptides to MHC class I molecules and that peptide is required for their maintenance at the cell surface. This was the first indication that MHC molecules are unstable in the absence of bound peptide.

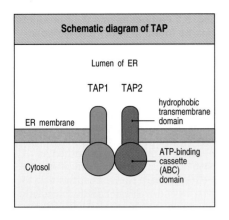

Fig. 5.3 TAP1 and TAP2 form a peptide transporter in the endoplasmic reticulum membrane. All transporters of the ATP-binding cassette (ABC) family are composed of four domains: two hydrophobic transmembrane domains that have multiple transmembrane regions, and two ATP-binding domains. TAP1 and TAP2 each encode one hydrophobic and one ATP-binding domain and assemble into a heterodimer to form a four-domain transporter. From similarities between the TAP molecules and other members of the ABC-transporter family, it is believed that the ATP-binding domains lie within the cytosol, whereas the hydrophobic domains project through the membrane into the lumen of the endoplasmic reticulum (ER).

Analysis of the DNA of the mutant cells showed that two genes encoding members of the ATP-binding cassette, or ABC, family of proteins are mutant or absent in these cells. ABC proteins mediate ATP-dependent transport of ions, sugars, amino acids, and peptides across membranes in many types of cells, including bacteria. The two ABC proteins missing in the mutant cells are normally associated with the endoplasmic reticulum membrane. Transfection of the mutant cells with both genes restores presentation of peptides by the cell's MHC class I molecules. These proteins are now called **Transporters associated with Antigen Processing-1 and -2 (TAP1 and TAP2)**. The two TAP proteins form a heterodimer (Fig. 5.3) and mutations in either TAP gene can prevent antigen presentation by MHC class I molecules. The genes *TAP1* and *TAP2* map within the MHC itself (see Section 5-9), and are inducible by interferons, which are produced in response to virus infection.

In *in vitro* assays using normal cell fractions, microsomal vesicles that mimic the endoplasmic reticulum will internalize peptides, which then bind to MHC class I molecules already present in the microsome lumen. Vesicles from TAP1 or TAP2 mutant cells do not transport peptides. Peptide transport into the normal microsomes requires ATP hydrolysis, proving that the TAP1:TAP2 complex is an ATP-dependent peptide transporter. Such experiments have also shown that the TAP complex has some specificity for the peptides it will transport. It prefers peptides of eight or more amino acids with hydrophobic or basic residues at the carboxy terminus—the exact features of peptides that bind MHC class I molecules. The TAP transporter provides the answer to the question of how viral peptides gain access to the lumen of the ER in order to bind to MHC class I molecules, but leaves open the question of how these peptides are generated.

5-3   Peptides for transport into the endoplasmic reticulum are generated in the cytosol.

Proteins in cells are continually being degraded and replaced with newly synthesized proteins. Much cytosolic protein degradation is carried out by a large, multicatalytic protease complex called the **proteasome** (Fig. 5.4). The proteasome is a large cylindrical complex of some 28 subunits, arranged in four stacked rings, each of seven subunits, and it has a hollow core lined by the active sites of the proteolytic subunits of the proteasome. Proteins to be degraded are introduced into the core of the proteasome and are there broken down into short peptides.

Various lines of evidence implicate the proteasome in the production of peptide ligands for MHC class I molecules. For example, the proteasome takes part in the ubiquitin-dependent degradation pathway for cytosolic proteins, and experimentally tagging proteins with ubiquitin also results in more efficient presentation of their peptides by MHC class I molecules. Moreover, inhibitors of the proteolytic activity of the proteasome also inhibit antigen presentation by MHC class I molecules. Whether the proteasome is the only cytosolic protease capable of generating peptides for transport into the endoplasmic reticulum is not known.

Two subunits of the proteasome, called LMP2 and LMP7, are encoded within the MHC near the *TAP1* and *TAP2* genes. Along with MHC class I and TAP molecules, their expression is induced by interferons, which are produced in response to viral infections. LMP2 and LMP7 substitute for two constitutively expressed subunits of the proteasome. A third subunit, MECL-1, which is not encoded within the MHC, is also induced by interferons and also displaces a constitutive proteasome subunit. These three inducible subunits and their constitutive counterparts are thought to be the active proteases of the proteasome. The replacement of the constitutive components by their

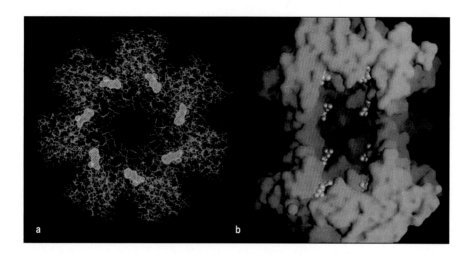

Fig. 5.4 The structure of the proteasome. Proteasomes are found throughout the eukaryotes and the archaebacteria, and their structure and function are highly conserved. The structure shown here is from an archae-bacterium, as the detailed structure of a mammalian proteasome has not yet been determined. The proteasome contains 28 subunits arranged to form a cylindrical structure composed of four rings, each of seven subunits. Panel a shows a horizontal cross-section through the proteasome, showing the arrangement of the seven subunits that comprise each ring; panel b shows a longitudinal section, in which the surface of the proteasome can be seen. The subunits that form the two central rings of the archaebacterial proteasome contain the proteolytic activity, and the active sites (seven in each ring) are indicated in green in panel a and in gold in panel b. Thus it is thought that proteins have to unfold and pass through the center of the cylinder in order to be degraded to peptides. It is not known exactly how the mammalian proteasome functions; it has six proteo-lytic sites, three in each of the two central rings, and these sites also lie in the center of the cylinder. It is therefore likely that its mechanism of action is very similar to that of the archae-bacterial proteasome. Photographs (× 667,000) courtesy of W. Baumeister.

interferon-inducible counterparts seems to change the specificity of the proteasome: in interferon-treated cells, there is increased cleavage of polypeptides after hydrophobic and basic residues, and reduced cleavage after acidic residues. This produces peptides with carboxy-terminal residues that are preferred anchor residues for binding to most MHC class I molecules and are also the preferred structures for transport by TAP.

MHC class I molecules also present peptides derived from membrane and secreted proteins, for example, the glycoproteins of viral envelopes. Membrane and secreted proteins are normally translocated into the lumen of the endoplasmic reticulum during their biosynthesis. Yet the peptides bound by MHC class I molecules bear evidence that such proteins are degraded in the cytosol. Asparagine-linked carbohydrate moieties commonly present on membrane-bound or secreted proteins can be removed in the cytosol by an enzyme reaction that changes the asparagine residue into aspartic acid, and this diagnostic sequence change can be seen in some peptides presented by MHC class I molecules. It now appears that endoplasmic reticulum proteins can be returned to the cytosol by the same translocation system that trans-ported them into the endoplasmic reticulum in the first place. This newly discovered mechanism, known as retrograde translocation, may be the normal mechanism by which proteins in the endoplasmic reticulum are turned over, and by which misfolded proteins in the endoplasmic reticulum are removed and degraded. Once in the cytosol, the polypeptides are degraded by the pro-teasome. The resulting peptides can then be transported back into the lumen of the endoplasmic reticulum via TAP and loaded onto MHC class I molecules.

5-4   Newly synthesized MHC class I molecules are retained in the endoplasmic reticulum until they bind peptide.

Binding of peptide is an important step in the assembly of stable MHC class I molecules. When the supply of peptides into the endoplasmic reticulum is disrupted, as in the *TAP* mutant cells, newly synthesized MHC class I mol-ecules are held in the endoplasmic reticulum in a partially folded state. This explains why cells with mutations in *TAP1* or *TAP2* fail to express MHC class I molecules at the cell surface. The folding and assembly of a complete MHC class I molecule (see Fig. 3.20) depends on the association of the MHC class I α chain first with β$_2$-microglobulin and then with peptide, and this process involves a number of accessory proteins with a chaperone-like function. Only after peptide has bound is the MHC class I molecule released from the endo-plasmic reticulum and allowed to reach the cell surface.

MHC Class I Deficiency

In humans, newly synthesized MHC class I α chains that enter the endoplasmic reticulum membranes bind to a chaperone protein, **calnexin**, which retains the MHC class I molecule in a partially folded state in the endoplasmic reticulum. Calnexin also associates with partially folded T-cell receptors, immunoglobulins, and MHC class II molecules, and so has a central role in the assembly of many immunological proteins. When β$_2$-microglobulin binds to the α chain, the partially folded α:β$_2$-microglobulin heterodimer dissociates from calnexin and now binds to a complex of proteins, one of which—**calreticulin**—is similar to calnexin and probably carries out a similar chaperone function. A second component of the complex is the TAP-associated protein **tapasin**, also encoded by a gene that lies within the MHC. Tapasin forms a bridge between MHC class I molecules and TAP1 and TAP2, allowing the partially folded α:β$_2$-microglobulin heterodimer to await the transport of a suitable peptide from the cytosol. A third component of this complex is the chaperone molecule **Erp57**, a protein disulfide isomerase that may have a role in breaking and reforming the disulfide bond in the MHC class I α$_2$ domain during peptide loading. Erp57 and calreticulin bind to a number of glycoproteins during their assembly in the endoplasmic reticulum and seem to be part of the cell's general quality-control mechanism. Finally, the binding of a peptide to the partially folded heterodimer releases it from the complex of TAP:tapasin:calreticulin:Erp57. The fully folded MHC class I molecule and its bound peptide are now able to leave the endoplasmic reticulum and are transported to the cell surface (Fig. 5.5).

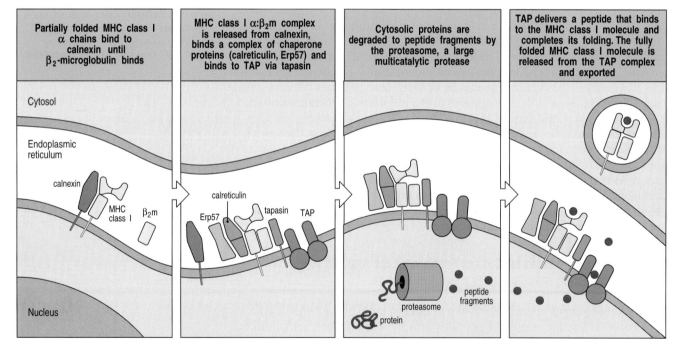

Fig. 5.5 MHC class I molecules do not leave the endoplasmic reticulum unless they bind peptides. Newly synthesized MHC class I α chains assemble in the endoplasmic reticulum with a membrane-bound protein, calnexin. When this complex binds β$_2$-microglobulin (β$_2$m), the MHC class I α:β$_2$m dimer dissociates from calnexin, and the partially folded MHC class I molecule then binds to the peptide transporter TAP by interacting with one molecule of the TAP-associated protein tapasin. The chaperone molecules calreticulin and Erp57 also bind to form part of this complex. The MHC class I molecule is retained within the endoplasmic reticulum until released by the binding of a peptide, which completes the folding of the MHC molecule. Peptides generated by the degradation of proteins in the cytoplasm are transported into the lumen of the endoplasmic reticulum by TAP. Once peptide has bound to the MHC molecule, the peptide:MHC complex leaves the endoplasmic reticulum and is transported through the Golgi apparatus to the cell surface.

Most of the peptides transported by TAP will not bind to the MHC molecules in that cell and are rapidly cleared out of the endoplasmic reticulum; there is evidence that they are transported back into the cytosol by an ATP-dependent transport mechanism distinct from that of TAP. It is not yet clear whether the TAP:tapasin complex directly loads peptides onto MHC class I molecules or whether binding to the TAP complex merely allows the MHC class I molecule to scan the transported peptides before they diffuse through the lumen of the endoplasmic reticulum and are transported back into the cytosol.

In cells with mutant TAP genes, the MHC class I molecules in the endoplasmic reticulum are unstable and are eventually translocated back into the cytosol, where they are degraded. Thus, the MHC class I molecule must bind a peptide to complete its folding and be transported onward from the endoplasmic reticulum. In uninfected cells, peptides derived from self proteins fill the peptide-binding cleft of the mature MHC class I molecules and are carried to the cell surface. In normal cells, MHC class I molecules are retained in the endoplasmic reticulum for some time, which suggests that they are present in excess of peptide. This is very important for the function of MHC class I molecules because they must be immediately available to transport viral peptides to the cell surface if the cell becomes infected. When a cell is infected by a virus, the presence of excess MHC class I molecules in the endoplasmic reticulum allows the rapid appearance of pathogen-derived peptides at the cell surface.

Because the presentation of viral peptides by MHC class I molecules signals CD8 T cells to kill the infected cell, some viruses have evolved ways of evading recognition by preventing the appearance of peptide:MHC class I complexes at the cell surface. The herpes simplex virus prevents the transport of viral peptides into the endoplasmic reticulum by producing a protein that binds to and inhibits TAP. Adenoviruses, on the other hand, encode a protein that binds to MHC class I molecules and retains them in the endoplasmic reticulum. Cytomegalovirus accelerates the retrograde translocation of MHC class I molecules back into the cytosol of the cell, where they are degraded. The advantage to a virus of blocking the recognition of infected cells is so great that it would not be surprising if other steps in the formation of MHC:peptide complexes, for example, the association of the MHC class I:chaperone complex with TAP, were found to be inhibited by some viruses.

5-5   Peptides presented by MHC class II molecules are generated in acidified endocytic vesicles.

Several classes of pathogen, including the protozoan parasite *Leishmania* and the mycobacteria that cause leprosy and tuberculosis, replicate inside intracellular vesicles in macrophages. Because they reside in membrane-bounded vesicles, the proteins of these pathogens are not accessible to proteasomes in the cytosol. Instead, after activation of the macrophage, proteins in vesicles are degraded by proteases within the vesicles into peptide fragments that bind to MHC class II molecules. In this way they are delivered to the cell surface where they can be recognized by CD4 T cells. Extracellular pathogens and proteins that are internalized into endocytic vesicles are also processed by this pathway and their peptides are presented to CD4 T cells (Fig. 5.6).

Most of what we know about protein processing in the endocytic pathway has come from experiments in which simple proteins are fed to macrophages and are taken up by endocytosis; in this way the processing of added antigen can be quantified. Proteins that bind to surface immunoglobulin on B cells and are internalized by receptor-mediated endocytosis are processed by the

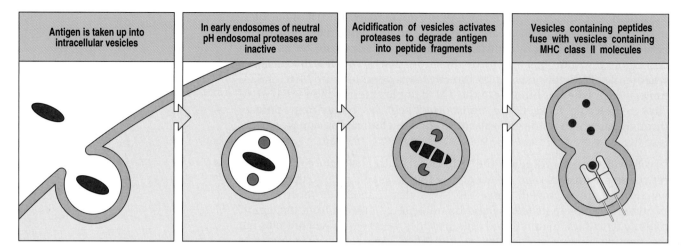

| Antigen is taken up into intracellular vesicles | In early endosomes of neutral pH endosomal proteases are inactive | Acidification of vesicles activates proteases to degrade antigen into peptide fragments | Vesicles containing peptides fuse with vesicles containing MHC class II molecules |

Fig. 5.6 Peptides that bind to MHC class II molecules are generated in acidified endocytic vesicles. In the case illustrated here, extracellular foreign antigens, such as bacteria or bacterial antigens, have been taken up by an antigen-presenting cell such as a macrophage or immature dendritic cell. In other cases, the source of the peptide antigen may be bacteria or parasites that have invaded the cell to replicate in intracellular vesicles. In both cases, the antigen processing pathway is the same. The pH of the endosomes containing the engulfed pathogens progressively decreases, activating proteases that reside within the vesicles to degrade the engulfed material. At some point on their pathway to the cell surface, newly synthesized MHC class II molecules pass through such acidified vesicles and bind peptide fragments of the antigen, transporting the peptides to the cell surface.

same pathway. Proteins that enter cells through endocytosis become enclosed in **endosomes**, which become increasingly acidic as they progress into the interior of the cell, eventually fusing with lysosomes. The endosomes and lysosomes contain proteases, known as acid proteases, which are activated at low pH and eventually degrade the protein antigens contained in the vesicles. Larger particulate material internalized by phagocytosis or macropinocytosis can also be handled by this pathway of antigen processing.

Drugs, such as chloroquine, that raise the pH of endosomes, making them less acid, inhibit the presentation of antigens that enter the cell in this way, suggesting that acid proteases are responsible for the processing of internalized antigen. Among these acid proteases are the cysteine proteases cathepsins B, D, S, and L, the last of which is the most active enzyme in this family. Antigen processing can be mimicked to some extent by digestion of proteins with these enzymes *in vitro* at acid pH. Cathepsins S and L may be the predominant proteases involved in the processing of vesicular antigens; mice that lack cathepsin B or cathepsin D show normal antigen processing, whereas mice with no cathepsin S cannot process antigen.

5-6 The invariant chain directs newly synthesized MHC class II molecules to acidified intracellular vesicles.

The function of MHC class II molecules is to bind peptides generated in the intracellular vesicles of macrophages, immature dendritic cells, B cells, and other antigen-presenting cells and to present these peptides to CD4 T cells. However, the biosynthetic pathway for MHC class II molecules, like that of other cell-surface glycoproteins, starts with their translocation into the endoplasmic reticulum, and they must therefore be prevented from binding prematurely to peptides transported into the endoplasmic reticulum lumen or to the cell's own newly synthesized polypeptides. As the endoplasmic reticulum is richly endowed with unfolded and partially folded polypeptide chains, a general mechanism is needed to prevent their binding in the open-ended MHC class II peptide-binding groove.

Binding is prevented by the assembly of newly synthesized MHC class II molecules with a protein known as the MHC class II-associated **invariant chain (Ii)**. The invariant chain forms trimers, with each subunit binding non-covalently to an MHC class II α:β heterodimer (Fig. 5.7). Ii binds to the MHC class II molecule with part of its polypeptide chain lying within the peptide-binding groove, thus blocking the groove and preventing the binding of either peptides or partially folded proteins. While this complex is being assembled in the endoplasmic reticulum, its component parts are associated with calnexin. Only when assembly is completed to produce a nine-chain complex is the complex released from calnexin for transport out of the endoplasmic reticulum. When it is part of the nine-chain complex, the MHC class II molecule cannot bind peptides or unfolded proteins, so that peptides present in the endoplasmic reticulum are not usually presented by MHC class II molecules. There is evidence that in the absence of invariant chains many MHC class II molecules are retained in the endoplasmic reticulum as complexes with misfolded proteins.

The invariant chain has a second function, which is to target delivery of the MHC class II molecules to a low-pH endosomal compartment where peptide loading can occur. The complex of MHC class II α:β heterodimers with invariant chain is retained for 2–4 hours in this compartment. During this time, the invariant chain is cleaved by acid proteases such as cathepsin S in several steps, as shown in Fig. 5.7. The initial cleavage events generate a truncated form of the invariant chain that remains bound to the MHC class II molecule and retains it within the proteolytic compartment. A subsequent cleavage releases the MHC class II molecule from the membrane-associated fragment of Ii, leaving a short fragment of Ii, called **CLIP** (for *cl*ass II-associated *i*nvariant-chain *p*eptide) still bound to the MHC class II molecule. MHC class II molecules associated with CLIP still cannot bind other peptides. CLIP must either dissociate or be displaced to allow a peptide to bind to the MHC molecule and enable the complex to be delivered to the cell surface. Cathepsin S cleaves Ii in most class II-positive cells, including antigen-presenting cells, whereas cathepsin L appears to substitute for cathepsin S in thymic cortical epithelial cells.

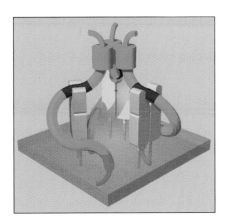
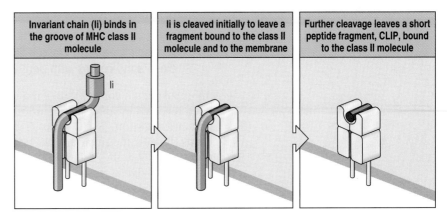

Invariant chain (Ii) binds in the groove of MHC class II molecule

Ii is cleaved initially to leave a fragment bound to the class II molecule and to the membrane

Further cleavage leaves a short peptide fragment, CLIP, bound to the class II molecule

Fig. 5.7 The invariant chain is cleaved to leave a peptide fragment, CLIP, bound to the MHC class II molecule. A model of the trimeric invariant chain bound to MHC class II α:β heterodimers is shown on the left. The CLIP portion is shown in red, the rest of the invariant chain in green, and the MHC class II molecule in yellow. In the endoplasmic reticulum, the invariant chain (Ii) binds to MHC class II molecules with the CLIP section of its polypeptide chain lying along the peptide-binding groove (model and left of three panels). After transport into an acidified vesicle, Ii is cleaved, initially just at one side of the class II molecule (center panel). The remaining portion of Ii (known as the leupeptin-induced peptide or LIP fragment) retains the transmembrane and cytoplasmic segments that contain the signals that target Ii:MHC class II complexes to the endosomal pathway. Subsequent cleavage (right panel) of LIP leaves only a short peptide still bound by the class II molecule; this peptide is the CLIP fragment. Model structure courtesy of P. Cresswell.

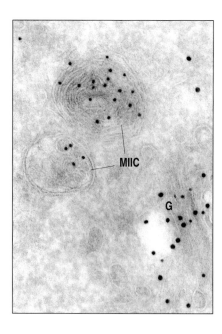

Fig. 5.8 MHC class II molecules are loaded with peptide in a specialized intracellular compartment. MHC class II molecules are transported from the Golgi apparatus (labeled G in this electron micrograph of an ultrathin section of a B cell) to the cell surface via specialized intracellular vesicles called the MHC class II compartment (MIIC). These have a complex morphology, showing internal vesicles and sheets of membrane. Antibodies labeled with different-sized gold particles identify the presence of both MHC class II molecules (small gold particles) and the invariant chain (large gold particles) in the Golgi, whereas only MHC class II molecules are detectable in the MIIC. This compartment is thus thought to be the one in which invariant chain is cleaved and peptide loading occurs. Magnification × 135,000. Photograph courtesy of Hans J. Geuze.

The endosomal compartment in which invariant chain is cleaved and MHC class II molecules encounter peptide is not yet clearly defined. Most newly synthesized MHC class II molecules are brought toward the cell surface in vesicles, which at some point fuse with incoming endosomes. However, there is also evidence that some MHC class II:Ii complexes are first transported to the cell surface and then re-internalized into endosomes. In either case, MHC class II:Ii complexes enter the endosomal pathway and there encounter and bind pathogen-derived peptides. Immunoelectron-microscopy using antibodies tagged with gold particles to localize Ii and MHC class II molecules within cells suggests that Ii is cleaved and peptides bind to MHC class II in a particular endosomal compartment called the **MIIC** (*MHC class II compartment*), late in the endosomal pathway (Fig. 5.8).

As with MHC class I molecules, MHC class II molecules in uninfected cells bind peptides derived from self proteins. MHC class II molecules that do not bind peptide after dissociation from the invariant chain are unstable; in the acidic pH of the endosome they aggregate and are rapidly degraded. It is therefore not surprising that peptides derived from MHC class II molecules form a large proportion of the self peptides presented by MHC class II molecules in normal uninfected cells. This suggests that, as for MHC class I molecules, MHC class II molecules are produced in excess. Thus, when a cell is infected by pathogens that proliferate in intracellular vesicles, when a phagocyte engulfs a pathogen, or when pathogen-derived proteins are internalized by B cells, the peptides generated from the pathogen proteins find plentiful empty MHC class II molecules to bind.

5-7   A specialized MHC class II-like molecule catalyzes loading of MHC class II molecules with peptides.

An unsuspected component of the vesicular antigen-processing pathway has been revealed by analysis of mutant human B-cell lines with a defect in antigen presentation. MHC class II molecules in these cell lines assemble correctly with the invariant chain and seem to follow the normal vesicular route. However, they fail to bind peptides derived from internalized proteins and often arrive at the cell surface with the CLIP peptide still bound.

The defect in these mutant cells lies in an MHC class II-like molecule called **HLA-DM** in humans (H-2M in mice). The HLA-DM genes are found near the TAP and LMP genes in the class II region of the MHC (see Fig. 5.10); they encode an α chain and a β chain that closely resemble those of other MHC class II molecules. The HLA-DM molecule is not expressed at the cell surface, however, but is found predominantly in the MIIC compartment and does not appear to require peptide for stabilization. HLA-DM binds to and stabilizes empty MHC class II molecules that would otherwise aggregate; in addition, it catalyzes both the release of the CLIP fragment from MHC class II:CLIP complexes and the binding of other peptides to the empty MHC class II molecule (Fig. 5.9).

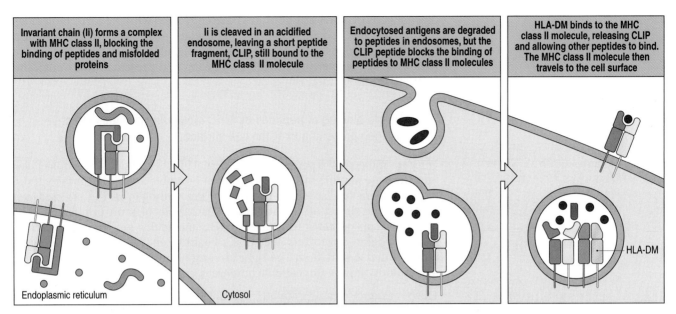

| Invariant chain (Ii) forms a complex with MHC class II, blocking the binding of peptides and misfolded proteins | Ii is cleaved in an acidified endosome, leaving a short peptide fragment, CLIP, still bound to the MHC class II molecule | Endocytosed antigens are degraded to peptides in endosomes, but the CLIP peptide blocks the binding of peptides to MHC class II molecules | HLA-DM binds to the MHC class II molecule, releasing CLIP and allowing other peptides to bind. The MHC class II molecule then travels to the cell surface |

Fig. 5.9 HLA-DM facilitates the loading of antigenic peptides onto class II molecules. The invariant chain binds to newly synthesized MHC class II molecules and blocks the binding of peptides and unfolded proteins in the endoplasmic reticulum and during the transport of the MHC class II molecule into acidified endocytic vesicles (first panel). In such vesicles, proteases cleave the invariant chain, leaving the CLIP peptide bound to the MHC class II molecule (second panel). Pathogens and their proteins are broken down into peptides within acidified endocytic vesicles, but these peptides cannot bind to MHC class II molecules that are occupied by CLIP (third panel). The class II-like molecule, HLA-DM, binds to MHC class II:CLIP complexes, catalyzing the release of CLIP and the binding of antigenic peptides (fourth panel).

HLA-DM also catalyzes the release of unstably bound peptides from MHC class II molecules. In the presence of a mixture of peptides capable of binding to MHC class II molecules, as occurs in the MIIC, HLA-DM will continuously bind and rebind to peptide:MHC class II complexes, removing weakly bound peptides and allowing other peptides to replace them. Antigens presented by MHC class II molecules may have to persist on the surface of antigen-presenting cells for some days before encountering T cells able to recognize them. The ability of HLA-DM to remove unstably bound peptides, sometimes called 'peptide editing,' ensures that the peptide:MHC class II complexes displayed on the surface of the antigen-presenting cell will survive long enough to stimulate the appropriate CD4 cells.

A second atypical MHC class II molecule, called HLA-DO (H-2O in mice), is produced in thymic epithelial cells and B cells. This molecule is a heterodimer of the HLA-DNα chain and the HLA-DOβ chain (see Fig. 5.10). HLA-DO is not present at the cell surface, being found only in intracellular vesicles, and it does not appear to bind peptides. Instead, it acts as a negative regulator of HLA-DM, binding to it and inhibiting both the HLA-DM-catalyzed release of CLIP from, and the binding of other peptides to, MHC class II molecules. Expression of the HLA-DOβ chain is not increased by interferon-γ (IFN-γ), whereas the expression of HLA-DM is. Thus, during inflammatory responses, in which IFN-γ is produced by T cells and NK cells (see Section 2-25), the increased expression of HLA-DM is able to overcome the inhibitory effects of HLA-DO. Why the antigen-presenting ability of thymic epithelial cells and of B cells should be regulated in this way is not known; in thymic epithelial cells the purpose may be to select developing CD4 T cells by using a repertoire of self peptides different from those to which they will be exposed as mature T cells.

The role of the HLA-DM molecule in facilitating the binding of peptides to MHC class II molecules parallels that of the TAP molecules in facilitating peptide binding to MHC class I molecules. Thus it seems likely that specialized mechanisms of delivering peptides have coevolved with the MHC molecules themselves. It is also likely that pathogens have evolved strategies to inhibit loading of peptides onto MHC class II molecules, much as viruses have found ways of subverting antigen processing and presentation through the MHC class I molecules.

5-8 Stable binding of peptides by MHC molecules provides effective antigen presentation at the cell surface.

For MHC molecules to perform their essential function of signaling intracellular infection, the peptide:MHC complex must be stable at the cell surface. If the complex were to dissociate too readily, the pathogen in the infected cell could escape detection. In addition, MHC molecules on uninfected cells could pick up peptides released by MHC molecules on infected cells and falsely signal to cytotoxic T cells that a healthy cell is infected, triggering its unwarranted destruction. The tight binding of peptides by MHC molecules makes both of these undesirable outcomes unlikely.

The persistence of a peptide:MHC complex on a cell can be measured by its ability to stimulate T cells, while the fate of the MHC molecules themselves can be directly followed by specific staining. In this way it can be shown that specific peptide:MHC complexes on living cells are lost from the surface and re-internalized as part of natural protein turnover at the same rate as the MHC molecules themselves, indicating that peptide binding is essentially irreversible. This stable binding enables even rare peptides to be transported efficiently to the cell surface by MHC molecules, and allows long-term display of these complexes on the surface of the infected cell. This fulfills the first of the requirements for effective antigen presentation.

The second requirement is that if a peptide should dissociate from a cell-surface MHC molecule, peptides from the surrounding extracellular fluid would not be able to bind to the empty peptide-binding site. In fact, removal of the peptide from a purified MHC class I molecule has been shown to require denaturation of the protein. When peptide dissociates from an MHC class I molecule at the surface of a living cell, the molecule changes conformation, the β_2-microglobulin dissociates, and the α chain is internalized and rapidly degraded. Thus, most empty MHC class I molecules are quickly lost from the cell surface.

At neutral pH, empty MHC class II molecules are more stable than empty MHC class I molecules, yet empty MHC class II molecules are also removed from the cell surface. Empty MHC class II molecules aggregate readily, and internalization of such aggregates may account for the loss of empty MHC class II molecules from the surface of cells. Moreover, peptide loss from MHC class II molecules is most likely when the molecules transit through acidified endosomes as part of the normal process of cell membrane recycling. At acidic pH, MHC class II molecules are able to bind peptides that are present in the vesicles, but those that fail to do so are rapidly degraded.

Thus, both MHC class I and class II molecules are effectively prevented from acquiring peptides from the surrounding extracellular fluid. This ensures that T cells act selectively on infected cells or on cells specialized for antigen uptake and display, while sparing surrounding healthy cells.

Summary.

The most distinctive feature of antigen recognition by T cells is the form of the ligand recognized by the T-cell receptor. This comprises a peptide derived from the foreign antigen and bound to an MHC molecule. MHC molecules are cell-surface glycoproteins with a peptide-binding groove that can bind a wide variety of different peptides. The MHC molecule binds the peptide in an intracellular location and delivers it to the cell surface, where the combined ligand can be recognized by a T cell. There are two classes of MHC molecule, MHC class I and MHC class II, which bind peptides from proteins degraded in different intracellular sites. MHC class I molecules bind peptides from proteins degraded in the cytosol, where a multicatalytic protease, the proteasome, degrades both cytosolic proteins and proteins that have been transported back into the cytosol from the endoplasmic reticulum. Peptides produced by proteasomes are transported into the endoplasmic reticulum by a heterodimeric ATP-binding protein called TAP, and are then available for binding by partially folded MHC class I molecules that are held tethered to TAP. Peptide binding is an integral part of MHC class I assembly, and must occur before the MHC class I molecule can complete its folding and leave the endoplasmic reticulum for the cell surface. By binding stably to peptides from proteins degraded in the cytosol, MHC class I molecules on the surface of cells infected with viruses or other cytosolic pathogens display peptides from these pathogens. In contrast, MHC class II molecules are prevented from binding to peptides in the endoplasmic reticulum by their early association with the invariant chain, which fills and blocks their peptide-binding groove. They are targeted by the invariant chain to an acidic endosomal compartment where, in the presence of active proteases, in particular cathepsin S, and with the help of a specialized MHC class II-like molecule, which catalyzes peptide loading, the invariant chain is released and other peptides are bound. MHC class II molecules thus bind peptides from proteins that are degraded in endosomes. They capture peptides from pathogens that enter the vesicular system of macrophages, or from antigens internalized by immature dendritic cells or the immunoglobulin receptors of B cells. Different types of T cell are activated on recognizing foreign peptides presented by the different classes of MHC molecule. The CD8 T cells that recognize MHC class I:peptide complexes are specialized to kill any cells displaying foreign peptides and so rid the body of cells infected with viruses and other cytosolic pathogens. The CD4 T cells that recognize MHC class II:peptide complexes are specialized to activate other effector cells of the immune system; macrophages, for example, are activated to kill the intravesicular pathogens they harbor and B cells to secrete immunoglobulins against foreign molecules. Thus, the two classes of MHC molecule deliver peptides from different cellular compartments to the cell surface, where they are recognized by T cells mediating distinct and appropriate effector functions.

The major histocompatibility complex and its functions.

The function of MHC molecules is to bind peptide fragments derived from pathogens and display them on the cell surface for recognition by the appropriate T cells. The consequences are almost always deleterious to the pathogen—virus-infected cells are killed, macrophages are activated to kill bacteria living in their intracellular vesicles, and B cells are activated to produce antibodies that eliminate or neutralize extracellular pathogens. Thus, there is strong selective pressure in favor of any pathogen that has mutated in such a way that it escapes presentation by an MHC molecule.

Two separate properties of the MHC make it difficult for pathogens to evade immune responses in this way. First, the MHC is **polygenic**: it contains several different MHC class I and MHC class II genes, so that every individual possesses a set of MHC molecules with different ranges of peptide-binding specificities. Second, the MHC is highly **polymorphic**; that is, there are multiple variants of each gene within the population as a whole. The MHC genes are, in fact, the most polymorphic genes known. In this section, we will describe the organization of the genes in the MHC and discuss how the variation in MHC molecules arises. We will also see how the effect of polygeny and polymorphism on the range of peptides that can be bound contributes to the ability of the immune system to respond to the multitude of different and rapidly evolving pathogens.

5-9 Many proteins involved in antigen processing and presentation are encoded by genes within the major histocompatibility complex.

The major histocompatibility complex is located on chromosome 6 in humans and chromosome 17 in the mouse and extends over some 4 centimorgans of DNA, about 4×10^6 base pairs. In humans it contains more than 200 genes. As work continues to define the genes within and around the MHC, both its extent and the number of genes are likely to grow; in fact, recent studies suggest that the MHC may span at least 7×10^6 base pairs. The genes encoding the α chains of MHC class I molecules and the α and β chains of MHC class II molecules are linked within the complex; the genes for β₂-microglobulin and the invariant chain are on different chromosomes (chromosomes 15 and 5, respectively, in humans and chromosomes 2 and 18 in the mouse). Figure 5.10 shows the general organization of the MHC class I and II genes in human and mouse. In humans these genes are called *H*uman *L*eukocyte *A*ntigen or **HLA** genes, as they were first discovered through antigenic differences between white blood cells from different individuals; in the mouse they are known as the **H-2** genes.

There are three class I α-chain genes in humans, called HLA-A, -B, and -C. There are also three pairs of MHC class II α- and β-chain genes, called HLA-DR, -DP, and -DQ. However, in many cases the HLA-DR cluster contains an extra β-chain gene whose product can pair with the DRα chain. This means that the three sets of genes can give rise to four types of MHC class II molecule. All the MHC class I and class II molecules can present peptides to

Fig. 5.10 The genetic organization of the major histocompatibility complex (MHC) in human and mouse. The organization of the principal MHC genes is shown for both humans (where the MHC is called HLA and is on chromosome 6) and mice (in which the MHC is called H-2 and is on chromosome 17). The organization of the MHC genes is similar in both species. There are separate clusters of MHC class I genes (shown in red) and MHC class II genes (shown in yellow), although in the mouse an MHC class I gene (H-2K) appears to have translocated relative to the human MHC so that the class I region in mice is split in two. In both species there are three main class I genes, which are called HLA-A, -B, and -C in humans, and H2-K, -D, and -L in the mouse. The gene for β₂-microglobulin, although it encodes part of the MHC class I molecule, is located on a different chromosome, chromosome 15 in humans and chromosome 2 in the mouse. The class II region includes the genes for the α and β chains of the antigen-presenting MHC class II molecules HLA-DR, -DP, and -DQ (H-2A and -E in the mouse). In addition, the genes for the TAP1:TAP2 peptide transporter, the LMP genes that encode proteasome subunits, the genes encoding the DMα and DMβ chains, the genes encoding the α and β chains of the DO molecule (DNα and DOβ, respectively), and the gene for tapasin (TAPBP) are also in the MHC class II region. The so-called class III genes encode various other proteins with functions in immunity (see Fig. 5.11).

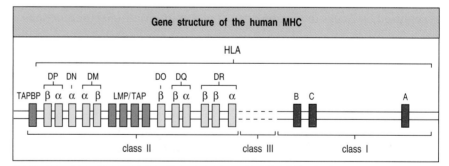

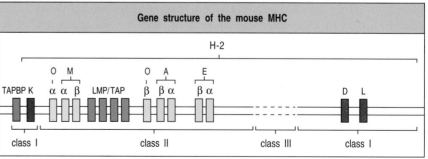

T cells, but each protein binds a different range of peptides (see Sections 3-16 and 3-17). Thus, the presence of several different genes of each MHC class means that any one individual is equipped to present a much broader range of peptides than if only one MHC molecule of each class were expressed at the cell surface.

The two TAP genes lie in the MHC class II region, in close association with the LMP genes that encode components of the proteasome, whereas the gene for tapasin, which binds to both TAP and empty MHC class I molecules, lies at the edge of the MHC nearest the centromere (see Fig. 5.10). The genetic linkage of the MHC class I genes, whose products deliver cytosolic peptides to the cell surface, with the TAP, tapasin, and proteasome genes, which encode the molecules that generate peptides in the cytosol and transport them into the endoplasmic reticulum, suggests that the entire MHC has been selected during evolution for antigen processing and presentation.

When cells are treated with the interferons IFN-α, -β, or -γ, there is a marked increase in transcription of MHC class I α-chain and β_2-microglobulin genes, and of the proteasome, tapasin, and TAP genes. Interferons are produced early in viral infections as part of the innate immune response, as described in Chapter 2, and so this effect increases the ability of cells to process viral proteins and present the resulting peptides at the cell surface. This helps to activate the appropriate T cells and initiate the adaptive immune response in response to the virus. The coordinated regulation of the genes encoding these components may be facilitated by the linkage of many of them in the MHC.

The HLA-DM genes, which encode the DM molecule whose function is to catalyze peptide binding to MHC class II molecules (see Section 5-7), are clearly related to the MHC class II genes. The DNα and DOβ genes, which encode the DO molecule, a negative regulator of DM, are also clearly related to the MHC class II genes. The classical MHC class II genes, along with the invariant-chain gene and the genes for DMα, β, and DNα, but not DOβ, are coordinately regulated. This distinct regulation of MHC class II genes by IFN-γ, which is made by activated T cells of T_H1 type as well as by activated CD8 and NK cells, allows T cells responding to bacterial infections to upregulate those molecules concerned in the processing and presentation of intravesicular antigens. Expression of all of these molecules is induced by IFN-γ (but not by IFN-α or -β), via the production of a transcriptional activator known as **MHC class II transactivator** (**CIITA**). An absence of CIITA causes severe immunodeficiency due to nonproduction of MHC class II molecules.

MHC Class II
Deficiency

5-10 A variety of genes with specialized functions in immunity are also encoded in the MHC.

Although the most important known function of the gene products of the MHC is the processing and presentation of antigens to T cells, many other genes map within this region; some of these are known to have other roles in the immune system, but many have yet to be characterized functionally. Figure 5.11 shows the detailed organization of the human MHC.

In addition to the highly polymorphic 'classical' MHC class I and class II genes, there are many genes encoding MHC class I-type molecules that show little polymorphism; most of these have yet to be assigned a function. They are linked to the class I region of the MHC and their exact number varies greatly between species and even between members of the same species. These genes have been termed **MHC class IB** genes; like MHC class I genes, they encode β_2-microglobulin-associated cell-surface molecules. Their expression on cells is variable, both in the amount expressed at the cell surface and in the tissue distribution.

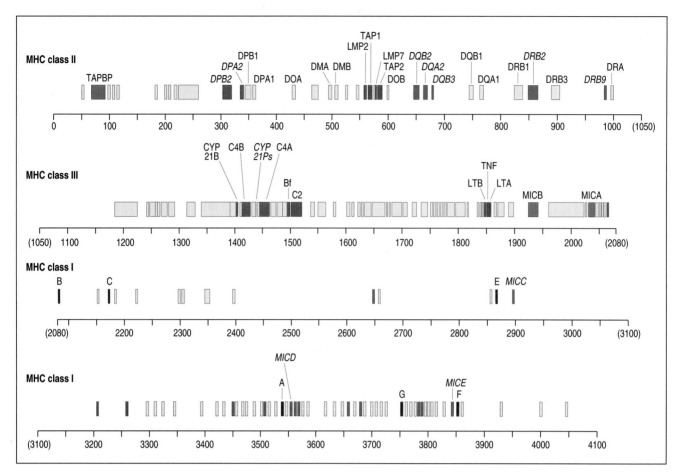

Fig. 5.11 Detailed map of the human MHC. The organization of the class I, class II, and class III regions of the human MHC are shown, with approximate genetic distances given in thousands of base pairs (kb). Most of the genes in the class I and class II regions are mentioned in the text. The additional genes indicated in the class I region (for example, E, F, and G) are class I-like genes, encoding class IB molecules; the additional class II genes are pseudogenes. The genes shown in the class III region encode the complement proteins C4 (two genes, shown as C4A and C4B), C2 and factor B (shown as Bf) as well as genes that encode the cytokines tumor necrosis factor-α (TNF) and lymphotoxin (LTA, LTB). Closely linked to the C4 genes is the gene encoding 21-hydroxylase (shown as CYP 21B), an enzyme involved in steroid synthesis. Genes shown in gray and named in italic are pseudogenes. The genes are colour coded, with the MHC class I genes being shown in red, except for the MIC genes, which are shown in blue; these are distinct from the other class I-like genes and are under different transcriptional control. The MHC class II genes are shown in yellow. Genes in the MHC region which have immune functions but are not related to the MHC class I and class II genes are shown in purple.

One of the mouse MHC class IB molecules, H2-M3, can present peptides with *N*-formylated amino termini, which is of interest because all bacteria initiate protein synthesis with *N*-formylmethionine. Cells infected with cytosolic bacteria can be killed by CD8 T cells that recognize *N*-formylated bacterial peptides bound to H2-M3. Whether an equivalent MHC class IB molecule exists in humans is not known. The large number of MHC class IB genes (50 or more in the mouse) means that many different MHC class IB molecules can exist in a single animal. They may, like the protein that presents *N*-formylmethionyl peptides, have specialized roles in antigen presentation. Some of them are known to be recognized by NK cell receptors, as we discuss further below.

Yet other MHC class IB genes have functions unrelated to the immune system. The HFe gene lies some 3×10^6 base pairs from HLA-A. Its product is expressed on cells in the intestinal tract, and has a function in iron metabolism, regulating the uptake of iron into the body, most likely through interactions with the transferrin receptor. Individuals defective for this gene have an iron-storage disease, hemochromatosis, in which an abnormally high level of iron is retained in the liver and other organs. Mice lacking β_2-microglobulin, and hence defective in the expression of all class I molecules, show a similar iron overload. Exactly how this gene product regulates the levels of iron within the body is not known, but it is unlikely to involve an immunological mechanism.

The other genes that map within the MHC include some that encode complement components (for example, C2, C4, and factor B) and some that encode cytokines—for example, tumor necrosis factor-α (TNF-α) and lymphotoxin (TNF-β)—all of which have important functions in immunity. These have been termed MHC class III genes, and are shown in Fig. 5.11. The functions of these genes are discussed in Chapters 2 and 9.

Many studies have established associations between susceptibility to certain diseases and particular alleles of MHC genes, and we now have considerable insight into how polymorphism in the classical MHC class I and class II genes can affect resistance or susceptibility. But although most of these MHC-influenced diseases are known or suspected to have an immune etiology, this is not true of all of them, and it is important to remember that there are many genes lying within the MHC that have no known or suspected immunological function. One of these is the enzyme 21-hydroxylase which, when deficient, causes congenital adrenal hyperplasia and, in severe cases, salt-wasting syndrome. Even where a disease-related gene is clearly homologous to immune system genes, as is the case with HFe, the disease mechanism may not be immune related. Disease associations mapping to the MHC must therefore be interpreted with caution, in the light of a detailed understanding of its genetic structure and the functions of its individual genes. Much remains to be learned about the latter and about the significance of all the genetic variation localized within the MHC. For instance, the C4 genes are highly polymorphic, but the adaptive significance of this genetic variability is not well understood.

5-11 Specialized MHC class I molecules act as ligands for activation and inhibition of NK cells.

Some class IB genes, for example the members of the **MIC** gene family, are under a different regulatory control from the classical MHC class I genes and are induced in response to cellular stress (such as heat shock). There are five MIC genes, but only two—*MICA* and *MICB*—are expressed and produce protein products. They are expressed in fibroblasts and epithelial cells, particularly in intestinal epithelial cells, and may play a part in innate immunity or in the induction of immune responses in circumstances where interferons are not produced. The MICA and MICB molecules are recognized by a receptor that is present on NK cells, $\gamma{:}\delta$ T cells and some CD8 T cells and is capable of activating these cells to kill MIC-expressing targets. The MIC receptor is composed of two chains. One is NKG2D, an 'activating' member of the NKG2 family of NK-cell receptors whose cytoplasmic domain lacks an inhibitory sequence motif found in other members of this family that act as inhibitory receptors (see Section 2-28); the other is a protein called DAP10, which transmits the signal into the interior of the cell by interacting with and activating intracellular protein tyrosine kinases.

Other MHC class IB molecules may inhibit cell killing by NK cells, as described in Chapter 2. Such a role has been suggested for the MHC class IB molecule HLA-G, which is expressed on fetus-derived placental cells that migrate into the uterine wall. These cells express no classical MHC class I molecules and cannot be recognized by CD8 T cells but, unlike other cells lacking classical MHC class I molecules, they are not killed by NK cells. This appears to be because HLA-G is recognized by an inhibitory receptor, ILT-2, on the NK cell, which prevents the NK cell killing the placental cell.

Another MHC class IB molecule, HLA-E, also has a specialized role in cell recognition by NK cells. HLA-E binds a very restricted subset of peptides, derived from the leader peptides of other HLA class I molecules. These peptide:HLA-E complexes can bind to the receptor NKG2A, which is present on NK cells in a complex with the cell-surface molecule CD94. NKG2A is an inhibitory member of the NKG2 family, and on stimulation inhibits the cytotoxic activity of the NK cell. Thus a cell that expresses either HLA-E or HLA-G is not killed by NK cells.

5-12   The protein products of MHC class I and class II genes are highly polymorphic.

Because of the polygeny of the MHC, every person will express at least three different antigen-presenting MHC class I molecules and three (or sometimes four) MHC class II molecules on his or her cells (see Section 5-9). In fact, the number of different MHC molecules expressed on the cells of most people is greater because of the extreme polymorphism of the MHC and the co-dominant expression of MHC gene products.

The term **polymorphism** comes from the Greek *poly*, meaning many, and *morphe*, meaning shape or structure. As used here, it means within-species variation at a gene locus, and thus in its protein product; the variant genes that can occupy the locus are termed **alleles**. There are more than 200 alleles of some human MHC class I and class II genes (Fig. 5.12), each allele being present at a relatively high frequency in the population. So there is only a small chance that the corresponding MHC locus on both the homologous chromosomes of an individual will have the same allele; most individuals will be **heterozygous** at MHC loci. The particular combination of MHC alleles found on a single chromosome is known as an **MHC haplotype**. Expression of MHC alleles is codominant, with the protein products of both the alleles at a locus being expressed in the cell, and both gene products being able to present antigens to T cells (Fig. 5.13). The extensive polymorphism at each locus thus has the potential to double the number of different MHC molecules expressed in an individual and thereby increases the diversity already available through polygeny (Fig. 5.14).

Fig. 5.12 Human MHC genes are highly polymorphic. With the notable exception of the DRα locus, which is functionally monomorphic, each locus has many alleles. The number of different alleles is shown in this figure by the height of the bars. The figures are the numbers of HLA alleles currently officially assigned by the WHO Nomenclature Committee for Factors of the HLA System as of August 2000.

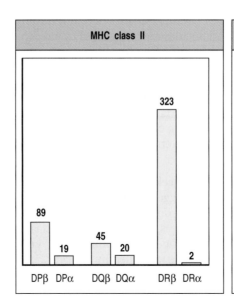
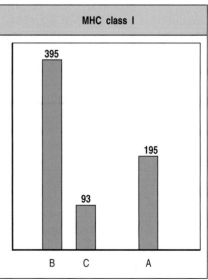

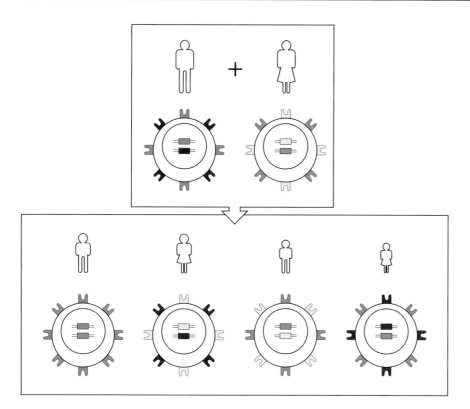

Fig. 5.13 Expression of MHC alleles is codominant. The MHC is so polymorphic that most individuals are likely to be heterozygous at each locus. Alleles are expressed from both MHC haplotypes in any one individual, and the products of all alleles are found on all expressing cells. In any mating, four possible combinations of haplotypes can be found in the offspring; thus siblings are also likely to differ in the MHC alleles they express, there being one chance in four that an individual will share both haplotypes with a sibling. One consequence of this is the difficulty of finding suitable donors for tissue transplantation.

Thus, with three MHC class I genes and a possible four sets of MHC class II genes on each chromosome 6, a human typically expresses six different MHC class I molecules and eight different MHC class II molecules on his or her cells. For the MHC class II genes, the number of different MHC molecules may be increased still further by the combination of α and β chains encoded by different chromosomes (so that two α chains and two β chains can give rise to four different proteins, for example). In mice it has been shown that not all combinations of α and β chains can form stable dimers and so, in practice, the exact number of different MHC class II molecules expressed depends on which alleles are present on each chromosome.

All MHC products are polymorphic to a greater or lesser extent, with the exception of the DRα chain and its mouse homologue Eα. These chains do not vary in sequence between different individuals and are said to be **monomorphic**. This might indicate a functional constraint that prevents variation in the DRα and Eα proteins, but no such special function has been found. Many mice, both domestic and wild, have a mutation in the Eα gene that prevents synthesis of the Eα protein. They thus lack cell-surface H-2E molecules, so if H-2E molecules do have a special function it is unlikely to be an essential one. All other MHC class I and class II genes are polymorphic.

Polymorphism

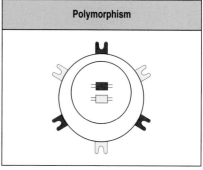

Polygeny

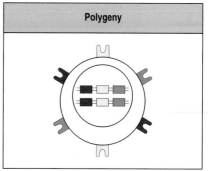

Polymorphism and polygeny

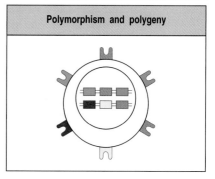

Fig. 5.14 Polymorphism and polygeny both contribute to the diversity of MHC molecules expressed by an individual. The high polymorphism of the classical MHC loci ensures a diversity in MHC gene expression in the population as a whole. However, no matter how polymorphic a gene, no individual can express more than two alleles at a single gene locus. Polygeny, the presence of several different related genes with similar functions, ensures that each individual produces a number of different MHC molecules. Polymorphism and polygeny combine to produce the diversity of MHC molecules seen both within an individual and in the population at large.

5-13 MHC polymorphism affects antigen recognition by T cells by influencing both peptide binding and the contacts between T-cell receptor and MHC molecule.

The products of individual MHC alleles can differ from one another by up to 20 amino acids, making each variant protein quite distinct. Most of the differences are localized to exposed surfaces of the outer domain of the molecule, and to the peptide-binding groove in particular (Fig. 5.15). The polymorphic residues that line the peptide-binding groove determine the peptide-binding properties of the different MHC molecules.

We have seen that peptides bind to MHC class I molecules through specific anchor residues (see Section 3-16), and that the amino acid side chains of these residues anchor the peptide by binding in pockets that line the peptide-binding groove. Polymorphism in MHC class I molecules affects which amino acids line these pockets and thus their binding specificity. In consequence, the anchor residues of peptides that bind to each allelic variant are different. The set of anchor residues that allows binding to a given MHC class I molecule is called a **sequence motif**. These sequence motifs make it possible to identify peptides within a protein that can potentially bind the appropriate MHC molecule, which may be very important in designing peptide vaccines. Different allelic variants of MHC class II molecules also bind different peptides,

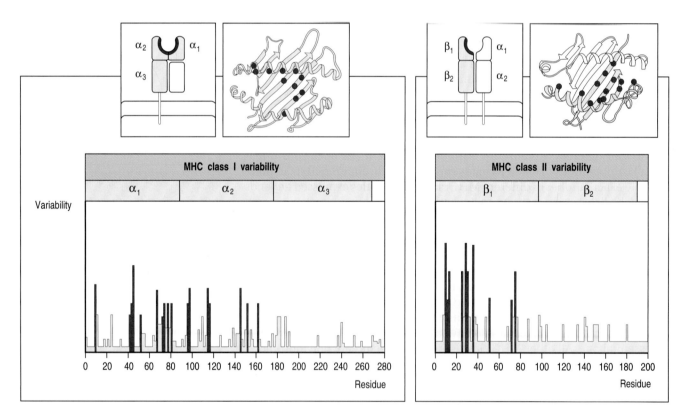

Fig. 5.15 Allelic variation occurs at specific sites within MHC molecules. Variability plots of the amino acid sequences of MHC molecules show that the variation arising from genetic polymorphism is restricted to the amino-terminal domains (α_1 and α_2 domains of class I molecules, and α_1 and β_1 domains of MHC class II molecules). These are the domains that form the peptide-binding cleft. Moreover, allelic variability is clustered in specific sites within the amino-terminal domains, lying in positions that line the peptide-binding cleft, either on the floor of the groove or directed inward from the walls. For the MHC class II molecule, the variability of the HLA-DR alleles is shown. For HLA-DR, and its homologues in other species, the α chain is essentially invariant and only the β chain shows significant polymorphism.

but the more open structure of the MHC class II peptide-binding groove and the greater length of the peptides bound in it allow greater flexibility in peptide binding (see Section 3-17). It is therefore more difficult to predict which peptides will bind to MHC class II molecules.

In rare cases, processing of a protein will not generate any peptides with a suitable motif for binding to any of the MHC molecules expressed by an individual. When this happens, the individual fails to respond to the antigen. Such failures in responsiveness to simple antigens were first reported in inbred animals, where they were called **immune response (Ir) gene defects**. These defects were identified and mapped to genes within the MHC long before the function of MHC molecules was understood. Indeed, they were the first clue to the antigen-presenting function of MHC molecules, although it was only much later that the 'Ir genes' were shown to encode MHC class II molecules. Ir gene defects are common in inbred strains of mice because the mice are homozygous at all their MHC loci and thus express only one type of MHC molecule from each gene locus. This limits the range of peptides they can present to T cells. Ordinarily, MHC polymorphism guarantees a sufficient number of different MHC molecules in a single individual to make this type of nonresponsiveness unlikely, even to relatively simple antigens such as small toxins. This has obvious importance for host defense.

Initially, the only evidence linking Ir gene defects to the MHC was genetic— mice of one MHC genotype could make antibody in response to a particular antigen, whereas mice of a different MHC genotype, but otherwise genetically identical, could not. The MHC genotype was somehow controlling the ability of the immune system to detect or respond to specific antigens, but it was not clear at the time that direct recognition of MHC molecules was involved.

Later experiments showed that the antigen specificity of T-cell recognition was controlled by MHC molecules. The immune responses affected by the Ir genes were known to be dependent on T cells, and this led to a series of experiments in mice to ascertain how MHC polymorphism might control T-cell responses. The earliest of these experiments showed that T cells could only be activated by macrophages or B cells that shared MHC alleles with the mouse in which the T cells originated. This was the first evidence that antigen recognition by T cells depends on the presence of specific MHC molecules in the antigen-presenting cell. The clearest example of this feature of T-cell recognition came, however, from studies of virus-specific cytotoxic T cells, for which Peter Doherty and Rolf Zinkernagel were awarded the Nobel Prize in 1996.

When mice are infected with a virus, they generate cytotoxic T cells that kill self cells infected with the virus, while sparing uninfected cells or cells infected with unrelated viruses. The cytotoxic T cells are thus virus-specific. A particularly striking outcome of these experiments was that the specificity of the cytotoxic T cells was also affected by the polymorphism of MHC molecules. Cytotoxic T cells induced by viral infection in mice of MHC genotype a (MHC^a) would kill any MHC^a cell infected with that virus but would not kill cells of MHC genotype b, or c, and so on, even if they were infected with the same virus. Because the MHC genotype restricts the antigen specificity of T cells, this effect is called **MHC restriction**. Together with the earlier studies on both B cells and macrophages, this work showed that MHC restriction is a critical feature of antigen recognition by all functional classes of T cells.

Because different MHC molecules bind different peptides, MHC restriction in immune responses to viruses and other complex antigens might be explained solely on this indirect basis. However, it can be seen from Fig. 5.15 that some of the polymorphic amino acids in MHC molecules are located in the α helices flanking the peptide-binding cleft in such a way that they are exposed on the outer surface of the peptide:MHC complex and can be directly

Fig. 5.16 T-cell recognition of antigens is MHC restricted. The antigen-specific T-cell receptor (TCR) recognizes a complex of antigenic peptide and MHC. One consequence of this is that a T cell specific for peptide x and a particular MHC allele, MHC^a (left panel), will not recognize the complex of peptide x with a different MHC allele, MHC^b (center panel), or the complex of peptide y with MHC^a (right panel). The co-recognition of peptide and MHC molecule is known as MHC restriction because the MHC molecule is said to restrict the ability of the T cell to recognize antigen. This restriction may either result from direct contact between MHC molecule and T-cell receptor or be an indirect effect of MHC polymorphism on the peptides that bind or on their bound conformation.

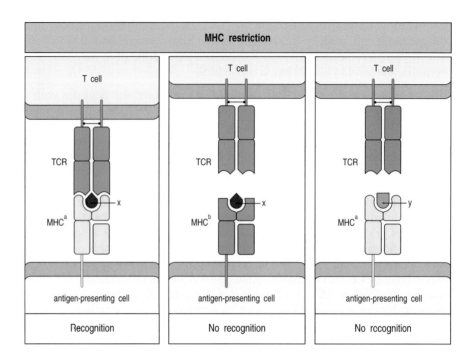

contacted by the T-cell receptor (see Fig. 3.27). It is therefore not surprising that when T cells are tested for their ability to recognize the same peptide bound to different MHC molecules, they readily distinguish the peptide bound to MHC^a from the same peptide bound to MHC^b. Thus, the specificity of a T-cell receptor is defined both by the peptide it recognizes and by the MHC molecule bound to it (Fig. 5.16). This restricted recognition may sometimes be caused by differences in the conformation of the bound peptide imposed by the different MHC molecules rather than by direct recognition of polymorphic amino acids in the MHC molecule itself. MHC restriction in antigen recognition therefore reflects the combined effect of differences in peptide binding and of direct contact between the MHC molecule and the T-cell receptor.

5-14   Nonself MHC molecules are recognized by 1–10% of T cells.

The discovery of MHC restriction, by revealing the physiological function of the MHC molecules, also helped explain the otherwise puzzling phenomenon of recognition of nonself MHC in the rejection of organs and tissues transplanted between members of the same species. Transplanted organs from donors bearing MHC molecules that differ from those of the recipient—even by as little as one amino acid—are invariably rejected. The rapid and very potent cell-mediated immune response to the transplanted tissue results from the presence in any individual of large numbers of T cells that are specifically reactive to nonself, or **allogeneic**, MHC molecules. Early studies on T-cell responses to allogeneic MHC molecules used the **mixed lymphocyte reaction**. In this reaction T cells from one individual are mixed with lymphocytes from a second individual. If the T cells of one individual recognize the other individual's MHC molecules as 'foreign,' the T cells will divide and proliferate. (The lymphocytes from the second individual are usually prevented from dividing by irradiation or treatment with the cytostatic drug mitomycin C.) Such studies have shown that roughly 1–10% of all T cells in an individual will respond to stimulation by cells from another, unrelated, member of the same species. This type of T-cell response is called **alloreactivity** because it represents the recognition of allelic polymorphism in allogeneic MHC molecules.

Before the role of the MHC molecules in antigen presentation was understood, it was a mystery why so many T cells should recognize nonself MHC molecules, as there is no reason the immune system should have evolved a defense against tissue transplants. However, once it was appreciated that T-cell receptors have evolved to recognize foreign peptides in combination with polymorphic MHC molecules, alloreactivity became easier to explain. From experiments in which T cells from animals lacking MHC class I and class II molecules have been artificially driven to mature, it has been shown that the ability to recognize MHC molecules is inherent in the genes that encode the T-cell receptor, rather than being dependent on selection for MHC recognition during T-cell development. The high frequency of alloreactive T cells clearly reflects the commitment of the T-cell receptor to the recognition of MHC molecules in general.

Mature T cells have, however, survived a stringent selection process for the ability to respond to foreign, but not self, peptides bound to self MHC molecules. It is therefore thought that the alloreactivity of mature T cells reflects the cross-reactivity of T-cell receptors normally specific for a variety of foreign peptides bound by self MHC molecules. Given a T-cell receptor that normally binds a foreign peptide displayed by a self MHC molecule (Fig. 5.17, left panel), there are two ways in which it may bind to nonself MHC molecules. In some cases, the peptide bound by the nonself MHC molecule interacts strongly with the T-cell receptor, and the T cells bearing this receptor are stimulated to respond. This type of cross-reactive recognition arises because the spectrum of peptides bound by nonself MHC molecules on the transplanted tissues differs from those bound by the host's own MHC, and it is known as peptide-dominant binding (Fig. 5.17, center panel). In a second type of cross-reactive recognition, known as MHC-dominant binding, alloreactive T cells respond because of direct binding of the T-cell receptor to distinctive features of the nonself MHC molecule (Fig. 5.17, right panel). In these cases the recognition is less dependent on the particular peptide bound; T-cell receptor binding to unique features of the nonself MHC molecule generates a strong signal because of the high concentration of the nonself MHC molecule on the surface of the presenting cell. Both these mechanisms may contribute to the high frequency of T cells that can respond to nonself MHC molecules on transplanted tissue.

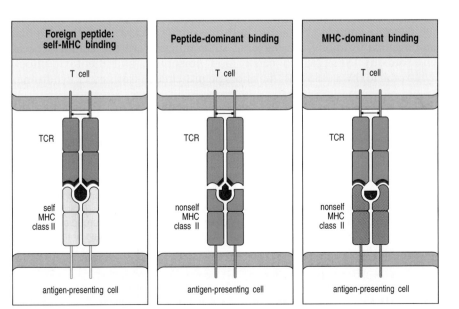

Fig. 5.17 Two modes of cross-reactive recognition that may explain alloreactivity. A T cell that is specific for one peptide:MHC combination (left panel) may cross-react with peptides presented by other, nonself (allogeneic), MHC molecules. This may come about in either of two ways. Most commonly, the peptides bound to the allogeneic MHC molecule fit well to the T-cell receptor (TCR), allowing binding even if there is not a good fit with the MHC molecule (center panel). Alternatively, but less often, the allogeneic MHC molecule may provide a better fit to the T-cell receptor, giving a tight binding that is thus less dependent on the peptide that is bound to the MHC molecule (right panel).

5-15   Many T cells respond to superantigens.

Superantigens are a distinct class of antigens that stimulate a primary T-cell response similar in magnitude to a response to allogeneic MHC. Such responses were first observed in mixed lymphocyte reactions using lymphocytes from strains of mice which were MHC identical but otherwise genetically distinct. The antigens provoking this reaction were originally designated **minor lymphocyte stimulating** (**Mls**) antigens, and it seemed reasonable to suppose that they might be functionally similar to the MHC molecules themselves. We now know that this is not the case, however. The Mls antigens found in these mice strains are encoded by retroviruses which have become stably integrated at various sites into the mouse chromosomes. They act as superantigens because they have a distinctive mode of binding to both MHC and T-cell receptor molecules that enables them to stimulate very large numbers of T cells. Superantigens are produced by many different pathogens, including bacteria, mycoplasmas, and viruses, and the responses they provoke are helpful to the pathogen rather than the host.

Superantigens are unlike other protein antigens, in that they are recognized by T cells without being processed into peptides that are captured by MHC molecules. Indeed, fragmentation of a superantigen destroys its biological activity, which depends on binding as an intact protein to the outside surface of an MHC class II molecule which has already bound peptide. In addition to binding MHC class II molecules, superantigens are able to bind the V_β region of many T-cell receptors (Fig. 5.18). Bacterial superantigens bind mainly to the V_β CDR2 loop, and to a smaller extent to the V_β CDR1 loop and an additional loop called the hypervariable 4 or HV4 loop. The HV4 loop is the predominant binding site for viral superantigens, at least for the Mls antigens encoded by the endogenous mouse mammary tumor viruses. Thus, the α-chain V region and the CDR3 of the β chain of the T-cell receptor have little effect on superantigen recognition, which is determined largely by the germline-encoded V sequences of the expressed β chain. Each superantigen is specific for one or a few of the different V_β gene segments, of which there are 20–50 in mice and humans; a superantigen can thus stimulate 2–20% of all T cells.

This mode of stimulation does not prime an adaptive immune response specific for the pathogen. Instead, it causes a massive production of cytokines by CD4 T cells, the predominant responding population of T cells. These cytokines have two effects on the host: systemic toxicity and suppression of the adaptive immune response. Both these effects contribute to microbial pathogenicity. Among the bacterial superantigens are the **staphylococcal enterotoxins** (**SEs**), which cause food poisoning, and the **toxic shock syndrome toxin-1** (**TSST-1**), the etiologic principle in toxic shock syndrome.

Toxic Shock Syndrome

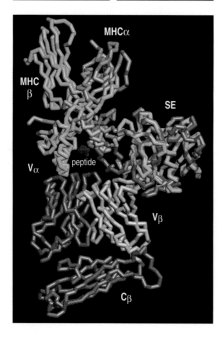

Fig. 5.18 Superantigens bind directly to T-cell receptors and to MHC molecules. Superantigens can bind independently to MHC class II molecules and to T-cell receptors, binding to the V_β domain of the T-cell receptor (TCR), away from the complementarity-determining regions, and to the outer faces of the MHC class II molecule, outside the peptide-binding site (top panels). The bottom panel shows a reconstruction of the interaction between a T-cell receptor, an MHC class II molecule and a staphylococcal enterotoxin (SE) superantigen, produced by superimposing separate structures of an enterotoxin:MHC class II complex onto an enterotoxin:T-cell receptor complex. The two enterotoxin molecules (actually SEC3 and SEB) are shown in turquoise and blue, binding to the α chain of the class II molecule (yellow) and to the β chain of the T-cell receptor (colored gray for the V_β domain and pink for the C_β domain). Molecular model courtesy of H.M. Li, B.A. Fields, and R.A. Mariuzza.

The role of viral superantigens in human disease is less clear. The T-cell responses to rabies virus and the Epstein–Barr virus indicate the existence of superantigens in these human pathogens but the genes encoding them have not yet been identified. The best characterized viral superantigens remain the mouse mammary tumor virus superantigens which are common as endogenous antigens in mice. We will see in Chapter 7 how these have made it possible to observe the deletion of self-reactive T cells as they develop in the thymus, and in Chapter 11 how the virus uses the response to its super-antigen to promote its own transmission.

5-16 MHC polymorphism extends the range of antigens to which the immune system can respond.

Most polymorphic genes encode proteins that vary by only one or a few amino acids, whereas the different allelic variants of MHC proteins differ by up to 20 amino acids. The extensive polymorphism of the MHC proteins has almost certainly evolved to outflank the evasive strategies of pathogens. Pathogens can avoid an immune response either by evading detection or by suppressing the ensuing response. The requirement that pathogen antigens must be presented by an MHC molecule provides two possible ways of evading detection. One is through mutations that eliminate from its proteins all peptides able to bind MHC molecules. The Epstein–Barr virus provides an example of this strategy. In regions of south-east China and in Papua New Guinea there are small isolated populations in which about 60% of individuals carry the HLA-A11 allele. Many isolates of the Epstein–Barr virus obtained from these populations have mutations in a dominant peptide epitope normally presented by HLA-A11; the mutant peptides no longer bind to HLA-A11 and cannot be recognized by HLA-A11-restricted T cells. This strategy is plainly much more difficult to follow if there are many different MHC molecules, and the presence of different loci encoding functionally related proteins may have been an evolutionary adaptation by hosts to this strategy by pathogens.

In large outbred populations, polymorphism at each locus can potentially double the number of different MHC molecules expressed by an individual, as most individuals will be heterozygotes. Polymorphism has the additional advantage that individuals in a population will differ in the combinations of MHC molecules they express and will therefore present different sets of peptides from each pathogen. This makes it unlikely that all individuals in a population will be equally susceptible to a given pathogen and its spread will therefore be limited. That exposure to pathogens over an evolutionary timescale can select for expression of particular MHC alleles is indicated by the strong association of the HLA-B53 allele with recovery from a potentially lethal form of malaria; this allele is very common in people from West Africa, where malaria is endemic, and rare elsewhere, where lethal malaria is uncommon.

Similar arguments apply to a second strategy for evading recognition. If pathogens can develop mechanisms to block the presentation of their peptides by MHC molecules, they can avoid the adaptive immune response. Adenoviruses encode a protein that binds to MHC class I molecules in the endoplasmic reticulum and prevents their transport to the cell surface, thus preventing the recognition of viral peptides by CD8 cytotoxic T cells. This MHC-binding protein must interact with a polymorphic region of the MHC class I molecule, as some allelic variants are retained in the endoplasmic reticulum by the adenoviral protein whereas others are not. Increasing the variety of MHC molecules expressed therefore reduces the likelihood that a pathogen will be able to block presentation by all of them and completely evade an immune response.

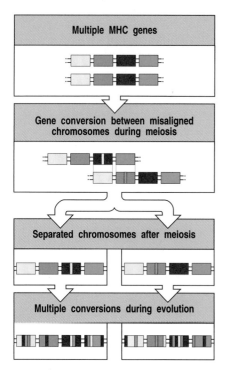

Fig. 5.19 Gene conversion can create new alleles by copying sequences from one MHC gene to another. Sequences can be transferred from one gene to a similar but different gene by a process known as gene conversion. For this to happen, the two genes must become apposed during meiosis. This can occur as a consequence of the misalignment of the two paired homologous chromosomes when there are many copies of similar genes arrayed in tandem—somewhat like buttoning in the wrong buttonhole. During the process of crossing-over and DNA recombination, a DNA sequence from one chromosome is sometimes copied to the other, replacing the original sequence. In this way several nucleotide changes can be inserted all at once into a gene and can cause several simultaneous amino acid changes between the new gene sequence and the original gene. Because of the similarity of the MHC genes to each other and their close linkage, gene conversion has occurred many times in the evolution of MHC alleles.

These arguments raise a question: if having three MHC class I loci is better than having one, why are there not far more MHC loci? The probable explanation is that each time a distinct MHC molecule is added to the MHC repertoire, all T cells that can recognize self peptides bound to that molecule must be removed in order to maintain self tolerance. It seems that the number of MHC loci present in humans and mice is about optimal to balance out the advantages of presenting an increased range of foreign peptides and the disadvantages of an increased loss of T cells from the repertoire.

5-17 Multiple genetic processes generate MHC polymorphism.

MHC polymorphism appears to have been strongly selected by evolutionary pressures. However, for selection to work efficiently in organisms that reproduce slowly, such as humans, there must also be powerful mechanisms for generating the variability on which selection can act. The generation of polymorphism in MHC molecules is an evolutionary problem not easily analyzed in the laboratory; however, it is clear that several genetic mechanisms contribute to the generation of new alleles. Some new alleles are the result of point mutations, but many arise from the combination of sequences from different alleles either by genetic recombination or by **gene conversion**, a process in which one sequence is replaced, in part, by another from a different gene (Fig. 5.19).

Evidence for gene conversion comes from studies of the sequences of different MHC alleles. These reveal that some changes involve clusters of several amino acids in the MHC molecule and require multiple nucleotide changes in a contiguous stretch of the gene. Even more significantly, the sequences that have been changed frequently derive from other MHC genes on the same chromosome, which is a typical signature of gene conversion. Genetic recombination between different alleles at the same locus may, however, have been more important than gene conversion in generating MHC polymorphism. A comparison of sequences of MHC alleles shows that many different alleles could represent recombination events between a relatively small set of hypothetical ancestral alleles (Fig. 5.20).

The effects of selective pressure in favor of polymorphism can be seen clearly in the pattern of point mutations in the MHC genes. Point mutations can be classified as replacement substitutions, which change an amino acid, or silent substitutions, which simply change the codon but leave the amino acid the same. Replacement substitutions occur within the MHC at a higher frequency relative to silent substitutions than would be expected, providing evidence that polymorphism has been actively selected for in the evolution of the MHC.

5-18 Some peptides and lipids generated in the endocytic pathway can be bound by MHC class I-like molecules that are encoded outside the MHC.

Some MHC class I-like genes map outside the MHC region. One family, called CD1, expressed on dendritic cells and monocytes as well as some thymocytes, functions in antigen presentation to T cells, but the molecules it encodes have two features that distinguish them from classical MHC class I molecules. The first is that the CD1 molecule, although similar to MHC class I molecules in its subunit organization and association with β_2-microglobulin, behaves like an MHC class II molecule. It is not retained within the endoplasmic reticulum by association with the TAP complex but is targeted to vesicles, where it binds its peptide ligand. The peptide antigens bound by

Fig. 5.20 Genetic recombination can create new MHC alleles by DNA exchange between different alleles of the same gene. Conventional meiotic recombination differs from gene conversion in that DNA segments are exchanged between alleles on the two homologous chromosomes rather than, as in gene conversion, being copied in one direction only. Analysis of many MHC allele sequences has shown that the swapping of segments of DNA has occurred many times in the evolution of MHC alleles. Closely related strains of mice have MHC genes where only one or two segments have been swapped between alleles (first panel), whereas more distantly related strains show a patchwork effect that results from the accumulation of many such recombination events (second panel). The variable parts of MHC molecules correspond to segments of the structure around the peptide-binding groove, such as the β strands or parts of the α helix shown in the bottom panels.

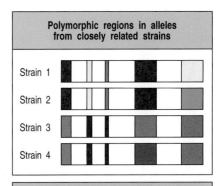

Polymorphic regions in alleles from closely related strains

Strain 1

Strain 2

Strain 3

Strain 4

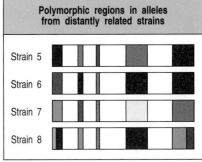

Polymorphic regions in alleles from distantly related strains

Strain 5

Strain 6

Strain 7

Strain 8

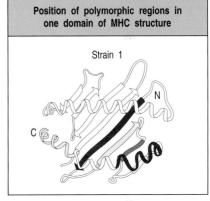

Position of polymorphic regions in one domain of MHC structure

Strain 1

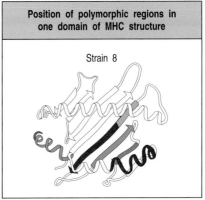

Position of polymorphic regions in one domain of MHC structure

Strain 8

CD1 are therefore derived from the breakdown of extracellular proteins within acidified endosomal compartments and, like the peptides that bind to MHC class II molecules, tend to be longer than the peptides that bind to classical MHC class I molecules.

The second unusual feature of CD1 molecules is that they are able to bind and present glycolipids, in particular the mycobacterial membrane components mycolic acid, glucose monomycolate, phosphoinositol mannosides, and lipoarabinomannan. These are derived either from internalized mycobacteria or from the uptake of lipoarabinomannans by the mannose receptor that is expressed by many phagocytic cells (see Section 2-15). These ligands will thus be delivered into the endocytic pathway, where they can be bound by CD1 molecules. The relationship between the peptide-binding and lipid-binding capacities of CD1 molecules is not clear. Structural studies show that the CD1 molecule has a deep binding groove into which the glycolipid antigens bind. Whether the peptide antigens also bind in this deep groove is not yet known; although CD1-binding peptides are predominantly hydrophobic in character, it is thought unlikely that they bind to the same site as the lipids. It appears that the CD1 genes have evolved as a separate lineage of antigen-presenting molecules able to present microbial lipids and glycolipids, as well as a subset of peptides, to T cells.

Summary.

The major histocompatibility complex (MHC) of genes consists of a linked set of genetic loci encoding many of the proteins involved in antigen presentation to T cells, most notably the MHC class I and class II glycoproteins (the MHC molecules) that present peptides to the T-cell receptor. The outstanding feature of the MHC molecules is their extensive polymorphism. This polymorphism is of critical importance in antigen recognition by T cells. A T cell recognizes antigen as a peptide bound by a particular allelic variant of an MHC molecule, and will not recognize the same peptide bound to other MHC molecules. This behavior of T cells is called MHC restriction. Most MHC alleles differ from one another by multiple amino acid substitutions, and these differences are focused on the peptide-binding site and adjacent regions that make direct contact with the T-cell receptor. At least three properties of MHC molecules are affected by MHC polymorphism: the range of peptides bound; the conformation of the bound peptide; and the direct interaction of the MHC molecule with the T-cell receptor. Thus the highly polymorphic nature of the MHC has functional consequences, and the evolutionary selection for this polymorphism suggests that it is critical to the role of the MHC molecules in the immune response. Powerful genetic mechanisms generate the variation that is seen among MHC alleles, and a compelling argument can be made that selective pressure to maintain a wide variety of MHC molecules in the population comes from infectious agents.

Summary to Chapter 5.

The antigen receptors on T cells recognize complexes of pathogen-derived peptides bound to MHC molecules on a target cell surface. There are two classes of MHC molecule—MHC class I molecules, which bind stably to peptides derived from proteins synthesized and degraded in the cytosol, and MHC class II molecules, which bind stably to peptides derived from proteins degraded in endocytic vesicles. In addition to being bound by the T-cell receptor, the two classes of MHC molecule are differentially recognized by the two co-receptor molecules, CD8 and CD4, which characterize the two major subsets of T cells. CD8 T cells recognize MHC class I:peptide complexes and are activated to kill cells displaying foreign peptides derived from cytosolic pathogens, such as viruses. CD4 T cells recognize MHC class II:peptide complexes and are specialized to activate other immune effector cells, for example B cells or macrophages, to act against the foreign antigens or pathogens that they have taken up. Thus, the two classes of MHC molecule deliver peptides from different cellular compartments to the cell surface, where they are recognized by different types of T cells that carry out the appropriate effector function. There are several genes for each class of MHC molecule, arranged in clusters within a larger region known as the major histocompatibility complex (MHC). Within the MHC, the genes for the MHC molecules are closely linked to genes involved in the degradation of proteins into peptides, the formation of the complex of peptide and MHC molecule, and the transport of these complexes to the cell surface. Because the several different genes for the MHC class I and class II molecules are highly polymorphic and are expressed in a codominant fashion, each individual expresses a number of different MHC class I and class II molecules. Each different MHC molecule can bind stably to a range of different peptides, and thus the MHC repertoire of each individual can recognize and bind many different peptide antigens. Because the T-cell receptor binds a combined peptide:MHC ligand, T cells show MHC-restricted antigen recognition, such that a given T cell is specific for a particular peptide bound to a particular MHC molecule. The genes encoding the T-cell receptors appear to have evolved to recognize MHC molecules, thus accounting for the high frequency of T cells that respond to allogeneic MHC molecules, such as those on an organ transplant from an unrelated donor.

General references.

Bodmer, J.G., Marsh, S.G.E., Albert, E.D., Bodmer, W.F., DuPont, B., Erlich, H.A., Mach, B., Mayr, W.R., Parham, P., Saszuki, T., et al.: **Nomenclature for factors of the HLA system, 1991.** *Tissue Antigens* 2000, **56**:289-290.

Germain, R.N.: **MHC-dependent antigen processing and peptide presentation: Providing ligands for T lymphocyte activation.** *Cell* 1994, **76**:287-299.

Klein, J.: *Natural History of the Major Histocompatibility Complex.* New York, J. Wiley & Sons, 1986.

Moller, G. (ed): **Origin of major histocompatibility complex diversity.** *Immunol. Rev.* 1995, **143**:5-292.

Section references.

5-1   **The MHC class I and class II molecules deliver peptides to the cell surface from two distinct intracellular compartments.**

Morrison, L.A., Lukacher, A.E., Braciale, V.L., Fan, D.P., and Braciale, T.J.: **Differences in antigen presentation to MHC class I- and class II-restricted influenza virus-specific cytolytic T-lymphocyte clones.** *J. Exp. Med.* 1986, **163**:903.

Song, R., and Harding C.V.: **Roles of proteasomes, transporter for antigen presentation (TAP), and β2-microglobulin in the processing of bacterial or particulate antigens via an alternate class I MHC processing pathway.** *J. Immunol.* 1996, **156**:4182-4190.

5-2   Peptides that bind to MHC class I molecules are actively transported from the cytosol to the endoplasmic reticulum.

Lankat-Buttgereit, B., and Tampe, R.: **The transporter associated with antigen processing TAP: structure and function.** *FEBS Lett.* 1999, **464**:108-112.

Townsend, A., Ohlen, C., Foster, L., Bastin, J., Lunggren, H.G., and Karre, K.: **A mutant cell in which association of class I heavy and light chains is induced by viral peptides.** *Cold Spring Harbor Symp. Quant. Biol.* 1989, **54**:299-308.

Uebel, S., and Tampe, R.: **Specificity of the proteasome and the TAP transporter.** *Curr. Opin. Immunol.* 1999, **11**:203-208.

5-3   Peptides for transport into the endoplasmic reticulum are generated in the cytosol.

Niedermann, G., Geier, E., Lucchiari-Hartz, M., Hitziger, N., Ramsperger, A., and Eichmann, K.: **The specificity of proteasomes: impact on MHC class I processing and presentation of antigens.** *Immunol. Rev.* 1999, **172**:29-48.

Stoltze, L., Nussbaum, A.K., Sijts, A., Emmerich, N.P., Kloetzel, P.M., and Schild, H.: **The function of the proteasome system in MHC class I antigen processing.** *Immunol. Today* 2000, **21**:317-319.

Yewdell, J., Anton, L. C., Bacik, I., Schubert, U., Snyder, H.L., and Bennink, J.R.: **Generating MHC class I ligands from viral gene products.** *Immunol. Rev.* 1999, **172**:97-108.

York, I.A., Goldberg, A.L., Mo, X.Y., and Rock, K.L.: **Proteolysis and class I major histocompatibility complex antigen presentation.** *Immunol. Rev.* 1999, **172**:49-66.

5-4   Newly synthesized MHC class I molecules are retained in the endoplasmic reticulum until they bind peptide.

van Endert, P.M.: **Genes regulating MHC class I processing of antigen.** *Curr. Opin. Immunol.* 1999, **11**:82-88.

Grandea, A.G., III, Golovina, T.N., Hamilton, S.E., Sriram,V., Spies, T., Brutkiewicz, R.R., Harty, J.T., Eisenlohr, L.C., and Van Kaer, L.: **Impaired assembly yet normal trafficking of MHC class I molecules in Tapasin mutant mice.** *Immunity* 2000, **13**:213-222.

Li, S., Paulsson, K.M., Chen, S., Sjogren, H.O., and Wang, P.: **Tapasin is required for efficient peptide binding to transporter associated with antigen processing.** *J. Biol. Chem.* 2000, **275**:1581-1586.

Miller, D.M., and Sedmak, D.D.: **Viral effects on antigen processing.** *Curr. Opin. Immunol.* 1999, **11**:94-99.

Pamer, E., and Cresswell, P.: **Mechanisms of MHC class I-restricted antigen processing.** *Annu. Rev. Immunol.* 1998, **16**:323-358.

Seliger, B., Maeurer, M.J., and Ferrone, S.: **Antigen-processing machinery breakdown and tumor growth.** *Immunol. Today* 2000, **21**:455-464.

5-5   Peptides presented by MHC class II molecules are generated in acidified endocytic vesicles.

Chapman, H.A.: **Endosomal proteolysis and MHC class II function.** *Curr. Opin. Immunol.* 1998, **10**:93-102.

Nakagawa, T.Y., and Rudensky, A.Y.: **The role of lysosomal proteinases in MHC class II-mediated antigen processing and presentation.** *Immunol. Rev.* 1999, **172**:121-129.

Pieters, J.: **MHC class II-restricted antigen processing and presentation.** *Adv. Immunol.* 2000, **75**:159-208.

5-6   The invariant chain directs newly synthesized MHC class II molecules to acidified intracellular vesicles.

Brachet, V., Raposo, G., Amigorena, S., and Mellman, I.: **Ii controls the transport of major histocompatibility class II molecules to and from lysosomes.** *J. Cell Biol.* 1997, **137**:51-55.

Jasanoff, A., Wagner, G., and Wiley, D.C.: **Structure of a trimeric domain of the MHC class II-associated chaperonin and targeting protein Ii.** *EMBO J.* 1998, **17**:6812-6818.

Kleijmeer, M.J., Morkowski, S., Griffith, J.M., Rudensky, A.Y., and Geuze, H.J.: **Major histocompatibility complex class II compartments in human and mouse B lymphoblasts represent conventional endocytic compartments.** *J. Cell Biol.* 1997, **139**:639-649.

5-7   A specialized MHC class II-like molecule catalyzes loading of MHC class II molecules with peptides.

Alfonso, C., Liljedahl, M., Winqvist, O., Surh, C.D., Peterson, P.A., Fung-Leung, W.P., and Karlsson, L.: **The role of H2-O and HLA-DO in major histocompatibility complex class II-restricted antigen processing and presentation.** *Immunol. Rev.* 1999, **172**:255-266.

Denzin, L.K., Sant'Angelo, D.B., Hammond, C., Surman, M.J., and Cresswell, P.: **Negative regulation by HLA-DO of MHC class II restricted antigen processing.** *Science* 1997, **278**:106-109.

Jensen, P.E., Weber, D.A., Thayer, W.P., Chen, X., and Dao, C.T.: **HLA-DM and the MHC class II antigen presentation pathway.** *Immunol. Res.* 1999, **20**:195-205.

Jensen, P.E., Weber, D.A., Thayer, W.P., Westerman, L.E., and Dao, C.T.: **Peptide exchange in MHC molecules.** *Immunol. Rev.* 1999, **172**:229-238.

Kropshofer, H., Arndt, S.O., Moldenhauer, G., Hammerling, G.J., and Vogt, A.B.: **HLA-DM acts as a molecular chaperone and rescues empty HLA-DR molecules at lysosomal pH.** *Immunity* 1997, **6**:293-302.

Kropshofer, H., Vogt, A.B., Moldenhauer, G.J.H., Blum, J.S., and Hammerling, G.J.: **Editing of the HLA-DR peptide repertoire by HLA-DM.** *EMBO J.* 1996, **15**:6145-6154.

Van Ham, S.M., Tjin, E.P.M., Lillemeier, B.F., Gruneberg, U., Van Meijgaarden, K.E., Pastoors, L., Verwoerd, D., Tulp, A., Canas, B., Rahman, D., Ottenhoff, T.H.M., Pappin, D.J.C., Trowsdale, J., and Neefjes, J.: **HLA-DO is a negative regulator of HLA-DM mediated MHC class II peptide loading.** *Curr. Biol.* 1997, **7**:950-957.

5-8   Stable binding of peptides by MHC molecules provides effective antigen presentation at the cell surface.

Lanzavecchia, A., Reid, P.A., and Watts, C.: **Irreversible association of peptides with class II MHC molecules in living cells.** *Nature* 1992, **357**:249-252.

5-9   Many proteins involved in antigen processing and presentation are encoded by genes within the major histocompatibility complex.

Beck, S., and Trowsdale, J.: **Sequence organisation of the class II region of the human MHC.** *Immunol. Rev.* 1999, **167**:201-210.

Herberg, J.A., Beck S., and Trowsdale, J.: **TAPASIN, DAXX, RGL2, HKE2 and four new genes (BING 1, 3 to 5) form a dense cluster at the centromeric end of the MHC.** *J. Mol. Biol.* 1998, **277**:839-857.

5-10   A variety of genes with specialized functions in immunity are also encoded in the MHC.

Alfonso, C., and Karlsson, L.: **Nonclassical MHC class II molecules.** *Annu. Rev. Immunol.* 2000, **18**:113-142.

Braud, V.M., Allan, D.S., and McMichael, A.J.: **Functions of nonclassical MHC and non-MHC-encoded class I molecules.** *Curr. Opin. Immunol.* 1999, **11**:100-108.

Ehrlich, R., and Lemonnier, F.A.: **HFE-A novel nonclassical class I molecule that is involved in iron metabolism.** *Immunity* 2000, **13**:585-588.

Maenaka, K., and Jones, E.Y.: **MHC superfamily structure and the immune system.** *Curr. Opin. Struct. Biol.* 1999, **9**:745-753.

5-11   Specialized MHC class I molecules act as ligands for activation and inhibition of NK cells.

Boyington, J.C., Riaz, A.N., Patamawenu, A., Coligan, J.E., Brooks, A.G., and

Sun, P.D.: **Structure of CD94 reveals a novel C-type lectin fold: implications for the NK cell-associated CD94/NKG2 receptors.** *Immunity* 1999, **10**:75-82.

Braud, V.M., and McMichael, A.J.: **Regulation of NK cell functions through interaction of the CD94/NKG2 receptors with the nonclassical class I molecule HLA-E.** *Curr. Top. Microbiol. Immunol.* 1999, **244**:85-95.

Lopez-Botet, M., and Bellon, T.: **Natural killer cell activation and inhibition by receptors for MHC class I.** *Curr. Opin. Immunol.* 1999, **11**:301-307.

Lopez-Botet, M., Bellon, T., Llano, M., Navarro, F., Garcia, P., and de Miguel, M.: **Paired inhibitory and triggering NK cell receptors for HLA class I molecules.** *Hum. Immunol.* 2000, **61**:7-17.

Lopez-Botet, M., Llano, M., Navarro, F., and Bellon, T.: **NK cell recognition of non-classical HLA class I molecules.** *Semin. Immunol.* 2000, **12**:109-119.

Navarro, F., Llano, M., Bellon, T., Colonna, M., Geraghty, D.E., and Lopez-Botet, M.: **The ILT2(LIR1) and CD94/NKG2A NK cell receptors respectively recognize HLA-G1 and HLA-E molecules co-expressed on target cells.** *Eur. J. Immunol.* 1999, **29**:277-283.

Vales-Gomez, M., Reyburn, H., and Strominger, J.: **Interaction between the human NK receptors and their ligands.** *Crit. Rev. Immunol.* 2000, **20**:223-244.

Vales-Gomez, M., Reyburn, H., and Strominger, J.: **Molecular analyses of the interactions between human NK receptors and their HLA ligands.** *Hum. Immunol.* 2000, **61**:28-38.

5-12 The protein products of MHC class I and class II genes are highly polymorphic.

Robinson, J., Malik, A., Parham, P., Bodmer, J.G., and Marsh, S.G.: **IMGT/HLA database—a sequence database for the human major histocompatibility complex.** *Tissue Antigens* 2000, **55**:280-287.

Ruiz, M., Giudicelli, V., Ginestoux, C., Stoehr, P., Robinson, J., Bodmer, J., Marsh, S. G., Bontrop, R., Lemaitre, M., Lefranc, G., Chaume, D., and Lefranc, M.P.: **IMGT, the international ImMunoGeneTics database.** *Nucleic Acids Res.* 2000, **28**:219-221.

Schreuder, G.M., Hurley, C.K., Marsh, S.G., Lau, M., Maiers, M., Kollman, C., and Noreen, H.: **The HLA dictionary 1999: a summary of HLA-A, -B, -C, -DRB1/3/4/5, -DQB1 alleles and their association with serologically defined HLA-A, -B, -C, -DR, and -DQ antigens.** *Hum. Immunol.* 1999, **60**:1157-1181.

5-13 MHC polymorphism affects antigen recognition by T cells by influencing both peptide binding and the contacts between T-cell receptor and MHC molecule.

Babbitt, B., Allen, P.M., Matsueda, G., Haber, P., and Unanue, E.R.: **Binding of immunogenic peptides to Ia histocompatibility molecules.** *Nature* 1985, **317**:359.

Hammer, J.: **New methods to predict MHC-binding sequences within protein antigens.** *Curr. Opin. Immunol.* 1995, **7**:263-269.

Katz, D.H., Hamaoka, T., Dorf, M.E., Maurer, P.H., and Benacerraf, B.: **Cell interactions between histoincompatible T and B lymphocytes. IV. Involvement of immune response (Ir) gene control of lymphocyte interaction controlled by the gene.** *J. Exp. Med.* 1973, **138**:734.

Rosenthal, A.S., and Shevach, E.M.: **Function of macrophages in antigen recognition by guinea pig T lymphocytes. I. Requirement for histocompatible macrophages and lymphocytes.** *J. Exp. Med.* 1973, **138**:1194.

Zinkernagel, R.M., and Doherty, P.C.: **Restriction of *in vivo* T-cell mediated cytotoxicity in lymphocytic choriomeningitis within a syngeneic or semiallogeneic system.** *Nature* 1974, **248**:701-702.

5-14 Nonself MHC molecules are recognized by 1–10% of T cells.

Geluk, A., van Meijgaarden, K.E., and Ottenhoff, T.H.: **Flexibility in T-cell receptor ligand repertoires depends on MHC and T-cell receptor clonotype.** *Immunology* 1997, **90**:370-375.

Merkenschlager, M., Graf, D., Lovatt, M., Bommhardt, U., Zamoyska, R., and Fisher, A.G.: **How many thymocytes audition for selection?** *J. Exp. Med.* 1997,

186:1149-1158.

O'Brien, D.P., Baecher-Allan, C.M., Burns, R.P., Jr., Shastri, N., and Barth, R.K.: **Elimination of T-cell-receptor beta-chain diversity in transgenic mice restricts antigen-specific but not alloreactive responses.** *Immunology* 1997, **91**:375-382.

Smith, P.A., Brunmark, A., Jackson, M.R., and Potter, T.A.: **Peptide-independent recognition by alloreactive cytotoxic T lymphocytes (CTL).** *J. Exp. Med.* 1997, **185**:1023-1033.

Speir, J.A., Garcia, K.C., Brunmark, A., Degano, M., Peterson, P.A., Teyton, L., and Wilson, I.A.: **Structural basis of 2C TCR allorecognition of H-2Ld peptide complexes.** *Immunity* 1998, **8**:553-562.

Zerrahn, J., Held, W., and Raulet, D.H.: **The MHC reactivity of the T cell repertoire prior to positive and negative selection.** *Cell* 1997, **88**:627-636.

5-15 Many T cells respond to superantigens.

Acha-Orbea, H., Finke, D., Attinger, A., Schmid, S., Wehrli, N., Vacheron, S., Xenarios, I., Scarpellino, L., Toellner, K.M., MacLennan, I.C., and Luther, S.A.: **Interplays between mouse mammary tumor virus and the cellular and humoral immune response.** *Immunol. Rev.* 1999, **168**:287-303.

Andersen, P.S., Lavoie, P.M., Sekaly, R.P., Churchill, H., Kranz, D.M., Schlievert, P.M., Karjalainen, K., and Mariuzza, R.A.: **Role of the T cell receptor alpha chain in stabilizing TCR-superantigen- MHC class II complexes.** *Immunity* 1999, **10**:473-483.

Fields, B.A., Malchiodi, E.L., Ysern, X., Stauffacher, C.V., Schlievert, P.M., Karjalainen, K., and Mariuzza, R.A.: **Crystal structure of a T cell receptor β chain complexed with a superantigen.** *Nature* 1996, **384**:188-192.

Macphail, S.: **Superantigens: mechanisms by which they may induce, exacerbate and control autoimmune diseases.** *Int. Rev. Immunol.* 1999, **18**:141-180.

Krakauer, T.: **Immune response to staphylococcal superantigens.** *Immunol. Res.* 1999, **20**:163-173.

Li, H., Llera, A., Tsuchiya, D., Leder, L., Ysern, X., Schlievert, P.M., Karjalainen, K., and Mariuzza, R.A.: **Three-dimensional structure of the complex between a T cell receptor beta chain and the superantigen staphylococcal enterotoxin B.** *Immunity* 1998, **9**:807-816.

5-16 MHC polymorphism extends the range of antigens to which the immune system can respond.

Franco, A., Ferrari, C., Sette, A., and Chisari, F.V.: **Viral mutations, TCR antagonism and escape from the immune response.** *Curr. Opin. Immunol.* 1995, **7**:525-531

Grosberg, R.K., and Hart, M.W.: **Mate selection and the evolution of highly polymorphic self/nonself recognition genes.** *Science* 2000, **289**:2111-2114.

Hill, A.V., Elvin, J., Willis, A.C., Aidoo, M., Allsopp, C.E.M., Gotch, F.M., Gao, X.M., Takiguchi, M., Greenwood, B.M., Townsend, A.R.M., McMichael, A.J., and Whittle, H.C.: **Molecular analysis of the association of B53 and resistance to severe malaria.** *Nature* 1992, **360**:435-440.

Potts, W.K., and Slev, P.R.: **Pathogen-based models favouring MHC genetic diversity.** *Immunol. Rev.* 1995, **143**:181-197.

5-17 Multiple genetic processes generate MHC polymorphism.

Gaur, L.K., and Nepom, G.T.: **Ancestral major histocompatibility complex DRB genes beget conserved patterns of localized polymorphisms.** *Proc. Natl. Acad. Sci. USA* 1996, **93**:5380-5383.

5-18 Some peptides and lipids generated in the endocytic pathway can be bound by MHC class I-like molecules that are encoded outside the MHC.

Brossay, L., and Kronenberg, M.: **Highly conserved antigen-presenting function of CD1d molecules.** *Immunogenetics* 1999, **50**:146-151.

Kronenberg, M., Brossay, L., Kurepa, Z., and Forman, J.: **Conserved lipid and peptide presentation functions of nonclassical class I molecules.** *Immunol. Today* 1999, **20**:515-521.

PART III

THE DEVELOPMENT OF MATURE LYMPHOCYTE RECEPTOR REPERTOIRES

Signaling Through Immune System Receptors

Cells communicate with their environment through a variety of cell-surface receptors that recognize and bind molecules present in the extracellular environment. The main function of T and B lymphocytes is to respond to antigen and so, in their case, the receptors for antigen are the most important and the best-studied. Binding of antigen to these receptors generates intracellular signals that alter the cells' behavior, and the mechanisms that bring this about will be the main topic of this chapter. Because of the diversity of antigen receptors in the normal lymphocyte population, most of our information on intracellular signaling in lymphocytes comes from tumor-derived lymphoid cell lines that are activated when their antigen receptors are stimulated by anti-receptor antibodies. However, more information is increasingly being derived from studies on normal cells from transgenic animals that express a single type of antigen receptor on their B or T cells. We will use this information to infer the signaling pathways generated when a mature naive lymphocyte binds its specific antigen and is activated to undergo clonal expansion followed by differentiation to a functional effector cell. We will, however, also consider how signaling via the antigen receptor and other lymphocyte receptors can lead to other responses such as inactivation or cell death, depending on the stage of development of the cell and the nature of the ligand.

The antigen receptors of B and T lymphocytes are present at the cell surface as multiprotein complexes. These are composed of clonally variable antigen-binding chains associated with invariant accessory chains that have signaling function. The B- and T-cell antigen receptors are formed from different proteins and have different recognition properties, which have been described in Chapters 3 and 4. After these receptors bind their ligand, however, the intracellular signaling pathways leading from the receptor are remarkably similar in B and T cells. In both cases they lead to the nucleus and to changes in gene expression that dictate the lymphocyte's response.

We will start this chapter by discussing some general principles of cell signaling and will introduce some of the common mechanisms used in intracellular signaling pathways, with particular reference to the antigen receptor signaling pathways. In the second part of the chapter we will outline the signaling pathways from the antigen receptors to the nucleus, and consider how these can be supplemented or inhibited by signals simultaneously received from other receptors. Other signals act through different receptors at different times to influence the development, survival, and responses of lymphocytes and other immune system cells, and these will be considered in the third and final part of the chapter.

General principles of transmembrane signaling.

The challenge that faces all cells that respond to external stimuli is how the recognition of a stimulus, usually by receptors on the outer surface of the cell, is able to effect changes within the cell. Extracellular signals are transmitted across the plasma membrane by receptor proteins, which are instrumental in converting extracellular ligand binding into an intracellular biochemical event. Conversion of a signal from one form into another is known as **signal transduction**, and in this part of the chapter we consider several different mechanisms of signal transduction in cell signaling. Cell-surface receptors activate intracellular signaling pathways and so convert an extracellular signal into an intracellular one that then transmits the signal onward. The signal is converted into different biochemical forms, distributed to different sites in the cell, and sustained and amplified as it proceeds toward its various destinations. One result of intracellular signaling may be changes in the cytoskeleton and secretory apparatus. This is seen in the activation of effector T cells, which direct the release of secretory vesicles to the site where the antigen receptor is bound to antigen on the target cell. The final destination of intracellular signaling is usually the nucleus, where the activation of transcription factors turns on new gene expression and cell division may be induced.

6-1   Binding of antigen leads to clustering of antigen receptors on lymphocytes.

All cell-surface receptors that have a signaling function are either transmembrane proteins themselves, or form parts of protein complexes that link the exterior and interior of the cell. Many receptors undergo a change in protein conformation on binding their ligand. In some types of receptor, this conformational change opens an ion channel into the cell and the resulting change in the concentration of important ions within the cell acts as the intracellular signal, which is then converted into an intracellular response. In other receptors, the conformational change affects the cytoplasmic portion of the receptor, enabling it to associate with and activate intracellular signaling proteins and enzymes.

The antigen receptors on lymphocytes transmit a signal when they bind a ligand that causes them to cluster together on the cell surface. The requirement for receptor clustering was first shown experimentally in somewhat artificial systems by using antibodies against the extracellular portion of the receptor to mimic antigen binding. Antibodies specific for the B-cell receptor or the T-cell receptor activate signaling by inducing clustering of the receptor complexes. This is a very convenient system for the analysis of early events after activation, as all the cells in a sample will be stimulated at the same time, making the course of the response easier to follow.

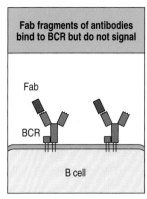

Fab fragments of antibodies bind to BCR but do not signal

Fab

BCR

B cell

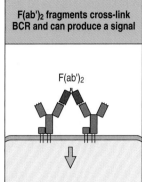

F(ab')₂ fragments cross-link BCR and can produce a signal

F(ab')₂

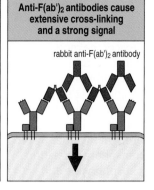

Anti-F(ab')₂ antibodies cause extensive cross-linking and a strong signal

rabbit anti-F(ab')₂ antibody

Fig. 6.1 Cross-linking of antigen receptors is the first step in lymphocyte activation. The requirement for receptor cross-linking is illustrated by the use of anti-immunoglobulin antibodies to activate the B-cell antigen receptor (BCR). As shown in the left panel, Fab fragments of an anti-immunoglobulin can bind to the receptors but cannot cross-link them; they also fail to activate B cells. F(ab')₂ fragments of the same anti-immunoglobulin, which have two binding sites, can bridge two receptors (center panel), and thus signal, albeit weakly, to the B cell. The most effective activation occurs when receptors are extensively cross-linked by first adding the F(ab')₂ fragments and then rabbit antibody molecules that bind and cross-link the bound F(ab')₂ fragments (right panel). The use of antibodies generally to stimulate receptors is described in Appendix I, Section A-19.

Antigen receptor clustering occurs when the receptors are cross-linked to each other. The importance of cross-linking was shown by comparing the response to stimulation with antibody F(ab')₂ fragments, which have two binding sites, and with Fab fragments, which have only one (see Fig. 3.3). On lymphocytes treated with Fab fragments the antigen receptors do not cluster and the cells make no response, whereas on lymphocytes treated with F(ab')₂ fragments the receptors become dimerized and the cells respond, although they may respond only weakly. The response is strongest when the F(ab')₂ cross-linked dimers are further clustered using anti-immunoglobulin sera directed against the F(ab')₂ fragments. The extensive cross-linking of the antigen receptors that then occurs delivers a very strong signal to the cell (Fig. 6.1).

How antigen receptors are clustered *in vivo* when B and T cells encounter their specific antigens is not yet completely understood. T-cell receptors are presumed to undergo clustering in response to contact with another cell surface bearing multiple copies of the specific MHC:peptide complex they recognize. As we will see in Chapter 8, the T-cell receptors become involved in an organized cluster with other cell-surface signaling molecules, but the details of this clustering remain poorly understood. B-cell receptors can be cross-linked by pathogens such as intact bacteria and viruses that have repetitive epitopes on their surfaces. Complex molecules that contain regularly repeated identical epitopes will have the same effect. However, it is still uncertain how B-cell receptors can be clustered by soluble monomeric antigens, such as most of the experimental antigens that immunologists use to study immune responses. An inability, or limited ability, of soluble monomeric antigens to induce receptor clustering may explain why the activation of naive B cells in response to these antigens depends on receiving activating signals from antigen-specific T cells. As we will see in Chapter 9, the binding of soluble monomeric antigen by the B-cell receptor triggers receptor-mediated endocytosis, but is not sufficient by itself to stimulate cell division and differentiation. However, receptor-mediated uptake allows the antigen to be processed and displayed as peptide fragments bound to MHC class II molecules at the cell surface. The B cell can then be recognized by an antigen-specific CD4 effector T cell, which can deliver activating signals that drive clonal expansion and differentiation.

Understanding how the binding of antigen leads to receptor clustering and signaling in lymphocytes is complicated by the diversity of antigen receptors and their ligands. In addition, as we will see in Section 6-8, co-receptors for antigen-linked molecules may also cluster with the receptor and contribute to the initiation of intracellular signaling. How ligand binding leads to receptor clustering and generates a signal is more clearly understood for some other simpler receptors, as we will see in the next section.

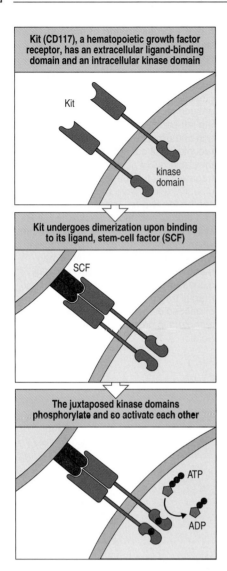

Kit (CD117), a hematopoietic growth factor receptor, has an extracellular ligand-binding domain and an intracellular kinase domain

Kit

kinase domain

Kit undergoes dimerization upon binding to its ligand, stem-cell factor (SCF)

SCF

The juxtaposed kinase domains phosphorylate and so activate each other

ATP

ADP

Fig. 6.2 Ligand binding to the growth factor receptor Kit induces receptor dimerization and transphosphorylation of its cytoplasmic tyrosine kinase domains. Kit (CD117) is a transmembrane protein with an external ligand-binding domain specific for stem-cell factor (SCF) and a cytoplasmic domain with intrinsic tyrosine kinase activity. In the unbound state, the kinase part of the receptor is inactive (top panel). When SCF binds to Kit, it causes the receptor proteins to dimerize; this allows the two tyrosine kinase domains to phosphorylate one another and so become activated. Transactivation of protein kinases by transphosphorylation is an important step in signaling from many cell-surface kinases.

6-2 Clustering of antigen receptors leads to activation of intracellular signal molecules.

Most of the receptors discussed in this chapter initiate intracellular signaling by the activation of **protein tyrosine kinases**, enzymes that affect the activity of other proteins by adding a phosphate group to certain tyrosine residues. The receptors for some growth factors provide the simplest example of this type of receptor. They have cytoplasmic domains that contain an intrinsic tyrosine kinase activity. These enzyme domains are normally inactive, but when brought together by receptor clustering they are able to activate each other by transphosphorylation (Fig. 6.2). Once activated, these tyrosine kinases can phosphorylate and activate other cytoplasmic signaling molecules.

The situation in the antigen receptors is somewhat more complex. As we will see later, they do not themselves have intrinsic tyrosine kinase activity. Instead, the cytoplasmic portions of some of the receptor components bind to intracellular protein tyrosine kinases, which are therefore known as **receptor-associated tyrosine kinases**. When the receptors cluster, these enzymes are brought together and act on each other and on the receptor cytoplasmic tails to initiate the signaling process as in the example above.

In the case of the antigen receptors, the first tyrosine kinases associated with the receptor are members of the Src (pronounced 'Sark') family of tyrosine kinases. The Src-family kinases are common components of signaling pathways concerned with the control of cell division and differentiation in vertebrates and other animals. The prototypic family member Src was initially discovered as the **oncogene** v-*src* which is responsible for the ability of the Rous sarcoma virus to produce tumors in chickens. This viral gene was subsequently shown to be a modified form of a normal cellular gene called c-*src* that the virus had picked up from its host cell at some time in the past. Several other common components of signaling pathways that regulate cell growth and division were also first discovered through their oncogenic action when mutated or removed from their normal controls.

The receptor-associated Src-family kinases play a key role in transducing signals across the lymphocyte membrane; their activation informs the cell interior that the receptor has encountered its antigen. But this is just the first step in a multistep process. When a cell is signaled by the binding of ligand to a kinase-coupled receptor, kinase activation initiates a cascade of intracellular signaling that transfers the signal to other molecules and eventually carries it to the nucleus.

6-3 Phosphorylation of receptor cytoplasmic tails by tyrosine kinases concentrates intracellular signaling molecules around the receptors.

Phosphorylation of enzymes and other proteins by protein kinases is a common general mechanism by which cells regulate their biochemical activity, and has many advantages as a control mechanism. It is rapid, not requiring new protein synthesis or protein degradation to change the biochemical activity of a cell. It can also be easily reversed by the action of **protein phosphatases**, which remove the phosphate group. Many enzymes become active when phosphorylated and inactive when dephosphorylated, or vice versa; the activity of many of the protein kinases involved in signaling is regulated in this way.

Another and equally important outcome of protein phosphorylation is the creation of a binding site for other proteins. This does not alter the intrinsic activity of the phosphorylated molecule. In this case, phosphorylation is used as a tag, allowing the recruitment of other proteins that bind to the phosphorylated site. For example, many kinases involved in signaling are associated with the inner surface of the cell membrane and can act only inefficiently upon their target proteins when these are free in the cytosol. Receptor activation and the phosphorylation of membrane-associated proteins can, however, create binding sites for these target proteins. Cytosolic proteins that bind to phosphorylated sites at the membrane are thus concentrated near to the kinase and can in their turn be phosphorylated and activated (Fig. 6.3). They can also, in some cases, be activated simply by binding to phosphotyrosine. This is an example of allosteric activation, as binding the phosphotyrosine leads to an alteration in their molecular conformation.

Proteins can be phosphorylated on three classes of amino acid—on tyrosine, on serine or threonine, or on histidine. Each of these requires a separate class of kinases to add phosphate groups; only the first two are currently known to be relevant to signaling within the immune system. As we have seen, the early events of signaling, associated with the clustering of the antigen receptors, predominantly involve protein tyrosine kinases; the later events also involve protein serine/threonine kinases.

In antigen receptor signaling, the phosphotyrosines generated by tyrosine kinase action form binding sites for a protein domain known as an **SH2 domain** (Src homology 2 domain). This is found in many intracellular signaling proteins including the Src-family kinases, in which SH2 domains were first discovered. Binding of SH2 domains to phosphotyrosines is a crucial mechanism for recruiting intracellular signaling molecules to an activated receptor. As well as the SH2 domain, Src-family kinases possess another binding domain known as SH3 or Src homology 3. This domain, which is also found in other proteins, binds to proline-rich regions in diverse proteins and can thus recruit these proteins into the signaling pathway, as we will see later. Src-family kinases are usually anchored to the cell membrane by a lipid moiety attached to their amino-terminal region. They are distributed over the inner surface of the cell membrane; during cell activation they become localized to sites of receptor signaling by binding to phosphotyrosine via their SH2 domains.

As a signaling mechanism, phosphorylation also has the advantage that it is easily and rapidly reversible by protein phosphatases that specifically remove the phosphate groups added by the protein kinases. It is crucial that components of signaling pathways can be readily returned to their unstimulated state; not only does this make the signaling pathway ready to receive another signal but it sets a limit on the time that any individual signal is active, preventing cellular responses from running out of control. It is not surprising, therefore, that the signaling pathways that link the cell surface to changes in gene expression use protein phosphorylation and dephosphorylation to regulate the activity of many of their components.

Fig. 6.3 Receptor activation recruits cytosolic proteins to the signaling pathway. Receptor-associated protein kinases are localized at the inner surface of the cell membrane and cannot activate their cytosolic targets efficiently unless these are brought to the membrane. However, another membrane-associated protein functions as an adaptor, being phosphorylated by the active kinase to create a binding site for the cytosolic target, bringing the target molecules to the membrane. These proteins can then be phosphorylated, and thus activated, by the membrane-associated kinase.

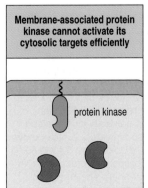

Membrane-associated protein kinase cannot activate its cytosolic targets efficiently

protein kinase

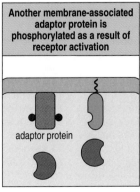

Another membrane-associated adaptor protein is phosphorylated as a result of receptor activation

adaptor protein

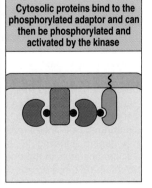

Cytosolic proteins bind to the phosphorylated adaptor and can then be phosphorylated and activated by the kinase

6-4 Intracellular signaling components recruited to activated receptors transmit the signal onward from the membrane and amplify it.

Several classes of protein are typically recruited to the activated receptors and participate in signal propagation. The enzyme **phospholipase C-γ (PLC-γ)** contains two SH2 domains through which it can bind to phosphotyrosine; it is thus recruited to the site of receptor-associated tyrosine kinase activity at the plasma membrane. PLC-γ has a crucial role in propagating the signal onward from the membrane and in amplifying it. Phosphorylation of a tyrosine residue in PLC-γ activates the enzyme, which then cleaves molecules of the membrane phospholipid **phosphatidylinositol bisphosphate (PIP₂)** into two parts, **inositol trisphosphate (IP₃)** and **diacylglycerol (DAG)** (Fig. 6.4). As one molecule of PLC-γ can generate many molecules of DAG and IP₃, this and similar enzymatic steps serve to amplify and sustain the signal. Production of DAG and IP₃ by activated PLC-γ is a common step in pathways from many types of receptor.

Interaction of IP₃ with its receptors in the endoplasmic reticulum causes the release of Ca²⁺ into the cytosol from this intracellular storage site, immediately raising intracellular free Ca²⁺ levels severalfold. Depletion of the endoplasmic reticulum calcium stores triggers the opening of calcium channels in the plasma membrane that let more Ca²⁺ into the cell, thus sustaining the signal (see Fig. 6.4). Increased intracellular free Ca²⁺ leads to the activation of the Ca²⁺-binding protein calmodulin, which in turn binds to and regulates the activity of several other proteins and enzymes in the cell, transmitting the signal onward along pathways that eventually converge on the nucleus. One protein that is regulated by the calcium pathway is the nuclear factor of activated T cells (NFAT), a transcription factor we will discuss further in Section 6-11.

The other product of PIP₂ cleavage is DAG, which remains associated with the inner surface of the plasma membrane. DAG helps to activate members of the **protein kinase C (PKC)** family (see Fig. 6.4). These are serine/threonine protein kinases that are thought to initiate several signaling pathways leading to the nucleus. Some isoforms of PKC are further activated by the Ca²⁺ released by IP₃ action. Thus, the two products of the cleavage of PIP₂ reinforce each other in activating PKC as well as having their own independent effects.

Many of the cellular processes that are activated in lymphocytes when antigen binds to its receptor are common to many cell types. For example, resting lymphocytes proliferate when exposed to antigen, whereas other cell types proliferate in response to particular growth factors; what differs in each case is the receptor that initiates the common response pathway in the different cell types. To link these different receptors to common intracellular signaling components, specialized **adaptor proteins** are needed.

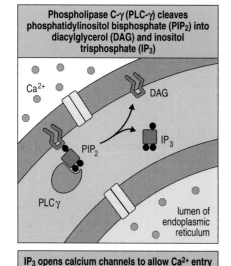

Phospholipase C-γ (PLC-γ) cleaves phosphatidylinositol bisphosphate (PIP₂) into diacylglycerol (DAG) and inositol trisphosphate (IP₃)

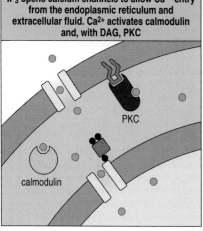

IP₃ opens calcium channels to allow Ca²⁺ entry from the endoplasmic reticulum and extracellular fluid. Ca²⁺ activates calmodulin and, with DAG, PKC

Fig. 6.4 The enzyme phospholipase C-γ cleaves inositol phospholipids to generate two important signaling molecules. Phosphatidylinositol bisphosphate (PIP₂) is a component of the inner leaflet of the plasma membrane. When phospholipase C-γ (PLC-γ) is activated, it cleaves PIP₂ into two parts, inositol trisphosphate (IP₃), which diffuses away from the membrane, and diacylglycerol (DAG), which stays in the membrane. Both these molecules are important in signaling. IP₃ binds to calcium channels in the endoplasmic reticulum (ER) membrane, opening the channels and allowing Ca²⁺ to enter the cytosol from stores in the ER. The depletion of Ca²⁺ from the ER in turn triggers the opening of channels in the plasma membrane that allows Ca²⁺ in from the external fluid. DAG attracts a protein kinase C (PKC) to the cell membrane where it is activated, often with the help of the increased level of Ca²⁺. The active forms of protein kinase C are serine/threonine kinases with several roles in cell activation. Raised Ca²⁺ levels also activate a protein known as calmodulin, a ubiquitous Ca²⁺-binding protein that is responsible for activating other Ca²⁺-dependent enzymes within the cell.

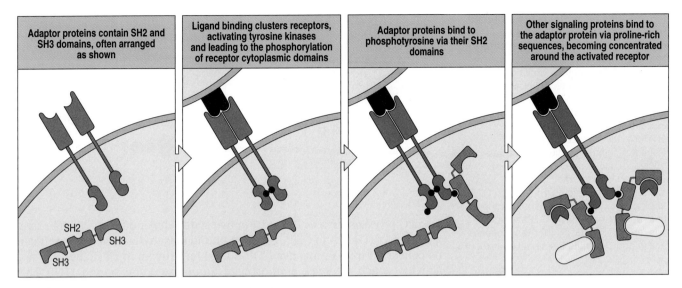

| Adaptor proteins contain SH2 and SH3 domains, often arranged as shown | Ligand binding clusters receptors, activating tyrosine kinases and leading to the phosphorylation of receptor cytoplasmic domains | Adaptor proteins bind to phosphotyrosine via their SH2 domains | Other signaling proteins bind to the adaptor protein via proline-rich sequences, becoming concentrated around the activated receptor |

Fig. 6.5 Signals are propagated from the receptor through adaptor proteins, which recruit other signaling proteins to the receptor. Adaptor proteins are specialized signaling molecules that usually have no enzymatic activity themselves. Instead, they allow other molecules to become associated with the activated receptors. Adaptors, as shown in the first panel, often contain SH2 domains flanked by SH3 domains. Once a receptor, in this case Kit, has been activated and trans-phosphorylated (second panel), adaptors can bind to the phosphotyrosines through their SH2 domains (third panel). Other molecules that contain proline-rich regions can now bind to the adaptors through the SH3 domains and be activated by the receptor-associated kinases (fourth panel).

In lymphocytes, the adaptor proteins that bind to the antigen receptors contain two or more domains, for example, SH2 and SH3 domains, that mediate protein–protein interactions. These proteins do not have kinase activity themselves and their function in general is to recruit other molecules to the activated receptor. Such a protein can bind to a phosphotyrosine residue via its SH2 domain and to other proteins containing proline-rich motifs via its SH3 domains (Fig. 6.5). Binding to an adaptor protein positions these other proteins at or near the cell membrane, where they can in turn be phosphorylated and activated by the tyrosine kinases associated with the receptor. One important family of proteins whose members are bound by adaptors and activated during signaling are the guanine-nucleotide exchange factors (GEFs), which, as we will see in the next section, pass the signal on to another common central component of many signaling pathways, the small G proteins.

6-5 Small G proteins activate a protein kinase cascade that transmits the signal to the nucleus.

Small GTP-binding proteins or **small G proteins** are another class of protein that serves to propagate signals from tyrosine kinase-associated receptors. The family of small single-chain GTP-binding proteins is distinct from the heterotrimeric G proteins that associate with seven-span transmembrane receptors such as the anaphylotoxin or chemokine receptors (see Section 6-16). The best-known small G protein is **Ras**. Like the Src-family kinases, Ras was discovered through its effects on cell growth. A gene encoding a mutated form of Ras was found in various animal retroviruses that cause tumors, and the corresponding cellular *Ras* gene (c-*Ras*) was subsequently found to be mutated in many different human tumors. The frequent discovery of mutant c-*Ras* in tumors indicated that the normal *Ras* gene had a critical role in the control of cell growth and focused attention on the physiological role of Ras. This protein has been highly conserved throughout evolution. Ras proteins are found in all eukaryotic cells and are activated in response to many different cell-activating ligands.

Fig. 6.6 Small G proteins are switched from inactive to active states by guanine-nucleotide exchange factors. Ras is a small GTP-binding protein with intrinsic GTPase activity. In its GTP-bound state, it is active (first panel), whereas in the GDP-bound state it is inactive. Most of the time, it is in the inactive state owing to its intrinsic GTPase activity (second panel). Receptor signaling activates guanine-nucleotide exchange factors (GEFs), which can bind to small G proteins such as Ras and displace GDP, allowing GTP to bind in its place (right panel). In the time before the intrinsic GTPase activity converts GTP to GDP, the Ras protein is active and transmits the signal onward.

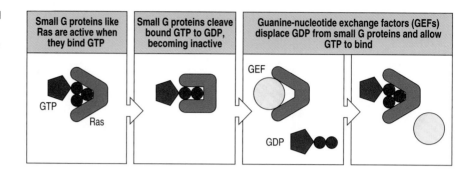

Small G proteins such as Ras exist in two states, depending on whether they are bound to GTP or to GDP. The GTP-bound form of Ras is active, but it can be converted into an inactive GDP-bound form by an intrinsic Ras-GTPase activity, which removes a phosphate group from the bound GTP. This reaction is accelerated by regulatory cofactors, and small G proteins do not normally stay permanently activated; instead they eventually turn themselves off. Thus the small G proteins are usually found in the inactive GDP-containing state, and they are activated only transiently in response to activating ligands. However, mutation at a single residue can lock them in the active state, causing them to become oncogenic.

The activation of small G proteins is mediated by **guanine-nucleotide exchange factors** (**GEFs**), which exchange GDP for GTP (Fig. 6.6). In lymphocytes, Ras and other small G proteins are recruited to the receptor site by adaptor proteins and are activated by GEFs bound to these adaptor proteins. Thus, G proteins can act as molecular switches, becoming switched on only when the cell-surface receptor is activated.

Once activated, small G proteins activate, among other things, a cascade of protein kinases known as the **mitogen-activated protein kinase** (**MAP kinase**) cascade. This kinase cascade is found in all multicellular animals and is responsible for many of the effects of activating ligands. The MAP kinase cascade leads directly to the phosphorylation and activation of transcription factors in the nucleus. In particular, the AP-1 family of transcription factors, which are heterodimers of the Fos and Jun proteins, the products of the oncogenes *fos* and *jun*, are activated through MAP kinase cascades. We will discuss these activation pathways in more detail in the next part of the chapter, where the structures of the lymphocyte antigen-receptor complexes are described and we look at the particular signals generated by the antigen receptors and the co-receptors that cluster with them.

Summary.

Lymphocyte antigen receptors signal for cell activation using signal transduction mechanisms common to many intracellular signaling pathways. On ligand binding, antigen receptor clustering leads to the activation of receptor-associated protein tyrosine kinases at the cytoplasmic face of the plasma membrane. These initiate intracellular signaling by phosphorylating tyrosine residues in the clustered receptor tails. The phosphorylated tyrosines act as binding sites for additional kinases and other signaling molecules that amplify the signal and transmit it onward. The enzyme phospholipase C-γ is recruited in this way and initiates two major pathways of intracellular signaling that are common to many other receptors. Cleavage of the membrane phospholipid PIP$_2$ by this enzyme produces the diffusible messenger inositol trisphosphate (IP$_3$) and membrane-bound diacylglycerol (DAG). IP$_3$ action leads to a sharp increase in the level of intracellular free Ca^{2+}, which activates various calcium-dependent enzymes. Together with Ca^{2+}, DAG initiates a second signaling pathway by activating members of the protein kinase C family. A third pathway involves small G proteins, proteins with GTPase activity that are activated by binding GTP, but then hydrolyze the GTP to GDP and become inactive. Small G proteins are recruited to the signaling pathway and activated by guanine-nucleotide exchange factors (GEFs), which catalyze the exchange of GDP for GTP. GEFs and other signaling molecules are linked to the activated receptors by adaptor proteins that bind to phosphorylated tyrosines through one protein domain, the SH2 domain, and to other signaling molecules through other domains including SH3. All these signaling pathways eventually converge on the nucleus to alter patterns of gene transcription.

Antigen receptor structure and signaling pathways.

The antigen receptors on B cells (the **B-cell receptor** or **BCR**) and T cells (the **T-cell receptor** or **TCR**) are multiprotein complexes made up of clonally variable antigen-binding chains—the heavy and light immunoglobulin chains in the B-cell receptor, and the TCRα and TCRβ chains in the T-cell receptor—that are associated with invariant accessory proteins. The invariant chains are required both for transport of the receptors to the cell surface and, most importantly, for initiating signaling when the receptors bind to an extracellular ligand. Antigen binding to the receptor generates signals that lead ultimately to the activation of nuclear transcription factors that turn on new gene expression and turn off genes typically expressed only in resting cells. In this part of the chapter we also see how clustering of the antigen receptors with co-receptors helps to generate these signals.

6-6 The variable chains of lymphocyte antigen receptors are associated with invariant accessory chains that carry out the signaling function of the receptor.

The antigen-binding portion of the B-cell receptor complex is a cell-surface immunoglobulin that has the same antigen specificity as the secreted antibodies that the B cell will eventually produce. Indeed, it is identical to a secreted monomeric immunoglobulin, except that it is attached to the membrane through the carboxy termini of the paired heavy chains. The mRNA for the cell-surface heavy chain is spliced in such a way that the carboxy terminus of the protein is made up of a transmembrane domain and a very short cytoplasmic tail (see Fig. 4.19). The heavy and light chains do not by themselves make up a complete cell-surface receptor, however. When cells were transfected with heavy- and light-chain cDNA derived from a cell expressing surface immunoglobulin, the immunoglobulin that was synthesized remained inside the transfected cell rather than appearing on the surface. This implied that other molecules were required for the immunoglobulin receptor to be expressed on the cell surface. Two proteins associated with heavy chains on the B-cell surface were subsequently identified and called **Igα** and **Igβ**. Transfection of Igα and Igβ cDNA along with that for the immunoglobulin chains results in the appearance of a B-cell receptor on the cell surface.

One Igα chain and one Igβ chain associates with each surface immunoglobulin molecule. Thus the complete B-cell receptor is thought to be a complex of six chains—two identical light chains, two identical heavy chains, one Igα, and one Igβ (Fig. 6.7). The Igα and Igβ genes are closely linked in the genome and

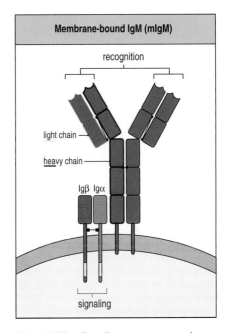

Membrane-bound IgM (mIgM)

recognition

light chain

heavy chain

Igβ Igα

signaling

Fig. 6.7 The B-cell receptor complex is made up of cell-surface immuno-globulin with one each of the invariant proteins Igα and Igβ. The immunoglobulin recognizes and binds antigen but cannot itself generate a signal. It is associated with antigen-nonspecific signaling molecules—Igα and Igβ. These each have a single immunoreceptor tyrosine-based activation motif (ITAM), shown as a yellow segment, in their cytosolic tails that enables them to signal when the B-cell receptor is ligated with antigen. Igα and Igβ are disulfide-linked and associated with the heavy chains, but it is not known which binds to the heavy chain.

encode proteins composed of a single amino-terminal immunoglobulin-like domain connected via a transmembrane domain to a cytoplasmic tail. Igα and Igβ provide the only substantial cytoplasmic domains present in the receptor complex and are crucial for signaling. The transmembrane form of the immunoglobulin heavy chain has a very short cytoplasmic tail and it was hard to understand how this could signal into the cell that the surface immunoglobulin was ligated. The discovery of Igα and Igβ solved this intellectual problem.

Signaling from the B-cell receptor complex depends on the presence in Igα and Igβ of amino acid sequences called **immunoreceptor tyrosine-based activation motifs** (**ITAMs**). These motifs were originally identified in the cytoplasmic tails of Igα and Igβ, but are now known to be present in the accessory chains involved in signaling from the T-cell receptor, and in the Fc receptors on mast cells, macrophages, monocytes, and natural killer (NK) cells that bind antibody constant regions. ITAMs are composed of two tyrosine residues separated by around 9–12 amino acids; the canonical ITAM sequence is ...YXX[L/V]X_{6-9}YXX[L/V]..., where Y is tyrosine, L is leucine, V is valine, and X represents any amino acid. Igα and Igβ each have a single ITAM in their cytosolic tails, giving the B-cell receptor a total of two ITAMs. When antigen binds, the tyrosines in these ITAMs become phosphorylated by receptor-associated Src-family tyrosine kinases Blk, Fyn, or Lyn. The ITAMs, by virtue of their two precisely spaced tyrosines, are then able to bind with high affinity to the tandem SH2 domains of members of a second family of protein tyrosine kinases. As we will see in Section 6-9, these kinases—Syk in B cells and ZAP-70 in T cells—are important in transmitting the signal onward.

The complete α:β T-cell antigen receptor complex contains several different accessory chains in addition to the highly variable TCRα and TCRβ chains which form heterodimers containing a single antigen-binding site (see Chapter 4). The invariant accessory chains are CD3γ, CD3δ, and CD3ε, which make up the **CD3 complex**, and the ζ chain, which is present as a largely intracytoplasmic homodimer. Although the exact stoichiometry of the T-cell receptor complex is not definitively established, it seems likely that two α:β heterodimers are associated at the cell surface with one CD3γ, one CD3δ, two CD3ε, and one ζ homodimer (Fig. 6.8, one α:β heterodimer omitted). The three CD3 proteins are encoded in adjacent genes that are regulated as a unit and are required for surface expression of the α:β heterodimers and for signaling via the receptor. Optimal expression and maximum signaling, however, also require the ζ chain, which is encoded elsewhere in the genome.

The CD3 proteins resemble Igα and Igβ in having an extracellular immunoglobulin-like domain and a single ITAM in their cytoplasmic tails. The ζ chain is distinct in having only a short extracellular domain, but it has three ITAMs in its cytoplasmic domain. The CD3 chains have negatively charged acidic residues in their transmembrane domains, which are able to interact with the positive charges of the α and β chains, as shown in Fig. 6.8. In total, the T-cell receptor complex is equipped with 10 ITAMs, which might give it greater flexibility in signaling compared with the B-cell receptor, as discussed later in this chapter.

Thus the antigen receptors of B and T lymphocytes are made from distinct sets of proteins but are similarly constructed. Both are molecular complexes made up of two types of functional component: variable chains that recognize the individual antigens and invariant chains that have a role both in the surface expression of the receptors and in transmitting signals to the cell's interior, enabling antigen recognition to be translated into action.

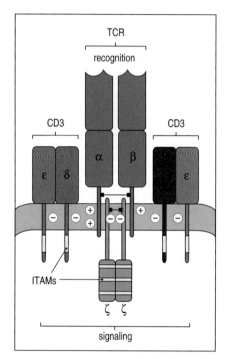

Fig. 6.8 The T-cell receptor complex is made up of antigen-recognition proteins and invariant signaling proteins. The T-cell receptor α:β heterodimer (TCR) recognizes and binds its peptide:MHC ligand, but cannot signal to the cell that antigen has bound. In the functional receptor complex, α:β heterodimers are associated with a complex of four other signaling chains (two ε, one δ, one γ) collectively called CD3, which are required for the cell-surface expression of the antigen-binding chains and for signaling. The cell-surface receptor complex is also associated with a homodimer of ζ chains, which signal to the interior of the cell upon antigen binding. Each CD3 chain has one ITAM (yellow segment), whereas each ζ chain has three. The transmembrane regions of each chain have either a net positive or negative charge as shown. It is thought that each complete receptor complex contains two α:β heterodimers associated with the six accessory chains shown in the figure, in which case the charges would balance.

6-7   The ITAMs associated with the B-cell and T-cell receptors are phosphorylated by protein tyrosine kinases of the Src family.

Phosphorylation of the tyrosines in ITAMs serves as the first intracellular signal indicating that the lymphocyte has detected its specific antigen. In B cells, three protein tyrosine kinases of the Src family—**Fyn**, **Blk**, and **Lyn**—are thought to be responsible for this. These Src-family kinases can associate with resting receptors through a low-affinity interaction with an unphosphorylated ITAM, to which they bind through a site in their amino-terminal domain. When the receptors cluster after antigen binding, the receptor-associated kinases phosphorylate and activate each other and are then thought to phosphorylate the tyrosine residues in the ITAMs of the cytoplasmic tails of Igβ and Igα. Phosphorylation of a single tyrosine in an ITAM allows the binding of a Src-family kinase through its SH2 domains, and kinases bound in this way are in turn activated to phosphorylate further ITAM tyrosine residues (Fig. 6.9).

The initial events in T-cell receptor signaling are also implemented by two Src-family kinases—**Lck**, which is constitutively associated with the cytoplasmic domain of the co-receptor molecules CD4 and CD8 (see Chapter 3), and Fyn, which associates with the cytoplasmic domains of the ζ and CD3ε chains upon receptor clustering. In the next section we discuss how CD4 or CD8 is clustered together with the antigen receptor when the receptor binds to its peptide:MHC ligand. Thus antigen recognition allows both Fyn and Lck to phosphorylate specific ITAMs on the accessory chains of the T-cell receptor complex.

The enzyme activity of the Src-family kinases is itself regulated by the phosphorylation status of the kinase domain and the carboxy-terminal region, each of which has regulatory tyrosine residues. Phosphorylation of the tyrosine in the kinase domain is activatory, while phosphorylation of a tyrosine at the carboxy terminus is inhibitory. Even after being phosphorylated at the activating tyrosine, Src-family kinases can be kept inactive by a protein tyrosine kinase called **Csk** (C-terminal Src kinase), which phosphorylates the inhibitory tyrosine (Fig. 6.10). As Csk activity is constitutive in resting cells, the Src proteins are generally inactive. An agent that counteracts the effects of Csk is the transmembrane protein tyrosine phosphatase **CD45**, also known as leukocyte common antigen, which is required for receptor signaling in lymphocytes and other cells. CD45 can remove the phosphate from phosphotyrosines, especially from the inhibitory tyrosine residue of Src-family kinases. Thus, the balance between Csk, which prevents the activation of Src-family kinases, and CD45, which restores their potential for being activated, determines the signaling activity of Src-family kinases in response to receptor aggregation. This is one way in which the threshold for initiating receptor signaling is regulated in lymphocytes. For example, as we will see in Chapter 10, activated effector T lymphocytes and memory T lymphocytes express an isoform of CD45 that associates with the T-cell receptor complex, thus facilitating signaling through the receptor. By contrast, naive T cells express a CD45 isoform that does not have this property.

A second method by which the activity of Src-family kinases is regulated is by controlling the level at which they are present in the cell. Src-family kinases can be covalently modified with ubiquitin, a signal that targets proteins for

Fig. 6.9 Src-family kinases are associated with antigen receptors and phosphorylate the tyrosines in ITAMs. The membrane-bound Src-family kinases Fyn, Blk, and Lyn associate with the B-cell antigen receptor by binding to ITAM motifs, either through their amino-terminal domains or, as shown in the figure, by binding a single phosphorylated tyrosine through their SH2 domains. After ligand binding and receptor clustering, they phosphorylate tyrosines in the ITAMs on the cytoplasmic tails of Igα and Igβ.

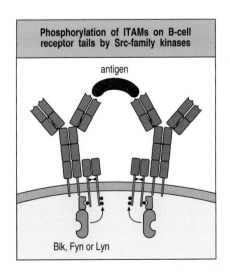

Phosphorylation of ITAMs on B-cell receptor tails by Src-family kinases

antigen

Blk, Fyn or Lyn

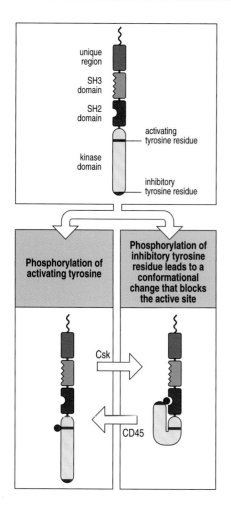

Fig. 6.10 Regulation of Src-family kinase activity. Src-family kinases contain two tyrosine residues (red bars) that are targets for phosphorylation. Phosphorylation of the tyrosine in the kinase domain (bottom left panel) stimulates kinase activity, and this tyrosine is a target for phosphorylation by receptor-associated tyrosine kinases. The second tyrosine lies near the carboxy terminus and has a regulatory function. When it has been phosphorylated, the kinase is inactive as a result of an interaction between the inhibitory phosphotyrosine and the SH2 domain, as shown in the lower right panel.

degradation by the proteasome, and this degradation pathway is controlled through association with a regulatory protein, Cbl. Cbl itself does not appear to add the ubiquitin to the Src-family kinases; rather, it acts as an adaptor between the kinases and ubiquitin ligase enzymes. This process may be used to set a maximum level of response by limiting the concentration of kinases within the cell. However, Cbl is itself a target of tyrosine phosphorylation after receptor aggregation, and it seems more likely that its role is to switch off cascades of activated Src-family kinases after the cell has become activated.

In signaling from the antigen receptor, the activation of Src-family kinases is the first step in a signaling process that passes the signal on to many different molecules. Before we consider how the signals generated by Src-family kinases are transmitted onward, we will look at how antigen binding by both B cells and T cells directly or indirectly activates co-receptor molecules that are essential for producing a strong and effective intracellular signal.

6-8 Antigen receptor signaling is enhanced by co-receptors that bind the same ligand.

Optimal signaling through the T-cell receptor complex occurs only when it clusters with the co-receptors CD4 or CD8. This is why most T cells with MHC class II-restricted T-cell receptors express CD4, which binds to the membrane-proximal domain of MHC class II molecules, whereas most T cells with MHC class I-restricted T-cell receptors express CD8, which binds MHC class I molecules. Expression of the appropriate co-receptor means that the peptide:MHC complex bound by the antigen receptor can be simultaneously bound by a co-receptor molecule. About 100 identical specific peptide:MHC complexes are required on a target cell to trigger a T cell expressing the appropriate co-receptor. In the absence of the co-receptor, 10,000 identical complexes (about 10% of all the MHC molecules on a cell) are required for optimal T-cell activation. This density is rarely, if ever, achieved *in vivo*.

Aggregation of the T-cell receptor with the appropriate co-receptor helps to activate the T cell by bringing the Lck tyrosine kinase associated with the cytoplasmic domain of the co-receptor together with the ITAMs and other targets associated with the cytoplasmic domains of the T-cell receptor complex (Fig. 6.11). One of the principal targets of Lck in T cells is another kinase, the ζ-chain-associated protein or ZAP-70, which is important in propagating the signal onward, as we will see in the next section.

B-cell receptor signaling is also enhanced by aggregation with a co-receptor. The **B-cell co-receptor** is expressed on mature B cells as a complex of the cell-surface molecules CD19, CD21, and CD81. One way in which this complex can be co-ligated with the B-cell receptor is through the recognition of an antigen that has activated complement. The CD21 molecule (also known as complement receptor 2, CR2) is a receptor for the C3d fragment of complement, so antigens, such as pathogens, that have activated complement directly (see Chapter 2) or through the activation of antibody can cross-link the B-cell receptor with CD21 and its associated proteins. This induces phosphorylation of the cytoplasmic tail of CD19 by B-cell receptor-associated tyrosine kinases, which in turn leads to the binding of Src-family kinases and

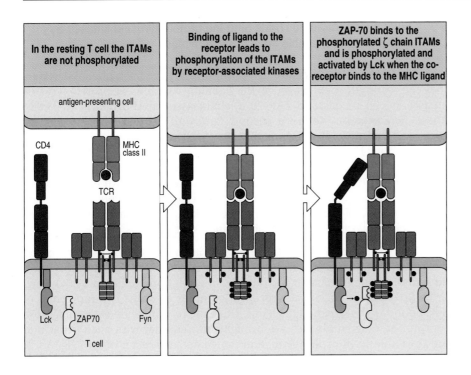

In the resting T cell the ITAMs are not phosphorylated

Binding of ligand to the receptor leads to phosphorylation of the ITAMs by receptor-associated kinases

ZAP-70 binds to the phosphorylated ζ chain ITAMs and is phosphorylated and activated by Lck when the co-receptor binds to the MHC ligand

Fig. 6.11 Clustering of the T-cell receptor and a co-receptor initiates signaling within the T cell. When T-cell receptors become clustered on binding MHC:peptide complexes on the surface of an antigen-presenting cell, activation of receptor-associated kinases such as Fyn leads to phosphorylation of the CD3γ, δ, and ε ITAMs as well as those on the ζ chain. The tyrosine kinase ZAP-70 binds to the phosphorylated ITAMs of the ζ chain, but is not activated until binding of the co-receptor to the MHC molecule on the antigen-presenting cell (here shown as CD4 binding to an MHC class II molecule) brings the kinase Lck into the complex. Lck then phospho-rylates and activates ZAP-70.

the recruitment of a lipid kinase called phosphatidylinositol 3-OH kinase (PI 3-kinase) (Fig. 6.12). Co-ligation of the B-cell receptor with its CD19/CD21/CD81 co-receptor increases signaling 1000- to 10,000-fold. The role of the third component of the B-cell receptor complex, CD81 (TAPA-1), is as yet unknown.

CD19 is expressed on all B cells from an early stage in their development, before CD21 and CD81 are expressed, and it appears to contribute to signaling through the B-cell receptor even in the absence of co-ligation through CD21. Thus B-cell activation that is experimentally induced by cross-linking the B-cell receptor with anti-receptor antibodies is accompanied by co-capping of CD19, the tyrosine phosphorylation of its cytoplasmic tail, and the activation of PI 3-kinase. By contrast, B cells from mice that lack CD19 fail to proliferate in response to B-cell receptor cross-linking and do not fully activate the intra-cellular signaling pathways normally generated when the B-cell receptor is cross-linked. These experiments suggest that CD19 can associate with the B-cell receptor, either constitutively or after receptor activation, and con-tribute to signaling even when the co-receptor has not been engaged through CD21 binding to complement. The physiological importance of CD19 is demonstrated by genetically engineered mice that lack this molecule; these mice make deficient B-cell responses to most antigens.

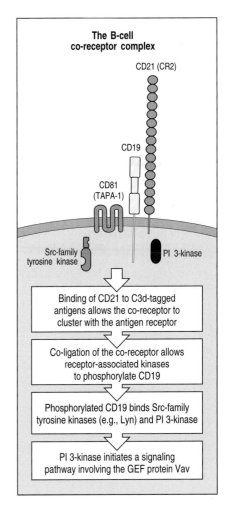

The B-cell co-receptor complex

CD21 (CR2)

CD19

CD81 (TAPA-1)

Src-family tyrosine kinase

PI 3-kinase

Binding of CD21 to C3d-tagged antigens allows the co-receptor to cluster with the antigen receptor

Co-ligation of the co-receptor allows receptor-associated kinases to phosphorylate CD19

Phosphorylated CD19 binds Src-family tyrosine kinases (e.g., Lyn) and PI 3-kinase

PI 3-kinase initiates a signaling pathway involving the GEF protein Vav

Fig. 6.12 B-cell antigen receptor signaling is modulated by a co-receptor complex of at least three cell-surface molecules, CD19, CD21, and CD81. Binding of the cleaved complement fragment C3d to antigen allows the tagged antigen to bind to both the B-cell receptor and the complement cell-surface protein CD21 (complement receptor 2, CR2), a component of the B-cell co-receptor complex. Cross-linking and clustering of the co-receptor with the antigen receptor results in phospho-rylation of tyrosine residues in the cytoplasmic domain of CD19 by protein kinases associated with the B-cell receptor; other Src-family kinases can bind to phosphorylated CD19 and so augment signaling through the B-cell receptor. Phosphorylated CD19 can also bind the enzyme phosphatidylinositol 3-OH kinase (PI 3-kinase), which is instrumental in recruiting the guanine-nucleotide exchange factor Vav to the receptor complex.

6-9 Fully phosphorylated ITAMs bind the protein tyrosine kinases Syk and ZAP-70 and enable them to be activated.

Once the ITAMs in the receptor cytoplasmic tails have been phosphorylated, they can recruit the next players in the signaling cascade, the protein tyrosine kinases **Syk** in B cells and **ZAP-70** (ζ-**chain-associated protein**) in T cells. These two proteins define a second family of protein tyrosine kinases expressed mainly in hematopoietic cells (Syk) or mainly in T lymphocytes (ZAP-70); they have two SH2 domains in their amino-terminal halves and a carboxy-terminal kinase domain. As each SH2 domain binds to one phosphotyrosine, these proteins preferentially bind to motifs with two phosphotyrosines spaced a precise distance apart; tyrosines spaced correctly are found in the ITAM motif. Thus, ZAP-70 or Syk molecules are recruited to the receptor complex upon full phosphorylation of the ITAMs.

Until Syk has bound to the doubly phosphorylated ITAM in the B-cell receptor, it is inactive enzymatically. To become active it must itself be phosphorylated, and this is thought to occur by transphosphorylation mediated by Syk itself or by Src kinases. Each B-cell receptor complex contains two molecules of Syk, bound to the Igα and Igβ chains. Once the receptors are clustered, these receptor-associated kinases are brought into contact with each other and are thus able to phosphorylate, and hence activate, each other (Fig. 6.13). Once activated, Syk phosphorylates target proteins to initiate a cascade of intracellular signaling molecules, which will be described in the next section.

ZAP-70 is not activated by transphosphorylation after binding to the ITAMs in the ζ chain; instead, it is activated by the co-receptor-associated Src kinase Lck (see Fig. 6.11). Once activated, ZAP-70 phosphorylates the substrate **LAT** (**linker of activation in T cells**) and the protein called SLP-76, a second linker or adaptor protein in T cells. LAT is a cytoplasmic protein containing multiple tyrosines and associates with the inside face of the plasma membrane through cysteine residues that are palmitoylated and thus become associated with membrane lipid rafts. The association of LAT with lipid rafts, small, cholesterol-rich areas in the membrane, is essential for recruitment of a signaling complex and the propagation of signals away from the membrane and on into the interior of the cell. By virtue of its many tyrosine residues, phosphorylated LAT is able to recruit proteins that bind through their SH2 domains and transmit the signal from the T-cell membrane to downstream targets. A protein serving the same function in B cells has recently been identified as SLP-65 or **BLNK** for **B**-cell **link**er protein. This, again, has multiple sites for tyrosine phosphorylation, and interacts with many of the same proteins as LAT and SLP-76, to which it has strong homology in sequence.

6-10 Downstream events are mediated by proteins that associate with the phosphorylated tyrosines and bind to and activate other proteins.

Once ZAP-70 tyrosine kinase has been activated in T cells, and LAT and SLP-76 have been phosphorylated, the next steps in the signaling pathway serve to propagate the signal at the cell membrane, and eventually to communicate it to the nucleus. A very similar signaling pathway exists in B cells, made up of activated Syk acting on BLNK, and again propagating the signal from the cell membrane into the nucleus. Signal propagation involves various proteins that bind to phosphotyrosine via SH2 domains, and proteins that are targets for protein tyrosine kinases. Several classes of protein participate in signal propagation. These include the enzyme PLC-γ (see Section 6-4), which initiates two of the main signaling pathways that lead to the nucleus. The third main pathway is generated by activation of the small G protein Ras. These

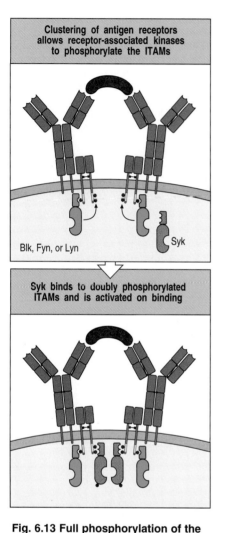

Clustering of antigen receptors allows receptor-associated kinases to phosphorylate the ITAMs

Blk, Fyn, or Lyn   Syk

Syk binds to doubly phosphorylated ITAMs and is activated on binding

Fig. 6.13 Full phosphorylation of the ITAMs on clustered Igα or β chains associated with the B-cell receptor creates binding sites for Syk and Syk activation via transphosphorylation. On clustering of the receptors, the receptor-associated tyrosine kinases Blk, Fyn, and Lyn phosphorylate the ITAMs on the cytoplasmic tails of Igα and Igβ (shown in blue and orange, respectively). Subsequently, Syk binds to the phosphorylated ITAMs of the Igβ chain. Because there are at least two receptor complexes in each cluster, Syk molecules become bound in close proximity and can activate each other by transphosphorylation, thus initiating further signaling.

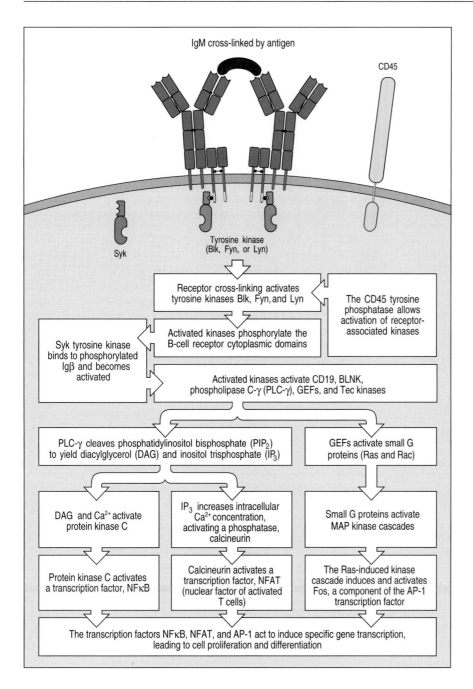

IgM cross-linked by antigen

CD45

Syk

Tyrosine kinase
(Blk, Fyn, or Lyn)

Receptor cross-linking activates
tyrosine kinases Blk, Fyn, and Lyn

The CD45 tyrosine
phosphatase allows
activation of receptor-
associated kinases

Syk tyrosine kinase
binds to phosphorylated
Igβ and becomes
activated

Activated kinases phosphorylate the
B-cell receptor cytoplasmic domains

Activated kinases activate CD19, BLNK,
phospholipase C-γ (PLC-γ), GEFs, and Tec kinases

PLC-γ cleaves phosphatidylinositol bisphosphate (PIP$_2$)
to yield diacylglycerol (DAG) and inositol trisphosphate (IP$_3$)

GEFs activate small G
proteins (Ras and Rac)

DAG and Ca^{2+} activate
protein kinase C

IP$_3$ increases intracellular
Ca^{2+} concentration,
activating a phosphatase,
calcineurin

Small G proteins activate
MAP kinase cascades

Protein kinase C activates
a transcription factor, NFκB

Calcineurin activates a
transcription factor, NFAT
(nuclear factor of activated
T cells)

The Ras-induced kinase
cascade induces and activates
Fos, a component of the AP-1
transcription factor

The transcription factors NFκB, NFAT, and AP-1 act to induce specific gene transcription,
leading to cell proliferation and differentiation

Fig. 6.14 Simplified outline of the intracellular signaling pathways initiated by cross-linking of B-cell receptors by antigen. Cross-linking of surface immunoglobulin molecules activates the receptor-associated Src-family protein tyrosine kinases Blk, Fyn, and Lyn. The CD45 phosphatase can remove an inhibitory phosphate from these kinases, thus allowing their activation. The receptor-associated kinases phosphorylate the ITAMs in the receptor complex, which bind and activate the cytosolic protein kinase Syk, whose activation has been described in Fig. 6.13. Syk then phosphorylates other targets, including the adaptor protein BLNK, which help to recruit Tec kinases that in turn phosphorylate and activate the enzyme phospholipase C-γ. PLC-γ cleaves the membrane phospholipid PIP$_2$ into IP$_3$ and DAG, thus initiating two of the three main signaling pathways to the nucleus. IP$_3$ releases Ca^{2+} from intracellular and extracellular sources, and Ca^{2+}-dependent enzymes are activated, whereas DAG activates protein kinase C with the help of Ca^{2+}. The third main signaling pathway is initiated by guanine-nucleotide exchange factors (GEFs) that become associated with the receptor and activate small GTP-binding proteins such as Ras. These in turn trigger protein kinase cascades (MAP kinase cascades) that lead to the activation of MAP kinases that move into the nucleus and phosphorylate proteins that regulate gene transcription. This scheme is a simplification of the events that actually occur during signaling, showing only the main events and pathways.

three pathways propagate the signal from the activated receptors at the plasma membrane and carry it to the nucleus, as illustrated in Fig. 6.14 for B cells and Fig. 6.15 for T cells. We will discuss in more detail below how each of these pathways is initiated and, in Section 6-11, how they lead to changes in gene expression in the nucleus. Activation of PLC-γ is linked to activation of the antigen receptor through a family of Src-like tyrosine kinases called **Tec kinases**, after a member of the family which is expressed in T and B cells. Other Tec kinases with an important role in lymphocytes are Btk, which is expressed in B lymphocytes and Itk, which is expressed in T lymphocytes. These kinases contain a domain, called a pleckstrin-homology (PH) domain, that allows them to interact with phosphorylated lipids on the inner face of the cell membrane. They also carry SH2 and SH3 domains that allow them to interact with the adaptor proteins LAT, SLP-76, and BLNK. These interactions

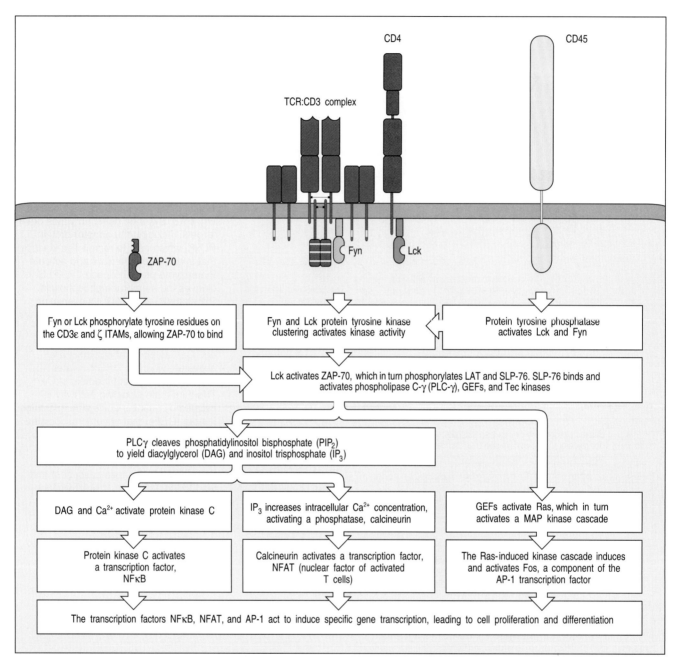

Fig. 6.15 Simplified outline of the intracellular signaling pathways initiated by the T-cell receptor complex and its co-receptor. The T-cell receptor complex and co-receptor (in this example the CD4 molecule) are associated with Src-family protein kinases, Fyn and Lck, respectively. It is thought that binding of a peptide:MHC ligand to the T-cell receptor and co-receptor brings together CD4, the T-cell receptor complex, and CD45. This allows the CD45 tyrosine phosphatase to remove inhibitory phosphate groups and thereby allow the activation of Lck and Fyn. Phosphorylation of the ζ chains enables them to bind the cytosolic tyrosine kinase ZAP-70. In freshly isolated T cells, inactive ZAP-70 is already bound to the ζ chain, so this step is thought to occur before stimulation with specific antigen.

The subsequent activation of bound ZAP-70 by phosphorylation leads to three important signaling pathways. ZAP-70 phosphorylates the adaptor proteins LAT and SLP-76, which in turn leads to the activation of PLC-γ by Tec kinases and the activation of Ras by guanine-nucleotide exchange factors. As illustrated, activated PLC-γ and Ras initiate three important signaling pathways that culminate in the activation of transcription factors in the nucleus. Together, NFκB, NFAT, and AP-1 act on the T-cell chromosomes, initiating new gene transcription that results in the differentiation, proliferation and effector actions of T cells. As with Fig. 6.14, this model is a simplified version showing the main events only.

allow Tec kinases to cluster around the activated antigen receptor where they are phosphorylated and activated by other Src-family tyrosine kinases. Once activated, Tec kinases phosphorylate and activate PLC-γ, which cleaves inositol phospholipids to generate the intracellular second messengers diacylglycerol and IP$_3$ (see Section 6-4). The importance of the Tec kinases in lymphocytes can be seen from inherited deficiencies in these enzymes. The human immunodeficiency disease X-linked agammaglobulinemia (XLA), in which B cells fail to mature (see Section 11-8), results from a mutation of Btk, while the same gene mutated in mice leads to a similar immunodeficiency called *xid*.

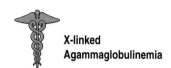

X-linked
Agammaglobulinemia

The activation of Ras is achieved by adaptor proteins and guanine-nucleotide exchange factors (see Section 6-5) recruited to the phosphorylated receptors. In B lymphocytes, the adaptor protein Shc binds to tyrosine residues that have been phosphorylated by the receptor-associated tyrosine kinases. Another adaptor protein, Grb2, which has an SH2 domain flanked on both sides by SH3 domains, forms a complex with Shc and this complex binds the guanine-nucleotide exchange factor SOS. SOS, in turn, is involved in activating Ras by the mechanism shown in Fig. 6.6. In T lymphocytes, the adaptor protein GADS, a homologue of Grb2, is recruited by phosphorylated LAT, which again uses SOS to recruit Ras to the pathway. Adaptor proteins thus form the scaffolding of a signaling complex, associated with lipid rafts, that links ligand binding by the antigen receptor at the cell surface to the activation of Ras, which then triggers further signaling events downstream.

Another small G protein is activated via the B-cell co-receptor complex (see Section 6-8). Phosphorylated CD19 binds a multifunctional intracellular signaling molecule called **Vav**; this is an adaptor protein that also contains guanine-nucleotide exchange factor activity. When Vav is activated, it can activate the small G protein **Rac**. Small G proteins such as Ras and Rac activate a cascade of protein kinases that leads directly to the phosphorylation and activation of transcription factors; we will discuss this in the next section. Vav and Rac can also influence changes in the actin cytoskeleton.

Thus, as we have seen so far, a signal originating from activated tyrosine kinases at the cell membrane can be propagated through several different pathways involving many intracellular proteins. We now turn to the question of how signals are transmitted to the nucleus, there to activate transcription factors that can regulate specific genes.

6-11   Antigen recognition leads ultimately to the induction of new gene synthesis by activating transcription factors.

The ultimate response of lymphocytes to extracellular signals is the induction of new gene expression. This is achieved through the activation of transcription factors—proteins that control the initiation of transcription by binding to regulatory sites in the DNA. Several important transcription factors involved in lymphocyte responses to antigen are activated as a consequence of phosphorylation by the MAP kinases (see Section 6-5). These kinases are themselves activated by phosphorylation; in the inactive, nonphosphorylated state they are resident in the cytoplasm, but when activated by phosphorylation they translocate into the nucleus.

Transcription factors are activated by MAP kinases by the general pathway shown in Fig. 6.16. In lymphocytes, the MAP kinases that are thought to activate transcription factors in response to antigen-receptor ligation are called Erk1 (for extracellular-regulated kinase-1) and Erk2. MAP kinases are unusual, in that their full activation requires phosphorylation on both a tyrosine and a threonine residue, which are separated in the protein by a single amino acid. This can be done only by MAP kinase kinases, enzymes with dual specificity for tyrosine and serine/threonine residues. In the context of antigen-receptor signaling, the MAP kinase kinases that activate Erk1 and Erk2 are called Mek1 and Mek2. MAP kinase kinases are themselves

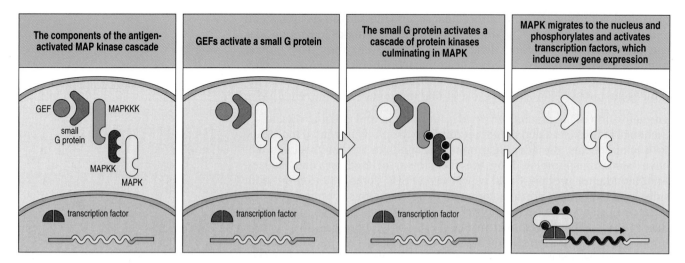

Fig. 6.16 MAP kinase cascades activate transcription factors. All MAP kinase cascades share the same general features. They are initiated by a small G protein, which is switched from an inactive state to an active state by a guanine-nucleotide exchange factor (GEF). The small G protein activates the first enzyme of the cascade, a protein kinase called a MAP kinase kinase kinase (MAPKKK). As expected from its name, the MAP kinase kinase kinase phosphorylates a second enzyme, MAP kinase kinase (MAPKK), which in its turn phosphorylates and activates the mitogen-activated protein kinase (MAP kinase, MAPK) on two sites, a tyrosine and a threonine separated by a single amino acid. This MAP kinase, when doubly phosphorylated, is both activated as a kinase and enabled to translocate from the cytosol into the nucleus, where it can phosphorylate and activate nuclear transcription factors.

activated by phosphorylation by a MAP kinase kinase kinase, which is the first kinase in the cascade. It is thought that in lymphocytes this MAP kinase kinase kinase is the serine/threonine kinase **Raf** (Fig. 6.17, left panel). The antigen receptor activates this cascade through the activation of Raf by the GTP-bound form of Ras.

The activation of transcription factors through a MAP kinase cascade is a critical part of many cell signaling pathways. How the signal is individualized for different stimuli is not yet known, although the existence of variant components at each level of the pathway may allow this to occur. For example, the MAP kinase cascade triggered by signals from the B-cell co-receptor through Vav and Rac uses different kinases from those in the antigen-receptor pathway, and activates different transcription factors (Fig. 6.17, right panel). The MAP kinase cascade activated by the antigen receptor activates the transcription factor Elk, which in turn upregulates the synthesis of the transcription factor Fos. By contrast, the MAP kinase pathway activated by co-ligation of the B-cell co-receptor activates the transcription factor Jun. These pathways can combine in their effects because heterodimers of Fos and Jun form AP-1 transcription factors, which regulate the expression of many genes involved in cell growth. It appears that both pathways are required to drive the clonal expansion of naive lymphocytes. In T cells, the MAP kinase pathway that activates Jun is activated through a cell-surface molecule known as CD28, which interacts with co-stimulatory molecules that are induced on the surface of antigen-presenting cells during the innate immune response to infection (see Section 2-18). Thus this second MAP kinase pathway appears to convey to the cell that the antigen is part of a pathogen that has been recognized by the innate immune system.

A transcription factor that is activated through antigen-receptor signaling, but by a different pathway, is **NFAT** (**nuclear factor of activated T cells**). This is something of a misnomer, as NFAT transcription factors are found not only in T cells but also in B cells, NK cells, mast cells, and monocytes, as well as in some nonhematopoietic cells. The function of NFAT is regulated by its intracellular location. NFAT contains amino acid sequence motifs specifying a nuclear localization signal, which enables it to be translocated into the nucleus, and a nuclear export signal, which directs the movement of NFAT back into the cytosol. In unstimulated cells, the nuclear localization signal is rendered inoperative by phosphorylation at serine/threonine residues; hence NFAT is retained in the cytosol after its synthesis. In addition, should any NFAT make its way to the nucleus, the nuclear export sequence also becomes phosphorylated by a kinase, glycogen synthase kinase 3 (GSK3), which resides primarily within the nucleus. The phosphorylated NFAT is then rapidly exported from the nucleus. The sum of these two effects is that all the NFAT in a resting cell is found in the cytosol.

NFAT is released from the cytosol by the action of the enzyme **calcineurin**, a serine/threonine protein phosphatase. Calcineurin is itself activated by the increase in intracellular free Ca^{2+} that accompanies lymphocyte activation (see Fig. 6.15). Once NFAT has been dephosphorylated by calcineurin, it enters the nucleus, where it can act as a transcriptional regulatory protein. However, activation of calcineurin alone is insufficient to allow NFAT to function within the nucleus. In the absence of other signals, the NFAT is phosphorylated by GSK3 and exported back out of the nucleus again. Other signals mediated through the T-cell receptor are required to downregulate the activity of GSK3 and permit NFAT to remain in the nucleus. These other signals, and signals generated through co-stimulation, may contribute to NFAT function in another way, as NFAT acts as a transcriptional regulator in combination with AP-1 transcription factors, which, as we have seen above, are dimers of Fos and Jun. In lymphocytes, NFAT interacts with Fos:Jun heterodimers that have been activated by the MAP kinase known as Jnk (or 'Junk', as it is commonly called) (see Fig. 6.17, right panel). This illustrates how complex the process of signal transduction can be, and how proteins that regulate transcription can integrate signals that come from different pathways.

The importance of NFAT in T-cell activation is illustrated by the effects of the selective inhibitors of NFAT called cyclosporin A and FK506 (tacrolimus). These drugs inhibit calcineurin and hence prevent the formation of active NFAT. T cells express low levels of calcineurin, so they are more sensitive to inhibition of this pathway than are many other cell types. Both cyclosporin A and FK506 thus act as effective inhibitors of T-cell activation with only limited side-effects (see Chapter 14). These drugs are used widely to prevent graft rejection, which they do by inhibiting the activation of alloreactive T cells.

Insulin-Dependent
Diabetes Mellitus

Fig. 6.17 Initiation of MAP kinase cascades by guanine-nucleotide exchange factors is involved in both antigen receptor and co-stimulatory signaling. Signaling through the antigen receptors of both B and T cells (left panel) leads to the activation of the guanine-nucleotide exchange factor SOS, which activates the small G protein Ras. Ras activates the MAP kinase cascade, where Raf, Mek, and Erk activate each other in turn. Erk finally phosphorylates and activates the transcription factor Elk, which enters the nucleus to initiate new gene transcription, particularly of the *Fos* gene. Signaling through the B-cell co-receptor or through the co-stimulatory receptor CD28 on T cells leads to the recruitment and activation of a different guanine-nucleotide exchange factor, Vav (right panel). Activation of Vav initiates a second MAP kinase cascade as Vav activates the small G protein Rac, which initiates the sequential activation of Mekk, Jnkk, and Jnk. Finally, Jnk (which stands for Jun N-terminal kinase) phosphorylates and activates Jun, which, together with Fos, forms the AP-1 transcription factor.

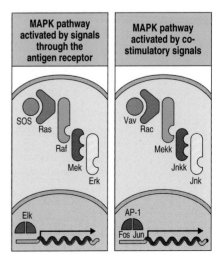

MAPK pathway activated by signals through the antigen receptor

SOS
Ras
Raf
Mek
Erk
Elk

MAPK pathway activated by co-stimulatory signals

Vav
Rac
Mekk
Jnkk
Jnk
AP-1
Fos Jun

6-12   Not all ligands for the T-cell receptor produce a similar response.

So far, we have assumed that all peptide:MHC complexes that are recognized by a given T-cell receptor will activate the T cell equally—that is, that the T-cell receptor is a binary switch with only two settings, 'on' and 'off.' Peptides that trigger the 'on' setting are called **agonist peptides**, by analogy with the agonist drugs that activate other types of receptor. However, experiments originally designed to explore the structural basis for antigen recognition by T cells unexpectedly showed that recognition does not necessarily lead to activation. Indeed, some variant peptide:MHC complexes that do not themselves activate a given T cell can actually inhibit its response to the agonist peptide:MHC complex. These peptides are usually called **antagonist peptides**, as they antagonize the action of the agonist peptide. Other peptides that can trigger only a part of the program activated by the agonist peptide are called **altered peptide ligands** or **partial agonists**. Such ligands can, for instance, induce the lymphocyte to secrete cytokines but not to proliferate. Most of the antagonist and partial agonist peptides that have been described are closely related in sequence to the agonist peptide, although this is not always the case. Several studies have suggested that the binding interactions between T-cell receptors and these variant peptide:MHC complexes are of a lower affinity than the binding interactions between a T-cell receptor and its agonist peptide:MHC ligand.

The extent to which antagonist peptides and altered peptide ligands influence physiological immune responses is not known, although they might contribute to the persistence of some viral infections. For example, mutant peptides of epitopes on cells infected with human immunodeficiency virus (HIV) can inhibit the activation of CD8 T cells specific for the original agonist epitope at a 1:100 ratio of mutant to agonist; this sort of effect could allow cells infected with a mutant virus, which arise commonly during the course of HIV infection, to inhibit virus-specific cytotoxic T cells and thereby promote the survival of all infected cells.

Differential signaling by variants of agonist peptides is thought to be important for the development of T cells in the thymus and their subsequent maintenance in the periphery. As we will discuss in Chapter 7, T cells are selected for their potential to be activated by foreign peptide antigens in the context of self MHC molecules through interactions with self peptide:self MHC complexes. This process seems to require the receipt of survival signals from self peptide:self MHC complexes that are related to, but not identical with, the foreign peptide:self MHC complexes that the mature T cell will recognize. We will discuss the evidence for survival signaling by altered peptide ligands further in Section 6-20.

The basis of the incomplete activating signal delivered by altered peptide ligands is unknown, but it has been shown that the recognition of these ligands leads to altered phosphorylation of CD3ε and ζ chains, and to the recruitment of the inactive form of ZAP-70 tyrosine kinase to the T-cell receptor. Cross-linking the T-cell receptor alone, without any co-receptor engagement, can also generate the same partial phosphorylation events within the cell. Thus, the incomplete signal generated by altered peptide ligands might reflect a failure to recruit the co-receptors or a failure of the T-cell receptor to interact with the co-receptor productively. Also, because the affinity of the T-cell receptor for the altered peptide ligand:MHC complex is lower than that for the activating complex, the T-cell receptor might dissociate too quickly from its ligand for a full activating signal to be delivered. Another possibility is that conformational changes in the T-cell receptor can contribute to the life-span of the signaling complex and its ability to recruit Lck, and that altered peptide ligands fail to trigger these conformational changes.

Antagonist signaling also recruits the tyrosine phosphatase SHP-1 to the T-cell receptor signaling complex, where it can dephosphorylate the signaling complex (see Section 6-14). This recruitment, which is seen late in activation by agonist peptides, occurs within one minute in T cells recognizing antagonist peptides.

6-13 Other receptors on leukocytes also use ITAMs to signal activation.

Although this chapter is focused on lymphocyte antigen receptors, which are the signaling machines that regulate adaptive immunity, other receptors on immune system cells also use the ITAM motif to transduce activating signals (Fig. 6.18). One example is FcγRIII (CD16); this is a receptor for IgG that triggers antibody-dependent cell-mediated cytotoxicity (ADCC) by NK cells, which we will learn about in Chapter 9; it is also found on macrophages and neutrophils, where it facilitates the uptake and destruction of antibody-bound pathogens. FcγRIII is associated with either an ITAM-containing ζ chain like those found in the T-cell receptor complex or with a second member of the same protein family known as the γ chain. The γ chain is also associated with another type of receptor—the Fcε receptor I (FcεRI) on mast cells. As we will discuss in Chapter 12, this receptor binds IgE antibodies and on cross-linking by allergens it triggers the degranulation of mast cells.

Other members of the ITAM-containing ζ-chain family have recently been discovered in NK cells. These proteins, called DAP10 and DAP12, each have only one ITAM, but in other respects they are similar to ζ chains and can form ζ-like homodimers in the membranes of NK cells. Both DAP10 and DAP12 associate with receptors called **killer activatory receptors (KARs)**. As we discussed in Section 2-27, these receptors can activate NK cells to kill infected or abnormal target cells. The killer activatory receptors signal through their associated ITAM-containing homodimer for the release of the cytotoxic granules by which NK cells kill their targets.

Several viral pathogens appear to have acquired ITAM-containing receptors from their hosts. These include the Epstein–Barr virus (EBV), whose *LMP2A* gene has a cytoplasmic tail which encodes an ITAM. This allows EBV to trigger B-cell proliferation by using the downstream signaling molecules we discussed in Sections 6-9 to 6-11. Other viruses also express genes that encode proteins with ITAMs in their transmembrane regions and also induce transformation. An example is the Kaposi sarcoma herpes virus (KSHV or HHV8).

6-14 Antigen-receptor signaling can be inhibited by receptors associated with ITIMs.

Both B and T cells receive signals that can counteract and modify the activating signals delivered through antigen receptors and co-receptors. These inhibitory signals usually block the response by raising the threshold at which signal transduction can occur. Most of these modifying signals are received through receptors that bear a distinct motif called an **immunoreceptor tyrosine-based inhibitory motif (ITIM)** in their cytoplasmic tails. In this motif, a large hydrophobic residue such as isoleucine or valine occurs two residues upstream of a tyrosine that is followed by two amino acids and a leucine to give the amino acid sequence ...[I/V]XYXXL... (Fig. 6.19).

The ITIM motif is found in several receptors that modulate activation signals in lymphocytes. It functions by recruiting one or other of the inhibitory phosphatases SHP-1, SHP-2 and SHIP. These phosphatases carry one or more SH2 domains that preferentially bind the phosphorylated tyrosines in the ITIM. SHP-1 and SHP-2 are protein tyrosine phosphatases and remove the phosphate

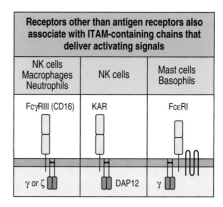

Receptors other than antigen receptors also associate with ITAM-containing chains that deliver activating signals		
NK cells Macrophages Neutrophils	NK cells	Mast cells Basophils
FcγRIII (CD16)	KAR	FcεRI
γ or ζ	DAP12	γ

Fig. 6.18 Other receptors that pair with ITAM-containing chains can deliver activating signals. Cells other than B and T cells have receptors that pair with accessory chains containing ITAMs, which are phosphorylated when the receptor is cross-linked. These receptors deliver activating signals. The Fcγ receptor III (CD16) is found on NK cells, macrophages, and neutrophils. Binding of IgG to this receptor activates the killing function of the NK cell, leading to the process known as antibody-dependent cell-mediated cytotoxicity (ADCC). Killer activatory receptors (KARs) are also found on NK cells. The Fcε receptor (FcεRI) is found on mast cells and basophils. It binds to IgE antibodies with very high affinity. When antigen subsequently binds to the IgE, the mast cell is triggered to release granules containing inflammatory mediators. The γ chain associated with the Fc receptors, and the DAP12, or similar DAP10, chain that associates with the NK killer activatory receptors contain one ITAM per chain and are present as homodimers.

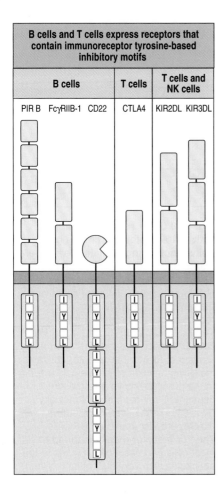

B cells and T cells express receptors that contain immunoreceptor tyrosine-based inhibitory motifs					
B cells			**T cells**	**T cells and NK cells**	
PIR B	FcγRIIB-1	CD22	CTLA4	KIR2DL	KIR3DL

Fig. 6.19 Some lymphocyte cell-surface receptors contain motifs involved in downregulating activation. Several receptors that transduce signals that inhibit lymphocyte or NK cell activation contain ITIMs (immunoreceptor tyrosine-based inhibitory motifs) in their cytoplasmic tails. These bind to various phosphatases that, when activated, inhibit signals derived from ITAM-containing receptors. Analysis of the mode of action of these receptors has implicated lipid and protein phosphatases that contain SH2 domains.

groups added by tyrosine kinases. SHIP is an inositol phosphatase and removes the 5′ phosphate from phosphatidylinositol trisphosphate (PI-3,4,5-P); the exact mechanism by which SHIP regulates signaling is not known, but it is thought to inhibit the activation of PLC-γ by inhibiting the recruitment of the Tec family of kinases, including Btk and Itk, and thus the production of DAG and IP$_3$ and the associated mobilization of calcium.

One example of an ITIM-containing receptor that inhibits lymphocyte activation is the B-cell Fc receptor for IgG known as FcγRIIB-1. It has long been known that the activation of naive B cells in response to antigen can be inhibited by soluble IgG antibodies that recognize the same antigen and therefore co-ligate the B-cell receptor with this Fc receptor. It was only recently, however, that the FcγRIIB-1 ITIM motif was defined and shown to function by drawing SHIP into a complex with the B-cell receptor.

Several other receptors on B and T cells contain the ITIM motif and inhibit cell activation when they are ligated along with the antigen receptors. One example is CD22, a B-cell transmembrane protein that inhibits B-cell signaling. PIR-B, the paired immunoglobulin-like receptor, also expressed on B cells, contains an ITIM that allows it to interact with the tyrosine phosphatase SHP-1. In T cells, the transmembrane protein CTLA-4 is induced by activation and then has a critical role in regulating T-cell signaling; it binds to the same co-stimulatory molecules as CD28 and inhibits signaling through the T-cell receptor by recruiting the tyrosine phosphatase SHP-2. Antibodies that block recognition by CTLA-4 of its ligands lead to massive increases in T-cell effector function. We will return to the therapeutic consequences of this in Chapter 14, when we look at how T cells can be induced to reject established tumors. The killer inhibitory receptors (KIR) on NK cells and T cells also have ITIM motifs; they recognize MHC class I molecules and transmit signals that inhibit the release of cytotoxic granules when NK cells recognize healthy uninfected cells (see Section 2-27). These examples illustrate how the ITIM motif is able to modify signals originating from the B-cell receptor, the T-cell receptor, Fc receptors, and NK cell receptors, even to the extent of completely negating them.

Summary.

Lymphocyte antigen receptors are multiprotein complexes made up of variable antigen-binding chains and invariant chains that transmit the signal that antigen has bound. The cytoplasmic tails of the invariant chains contain amino acid motifs called ITAMs, each possessing two tyrosine residues, that are targeted by receptor-associated protein tyrosine kinases of the Src family upon receptor aggregation. The B-cell receptor complex is associated with two such ITAMs, whereas the T-cell receptor complex is associated with 10, the larger number allowing the T-cell receptor a greater flexibility in signaling. Once the ITAMs have been phosphorylated by an Src-family kinase, the Syk-family kinases—Syk in B cells and ZAP-70 in T cells—bind and become activated. Linker and adaptor proteins are subsequently phosphorylated and serve to recruit enzymes that are activated by relocalization to the plasma membrane, by phosphorylation, or by both. Receptor phosphorylation initiates several signaling pathways, including those propagated through phospholipase C-γ and the small G proteins. These pathways converge on the nucleus and result in new patterns of gene expression. The small G proteins activate a cascade of serine/threonine protein kinases known as a MAP kinase cascade, which leads to the phosphorylation and activation of transcription factors. Signaling through the antigen receptors can be enhanced by signaling through the B-cell co-receptor on B cells or the CD4 and CD8 co-receptor molecules on T cells; signaling through the co-stimulatory receptor CD28 on T cells also contributes to activating naive T cells. Activating signals can be modulated or inhibited by signals from inhibitory receptors that are associated with chains containing a different motif, ITIM, in their cytoplasmic tails. This provides a mechanism for tuning the on/off threshold according to external stimuli or the state of development of the cell, allowing the modulation of the adaptive immune response. Signaling through the T-cell receptor also occurs in response to altered peptide or antagonist ligands, leading to a state of partial activation that can affect cell survival and the response to agonist ligands.

Other signaling pathways that contribute to lymphocyte behavior.

Lymphocytes are normally studied in terms of their responsiveness to antigen. However, they bear numerous other receptors that make them aware of events occurring both in their immediate neighborhood and at distant sites. Among these are receptors that detect the presence of infection and receptors that bind cytokines produced by the cell itself or reaching it from elsewhere. In the absence of infection, lymphocyte populations are also kept remarkably constant in numbers. This **homeostasis** is achieved by a host of extracellular factors that interact with receptors on lymphocytes, the most important of which is the antigen receptor. Other receptors and ligands that come into play include Fas and its ligand, various cytokine receptors and their ligands, and intracellular proteins such as Bcl-2 that modulate survival.

6-15   Microbes and their products release NFκB from its site in the cytosol through an ancient pathway of host defense against infection.

As we saw in Chapter 2, pathogens that infect the body are detected by the germline-encoded receptors of the innate immune system. One family of receptors that signals the presence of infection by microbes is known as the Toll family of receptors. The Toll receptor was originally identified in the fruit fly *Drosophila melanogaster* on account of its role in determining dorsoventral body pattern during fly embryogenesis. Later, it was found to participate in signaling in response to infection in adult flies. A close homologue of Toll has been identified in mammals, and similar proteins are also used by plants in their defense against viruses, indicating that the Toll pathway is an ancient signaling pathway that is used in innate defenses in most multicellular organisms.

In mammals, the Toll pathway leads to the activation of a transcription factor known as NFκB. Microbial substances such as lipopolysaccharide (LPS), a component of the cell wall of gram-negative bacteria, and many other microbes and their products can activate NFκB in lymphocytes and other cells through this pathway. LPS signals through a mammalian Toll-like receptor known as TLR-4, with which it interacts indirectly, through association with another cellular receptor (CD14) for LPS (see Section 2-17). Other microbial substances can interact directly with other Toll-like receptors; gram-positive bacteria, for example, interact with TLR-2. Toll-like receptors signal for NFκB activation through a pathway that seems to operate in most multicellular organisms. The cytoplasmic domain of the Toll receptor is known as a TIR domain because it is also found in the cytoplasmic tail of the receptor for the cytokine interleukin-1 (IL-1R). Ligand binding to the extracellular portion of the Toll receptor induces the TIR domain to bind and activate an adaptor protein known as MyD88. The IL-1 receptor also binds and activates MyD88 on binding IL-1. As the ligands recognized by the various Toll-like receptors and their downstream connections are now fully worked out, we illustrate this pathway in Fig. 6.20 and describe it below; the pathway from the IL-1R is very similar.

At one end, the MyD88 protein also has a TIR domain through which it interacts with the TIR domain of Toll or IL-1R. At the other end of the MyD88 protein is another domain that mediates protein–protein interactions. This is known as a **death domain** because it was first found in proteins involved in programmed cell death. Through this death domain, the bound adaptor

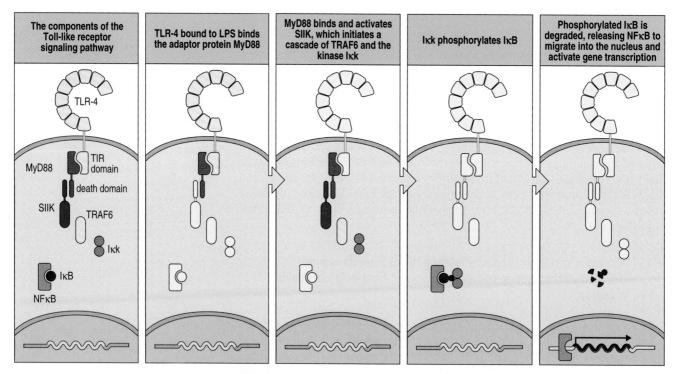

Fig. 6.20 The transcription factor NFκB is activated by signals from receptors of the Toll-like receptor (TLR) family. The cytoplasmic domain of TLR-4 is homologous to that of the receptor for the cytokine IL-1 and is called a TIR (Toll/IL-1R) domain. This domain can interact with other TIR domains, including that found in the adaptor protein MyD88. MyD88 also contains a death domain through which it interacts with the death domain in a serine/threonine innate immunity kinase (SIIK), which in turn activates TRAF6. TRAF6 activates Iκ kinase-α and Iκ kinase-β, which form the dimer Iκk. Iκk phosphorylates IκB, a cytosolic protein that is retaining the transcription factor NFκB in the cytoplasm. Once phosphorylated, IκB dissociates from its complex with NFκB, allowing NFκB to enter the nucleus and activate genes involved in host defense against infection.

protein interacts with another death domain on a serine/threonine innate immunity kinase (SIIK) known as IRAK, or the IL-1R-associated kinase. This initiates a kinase activation cascade through which two kinases known as Iκkα and Iκkβ are activated to form a dimer (Iκk) that phosphorylates an inhibitory protein known as IκB. This protein is bound in a complex in the cytosol with the transcription factor NFκB and inhibits its action by retaining it there. When IκB is phosphorylated, it dissociates from the complex and is rapidly degraded by proteasomes. After removal of IκB, NFκB enters the nucleus and binds to various promoters, activating genes that contribute to adaptive immunity and the secretion of pro-inflammatory cytokines. Also activated is the gene for IκB itself, which is rapidly synthesized and inactivates the NFκB signal.

The responses activated by this pathway depend upon the cell type and organism concerned: in adult *Drosophila melanogaster*, Toll-family receptors induce the expression of antimicrobial peptides in response to bacterial or fungal infection. In mammals, as we saw in Chapter 2, the Toll signaling pathway contributes to the initiation of an adaptive immune response by inducing the expression of co-stimulatory molecules on tissue dendritic cells and the migration of these cells from a site of infection to a local lymph node. Here they can function as antigen-presenting cells and provide a naive lymphocyte that recognizes its antigen with a co-stimulatory signal that combines with the antigen-receptor signal to drive clonal expansion and differentiation.

6-16   Bacterial peptides, mediators of inflammatory responses, and chemokines signal through members of the seven-transmembrane-domain, trimeric G protein-coupled receptor family.

Another way in which cells in the innate immune system are able to detect the presence of infection is by binding bacterial peptides containing *N*-formylmethionine, or fMet, a modified amino acid that initiates all proteins synthesized in prokaryotes. The receptor that recognizes these peptides is known as the fMet-Leu-Phe (fMLP) receptor, after a tripeptide for which it has a high affinity, though it is not restricted to binding just this tripeptide. The fMLP receptor belongs to an ancient and widely distributed family of receptors that have seven membrane-spanning segments; the best-characterized members of this family are the photoreceptors rhodopsin and bacteriorhodopsin. In the immune system, members of this family of receptors have a number of essential roles; the receptors for the anaphylotoxins (see Section 2-12) and for chemokines (see Section 2-20) belong to this family.

All receptors of this family use the same mechanism of signaling; ligand binding activates a member of a class of GTP-binding proteins called **G proteins**. These are sometimes called large G proteins, to distinguish them from the smaller Ras-like family of GTP-binding proteins, or heterotrimeric G proteins, as each is made up of three subunits—Gα, Gβ, and Gγ. Roughly 20 different large G proteins are known, each interacting with different cell-surface receptors and transmitting signals to different intracellular pathways. In the resting state, the trimeric G protein is inactive, not associated with the receptor and has a molecule of GDP bound to the α subunit. When the receptor binds its ligand, a conformational change in the receptor allows it now to bind the G protein, displacing the GDP molecule from the G protein and replacing it with a molecule of GTP. The G protein now dissociates into two components, the α subunit and the combined βγ subunits; each of these components is capable of interacting with other cellular components to transmit and amplify the signal. Binding of the α subunit to its ligand activates the GTPase activity of this subunit, cleaving the molecule of GTP to GDP, and thus allowing the α and βγ subunits to reassociate (Fig. 6.21).

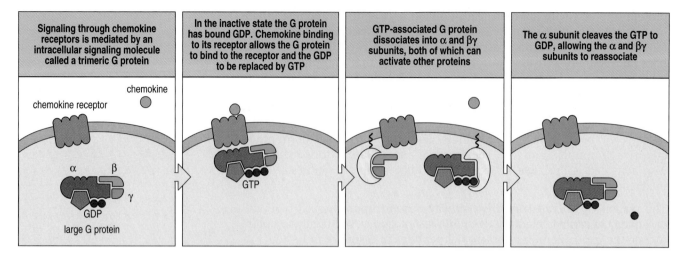

| Signaling through chemokine receptors is mediated by an intracellular signaling molecule called a trimeric G protein | In the inactive state the G protein has bound GDP. Chemokine binding to its receptor allows the G protein to bind to the receptor and the GDP to be replaced by GTP | GTP-associated G protein dissociates into α and βγ subunits, both of which can activate other proteins | The α subunit cleaves the GTP to GDP, allowing the α and βγ subunits to reassociate |

Fig. 6.21 Seven-transmembrane-domain receptors signal by coupling with trimeric GTP-binding proteins. Seven-transmembrane-domain receptors such as the chemokine receptors signal through trimeric GTP-binding proteins known as large G proteins. In the inactive state, the G protein is bound to GDP. When the receptor binds its ligand, the G protein binds to the receptor and the GDP is replaced by GTP. This triggers dissociation of the trimeric G protein into α and βγ subunits, both of which can activate other proteins at the inner surface of the cell membrane. The activated response ceases when the α subunit, which contains intrinsic GTPase activity, cleaves the GTP to GDP, allowing the α and βγ subunits to reassociate.

Important targets for the activated G protein subunits are adenylate cyclase and phospholipase C, whose activation gives rise to the second messengers cyclic AMP, IP_3 and Ca^{2+}. These in turn activate a variety of intracellular pathways that affect cell metabolism, motility, gene expression, and cell division. Thus activation of G protein-coupled receptors can have a wide variety of effects depending on the exact nature of the receptor and the G proteins that it interacts with, as well as the different downstream pathways that are activated in different cell types.

6-17 Cytokines signal lymphocytes by binding to cytokine receptors and triggering Janus kinases to phosphorylate and activate STAT proteins.

Cytokines, which we encountered in Chapter 2, are small proteins (of ~20 kDa) that each act on a specific receptor. They are secreted by a variety of cells, usually in response to an external stimulus, and they can then act on the cells that produce them (autocrine action), on other cells in the immediate vicinity (paracrine action), or on cells at a distance (endocrine action) after being carried in blood or tissue fluids. Cytokines affect cell behavior in a variety of ways and, as we will see in subsequent chapters, they play key roles in controlling the growth, development, and functional differentiation of lymphocytes, and as effector molecules of activated T cells. Many cytokines bind to receptors that use a particularly rapid and direct signaling pathway to effect changes in gene expression in the nucleus.

Cytokine binding to such receptors activates receptor-associated tyrosine kinases of the **Janus kinase** family (**JAKs**), so-called because they have two symmetrical kinase-like domains, and thus resemble the two-headed mythical Roman god Janus. These kinases then phosphorylate cytosolic proteins called **signal transducers and activators of transcription** (**STATs**). Phosphorylation of STAT proteins leads to their homo- and heterodimerization through interactions involving their SH2 domains; STAT dimers can then translocate to the nucleus, where they activate various genes (Fig. 6.22). The proteins encoded by these genes contribute to the growth and differentiation of particular subsets of lymphocytes.

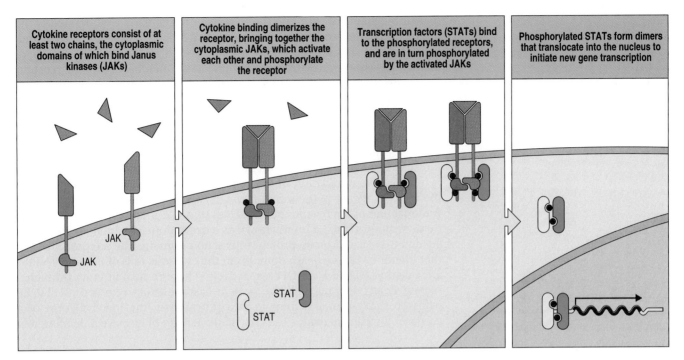

| Cytokine receptors consist of at least two chains, the cytoplasmic domains of which bind Janus kinases (JAKs) | Cytokine binding dimerizes the receptor, bringing together the cytoplasmic JAKs, which activate each other and phosphorylate the receptor | Transcription factors (STATs) bind to the phosphorylated receptors, and are in turn phosphorylated by the activated JAKs | Phosphorylated STATs form dimers that translocate into the nucleus to initiate new gene transcription |

Fig. 6.22 Many cytokine receptors signal by a rapid pathway using receptor-associated kinases to activate specific transcription factors. Many cytokines act via receptors that are associated with cytoplasmic Janus kinases (JAKs). The receptor consists of at least two chains, each associated with a specific JAK (left panel). Ligand binding and dimerization of the receptor chains brings together the JAKs, which can transactivate each other, subsequently phosphorylating tyrosines in the receptor tails (second panel). Members of the STAT (signal transducer and activators of transcription) family of proteins bind to the phosphorylated receptors and are themselves phosphorylated by the JAKs (third panel). On phosphorylation, STAT proteins dimerize by binding phosphotyrosine residues in SH2 pockets and go rapidly to the nucleus, where they bind to and activate transcription of a variety of genes important for adaptive immunity.

In this pathway, gene transcription is activated very soon after the cytokine binds to its receptor, and specificity of signaling in response to different cytokines is achieved by using different combinations of JAKs and STATs. This signaling pathway is used by most of the cytokines that are released by T cells in response to antigen. Although cytokines are not in themselves antigen specific, their effects can be targeted in an antigen-specific manner by their directed release in antigen-specific cell–cell interactions and their selective action on the cell that triggers their production, as we will see in Chapter 8.

6-18   Programmed cell death of activated lymphocytes is triggered mainly through the receptor Fas.

When antigen-specific lymphocytes are activated through their antigen receptors in an adaptive immune response, they first undergo blast transformation and begin to increase their numbers exponentially by cell division. This clonal expansion can continue for up to 7 or 8 days, so that lymphocytes specific for the infecting pathogen increase vastly in numbers and can come to predominate in the population. In the response to certain viruses, nearly 50% of the CD8 T cells at the peak of the response are specific for a single virus-derived peptide:MHC class I complex. After clonal expansion, the activated T cells undergo their final differentiation into effector cells; these remove the pathogen from the body, which terminates the antigenic stimulus.

When the infection has terminated, the activated effector T cells are no longer needed and cessation of the antigenic stimulus prompts them to undergo programmed cell death or apoptosis. Apoptosis can probably be induced by several mechanisms, but one that has been particularly well defined is the interaction of the receptor molecule **Fas** on T cells with its ligand, **Fas ligand**. Fas ligand is a member of the tumor necrosis factor (TNF) family of membrane-associated cytokines, whereas Fas is a member of the TNF receptor family. Both Fas and its ligand are normally induced during the course of an adaptive immune response. TNF and its receptor TNFR-I can act in a similar way to Fas ligand and Fas but their actions are far less significant.

All pathways inducing apoptosis lead to the activation of a series of cysteine proteases that cleave protein chains after aspartic acid residues and have therefore been called **caspases**. In the case of activated lymphocytes, apoptosis is initiated by stimulation of the receptors Fas or TNFR-I. The ligands for these receptors are in the form of trimers, and when they bind, they induce trimerization of the receptors themselves (Fig. 6.23). The cytoplasmic tails of these receptors share a motif known as a death domain which, as we saw in Section 6-15, is a protein–protein interaction domain. The adaptor proteins that interact with the death domains in the cytosolic tails of Fas and TNFR-I are called FADD and TRADD respectively. These in turn interact through a second death domain with the protein caspase 8 (also known as FLICE), whose carboxy-terminal domain is a procaspase (the inactive form of a caspase). Binding activates the enzymatic activity of caspase 8, leading to a

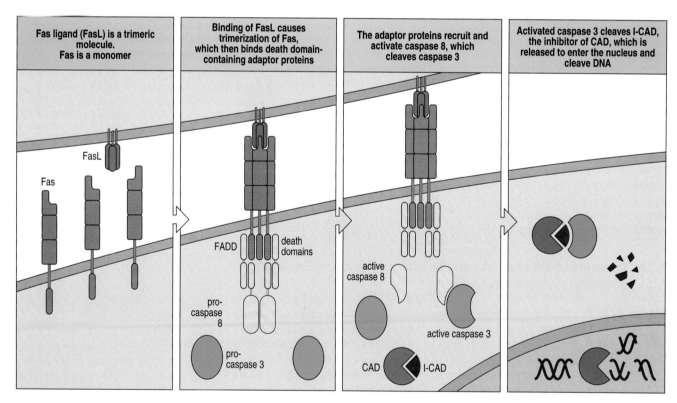

Fig. 6.23 Binding of Fas ligand to Fas initiates the process of apoptosis. The Fas ligand (FasL) recognized by Fas is a homotrimer, and when it binds it induces the trimerization of Fas. This brings the death domains in the Fas cytoplasmic tails together. A number of adaptor proteins containing death domains bind to the death domains of Fas, in particular the protein FADD, which in turn interacts through a second death domain with the protease caspase 8. Clustered caspase 8 can transactivate, cleaving caspase 8 itself to release an active caspase domain that in turn can activate other caspases. The ensuing caspase cascade culminates in the activation of the caspase-activatable DNase (CAD), which is present in all cells in an inactive cytoplasmic form bound to an inhibitory protein called I-CAD. When I-CAD is broken down by caspases, CAD can enter the nucleus where it cleaves DNA into the 200-base-pair fragments that are characteristic of apoptosis.

protease cascade in which activated caspases cleave and activate a succession of downstream caspases. At the end of this pathway a caspase-activated DNase (CAD) enters the nucleus and cleaves DNA to produce the DNA fragments characteristic of an apoptotic cell. These fragments can be labeled by a procedure known as TUNEL (see Appendix I, Section A-32) that selectively stains and detects only those cells that have undergone apoptosis.

Mutations in the genes encoding Fas or Fas ligand have now been identified in both mice and humans. These mutations are associated with an excessive accumulation of abnormal T cells that lack both co-receptor proteins and express the CD45 isoform usually expressed by B cells. It is thought that these cells have been activated but subsequently failed to die. The mutations that cause this phenotype are mostly recessive; that is, both copies of the gene for either Fas or Fas ligand must be defective to produce an effect. However, in some cases in humans, the mutant phenotype is seen in heterozygous individuals. The production of an effect in heterozygotes is likely to reflect the need for trimerization of Fas for efficient operation of the Fas–Fas ligand interaction, and the fact that if one of the members of the trimer is mutant, the trimer cannot transduce a signal.

Autoimmune
Lymphoproliferative
Syndrome

6-19 Lymphocyte survival is maintained by a balance between death-promoting and death-inhibiting members of the Bcl-2 family of proteins.

Apoptosis plays a major role in the development and maintenance of all multicellular organisms. The apoptotic program is present in all cells and may be triggered by an absence of appropriate survival signals as well as by external stimuli as described for Fas-induced cell death above. Thus it is not surprising that all cells also possess a separate set of proteins that can inhibit programmed cell death.

The first member of this family of proteins was discovered as an oncogene in B cells. When tumors form in the B-cell lineage, they are frequently associated with chromosomal translocations in which the chromosomal DNA is broken and an active immunoglobulin locus is joined to a gene that affects cell growth, usually activating that gene in the process. By cloning the DNA breakpoints, one can isolate the gene that has been activated by the translocation. One such gene is *bcl-2*, which was isolated from the second B-cell lymphoma to have its breakpoint identified. *bcl-2* is homologous with the *Caenorhabditis elegans* gene *ced-9*, which is a cell death-inhibitory gene. In cultured B cells and in transgenic mice, the expression of *bcl-2* protects against cell death. *bcl-2* is a member of a small family of closely related genes, some of which inhibit cell death whereas others promote it. These genes can be divided into death-inhibiting genes, such as *bcl-2* and *bcl-X_L*, and death-promoting genes, such as *Bax* and *Bad*. The proteins encoded by these genes act as dimers, and as Bcl-2 and Bax proteins can dimerize with each other to form heterodimers, the more abundant protein determines whether the cell lives or dies. One way in which Bcl-2 acts to prevent cell death is shown in Fig. 6.24.

The balance between death-promoting and death-inhibiting gene expression is critically important in lymphocytes, because lymphocyte populations are regulated so that a person will, in the absence of infection, maintain a constant level of T and B cells despite the production and death of many lymphocytes each day. The fate of individual lymphocytes is set by signals delivered mainly or entirely through the antigen-specific receptors, as we will learn in Chapter 7, which deals with the production of the mature repertoire of receptors on B and T lymphocytes. In the last section of this chapter we will look at the evidence for a continued role of antigen-receptor signaling in maintaining the survival of mature T and B cells.

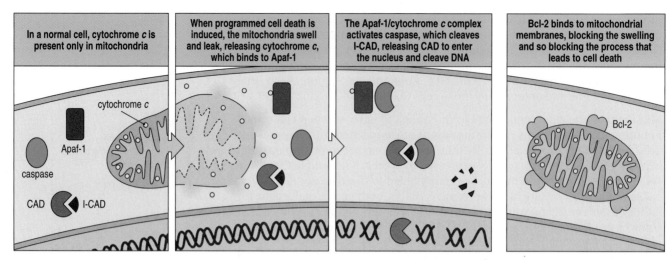

Fig. 6.24 Bcl-2 inhibits the processes that lead to programmed cell death. In normal cells, cytochrome *c* is confined to the mitochondria (first panel). However, during apoptosis the mitochondria swell, allowing the cytochrome *c* to leak out into the cytosol (second panel). There it interacts with the protein Apaf-1, forming a cytochrome *c*:Apaf-1 complex that can activate caspases. An activated caspase cleaves I-CAD (see Fig. 6.23), which leads to DNA fragmentation (third panel). Bcl-2 interacts with the mitochondrial outer membrane and blocks the mitochondrial swelling that leads to cytochrome *c* release (last panel).

6-20 Homeostasis of lymphocyte populations is maintained by signals that lymphocytes are continually receiving through their antigen receptors.

Most of what we know about signaling through the B-cell receptor and the T-cell receptor derives from observations in cultured B-cell and T-cell lines. In these cells, as discussed in Section 6-9, the tyrosine kinases Syk and ZAP-70 are recruited as a consequence of receptor clustering and the subsequent phosphorylation of ITAMs in the receptor subunit tails. However, in T cells isolated directly from lymph nodes or the thymus, ZAP-70 is already bound to partly phosphorylated ζ chains but is not yet activated. The same state can be produced in both normal naive T cells and cultured T cells by exposure to altered peptide ligands (see Section 6-12). It therefore seems likely that the phosphorylated ITAMs and bound ZAP-70 seen in mature naive T cells *in vivo* reflect the receipt of signals from self MHC molecules bound to self peptides that behave like altered peptide ligands.

As we will see in Chapter 7, the ability to interact with self peptide:self MHC ligands is a criterion for survival during T-cell development in the thymus. Developing T cells are subject to stringent testing once the α:β T-cell receptor is expressed. This selection process retains those cells whose receptors interact effectively with various self peptide:self MHC ligands (positive selection), and removes those T cells that either cannot participate in such interactions (death by neglect) or recognize a self peptide:self MHC complex so well that they could damage host cells if allowed to mature; such cells are removed by clonal deletion (negative selection). Those T cells that mature and emerge into the periphery have therefore been selected for their ability to recognize self MHC:self peptide complexes without being fully activated by them.

A number of studies have now shown that naive T cells in the peripheral lymphoid tissues continue to receive signals through interactions with self MHC:self peptide complexes. These signals enable the cells to survive and are delivered most effectively by the cells that are most capable of T-cell activation, namely the dendritic cells. Thus, the T cells in mice transgenic for a

T-cell receptor that is positively selected on a particular self MHC:self peptide complex can survive only if they receive similar signals from self MHC:self peptide complexes in the periphery. It seems as though the peripheral T-cell repertoire is maintained by a continuous dialogue between dendritic cells presenting self peptides on self MHC molecules and recirculating naive T cells that were positively selected by signals from the same self peptide:self MHC molecules during development. This dialogue is likely to account for the state of partial phosphorylation of the ζ chains discussed above. A requirement for repeated interaction with dendritic cells resident in peripheral lymphoid tissue could also account for the tight regulation of the numbers of naive CD4 and CD8 cells. For memory T cells, the factors that govern survival are less well defined. It is clear that self MHC:self peptide recognition is not required for the survival of memory T cells; in fact, the presence of MHC molecules does not seem to be required at all. Rather, it is likely that the levels of particular cytokines serve to maintain the memory T-cell pool.

In B cells, it is also clear that signaling through the antigen receptor determines cell survival from the time that it is first expressed on the cell surface. As we will see in Chapter 7, autoreactive B cells are induced to die on binding antigen. However, the expression of a functional B-cell receptor at the cell surface is also essential for cell maturation and survival. The role of the B-cell receptor in signaling for survival in the periphery was demonstrated quite dramatically using a conditional gene knockout strategy (see Appendix I, Section A-47). Mice were made transgenic for a rearranged immunoglobulin V_H gene flanked by *loxP* sites. This construct was 'knocked-in' to the site at which the rearranged V_H gene normally occurs in such a way that only the transgene could produce the heavy-chain V region. The animals were also made transgenic for the enzyme Cre recombinase, which can excise *loxP*-flanked genes; the transgene encoding the Cre recombinase was made inducible by interferon. After treatment with interferon, the induced Cre recombinase removed the transgenic rearranged V_H gene, preventing the production of a heavy chain. Most of the B cells in the treated animals lost their receptors; these receptor-negative B cells rapidly disappeared. Thus, the B-cell receptor is clearly required to keep recirculating B cells alive, and must have a role in perceiving or transmitting survival signals to each B cell. Evidence also shows that receptor specificity can affect this process. However, the ligand or ligands responsible for signaling for B-cell survival are not yet known.

The identity of the ligands responsible for delivering survival signals to T and B cells through their antigen receptors remains an important question in immunology. There is also much to learn about the signaling process itself. Survival signals are likely to regulate the level of proteins in the Bcl-2 family because, as we discussed in Section 6-18, increased levels of Bcl-2 and Bcl-X_L promote the survival of lymphocytes, whereas increased levels of Bax and Bad have an opposing effect. As yet, however, a direct link between antigen-receptor signaling for survival and the regulation of the Bcl-2 family has not been shown.

Summary.

Many different signals govern lymphocyte behavior, only some of which are delivered via the antigen receptor. Lymphocyte development, activation, and longevity are clearly influenced by the antigen receptor, but these processes are also regulated by other extracellular signals. Other signals are delivered in a variety of ways. An ancient signaling pathway with a role in host defense leads rapidly from the IL-1 receptor or a similar receptor called Toll to initiate the detachment and degradation of the inhibitory protein IκB from the transcription factor NFκB, which can then enter the nucleus and activate the transcription of many genes. Many cytokines signal through an express pathway that links receptor-associated JAK kinases to preformed STAT proteins, which after phosphorylation dimerize through their SH2 domains and head for the nucleus. Activated lymphocytes are programmed to die when the Fas receptor that they express binds the Fas ligand. This transmits a death signal, which activates a protease cascade that triggers apoptosis. Lymphocyte apoptosis is inhibited by some members of the intracellular Bcl-2 family and promoted by others. Working out the complete picture of the signals processed by lymphocytes as they develop, circulate, respond to antigen, and die is an immense and exciting prospect.

Summary to Chapter 6.

In adaptive immune responses, signaling through the antigen receptors of naive mature lymphocytes induces the clonal expansion and differentiation of antigen-specific lymphocytes after engagement with foreign antigen, provided that, in the case of T cells, they receive a co-stimulatory signal through CD28. Lymphocyte antigen receptors belong to the general class of receptors that are associated with cytoplasmic protein tyrosine kinases. The antigen-binding chains of the receptors are associated on the cell surface with invariant chains that are responsible for generating an intracellular signal indicating that antigen has bound. These chains contain sequence motifs called ITAMs, which after phosphorylation by receptor-associated tyrosine kinases recruit intracellular signaling molecules to the activated receptor. The intracellular signaling pathway leading from the antigen receptors results in gene activation, new protein synthesis, and the stimulation of cell division. This pathway is subject to regulation at most of its steps; these control points form important checkpoints in the pathways leading to lymphocyte activation and the clonal expansion and differentiation of antigen-specific lymphocytes that occurs during an adaptive immune response. As well as enabling lymphocytes to respond to foreign antigens in an adaptive immune response, signals delivered through the antigen receptors are important in selecting lymphocytes for removal or survival during lymphocyte development in the primary lymphoid organs as well as survival later on in the periphery. The ligands responsible for providing survival signals appear similar to altered peptide ligands in the case of T cells but are unknown in the case of B cells: their identity remains a central question in immunology. Lymphocytes also carry receptors for many other extracellular signals, such as cytokines and Fas ligand. The latter, by interacting with the cell-surface receptor Fas on activated lymphocytes, induces apoptosis and is involved in controlling lymphocyte numbers and in removing activated lymphocytes once an infection has been cleared.

General references.

DeFranco, A.L., and Weiss, A.: **Lymphocyte activation and effector functions.** *Curr. Opin. Immunol.* 1998, **10**:243-367.

Healy, J.L., and Goodnow, C.C.: **Positive versus negative signaling by lymphocyte antigen receptors.** *Annu. Rev. Immunol.* 1998, **16**:645-670.

Weiss, A., and Littman, D.R.: **Signal transduction by lymphocyte antigen receptors.** *Cell* 1994, **76**:263-274.

Section references.

6-1 **Binding of antigen leads to clustering of antigen receptors on lymphocytes.**

Deller, M.C., and Yvonne Jones, E.: **Cell surface receptors.** *Curr. Opin. Struct. Biol.* 2000, **10**:213-219.

Metzger, H., Chen, H., Goldstein, B., Haleem-Smith, H., Inman, J.K., Peirce, M., Torigoe, C., Vonakis, B., and Wofsy, C.: **A quantitative approach to signal** transduction. *Immunol. Lett.* 1999, **68**:53-57.

Monks, C.R.F., Kupfer, H., Tamir, I., Barlow, A., and Kupfer, A.: **Selective modulation of protein kinase C-t during T cell activation.** *Nature* 1997, **385**:83-86.

6-2   Clustering of antigen receptors leads to activation of intracellular signal molecules.

Germain, R.N.: **T-cell signaling: the importance of receptor clustering.** *Curr. Biol.* 1997, **7**:R640-R644.

Klemm, J.D., Schreiber, S.L., and Crabtree, G.R.: **Dimerization as a regulatory mechanism in signal transduction.** *Annu. Rev. Immunol.* 1998, **16**:569-592.

6-3   Phosphorylation of receptor cytoplasmic tails by tyrosine kinases concentrates intracellular signaling molecules around the receptors.

Buday, L.: **Membrane-targeting of signalling molecules by SH2/SH3 domain-containing adaptor proteins.** *Biochim. Biophys. Acta* 1999, **1422**:187-204.

Fisher, M.J., Paton, R.C., and Matsuno, K.: **Intracellular signalling proteins as smart' agents in parallel distributed processes.** *Biosystems* 1999, **50**:159-171.

Kholodenko, B.N., Hoek, J.B., and Westerhoff, H.V.: **Why cytoplasmic signalling proteins should be recruited to cell membranes.** *Trends Cell Biol.* 2000, **10**:173-178.

6-4   Intracellular signaling components recruited to activated receptors transmit the signal onward from the membrane and amplify it.

Pawson, T., and Scott, J.D.: **Signaling through scaffold, anchoring and adaptor proteins.** *Science* 1997, **278**:2075-2080.

Peterson, E.J., Clements, J.L., Fang, N., and Koretzky, G.A.: **Adaptor proteins in lymphocyte antigen receptor signaling.** *Curr. Opin. Immunol.* 1998, **10**:337-344.

6-5   Small G proteins activate a protein kinase cascade that transmits the signal to the nucleus.

Caffrey, D.R., O'Neill, L.A., and Shields, D.C.: **The evolution of the MAP kinase pathways: coduplication of interacting proteins leads to new signaling cascades.** *J. Mol. Evol.* 1999, **49**:567-582.

Cantrell, D.: **T cell antigen receptor signal transduction pathways.** *Annu. Rev. Immunol.* 1996, **14**:259-274.

Henning, S.W., and Cantrell, D.A.: **GTPases in antigen receptor signaling.** *Curr. Opin. Immunol.* 1998, **10**:322-329.

Kyriakis, J.M.: **Making the connection: coupling of stress-activated ERK/MAPK (extracellular-signal-regulated kinase/mitogen-activated protein kinase) core signalling modules to extracellular stimuli and biological responses.** *Biochem. Soc. Symp.* 1999, **64**:29-48.

Mulder, K.M.: **Role of Ras and Mapks in TGF-beta signaling.** *Cytokine Growth Factor Rev.* 2000, **11**:23-35.

Yamamoto, H., Atsuchi, N., Tanaka, H., Ogawa, W., Abe, M., Takeshita, A., and Ueno, H.: **Separate roles for H-Ras and Rac in signaling by transforming growth factor (TGF)-beta. H-Ras is essential for activation of MAP kinase, partially required for transcriptional activation by TGF-beta, but not required for signaling of growth suppression by TGF-beta.** *Eur. J. Biochem.* 1999. **264**:110-119.

6-6   The variable chains of lymphocyte antigen receptors are associated with invariant accessory chains that carry out the signaling function of the receptor.

Borst, J., Jacobs, H., and Brouns, G.: **Composition and function of T-cell receptor and B-cell receptor complexes on precursor lymphocytes.** *Curr. Opin. Immunol.* 1996, **8**:181-190.

Gobel, T.W., and Bolliger, L.: **Evolution of the T cell receptor signal transduction units.** *Curr. Top. Microbiol. Immunol.* 2000, **248**:303-320.

Malissen, B., Ardouin, L., Lin, S.Y., Gillet, A., and Malissen, M.: **Function of the CD3 subunits of the pre-TCR and TCR complexes during T cell development.** *Adv. Immunol.* 1999, **72**:103-148.

6-7   The ITAMs associated with the B-cell and T-cell receptors are phosphorylated by protein tyrosine kinases of the Src family.

Lin, J., and Weiss, A.: **T cell receptor signalling.** *J. Cell Sci.* 2001, **114**:243-244.

Hegedus, Z., Chitu, V., Toth, G.K., Finta, C., Varadi, G., Ando, I., and Monostori, E.: **Contribution of kinases and the CD45 phosphatase to the generation of tyrosine phosphorylation patterns in the T-cell receptor complex zeta chain.** *Immunol. Lett.* 1999, **67**:31-39.

Pao, L.I., Famiglietti, S.J., and Cambier, J.C.: **Asymmetrical phosphorylation and function of immunoreceptor tyrosine-based activation motif tyrosines in B cell antigen receptor signal transduction.** *J. Immunol.* 1998, **160**:3305-3314.

Turner, M., Schweighoffer, E., Colucci, F., Di Santo, J.P., and Tybulewicz, V.L.: **Tyrosine kinase SYK: essential functions for immunoreceptor signalling.** *Immunol. Today* 2000, **21**:148-154.

6-8   Antigen receptor signaling is enhanced by co-receptors that bind the same ligand.

Denny, M.F., Patai, B., and Straus, D.B.: **Differential T-cell antigen receptor signaling mediated by the Src family kinases Lck and Fyn.** *Mol. Cell. Biol.* 2000, **20**:1426-1435.

Garcia, K.C.: **Molecular interactions between extracellular components of the T-cell receptor signaling complex.** *Immunol. Rev.* 1999, **172**:73-85.

6-9   Fully phosphorylated ITAMs bind the protein tyrosine kinases Syk and ZAP-70 and enable them to be activated.

Pao, L.I., and Cambier, J.C.: **Syk, but not Lyn, recruitment to B cell antigen receptor and activation following stimulation of CD45 B cells.** *J. Immunol.* 1997, **158**:2663-2669.

Visco, C., Magistrelli, G., Bosotti, R., Perego, R., Rusconi, L., Toma, S., Zamai, M., Acuto, O., and Isacchi, A.: **Activation of Zap-70 tyrosine kinase due to a structural rearrangement induced by tyrosine phosphorylation and/or ITAM binding.** *Biochemistry* 2000, **39**:2784-2791.

Zoller, K.E., MacNeil, I.A., and Brugge, J.S.: **Protein tyrosine kinases Syk and ZAP-70 display distinct requirements for Src family kinases in immune response receptor signal transduction.** *J. Immunol.* 1997, **158**:1650-1659.

6-10   Downstream events are mediated by proteins that associate with the phosphorylated tyrosines and bind to and activate other proteins.

Fusaki, N., Tomita, S., Wu, Y., Okamoto, N., Goitsuka, R., Kitamura, D., and Hozumi, N.: **BLNK is associated with the CD72/SHP-1/Grb2 complex in the WEHI231 cell line after membrane IgM cross-linking.** *Eur. J. Immunol.* 2000, **30**:1326-1330.

Schraven, B., Marie-Cardine, A., Hubener, C., Bruyns, E., and Ding, I.: **Integration of receptor-mediated signals in T cells by transmembrane adaptor proteins.** *Immunol. Today* 1999, **20**:431-434.

van Oers, N.S.: **T cell receptor-mediated signs and signals governing T cell development.** *Semin. Immunol.* 1999, **11**:227-237.

Wienands, J.: **The B-cell antigen receptor: formation of signaling complexes and the function of adaptor proteins.** *Curr. Top. Microbiol. Immunol.* 2000, **245**:53-76.

Zhang, W., and Samelson, L.E.: **The role of membrane-associated adaptors in T cell receptor signalling.** *Semin. Immunol.* 2000. **12**:35-41.

6-11   Antigen recognition leads ultimately to the induction of new gene synthesis by activating transcription factors.

Jacinto, E., Werlin, G., and Karin, M.: **Cooperation between Syk and Rac1** leads to synergistic JNK activation in T lymphocytes. *Immunity* 1998, **8**:31-41.

Yoshida, H., Nisina, H., Takimoto, H., Marengere, L.E.M., Wakeham, A.C., Bouchard, D., Kong, Y.-Y., Ohteki, T., Shahinian, A., Bachmann, M., Ohashi, P.S., Penninger, J., Crabtree, G.R., and Mak, T.W.: **The transcription factor NF-ATc1 regulates lymphocyte proliferation and Th2 cytokine production.** *Immunity* 1998, **8**:115-124.

6-12 Not all ligands for the T-cell receptor produce a similar response.

Alam, S.M., Davies, G.M., Lin, C.M., Zal, T., Nasholds, W., Jameson, S.C., Hogquist, K.A., Gascoigne, N.R., and Travers, P.J.: **Qualitative and quantitative differences in T cell receptor binding of agonist and antagonist ligands.** *Immunity* 1999, **10**:227-237.

Germain, R.N., and Stefanova, I.: **The dynamics of T cell receptor signaling: complex orchestration and the key roles of tempo and cooperation.** *Annu. Rev. Immunol.* 1999, **17**:467-522.

Itoh, Y., Hemmer, B., Martin, R., and Germain, R.N.: **Serial TCR engagement and down-modulation by peptide:MHC molecule ligands: relationship to the quality of individual TCR signaling events.** *J. Immunol.* 1999, **162**:2073-2080.

Madrenas, J.: **Differential signalling by variant ligands of the T cell receptor and the kinetic model of T cell activation.** *Life Sci.* 1999, **64**:717-731.

Sykulev, Y., Vugmeyster, Y., Brunmark, A., Ploegh, H.L., and Eisen, H.N.: **Peptide antagonism and T cell receptor interactions with peptide-MHC complexes.** *Immunity* 1998, **9**:475-483.

6-13 Other receptors on leukocytes also use ITAMs to signal activation.

Amigorena, S., and Bonnerot, C.: **Fc receptor signaling and trafficking: a connection for antigen processing.** *Immunol. Hev.* 1999, **172**:279-284.

Greenberg, S.: **Modular components of phagocytosis.** *J. Leukoc. Biol.* 1999, **66**:712-717

Lanier, L.L., and Bakker, A.B.: **The ITAM-bearing transmembrane adaptor DAP12 in lymphoid and myeloid cell function.** *Immunol. Today* 2000, **21**:611-614.

Lopez-Botet, M., Bellon, T., Llano, M., Navarro, F., Garcia, P., and de Miguel, M.: **Paired inhibitory and triggering NK cell receptors for HLA class I molecules.** *Hum. Immunol.* 2000, **61**:7-17.

6-14 Antigen-receptor signaling can be inhibited by receptors associated with ITIMs.

Bolland, S., and Ravetch, J.V.: **Inhibitory pathways triggered by ITIM-containing receptors.** *Adv. Immunol.* 1999, **72**:149-177.

Bruhns, P., Vely, F., Malbec, O., Fridman, W.H., Vivier, E., and Daeron, M.: **Molecular basis of the recruitment of the SH2 domain-containing inositol 5-phosphatases SHIP1 and SHIP2 by fcgamma RIIB.** *J. Biol. Chem.* 2000, **275**:37357-37364.

Moretta, A., Biassoni, R., Bottino, C., and Moretta, L.: **Surface receptors delivering opposite signals regulate the function of human NK cells.** *Semin. Immunol.* 2000, **12**:129-138.

Sinclair, N.R.: **Immunoreceptor tyrosine-based inhibitory motifs on activating molecules.** *Crit. Rev. Immunol.* 2000, **20**:89-102.

6-15 Microbes and their products release NFκB from its site in the cytosol through an ancient pathway of host defense against infection.

Hacker, H.: **Signal transduction pathways activated by CpG-DNA.** *Curr. Top. Microbiol. Immunol.* 2000, **247**:77-92.

Medzhitov, R., and Janeway, C., Jr.: **The toll receptor family and microbial recognition.** *Trends Microbiol.* 2000, **8**:452-456.

Muzio, M., Polentarutti, N., Bosisio, D., Manoj Kumar, P.P., and Mantovani, A.: **Toll-like receptor family and signalling pathway.** *Biochem. Soc. Trans.* 2000, **28**:563-566.

6-16 Bacterial peptides, mediators of inflammatory responses, and chemokines signal through members of the seven-transmembrane-domain, trimeric G protein-coupled receptor family.

Pan, Z.K.: **Anaphylatoxins C5a and C3a induce nuclear factor kappaB activation in human peripheral blood monocytes.** *Biochim. Biophys. Acta* 1998, **1443**:90-98.

Rojo, D., Suetomi, K., and Navarro, J.: **Structural biology of chemokine receptors.** *Biol. Res.* 1999, **32**:263-272.

Slupsky, J.R., Quitterer, U., Weber, C.K., Gierschik, P., Lohse, M.J., and Rapp, U.R.: **Binding of Gβγ subunits to cRaf1 downregulates G-protein-coupled receptor signalling.** *Curr. Biol.* 1999, **9**:971-974.

6-17 Cytokines signal lymphocytes by binding to cytokine receptors and triggering Janus kinases to phosphorylate and activate STAT proteins.

Leonard, W.J., and O'Shea, J.J.: **Jaks and STATs: biological implications.** *Annu. Rev. Immunol.* 1998, **16**:293-322.

Liu, K.D., Gaffen, S.L., and Goldsmith, M.A.: **JAK/STAT signaling by cytokine receptors.** *Curr. Opin. Immunol.* 1998, **10**:271-278.

O'Shea, J.J.: **Jaks, STATs, cytokine signal transduction, and immunoregulation: are we there yet?** *Immunity* 1997, **7**:1-11.

6-18 Programmed cell death of activated lymphocytes is triggered mainly through the receptor Fas.

Janssen, O., Sanzenbacher, R., and Kabelitz, D.: **Regulation of activation-induced cell death of mature T-lymphocyte populations.** *Cell Tissue Res.* 2000, **301**:85-99.

Ju, S.T., Matsui, K., and Ozdemirli, M.: **Molecular and cellular mechanisms regulating T and B cell apoptosis through Fas/FasL interaction.** *Int. Rev. Immunol.* 1999, **18**:485-513.

Krammer, P.H.: **CD95's deadly mission in the immune system.** *Nature* 2000, **407**:789-795.

Lenardo, M., Chan, K.M., Hornung, F., McFarland, H., Siegel, R., Wang, J., and Zheng, L.: **Mature T lymphocyte apoptosis—immune regulation in a dynamic and unpredictable antigenic environment.** *Annu. Rev. Immunol.* 1999, **17**:221-253.

6-19 Lymphocyte survival is maintained by a balance between death-promoting and death-inhibiting members of the Bcl-2 family of proteins.

Basu, A., and Haldar, S.: **The relationship between Bcl2, Bax and p53: consequences for cell cycle progression and cell death.** *Mol. Hum. Reprod.* 1998, **4**:1099-1109.

Chao, D.T., and Korsmeyer, S.J.: **BCL-2 family: regulators of cell death.** *Annu. Rev. Immunol.* 1998, **16**:395-419.

McDonnell, J.M., Fushman, D., Milliman, C.L., Korsmeyer, S.J., and Cowburn, D.: **Solution structure of the proapoptotic molecule BID: a structural basis for apoptotic agonists and antagonists.** *Cell* 1999, **96**:625-634.

6-20 Homeostasis of lymphocyte populations is maintained by signals that lymphocytes are continually receiving through their antigen receptors.

Agenes, F., Rosado, M.M., and Freitas, A.A.: **Considerations on B cell homeostasis.** *Curr. Top. Microbiol. Immunol.* 2000, **252**:68-75.

Freitas, A.A., and Rocha, B.: **Population biology of lymphocytes: the flight for survival.** *Annu. Rev. Immunol.* 2000, **18**:83-111.

Goldrath, A.W., and Bevan, M.J.: **Selecting and maintaining a diverse T-cell repertoire.** *Nature* 1999, **402**:255-262.

Ku, C.C., Murakami, M., Sakamoto, A., Kappler, J., and Marrack, P.: **Control of homeostasis of CD8+ memory T cells by opposing cytokines.** *Science* 2000, **288**:675-678.

The Development and Survival of Lymphocytes

7

As described in Chapters 3 and 4, the antigen receptors carried by B and T lymphocytes are immensely variable in their antigen specificity, enabling an individual to make immune responses against the wide range of pathogens encountered during a lifetime. This diverse repertoire of **B-cell receptors** and **T-cell receptors** is generated during the development of B cells and T cells, respectively, from their uncommitted precursors. The production of new lymphocytes, or **lymphopoiesis**, takes place in specialized lymphoid tissues— the **central lymphoid tissues**—which are the bone marrow in the case of B cells and the thymus for T cells. Like all hematopoietic cells, lymphocyte precursors originate in the bone marrow, but while B cells complete most of their development within the bone marrow, T cells are generated in the thymus from precursor cells that migrate from the bone marrow.

The antigen specificity of an individual lymphocyte is determined early in its differentiation, when the DNA sequences encoding the variable regions of immunoglobulins, in B cells, and T-cell receptors, in T cells, are assembled from gene segments, as described in Chapter 4. Because of this requirement for gene rearrangement, the early stages of development of B cells and T cells proceed along broadly similar lines. In both B cells and T cells this aspect of development is regulated in similar ways to ensure both the diversity of the lymphocyte repertoire as a whole and the unique antigen specificity of the individual lymphocyte.

The expression of an antigen receptor on the surface of a lymphocyte marks a watershed in its development, as it can now detect ligands that bind to this receptor. In the next phase of lymphocyte development, the receptor is tested for its antigen-recognition properties against molecules present in the immediate environment. The specificity and affinity of the receptor for these ligands determines the fate of the immature lymphocyte; that is, whether the cell is selected to survive and develop further, or whether it dies without reaching maturity.

In general, it appears that developing lymphocytes whose receptors interact weakly with self antigens, or bind self antigens in a particular way, receive a signal that enables them to survive; this type of selection is known as **positive selection**. Positive selection is particularly critical in the development of $\alpha{:}\beta$ T cells, which recognize antigen as peptides bound to MHC molecules (see Chapter 3). Since T cells that cannot recognize the body's own MHC molecules would not be able to mount an immune response against any antigen, this process of positive selection ensures that an individual will have a repertoire of T cells that are capable of responding to foreign antigenic peptides when they are bound to his or her own MHC molecules.

Lymphocytes whose receptors bind strongly to self antigens, on the other hand, receive signals that lead to their death; this is termed **negative selection**. Strongly self-reactive lymphocytes are therefore removed from the repertoire before they become fully mature and might initiate damaging autoimmune reactions. In this way immunological **tolerance** is established to ubiquitous self antigens. The default fate of developing lymphocytes, in the absence of any signal being received from the receptor, is death and, as we will see, the vast majority of developing lymphocytes die either before emerging from the central lymphoid organs or before maturing in the peripheral lymphoid organs.

The lymphocytes that survive to form the mature lymphocyte population are thus only a small fraction of those generated in the bone marrow or thymus. Nonetheless, these cells express a large repertoire of receptors capable of responding to a virtually unlimited variety of nonself structures. This repertoire provides the raw material on which clonal selection acts in an adaptive immune response.

In this chapter we will describe the different stages of the development of B cells and T cells in mice and humans, from the uncommitted stem cell up to the mature, functionally specialized, lymphocyte with its unique antigen receptor, ready to respond to a foreign antigen. In the first two parts of the chapter we define the stages through which lymphocytes develop in the central lymphoid organs and how the unselected primary receptor repertoire is generated. We then discuss what is known of the mechanisms by which positive selection and tolerance to self occur once a cell expresses an antigen receptor at the surface.

In the last part of the chapter we will follow the fate of newly generated lymphocytes as they leave the central lymphoid organs and migrate to the peripheral lymphoid tissues, where some further maturation occurs. Mature lymphocytes continually recirculate between the blood and peripheral lymphoid tissues and, in the absence of infection, their total number remains relatively constant, despite the continual production of new ones. We look at the factors that govern the survival of naive lymphocytes in the peripheral lymphoid organs, and thus the maintenance of lymphocyte homeostasis.

The main phases of a lymphocyte's life history are shown for B cells in Fig. 7.1 and for T cells in Fig. 7.2. In this chapter we describe the stages of development that lead to a cell gaining a place among the population of mature lymphocytes in the periphery. The final stages in the life history of a mature lymphocyte, in which encounter with foreign antigen activates it to become

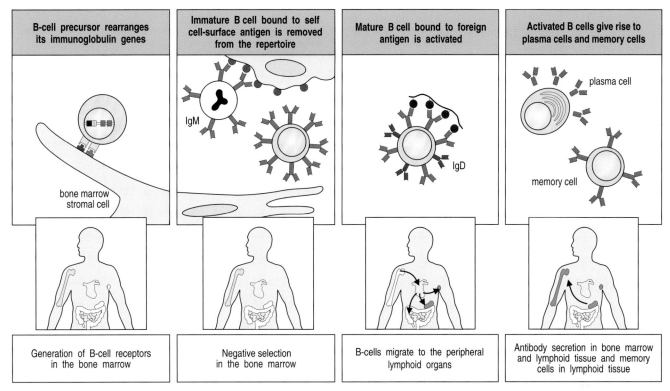

B-cell precursor rearranges its immunoglobulin genes	Immature B cell bound to self cell-surface antigen is removed from the repertoire	Mature B cell bound to foreign antigen is activated	Activated B cells give rise to plasma cells and memory cells
Generation of B-cell receptors in the bone marrow	Negative selection in the bone marrow	B-cells migrate to the peripheral lymphoid organs	Antibody secretion in bone marrow and lymphoid tissue and memory cells in lymphoid tissue

Fig. 7.1 B cells originate from a lymphoid progenitor in the bone marrow. In the first phase of development, progenitor B cells in the bone marrow rearrange their immunoglobulin genes. This phase is independent of antigen but is dependent on interactions with bone marrow stromal cells (first panels). It ends in an immature B cell that carries an antigen receptor in the form of cell-surface IgM and can now interact with antigens in its environment. Immature B cells that are strongly stimulated by antigen at this stage either die or are inactivated in a process of negative selection, thus removing many self-reactive B cells from the repertoire (second panels). In the third phase of development, the surviving immature B cells emerge into the periphery and mature to express IgD as well as IgM. They can now be activated by encounter with their specific foreign antigen in a secondary lymphoid organ (third panels). Activated B cells proliferate, and differentiate into antibody-secreting plasma cells and long-lived memory cells (fourth panels).

an effector cell or a memory cell, are discussed in Chapters 8–10. The last part of this chapter includes a discussion of the lymphoid tumors; these represent cells that have escaped from the normal controls on cell proliferation and are also of interest because they capture features of the different developmental stages of B cells and T cells.

Generation of lymphocytes in bone marrow and thymus.

The greater part of lymphocyte development in mammals occurs in the specialized environments of the central lymphoid organs—the bone marrow (and the liver in the fetus) for B cells and the thymus for T cells. In the fetus and the juvenile, these tissues are the source of large numbers of new lymphocytes, which migrate to populate the peripheral lymphoid tissues. In mature individuals, development of new T cells in the thymus slows down and T-cell numbers are maintained through division of mature T cells outside of the central lymphoid organs. New B cells, on the other hand, are continually produced from the bone marrow, even in adults.

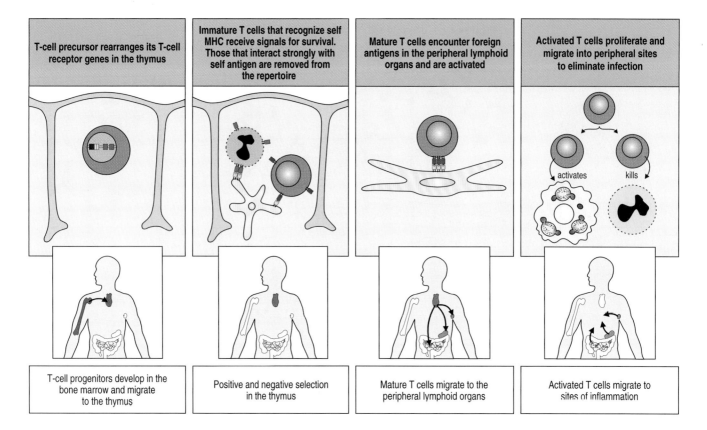

T-cell precursor rearranges its T-cell receptor genes in the thymus	Immature T cells that recognize self MHC receive signals for survival. Those that interact strongly with self antigen are removed from the repertoire	Mature T cells encounter foreign antigens in the peripheral lymphoid organs and are activated	Activated T cells proliferate and migrate into peripheral sites to eliminate infection
T-cell progenitors develop in the bone marrow and migrate to the thymus	Positive and negative selection in the thymus	Mature T cells migrate to the peripheral lymphoid organs	Activated T cells migrate to sites of inflammation

Fig. 7.2 The development of T cells. T-cell development follows broadly similar lines to that of B cells. T-cell precursors migrate from the bone marrow to the thymus where the T-cell receptor genes are rearranged (first panels), α:β T-cell receptors that are compatible with self MHC molecules transmit a survival signal on interacting with thymic epithelium, leading to positive selection of the cells that bear them. Self-reactive receptors transmit a signal that leads to cell death and are thus removed from the repertoire in a process of negative selection (second panels). T cells that survive selection mature and leave the thymus to circulate in the periphery; they repeatedly leave the blood to migrate through the peripheral lymphoid organs where they may encounter their specific foreign antigen and become activated (third panels). Activation leads to clonal expansion and differentiation into effector T cells. These are attracted to sites of infection where they can kill infected cells or activate macrophages (fourth panels); others are attracted into B-cell areas where they help to activate an antibody response (not shown).

In this part of the chapter we describe the nature of these primary lymphopoietic environments and the developmental stages through which lymphocytes pass. These stages are defined mainly by the various steps in the assembly and expression of functional antigen receptor genes, and by the appearance of features that distinguish the different functional types of B and T cells. At each step of lymphocyte development, the progress of gene rearrangement is monitored. A successful gene rearrangement that leads to the production of a protein chain serves as a signal for the cell to progress to the next stage of development.

T-cell development is more complicated than B-cell development as it has to accommodate the production of two distinct lineages of T cells with different types of T-cell receptor—α:β and γ:δ (see Section 4-13). The α:β and γ:δ T cells diverge early in T-cell development. There is a further division of the α:β T cells into CD4 and CD8 T cells (see Section 3-12) that occurs in immature T cells after the T-cell receptor genes have been assembled and expressed.

7-1 Lymphocyte development occurs in specialized environments and is regulated by the somatic rearrangement of the antigen-receptor genes.

Certain basic principles apply to the process by which precursor cells develop into committed B or T cells expressing antigen-specific receptors—the immunoglobulins and T-cell receptors, respectively. In both humans and mice this process of lymphocyte differentiation occurs in ordered stages, and these are marked by successive steps in the rearrangement of the antigen-receptor genes and expression of their protein products, as well as by changes in the expression of other cell-surface and intracellular proteins. The successful execution of this intrinsic developmental program requires signals from the specialized microenvironments in which lymphocytes develop—bone marrow and fetal liver for B cells and thymus for T cells. These tissues provide a network of specialized nonlymphoid **stromal cells** that interact intimately with the developing lymphocytes, providing signals through secreted growth factors and cell-surface molecules that bind receptors on the lymphocyte precursor cells.

The antigen specificity of each individual lymphocyte is determined through the assembly of V, D, and J gene segments to generate rearranged V genes encoding the antigen-receptor variable (V) region (see Sections 4-2 and 4-11). The expression of a complete antigen receptor requires the successful rearrangement of two different genetic loci in order to produce the protein chains of the antigen receptor—the heavy and light chains of immunoglobulins or the α and β chains (or γ and δ chains) of the T-cell receptor. Not all gene rearrangement events are successful, however. Because of the imprecision in the recombination process, not all rearrangements produce a complete in-frame DNA sequence that can be translated into protein. The successful assembly of a V region is monitored for each locus, and defines the different stages of lymphocyte development. A successful gene rearrangement, termed a **productive rearrangement**, leads to the synthesis of the protein product; this is the signal for the cell to progress to the next stage of development.

In the case of B cells, for example, the assembly of an in-frame heavy-chain VDJ sequence leads to expression of the heavy chain; this is sensed by the cell as a signal to stop heavy-chain locus rearrangement, to divide several times, and then to commence rearrangement of a light-chain locus. Unsuccessful rearrangements that do not result in a protein are termed **nonproductive rearrangements**, and if no further rearrangement can occur to rescue the situation, the lymphocyte dies.

To survive these first stages of development, therefore, B cells must make a productive rearrangement at the immunoglobulin heavy-chain locus and at one of the two light-chain loci, kappa or lambda. T cells must make productive rearrangements at either the α-chain locus and the β-chain locus, which produces an α:β T cell, or at the γ-chain and δ-chain loci, which produces a γ:δ T cell. Those cells that fail to make the necessary productive rearrangements die *in situ* by apoptosis.

The events that lead to a B cell with an immunoglobulin B-cell receptor and a T cell with an α:β T-cell receptor have a very close correspondence, and we will discuss these here, leaving aside the complicating factor of rearrangements at the γ and δ loci that are also occurring in the earliest T-cell precursors; these will be returned to later in the chapter. Only one gene locus is rearranged at a time, and the loci are each rearranged in a fixed sequence. Both B cells and T cells rearrange the locus that contains D gene segments first; in the case of B cells this is the immunoglobulin heavy-chain locus; for T cells it is the T-cell receptor β-chain locus. Only if a productive rearrangement is made do developing B cells go on to rearrange a light-chain locus, and T cells go on to rearrange the α-chain locus.

The protein product of each antigen-receptor locus is intended to be expressed paired with another chain; an α and a β chain make up the T-cell receptor, for example (see Fig. 3.12). How, therefore, can a developing T cell test whether a productive rearrangement of a β chain-gene has occurred when it has not yet rearranged the α-chain locus? The solution is that both B cells and T cells produce invariant 'surrogate' partner chains at this stage in development. These surrogates pair with the heavy chain or the β chain to produce 'receptors' that can be expressed at the cell surface. Formation of these receptors generates signals that result in the cessation of VDJ rearrangement. This is followed by several rounds of cell division before the cells proceed to the next stage of development, which is V to J rearrangement at a light-chain locus in B cells and at the α-chain locus in T cells. If this rearrangement is productive, the lymphocyte can then express a *bona fide* immunoglobulin or T-cell receptor at the surface.

7-2   B cells develop in the bone marrow with the help of stromal cells and achieve maturity in peripheral lymphoid organs.

B-cell development is dependent on the nonlymphoid stromal cells of the bone marrow; stem cells isolated from the bone marrow and grown in culture fail to differentiate into B cells unless bone marrow stromal cells are also present. The stroma, whose name derives from the Greek word for a mattress, thus provides a necessary support for B-cell development. The contribution of the stromal cells is twofold. First, they form specific adhesive contacts with the developing B-lineage cells by interactions between cell-adhesion molecules and their ligands. Second, they provide growth factors that stimulate lymphocyte differentiation and proliferation (Fig. 7.3).

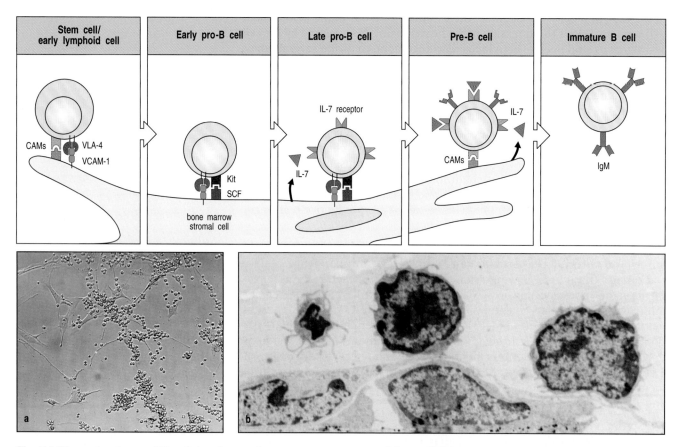

Fig. 7.3 The early stages of B-cell development are dependent on bone marrow stromal cells. The upper panels show the interactions between precursor B cells and stromal cells that are required for the development to the immature B-cell stage. The designations pro-B cell and pre-B cell refer to defined phases of B-cell development as described in Fig. 7.5. Lymphoid progenitor cells and early pro-B cells bind to the adhesion molecule VCAM-1 on stromal cells through the integrin VLA-4 and also interact through other cell-adhesion molecules (CAMs). These adhesive interactions promote the binding of the receptor tyrosine kinase Kit on the surface of the pro-B cell to stem-cell factor (SCF) on the stromal cell, which activates the kinase and induces the proliferation of the B-cell progenitors. Later stages require interleukin-7 (IL-7) for proliferation and further development. Panel a: light micrograph showing small round cells, which are the B-lymphoid progenitors, in intimate contact with cultured stromal cells, which have extended processes fastening them to the plastic dish on which they are grown. Panel b: high-magnification electron micrograph of a similar cell culture in which two lymphoid cells are seen adhering to a flattened stromal cell. Photographs courtesy of A. Rolink (a); P. Kincade and P.L. Witte (b).

Fig. 7.4 B-lineage cells move within the bone marrow toward its central axis as they mature. Part of a transverse section of a rat femur photographed through a fluorescence microscope to identify cells (green) stained for the enzyme terminal deoxynucleotidyl transferase (TdT), which marks the pro-B stage (see Fig. 7.5). Early pro-B cells expressing TdT are concentrated near the endosteum (the inner bone surface), as seen toward the upper right of the picture. As development proceeds, cells of the B-cell lineage lose TdT and pass toward the center of the marrow cavity (lower left), where they wait in sinuses ready for export. Photograph courtesy of D. Opstelten and M. Hermans.

A number of bone marrow growth factors have been identified and their functions elucidated. The growth of early B-lineage cells is stimulated by stem-cell factor (SCF), a membrane-bound cytokine present on stromal cells, which interacts with the cell-surface receptor tyrosine kinase Kit on B-cell precursors. Developing B cells at later stages require the secreted cytokine interleukin-7 (IL-7). The chemokine stromal cell-derived factor 1 or pre-B cell growth-stimulating factor (SDF-1/PBSF) has an important role in the early stages of B-cell development, as shown by the failure of B-cell development in mice lacking the gene for this molecule. SDF-1 is produced constitutively by bone marrow stromal cells and one of its roles may be to retain developing B-cell precursors in the marrow microenvironment. Other adhesion molecules and growth factors produced by stromal cells are known to have roles in B-cell development; this is an active area of research and a full understanding of the factors that regulate B-cell differentiation has yet to be achieved.

As B-lineage cells mature, they migrate within the marrow, remaining in contact with the stromal cells. The earliest stem cells lie in a region called the subendosteum, which is adjacent to the inner bone surface. As maturation proceeds, B-lineage cells move toward the central axis of the marrow cavity (Fig. 7.4). Later stages of maturation become less dependent on contact with stromal cells, and the final stages of development of immature B cells into mature B cells occur in peripheral lymphoid organs such as the spleen.

7-3   Stages in B-cell development are distinguished by the expression of immunoglobulin chains and particular cell-surface proteins.

The stages in primary B-cell development are defined by the sequential rearrangement and expression of heavy- and light-chain immunoglobulin genes (Fig. 7.5). The first classification of B-lineage cells into different developmental stages was made according to whether they expressed no immunoglobulin chains, or the immunoglobulin heavy chain only, or both heavy and light chains. Subsequently, intermediate differentiation stages have been distinguished on the basis of the expression of other cell-surface proteins, together with direct DNA analysis of the state of the immunoglobulin gene loci.

The earliest B-lineage cells are known as **pro-B cells**, as they are progenitor cells with limited self-renewal capacity. They are derived from pluripotent hematopoietic stem cells and are identified by the appearance of cell-surface proteins characteristic of early B-lineage cells. Rearrangement of the immunoglobulin heavy-chain locus takes place in pro-B cells; D_H to J_H joining at the **early pro-B cell** stage is followed by V_H to DJ_H joining at the **late pro-B cell** stage.

Productive VDJ_H joining leads to the expression of an intact μ heavy chain, which is the hallmark of the next main stage of development, the **pre-B cell** stage. The μ chain in **large pre-B cells** is expressed intracellularly and possibly

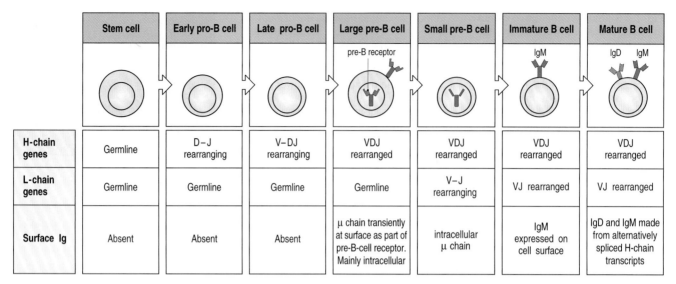

	Stem cell	Early pro-B cell	Late pro-B cell	Large pre-B cell	Small pre-B cell	Immature B cell	Mature B cell
H-chain genes	Germline	D–J rearranging	V–DJ rearranging	VDJ rearranged	VDJ rearranged	VDJ rearranged	VDJ rearranged
L-chain genes	Germline	Germline	Germline	Germline	V–J rearranging	VJ rearranged	VJ rearranged
Surface Ig	Absent	Absent	Absent	μ chain transiently at surface as part of pre-B-cell receptor. Mainly intracellular	intracellular μ chain	IgM expressed on cell surface	IgD and IgM made from alternatively spliced H-chain transcripts

Fig. 7.5 The development of a B-lineage cell proceeds through several stages marked by the rearrangement and expression of the immunoglobulin genes. The stem cell has not yet begun to rearrange its immunoglobulin (Ig) gene segments; they are in the germline configuration as found in all nonlymphoid cells. The heavy-chain (H-chain) locus rearranges first. Rearrangement of a D gene segment to a J_H gene segment occurs in early pro-B cells, generating late pro-B cells in which V_H to DJ_H rearrangement occurs. A successful VDJ_H rearrangement leads to the expression of a complete immunoglobulin heavy chain as part of the pre-B-cell receptor, which is found mainly in the cytoplasm and to some degree on the surface of the cell. Once this occurs the cell is stimulated to become a large pre-B cell which actively divides. Large pre-B cells then cease dividing and become small resting pre-B cells, at which point they cease expression of the surrogate light chains and express the μ heavy chain alone in the cytoplasm. When the cells are again small, they reexpress the RAG proteins and start to rearrange the light-chain (L-chain) genes. Upon successfully assembling a light-chain gene, a cell becomes an immature B cell that expresses a complete IgM molecule at the cell surface. Mature B cells produce a δ heavy chain as well as a μ heavy chain, by a mechanism of alternative mRNA splicing, and are marked by the additional appearance of IgD on the cell surface.

in small amounts at the cell surface, in combination with a surrogate light chain, to form the **pre-B-cell receptor**. Expression of the pre-B-cell receptor signals the cell to halt heavy-chain locus rearrangement and production of the surrogate light chain, and to divide several times before giving rise to **small pre-B cells**, in which light-chain rearrangements begin. Once a light-chain gene is assembled and a complete IgM molecule is expressed on the cell surface, the cell is defined as an **immature B cell**. The expression of the immunoglobulin heavy and light chains are key milestones in this differentiation pathway. These events do more than simply delineate stages of the pathway; the expression of an intact heavy chain, and later of a complete immunoglobulin molecule, actively regulates progression from one stage to the next.

All development up to this point has taken place in the bone marrow and is independent of antigen. Immature B cells now undergo selection for self-tolerance and subsequently for the ability to survive in the peripheral lymphoid tissues. B cells that survive in the periphery undergo further differentiation to become **mature B cells** that express IgD in addition to IgM. These cells, also called **naive B cells** until they encounter their specific antigen, recirculate through peripheral lymphoid tissues, where they may encounter and be activated by the appropriate foreign antigen.

As B cells develop from pro-B cells to mature B cells, they express proteins other than immunoglobulin that are characteristic of each stage. Many of these proteins are expressed on the cell surface and are useful markers for B-lineage cells at different developmental stages. Fig. 7.6 summarizes their expression patterns. The functions of some of these proteins are understood, whereas others serve at present simply as useful signposts for the study of B-cell development. One of the first identifiable proteins expressed on the surface of B-lineage cells is CD45R (known in mice as B220). This is a B-cell-specific form of the CD45 protein originally known as the common leukocyte antigen; T cells, monocytes, and neutrophils express other variants of this protein. CD45R, which is expressed throughout B-cell development from pro-B cells right up to the antibody-secreting plasma cells, is a protein tyrosine phosphatase that functions in B-cell receptor signaling (see Section 6-7). Another protein that is expressed throughout B-cell development is CD19, which also participates in B-cell receptor signaling. Since signaling through the pre-B and B-cell receptors guides B-cell development, the earliest B-cell precursors begin to assemble the signaling components of the receptor complex including CD45R and CD19. The earliest B-cell precursors also express the receptor for IL-7, which is an essential growth factor for both developing B cells and T cells. Blocking signaling by infusion of an anti-IL-7 antibody will halt B-cell development, as will mutations that inactivate either IL-7 or its receptor.

Another cell-surface protein first detected at the pro-B cell stage is CD43 (the mucin leukosialin), but this is lost as cells progress to become immature B cells. CD43 functions both as an adhesion molecule that may guide cell–cell interactions—for example those of B-cell precursors with stromal cells—and also as a signaling molecule, although not as part of the B-cell receptor complex.

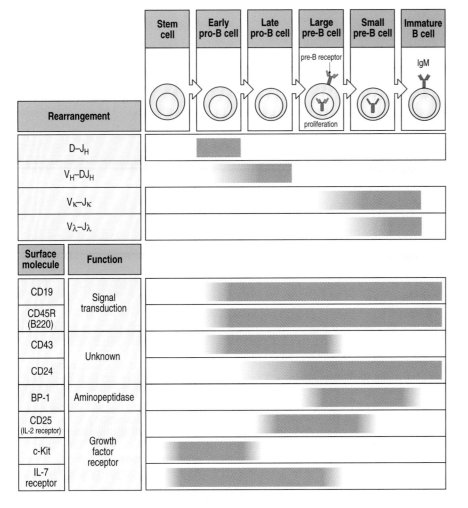

Fig. 7.6 The correlation of the stages of B-cell development with immunoglobulin gene segment rearrangement and expression of cell-surface proteins. Stages of B-cell development are defined by which gene segments are undergoing rearrangement as well as by cell-surface proteins. The earliest B-lineage surface markers are CD19 and CD45R (B220 in the mouse). The expression of these persists throughout B-cell development. A pro-B cell is also distinguished by expression of CD43, c-Kit, and the IL-7 receptor. An early pro-B cell is rearranging D to J_H and a late pro-B cell V_H to DJ_H. A late pro-B cell starts to express CD24 and CD25. When a late pro-B cell makes a successful VDJ join, it can then express an immunoglobulin heavy chain, which defines it as a pre-B cell. A pre-B cell is phenotypically distinguished by expression of BP-1, whereas c-Kit and the IL-7 receptor are no longer expressed. Before going on to rearrange a light-chain locus, a pre-B cell enlarges and undergoes several cycles of cell division. It then becomes a small pre-B cell, which can rearrange light-chain gene segments, first at the κ locus and, if not successful, then at the λ locus. A cell that has made a productive light chain becomes an immature B cell expressing surface IgM, and has completed the antigen-independent phase of B-cell development.

Exactly why CD43 is expressed at these early stages of B-cell development, but not later, is not yet known. Other cell-surface molecules expressed during early stages of B-cell development include the heat-stable antigen (HSA, CD24) and the aminopeptidase BP-1. The functions of these molecules in B-cell development are unknown, although B cells apparently develop normally in mice lacking BP-1. At the late pro-B cell stage, Kit, the receptor for SCF-1 mutates. Kit is a factor that stimulates both lymphoid and myeloid development in the bone marrow. After Kit turns off, at the large pre-B cell stage, the low affinity IL-2 receptor, CD25, is expressed. Signaling through Kit, in concert with IL-7 and other stromal-derived signals, promotes pro- and pre-B cell proliferation (see Fig. 7.3).

7-4 T cells also originate in the bone marrow, but all the important events in their development occur in the thymus.

T lymphocytes develop from a common lymphoid progenitor in the bone marrow that also gives rise to B lymphocytes, but those progeny destined to give rise to T cells leave the bone marrow and migrate to the thymus (see Fig. 7.2). This is the reason they are called thymus-dependent (T) lymphocytes or T cells. The thymus is situated in the upper anterior thorax, just above the heart. It consists of numerous lobules, each clearly differentiated into an outer cortical region—the **thymic cortex**—and an inner **medulla** (Fig. 7.7). In

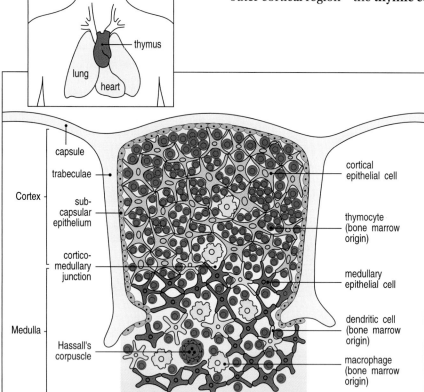

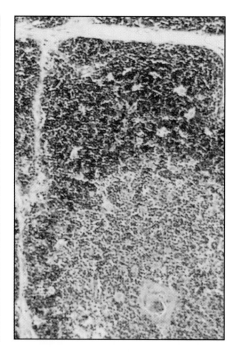

Fig. 7.7 The cellular organization of the human thymus. The thymus, which lies in the midline of the body, above the heart, is made up of several lobules, each of which contains discrete cortical (outer) and medullary (central) regions. As shown in the diagram on the left, the cortex consists of immature thymocytes (dark blue), branched cortical epithelial cells (pale blue), with which the immature cortical thymocytes are closely associated, and scattered macrophages (yellow), which are involved in clearing apoptotic thymocytes. The medulla consists of mature thymocytes (dark blue), and medullary epithelial cells (orange), along with macrophages (yellow) and dendritic cells (yellow) of bone marrow origin. Hassall's corpuscles are probably also sites of cell destruction. The thymocytes in the outer cortical cell layer are proliferating immature cells, whereas the deeper cortical thymocytes are mainly immature T cells undergoing thymic selection. The photograph shows the equivalent section of a human thymus, stained with hematoxylin and eosin. The cortex is darkly staining; the medulla is lightly stained. The large body in the medulla is a Hassall's corpuscle. Photograph courtesy of C.J. Howe.

young individuals, the thymus contains large numbers of developing T-cell precursors embedded in a network of epithelia known as the **thymic stroma**, which provides a unique microenvironment for T-cell development analogous to that provided by the stromal cells of the bone marrow.

The thymic stroma arises early in embryonic development from the endo dermal and ectodermal layers of embryonic structures known as the third pharyngeal pouch and third branchial cleft. Together these epithelial tissues form a rudimentary thymus, or **thymic anlage**. The thymic anlage then attracts cells of hematopoietic origin, which colonize it; these give rise to large numbers of **thymocytes**, which are committed to the T-cell lineage, and the **intrathymic dendritic cells**. The thymocytes are not simply passengers within the thymus; they influence the arrangement of the thymic epithelial cells on which they depend for survival, inducing the formation of a reticular epithelial structure that surrounds the developing thymocytes (Fig. 7.8). The thymus is independently colonized by numerous macrophages, also of bone marrow origin.

The cellular architecture of the human thymus is illustrated in Fig. 7.7. Bone marrow derived cells are differentially distributed between the thymic cortex and medulla; the cortex contains only immature thymocytes and scattered macrophages, whereas more mature thymocytes, along with dendritic cells and macrophages, are found in the medulla. This reflects the different developmental events that occur within these two compartments, as we will discuss further in Section 7-7.

The importance of the thymus in immunity was first discovered through experiments on mice, and indeed, most of our knowledge of T-cell development within the thymus comes from the mouse. It was found that surgical removal of the thymus (**thymectomy**) at birth resulted in immunodeficient mice, focusing interest on this organ at a time when the difference between T and B lymphocytes in mammals had not yet been defined. Much evidence has accumulated since to establish the importance of the thymus in T-cell development, including observations of immunodeficient children. Thus, for example, in **DiGeorge's syndrome** in humans, and in mice with the *nude* mutation (which also causes hairlessness), the thymus fails to form and the affected individual produces B lymphocytes but few T lymphocytes.

The crucial role of the thymic stroma in inducing the differentiation of bone marrow-derived precursor cells can be demonstrated by tissue grafts between two mutant mice, each lacking mature T cells for a different reason. In *nude* mice the thymic epithelium fails to differentiate, whereas in *scid* mice B and T lymphocytes fail to develop because of a defect in T-cell receptor gene rearrangement (see Section 4-5). Reciprocal grafts of thymus and bone marrow between these immunodeficient strains show that *nude* bone marrow precursors develop normally in a *scid* thymus (Fig. 7.9). Thus, the defect in *nude* mice is in the thymic stromal cells. Transplanting a *scid* thymus into *nude* mice leads to T-cell development. However, *scid* bone marrow cannot develop T cells even in a wild-type recipient.

In mice, the thymus continues to develop for 3 to 4 weeks after birth, whereas in humans it is fully developed at birth. The rate of T-cell production by the thymus is greatest before puberty. After puberty, the thymus begins to shrink and the production of new T cells in adults is lower, although it does continue throughout life. In both mice and humans, removal of the thymus after puberty is not accompanied by any notable loss of T-cell function. Thus, it seems that once the T-cell repertoire is established, immunity can be sustained without the production of significant numbers of new T cells; the pool of peripheral T cells is instead maintained by the division of mature T cells.

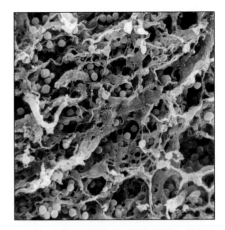

Fig. 7.8 The epithelial cells of the thymus form a network surrounding developing thymocytes. In this scanning electron micrograph of the thymus, the developing thymocytes (the spherical cells) occupy the interstices of an extensive network of epithelial cells. Photograph courtesy of W. van Ewijk.

Fig. 7.9 The thymus is critical for the maturation of bone marrow-derived cells into T cells. Mice with the *scid* mutation (upper left photograph) have a defect that prevents lymphocyte maturation, whereas mice with the *nude* mutation (upper right photograph) have a defect that affects the development of the cortical epithelium of the thymus. T cells do not develop in either strain of mouse: this can be demonstrated by staining spleen cells with antibodies specific for mature T cells and analyzing them in a flow cytometer (see Appendix I, Section A-22), as represented by the blue line in the graphs in the bottom panels. Bone marrow cells from *nude* mice can restore T cells to *scid* mice (red line in graph on left), showing that, in the right environment, the *nude* bone marrow cells are intrinsically normal, and capable of producing T cells. Thymic epithelial cells from *scid* mice can induce the maturation of T cells in *nude* mice (red line in graph on right), demonstrating that the thymus provides the essential microenvironment for T-cell development.

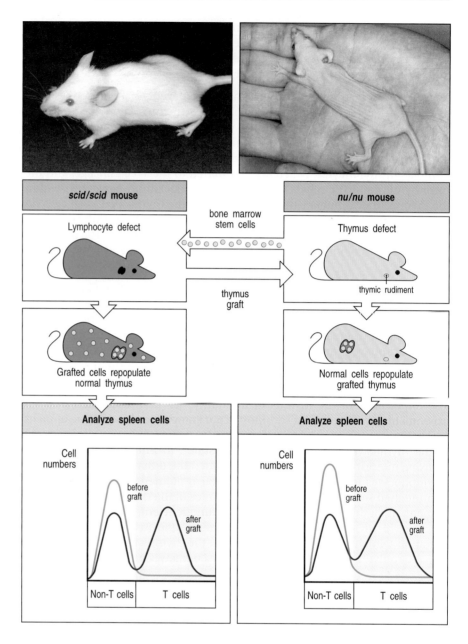

7-5   Most developing T cells die in the thymus.

T-cell precursors arriving in the thymus from the bone marrow spend up to a week differentiating there before they enter a phase of intense proliferation. In a young adult mouse the thymus contains around 10^8 to 2×10^8 thymocytes. About 5×10^7 new cells are generated each day; however, only about 10^6 to 2×10^6 (roughly 2–4%) of these will leave the thymus each day as mature T cells. Despite the disparity between the numbers of T cells generated daily in the thymus and the number leaving, the thymus does not continue to grow in size or cell number. This is because approximately 98% of the thymocytes that develop in the thymus also die within the thymus. No widespread damage is seen, indicating that death is occurring by apoptosis rather than by necrosis (see Section 1-11).

Changes in the plasma membrane of cells undergoing apoptosis lead to their rapid phagocytosis, and apoptotic bodies, which are the residual condensed chromatin of apoptotic cells, are seen inside macrophages throughout the

Fig. 7.10 Developing T cells that undergo apoptosis are ingested by macrophages in the thymic cortex. Panel a shows a section through the thymic cortex and part of the medulla in which cells have been stained for apoptosis with a red dye. Thymic cortex is to the right of the photograph. Apoptotic cells are scattered throughout the cortex but are rare in the medulla. Panel b shows a section of thymic cortex at higher magnification that has been stained red for apoptotic cells and blue for macrophages. The apoptotic cells can be seen within macrophages. Magnifications: panel a, × 45; panel b, × 164. Photographs courtesy of J. Sprent and C. Suhr.

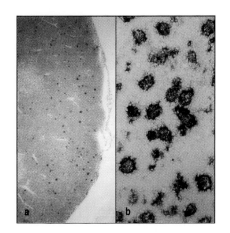

thymic cortex (Fig. 7.10). This apparently profligate waste of thymocytes is a crucial part of T-cell development as it reflects the intensive screening that each new thymocyte undergoes for the ability to recognize self MHC and for self tolerance.

7-6   Successive stages in the development of thymocytes are marked by changes in cell-surface molecules.

Developing thymocytes pass through a series of distinct phases that are marked by changes in the status of T-cell receptor genes and in the expression of the T-cell receptor, and by changes in expression of cell-surface proteins such as the CD3 complex and the co-receptor proteins CD4 and CD8. These surface changes reflect the state of functional maturation of the cell. Particular combinations of cell-surface proteins can thus be used as markers for T cells at different stages of differentiation. The principal stages are summarized in Fig. 7.11. Two distinct lineages of T cells—α:β and γ:δ, which have different types of T-cell receptor—are produced early in T-cell development. Later, α:β T cells develop into two distinct functional subsets, CD4 and CD8 T cells.

When progenitor cells first enter the thymus from the bone marrow, they lack most of the surface molecules characteristic of mature T cells and their receptor genes are unrearranged. These cells give rise to the major population of α:β T cells and the minor population of γ:δ T cells. If injected into the peripheral circulation, these lymphoid progenitors can even give rise to B cells and NK cells (see Section 1-1). Interactions with the thymic stroma trigger an initial

Fig. 7.11 Changes in cell-surface molecules allow thymocyte populations at different stages of maturation to be distinguished. The most important cell-surface molecules for identifying thymocyte subpopulations have been CD4, CD8, and T-cell receptor complex molecules (CD3, and the T-cell receptor α and β chains). The earliest cell population in the thymus does not express any of these. As these cells do not express CD4 or CD8, they are called 'double-negative' thymocytes. These precursor cells give rise to two T-cell lineages, the minority population of γ:δ T cells (which lack CD4 or CD8 even when mature), and the majority α:β T-cell lineage. Development of prospective α:β T cells proceeds through stages where both CD4 and CD8 are expressed by the same cell; these are known as 'double-positive' thymocytes. At first, double-positive cells express the pre-T-cell receptor (pTα:β). These cells enlarge and divide. Later, they become small resting double-positive cells in which low levels of the T-cell receptor (α:β) itself are expressed. Most thymocytes (~97%) die within the thymus after becoming small double-positive cells. Those cells whose receptors can interact with self MHC molecules lose expression of either CD4 or CD8 and increase the level of expression of the T-cell receptor. The outcome of this process are the 'single-positive' thymocytes, which, after maturation, are exported from the thymus as mature single-positive CD4 or CD8 T cells.

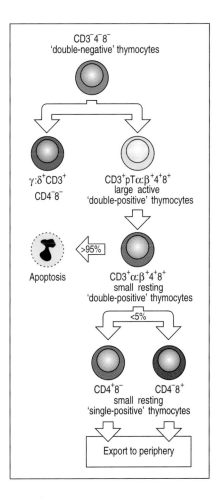

CD3⁻4⁻8⁻
'double-negative' thymocytes

γ:δ⁺CD3⁺
CD4⁻8⁻

CD3⁺pTα:β⁺4⁺8⁺
large active
'double-positive' thymocytes

Apoptosis   >95%

CD3⁺α:β⁺4⁺8⁺
small resting
'double-positive' thymocytes

<5%

CD4⁺8⁻   CD4⁻8⁺
small resting
'single-positive' thymocytes

Export to periphery

phase of differentiation along the T-cell lineage pathway followed by cell proliferation, and the expression of the first cell-surface molecules specific for T cells, for example CD2 and (in mice) Thy-1. At the end of this phase, which can last about a week, the thymocytes bear distinctive markers of the T-cell lineage, but they do not express any of the three cell-surface markers that define mature T cells. These are the CD3:T-cell receptor complex and the co-receptors CD4 or CD8. Because of the absence of CD4 and CD8 such cells are called **'double-negative' thymocytes** (see Fig. 7.11).

In the fully developed thymus, these immature double-negative T cells form approximately 60% of a small, highly heterogeneous pool of CD4$^-$ CD8$^-$ cells (about 5% of total thymocytes), which also includes two populations of more mature T cells that belong to minority lineages. One of these, representing about 20% of all the double-negative cells in the thymus, or 1% of total thymocytes, comprises cells that have rearranged and are expressing the genes encoding the γ:δ T-cell receptor; we will return to these cells in Section 7-13. The second, representing another 20% of all double negatives, includes cells bearing α:β T-cell receptors of a very limited diversity; these cells also express the NK1.1 receptor commonly found on NK cells, hence they are known as 'NK1.1$^+$ T cells' (sometimes simply called NK T cells). NK T cells are activated as part of the early response to many infections; they differ from the major lineage of α:β T cells in recognizing CD1 molecules rather than MHC class I or MHC class II molecules (see Section 5-18) and they are not shown in Fig. 7.11. In this and subsequent discussions, we will reserve the term double-negative thymocytes for the immature thymocytes that do not yet express a complete T-cell receptor molecule. These cells give rise to both γ:δ and α:β T cells (see Fig. 7.11). Most of them develop along the α:β pathway.

The α:β pathway is shown in more detail in Fig. 7.12. The double-negative stage can be subdivided on the basis of expression of the adhesion molecule CD44, CD25 (the α chain of the IL-2 receptor), and c-Kit, the receptor for the hematopoietic cytokine, stem cell factor. At first, double-negative thymocytes express c-Kit and CD44 but not CD25; in these cells, the genes encoding both chains of the T-cell receptor are in the germline configuration. As the thymocytes mature further, they begin to express CD25 on their surface and, later still, expression of CD44 and c-Kit is reduced. In these latter cells, which are known as CD44^{low} CD25$^+$ cells, rearrangement of the T-cell receptor β-chain locus occurs. Cells that fail to make a successful rearrangement of the β locus remain in the CD44^{low} CD25$^+$ stage and soon die, whereas cells that make productive β-chain gene rearrangements and express the β chain lose expression of CD25 once again. The functional significance of the transient expression of CD25 is unclear; T cells develop normally in mice in which the IL-2 gene has been deleted by gene knockout (see Appendix I, Section A-47); however, another cytokine, IL-15, also binds to the IL-2 receptor and may be able to compensate for the loss of IL-2. By contrast, c-Kit is quite important for the development of the earliest double-negative thymocytes in that mice lacking c-Kit have a markedly reduced number of double-negative T cells. In addition, the IL-7 receptor is also essential for early T-cell development, as T cells do not develop in either humans or mice when this receptor is defective.

The β chains expressed by CD44^{low} CD25$^+$ thymocytes pair with a surrogate α chain called **pTα** (pre-T-cell α), which allows them to assemble a **pre-T-cell receptor** that is analogous in structure and function to the pre-B-cell receptor. The pre-T-cell receptor is expressed on the cell surface as a complex with the CD3 molecules that provide the signaling components of T-cell receptors (see Section 6-6). The assembly of the CD3:pre-T-cell receptor complex leads to cell proliferation, the arrest of further β-chain gene rearrangements, and the expression of CD8 and CD4. These double-positive thymocytes comprise the vast majority of thymocytes. Once the

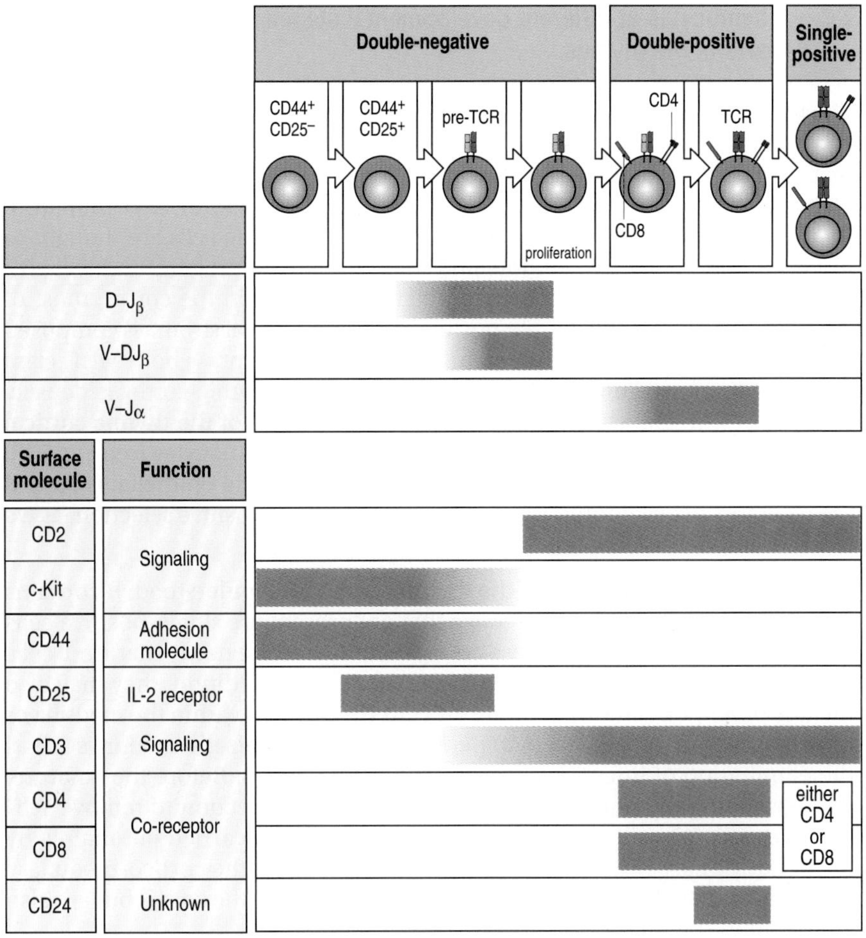

Fig. 7.12 The correlation of stages of α:β T-cell development with T-cell receptor gene rearrangement and expression of cell-surface proteins. Lymphoid precursors are triggered to proliferate and become thymocytes committed to the T-cell lineage through interactions with the thymic stroma. These negative cells express CD44 and c-Kit and, at a later stage, the α chain of the IL-2 receptor, CD25. After this, the CD44⁺ CD25⁺ cells begin to rearrange the β-chain locus, becoming $CD44^{low}$ and $c\text{-}Kit^{low}$ as this occurs. The cells are arrested in the $CD44^{low}$ $CD25^{+}$ stage until they productively rearrange the β-chain locus; the in-frame β chain then pairs with the surrogate pTα chain and is expressed on the cell surface, which triggers entry into the cell cycle. Expression of pTα:β on the cell surface is associated with small amounts of CD3, and causes the loss of CD25, cessation of β-chain gene rearrangement, cell proliferation, and the expression of CD4 and CD8. After the cells cease proliferating and revert to small $CD4^{+}$ $CD8^{+}$ double-positive cells, they begin rearrangement at the α-chain locus. The cells then express low levels of an α:β T-cell receptor and the associated CD3 complex and are ready for selection. Most cells die by failing to be positively selected or as a consequence of negative selection, but some are selected to mature into CD4 or CD8 single-positive cells and eventually to leave the thymus.

large double-positive thymocytes cease to proliferate and become small double-positive cells, the α-chain locus begins to rearrange. As we will see later in this chapter, the structure of the α locus allows multiple successive rearrangement attempts, so that a successful rearrangement at the α-chain locus is achieved in most developing thymocytes. Thus most double-positive cells produce an α:β T-cell receptor.

Small **double-positive thymocytes** initially express low levels of the T-cell receptor. Most of these cells bear receptors that cannot recognize self MHC molecules; they are destined to fail positive selection and die. On the other hand, those double-positive cells that recognize self MHC and can therefore undergo positive selection, go on to mature and express high levels of the T-cell receptor. Concurrently, they cease to express one or other of the two co-receptor molecules, becoming either CD4 or CD8 **single-positive thymocytes**. Thymocytes also undergo negative selection during and after the double-positive stage in development, which eliminates those cells capable of responding to self antigens. Approximately 2% of the double positives survive this dual screening and mature as single-positive T cells that are gradually exported from the thymus to form the peripheral T-cell repertoire. The time between the entry of a T-cell progenitor into the thymus and the export of its mature progeny is estimated to be around 3 weeks in the mouse.

7-7   Thymocytes at different developmental stages are found in distinct parts of the thymus.

The thymus is divided into two main regions, a peripheral cortex and a central medulla (see Fig. 7.7). Most T-cell development takes place in the cortex; only mature single-positive thymocytes are seen in the medulla. At the outer edge of the cortex, in the subcapsular region of the thymus (Fig. 7.13), large immature double-negative thymocytes proliferate vigorously; these cells are thought to represent the thymic progenitors and their immediate progeny and will give rise to all subsequent thymocyte populations. Deeper in the cortex, most of the thymocytes are small double-positive cells. The cortical stroma is composed of epithelial cells with long branching processes that express both MHC class II and MHC class I molecules on their surface. The thymic cortex is densely packed with thymocytes, and the branching processes of the thymic cortical epithelial cells make contact with almost all cortical thymocytes (see Fig. 7.8). Contact between the MHC molecules on thymic cortical epithelial cells and the receptors of developing T cells has a crucial role in positive selection, as we will see later in this chapter.

The function of the medulla of the thymus is less well understood. It contains relatively few thymocytes, and those that are present are single-positive cells resembling mature T cells. These cells probably include newly mature T cells that are leaving the thymus through the medulla. In addition, they may also include other populations of mature T cells that remain within the medulla or return to it from the periphery to perform some specialized function, such as the elimination of infectious agents within the thymus. Before they mature, the developing thymocytes must undergo negative selection to remove self-reactive cells. We will see that this selective process is carried out mainly by the dendritic cells, which are particularly numerous at the cortico-medullary junction, and by the macrophages that are scattered in the cortex but are also abundant in the thymic medulla.

Summary.

B cells are generated and develop in the specialized microenvironment of the bone marrow, while the thymus provides a specialized and architecturally organized microenvironment for the development of T cells. As B cells

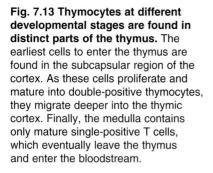

Fig. 7.13 Thymocytes at different developmental stages are found in distinct parts of the thymus. The earliest cells to enter the thymus are found in the subcapsular region of the cortex. As these cells proliferate and mature into double-positive thymocytes, they migrate deeper into the thymic cortex. Finally, the medulla contains only mature single-positive T cells, which eventually leave the thymus and enter the bloodstream.

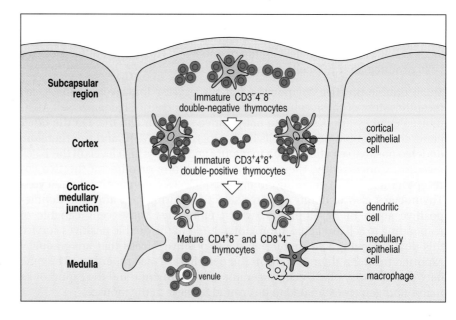

differentiate from primitive stem cells, they proceed through stages that are marked by the sequential rearrangement of immunoglobulin gene segments to generate a diverse repertoire of antigen receptors. This developmental program also involves changes in the expression of other cellular proteins. Precursors of T cells migrate from the bone marrow and mature in the thymus. This process is similar to that for B cells, including the sequential rearrangement of antigen receptor gene segments. Developing T cells pass through a series of stages that can be distinguished by the differential expression of CD44 and CD25, the CD3:T-cell receptor complex proteins, and the co-receptor proteins CD4 and CD8. The development of both T and B cells is guided by the environment, particularly by stromal cells that provide contact-dependent signals and growth factors for developing lymphocytes. In the case of T cells, development is compartmentalized, with different types of stromal cells in the thymic cortex and medulla. Most steps in T-cell differentiation occur in the cortex of the thymus. The thymic medulla contains mainly mature T cells. Lymphocyte development is accompanied by extensive cell death, reflecting intense selection and the elimination of those cells with inappropriate receptor specificities. B cells are produced throughout life, whereas T-cell production from the thymus slows down after puberty.

The rearrangement of antigen-receptor gene segments controls lymphocyte development.

In this part of the chapter we look in more detail at the steps that lead to a mature lymphocyte expressing a unique antigen receptor on its surface. The binding site of this receptor is formed from the variable regions of two different receptor chains and, in general, lymphocyte development is regulated so that each mature cell produces only one of each of these (for example, one immunoglobulin heavy chain and one light chain in B cells), and thus bears receptors of a single specificity. Production of a complete antigen receptor thus entails two series of gene segment rearrangements, one for each receptor-chain locus. Each series of rearrangements continues until a protein product is made, at which point the cell moves on to the next stage of development. This process is guided by signals that regulate the expression of the transcription factors and enzymes that control the rearrangement process. Such signals are generated by expression of a complete antigen receptor and also by expression of a pre-B- or pre-T-cell receptor in which the first receptor chain to be successfully rearranged is combined with a surrogate second chain; these receptors form complexes with accessory chains that have a signaling function. T-cell precursors develop to express either of two mutually exclusive types of T-cell receptor, the $\alpha{:}\beta$ receptor or the $\gamma{:}\delta$ receptor, and we will describe how this is thought to occur.

7-8   B cells undergo a strictly programmed series of gene rearrangements in the bone marrow.

As we saw in Chapter 4, the genes that encode immunoglobulin V regions are initially organized as an array of separate gene segments. These must be rearranged and joined together in the developing B cell to produce a complete V-region sequence. The recombination process is imprecise, however, with the random addition of nucleotides at the joins between gene segments (see Section 4-8). This means that it is a matter of chance whether the juxtaposed

J sequence, for example, and the μ constant (C)-region sequence downstream (see Fig. 4.8), can be read in the correct reading frame; each time a V gene segment undergoes rearrangement to a J segment, or to an already rearranged DJ sequence, there is a roughly two in three chance of generating an out-of-frame sequence downstream from the join. Thus, B-cell development has evolved to preserve and multiply those B cells that have made productive joins and to eliminate cells that have not.

In addition, because there are two alleles for each immunoglobulin locus in the diploid genome, each of which can rearrange, the cell must prevent both alleles from making productive joins, lest the cell express two or more receptors of different antigen specificities. This is accomplished by checking for productive joins as soon as an allele has rearranged. When a productive join is made, it generally signals the cell to cease the current phase of rearrangement and progress to the next stage. Assembly of the genes for a complete receptor requires three separate recombination events, which occur at different stages of B-cell development. These are, in the order that they occur: the joining of D to J_H and V_H to DJ_H to produce the functional heavy-chain gene, and the joining of V_L to J_L to produce the functional light-chain gene. The kappa chain locus is generally rearranged before the lambda chain locus, the latter only initiating rearrangement if the kappa locus rearrangements have failed to generate a productive join.

Because only about one in three joins will be successful, and three successful joins are required to express a complete immunoglobulin molecule, a large number of developing B cells are lost because they fail to make a productive rearrangement at one of these stages. The sequence of immunoglobulin gene rearrangements and the points at which nonproductive rearrangements can lead to cell loss are shown in Fig. 7.14. Far fewer cells are expected to be lost

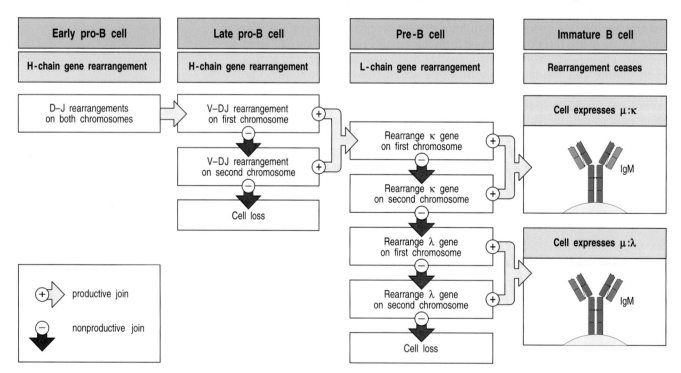

Fig. 7.14 The steps in immunoglobulin gene rearrangement at which cells can be lost. The developmental program usually rearranges the heavy-chain (H-chain) locus first and then the light-chain (L-chain) loci. Cells are allowed to progress to the next stage when a productive rearrangement has been achieved. Each rearrangement has about a one in three chance of being successful, but if the first attempt is nonproductive, development is suspended and there is a chance for one or more further attempts. The scope for repeated rearrangements is greater at the light-chain loci (see Fig. 7.16), so that fewer cells are lost between the pre-B and immature B cell stages than in the pro-B to pre-B transition.

because of failure to make productive light-chain gene rearrangements than are lost at the stage of heavy-chain gene rearrangement. This is partly because there are two light-chain loci—κ and λ—that can be rearranged, and partly because the opportunity for successive rearrangement attempts is much greater at each light-chain locus.

7-9 Successful rearrangement of heavy-chain immunoglobulin gene segments leads to the formation of a pre-B-cell receptor that halts further V_H to DJ_H rearrangement and triggers the cell to divide.

Immunoglobulin heavy-chain gene rearrangement begins in early pro-B cells with D to J_H joining. This typically occurs at both alleles of the heavy-chain locus, at which point the cell becomes a late pro-B cell. The cell then proceeds to rearrange a V_H gene segment to the DJ_H sequence. Most D to J_H joins in humans are potentially useful, as almost all human D gene segments can be translated in all three reading frames without encountering a stop codon. Thus, there is no need of a special mechanism for distinguishing successful D–J_H joins, and at this early stage there is also no need to ensure that only one allele undergoes rearrangement. Indeed, given the likely rate of failure at later stages, starting off with two successfully rearranged D–J sequences is an advantage.

V_H to DJ_H rearrangement occurs first on only one chromosome. A successful rearrangement means that intact μ chains are produced, V_H to DJ_H rearrangement ceases, and the cell progresses to become a pre-B cell. In at least two out of three cases, however, the first rearrangement is nonproductive, and V_H to DJ_H rearrangement continues on the other chromosome, again with a theoretical one in three chance of being productive. A rough estimate of the chance of generating a pre-B cell is thus something less than 55% $(1/3 + (2/3 \times 1/3) = 0.55)$.

The large pre-B cell in which a successful heavy-chain gene rearrangement has just occurred stops V_H to DJ_H rearrangement and begins to divide. This change in the cell's state is thought to occur after the transient expression of the rearranged heavy chain as part of a pre-B-cell receptor, which in some way provides the cell with a signal to halt rearrangement and to start proliferation. Pro-B cells in which rearrangements at both heavy-chain alleles are non-productive are unable to receive this signal, and are eliminated. A considerable proportion of pro-B cells is therefore lost at this stage.

The **pre-B-cell receptor** is formed by an association between the μ heavy chain and two proteins that are made in pro-B cells, which pair noncovalently to form a surrogate light chain (Fig. 7.15). One of these is called $\lambda 5$ because of its close similarity to the C domain of the λ light chain, and the other, called **VpreB**, resembles a light-chain V domain but has an extra amino-terminal region. The signaling capability of the pre-B-cell receptor depends on its further association with Igα and Igβ, two invariant accessory chains that also attend the mature B-cell receptor (see Section 6-6). Thus $\lambda 5$, VpreB, the μ heavy chain, and the attendant Igα and Igβ chains together form a **pre-B-cell receptor complex**, which structurally resembles a mature B-cell receptor complex.

The pre-B-cell receptor complex is expressed only transiently, perhaps because the production of $\lambda 5$ stops as soon as pre-B-cell receptors begin to be formed. Although the pre-B-cell receptor is expressed at low levels on the surface of pre-B cells, it is not clear whether it interacts with an external ligand. It may simply be the assembly of the receptor, or even its insertion into the endoplasmic reticulum membrane, where most pre-B-cell receptor molecules are found, that generates the signals required for further development.

Fig. 7.15 A productively rearranged immunoglobulin gene is expressed immediately as a protein by the developing B cell. In early pro-B cells, heavy-chain gene rearrangement is not yet complete and no functional μ protein is expressed, as shown in the top panel. As soon as a productive heavy-chain gene rearrangement has taken place, μ chains are expressed by the cell in a complex with two other chains, λ5 and VpreB, which together make up a surrogate light chain. The whole immunoglobulin-like complex is known as the pre-B-cell receptor (second panel). It is also associated with two other protein chains, Igα (CD79α) and Igβ (CD79β), in the cell. These associated chains signal the B cell to halt heavy-chain gene rearrangement, and drive the transition to the large pre-B cell stage by inducing proliferation. The progeny of large pre-B cells stop dividing and become small pre-B cells, in which light-chain gene rearrangements commence. Successful light-chain gene rearrangement results in the production of a light chain that binds the μ chain to form a complete IgM molecule, which is expressed together with Igα and Igβ at the cell surface, as shown in the third panel. Signaling via these surface IgM molecules is thought to trigger the cessation of light-chain gene rearrangement.

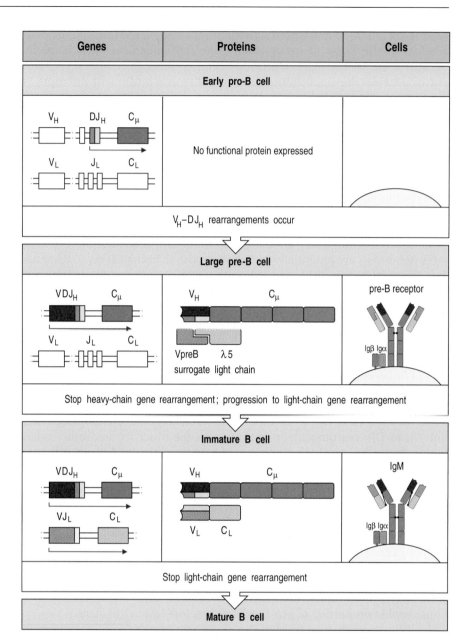

Nevertheless, its formation is an important checkpoint in B-cell development. In mice that either lack λ5 or possess mutant heavy-chain genes that cannot produce transmembrane heavy chains, the pre-B-cell receptor cannot be formed and B-cell development is blocked after heavy-chain gene rearrangement. Because the normal signals that halt V_H to DJ_H rearrangement are not given, λ5 knockout mice have rearrangements of the heavy-chain genes on both chromosomes in all pre-B cells, so that about 10% of the cells have two productive VDJ_H rearrangements.

In normal mice, the appearance of the pre-B-cell receptor coincides with inactivation of one of the recombinase subunits, RAG-2, by phosphorylation, which targets it for degradation. In addition, synthesis of the mRNAs for both RAG-2 and the other recombinase subunit, RAG-1, is suppressed, suggesting that this is the mechanism by which further rearrangement at the heavy-chain locus is blocked. Expression of the pre-B-cell receptor is also associated with cell enlargement, followed by a burst of proliferation. These cells then undergo the transition to small resting pre-B cells, in which the RAG-1:RAG-2 recombinase is again produced and the light-chain locus can be rearranged. The pre-B-cell receptor therefore appears to signal to the cell that a complete heavy-chain gene has been formed, that further rearrangements at this locus should be suppressed, and that development can proceed to the next stage. The intracellular tyrosine kinase Btk (see Section 6-10) is thought to play a part in transducing this signal, as in its absence B-cell development is blocked at the pre-B cell stage.

In the mouse, the large pre-B cells divide several times, expanding the population of cells with successful in-frame joins by approximately thirty to sixtyfold before they become resting small pre-B cells. A large pre-B cell with a particular rearranged heavy-chain gene therefore gives rise to numerous progeny. Upon reaching the small pre-B-cell stage, each of these progeny can make a different rearranged light-chain gene. Thus cells with many different antigen specificities are generated from a single pre-B cell. This makes an important contribution to B-cell receptor diversity. A similar course of events occurs in the thymus, allowing a diversity of T-cell receptor α chains to be expressed with a successfully rearranged β chain during T-cell development.

X-linked
Agammaglobulinemia

7-10 Rearrangement at the immunoglobulin light-chain locus leads to cell-surface expression of the B-cell receptor.

In mice and humans, the κ light-chain locus tends to rearrange before the λ locus. This was first deduced from the observation that myeloma cells secreting λ light chains generally have both their κ and λ light-chain genes rearranged, whereas in myelomas secreting κ light chains generally only the κ genes are rearranged. This order is occasionally reversed, however, and λ gene rearrangement does not absolutely require the prior rearrangement of the κ genes.

As with the heavy-chain locus, rearrangements at the light-chain locus generally take place at only one allele at a time. Unlike the case of the heavy-chain genes, however, there is scope for repeated rearrangements of unused V and J gene segments at each allele (Fig. 7.16). Several successive attempts at productive rearrangement of a light-chain gene can therefore be made on one chromosome before initiating any rearrangements on the second chromosome.

The chances of eventually generating an intact light chain are greatly increased by the potential for multiple successive rearrangement events at each allele and by the chance to successfully rearrange either of the two light-chain loci. As a result, most cells that reach the pre-B cell stage succeed in generating progeny that bear intact IgM molecules and can be classified as immature B cells.

Once a light-chain gene has been rearranged successfully, light chains are synthesized and combine with the heavy chain to form intact IgM (see Fig. 7.15). IgM appears at the cell surface together with Igα and Igβ to form the functional B-cell receptor complex. If the newly expressed receptor encounters a strongly cross-linking antigen—that is, if the B cell is strongly self-reactive—development is halted and the cell will not mature further. This is the first negative selection process that B cells undergo. On the other hand, if the IgM is not self-reactive, the cell continues to mature. It is not yet clear how, in the absence of binding to a specific antigen, the B cell senses that a functional immunoglobulin receptor has been expressed, and thereby receives signals for further maturation. A role for Igα in signaling at this stage is indicated by a reduction in B-lineage cells in mice that express Igα with a truncated cytoplasmic domain, which therefore cannot signal to the interior of the cell. In these mice the number of immature B cells in the marrow is reduced four-fold, and the number of peripheral B cells is reduced one hundredfold. This

Fig. 7.16 Nonproductive light-chain gene rearrangements can be rescued by further gene rearrangement. The organization of the light-chain loci in mice and humans offers many opportunities for rescue of pre-B cells that initially make an out-of-frame light-chain gene rearrangement. Light-chain rescue is illustrated for the human κ locus. If the first rearrangement is nonproductive, a 5′ V_κ gene segment can recombine with a 3′ J_κ gene segment to remove the out-of-frame join and replace it. In principle, this can happen up to five times on each chromosome, because there are five functional J_κ gene segments in humans. If all rearrangements of κ-chain genes fail to yield a productive light-chain join, λ-chain gene rearrangement may succeed (see Fig. 7.14).

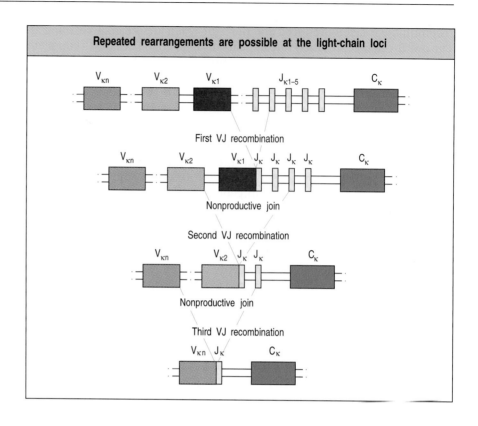

shows that an ability to signal through Igα is particularly important in dictating emigration of B cells from the bone marrow and/or their survival in the periphery once a complete immunoglobulin molecule is expressed.

Thus, by signaling the completion of a productive rearrangement, the synthesis of an immunoglobulin heavy chain or light chain and its assembly into a receptor (either the pre-B-cell receptor or IgM) leads to the further maturation of the B cell and ultimately to the cessation of further rearrangements. The shutdown of rearrangement once a productive join is made is the mechanism underlying **allelic exclusion**, a term that signifies the expression of only one of the two alleles of a given gene in a diploid cell (Fig. 7.17). Allelic exclusion occurs at both the heavy-chain locus and the light-chain loci. The phenomenon was discovered over 30 years ago, when it provided one of the original pieces of experimental support for the clonal selection theory. A similar mechanism underlies **isotypic exclusion**, production of a light chain from only one of the two light-chain loci—κ or λ—in each individual cell.

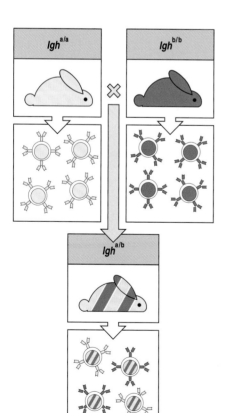

Fig. 7.17 Allelic exclusion in individual B cells. Most species have genetic polymorphisms of the constant regions of their immunoglobulin heavy- and light-chain genes; these are known as allotypes (see Section 4-20). In rabbits, for example, all of the B cells in an individual homozygous for the *a* allele of the immunoglobulin heavy-chain locus (*Igh*) will express immunoglobulin of allotype a, whereas in an individual homozygous for the *b* allele all the B cells make immunoglobulin of allotype b. In a heterozygous animal, which carries the *a* allele on one of the *Igh* chromosomes and the *b* allele on the other, individual B cells can be shown to carry either a-type or b-type immunoglobulin, but not both. This allelic exclusion reflects productive rearrangement of only one of the two parental *Igh* alleles.

Allelic exclusion seems to operate without substantial allelic preference, as the alleles at each locus are generally expressed at roughly equal frequencies; chance probably determines which allele rearranges first. In light-chain isotypic exclusion, however, there is a decided preference for which locus is rearranged first, and the ratios of κ-expressing versus λ-expressing mature B cells vary from one extreme to the other in different species. In mice and rats it is 95% κ to 5% λ; in humans it is typically 65:35, and in cats it is 5:95, the opposite of that in mice. These ratios correlate most strongly with the number of functional V_κ and V_λ gene segments in the species genome. They also reflect the kinetics and efficiency of gene segment rearrangements. The κ:λ ratio in the mature lymphocyte population is useful in clinical diagnostics, as an aberrant κ:λ ratio indicates the dominance of one clone and the presence of a lymphoproliferative disorder, which may be malignant.

7-11 The expression of proteins regulating immunoglobulin gene rearrangement and function is developmentally programmed.

A variety of proteins have a role in the development of B cells, and many of these contribute to regulating and executing the sequential steps in immunoglobulin gene segment rearrangement that mark the different stages in B-cell differentiation. Figure 7.18 lists some of these proteins, and shows how their expression is regulated through the different stages of B-cell development.

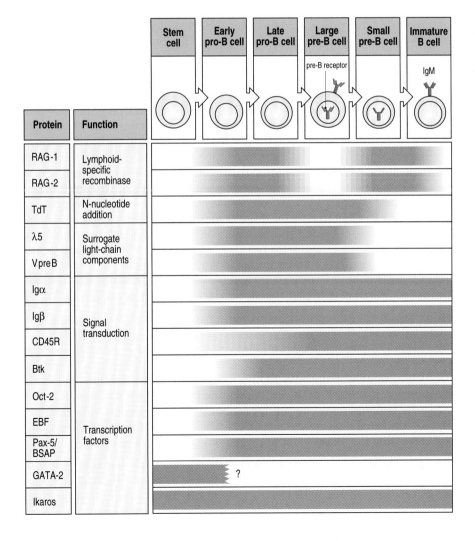

Fig. 7.18 The temporal expression of several cellular proteins known to be important for B-cell development. The proteins listed here are a selection of those known to be associated with early B-lineage development, and have been included because of their proven importance in the developmental sequence, largely on the basis of studies in mice. Their individual contributions to B-cell development are discussed in the text, with the exception of the Octamer Transcription Factor, Oct-2, which binds the octamer ATGCAAAT found in the heavy-chain promoter and elsewhere, and GATA-2, which is one example of the many transcription factors that are active in several hematopoietic lineages. The *pax-5* gene product, known as B-lineage-specific activator protein (BSAP), is involved in regulating the expression of several of the other proteins listed. The tight temporal regulation of the expression of these proteins, and of the immunoglobulin genes themselves, would be expected to impose a strict sequence on the events of B-cell differentiation.

Both immunoglobulin gene rearrangement and T-cell receptor gene rearrangement depend on the proteins **RAG-1** and **RAG-2**, products of the recombination-activation genes *RAG-1* and *RAG-2*. These proteins are unique to vertebrates, and thus are at least part of the reason that only vertebrates have rearranging antigen receptor genes and can mount an adaptive immune response. The RAG-1:RAG-2 dimer is a component of the V(D)J recombinase (see Section 4-5) which is active at very early stages of lymphoid development. In both B- and T-cell lineages there is a later and temporary suppression of *RAG* gene expression after successful rearrangements have been completed at the first of the two rearranging loci (the immunoglobulin heavy-chain locus in B cells and the TCR-β locus in T cells). Thus the RAG proteins are inactive during the burst of cell proliferation that follows these first successful rearrangements (see Sections 7-9 and 7-15), but they are resynthesized later when the cells cease dividing and go on to make rearrangements at the second locus.

Another enzyme, **terminal deoxynucleotidyl transferase (TdT)**, contributes to the diversity of both B-cell and T-cell antigen receptor repertoires by adding N-nucleotides at the joints between rearranged gene segments (see Section 4-8). As with the RAG proteins, TdT is expressed in early lymphoid progenitors but, unlike the RAG-1:RAG-2 enzyme, it is not essential to the rearrangement process. Indeed, at the time in fetal development when the peripheral immune system is first being supplied with T and B lymphocytes, TdT is expressed at low levels, if at all. In adult humans, it is expressed in pro-B cells but its expression declines at the pre-B cell stage, when heavy-chain gene rearrangement is complete and light-chain gene rearrangement has commenced. This timing explains why N-nucleotides are found in the V–D and D–J joints of heavy-chain genes but only in about a quarter of human light-chain joints. N-Nucleotides are rarely found in mouse light-chain V–J joints, showing that TdT is switched off slightly earlier in the development of mouse B cells.

Proteins required for formation of a functional receptor complex are also essential for B-cell development. The invariant proteins Igα (CD79α) and Igβ (CD79β) are components of both the pre-B-cell receptor and the B-cell receptor complexes on the cell surface (see Fig. 7.15). As well as enabling immunoglobulins to be transported to the cell surface, Igα and Igβ transduce signals from these receptors by interacting with intracellular tyrosine kinases through their cytoplasmic tails (see Section 6-6). Igα and Igβ are expressed from the pro-B cell stage until the death of the cell or its terminal differentiation into an antibody-secreting plasma cell. Curiously, mice lacking Igβ have a block in B-cell development at the pro-B cell stage, before VDJ$_H$ rearrangements are complete. The requirement for Igβ is thought to be due to the assembly of an Igα:Igβ complex with the chaperone protein calnexin in the normal pro-B cell, which may trigger signals needed for further development.

X-linked Agammaglobulinemia

Other signal transduction proteins with a role in B-cell development are included in Fig. 7.18. Bruton's tyrosine kinase (Btk), a Tec-family kinase (see Section 6-10), has received intense scrutiny because mutations in the *Btk* gene cause a profound B-lineage-specific immune deficiency, Bruton's X-linked agammaglobulinemia (XLA), in which no mature B cells are produced. In humans, the block in B-cell development caused by mutations at the *XLA* locus is almost total, interrupting the transition from pre-B cell to immature B cell. A similar, though less severe, defect called X-linked immunodeficiency or **xid** arises from mutations in the corresponding gene in mice.

Finally, several transcription factors or gene-regulatory proteins are essential for B-cell development, as shown by deficiencies of the B-cell lineage in genetically engineered mice lacking these proteins. At least 10 transcription factors necessary for normal B-lineage development have been described, and there are likely to be others. Many of these, like Ikaros, the absence of which leads to a lack of B cells, are also required for the development of other hematopoietic lineages as well. Essential transcription factors for B-cell development include the products of the *E2A* gene. These products, which are derived by alternative splicing, induce another transcription factor, the early B-cell factor (EBF), which in turn regulates the transcription of the gene for Igα. In the absence of E2A or EBF even the earliest identifiable stage in B-cell development, D–J$_H$ joining, fails to occur. Another important transcription factor is the *pax-5* gene product, one isoform of which is the B-lineage-specific activator protein (BSAP). This protein is active in late pro-B cells and enables the proper functioning of the heavy-chain enhancers located 3′ of the heavy-chain C-region genes. It also binds to regulatory sites in the genes for λ5, VpreB, and other B-cell specific proteins. In the absence of BSAP, pro-B cells fail to develop further down the B-cell pathway but can be induced to give rise to T cells and various myeloid cell types. Thus BSAP is required for the commitment of the pro-B cell to the B-cell lineage.

It seems likely that these regulatory proteins and others like them together direct the developmental program of B-lineage cells. In particular, the proteins involved in the tissue-specific transcriptional regulation of immunoglobulin genes are likely to be important in regulating the order of events in gene rearrangement. The V(D)J recombinase system is not in itself lineage-specific; it operates in both B- and T-lineage cells and uses the same core enzymes, RAG-1:RAG-2, which recognize the same conserved recombination signal sequences in both immunoglobulin and T-cell receptor genes. Yet rearrangements of T-cell receptor genes do not occur in B-lineage cells, nor do complete rearrangements of immunoglobulin genes occur in T cells. The ordered rearrangement events that do occur are associated with low-level transcription of the gene segments about to be joined. This is probably because key transcription factors such as BSAP bind to the DNA and 'open' the chromatin, making it accessible to the recombinase enzymes (Fig. 7.19).

As a consequence of immunoglobulin gene rearrangement, the promoter upstream of the V gene segments is brought nearer to the enhancers associated with the C gene segments. This in turn brings transcription factors that have bound the promoter and enhancer into proximity, resulting in a dramatic increase in transcription of the rearranged segments. Thus, gene rearrangement can be viewed as a powerful mechanism for regulating gene expression, as well as for generating receptor diversity. Several cases of gene rearrangement that brings the rearranged genes under the control of a new promoter are known from prokaryotes and single-celled eukaryotes, but in vertebrates only the immunoglobulin and T-cell receptor genes are known to use gene rearrangement to regulate gene expression.

7-12 T cells in the thymus undergo a series of gene segment rearrangements similar to those of B cells.

Developing T cells face a similar challenge to developing B cells. They must assemble a functional gene for each T-cell receptor chain while at the same time ensuring that each T cell expresses receptors of only one specificity. Not surprisingly, T cells follow an almost identical strategy to B cells, in that the receptor is assembled in stages, with each stage being checked for correct assembly. Moreover, as in B cells, a productively rearranged gene is expressed as soon as it is made, and the product(s) are assembled into a receptor complex, in this case either a pre-T-cell receptor or a *bona fide* T-cell receptor. Expression of this receptor is instrumental in promoting further development, which ultimately results in shutting down further rearrangement at the locus that has just been active.

Fig. 7.19 Proteins binding to promoter and enhancer elements contribute to the sequence of gene rearrangement and regulate the level of RNA transcription. The immunoglobulin heavy-chain locus is illustrated. First panel: in germline DNA, stem cells, and nonlymphoid cells, the chromatin containing the immunoglobulin genes is in a closed conformation. Second panel: in the early pro-B cell, lineage-specific proteins bind to the Ig enhancer elements (e); for the heavy-chain locus these are in the J–C intron and 3′ to the C exons. The DNA is now in an open conformation, and low-level transcription from promoters (P) upstream of the D and J gene segments occurs. Third panel: the rearrangement of D to J that follows the initiation of transcription of the D and J gene segments leads to a low level of transcription from promoters located upstream of the D gene segments. For some D–J joins in the mouse this may result in the expression of a truncated heavy-chain (D_μ) at levels sufficient to abort further development, but in most cases it is followed by initiation of low-level transcription of an upstream V gene segment. Fourth panel: subsequent rearrangement of a V gene segment brings its promoter under the influence of the heavy-chain enhancers, leading to enhanced production of a μ heavy-chain mRNA in large pre-B cells and their progeny. S_μ represents the switch signal sequence for isotype switching (see Section 4-16).

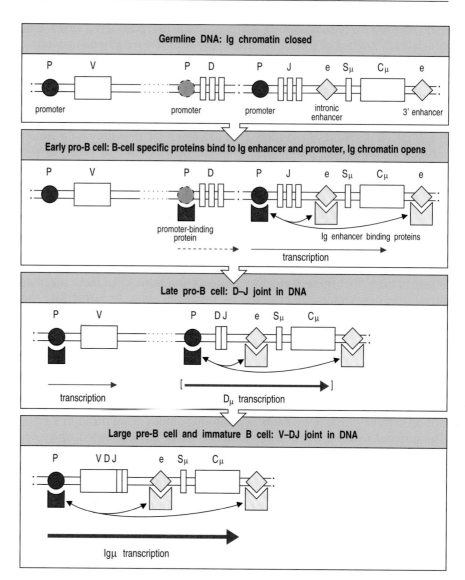

Despite the similarities with B-cell development, the control of antigen-receptor assembly during development is more complicated for T cells because there are two different kinds of T cells that could be generated from an undifferentiated precursor—α:β T cells or γ:δ T cells. These two types are distinguished by the different genetic loci that are used to make their T-cell receptors, as described in Section 4-13. Thus, the T-cell developmental program must control to which of the two lineages a precursor commits and must also ensure that a fully developed T cell only expresses receptor components of one or the other lineage. Another key difference between B and T cells is that the final assembly of an immunoglobulin leads to cessation of gene rearrangement and initiates the further differentiation of the B cell, whereas in the case of T cells, rearrangement of the V_α gene segments continues unless there is signaling to positively select the receptor.

7-13　T cells with α:β or γ:δ receptors arise from a common progenitor.

T cells bearing γ:δ receptors differ from α:β T cells in the types of antigen they recognize, in the pattern of expression of the CD4 and CD8 co-receptors, and in their anatomical distribution in the periphery. The two types of T cell also differ in function, although relatively little is known about the function of γ:δ T cells (see Sections 2-28 and 3-19). The gene rearrangements found in thymocytes and in mature γ:δ and α:β T cells suggest that these two cell lineages diverge from a common precursor after certain gene rearrangements have already occurred (Fig. 7.20). Mature γ:δ T cells can have productively rearranged β-chain genes, and mature α:β T cells often contain rearranged, but mostly (about 80%) out-of-frame, γ-chain genes.

The β, γ, and δ loci undergo rearrangement almost simultaneously in developing thymocytes. At present, the factors that regulate the lineage commitment of these cells are not known; indeed, commitment to the γ:δ lineage might simply depend on whether productive rearrangements at a γ and a δ gene have occurred in the same cell. In this view, successful rearrangement of a γ and a δ gene leads to the expression of a functional γ:δ T-cell receptor that signals the cell to differentiate along the γ:δ lineage. In most precursors, however, there is a successful rearrangement of a β-chain gene, which results in the production of a functional β-chain protein, before successful rearrangement of *both* γ and δ has occurred. The β chain pairs with the surrogate α chain, pTα, to create a pre-T-cell receptor (β:pTα),

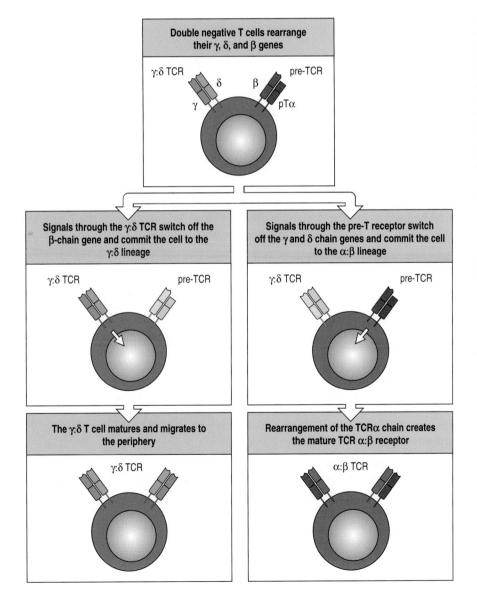

Fig. 7.20 Signals through the γ:δ TCR and the pre-T-cell receptor compete to determine the fate of thymocytes. During differentiation of T cells in the thymus, at the double-negative stage where they express neither CD4 nor CD8, the developing thymocytes express both a γ:δ T-cell receptor and the pre-T-cell receptor composed of the TCRβ chain and the pre-Tα chain (pTα—top panel). At this stage, if the thymocyte receives signals through the γ:δ receptor, the cell commits to the γ:δ lineage, switching off expression of the β-chain gene and thus the pre-Tα receptor (middle left panel). This cell then matures into a γ:δ T cell and migrates out of the thymus into the peripheral circulation (bottom left panel). On the other hand, if the developing thymocyte receives a signal through the pre-TCR (middle right panel) then the cell switches off expression of the γ:δ TCR, deletes the δ chain genes as a preliminary to rearranging the TCRα-chain locus, and goes on to express a mature α:β T-cell receptor (bottom right panel).

thus arresting further gene rearrangement and signaling the thymocyte to proliferate, to express its co-receptor genes, and eventually to start rearranging the α-chain genes. It is not known whether the TCRβ:pTα receptor recognizes a specific ligand, but it is known that its expression leads to signaling via the cytoplasmic tyrosine kinase Lck and that this is crucial for the further development of an α:β T cell (see Section 7-15). It seems likely that signals through the pre-T receptor also commit the cell to the α:β lineage (see Fig. 7.20). According to this view, the mature γ:δ cells that carry productive β-gene rearrangements committed to the γ:δ rather than α:β lineage because they have received a signal from an assembled γ:δ receptor before having assembled a functional pre-T-cell receptor. An alternative theory is that these cells were diverted into becoming γ:δ cells at a later stage, after receiving signals from a pre-T receptor and progressing to the second phase of RAG enzyme activity. However, it remains uncertain whether further rearrangements at the γ and δ loci can occur at this point. Normally, α-gene rearrangements ensue and these delete the intervening δ-chain gene segments as an extrachromosomal circle (see Section 4-13 and Fig. 4.15). The subsequent maturation of α:β T cells depends on rearrangements at the α locus leading to a functional α:β receptor that is positively selected for its ability to recognize self MHC molecules. The signals that drive a γ:δ T cell to mature are also unknown. Certain γ:δ T cells seem able to develop in the absence of a functioning thymus and are present, for example, in *nude* mice.

7-14   T cells expressing particular γ- and δ-chain V regions arise in an ordered sequence early in life.

We have just described how a precursor T cell that can become either an α:β or a γ:δ T cell becomes directed to one or the other lineage. During the development of the organism, these pathways are not equally used; instead, the generation of the various types of T cells—even the particular V region assembled in γ:δ cells—are developmentally controlled. The first T cells to appear during embryonic development carry γ:δ T-cell receptors (Fig. 7.21). In the mouse, where the development of the immune system can be studied in detail, γ:δ T cells first appear in discrete waves or bursts, with the T cells in each wave populating distinct sites in the adult animal.

The first wave of γ:δ T cells populates the epidermis; the T cells become wedged among the keratinocytes and adopt a dendritic-like form which has given them the name of **dendritic epidermal T cells** (**dETCs**). The second wave homes to the epithelia of the reproductive tract. Remarkably, given the large number of theoretically possible rearrangements, the receptors expressed by these early waves of γ:δ T cells are essentially homogeneous. All the cells in each wave assemble the same V_γ and V_δ regions. Each different wave, however, uses a different set of V, D, and J gene segments. Thus, certain V, D, and J gene segments are selected for rearrangement at particular times during embryonic development; the reasons for this limitation are poorly understood. There are no N-nucleotides contributing additional diversity at the junctions between V, D, and J gene segments, reflecting the absence of the enzyme TdT from these fetal T cells.

After these initial waves, T cells are produced continuously rather than in bursts, and α:β T cells predominate, making up more than 95% of thymocytes. The γ:δ T cells produced at this stage are different from those of the early waves. They have a considerably more diverse receptor repertoire, for which several different V gene segments have been used, and the receptor sequences have abundant N-nucleotide additions. Most of these γ:δ T cells, like α:β T cells, are found in peripheral lymphoid tissues rather than in the epithelial sites populated by the early γ:δ T cells.

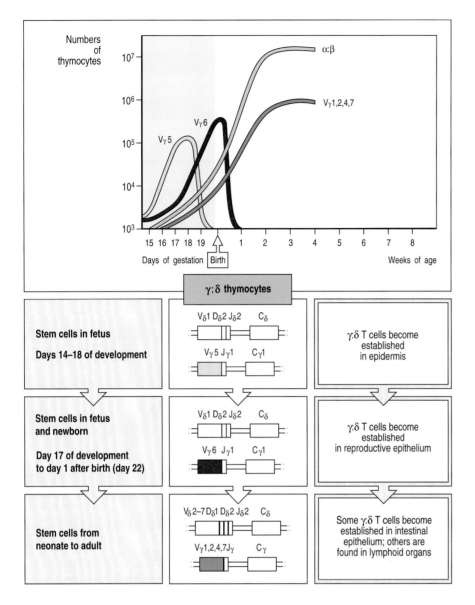

Fig. 7.21 The rearrangement of T-cell receptor γ and δ genes in the mouse proceeds in waves of cells expressing different Vγ and Vδ gene segments. At about 2 weeks of gestation, the Cγ1 locus is expressed with its closest V gene (Vγ5; also known as Vγ3). After a few days Vγ5-bearing cells decline (upper panel) and are replaced by cells expressing the next most proximal gene, Vγ6. Both these rearranged γ chains are expressed with the same rearranged δ-chain gene, as shown in the lower panels, and there is little junctional diversity in either the Vγ or the Vδ chain. As a consequence, most of the γ:δ T cells produced in each of these early waves have the same specificity, although the antigen recognized in each case is not known. The Vγ5-bearing cells become established selectively in the epidermis, whereas the Vγ6-bearing cells become established in the epithelium of the reproductive tract. After birth, the α:β T-cell lineage becomes dominant and, although γ:δ T cells are still produced, they are a much more heterogeneous population, bearing receptors with a great deal of junctional diversity.

The developmental changes in V gene segment usage and N-nucleotide addition in murine γ:δ T cells parallel changes in B-cell populations during fetal development, which will be discussed later (see Section 7-28). Their functional significance is unclear, however, and not all of these changes in the pattern of receptors expressed by γ:δ T cells occur in humans. Certainly, the γ:δ T cells that home to the skin of mice, the dendritic epidermal T cells (dETCs), do not seem to have exact human counterparts, although there are γ:δ T cells in the human reproductive and gastrointestinal tracts. The mouse dETCs may serve as sentinel cells which are activated upon local tissue damage or as cells that regulate inflammatory processes (see Section 2-28).

7-15 Rearrangement of the β-chain locus and production of a β chain trigger several events in developing thymocytes.

T cells expressing α:β receptors first appear a few days after the earliest γ:δ T cells and rapidly become the most abundant type of thymocyte (see Fig. 7.21). The rearrangement of the β- and α-chain loci during T-cell development follows a sequence that closely parallels the rearrangement of immunoglobulin heavy- and light-chain loci during B-cell development (see Section 7-9). As shown in Fig. 7.22, the β-chain genes rearrange first, with the $D_β$ gene segments rearranging to $J_β$ gene segments, and this is followed by $V_β$ to $DJ_β$ gene rearrangement. If no functional β chain can be synthesized from these rearrangements, the cell will not be able to produce a pre-T-cell receptor and will die unless it makes successful rearrangements at both the γ and δ loci (see Section 7-13). However, unlike B cells with nonproductive

Fig. 7.22 The stages of gene rearrangement in α:β T cells. The sequence of gene rearrangements is shown, together with an indication of the stage at which the events take place and the nature of the cell-surface receptor molecules expressed at each stage. The β-chain locus rearranges first, in CD4⁻ CD8⁻ double-negative thymocytes expressing CD25 and low levels of CD44. As with immunoglobulin heavy-chain genes, D to J gene segments rearrange before V gene segments rearrange to DJ (second and third panels). It is possible to make up to four attempts to generate a productive rearrangement at the β-chain locus, as there are four D gene segments and two sets of J gene segments (not shown). The productively rearranged gene is expressed initially within the cell and then at low levels on the cell surface. It associates with pTα, a surrogate 33 kDa α chain that is equivalent to λ5 in B-cell development, and this pTα:β heterodimer forms a complex with the CD3 chains (fourth panel). The expression of the pre-T-cell receptor signals the developing thymocytes via the tyrosine kinase Lck to halt β-chain gene rearrangement, and to undergo multiple cycles of division. At the end of this proliferative burst, the CD4 and CD8 molecules are expressed, the cell ceases cycling, and the α chain is now able to undergo rearrangement. The first α-chain gene rearrangement deletes all δ D, J, and C gene segments on that chromosome, although these are retained as a circular DNA, proving that these are nondividing cells (bottom panel). This permanently inactivates the δ-chain gene. Rearrangements at the α-chain locus can proceed through several cycles, because of the large number of $V_α$ and $J_α$ gene segments, so that productive rearrangements almost always occur. When a functional α chain is produced that pairs efficiently with the β chain, the CD3^low CD4⁺ CD8⁺ thymocyte is ready to undergo selection for its ability to recognize self peptides in association with self MHC molecules.

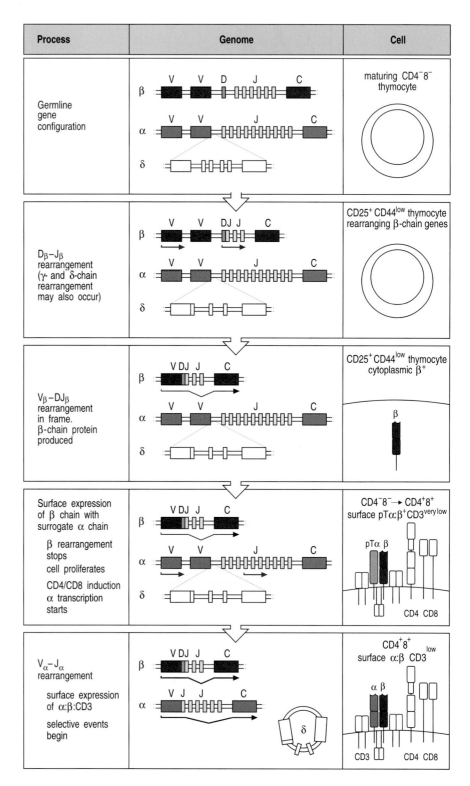

immunoglobulin heavy-chain gene rearrangements, thymocytes with non-productive β-chain VDJ rearrangements can be rescued by subsequent rearrangements, which are possible owing to the presence of two clusters of D_β and J_β gene segments upstream of two C_β genes (see Fig. 4.11). This increases the likelihood of a productive VDJ join from 55% for immunoglobulin heavy-chain genes to more than 80% for T-cell receptor β-chain genes.

Once a productive β-chain gene rearrangement has occurred, the β chain is expressed together with the invariant partner chain pTα and the CD3 molecules (see Fig. 7.22) and is transported to the cell surface. The β:pTα complex is a functional pre-T-cell receptor analogous to the μ:VpreB:λ5 pre-B-cell receptor complex in B-cell development (see Section 7-9). Expression of the pre-T-cell receptor triggers the phosphorylation and degradation of RAG-2, halting β-chain gene rearrangement and thus ensuring allelic exclusion at the β locus. It also induces rapid cell proliferation and eventually the expression of the co-receptor proteins CD4 and CD8. All these events require the cytoplasmic protein tyrosine kinase Lck, which subsequently associates with the co-receptor proteins. In mice genetically deficient in Lck, T-cell development is arrested before the $CD4^+$ $CD8^+$ double-positive stage and no α-chain gene rearrangements are made.

The role of the expressed β chain in suppressing further β-chain locus rearrangement can be demonstrated in transgenic mice containing a rearranged T-cell receptor β-chain transgene: these mice express the transgenic β chain on virtually 100% of their T cells, showing that rearrangement of the endogenous β-chain genes is strongly suppressed. The importance of pre-Tα has been shown by the hundredfold decrease in α:β T cells and by the absence of allelic exclusion at the β locus in mice deficient in pre-Tα.

During the proliferative phase triggered by expression of the pre-T-cell receptor, the *RAG-1* and *RAG-2* genes are also repressed. Hence, no rearrangement of the α-chain locus occurs until the proliferative phase ends, when *RAG-1* and *RAG-2* genes are transcribed again, and the functional RAG-1:RAG-2 heterodimer accumulates. This ensures that each cell in which a β-chain gene has been successfully rearranged gives rise to many $CD4^+$ $CD8^+$ double-positive thymocytes. Once the cells stop dividing, each of these can independently rearrange its α-chain genes, so that a single functional β chain can be associated with many different α chains in the progeny cells. During the period of α-chain gene rearrangement, α:β T-cell receptors are first expressed, and selection by peptide:MHC complexes in the thymus can begin.

Along the progression of T cells from the double-negative to the double-positive and finally single-positive stage, there is a distinct pattern of expression of molecules involved in rearrangement, signaling, and also transcription factors that most likely control developmental fates as well as the expression of important T-cell loci like those of the T-cell receptor itself. The expression of these molecules through T-cell development is illustrated in Fig. 7.23. The expression of the RAG proteins has already been described. TdT, the protein responsible for insertion of N-nucleotides at gene segment junctions in both B and T cells, is expressed throughout the developmental period in which thymocytes are rearranging T-cell receptor gene segments; N-nucleotides can be found at the junctions of all rearranged α and β genes. Lck, a Src-family tyrosine kinase, and ZAP-70, another tyrosine kinase, are both expressed from an early stage in thymocyte development. Lck plays a key role in signaling of thymocytes at the pre-TCR$^+$ stage, since mice lacking Lck have a profound loss of cells beyond this stage. Lck is also important for γ:δ T-cell development. On the other hand, gene knockout studies (see Appendix I, Section A-47) show that ZAP-70, though expressed from the double-negative stage onward, plays a later role in promoting the development of single-positive thymocytes from double-positive thymocytes. Fyn is expressed at increasing levels from the

Fig. 7.23 The temporal expression of several cellular proteins known to be important for early T-cell development. The expression of a set of proteins is depicted with respect to the stages of thymocyte development as determined by cell-surface marker expression. The proteins listed here are a selection of those known to be associated with early T-lineage development, and have been included because of their proven importance in the developmental sequence, largely on the basis of studies in mice. Their individual contributions to T-cell development are discussed in the text.

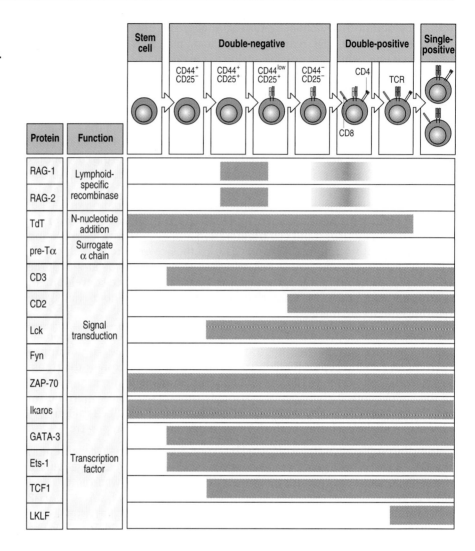

double-positive stage onward, but is not absolutely essential for thymocyte development as long as Lck is present. The roles these molecules play in mature T cell signaling are described in Chapter 6. Finally, a set of transcription factors guides the development of thymocytes from one stage to the next. A number of important factors have been identified; in some cases gene knockout studies have shown at which stage of development they play essential roles. Ikaros and GATA-3 are expressed in early T-cell progenitors, and in the absence of either, T-cell development is generally disrupted; moreover, these molecules also have roles in the normal functioning of mature T cells. In contrast, Ets-1, though also expressed in early progenitors, does not have an essential role for T-cell development, although mice lacking this factor do not make NK cells. TCF1 (T-cell factor-1) is first expressed during the double-negative stage and in its absence double-negative T cells that make productive β rearrangements do not undergo proliferation as usual in response to the pre-TCR signal. Thus, TCF1-deficient mice fail to make double-positive thymocytes efficiently. LKLF (lung Kruppel-like factor) is first expressed late in thymocyte development at the single-positive stage; if LKLF is absent, single-positive cells do develop, but they are abnormal, displaying a partially activated phenotype. Thus, transcription factors are turned on at various developmental stages and are responsible for normal development through those stages by controlling the expression of important genes. Transcription factors mediate the response to developmental signals by turning on the set of appropriate genes when the factors are activated to bind DNA.

7-16   T-cell α-chain genes undergo successive rearrangements until positive selection or cell death intervenes.

The T-cell receptor α-chain genes are comparable to the immunoglobulin κ and λ light-chain genes in that they do not have D gene segments and are rearranged only after their partner receptor-chain gene has been expressed. As with the immunoglobulin light-chain genes, repeated attempts at rearrangement are possible. Indeed, the presence of multiple V_α gene segments, and around 60 J_α gene segments spread over some 80 kb of DNA allows many successive V_α to J_α rearrangements. This means that T cells with an initial nonproductive α gene rearrangement are much more likely to be rescued by a subsequent rearrangement than are B cells with a nonproductive light-chain gene rearrangement (Fig. 7.24).

The potential for many successive rearrangements at both alleles of the α-chain locus virtually guarantees that a functional α chain will be produced in every developing T cell. Moreover, many T cells have in-frame rearrangements on both chromosomes and thus can produce two types of α chain. This is possible because, unlike the situation in B cells, expression of the T-cell receptor is not in itself sufficient to shut off gene rearrangement. Thus, rearrangement of α-chain genes continues even after production of a T-cell receptor at the cell surface. Continued rearrangements can allow several different α chains to be produced successively in each developing T cell and to be tested for self MHC recognition in partnership with the same β chain. This phase of gene rearrangement lasts for 3 or 4 days in the mouse and only ceases when positive selection occurs as a consequence of receptor engagement, or when the cell dies. Thus, in the strict sense, T-cell receptor α-chain

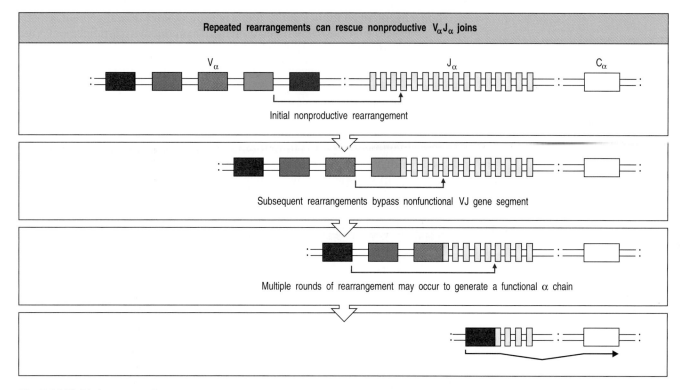

Fig. 7.24 Multiple successive rearrangement events can rescue nonproductive T-cell receptor α-chain gene rearrangements. The multiplicity of V and J gene segments at the α-chain locus allows successive rearrangement events to 'leapfrog' over previously rearranged VJ segments, deleting any intervening gene segments. The α-chain rescue pathway resembles that of the immunoglobulin κ light-chain genes (see Section 7-10), but the number of possible successive rearrangements is greater. α-chain Gene rearrangement continues until either a productive rearrangement leads to positive selection or the cell dies.

genes are not subject to allelic exclusion (see Section 7-10). However, as we will see in the next part of this chapter, only T-cell receptors that are positively selected for self MHC recognition can function in self MHC-restricted responses. The regulation of α-chain gene rearrangement by positive selection therefore ensures that each T cell has only a single functional specificity, even if two different α chains are expressed.

Clearly, the engagement of any particular T-cell receptor with a self MHC:self peptide ligand will depend on the receptor's specificity. Thus, the phase of α-chain gene rearrangement marks an important change in the forces shaping the destiny of the T cell. Up to this point, the development of the thymocyte has been independent of antigen; from this point on, developmental decisions depend on the interaction of the T-cell receptor with its peptide:MHC ligands, as described in the next part of the chapter.

Summary.

As lymphocytes differentiate from primitive stem cells, they proceed through stages that are marked by the sequential rearrangement of the antigen-receptor gene segments at the different genetic loci. As each complete receptor-chain gene is generated, the protein it encodes is expressed as part of a receptor, and this signals the developing cell to progress to the next developmental step. For B cells, the heavy-chain locus is rearranged first. If rearrangement is successful and a pre-B-cell receptor is made, heavy-chain gene rearrangement ceases and the resulting pre-B cells proliferate, followed by the start of rearrangement at a light chain locus. If the initial light-chain gene rearrangement is productive, a complete immunoglobulin B-cell receptor is formed, gene rearrangement ceases, and B-cell development proceeds. If it is not, light-chain gene rearrangement continues until either a new productive rearrangement is made or all available J regions have been used up. If a productive rearrangement is not made, the developing B cell dies.

In developing T cells, receptor loci rearrange according to a defined program similar to that in B cells. There is, however, the added complication that individual precursor T cells can follow one of two distinct lines of development. There are four T-cell receptor gene loci—γ, δ, α, and β. Rearrangement at these loci leads to cells bearing either γ:δ T-cell receptors or α:β T-cell receptors. Early in ontogeny, γ:δ T cells predominate, but from birth onward more than 90% of thymocytes express T-cell receptors encoded by rearranged α and β genes. In developing thymocytes, the γ, δ, and β loci rearrange first, and start rearranging virtually simultaneously. Productive rearrangements of both a γ and a δ gene may lead to the production of a functional γ:δ T-cell receptor and the development of a γ:δ T cell. Most cells, however, enter the α:β lineage. In these cells the generation of a β-chain gene before both γ and δ have rearranged leads to the expression of a functional β chain and a pre-T-cell receptor. This receptor, like the pre-B-cell receptor, signals the developing cell to proliferate, to arrest β-chain gene rearrangement, to express CD4 and CD8, and eventually to rearrange the α-chain gene. Rearrangement of the α-chain genes continues in these CD4+ CD8+ double-positive thymocytes until positive selection allows the maturation of a single-positive α:β cell; almost all cells will make productive α-gene rearrangements but the vast majority of thymocytes die at the CD4 CD8 double-positive stage because they are not positively selected.

Thus gene rearrangement follows an ordered sequence in both B and T lineages to produce immature lymphocytes that each bear antigen receptors of a single specificity on their surface. These antigen receptors can now be tested for their antigen-recognition properties and the cells selected accordingly; these selection processes are described in the next part of this chapter.

Interaction with self antigens selects some lymphocytes for survival but eliminates others.

We have followed the development of a lymphocyte from a committed precursor to the point at which a complete antigen receptor is expressed on the cell surface. Development up to this point has been focused on testing for productive gene rearrangements and multiplying those cells that are successful, while at the same time controlling the rearrangement process so that cells express only one receptor. Once the cell can express that receptor, however, the focus shifts. Now the fate of the immature lymphocyte will be determined by the specificity of its antigen receptor. Most obviously, lymphocytes with strongly self-reactive receptors should be eliminated to prevent autoimmune reactions; this negative selection is one of the ways in which the immune system is made self-tolerant. In addition, given the incredible diversity of receptors that the rearrangement process can generate, it is important that those lymphocytes that mature are likely to be useful in recognizing and responding to foreign antigens, especially as an individual can only express a small fraction of the total possible receptor repertoire in his or her lifetime. Indeed, certainly for T cells and probably for B cells, a process of positive selection identifies and preserves lymphocytes that are likely to be able to respond to foreign antigens; those that do not pass this test die by neglect. In this part of the chapter we will describe how the processes of positive and negative selection shape the mature lymphocyte repertoire.

7-17   Immature B cells that bind self antigens undergo further receptor rearrangement, or die, or are inactivated.

Once an immature B cell expresses IgM on its surface (sIgM), its fate is guided by the nature of the signals it receives through its antigen receptor. This was first demonstrated by experiments in which antigen receptors on immature B cells were experimentally stimulated *in vivo* using anti-μ chain antibodies (see Appendix I, Section A-10); the outcome was elimination of the immature B cells. More recent experiments using mice expressing B-cell receptor transgenes have confirmed these earlier findings, but have also shown that immediate elimination is not the only possible outcome of binding to a self antigen. Instead, there are four possible fates for self-reactive immature B cells, depending on the nature of the ligand to which they are capable of binding (Fig. 7.25). These fates are cell death by apoptosis; the production of a new receptor by receptor editing (Fig. 7.26); the induction of a permanent state of unresponsiveness to antigen; and ignorance. An ignorant cell is defined as one that has affinity for a self antigen but that does not sense the self antigen, either because it is sequestered, is in low concentration, or does not cross-link the B-cell receptor.

Apoptosis and elimination of immature self-reactive B cells seems to predominate when the interacting self antigen is multivalent, for example the multiple copies of an MHC molecule on a cell surface. This antigen-induced loss of cells from the B-cell population is known as **clonal deletion**. The effect of encounter with a multivalent antigen was tested in mice transgenic for both chains of an immunoglobulin specific for H-2K^b MHC class I molecules. In such mice nearly all the B cells that develop bear the anti-MHC immunoglobulin as sIgM, because the presence of the already rearranged transgenes inhibits rearrangement of the endogenous immunoglobulin genes (see Sections 7-9 and 7-10). If the transgenic mouse does not express

Fig. 7.25 Binding to self molecules in the bone marrow can lead to the death or inactivation of immature B cells. Left panels: when developing B cells express receptors that recognize multivalent ligands, for example, ubiquitous self cell-surface molecules such as those of the MHC, they are deleted from the repertoire (clonal deletion). These B cells either undergo receptor editing, so that the self-reactive receptor specificity is deleted, or the cells themselves undergo programmed cell death or apoptosis. Center left panels: immature B cells that bind soluble self antigens able to cross-link the B-cell receptor are rendered unresponsive to the antigen (anergic) and bear little surface IgM. They migrate to the periphery where they express IgD but remain anergic; if in competition with other B cells in the periphery, they are rapidly lost. Center right panels: immature B cells that bind soluble self antigens with low affinity or that bind monovalent antigens do not receive any signal as a result of this interaction and mature normally to express both IgM and IgD at the cell surface. Such cells are potentially self-reactive, and they are said to be clonally ignorant as their ligand is present but is not able to activate them. Right panels: immature B cells that do not encounter antigen mature normally; they migrate from the bone marrow to the peripheral lymphoid tissues where they may become mature recirculating B cells bearing both IgM and IgD on their surface.

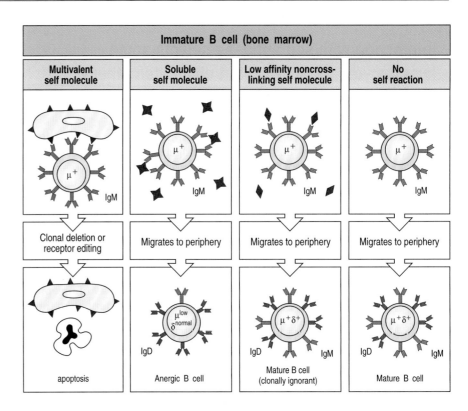

H-$2K^b$, normal numbers of B cells develop, all bearing transgene-encoded anti-H-$2K^b$ receptors. However, in mice expressing both H-$2K^b$ and the immuno-globulin transgenes, B-cell development is blocked. Normal numbers of pre-B cells and immature B cells are found, but B cells expressing the anti-H-$2K^b$ immunoglobulin as sIgM never mature to populate the spleen and lymph nodes; instead, most of these immature B cells die in the bone marrow by apoptosis. This is because the anti-H-$2K^b$ immunoglobulin on the immature B cells interacts strongly with the H-$2K^b$ molecules on the bone marrow stromal cells.

Closer analysis of this experimental system and others like it revealed the surprising finding that clonal deletion was not the only outcome in these circumstances. There was, in fact, an interval before cell death during which the self-reactive B cell might be rescued by further gene rearrangements that replaced the self-reactive receptor with a new receptor that was not autoreactive (see Fig. 7.26). This mechanism for replacing receptors, termed **receptor editing**, works as follows. When an immature B cell first expresses a light chain, and produces sIgM, the *RAG* genes are still being expressed and RAG protein is being made. If the B cell does not interact with self antigen and there is no strong cross-linking of sIgM, gene rearrangement ceases and the B cell continues its development; RAG protein begins to decline but does not completely disappear until the B cell reaches full maturity in the spleen. However, in self-reactive B cells, strong cross-linking of sIgM, as a result of encounter with the self antigen, halts further development, and *RAG* gene expression continues at levels similar to that in pre-B cells undergoing light-chain gene rearrangement. As a consequence of the continued presence of recombinase, light-chain gene rearrangement continues, even though the cell has already made one productive rearrangement at this locus. The light-chain loci are able to make numerous successive rearrangements (see Fig. 7.16), and these rearrangements can rescue immature self-reactive B cells by deleting or displacing the rearrangement that encodes the self-reactive receptor and replacing it with another sequence.

This continuation of light-chain gene rearrangement has parallels with the continuation of α-gene rearrangement in developing T cells, but it should be emphasized that in B cells it only occurs if the receptor encounters a strongly cross-linking antigen. In T cells, in contrast, α-gene rearrangement and 'receptor editing' continue as part of the normal developmental program until the cell is positively selected or dies.

More recently, receptor editing has been shown unambiguously in mice bearing transgenes for autoantibody heavy and light chains that have been placed within the immunoglobulin loci by the homologous recombination method explained in Appendix I, Section A-47. The transgene imitates a primary gene rearrangement and is surrounded by unused endogenous gene segments. In mice that express the antigen recognized by the transgene-encoded receptor, the few mature B cells that emerge into the periphery have used these surrounding gene segments for further rearrangements that replace the autoreactive light-chain transgene with a nonautoreactive rearranged gene.

At a light-chain locus, the multiplicity of V and J segments allows the unused V and J gene segments to be selected for multiple further rearrangements (see Fig. 7.16). In addition, cells that have exhausted the J_κ regions can rearrange the λ locus. It is not clear whether receptor editing occurs at the heavy-chain locus. There are no available D segments at a rearranged heavy-chain locus, so a new rearrangement cannot simply occur by the normal mechanism and at the same time remove the preexisting one. Instead a process of V_H replacement may use embedded recombination signal sequences in a recombination event that displaces the V gene segment from the self-reactive rearrangement and replaces it with a new V gene segment. This has been observed in some B-cell tumors, but whether it occurs during normal B-cell development in response to signals from autoreactive B-cell receptors is not known.

It was originally thought that successful production of a heavy chain and a light chain caused the almost instantaneous shut down of further light-chain locus rearrangement and that this ensured both allelic and isotypic exclusion (see Section 7-10). The unexpected ability of self-reactive B cells to continue to rearrange their light-chain genes, even after having made a productive rearrangement, has raised questions about this supposed mechanism of allelic exclusion.

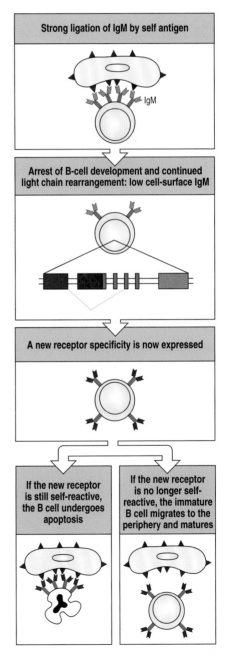

Fig. 7.26 Replacement of light chains by receptor editing can rescue some self-reactive B cells by changing their antigen specificity. When a developing B cell expresses antigen receptors that are strongly cross-linked by multivalent self antigens such as MHC molecules on cell surfaces (top panel), the B cell undergoes developmental arrest. The cell lowers surface expression of IgM and does not turn off the *RAG* genes (second panel). Continued synthesis of RAG proteins allows the cell to continue light-chain gene rearrangement. This usually leads to a new productive rearrangement and expression of a new light chain, which combines with the previous heavy chain to form a new receptor (receptor editing; third panel). If this new receptor is not self-reactive, the cell is 'rescued' and continues normal development much like a cell that had never reacted with self (bottom right panel). If the cell remains self-reactive, it may be rescued by another cycle of rearrangement, but if it continues to react strongly with self it will undergo programmed cell death or apoptosis and be deleted from the repertoire (clonal deletion; bottom left panel).

Undoubtedly, however, the decline in the level of RAG protein that occurs after a successful nonself-reactive rearrangement is crucial to maintaining allelic exclusion, as it will reduce the chance of a subsequent rearrangement. Furthermore, any additional productive rearrangement that did still occur would not necessarily breach allelic exclusion: if it occurred on the same chromosome it would eliminate the existing productive rearrangement, while if it occurred on the other chromosome it would be nonproductive in two out of three cases. Thus, the observed fall in the level of RAG protein may be sufficient to ensure that a second productive rearrangement is rarely made, and could be the principal, if not sole, mechanism behind allelic exclusion. Consistent with this idea, it appears that allelic exclusion is not absolute, since there are rare B cells that express two light chains.

We have so far discussed the fate of newly formed B cells that undergo multivalent cross-linking of their sIgM. Those immature B cells that encounter more weakly cross-linking self antigens of low valence, such as small soluble proteins, respond differently. In this situation, the self-reactive B cells tend to be inactivated and enter a state of permanent unresponsiveness, or **anergy**, but do not immediately die (see Fig. 7.25). **Anergic** B cells cannot be activated by their specific antigen even with help from antigen-specific T cells (see Section 1-15). Again, this phenomenon was elucidated using transgenic mice. When hen egg lysozyme (HEL) is expressed in soluble form from a transgene in mice that are also transgenic for high-affinity anti-HEL immunoglobulin, the HEL-specific B cells mature but are unable to respond to antigen. The anergic cells retain their IgM within the cell and transport little to the surface. In addition, they develop a partial block in signal transduction so that, despite normal levels of HEL-binding sIgD, the cells cannot be stimulated by cross-linking this receptor. It seems that signal transduction is blocked at a step before the phosphorylation of the Igα and Igβ chains (see Section 6-6), although the exact step is not yet known. The signaling defect may involve the inability of B-cell receptor molecules on tolerant B cells to enter regions of the cell in which important other signaling molecules normally segregate in order to transmit a complete signal subsequent to antigen binding. Furthermore, cells that have received an anergizing signal may upregulate molecules that inhibit signaling and transcription.

Anergic B cells are not only impaired in signal transduction and expression of sIgM. Their migration within peripheral lymphoid organs is altered and their life-span and ability to compete with immunocompetent B cells are compromised. Mature B cells recirculate through lymphoid follicles during their life-span. In normal circumstances, in which B cells binding soluble self antigen are in a minority, anergic B cells are detained in the T-cell areas of the peripheral lymphoid tissue and are excluded from lymphoid follicles. As anergic B cells cannot be activated by T cells, and T-cell help will in any case not be available for self antigens, to which the T cells will be tolerant, these self-reactive B cells will not be activated to secrete antibody. Instead they die relatively soon, presumably because they fail to get survival signals from T cells, thus ensuring that the long-lived pool of peripheral B cells is purged of potentially self-reactive cells.

The fourth potential fate of self-reactive immature B cells is that nothing happens to them; they remain in a state of immunological 'ignorance' of their self antigen (see Fig. 7.25). It is clear that some B cells with a weak but definite affinity for a self antigen, mature as if they were not self-reactive at all. Such B cells do not respond to the presence of their self-antigen because it interacts so weakly with the receptor that little if any, intracellular signal is generated when it binds. Alternatively, some self-reactive B cells may not encounter their antigen at this stage because it is not accessible to B cells developing in the bone marrow and spleen. The maturation of these B cells reflects a balance that the immune system strikes between purging all self-reactivity and maintaining the ability to respond to pathogens. If elimination of self-reactive cells were too efficient, the receptor repertoire might become too limited and thus not be able to recognize a wide variety of pathogens. Some autoimmune disease might be the price of this balance, as it is very likely that these low-affinity self-reactive lymphocytes can be activated and cause disease under certain circumstances. Thus these cells can be thought of as the seeds of autoimmune disease. Normally, however, ignorant B cells will be held in check by a lack of T-cell help, the continued inaccessibility of the self antigen, or the tolerance that can be induced in mature B cells, as described in the next section.

7-18   Mature B cells can also be rendered self-tolerant.

It is likely that most self-reactive B cells will encounter their antigens while still immature, as many self antigens circulate through tissues in soluble form or are expressed by many different cell types, including those in the bone marrow. Nonetheless, some self antigens are not present in the tissues through which immature B cells pass, and thus B cells expressing receptors specific for these antigens will survive to become mature. Lack of T-cell help, or an inability of the antigen to strongly cross-link the B-cell receptor will usually prevent these cells from becoming activated and causing problems. However, mechanisms for eliminating or inactivating mature self-reactive B cells, and even activated self-reactive B cells, also exist.

For example, B cells expressing immunoglobulin specific for H-2K^b MHC class I molecules are deleted even when, in transgenic animals, the expression of the H-2K^b molecule is restricted to the liver by use of a liver-specific gene promoter. B cells that encounter strongly cross-linking antigens in the periphery undergo apoptosis directly, unlike their counterparts in the bone marrow, which attempt further receptor rearrangements instead. The different outcomes may be due to the fact that the B cells in the periphery are more mature and can no longer rearrange their light-chain loci. On the other hand, stromal signals may be important; recent data suggest that stromal cells in the spleen instruct B cells to respond in a different way to signals delivered through the antigen receptor compared with the stromal cells of the bone marrow.

Mature B cells that encounter and bind an abundant soluble antigen become anergized. This was demonstrated in mice by placing the HEL transgene under the control of an inducible promoter that can be regulated by changes in the diet. It is thus possible to induce the production of lysozyme at any time and thereby study its effects on HEL-specific B cells at different stages of maturation. These experiments have shown that both mature and immature B cells are inactivated when they are chronically exposed to soluble antigen.

Finally, there may be a mechanism to eliminate activated B cells if they mutate to self-reactivity while they are proliferating and undergoing somatic hypermutation in germinal centers of peripheral lymphoid tissues (see Section 4-9). This possibility has been demonstrated by first stimulating a germinal center response by immunizing with foreign antigen, and then infusing large quantities of this antigen in soluble form, a procedure that induced apoptosis in the responding B cells. This experiment was designed to mimic the situation in which a germinal center B cell alters its specificity to express a mutated immunoglobulin with high affinity for self antigen; however, whether this form of self-tolerance operates under normal conditions is not yet clear.

7-19 Only thymocytes whose receptors can interact with self MHC:self peptide complexes can survive and mature.

As we learned in Chapters 3 and 5, α:β T-cell receptors recognize intracellular antigens after they have been processed into peptides that can be complexed with self MHC molecules. These antigen-presenting MHC molecules are encoded by a set of highly polymorphic genes found at the MHC locus, and their variability extends the range of peptides that can be presented to T cells in any one individual. However, MHC polymorphism also affects antigen recognition by T cells, since the T-cell receptor contacts both the bound peptide and the surrounding polymorphic surface of the MHC molecule itself. Each individual T-cell receptor is specific for a particular combination of MHC molecule and peptide antigen and is thus said to be MHC-restricted for antigen recognition. Not all the T-cell receptors generated by gene rearrangement will be able to recognize self MHC molecules and function in self MHC-restricted responses to foreign antigens; those that have this capability are selected for survival by a process of positive selection that occurs in the thymus.

Positive selection was demonstrated in experiments using mice whose bone marrow had been completely replaced by bone marrow from a mouse of different MHC genotype that was genetically identical except in the MHC gene region. These mice are known as **bone marrow chimeras** (see Appendix I, Section A-43). The recipient mouse is first irradiated in order to destroy all its own lymphocytes and bone marrow progenitor cells; after bone marrow transplantation, all bone marrow-derived cells will be of the donor genotype. These will include all lymphocytes, as well as the antigen-presenting cells they interact with. The rest of the animal's tissues, including the nonlymphoid stromal cells of the thymus will, however, be of the recipient MHC genotype.

In the experiments that demonstrated positive selection, the donor mice were F_1 hybrids derived from MHC^a and MHC^b parents, and thus were of the $MHC^{a\times b}$ genotype. The irradiated recipients were one of the parental strains, either MHC^a or MHC^b. Because of MHC restriction, individual T cells recognize either MHC^a or MHC^b, but not both. In normal circumstances, roughly equal numbers of the $MHC^{a\times b}$ T cells from $MHC^{a\times b}$ F_1 hybrid mice will recognize antigen presented by MHC^a or MHC^b. But in bone marrow chimeras in which T cells of $MHC^{a\times b}$ genotype develop in an MHC^a thymus, the T cells turn out to recognize antigen mainly, if not exclusively, when it is presented by MHC^a molecules, even though the antigen-presenting cells display antigen bound to both MHC^a and MHC^b (Fig. 7.27). These experiments showed clearly that the MHC molecules present in the environment in which T cells develop determine the MHC restriction of the mature T-cell receptor repertoire, and provided the first evidence for positive selection of developing T cells.

A further experiment involving grafts of thymus tissue demonstrated that the thymic component responsible for positive selection is the thymic stroma. For this experiment, the recipient animals were athymic *nude* or thymectomized mice of $MHC^{a\times b}$ genotype who were given thymic stromal grafts of MHC^a genotype. Thus, all of their cells carried both MHC^a and MHC^b, except those of the thymic stroma. The $MHC^{a\times b}$ bone marrow cells of these mice also mature into T cells that recognize antigens presented by MHC^a but not antigens presented by MHC^b. Thus, what mature T cells consider to be self MHC is determined by the MHC molecules expressed by the thymic stromal cells that they encounter during intrathymic development.

The chimeric mice used to demonstrate positive selection produce normal T-cell responses to foreign antigens. By contrast, chimeras made by injecting MHC^a bone marrow cells into MHC^b animals cannot make normal T-cell responses. This is because the T cells in these animals have been selected to recognize peptides when they are presented by MHC^b, whereas

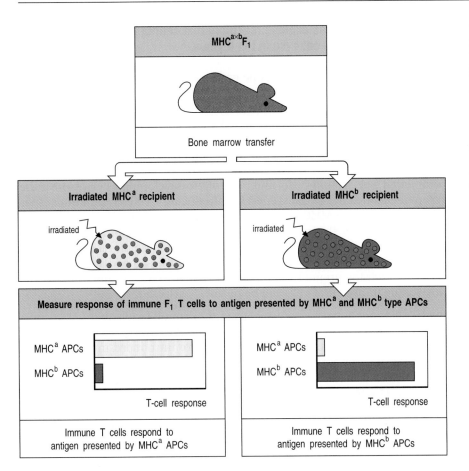

Fig. 7.27 Positive selection is revealed by bone marrow chimeric mice. As shown in the top two sets of panels, bone marrow from an MHC$^{a\times b}$ F$_1$ hybrid mouse is transferred to a lethally irradiated recipient mouse of either parental MHC type (MHC^a or MHC^b). When these chimeric mice are immunized with antigen, the antigen can be presented by the bone marrow-derived MHC$^{a\times b}$ APCs in association with both MHC^a and MHC^b molecules. The T cells from an MHC$^{a\times b}$ F$_1$ mouse include cells that respond to antigen presented by antigen-presenting cells (APCs) from MHC^a mice and cells that respond to APCs from MHC^b mice (not shown). But when T cells from the chimeric animals were tested *in vitro* with APCs bearing MHC^a or MHC^b only, they respond far better to antigen presented by the MHC molecules of the recipient MHC type, as shown in the bottom panels. This shows that the T cells have undergone positive selection for MHC restriction in the recipient thymus.

the antigen-presenting cells that they encounter as mature T cells in the periphery are bone marrow-derived MHC^a cells. The T cells will therefore fail to recognize antigen presented by antigen-presenting cells of their own MHC type, and T cells can be activated in these animals only if antigen-presenting cells of the MHC^b type are injected together with the antigen. Thus, for a bone marrow graft to reconstitute T-cell immunity, there must be at least one MHC molecule in common between donor and recipient (Fig. 7.28).

Bone marrow donor	Recipient	Mice contain APC of type:	Secondary T-cell responses to antigen presented *in vitro* by APC of type:	
			MHC^a APC	MHC^b APC
MHC$^{a\times b}$	MHC^a	MHC$^{a\times b}$	Yes	No
MHC$^{a\times b}$	MHC^b	MHC$^{a\times b}$	No	Yes
MHC^a	MHC^b	MHC^a	No	No
MHC^a	MHC^b + MHC^b APC	MHC^a + MHC^b	No	Yes

Fig. 7.28 Summary of T-cell responses to immunization in bone marrow chimeric mice. A set of bone marrow chimeric mice with different combinations of donor and recipient MHC types were made. These mice were then immunized and their T cells were isolated. These were then tested *in vitro* for a secondary immune reaction using MHC^a or MHC^b antigen-presenting cells (APCs). The results are indicated in the last two columns. T cells can make antigen-specific immune responses far better if the APCs present in the host at the time of priming share at least one MHC molecule with the thymus in which the T cells developed.

7-20   Most thymocytes express receptors that cannot interact with self MHC and these cells die in the thymus.

Bone marrow chimeras and thymic grafting provided the first evidence for the central importance of the thymus in positive selection, but more detailed investigation of the process has used mice transgenic for rearranged T-cell receptor genes. The rearranged α- and β-chain genes are molecularly cloned from a T-cell line or clone (see Appendix I, Section A-24) whose origin, antigen specificity, and MHC restriction are well characterized. When such genes are introduced into the mouse genome, rearrangement of the endogenous genes is inhibited (see Section 7-15), and most of the developing T cells express the receptor encoded by the transgenes. By introducing T-cell receptor transgenes into mice of known MHC genotype, the effect of known MHC molecules on the maturation of thymocytes with known recognition properties can be studied. It is routinely found that T cells bearing a transgenic receptor develop to maturity only within a thymus that expresses the same MHC molecule as that on which the original T-cell receptor was selected. In this situation the developing transgenic T cells can be positively selected.

Such experiments have also established the fate of T cells that fail positive selection. Rearranged receptor genes from a mature T cell specific for a peptide presented by a particular MHC molecule were introduced into a recipient mouse lacking that MHC molecule, and the fate of the thymocytes was investigated by staining with antibodies specific for the transgenic receptor. Antibodies to other molecules such as CD4 and CD8 were used at the same time to mark the stages of T-cell development. It was found that cells that fail to recognize the MHC molecules present on the thymic epithelium never progress further than the double-positive stage and die in the thymus within 3 or 4 days of their last division.

In a normal thymus, the fate of each thymocyte depends on the specificity of the receptor it expresses and, as we saw in Section 7-16, the specificity can undergo several changes as the α-chain genes continue to rearrange. The ability of a single developing thymocyte to express several different rearranged α-chain genes during the time that it is susceptible to positive selection must increase the yield of useful T cells significantly; without this mechanism many more thymocytes would fail positive selection and die. However, this continued rearrangement of α-chain genes also makes it likely that a significant percentage of T cells will express two receptors, sharing a β chain but differing in their α chains. Indeed, one can predict that if the frequency of positive selection is sufficiently low, roughly one in three mature T cells will have two α chains at the cell surface. This was confirmed recently for both human and mouse T cells.

T cells with dual specificity might be expected to give rise to inappropriate immune responses if the cell is activated through one receptor yet can act upon target cells recognized by the second receptor. However, only one of the two receptors is likely to be able to recognize peptide presented by a self MHC molecule as, once the cell has been positively selected, α-chain gene rearrangement ceases. Thus, the existence of cells with two α-chain genes productively rearranged and two α chains expressed at the cell surface does not seem to challenge the importance of clonal selection, which depends on a single functional specificity being expressed by each cell.

7-21   Positive selection acts on a repertoire of receptors with inherent specificity for MHC molecules.

Positive selection acts on a repertoire of receptors whose specificity is determined by a combination of germline gene segments and junctional regions whose diversity is randomly created as the genes rearrange (see Section 4-11). The unselected receptor repertoire must be capable of recognizing all of the hundreds of different allelic variants of MHC molecules present in the population, as the genes for the T-cell receptor α and β chains seregate in the population independently from those of the MHC. If the binding specificity of the unselected repertoire were completely random, only a very small proportion of thymocytes would be expected to recognize an MHC molecule. However, it seems that the variable CDR1 and CDR2 loops of both chains of the T-cell receptor, which are encoded within the germline V gene segments (see Section 4-12), give the T-cell receptor an intrinsic specificity for MHC molecules. This is evident from the way these two regions contact MHC molecules in crystal structures (see Section 3-18). An inherent specificity for MHC molecules has also been shown by examining mature T cells that represent an unselected repertoire of receptors. Such T cells can be generated in fetal thymic organ cultures, using thymuses that do not express either MHC class I or MHC class II molecules, by substituting binding of anti-β-chain antibodies and anti-CD4 antibodies for the receptor engagement responsible for normal positive selection. When the reactivity of these antibody-selected CD4 T cells is tested, roughly 5% are able to respond to any one MHC class II genotype and, because they developed without selection by MHC molecules, this must reflect a specificity inherent in the germline V gene segments. This germline-encoded specificity for MHC should significantly increase the proportion of receptors that can be positively selected in any one individual.

7-22 Positive selection coordinates the expression of CD4 or CD8 with the specificity of the T-cell receptor and the potential effector functions of the cell.

At the time of positive selection, the thymocyte expresses both CD4 and CD8 co-receptor molecules. At the end of the selection process, mature thymocytes ready for export to the periphery express only one of these co-receptors. Moreover, almost all mature T cells that express CD4 have receptors that recognize peptides bound to self MHC class II molecules and are programmed to become cytokine-secreting cells. In contrast, most of the cells that express CD8 have receptors that recognize peptides bound to self MHC class I molecules and are programmed to become cytotoxic effector cells. Thus, positive selection also determines the cell-surface phenotype and functional potential of the mature T cell, selecting the appropriate co-receptor for efficient antigen recognition and the appropriate program for the T cell's eventual functional differentiation in an immune response.

Experiments with mice made transgenic for rearranged T-cell receptor genes show clearly that it is the specificity of the T-cell receptor for self MHC molecules that determines which co-receptor a mature T cell will express. If the T-cell receptor transgenes encode a receptor specific for antigen presented by self MHC class I molecules, mature T cells that express the transgenic receptor are CD8 T cells. Similarly, in mice made transgenic for a receptor that recognizes antigen with self MHC class II molecules, mature T cells that express the transgenic receptor are CD4 T cells (Fig. 7.29).

The importance of MHC molecules in this selection is illustrated by the class of human immunodeficiency diseases known as **bare lymphocyte syndromes**, which are caused by mutations that lead to an absence of MHC molecules on lymphocytes and thymic epithelial cells. People who lack MHC class II molecules have CD8 T cells but only a few, highly abnormal, CD4 T cells; a similar result has been obtained in mice in which MHC class II expression has been eliminated by targeted gene disruption (see Appendix I, Section A-47).

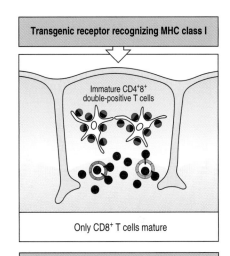

Transgenic receptor recognizing MHC class I

Immature CD4⁺8⁺ double-positive T cells

Only CD8⁺ T cells mature

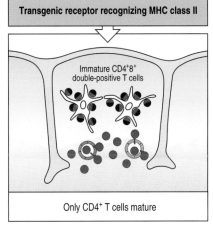

Transgenic receptor recognizing MHC class II

Immature CD4⁺8⁺ double-positive T cells

Only CD4⁺ T cells mature

Fig. 7.29 Positive selection determines co-receptor specificity. In mice transgenic for T-cell receptors restricted by an MHC class I molecule (top panel), the only mature T cells to develop have the CD8 (red) phenotype. In mice transgenic for receptors restricted by an MHC class II molecule (bottom panel), all the mature T cells have the CD4 (blue) phenotype. In both cases, normal numbers of immature, double-positive thymocytes are found. The specificity of the T-cell receptor determines the outcome of the developmental pathway, ensuring that the only T cells that mature are those equipped with a co-receptor that is able to bind the same self MHC molecule as the T-cell receptor.

MHC Class I
Deficiency

Likewise, mice and humans that lack MHC class I molecules lack CD8 T cells. Thus, MHC class II molecules are required for CD4 T-cell development, whereas MHC class I molecules are required for CD8 T-cell development.

In mature T cells, the co-receptor functions of CD8 and CD4 depend on their respective abilities to bind invariant sites on MHC class I and MHC class II molecules (see Section 3-12). Co-receptor binding to an MHC molecule is also required for normal positive selection. as shown for CD4 in the experiment discussed below (see Section 7-23). Thus positive selection depends on engagement of both the antigen receptor and co-receptor with an MHC molecule, and determines the survival of single-positive cells that express only the appropriate co-receptor. However, the exact mechanism whereby lineage commitment is coordinated with receptor specificity remains to be established. At present, it seems as though the developing thymocyte integrates the signals that it gets from both the antigen receptor and the co-receptor in order to determine its fate. Co-receptor-associated Lck signals are most effectively delivered when CD4 is engaged as a co-receptor rather than CD8, and these Lck signals play a large role in the decision to become a CD4 mature cell. Other signaling molecules are clearly important for thymocyte survival and development, but appear to be common to the commitment of both lineages. A few of these may, nonetheless, preferentially influence the commitment of thymocytes to one lineage over the other: for example, the MAPK signaling pathway initiated by the antigen receptor (see Section 6-11) biases cells toward the CD4 versus the CD8 lineage. It is a general principle of lineage commitment that different signals must be created in ordcr to activate lineage-specific factors and generate a divergence of developmental programming. While much remains to be discovered about this process in developing α:β thymocytes, it is clear that the different signals that are created result in a divergence of functional programming, so that the ability to express genes involved in the killing of target cells, for example, develops in CD8 T cells, whereas the potential to express various cytokine genes develops in CD4 T cells.

7-23 Thymic cortical epithelial cells mediate positive selection of developing thymocytes.

The thymic cortical epithelial cell is the stromal cell type critical for positive selection. These cells form a web of cell processes that make close contacts with the double-positive T cells undergoing positive selection (see Fig. 7.8) and T-cell receptors can be seen clustering with MHC molecules at the sites of contact.

Direct evidence that thymic cortical epithelial cells mediate positive selection comes from an ingenious manipulation of mice whose MHC class II genes have been eliminated by targeted gene disruption (Fig. 7.30). Mutant mice that lack MHC class II molecules do not normally produce CD4 T cells. To test the role of the thymic epithelium in positive selection, an MHC class II gene was placed under the control of a promoter that restricted its expression to thymic cortical epithelial cells, and was introduced as a transgene into these mutant mice. CD4 T cells then developed. A further variant of this experiment shows that, in order to promote the development of CD4 T cells, the MHC class II molecule on the thymic epithelium must be able to interact effectively with CD4. Thus, when the MHC class II transgene expressed in the thymus contains a mutation that prevents its binding to CD4, very few CD4 T cells develop. Equivalent studies of CD8 interaction with MHC class I molecules show that co-receptor binding is necessary for normal positive selection of CD8 cells as well.

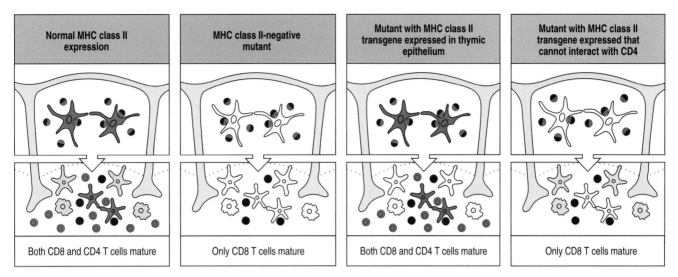

Normal MHC class II expression	MHC class II-negative mutant	Mutant with MHC class II transgene expressed in thymic epithelium	Mutant with MHC class II transgene expressed that cannot interact with CD4
Both CD8 and CD4 T cells mature	Only CD8 T cells mature	Both CD8 and CD4 T cells mature	Only CD8 T cells mature

Fig. 7.30 Thymic cortical epithelial cells mediate positive selection. The expression of MHC class II molecules in the thymus of normal and mutant strains of mice is shown by coloring the stromal cells only if they are expressing MHC class II molecules. In the thymus of normal mice (first panels), which express MHC class II molecules on epithelial cells in the thymic cortex (blue) as well as on medullary epithelial cells (orange) and bone marrow-derived cells (yellow), both CD4 (blue) and CD8 (red) T cells mature. Double-positive thymocytes are shown as half red/half blue. The second panels represent mutant mice in which MHC class II expression has been eliminated by targeted gene disruption; in these mice, few CD4 T cells develop, although CD8 T cells develop normally. In MHC class II-negative mice containing an MHC class II transgene engineered so that it is expressed only on the epithelial cells of the thymic cortex (third panels), normal numbers of CD4 T cells mature. The MHC class II molecule needs to be able to interact with the CD4 protein, as mutant MHC class II molecules with a defective CD4 binding site do not allow the positive selection of CD4 T cells (fourth panels). Thus, the cortical epithelial cells are the critical cell type mediating positive selection.

The critical role of the thymic epithelium in positive selection raises the question whether there is anything distinctive about the antigen-presenting properties of these cells. This is not clear at present; however, thymic epithelium may differ from other tissues in the proteases used to degrade the invariant chain (Ii) (see Section 5-6). The protease cathepsin L dominates in thymic cortical epithelium, whereas cathepsin S seems to be most important in other tissues. Consequently, CD4 T-cell development is severely impaired in cathepsin L knockout mice. Thymic epithelial cells do seem to bear on their cell surfaces a relatively high density of MHC class II molecules that retain the invariant chain-associated peptide (CLIP) (see Fig. 5.7). However, they also present a range of other peptides, and it remains to be seen whether the MHC:peptide complexes presented by these cells have any special characteristics that are important for positive selection.

Artificial manipulation of the peptides presented by thymic epithelium has a profound effect on positive selection. The effect of selecting thymocytes in the presence of a single peptide:MHC class II complex has been demonstrated in experiments in which the α chain of H-2M, the mouse homologue of human HLA-DM, is disrupted (Fig. 7.31). In mice lacking a functional H-2Mα chain, the CLIP fragment of the invariant chain is not released from the newly synthesized MHC class II molecules (see Section 5-7). These CLIP-associated MHC class II molecules are unable to bind the self peptides present in the endosomes, and the main self peptide presented by MHC class II molecules on the surface of the thymic epithelium is therefore the invariant CLIP peptide. In these mice, the total number of CD4 T cells is reduced twofold to threefold. A diversity of T-cell receptor β chains is expressed in these cells, but of seven T-cell receptor transgenes examined so far, none is positively selected in H-2Mα knockout mice despite the presence of typical levels of the MHC class II molecule that normally selects them.

Fig. 7.31 The peptides bound to MHC class II molecules can affect the T-cell receptor repertoire. The left panels show the normal situation, in which a range of peptides is presented by antigen-presenting cells (APCs) to immature T cells in the thymus, with the consequent deletion of self-reactive T cells. The right panels show a case in which a single peptide predominates. Mice with their H-2Mα gene disrupted by targeted mutagenesis express MHC class II molecules that predominantly carry the CLIP peptide of the invariant chain (top right panel). Their MHC class II molecules thus present only the CLIP peptide to T cells maturing in the thymus. CD4 T cells mature in the presence of this single dominant peptide: MHC complex but they are reduced in number by twofold to threefold, even though negative selection of MHC class II-restricted T cells will only remove T cells specific for the CLIP peptide (middle right panel). Mature T cells from such mice respond strongly to MHC-identical APCs, which express the normal array of self peptides (bottom right panel). This shows that a majority of the T cells positively selected by this single dominant MHC:peptide complex are reactive with other self peptides complexed with the same MHC molecule. In normal mice, the T cells bearing these receptors would be deleted by negative selection (middle and bottom left panels).

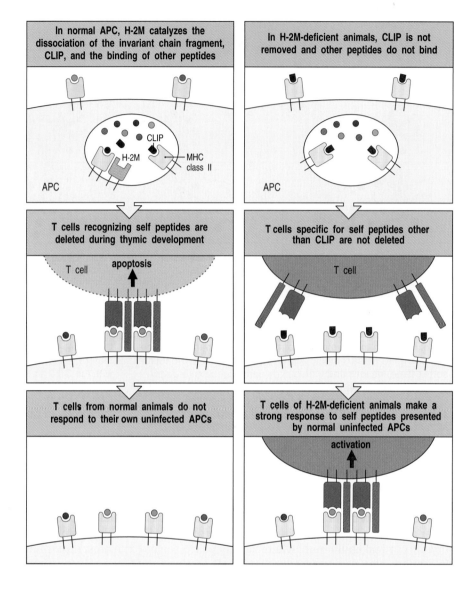

Thus it seems that a significantly reduced repertoire of T cells is selected in the presence of a single predominant peptide:MHC class II complex. In addition, a high proportion of the T cells that are positively selected in H2-M-deficient mice and reach maturity are self-reactive, as shown by their activation by antigen-presenting cells from wild-type mice of the same MHC genotype. These T cells would have been eliminated by negative selection in a nonmutant mouse (see Fig. 7.31). These self-reactive cells might have been positively selected through an interaction dominated by contacts between the T-cell receptor and MHC class II molecule, to which the bound peptide made a minimal contribution. Such T cells would thus be more likely to react with the same MHC molecule complexed with a different self peptide than would cells that were positively selected through an interaction dominated by contacts with the bound peptide.

The influence of peptide diversity on positive selection is seen even more clearly in experiments using H2-M-deficient mice that have been bred to carry a rearranged T-cell receptor β-chain transgene. Endogenous β-gene rearrangements are suppressed in such mice and this limits the range of antigen receptor specificities on which positive selection can act. In control mice that are MHC-identical and carry the same transgene but are not H2-M-deficient, positive selection is able to generate mature T cells that express a large repertoire of α-chain genes. By contrast, mature T cells in the H-2M-deficient animals have a highly restricted repertoire of endogenous T-cell receptor α chains, and most of the selected T cells react against self MHC molecules when these are combined with the normal range of self peptides. These experiments illustrate the results of positive selection for α-gene rearrangements when there is only a single β-chain rearrangement with which they can combine, and when selection occurs on only a small set of self peptide:self MHC complexes. They therefore suggest that a diversity of bound peptides is important for the positive selection of a diverse repertoire of self MHC-restricted T cells.

7-24   T cells that react strongly with ubiquitous self antigens are deleted in the thymus.

When the T-cell receptor of a mature naive T cell is ligated by peptide plus MHC antigen displayed by a professional antigen-presenting cell in a peripheral lymphoid organ, the T cell is activated to proliferate and produce effector T cells (see Section 1-12). In contrast, when the T-cell receptor of a developing thymocyte is similarly ligated by antigen on stromal or bone marrow-derived cells in the thymus, it dies by apoptosis. The response of immature T cells to stimulation by antigen is the basis of negative selection. Elimination of these T cells in the thymus prevents their potentially harmful activation later on, if they should encounter the same peptides as mature T cells. Negative selection has been demonstrated in mice that express a transgenic T-cell receptor specific for a peptide of ovalbumin bound to an MHC class II molecule. As explained in Section 7-15, all the T cells developing in such a mouse will express the transgenic receptor. When these mice are injected with the ovalbumin peptide, the mature CD4 T cells in the periphery become activated, but most of the intrathymic T cells die (Fig. 7.32). Similar results can be obtained in thymic organ culture with T cells from both normal

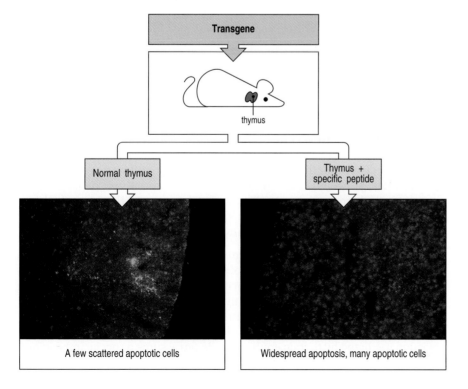

Fig. 7.32 T cells specific for self antigens are deleted in the thymus. In mice transgenic for a T-cell receptor that recognizes a known peptide antigen complexed with self MHC, all of the T cells have the same specificity; in the absence of the peptide, most thymocytes mature and migrate to the periphery. This can be seen in the bottom left panel, where a normal thymus is stained with antibody to identify the medulla (in green), and by the TUNEL technique (see Appendix I, Section A-32) to identify apoptotic cells (in red). If the mice are injected with the peptide that is recognized by the transgenic T-cell receptor, then massive cell death occurs in the thymus, as shown by the increased numbers of apoptotic cells in the right-hand bottom panel. Photographs courtesy of A. Wack and D. Kioussis.

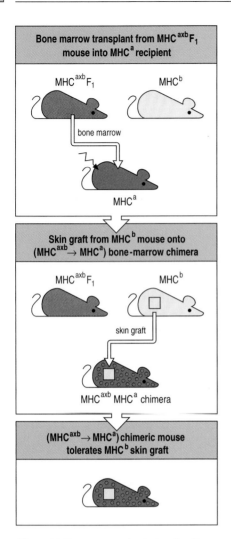

Fig. 7.33 Bone marrow-derived cells mediate negative selection in the thymus. When MHC^{axb} F$_1$ bone marrow is injected into an irradiated MHC^a mouse, the T cells mature on thymic epithelium expressing only MHC^a molecules. Nevertheless, the chimeric mice are tolerant to skin grafts expressing MHC^b molecules (provided that these grafts do not present skin-specific peptides that differ between strains a and b). This implies that the T cells whose receptors recognize self antigens presented by MHC^b have been eliminated in the thymus. As the transplanted MHC^{axb} F$_1$ bone marrow cells are the only source of MHC^b molecules in the thymus, bone marrow-derived cells must be able to induce negative selection.

and transgenic mice, showing that secondary effects caused by the induction of cytokines or corticosteroids due to simultaneous activation of peripheral T cells *in vivo* cannot account for this cell death of immature thymocytes.

The deletion of developing T cells that recognize self peptides synthesized naturally in the thymus has also been demonstrated experimentally. The negative selection of such thymocytes was observed in mice made transgenic for rearranged genes encoding T-cell receptors specific for self peptides expressed only in male mice. Thymocytes bearing these receptors disappear from the developing T-cell population in male mice at the CD4$^+$ CD8$^+$ double-positive stage of development, and no single-positive cells bearing the transgenic receptors mature. By contrast, in female mice, which lack the male-specific peptide, the transgenic T cells mature normally. This initial observation has been confirmed using T-cell receptor transgenes that recognize other antigens, with similar results.

These experiments illustrate the principle that self peptide:self MHC complexes encountered in the thymus purge the T-cell repertoire of immature T cells bearing self-reactive receptors. Not all self proteins are expressed in the thymus, however, and those that appear in other tissues, or are expressed at different stages in development, such as after puberty, will encounter mature T cells with the potential to respond to them. However, there are mechanisms that prevent mature T cells from responding to such antigens, and these will be discussed in Chapter 13, when we consider the problem of autoimmune responses and their avoidance.

7-25   Negative selection is driven most efficiently by bone marrow-derived antigen-presenting cells.

Negative selection in the thymus can be mediated by several different cell types. The most important are the bone marrow-derived dendritic cells and macrophages. These are professional antigen-presenting cell types that also activate mature T cells in peripheral lymphoid tissues. The self antigens presented by these cells are therefore the most important source of potential autoimmune responses, and T cells responding to such self peptides must be eliminated in the thymus.

Experiments using bone marrow chimeric mice have shown clearly the role of thymic macrophages and dendritic cells in negative selection. In these experiments, MHC^{axb} F$_1$ bone marrow is grafted into one of the parental strains (MHC^a in Fig. 7.33). The MHC^{axb} T cells developing in the grafted animals are thus exposed to the thymic epithelium of the MHC^a host strain. Bone marrow-derived dendritic cells and macrophages will, however, express both MHC^a and MHC^b. The bone marrow chimeras will tolerate skin grafts from either MHC^a or MHC^b animals (see Fig. 7.33), and from the acceptance of both types of skin grafts we can infer that the developing T cells are not self-reactive for either of the two MHC antigens. The only cells that could present self peptide:MHC^b complexes to thymocytes, and thus induce tolerance to MHC^b, are the bone marrow-derived cells. The dendritic cells and macrophages are therefore assumed to have a crucial role in negative selection.

In addition, both the thymocytes themselves and thymic epithelial cells also have the ability to cause the deletion of self-reactive cells. Such reactions may normally be of secondary significance compared to the dominant role of bone marrow-derived cells in negative selection. In patients undergoing bone marrow transplantation from an unrelated donor, however, where all the thymic macrophages and dendritic cells are of donor type, negative selection mediated by thymic epithelial cells can assume a special importance in maintaining tolerance to the recipient's own antigens.

7-26   Endogenous superantigens mediate negative selection of T-cell receptors derived from particular V_β gene segments.

It is virtually impossible to demonstrate directly the negative selection of T cells specific for any particular self antigen in the normal thymus because such T cells will be too rare to detect. There is, however, one case in which negative selection can be seen on a large scale in normal mice and the point at which it occurs in T-cell development can be identified. In the most striking examples, T cells expressing receptors encoded by particular V_β gene segments are virtually eliminated in the affected mouse strains. This occurs as the consequence of the interaction of immature thymocytes with endogenous superantigens present in those strains. We learned in Chapter 5 (see Section 5-15) that superantigens are viral or bacterial proteins that bind tightly to both MHC class II molecules and particular V_β domains, irrespective of the antigen specificity of the receptor and the peptide bound by the MHC molecule (see Fig. 5.18).

The endogenous superantigens of mice are encoded by mouse mammary tumor virus (MMTV) genomes that have become integrated at various sites into the mouse chromosomes. Different mouse strains have different complements of inherited MMTV genomes, and therefore express different viral antigens. Like the bacterial superantigens, these MMTV superantigens induce strong T-cell responses; indeed, they were originally designated **minor lymphocyte-stimulating (Mls)** antigens (see Section 5-15). Mice that carry these endogenous superantigens are said to be Mls$^+$, and a series of Mls antigens (Mls-1^a, Mls-1^b, ...) have been identified by their ability to stimulate primary T-cell responses when T cells from a strain lacking the superantigen are mixed with B cells from MHC-identical mice that express it.

In Mls$^+$ strains, T cells bearing V_β regions to which the Mls proteins bind, die by apoptosis during intrathymic maturation. For example, one variant of the Mls antigen (Mls-1^a or MTV7) deletes all thymocytes expressing the $V_\beta 6$ V gene segment (and also those expressing $V_\beta 8.1$ and $V_\beta 9$), whereas such cells are not deleted in mice that lack Mls-1^a. Thus, the expression of endogenous superantigens in mice has a profound impact on the T-cell receptor repertoire. This sort of deletion has not yet been seen in any other species, including humans, despite the presence of endogenous retroviral sequences in many mammals.

In mice that express the superantigen and thus are tolerant to it, cells expressing receptors responsive to superantigens are found among the double-positive thymocytes and are abundant in thymic cortex. They are, however, absent from the thymic medulla and tissues outside the thymus. This suggests that superantigens might delete relatively mature T cells as they migrate out of the cortex into the medulla, where a particularly dense network of dendritic cells marks the cortico-medullary junction (Fig. 7.34). Although clonal deletion by superantigens is a powerful tool for examining negative selection in normal mice, superantigen-driven clonal deletion might not be representative of clonal deletion by self peptide:self MHC complexes. What is clear is that clonal

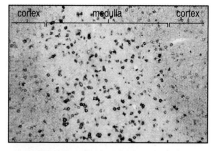

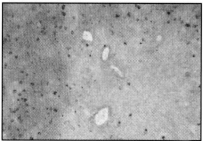

Fig. 7.34 Clonal deletion by Mls-1^a occurs late in the development of thymocytes. T cells with Mls-1^a-responsive receptors encoded by $V_\beta 6$ are seen in both the cortex and medulla of Mls-1^b mice (top panel, cells stained brown with anti-$V_\beta 6$ antibody). Note that the mature cells in the medulla express higher levels of the receptor and thus stain more darkly than the immature cells in the cortex. In Mls-1^a mice (lower panel) there is no obvious reduction in the number of immature cortical T cells expressing the $V_\beta 6$ receptor, but the mature cells are not found. Photographs courtesy of H. Hengartner.

deletion by either superantigens or self peptide:self MHC complexes generates a repertoire of T cells that does not respond to the self antigens expressed by its own professional antigen-presenting cells.

7-27   The specificity and strength of signals for negative and positive selection must differ.

We have described some of the experiments that contributed to the large body of evidence that T cells are selected for both self MHC restriction and self tolerance by MHC molecules expressed on stromal cells in the thymus. We now turn to the central question posed by positive and negative selection: how can engagement of the receptor by self MHC:self peptide complexes lead both to further maturation of thymocytes during positive selection and to cell death during negative selection? The answers to these questions are still not known for certain, but two possible mechanisms have been suggested. We will briefly describe each, before going on to discuss some recent experiments that bear on the mechanism of positive selection.

There are two issues to be resolved. First, the interactions that lead to positive selection must include more receptor specificities than those that lead to negative selection. Otherwise, all the cells that were positively selected in the thymic cortex would be eliminated by negative selection, and no T cells would ever leave the thymus (Fig. 7.35). Second, the consequences of the

Fig. 7.35 The specificity or affinity of positive selection must differ from that of negative selection. Immature T cells are positively selected in such a way that only those thymocytes whose receptors can engage the peptide:MHC complexes on thymic epithelium mature, giving rise to a population of thymocytes restricted for self MHC. Negative selection removes those thymocytes whose receptors engage with self peptides complexed with self MHC molecules, giving a self-tolerant population of thymocytes. If the specificity and/or avidity of positive and negative selection were the same (left panels), all the T cells that survive positive selection would be deleted during negative selection. Only if the specificity and/or avidity of negative selection is different from that of positive selection (right panels) can thymocytes mature into T cells.

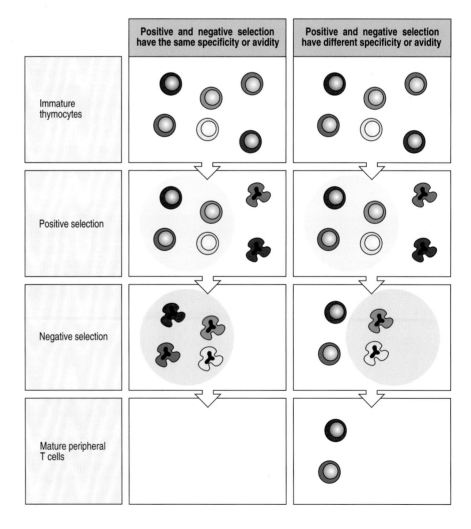

interactions that lead to positive and negative selection must differ; cells that recognize self peptide:self MHC complexes on cortical epithelial cells are induced to mature, whereas those whose receptors might confer strong and potentially damaging autoreactivity are induced to die.

Two main hypotheses have been proposed to account for these differences between positive and negative selection. The first is the **avidity hypothesis**. This states that the outcome of MHC:peptide binding by thymocyte T-cell receptors depends on the strength of the signal delivered by the receptor on binding, and that this will, in turn, depend upon both the affinity of the T-cell receptor for the MHC:peptide complex and the density of the complex on a thymic cortical epithelial cell. Thymocytes that are signaled weakly are rescued from apoptosis and are thus positively selected, whereas thymocytes that are signaled strongly are driven to apoptosis and are thus negatively selected. Because more complexes are likely to bind weakly rather than strongly, this will result in the positive selection of a larger repertoire of cells than are negatively selected.

Alternatively, the delivery of incomplete activating signals by self peptides could account for positive selection: we will call this the **differential signaling hypothesis**. Under this hypothesis, it is the nature of the signal delivered by the receptor, not just the number of receptors engaged, that distinguishes positive from negative selection. According to the avidity hypothesis, the same MHC:peptide complex could drive positive or negative selection of the same receptor, depending on its density on the cell surface. This could not occur under the differential signaling hypothesis, because it proposes that the signals leading to positive and negative selection are qualitatively different.

A new approach to testing these hypotheses has opened up with the recent description of **antagonist peptides**. These are known as antagonist peptides because they inhibit the response of mature T cells to their normal stimulatory, or **agonist**, peptide. On binding to the antigen receptor of a mature T cell, antagonist peptide:MHC complexes generate some, but not all, of the intracellular signaling events that are associated with full agonist-driven T-cell activation (see Section 6-12). Recognition of antagonist peptides by thymocytes has been shown to induce positive selection, whereas recognition of agonist peptides induces negative selection. Thus the differences between the ways agonist and antagonist peptides interact with the receptor and signal to the cell are likely to be relevant to the issue of how positive selection works.

The experiments shown in Fig. 7.36 used a T-cell receptor of known specificity whose *in vivo* behavior with different variants of its peptide antigen had been characterized. The affinities of this T-cell receptor for the agonist peptide and for its antagonist variants (all bound to the appropriate MHC class I molecule) were measured using an affinity sensor (see Appendix I, Section A-30). Initially, affinity measurements were carried out at room temperature and revealed very slight differences in affinity for the agonist versus antagonist peptides. At physiological temperatures, the differences were larger, with the antagonist peptide:MHC complexes having lower affinities for the T-cell receptor. However, the affinities were still broadly similar and this parameter failed to capture a more radical difference in the binding of the T-cell receptor that was observed under these conditions. The complexes of MHC class I molecules with agonist peptides induced dimerization of the T-cell receptors, whereas the complexes with antagonist peptides did not. This suggests that *in vivo* the agonist peptides induce the receptor clustering required for generating the signals that lead to activation (see Section 6-2) and, during thymocyte development, to deletion. The antagonist peptides, in contrast, bind but fail to induce receptor clustering, and therefore deliver a qualitatively

Fig. 7.36 The differences between negative and positive selection may be due to differences in the aggregation of T-cell receptors upon ligand binding. Agonist, antagonist, and null peptides can be identified for a particular T-cell receptor by assaying the responses of mature T-cell clones. Their effects on mature T cells *in vivo* can be tested in mice made transgenic for this T-cell receptor (left-hand panels). By using thymic lobe organ cultures from mice that have also been genetically engineered to express only the MHC class I molecule recognized by the receptor, it is possible to test for the effects of these peptides on thymocyte development (right-hand panels). Binding to an agonist peptide (top panels) induces organized aggregation of T-cell receptors on the cell surface and effective signaling through the T-cell receptor. Agonist peptides trigger the full activation of mature T cells, however if they are encountered as self peptides by developing thymocytes they induce apoptosis and hence negative selection. Antagonist peptides (center panels) bind the T-cell receptor but fail to induce organized receptor aggregation. These peptides inhibit mature T cells but can drive the positive selection of developing thymocytes. Irrelevant peptides fail to engage the T-cell receptor at all and are ignored by mature T cells and developing thymocytes. If only irrelevant peptides are presented, the thymocyte is not positively selected and it dies by neglect.

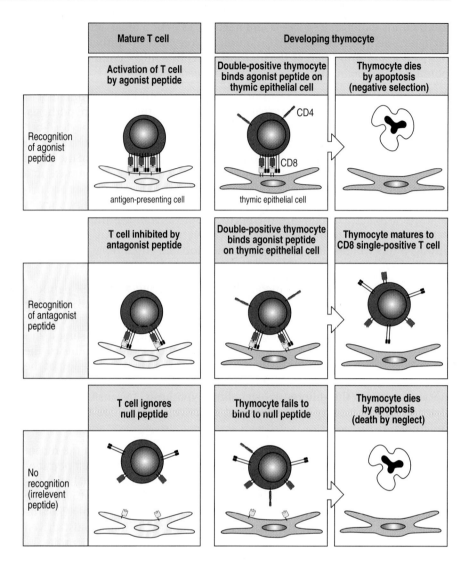

different signal; during thymocyte development, recognition of the antagonist peptide delivers a survival signal to the cell. As the agonist and antagonist peptides bind the T-cell receptor with similar affinity, but only the antagonist induces positive selection, the implication is that the signals delivered by the agonist peptides that cause deletion and by the antagonist peptides that mediate positive selection are fundamentally different. This suggests that the differential signaling hypothesis of T-cell selection is correct. These experiments have, however, only been carried out in a single laboratory, so one should keep an open mind until other laboratories produce similar results in other systems.

7-28   The B-1 subset of B cells has a distinct developmental history and expresses a distinctive repertoire of receptors.

A minority subset of B cells (comprising about 5%) in mice and humans, and the major population in rabbits, arises during fetal development and has a restricted receptor repertoire (Fig. 7.37). The B cells belonging to this subset were first identified by surface expression of the protein CD5 and are also characterized by high levels of sIgM with little sIgD, even when mature. They are termed **B-1 cells**, because their development precedes that of the conventional B cells whose development has been discussed up to now and which are sometimes termed **B-2 cells**. B-1 cells are also known as **CD5 B cells**, although CD5 itself cannot be essential for their function because cells that have similar traits develop normally in mice lacking the CD5 gene, and B-1 cells in rats do not display CD5. B-1 cells are of interest to clinicians, as they are the origin of the common B-cell tumor chronic lymphocytic leukemia (CLL). CLL cells often display CD5, which is a useful diagnostic clue.

There is great debate about the origin of B-1 cells, and it is not yet clear whether they arise as a distinct lineage from a unique precursor cell, or instead differentiate to the B-1 phenotype from a precursor cell that could also give rise to B-2 cells. In the mouse, fetal liver mainly produces B-1 cells, whereas adult bone marrow generates predominantly B-2 cells, and this has been interpreted as support for the unique precursor hypothesis. However, the weight of evidence favors the idea that the antigen specificity of the B-cell receptor determines whether the precursor becomes a B-1 or a B-2 cell. Under this hypothesis, commitment to the B-1 or B-2 subset would be due to a selection step, rather than being a distinct lineage difference like that between $\gamma{:}\delta$ and $\alpha{:}\beta$ T cells. It is, however, difficult to rule out the idea that cells are committed before this but only survive if their receptor specificity matches the predetermined fate.

The two hypotheses of B-cell commitment might be reconciled by proposing that fetal cells tend to generate B-1 cells because of the particular antigen specificities their immunoglobulin gene rearrangements tend to generate. Indeed, fetal B-cell precursors preferentially rearrange certain V_H gene segments, and also do not incorporate N-nucleotides; thus their receptors have a restricted and particular range of specificities, analogous to the T-cell receptors of the early waves of $\gamma{:}\delta$ T cells (see Section 7-14). The V gene

Fig. 7.37 A comparison of the properties of B-1 cells and conventional B cells (B-2 cells). B-1 cells can develop in unusual sites in the fetus, such as the omentum, in addition to the liver. B-1 cells predominate in the young animal although they probably can be produced throughout life. Being mainly produced during fetal and neonatal life, their rearranged variable-region sequences contain few N-nucleotides. B-1 cells are best thought of as a partially activated self-renewing pool of lymphocytes that are selected by ubiquitous self and foreign antigens. Because of this selection, and possibly because the cells are produced early in life, the B-1 cells have a restricted repertoire of variable regions and antigen-binding specificities. B-1 cells seem to be the major population of B cells in certain body cavities, most probably because of exposure at these sites to antigens that drive B-1 cell proliferation. Partial activation also leads to secretion of mainly IgM antibody; B-1 cells contribute much of the IgM that circulates in the blood. The limited diversity of the B-1 cell repertoire and the propensity of B-1 cells to react with common bacterial carbohydrate antigens suggest that they carry out a more primitive, less adaptive, immune response than conventional B cells (B-2 cells). In this regard, they are comparable to $\gamma{:}\delta$ T cells.

Property	B-1 cells	Conventional B-2 cells
When first produced	Fetus	After birth
N-regions in VDJ junctions	Few	Extensive
V-region repertoire	Restricted	Diverse
Primary location	Body cavities (peritoneal, pleural)	Secondary lymphoid organs
Mode of renewal	Self-renewing	Replaced from bone marrow
Spontaneous production of immunoglobulin	High	Low
Isotypes secreted	IgM >> IgG	IgG > IgM
Response to carbohydrate antigen	Yes	Maybe
Response to protein antigen	Maybe	Yes
Requirement for T-cell help	No	Yes
Somatic hypermutation	Low–none	High
Memory development	Little or none	Yes

segments that are commonly used to encode the receptors of B-1 cells seem to have evolved to recognize common bacterial and self antigens. These specificities may then be positively selected so that the cell matures with the phenotype of a B-1 cell.

Regardless of how B-1 cells originate, they are certainly expanded and maintained by interaction with self antigens or nonself antigens normally present in the body, such as those of the bacterial gut flora. Expansion by a relatively small number of ubiquitous antigens is a form of positive selection that would also tend to restrict the receptor repertoire of the B-1 cell population. In adult animals, the population of B-1 cells is maintained by continued division in peripheral sites such as the peritoneal and pleural cavities, a process that requires the cytokine IL-10 in addition to stimulation through the B-cell receptor. Some of the antigens that drive B-1 cell expansion, such as the phospholipid phosphatidylcholine, are encountered on the surface of bacteria that colonize the gut. Interestingly, a self antigen that can drive the expansion of B-1 cells has recently been identified on the surface of thymocytes. Why a ligand for B-1 cells should be expressed on T cells is unknown at present.

Little is yet known about the functions of B-1 cells. Their location suggests a role in defending the body cavities, while their restricted repertoire of receptors appears to equip them for a function in the early, nonadaptive phase of an immune response (see Section 2-28). Indeed, the V gene segments that are used to encode the receptors of B-1 cells might have evolved by natural selection to recognize common bacterial antigens, thus allowing them to contribute to the very early phases of the adaptive immune response. In practice, it is found that B-1 cells make little contribution to adaptive immune responses to most protein antigens but contribute strongly to some antibody responses against carbohydrate antigens. Moreover, a large proportion of the IgM that normally circulates in the blood of nonimmunized mice derives from B-1 cells. The existence of these so-called 'natural antibodies,' which are highly cross-reactive, and bind with low affinity to both microbial and self antigens, supports the view that B-1 cells are partially activated as they are selected for self-renewal by ubiquitous self and foreign antigens.

Summary.

Initially, random receptor rearrangements and junctional diversity create a broad repertoire of antigen receptors. Only some of these will not be dangerously self-reactive and yet be useful to the immune system. Cells meeting these criteria are selected, a process that begins as soon as immature lymphocytes express antigen receptors. Lymphocyte development involves both negative and positive selection. Negative selection includes deletion from the repertoire, receptor editing, and anergy, which most often are imposed on immature self-reactive lymphocytes in the thymus or bone marrow. However, even mature lymphocytes can be subject to negative selection when presented with a strong antigen signal without the usual co-stimulation needed for activation. Positive selection is best defined for T cells and B-1 cells; as we will discuss in the next part of this chapter, B-2 cells may also be positively selected as they mature to populate the periphery, although this is less well-established. For developing T cells, recognition of self MHC:self peptide complexes on thymic epithelial cells provides an as yet poorly defined positive survival signal. T-cell receptors capable of responding to foreign peptides presented by self MHC molecules are thus selected from a primary repertoire that has an inherent specificity for all the different MHC molecules in the population. Positive selection also ensures the functional matching of receptor, co-receptor, and the class of MHC molecule recognized. The paradox that recognition of self MHC:self peptide ligands by the

T-cell receptor can lead to two opposing effects, namely positive and negative selection, is one of the central mysteries of immunology. Its solution will come from a full understanding of the ligand–receptor interactions, the signal transduction mechanisms, and the physiology of each step of the process. An antigen-driven selection process during development may also lead to the generation of B-1 cells, a minor subset of B cells. These cells originate early in fetal life and are self-renewing in the periphery. They reside mainly in the pleural and peritoneal cavities and have B-cell receptors with a limited range of antigen specificities compared with conventional B-2 cells.

Survival and maturation of lymphocytes in peripheral lymphoid tissues.

Once lymphocytes have left the central lymphoid tissues, they are carried in the blood to the peripheral lymphoid tissues. These have a highly organized architecture, with distinct areas of B cells and T cells. Their organization, and the survival of newly formed lymphocytes, is determined by interactions between the lymphocytes and the other cell types that make up the lymphoid tissues. Before considering the factors governing the survival of newly formed lymphocytes in the periphery, we therefore look briefly at the organization and development of these tissues and the signals that guide lymphocytes to their correct locations within them. If a lymphocyte does not encounter its specific antigen and become activated within a peripheral lymphoid tissue, it leaves the tissue and recirculates via lymph and blood (see Section 1-4), continually reentering lymphoid tissues until either antigen is encountered or the lymphocyte dies. Its place is then taken by a newly formed lymphocyte; this enables a turnover of the receptor repertoire, and ensures that lymphocyte numbers remain constant.

7-29 Newly formed lymphocytes home to particular locations in peripheral lymphoid tissues.

As we saw in Chapter 1, the various peripheral lymphoid organs are organized roughly along the same lines, with distinct areas of B cells and T cells. As well as lymphocytes, peripheral lymphoid tissues are composed of several other types of leukocyte, principally macrophages and dendritic cells, and nonleukocyte stromal cells. The lymphoid tissue of the spleen is the white pulp, whose overall design is illustrated in Fig. 1.9. Each area of white pulp is demarcated by a marginal sinus, a vascular network that branches from the central arteriole.

Newly formed lymphocytes enter the spleen via the blood, from which they migrate to the appropriate areas of the white pulp. Lymphocytes that survive their passage through the spleen leave via the marginal sinus. Within the white pulp are clearly separated areas of T cells and B cells. T cells are clustered around the central arteriole, with the B-cell areas or follicles located farther out. Some follicles contain germinal centers, areas in which B cells involved in an immune response are proliferating and undergoing somatic hypermutation (see Section 4-9). In follicles with germinal centers, the resting B cells that are not part of the immune response are pushed outward to make up the **mantle zone** or **corona** around the proliferating lymphocytes. The antigen-driven production of germinal centers will be described in detail when we consider B-cell responses in Chapter 9.

The **marginal zone** of the white pulp (see Fig. 1.9), which is at the border of the white pulp, is a highly organized region whose function is poorly understood. It contains a unique population of B cells, the **marginal zone B cells**, which do not recirculate. These appear to be resting mature B cells, yet they have a different set of surface proteins from the major follicular population of B cells. For example, they express lower levels of CD23, a C-type lectin, and high levels of both the MHC class I-like molecule CD1 (see Section 5-18) and the complement C3-fragment receptor CD21/35.

Marginal zone B cells may have restricted antigen specificities, biased toward common environmental and even self antigens, and may be adapted to provide a quick response if such antigens enter the bloodstream; they may not, for example, require T-cell help to become activated. Both functionally and phenotypically, marginal zone B cells resemble B-1 cells; recent data suggest they may even be positively selected much as B-1 cells are. However, they are distinct both in location and in surface protein expression; for example, marginal zone B cells do not express high levels of CD5. The marginal zone also contains dendritic cells as well as two different types of macrophage, one of which lines the marginal sinus (the so-called 'metalophilic macrophages') whereas the other is scattered throughout the marginal zone.

B cells and T cells are not the only cells in their respective areas. The B-cell zone contains a network of **follicular dendritic cells** (**FDCs**), which are mainly concentrated in the area of the follicle most distant from the central arteriole. Follicular dendritic cells have long processes, from which they get their name, and are in contact with B cells. They differ, however, from the dendritic cells we have encountered previously (see Section 1-6) in that they are not leukocytes and are not derived from bone marrow precursors; in addition, they are not phagocytic and do not express MHC class II proteins. Follicular dendritic cells seem to be specialized to capture antigen in the form of immune complexes—complexes of antigen, antibody, and complement. The immune complexes are not internalized but remain intact on the follicular dendritic cell surface, where the antigen can be recognized by B cells. Follicular dendritic cells are also important in the development of B-cell follicles. Beyond this, their functions are not entirely clear.

T-cell zones contain a network of bone marrow-derived dendritic cells, sometimes known as **interdigitating dendritic cells** from the way their processes interweave among the T cells. There are two subtypes of these dendritic cells, distinguished by characteristic cell-surface proteins. The so-called **lymphoid dendritic cells** express the α chain of CD8 whereas the **myeloid dendritic cells** do not express CD8, but express CD11b, an integrin that is also expressed by macrophages. Both types of dendritic cells appear to derive from a common myeloid precursor. Whether they subsequently represent two separate lineages of dendritic cells with discrete functions, two different developmental stages within the same lineage, or two alternative fates for a dendritic cell depending on environmental stimuli is not known, and is an area of active research.

Lymph nodes have a very similar architecture to the spleen (see Fig. 1.8). B-cell follicles with similar structure and composition to those in the spleen are located just under the outer capsule. T-cell zones surround the follicles in the paracortical areas. Unlike the spleen, lymph nodes have no marginal sinus surrounding the lymphocyte areas and thus there is no marginal zone. Also unlike the spleen, lymph nodes collect lymph, which enters in the subcapsular space. Newly formed lymphocytes enter the lymph node from the blood through the walls of specialized blood vessels, the high endothelial venules (HEVs), which are located within the T-cell zones. Naive B cells migrate through the T-cell area to the follicle where, unless they encounter their specific antigen and become activated, they remain for about a day. B cells and T cells leave in the lymph via the efferent lymphatic, which returns them eventually to the blood.

Lymphoid tissues are also associated with epithelial surfaces that provide physical barriers against infection. Peyer's patches are lymph node-like structures that are interspersed at intervals just beneath the gut epithelium (see Fig. 1.10). They comprise B-cell follicles located immediately under specialized areas, or 'domes,' of epithelium known as follicular-associated epithelium, with T-cell zones that lie more deeply. Follicular-associated epithelium is composed of cells that lack the typical brush border. Instead, these so-called M cells are adapted to channel antigens and pathogens from the gut lumen to the Peyer's patch. Peyer's patches and similar tissue present in the tonsils provide specialized sites where B cells can become committed to synthesizing IgA. The mucosal epithelial surfaces lining the mouth, respiratory tract, and reproductive tract have their own specialized immune systems called collectively the mucosal-associated lymphoid tissues (MALT) (see Section 1-3). The epithelial cells of these surfaces as well as the skin are interspersed with a dense network of dendritic cells, which in the skin are called Langerhans' cells. The stromal cells of the MALT, and in particular the gut-associated lymphoid tissues (GALT), a part of the overall MALT, secrete the cytokine TGF-β, which induces IgA secretion in the culture system described in Section 9-4. In addition, as discussed in Section 7-14, during fetal development, waves of γ:δ T cells with specific γ- and δ-gene rearrangements leave the thymus and migrate to these epithelial barriers.

7-30   The development and organization of peripheral lymphoid tissues is controlled by cytokines and chemokines.

Surprisingly, members of the TNF/TNFR family—which had been thought to be involved in inflammation and cell death—are also critical for normal lymphoid development. This has been best demonstrated by a series of knockout mice in which either the ligand or receptor has been inactivated (Fig. 7.38). As shown, these knockouts have complicated phenotypes. This is partly due to

		Effects seen in knockout (KO) mice				
Receptor	Ligands	Spleen	Peripheral lymph node	Mesenteric lymph node	Peyer's patch	Follicular dendritic cells
TNFR-I	TNF-α LT-α_3	Distorted architecture	Present in TNF-α KO Absent in LT-α KO due to lack of LT-β signals	Present	Reduced	Absent
LT-β receptor	TNF-α LT-α_2/β_1 LIGHT	Distorted No marginal zones	Absent	Present in LT-β KO Absent in LT-β receptor KO	Absent	Absent
HVEM	LT-α_3 LIGHT	Although both LT-α and LIGHT can bind HVEM, there is no known role of HVEM signaling in organogenesis				

Fig. 7.38 TNF family members and their receptors play important roles in lymphoid development and the architecture of the secondary lymphoid organs. These roles have been deduced mainly from the study of knockout mice deficient in one or more TNF family member (ligand) and/or receptor. Since some receptors bind more than one ligand, and some ligands bind more than one receptor, the roles are complex, and have been determined by analyzing a variety of mutant mice. Here we organize the defects according to the two major receptors, TNFR-I and the LT-β receptor, along with a relatively new receptor, the herpes virus entry mediator (HVEM) which may also play a role. Note that the receptors were named for the first known ligand that binds them, even though we now know that additional ligands can also bind. In some cases, the loss of one ligand leads to a different phenotype than the loss of another; this is due to the ability of the ligand to bind another receptor, and is indicated in the figure. In addition the LT-α protein chain contributes to two distinct ligands, LT-α_3 and LT$\alpha_2\beta_1$, each of which has a distinct receptor. In general, signaling through the LT-β receptor is required for lymph node and follicular dendritic cell development and normal splenic architecture, while signaling through the TNFR-I receptor is also required for follicular dendritic cells and normal splenic architecture but not lymph node development.

the fact that individual TNF family ligands can bind to multiple receptors and, conversely, many receptors can bind more than one ligand. In addition, it seems clear that there is some overlapping function or cooperation between ligands. Nonetheless, some general conclusions can be drawn.

Lymph node development is dependent on the expression of lymphotoxin (LT) family members. Interestingly, not all lymph nodes are dependent on the same signals from these family members. LT-α_3, a soluble homotrimer of the LT-α protein chain, supports the development of cervical and mesenteric lymph nodes, and possibly lumbar and sacral lymph nodes. All of these lymph nodes drain mucosal sites. LT-α_3 probably exerts these effects by binding to the TNFR-I and possibly also another TNFR family member called HVEM. The membrane-bound heterotrimer comprised of LT-α and the distinct protein chain LT-β (LT-$\alpha_2\beta_1$) binds only to the LT-β receptor and supports the development of all the other lymph nodes. In addition, Peyer's patches do not form in the absence of the membrane-bound LT heterotrimer. These effects are not reversible in adult animals and, in fact, it can be shown that there are certain critical developmental periods during which the absence or inhibition of these LT family members will irrevocably prevent the development of lymph nodes and Peyer's patches.

The spleen develops in all of the mice deficient in various TNF or TNFR family members that have been studied. However, its architecture is very abnormal in many of these knockouts (see Fig. 7.38). LT (most likely the membrane-bound heterotrimer) is required for the normal segregation of T cell and B cell zones. TNF-α, binding to the TNFR-I, also contributes to the organization of the white pulp, in that when TNF-α signals are disrupted, B cells surround T-cell zones in a ring rather than in discrete follicles. In addition, the marginal zones are not well defined when TNF-α or its receptor is absent. Perhaps most importantly, follicular dendritic cells are not found in mice that lack TNF-α or TNFR-I. These mice, which do have lymph nodes since they do express LT family members, nonetheless lack follicular dendritic cells in those lymph nodes and in their Peyer's patches. Similarly, mice that cannot form or signal via the membrane-bound LT-$\alpha_2\beta_1$ heterotrimer also lack normal follicular dendritic cells in spleen and in any residual lymph nodes. Unlike the situation in lymph node development, which is irreversible, disorganized lymphoid structure is reversible when the missing TNF family member is restored. B cells are the likely source for the membrane-bound LT since normal B cells can restore follicular dendritic cells and follicles when transferred to RAG-deficient recipients (which lack lymphocytes). Very recently, a similar role for B cells in the development of M cells in Peyer's patches was discovered. In this case it appears that signals independent of LT-α are required since LT-α-deficient B cells will still restore the development of M cells in Peyer's patches.

The precise location of B cells, T cells, macrophages, and dendritic cells in peripheral lymphoid tissue is controlled by chemoattractant cytokines, or **chemokines**, which are produced by both stromal cells and bone marrow-derived cells (Fig. 7.39). B cells are attracted to the follicles by **B-lymphocyte chemokine (BLC)**, for which B cells constitutively express the receptor, CXCR5. The most likely source of BLC is the follicular dendritic cell (FDC). As mentioned, B cells in turn are the source of LT that is required for the development of FDCs. This reciprocal dependence of B cells and FDCs illustrates the complex web of interactions that organizes secondary lymphoid tissues. T cells can also express CXCR5, albeit at a lower level, and this may explain how T cells are able to enter B-cell follicles, which they do on activation, to participate in the formation of the germinal center.

Two chemokines, MIP-3β and secondary lymphoid chemokine (SLC) account for T-cell localization to the T zones. They both bind to the receptor

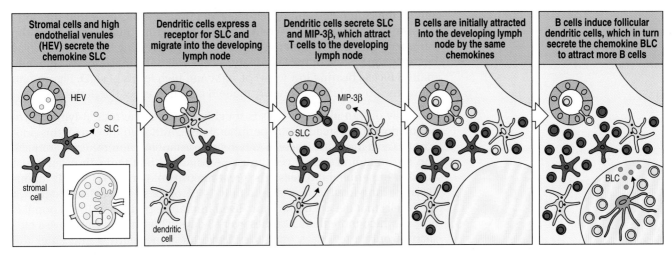

| Stromal cells and high endothelial venules (HEV) secrete the chemokine SLC | Dendritic cells express a receptor for SLC and migrate into the developing lymph node | Dendritic cells secrete SLC and MIP-3β, which attract T cells to the developing lymph node | B cells are initially attracted into the developing lymph node by the same chemokines | B cells induce follicular dendritic cells, which in turn secrete the chemokine BLC to attract more B cells |

Fig. 7.39 The organization of a lymphoid organ is orchestrated by chemokines. The cellular organization of a lymphoid organ is initiated by stromal cells and by vascular endothelial cells, which express the chemokine SLC (first panel). Dendritic cells express a receptor for SLC and are attracted by it to the site of the developing lymph node (second panel). There the dendritic cells also secrete SLC, augmenting the expression by stromal cells and vascular endothelial cells. The dendritic cells also express a second chemokine, MIP-3β, and the combination of SLC and MIP-3β attracts T cells to the developing lymph node (third panel). The same combination of chemokines also attracts B cells into the developing lymph node (fourth panel). The B cells are able to either induce the differentiation of follicular dendritic cells or direct their recruitment into the lymph node. Once present, the follicular dendritic cells secrete a chemokine, BLC, which is a chemoattractant for B cells. The production of BLC drives the organization of B cells into discrete B-cell areas (follicles) around the follicular dendritic cells and contributes to the further recruitment of B cells from the circulation into the lymph node (fifth panel).

CCR7, which is present on T cells; mice that lack CCR7 do not form normal T zones and have markedly impaired primary immune responses. SLC is produced by stromal cells of the T zone in spleen, and by the endothelial cells of HEVs in lymph nodes and Peyer's patches. Another source of MIP-3β and SLC is the interdigitating dendritic cells, which are also prominent in the T zones. Indeed, dendritic cells themselves express CCR7 and will localize to T zones even in RAG-deficient mice, which lack lymphocytes. Thus, the T zone might be organized first through the attraction of dendritic cells and T cells by SLC produced by stromal cells; this organization would then be reinforced by SLC and MIP-3β secreted by resident mature dendritic cells, which in turn attract more T cells and dendritic cells.

B cells—particularly activated ones—also express CCR7 but at lower levels than do T cells or dendritic cells. This may account for their characteristic migration pattern, which is first through the T zone (where they may linger if activated) and then to the follicle. Although the cellular organization of T-cell and B-cell areas in lymph nodes and Peyer's patches has been less well studied, it seems likely that it is controlled by similar, if not identical, chemokines and receptors.

7-31   Only a small fraction of immature B cells mature and survive in peripheral lymphoid tissues.

Our knowledge of the dynamics of B-cell populations comes from labeling immature B cells in the bone marrow of normal young adult mice with the thymidine analogue bromodeoxyuridine. These experiments showed that about 35×10^6 large pre-B cells enter mitosis each day. However, only $10–20 \times 10^6$ new B lymphocytes emerge as the end products of primary development in the bone marrow. The loss of more than half the initial pre-B cells may be due to failure to make a productive light-chain gene rearrangement and by the clonal deletion of self-reactive immature B cells within the bone marrow (see Section 7-17). When B cells emerge from bone marrow into the periphery, they are still functionally immature, expressing high levels of sIgM but little sIgD. Most of these immature cells will not survive to become fully mature B cells bearing low levels of sIgM and high levels of sIgD. Figure 7.40 shows the possible fates of newly made B cells that enter the periphery.

The daily output of new B cells is roughly 5–10% of the total B-lymphocyte population in the steady-state peripheral pool. Although the size of this pool is not easy to measure, it seems to remain constant in unimmunized animals, and the stream of new B cells must be balanced by the death of an equal number of peripheral B cells. The great majority of peripheral B cells (about 90%) are long-lived, however, and only 1–2% of these die each day. Most of the B cells that die are in the short-lived immature peripheral B-cell population, of which more than 50% dies every 3 days. Thus most B cells have a lifespan of only a few days once they leave the bone marrow and enter the periphery; only a small number of newly made B cells survive to become part of the pool of relatively long-lived peripheral B cells.

The failure of most B cells to survive for more than a few days in the periphery appears to be mainly due to competition between peripheral B cells for access to the follicles in the peripheral lymphoid tissues. The follicles appear to provide essential survival and possibly maturation signals for naive B cells. If newly made immature B cells do not enter a follicle, therefore, their passage through the periphery is halted and they die. The limited number of lymphoid follicles cannot accommodate the very large numbers of naive B cells poured out into the periphery each day and so there is continual competition for entry.

The population dynamics indicate that competition for follicle entry favors mature B cells that are already established in the relatively long-lived and stable peripheral B-cell pool. Why this should be is not entirely clear. Mature B cells have undergone phenotypic changes that might make their access to the follicles easier; for example, they express the receptor, CXCR5, for the chemoattractant BLC, expressed by follicular dendritic cells. They also have increased expression of the B-cell co-receptor component CR2 (CD21), which affects the signaling capacity of the B cell.

Fig. 7.40 Proposed population dynamics of conventional B cells. B cells are produced as receptor-positive immature B cells in the bone marrow. The most avidly self-reactive B cells are removed at this stage. B cells then migrate to the periphery where they enter the secondary lymphoid tissues. It is estimated that $10–20 \times 10^6$ B cells are produced by the bone marrow and exported each day in a mouse, and an equal number is lost from the periphery. There seem to be two classes of peripheral B cell: long-lived B cells and short-lived B cells. The short-lived B cells are, by definition, recently formed B cells. Most of the turnover of short-lived B cells might result from B cells that fail to enter lymphoid follicles. In some cases this is a consequence of being rendered anergic by binding to soluble self antigen; for the remaining immature B cells, entry into lymphoid follicles is thought to entail some form of positive selection. Thus the remainder of the short-lived B cells fail to join the long-lived pool because they are not positively selected. About 90% of all peripheral B cells are relatively long-lived mature B cells that appear to have undergone positive selection in the periphery. These mature naive B cells recirculate through peripheral lymphoid tissues and have a half-life of 6–8 weeks in mice. Memory B cells, which have been activated previously by antigen and T cells, are thought to have a longer life.

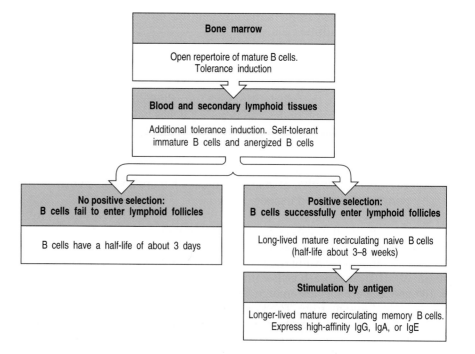

What determines the fate of an immature B cell? Is its survival simply a matter of chance, or are survival and further development determined by positive selection involving the specificity of its antigen receptor, as in the positive selection of T cells in the thymus? Several experiments indicate that signaling through the B-cell receptor has a positive role in the maturation and continued recirculation of peripheral B cells. A clever method of inactivating the B-cell receptor in mature B cells by conditional gene deletion has demonstrated that continuous expression of the B-cell receptor is required for B-cell survival. Mice that lack the tyrosine kinase Syk, which is involved in signaling from the B-cell receptor (see Section 6-10), fail to develop mature B cells although they do have immature B cells. Thus, a Syk-transduced signal may be required for final B-cell maturation or for the survival of mature B cells. Although each B-cell receptor has a unique specificity, such signaling need not depend on antigen-specific interactions; the receptor could, for example, be responsible for 'tonic' signaling, in which a weak but significant signal is generated by the assembly of the receptor complex and infrequently triggers some or all of the downstream signaling events.

Nonetheless, receptor specificity might also have a part in selecting a peripheral B-cell pool that continues to do well in the ongoing competition to survive. We know that some of the short-lived peripheral B cells fail to survive because they have bound soluble self antigen and become anergic (see Section 7-17), and are excluded from the lymphoid follicles for this reason. In addition, recent evidence shows that the B-cell receptor repertoire of surviving mature B cells is enriched for certain antigen specificities compared with the immature population, strongly favoring the idea that there is positive selection of B cells for maturation. However, the extent to which recruitment of immature B cells into the long-lived pool of recirculating mature B cells is ligand-mediated and governed by receptor specificity is not known.

Even among long-lived peripheral B cells the labeling studies in mice reveal a broad distribution of life-spans. One part of this population is composed of relatively long-lived mature naive B cells with a half-life of 1–2 months. As discussed above, these cells require a B-cell receptor for survival; the nature of other signals required for their maintenance is not clear. They might obtain a variety of survival signals from their normal environment, the B-cell follicle, through which they recirculate every few days. Peripheral B cells also include nondividing B cells that are very long-lived. These include the memory B cells that differentiate from mature B cells after their first encounter with antigen (see Section 1-12). Memory B cells persist for extended periods after antigenic stimulation and may require intermittent follicular signals for their survival; we will return to B-cell memory in Chapter 10.

The continual production and loss of new B cells ensures that the receptor repertoire is continually changing in order to meet new antigenic challenges. On the other hand, the persistence of memory cells—the progeny of cells that have been activated—ensures that those cells proven to recognize pathogens are retained to combat reinfection.

7-32 The life-span of naive T cells in the periphery is determined by ongoing contact with self peptide:self MHC complexes similar to those that initially selected them.

When T cells have expressed their receptors and co-receptors, and matured within the thymus for a further week or so, they emigrate to the periphery. Unlike B cells emigrating from bone marrow, only small numbers of T cells are exported from the thymus, roughly $1–2 \times 10^6$ per day in the mouse. Mature naive CD8 and CD4 T cells outside the thymus seem to be sustained by repeated contact with MHC:self peptide complexes similar or identical to those that originally positively selected them. This was shown by transfer of

T cells from their normal environment to recipients that lacked MHC molecules, or lacked the 'correct' MHC molecules that originally selected the T cells. The T cells did not survive as long in such environments as did T cells transferred into recipients that had the correct MHC molecules. Contact with the appropriate self peptide:self MHC complex leads to infrequent division of mature naive T cells being maintained in the peripheral lymphoid organs; this slow increase in T-cell numbers must be balanced by a slow loss of T cells, such that the number of T cells remains roughly constant. Most likely, this loss involves the daughters of such dividing naive cells.

Where do the CD4 and CD8 T cells encounter their positively selecting ligands in the periphery? Current evidence favors the idea that these ligands are encountered on lymphoid dendritic cells in T-cell zones (see Section 7-29). These cells are similar to the dendritic cells that migrate to the lymph nodes from peripheral sites (see Section 1-6), but lack sufficient co-stimulatory potential to induce full T-cell activation. However, the study of peripheral positive selection is still in its infancy, and a clear picture has yet to emerge.

Memory T cells also persist, like memory B cells, following immunization. Paradoxically, their turnover is faster than that of the naive T cells although expanded clones of memory cells are maintained at a relatively constant size by balanced proliferation and cell death. Unlike the naive cells, memory T cells do not appear to require continual stimulation by self peptide:self MHC ligands for this turnover to take place. Memory T cells transferred into mice that lack both MHC class I and MHC class II molecules are able to survive and to proliferate; these cells appear dependent on cytokine stimulation rather than upon antigenic stimulation.

7-33   B-cell tumors often occupy the same site as their normal counterparts.

Tumors retain many of the characteristics of the cell type from which they arose, especially when the tumor is relatively differentiated and slow growing. This is clearly illustrated in the case of B-cell tumors. Tumors corresponding to essentially all stages of B-cell development have been found in humans, from the earliest stages to the myelomas that represent malignant outgrowths of plasma cells (Fig. 7.41).

Furthermore, each type of tumor retains its characteristic homing properties. Thus, a tumor that resembles mature, germinal center or memory cells homes to follicles in lymph nodes and spleen, giving rise to a **follicular center cell lymphoma**, whereas a tumor of plasma cells usually disperses to many different sites in the bone marrow much as normal plasma cells do, from which comes the clinical name of **multiple myeloma** (tumor of bone marrow). These similarities mean that it is possible to use tumor cells, which are available in large quantities, to study the cell-surface molecules and signaling pathways responsible for homing behavior.

Multiple Myeloma

The status of the immunoglobulin genes in a B-cell tumor provides important information on its origin (see Fig. 7.41); in particular it tells us whether the B cell that gave rise to it has been through a germinal center. Tumors arising from lymphoid precursors, pre-B cells, and B-1 cells (the last of which is responsible for the vast majority of **chronic lymphocytic leukemias (CLL)**) have V genes that have not undergone somatic hypermutation. By contrast, tumors of mature B cells, such as follicular lymphoma or **Burkitt's lymphoma**, the latter arising from germinal center B cells, express mutated V genes. If the V genes from several different Burkitt's lymphoma lines from the same patient are sequenced, minor variations (intraclonal variations) are seen because somatic hypermutation is ongoing in the tumor cells. Later-stage B-cell tumors such as multiple myelomas contain mutated genes but do not display intraclonal variation; this is because by this stage in B-cell development somatic hypermutation has ceased.

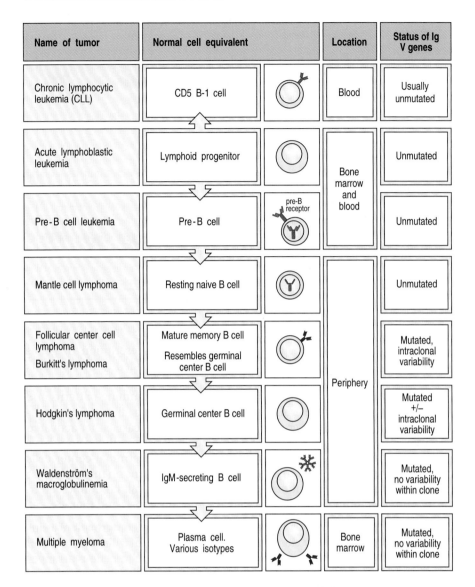

Name of tumor	Normal cell equivalent		Location	Status of Ig V genes
Chronic lymphocytic leukemia (CLL)	CD5 B-1 cell		Blood	Usually unmutated
Acute lymphoblastic leukemia	Lymphoid progenitor		Bone marrow and blood	Unmutated
Pre-B cell leukemia	Pre-B cell	pre-B receptor		Unmutated
Mantle cell lymphoma	Resting naive B cell			Unmutated
Follicular center cell lymphoma / Burkitt's lymphoma	Mature memory B cell / Resembles germinal center B cell		Periphery	Mutated, intraclonal variability
Hodgkin's lymphoma	Germinal center B cell			Mutated +/– intraclonal variability
Waldenström's macroglobulinemia	IgM-secreting B cell			Mutated, no variability within clone
Multiple myeloma	Plasma cell. Various isotypes		Bone marrow	Mutated, no variability within clone

Fig. 7.41 B-cell tumors represent clonal outgrowths of B cells at various stages of development. Each type of tumor cell has a normal B-cell equivalent, homes to similar sites, and has behavior similar to that cell. Thus, myeloma cells look much like the plasma cells from which they derive, they secrete immunoglobulin, and they are found predominantly in the bone marrow. The most enigmatic of B-cell tumors is Hodgkin's disease, which consists of two cell phenotypes: a lymphoid cell and a large, odd-looking cell known as a Reed–Sternberg (RS) cell. The RS cell appears to derive from a germinal center B cell that has decreased expression of surface immunoglobulin, possibly due to somatic mutation. Many lymphomas and myelomas can go through a preliminary, less aggressive, lymphoproliferative phase, and some mild lympho-proliferations seem to be benign.

Recently, a great mystery about the cellular origin of a class of tumors called **Hodgkin's disease** has been settled. The bizarre-looking cell that is characteristic of Hodgkin's disease, known as a **Reed–Sternberg (RS) cell**, was previously thought to be of T-cell or dendritic cell origin. DNA analysis has now shown that these cells have rearranged immunoglobulin genes, classifying them as outgrowths of a single B cell. In nearly all cases, Reed–Sternberg cells lack expression of surface immunoglobulin, which had prevented their earlier recognition as of B-cell origin. In many cases, the reason for loss of surface immunoglobulin is a somatic mutation that inactivates one of the Ig V-region genes. How the originally transformed B cell changes morphology to become an RS cell is not known. Curiously, in Hodgkin's disease, RS cells are sometimes a minority population; the more numerous surrounding cells are usually polyclonal T and B cells that may be reacting to the RS cells or to a soluble factor they secrete. In some cases, the surrounding cells are also abnormal, and it was recently shown that these are B-cell lymphomas that share the same rearrangement as the RS cells in the tumor, thus providing a clue as to the origin of RS cells. The main lesson from our evolving understanding of Hodgkin's disease is that by using molecular markers of cell lineage it is possible to learn a great deal about particular tumors, so that one might better design therapies and also monitor their progress.

Fig. 7.42 Clonal analysis of B-cell and T-cell tumors. DNA analysis of tumor cells by Southern blotting techniques can detect and monitor lymphoid malignancy. Left panel: B-cell tumor analysis. In a sample from a healthy person (left lane), immunoglobulin genes are in the germline configuration in non-B cells, so a digest of their DNA with a suitable restriction endonuclease yields a single germline DNA fragment when probed with an immunoglobulin heavy-chain J-region probe (J_H). Normal B cells present in this sample make many different rearrangements to J_H, producing a spectrum of 'bands' each so faint that it is undetectable. By contrast, in samples from patients with B-cell malignancies (patient 1 and patient 2), where a single cell has given rise to all the tumor cells in the sample, two extra bands are seen with the J_H probe. These bands are characteristic of each patient's tumor and result from the rearrangement of both alleles of the J_H gene in the original tumor cells. The intensity of the bands compared with that of the germline band gives an indication of the abundance of tumor cells in the sample. After antitumor treatment (see patient 1), the intensity of the tumor-specific bands can be seen to diminish. Right panel: the unique rearrangement events in each T cell can be similarly used to identify tumors of T cells by Southern blotting. The probe used in this case was for the T-cell receptor β-chain constant regions ($C_\beta 1$ and $C_\beta 2$). DNA from a placenta (P), a tissue in which the T-cell receptor genes are not rearranged, shows one prominent band for each region. DNA from peripheral blood lymphocytes from two patients suffering from T-cell tumors (T_1 and T_2) gives additional bands that correspond to specific rearrangements (arrowed) that are present in a large number of cells (the tumor). As with B cells, no bands deriving from rearranged genes in normal T cells also present in the patients' samples can be seen, as no one rearranged band is present at sufficient concentration to be detected in this assay. Left panel: photograph courtesy of T.J. Vulliamy and L. Luzzatto. Right panel: photograph courtesy of T. Diss.

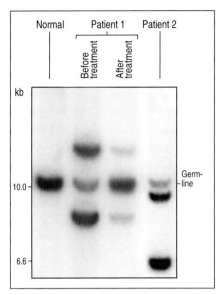

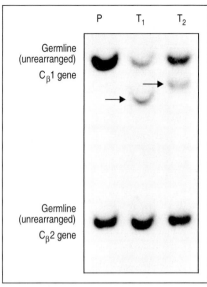

Recently, techniques that allow the comprehensive description of genes expressed in tumor cells and in normal cells have provided further insights into how tumors relate to normal tissues (see Appendix I, Section A-35). This effort has confirmed previous classifications based on homing patterns and has even allowed further subdivision of tumor types, like diffuse non-Hodgkin's lymphoma, into groups that resemble in this case either activated B cells or germinal center B cells. These molecularly based subdivisions have prognostic significance in that tumors that resemble germinal center cells respond much better to therapy. Expansions and variations on this molecular approach to classifying tumors of the immune system will no doubt lead to further understanding of tumors as well as better prognosis and therapy.

The general conclusion that a tumor represents the clonal outgrowth of a single transformed cell is very clearly illustrated by tumors of B-lineage cells. All the cells in a B-cell tumor have identical immunoglobulin gene rearrangements, decisively documenting their origin from one cell. This is useful for clinical diagnosis, as tumor cells can be detected by sensitive assays for these homogeneous rearrangements (Fig. 7.42). Only rarely are biclonal tumors, with two patterns of rearrangement, found.

7-34   A range of tumors of immune system cells throws light on different stages of T-cell development.

We saw in the preceding section that tumors of lymphoid cells corresponding in phenotype to intermediate stages in the development of the B cell provide invaluable tools in the analysis of B-cell differentiation. Tumors of T cells and other cells involved in T-cell development have been identified but, unlike the malignancies of B cells, few that correspond to intermediate stages in T-cell development have been identified in humans. Instead, the tumors resemble either mature T cells or, in **acute lymphoblastic leukemia**, the earliest type of lymphoid progenitor (Fig. 7.43). One possible reason for the rarity of tumors corresponding to intermediate stages is that immature T cells are programmed to die unless rescued within a very narrow time window by positive selection (see Section 7-21). Thymocytes might simply not linger long enough at the intermediate stages of their development to provide an opportunity for malignant transformation. Thus, only cells that are already transformed at earlier stages, or that do not become transformed until the T cell has matured, are ever seen as tumors.

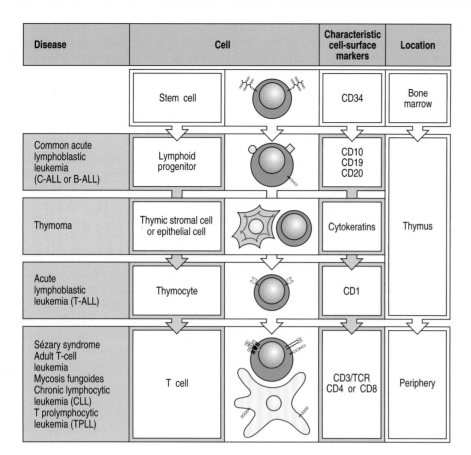

Disease	Cell		Characteristic cell-surface markers	Location
	Stem cell		CD34	Bone marrow
Common acute lymphoblastic leukemia (C-ALL or B-ALL)	Lymphoid progenitor		CD10 CD19 CD20	
Thymoma	Thymic stromal cell or epithelial cell		Cytokeratins	Thymus
Acute lymphoblastic leukemia (T-ALL)	Thymocyte		CD1	
Sézary syndrome Adult T-cell leukemia Mycosis fungoides Chronic lymphocytic leukemia (CLL) T prolymphocytic leukemia (TPLL)	T cell		CD3/TCR CD4 or CD8	Periphery

Fig. 7.43 T-cell tumors represent monoclonal outgrowths of normal cell populations. Each distinct T-cell tumor has a normal equivalent, as also seen with B-cell tumors, and retains many of the properties of the cell from which it develops. However, unlike B-cell tumors, tumors of T cells lack the intermediates in the T-cell developmental pathway. Some of these tumors represent massive outgrowth of a rare cell type. For example, acute lymphoblastic leukemia is derived from the lymphoid progenitor cell. One T-cell-related tumor is also included. Thymomas derive from thymic stromal or epithelial cells. Some characteristic cell-surface markers for each stage are also shown. For example, CD10 (acute lymphoblastic leukemia antigen or CALLA) is a widely used marker for acute lymphoblastic leukemia. Note that T-cell chronic lymphocytic leukemia (CLL) cells express CD8, whereas the other T-cell tumors mentioned express CD4. Adult T-cell leukemia is caused by the retrovirus HTLV-1.

The behavior of T-cell and other lymphoid tumors has provided insight into different aspects of T-cell biology, and vice versa. T-cell tumors provide valuable information about the phenotype, homing properties, and receptor gene rearrangements of normal T-cell types. For example, **cutaneous T-cell lymphomas**, which home to the skin and proliferate slowly, are clonal outgrowths of a CD4 T cell that, when activated, homes to the skin. Finally, a tumor of thymic stroma, called a **thymoma**, is frequently present in certain types of autoimmune disease, and removal of these tumors often ameliorates the disease. The reasons for this are not yet known.

7-35 Malignant lymphocyte tumors frequently carry chromosomal translocations that join immunoglobulin loci to genes regulating cell growth.

The unregulated accumulation of cells of a single clone, which is the most striking characteristic of tumors, is caused by mutations that release the founder cell from the normal restraints on its growth or prevent its normal programmed death. In B-cell tumors, the disruption of normal cellular homeostatic controls is often associated with an aberrant immunoglobulin gene rearrangement, in which one of the immunoglobulin loci is joined to a gene on another chromosome. This genetic fusion with another chromosome is known as a **translocation**, and in B-cell tumors such translocations are found to disrupt the expression and function of genes important for controlling cell growth. Cellular genes that cause cancer when their function or expression is disrupted are termed **oncogenes**.

Translocations give rise to chromosomal abnormalities that are visible microscopically in metaphase. Characteristic translocations are seen in different B-cell tumors and reflect the involvement of a particular oncogene in

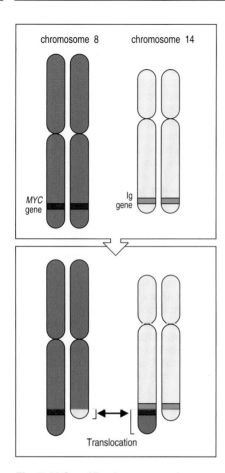

Fig. 7.44 Specific chromosomal rearrangements are found in some lymphoid tumors. If a chromosomal rearrangement joins one of the immunoglobulin genes to a cellular oncogene, it can result in the aberrant expression of the oncogene under the control of the immunoglobulin regulatory sequences. Such chromosomal rearrangements are frequently associated with B-cell tumors. In the example shown, from Burkitt's lymphoma, the translocation of the oncogene *MYC* from chromosome 8 to the immunoglobulin heavy-chain locus on chromosome 14 results in the deregulated expression of *MYC* and the unregulated growth of the B cell. The immunoglobulin gene located on the normal chromosome 14 is usually productively rearranged and the tumors that result from such translocations generally have a mature B-cell phenotype and express immunoglobulin.

each tumor type. Characteristic translocations, involving the T-cell receptor loci, are also seen in T-cell tumors. Immunoglobulin and T-cell receptor loci are sites at which double-stranded DNA breaks occur during gene rearrangement isotype switching, and in B cells, somatic hypermutation, so it is not surprising that they are especially likely to be sites of chromosomal translocation in T and B cells.

The analysis of chromosomal abnormalities has revealed much about the regulation of B-cell growth and the disruption of growth control in tumor cells. In Burkitt's lymphoma cells, the *MYC* oncogene on chromosome 8 is recombined with an immunoglobulin locus by translocations that involve either chromosome 14 (heavy chain) (Fig. 7.44), chromosome 2 (κ light chain), or chromosome 22 (λ light chain). The Myc protein is known to be involved in the control of the cell cycle in normal cells. The translocation deregulates expression of the Myc protein, which leads to increased proliferation of B cells, although other mutations elsewhere in the genome are also needed before a B-cell tumor results.

Other B-cell lymphomas bear a chromosomal translocation of immunoglobulin genes to the oncogene *bcl-2*, increasing the production of Bcl-2 protein. The Bcl-2 protein prevents programmed cell death in B-lineage cells, so its abnormal expression allows some B cells to survive and accumulate beyond their normal life-span. During this time, further genetic changes can occur that lead to malignant transformation. Mice carrying an expressed *bcl-2* transgene tend to develop B-cell lymphomas in later life.

Summary.

Lymphocytes that survive the selective processes that operate in the central lymphoid organs are exported to the peripheral lymphoid organs. Here, their fate is still controlled by their antigen receptors. In the absence of encounter with their specific foreign antigen, naive lymphocytes require at least weak engagement of their antigen receptors for long-term survival. Experimental evidence suggests that this occurs primarily by contact with the same ligand that initiated the positive selection of the lymphocyte in the first place, or one very similar. The evidence for such receptor-mediated survival signals is clearest in the case of T cells, but they also seem to be needed for B-1 cells and marginal zone B cells, in which case they may promote expansion and survival, and most likely also for B-2 cells, in which case they promote survival without expansion. Although in a few cases for B-1 and marginal zone B cells, ligands that select B cells are known, in general the ligands involved are unknown.

The organization of the peripheral lymphoid tissues and the homing of B and T cells to distinct areas are controlled by cytokines and chemokines. Large numbers of immature B cells leave the bone marrow daily. Most of these emigrants die soon after their arrival in the periphery, thus keeping the number of circulating B cells fairly constant. A small number acquire the signs of maturity and become longer-lived naive B cells. The lymphoid follicle, through which B cells must circulate in order to survive, seems to provide signals for maturation and survival. T cells leave the thymus as fully mature cells and in smaller numbers than B cells. They are generally long-lived, and are thought to be slowly self-renewing in the peripheral lymphoid tissues, being maintained by repeated contacts with self peptide:self MHC complexes that can be recognized by the T-cell receptor but that do not cause T-cell activation. The distinct minority subpopulations of lymphocytes, such as the B-1 cells (see Section 7-28), marginal zone B cells (see Section 7-29), γ:δ T cells (see Section 7-14), and the double-negative T cells with α:β receptors of very limited diversity (see Section 7-6), have different developmental histories and functional properties from conventional B-2 cells and α:β T cells, and are likely to be regulated independently of these majority B- and T-cell populations.

Very rarely, an individual B cell or T cell undergoes mutation and gives rise to a leukemia or lymphoma. The properties of the different lymphoid tumors reflect the developmental stage of the cell from which the tumor derives. All lymphoid tumors except those derived from very early uncommitted cells have characteristic gene rearrangements that allow their placement in the B or T lineage. These rearrangements are frequently accompanied by chromosomal translocations, often between a locus involved in generating the antigen receptor and a cellular proto-oncogene.

Summary to Chapter 7.

In this chapter we have learned about the formation of the B- and T-cell lineages from a primitive lymphoid progenitor. The somatic gene rearrangements that generate the highly diverse repertoire of antigen receptors—immunoglobulin for B cells and the T-cell receptor for T cells—occur in the early stages of the development of T cells and B cells from the common lymphoid progenitor. Mammalian B-cell development takes place in fetal liver and, after birth, in the bone marrow; T cells also originate in the bone marrow but undergo most of their development in the thymus. B-cell specific and T-cell specific transcription factors and other proteins ensure that the immunoglobulin genes are rearranged in developing B cells and the T-cell receptor genes in developing T cells, but much of the somatic recombination machinery, including the RAG proteins that are an essential part of the V(D)J recombinase, is common to both. Gene rearrangement proceeds successively at each gene locus, and the cell is only allowed to proceed to the next stage of development if the rearrangement has been successful and has produced an in-frame sequence that can be translated into a protein chain. Cells that do not generate successful rearrangements for both receptor chains die by apoptosis. The course of B-cell development is summarized in Fig. 7.45, and that of α:β T cells in Fig. 7.46.

Once a functional antigen receptor has appeared on the cell surface, the newly minted lymphocyte is ready for stringent tests of its potential usefulness on the one hand and its autoreactivity on the other. Positive selection selects potentially useful cells for survival, while negative selection removes self-reactive cells from the lymphocyte repertoire, rendering it tolerant to the antigens of the body. Tolerance is enforced at different stages throughout the development of both B and T cells, whereas positive selection appears to happen at a particular point. Positive selection is particularly crucial for T cells, as it ensures that only cells bearing T-cell receptors that can recognize antigen in combination with self MHC molecules will continue to mature. It occurs in the thymus, in thymocytes that have just started to produce a T-cell receptor and are still expressing both of the co-receptors CD4 and CD8. Though signals through co-receptors are clearly important, precisely how the thymocyte matches up the specificity of its T-cell receptor, the specificity of the co-receptor for MHC molecules, and the appropriate differentiated cellular function is still an open question. In the case of B cells, positive selection seems to occur at the final transition from immature to mature B cells, which occurs in peripheral lymphoid tissues. How it is effected is not known at present.

Occasionally, B cells and T cells undergo malignant transformation, giving rise to tumors that have escaped normal growth controls while retaining most features of the parent cell, including its characteristic homing pattern. These tumors frequently carry translocations involving the antigen-receptor loci and other genes that are intimately involved in lymphocyte growth regulation or cell death; thus these translocations are a goldmine of information about the genes and proteins that regulate lymphocyte homeostasis.

Fig. 7.45 A summary of the development of human conventional B-lineage cells. The state of the immunoglobulin genes, the expression of some essential intracellular proteins, and the expression of some cell-surface molecules are shown for successive stages of B-2-cell development. Note that since the immunoglobulin genes undergo further changes during antigen-driven development, such as isotype switch and somatic hypermutation (see Chapter 4), these are chronicled throughout development.

B cells		Heavy-chain genes	Light-chain genes	Intra-cellular proteins	Surface marker proteins
Stem cell		Germline	Germline		CD34, CD45
Early pro-B cell		D–J rearranged	Germline	RAG-1 RAG-2 TdT λ5, VpreB	CD34, CD45 MHC class II CD10, CD19 CD38
Late pro-B cell		V–DJ rearranged	Germline	TdT λ5, VpreB	CD45R MHC class II CD10, CD19 CD38, CD20 CD40
Large pre-B cell	pre-B receptor	VDJ rearranged	Germline	λ5, VpreB	CD45R MHC class II pre-B-R CD19, CD38 CD20, CD40
Small pre-B cell	cytoplasmic μ	VDJ rearranged	V–J rearrangement	μ RAG-1 RAG-2	CD45R MHC class II CD19, CD38 CD20, CD40
Immature B cell	IgM	VDJ rearranged. μ heavy chain produced in membrane form	VJ rearranged		CD45R MHC class II IgM CD19, CD20 CD40
Mature naive B cell	IgD IgM	VDJ rearranged. μ chain produced in membrane form. Alternative splicing yields μ + δ mRNA	VJ rearranged		CD45R MHC class II IgM, IgD CD19, CD20 CD21, CD40
Lympho-blast	IgM	VDJ rearranged. Alternative splicing yields secreted μ chains	VJ rearranged	Ig	CD45R MHC class II CD19, CD20 CD21, CD40
Memory B cell	IgG	Isotype switch to Cγ, Cα, or Cε. Somatic hypermutation	VJ rearranged. Somatic hypermutation		CD45R MHC class II IgG, IgA CD19, CD20 CD21, CD40
Plasma cell	IgG	Isotype switch and alternative splicing yields secreted γ, α, or ε chains	VJ rearranged	Ig	Plasma cell antigen-1 CD38

T cells		β-chain gene rearrangements	α-chain gene rearrangements	Intra-cellular proteins	Surface marker proteins	
Stem cell		Germline	Germline		CD34?	BONE MARROW
Early double-negative thymocyte		D–J rearranged	Germline	RAG-1 RAG-2 TdT Lck ZAP-70	CD2 HSA CD44hi	THYMUS
Late double-negative thymocyte		V–DJ rearranged	Germline	RAG-1 RAG-2 TdT Lck ZAP-70	CD25 CD44lo HSA	
Early double-positive thymocyte	pre-T receptor		V–J rearranged	RAG-1 RAG-2	Pre-Tα CD4 CD8 HSA	
Late double-positive thymocyte	T-cell receptor			Lck ZAP-70	CD69 CD4 CD8 HSA	
Naive CD4 T cell				Lck ZAP-70 LKLF	CD4 CD62L CD45RA CD5	PERIPHERY
Memory CD4 T cell				Lck ZAP-70	CD4 CD45RO CD44	
Effector CD4 T cell type 2				Type 1: IL-5 IL-4 IL-10 Type 2: IL2 IFN-γ	CD4 CD45RO CD44hi Fas FasL (type 1)	
Naive CD8 T cell					CD8 CD45RA	
Memory CD8 T cell					CD8 CD45RO CD44	
Effector CD8 T cell				IFN-γ granzyme perforin	FasL Fas CD8 CD44hi	

Row-group labels (left margin, top to bottom): ANTIGEN INDEPENDENT; ANTIGEN DEPENDENT; TERMINAL DIFFERENTIATION; ANTIGEN DEPENDENT; TERMINAL DIFFERENTIATION.

Fig. 7.46 A summary of the development of human α:β T cells. The state of the T-cell receptor genes, the expression of some essential intracellular proteins, and the expression of some cell-surface molecules are shown for successive stages of α:β T-cell development. Note that since the TCR genes do not undergo further changes during antigen-driven development, only the phases during which they are actively undergoing rearrangement in the thymus are indicated. The antigen-dependent phases of CD4 and CD8 cells are depicted separately.

General references.

Alt, F., and Marrack, P. (eds): *Current Opinion in Immunology*, Vol. 10. London, Current Biology, 1998.

von Boehmer, H.: **Positive selection of lymphocytes.** *Cell* 1994, **76**:219-228.

von Boehmer, H.: **The developmental biology of T lymphocytes.** *Annu. Rev. Immunol.* 1993, **6**:309-326.

Casali, P., and Silberstein, L.E.S. (eds): **Immunoglobulin gene expression in development and disease.** *Ann. N.Y. Acad. Sci.* 1995, **764**.

Hardy, R.R., and Hayakawa, K.: **B-lineage differentiation stages resolved by multiparameter flow-cytometry.** *Ann. N.Y. Acad. Sci.* 1995, **764**:19-24.

Kee, B.L., and Paige, C.J.: **Murine B-cell development—commitment and progression from multipotential progenitors to mature B-lymphocytes.** *Intl. Rev. Cytol.* 1995, **157**:129-179.

Linette, G.P., and Korsmeyer, S.J.: **Differentiation and cell-death—lessons from the immune-system.** *Curr. Opin. Cell Biol.* 1994, **6**:809-815.

Loffert, D., Schaal, S., Ehlich, A., Hardy, R.R., Zou, Y.R., Muller, W., and Rajewsky, K.: **Early B-cell development in the mouse—insights from mutations introduced by gene targeting.** *Immunol. Rev.* 1994, **137**:135-153.

Melchers, F., ten Boekel, E., Seidl, T., Kong, X.C., Yamagami, T., Onishi, K., Shimizu, T., Rolink, A.G., and Andersson, J.: **Repertoire selection by pre-B-cell receptors and B-cell receptors, and genetic control of B-cell development from immature to mature B cells.** *Immunol. Rev.* 2000, **175**:33-46.

Moller, G. (ed): **Positive T-cell selection in the thymus.** *Immunol. Rev.* 1993, **135**:5-242.

Nossal, G.J.V.: **Negative selection of lymphocytes.** *Cell* 1994, **76**:229-239.

Rajewsky, K.: **Clonal selection and learning in the antibody system.** *Nature* 1996, **381**:751-758.

Rolink, A., and Melchers, F.: **B-Lymphopoiesis in the mouse.** *Adv. Immunol.* 1993, **53**:123-156.

Spangrude, G.J.: **Biological and clinical aspects of hematopoietic stem-cells.** *Annu. Rev. Med.* 1994, **45**:93-104.

Section references.

7-1 Lymphocyte development occurs in specialized environments and is regulated by the somatic rearrangement of the antigen-receptor genes.

van Ewijk, W.: **T-cell differentiation is influenced by thymic microenvironments.** *Annu. Rev. Immunol.* 1991, **9**:591-615.

van Ewijk, W., Hollander, G., Terhorst, C., and Wang, B.: **Stepwise development of thymic microenvironments *in vivo* is regulated by thymocyte subsets.** *Development* 2000, **127**:1583-1591.

Funk, P.E., Kincade, P.W., and Witte, P.L.: **Native associations of early hematopoietic stem-cells and stromal cells isolated in bone-marrow cell aggregates.** *Blood* 1994, **83**:361-369.

7-2 B cells develop in the bone marrow with the help of stromal cells and achieve maturity in peripheral lymphoid organs.

Jacobsen, K., Kravitz, J., Kincade, P.W., and Osmond, D.G.: **Adhesion receptors on bone-marrow stromal cells—*in vivo* expression of vascular cell adhesion molecule-1 by reticular cells and sinusoidal endothelium in normal and γ-irradiated mice.** *Blood* 1996, **87**:73-82.

Nagasawa, T., Hirota, S., Tachibana, V., Takakura, N., Nishikawa, S., Kitamura, Y., Yoshida, V., Kikutani, H., and Kishimoto, T.: **Defects of B-cell lymphopoiesis and bone-marrow myelopoiesis in mice lacking the CXC chemokine PBSF/SDF-1.** *Nature* 1996, **382**:635-638.

Rosenberg, N., and Kincade, P.W.: **B-lineage differentiation in normal and transformed-cells and the microenvironment that supports it.** *Curr. Opin. Immunol.* 1994, **6**:203-211.

7-3 Stages in B-cell development are distinguished by the expression of immunoglobulin chains and particular cell-surface proteins.

Hardy, R.R., Carmack, C.E., Shinton, S.A., Kemp, J.D., and Hayakawa, K.: **Resolution and characterization of pro-B and pre-pro-B cell stages in normal mouse bone marrow.** *J. Exp. Med.* 1991, **173**:1213.

Osmond, D.G., Rolink, A., and Melchers, F.: **Murine B lymphopoiesis: towards a unified model.** *Immunol. Today* 1998, **19**:65-68.

7-4 T cells also originate in the bone marrow, but all the important events in their development occur in the thymus.

Anderson, G., Moore, N.C., Owen, J.J.T., and Jenkinson, E.J.: **Cellular interactions in thymocyte development.** *Annu. Rev. Immunol.* 1996, **14**:73-99.

Carlyle, J.R., and Zúñiga-Pflücker, J.C.: **Requirement for the thymus in alpha-beta T lymphocyte lineage commitment.** *Immunity* 1998, **9**:187-197.

Cordier, A.C., and Haumont, S.M.: **Development of thymus, parathyroids, and ultimobranchial bodies in NMRI and *nude* mice.** *Am. J. Ana.* 1980, **157**:227.

Nehls, M., Kyewski, B., Messerle, M., Waldschütz, R., Schüddekopf, K., Smith, A.J.H., and Boehm, T.: **Two genetically separable steps in the differentiation of thymic epithelium.** *Science* 1996, **272**:886-889.

Zúñiga-Pflücker, J.C., and Lenardo, M.J.: **Regulation of thymocyte development from immature progenitors.** *Curr. Opin. Immunol.* 1996, **8**:215-224

7-5 Most developing T cells die in the thymus.

Shortman, K., Egerton, M., Spangrude, G.J., and Scollay, R.: **The generation and fate of thymocytes.** *Semin. Immunol.* 1990, **2**:3-12.

Strasser, A.: **Life and death during lymphocyte development and function: evidence for two distinct killing mechanisms.** *Curr. Opin. Immunol.* 1995, **7**:228-234.

Surh, C.D., and Sprent, J.: **T-cell apoptosis detected *in situ* during positive and negative selection in the thymus.** *Nature* 1994, **372**:100-103.

7-6 Successive stages in the development of thymocytes are marked by changes in cell-surface molecules.

Petrie, H.T., Hugo, P., Scollay, R., and Shortman, K.: **Lineage relationships and developmental kinetics of immature thymocytes: CD3, CD4, and CD8 acquisition *in vivo* and *in vitro*.** *J. Exp. Med.* 1990, **172**:1583-1588.

Saint-Ruf, C., Ungewiss, K., Groettrup, M., Bruno, L., Fehling, H.J., and von Boehmer, H.: **Analysis and expression of a cloned pre-T-cell receptor gene.** *Science* 1994, **266**:1208.

Shortman, K., and Wu, L.: **Early T lymphocyte progenitors.** *Annu. Rev. Immunol.* 1996, **14**:29-47.

7-7 Thymocytes at different developmental stages are found in distinct parts of the thymus.

Picker, L.J., and Siegelman, M.H.: Lymphoid tissues and organs, in Paul, W.E. (ed): *Fundamental Immunology*, 3rd edn. New York, Raven Press, 1993.

7-8 B cells undergo a strictly programmed series of gene rearrangements in the bone marrow.

Ehrlich, A., and Kuppers, R.: **Analysis of immunoglobulin gene rearrangements in single B cells.** *Curr. Opin. Immunol.* 1995, **7**:281-284.

Ten-Boekel, E., Melchers, F., and Rolink, A.: **The status of Ig loci rearrangements in single cells from different stages of B-cell development.** *Intl. Immunol.* 1995, **7**:1013-1019.

7-9 Successful rearrangement of heavy-chain immunoglobulin gene segments leads to the formation of a pre-B-cell receptor that halts further V_H to DJ_H rearrangement and triggers the cell to divide.

Grawunder, U., Leu, T.M.J., Schatz, D.G., Werner, A., Rolink, A.G., Melchers, F., and Winkler, T.H.: **Down-regulation of Rag1 and Rag2 gene expression in pre-B cells after functional immunoglobulin heavy-chain rearrangement.** *Immunity* 1995, **3**:601-608.

Loffert, D., Ehlich, A., Muller, W., and Rajewsky, K.: **Surrogate light-chain expression is required to establish immunoglobulin heavy-chain allelic exclusion during early B-cell development.** *Immunity* 1996, **4**:133-144.

Melchers, F., ten Boekel, E., Yamagami, T., Andersson, J., and Rolink, A.: **The roles of preB and B cell receptors in the stepwise allelic exclusion of mouse IgH and L chain gene loci.** *Semin. Immunol.* 1999, **11**:307-317.

7-10 Rearrangement at the immunoglobulin light-chain locus leads to cell-surface expression of the B-cell receptor.

Arakawa, H., Shimizu, T., and Takeda, S.: **Reevaluation of the probabilities for productive rearrangements on the κ-loci and λ-loci.** *Intl. Immunol.* 1996, **8**:91-99.

Gorman, J.R., van der Stoep, N., Monroe, R., Cogne, M., Davidson, L., and Alt, F.W.: **The Igκ 3′ enhancer influences the ratio of Igκ versus Igλ B lymphocytes.** *Immunity* 1996, **5**:241-252.

Takeda, S., Sonoda, E., and Arakawa, H.: **The κ–λ ratio of immature B cells.** *Immunol. Today* 1996, **17**:200-201.

7-11 The expression of proteins regulating immunoglobulin gene rearrangement and function is developmentally programmed.

Desiderio, S.: **Lymphopoiesis—transcription factors controlling B-cell development.** *Curr. Biol.* 1995, **5**:605-608.

Khan, W.N., Alt, F.W., Gerstein, R.M., Malynn, B.A., Larsson, I., Rathbun, G., Davidson, L., Muller, S., Kantor, A.B., Herzenberg, L.A., Rosen, F.S., and Sideras, P.: **Defective B-cell development and function in btk-deficient mice.** *Immunity* 1995, **3**:283-299.

Neurath, M.F., Stuber, E.R., and Strober, W.: **BSAP—a key regulator of B-cell development and differentiation.** *Immunol. Today* 1995, **16**:564-569.

Opstelten, D.: **B lymphocyte development and transcription regulation *in vivo*.** *Adv. Immunol.* 1996, **63**:197-268.

Sleckman, B.P., Gorman, J.R., and Alt, F.W.: **Accessibility control of antigen receptor variable region gene assembly—role of *cis*-acting elements.** *Annu. Rev. Immunol.* 1996, **14**:459-481.

7-12 T cells in the thymus undergo a series of gene segment rearrangements similar to those of B cells.

Lauzurica, P., and Krangel, M.S.: **Temporal and lineage-specific control of T-cell receptor α/δ gene rearrangement by T-cell receptor α and δ enhancers.** *J. Exp. Med.* 1994, **179**:1913-1921.

Livak, F., Petrie, H.T., Crispe, I.N., and Schatz, D.G.: **In-frame TCR δ gene rearrangements play a critical role in the αβ/γδ T cell lineage decision.** *Immunity* 1995, **2**:617-627.

7-13 T cells with α:β or γ:δ receptors arise from a common progenitor.

Fehling, H.J., Gilfillan, S., and Ceredig, R.: **Alpha beta/gamma delta lineage commitment in the thymus of normal and genetically manipulated mice.** *Adv. Immunol.* 1999, **71**:1-76.

Hayday, A.C., Barber, D.F., Douglas, N., and Hoffman, E.S.: **Signals involved in gamma/delta T cell versus alpha/beta T cell lineage commitment.** *Semin. Immunol.* 1999, **11**:239-249.

Kang, J., and Raulet, D.H.: **Events that regulate differentiation of α β TCR⁺**

and γ δ TCR⁺ T cells from a common precursor. *Semin. Immunol.* 1997, **9**:171-179.

Kang, J., Coles, M., Cado, D., and Raulet, D.H.: **The developmental fate of T cells is critically influenced by TCRγδ expression.** *Immunity* 1998, **8**:427-438.

7-14 T cells expressing particular γ- and δ-chain V regions arise in an ordered sequence early in life.

Dunon, D., Courtois, D., Vainio, O., Six, A., Chen, C.H., Cooper, M.D., Dangy, J.P., and Imhof, B.A.: **Ontogeny of the immune system: γδ and αβ T cells migrate from thymus to the periphery in alternating waves.** *J. Exp. Med.* 1997, **186**:977-988.

Havran, W.L., and Boismenu, R.: **Activation and function of γδ T cells.** *Curr. Opin. Immunol.* 1994, **6**:442-446.

Itohara, S., Nakanishi, N., Kanagawa, O., Kubo, R., and Tonegawa, S.: **Monoclonal antibodies specific to native murine T cell receptor γδ analysis of γδ T cells in thymic ontogeny and peripheral lymphoid organs.** *Proc. Natl. Acad. Sci. USA* 1989, **86**:5094-5098.

7-15 Rearrangement of the β-chain locus and production of a β chain trigger several events in developing thymocytes.

Dudley, E.C., Petrie, H.T., Shah, L.M., Owen, M.J., and Hayday, A.C.: **T-cell receptor β chain gene rearrangement and selection during thymocyte development in adult mice.** *Immunity* 1994, **1**:83-93.

Philpott, K.I., Viney, J.L., Kay, G., Rastan, S., Gardiner, E.M., Chae, S., Hayday, A.C., and Owen, M.J.: **Lymphoid development in mice congenitally lacking T cell receptor α β-expressing cells.** *Science* 1992, **256**:1448-1453.

von Boehmer, H., and Fehling, H.J.: **Structure and function of the pre-T cell receptor.** *Annu. Rev. Immunol.* 1997, **15**:433-452.

7-16 T-cell α-chain genes undergo successive rearrangements until positive selection or cell death intervenes.

Hardardottir, F., Baron, J.L., and Janeway, C.A., Jr.: **T cells with two functional antigen-specific receptors.** *Proc. Natl. Acad. Sci. USA* 1995, **92**:354-358.

Marrack, P., and Kappler, J.: **Positive selection of thymocytes bearing alpha beta T cell receptors.** *Curr. Opin. Immunol.* 1997, **9**:250-255.

Padovan, E., Casorati, G., Dellabona, P., Meyer, S., Brockhaus, M., and Lanzavecchia, A.: **Expression of two T-cell receptor α chains: dual receptor T cells.** *Science* 1993, **262**:422-424.

Petrie, H.T., Livak, F., Schatz, D.G., Strasser, A., Crispe, I.N., and Shortman, K.: **Multiple rearrangements in T-cell receptor α-chain genes maximize the production of useful thymocytes.** *J. Exp. Med.* 1993, **178**:615-622.

7-17 Immature B cells that bind self antigens undergo further receptor rearrangement, or die, or are inactivated.

Chen, C., Nagy, Z., Radic, M.Z., Hardy, R.R., Huszar, D., Camper, S.A., and Weigert, M.: **The site and stage of anti-DNA B-cell deletion.** *Nature* 1995, **373**:252-255.

Cornall, R.J., Goodnow, C.C., and Cyster, J.G.: **The regulation of self-reactive B cells.** *Curr. Opin. Immunol.* 1995, **7**:804-811.

Melamed, D., Benschop, R.J., Cambier, J.C., and Nemazee, D.: **Developmental regulation of B lymphocyte immune tolerance compartmentalizes clonal selection from receptor selection.** *Cell* 1998, **92**:173-182.

Nemazee, D., Russell, D., Arnold, B., Haemmerling, G., Allison, J., Miller, J.F.A.P., Morahan, G., and Beurki, K.: **Clonal deletion of autospecific B lymphocytes.** *Immunol. Rev.* 1991, **122**:117-132.

Prak, E.L., and Weigert, M.: **Light-chain replacement—a new model for antibody gene rearrangement.** *J. Exp. Med.* 1995, **182**:541-548.

Tiegs, S.L., Russell, D.M., and Nemazee, D.: **Receptor editing in self-reactive bone marrow B cells.** *J. Exp. Med.* 1993, **177**:1009-1020.

7-18 Mature B cells can also be rendered self-tolerant.

Goodnow, C.C., Crosbie, J., Jorgensen, H., Brink, R.A., and Basten, A.: **Induction of self-tolerance in mature peripheral B lymphocytes.** *Nature* 1989, **342**:385-391.

Russell, D.M., Dembic, Z., Morahan, G., Miller, J.F.A.P., Burki, K., and Nemazee, D.: **Peripheral deletion of self-reactive B cells.** *Nature* 1991, **354**:308-311.

7-19 Only thymocytes whose receptors can interact with self MHC:self peptide complexes can survive and mature.

Fink, P.J., and Bevan, M.J.: **H-2 antigens of the thymus determine lymphocyte specificity.** *J. Exp. Med.* 1978, **148**:766-775.

Hogquist, K.A., Tomlinson, A.J., Kieper, W.C., McGargill, M.A., Hart, M.C., Naylor, S., and Jameson, S.C.: **Identification of a naturally occurring ligand for thymic positive selection.** *Immunity* 1997, **6**:389-399.

Ignatowicz, L., Kappler, J., and Marrack, P.: **The repertoire of T cells shaped by a single MHC/peptide ligand.** *Cell* 1996, **84**:521-529.

Zinkernagel, R.M., Callahan, G.N., Klein, J., and Dennert, G.: **Cytotoxic T cells learn specificity for self H-2 during differentiation in the thymus.** *Nature* 1978, **271**:251-253.

7-20 Most thymocytes express receptors that cannot interact with self MHC and these cells die in the thymus.

Huessman, M., Scott, B., Kisielow, P., and von Boehmer, H.: **Kinetics and efficacy of positive selection in the thymus of normal and T-cell receptor transgenic mice.** *Cell* 1991, **66**:533-562.

Zerrahn, J., Held, W., and Raulet, D.H.: **The MHC reactivity of the T cell repertoire prior to positive and negative selection.** *Cell* 1997, **88**:627-636.

7-21 Positive selection acts on a repertoire of receptors with inherent specificity for MHC molecules.

Merkenschlager, M., Graf, D., Lovatt, M., Bommhardt, U., Zamoyska, R., and Fisher, A.G.: **How many thymocytes audition for selection?** *J. Exp. Med.* 1997, **186**:1149-1158.

7-22 Positive selection coordinates the expression of CD4 or CD8 with the specificity of the T-cell receptor and the potential effector functions of the cell.

Albert Basson, M., Bommhardt, U., Cole, M.S., Tso, J.Y., and Zamoyska, R.: **CD3 ligation on immature thymocytes generates antagonist-like signals appropriate for CD8 lineage commitment, independently of T cell receptor specificity.** *J. Exp. Med.* 1998, **187**:1249-1260.

von Boehmer, H., Kisielow, P., Lishi, H., Scott, B., Borgulya, P., and Teh, H.S.: **The expression of CD4 and CD8 accessory molecules on mature T cells is not random but correlates with the specificity of the αβ receptor for antigen.** *Immunol. Rev.* 1989, **109**:143-151.

Bommhardt, U., Cole, M.S., Tso, J.Y., and Zamoyska, R.: **Signals through CD8 or CD4 can induce commitment to the CD4 lineage in the thymus.** *Eur. J. Immunol.* 1997, **27**:1152-1163.

Kaye, J., Hsu, M.L., Sauvon, M.E., Jameson, S.C., Gascoigne, N.R.J., and Hedrick, S.M.: **Selective development of CD4+ T cells in transgenic mice expressing a class II MHC-restricted antigen receptor.** *Nature* 1989, **341**:746-748.

Lundberg, K., Heath, W., Kontgen, F., Carbone, F.R., and Shortman, K.: **Intermediate steps in positive selection: differentiation of CD4+8int TCRint thymocytes into CD4-8+TCRhi thymocytes.** *J. Exp. Med.* 1995, **181**:1643-1651.

Singer, A., Bosselut, R., and Bhandoola, A.: **Signals involved in CD4/CD8 lineage commitment: current concepts and potential mechanisms.** *Semin. Immunol.* 1999, **11**:273-281.

7-23 Thymic cortical epithelial cells mediate positive selection of developing thymocytes.

Cosgrove, D., Chan, S.H., Waltzinger, C., Benoist, C., and Mathis, D.: **The thymic compartment responsible for positive selection of CD4+ T cells.** *Intl. Immunol.* 1992, **4**:707-710.

Ernst, B.B., Surh, C.D., and Sprent, J.: **Bone marrow-derived cells fail to induce positive selection in thymus reaggregation cultures.** *J. Exp. Med.* 1996, **183**:1235-1240.

Fowlkes, B.J., and Schweighoffer, E.: **Positive selection of T cells.** *Curr. Opin. Immunol.* 1995, **7**:188-195.

7-24 T cells that react strongly with ubiquitous self antigens are deleted in the thymus.

Kishimoto, H., and Sprent, J.: **Negative selection in the thymus includes semimature T cells.** *J. Exp. Med.* 1997, **185**:263-271.

Zal, T., Volkmann, A., and Stockinger, B.: **Mechanisms of tolerance induction in major histocompatibility complex class II-restricted T cell specific for a blood-borne self antigen.** *J. Exp. Med.* 1994, **180**:2089-2099.

7-25 Negative selection is driven most efficiently by bone marrow-derived antigen-presenting cells.

Matzinger, P., and Guerder, S.: **Does T cell tolerance require a dedicated antigen-presenting cell?** *Nature* 1989, **338**:74-76.

Sprent, J., and Webb, S.R.: **Intrathymic and extrathymic clonal deletion of T cells.** *Curr. Opin. Immunol.* 1995, **7**:196-205.

Webb, S.R., and Sprent, J.: **Tolerogenicity of thymic epithelium.** *Eur. J. Immunol.* 1990, **20**:2525-2528.

7-26 Endogenous superantigens mediate negative selection of T-cell receptors derived from particular Vβ gene segments.

Kappler, J.W., Roehm, N., and Marrack, P.: **T-cell tolerance by clonal elimination in the thymus.** *Cell* 1987, **49**:273-280.

MacDonald, H.R., Schneider, R., Lees, R.K., Howe, R.C., Acha-Orbea, H., Festenstein, H., Zinkernagel, R.M., and Hengartner, H.: **T-cell receptor Vβ use predicts reactivity and tolerance to Mlsa-encoded antigens.** *Nature* 1988, **332**:40-45.

7-27 The specificity and strength of signals for negative and positive selection must differ.

Alberola-Ila, J., Hogquist, K.A., Swan, K.A., Bevan, M.J., and Perlmutter, R.M.: **Positive and negative selection invoke distinct signaling pathways.** *J. Exp. Med.* 1996, **184**:9-18.

Ashton-Rickardt, P.G., Bandeira, A., Delaney, J.R., Van Kaer, L., Pircher, H.P., Zinkernagel, R.M., and Tonegawa, S.: **Evidence for a differential avidity model of T-cell selection in the thymus.** *Cell* 1994, **74**:577.

Bommhardt, U., Basson, M.A., Krummrei, U., and Zamoyska, R.: **Activation of the extracellular signal-related kinase/mitogen-activated protein kinase pathway discriminates CD4 versus CD8 lineage commitment in the thymus.** *J. Immunol.* 1999, **163**:715-722.

Bommhardt, U., Scheuring, Y., Bickel, C., Zamoyska, R., and Hunig, T.: **MEK activity regulates negative selection of immature CD4+CD8+ thymocytes.** *J. Immunol.* 2000, **164**:2326-2337.

Hogquist, K.A., Jameson, S.C., Heath, W.R., Howard, J.L., Bevan, M.J., and Carbane, F.R.: **T-cell receptor antagonist peptides induce positive selection.** *Cell* 1994, **76**:17-27.

7-28 The B-1 subset of B cells has a distinct developmental history and expresses a distinctive repertoire of receptors.

Clarke, S.H., and Arnold, L.W.: **B-1 cell development: evidence for an uncommitted immunoglobulin (Ig)M+ B cell precursor in B-1 cell differentiation.** *J. Exp. Med.* 1998, **187**:1325-1334.

Hardy, R.R., and Hayakawa, K.: **A developmental switch in B lymphopoiesis.** *Proc. Natl. Acad. Sci. USA* 1991, **88**:11550-11554.

Hayakawa, K., Asano, M., Shinton, S.A., Gui, M., Allman, D., Stewart, C.L., Silver, J., and Hardy, R.R.: **Positive selection of natural autoreactive B cells.** *Science* 1999, **285**:113-116.

Murakami, M., and Honjo, T.: **Involvement of B-1 cells in mucosal immunity and autoimmunity.** *Immunol. Today* 1995, **16**:534-539.

7-29 Newly formed lymphocytes home to particular locations in peripheral lymphoid tissues.

Liu, Y.J.: **Sites of B lymphocyte selection, activation, and tolerance in spleen.** *J. Exp. Med.* 1997, **186**:625-629.

Loder, F., Mutschler, B., Ray, R.J., Paige, C.J., Sideras, P., Torres, R., Lamers, M.C., and Carsetti, R.: **B cell development in the spleen takes place in discrete steps and is determined by the quality of B cell receptor-derived signals.** *J. Exp. Med.* 1999, **190**:75-89.

7-30 The development and organization of peripheral lymphoid tissues is controlled by cytokines and chemokines.

Cyster, J.G.: **Chemokines and cell migration in secondary lymphoid organs.** *Science* 1999, **286**:2098-2102.

Cyster, J.G.: **Leukocyte migration: scent of the T zone.** *Curr. Biol.* 2000, **10**:R30-R33.

Cyster, J.G., Ngo, V.N., Ekland, E.H., Gunn, M.D., Sedgwick, J.D., and Ansel, K.M.: **Chemokines and B-cell homing to follicles.** *Curr. Top. Microbiol. Immunol.* 1999, **246**:87-92.

Douni, E., Akassoglou, K., Alexopoulou, L., Georgopoulos, S., Haralambous, S., Hill, S., Kassiotis, G., Kontoyiannis, D., Pasparakis, M., Plows, D., Probert, L., and Kollias, G.: **Transgenic and knockout analysis of the role of TNF in immune regulation and disease pathogenesis.** *J. Inflamm.* 1996, **47**:27-38.

Forster, R., Mattis, A.E., Kremmer, E., Wolf, E., Brem, G., and Lipp, M.: **A putative chemokine receptor, BLR1, directs B cell migration to defined lymphoid organs and specific anatomic compartments of the spleen.** *Cell* 1996, **87**:1037-1047.

Fu, Y.X., and Chaplin, D.D.: **Development and maturation of secondary lymphoid tissues.** *Annu. Rev. Immunol.* 1999, **17**:399-433.

Mariathasan, S., Matsumoto, M., Baranyay, F., Nahm, M.H., Kanagawa, O., and Chaplin, D.D.: **Absence of lymph nodes in lymphotoxin-α (LTα)-deficient mice is due to abnormal organ development, not defective lymphocyte migration.** *J. Inflamm.* 1995, **45**:72-78.

7-31 Only a small fraction of immature B cells mature and survive in peripheral lymphoid tissues

Allman, D.M., Ferguson, S.E., Lentz, V.M., and Cancro, M.P.: **Peripheral B cell maturation. II. Heat-stable antigen(hi) splenic B cells are an immature developmental intermediate in the production of long-lived marrow-derived B cells.** *J. Immunol.* 1993, **151**:4431-4444.

Cyster, J.G., Hartley, S.B., and Goodnow, C.C.: **Competition for follicular niches excludes self-reactive cells from the recirculating B-cell repertoire.** *Nature* 1994, **371**:389-395.

Fulcher, D.A., and Basten, A.: **Reduced lifespan of anergic self-reactive B cells in a double-transgenic model.** *J. Exp. Med.* 1994, **179**:125-134.

Lam, K.P., Kuhn, R., and Rajewsky, K.: *In vivo* **ablation of surface immunoglobulin on mature B cells by inducible gene targeting results in rapid cell death.** *Cell* 1997, **90**:1073-1083.

Levine, M.H., Haberman, A.M., Sant'Angelo, D.B., Hannum, L.G., Cancro, M.P., Janeway, C.A., Jr., and Shlomchik, M.J.: **A B-cell receptor-specific selection step governs immature to mature B cell differentiation.** *Proc. Natl. Acad. Sci. USA* 2000, **97**:2743-2748.

Osmond, D.G.: **The turnover of B-cell populations.** *Immunol. Today* 1993, **14**:34-37.

Turner, M., Gulbranson-Judge, A., Quinn, M.E., Walters, A.E., MacLennan, I.C., and Tybulewicz, V.L.: **Syk tyrosine kinase is required for the positive selection of immature B cells into the recirculating B cell pool.** *J. Exp. Med.* 1997, **186**:2013-2021.

7-32 The life-span of naive T cells in the periphery is determined by ongoing contact with self peptide:self MHC complexes similar to those that initially selected them.

Freitas, A.A., and Rocha, B.: **Peripheral T cell survival.** *Curr. Opin. Immunol.* 1999, **11**:152-156.

Murali-Krishna, K., Lau, L.L., Sambhara, S., Lemonnier, F., Altman, J., and Ahmed, R.: **Persistence of memory CD8 T cells in MHC class I-deficient mice.** *Science* 1999, **286**:1377-1381.

7-33 B-cell tumors often occupy the same site as their normal counterparts.

Alizadeh, A.A., and Staudt, L.M.: **Genomic-scale gene expression profiling of normal and malignant immune cells.** *Curr. Opin. Immunol.* 2000, **12**:219-225.

Cotran, R.S., Kumar, V., and Robbins, S.L.: *Diseases of white cells, lymph nodes, and spleen. Pathologic basis of disease,* 5th edn, 629-672. Philadelphia, W.B. Saunders, 1994.

7-34 A range of tumors of immune system cells throws light on different stages of T-cell development.

Hwang, L.Y., and Baer, R.J.: **The role of chromosome translocations in T cell acute leukemia.** *Curr. Opin. Immunol.* 1995, **7**:659-664.

Rabbitts, T.H.: **Chromosomal translocations in human cancer.** *Nature* 1994, **372**:143-149.

7-35 Malignant lymphocyte tumors frequently carry chromosomal translocations that join immunoglobulin loci to genes regulating cell growth.

Cory, S.: **Regulation of lymphocyte survival by the Bcl-2 gene family.** *Annu. Rev. Immunol.* 1995, **13**:513-543.

Rabbitts, T.H.: **Chromosomal translocations in human cancer.** *Nature* 1994, **372**:143-149.

Yang, E., and Korsmeyer, S.J.: **Molecular thanatopsis—a discourse on the Bcl-2 family and cell death.** *Blood* 1996, **88**:386-401.

PART IV

THE ADAPTIVE IMMUNE RESPONSE

T Cell-Mediated Immunity

8

Once they have completed their development in the thymus, T cells enter the bloodstream and are carried by the circulation. On reaching a peripheral lymphoid organ they leave the blood again to migrate through the lymphoid tissue, returning to the bloodstream to recirculate between blood and peripheral lymphoid tissue until they encounter their specific antigen. Mature recirculating T cells that have not yet encountered their antigens are known as **naive T cells**. To participate in an adaptive immune response, a naive T cell must first encounter antigen, and then be induced to proliferate and differentiate into cells capable of contributing to the removal of the antigen. We will term such cells **armed effector T cells** because they act very rapidly when they encounter their specific antigen on other cells. The cells on which armed effector T cells act will be referred to as their **target cells**.

In this chapter, we will see how naive T cells are activated to produce armed effector T cells the first time they encounter their specific antigen in the form of a peptide:MHC complex on the surface of an activated **antigen-presenting cell** (**APC**). The most important antigen-presenting cells are the highly specialized **dendritic cells**, whose only known function is to ingest and present antigen. Tissue dendritic cells ingest antigen at sites of infection and are activated as part of the innate immune response. This induces their migration to local lymphoid tissue and their maturation into cells that are highly effective at presenting antigen to recirculating T cells. These mature dendritic cells are distinguished by surface molecules, known as **co-stimulatory molecules**, that synergize with antigen in the activation of naive T cells. **Macrophages**, which we described in Chapter 2 as phagocytic cells that provide a first line of defense against infection, can also be activated to express co-stimulatory and MHC class II molecules. This enables them to act as antigen-presenting cells, although they are less powerful than dendritic cells at activating naive T cells. B cells can also serve as antigen-presenting cells in some circumstances. Once a T-cell response has been initiated, macrophages and B cells that have taken up specific antigen also become targets for armed effector T cells. Dendritic cells, macrophages, and B cells are often known as professional antigen-presenting cells.

Effector T cells, as we learned in Chapter 5, fall into three functional classes that detect peptide antigens derived from different types of pathogen. Peptides from intracellular pathogens that multiply in the cytoplasm are carried to the cell surface by MHC class I molecules and presented to CD8 T cells. These differentiate into **cytotoxic T cells** that kill infected target cells. Peptide antigens from pathogens multiplying in intracellular vesicles, and those derived from ingested extracellular bacteria and toxins, are carried to the cell surface by MHC class II molecules and presented to CD4 T cells. These can differentiate into two types of effector T cell, called T_H1 and T_H2. Pathogens that accumulate in large numbers inside macrophage and dendritic cell vesicles tend to stimulate the differentiation of T_H1 cells, whereas extracellular antigens tend to stimulate the production of T_H2 cells. T_H1 cells activate the microbicidal properties of macrophages, and induce B cells to make IgG antibodies that are very effective at opsonizing extracellular pathogens for uptake by phagocytic cells. T_H2 cells initiate the humoral immune response by activating naive antigen-specific B cells to produce IgM antibodies. These T_H2 cells can subsequently stimulate the production of different isotypes, including IgA and IgE, as well as neutralizing and/or weakly opsonizing subtypes of IgG. Fig. 8.1 shows the involvement of the different effector T cells in the immune responses to different classes of pathogens.

The activation of naive T cells in response to antigen, and their subsequent proliferation and differentiation, constitutes a **primary immune response**. At the same time as providing armed effector T cells, this response generates

Fig. 8.1 The role of effector T cells in cell-mediated and humoral immune responses to representative pathogens. Cell-mediated immune responses involve the destruction of infected cells by cytotoxic T cells, or the destruction of intracellular pathogens by macrophages activated by T_H1 cells, and are directed principally at intracellular pathogens. However, T_H1 cells can also contribute to humoral immunity by inducing the production of strongly opsonizing antibodies, whereas T_H2 cells initiate the humoral response by activating naive B cells to secrete IgM, and induce the production of other antibody isotypes including weakly opsonizing antibodies such as IgG1 and IgG3 (mouse) and IgG2 and IgG4 (human) as well as IgA and IgE (mouse and human). All types of antibody contribute to humoral immunity, which is directed principally at extracellular pathogens. Note, however, that both cell-mediated and humoral immunity are involved in many infections, such as the response to *Pneumocystis carinii*, which requires antibody for ingestion by phagocytes and macrophage activation for effective destruction of the ingested pathogen.

	Cell-mediated immunity		Humoral immunity
Typical pathogens	Vaccinia virus Influenza virus Rabies virus *Listeria*	*Mycobacterium tuberculosis* *Mycobacterium leprae* *Leishmania donovani* *Pneumocystis carinii*	*Clostridium tetani* *Staphylococcus aureus* *Streptococcus pneumoniae* Polio virus *Pneumocystis carinii*
Location	Cytosol	Macrophage vesicles	Extracellular fluid
Effector T cell	Cytotoxic CD8 T cell	T_H1 cell	T_H2/T_H1 cell
Antigen recognition	Peptide:MHC class I on infected cell	Peptide:MHC class II on infected macrophage	Peptide:MHC class II on antigen-specific B cell
Effector action	Killing of infected cell	Activation of infected macrophages	Activation of specific B cell to make antibody

immunological memory, which gives protection from subsequent challenge by the same pathogen. The generation of memory T cells, long-lived cells that give an accelerated response to antigen, is much less well understood than the generation of effector T cells and will be dealt with in Chapter 10. Memory T cells differ in several ways from naive T cells, but like naive T cells they are quiescent and require activation by antigen-presenting cells with co-stimulatory activity in order to regenerate effector T cells.

Armed effector T cells differ in many ways from their naive precursors, and these changes equip them to respond quickly and efficiently when they encounter specific antigen on target cells. In the last two sections of this chapter we will describe the specialized mechanisms of T cell-mediated cytotoxicity and of macrophage activation by armed effector T cells, the major components of **cell-mediated immunity**. We will leave the activation of B cells by helper T cells until Chapter 9, where the humoral, or antibody-mediated, immune response is discussed.

The production of armed effector T cells.

In order to be activated, a naive T cell must recognize a foreign peptide bound to a self MHC molecule. But this is not, on its own, sufficient for activation. That requires the simultaneous delivery of a **co-stimulatory signal** by a specialized antigen-presenting cell. Only dendritic cells, macrophages, and B cells are able to express both classes of MHC molecule as well as the co-stimulatory cell-surface molecules that drive the clonal expansion of naive T cells and their differentiation into armed effector T cells.

The most potent activators of naive T cells are mature dendritic cells and these are thought to initiate most, perhaps all, T-cell responses *in vivo*. As we will describe in this part of the chapter, immature dendritic cells in the tissues take up antigen at sites of infection and are activated to travel to local lymphoid tissue. Here they mature into cells that express high levels of co-stimulatory molecules and the adhesion molecules that mediate interactions with the naive T cells continually recirculating through these tissues. The activation and clonal expansion of a naive T cell on initial encounter with antigen on the surface of an antigen-presenting cell is often called **priming**, to distinguish it from the responses of armed effector T cells to antigen on their target cells, and the responses of primed memory T cells.

8-1   T-cell responses are initiated in peripheral lymphoid organs by activated antigen-presenting cells.

Adaptive immune responses are not initiated at the site where a pathogen first establishes a focus of infection. They occur in the organized peripheral lymphoid tissues through which naive T cells are continually migrating. Pathogens or their products are transported to lymphoid tissue in the lymph that drains the infected tissue, or, more rarely, by the blood. Pathogens infecting mucosal surfaces accumulate in lymphoid tissues such as the Peyer's patches of the gut or the tonsils; those that enter the blood are trapped in the spleen; while those infecting peripheral sites are trapped in the lymph nodes directly downstream of the site of infection (see Section 1-3). All these lymphoid organs contain cells specialized for capturing antigen and presenting it to T cells. The most important of these are the dendritic cells, which capture antigen at the site of infection and then migrate to the down-stream lymph node.

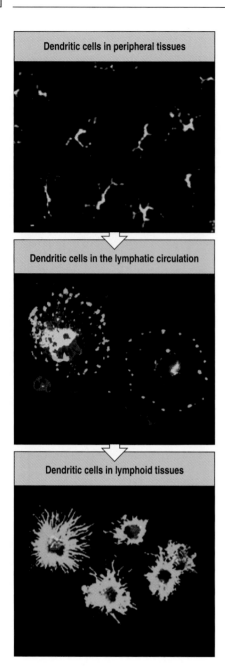

Dendritic cells in peripheral tissues

Dendritic cells in the lymphatic circulation

Dendritic cells in lymphoid tissues

Fig. 8.2 Immature dendritic cells take up antigen in the tissues. Immature dendritic cells in the tissues have a very dendritic morphology, with many long processes, as shown in the top panel, where the tissue is stained for MHC class II molecules in green and for a lysosomal protein in red. The cell bodies of these immature dendritic cells are difficult to distinguish in this figure, but what you can see is that the cell contains many endocytic vesicles that stain both for MHC class II molecules and for the lysosomal protein; when these two colors overlap they give rise to a yellow fluorescence. These immature cells are activated and leave the tissues to migrate through the lymphatics to secondary lymphoid tissues. During this migration their morphology changes, as shown in the middle panel. The dendritic cells also stop phagocytosing antigen, and in this panel you can start to see that the staining for lysosomal proteins in red is distinct from that for MHC class II molecules in green. Finally, in the lymph nodes (bottom panel), they become mature dendritic cells that express high levels of peptide:MHC complexes and co-stimulatory molecules and are very good at stimulating naive CD4 T cells. Here the cells do not phagocytose and again the red staining of lysosomal proteins is quite distinct from the green-stained MHC class II molecules displayed at high density on many dendritic processes. Photographs courtesy of I. Mellman, P. Pierre, and S. Turley.

The delivery of antigen from a site of infection to downstream lymphoid tissue and its subsequent presentation to naive T cells is actively aided by the innate immune response to infection. As discussed in Chapter 2, this is rapidly triggered at the site of infection by nonclonotypic receptors that recognize molecular patterns that are associated with pathogens but not host cells. One of the induced responses of innate immunity is an inflammatory reaction that increases the entry of plasma into the infected tissues and the consequent drainage of tissue fluids into the lymph. Another is the induced maturation of tissue dendritic cells that have been taking up particulate and soluble antigens at the site of infection (Fig. 8.2). These cells are activated through receptors that signal the presence of pathogen components bound by dendritic cell receptors, or by cytokines produced during the inflammatory response. The dendritic cells respond by migrating to the lymph node and expressing the co-stimulatory molecules that are required, in addition to antigen, for the activation of naive T cells. Macrophages, which are phagocytic cells found in the tissues and scattered throughout lymphoid tissue, and B cells, which bind pathogen components, may be similarly induced through nonspecific receptors to express co-stimulatory molecules and act as antigen-presenting cells. Thus the innate immune response to infection hastens the transport of antigens to the local lymphoid tissue, and enables those cells that have taken up antigen to present it effectively to the naive T cells that migrate through this tissue.

The distribution of dendritic cells, macrophages, and B cells in a lymph node is shown in Fig. 8.3. Dendritic cells are present mainly in the T-cell areas. These cells are named after their fingerlike processes, which form a network of branches among the T cells. By the time they arrive in the lymph nodes, dendritic cells have lost their ability to capture new antigen. They are, however, able to present the antigens they ingested at the site of infection and in their mature, activated form they are the most potent antigen-presenting cells for naive T cells.

Macrophages are found in many areas of the lymph node, especially in the marginal sinus, where the afferent lymph enters the lymphoid tissue, and in the medullary cords, where the efferent lymph collects before flowing into the blood. Here they can actively ingest microbes and particulate antigens and so prevent them from entering the blood. As most pathogens are particulate, macrophages in the T-cell areas may stimulate immune responses to many sources of infection.

The production of armed effector T cells

Fig. 8.3 Antigen-presenting cells are distributed differentially in the lymph node. Dendritic cells are found throughout the cortex of the lymph node in the T-cell areas. Macrophages are distributed throughout but are mainly found in the marginal sinus, where the afferent lymph collects before percolating through the lymphoid tissue, and also in the medullary cords, where the efferent lymph collects before passing via the efferent lymphatics into the blood. B cells are found mainly in the follicles. The three types of antigen-presenting cell are thought to be adapted to present different types of pathogen or products of pathogens, but mature dendritic cells are by far the strongest activators of naive T cells.

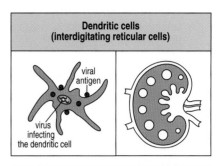

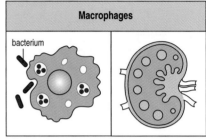

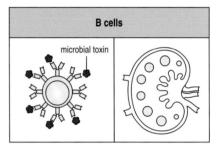

Finally, the B cells, which recirculate through the lymphoid tissues and concentrate in the lymphoid follicles, are particularly efficient at taking up soluble antigens such as bacterial toxins by the specific binding of antigen to the B-cell surface immunoglobulin. The antigen:receptor complex is internalized by receptor-mediated phagocytosis, and degraded fragments of the antigen can return to the B-cell surface complexed with MHC class II molecules. Antigen-specific B cells can thus activate naive CD4 T cells if the B cells are also induced to express co-stimulatory molecules. B cells are, however, very inefficient at initiating adaptive immune responses. This is because only those with the appropriate receptor specificity can internalize and present a particular antigen at high frequency, and these will be very scarce. Thus, the probability of their encountering a naive T cell specific for the same antigen is very low.

The antigen-presenting function of dendritic cells, macrophages, and B cells will be discussed in more detail in Sections 8-5 to 8-7. Only these three cell types express the specialized co-stimulatory molecules required to activate naive T cells; furthermore, all of these cell types express these molecules only when suitably activated in the context of a response to infection. Dendritic cells can take up, process, and present a wide variety of pathogens and antigens and appear to be the most important activators of naive T cells, whereas macrophages and B cells specialize in processing and presenting antigens from ingested pathogens and soluble antigens, respectively, and are also the targets of subsequent actions of armed effector CD4 T cells.

8-2   Naive T cells sample the MHC:peptide complexes on the surface of antigen-presenting cells as they migrate through peripheral lymphoid tissue.

Naive T cells enter lymphoid tissue by crossing the walls of specialized venules known as **high endothelial venules** (HEV). They circulate continuously from the bloodstream to the lymphoid organs and back to the blood, making contact with many thousands of antigen-presenting cells in the lymphoid tissues every day. These contacts allow the sampling of MHC:peptide complexes on the surface of these antigen-presenting cells, which is important for two reasons. One is that it appears to reinforce the process of positive selection for self MHC recognition that occurred during T-cell development. As we discussed in Chapter 7, T-cell receptors are selected for their ability to interact with self MHC:self peptide complexes during T-cell development. In this way, a repertoire of mature T cells is selected that can be activated by nonself peptides bound to the same MHC molecules. Recent experiments show that T-cell survival in the periphery also depends on contact with self MHC:self peptide ligands (see Section 7-32), and that the signals required for survival are delivered effectively through interactions with MHC:peptide complexes on dendritic cells. Thus, as naive T cells migrate through peripheral lymphoid tissue, they receive survival signals through their interactions with dendritic cells. At the same time, the sampling of MHC:peptide ligands ensures that each T cell has a high probability of encountering antigens derived from

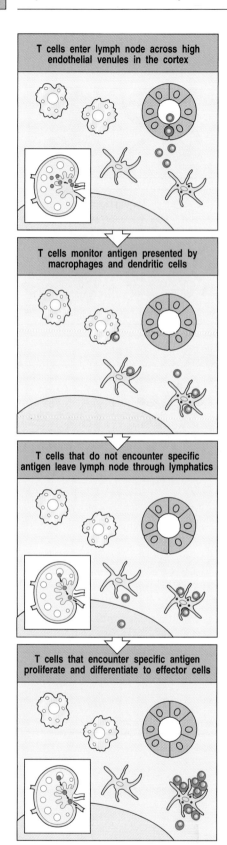

T cells enter lymph node across high endothelial venules in the cortex

T cells monitor antigen presented by macrophages and dendritic cells

T cells that do not encounter specific antigen leave lymph node through lymphatics

T cells that encounter specific antigen proliferate and differentiate to effector cells

Fig. 8.4 Naive T cells encounter antigen during their recirculation through peripheral lymphoid organs. Naive T cells recirculate through peripheral lymphoid organs, such as the lymph node shown here, entering through specialized regions of vascular endothelium called high endothelial venules. On leaving the blood vessel, the T cells enter the deep cortex of the lymph node, where they encounter mature dendritic cells. Those T cells shown in green do not encounter their specific antigen. They receive a survival signal through their interaction with self MHC:self peptide complexes and leave the lymph node through the lymphatics to return to the circulation. T cells shown in blue encounter their specific antigen on the surface of an antigen-presenting cell and are activated to proliferate and to differentiate into armed effector T cells. These antigen-specific armed effector T cells, now increased a hundred-fold to a thousandfold in number, also leave the lymph node via the efferent lymphatics and enter the circulation.

pathogens at any site of infection. This is crucial for the initiation of an adaptive immune response, as only one naive T cell in 10^4–10^6 is likely to be specific for a particular antigen, and adaptive immunity depends on the activation and expansion of such rare antigen-specific T cells (Fig. 8.4). The T cells that do not encounter their antigen eventually reach the medulla of the lymph node, from where they are carried by the efferent lymphatics back to the blood to continue recirculating through other lymphoid organs. Naive T cells that recognize their antigen on the surface of a dendritic cell cease to migrate, and embark on the steps that generate armed effector cells. The generation of effector cells from a naive T cell takes several days. At the end of this period, the armed effector T cells leave the lymphoid organ and reenter the blood-stream to migrate to sites of infection.

8-3 Lymphocyte migration, activation, and effector function depend on cell–cell interactions mediated by cell-adhesion molecules.

The migration of naive T cells through the lymph nodes, and their initial interactions with antigen-presenting cells, depend on cells binding to each other through interactions that are not antigen-specific. Similar interactions eventually guide the effector T cells into the peripheral tissues and play an important part in their interaction with target cells. Binding of T cells to other cells is controlled by an array of adhesion molecules on the surface of the T lymphocyte that recognize a complementary array of adhesion molecules on the surface of the interacting cell. The main classes of adhesion molecule involved in lymphocyte interactions are the selectins, the integrins, members of the immunoglobulin superfamily, and some mucinlike molecules. We have already encountered members of the first three classes in the recruitment of neutrophils and monocytes to sites of infection during an innate immune response (see Section 2-22). Most adhesion molecules play fairly broad roles in the generation of immune responses. Many that are involved in lymphocyte migration and the interactions of armed effector T cells with their targets are also involved in interactions between other leukocytes. Adhesion molecules are important in getting lymphocytes together in adaptive immune responses that involve T-cell–B-cell interactions, and we will describe these in Chapter 10, where we present an integrated view of the immune response.

The selectins (Fig. 8.5) are particularly important for leukocyte homing to particular tissues, and can be expressed either on leukocytes (**L-selectin**, CD62L) or on vascular endothelium (**P-selectin**, CD62P, and **E-selectin**, CD62E). L-Selectin is expressed on naive T cells and guides their exit from the blood into peripheral lymphoid tissues. P-Selectin and E-selectin are expressed on the vascular endothelium at sites of infection and serve to recruit effector cells into the tissues at these sites (see Sections 2-21 and 2-22).

Selectins are cell-surface molecules with a common core structure, distinguished from each other by the presence of different lectinlike domains in their extracellular portion (see Fig. 2.34). The lectin domains bind to particular sugar groups, and each selectin binds to a cell-surface carbohydrate. L-Selectin binds to the carbohydrate moiety, sulfated sialyl-Lewis[x], of mucinlike molecules called **vascular addressins**, which are expressed on the surface of vascular endothelial cells. Two of these addressins, **CD34** and **GlyCAM-1**, are expressed as sulfated sialyl-Lewis[x] molecules on high endothelial venules in lymph nodes. A third, **MAdCAM-1**, is expressed on endothelium in mucosa, and guides lymphocyte entry into mucosal lymphoid tissue such as that of the gut.

The interaction between L-selectin and the vascular addressins is responsible for the specific homing of naive T cells to lymphoid organs but does not, on its own, enable the cell to cross the endothelial barrier into the lymphoid tissue. For this, proteins from two other families—the integrins and the immunoglobulin superfamily—are required. These proteins also play a critical part in the subsequent interactions of lymphocytes with antigen-presenting cells and later with their target cells.

The integrins comprise a large family of cell-surface proteins that mediate adhesion between cells, and between cells and the extracellular matrix, in normal development as well as in immune and inflammatory responses. Integrins bind tightly to their ligands after receiving signals that induce a change in conformation. For example, as we saw in Chapter 2, signaling by chemokines activates integrins on leukocytes to bind tightly to the vascular surface during migration of leukocytes into sites of inflammation. Chemokines similarly activate T-cell integrins during the migration of T lymphocytes into lymphoid organs and in the migration of activated T lymphocytes to sites of infection.

The migration of naive T cells into lymphoid tissues is mediated by the chemokine **SLC** (**secondary lymphoid tissue chemokine**). This is expressed by the high vascular endothelium, stromal cells, and dendritic cells in lymphoid tissue, and binds to the CCR7 chemokine receptor on naive T cells. This interaction, by a mechanism as yet unknown, is able to increase the affinity of integrin binding, arresting the T cell's progress through the blood and enabling it to enter the lymphoid tissue. Similar interactions with chemokines produced at sites of inflammation direct effector T-cell migration into the tissues; this will be discussed in more detail when we describe the functions of effector T cells in Chapter 10. Chemokines are not the only molecules able to signal the upregulation of integrin affinity; later in this chapter we will see how signaling through the T-cell receptor also triggers T-cell integrins to adhere tightly to their ligands on the antigen-presenting cell.

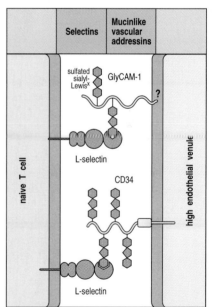

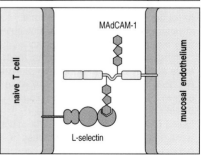

Fig. 8.5 L-Selectin and the mucinlike vascular addressins direct naive lymphocyte homing to lymphoid tissues. L-Selectin is expressed on naive T cells, which bind to sulfated sialyl-Lewis[x] moieties on the vascular addressins CD34 and GlyCAM-1 on high endothelial venules in order to enter lymph nodes. The relative importance of CD34 and GlyCAM-1 in this interaction is unclear. GlyCAM-1 is expressed exclusively on high endothelial venules but has no transmembrane region and it is unclear how it is attached to the membrane; CD34 has a transmembrane anchor and is expressed in appropriately glycosylated form only on high endothelial venule cells, although it is found in other forms on other endothelial cells. The addressin MAdCAM-1 is expressed on mucosal endothelium and guides entry into mucosal lymphoid tissue. The icon shown represents mouse MadCAM-1, which contains an IgA-like domain closest to the cell membrane; human MadCAM-1 has an elongated mucinlike domain and lacks the IgA-like domain. L-Selectin recognizes the carbohydrate moieties on the vascular addressins.

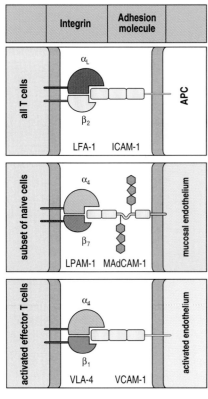

Leukocyte
Adhesion
Deficiency

Fig. 8.6 Integrins are important in T-lymphocyte adhesion. Integrins are heterodimeric proteins containing a β chain, which defines the class of integrin, and an α chain, which defines the different integrins within a class. The α chain is larger than the β chain and contains binding sites for divalent cations that may be important in signaling. LFA-1 is a $β_2$ integrin which is expressed on all T cells and indeed on all leukocytes. It binds ICAMs and is important in the adhesive interactions that mediate cell migration and in the interactions of T cells with antigen-presenting or target cells; the level of its expression is increased on armed effector T cells. Lymphocyte Peyer's patch adhesion molecule, LPAM-1 (integrin $α_4{:}β_7$) is expressed by a subset of naive T cells and contributes to mucosal homing by supporting adhesion through interactions with MAdCAM-1. VLA-4 is a $β_1$ integrin which is upregulated following T-cell activation. It binds to VCAM-1 on activated endothelium and, as we will discuss further in Chapter 10, is important for recruiting armed effector T cells into sites of infection.

The integrins were introduced in Chapter 2, so we will just review their most important characteristics here. An integrin molecule consists of a large α chain that pairs noncovalently with a smaller β chain. There are several subfamilies of integrins, broadly defined by their common β chains. We will be concerned chiefly with the **leukocyte integrins**, which have a common $β_2$ chain with distinct α chains (Fig. 8.6). All T cells express a $β_2$ integrin known as **lymphocyte function-associated antigen-1 (LFA-1)**. This leukocyte integrin is also found on macrophages and neutrophils, and is involved in their recruitment to sites of infection (see Sections 2-21 and 2-22). LFA-1 plays a similar role in the migration of both naive and effector T cells out of the blood. In addition, it is thought to be the most important adhesion molecule for T-lymphocyte activation, because antibodies to LFA-1 effectively inhibit the activation of both naive and armed effector T cells

Surprisingly, T-cell responses can be normal in patients lacking the $β_2$ integrin chain and hence all integrins that contain $β_2$, such as LFA-1. This is probably because T cells also express other adhesion molecules, including CD2 and $β_1$ integrins, which may be able to compensate for the absence of LFA-1. Expression of the $β_1$ integrins increases significantly at a late stage in T-cell activation, and they are thus often called **VLAs**, for **very late activation antigens**; we will see in Chapter 10 that they play an important part in directing armed effector T cells to their inflamed target tissues.

Many cell-surface adhesion molecules are members of the immunoglobulin superfamily, which also includes the antigen receptors of T and B cells, the co-receptors CD4, CD8, and CD19, and the invariant domains of MHC molecules. At least five adhesion molecules of the immunoglobulin superfamily are especially important in T-cell activation (Fig. 8.7). Three very similar **intercellular adhesion molecules (ICAMs)—ICAM-1, ICAM-2, and ICAM-3—** all bind to the T-cell integrin LFA-1. ICAM-1 and ICAM-2 are expressed on endothelium as well as on antigen-presenting cells; binding to these molecules enables lymphocytes to migrate through blood vessel walls. ICAM-3 is

Fig. 8.7 Adhesion molecules involved in leukocyte interactions. Several structural families of adhesion molecules play a part in lymphocyte migration, homing, and cell–cell interactions; most have already been introduced in Fig. 2.34. One new member, described only in 2000, is the ICAM-3-binding protein made by dendritic cells and called DC-SIGN. It is suspected to have a major role in interactions between dendritic cells and T cells.

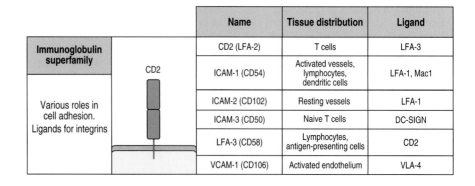

	Name	Tissue distribution	Ligand
Immunoglobulin superfamily CD2 Various roles in cell adhesion. Ligands for integrins	CD2 (LFA-2)	T cells	LFA-3
	ICAM-1 (CD54)	Activated vessels, lymphocytes, dendritic cells	LFA-1, Mac1
	ICAM-2 (CD102)	Resting vessels	LFA-1
	ICAM-3 (CD50)	Naive T cells	DC-SIGN
	LFA-3 (CD58)	Lymphocytes, antigen-presenting cells	CD2
	VCAM-1 (CD106)	Activated endothelium	VLA-4

expressed only on leukocytes and is thought to play an important part in adhesion between T cells and antigen-presenting cells, particularly dendritic cells. In addition to binding LFA-1, ICAM-3 binds with high affinity to a recently discovered lectin called DC-SIGN, which is found only on dendritic cells. Another interaction involving immunoglobulin superfamily molecules is mediated by LFA-3 on the antigen-presenting cell binding to CD2 on the T cell; this interaction synergizes with that between LFA-1 and ICAM-1 and ICAM-2.

8-4 The initial interaction of T cells with antigen-presenting cells is mediated by cell-adhesion molecules.

As they migrate through the cortical region of the lymph node, naive T cells bind transiently to each antigen-presenting cell they encounter. Antigen-presenting cells, and dendritic cells in particular, bind naive T cells very efficiently through interactions between LFA-1, CD2, and ICAM-3 on the T cell, and ICAM-1, ICAM-2, LFA-3, and DC-SIGN on the antigen-presenting cell (Fig. 8.8). The binding of ICAM-3 to DC-SIGN is unique to the interaction between dendritic cells and T cells, while the other molecules synergize in the binding of lymphocytes to all three types of antigen-presenting cell. Perhaps because of this synergy, the precise role of each adhesion molecule has been difficult to distinguish. People lacking LFA-1 can have normal T-cell responses, and this also seems to be the case for genetically engineered mice lacking CD2. It would not be surprising if there were enough redundancy in the molecules mediating T-cell adhesive interactions to enable immune responses to occur in the absence of any one of them; such molecular redundancy has been observed in other complex biological processes.

The transient binding of naive T cells to antigen-presenting cells is crucial in providing time for T cells to sample large numbers of MHC molecules on each antigen-presenting cell for the presence of specific peptide. In those rare cases in which a naive T cell recognizes its peptide:MHC ligand, signaling through the T-cell receptor induces a conformational change in LFA-1, which greatly increases its affinity for ICAM-1 and ICAM-2. This conformational change is the same as that induced by signaling through chemokine receptors during the migration of leukocytes to sites of infection (see Section 2-20), although its mechanism is not known. The change in LFA-1 stabilizes the association between the antigen-specific T cell and the antigen-presenting cell (Fig. 8.9). The association can persist for several days, during which time the naive T cell proliferates and its progeny, which also adhere to the antigen-presenting cell, differentiate into armed effector T cells.

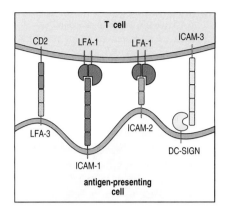

Fig. 8.8 Cell-surface molecules of the immunoglobulin superfamily are important in the interactions of lymphocytes with antigen-presenting cells. In the initial encounter of T cells with antigen-presenting cells, CD2 binding to LFA-3 on the antigen-presenting cell synergizes with LFA-1 binding to ICAM-1 and ICAM-2. One interaction that appears to be exclusive to the interaction of naive T cells with dendritic cells is that between ICAM-3 on the naive T cell and a recently identified molecule specific to dendritic cells and known as DC-SIGN. DC-SIGN is a C-type lectin that binds ICAM-3 with high affinity. LFA-1 is the $\alpha_L:\beta_2$ integrin heterodimer CD11a:CD18. LFA-3 is also known as CD58, and ICAM-1, -2, and -3 are CD54, CD102, and CD50, respectively.

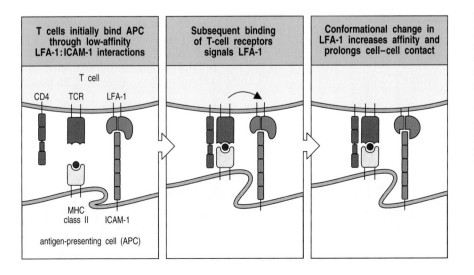

Fig. 8.9 Transient adhesive interactions between T cells and antigen-presenting cells are stabilized by specific antigen recognition. When a T cell binds to its specific ligand on an antigen-presenting cell, intracellular signaling through the T-cell receptor (TCR) induces a conformational change in LFA-1 that causes it to bind with higher affinity to ICAMs on the antigen-presenting cell. The T cell shown here is a CD4 T cell.

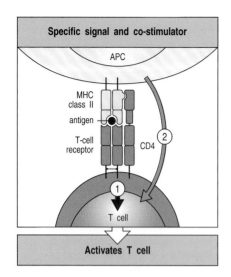

Fig. 8.10 Activation of naive T cells requires two independent signals. Binding of the peptide:MHC complex by the T-cell receptor and, in this example, the CD4 co-receptor, transmits a signal (arrow 1) to the T cell that antigen has been encountered. Activation of naive T cells requires a second signal (arrow 2), the co-stimulatory signal, to be delivered by the same antigen-presenting cell.

Most T-cell encounters with antigen-presenting cells do not, however, result in recognition of an antigen. In these encounters, the T cells must be able to separate efficiently from the antigen-presenting cells so that they can continue to migrate through the lymph node, eventually leaving via the efferent lymphatic vessels to reenter the blood and continue circulating. Dissociation, like stable binding, may also involve signaling between the T cell and the antigen-presenting cells, but little is known of its mechanism.

8-5 Both specific ligand and co-stimulatory signals provided by an antigen-presenting cell are required for the clonal expansion of naive T cells.

We saw in Chapter 3 that armed effector T cells are triggered when their antigen-specific receptors and either the CD4 or CD8 co-receptors bind to peptide:MHC complexes. By contrast, ligation of the T-cell receptor and co-receptor does not, on its own, stimulate naive T cells to proliferate and differentiate into armed effector T cells. The antigen-specific clonal expansion of naive T cells requires a second, or co-stimulatory, signal (Fig. 8.10), which must be delivered by the same antigen-presenting cell on which the T cell recognizes its antigen. CD8 T cells appear to require a stronger co-stimulatory signal than CD4 cells and, as we will see later, their clonal expansion is aided by CD4 cells interacting with the same antigen-presenting cell.

The best-characterized co-stimulatory molecules are the structurally related glycoproteins **B7.1** (CD80) and **B7.2** (CD86). We will call them the **B7 molecules** from here on, as functional differences between the two have yet to be defined. The B7 molecules are homodimeric members of the immunoglobulin superfamily that are found exclusively on the surfaces of cells that can stimulate T-cell proliferation. Their role in co-stimulation has been demonstrated by transfecting fibroblasts that express a T-cell ligand with genes encoding B7 molecules and showing that the fibroblasts could then stimulate the clonal expansion of naive T cells. The receptor for B7 molecules on the T cell is **CD28**, yet another member of the immunoglobulin superfamily (Fig. 8.11). Ligation of CD28 by B7 molecules or by anti-CD28 antibodies co-stimulates the clonal expansion of naive T cells, whereas anti-B7 antibodies, which inhibit the binding of B7 molecules to CD28, inhibit T-cell responses. Although other molecules have been reported to co-stimulate naive T cells, so far only the B7 molecules have been shown definitively to provide co-stimulatory signals for naive T cells in normal immune responses.

Once a naive T cell is activated, however, it expresses a number of proteins that contribute to sustaining or modifying the co-stimulatory signal that drives clonal expansion and differentiation. One such protein is **CD40 ligand**, so-called because it binds to **CD40** on antigen-presenting cells. Binding of CD40 ligand by CD40 transmits activating signals to the T cell and also activates the antigen-presenting cell to express B7 molecules, thus stimulating further T-cell proliferation. CD40 and CD40 ligand belong to the TNF family

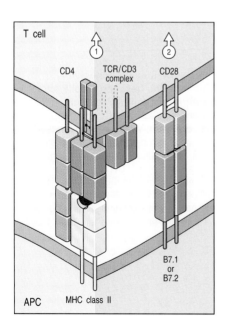

Fig. 8.11 The principal co-stimulatory molecules expressed on antigen-presenting cells are B7 molecules, which bind the T-cell protein CD28. Binding of the T-cell receptor (TCR) and its co-receptor CD4 to the peptide:MHC class II complex on the antigen-presenting cell (APC) delivers a signal (arrow 1) that can induce the clonal expansion of T cells only when the co-stimulatory signal (arrow 2) is given by binding of CD28 to B7 molecules. Both CD28 and B7 molecules are members of the immunoglobulin superfamily. B7.1 (CD80) and B7.2 (CD86) are homo-dimers, each of whose chains has one immunoglobulin V-like domain and one C-like domain. CD28 is a disulfide-linked homodimer in which each chain has one V-like domain.

of receptors and ligands and, as we will describe later in this chapter, have a central role in the effector function of fully differentiated T cells. Their earlier role in sustaining the development of a T-cell response is demonstrated by mice lacking CD40 ligand; when these mice are immunized, the clonal expansion of responding T cells is curtailed at an early stage. Another pair of TNF family molecules that appear to contribute to co-stimulation of T cells are the T-cell molecule **4-1BB** (CD137) and its ligand **4-1BBL**, which is expressed on activated dendritic cells, macrophages, and B cells. As with CD40L and CD40, the effects of this receptor–ligand interaction are bidirectional, with both the T cell and the antigen-presenting cell receiving activating signals; this process is sometimes referred to as the T-cell/antigen-presenting cell dialogue.

CD28-related proteins are also induced on activated T cells and serve to modify the co-stimulatory signal as the T-cell response develops. One is **CTLA-4** (CD152), an additional receptor for B7 molecules. CTLA-4 closely resembles CD28 in sequence, and the two proteins are encoded by closely linked genes. However, CTLA-4 binds B7 molecules about 20 times more avidly than does CD28 and delivers an inhibitory signal to the activated T cell (Fig. 8.12). This makes the activated progeny of a naive T cell less sensitive to stimulation by the antigen-presenting cell and limits the amount of an autocrine T-cell growth factor, interleukin-2 (IL-2), that is produced. Thus, binding of CTLA-4 to B7 molecules is essential for limiting the proliferative response of activated T cells to antigen and B7. This was confirmed by producing mice with a disrupted CTLA-4 gene; such mice develop a fatal disorder characterized by massive lymphocyte proliferation.

A third CD28-related protein is induced on activated T cells and can enhance T-cell responses; this inducible co-stimulator, or **ICOS**, binds a ligand known as **LICOS**, the ligand of ICOS, which is distinct from B7.1 and B7.2. LICOS is produced on activated dendritic cells, monocytes, and B cells, but its contribution to immune responses has not yet been clearly defined. Although it resembles CD28 in driving T-cell growth, it differs from CD28 in not inducing IL-2; instead it induces IL-10 (see Fig. 8.32).

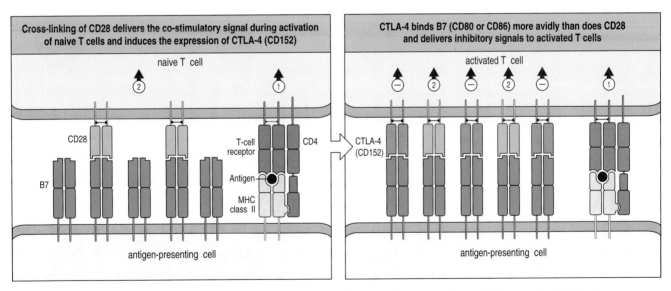

Fig. 8.12 T-cell activation through the T-cell receptor and CD28 leads to the increased expression of CTLA-4, an inhibitory receptor for B7 molecules. Naive T cells express CD28, which delivers a co-stimulatory signal on binding B7 molecules (left panel), thereby driving the activation and expansion of T cells that encounter specific antigen presented by an antigen-presenting cell. Once activated, T cells express increased levels of CTLA-4 (CD152; right panel). CTLA-4 has a higher affinity for B7 molecules than does CD28 and thus binds most or all of the B7 molecules, effectively shutting down the proliferative phase of the response.

Thus antigen-presenting cells engage in a co-stimulatory dialogue with T cells that recognize the antigens they display. This dialogue involves the delivery and receipt of signals through a number of different molecules, but appears to be initiated through the binding of B7 molecules to CD28 on a naive T cell. Antigen-presenting cells are activated to express B7 molecules on detecting the presence of infection through receptors of the innate immune system. The requirement for the simultaneous delivery of antigen-specific and co-stimulatory signals by one cell in the activation of naive T cells means that only such activated antigen-presenting cells, principally the dendritic cells that migrate into lymphoid tissue after being activated by binding and ingesting pathogens, can initiate T-cell responses. This is important, because not all potentially self-reactive T cells are deleted in the thymus; peptides derived from proteins made only in specialized cells in peripheral tissues might not be encountered during negative selection of thymocytes. Self-tolerance could be broken if naive autoreactive T cells could recognize self antigens on tissue cells and then be co-stimulated by an antigen-presenting cell, either locally or at a distant site. Thus, the requirement that the same cell presents both the specific antigen and the co-stimulatory signal is important in preventing destructive immune responses to self tissues. Indeed, antigen binding to the T-cell receptor in the absence of co-stimulation not only fails to activate the cell, it instead leads to a state called **anergy**, in which the T cell becomes refractory to activation by specific antigen even when the antigen is subsequently presented to it by a professional antigen-presenting cell (Fig. 8.13).

Fig. 8.13 The requirement for one cell to deliver both the antigen-specific signal and the co-stimulatory signal is crucial in preventing immune responses to self antigens. In the upper panels, a T cell recognizes a viral peptide on the surface of an antigen-presenting cell and is activated to proliferate and differentiate into an effector cell capable of eliminating any virus-infected cell. However, naive T cells that recognize antigen on cells that cannot provide co-stimulation become anergic, as when a T cell recognizes a self antigen expressed by an uninfected epithelial cell (lower panels). This T cell does not differentiate into an armed effector cell, and cannot be stimulated further by an antigen-presenting cell presenting that antigen.

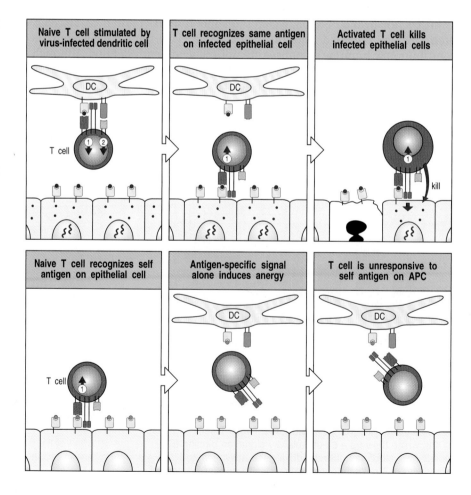

Now that we have discussed the molecular interactions that allow naive T cells to adhere transiently to antigen-presenting cells and scan their MHC:peptide complexes, and also the adhesion and co-stimulatory molecules that contribute to T-cell activation once a specific antigen is encountered, we will look more closely at the properties of the three types of antigen-presenting cell. Dendritic cells, macrophages, and B cells differ in their selectivity of antigen uptake, their antigen-processing properties, and their co-stimulatory and migratory behavior, and thus have distinctive functions in initiating T-cell responses.

8-6   Dendritic cells specialize in taking up antigen and activating naive T cells.

The only known function of dendritic cells is to present antigen to T cells, and the mature dendritic cells found in lymphoid tissues are by far the most potent stimulators of naive T cells. This ability is not shared, however, by the immature dendritic cells found under most surface epithelia and in most solid organs such as the heart and kidneys. Dendritic cells arise from myeloid progenitors within the bone marrow, and emerge from the bone marrow to migrate in the blood to peripheral tissues. In these tissues, they have an immature phenotype that is associated with low levels of MHC proteins, and they lack co-stimulatory B7 molecules (Fig 8.14, top panel). They are not yet equipped to stimulate naive T cells. However, they share with their close relatives the macrophages, the ability to recognize and ingest pathogens through receptors that recognize features common to microbial surfaces, and they are very active in taking up antigens by phagocytosis using receptors such as DEC 205. Other extracellular antigens are taken up nonspecifically by **macropinocytosis**, in which large volumes of surrounding fluid are engulfed.

Typical of immature dendritic cells are the **Langerhans' cells** of the skin. These are actively phagocytic and contain large granules, known as Birbeck granules, which may be a type of phagosome. An infection triggers the migration of Langerhans' cells to the regional lymph nodes (Figs 8.15 and 8.2). Here, they rapidly lose the ability to take up and process antigen, but synthesize new MHC molecules that present peptides of pathogens at a high level. On arriving in the regional lymph node, they also express B7 molecules, which can co-stimulate naive T cells, and also large numbers of adhesion molecules, which enable them to interact with antigen-specific T cells. In this way the Langerhans' cells capture antigens from invading pathogens and differentiate into mature dendritic cells that are uniquely fitted for presenting these antigens and activating naive T cells.

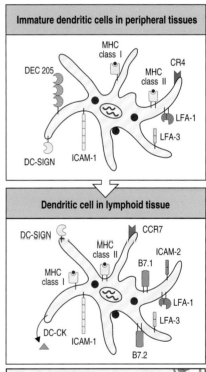

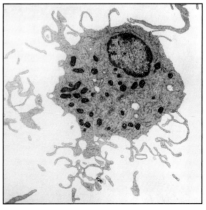

Fig. 8.14 Dendritic cells mature through at least two definable stages to become potent antigen-presenting cells in lymphoid tissue. Dendritic cells arise from bone marrow progenitors and migrate via the blood to peripheral tissues and organs, where they are highly phagocytic via receptors such as DEC 205 and are actively macro-pinocytic but do not express co-stimulatory molecules (top panel). At sites of infection they pick up antigen and are induced to migrate via the afferent lymphatic vessels to the regional lymph node (see Fig. 8.15). Here they exhibit high levels of T-cell-activating potential but are no longer phagocytic. Dendritic cells in lymphoid tissue express B7.1, B7.2, and high levels of MHC class I and class II molecules, as well as high levels of the adhesion molecules ICAM-1, ICAM-2, LFA-1, and LFA-3 (center panel). They also express high levels of the dendritic-cell-specific adhesion molecule DC-SIGN, which binds ICAM-3 with high affinity. The photograph shows a mature dendritic cell. Photograph courtesy of J. Barker.

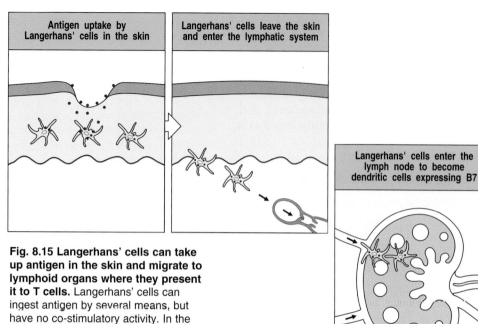

Fig. 8.15 Langerhans' cells can take up antigen in the skin and migrate to lymphoid organs where they present it to T cells. Langerhans' cells can ingest antigen by several means, but have no co-stimulatory activity. In the presence of infection, they take up antigen locally in the skin and then migrate to the lymph nodes. There they differentiate into dendritic cells that can no longer ingest antigen but now have co-stimulatory activity.

Immature dendritic cells persist in the peripheral tissues for variable lengths of time. When an infection occurs, they are stimulated to migrate via the lymphatics to the local lymphoid tissues, where they have a completely different phenotype. The dendritic cells in lymphoid tissue are no longer able to engulf antigens by phagocytosis or by macropinocytosis. However, they now express very high levels of long-lived MHC class I and MHC class II molecules; this enables them to stably present peptides from proteins acquired from the infecting pathogens. They also express very high levels of adhesion molecules, including DC-SIGN, as well as high levels of B7 molecules (Fig. 8.14, center panel). They also secrete a chemokine that specifically attracts naive T cells; this chemokine, called DC-CK, is expressed only in dendritic cells in lymphoid tissues. These properties help to explain dendritic cells' ability to stimulate strong naive T-cell responses.

Although activated mature dendritic cells will also present some self peptides, the T-cell receptor repertoire has been purged in the thymus of receptors that recognize self peptides presented by dendritic cells (see Chapter 7), and thus T-cell responses against ubiquitous self antigens are avoided. In addition, tissue dendritic cells reaching the end of their life-span without having been activated by infection also travel via the lymphatics to local lymphoid tissue. Because they do not express the appropriate co-stimulatory molecules, these cells induce tolerance to any self antigens derived from peripheral tissues that they display.

The signals that activate tissue dendritic cells to migrate and mature after taking up antigen are clearly of key importance in determining whether an adaptive immune response will be initiated. These signals can be generated through direct interactions with pathogens or by cytokine stimulation, but in both cases they are thought to be a consequence of the recognition of invading pathogens by nonclonotypic receptors of the innate immune system. The best-understood example is the response to gram-negative bacteria, whose cell walls contain lipopolysaccharide (LPS). Receptors that recognize LPS are found on dendritic cells and macrophages, and these associate with the Toll-like signaling receptor TLR-4, which then activates the transcription factor NFκB (see Sections 2-17 and 6-15). Signaling through this pathway induces the expression of B7 molecules, and of cytokines such as TNF-α, which stimulate the migration of tissue dendritic cells. Thus an immature tissue dendritic cell that binds and internalizes a gram-negative bacterium is induced to migrate to local lymphoid tissue and present bacterium-derived peptide antigens to naive T cells. Other members of the TLR family are expressed on tissue dendritic cells, and are thought to be involved in detecting and signaling the presence of other classes of pathogen. Other types of receptor that can bind pathogens, such as receptors for complement, or phagocytic receptors such as the mannose receptor, are also expressed on dendritic cells and may contribute to their activation.

Pathogens that have evolved to escape recognition by phagocytic receptors are taken up by tissue dendritic cells through the process of macropinocytosis, and can then be presented to T cells. This is thought to occur after intracellular degradation of the pathogen to reveal components that trigger activation of the dendritic cell. Bacterial DNA containing unmethylated CpG dinucleotide motifs induces the rapid activation of dendritic cells. This probably occurs after recognition of the DNA by an intracellular receptor called TLR-9. Exposure to bacterial DNA activates NFκB and mitogen-activated protein kinase (MAP kinase) signaling pathways, leading to the production of cytokines such as IL-6, IL-12, IL-18, and interferon (IFN)-α and IFN-γ. In turn, these induce and augment the expression of co-stimulatory molecules. Bacterial heat-shock proteins are another internal bacterial constituent that can activate the antigen-presenting function of dendritic cells. Some viruses are thought to be recognized inside the dendritic cell, as a consequence of the production of double-stranded RNA in the course of their replication. As discussed in Section 2-25, viral infection also induces the production of IFN-α by infected cells. IFN-α is one of the cytokines that can activate dendritic cells to express co-stimulatory molecules.

Dendritic cells are likely to be particularly important in stimulating T-cell responses to viruses, which fail to induce co-stimulatory activity in other types of antigen-presenting cell. Viruses may infect dendritic cells by binding to any of several molecules on the cell surface, or after being engulfed but not destroyed by immature dendritic cells. Such viruses synthesize their proteins using the dendritic cell's own protein synthesis machinery, leading to surface expression of viral peptides by MHC class I molecules just as in other types of infected cell. Viral peptides will also be presented on both MHC class I and MHC class II molecules as a result of uptake of viral particles by phagocytic receptors such as the mannose receptor, which can recognize many viruses, or through macropinocytosis. The mechanism by which peptides generated by degradation of viral proteins in the endosomal pathway can be presented by MHC class I molecules is not known, nor, in fact, whether there is only one such mechanism. Nevertheless, it is clear that extracellular proteins taken up by dendritic cells can give rise to peptides presented by MHC class I molecules. In this way, viruses that are not able to infect dendritic cells are still able to stimulate effective immune responses. Thus, any virus-infected cell is able to activate naive CD8 T cells, generating cytotoxic CD8 effector T cells that can kill infected cells, and also to activate CD4 T cells that can stimulate the production of antibodies.

Dendritic cells are believed to present antigens from fungal as well as viral and bacterial pathogens. Indeed, they are thought to initiate immune responses to a wide range of pathogens, and to be able to distinguish between different classes of pathogen. This is reflected in the synthesis of different effector molecules by the activated dendritic cells, which in turn influence the differentiation of the responding T cells into different subclasses, which is discussed further in Section 10-5. In addition to pathogen-associated antigens, dendritic cells are thought to present protein antigens from environmental sources that trigger allergic reactions upon inhalation (see Chapter 12), and alloantigens deriving from a transplanted organ, which form the basis for graft rejection (see Chapter 13). In principle, any nonself antigen will be immunogenic if it is taken up and presented by a dendritic cell that is activated to migrate to nearby lymphoid tissues and mature. The normal physiology of dendritic cells is to migrate, and this is increased by stimuli that activate the linings of the lymphatics, like transplantation, which is why dendritic cells are so potent at stimulating allograft reactions.

8-7 Macrophages are scavenger cells that can be induced by pathogens to present foreign antigens to naive T cells.

As we learned in Chapter 2, many of the microorganisms that enter the body are engulfed and destroyed by phagocytes, which provide an innate, antigen-nonspecific first line of defense against infection. Microorganisms that are destroyed by phagocytes without additional help from T cells do not cause disease and do not require an adaptive immune response. Pathogens, by definition, have developed mechanisms to avoid elimination by innate immunity, and the targeting and removal of such pathogens is the function of the adaptive immune response. Mononuclear phagocytes or macrophages that have bound and ingested microorganisms but have failed to destroy them, contribute to the adaptive immune response by acting as antigen-presenting cells. As we will see later in this chapter and in Chapter 10, the adaptive immune response is in turn able to stimulate the microbicidal and phagocytic capacities of these cells.

Resting macrophages have few or no MHC class II molecules on their surface, and do not express B7 molecules. The expression of both MHC class II and B7 molecules is induced by the ingestion of microorganisms and recognition of their foreign molecular patterns. Macrophages, like tissue dendritic cells, have a variety of receptors that recognize microbial surface components, including the mannose receptor, the scavenger receptor, complement receptors, and several Toll-like receptors (see Chapter 2). These receptors function in the innate immune defense mediated by macrophages; they are involved in the uptake of microorganisms by phagocytosis and in signaling for the secretion of pro-inflammatory cytokines that recruit and activate more phagocytes. In addition, they can play the same role as tissue dendritic cells, and allow the macrophage to function as an antigen-presenting cell. Once bound, microorganisms are engulfed and degraded in the endosomes and lysosomes, generating peptides that can be presented by MHC class II molecules. At the same time, the receptors recognizing these microorganisms transmit a signal that leads to expression of MHC class II molecules and B7 molecules.

Thus the induction of co-stimulatory activity by common microbial constituents occurs in both dendritic cells and macrophages. This is believed to allow the immune system to distinguish antigens borne by infectious agents from antigens associated with innocuous proteins, including self proteins. Indeed, many foreign proteins do not induce an immune response when injected on their own, presumably because they fail to induce co-stimulatory activity in antigen-presenting cells. When such protein antigens are mixed with bacteria, however, they become immunogenic, because the bacteria induce the essential co-stimulatory activity in cells that ingest the protein (Fig. 8.16). Bacteria used in this way are known as **adjuvants** (see Appendix I, Section A-4). We will see in Chapter 13 how self tissue proteins

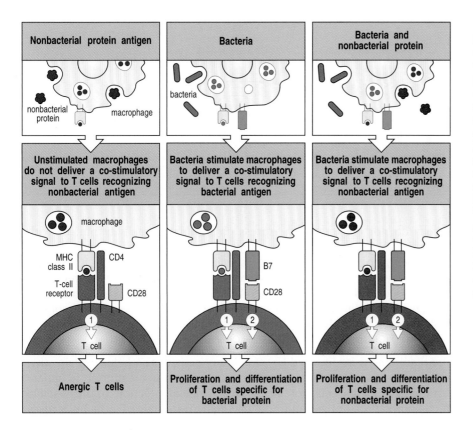

Fig. 8.16 Microbial substances can induce co-stimulatory activity in macrophages. If protein antigens are taken up and presented by macrophages in the absence of bacterial components that induce co-stimulatory activity in the macrophage, T cells specific for the antigen will become anergic (refractory to activation). Many bacteria induce the expression of co-stimulators by antigen-presenting cells, and macrophages presenting peptide antigens derived by degradation of such bacteria can activate naive T cells. When bacteria are mixed with protein antigens, the protein antigens are rendered immunogenic because the bacteria induce co-stimulatory B7 molecules in the antigen-presenting cells. Such added bacteria act as adjuvants (see Appendix I, Section A-4).

mixed with bacterial adjuvants can induce autoimmune diseases, illustrating the crucial importance of the regulation of co-stimulatory activity in discrimination of self from nonself.

As macrophages continuously scavenge dead or dying cells, which are rich sources of self antigens, it is particularly important that they should not activate T cells in the absence of microbial infection. The Kupffer cells of the liver sinusoids and the macrophages of the splenic red pulp, in particular, remove large numbers of dying cells from the blood daily. Kupffer cells express little MHC class II and no TLR-4, the Toll-like receptor on human cells that signals the presence of LPS. Thus, although they generate large amounts of self peptides in their endosomes and lysosomes, these macrophages are not likely to elicit an autoimmune response.

8-8   B cells are highly efficient at presenting antigens that bind to their surface immunoglobulin.

Macrophages cannot take up soluble antigens efficiently, whereas immature dendritic cells can take up large amounts of antigen from extracellular fluid by macropinocytosis. B cells, by contrast, are uniquely adapted to bind specific soluble molecules through their cell-surface immunoglobulin. B cells internalize the antigens bound by their surface immunoglobulin receptors and then display peptide fragments of antigen as peptide:MHC class II complexes. Because this mechanism of antigen uptake is highly efficient, and B cells constitutively express high levels of MHC class II molecules, high levels of specific peptide:self MHC class II complexes are generated at the B-cell surface (Fig. 8.17). This pathway of antigen presentation allows B cells to be targeted by antigen-specific CD4 T cells, which drive their differentiation, as we will see in Chapter 9. In circumstances in which the presenting B cells are induced to express co-stimulatory activity, it also allows B cells to activate naive T cells.

Fig. 8.17 B cells can use their immunoglobulin receptor to present specific antigen very efficiently to T cells. Surface immunoglobulin allows B cells to bind and internalize specific antigen very efficiently. The internalized antigen is processed in intracellular vesicles where it binds to MHC class II molecules. These vesicles are then transported to the cell surface where the MHC class II:antigen complex can be recognized by T cells. When the protein antigen is not recognized specifically by the B cell, its internalization is inefficient and only a low density of fragments of such proteins are subsequently presented at the B-cell surface (not shown).

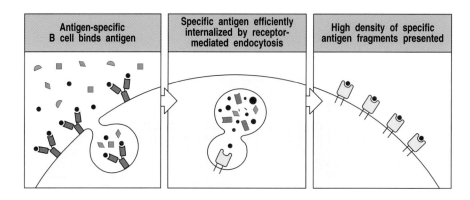

| Antigen-specific B cell binds antigen | Specific antigen efficiently internalized by receptor-mediated endocytosis | High density of specific antigen fragments presented |

B cells do not constitutively express co-stimulatory activity but, as with dendritic cells and macrophages, they can be induced by various microbial constituents to express B7.1 and especially B7.2. Indeed, B7.1 was first identified as a molecule expressed on B cells activated by microbial lipopolysaccharide. These observations help explain why it is essential to co-inject bacterial adjuvants in order to produce an immune response to soluble proteins such as ovalbumin, hen egg-white lysozyme, and cytochrome *c*, which may require B cells as antigen-presenting cells.

The requirement for induced co-stimulatory activity also helps explain why, although B cells present soluble proteins efficiently, they are unlikely to initiate responses to soluble self proteins in the absence of infection. In the absence of co-stimulation, antigen not only fails to activate naive T cells but causes them to become anergic, or nonresponsive (see Fig. 8.13). This provides an additional safeguard to the mechanisms discussed in Chapter 7 whereby potentially self-reactive T and B cells are eliminated or inactivated as they develop in the thymus and bone marrow.

Although much of what we know about the immune system in general, and about T-cell responses in particular, has been learned from the study of immune responses to soluble protein immunogens presented by B cells, it is not clear how important B cells are in priming naive T cells in natural immune responses. Soluble protein antigens are not abundant during natural infections; most natural antigens, such as bacteria and viruses, are particulate, whereas soluble bacterial toxins act by binding to cell surfaces and so are present only at low concentrations in solution. Some natural immunogens enter the body as soluble molecules; examples are insect toxins, anticoagulants injected by blood-sucking insects, snake venoms, and many allergens. However, tissue dendritic cells could also be responsible for activating naive T cells that recognize these antigens. Although tissue dendritic cells could not concentrate these antigens in the same way as antigen-specific B cells, they may be more likely to encounter a naive T cell with the appropriate antigen specificity than the limited number of B cells able to bind and concentrate a particular antigen. The chances of a B cell encountering a T cell that can recognize the peptide antigens it displays is greatly increased once a naive T cell has been detained in lymphoid tissue by finding its antigen on the surface of a dendritic cell.

T-cell responses can thus be primed by three distinct types of antigen-presenting cell. Dendritic cells are optimally equipped to present a wide variety of antigens to naive T cells, while macrophages stimulate T-cell responses to the pathogens they take up but are unable to eliminate, and B cells specialize in presenting fragments of the antigen to which their surface immunoglobulin binds (Fig. 8.18). In each of these cell types, as we saw in Chapter 2, the expression of co-stimulatory activity is controlled so as to provoke responses against pathogens while avoiding immunization against self.

	Dendritic cells	Macrophages	B cells
Antigen uptake	+++ Macropinocytosis and phagocytosis by tissue dendritic cells Viral infection	Phagocytosis +++	Antigen-specific receptor (Ig) ++++
MHC expression	Low on tissue dendritic cells High on dendritic cells in lymphoid tissues	Inducible by bacteria and cytokines − to +++	Constitutive Increases on activation +++ to ++++
Co-stimulator delivery	Constitutive by mature, nonphagocytic lymphoid dendritic cells ++++	Inducible − to +++	Inducible − to +++
Antigen presented	Peptides Viral antigens Allergens	Particulate antigens Intracellular and extracellular pathogens	Soluble antigens Toxins Viruses
Location	Lymphoid tissue Connective tissue Epithelia	Lymphoid tissue Connective tissue Body cavities	Lymphoid tissue Peripheral blood

Fig. 8.18 The properties of the various antigen-presenting cells. Dendritic cells, macrophages, and B cells are the main cell types involved in the initial presentation of foreign antigens to naive T cells. These cells vary in their means of antigen uptake, MHC class II expression, co-stimulator expression, the type of antigen they present effectively, their locations in the body, and their surface adhesion molecules (not shown).

8-9 Activated T cells synthesize the T-cell growth factor interleukin-2 and its receptor.

Naive T cells can live for many years without dividing. These small resting cells have condensed chromatin and a scanty cytoplasm and synthesize little RNA or protein. On activation, they must reenter the cell cycle and divide rapidly to produce the large numbers of progeny that will differentiate into armed effector T cells. Their proliferation and differentiation are driven by a cytokine called **interleukin-2 (IL-2)**, which is produced by the activated T cell itself.

The initial encounter with specific antigen in the presence of the required co-stimulatory signal triggers entry of the T cell into the G_1 phase of the cell cycle; at the same time, it also induces the synthesis of IL-2 along with the α chain of the IL-2 receptor. The IL-2 receptor has three chains: α, β, and γ (Fig. 8.19). Resting T cells express a form of this receptor composed of β and γ chains which binds IL-2 with moderate affinity, allowing resting T cells to respond to very high concentrations of IL-2. Association of the α chain with

Fig. 8.19 High-affinity IL-2 receptors are three-chain structures that are produced only on activated T cells. On resting T cells, the β and γ chains are expressed constitutively. They bind IL-2 with moderate affinity. Activation of T cells induces the synthesis of the α chain and the formation of the high-affinity heterotrimeric receptor. The β and γ chains show similarities in amino acid sequence to cell-surface receptors for growth hormone and prolactin, both of which also regulate cell growth and differentiation.

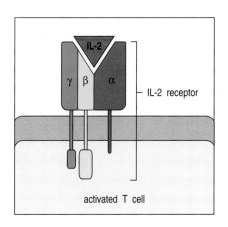

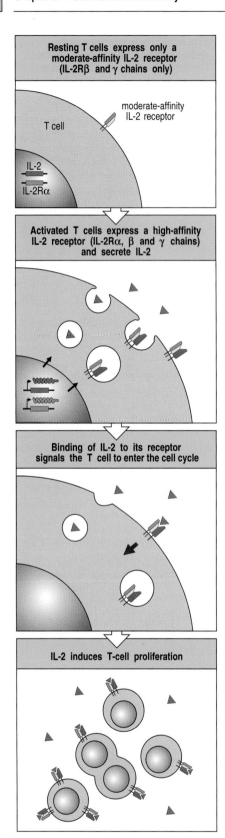

Resting T cells express only a moderate-affinity IL-2 receptor (IL-2Rβ and γ chains only)

T cell

moderate-affinity IL-2 receptor

IL-2

IL-2Rα

Activated T cells express a high-affinity IL-2 receptor (IL-2Rα, β and γ chains) and secrete IL-2

Binding of IL-2 to its receptor signals the T cell to enter the cell cycle

IL-2 induces T-cell proliferation

Fig. 8.20 Activated T cells secrete and respond to IL-2. Activation of naive T cells by the recognition of a peptide:MHC complex accompanied by co-stimulation induces expression and secretion of IL-2 and the expression of high-affinity IL-2 receptors. IL-2 binds to the high-affinity IL-2 receptors to promote T-cell growth in an autocrine fashion.

the β and γ chains creates a receptor with a much higher affinity for IL-2, allowing the cell to respond to very low concentrations of IL-2. Binding of IL-2 to the high-affinity receptor then triggers progression through the rest of the cell cycle. T cells activated in this way can divide two to three times a day for several days, allowing one cell to give rise to a clone composed of thousands of progeny that all bear the same receptor for antigen (Fig. 8.20). IL-2 also promotes the differentiation of these cells into armed effector T cells.

8-10 The co-stimulatory signal is necessary for the synthesis and secretion of IL-2.

The production of IL-2 determines whether a T cell will proliferate and become an armed effector cell, and the most important function of the co-stimulatory signal is to promote the synthesis of IL-2. Antigen recognition by the T-cell receptor ultimately induces the synthesis of several transcription factors (see Chapter 6). One of these factors, **NFAT** (**nuclear factor of activated T cells**), binds to the promoter region of the IL-2 gene and is needed to activate its transcription. IL-2 gene transcription on its own, however, does not lead to the production of IL-2, which additionally requires CD28 ligation by B7. One effect of signaling through CD28 is thought to be the stabilization of IL-2 mRNA. Cytokine mRNAs are very short-lived because of an 'instability' sequence in their 3′ untranslated region. This instability prevents sustained cytokine production and release, and enables cytokine activity to be tightly regulated. The stabilization of IL-2 mRNA increases IL-2 synthesis by 20- to 30-fold. A second effect of CD28 ligation is to activate transcription factors (AP-1 and NFκB) that increase transcription of IL-2 mRNA by about three-fold. These two effects together increase IL-2 protein production by as much as 100-fold. When a T cell recognizes specific antigen in the absence of co-stimulation through its CD28 molecule, little IL-2 is produced and the T cell does not proliferate.

The central importance of IL-2 in initiating adaptive immune responses is well illustrated by the drugs that are most commonly used to suppress undesirable immune responses such as transplant rejection. The immunosuppressive drugs cyclosporin A and FK506 (tacrolimus) inhibit IL-2 production by disrupting signaling through the T-cell receptor, whereas rapamycin (sirolimus) inhibits signaling through the IL-2 receptor. Cyclosporin A and rapamycin act synergistically to inhibit immune responses by preventing the IL-2-driven clonal expansion of T cells. The mode of action of these drugs will be considered in detail in Chapter 14.

8-11 Antigen recognition in the absence of co-stimulation leads to T-cell tolerance.

Antigen recognition in the absence of co-stimulation inactivates naive T cells, inducing a state known as anergy. The most important change in anergic T cells is their inability to produce IL-2. This prevents them from proliferating and differentiating into effector cells when they encounter antigen, even if the antigen is subsequently presented by antigen-presenting cells. This helps to ensure the tolerance of T cells to self tissue antigens. Although anergy has

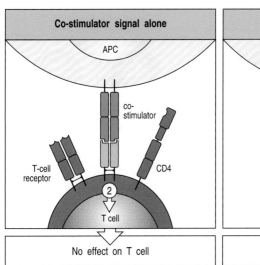
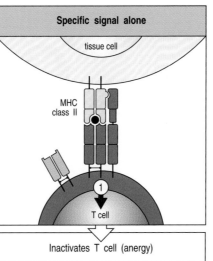

Fig. 8.21 T-cell tolerance to antigens expressed on tissue cells results from antigen recognition in the absence of co-stimulation. An antigen-presenting cell (APC) will neither activate nor inactivate a T cell if the appropriate antigen is not present on the APC surface, even if it expresses a co-stimulatory molecule and can deliver signal 2 (left panel). However, when a T cell recognizes antigen in the absence of co-stimulatory molecules, it receives signal 1 alone and is inactivated (right panel). This allows self antigens expressed on tissue cells to induce tolerance in the peripheral T-cell population.

only been demonstrated formally *in vitro*, there is sufficiently compelling evidence from studies *in vivo* showing peripheral tolerance to various antigens to assume that it happens in this setting as well.

As we saw in Section 7-24, any protein synthesized by all cells will be presented by antigen-presenting cells in the thymus and will cause clonal deletion of the T cells reactive to these ubiquitous self proteins. However, many proteins have specialized functions and are made only by the cells of certain tissues. Because MHC class I molecules present only peptides derived from proteins synthesized within the cell, such tissue-specific peptides will not be displayed on the MHC molecules of thymic cells, and T cells recognizing them are unlikely to be deleted in the thymus. An important factor in avoiding autoimmune responses to such tissue-specific proteins is the absence of co-stimulatory activity on tissue cells. Naive T cells recognizing self peptides on tissue cells are not activated; instead they may be induced to enter a state of anergy (Fig. 8.21).

Although the deletion of potentially autoreactive T cells is readily understood as a simple way to maintain self tolerance, the retention of anergic T cells specific for tissue antigens is less easy to understand. It would seem more economical and efficient to eliminate such cells; indeed, binding of the T-cell receptor on peripheral T cells in the absence of co-stimulators can lead to programmed cell death as well as to anergy. Nevertheless, some T cells persist in an anergic state *in vivo*. One possible explanation for this is that such anergic T cells have a role in preventing responses by naive, nonanergic T cells to foreign antigens that mimic self peptide:self MHC complexes. The persisting anergic T cells could recognize and bind to such peptide:MHC complexes on antigen-presenting cells without responding, and thus could compete with naive, potentially autoreactive cells of the same specificity. In this way, anergic T cells could serve to prevent the accidental activation of autoreactive T cells by infectious agents, thus actively contributing to tolerance.

8-12   Proliferating T cells differentiate into armed effector T cells that do not require co-stimulation to act.

Late in the proliferative phase of the T-cell response induced by IL-2, after 4–5 days of rapid growth, activated T cells differentiate into armed effector T cells that can synthesize all the effector molecules required for their specialized functions as helper or cytotoxic T cells. In addition, all classes of armed effector

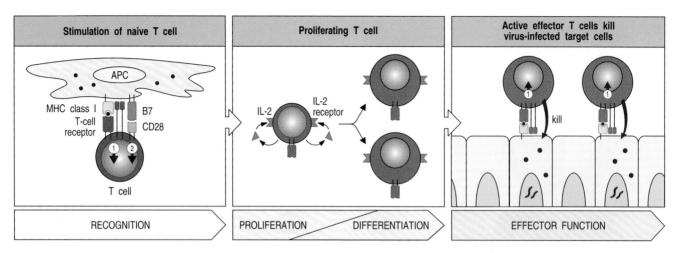

Fig. 8.22 Armed effector T cells can respond to their target cells without co-stimulation. A naive T cell that recognizes antigen on the surface of an antigen-presenting cell and receives the required two signals (arrows 1 and 2, left panel) becomes activated, and both secretes and responds to IL-2. IL-2-driven clonal expansion (center panel) is followed by the differentiation of the T cells to armed effector cell status. Once the cells have differentiated into effector T cells, any encounter with specific antigen triggers their effector actions without the need for co-stimulation. Thus, as illustrated here, a cytotoxic T cell can kill targets that express only the peptide:MHC ligand and not co-stimulatory signals (right panel).

T cells have undergone changes that distinguish them from naive T cells. One of the most critical is in their activation requirements: once a T cell has differentiated into an armed effector cell, encounter with its specific antigen results in immune attack without the need for co-stimulation (Fig. 8.22).

This applies to all classes of armed effector T cells. Its importance is particularly easy to understand in the case of cytotoxic CD8 T cells, which must be able to act on any cell infected with a virus, whether or not the infected cell can express co-stimulatory molecules. However, it is also important for the effector function of CD4 cells, as armed effector CD4 T cells must be able to activate B cells and macrophages that have taken up antigen, even if, as is often the case, they have too little co-stimulatory activity to activate a naive CD4 T cell.

Changes are also seen in the cell-adhesion molecules expressed by armed effector T cells. They express higher levels of LFA-1 and CD2, but lose cell-surface L-selectin and thus cease to recirculate through lymph nodes. Instead, they express the integrin VLA-4, which allows them to bind to vascular endothelium at sites of inflammation. This allows the armed effector T cells to enter sites of infection and put their armory of effector proteins to good use. These changes in the T-cell surface are summarized in Fig. 8.23.

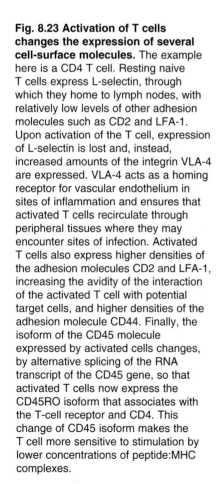

Fig. 8.23 Activation of T cells changes the expression of several cell-surface molecules. The example here is a CD4 T cell. Resting naive T cells express L-selectin, through which they home to lymph nodes, with relatively low levels of other adhesion molecules such as CD2 and LFA-1. Upon activation of the T cell, expression of L-selectin is lost and, instead, increased amounts of the integrin VLA-4 are expressed. VLA-4 acts as a homing receptor for vascular endothelium in sites of inflammation and ensures that activated T cells recirculate through peripheral tissues where they may encounter sites of infection. Activated T cells also express higher densities of the adhesion molecules CD2 and LFA-1, increasing the avidity of the interaction of the activated T cell with potential target cells, and higher densities of the adhesion molecule CD44. Finally, the isoform of the CD45 molecule expressed by activated cells changes, by alternative splicing of the RNA transcript of the CD45 gene, so that activated T cells now express the CD45RO isoform that associates with the T-cell receptor and CD4. This change of CD45 isoform makes the T cell more sensitive to stimulation by lower concentrations of peptide:MHC complexes.

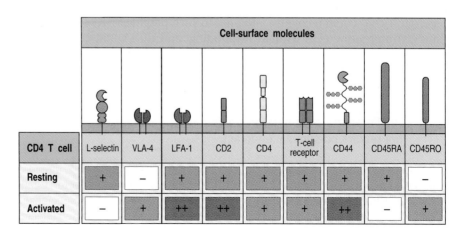

Cell-surface molecules								
CD4 T cell	L-selectin	VLA-4	LFA-1	CD2	CD4	T-cell receptor	CD44	CD45RA	CD45RO
Resting	+	–	+	+	+	+	+	+	–
Activated	–	+	++	++	+	+	++	–	+

8-13   The differentiation of CD4 T cells into T$_H$1 or T$_H$2 cells determines whether humoral or cell-mediated immunity will predominate.

Naive CD8 T cells emerging from the thymus are already predestined to become cytotoxic cells, even though they are not yet expressing any of the differentiated functions of armed effector cells. The case of CD4 T cells, however, is more complex. Naive CD4 T cells can differentiate upon activation into either T$_H$1 or T$_H$2 cells, which differ in the cytokines they produce and thus in their function. The decision on which fate the progeny of a naive CD4 T cell will follow is made during the clonal expansion that takes place after the first encounter with antigen (Fig. 8.24).

The factors that determine whether a proliferating CD4 T cell will differentiate into a T$_H$1 or a T$_H$2 cell are not fully understood. The cytokines elicited by infectious agents (principally IFN-γ, IL-12, and IL-4), the co-stimulators used to drive the response, and the nature of the peptide:MHC ligand all have an effect. In particular, because the decision to differentiate into T$_H$1 versus T$_H$2 cells occurs early in the immune response, the cytokines produced in response to pathogens by cells of the innate immune system play an important part in shaping the subsequent adaptive response; we will learn more about this in Chapter 10.

The consequences of inducing T$_H$1 versus T$_H$2 cells are profound: the selective production of T$_H$1 cells leads to cell-mediated immunity, whereas the production of predominantly T$_H$2 cells provides humoral immunity. A striking example of the difference this can make to the outcome of infection is seen in leprosy, a disease caused by infection with *Mycobacterium leprae*. *M. leprae*, like *M. tuberculosis*, grows in macrophage vesicles, and effective host defense requires macrophage activation by T$_H$1 cells. In patients with tuberculoid leprosy, in which T$_H$1 cells are preferentially induced, few live bacteria are found, little antibody is produced, and, although skin and peripheral nerves are damaged by the inflammatory responses associated with macrophage activation, the disease progresses slowly and the patient usually survives. However, when T$_H$2 cells are preferentially induced, the main response is humoral, the antibodies produced cannot reach the intracellular bacteria, and the patients develop lepromatous leprosy, in which *M. leprae* grows abundantly in macrophages, causing gross tissue destruction that is eventually fatal.

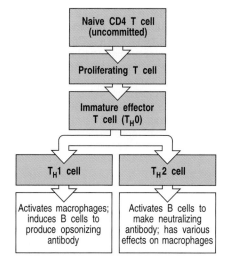

Fig. 8.24 The stages of activation of CD4 T cells. Naive CD4 T cells first respond to their specific peptide:MHC class II complexes by making IL-2 and proliferating. These cells then differentiate into a cell type known as T$_H$0, which has some of the effector functions characteristic of T$_H$1 and T$_H$2 cells. The T$_H$0 cell has the potential to become either a T$_H$1 cell or a T$_H$2 cell.

Lepromatous Leprosy

8-14   Naive CD8 T cells can be activated in different ways to become armed cytotoxic effector cells.

Naive CD8 T cells differentiate into cytotoxic cells, and perhaps because the effector actions of these cells are so destructive, naive CD8 T cells require more co-stimulatory activity to drive them to become armed effector cells than do naive CD4 T cells. This requirement can be met in two ways. The simplest is activation by dendritic cells, which have high intrinsic co-stimulatory activity. These cells can directly stimulate CD8 T cells to synthesize the IL-2 that drives their own proliferation and differentiation (Fig. 8.25). This has been exploited to generate cytotoxic T-cell responses against tumors, as we will see in Chapter 14.

Cytotoxic T-cell responses to some viruses and tissue grafts, however, seem to require the presence of CD4 T cells during the priming of the naive CD8 T cell. In these responses, both the naive CD8 T cell and the CD4 T cell must recognize related antigens on the surface of the same antigen-presenting cell. In this case, it is thought that the actions of the CD4 T cell may be needed to compensate for inadequate co-stimulation of naive CD8 T cells by the antigen-presenting cell. This compensatory effect is currently thought to occur by the

Fig. 8.25 Naive CD8 T cells can be activated directly by potent antigen-presenting cells. Naive CD8 T cells that encounter peptide:MHC class I complexes on the surface of dendritic cells, which express high levels of co-stimulatory molecules (left panel), are activated to produce IL-2 (right panel) and proliferate in response to it, eventually differentiating into armed cytotoxic CD8 T cells (not shown).

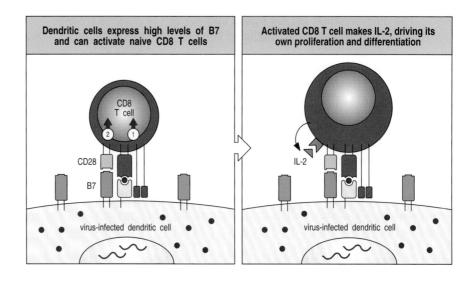

Dendritic cells express high levels of B7 and can activate naive CD8 T cells

Activated CD8 T cell makes IL-2, driving its own proliferation and differentiation

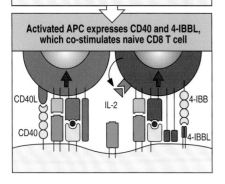

APC stimulates effector CD4 T cell, which in turn activates the APC

Activated APC expresses CD40 and 4-IBBL, which co-stimulates naive CD8 T cell

Fig. 8.26 Some CD8 T-cell responses require CD4 T cells. CD8 T cells recognizing antigen on weakly co-stimulating cells may become activated only in the presence of CD4 T cells bound to the same antigen-presenting cell. This happens mainly by an effector CD4 T cell recognizing antigen on the antigen-presenting cell and being triggered to induce increased levels of co-stimulatory activity on the antigen-presenting cell, which in turn activates the CD8 T cell to make its own IL-2.

recruitment of an armed effector CD4 T cell that activates the antigen-presenting cell to express higher levels of co-stimulatory activity. We have seen that this is one of the actions of the CD40 ligand, which is expressed once T cells have been activated. Binding of CD40 ligand on the CD4 T cell to CD40 on the antigen-presenting cell induces B7 expression and enables the antigen-presenting cell to co-stimulate the CD8 T cell directly (Fig. 8.26).

Summary.

The crucial first step in adaptive immunity is the activation of naive antigen-specific T cells by antigen-presenting cells. This occurs in the lymphoid tissues and organs through which naive T cells are constantly passing. The most distinctive feature of antigen-presenting cells is the expression of co-stimulatory molecules, of which the B7.1 and B7.2 molecules are the best characterized. Naive T cells will respond to antigen only when one cell presents both specific antigen to the T-cell receptor and a B7 molecule to CD28, the receptor for B7 on the T cell. The three cell types that can serve as antigen-presenting cells are dendritic cells, macrophages, and B cells. Each of these cells has a distinct function in eliciting immune responses. Tissue dendritic cells take up antigens by phagocytosis and macropinocytosis and are stimulated by infection to migrate to the local lymphoid tissue, where they differentiate into mature dendritic cells expressing co-stimulatory activity. They serve as the most potent activators of naive T-cell responses. Macrophages efficiently ingest particulate antigens such as bacteria and are induced by infectious agents to express MHC class II molecules and co-stimulatory activity. The unique ability of B cells to bind and internalize soluble protein antigens via their receptors may be important in activating T cells to this class of antigen, provided that co-stimulatory molecules are also induced on the B cell. In all three types of antigen-presenting cell, the expression of co-stimulatory molecules is activated in response to signals from receptors that also function in innate immunity to signal the presence of infectious agents (see Chapter 2).

The activation of T cells by antigen-presenting cells leads to their proliferation and the differentiation of their progeny into armed effector T cells. This depends on the production of cytokines, in particular the T-cell growth factor IL-2, which binds to a high-affinity receptor on the activated T cell. T cells whose antigen receptors are ligated in the absence of co-stimulatory signals fail to make IL-2 and instead become anergic or die. This dual requirement for both receptor ligation and co-stimulation helps to prevent naive T cells from responding to self antigens on tissue cells, which lack co-stimulatory activity. Proliferating T cells develop into armed effector T cells, the critical event in most adaptive immune responses. Once an expanded clone of T cells achieves effector function, its armed effector T-cell progeny can act on any target cell that displays antigen on its surface. Effector T cells can mediate a variety of functions. Their most important functions are the killing of infected cells by CD8 cytotoxic T cells and the activation of macrophages by T_H1 cells, which together make up cell-mediated immunity, and the activation of B cells by both T_H2 and T_H1 cells to produce different classes of antibody, thus driving the humoral immune response.

General properties of armed effector T cells.

All T-cell effector functions involve the interaction of an armed effector T cell with a target cell displaying specific antigen. The effector proteins released by these T cells are focused on the appropriate target cell by mechanisms that are activated by recognition of antigen on the target cell. The focusing mechanism is common to all types of effector T cells, whereas their effector actions depend on the array of membrane and secreted proteins they express or release upon receptor ligation. The different types of effector T cell are specialized to deal with different types of pathogen, and the effector molecules they are programmed to produce cause distinct and appropriate effects on the target cell (Fig. 8.27).

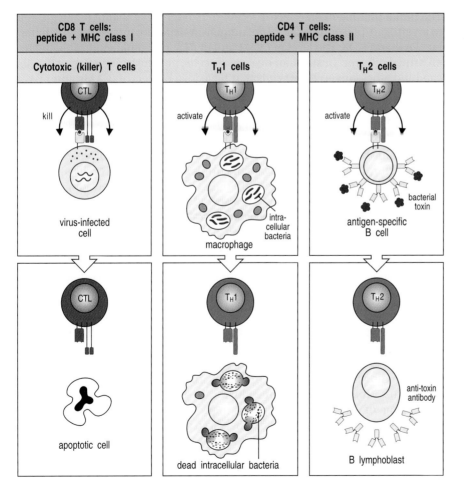

Fig. 8.27 There are three classes of effector T cell, specialized to deal with three classes of pathogen. CD8 cytotoxic cells (left panels) kill target cells that display peptide fragments of cytosolic pathogens, most notably viruses, bound to MHC class I molecules at the cell surface. T_H1 cells (middle panels) and T_H2 cells (right panels) both express the CD4 co-receptor and recognize fragments of antigens degraded within intracellular vesicles, displayed at the cell surface by MHC class II molecules. T_H1 cells activate macrophages, enabling them to destroy intracellular microorganisms more efficiently; they can also activate B cells to produce strongly opsonizing antibodies belonging to certain IgG subclasses (IgG1 and IgG3 in humans, and their homologues IgG2a and IgG2b in the mouse). T_H2 cells, on the other hand, drive B cells to differentiate and produce immunoglobulins of all other types, and are responsible for initiating B-cell responses by activating naive B cells to proliferate and secrete IgM. The various types of immunoglobulin together make up the effector molecules of the humoral immune response.

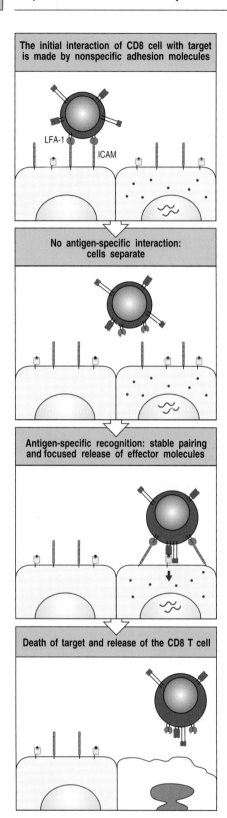

The initial interaction of CD8 cell with target is made by nonspecific adhesion molecules

LFA-1

ICAM

No antigen-specific interaction: cells separate

Antigen-specific recognition: stable pairing and focused release of effector molecules

Death of target and release of the CD8 T cell

Fig. 8.28 Interactions of T cells with their targets initially involve nonspecific adhesion molecules. The major initial interaction is between LFA-1 on the T cell, illustrated here as a cytotoxic CD8 T cell, and ICAM-1 or ICAM-2 on the target cell (top panel). This binding allows the T cell to remain in contact with the target cell and to scan its surface for the presence of specific peptide:MHC complexes. If the target cell does not carry the specific antigen, the T cell disengages (second panel) and can scan other potential targets until it finds the specific antigen (third panel). Signaling through the T-cell receptor increases the strength of the adhesive interactions, prolonging the contact between the two cells and stimulating the T cell to deliver its effector molecules. The T cell then disengages (bottom panel).

8-15 Effector T-cell interactions with target cells are initiated by antigen-nonspecific cell-adhesion molecules.

Once an effector T cell has completed its differentiation in the lymphoid tissue it must find target cells that are displaying the MHC:peptide complex that it recognizes. Some T$_H$2 cells encounter their B-cell targets without leaving the lymphoid tissue, as we discuss further in Chapter 9. However, most of the armed effector T cells emigrate from their site of activation in lymphoid tissues and enter the blood via the thoracic duct. Because of the cell-surface changes that have occurred during differentiation, they can now migrate into tissues, particularly at sites of infection. They are guided to these sites by changes in the adhesion molecules expressed on the endothelium of the local blood vessels as a result of infection, and by local chemotactic factors, as we will see in Chapter 10.

The initial binding of an effector T cell to its target, like that of a naive T cell to an antigen-presenting cell, is an antigen-nonspecific interaction mediated by LFA-1 and CD2. The level of LFA-1 and of CD2 is twofold to fourfold higher on armed effector T cells than on naive T cells, and so armed effector T cells can bind efficiently to target cells that have lower levels of ICAMs and LFA-3 on their surface than do the professional antigen-presenting cells. This interaction is normally transient unless recognition of antigen on the target cell through the T-cell receptor triggers an increase in the affinity of the T-cell's LFA-1 for its ligands on the target cell. The T cell binds more tightly to its target and remains bound for long enough to release its effector molecules. Armed CD4 effector T cells, which activate macrophages or induce B cells to secrete antibody, must maintain contact with their targets for relatively long periods. Cytotoxic T cells, by contrast, can be observed under the microscope attaching to and dissociating from successive targets relatively rapidly as they kill them (Fig. 8.28). Killing of the target, or some local change in the T cell, then allows the effector T cell to detach and address new targets. How armed CD4 effector T cells disengage from their antigen-negative targets is not known, although current evidence suggests that CD4 binding directly to MHC class II molecules on target cells that are not displaying specific antigen, signals the cell to detach.

8-16 Binding of the T-cell receptor complex directs the release of effector molecules and focuses them on the target cell.

When binding to peptide:MHC complexes, the T-cell receptor molecules and their cross-linked co-receptors cluster at the site of cell–cell contact. Clustering of the T-cell receptors then signals a reorientation of the cytoskeleton that polarizes the effector cell so as to focus the release of effector molecules at the site of contact with the target cell, as illustrated for a cytotoxic T cell in Fig. 8.29. Polarization of the cell starts with the local reorganization of the cortical actin cytoskeleton at the site of contact; this in turn leads to the

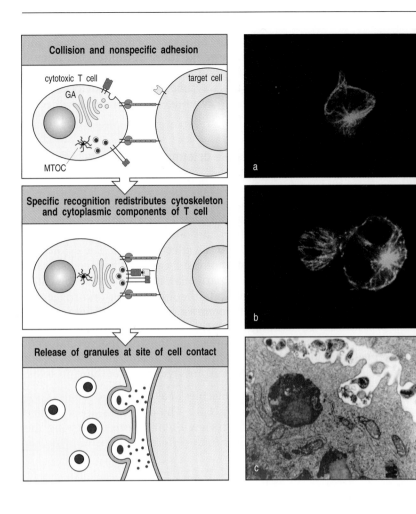

Collision and nonspecific adhesion

cytotoxic T cell

GA

MTOC

target cell

Specific recognition redistributes cytoskeleton and cytoplasmic components of T cell

Release of granules at site of cell contact

a

b

c

Fig. 8.29 The polarization of T cells during specific antigen recognition allows effector molecules to be focused on the antigen-bearing target cell. The example illustrated here is a CD8 cytotoxic T cell. Cytotoxic CD8 cells contain specialized lysosomes called lytic granules, which contain cytotoxic proteins. Initial binding to a target cell through adhesion molecules does not have any effect on the location of the lytic granules. Binding of the T-cell receptor causes the T cell to become polarized: reorganization within the cortical actin cytoskeleton at the site of contact has the effect of aligning the microtubule-organizing center (MTOC), which in turn aligns the secretory apparatus, including the Golgi apparatus (GA), towards the target cell. Proteins stored in lytic granules derived from the Golgi are then directed specifically onto the target cell. The photomicrograph in panel a shows an unbound, isolated cytotoxic T cell. The microtubule cytoskeleton is stained in green and the lytic granules in red. Note how the lytic granules are dispersed throughout the T cell. Panel b depicts a cytotoxic T cell bound to a (larger) target cell. The lytic granules are now clustered at the site of cell–cell contact in the bound T cell. The electron micrograph in panel c shows the release of granules from a cytotoxic T cell. Panels a and b courtesy of G. Griffiths. Panel c courtesy of E.R. Podack.

reorientation of the microtubule-organizing center (MTOC), the center from which the microtubule cytoskeleton is produced, and of the Golgi apparatus (GA), through which most proteins destined for secretion travel. In the cytotoxic T cell, the cytoskeletal reorientation focuses exocytosis of the preformed lytic granules at the site of contact with its target cell.

The polarization of a T cell also focuses the secretion of soluble effector molecules whose synthesis is induced *de novo* by ligation of the T-cell receptor. For example, the secreted cytokine IL-4, which is the principal effector molecule of T_H2 cells, is confined and concentrated at the site of contact with the target cell (see Fig. 9.6). It has been shown that the enhanced binding of LFA-1 to ICAM-1 creates a molecular seal surrounding the clustered T-cell receptors, CD4 co-receptors, and CD28 molecules (Fig. 8.30).

Thus, the antigen-specific T-cell receptor controls the delivery of effector signals in three ways: it induces the stable binding of effector cells to their specific target cells to create a tightly held, narrow space in which effector molecules can be concentrated; it focuses their delivery at the site of contact

Outer ring (red)	Inner circle (green)
LFA-1:ICAM-1	TCR, CD4, CD28

Fig. 8.30 Tight junctions are formed between armed effector T cells and their targets. Confocal fluorescence micrograph of the area of contact between a T cell and a B cell (as viewed through one of the cells). The outer red ring is made up of LFA-1 on the T cell and its counterreceptors on the target cell, whereas molecules that cluster in the center of the ring (bright green) include the T-cell receptor complex, the co-receptor CD4, and CD28. Photograph courtesy of A. Kupfer.

by inducing a reorientation of the secretory apparatus of the effector cell; and it triggers their synthesis and/or release. All these receptor-coordinated mechanisms contribute to the selective action of effector molecules on the target cell bearing specific antigen. In this way, effector T-cell activity is highly selective for those target cells that display antigen, although the effector molecules themselves are not antigen-specific.

8-17   The effector functions of T cells are determined by the array of effector molecules they produce.

The effector molecules produced by armed effector T cells fall into two broad classes: **cytotoxins**, which are stored in specialized lytic granules and released by cytotoxic CD8 T cells, and **cytokines** and related membrane-associated proteins, which are synthesized *de novo* by all effector T cells. The cytotoxins are the principal effector molecules of cytotoxic T cells and will be discussed further in Section 8-22. Their release, in particular, must be tightly regulated as they are not specific: they can penetrate the lipid bilayer and trigger an intrinsic death program in any cell. By contrast, cytokines and membrane-associated proteins act by binding to specific receptors on the target cell. Cytokines and membrane-associated proteins are the principal mediators of CD4 T-cell effector actions, and the main effector actions of CD4 cells are therefore directed at specialized cells that express receptors for these proteins.

The effector actions and main effector molecules of all three functional classes of effector T cell are summarized in Fig. 8.31. The cytokines are a diverse group of proteins and we will briefly review them before discussing the T-cell cytokines and their contributions to the effector actions of cytotoxic

Fig. 8.31 The three main types of armed effector T cell produce distinct sets of effector molecules. CD8 T cells are predominantly killer T cells that recognize pathogen-derived peptides bound to MHC class I molecules. They release perforin (which creates holes in the target cell membrane), granzymes (which are proteases that act intra-cellularly to trigger apoptosis), and often the cytokine IFN-γ. A membrane-bound effector molecule expressed on CD8 T cells is Fas ligand. When this binds to Fas on a target cell it activates apoptosis in the Fas-bearing cell. CD4 T cells recognize peptides bound to MHC class II molecules and are of two functional types: T$_H$1 and T$_H$2. T$_H$1 cells are specialized to activate macrophages that are infected by or have ingested pathogens; they secrete IFN-γ as well as other effector molecules, and express membrane-bound CD40 ligand and/or Fas ligand. These are both members of the TNF family but CD40 ligand triggers the activation of the target cell, whereas Fas ligand triggers the death of Fas-expressing cells, so their pattern of expression has a strong influence on their function. T$_H$2 cells are specialized for B-cell activation; they secrete the B-cell growth factors IL-4 and IL-5. The principal membrane-bound effector molecule expressed by T$_H$2 cells is CD40 ligand, which binds to CD40 on the B cell and induces B-cell proliferation.

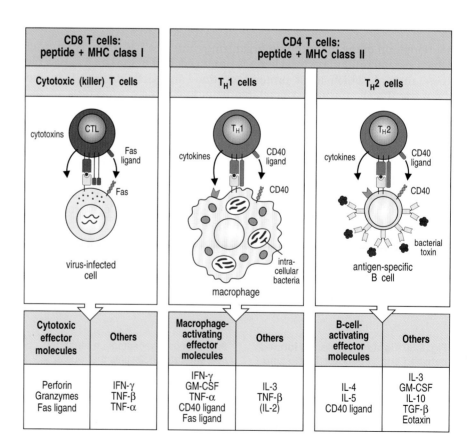

CD8 T cells: peptide + MHC class I		CD4 T cells: peptide + MHC class II			
Cytotoxic (killer) T cells		T$_H$1 cells		T$_H$2 cells	
Cytotoxic effector molecules	Others	Macrophage-activating effector molecules	Others	B-cell-activating effector molecules	Others
Perforin Granzymes Fas ligand	IFN-γ TNF-β TNF-α	IFN-γ GM-CSF TNF-α CD40 ligand Fas ligand	IL-3 TNF-β (IL-2)	IL-4 IL-5 CD40 ligand	IL-3 GM-CSF IL-10 TGF-β Eotaxin

CD8 T cells, T_H1 cells, and T_H2 cells. As we will see, soluble cytokines and membrane-associated molecules often act in combination to mediate the effects of T cells on their target cells.

The membrane-associated effector molecules, which we will discuss further in Section 8-20, are all structurally related to **tumor necrosis factor** (**TNF**), and their receptors on target cells are members of the **TNF receptor** (**TNFR**) family. All three classes of effector T cell express one or more members of the TNF family upon recognizing their specific antigen on the target cell. The membrane-bound TNF family member CD40 ligand is of particular importance for CD4 T-cell effector function; it is induced on T_H1 and T_H2 cells, and delivers activating signals to B cells and macrophages through the TNFR protein CD40. TNF-α is made by T_H1 cells, some T_H2 cells, and by cytotoxic T cells in soluble and membrane-associated forms, and can also deliver activating signals to macrophages. Some members of the TNF family can stimulate death by apoptosis. Thus **Fas ligand** (CD95L), the principal membrane-associated TNF-related molecule expressed by cytotoxic T cells, can trigger death by apoptosis in target cells bearing the receptor protein **Fas** (CD95); some T_H1 cells also express Fas ligand and can kill Fas-bearing cells with which they interact. Death by this mechanism appears to be important for removing activated Fas-bearing lymphocytes; if it fails, a lymphoproliferative disease associated with severe autoimmunity results.

8-18 Cytokines can act locally or at a distance.

Cytokines are small soluble proteins secreted by one cell that can alter the behavior or properties of the cell itself or of another cell. They are released by many cells in addition to those of the immune system. We have already discussed the cytokines released by phagocytic cells in Chapter 2, where we dealt with the inflammatory reactions that play an important part in innate immunity; here we are concerned mainly with the cytokines that mediate the effector functions of T cells. Cytokines produced by lymphocytes are often called **lymphokines**, but this nomenclature can be confusing because some lymphokines are also secreted by nonlymphoid cells; we will therefore use the generic term 'cytokine' for all of them. Most cytokines produced by T cells are given the name **interleukin** (**IL**) followed by a number: we have encountered several interleukins already in this chapter. Cytokines of immunological interest are listed in Appendix III.

Most cytokines have a multitude of different biological effects when tested at high concentration in biological assays *in vitro* but targeted disruption of genes for cytokines and cytokine receptors in knockout mice (see Appendix I, Section A-47) has helped to clarify their physiological roles. The major actions of the cytokines produced by effector T cells are given in Fig. 8.32. As the effect of a cytokine varies depending on the target cell, the actions are listed according to the major target cell types—B cells, T cells, macrophages, hematopoietic cells, and tissue cells.

The main cytokine released by CD8 effector T cells is IFN-γ, which can block viral replication or even lead to the elimination of virus from infected cells without killing them. T_H1 cells and T_H2 cells release different, but overlapping, sets of cytokines, which define their distinct actions in immunity. T_H2 cells secrete IL-4 and IL-5, which activate B cells, and IL-10, which inhibits macrophage activation. T_H1 cells secrete IFN-γ, which is the main macrophage-activating cytokine, and lymphotoxin (LT-α or TNF-β), which activates macrophages, inhibits B cells and is directly cytotoxic for some cells. The T_H0 cells from which both of these functional classes derive (see Fig. 8.24) also secrete cytokines, including IL-2, IL-4, and IFN-γ, and may therefore have a distinctive effector function.

We have already discussed in Section 8-16 how the T-cell receptor can orchestrate the polarized release of these cytokines so that they are concentrated at the site of contact with the target cell. Furthermore, most of the soluble cytokines have local actions that synergize with those of the membrane-bound effector molecules. The effect of all these molecules is therefore combinatorial, and as the membrane-bound effectors can only bind to receptors on an interacting cell, this is another mechanism by which selective effects of cytokines are focused on the target cell. The effects of

Cytokine	T-cell source	Effects on					Effect of gene knockout
		B cells	T cells	Macrophages	Hematopoietic cells	Other somatic cells	
Interleukin-2 (IL-2)	T_H0, T_H1, some CTL	Stimulates growth and J-chain synthesis	Growth	–	Stimulates NK cell growth	–	↓ T-cell responses IBD
Interferon-γ (IFN-γ)	T_H1, CTL	Differentiation IgG2a synthesis	Inhibits T_H2 cell growth	Activation, ↑ MHC class I and class II	Activates NK cells	Antiviral ↑ MHC class I and class II	Susceptible to mycobacteria
Lymphotoxin (LT, TNF-β)	T_H1, some CTL	Inhibits	Kills	Activates, induces NO production	Activates neutrophils	Kills fibroblasts and tumor cells	Absence of lymph nodes Disorganized spleen
Interleukin-4 (IL-4)	T_H2	Activation, growth IgG1, IgE ↑ MHC class II induction	Growth, survival	Inhibits macrophage activation	↑Growth of mast cells	–	No T_H2
Interleukin-5 (IL-5)	T_H2	Differentiation IgA synthesis	–	–	↑ Eosinophil growth and differentiation		
Interleukin-10 (IL-10)	T_H2	↑ MHC class II	Inhibits T_H1	Inhibits cytokine release	Co-stimulates mast cell growth		IBD
Interleukin-3 (IL-3)	T_H1, T_H2, some CTL	–	–	–	Growth factor for progenitor hematopoietic cells (multi-CSF)	–	–
Tumor necrosis factor-α (TNF-α)	T_H1, some T_H2, some CTL	–	–	Activates, induces NO production	–	Activates microvascular endothelium	Resistance to gram –ve sepsis
Granulocyte-macrophage colony-stimulating factor (GM-CSF)	T_H1, some T_H2, some CTL	Differentiation	Inhibits growth	Activation Differentiation to dendritic cells	↑ Production of granulocytes and macrophages (myelopoiesis) and dendritic cells	–	–
Transforming growth factor-β (TGF-β)	CD4 T cells	Inhibits growth IgA switch factor	–	Inhibits activation	Activates neutrophils	Inhibits/ stimulates cell growth	Death at ~10 weeks

Fig. 8.32 The nomenclature and functions of well-defined T-cell cytokines. The major actions are noted in boxes. Each cytokine has multiple activities on different cell types. The mixture of cytokines secreted by a given cell type produces many effects through what is called a 'cytokine network.' Major activities of effector cytokines are highlighted in red. ↑, increase; ↓, decrease; CTL, cytotoxic lymphocyte; NK, natural killer cell; CSF, colony-stimulating factor; IBD, inflammatory bowel disease; NO, nitric oxide.

some cytokines are further confined to target cells by tight regulation of their synthesis: as we will see later, the synthesis of cytokines such as IL-2, IL-4, and IFN-γ is controlled, so that secretion from T cells does not continue after the interaction with a target cell ends.

Some cytokines, however, have more distant effects. IL-3 and GM-CSF (see Fig. 8.32), for example, which are released by both types of CD4 effector T cell, act on bone marrow cells to stimulate the production of macrophages and granulocytes, both of which are important nonspecific effector cells in both humoral and cell-mediated immunity. IL-3 and GM-CSF also stimulate the production of dendritic cells from bone marrow precursors. IL-5, produced by T$_H$2 cells, can increase the production of eosinophils, which contribute to the late phase of allergic reactions, in which there is a predominant activation of T$_H$2 cells (see Chapter 12). Whether a cytokine effect is local or more distant is likely to reflect the amounts released, the degree to which this release is focused on the target cell, and the stability of the cytokine *in vivo* but, for most of the cytokines, in particular those with more distant effects, these factors are not yet known.

8-19   Cytokines and their receptors fall into distinct families of structurally related proteins.

Cytokines can be grouped by structure into families—the hematopoietins, the interferons, and the TNF family (Fig. 8.33)—and their receptors can likewise be grouped (Fig. 8.34). We have already encountered members of all of these families in Chapter 2, and have given an overview of the chemokine family

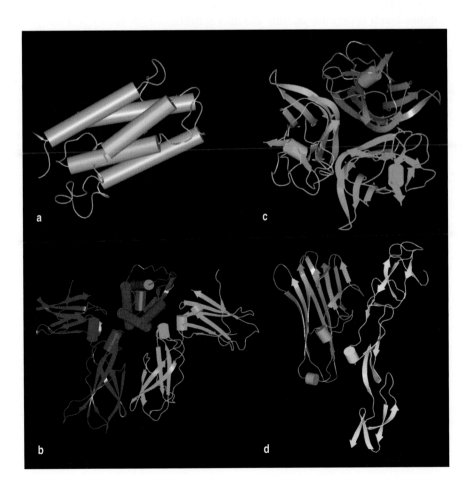

Fig. 8.33 Cytokines and their receptors can be grouped into a small number of structural families. Representatives of the hematopoietin and TNF families are shown here, as most of the cytokines made by effector T cells belong to one or other of these families. Cytokines are in the top row with their receptors below. The hematopoietins are represented by IL-4 (a). They are small single-chain proteins. A hypothetical model of the dimeric IL-4 receptor structure (based on the known structure of the related human growth hormone receptor) is shown in b, with bound IL-4 in red. Tumor necrosis factor (TNF) and its related molecules occur as trimers, as shown in c. The structure of one subunit of a TNF receptor binding a monomeric TNF is shown in d. The other structural families of immunological interest are the interferons and their receptors (see Fig. 8.34), and the chemokines and their receptors (see Fig. 2.32).

Fig. 8.34 Cytokine receptors belong to families of receptor proteins, each with a distinctive structure. There is a large family of cytokine receptors, which are divided into two subsets on the basis of the presence or absence of particular sequence motifs. Many cytokine receptors are members of the hematopoietin-receptor family, also called the class I cytokine receptor family. This family is named after the first of its members to be defined, the hematopoietin receptor. A smaller number of receptors fall into the class II cytokine receptor superfamily; many of these are receptors for interferons or interferon-like cytokines. Other super-families of cytokine receptors are the tumor necrosis factor-receptor (TNFR) family, and the chemokine-receptor family, which are part of a very large family of large G protein-coupled receptors. Each family member is a variant with a distinct specificity, performing a particular function on the cell that expresses it. In the hemato-poietin-receptor family, the α chain often defines the ligand specificity of the receptor, whereas the β or γ chain confers the intracellular signaling function. For the TNFR family, the ligands act as trimers and may be associated with the cell membrane rather than being secreted. Of the receptors listed here, some have been mentioned already in this book, some will occur in later chapters, and some are important examples from other biological systems. The diagrams indicate the representations of these receptors that you will encounter throughout this book.

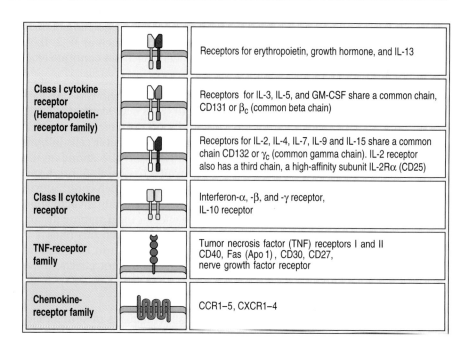

Class I cytokine receptor (Hematopoietin-receptor family)		Receptors for erythropoietin, growth hormone, and IL-13
		Receptors for IL-3, IL-5, and GM-CSF share a common chain, CD131 or β$_c$ (common beta chain)
		Receptors for IL-2, IL-4, IL-7, IL-9 and IL-15 share a common chain CD132 or γ$_c$ (common gamma chain). IL-2 receptor also has a third chain, a high-affinity subunit IL-2Rα (CD25)
Class II cytokine receptor		Interferon-α, -β, and -γ receptor, IL-10 receptor
TNF-receptor family		Tumor necrosis factor (TNF) receptors I and II CD40, Fas (Apo 1), CD30, CD27, nerve growth factor receptor
Chemokine-receptor family		CCR1–5, CXCR1–4

there (see Section 2-20). We will focus here on the receptors for the hematopoietins, the TNF family, and IFN-γ on account of their role in T-cell effector function. Members of the TNF family act as trimers, most of which are membrane-bound and so are quite distinct in their properties from the other cytokines. Nevertheless, they share some important properties with the soluble T-cell cytokines, as they are also synthesized *de novo* upon antigen recognition by T cells, and affect the behavior of the target cell.

Many of the soluble cytokines made by effector T cells are members of the hematopoietin family. These cytokines and their receptors can be further divided into subfamilies characterized by functional similarities and genetic linkage. For instance, IL-3, IL-4, IL-5, IL-13, and GM-CSF are related structurally, their genes are closely linked in the genome, and all are major cytokines produced by T$_H$2 cells. In addition, they bind to closely related receptors, which form the family of class I cytokine receptors. The IL-3, IL-5, and GM-CSF receptors share a common β chain. Another subgroup of class I cytokine receptors is defined by their use of the γ chain of the IL-2 receptor; this is shared by receptors for the cytokines IL-2, IL-4, IL-7, IL-9, and IL-15 and is now called the **common γ chain** (γ$_c$). More distantly related, the receptor for IFN-γ is a member of a small family of cytokine receptors with some similarities to the hematopoietin receptor family. These so-called class II cytokine receptors include the receptor for IFN-α and IFN-β, and the IL-10 receptor. Overall, the structural, functional, and genetic relations between the cytokines and their receptors suggest that they may have diversified in parallel during the evolution of increasingly specialized effector functions.

These specific functional effects depend on intracellular signaling events that are triggered by the cytokines binding to their specific receptors. The hematopoietin and interferon receptors all signal through a similar pathway, which is described in Chapter 6. The key signaling molecules of this pathway are members of the Janus family of cytoplasmic tyrosine kinases (JAKs) and their targets the signal transducing activators of transcription (STATs), which enter the nucleus to activate specific genes. As the JAKs and STATs are present as families of related molecules, different members may be activated to achieve different effects.

8-20   The TNF family of cytokines are trimeric proteins that are often associated with the cell surface.

TNF-α is made by T cells in soluble and membrane-associated forms, both of which are made up of three identical protein chains (a homotrimer, see Fig. 8.33). TNF-β (LT-α) can be produced as a secreted homotrimer, but is usually linked to the cell surface by forming heterotrimers with a third, membrane-associated, member of this family called LT-β. The receptors for these molecules, TNFR-I and TNFR-II, form homotrimers when bound to either TNF-α or LT. The trimeric structure is characteristic of all members of the TNF family, and the ligand-induced trimerization of their receptors seems to be the critical event in initiating signaling.

Most effector T cells express members of the TNF protein family as cell-surface molecules. The most important TNF-family proteins in T-cell effector function are TNF-α, TNF-β, Fas ligand, and CD40 ligand, the latter two always being cell-surface associated. These molecules all bind receptors that are members of the TNFR family; TNFR-I and II can each interact with either TNF-α or TNF-β, whereas Fas ligand and CD40 ligand bind respectively to the transmembrane protein Fas and to the transmembrane protein CD40 on target cells

Fas is expressed on many cells, especially on activated lymphocytes. Activation of Fas by the Fas ligand has profound consequences for the cell as Fas contains a 'death domain' in its cytoplasmic tail, which can initiate an activation cascade of cellular proteases called caspases that leads to apoptotic cell death (see Fig. 6.23). Fas is important in maintaining lymphocyte homeostasis, as can be seen from the effects of mutations in the Fas or Fas ligand genes. Mice and humans with a mutant form of Fas develop a lymphoproliferative disease associated with severe autoimmunity. A mutation in the gene encoding the Fas ligand in another mouse strain creates a nearly identical phenotype. These mutant phenotypes represent the best-characterized examples of generalized autoimmunity caused by single-gene defects. Other TNFR family members, including TNFR-I, are also associated with death domains and can also induce programmed cell death. Thus, TNF-α and TNF-β can induce programmed cell death by binding to TNFR-I.

The cytoplasmic tail of CD40 lacks a death domain; instead, it appears to be linked to proteins called **TRAFs** (TNF-receptor-associated factors), about which little is known. CD40 is involved in macrophage and B-cell activation; the ligation of CD40 on B cells promotes growth and isotype switching, whereas CD40 ligation on macrophages induces them to secrete TNF-α and to become receptive to much lower concentrations of IFN-γ. Deficiency in CD40 ligand expression is associated with immunodeficiency, as we will learn in Chapters 9 and 11.

Hyper IgM
Immunodeficiency

Summary.

Interactions between armed effector T cells and their targets are initiated by transient nonspecific adhesion between the cells. T-cell effector functions are elicited only when peptide:MHC complexes on the surface of the target cell are recognized by the receptor on an armed effector T cell. This recognition event triggers the armed effector T cell to adhere more strongly to the antigen-bearing target cell and to release its effector molecules directly at the target cell, leading to the activation or death of the target. The consequences of antigen recognition by an armed effector T cell are determined largely by the set of effector molecules it produces on binding a specific target cell. CD8 cytotoxic T cells store preformed cytotoxins in specialized lytic granules whose release can be tightly focused at the site of contact with the infected target cell. Cytokines, and one or more members of the TNF family of membrane-associated effector proteins, are synthesized *de novo* by all three types of effector T cell. T_H2 cells express B-cell activating effector molecules, whereas T_H1 cells express effector molecules that activate macrophages. CD8 T cells express membrane-associated Fas ligand that induces programmed cell death in cells bearing Fas; they also release IFN-γ. Membrane-associated effector molecules can deliver signals only to an interacting cell bearing the appropriate receptor, whereas soluble cytokines can act on cytokine receptors expressed locally on the target cell, or on hematopoietic cells at a distance. The actions of cytokines and membrane-associated effector molecules through their specific receptors, together with the effects of cytotoxins released by CD8 cells, account for most of the effector functions of T cells.

T cell-mediated cytotoxicity.

All viruses, and some bacteria, multiply in the cytoplasm of infected cells; indeed, the virus is a highly sophisticated parasite that has no biosynthetic or metabolic apparatus of its own and, in consequence, can replicate only inside cells. Once inside cells, these pathogens are not accessible to antibodies and can be eliminated only by the destruction or modification of the infected cells on which they depend. This role in host defense is fulfilled by cytotoxic CD8 T cells. The critical role of cytotoxic T cells in limiting such infections is seen in the increased susceptibility of animals artificially depleted of these T cells, or of mice or humans that lack the MHC class I molecules that present antigen to CD8 T cells. As well as controlling infection by viruses and cytoplasmic bacteria, CD8 T cells are important in controlling some protozoan infections and are crucial, for example, in host defense against the protozoan *Toxoplasma gondii*, a vesicular parasite that exports peptides from the infected vesicles to the cytosol, from which they enter the MHC class I processing pathway. The elimination of infected cells without the destruction of healthy tissue requires the cytotoxic mechanisms of CD8 T cells to be both powerful and accurately targeted.

8-21 Cytotoxic T cells can induce target cells to undergo programmed cell death.

Cells can die in either of two ways. Physical or chemical injury, such as the deprivation of oxygen that occurs in heart muscle during a heart attack, or membrane damage with antibody and complement, leads to cell disintegration or **necrosis**. The dead or necrotic tissue is taken up and degraded by phagocytic cells, which eventually clear the damaged tissue and heal the wound. The other form of cell death is known as **programmed cell death** or **apoptosis**. Apoptosis is a normal cellular response that is crucial in the tissue remodeling that occurs during development and metamorphosis in all multicellular animals. As we saw in Chapter 7, most thymocytes die an apoptotic death when they fail positive selection or, much less often, are negatively selected as a result of recognizing self antigens. Early changes seen in apoptosis are nuclear blebbing, alteration in cell morphology, and, eventually, fragmentation of the DNA. The cell then destroys itself from within, shrinking by shedding membrane-bound vesicles, and degrading itself until little is left. A hallmark of this type of cell death is the fragmentation of nuclear DNA into 200-base-pair (bp) pieces through the activation of endogenous nucleases that cleave the DNA between nucleosomes, each of which contains about 200 bp of DNA.

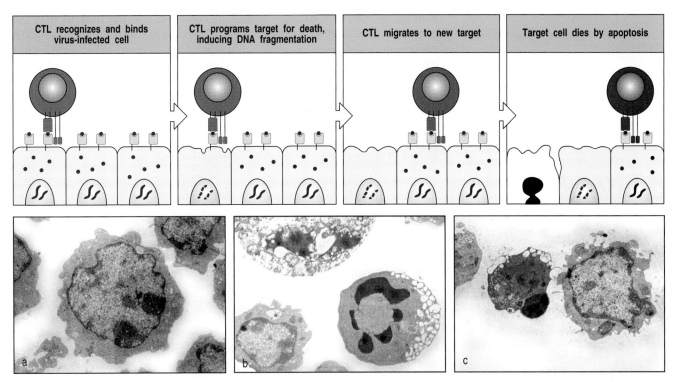

| CTL recognizes and binds virus-infected cell | CTL programs target for death, inducing DNA fragmentation | CTL migrates to new target | Target cell dies by apoptosis |

Fig. 8.35 Cytotoxic CD8 T cells can induce apoptosis in target cells. Specific recognition of peptide:MHC complexes on a target cell (top panels) by a cytotoxic CD8 T cell (CTL) leads to the death of the target cell by apoptosis. Cytotoxic T cells can recycle to kill multiple targets. Each killing requires the same series of steps, including receptor binding and directed release of cytotoxic proteins stored in lytic granules. The process of apoptosis is shown in the micrographs (bottom panels), where panel a shows a healthy cell with a normal nucleus. Early in apoptosis (panel b) the chromatin becomes condensed (red) and, although the cell sheds membrane vesicles, the integrity of the cell membrane is retained, in contrast to the necrotic cell in the upper part of the same field. In late stages of apoptosis (panel c), the cell nucleus (middle cell) is very condensed, no mitochondria are visible, and the cell has lost much of its cytoplasm and membrane through the shedding of vesicles. Photographs (× 3500) courtesy of R. Windsor and E. Hirst.

Cytotoxic T cells kill their targets by programming them to undergo apoptosis (Fig. 8.35). When cytotoxic T cells are mixed with target cells and rapidly brought into contact by centrifugation, they can program antigen-specific target cells to die within 5 minutes, although death may take hours to become fully evident. The short period required by cytotoxic T cells to program their targets to die reflects the release of preformed effector molecules, which activate an endogenous apoptotic pathway within the target cell.

As well as killing the host cell, the apoptotic mechanism may also act directly on cytosolic pathogens. For example, the nucleases that are activated in apoptosis to destroy cellular DNA can also degrade viral DNA. This prevents the assembly of virions and thus the release of infectious virus, which could otherwise infect nearby cells. Other enzymes activated in the course of apoptosis may destroy nonviral cytosolic pathogens. Apoptosis is therefore preferable to necrosis as a means of killing infected cells; in necrosis, intact pathogens are released from the dead cell and these can continue to infect healthy cells, or can parasitize the macrophages that ingest them.

8-22 Cytotoxic effector proteins that trigger apoptosis are contained in the granules of CD8 cytotoxic T cells.

The principal mechanism through which cytotoxic T cells act is by the calcium-dependent release of specialized **lytic granules** upon recognition of antigen on the surface of a target cell. These granules are modified lysosomes that

Protein in lytic granules of cytotoxic T cells	Actions on target cells
Perforin	Polymerizes to form a pore in target membrane
Granzymes	Serine proteases, which activate apoptosis once in the cytoplasm of the target cell

Fig. 8.36 Cytotoxic effector proteins released by cytotoxic T cells.

contain at least two distinct classes of cytotoxic effector protein that are expressed selectively in cytotoxic T cells (Fig. 8.36). Such proteins are stored in the lytic granules in an active form, but conditions within the granules prevent them from functioning until after their release. One of these cytotoxic proteins, known as **perforin**, polymerizes to form transmembrane pores in target cell membranes. The other class of cytotoxic proteins comprises at least three proteases called **granzymes**, which belong to the same family of enzymes—the serine proteases—as the digestive enzymes trypsin and chymotrypsin. Granules that store perforin and granzymes can be seen in armed CD8 cytotoxic effector cells in tissue lesions.

When purified granules from cytotoxic T cells are added to target cells *in vitro*, they lyse the cells by creating pores in the lipid bilayer. The pores consist of polymers of perforin, which is a major constituent of these granules. On release from the granule, perforin forms a cylindrical structure that is lipophilic on the outside and hydrophilic down a hollow center with an inner diameter of 16 nm (Fig. 8.37). It is not known whether this structure is first formed and then inserted into the lipid bilayer of the target cell membrane, or whether it is formed in the bilayer itself. The pore that is formed allows water and salts to pass rapidly into the cell. With the integrity of the cell membrane destroyed, the cells die rapidly. Large numbers of purified granules can kill target cells *in vitro* without inducing fragmentation of cellular DNA, but this lytic mechanism of cell killing probably occurs only at artificially high levels of perforin that do not reflect the physiological activity of cytotoxic T cells.

Both perforin and granzymes are required for effective cell killing. The separate roles of perforin and granzymes have been investigated in a cell system that relies upon similarities between the lytic granules of T cells and the granules of mast cells. Release of mast cell granules occurs on cross-linking of the Fcε

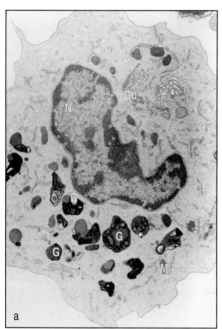
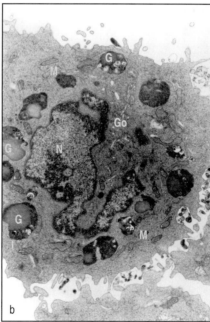
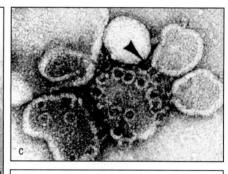
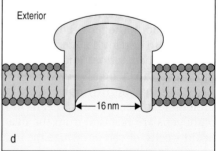

Fig. 8.37 Perforin released from the lytic granules of cytotoxic T cells can insert into the target cell membrane to form pores. Perforin molecules, as well as several other effector molecules, are contained in the granules of cytotoxic T cells (panel a). When a CD8 cytotoxic T cell recognizes its target, the granule contents are released onto the target cell (panel b, bottom right quadrant). The perforin molecules released from the granules polymerize in the membrane of the target cell to form pores. The structure of these pores is best seen when purified perforin is added to synthetic lipid vesicles (panel c: pores are seen both end on, as circles, and sideways on, arrow). The pores span the target cell membrane (panel d). G, granule; N, nucleus; M, mitochondrion; Go, Golgi apparatus. Photographs courtesy of E. Podack.

receptor (see Chapter 9), just as release of lytic granules from CD8 T cells occurs on cross-linking of the T-cell receptor. The mechanism of signaling for granule release is thought to be the same or similar in both cases, as both the Fcε receptor and the T-cell receptor have ITAM motifs in their cytoplasmic domains, and cross-linking leads to tyrosine phosphorylation of the ITAMs (see Chapter 6).

When a mast-cell line is transfected with the gene for perforin or for granzyme, the gene products are stored in mast cell granules, and when the cell is activated through its Fcε receptor, these granules are released. When transfected with the gene for perforin alone, mast cells can kill other cells, but large numbers of the transfected cells are needed as the killing is not very efficient. By contrast, mast cells transfected with the gene for granzyme B alone are unable to kill other cells. However, when perforin-transfected mast cells are also transfected with the gene encoding granzyme B, the cells or their purified granules become as effective at killing targets as granules from cytotoxic cells, and granules from both types of cell induce DNA fragmentation. This suggests that perforin makes pores through which the granzymes can move into the target cell.

The granzymes are proteases, so although they have a role in triggering apoptosis in the target cell, they cannot act directly to fragment the DNA. Rather, they must activate an enzyme, or more probably an enzyme cascade, in the target cell. Granzyme B can cleave the ubiquitous cellular enzyme CPP-32, which is believed to have a key role in programmed cell death in all cells. CPP-32 is a caspase and activates a nuclease, called caspase-activated deoxyribonuclease or CAD, by cleaving an inhibitory protein (ICAD) that binds to and inactivates CAD. This enzyme is believed to be the final effector of DNA degradation in apoptosis.

Cells undergoing programmed cell death are rapidly ingested by nearby phagocytic cells. The phagocytes recognize some change in the cell membrane, most probably the exposure of phosphatidylserine, which is normally found only in the inner leaflet of the membrane. The ingested cell is then completely broken down and digested by the phagocyte without the induction of co-stimulatory proteins. Thus, apoptosis is normally an immunologically 'quiet' process; that is, apoptotic cells do not normally contribute to or stimulate immune responses.

The importance of perforin in this process is well illustrated in mice that have had their perforin gene knocked out. Such mice are severely defective in their ability to mount a cytotoxic T-cell response to many but not all viruses, whereas mice that are defective in the granzyme B gene have a less profound defect, probably because there are several genes coding for granzymes.

8-23   Activated CD8 T cells and some CD4 effector T cells express Fas ligand, which can also activate apoptosis.

The release of granule contents accounts for most of the cytotoxic activity of CD8 effector T cells, as shown by the loss of most killing activity in perforin knockout mice. This granule-mediated killing is strictly calcium-dependent, yet some cytotoxic actions of CD8 T cells survive calcium depletion. Moreover, some CD4 T cells can also kill other cells, yet do not contain granules and make neither perforin nor granzymes. These observations imply that there must be a second perforin-independent mechanism of cytotoxicity. This mechanism involves the binding of Fas in the target cell membrane by the Fas ligand, which is present in the membranes of activated cytotoxic T cells and T_H1 cells. Ligation of Fas leads to activation of caspases, which induce apoptosis in the target cell (see Fig. 6.23). As discussed in Section 8-20, the lymphoproliferative and autoimmune disorders seen in mice and

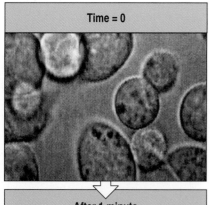

Time = 0

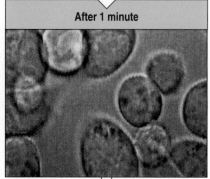

After 1 minute

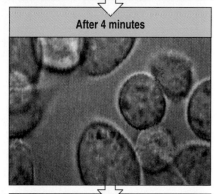

After 4 minutes

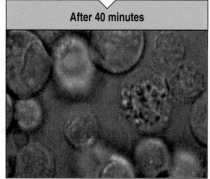

After 40 minutes

Fig. 8.38 Effector molecules are released from T-cell granules in a highly polar fashion. The granules of cytotoxic T cells can be labeled with fluorescent dyes, allowing them to be seen under the microscope, and their movements followed by time-lapse photography. Here we show a series of pictures taken during the interaction of a cytotoxic T cell with a target cell, which is eventually killed. In the top panel, at time 0, the T cell (upper right) has just made contact with a target cell (diagonally below). At this time, the granules of the T cell, labeled with a red fluorescent dye, are distant from the point of contact. In the second panel, after 1 minute has elapsed, the granules have begun to move towards the target cell, a move that has essentially been completed in the third panel, after 4 minutes. After 40 minutes, in the last panel, the granule contents have been released into the space between the T cell and the target, which has begun to undergo apoptosis (note the fragmented nucleus). The T cell will now disengage from the target cell and can recognize and kill other targets. Photographs courtesy of G. Griffiths.

humans with mutations in genes for either Fas or Fas ligand imply that this pathway of killing is very important in regulating peripheral immune responses. Fas is expressed on activated lymphocytes and Fas–Fas ligand interactions are important in terminating lymphocyte growth after the removal of the initiating pathogen.

8-24 Cytotoxic T cells are selective and serial killers of targets expressing specific antigen.

When cytotoxic T cells are offered a mixture of equal amounts of two target cells, one bearing specific antigen and the other not, they kill only the target cell bearing the specific antigen. The 'innocent bystander' cells and the cytotoxic T cells themselves are not killed, despite the fact that cloned cytotoxic T cells can be recognized and killed by other cytotoxic T cells just like any tissue cell. At first sight this may seem surprising, because the effector molecules released by cytotoxic T cells lack any specificity for antigen. The explanation probably lies in the highly polar release of the effector molecules. As we saw in Fig. 8.29, cytotoxic T cells orient their Golgi apparatus and microtubule-organizing center to focus secretion on the point of contact with a target cell. Granule movement toward the point of contact is shown in Fig. 8.38. Cytotoxic T cells attached to several different target cells reorient their secretory apparatus toward each cell in turn and kill them one by one, strongly suggesting that the mechanism whereby cytotoxic mediators are released allows attack at only one point of contact at any one time. The narrowly focused action of cytotoxic CD8 T cells allows them to kill single infected cells in a tissue without creating widespread tissue damage (Fig. 8.39) and is of critical importance in tissues where cell regeneration does not occur, as in neurons of the central nervous system, or is very limited, as in the pancreatic islets.

Cytotoxic T cells can kill their targets rapidly because they store preformed cytotoxic proteins in forms that are inactive in the environment of the lytic granule. Cytotoxic proteins are synthesized and loaded into the lytic granules during the first encounter of a naive cytotoxic precursor T cell with its specific antigen. Ligation of the T-cell receptor similarly induces *de novo* synthesis of perforin and granzymes in armed effector CD8 T cells, so that the supply of lytic granules is replenished. This makes it possible for a single CD8 T cell to kill many targets in succession.

8-25 Cytotoxic T cells also act by releasing cytokines.

Although the secretion of perforin and granzymes is the main way by which cytotoxic CD8 T cells eliminate infection, with the expression of Fas ligand playing a lesser role, most cytotoxic CD8 T cells also release the cytokines IFN-γ, TNF-α, and TNF-β, which contribute to host defense in several ways. IFN-γ directly inhibits viral replication, and also induces the increased expression of MHC class I and other molecules involved in peptide loading of the newly synthesized MHC class I proteins in infected cells. This increases the chance that infected cells will be recognized as target cells for cytotoxic attack. IFN-γ also activates macrophages, recruiting them to sites of infection both as effector cells and as antigen-presenting cells. The activation of macrophages by IFN-γ is a critical component of the host immune response to intracellular protozoan pathogens such as *Toxoplasma gondii*. IFN-γ also has a secondary role in decreasing the tryptophan concentration within responsive cells and thus can kill intracellular parasites, effectively by starvation. TNF-α or TNF-β can synergize with IFN-γ in macrophage activation, and in killing some target cells through their interaction with TNFR-I. Thus, armed effector cytotoxic CD8 T cells act in a variety of ways to limit the spread of cytosolic pathogens. The relative importance of each of these mechanisms remains to be determined.

Summary.

Armed effector cytotoxic CD8 T cells are essential in host defense against pathogens that live in the cytosol, the commonest of which are viruses. These cytotoxic T cells can kill any cell harboring such pathogens by recognizing foreign peptides that are transported to the cell surface bound to MHC class I molecules. Cytotoxic CD8 T cells carry out their killing function by releasing two types of preformed cytotoxic protein: the granzymes, which seem able to induce apoptosis in any type of target cell, and the pore-forming protein perforin, which punches holes in the target-cell membrane through which the granzymes can enter. These properties allow the cytotoxic T cell to attack and destroy virtually any cell that is infected with a cytosolic pathogen. A membrane-bound molecule, the Fas ligand, expressed by CD8 and some CD4 T cells, is also capable of inducing apoptosis by binding to Fas expressed by some target cells. Cytotoxic CD8 T cells also produce IFN-γ, which is an inhibitor of viral replication and is an important inducer of MHC class I expression and macrophage activation. Cytotoxic T cells kill infected targets with great precision, sparing adjacent normal cells. This precision is critical in minimizing tissue damage while allowing the eradication of infected cells.

Macrophage activation by armed CD4 T$_H$1 cells.

Some microorganisms such as mycobacteria, the causative agents of tuberculosis and leprosy, are intracellular pathogens that grow primarily in phagolysosomes of macrophages. There they are shielded from the effects of both antibodies and cytotoxic T cells. These microbes maintain themselves in the usually hostile environment of the phagocyte by inhibiting the fusion of lysosomes to the phagosomes in which they grow, or by preventing the acidification of these vesicles that is required to activate lysosomal proteases. Such microorganisms can be eliminated when the macrophage is activated by a T$_H$1 cell. Armed T$_H$1 cells act by synthesizing membrane-associated proteins and a range of soluble cytokines whose local and distant actions coordinate the immune response to these intracellular pathogens. Armed T$_H$1 effector cells can also activate macrophages to kill recently ingested pathogens.

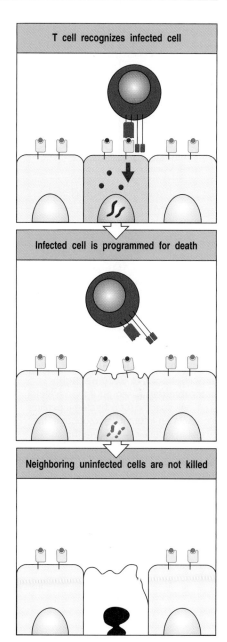

Fig. 8.39 Cytotoxic T cells kill target cells bearing specific antigen while sparing neighboring uninfected cells. All the cells in a tissue are susceptible to lysis by the cytotoxic proteins of armed effector CD8 T cells, but only infected cells are killed. Specific recognition by the T-cell receptor identifies which target cell to kill, and the polarized release of granules (not shown) ensures that neighboring cells are spared.

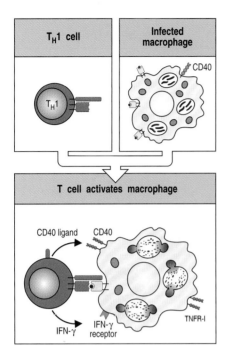

Fig. 8.40 T$_H$1 cells activate macrophages to become highly microbicidal. When an armed effector T$_H$1 cell specific for a bacterial peptide contacts an infected macrophage, the T cell is induced to secrete the macrophage-activating factor IFN-γ and to express CD40 ligand. Together, these newly synthesized T$_H$1 proteins activate the macrophage.

8-26 Armed T$_H$1 cells have a central role in macrophage activation.

A number of important pathogens live within macrophages, whereas many others are ingested by macrophages from the extracellular fluid. In many cases, the macrophage is able to destroy such pathogens without the need for T-cell activation, as we have seen in Chapter 2, but in several clinically important infections CD4 T cells are needed to provide activating signals for macrophages. The induction of antimicrobial mechanisms in macrophages is known as **macrophage activation** and is the principal effector action of T$_H$1 cells. Among the extracellular pathogens that are killed when macrophages are activated is *Pneumocystis carinii*, which, because of a deficiency of CD4 T cells, is a common cause of death in people with AIDS. Macrophage activation can be measured by the ability of activated macrophages to damage a broad spectrum of microbes as well as certain tumor cells. This ability to act on extracellular targets extends to healthy self cells, which means that macrophages must normally be maintained in a nonactivated state.

Macrophages require two signals for activation. One of these is provided by IFN-γ; the other can be provided by a variety of means, and is needed to sensitize the macrophage to respond to IFN-γ. Armed effector T$_H$1 cells can deliver both signals. IFN-γ is the most characteristic cytokine produced by armed T$_H$1 cells on interacting with their specific target cells, whereas the CD40 ligand expressed by the T$_H$1 cell delivers the sensitizing signal by contacting CD40 on the macrophage (Fig. 8.40). CD8 T cells are also an important source of IFN-γ and can activate macrophages presenting antigens derived from cytosolic proteins; mice lacking MHC class I molecules, and which thus have no CD8 T cells, show increased susceptibility to some parasitic infections. Macrophages can be made more sensitive to IFN-γ by very small amounts of bacterial lipopolysaccharide, and this latter pathway may be particularly important when CD8 T cells are the primary source of the IFN-γ. It is also possible that membrane-associated TNF-α or TNF-β can substitute for CD40 ligand in macrophage activation. These cell-associated molecules apparently stimulate the macrophage to secrete TNF-α, and antibody to TNF-α can inhibit macrophage activation. T$_H$2 cells are inefficient macrophage activators because they produce IL-10, a cytokine that can deactivate macrophages, and they do not produce IFN-γ. They do express CD40 ligand, however, and can deliver the contact-dependent signal required to activate macrophages to respond to IFN-γ.

8-27 The production of cytokines and membrane-associated molecules by armed CD4 T$_H$1 cells requires new RNA and protein synthesis.

Within minutes of the recognition of specific antigen by armed effector cytotoxic CD8 T cells, directed exocytosis of preformed perforins and granzymes programs the target cell to die via apoptosis. In contrast, when armed T$_H$1 cells encounter their specific ligand, they must synthesize *de novo* the cytokines and cell-surface molecules that mediate their effects. This process requires hours rather than minutes, so T$_H$1 cells must adhere to their target cells for far longer than cytotoxic T cells.

Recognition of its target by a T$_H$1 cell rapidly induces transcription of cytokine genes and new protein synthesis begins within an hour of receptor triggering. The newly synthesized cytokines are then delivered directly through microvesicles of the constitutive secretory pathway to the site of contact between the T-cell membrane and the macrophage. It is thought that the newly synthesized cell-surface CD40 ligand is also expressed in this polarized fashion. This means that, although all macrophages have receptors for IFN-γ, the macrophage actually displaying antigen to the armed T$_H$1 cell is far more likely to become activated by it than are neighboring uninfected macrophages.

8-28   Activation of macrophages by armed T$_H$1 cells promotes microbial killing and must be tightly regulated to avoid tissue damage.

T$_H$1 cells activate infected macrophages through cell contact and the focal secretion of IFN-γ. This generates a series of biochemical responses that converts the macrophage into a potent antimicrobial effector cell (Fig. 8.41). Activated macrophages fuse their lysosomes more efficiently to phagosomes, exposing intracellular or recently ingested extracellular microbes to a variety of microbicidal lysosomal enzymes. Activated macrophages also make oxygen radicals and nitric oxide (NO), both of which have potent antimicrobial activity, as well as synthesizing antimicrobial peptides and proteases that can be released to attack extracellular parasites.

Additional changes in the activated macrophage help to amplify the immune response. The number of MHC class II molecules, B7 molecules, CD40, and TNF receptors on the macrophage surface increases, making the cell both more effective at presenting antigen to fresh T cells, which may thereby be recruited as effector cells, and more responsive to CD40 ligand and to TNF-α. TNF-α synergizes with IFN-γ in macrophage activation, particularly in the induction of the reactive nitrogen metabolite NO, which has broad antimicrobial activity. The NO is produced by the enzyme **inducible NO synthase (iNOS)**, and mice that have had the gene for iNOS knocked out are highly susceptible to infection with several intracellular pathogens. Activated macrophages secrete IL-12, which directs the differentiation of activated naive CD4 T cells into T$_H$1 effector cells, as we will learn in Chapter 10. These and many other surface and secreted molecules of activated macrophages are instrumental in the effector actions of macrophages in cell-mediated responses, and they are also important effectors in humoral immune responses, which we will discuss in Chapter 9, and in recruiting other immune cells to sites of infection, a function to which we will return in Chapter 10.

Because activated macrophages are extremely effective in destroying pathogens, one may ask why macrophages are not simply maintained in a state of constant activation. Besides the fact that macrophages consume huge quantities of energy to maintain the activated state, macrophage activation *in vivo* is usually associated with localized tissue destruction that apparently results from the release of antimicrobial mediators such as oxygen radicals, NO, and proteases, which are also toxic to host cells. The ability of activated macrophages to release toxic mediators is important in host defense because it enables them to attack large extracellular pathogens that they cannot ingest, such as parasitic worms. This can only be achieved, however, at the expense of tissue damage. Tight regulation of the activity of macrophages by T$_H$1 cells thus allows the specific and effective deployment of this potent means of host defense while minimizing local tissue damage and energy consumption.

Macrophage activation is contained by mechanisms that control IFN-γ synthesis by activated effector T cells. This seems to be achieved by regulating the half-life of the mRNA encoding IFN-γ. IFN-γ mRNA, like that encoding a variety of other cytokines, contains a sequence (AUUUA)$_n$ in its 3' untranslated region that greatly reduces its half-life, and this serves to limit the period of cytokine production. Activation of the T cell appears to induce the production of a new protein that promotes cytokine mRNA degradation: treatment of activated effector T cells with the protein synthesis inhibitor cycloheximide greatly increases the level of cytokine mRNA. The rapid destruction of cytokine mRNA, together with the focal delivery of IFN-γ at the point of contact between the activated T$_H$1 cell and its macrophage target, thus limits the action of the effector T cell to the infected macrophage. We will see in Chapter 9, when we consider the activation of

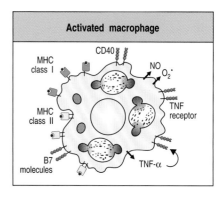

Fig. 8.41 Activated macrophages undergo changes that greatly increase their antimicrobial effectiveness and amplify the immune response. Activated macrophages increase their expression of CD40 and of TNF receptors, and secrete TNF-α. This autocrine stimulus synergizes with IFN-γ secreted by T$_H$1 cells to increase the antimicrobial action of the macrophage, in particular by inducing the production of nitric oxide (NO) and oxygen radicals (O$_2^{\bullet}$). The macrophage also upregulates its B7 molecules in response to binding to CD40 ligand on the T cell, and increases its expression of MHC class II molecules, thus allowing further activation of resting CD4 T cells.

B cells by T_H2 cells, that the same mechanisms direct and limit T-cell help to the specific antigen-binding B cell. In addition, macrophage activation itself is markedly inhibited by cytokines such as transforming growth factor-β (TGF-β), IL-4, IL-10, and IL-13. Because several of these inhibitory cytokines are produced by T_H2 cells, the induction of CD4 T cells belonging to the T_H2 subset represents an important pathway for controlling the effector functions of activated macrophages.

8-29 T_H1 cells coordinate the host response to intracellular pathogens.

The activation of macrophages by armed T_H1 cells expressing CD40 ligand and secreting IFN-γ is central to the host response to pathogens that proliferate in macrophage vesicles. In mice in which the IFN-γ gene or the CD40 ligand gene has been destroyed by targeted gene disruption, production of anti-microbial agents by macrophages is impaired, and the animals succumb to sublethal doses of *Mycobacterium* species and *Leishmania* species. Macrophage activation is also critical in controlling vaccinia virus. Mice lacking TNF receptors also show increased susceptibility to these pathogens. However, although IFN-γ and CD40 ligand are probably the most important effector molecules synthesized by T_H1 cells, the immune response to pathogens that proliferate in macrophage vesicles is complex, and other cytokines secreted by T_H1 cells have a crucial role in coordinating these responses (Fig. 8.42). For example, macrophages that are chronically infected with intracellular bacteria may lose the ability to become activated. Such cells

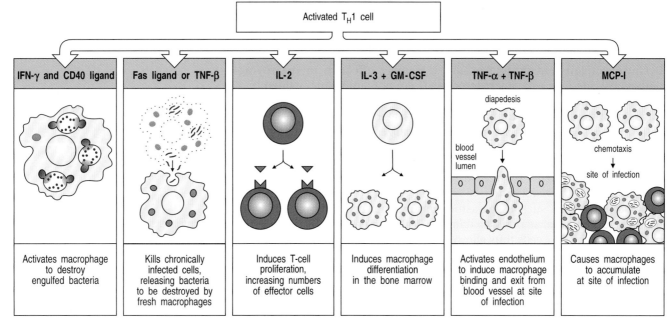

IFN-γ and CD40 ligand	Fas ligand or TNF-β	IL-2	IL-3 + GM-CSF	TNF-α + TNF-β	MCP-I
Activates macrophage to destroy engulfed bacteria	Kills chronically infected cells, releasing bacteria to be destroyed by fresh macrophages	Induces T-cell proliferation, increasing numbers of effector cells	Induces macrophage differentiation in the bone marrow	Activates endothelium to induce macrophage binding and exit from blood vessel at site of infection	Causes macrophages to accumulate at site of infection

Fig. 8.42 The immune response to intracellular bacteria is coordinated by activated T_H1 cells. The activation of T_H1 cells by infected macrophages results in the synthesis of cytokines that both activate the macrophage and coordinate the immune response to intracellular pathogens. IFN-γ and CD40 ligand synergize in activating the macrophage, which allows it to kill engulfed pathogens. Chronically infected macrophages lose the ability to kill intracellular bacteria, and Fas ligand or TNF-β produced by the T_H1 cell can kill these macrophages, releasing the engulfed bacteria, which are taken up and killed by fresh macrophages. In this way, IFN-γ and TNF-β synergize in the removal of intracellular bacteria. IL-2 produced by T_H1 cells induces T-cell proliferation and potentiates the release of other cytokines. IL-3 and GM-CSF stimulate the production of new macrophages by acting on hematopoietic stem cells in the bone marrow. New macrophages are recruited to the site of infection by the action of TNF-α and TNF-β (and other cytokines) on vascular endothelium, which signal macrophages to leave the bloodstream and enter the tissues. A chemokine with macrophage chemotactic activity (MCP-1) signals macrophages to migrate into sites of infection and accumulate there. Thus, the T_H1 cell coordinates a macrophage response that is highly effective in destroying intracellular infectious agents.

could provide a reservoir of infection that is shielded from immune attack. Activated T$_H$1 cells can express Fas ligand and thus kill a limited range of target cells that express Fas, including macrophages, thereby destroying these infected cells.

Whereas some intravesicular bacteria pose a hazard by incapacitating chronically infected macrophages, others, including some mycobacteria and *Listeria monocytogenes,* can escape from cell vesicles and enter the cytoplasm, where they are not susceptible to macrophage activation. Their presence can, however, be detected by cytotoxic CD8 T cells, which can release them by killing the cell. The pathogens released when macrophages are killed either by T$_H$1 cells or by cytotoxic CD8 T cells can be taken up by freshly recruited macrophages still capable of activation to antimicrobial activity.

Another very important function of T$_H$1 cells is the recruitment of phagocytic cells to sites of infection. T$_H$1 cells recruit macrophages by two mechanisms. First, they make the hematopoietic growth factors IL-3 and GM-CSF, which stimulate the production of new phagocytic cells in the bone marrow. Second, TNF-α and TNF-β, which are secreted by T$_H$1 cells at sites of infection, change the surface properties of endothelial cells so that phagocytes adhere to them, while chemokines such as macrophage chemotactic protein (MCP-1), produced by T$_H$1 cells in the inflammatory response, serve to direct the migration of these phagocytic cells through the vascular endothelium to the site of the infection (see Section 10-8).

When microbes effectively resist the microbicidal effects of activated macrophages, chronic infection with inflammation can develop. Often, this has a characteristic pattern, consisting of a central area of macrophages surrounded by activated lymphocytes. This pathological pattern is called a **granuloma** (Fig. 8.43). Giant cells consisting of fused macrophages usually form the center of these granulomas. This serves to 'wall-off' pathogens that resist destruction. T$_H$2 cells seem to participate in granulomas along with T$_H$1 cells, perhaps by regulating their activity and preventing widespread tissue damage. In tuberculosis, the center of the large granulomas can become isolated and the cells there die, probably from a combination of lack of oxygen and the cytotoxic effects of activated macrophages. As the dead tissue in the center resembles cheese, this process is called **caseation necrosis**. Thus, the activation of T$_H$1 cells can cause significant pathology. Their nonactivation, however, leads to the more serious consequence of death from disseminated infection, which is now seen frequently in patients with AIDS and concomitant mycobacterial infection.

Summary.

CD4 T cells that can activate macrophages have a critical role in host defense against those intracellular and extracellular pathogens that resist killing after being engulfed by macrophages. Macrophages are activated by membrane-bound signals delivered by activated T$_H$1 cells as well as by the potent macrophage-activating cytokine IFN-γ, which is secreted by activated T cells. Once activated, the macrophage can kill intracellular and ingested bacteria. Activated macrophages can also cause local tissue damage, which explains why this activity must be strictly regulated by antigen-specific T cells. T$_H$1 cells produce a range of cytokines and surface molecules that not only activate infected macrophages but can also kill chronically infected senescent macrophages, stimulate the production of new macrophages in bone marrow, and recruit fresh macrophages to sites of infection. Thus, T$_H$1 cells have a central role in controlling and coordinating host defense against certain intracellular pathogens. It is likely that the absence of this function explains the preponderance of infections with intracellular pathogens in adult AIDS patients.

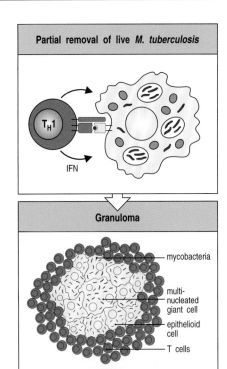

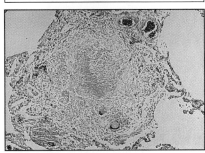

Fig. 8.43 Granulomas form when an intracellular pathogen or its constituents cannot be totally eliminated. When mycobacteria (red) resist the effects of macrophage activation, a characteristic localized inflammatory response called a granuloma develops. This consists of a central core of infected macrophages. The core may include multinucleated giant cells, which are fused macrophages, surrounded by large macrophages often called epithelioid cells. Mycobacteria can persist in the cells of the granuloma. The central core is surrounded by T cells, many of which are CD4-positive. The exact mechanisms by which this balance is achieved, and how it breaks down, are unknown. Granulomas, as seen in the bottom panel, also form in the lungs and elsewhere in a disease known as sarcoidosis, which may be caused by occult mycobacterial infection. Photograph courtesy of J. Orrell.

Summary to Chapter 8.

Armed effector T cells are crucial to almost all adaptive immune responses. Adaptive immune responses are initiated when naive T cells encounter specific antigen on the surface of an antigen-presenting cell that also expresses the co-stimulatory molecules B7.1 and B7.2. In most cases, these first encounters with antigen are thought to occur with a dendritic cell that has taken up antigen at a site of infection, migrated to local lymphoid tissue and matured to become a potent activator of naive T cells. The activated T cells produce IL-2, which drives them to proliferate and differentiate into armed effector T cells. All T-cell effector functions involve cell–cell interactions. When armed effector T cells recognize specific antigen on target cells, they release mediators that act directly on the target cell, altering its behavior. The triggering of armed effector T cells by peptide:MHC complexes is independent of co-stimulation, so that any infected target cell can be activated or destroyed by an armed effector T cell. CD8 cytotoxic T cells kill target cells infected with cytosolic pathogens, removing sites of pathogen replication. CD4 T_H1 cells activate macrophages to kill intracellular parasites. CD4 T_H2 cells are essential in the activation of B cells to secrete the antibodies that mediate humoral immune responses directed against extracellular pathogens, as will be seen in Chapter 9. Thus, effector T cells control virtually all known effector mechanisms of the adaptive immune response.

General references.

Ihle, J.N.: **Cytokine receptor signaling.** *Nature* 1995, **377**:591-594.

Janeway, C.A., and Bottomly, K.: **Signals and signs for lymphocyte responses.** *Cell* 1994, **76**:275-285.

Lenschow, D.J., Walunas, T.L., and Bluestone, J.A.: **CD28/B7 system of T cell costimulation.** *Annu. Rev. Immunol.* 1996, **14**:233-258.

Mosmann, T.R., and Coffman, R.L.: **T_H1 and T_H2 cells: different patterns of lymphokine secretion lead to different functional properties.** *Annu. Rev. Immunol.* 1989, **7**:145-173.

Springer, T.A.: **Traffic signals for lymphocyte recirculation and leukocyte emigration: the multistep paradigm.** *Cell* 1994, **76**:301-314.

Section references.

8-1 **T-cell responses are initiated in peripheral lymphoid organs by activated antigen-presenting cells.**

Picker, L.J., and Butcher, E.C.: **Physiological and molecular mechanisms of lymphocyte homing.** *Annu. Rev. Immunol.* 1993, **10**:561-591.

Steptoe, R.J., Li, W., Fu, F., O'Connell, P.J., and Thomson, A.W.: **Trafficking of APC from liver allografts of Flt3L-treated donors: augmentation of potent allostimulatory cells in recipient lymphoid tissue is associated with a switch from tolerance to rejection.** *Transpl. Immunol.* 1999, **7**:51-57.

8-2 **Naive T cells sample the MHC:peptide complexes on the surface of antigen-presenting cells as they migrate through peripheral lymphoid tissue.**

Schlienger, K., Craighead, N., Lee, K.P., Levine, B.L., and June, C.H.: **Efficient priming of protein antigen-specific human CD4($^+$) T cells by monocyte-derived dendritic cells.** *Blood* 2000, **96**:3490-3498.

Thery, C., and Amigorena, S.: **The cell biology of antigen presentation in dendritic cells.** *Curr. Opin. Immunol.* 2001, **13**:45-51.

8-3 **Lymphocyte migration, activation, and effector function depend on cell–cell interactions mediated by cell-adhesion molecules.**

Madri, J.A., and Graesser, D.: **Cell migration in the immune system: the evolving inter-related roles of adhesion molecules and proteinases.** *Dev. Immunol.* 2000, **7**:103-116.

Picker, L.J.: **Control of lymphocyte homing.** *Curr. Opin. Immunol.* 1994, **6**:394-406.

8-4 **The initial interaction of T cells with antigen-presenting cells is mediated by cell-adhesion molecules.**

Ganpule, G., Knorr, R., Miller, J.M., Carron, C.P., and Dustin, M.L.: **Low affinity of cell surface lymphocyte function-associated antigen-1 (LFA-1) generates selectivity for cell-cell interactions.** *J. Immunol.* 1997, **159**:2685-2692.

Grakoui, A., Bromley, S.K., Sumen, C., Davis, M.M., Shaw, A.S., Allen, P.M., and Dustin, M.L.: **The immunological synapse: a molecular machine controlling T cell activation.** *Science* 1999, **285**:221-227.

Gunzer, M., Schafer, A., Borgmann, S., Grabbe, S., Zanker, K.S., Brocker, E.B., Kampgen, E., and Friedl, P.: **Antigen presentation in extracellular matrix: interactions of T cells with dendritic cells are dynaic, short lived, and sequential.** *Immunity* 2000, **13**:323-332.

8-5 **Both specific ligand and co-stimulatory signals provided by an antigen-presenting cell are required for the clonal expansion of naive T cells.**

Gonzalo, J.A., Delaney, T., Corcoran, J., Goodearl, A., Gutierrez-Ramos, J.C., and Coyle, A.J.: **Cutting edge: the related molecules CD28 and inducible costimulator deliver both unique and complementary signals required for optimal T cell activation.** *J. Immunol.* 2001, **166**:1-5.

Henry, J., Miller, M.M., and Pontarotti, P.: **Structure and evolution of the extended B7 family.** *Immunol. Today* 1999, **20**:285-288.

Wang, S., Zhu, G., Chapoval, A.I., Dong, H., Tamada, K., Ni, J., and Chen, L.:

Costimulation of T cells by B7-H2, a B7-like molecule that binds ICOS. *Blood* 2000, **96**:2808-2813.

8-6 Dendritic cells specialize in taking up antigen and activating naive T cells.

Banchereau, J., and Steinman, R.M.: **Dendritic cells and the control of immunity.** *Nature* 1998, **392**:245-252.

Blanas, E., Davey, G.M., Carbone, F.R., and Heath, W.R.: **A bone marrow-derived APC in the gut-associated lymphoid tissue captures oral antigens and presents them to both CD4⁺ and CD8⁺ T cells.** *J. Immunol.* 2000, **164**:2890-2896.

8-7 Macrophages are scavenger cells that can be induced by pathogens to present foreign antigens to naive T cells.

Goerdt, S., Politz, O., Schledzewski, K., Birk, R., Gratchev, A., Guillot, P., Hakiy, N., Klemke, C.D., Dippel, E., Kodelja, V., and Orfanos, C.E.: **Alternative versus classical activation of macrophages.** *Pathobiology* 1999, **67**:222-226.

Mitchell, D.A., Nair, S.K., and Gilboa, E.: **Dendritic cell/macrophage precursors capture exogenous antigen for MHC class I presentation by dendritic cells.** *Eur. J. Immunol.* 1998, **28**:1923-1933.

Underhill, D.M., Bassetti, M., Rudensky, A., and Aderem, A.: **Dynamic interactions of macrophages with T cells during antigen presentation.** *J. Exp. Med.* 1999, **190**:1909-1914.

8-8 B cells are highly efficient at presenting antigens that bind to their surface immunoglobulin.

Guermonprez, P., England, P., Bedouelle, H., and Leclerc, C.: **The rate of dissociation between antibody and antigen determines the efficiency of antibody-mediated antigen presentation to T cells.** *J. Immunol.* 1998, **161**:4542-4548.

Lanzavecchia, A.: **Receptor-mediated antigen uptake and its effect on antigen presentation to class II-restricted T lymphocytes.** *Annu. Rev. Immunol.* 1993, **8**:773-793.

8-9 Activated T cells synthesize the T-cell growth factor interleukin-2 and its receptor.

&

8-10 The co-stimulatory signal is necessary for the synthesis and secretion of IL-2.

Cerdan, C., Martin, Y., Courcoul, M., Mawas, C., Birg, F., and Olive, D.: **CD28 costimulation regulates long-term expression of the three genes (alpha, beta, gamma) encoding the high-affinity IL2 receptor.** *Res. Immunol.* 1995, **146**:164-168.

Jain, J., Loh, C., and Rao, A.: **Transcriptional regulation of the IL-2 gene.** *Curr. Opin. Immunol.* 1995, **7**:333-342.

Minami, Y., Kono, T., Miyazaki, T., and Taniguchi, T.: **The IL-2 receptor complex: its structure, function, and target genes.** *Annu. Rev. Immunol.* 1993, **11**:245-267.

8-11 Antigen recognition in the absence of co-stimulation leads to T-cell tolerance.

Chai, J.G., Vendetti, S., Bartok, I., Schoendorf, D., Takacs, K., Elliott, J., Lechler, R., and Dyson, J.: **Critical role of costimulation in the activation of naive antigen-specific TCR transgenic CD8⁺ T cells in vitro.** *J. Immunol.* 1999, **163**:1298-1305.

Bachmann, M.F., McKall-Faienza, K., Schmits, R., Bouchard, D., Beach, J., Speiser, D.E., Mak, T.W., and Ohashi, P.S.: **Distinct roles for LFA-1 and CD28 during activation of naive T cells: adhesion versus costimulation.** *Immunity* 1997, **7**:549-557.

Greenfield, E.A., Nguyen, K.A., and Kuchroo, V.K.: **CD28/B7 costimulation: a review.** *Crit. Rev. Immunol.* 1998, **18**:389-418.

8-12 Proliferating T cells differentiate into armed effector T cells that do not require co-stimulation to act.

Gudmundsdottir, H., Wells, A.D., and Turka, L.A.: **Dynamics and requirements of T cell clonal expansion in vivo at the single-cell level: effector function is linked to proliferative capacity.** *J. Immunol.* 1999, **162**:5212-5223.

London, C.A., Lodge, M.P., and Abbas, A.K.: **Functional responses and co-stimulator dependence of memory CD4⁺ T cells.** *J. Immunol.* 2000, **164**:265-272.

Schweitzer, A.N., and Sharpe, A.H.: **Studies using antigen-presenting cells lacking expression of both B7-1 (CD80) and B7-2 (CD86) show distinct requirements for B7 molecules during priming versus restimulation of Th2 but not Th1 cytokine production.** *J. Immunol.* 1998, **161**:2762-2771.

8-13 The differentiation of CD4 T cells into T_H1 or T_H2 cells determines whether humoral or cell-mediated immunity will predominate.

Bradley, L.M., Harbertson, J., Freschi, G.C., Kondrack, R., and Linton, P.J.: **Regulation of development and function of memory CD4 subsets.** *Immunol. Res.* 2000, **21**:149-158.

Nath, I., Vemuri, N., Reddi, A.L., Jain, S., Brooks, P., Colston, M.J., Misra, R.S., and Ramesh, V.: **The effect of antigen presenting cells on the cytokine profiles of stable and reactional lepromatous leprosy patients.** *Immunol. Lett.* 2000, **75**:69-76.

O'Garra, A., and Arai, N.: **The molecular basis of T helper 1 and T helper 2 cell differentiation.** *Trends Cell Biol.* 2000, **10**:542-550.

8-14 Naive CD8 T cells can be activated in different ways to become armed cytotoxic effector cells.

Andreasen, S.O., Christensen, J.E., Marker, O., and Thomsen, A.R.: **Role of CD40 ligand and CD28 in induction and maintenance of antiviral CD8⁺ effector T cell responses.** *J. Immunol.* 2000, **164**:3689-3697.

Mackey, M.F., Barth, R.J., and Noelle, R.J.: **The role of CD40/CD154 interactions in the priming, differentiation, and effector function of helper and cytotoxic T cells.** *J. Leukoc. Biol.* 1998, **63**:418-428.

Thomsen, A.R., Nansen, A., Christensen, J.P., Andreasen, S.O., and Marker, O.: **CD40 ligand is pivotal to efficient control of virus replication in mice infected with lymphocytic choriomeningitis virus.** *J. Immunol.* 1998, **161**:4583-4590.

8-15 Effector T-cell interactions with target cells are initiated by antigen-nonspecific cell-adhesion molecules.

O'Rourke, A.M., and Mescher, M.F.: **Cytotoxic T lymphocyte activation involves a cascade of signaling and adhesion events.** *Nature* 1992, **358**:253-255.

Rodrigues, M., Nussezwieg, R.S., Romero, P., and Zavala, F.: **The in vivo cytotoxic activity of CD8⁺ T-cell clones correlates with their levels of expression of adhesion molecules.** *J. Exp. Med.* 1992, **175**:895-905.

van Seventer, G.A., Simuzi, Y., and Shaw, S.: **Roles of multiple accessory molecules in T-cell activation.** *Curr. Opin. Immunol.* 1991, **3**:294-303.

8-16 Binding of the T-cell receptor complex directs the release of effector molecules and focuses them on the target cell.

Griffiths, G.M.: **The cell biology of CTL killing.** *Curr. Opin. Immunol.* 1995, **7**:343-348.

Kupfer, H., Monks, C.R., and Kupfer, A.: **Small splenic B cells that bind to antigen-specific T helper (Th) cells and face the site of cytokine production in the Th cells selectively proliferate: immunofluorescence microscopic studies of Th-B antigen-presenting cell interactions.** *J. Exp. Med.* 1994, **179**:1507-1515.

Monks, C.R., Freiberg, B.A., Kupfer, H., Sciaky, N., and Kupfer, A.: **Three-dimensional segregation of supramolecular activation clusters in T cells.** *Nature* 1998, **395**:82-86.

8-17 The effector functions of T cells are determined by the array of effector molecules they produce.

&

8-18 Cytokines can act locally or at a distance.

Guidotti, L.G., and Chisari, F.V.: **Cytokine-mediated control of viral infections.** *Virology* 2000, **273**:221-227.

Harty, J.T., Tvinnereim, A.R., and White, D.W.: **CD8⁺ T cell effector mechanisms in resistance to infection.** *Annu. Rev. Immunol.* 2000, **18**:275-308.

Romagnani, S.: **Th1/Th2 cells.** *Inflamm. Bowel Dis.* 1999, **5**:285-294.

Hunter, C.A., and Reiner, S.L.: **Cytokines and T cells in host defense.** *Curr. Opin. Immunol.* 2000, **12**:413-418.

Arai, K., Lee, F., Miyajima, A., Miyatake, S., Arai, N., and Yokota, T.: **Cytokines: co-ordinators of immune and inflammatory responses.** *Annu. Rev. Biochem.* 1990, **59**:783.

8-19 Cytokines and their receptors fall into distinct families of structurally related proteins.

Balkwill, F.: **The molecular and cellular biology of the chemokines.** *J. Viral Hepat.* 1998, **5**:1-14.

Bravo, J., and Heath, J.K.: **Receptor recognition by gp130 cytokines.** *EMBO J.* 2000, **19**:2399-2411.

Schindler, C., and Brutsaert, S.: **Interferons as a paradigm for cytokine signal transduction.** *Cell Mol. Life Sci.* 1999, **55**:1509-1522.

Thompson, A.: *The Cytokine Handbook.* 2nd edn. Academic Press, San Diego, 1994.

Wajant, H., Grell, M., and Scheurich, P.: **TNF receptor associated factors in cytokine signaling.** *Cytokine Growth Factor Rev.* 1999, **10**:15-26.

8-20 The TNF family of cytokines are trimeric proteins that are often associated with the cell surface.

Matsumoto, M.: **Role of TNF ligand and receptor family in the lymphoid organogenesis defined by gene targeting.** *J. Med. Invest.* 1999, **46**:141-150.

Mullberg, J., Althoff, K., Jostock, T., and Rose-John, S.: **The importance of shedding of membrane proteins for cytokine biology.** *Eur. Cytokine Netw.* 2000, **11**:27-38.

Orlinick, J.R., and Chao, M.V.: **TNF-related ligands and their receptors.** *Cell Signal.* 1998, **10**:543-551.

Screaton, G., and Xu, X.N.: **T cell life and death signalling via TNF-receptor family members.** *Curr. Opin. Immunol.* 2000, **12**:316-322.

8-21 Cytotoxic T cells can induce target cells to undergo programmed cell death.

Henkart, P.A.: **Lymphocyte-mediated cytotoxicology: two pathways and multiple effector molecules.** *Immunity* 1994, **1**:343-346.

Squier, M.K.T., and Cohen, J.J.: **Cell-mediated cytotoxic mechanisms.** *Curr. Opin. Immunol.* 1994, **6**:447-452.

8-22 Cytotoxic effector proteins that trigger apoptosis are contained in the granules of CD8 cytotoxic T cells.

Barry, M., Heibein, J.A., Pinkoski, M.J., Lee, S.F., Moyer, R.W., Green, D.R., and Bleackley, R.C.: **Granzyme B short-circuits the need for caspase 8 activity during granule-mediated cytotoxic T-lymphocyte killing by directly cleaving Bid.** *Mol. Cell Biol.* 2000, **20**:3781-3794.

Edwards, K.M., Davis, J.E., Browne, K.A., Sutton, V.R., and Trapani, J.A.: **Antiviral strategies of cytotoxic T lymphocytes are manifested through a variety of granule-bound pathways of apoptosis induction.** *Immunol. Cell Biol.* 1999, **77**:76-89.

Kägi, B., Ledermann, K., and Bürki, R.: **Molecular mechanisms of lymphocyte-mediated cytotoxicity and their role in immunological protection and pathogenesis *in vivo*.** *Annu. Rev. Immunol.* 1994, **12**:207-232.

8-23 Activated CD8 T cells and some CD4 effector T cells express Fas ligand, which can also activate apoptosis.

Medana, I.M., Gallimore, A., Oxenius, A., Martinic, M.M., Wekerle, H., and Neumann, H.: **MHC class I-restricted killing of neurons by virus-specific CD8+ T lymphocytes is effected through the Fas/FasL, but not the perforin pathway.** *Eur. J. Immunol.* 2000, **30**:3623-3633.

Suda, T., Takahashi, T., Goldstein P., and Nagata, S.: **Molecular cloning and expression of the Fas ligand, a novel member of the tumor necrosis factor family.** *Cell* 1993, **75**:1169-1178.

Watanbe, F.R., Branna, C.I., Copeland, N.G., Jenkins, N.A., and Nagata, S.: **Lymphoproliferation disorder in mice explained by defects in Fas antigen that mediates apoptosis.** *Nature* 1992, **356**:314-317.

8-24 Cytotoxic T cells are selective and serial killers of targets expressing specific antigen.

Kuppers, R.C., and Henney, C.S.: **Studies on the mechanism of lymphocyte-mediated cytolysis. IX. Relationships between antigen recognition and lytic expression in killer T cells.** *J. Immunol.* 1977, **118**:71-76.

8-25 Cytotoxic T cells also act by releasing cytokines.

Callan, M.F., Fazou, C., Yang, H., Rostron, T., Poon, K., Hatton, C., and McMichael, A.J.: **CD8(+) T-cell selection, function, and death in the primary immune response *in vivo*.** *J. Clin. Invest.* 2000, **106**:1251-1261.

Fowler, D.H., and Gress, R.E.: **Th2 and Tc2 cells in the regulation of GVHD, GVL, and graft rejection: considerations for the allogeneic transplantation therapy of leukemia and lymphoma.** *Leuk. Lymphoma* 2000, **38**:221-234.

Vukmanovic-Stejic, M., Vyas, B., Gorak-Stolinska, P., Noble, A., and Kemeny, D.M.: **Human Tc1 and Tc2/Tc0 CD8 T-cell clones display distinct cell surface and functional phenotypes.** *Blood* 2000, **95**:231-240.

Wang, B., Fujisawa, H., Zhuang, L., Freed, I., Howell, B.G., Shahid, S., Shivji, G.M., Mak, T.W., and Sauder, D.N.: **CD4+ Th1 and CD8+ type 1 cytotoxic T cells both play a crucial role in the full development of contact hypersensitivity.** *J. Immunol.* 2000, **165**:6783-6790.

8-26 Armed T$_H$1 cells have a central role in macrophage activation.

Munoz Fernandez, M.A., Fernandez, M.A., and Fresno, M.: **Synergism between tumor necrosis factor-α and interferon-γ on macrophage activation for the killing of intracellular *Trypanosoma crusi* through a nitric oxide-dependent mechanism.** *Eur. J. Immunol.* 1992, **22**:301-307.

Stout, R., and Bottomly, K.: **Antigen-specific activation of effector macrophages by interferon-γ producing (T$_H$1) T-cell clones: failure of IL-4 producing (T$_H$2) T-cell clones to activate effector functions in macrophages.** *J. Immunol.* 1989, **142**:760.

8-27 The production of cytokines and membrane-associated molecules by armed CD4 T$_H$1 cells requires new RNA and protein synthesis.

Shaw, G., and Karmen, R.: **A conserved UAU sequence from the 3′ untranslated region of GM-CSF mRNA mediates selective mRNA degradation.** *Cell* 1986, **46**:659.

8-28 Activation of macrophages by armed T$_H$1 cells promotes microbial killing and must be tightly regulated to avoid tissue damage.

James, D.G.: **A clinicopathological classification of granulomatous disorders.** *Postgrad. Med. J.* 2000, **76**:457-465.

Paulnock, D.M.: **Macrophage activation by T cells.** *Curr. Opin. Immunol.* 1992, **4**:344-349.

8-29 T$_H$1 cells coordinate the host response to intracellular pathogens.

Alexander, J., Satoskar, A.R., and Russell, D.G.: **Leishmania species: models of intracellular parasitism.** *J. Cell. Sci.* 1999, **112**:2993-3002.

Denkers, E.Y., and Gazzinelli, R.T.: **Regulation and function of T-cell-mediated immunity during Toxoplasma gondii infection.** *Clin. Microbiol. Rev.* 1998, **11**:569-588.

Yamamura, M., Uyemura, K., Deans, R,J., Weinberg, K., Rea, T.H., Bloom, B.R., and Modlin, R.L.: **Defining protective responses to pathogens: cytokine profiles in leprosy lesions.** *Science* 1991, **254**:277-279.

The Humoral Immune Response

9

Many of the bacteria that cause infectious disease in humans multiply in the extracellular spaces of the body, and most intracellular pathogens spread by moving from cell to cell through the extracellular fluids. The extracellular spaces are protected by the **humoral immune response**, in which antibodies produced by B cells cause the destruction of extracellular microorganisms and prevent the spread of intracellular infections. The activation of B cells and their differentiation into antibody-secreting plasma cells (Fig. 9.1) is triggered by antigen and usually requires helper T cells. The term 'helper T cell' is often used to mean a cell from the T_H2 class of CD4 T cells (see Chapter 8), but a subset of T_H1 cells can also help in B-cell activation. In this chapter we will therefore use the term **helper T cell** to mean any armed effector CD4 T cell that can activate a B cell. Helper T cells also control isotype switching and have a role in initiating somatic hypermutation of antibody variable V-region genes, molecular processes that were described in Chapter 4.

Antibodies contribute to immunity in three main ways (see Fig. 9.1). To enter cells, viruses and intracellular bacteria bind to specific molecules on the target cell surface. Antibodies that bind to the pathogen can prevent this and are said to **neutralize** the pathogen. Neutralization by antibodies is also important in preventing bacterial toxins from entering cells. Antibodies protect against bacteria that multiply outside cells mainly by facilitating uptake of the pathogen by phagocytic cells that are specialized to destroy ingested bacteria. Antibodies do this in either of two ways. In the first, bound antibodies coating the pathogen are recognized by Fc receptors on phagocytic cells that bind to

Fig. 9.1 The humoral immune response is mediated by antibody molecules that are secreted by plasma cells. Antigen that binds to the B-cell antigen receptor signals B cells and is, at the same time, internalized and processed into peptides that activate armed helper T cells. Signals from the bound antigen and from the helper T cell induce the B cell to proliferate and differentiate into a plasma cell secreting specific antibody (top two panels). These antibodies protect the host from infection in three main ways. They can inhibit the toxic effects or infectivity of pathogens by binding to them: this is termed neutralization (bottom left panel). By coating the pathogens, they can enable accessory cells that recognize the Fc portions of arrays of antibodies to ingest and kill the pathogen, a process called opsonization (bottom center panel). Antibodies can also trigger activation of the complement system. Complement proteins can strongly enhance opsonization, and can directly kill some bacterial cells (bottom right panel).

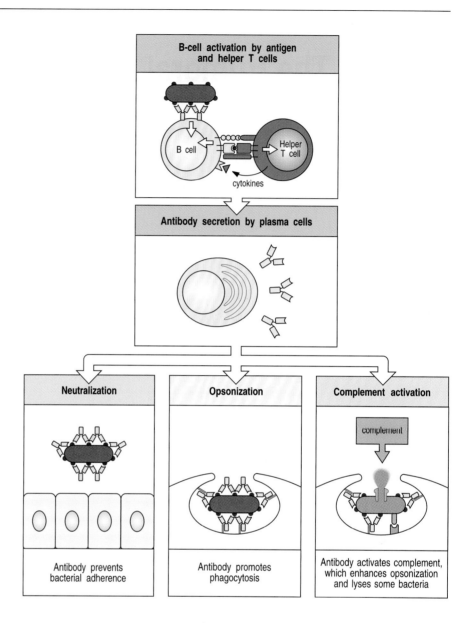

the antibody constant C region (see Section 4-18). Coating the surface of a pathogen to enhance phagocytosis is called **opsonization**. Alternatively, antibodies binding to the surface of a pathogen can activate the proteins of the complement system, which was described in Chapter 2. **Complement activation** results in complement proteins being bound to the pathogen surface, and these opsonize the pathogen by binding complement receptors on phagocytes. Other complement components recruit phagocytic cells to the site of infection, and the terminal components of complement can lyse certain microorganisms directly by forming pores in their membranes. Which effector mechanisms are engaged in a particular response is determined by the isotype or class of the antibodies produced.

In the first part of this chapter we will describe the interactions of B cells with helper T cells that lead to the production of antibodies, the affinity maturation of this antibody response, the isotype switching that confers functional diversity, and the generation of memory B cells that provide long-lasting immunity to reinfection. In the rest of the chapter we will discuss in detail the mechanisms whereby antibodies contain and eliminate infections.

B-cell activation by armed helper T cells.

The surface immunoglobulin that serves as the **B-cell antigen receptor (BCR)** has two roles in B-cell activation. First, like the antigen receptor on T cells, it transmits signals directly to the cell's interior when it binds antigen (see Section 6-1). Second, the B-cell antigen receptor delivers the antigen to intracellular sites where it is degraded and returned to the B-cell surface as peptides bound to MHC class II molecules (see Chapter 5). The peptide:MHC class II complex can be recognized by antigen-specific armed helper T cells, stimulating them to make proteins that, in turn, cause the B cell to proliferate and its progeny to differentiate into antibody-secreting cells. Some microbial antigens can activate B cells directly in the absence of T-cell help. The ability of B cells to respond directly to these antigens provides a rapid response to many important bacterial pathogens. However, somatic hypermutation and switching to certain immunoglobulin isotypes depend on the interaction of antigen-stimulated B cells with helper T cells and other cells in the peripheral lymphoid organs. Antibodies induced by microbial antigens alone are therefore less variable and less functionally versatile than those induced with T-cell help.

9-1 The humoral immune response is initiated when B cells that bind antigen are signaled by helper T cells or by certain microbial antigens alone.

It is a general rule in adaptive immunity that naive antigen-specific lymphocytes are difficult to activate by antigen alone. Naive T cells require a co-stimulatory signal from professional antigen-presenting cells; naive B cells require accessory signals that can come either from an armed helper T cell or, in some cases, directly from microbial constituents.

Antibody responses to protein antigens require antigen-specific T-cell help. B cells can receive help from armed helper T cells when antigen bound by surface immunoglobulin is internalized and returned to the cell surface as peptides bound to MHC class II molecules. Armed helper T cells that recognize the peptide:MHC complex then deliver activating signals to the B cell. Thus, protein antigens binding to B cells both provide a specific signal to the B cell by cross-linking its antigen receptors and allow the B cell to attract antigen-specific T-cell help. These antigens are unable to induce antibody responses in animals or humans who lack T cells, and they are therefore known as **thymus-dependent** or **TD antigens** (Fig. 9.2, top two panels).

The **B-cell co-receptor** complex of CD19:CD21:CD81 (see Section 6-8) can greatly enhance B-cell responsiveness to antigen. CD21 (also known as complement receptor 2, CR2) is a receptor for the complement fragment C3d (see Section 2-11). When mice are immunized with hen egg lysozyme coupled to three linked molecules of the complement fragment C3dg, the modified lysozyme induces antibody without added adjuvant at doses up to 10,000 times smaller than unmodified hen egg lysozyme. Whether binding of CD21 enhances B-cell responsiveness by increasing B-cell signaling, by inducing co-stimulatory molecules on the B cell, or by increasing the receptor-mediated uptake of antigen, is not yet known. As we will see later in this chapter, antibodies already bound to antigens can activate the complement system, thus coating the antigen with C3d and producing a more potent antigen, which in turn leads to more efficient B-cell activation and antibody production.

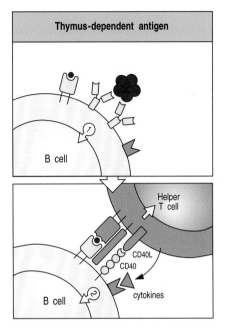

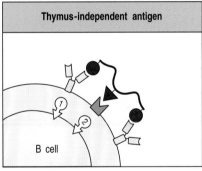

Fig. 9.2 A second signal is required for B-cell activation by either thymus-dependent or thymus-independent antigens. The first signal required for B-cell activation is delivered through its antigen receptor (top panel). For thymus-dependent antigens, the second signal is delivered by a helper T cell that recognizes degraded fragments of the antigen as peptides bound to MHC class II molecules on the B-cell surface (center panel); the interaction between CD40 ligand (CD40L) on the T cell and CD40 on the B cell contributes an essential part of this second signal. For thymus-independent antigens, the second signal can be delivered by the antigen itself (lower panel), or by non-thymus-derived accessory cells (not shown).

Although armed peptide-specific helper T cells are required for B-cell responses to protein antigens, many microbial constituents, such as bacterial polysaccharides, can induce antibody production in the absence of helper T cells. These microbial antigens are known as **thymus-independent** or **TI antigens** because they induce antibody responses in individuals who have no T lymphocytes. The second signal required to activate antibody production to TI antigens is either provided directly by recognition of a common microbial constituent (see Fig. 9.2, bottom panel) or by a nonthymus-derived accessory cell in conjunction with massive cross-linking of B-cell receptors, which would occur when a B cell binds repeating epitopes on the bacterial cell. Thymus-independent antibody responses provide some protection against extracellular bacteria, and we will return to them later.

9-2   Armed helper T cells activate B cells that recognize the same antigen.

T-cell dependent antibody responses require the activation of B cells by helper T cells that respond to the same antigen; this is called **linked recognition**. This means that before B cells can be induced to make antibody to an infecting pathogen, a CD4 T cell specific for peptides from this pathogen must first be activated to produce the appropriate armed helper T cells. This presumably occurs by interaction with an antigen-presenting dendritic cell (see Section 8-1). Although the epitope recognized by the armed helper T cell must therefore be linked to that recognized by the B cell, the two cells need not recognize identical epitopes. Indeed, we saw in Chapter 5 that T cells can recognize internal peptides that are quite distinct from the surface epitopes on the same protein recognized by B cells. For more complex natural antigens, such as viruses, the T cell and the B cell might not even recognize the same protein. It is, however, crucial that the peptide recognized by the T cell be a physical part of the antigen recognized by the B cell, which can thus produce the appropriate peptide after internalization of the antigen bound to its B-cell receptors.

For example, by recognizing an epitope on a viral protein coat, a B cell can internalize a complete virus particle. After internalization, the virus particle is degraded and peptides from internal viral proteins as well as coat proteins can be displayed by MHC class II molecules on the B-cell surface. Helper T cells that have been primed earlier in an infection by macrophages or dendritic cells presenting these internal peptides can then activate the B cell to make antibodies that recognize the coat protein (Fig. 9.3).

The specific activation of the B cell by a T cell sensitized to the same antigen or pathogen depends on the ability of the antigen-specific B cell to concentrate the appropriate peptide on its surface MHC class II molecules. B cells that bind a particular antigen are up to 10,000 times more efficient at displaying peptide fragments of that antigen on their MHC class II molecules than are B cells that do not bind the antigen. Armed helper T cells will thus help only those B cells whose receptors bind an antigen containing the peptide they recognize.

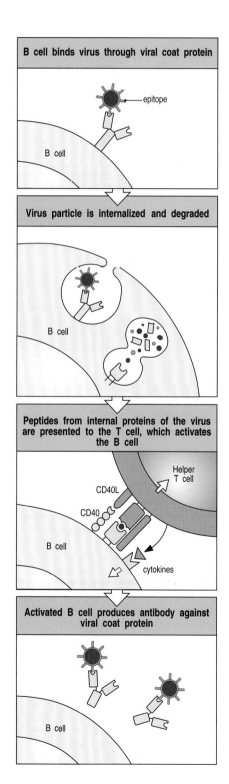

Fig. 9.3 B cells and helper T cells must recognize epitopes of the same molecular complex in order to interact. An epitope on a viral coat protein is recognized by the surface immunoglobulin on a B cell and the virus is internalized and degraded. Peptides derived from viral proteins, including internal proteins, are returned to the B-cell surface bound to MHC class II molecules (see Chapter 5). Here, these complexes are recognized by helper T cells, which help to activate the B cells to produce antibody against the coat protein.

Fig. 9.4 Protein antigens attached to polysaccharide antigens allow T cells to help polysaccharide-specific B cells. *Haemophilus influenzae* type B vaccine is a conjugate of bacterial polysaccharide and the tetanus toxoid protein. The B cell recognizes and binds the polysaccharide, internalizes and degrades the whole conjugate and then displays toxoid-derived peptides on surface MHC class II molecules. Helper T cells generated in response to earlier vaccination against the toxoid recognize the complex on the B-cell surface and activate the B cell to produce anti-polysaccharide antibody. This antibody can then protect against infection with *H. influenzae* type B.

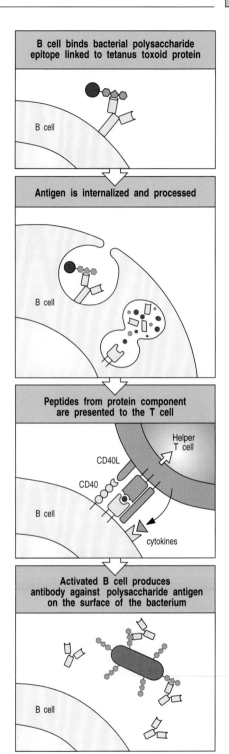

The requirement for linked recognition has important consequences for the regulation and manipulation of the humoral immune response. One is that linked recognition helps ensure self tolerance, as will be described in Chapter 13. An important application of linked recognition is in the design of vaccines, such as that used to immunize infants against *Haemophilus influenzae* type B. This bacterial pathogen can infect the lining of the brain, called the meninges, causing meningitis and, in severe cases, neurological damage or death. Protective immunity to this pathogen is mediated by antibodies against its capsular polysaccharide. Although adults make very effective thymus-independent responses to these polysaccharide antigens, such responses are weak in the immature immune system of the infant. To make an effective vaccine for use in infants, therefore, the polysaccharide is linked chemically to tetanus toxoid, a foreign protein against which infants are routinely and successfully vaccinated (see Chapter 14). B cells that bind the polysaccharide component of the vaccine can be activated by helper T cells specific for peptides of the linked toxoid (Fig. 9.4).

Linked recognition was originally discovered through studies of the production of antibodies to haptens (see Appendix I, Section A-1). Haptens are small chemical groups that cannot elicit antibody responses on their own because they cannot cross-link B-cell receptors and they cannot recruit T-cell help. When coupled at high density to a carrier protein, however, they become immunogenic, because the protein will carry multiple hapten groups that can now cross-link B-cell receptors. In addition, T-cell dependent responses are possible because T cells can be primed to peptides derived from the protein. Coupling of a hapten to a protein is responsible for the allergic responses shown by many people to the antibiotic penicillin, which reacts with host proteins to form a coupled hapten that can stimulate an antibody response, as we will learn in Chapter 12.

9-3 Antigenic peptides bound to self MHC class II molecules trigger armed helper T cells to make membrane-bound and secreted molecules that can activate a B cell.

Armed helper T cells activate B cells when they recognize the appropriate peptide:MHC class II complex on the B-cell surface (Fig. 9.5). As with armed T_H1 cells acting on macrophages, recognition of peptide:MHC class II complexes on B cells triggers armed helper T cells to synthesize both cell-bound and secreted effector molecules that synergize in activating the B cell. One particularly important T-cell effector molecule is a membrane-bound molecule of the tumor necrosis factor (TNF) family known as **CD40 ligand** (**CD40L**, also known as **CD154**) because it binds to the B-cell surface molecule **CD40**. CD40 is a member of the TNF-receptor family of cytokine receptors (see Section 8-20) however, it does not contain a 'death domain.' It is involved in directing all phases of the B-cell response. Binding of CD40 by CD40L helps to drive the resting B cell into the cell cycle and is essential for B-cell responses to thymus-dependent antigens.

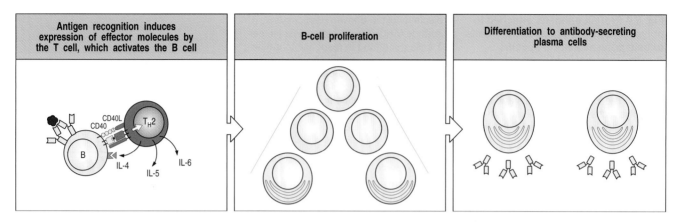

Fig. 9.5 Armed helper T cells stimulate the proliferation and then the differentiation of antigen-binding B cells. The specific interaction of an antigen-binding B cell with an armed helper T cell leads to the expression of the B-cell stimulatory molecule CD40 ligand (CD40L) on the helper T-cell surface and to the secretion of the B-cell stimulatory cytokines IL-4, IL-5, and IL-6, which drive the proliferation and differentiation of the B cell into antibody-secreting plasma cells.

B cells are stimulated to proliferate *in vitro* when they are exposed to a mixture of artificially synthesized CD40L and the cytokine interleukin-4 (IL-4). IL-4 is also made by armed T_H2 cells when they recognize their specific ligand on the B-cell surface, and IL-4 and CD40L are thought to synergize in driving the clonal expansion that precedes antibody production *in vivo*. IL-4 is secreted in a polar fashion by the T_H2 cell and is directed at the site of contact with the B cell (Fig. 9.6) so that it acts selectively on the antigen-specific target B cell.

Fig. 9.6 When an armed helper T cell encounters an antigen-binding B cell, it becomes polarized and secretes IL-4 and other cytokines at the point of cell–cell contact. On binding antigen on the B cell through its T-cell receptor, the helper T cell is induced to express CD40 ligand (CD40L), which binds to CD40 on the B cell. As shown in the top left panel, the tight junction formed between the cells upon antigen-specific binding seems to be sealed by a ring of adhesion molecules, with LFA-1 on the T cell interacting with ICAM-1 on the B cell (see Fig. 8.30). The cytoskeleton becomes polarized, as revealed by the relocation of the cytoskeletal protein talin (stained red in right center panel), to the point of cell–cell contact, and the secretory apparatus (the Golgi apparatus) is reoriented by the cytoskeleton toward the point of contact with the B cell. As shown in the bottom panels, cytokines are released at the point of contact. The bottom right panel shows IL-4 (stained green) confined to the space between the B cell and the helper T cell. MTOC, microtubule-organizing center. Photographs courtesy of A. Kupfer.

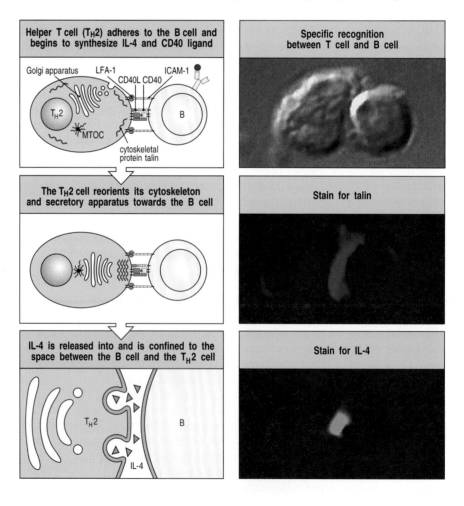

The combination of B-cell receptor and CD40 ligation, along with IL-4 and other signals derived from direct T-cell contact, leads to B-cell proliferation. Some of these contact signals have recently been elucidated. They involve other TNF/TNF-receptor family members, including **CD30** and **CD30 ligand** and **BLyS** (**B lymphocyte stimulator**) and its receptor on B cells, **TACI**. After several rounds of proliferation, B cells can further differentiate into antibody-secreting plasma cells. Two additional cytokines, IL-5 and IL-6, both secreted by helper T cells, contribute to these later stages of B-cell activation.

9-4 Isotype switching requires expression of CD40L by the helper T cell and is directed by cytokines.

Antibodies are remarkable not only for the diversity of their antigen-binding sites but also for their versatility as effector molecules. The specificity of an antibody response is determined by the antigen-binding site, which consists of the two variable V domains, V_H and V_L; however, the effector action of the antibody is determined by the isotype of its heavy-chain C region (see Section 4-15). A given heavy-chain V domain can become associated with the C region of any isotype through the process of isotype switching (see Section 4-16). We will see later in this chapter how antibodies of each isotype contribute to the elimination of pathogens. The DNA rearrangements that underlie isotype switching and confer this functional diversity on the humoral immune response are directed by cytokines, especially those released by armed effector CD4 T cells.

All naive B cells express cell-surface IgM and IgD, yet IgM makes up less than 10% of the immunoglobulin found in plasma, where the most abundant isotype is IgG. Much of the antibody in plasma has therefore been produced by B cells that have undergone isotype switching. Little IgD antibody is produced at any time, so the early stages of the antibody response are dominated by IgM antibodies. Later, IgG and IgA are the predominant isotypes, with IgE contributing a small but biologically important part of the response. The overall predominance of IgG results, in part, from its longer lifetime in the plasma (see Fig. 4.16).

Isotype switching does not occur in individuals who lack functional CD40L, which is necessary for productive interactions between B cells and helper T cells, such individuals make only small amounts of IgM antibodies in response to thymus-dependent antigens and have abnormally high levels of IgM in their plasma. These IgM antibodies may be induced by thymus-independent antigens expressed by the pathogens that chronically infect these patients, who suffer from severe humoral immunodeficiency, as we will see in Chapter 11.

Hyper IgM
Immunodeficiency

Most of what is known about the regulation of isotype switching by helper T cells has come from experiments in which mouse B cells are stimulated with bacterial lipopolysaccharide (LPS) and purified cytokines *in vitro*. These experiments show that different cytokines preferentially induce switching to different isotypes. Some of these cytokines are the same as those that drive B-cell proliferation in the initiation of a B-cell response. In the mouse, IL-4 preferentially induces switching to IgG1 and IgE, whereas transforming growth factor (TGF)-β induces switching to IgG2b and IgA. T_H2 cells make both of these cytokines as well as IL-5, which induces IgA secretion by cells that have already undergone switching. Although T_H1 cells are relatively poor initiators of antibody responses, they participate in isotype switching by releasing interferon (IFN)-γ, which preferentially induces switching to IgG2a and IgG3. The role of cytokines in directing B cells to make the different antibody isotypes is summarized in Fig. 9.7.

Fig. 9.7 Different cytokines induce switching to different isotypes. The individual cytokines induce (violet) or inhibit (red) production of certain isotypes. Much of the inhibitory effect is probably the result of directed switching to a different isotype. These data are drawn from experiments with mouse cells.

Role of cytokines in regulating Ig isotype expression							
Cytokines	IgM	IgG3	IgG1	IgG2b	IgG2a	IgE	IgA
IL-4	Inhibits	Inhibits	Induces		Inhibits	Induces	
IL-5							Augments production
IFN-γ	Inhibits	Induces	Inhibits		Induces	Inhibits	
TGF-β	Inhibits	Inhibits		Induces			Induces

Cytokines induce isotype switching by stimulating the formation and splicing of mRNA transcribed from the switch recombination sites that lie 5′ to each heavy-chain C gene (see Fig. 4.20). When activated B cells are exposed to IL-4, for example, transcription from a site upstream of the switch regions of $C_{\gamma 1}$ and C_ε can be detected a day or two before switching occurs (Fig. 9.8). Recent data suggest that the production of a spliced switch transcript has a role in directing switching, but the mechanism is not yet clear. Each of the cytokines that induces switching seems to induce transcription from the switch regions of two different heavy-chain C genes, promoting specific recombination to one or other of these genes only. Such a directed mechanism is supported by the observation that individual B cells frequently undergo switching to the same C gene on both chromosomes, even though the antibody heavy chain is only being expressed from one of the chromosomes. Thus, helper T cells regulate both the production of antibody by B cells and the isotype that determines the effector function of the antibody.

Fig. 9.8 Isotype switching is preceded by transcriptional activation of heavy-chain C-region genes. Resting naive B cells transcribe the μ and δ genes at a low rate, giving rise to surface IgM and IgD. Bacterial lipopolysaccharide (LPS), which can activate B cells independently of antigen, induces IgM secretion. In the presence of IL-4, however, $C_{\gamma 1}$ and C_ε are transcribed at a low rate, presaging switches to IgG1 and IgE production. The transcripts originate before the 5′ end of the region to which switching occurs, and do not code for protein. Similarly, TGF-β gives rise to $C_{\gamma 2b}$ and C_α transcripts and drives switching to IgG2b and IgA. It is not known what determines which of the two transcriptionally activated heavy-chain C genes undergoes switching. Arrows indicate transcription. The figure shows isotype switching in the mouse.

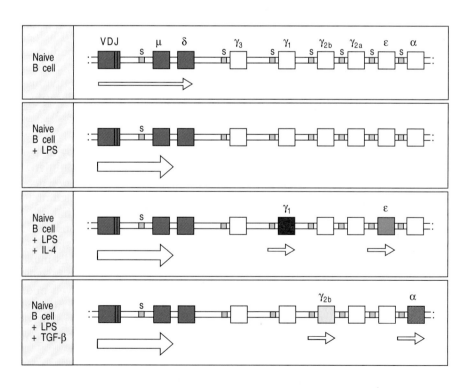

9-5   Antigen-binding B cells are trapped in the T-cell zone of secondary lymphoid tissues and are activated by encounter with armed helper T cells.

One of the most puzzling features of the antibody response is how an antigen-specific B cell manages to encounter a helper T cell with an appropriate antigen specificity. This question arises because the frequency of naive lymphocytes specific for any given antigen is estimated to be between 1 in 10,000 and 1 in 1,000,000. Thus, the chance of an encounter between a T lymphocyte and a B lymphocyte that recognize the same antigen should be between 1 in 10^8 and 1 in 10^{12}. Achieving such an encounter is a far more difficult challenge than getting effector T cells activated, because, in the latter case, only one of the two cells involved has specific receptors. Moreover, T cells and B cells mostly occupy quite distinct zones in peripheral lymphoid tissue (see Fig. 1.8). As in naive T-cell activation (see Chapter 8), the answer seems to lie in the antigen-specific trapping of migrating lymphocytes.

When an antigen is introduced into an animal, it is captured and processed by professional antigen-presenting cells, especially the dendritic cells that migrate from the tissues into the T-cell zones of local lymph nodes. Recirculating naive T cells pass by such cells continuously and those rare T cells whose receptors bind peptides derived from the antigen are trapped very efficiently. This trapping clearly involves the specific antigen receptor on the T cell, although it is stabilized by the activation of adhesion molecules and chemokines as we learned in Sections 8-3 and 8-4. Ingenious experiments using mice transgenic for rearranged immunoglobulin genes show that, in the presence of the appropriate antigen, B cells with antigen-specific receptors are also trapped in the T-cell zones of lymphoid tissue by a similar mechanism. On encountering antigen, migrating antigen-binding B cells are arrested by the activation of adhesion molecules and the engagement of chemokine receptors such as CCR7, a receptor for MIP-3β and SLC.

Trapping of B cells in the T-cell zones provides an elegant solution to the problem posed at the beginning of this section. T cells are themselves trapped and activated to helper status in the T-cell zones, and when B cells migrate into lymphoid tissue through high endothelial venules they first enter these same T-cell zones. Most of the B cells move quickly through the T-cell zone into the B-cell zone (the primary follicle), but those B cells that have bound antigen are trapped. Thus, antigen-binding B cells are selectively trapped in precisely the correct location to maximize the chance of encountering a helper T cell that can activate them. Interaction with armed helper T cells activates the B cell to establish a **primary focus** of clonal expansion (Fig. 9.9). Here, at the border between T-cell and B-cell zones, both types of lymphocyte will proliferate for several days to constitute the first phase of the primary humoral immune response.

After several days, the primary focus of proliferation begins to involute. Many of the lymphocytes comprising the focus undergo apoptosis. However, some of the proliferating B cells differentiate into antibody-synthesizing **plasma cells** and migrate to the red pulp of the spleen or the medullary cords of the lymph node. The differentiation of a B cell into a plasma cell is accompanied by many morphological changes that reflect its commitment to the production of large amounts of secreted antibody. The properties of resting B cells and plasma cells are compared in Fig. 9.10. Plasma cells have abundant cytoplasm dominated by multiple layers of rough endoplasmic reticulum (see Fig. 1.19). The nucleus shows a characteristic pattern of peripheral chromatin condensation, a prominent perinuclear Golgi apparatus is visible, and the cisternae of the endoplasmic reticulum are rich in immunoglobulin, which makes up 10–20% of all the protein synthesized. MHC class II molecules are not

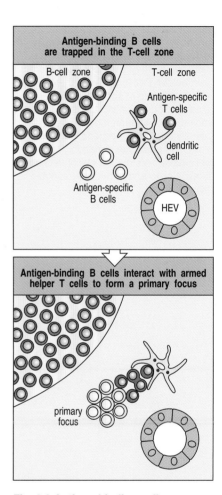

Fig. 9.9 Antigen-binding cells are trapped in the T-cell zone. Upon entry into lymphoid tissues through a high endothelial venule (HEV), T cells and B cells home to different regions, as described in Chapter 7. Antigen-specific T cells remain in the T-cell zone provided that they encounter antigen on the surface of a antigen-presenting cell such as a dendritic cell. B cells normally move rapidly through the T-cell zone, unless they bind specific antigen, in which case they are trapped before leaving the T-cell zone and thus can interact with antigen-specific armed helper T cells. This interaction gives rise to a primary focus of B cells and T cells near the border between B-cell and T-cell zones.

Fig. 9.10 Plasma cells secrete antibody at a high rate but can no longer respond to antigen or helper T cells. Resting naive B cells carry surface immunoglobulin (usually IgM and IgD) and MHC class II molecules on their surface. Their V genes do not carry somatic mutations. They can take up antigen and present it to helper T cells, which then induce the B cells to proliferate, switch isotype, and undergo somatic hypermutation; however, B cells do not secrete significant amounts of antibody. Plasma cells are terminally differentiated B cells that secrete antibodies. They can no longer interact with helper T cells because they have very low levels of surface immunoglobulin and lack MHC class II molecules, although they have usually already undergone isotype switching and hypermutation. Plasma cells have also lost the ability to change isotype or to undergo further somatic hypermutation.

	Property					
	Intrinsic			Inducible		
B-lineage cell	Surface Ig	Surface MHC class II	High-rate Ig secretion	Growth	Somatic hyper-mutation	Isotype switch
Resting B cell	High	Yes	No	Yes	Yes	Yes
Plasma cell	Low	No	Yes	No	No	No

expressed, so plasma cells can no longer present antigen to helper T cells, although these T cells may still provide important signals for plasma cell differentiation and survival, like IL-6 and CD40L. Surface immunoglobulin is still expressed on plasma cells at low levels, and recent evidence suggests that the survival of plasma cells may be determined in part by their ability to continue to bind antigen. Plasma cells have a range of life-spans. Some survive for only days to a few weeks after their final differentiation, whereas others are very long-lived and account for the persistence of antibody responses.

9-6   The second phase of the primary B-cell immune response occurs when activated B cells migrate to follicles and proliferate to form germinal centers.

There is another fate for some of the B cells and T cells that proliferate in the primary focus. Some of these cells migrate into a **primary lymphoid follicle** (Fig. 9.11) where they continue to proliferate and ultimately form a **germinal center** (Fig. 9.12). Germinal centers are composed mainly of proliferating B cells, but antigen-specific T cells make up about 10% of germinal center lymphocytes and provide indispensable help to the B cells. The germinal center is essentially an island of cell division that sets up amidst a sea of resting B cells in the primary follicles; germinal center B cells displace the resting B cells toward the periphery of the follicle, forming a **mantle zone** of resting cells around the center. Primary follicles contain resting B cells clustered around a dense network of processes extending from a specialized cell type, the **follicular dendritic cell** (**FDC**). Follicular dendritic cells attract both naive and activated B cells into the follicles by secreting the chemokine BLC (see Section 7-30).

The early events in the primary focus lead to the prompt secretion of specific antibody that serves as immediate protection to the infected individual. The germinal center reaction, on the other hand, provides for a more effective later response, should the pathogen establish a chronic infection or the host become reinfected. To this end, B cells undergo a number of important modifications in the germinal center These include **somatic hypermutation** (see Chapter 4), which alters the V regions of B cells, **affinity maturation**, which selects for survival of B cells with high affinity for the antigen, and **isotype switching** (see Sections 9-4 and 4-16), which allows these selected B cells to express a variety of effector functions in the form of antibodies of different isotypes. The selected B cells will either differentiate into memory B cells, the function of which will be described in Chapter 10, or into plasma cells, which will begin to secrete higher-affinity and isotype-switched antibody during the latter part of the primary immune response.

Fig. 9.11 Activated B cells form germinal centers in lymphoid follicles. Some B cells activated in the primary focus migrate to form a germinal center within a primary follicle. Germinal centers are sites of rapid B-cell proliferation and differentiation. Follicles in which germinal centers have formed are known as secondary follicles. Within the germinal center, B cells commence their differentiation into either antibody- secreting plasma cells or memory B cells. Plasma cells leave the germinal center and migrate to the medullary cords or leave the lymph node altogether via the efferent lymphatics and migrate to the bone marrow. Memory B cells continue to recirculate through the B-cell zones of secondary lymphoid tissue (not shown) and some may preferentially reside in the splenic marginal zone as described in Chapter 7.

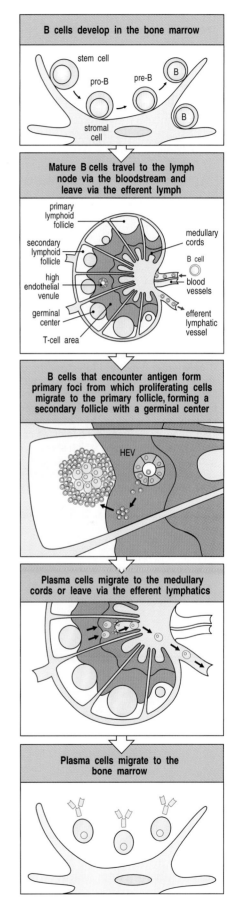

The germinal center is a site of intense cell proliferation, with B cells dividing every 6 to 8 hours. Initially, these rapidly proliferating B cells dramatically reduce their expression of surface immunoglobulin, particularly of IgD. These B cells are termed **centroblasts**. As time goes on, some B cells reduce their rate of division and begin to express higher levels of surface immunoglobulin. These are termed **centrocytes**. The centroblasts at first proliferate in the **dark zone** of the germinal center (see Fig. 9.12), so called because the proliferating cells are densely packed. With further development, B cells begin to fill the **light zone** of the germinal center, an area of the follicle that is more richly supplied with follicular dendritic cells and less densely packed with cells. It was thought originally that only the centroblasts in the dark zone proliferated, whereas centrocytes in the light zone did not divide. Indeed, this may be the case in chronic germinal centers found in inflamed tonsils that have been surgically removed. However, in newly forming germinal centers in mice, it is now apparent that proliferation can occur in both light and dark zones, and that proliferative cells in the dark zone can express moderate amounts of immunoglobulin on their surface. So the distinction between dark and light zones as areas of B-cell proliferation or quiescence does not strictly apply to primary germinal centers, at least in mice. Follicular dendritic cells, which originally were most prominent in the light zone, appear to react to germinal center formation and begin to extend more prominently throughout the germinal center as it develops. The result is that a mature germinal center at day 15 after immunization more resembles a light zone, with few of the classic dark zone characteristics. This view of germinal center evolution may help to explain how B cells with high affinity for immunizing antigen are selected, as we now discuss.

9-7 Germinal center B cells undergo V-region somatic hypermutation and cells with mutations that improve affinity for antigen are selected.

The process of somatic hypermutation, as one of the four mechanisms that create immunoglobulin diversity, was described in Chapter 4. Here we describe the signals that initiate hypermutation and the biological conse- quences of mutation for those cells. Somatic hypermutation is normally restricted to B cells that are proliferating in germinal centers. This was first shown by FACS sorting of germinal center B cells (see Appendix I, Section A-22) and sequencing of the V genes of cell lines derived from them; later, it was shown more directly by sequencing the V genes that were amplified by PCR of DNA isolated from germinal center B cells that had been micro- dissected from histologic sections. However, *in vitro* studies have shown that B cells can be induced to undergo hypermutation outside of germinal centers when their B-cell receptors are cross-linked and they receive help, including cytokines and CD40L stimulation, from activated T cells. In fact, mice that lack germinal centers owing to a mutation in the lymphotoxin-α gene (see Section 7-30) still support B-cell hypermutation, although where this takes place is unknown.

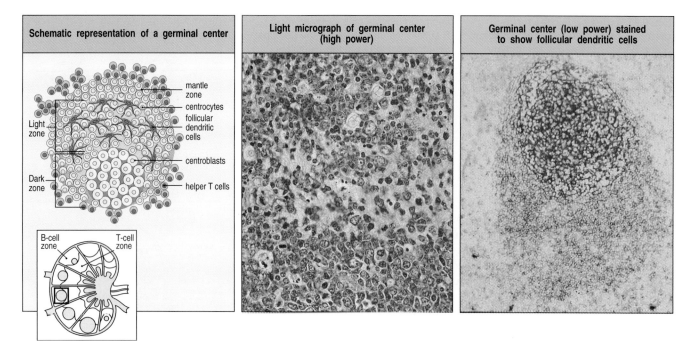

Schematic representation of a germinal center	Light micrograph of germinal center (high power)	Germinal center (low power) stained to show follicular dendritic cells

Fig. 9.12 Germinal centers are formed when activated B cells enter lymphoid follicles. The germinal center is a specialized microenvironment in which B-cell proliferation, somatic hypermutation, and selection for antigen binding all occur. Rapidly proliferating B cells in germinal centers are called centroblasts. Closely packed centroblasts form the so-called 'dark zone' of the germinal center, as can be seen in the lower part of the center panel, which shows a section through a germinal center. As these cells mature, they become small centrocytes, moving out into an area of the germinal center called the 'light zone' (the upper part of the center panel), where the centrocytes make contact with a dense network of follicular dendritic cell (FDC) processes. The FDCs are not stained in the center panel but can be seen clearly in the right panel, where both FDCs (stained blue with an antibody against Bu10, an FDC-specific marker) in the germinal center and also the mature B cells in the mantle zone (stained brown with an antibody against IgD) can be seen. The plane of this section chiefly reveals the dense network of FDCs in the light zone, although the less dense network in the dark zone can just be seen at the bottom half of the figure below the intensely stained area. Photographs courtesy of I. MacLennan.

Unlike the other mechanisms of immunoglobulin diversification (see Section 4-6), which generate B cells with radically differing B-cell receptors, somatic hypermutation has the potential to create a series of related B cells that differ subtly in their specificity and affinity for antigen. This is because somatic hypermutation generally involves individual point mutations that change only a single amino acid. Immunoglobulin V-region genes accumulate mutations at a rate of about one base pair change per 10^3 base pairs per cell division. The mutation rates of all other somatic cell DNA are much lower: around one base pair change per 10^{10} base pairs per cell division. As each of the expressed heavy- and light-chain V-region genes is encoded by about 360 base pairs, and about three out of every four base changes results in an altered amino acid, every second B cell will acquire a mutation in its receptor at each division. These mutations also affect some DNA flanking the rearranged V gene but they generally do not extend into the C-region exons. Thus, random point mutations are somehow targeted to the rearranged V genes in a B cell.

The point mutations accumulate in a stepwise manner as B-cell clones expand in the germinal center. Generally, a B cell will not acquire more than one or two new mutations in each generation. Mutations can affect the ability of a B cell to bind antigen and thus will affect the fate of the B cell in the germinal center, as diagrammed in Fig. 9.13. Most mutations have a negative impact on the ability of the B-cell receptor to bind the original antigen. For example, some mutations will abolish receptor function altogether by introducing a stop codon that prevents proper translation; other deleterious

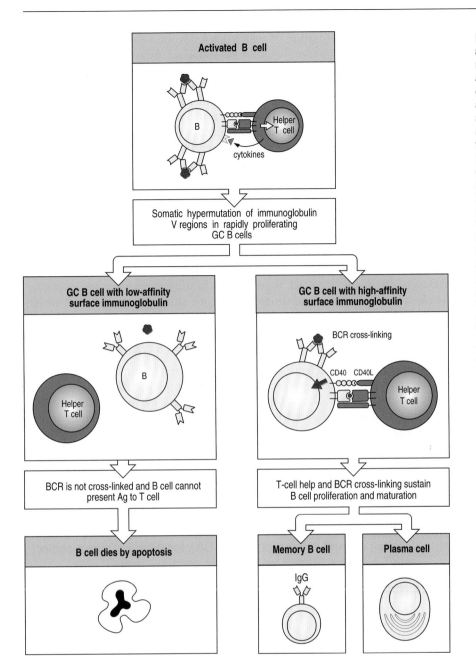

Fig. 9.13 After T-cell-dependent activation, B cells undergo rounds of mutation and selection for higher-affinity mutants in the germinal center, ultimately resulting in high-affinity memory B cells and antibody secreted from plasma cells. B cells are first activated outside of follicles by the combination of antigen and T cells (top panel). They migrate to germinal centers (GCs; not shown), where the remaining events occur. Somatic hypermutation can result in amino acid replacements in immunoglobulin V regions that affect the fate of the B cell. Mutations that result in a B-cell receptor (BCR) of lower affinity for the antigen (left panels) will prevent the B cell from being activated as efficiently, as both B-cell receptor cross-linking and the ability of the B cell to present peptide antigen to T cells are reduced. This results in the B cell dying by apoptosis. In this way, low-affinity cells are purged from the germinal center. Most mutations are either negative or neutral (not shown) and thus the germinal center is a site of massive B-cell death as well as of proliferation. Some mutations, however, will improve the ability of the B-cell receptor to bind antigen. This increases the B cell's chance of interacting with T cells, and thus of proliferating and surviving (right panels). Surviving cells undergo repeated cycles of mutation and selection during which some of the progeny B cells undergo differentiation to either memory B cells or plasma cells (bottom right panels) and leave the germinal center. The signals that control these differentiation decisions are unknown.

mutations alter framework region amino acids that are essential for correct immunoglobulin folding; and still others alter amino acids in the complementarity-determining regions that are responsible for contacting antigen. These deleterious mutations are disastrous for the cells that harbor them; these cells are eliminated by apoptosis either because they can no longer make a B-cell receptor or because they cannot compete with sibling cells that bind antigen more strongly. Deleterious mutation is evidently a frequent event, as germinal centers are filled with apoptotic B cells that are quickly engulfed by macrophages, resulting in **tingible body macrophages**, which contain dark-staining nuclear debris in their cytoplasm and are a long-recognized histologic feature of germinal centers.

More rarely, mutations will improve the affinity of a B-cell receptor for antigen. Cells that harbor these mutations are efficiently selected and expanded. Whether this is due to prevention of cell death and/or enhancement of cell division is still unclear. In either case, it is clear that selection is incremental.

After each round of mutation, B cells begin to express the new receptor, and it determines the cell's fate, whether favorable or unfavorable. If favorable, the cell undergoes another round of division and mutation and the expression and selection process is repeated. In this way, the affinity and specificity of positively selected B cells is continually refined during the germinal center response. The fact that both centroblasts and centrocytes proliferate and can express immunoglobulin explains how mutation and positive selection can take place simultaneously throughout the germinal center without the need for migration back and forth between the dark and light zones. Evidence of positive and negative selection is seen in the pattern of somatic hyper-mutations in V regions of B cells that have survived passage through the germinal center (see Section 4-9). The existence of negative selection is shown by the relative scarcity of amino acid replacements in the framework regions, reflecting the loss of cells that had mutated any one of the many residues that are critical for immunoglobulin V-region folding. Negative selection is an important force in the germinal center, most likely eliminating about one in every two cells. Were it not for substantial negative selection, B cells dividing three to four times per day in a single germinal center would quickly create enough progeny to overwhelm the entire organism; more than a billion cells could be created in 10 days in a single germinal center. Instead, a germinal center actually contains a few thousand B cells at its peak.

The mark of positive selection, on the other hand, is an accumulation of numerous amino acid replacements in the complementarity-determining regions (see Fig. 4.9). The consequence of these cycles of proliferation, mutation, and selection, which all happen within the germinal center, is that the average affinity of the population of responding B cells for its antigen increases over time, largely explaining the observed phenomenon of affinity maturation of the antibody response. The selection process can be quite stringent: although 50 to 100 B cells may seed the germinal center, most of these leave no progeny, and by the time the germinal center reaches maximum size, it is typically composed of the descendants of only one or a few B cells.

9-8 Ligation of the B-cell receptor and CD40, together with direct contact with T cells, are all required to sustain germinal center B cells.

Germinal center B cells are inherently prone to die and, in order to survive, they must receive specific signals. It was originally discovered *in vitro* that germinal center B cells could be kept alive by simultaneously cross-linking their B-cell receptors and ligating their cell-surface CD40. *In vivo*, these signals are delivered by antigen and T cells, respectively. Additional signals are also required for survival, which are delivered by direct contact with T cells. The nature of these signals is still obscure, but one signaling system involving the TNF-family member BLyS (the T-cell signal) and TACI (its receptor on B cells) has recently been found to be essential for the maintenance of germinal centers.

The source of antigen in the germinal center has been the matter of some controversy. Antigen can be trapped and stored for long periods of time in the form of immune complexes on follicular dendritic cells (Figs 9.14 and 9.15) and it was therefore assumed that this was the antigen that sustained germinal center B-cell proliferation. While this may be true under certain circumstances, there is now evidence that antigen on follicular dendritic cells is not required to sustain a normal germinal center response. Indeed, the role of the antigen depot on these cells is unknown, although it could be to maintain long-lived plasma cells. Where does the antigen that sustains the germinal center come from? Under normal circumstances, it is most likely that live pathogens carried to the lymphoid tissues and multiplying there will continue to provide antigens until they are eliminated by the immune response, after

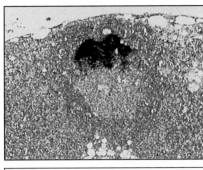

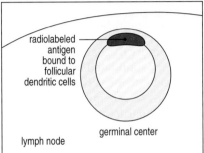

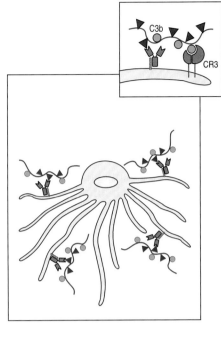

Fig. 9.14 Immune complexes bind to the surface of follicular dendritic cells. Radiolabeled antigen localizes to, and persists in, lymphoid follicles of draining lymph nodes (see light micrograph and the schematic representation below, showing a germinal center in a lymph node). Radiolabeled antigen has been injected 3 days previously and its localization in the germinal center is shown by the intense dark staining. The antigen is in the form of antigen:antibody:complement complexes bound to Fc and complement receptors on the surface of the follicular dendritic cell. These complexes are not internalized, as depicted schematically for immune complexes bound to both Fc and CR3 receptors in the right panel and insert. Antigen can persist in this form for long periods. Photograph courtesy of J. Tew.

which the germinal center decays. Immunizations with protein antigens are usually given in a form that slowly releases the antigen over time, which mimics the situation with live pathogens. Indeed, it is difficult to stimulate germinal center formation by immunization without either a live replicating pathogen or a sustained release of antigen in adjuvant (see Appendix I, Section A-4).

How the various signals that maintain the germinal center exert their effects on B cells is not completely understood. The combined signals from the B-cell receptor and CD40 seem to upregulate a protein called Bcl-X_L, a relative of Bcl-2, which promotes B-cell survival (see Chapter 6). There are doubtless many other signals yet to be discovered that promote B-cell differentiation.

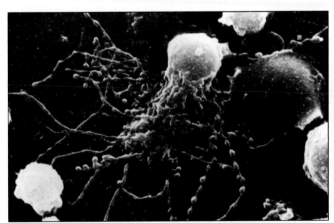

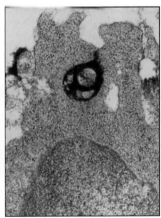

Fig. 9.15 Immune complexes bound to follicular dendritic cells form iccosomes, which are released and can be taken up by B cells in the germinal center. Follicular dendritic cells have a prominent cell body and many dendritic processes. Immune complexes, bound to complement and Fc receptors on the follicular dendritic cell surface, become clustered, forming prominent 'beads' along the dendrites. An intermediate form of follicular dendritic cell is shown (left panel) with both straight filiform dendrites and those that are becoming beaded. These beads are shed from the cell as iccosomes (immune complex-coated bodies), which can bind (center panel) and be taken up by B cells in the germinal center (right panel). In the center and right panels, the iccosome has been formed with immune complexes containing horseradish peroxidase, which is electron-dense and thus appears dark in the transmission electron micrographs. Photographs courtesy of A.K. Szakal.

9-9 Surviving germinal center B cells differentiate into either plasma cells or memory cells.

The purpose of the germinal center reaction is to enhance the later part of the primary immune response. Some germinal center cells differentiate first into plasmablasts and then into plasma cells. Plasmablasts continue to divide rapidly but have begun to specialize to secrete antibody at a high rate; they are destined to become nondividing, terminally differentiated plasma cells and thus represent an intermediate stage of differentiation. These plasma cells will migrate to the bone marrow, where a subset of them will live for a long period of time. Plasma cells obtain signals from bone marrow stromal cells that are essential for their survival. These plasma cells provide a source of long-lasting high-affinity antibody.

Other germinal center cells differentiate into **memory B cells**. Memory B cells are long-lived descendents of cells that were once stimulated by antigen and had proliferated in the germinal center. These cells divide very slowly if at all; they express surface immunoglobulin, but do not secrete antibody at a high rate. Since the precursors of memory B cells once participated in a germinal center reaction, memory B cells inherit the genetic changes that occurred in germinal center cells, including somatic mutations and the gene rearrangements that result in isotype switch (see Sections 4-9 and 4-16). The signals that control which differentiation path a B cell takes, and even whether at any given point the B cell continues to divide instead of differentiating, are unclear.

It has been proposed that signals from follicular dendritic cells (FDCs) are important in stimulating a B cell to become a memory cell. However, memory cells can develop in mutant mice lacking FDCs, albeit with reduced efficiency, so there may be other sources of signals. Another possibility is that affinity for antigen controls B-cell differentiation, with high-affinity cells perhaps being preferentially stimulated to become memory cells while the lower-affinity cells are allowed to undergo further cycles of proliferation, mutation, and selection. This is just one of the mysteries of the germinal center that immunologists have yet to solve. Immunological memory is discussed in detail in Chapter 10.

9-10 B-cell responses to bacterial antigens with intrinsic ability to activate B cells do not require T-cell help.

Although antibody responses to most protein antigens are dependent on helper T cells, humans and mice with T-cell deficiencies nevertheless make antibodies to many bacterial antigens. This is because the special properties of some bacterial polysaccharides, polymeric proteins, and lipopolysaccharides enable them to stimulate naive B cells in the absence of peptide-specific T-cell help. These antigens are known as thymus-independent antigens (TI antigens) because they stimulate strong antibody responses in athymic individuals. These nonprotein bacterial products cannot elicit classical T-cell responses, yet they induce antibody responses in normal individuals. However, B-cell responses to these TI antigens are influenced by the presence of T cells, perhaps indirectly through cytokines such as IL-5 since they are greatly diminished in animals that have no T cells at all.

Thymus-independent antigens fall into two classes that activate B cells by two different mechanisms. **TI-1 antigens** possess an intrinsic activity that can directly induce B-cell division. At high concentration, these molecules cause the proliferation and differentiation of most B cells regardless of their antigen specificity; this is known as **polyclonal activation** (Fig. 9.16, top two panels). TI-1 antigens are thus often called **B-cell mitogens**, a mitogen being a substance that induces cells to undergo mitosis. An example of a B-cell mitogen and TI-1 antigen is LPS, which binds to LPS-binding protein and CD14 (see Chapter 2), which then associate with the receptor TLR-4 on B cells. LPS activates B cells only at doses at least 100 times greater than those needed to activate dendritic cells. Thus, when B cells are exposed to concentrations of TI-1 antigens that are 10^3–10^5 times lower than those used for polyclonal activation, only those B cells whose B-cell receptors also specifically bind the TI-1 molecules become activated. At these low antigen concentrations, sufficient amounts of TI-1 for B-cell activation can only be concentrated on the B-cell surface with the aid of this specific binding (Fig. 9.16, bottom two panels). In the presence of large amounts of the TI-1 antigen, this concentrating effect is not required, and all B cells can be stimulated.

It is likely that, as with any pathogen antigen, concentrations of TI-1 antigens are low during the early stages of infections *in vivo*; thus, only antigen-specific B cells are likely to be activated and these will produce antibodies specific for the TI-1 antigen. Such responses have an important role in defense against several extracellular pathogens, as they arise earlier than thymus-dependent responses since they do not require prior priming and clonal expansion of helper T cells. However, TI-1 antigens are inefficient inducers of isotype switching, affinity maturation, or memory B cells, all of which require specific T-cell help.

9-11 B-cell responses to bacterial polysaccharides do not require peptide-specific T-cell help.

The second class of thymus-independent antigens consist of molecules such as bacterial capsular polysaccharides that have highly repetitive structures. These thymus-independent antigens, called **TI-2 antigens**, contain no intrinsic B-cell-stimulating activity. Whereas TI-1 antigens can activate both immature and mature B cells, TI-2 antigens can activate only mature B cells; immature B cells, as we saw in Chapter 7, are inactivated by repetitive epitopes. This might be why infants do not make antibodies to polysaccharide antigens efficiently; most of their B cells are immature. Responses to several TI-2 antigens are prominent among B-1 cells (also known as CD5 B cells), which comprise an autonomously replicating subpopulation of B cells, and among marginal zone B cells, another unique subset of nonrecirculating B cells that line the border of the splenic white pulp (see Chapter 7). Although B-1 cells arise early in development, young children do not make a fully effective response to carbohydrate antigens until about 5 years of age. On the other hand, marginal zone B cells are rare at birth and accumulate with age; they may thus be responsible for most physiological TI-2 responses, which also increase with age.

TI-2 antigens most probably act by extensively cross-linking the B-cell receptors of mature B cells specific for the antigen (Fig. 9.17, left panels). Excessive receptor cross-linking, however, renders mature B cells unresponsive or anergic, just as it does immature B cells. Thus, epitope density seems to be critical in the activation of B cells by TI-2 antigens: at too low a density, receptor cross-linking is insufficient to activate the cell; at too high a density, the B cell becomes anergic.

Although responses to TI-2 antigens can occur in nude mice (which lack a thymus), depletion of all T cells by knocking out the TCRβ and TCRδ loci eliminates responses to TI-2 antigens. Moreover, responses to TI-2 antigens can be augmented *in vivo* by transferring small numbers of T cells to these T-cell deficient mice. How T cells contribute to TI-2 responses is not clear. One possibility is that T cells can recognize and become activated by TI-2 antigens through cell-surface molecules shared by all T cells (Fig. 9.17, right panels). Alternatively, the help might come from γ:δ T cells or from CD4 CD8 double-negative α:β T cells. The T-cell receptors on these cells recognize certain

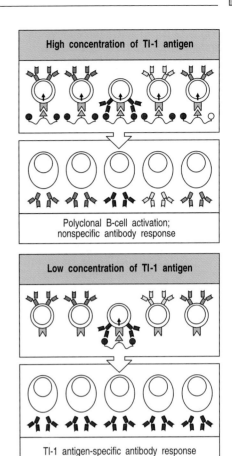

Fig. 9.16 Thymus-independent type 1 antigens (TI-1 antigens) are polyclonal B-cell activators at high concentrations, whereas at low concentrations they induce an antigen-specific antibody response. At high concentrations, the signal delivered by the B-cell-activating moiety of TI-1 antigens is sufficient to induce proliferation and antibody secretion by B cells in the absence of specific antigen binding to surface immunoglobulin. Thus, all B cells respond (top panels). At low concentrations, only B cells specific for the TI-1 antigen bind enough of it to focus its B-cell activating properties onto the B cell; this gives a specific antibody response to epitopes on the TI-1 antigen (lower panels).

Fig. 9.17 B-cell activation by thymus-independent type 2 antigens (TI-2 antigens) requires, or is greatly enhanced by, cytokines. Multiple cross-linking of the B-cell receptor by TI-2 antigens can lead to IgM antibody production (left panels), but there is evidence that helper T cells greatly augment these responses and lead to isotype switching as well (right panels). It is not clear how T cells are activated in this case, because polysaccharide antigens cannot produce peptide fragments that might be recognized by T cells on the B-cell surface. One possibility is that a component of the antigen binds to a cell-surface molecule common to all helper T cells, as shown in the figure. Another possibility (not shown) is that certain γ:δ T cells or CD4 CD8 double-negative α:β T cells can provide help, as some of these cells have T-cell receptors that recognize certain polysaccharrides bound to unconventional MHC molecules such as CD1.

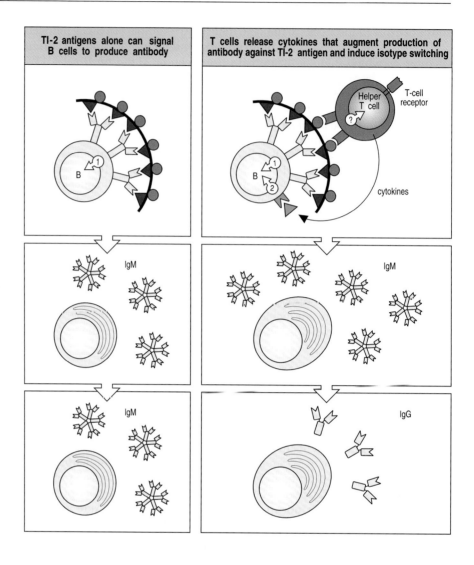

polysaccharides bound to unconventional MHC class I or class I-like molecules such as CD1. Such T cells can develop outside the thymus, principally in the gut.

B-cell responses to TI-2 antigens provide a prompt and specific response to an important class of pathogen. Many common extracellular bacterial pathogens are surrounded by a polysaccharide capsule that enables them to resist ingestion by phagocytes. The bacteria not only escape direct destruction by phagocytes but also avoid stimulating T-cell responses through the presentation of bacterial peptides by macrophages. Antibody that is produced rapidly in response to this polysaccharide capsule without the help of peptide-specific T cells can coat these bacteria, promoting their ingestion and destruction by phagocytes by mechanisms we will describe later in this chapter. The common encapsulated extracellular bacteria are often known as **pyogenic bacteria**, as they typically cause the formation of abundant pus, which consists chiefly of dead and dying neutrophils that have been recruited to the site of infection. Both IgM and IgG antibodies are induced by TI-2 antigens and are likely to be an important part of the humoral immune response in many bacterial infections. We mentioned earlier the importance of antibodies to the capsular polysaccharide of *Haemophilus influenzae* type B, a TI-2 antigen, in protective immunity to this bacterium. A further example of the importance of TI-2 responses can be seen in patients with an immuno-deficiency disease known as the **Wiskott–Aldrich syndrome**. These patients can respond, although poorly, to protein antigens but fail to make antibody

Wiskott–Aldrich Syndrome

against polysaccharide antigens and are highly susceptible to infection with encapsulated bacteria. Thus, the TI responses are important components of the humoral immune response to nonprotein antigens that do not engage peptide-specific T-cell help; the distinguishing features of thymus-dependent, TI-1, and TI-2 antibody responses are summarized in Fig. 9.18.

Summary.

B-cell activation by many antigens, especially monomeric proteins, requires both binding of the antigen by the B-cell surface immunoglobulin—the B-cell receptor—and interaction of the B cell with antigen-specific helper T cells. Helper T cells recognize peptide fragments derived from the antigen internalized by the B cell and displayed by the B cells as peptide:MHC class II complexes. Helper T cells stimulate the B cell through the binding of CD40L on the T cell to CD40 on the B cell, through interaction of other TNF–TNF-receptor family ligand pairs, and by the directed release of cytokines. The initial interaction occurs in the T-cell area of secondary lymphoid tissue, where both antigen-specific and helper T cells and antigen-specific B cells are trapped as a consequence of binding antigen; further interactions between T cells and B cells occur after migration into the B-cell zone or follicle, and formation of a germinal center. Helper T cells induce a phase of vigorous B-cell proliferation, and direct the differentiation of the clonally expanded progeny of the naive B cells into either antibody-secreting plasma cells or memory B cells. During the differentiation of activated B cells, the antibody isotype can change in response to cytokines released by helper T cells, and the antigen-binding properties of the antibody can change by somatic hyper-mutation of V-region genes. Somatic hypermutation and selection for high-affinity binding occur in the germinal centers. Helper T cells control these processes by selectively activating cells that have retained their specificity for the antigen and by inducing proliferation and differentiation into plasma cells and memory B cells. Some nonprotein antigens stimulate B cells in the

	TD antigen	TI-1 antigen	TI-2 antigen
Antibody response in infants	Yes	Yes	No
Antibody production in congenitally athymic individual	No	Yes	Yes
Antibody response in absence of all T cells	No	Yes	No
Primes T cells	Yes	No	No
Polyclonal B-cell activation	No	Yes	No
Requires repeating epitopes	No	No	Yes
Examples of antigen	Diphtheria toxin Viral hemagglutinin Purified protein derivative (PPD) of *Mycobacterium tuberculosis*	Bacterial lipopoly-saccharide *Brucella abortus*	Pneumococcal poly-saccharide *Salmonella* polymerized flagellin Dextran Hapten-conjugated Ficoll (polysucrose)

Fig. 9.18 Properties of different classes of antigen that elicit antibody responses.

absence of linked recognition by peptide-specific helper T cells. These thymus-independent antigens induce only limited isotype switching and do not induce memory B cells. However, responses to these antigens have a critical role in host defense against pathogens whose surface antigens cannot elicit peptide-specific T-cell responses.

The distribution and functions of immunoglobulin isotypes.

Extracellular pathogens can find their way to most sites in the body and antibodies must be equally widely distributed to combat them. Most classes of antibody are distributed by diffusion from their site of synthesis, but specialized transport mechanisms are required to deliver antibodies to lumenal epithelial surfaces, such as those of the lung and intestine. The distribution of antibodies is determined by their isotype, which can limit their diffusion or enable them to engage specific transporters that deliver them across epithelia. In this part of the chapter we will describe the mechanisms by which antibodies of different isotypes are directed to the compartments of the body in which their particular effector functions are appropriate, and discuss the protective functions of antibodies that result solely from their binding to pathogens. In the last part of the chapter we will discuss the effector cells and molecules that are specifically engaged by different isotypes.

9-12   Antibodies of different isotype operate in distinct places and have distinct effector functions.

Pathogens most commonly enter the body across the epithelial barriers of the mucosa lining the respiratory, digestive, and urogenital tracts, or through damaged skin, and can then establish infections in the tissues. Less often, insects, wounds, or hypodermic needles introduce microorganisms directly into the blood. The body's mucosal surfaces, tissues, and blood are all protected by antibodies from such infections; these antibodies serve to neutralize the pathogen or promote its elimination before it can establish a significant infection. Antibodies of different isotypes are adapted to function in different compartments of the body. Because a given V region can become associated with any C region through isotype switching (see Section 4-16), the progeny of a single B cell can produce antibodies, all specific for the same eliciting antigen, that provide all of the protective functions appropriate for each body compartment.

The first antibodies to be produced in a humoral immune response are always IgM, because IgM can be expressed without isotype switching (see Figs 4.20 and 9.8). These early IgM antibodies are produced before B cells have undergone somatic hypermutation and therefore tend to be of low affinity. IgM molecules, however, form pentamers whose 10 antigen-binding sites can bind simultaneously to multivalent antigens such as bacterial capsular polysaccharides. This compensates for the relatively low affinity of the IgM monomers by multipoint binding that confers high overall avidity. As a result of the large size of the pentamers, IgM is mainly found in the blood and, to a lesser extent, the lymph. The pentameric structure of IgM makes it especially effective in activating the complement system, as we will see in the last part of this chapter. Infection of the bloodstream has serious consequences unless it is controlled quickly, and the rapid production of IgM and its efficient activation of the complement system are important in controlling such infections. Some IgM is also produced in secondary and subsequent responses, and after somatic hypermutation, although other isotypes dominate the later phases of the antibody response.

Antibodies of the other isotypes—IgG, IgA, and IgE—are smaller in size and diffuse easily out of the blood into the tissues. Although IgA can form dimers, as we saw in Chapter 4, IgG and IgE are always monomeric. The affinity of the individual antigen-binding sites for their antigen is therefore critical for the effectiveness of these antibodies, and most of the B cells expressing these isotypes have been selected for increased affinity of antigen-binding in germinal centers. IgG is the principal isotype in the blood and extracellular fluid, whereas IgA is the principal isotype in secretions, the most important being those of the mucus epithelium of the intestinal and respiratory tracts. Whereas IgG efficiently opsonizes pathogens for engulfment by phagocytes and activates the complement system, IgA is a less potent opsonin and a weak activator of complement. This distinction is not surprising, as IgG operates mainly in the body tissues, where accessory cells and molecules are available, whereas IgA operates mainly on epithelial surfaces where complement and phagocytes are not normally present, and therefore functions chiefly as a neutralizing antibody.

Finally, IgE antibody is present only at very low levels in blood or extracellular fluid, but is bound avidly by receptors on mast cells that are found just beneath the skin and mucosa, and along blood vessels in connective tissue. Antigen binding to this IgE triggers mast cells to release powerful chemical mediators that induce reactions, such as coughing, sneezing, and vomiting, that can expel infectious agents, as will be discussed below when we describe the receptors that bind immunoglobulin C regions and engage effector functions. The distribution and main functions of antibodies of the different isotypes are summarized in Fig. 9.19.

Functional activity	IgM	IgD	IgG1	IgG2	IgG3	IgG4	IgA	IgE
Neutralization	+	–	++	++	++	++	++	–
Opsonization	–	–	+++	*	++	+	+	–
Sensitization for killing by NK cells	–	–	++	–	++	–	–	–
Sensitization of mast cells	–	–	+	–	+	–	–	+++
Activates complement system	+++	–	++	+	+++	–	+	–

Distribution	IgM	IgD	IgG1	IgG2	IgG3	IgG4	IgA	IgE
Transport across epithelium	+	–	–	–	–	–	+++ (dimer)	–
Transport across placenta	–	–	+++	+	++	+/–	–	–
Diffusion into extravascular sites	+/–	–	+++	+++	+++	+++	++ (monomer)	+
Mean serum level (mg ml^{-1})	1.5	0.04	9	3	1	0.5	2.1	3×10^{-5}

Fig. 9.19 Each human immunoglobulin isotype has specialized functions and a unique distribution. The major effector functions of each isotype (+++) are shaded in dark red, whereas lesser functions (++) are shown in dark pink, and very minor functions (+) in pale pink. The distributions are marked similarly, with actual average levels in serum being shown in the bottom row. *IgG2 can act as an opsonin in the presence of Fc receptors of a particular allotype, found in about 50% of white people.

9-13 Transport proteins that bind to the Fc regions of antibodies carry particular isotypes across epithelial barriers.

IgA-secreting plasma cells are found predominantly in the connective tissue called the lamina propria, which lies immediately below the basement membrane of many surface epithelia. From there, the IgA antibodies can be transported across the epithelium to its external surface, for example, to the lumen of the gut or the bronchi. IgA antibody synthesized in the lamina propria is secreted as a dimeric IgA molecule associated with a single J chain (see Fig. 4.23). This polymeric form of IgA binds specifically to the poly-Ig receptor, which is present on the basolateral surfaces of the overlying epithelial cells (Fig. 9.20). When the poly-Ig receptor has bound a molecule of dimeric IgA, the complex is internalized and carried through the cytoplasm of the epithelial cell in a transport vesicle to its luminal surface. This process is called **transcytosis**. At the apical or luminal surface of the epithelial cell, the poly-Ig receptor is cleaved enzymatically, releasing the extracellular portion of the receptor still attached to the Fc region of the dimeric IgA. This fragment of receptor, called the **secretory component**, may help to protect the IgA dimer from proteolytic cleavage. Some molecules of dimeric IgA diffuse from the lamina propria into the extracellular spaces of the tissues, draining into the bloodstream before being excreted into the gut via the bile. Therefore, it is not surprising that patients with obstructive jaundice, a condition in which bile is not excreted, show a marked increase in dimeric IgA in the plasma.

The principal sites of IgA synthesis and secretion are the gut, the respiratory epithelium, the lactating breast, and various other exocrine glands such as the salivary and tear glands. It is believed that the primary functional role of IgA antibodies is to protect epithelial surfaces from infectious agents, just as

Fig. 9.20 Transcytosis of IgA antibody across epithelia is mediated by the poly-Ig receptor, a specialized transport protein. Most IgA antibody is synthesized in plasma cells lying just beneath epithelial basement membranes of the gut, the respiratory epithelia, the tear and salivary glands, and the lactating mammary gland. The IgA dimer bound to a J chain diffuses across the basement membrane and is bound by the poly-Ig receptor on the basolateral surface of the epithelial cell. The bound complex undergoes transcytosis in which it is transported in a vesicle across the cell to the apical surface, where the poly-Ig receptor is cleaved to leave the extracellular IgA-binding component bound to the IgA molecule as the so-called secretory component. The residual piece of the poly-Ig receptor is nonfunctional and is degraded. In this way, IgA is transported across epithelia into the lumens of several organs that are in contact with the external environment.

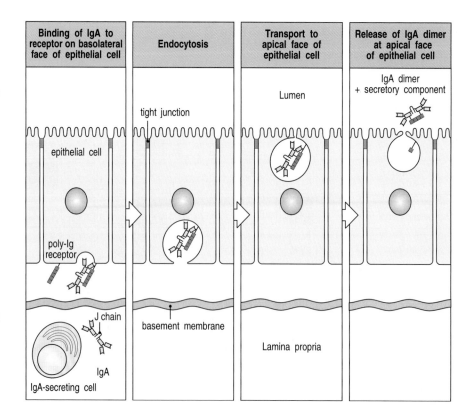

Fig. 9.21 FcRn binds to the Fc portion of IgG. The structure of a molecule of FcRn (white) bound to one chain of the Fc portion of IgG (blue) is shown. FcRn transports IgG molecules across the placenta in humans and also across the gut in rats and mice. It also plays a role in the homeostasis of IgG in adults. Although only one molecule of FcRn is shown binding to the Fc portion, it is thought that it takes two molecules of FcRn to capture one molecule of IgG. Photograph courtesy of P. Björkman.

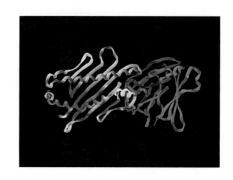

IgG antibodies protect the extracellular spaces of the internal tissues. IgA antibodies prevent the attachment of bacteria or toxins to epithelial cells and the absorption of foreign substances, and provide the first line of defense against a wide variety of pathogens. Newborn infants are especially vulnerable to infection, having had no prior exposure to the microbes in the environment they enter at birth. IgA antibodies are secreted in breast milk and are thereby transferred to the gut of the newborn infant, where they provide protection from newly encountered bacteria until the infant can synthesize its own protective antibody.

IgA is not the only protective antibody a mother passes on to her baby. Maternal IgG is transported across the placenta directly into the bloodstream of the fetus during intrauterine life; human babies at birth have as high a level of plasma IgG as their mothers, and with the same range of antigen specificities. The selective transport of IgG from mother to fetus is due to an IgG transport protein in the placenta, FcRn, which is closely related in structure to MHC class I molecules. Despite this similarity, FcRn binds IgG quite differently from the binding of peptide to MHC class I, as its peptide-binding groove is occluded. It binds to the Fc portion of IgG molecules (Fig. 9.21). Two molecules of FcRn bind one molecule of IgG, bearing it across the placenta. In some rodents, FcRn also delivers IgG to the circulation of the neonate from the gut lumen. Maternal IgG is ingested by the newborn animal in its mother's milk and colostrum, the protein-rich fluid secreted by the early postnatal mammary gland. In this case, FcRn transports the IgG from the lumen of the neonate gut into the blood and tissues. Interestingly, FcRn is also found in adults in the gut and liver and on endothelial cells. Its function in adults is to regulate the levels of IgG in serum and other body fluids, which it does by binding circulating antibody, endocytosing it, and then recycling to the cell surface.

By means of these specialized transport systems, mammals are supplied from birth with antibodies against pathogens common in their environments. As they mature and make their own antibodies of all isotypes, these are distributed selectively to different sites in the body (Fig. 9.22). Thus, throughout life, isotype switching and the distribution of isotypes through the body provide effective protection against infection in extracellular spaces.

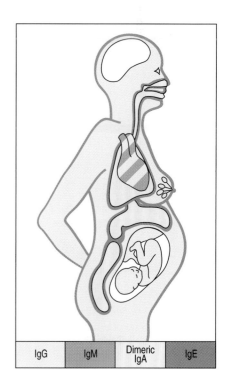

| IgG | IgM | Dimeric IgA | IgE |

Fig. 9.22 Immunoglobulin isotypes are selectively distributed in the body. IgG and IgM predominate in plasma, whereas IgG and monomeric IgA are the major isotypes in extracellular fluid within the body. Dimeric IgA predominates in secretions across epithelia, including breast milk. The fetus receives IgG from the mother by transplacental transport. IgE is found mainly associated with mast cells just beneath epithelial surfaces (especially of the respiratory tract, gastro-intestinal tract, and skin). The brain is normally devoid of immunoglobulin.

9-14   High-affinity IgG and IgA antibodies can neutralize bacterial toxins.

Many bacteria cause disease by secreting proteins called toxins, which damage or disrupt the function of the host's cells (Fig. 9.23). To have an effect, a toxin must interact specifically with a molecule that serves as a receptor on the surface of the target cell. In many toxins, the receptor-binding domain is on one polypeptide chain whereas the toxic function is carried by a second chain. Antibodies that bind to the receptor-binding site on the toxin molecule can prevent the toxin from binding to the cell and thus protect the cell from attack (Fig. 9.24). Antibodies that act in this way to neutralize toxins are referred to as **neutralizing antibodies**.

Most toxins are active at nanomolar concentrations: a single molecule of diphtheria toxin can kill a cell. To neutralize toxins, therefore, antibodies must be able to diffuse into the tissues and bind the toxin rapidly and with high affinity. The ability of IgG antibodies to diffuse easily throughout the extracellular fluid and their high affinity make these the principal neutralizing antibodies for toxins found in tissues. IgA antibodies similarly neutralize toxins at the mucosal surfaces of the body.

Diphtheria and tetanus toxins are two bacterial toxins in which the toxic and receptor-binding functions are on separate protein chains. It is therefore possible to immunize individuals, usually as infants, with modified toxin

Fig. 9.23 Many common diseases are caused by bacterial toxins. These toxins are all exotoxins—proteins secreted by the bacteria. High-affinity IgG and IgA antibodies protect against these toxins. Bacteria also have nonsecreted endotoxins, such as lipopolysaccharide, which are released when the bacterium dies. The endotoxins are also important in the pathogenesis of disease, but there the host response is more complex because the innate immune system has receptors for some of these (see Chapters 2 and 10).

Disease	Organism	Toxin	Effects *in vivo*
Tetanus	*Clostridium tetani*	Tetanus toxin	Blocks inhibitory neuron action leading to chronic muscle contraction
Diphtheria	*Corynebacterium diphtheriae*	Diphtheria toxin	Inhibits protein synthesis leading to epithelial cell damage and myocarditis
Gas gangrene	*Clostridium perfringens*	Clostridial toxin	Phospholipase activation leading to cell death
Cholera	*Vibrio cholerae*	Cholera toxin	Activates adenylate cyclase, elevates cAMP in cells, leading to changes in intestinal epithelial cells that cause loss of water and electrolytes
Anthrax	*Bacillus anthracis*	Anthrax toxic complex	Increases vascular permeability leading to edema, hemorrhage, and circulatory collapse
Botulism	*Clostridium botulinum*	Botulinum toxin	Blocks release of acetylcholine leading to paralysis
Whooping cough	*Bordetella pertussis*	Pertussis toxin	ADP-ribosylation of G proteins leading to lymphoproliferation
		Tracheal cytotoxin	Inhibits cilia and causes epithelial cell loss
Scarlet fever	*Streptococcus pyogenes*	Erythrogenic toxin	Vasodilation leading to scarlet fever rash
		Leukocidin Streptolysins	Kill phagocytes, allowing bacterial survival
Food poisoning	*Staphylococcus aureus*	Staphylococcal enterotoxin	Acts on intestinal neurons to induce vomiting. Also a potent T-cell mitogen (SE superantigen)
Toxic-shock syndrome	*Staphylococcus aureus*	Toxic-shock syndrome toxin	Causes hypotension and skin loss. Also a potent T-cell mitogen (TSST-1 superantigen)

Toxic Shock
Syndrome

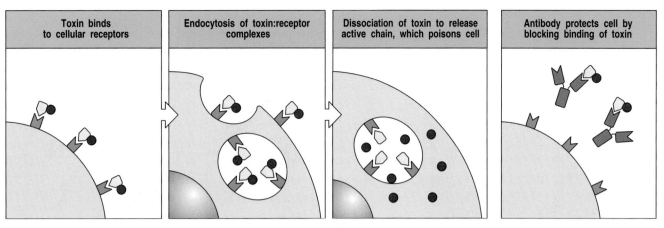

Toxin binds to cellular receptors	Endocytosis of toxin:receptor complexes	Dissociation of toxin to release active chain, which poisons cell	Antibody protects cell by blocking binding of toxin

Fig. 9.24 Neutralization of toxins by IgG antibodies protects cells from their damaging action. Many bacteria (as well as venomous insects and snakes) cause their damaging effects by elaborating toxic proteins (see Fig. 9.23). These toxins are usually composed of several distinct moieties. One part of the toxin molecule binds a cellular receptor, which enables the molecule to be internalized. Another part of the toxin molecule then enters the cytoplasm and poisons the cell. Antibodies that inhibit toxin binding can prevent, or neutralize, these effects.

molecules in which the toxic chain has been denatured. These modified toxins, called toxoids, lack toxic activity but retain the receptor-binding site. Thus, immunization with the toxoid induces neutralizing antibodies that protect against the native toxin.

With some insect or animal venoms that are so toxic that a single exposure can cause severe tissue damage or death, the adaptive immune response is too slow to be protective. Exposure to these venoms is a rare event and protective vaccines have not been developed for use in humans. Instead, neutralizing antibodies are generated by immunizing other species, such as horses, with insect and snake venoms to produce anti-venom antibodies (antivenins) for use in protecting humans. Transfer of antibodies in this way is known as **passive immunization** (see Appendix I, Section A-37).

9-15 High-affinity IgG and IgA antibodies can inhibit the infectivity of viruses.

Animal viruses infect cells by binding to a particular cell-surface receptor, often a cell-type-specific protein that determines which cells they can infect. The **hemagglutinin** of influenza virus, for example, binds to terminal sialic acid residues on the carbohydrates of glycoproteins present on epithelial cells of the respiratory tract. It is known as hemagglutinin because it recognizes and binds to similar sialic acid residues on chicken red blood cells and agglutinates these red blood cells. Antibodies to the hemagglutinin can prevent infection by the influenza virus. Such antibodies are called virus-neutralizing antibodies and, as with the neutralization of toxins, high-affinity IgA and IgG antibodies are particularly important.

Many antibodies that neutralize viruses do so by directly blocking viral binding to surface receptors (Fig. 9.25). However, viruses are sometimes successfully neutralized when only a single molecule of antibody is bound to a virus particle that has many receptor-binding proteins on its surface. In these cases, the antibody must cause some change in the virus that disrupts its structure and either prevents it from interacting with its receptors or interferes with the fusion of the virus membrane with the cell surface after the virus has engaged its surface receptor.

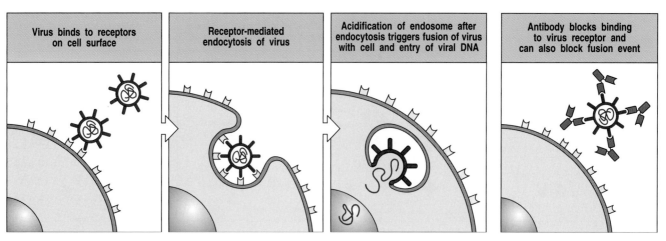

| Virus binds to receptors on cell surface | Receptor-mediated endocytosis of virus | Acidification of endosome after endocytosis triggers fusion of virus with cell and entry of viral DNA | Antibody blocks binding to virus receptor and can also block fusion event |

Fig. 9.25 Viral infection of cells can be blocked by neutralizing antibodies. For a virus to multiply within a cell, it must introduce its genes into the cell. The first step in entry is usually the binding of the virus to a receptor on the cell surface. For enveloped viruses, as shown in the figure, entry into the cytoplasm requires fusion of the viral envelope and the cell membrane. For some viruses, this fusion event takes place on the cell surface (not shown); for others it can occur only within the more acidic environment of endosomes, as shown here. Nonenveloped viruses must also bind to receptors on cell surfaces but they enter the cytoplasm by disrupting endosomes. Antibodies bound to viral-surface proteins neutralize the virus, inhibiting either its initial binding to the cell or its subsequent entry.

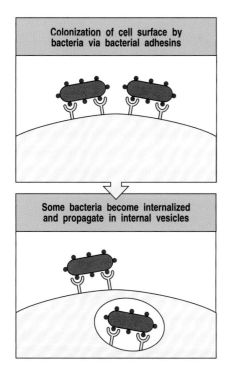

| Colonization of cell surface by bacteria via bacterial adhesins |

| Some bacteria become internalized and propagate in internal vesicles |

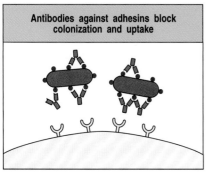

| Antibodies against adhesins block colonization and uptake |

9-16   Antibodies can block the adherence of bacteria to host cells.

Many bacteria have cell-surface molecules called adhesins that enable them to bind to the surface of host cells. This adherence is critical to the ability of these bacteria to cause disease, whether they subsequently enter the cell, as do some pathogens such as *Salmonella* species, or remain attached to the cell surface as extracellular pathogens (Fig. 9.26). *Neisseria gonorrhoeae*, the causative agent of the sexually transmitted disease gonorrhea, has a cell-surface protein known as pilin. Pilin enables the bacterium to adhere to the epithelial cells of the urinary and reproductive tracts and is essential to its infectivity. Antibodies against pilin can inhibit this adhesive reaction and prevent infection.

IgA antibodies secreted onto the mucosal surfaces of the intestinal, respiratory, and reproductive tracts are particularly important in preventing infection by preventing the adhesion of bacteria, viruses, or other pathogens to the epithelial cells lining these surfaces. The adhesion of bacteria to cells within tissues can also contribute to pathogenesis, and IgG antibodies against adhesins can protect from damage much as IgA antibodies protect at mucosal surfaces.

Fig. 9.26 Antibodies can prevent attachment of bacteria to cell surfaces. Many bacterial infections require an interaction between the bacterium and a cell-surface receptor. This is particularly true for infections of mucosal surfaces. The attachment process involves very specific molecular interactions between bacterial adhesins and their receptors on host cells; antibodies against bacterial adhesins can block such infections.

9-17   Antibody:antigen complexes activate the classical pathway of complement by binding to C1q.

Another way in which antibodies can protect against infection is by activation of the cascade of complement proteins. We have described these proteins in Chapter 2, as they can also be activated on pathogen surfaces in the absence of antibody, as part of the innate immune response. Complement activation proceeds via a series of proteolytic cleavage reactions, in which inactive components, present in plasma, are cleaved to form proteolytic enzymes that attach covalently to the pathogen surface. All known pathways of complement activation converge to generate the same set of effector actions: the pathogen surface or immune complex is coated with covalently attached fragments (principally C3b) that act as opsonins to promote uptake and removal by phagocytes. At the same time, small peptides with inflammatory and chemotactic activity are released (principally C5a) so that phagocytes are recruited to the site. In addition, the terminal complement components can form a membrane-attack complex that damages some bacteria.

Antibodies initiate complement activation by a pathway known as the classical pathway because it was the first pathway of complement activation to be discovered. The full details of this pathway, and of the other two known pathways of complement activation, are given in Chapter 2, but we will describe here how antibody is able to initiate the classical pathway after binding to pathogen, or after forming immune complexes.

The first component of the classical pathway of complement activation is C1, which is a complex of three proteins called C1q, C1r, and C1s. Two molecules each of C1r and C1s are bound to each molecule of C1q (see Fig. 2.10). Complement activation is initiated when antibodies attached to the surface of a pathogen bind C1q. C1q can be bound by either IgM or IgG antibodies but, because of the structural requirements of binding to C1q, neither of these antibody isotypes can activate complement in solution; the cascade is initiated only when the antibodies are bound to multiple sites on a cell surface, normally that of a pathogen.

The C1q molecule has six globular heads joined to a common stem by long, filamentous domains that resemble collagen molecules; the whole C1q complex has been likened to a bunch of six tulips held together by the stems. Each globular head can bind to one Fc domain, and binding of two or more globular heads activates the C1q molecule. In plasma, the pentameric IgM molecule has a planar conformation that does not bind C1q (Fig. 9.27, left panel); however, binding to the surface of a pathogen deforms the IgM pentamer so that it looks like a staple (see Fig. 9.27, right panel), and this

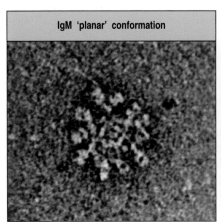

IgM 'planar' conformation

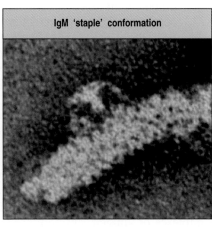

IgM 'staple' conformation

Fig. 9.27 The two conformations of IgM. The left panel shows the planar conformation of soluble IgM; the right panel shows the staple conformation of IgM bound to a bacterial flagellum. Photographs ($\times$ 760,000) courtesy of K.H. Roux.

Fig. 9.28 The classical pathway of complement activation is initiated by binding of C1q to antibody on a surface such as a bacterial surface. In the left panels, one molecule of IgM, bent into the 'staple' conformation by binding several identical epitopes on a pathogen surface, allows the globular heads of C1q to bind to its Fc pieces on the surface of the pathogen. In the right panels, multiple molecules of IgG bound on the surface of a pathogen allow the binding of a single molecule of C1q to two or more Fc pieces. In both cases, the binding of C1q activates the associated C1r, which becomes an active enzyme that cleaves the pro-enzyme C1s, generating a serine protease that initiates the classical complement cascade (not illustrated).

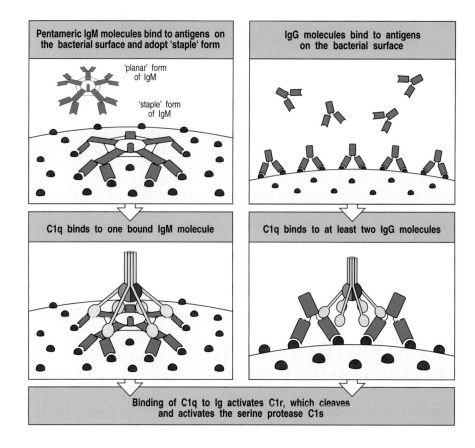

distortion exposes binding sites for the C1q heads. Although C1q binds with low affinity to some subclasses of IgG in solution, the binding energy required for C1q activation is achieved only when a single molecule of C1q can bind two or more IgG molecules that are held within 30–40 nm of each other as a result of binding antigen. This requires many molecules of IgG to be bound to a single pathogen. For this reason, IgM is much more efficient in activating complement than is IgG. The binding of C1q to a single bound IgM molecule, or to two or more bound IgG molecules, leads to the activation of an enzymatic activity in C1r, triggering the complement cascade as shown schematically in Fig. 9.28. This translates antibody binding into the activation of the complement cascade, which, as we learned in Chapter 2, can also be triggered by direct binding of C1q to the pathogen surface.

9-18 Complement receptors are important in the removal of immune complexes from the circulation.

Many small soluble antigens form antibody:antigen complexes known as **immune complexes** that contain too few molecules of IgG to be readily bound to the Fcγ receptors we will discuss in the next part of the chapter. These antigens include toxins bound by neutralizing antibodies and debris from dead microorganisms. Such immune complexes are found after most infections and are removed from the circulation through the action of complement. The soluble immune complexes trigger their own removal by activating complement, again through the binding of C1q, leading to the covalent binding of the activated components C4b and C3b to the complex, which is then cleared from the circulation by the binding of C4b and C3b to CR1 on the surface of erythrocytes. The erythrocytes transport the bound complexes of antigen, antibody, and complement to the liver and spleen.

Fig. 9.29 Erythrocyte CR1 helps to clear immune complexes from the circulation. CR1 on the erythrocyte surface has an important role in the clearance of immune complexes from the circulation. Immune complexes bind to CR1 on erythrocytes, which transport them to the liver and spleen, where they are removed by macrophages expressing receptors for both Fc and bound complement components.

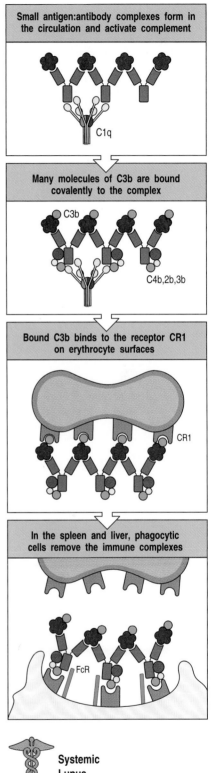

Here, macrophages bearing CR1 and Fc receptors remove the complexes from the erythrocyte surface without destroying the cell, and then degrade them (Fig. 9.29). Even larger aggregates of particulate antigen and antibody can be made soluble by activation of the classical complement pathway, and then removed by binding to complement receptors.

Immune complexes that are not removed tend to deposit in the basement membranes of small blood vessels, most notably those of the renal glomerulus where the blood is filtered to form urine. Immune complexes that pass through the basement membrane of the glomerulus bind to the complement receptor CR1 on the renal podocytes, cells that lie beneath the basement membrane. The functional significance of these receptors in the kidney is unknown; however, they play an important part in the pathology of some autoimmune diseases.

In the autoimmune disease systemic lupus erythematosus, which we will describe in Chapter 13, excessive levels of circulating immune complexes cause huge deposits of antigen, antibody, and complement on the podocytes, damaging the glomerulus; kidney failure is the principal danger in this disease. Immune complexes can also be a cause of pathology in patients with deficiencies in the early components of complement. Such patients do not clear immune complexes effectively and they also suffer tissue damage, especially in the kidneys, in a similar way.

Summary.

The T-cell dependent antibody response begins with IgM secretion but quickly progresses to the production of all the different isotypes. Each isotype is specialized both in its localization in the body and in the functions it can perform. IgM antibodies are found mainly in blood; they are pentameric in structure. IgM is specialized to activate complement efficiently upon binding antigen. IgG antibodies are usually of higher affinity and are found in blood and in extracellular fluid, where they can neutralize toxins, viruses, and bacteria, opsonize them for phagocytosis, and activate the complement system. IgA antibodies are synthesized as monomers, which enter blood and extra-cellular fluids, or as dimeric molecules in the lamina propria of various epithelia. IgA dimers are selectively transported across these epithelia into sites such as the lumen of the gut, where they neutralize toxins and viruses and block the entry of bacteria across the intestinal epithelium. Most IgE antibody is bound to the surface of mast cells that reside mainly just below body surfaces; antigen binding to this IgE triggers local defense reactions. Thus, each of these isotypes occupies a particular site in the body and has a particular role in defending the body against extracellular pathogens and their toxic products. Antibodies can accomplish this by direct interactions with pathogens or their products, for example by binding to active sites of toxins and neutralizing them or by blocking their ability to bind to host cells through specific receptors. When antibodies of the appropriate isotype bind to antigens, they can activate the classical pathway of complement, which leads to the elimination of the pathogen by the various mechanisms described in Chapter 2. Soluble immune complexes of antigen and antibody also fix complement and are cleared from the circulation via complement receptors on red blood cells.

Systemic Lupus Erythematosus

The destruction of antibody-coated pathogens via Fc receptors.

The ability of high-affinity antibodies to neutralize toxins, viruses, or bacteria can protect against infection but does not, on its own, solve the problem of how to remove the pathogens and their products from the body. Moreover, many pathogens cannot be neutralized by antibody and must be destroyed by other means. Many pathogen-specific antibodies do not bind to neutralizing targets on pathogen surfaces and thus need to be linked to other effector mechanisms in order to play their part in host defense. We have already seen how antibody binding to antigen can activate complement. Another important defense mechanism is the activation of a variety of **accessory effector cells** bearing receptors called **Fc receptors** because they are specific for the Fc portion of antibodies of a particular isotype. Through these receptors, accessory cells dispose of neutralized microorganisms and attack resistant extracellular pathogens. This mechanism maximizes the effectiveness of all antibodies regardless of where they bind. Accessory cells include the phagocytic cells (macrophages and neutrophils), which ingest antibody-coated bacteria and kill them, and other cells—natural killer (NK) cells, eosinophils, basophils, and mast cells (see Fig. 1.4)—which are triggered to secrete stored mediators when their Fc receptors are engaged. Accessory cells are activated when their Fc receptors are aggregated by binding to the multiple Fc regions of antibody molecules coating a pathogen. They can also be activated by soluble mediators, which include products of the complement cascade, which can itself be activated by antibody.

9-19   The Fc receptors of accessory cells are signaling receptors specific for immunoglobulins of different isotypes.

The Fc receptors are a family of cell-surface molecules that bind the Fc portion of immunoglobulins. Each member of the family recognizes immunoglobulin of one isotype or a few closely related isotypes through a recognition domain on the α chain of the Fc receptor. Fc receptors are themselves members of the immunoglobulin superfamily. Different accessory cells bear Fc receptors for antibodies of different isotypes, and the isotype of the antibody thus determines which accessory cell will be engaged in a given response. The different Fc receptors, the cells that express them, and their isotype specificity are shown in Fig. 9.30.

Most Fc receptors function as part of a multisubunit complex. Only the α chain is required for specific recognition; the other chains are required for transport to the cell surface and for signal transduction when an Fc region is bound. Signal transduction by many of these Fc receptors is mediated by the γ chain, which is closely related to the ζ chain of the T-cell receptor complex. Some Fcγ receptors, the Fcα receptor, and the high-affinity receptor for IgE use a γ chain for signaling; an exception is human FcγRII-A, a single-chain receptor in which the cytoplasmic domain of the α chain replaces the function of the γ chain. FcγRII-B1 and FcγRII-B2 are also single-chain receptors but function as inhibitory receptors as they contain an ITIM that engages the inositol 5′-phosphatase SHIP (see Section 6-14). Although the most prominent function of Fc receptors is the activation of accessory cells to attack pathogens, they can also contribute in other ways to immune responses. For example, the FcγRII-B receptor negatively regulates B cells,

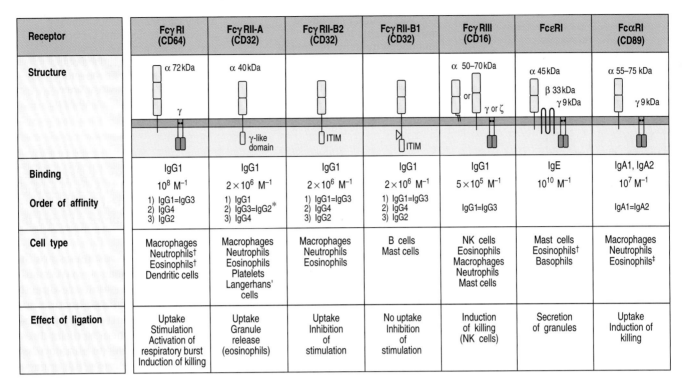

Receptor	FcγRI (CD64)	FcγRII-A (CD32)	FcγRII-B2 (CD32)	FcγRII-B1 (CD32)	FcγRIII (CD16)	FcεRI	FcαRI (CD89)
Structure	α 72 kDa; γ	α 40 kDa			α 50–70 kDa; γ or ζ	α 45 kDa; β 33 kDa; γ 9 kDa	α 55–75 kDa; γ 9 kDa
		γ-like domain	ITIM	ITIM			
Binding	IgG1 10^8 M^{-1}	IgG1 2×10^6 M^{-1}	IgG1 2×10^6 M^{-1}	IgG1 2×10^6 M^{-1}	IgG1 5×10^5 M^{-1}	IgE 10^{10} M^{-1}	IgA1, IgA2 10^7 M^{-1}
Order of affinity	1) IgG1=IgG3 2) IgG4 3) IgG2	1) IgG1 2) IgG3=IgG2* 3) IgG4	1) IgG1=IgG3 2) IgG4 3) IgG2	1) IgG1=IgG3 2) IgG4 3) IgG2	IgG1=IgG3		IgA1=IgA2
Cell type	Macrophages Neutrophils† Eosinophils† Dendritic cells	Macrophages Neutrophils Eosinophils Platelets Langerhans' cells	Macrophages Neutrophils Eosinophils	B cells Mast cells	NK cells Eosinophils Macrophages Neutrophils Mast cells	Mast cells Eosinophils† Basophils	Macrophages Neutrophils Eosinophils‡
Effect of ligation	Uptake Stimulation Activation of respiratory burst Induction of killing	Uptake Granule release (eosinophils)	Uptake Inhibition of stimulation	No uptake Inhibition of stimulation	Induction of killing (NK cells)	Secretion of granules	Uptake Induction of killing

Fig. 9.30 Distinct receptors for the Fc region of the different immunoglobulin isotypes are expressed on different accessory cells. The subunit structure and binding properties of these receptors and the cell types expressing them are shown. The complete multimolecular structure of most receptors is not yet known but they might all be multichain molecular complexes similar to the Fcε receptor I (FcεRI). The exact chain composition of any receptor can vary from one cell type to another. For example, FcγRIII in neutrophils is expressed as a molecule with a glycophosphoinositol membrane anchor, without γ chains, whereas in NK cells it is a transmembrane molecule associated with γ chains as shown. The FcγRII-B1 differs from the FcγRII-B2 by the presence of an additional exon in the intracellular region. This exon prevents the FcγRII-B1 from being internalized upon cross-linking. The binding affinities are taken from data on human receptors. *Only some allotypes of FcγRII-A bind IgG2. †In these cases Fc receptor expression is inducible rather than constitutive. ‡In eosinophils, the molecular weight of CD89α is 70–100 kDa.

mast cells, macrophages, and neutrophils by adjusting the threshold at which immune complexes will activate these cells. Fc receptors expressed by dendritic cells enable them to ingest antigen:antibody complexes and present antigenic peptides to T cells.

9-20   Fc receptors on phagocytes are activated by antibodies bound to the surface of pathogens and enable the phagocytes to ingest and destroy pathogens.

Phagocytes are activated by IgG antibodies, especially IgG1 and IgG3, that bind to specific Fcγ receptors on the phagocyte surface (see Fig. 9.30). As phagocyte activation can initiate an inflammatory response and cause tissue damage, it is essential that the Fc receptors on phagocytes are able to distinguish antibody molecules bound to a pathogen from the much larger number of free antibody molecules that are not bound to anything. This distinction is made possible by the aggregation or multimerization of antibodies that occurs when they bind to multimeric antigens or to multivalent antigenic particles such as viruses and bacteria. Fc receptors on the surface of an accessory cell bind antibody-coated particles with higher avidity than immunoglobulin monomers, and this is probably the principal mechanism by which bound antibodies are distinguished from free immunoglobulin

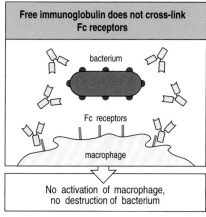

Free immunoglobulin does not cross-link Fc receptors

bacterium

Fc receptors

macrophage

No activation of macrophage, no destruction of bacterium

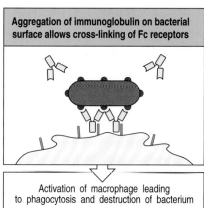

Aggregation of immunoglobulin on bacterial surface allows cross-linking of Fc receptors

Activation of macrophage leading to phagocytosis and destruction of bacterium

Fig. 9.31 Bound antibody is distinguishable from free immunoglobulin by its state of aggregation. Free immunoglobulin molecules bind most Fc receptors with very low affinity and can not cross-link Fc receptors. Antigen-bound immunoglobulin, however, can bind effectively to Fc receptors in a high-avidity interaction because several antibody molecules that are bound to the same surface bind to multiple Fc receptors on the surface of the accessory cell. This Fc receptor cross-linking sends a signal to activate (or sometimes inhibit, not shown) the cell bearing it.

(Fig. 9.31). The result is that Fc receptors enable accessory cells to detect pathogens through bound antibody molecules. Thus, specific antibody together with Fc receptors gives accessory cells that lack intrinsic specificity the ability to identify and remove pathogens and their products from the extracellular spaces.

The most important accessory cells in humoral immune responses are the phagocytic cells of the monocytic and myelocytic lineages, particularly macrophages and neutrophils (see Chapter 2). Many bacteria are directly recognized, ingested, and destroyed by phagocytes, and these bacteria are not pathogenic in normal individuals (see Chapter 2). Bacterial pathogens, however, often have polysaccharide capsules that allow them to resist direct engulfment by phagocytes. These bacteria become susceptible to phagocytosis, however, when they are coated with antibody and complement that engages the Fcγ or Fcα receptors and CR1 on phagocytic cells, triggering bacterial uptake (Fig. 9.32). Phagocytosis by binding to complement receptors is particularly important early in the immune response, before isotype-switched antibodies have been made. Capsular polysaccharides belong to the TI-2 class of thymus-independent antigens (see Section 9-11) and therefore can stimulate the early production of IgM antibodies. IgM is not an opsonizing antibody in itself, as there are no Fc receptors for IgM, but it is effective at activating the complement system. IgM binding to encapsulated bacteria thus triggers opsonization of these bacteria by complement and their prompt ingestion and destruction by phagocytes bearing complement receptors.

Both the internalization and destruction of microorganisms are greatly enhanced by interactions between the molecules coating an opsonized microorganism and their receptors on the phagocyte surface. When an antibody-coated pathogen binds to Fcγ receptors on the surface of a phago-cyte, for example, the cell surface extends around the surface of the particle through successive binding of Fcγ receptors to the antibody Fc regions bound to the pathogen surface. This is an active process triggered by the stimulation of Fcγ receptors. Endocytosis leads to enclosure of the particle in an acidified cytoplasmic vesicle called a phagosome. The phagosome then fuses with one or more lysosomes to generate a phagolysosome, releasing the lysosomal enzymes into the phagosome interior where they destroy the bacterium (see Fig. 9.32). The process of bacterial destruction in the phagolysosome was described in detail in Section 2-3.

Some particles are too large for a phagocyte to ingest; parasitic worms are one example. In this case, the phagocyte attaches to the surface of the antibody-coated parasite via its Fcγ, Fcα, or Fcε receptors, and the lysosomes fuse with the attached surface membrane. This reaction discharges the contents of the lysosome onto the surface of the parasite, damaging it directly in the extra-cellular space. While the principal phagocytes in the destruction of bacteria are macrophages and neutrophils, large parasites such as helminths are more usually attacked by eosinophils (Fig. 9.33). Thus, Fcγ and Fcα receptors can trigger the internalization of external particles by phagocytosis, or the externalization of internal vesicles by exocytosis. Cross-linking of IgE bound

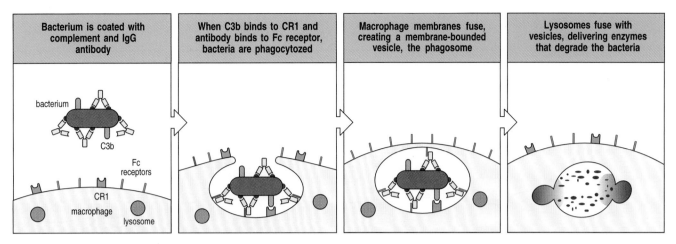

| Bacterium is coated with complement and IgG antibody | When C3b binds to CR1 and antibody binds to Fc receptor, bacteria are phagocytozed | Macrophage membranes fuse, creating a membrane-bounded vesicle, the phagosome | Lysosomes fuse with vesicles, delivering enzymes that degrade the bacteria |

Fig. 9.32 Fc and complement receptors on phagocytes trigger the uptake and degradation of antibody-coated bacteria. Many bacteria resist phagocytosis by macrophages and neutrophils. Antibodies bound to these bacteria, however, enable them to be ingested and degraded through interaction of the multiple Fc domains arrayed on the bacterial surface with Fc receptors on the phagocyte surface. Antibody coating also induces activation of the complement system and the binding of complement components to the bacterial surface. These can interact with complement receptors (for example CR1) on the phagocyte. Fc receptors and complement receptors synergize in inducing phagocytosis. Bacteria coated with IgG antibody and complement are therefore more readily ingested than those coated with IgG alone. Binding of Fc and complement receptors signals the phagocyte to increase the rate of phagocytosis, fuse lysosomes with phagosomes, and increase its bactericidal activity.

to the high-affinity FcεRI usually results in exocytosis. We will see in the next three sections that natural killer cells and mast cells also release mediators stored in their vesicles when their Fc receptors are aggregated.

9-21 Fc receptors activate natural killer cells to destroy antibody-coated targets.

Infected cells are usually destroyed by T cells alerted by foreign peptides bound to cell-surface MHC molecules. However, virus-infected cells can also signal the presence of intracellular infection by expressing on their surfaces viral proteins that can be recognized by antibodies. Cells bound by such antibodies can then be killed by a specialized non-T, non-B lymphoid cell called a **natural killer cell** (**NK cell**), which we met earlier in Chapter 2. NK cells are large lymphoid cells with prominent intracellular granules; they make up a small fraction of peripheral blood lymphoid cells. They bear no known antigen-specific receptors but are able to recognize and kill a limited range of abnormal cells. They were first discovered because of their ability to kill some tumor cells but are now known to have an important role in innate immunity.

The destruction of antibody-coated target cells by NK cells is called **antibody-dependent cell-mediated cytotoxicity** (**ADCC**) and is triggered when antibody bound to the surface of a cell interacts with Fc receptors on the NK cell (Fig. 9.34). NK cells express the receptor FcγRIII (CD16), which recognizes the IgG1 and IgG3 subclasses and triggers cytotoxic attack by the NK cell on antibody-coated target cells. The mechanism of attack is exactly analogous to that of cytotoxic T cells, involving the release of cytoplasmic granules containing perforin and granzymes (see Section 8-22). The importance of ADCC in defense against infection with bacteria or viruses has not yet been fully established. However, ADCC represents yet another mechanism by which, through engaging an Fc receptor, antibodies can direct an antigen-specific attack by an effector cell that itself lacks specificity for antigen.

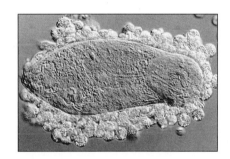

Fig. 9.33 Eosinophils attacking a schistosome larva in the presence of serum from an infected patient. Large parasites, such as worms, cannot be ingested by phagocytes; however, when the worm is coated with antibody, especially IgE, eosinophils can attack it through their binding to the high-affinity FcεRI. Similar attacks can be mounted by other Fc receptor-bearing cells on various large targets. These cells will release toxic contents of their granules directly onto the target, a process known as exocytosis. Photograph courtesy of A. Butterworth.

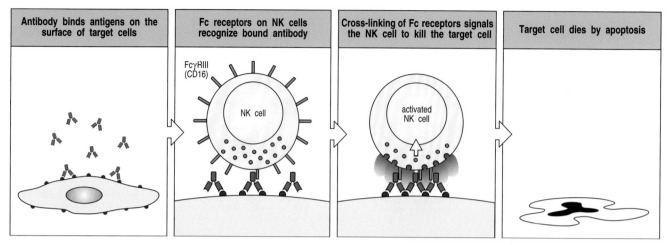

| Antibody binds antigens on the surface of target cells | Fc receptors on NK cells recognize bound antibody | Cross-linking of Fc receptors signals the NK cell to kill the target cell | Target cell dies by apoptosis |

Fig. 9.34 Antibody-coated target cells can be killed by NK cells in antibody-dependent cell-mediated cytotoxicity (ADCC). NK cells (see Chapter 2) are large granular non-T, non-B lymphoid cells that have FcγRIII (CD16) on their surface. When these cells encounter cells coated with IgG antibody, they rapidly kill the target cell. The importance of ADCC in host defense or tissue damage is still controversial.

9-22   Mast cells, basophils, and activated eosinophils bind IgE antibody via the high-affinity Fcε receptor.

When pathogens cross epithelial barriers and establish a local focus of infection, the host must mobilize its defenses and direct them to the site of pathogen growth. One mechanism by which this is achieved is to activate a specialized cell type known as a **mast cell**. Mast cells are large cells containing distinctive cytoplasmic granules that contain a mixture of chemical mediators, including histamine, that act rapidly to make local blood vessels more permeable. Mast cells have a distinctive appearance after staining with the dye toluidine blue that makes them readily identifiable in tissues (see Fig. 1.4). They are found in particularly high concentrations in vascularized connective tissues just beneath body epithelial surfaces, including the submucosal tissues of the gastrointestinal and respiratory tracts and the dermis that lies just below the surface of the skin.

Mast cells can be activated to release their granules, and to secrete lipid inflammatory mediators and cytokines, via antibody bound to Fc receptors specific for IgE (FcεRI) and IgG (FcγRIII). We have seen earlier that most Fc receptors bind stably to the Fc region of antibodies only when these are bound to antigen. By contrast, FcεRI binds monomeric IgE antibodies with a very high affinity, measured at approximately 10^{10} M^{-1}. Thus, even at the low levels of IgE found circulating in normal individuals, a substantial portion of the total IgE is bound to the FcεRI on mast cells and on circulating basophilic granulocytes or **basophils**. Eosinophils can also express Fc receptors, but only express FcεRI when activated and recruited to an inflammatory site.

Although mast cells are usually stably associated with bound IgE, they are not activated simply by the binding of monomeric antigens to this IgE. Mast-cell activation only occurs when the bound IgE is cross-linked by multivalent antigen. This signal activates the mast cell to release the contents of its granules, which occurs in seconds (Fig. 9.35), and to synthesize and release lipid mediators such as prostaglandin D_2 and leukotriene C4, and to secrete cytokines such as TNF-α, thereby initiating a local inflammatory response. Degranulation releases the stored histamine, causing a local increase in blood flow and vascular permeability that quickly leads to accumulation of fluid and blood proteins, including antibodies, in the surrounding tissue.

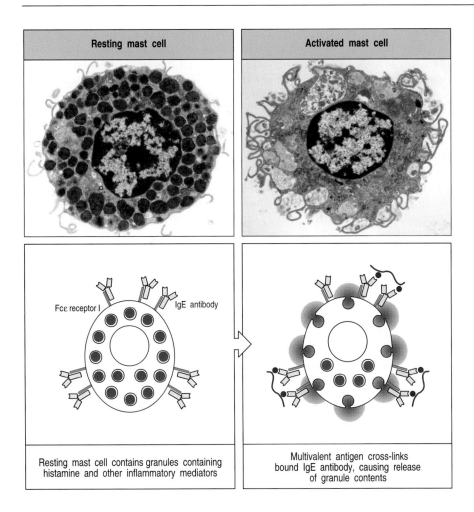

Resting mast cell	Activated mast cell

Fcε receptor I   IgE antibody

Resting mast cell contains granules containing histamine and other inflammatory mediators

Multivalent antigen cross-links bound IgE antibody, causing release of granule contents

Fig. 9.35 IgE antibody cross-linking on mast-cell surfaces leads to a rapid release of inflammatory mediators. Mast cells are large cells found in connective tissue that can be distinguished by secretory granules containing many inflammatory mediators. They bind stably to monomeric IgE antibodies through the very high-affinity Fcε receptor I. Antigen cross-linking of the bound IgE antibody molecules triggers rapid degranulation, releasing inflammatory mediators into the surrounding tissue. These mediators trigger local inflammation, which recruits cells and proteins required for host defense to sites of infection. These cells are also triggered during allergic reactions when allergens bind to IgE on mast cells. Photographs courtesy of A.M. Dvorak.

Shortly afterwards, there is an influx of blood-borne cells such as polymorphonuclear leukocytes and later macrophages, eosinophils, and effector lymphocytes. This influx can last a few minutes to a few hours and produces an inflammatory response at the site of infection. Thus, mast cells are part of the front-line host defenses against pathogens that enter the body across epithelial barriers.

9-23 IgE-mediated activation of accessory cells has an important role in resistance to parasite infection.

Mast cells are thought to serve at least three important functions in host defense. First, their location near body surfaces allows them to recruit both specific and nonspecific effector elements to sites where infectious agents are most likely to enter the internal milieu. Second, they also increase the flow of lymph from sites of antigen deposition to the regional lymph nodes, where naive lymphocytes are first activated. Third, their ability to trigger muscular contraction can contribute to the physical expulsion of pathogens from the lungs or the gut. Mast cells respond rapidly to the binding of antigen to surface-bound IgE antibodies, and their activation leads to the recruitment and activation of basophils and eosinophils, which contribute further to the IgE-mediated response. There is increasing evidence that such IgE-mediated responses are crucial to defense against parasite infestation.

A role for mast cells in the clearance of parasites is suggested by accumulation of mast cells in the intestine, known as **mastocytosis**, that accompanies helminth infection, and by observations in W/W^V mutant mice, which have a profound mast-cell deficiency caused by mutation of the gene c-*kit*. These mutant mice show impaired clearance of the intestinal nematodes *Trichinella spiralis* and *Strongyloides* species. Clearance of *Strongyloides* is even more impaired in W/W^V mice that lack IL-3 and therefore, in addition to lacking mast cells, fail to produce basophils. Thus both mast cells and basophils seem to contribute to defense against these helminth parasites. Other evidence also points to the importance of IgE antibodies and eosinophils in defense against parasites. Infections by certain classes of parasite, particularly helminths, are strongly associated with the production of IgE antibodies and the presence of an abnormally large number of eosinophils (**eosinophilia**) in blood and tissues. Furthermore, experiments in mice show that depletion of eosinophils by using polyclonal anti-eosinophil antisera increases the severity of infection by the parasitic helminth *Schistosoma mansoni*. Eosinophils seem to be directly responsible for helminth destruction; examination of infected tissues shows degranulated eosinophils adhering to helminths, and experiments *in vitro* have shown that eosinophils can kill *Schistosoma mansoni* in the presence of specific IgE (see Fig. 9.33), IgG, or IgA anti-schistosome antibodies.

The role of IgE, mast cells, basophils, and eosinophils can also be seen in resistance to the feeding of blood-sucking ixodid ticks. Normal skin at the site of a tick bite shows degranulated mast cells, and an accumulation of basophils and eosinophils that are degranulated, an indicator of recent activation. Resistance to subsequent feeding by these ticks develops after the first exposure, suggesting a specific immunological mechanism. Mast-cell deficient mice show no such acquired resistance to tick species, and in guinea pigs the depletion of either basophils or eosinophils by specific polyclonal antibodies also reduces resistance to tick feeding. Finally, recent experiments have shown that resistance to ticks in mice is mediated by specific IgE antibody.

Thus, many clinical studies and experiments support a role for this system of IgE binding to the high-affinity FcεRI in host resistance to pathogens that enter across epithelia. We will see later, in Chapter 12, that the same system accounts for many of the symptoms in allergic diseases such as asthma, hay-fever, and the life-threatening response known as systemic anaphylaxis.

Summary.

Antibody-coated pathogens are recognized by accessory effector cells through Fc receptors that bind to the multiple constant regions (Fc portions) provided by the bound antibodies. Binding activates the accessory cell and triggers destruction of the pathogen. Fc receptors comprise a family of proteins, each of which recognizes immunoglobulins of particular isotypes. Fc receptors on macrophages and neutrophils recognize the constant regions of IgG or IgA antibodies bound to a pathogen and trigger the engulfment and destruction of IgG- or IgA-coated bacteria. Binding to the Fc receptor also induces the production of microbicidal agents in the intracellular vesicles of the phagocyte. Eosinophils are important in the elimination of parasites too large to be engulfed; they bear Fc receptors specific for the constant region of IgG, as well as high-affinity receptors for IgE; aggregation of these receptors triggers the release of toxic substances onto the surface of the parasite. NK cells, tissue mast cells, and blood basophils also release their granule contents when their Fc receptors are engaged. The high-affinity receptor for IgE is expressed constitutively by mast cells and basophils, and is induced in activated eosinophils. It differs from other Fc receptors in that it can bind free monomeric antibody, thus enabling an immediate response to pathogens at their site of first entry into the tissues. When IgE bound to the surface of a mast cell is aggregated by binding to antigen, it triggers the release of histamine and many other mediators that increase the blood flow to sites of infection; it thereby recruits antibodies and effector cells to these sites. Mast cells are found principally below epithelial surfaces of the skin and the digestive and respiratory tracts, and their activation by innocuous substances is responsible for many of the symptoms of acute allergic reactions, as will be described in Chapter 12.

Summary to Chapter 9.

The humoral immune response to infection involves the production of antibody by plasma cells derived from B lymphocytes, the binding of this antibody to the pathogen, and the elimination of the pathogen by accessory cells and molecules of the humoral immune system. The production of antibody usually requires the action of helper T cells specific for a peptide fragment of the antigen recognized by the B cell. The B cell then proliferates and differentiates, first at the T zone–B zone boundary in secondary lymphoid tissues and later in the germinal center, where somatic hypermutation diversifies the B-cell receptors expressed by a clone of B cells. The B cells that bind antigen most avidly are selected for further differentiation by the continual requirement for contact with antigen and the requirement to present antigen-derived peptides to germinal center helper T cells. These events allow the affinity of antibodies to increase over the course of an immune response, especially in repeated responses to the same antigen. Helper T cells also direct isotype switching, leading to the production of antibody of various isotypes that can be distributed to various body compartments.

IgM is produced early in the response and has a major role in protecting against infection in the bloodstream, whereas more mature isotypes such as IgG diffuse into the tissues. Certain pathogens that both have highly repeating antigenic determinants and express mitogens that intrinsically stimulate B cells can elicit IgM and some IgG independently of T-cell help. Such antigens are called TI antigens, and the antibody elicited by these antigens can provide an early protective immune response to bacteria. Multimeric IgA is produced in the lamina propria and transported across epithelial surfaces, whereas IgE is made in small amounts and binds avidly to the surface of mast cells. Antibodies that bind with high affinity to critical sites on toxins, viruses, and bacteria can neutralize them. However, pathogens and their products are destroyed and removed from the body largely through uptake into phagocytes and degradation inside these cells. Antibodies that coat pathogens bind to Fc receptors on phagocytes, which are thereby triggered to engulf and destroy the pathogen. Binding of antibody C regions to Fc receptors on other cells leads to exocytosis of stored mediators, and this is particularly important in parasite infections, where Fcε-expressing mast cells and activated eosinophils are triggered by antigen binding to IgE antibody to release inflammatory mediators directly onto parasite surfaces. This same mechanism for activating mast cells can cause immunopathology, when IgE on mast cells binds to innocuous substances, resulting in allergic reactions, as we will see in Chapter 12. Antibodies can also initiate the destruction of pathogens by activating the complement system. Complement components can opsonize pathogens for uptake by phagocytes, can recruit phagocytes to sites of infection, and can directly destroy pathogens by creating pores in their cell membrane. Receptors for complement components and Fc receptors often synergize in activating the uptake and destruction of pathogens and immune complexes. Thus, the humoral immune response is targeted to the infecting pathogen through the production of specific antibody; however, the effector actions of that antibody are determined by its isotype and are the same for all pathogens bound by antibody of a particular isotype.

General references.

Liu, Y.J., Zhang, J., Lane, P.J., Chan, E.Y., and MacLennan, I.C.: **Sites of specific B cell activation in primary and secondary responses to T cell-dependent and T cell-independent antigens.** *Eur. J. Immunol.* 1991, **21**:2951-2962.

Metzger, H. (ed): *Fc Receptors and the Action of Antibodies*, 1st edn. Washington, DC, American Society for Microbiology, 1990.

Rajewsky, K.: **Clonal selection and learning in the antibody system.** *Nature* 1996, **381**:751-758.

Section references.

9-1   The humoral immune response is initiated when B cells that bind antigen are signaled by helper T cells or by certain microbial antigens alone.

Gulbranson-Judge, A., and MacLennan, I.: **Sequential antigen-specific growth of T cells in the T zones and follicles in response to pigeon cytochrome c.** *Eur. J. Immunol.* 1996, **26**:1830-1837.

O'Rourke, L., Tooze, R., and Fearon, D.T.: **Co-receptors of B lymphocytes.** *Curr. Opin. Immunol.* 1997, **9**:324-329.

9-2   Armed helper T cells activate B cells that recognize the same antigen.

Lanzavecchia, A.: **Receptor-mediated antigen uptake and its effect on antigen presentation to class II-restricted T lymphocytes.** *Annu. Rev. Immunol.* 1990, **8**:773-793.

MacLennan, I.C.M., Gulbranson-Judge, A., Toellner, K.M., Casamayor-Palleja, M., Chan, E., Sze, D.M.Y., Luther, S.A., and Orbea, H.A.: **The changing preference of T and B cells for partners as T-dependent antibody responses develop.** *Immunol. Rev.* 1997, **156**:53-66.

Parker, D.C.: **T cell-dependent B-cell activation.** *Annu. Rev. Immunol.* 1993, **11**:331-340.

9-3   Antigenic peptides bound to self MHC class II molecules trigger armed helper T cells to make membrane-bound and secreted molecules that can activate a B cell.

Croft, M., and Swain, S.L.: **B cell response to fresh and effector T helper cells. Role of cognate T-B interaction and the cytokines IL-2, IL-4, and IL-6.** *J. Immunol.* 1991, **146**:4055-4064.

Jaiswal, A.I., and Croft, M.: **CD40 ligand induction on T cell subsets by peptide-presenting B cells.** *J. Immunol.* 1997, **159**:2282-2291.

Lane, P., Traunecker, A., Hubele, S., Inui, S., Lanzavecchia, A., and Gray, D.: **Activated human T cells express a ligand for the human B cell-associated antigen CD40 which participates in T cell-dependent activation of B lymphocytes.** *Eur. J. Immunol.* 1992, **22**:2573-2578.

Lohoff, M., Koch, A., and Rollinghoff, M.: **Two signals are involved in polyclonal B cell stimulation by T helper type 2 cells: a role for LFA-1 molecules and interleukin 4.** *Eur. J. Immunol.* 1992, **22**:599-602.

Noelle, R.J., Roy, M., Shepherd, D.M., Stamenkovic, I., Ledbetter, J.A., and Aruffo, A.: **A novel ligand on activated T helper cells binds CD40 and transduces the signal for the cognate activation of B cells.** *Proc. Natl. Acad. Sci. USA* 1992, **89**:6550.

Shanebeck, K.D., Maliszewski, C.R., Kennedy, M.K., Picha, K.S., Smith, C.A., Goodwin, R.G., and Grabstein, K.H.: **Regulation of murine B cell growth and differentiation by CD30 ligand.** *Eur. J. Immunol.* 1995, **25**:2147-2153.

Valle, A., Zuber, C.E., Defrance, T., Djossou, O., De, R.M., and Banchereau, J.: **Activation of human B lymphocytes through CD40 and interleukin 4.** *Eur. J. Immunol.* 1989, **19**:1463-1467.

Yan, M., Marsters, S.A., Grewal, I.S., Wan, H., Ashkenazi, A., and Dixit, V.M.: **Identification of a receptor for BLyS demonstrates a crucial role in humoral immunity.** *Nat. Immunol.* 2000, **1**:37-41.

9-4   Isotype switching requires expression of CD40L by the helper T cell and is directed by cytokines.

Francke, U., and Ochs, H.D.: **The CD40 ligand, gp39, is defective in activated T cells from patients with X-linked hyper-IgM syndrome.** *Cell* 1993, **72**:291-300.

Jumper, M., Splawski, J., Lipsky, P., and Meek, K.: **Ligation of CD40 induces sterile transcripts of multiple Ig H chain isotypes in human B cells.** *J. Immunol.* 1994, **152**:438-445.

Snapper, C.M., and Paul, W.E.: **Interferon γ and B cell stimulatory factor-1 reciprocally regulate Ig isotype production.** *Science* 1987, **236**:944-947.

Snapper, C.M., Kehry, M.R., Castle, B.E., and Mond, J.J.: **Multivalent, but not divalent, antigen receptor cross-linkers synergize with CD40 ligand for induction of Ig synthesis and class switching in normal murine B cells.** *J. Immunol.* 1995, **154**:1177-1187.

Stavnezer, J.: **Immunoglobulin class switching.** *Curr. Opin. Immunol.* 1996, **8**:199-205.

9-5   Antigen-binding B cells are trapped in the T-cell zone of secondary lymphoid tissues and are activated by encounter with armed helper T cells.

Garside, P., Ingulli, E., Merica, R.R., Johnson, J.G., Noelle, R.J., and Jenkins, M.K.: **Visualization of specific B and T lymphocyte interactions in the lymph node.** *Science* 1998, **281**:96-99.

Jacob, J., Kassir, R., and Kelsoe, G.: ***In situ* studies of the primary immune response to (4-hydroxy-3-nitrophenyl)acetyl. I. The architecture and dynamics of responding cell population.** *J. Exp. Med.* 1991, **173**:1165-1175.

9-6   The second phase of the primary B-cell immune response occurs when activated B cells migrate to follicles and proliferate to form germinal centers.

Brachtel, E.F., Washiyama, M., Johnson, G.D., Tenner-Racz, K., Racz, P., and MacLennan, I.C.: **Differences in the germinal centres of palatine tonsils and lymph nodes.** *Scand. J. Immunol.* 1996, **43**:239-247.

Camacho, S.A., Kosco-Vilbois, M.H., and Berek, C.: **The dynamic structure of the germinal center.** *Immunol. Today* 1998, **19**:511-514.

Jacob, J., and Kelsoe, G.: ***In situ* studies of the primary immune response to (4-hydroxy-3-nitrophenyl)acetyl. II. A common clonal origin for periarteriolar lymphoid sheath-associated foci and germinal centers.** *J. Exp. Med.* 1992, **176**:679-687.

Jacob, J., Przylepa, J., Miller, C., and Kelsoe, G.: ***In situ* studies of the primary immune response to (4-hydroxy-3-nitrophenyl)acetyl. III. The kinetics of V region mutation and selection in germinal center B cells.** *J. Exp. Med.* 1993, **178**:1293-1307.

Kelsoe, G.: **The germinal center: a crucible for lymphocyte selection.** *Sem. Immunol.* 1996, **8**:179-184.

MacLennan, I.C.M.: **Germinal centers.** *Annu. Rev. Immunol.* 1994, **12**:117-139.

9-7   Germinal center B cells undergo V-region somatic hypermutation and cells with mutations that improve affinity for antigen are selected.

Clarke, S.H., Huppi, K., Ruezinsky, D., Staudt, L., Gerhard, W., and Weigert, M.: **Inter- and intraclonal diversity in the antibody response to influenza hemagglutinin.** *J. Exp. Med.* 1985, **161**:687-704.

Jacob, J., Kelsoe, G., Rajewsky, K., and Weiss, U.: **Intraclonal generation of antibody mutants in germinal centres.** *Nature* 1991, **354**:389-392.

Shlomchik, M.J., Litwin, S., and Weigert, M.: **The influence of somatic mutation on clonal expansion.** *Prog. Immunol. Proc. 7th Int. Cong. Immunol.* 1990, **7**:415-423.

Ziegner, M., Steinhauser, G., and Berek, C.: **Development of antibody diversity** in single germinal centers: selective expansion of high-affinity variants. *Eur. J. Immunol.* 1994, **24**:2393-2400.

9-8 Ligation of the B-cell receptor and CD40, together with direct contact with T cells, are all required to sustain germinal center B cells.

Humphrey, J.H., Grennan, D., and Sundaram, V.: The origin of follicular dendritic cells in the mouse and the mechanism of trapping of immune complexes on them. *Eur. J. Immunol.* 1984, **14**:1859.

Liu, Y.J., Joshua, D.E., Williams, G.T., Smith, C.A., Gordon, J., and MacLennan, I.C.M.: Mechanism of antigen-driven selection in germinal centres. *Nature* 1989, **342**:929-931.

Wang, Z., Karras, J.G., Howard, R.G., and Rothstein, T.L.: Induction of bcl-x by CD40 engagement rescues sIg-induced apoptosis in murine B cells. *J. Immunol.* 1995, **155**:3722-3725.

9-9 Surviving germinal center B cells differentiate into either plasma cells or memory cells.

Coico, R.F., Bhogal, B.S., and Thorbecke, G.J.: Relationship of germinal centers in lymphoid tissue to immunolgic memory. IV. Transfer of B cell memory with lymph node cells fractionated according to their receptors for peanut agglutinin. *J. Immunol.* 1983, **131**:2254.

Koni, P.A., Sacca, R., Lawton, P., Browning, J.L., Ruddle, N.H., and Flavell, R.A.: Distinct roles in lymphoid organogenesis for lymphotoxins alpha and beta revealed in lymphotoxin beta-deficient mice. *Immunity* 1997, **6**:491-500.

Kosco, M.H., Burton, G.F., Kapasi, Z.F., Szakal, A.K., and Tew, J.G.: Antibody-forming cell induction during an early phase of germinal centre development and its delay with ageing. *Immunology* 1989, **68**:312-316.

Matsumoto, M., Lo, S.F., Carruthers, C.J.L., Min, J., Mariathasan, S., Huang, G., Plas, D.R., Martin, S.M., Geha, R.S., Nahm, M.H., and Chaplin, D.D.: Affinity maturation without germinal centres in lymphotoxin-α-deficient mice. *Nature* 1996, **382**:462-466.

Tew, J.G., DiLosa, R.M., Burton, G.F., Kosco, M.H., Kupp, L.I., Masuda, A., and Szakal, A.K.: Germinal centers and antibody production in bone marrow. *Immunol. Rev.* 1992, **126**:99-112.

9-10 B-cell responses to bacterial antigens with intrinsic ability to activate B cells do not require T-cell help.

Anderson, J., Coutinho, A., Lernhardt, W., and Melchers, F.: Clonal growth and maturation to immunoglobulin secretion *in vitro* of every growth-inducible B lymphocyte. *Cell* 1977, **10**:27-34.

9-11 B-cell responses to bacterial polysaccharides do not require peptide-specific T-cell help.

Mond, J.J., Lees, A., and Snapper, C.M.: T cell-independent antigens type 2. *Annu. Rev. Immunol.* 1995, **13**:655-692.

9-12 Antibodies of different isotype operate in distinct places and have distinct effector functions.

Clark, M.R.: IgG effector mechanisms. *Chem. Immunol.* 1997, **65**:88-110.
Herrod, H.G.: IgG subclass deficiency. *Allergy Proc.* 1992, **13**:299-302.
Janeway, C.A., Rosen, F.S., Merler, E., and Alper, C.A.: *The Gamma Globulins*, 2nd edn. Boston, Little Brown and Co., 1967.
Ward, E.S., and Ghetie, V.: The effector functions of immunoglobulins: implications for therapy. *Thera. Immunol.* 1995, **2**:77-94.

9-13 Transport proteins that bind to the Fc regions of antibodies carry particular isotypes across epithelial barriers.

Burmeister, W.P., Gastinel, L.N., Simister, N.E., Blum, M.L., and Bjorkman, P.J.: Crystal structure at 2.2 Å resolution of the MHC-related neonatal Fc receptor. *Nature* 1994, **372**:336-343.

Corthesy, B., and Kraehenbuhl, J.P.: Antibody-mediated protection of mucosal surfaces. *Curr. Top. Microbiol. Immunol.* 1999, **236**:93-111.

Ghetie, V., and Ward, E.S.: Multiple roles for the major histocompatibility complex class I-related receptor FcRn. *Annu. Rev. Immunol.* 2000, **18**:739-766.

Lamm, M.E.: Current concepts in mucosal immunity. IV. How epithelial transport of IgA antibodies relates to host defense. *Am. J. Physiol.* 1998, **274**:G614-G617.

Mostov, K.E.: Transepithelial transport of immunoglobulins. *Annu. Rev. Immunol.* 1994, **12**:63-84.

Simister, N.E., and Mostov, K.E.: An Fc receptor structurally related to MHC class I antigens. *Nature* 1989, **337**:184-187.

9-14 High-affinity IgG and IgA antibodies can neutralize bacterial toxins.

Robbins, F.C., and Robbins, J.B.: Current status and prospects for some improved and new bacterial vaccines. *Am. J. Pub. Health* 1986, **7**:105-125.

9-15 High-affinity IgG and IgA antibodies can inhibit the infectivity of viruses.

Mandel, B.: Neutralization of polio virus: a hypothesis to explain the mechanism and the one hit character of the neutralization reaction. *Virology* 1976, **69**:500-510.

Possee, R.D., Schild, G.C., and Dimmock, N.J.: Studies on the mechanism of neutralization of influenza virus by antibody: evidence that neutralizing antibody (anti-hemaglutanin) inactivates influenza virus *in vivo* by inhibiting virion transcriptase activity. *J. Gen. Virol.* 1982, **58**:373-386.

Roost, H.P., Bachmann, M.F., Haag, A., Kalinke, U., Pliska, V., Hengartner, H., and Zinkernagel, R.M.: Early high-affinity neutralizing anti-viral IgG responses without further overall improvements of affinity. *Proc. Natl. Acad. Sci. USA* 1995, **92**:1257-1261.

9-16 Antibodies can block the adherence of bacteria to host cells.

Fischetti, V.A., and Bessen, D.: Effect of mucosal antibodies to M protein in colonization by group A streptococci, in Switalski, L., Hook, M., and Beachery, E. (eds): *Molecular Mechanisms of Microbial Adhesion*. New York, Springer, 1989.

9-17 Antibody:antigen complexes activate the classical pathway of complement by binding to C1q.

Cooper, N.R.: The classical complement pathway. Activation and regulation of the first complement component. *Adv. Immunol.* 1985, **37**:151-216.

Perkins, S.J., and Nealis, A.S.: The quaternary structure in solution of human complement subcomponent C1r$_2$C1s$_2$. *Biochem. J.* 1989, **263**:463-469.

9-18 Complement receptors are important in the removal of immune complexes from the circulation.

Schifferli, J.A., and Taylor, J.P.: Physiologic and pathologic aspects of circulating immune complexes. *Kidney Intl.* 1989, **35**:993-1003.

Schifferli, J.A., Ng, Y.C., and Peters, D.K.: The role of complement and its receptor in the elimination of immune complexes. *N. Engl. J. Med.* 1986, **315**:488-495.

9-19 The Fc receptors of accessory cells are signaling receptors specific for immunoglobulins of different isotypes.

Ravetch, J.V.: Fc receptors. *Curr. Opin. Immunol.* 1997, **9**:121-125.
Ravetch, J.V., and Clynes, R.A.: Divergent roles for Fc receptors and complement *in vivo*. *Annu. Rev. Immunol.* 1998, **16**:421-432.

9-20 Fc receptors on phagocytes are activated by antibodies bound to the surface of pathogens and enable the phagocytes to ingest and destroy pathogens.

Gounni, A.S., Lamkhioued, B., Ochiai, K., Tanaka, Y., Delaporte, E., Capron, A., Kinet, J.P., and Capron, M.: **High-affinity IgE receptor on eosinophils is involved in defence against parasites**. *Nature* 1994, **367**:183-186.

Karakawa, W.W., Sutton, A., Schneerson, R., Karpas, A., and Vann, W.F.: **Capsular antibodies induce type-specific phagocytosis of capsulated** *Staphylococcus aureus* **by human polymorphonuclear leukocytes**. *Infect. Immun.* 1986, **56**:1090-1095.

9-21 Fc receptors activate natural killer cells to destroy antibody-coated targets.

Lanier, L.L., and Phillips, J.H.: **Evidence for three types of human cytotoxic lymphocyte**. *Immunol. Today* 1986, **7**:132.

Lanier, L.L., Ruitenberg, J.J., and Phillips, J.H.: **Functional and biochemical analysis of CD16 antigen on natural killer cells and granulocytes**. *J. Immunol.* 1988, **141**:3487-3485.

9-22 Mast cells, basophils, and activated eosinophils bind IgE antibody via the high-affinity Fcε receptor.

Beaven, M.A., and Metzger, H.: **Signal transduction by Fc receptors: the FcεRI case**. *Immunol. Today* 1993, **14**:222-226.

Sutton, B.J., and Gould, H.J.: **The human IgE network**. *Nature* 1993, **366**:421-428.

9-23 IgE-mediated activation of accessory cells has an important role in resistance to parasite infection.

Capron, A., and Dessaint, J.P.: **Immunologic aspects of schistosomiasis**. *Annu. Rev. Med.* 1992, **43**:209-218.

Grencis, R.K., Else, K.J., Huntley, J.F., and Nishikawa, S.I.: **The** *in vivo* **role of stem cell factor (c-kit ligand) on mastocytosis and host protective immunity to the intestinal nematode** *Trichinella spiralis* **in mice**. *Parasite Immunol.* 1993, **15**:55-59.

Kasugai, T., Tei, H., Okada, M., Hirota, S., Morimoto, M., Yamada, M., Nakama, A., Arizono, N., and Kitamura, Y.: **Infection with** *Nippostrongylus brasiliensis* **induces invasion of mast cell precursors from peripheral blood to small intestine**. *Blood* 1995, **85**:1334-1340.

Ushio, H., Watanabe, N., Kiso, Y., Higuchi, S., and Matsuda, H.: **Protective immunity and mast cell and eosinophil responses in mice infested with larval** *Haemaphysalis longicornis* **ticks**. *Parasite Immunol.* 1993, **15**:209-214.

Adaptive Immunity to Infection

Throughout this book we have examined the individual mechanisms by which both the innate and the adaptive immune responses function to protect the individual from invading microorganisms. In this chapter, we consider how the cells and molecules of the immune system work as an integrated defense system to eliminate or control the infectious agent and how the adaptive immune system provides long-lasting protective immunity. This is the first of several chapters that consider how the immune system functions as a whole in health and disease. Subsequent chapters will examine how failures of immune defense and unwanted immune responses occur, and how the immune response can be manipulated to benefit the individual.

In the first part of this chapter, we briefly consider the diversity of pathogens the immune system can encounter and outline the general course of an infection. The mechanisms of innate immunity, which we discussed in detail in Chapter 2, are brought into play in the earliest phases of the infection and may succeed in repelling it. Pathogens, however, have developed strategies that allow them, at least on occasion, to elude or overcome the mechanisms of innate immune defense and establish a focus of infection from which they can spread. In these circumstances, the innate immune response sets the scene for the induction of an adaptive immune response, the focus of the second part of the chapter. Several days are required for the clonal expansion and differentiation of naive lymphocytes into effector T cells and antibody-secreting B cells that, in most cases, effectively target the pathogen for elimination (Fig. 10.1). During this period, specific immunological memory is also established. This ensures a rapid reinduction of antigen-specific antibody and armed effector T cells on subsequent encounters with the same pathogen, thus providing long-lasting protection against reinfection.

Innate immunity is an essential prerequisite for the adaptive immune response, as the antigen-specific lymphocytes of the adaptive immune response are activated by co-stimulatory molecules that are induced on cells of the innate immune system during their interaction with micoorganisms. The cytokines produced during these early phases also play an important part in stimulating the subsequent adaptive immune response and shaping its development; determining, for example, whether the response is predominantly T cell-mediated or predominantly humoral. We have already described the generation and function of effector T cells and antibodies in Chapters 8 and 9. In this chapter, we will discuss how the different phases of host defense are orchestrated in space and time, and how changes in specialized cell-surface molecules and chemokines guide lymphocytes to the appropriate site of action at different stages of the adaptive immune response.

The most frequent site of encounter between the body and microorganisms and other antigens is the mucosal immune system. This lines the airways, gastrointestinal tract, and urogenital system and is the most extensive compartment of the immune system. The third part of the chapter describes

Fig. 10.1 The course of a typical acute infection. 1. The level of infectious agent increases as the pathogen replicates. 2. When numbers of the pathogen exceed the threshold dose of antigen required for an adaptive response, the response is initiated; the pathogen continues to grow, retarded only by the innate and nonadaptive responses. At this stage, immunological memory also starts to be induced. 3. After 4–5 days, effector cells and molecules of the adaptive response start to clear the infection. 4. When the infection is cleared and the dose of antigen falls below the response threshold, the response ceases, but antibody, residual effector cells, and also immunological memory provide lasting protection against reinfection in most cases.

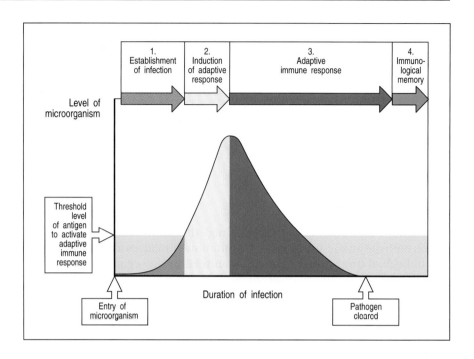

the functions and properties of the adaptive immune responses mounted by the mucosal immune system. These not only protect the body from infection, but are also designed to stop the immune system from responding inappropriately to the many environmental antigens and potential allergens with which the mucosal lymphoid tissues come into contact, most particularly the foods we eat every day.

We return to immunological memory that provides long-lasting and sometimes life-long protection against reinfection by many pathogens in the last part of the chapter. Memory responses differ in several ways from primary responses and we will discuss the reasons for this, and what is known of how immunological memory is maintained.

Infectious agents and how they cause disease.

Infectious disease can be devastating, and sometimes fatal, to the host. In this part of the chapter we will briefly examine the stages of infection, and the various types of infectious agents.

10-1   The course of an infection can be divided into several distinct phases.

The process of infection can be broken down into stages, each of which can be blocked by different defense mechanisms. In the first stage, a new host is exposed to infectious particles shed by an infected individual. The number, route, mode of transmission, and stability of an infectious agent outside the host determines its infectivity. Some pathogens, such as anthrax, are spread by spores that are highly resistant to heat and drying, while others, such as the human immunodeficiency virus (HIV), are spread only by the exchange of bodily fluids or tissues because they are unable to survive as infectious agents outside the body.

The first contact with a new host occurs through an epithelial surface. This may be the skin or the internal mucosal surfaces of the respiratory, gastro-intestinal, and urogenital tracts. After making contact, an infectious agent must establish a focus of infection. This involves adhering to the epithelial surface, and then colonizing it, or penetrating it to replicate in the tissues (Fig. 10.2, left-hand panels). Many microorganisms are repelled at this stage by innate immunity. We have discussed the innate immune defense mediated by epithelia and by phagocytes and complement in the underlying tissues in Chapter 2. Chapter 2 also discusses how NK cells are activated in response to intracellular infections, and how a local inflammatory response and induced cytokines and chemokines can bring more effector cells and molecules to the site of an infection while preventing pathogen spread into the blood. These innate immune responses use a variety of germline-encoded receptors to discriminate between microbial and host cell surfaces, or infected and normal cells. They are not as effective as adaptive immune responses, which can afford to be more powerful on account of their antigen specificity. However, they can prevent an infection being established, or failing that, contain it while an adaptive immune response develops.

Only when a microorganism has successfully established a site of infection in the host does disease occur, and little damage will be caused unless the agent is able to spread from the original site of infection or can secrete toxins that can spread to other parts of the body. Extracellular pathogens spread by direct extension of the focus of infection through the lymphatics or the bloodstream. Usually, spread by the bloodstream occurs only after the lymphatic system has been overwhelmed by the burden of infectious agent. Obligate intracellular pathogens must spread from cell to cell; they do so either by direct transmission from one cell to the next or by release into the extracellular fluid and reinfection of both adjacent and distant cells. Many

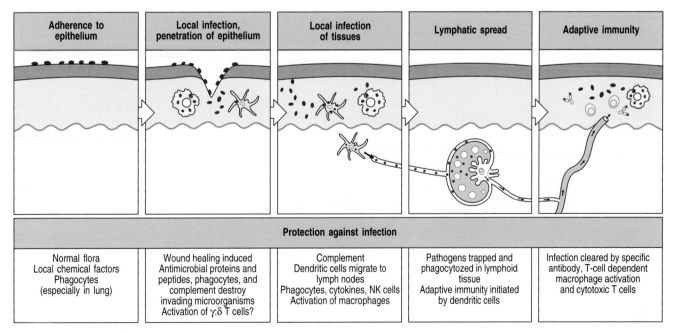

Adherence to epithelium	Local infection, penetration of epithelium	Local infection of tissues	Lymphatic spread	Adaptive immunity

Protection against infection				
Normal flora Local chemical factors Phagocytes (especially in lung)	Wound healing induced Antimicrobial proteins and peptides, phagocytes, and complement destroy invading microorganisms Activation of γ:δ T cells?	Complement Dendritic cells migrate to lymph nodes Phagocytes, cytokines, NK cells Activation of macrophages	Pathogens trapped and phagocytozed in lymphoid tissue Adaptive immunity initiated by dendritic cells	Infection cleared by specific antibody, T-cell dependent macrophage activation and cytotoxic T cells

Fig. 10.2 Infections and the responses to them can be divided into a series of stages. These are illustrated here for an infectious microorganism entering across an epithelium, the commonest route of entry. The infectious organism must first adhere to epithelial cells and then cross the epithelium. A local nonadaptive response helps contain the infection and delivers antigen to local lymph nodes, leading to adaptive immunity and clearance of the infection. The role of γ:δ T cells is uncertain, as indicated by the question mark.

common food poisoning organisms cause pathology without spreading into the tissues. They establish a site of infection on the epithelial surface in the lumen of the gut and cause no direct pathology themselves, but they secrete toxins that cause damage either *in situ* or after crossing the epithelial barrier and entering the circulation.

Most infectious agents show a significant degree of host specificity, causing disease only in one or a few related species. What determines host specificity for every agent is not known, but the requirement for attachment to a particular cell-surface molecule is one critical factor. As other interactions with host cells are also commonly needed to support replication, most pathogens have a limited host range. The molecular mechanisms of host specificity comprise an area of research known as molecular pathogenesis, which falls outside the scope of this book.

While most microorganisms are repelled by innate host defenses, an initial infection, once established, generally leads to perceptible disease followed by an effective host adaptive immune response. This is initiated in the local lymphoid tissue, in response to antigens presented by dendritic cells activated during the course of the innate immune response (Fig. 10.2, third and fourth panels). Antigen-specific effector T cells and antibody-secreting B cells are generated by clonal expansion and differentiation over the course of several days, during which time the induced responses of innate immunity continue to function. Eventually, antigen-specific T cells and then antibodies are released into the blood and recruited to the site of infection (Fig. 10.2, last panel). A cure involves the clearance of extracellular infectious particles by antibodies and the clearance of intracellular residues of infection through the actions of effector T cells.

After many types of infection there is little or no residual pathology following an effective primary response. In some cases, however, the infection or the response to it causes significant tissue damage. In other cases, such as infection with cytomegalovirus or *Mycobacterium tuberculosis*, the infection is contained but not eliminated and can persist in a latent form. If the adaptive immune response is later weakened, as it is in acquired immune deficiency syndrome (AIDS), these diseases reappear as virulent systemic infections. We will focus on the strategies used by certain pathogens to evade or subvert adaptive immunity and thereby establish a persistent infection in the first part of Chapter 11.

In addition to clearing the infectious agent, an effective adaptive immune response prevents reinfection. For some infectious agents, this protection is essentially absolute, while for others infection is reduced or attenuated upon reexposure.

10-2 Infectious diseases are caused by diverse living agents that replicate in their hosts.

The agents that cause disease fall into five groups: viruses, bacteria, fungi, protozoa, and helminths (worms). Protozoa and worms are usually grouped together as parasites, and are the subject of the discipline of parasitology, whereas viruses, bacteria, and fungi are the subject of microbiology. In Fig. 10.3, the classes of microorganisms and parasites that cause disease are listed, with typical examples of each. The remarkable variety of these pathogens has caused the natural selection of two crucial features of adaptive immunity. First, the advantage of being able to recognize a wide range of different pathogens has driven the development of receptors on B and T cells of equal or greater diversity. Second, the distinct habitats and life cycles of pathogens have to be countered by a range of distinct effector mechanisms. The characteristic features of each pathogen are its mode of transmission, its mechanism of

Some common causes of disease in humans			
Viruses	DNA viruses	Adenoviruses	Human adenoviruses (e.g., types 3, 4, and 7)
		Herpesviruses	Herpes simplex, varicella zoster, Epstein–Barr virus, cytomegalovirus, Kaposi's sarcoma
		Poxviruses	Vaccinia virus
		Parvoviruses	Human parvovirus
		Papovaviruses	Papilloma virus
		Hepadnaviruses	Hepatitis B virus
	RNA viruses	Orthomyxoviruses	Influenza virus
		Paramyxoviruses	Mumps, measles, respiratory syncytial virus
		Coronaviruses	Common cold viruses
		Picornaviruses	Polio, coxsackie, hepatitis A, rhinovirus
		Reoviruses	Rotavirus, reovirus
		Togaviruses	Rubella, arthropod-borne encephalitis
		Flaviviruses	Arthropod-borne viruses, (yellow fever, dengue fever)
		Arenaviruses	Lymphocytic choriomeningitis, Lassa fever
		Rhabdoviruses	Rabies
		Retroviruses	Human T-cell leukemia virus, HIV
Bacteria	Gram +ve cocci	Staphylococci	*Staphylococcus aureus*
		Streptococci	*Streptococcus pneumoniae, S. pyogenes*
	Gram –ve cocci	Neisseriae	*Neisseria gonorrhoeae, N. meningitidis*
	Gram +ve bacilli		*Corynebacteria, Bacillus anthracis, Listeria monocytogenes*
	Gram –ve bacilli		*Salmonella, Shigella, Campylobacter, Vibrio, Yersinia, Pasteurella, Pseudomonas, Brucella, Haemophilus, Legionella, Bordetella*
	Anaerobic bacteria	Clostridia	*Clostridium tetani, C. botulinum, C. perfringens*
	Spirochetes		*Treponema pallidum, Borrelia burgdorferi, Leptospira interrogans*
	Mycobacteria		*Mycobacterium tuberculosis, M. leprae, M. avium*
	Rickettsias		*Rickettsia prowazeki*
	Chlamydias		*Chlamydia trachomatis*
	Mycoplasmas		*Mycoplasma pneumoniae*
Fungi			*Candida albicans, Cryptococcus neoformans, Aspergillus, Histoplasma capsulatum, Coccidioides immitis, Pneumocystis carinii*
Protozoa			*Entamoeba histolytica, Giardia, Leishmania, Plasmodium, Trypanosoma, Toxoplasma gondii, Cryptosporidium*
Worms	Intestinal		*Trichuris trichura, Trichinella spiralis, Enterobius vermicularis, Ascaris lumbricoides, Ancylostoma, Strongyloides*
	Tissues		*Filaria, Onchocerca volvulus, Loa loa, Dracuncula medinensis*
	Blood, liver		*Schistosoma, Clonorchis sinensis*

Fig. 10.3 A variety of microorganisms can cause disease. Pathogenic organisms are of five main types: viruses, bacteria, fungi, protozoa, and worms. Some common pathogens in each group are listed in the column on the right.

replication, its pathogenesis or the means by which it causes disease, and the response it elicits. We will focus here on the immune responses to these pathogens.

Infectious agents can grow in various body compartments, as shown schematically in Fig. 10.4. We have already seen that two major compartments can be defined—intracellular and extracellular. Intracellular pathogens must invade host cells in order to replicate, and so must either be prevented from entering cells or be detected and eliminated once they have done so. Such pathogens can be subdivided further into those that replicate freely in the cell, such as viruses and certain bacteria (species of *Chlamydia* and *Rickettsia* as well as *Listeria*), and those, such as the mycobacteria, that replicate in cellular vesicles. Viruses can be prevented from entering cells by neutralizing antibodies whose production relies on T_H2 cells (see Section 9-14), while once within cells they are dealt with by virus-specific cytotoxic T cells, which recognize and kill the infected cell (see Section 8-21). Intravesicular pathogens, on the other hand, mainly infect macrophages and can be eliminated with the aid of pathogen-specific T_H1 cells, which activate infected macrophages to destroy the pathogen (see Section 8-26).

Many microorganisms replicate in extracellular spaces, either within the body or on the surface of epithelia. Extracellular bacteria are usually susceptible to killing by phagocytes and thus pathogenic species have developed means of resisting engulfment. The encapsulated gram-positive cocci, for instance, grow in extracellular spaces and resist phagocytosis by means of their polysaccharide capsule. This means they are not immediately eliminated by tissue phagocytes on infecting a previously unexposed host. However, if this mechanism of resistance is overcome by opsonization by complement and specific antibody, they are readily killed after ingestion by phagocytes. Thus, these extracellular bacteria are cleared by means of the humoral immune response (see Chapter 9).

Fig. 10.4 Pathogens can be found in various compartments of the body, where they must be combated by different host defense mechanisms. Virtually all pathogens have an extracellular phase where they are vulnerable to antibody-mediated effector mechanisms. However, intracellular phases are not accessible to antibody, and these are attacked by T cells.

	Extracellular		Intracellular	
	Interstitial spaces, blood, lymph	Epithelial surfaces	Cytoplasmic	Vesicular
Site of infection		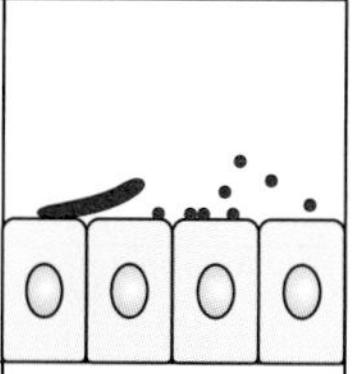	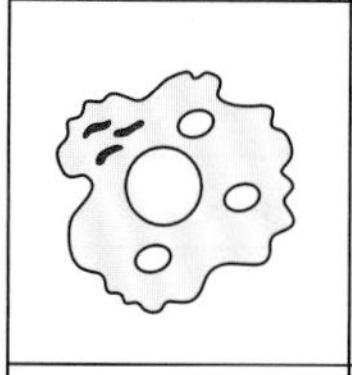	
Organisms	Viruses Bacteria Protozoa Fungi Worms	*Neisseria gonorrhoeae* Worms Mycoplasma *Streptococcus pneumoniae* *Vibrio cholerae* *Escherichia coli* *Candida albicans* *Helicobacter pylori*	Viruses *Chlamydia* spp. *Rickettsia* spp. Listeria *monocytogenes* Protozoa	Mycobacteria *Salmonella typhimurium* *Leishmania* spp. *Listeria* spp. *Trypanosoma* spp. *Legionella pneumophila* *Cryptococcus neoformans* *Histoplasma* *Yersinia pestis*
Protective immunity	Antibodies Complement Phagocytosis Neutralization	Antibodies, especially IgA Antimicrobial peptides	Cytotoxic T cells NK cells	T-cell and NK-cell dependent macrophage activation

Direct mechanisms of tissue damage by pathogens			Indirect mechanisms of tissue damage by pathogens		
Exotoxin production	Endotoxin	Direct cytopathic effect	Immune complexes	Anti-host antibody	Cell-mediated immunity

Pathogenic mechanism (row)

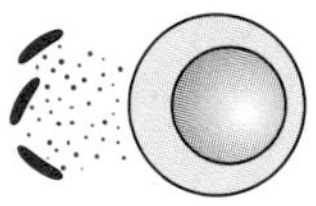

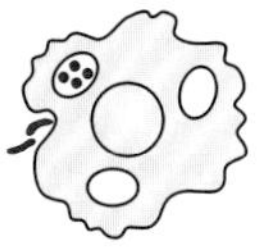

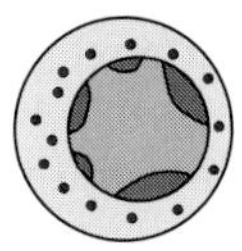

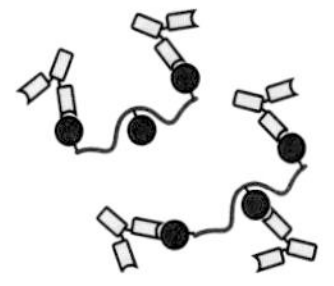

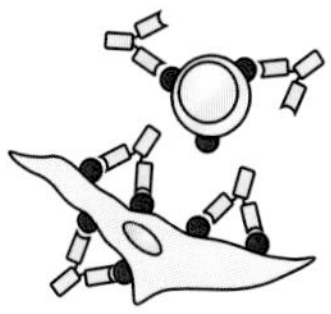

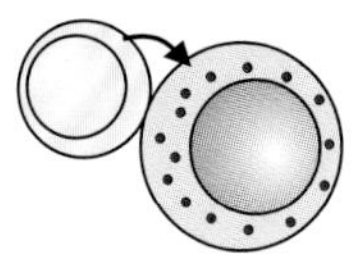

Exotoxin production	Endotoxin	Direct cytopathic effect	Immune complexes	Anti-host antibody	Cell-mediated immunity
Streptococcus pyogenes *Staphylococcus aureus* *Corynebacterium diphtheriae* *Clostridium tetani* *Vibrio cholerae*	*Escherichia coli* *Haemophilus influenzae* *Salmonella typhi* *Shigella* *Pseudomonas aeruginosa* *Yersinia pestis*	Variola Varicella-zoster Hepatitis B virus Polio virus Measles virus Influenza virus Herpes simplex virus	Hepatitis B virus Malaria *Streptococcus pyogenes* *Treponema pallidum* Most acute infections	*Streptococcus pyogenes* *Mycoplasma pneumoniae*	*Mycobacterium tuberculosis* *Mycobacterium leprae* Lymphocytic choriomeningitis virus *Borrelia burgdorferi* *Schistosoma mansoni* Herpes simplex virus
Tonsilitis, scarlet fever Boils, toxic shock syndrome, food poisoning Diphtheria Tetanus Cholera	Gram-negative sepsis Meningitis, pneumonia Typhoid Bacillary dysentery Wound infection Plague	Smallpox Chickenpox, shingles Hepatitis Poliomyelitis Measles, subacute sclerosing panencephalitis Influenza Cold sores	Kidney disease Vascular deposits Glomerulonephritis Kidney damage in secondary syphilis Transient renal deposits	Rheumatic fever Hemolytic anemia	Tuberculosis Tuberculoid leprosy Aseptic meningitis Lyme arthritis Schistosomiasis Herpes stromal keratitis

The leftmost column labels, top to bottom: **Pathogenic mechanism**, **Infectious agent**, **Disease**.

Fig. 10.5 Pathogens can damage tissues in a variety of different ways. The mechanisms of damage, representative infectious agents, and the common names of the diseases associated with each are shown. Exotoxins are released by microorganisms and act at the surface of host cells, for example, by binding to receptors. Endotoxins, which are intrinsic components of microbial structure, trigger phagocytes to release cytokines that produce local or systemic symptoms. Many pathogens are cytopathic, directly damaging the cells they infect. Finally, adaptive immune response to the pathogen can generate antigen:antibody complexes that can activate neutrophils and macrophages, antibodies that can cross-react with host tissues, or T cells that kill infected cells. All of these have some potential to damage the host's tissues. In addition, neutrophils, the most abundant cells early in infection, release many proteins and small-molecule inflammatory mediators that both control infection and cause tissue damage (not shown).

Different infectious agents cause markedly different diseases, reflecting the diverse processes by which they damage tissues (Fig. 10.5). Many extracellular pathogens cause disease by releasing specific toxic products or protein toxins (see Fig. 9.23), which can induce the production of neutralizing antibodies (see Section 9-14). Intracellular infectious agents frequently cause disease by damaging the cells that house them. The specific killing of virus-infected cells by cytotoxic T cells thus not only prevents virus spread but removes damaged cells. The immune response to the infectious agent can itself be a major cause of pathology in several diseases (see Fig. 10.5). The pathology caused by a particular infectious agent also depends on the site in which it grows; *Streptococcus pneumoniae* in the lung causes pneumonia, whereas in the blood it causes a rapidly fatal systemic illness.

As we learned in Chapter 2, for a pathogen to invade the body, it must first bind to or cross the surface of an epithelium. When the infection is due to intestinal pathogens such as *Salmonella typhi*, the causal agent of typhoid fever, or *Vibrio cholerae*, which causes cholera, the adaptive immune response occurs in the specialized mucosal immune system associated with the gastrointestinal tract, as described later in this chapter. Some intestinal pathogens even target the M cells of the gut mucosal immune system, which are specialized to transport antigens across the epithelium, as a means of entry.

Many pathogens cannot be entirely eliminated by the immune response. But neither are most pathogens universally lethal. Those pathogens that have persisted for many thousands of years in the human population are highly evolved to exploit their human hosts, and cannot alter their pathogenicity without upsetting the compromise they have achieved with the human immune system. Rapidly killing every host it infects is no better for the long-term survival of a pathogen than being wiped out by the immune response before it has had time to infect another individual. In short, we have learned to live with our enemies, and they with us. However, we must be on the alert at all times for new pathogens and new threats to health. The human immunodeficiency virus that causes AIDS serves as a warning to mankind that we remain constantly vulnerable to the emergence of new infectious agents.

Summary.

The mammalian body is susceptible to infection by many pathogens, which must first make contact with the host and then establish a focus of infection in order to cause infectious disease. To establish an infection, the pathogen must first colonize the skin or the internal mucosal surfaces of the respiratory, gastrointestinal, or urogenital tracts and then overcome or bypass the innate immune defenses associated with the epithelia and underlying tissues. If it succeeds in doing this, it will provoke an adaptive immune response that will take effect after several days and will usually clear the infection. Pathogens differ greatly in their lifestyles and means of pathogenesis, requiring an equally diverse set of defensive responses from the host immune system.

The course of the adaptive response to infection.

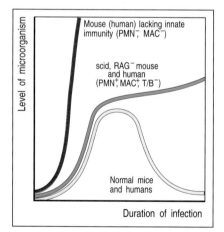

Fig. 10.6 The time course of infection in normal and immunodeficient mice and humans. The red curve shows the rapid growth of microorganisms in the absence of innate immunity, when macrophages (MAC) and polymorphonuclear leukocytes (PMN) are lacking. The green curve shows the course of infection in mice and humans that have innate immunity but have no T or B lymphocytes and so lack adaptive immunity. The yellow curve shows the normal course of an infection in immunocompetent mice or humans.

It is not known how many infections are dealt with solely by the nonadaptive mechanisms of innate immunity discussed in Chapter 2; this is because such infections are eliminated early and produce little in the way of symptoms or pathology. Moreover, deficiencies in nonadaptive defenses are rare, so it has seldom been possible to study their consequences. Innate immunity does, however, appear to be essential for effective host defense, as shown by the progression of infection in mice that lack components of innate immunity but have an intact adaptive immune system (Fig. 10.6). Adaptive immunity is also essential, as shown by the immunodeficiency syndromes associated with failure of one or the other arm of the adaptive immune response (see Chapter 11). Adaptive immunity is triggered when an infection eludes the innate defense mechanisms and generates a threshold dose of antigen (see Fig. 10.1). This antigen then initiates an adaptive immune response, which becomes effective only after several days, the time required for antigen-specific T cells and B cells to locate their specific foreign antigen, to proliferate, and to differentiate into armed effector cells. In the earlier chapters of this book, we discussed the cells and molecules that mediate the adaptive immune response, and the interactions between them. We are now ready to see how each cell type is recruited in turn in the course of a primary adaptive immune response to a pathogen, and how the effector lymphocytes and antibodies that are generated in response to antigen are dispersed to their sites of action. These clear the infection and protect against reinfection in the short term. A primary adaptive response also establishes a state of long-lasting protective immunity that is ultimately mediated by long-lived resting memory cells, to which we will return in the last part of this chapter.

10-3   The nonspecific responses of innate immunity are necessary for an adaptive immune response to be initiated.

The establishment of a focus of infection in tissues and the response of the innate immune system to it produce changes in the immediate environment of the infection. In a bacterial infection, the first thing that usually happens is that the infected tissue becomes inflamed (see Fig. 1.12). As we learnt in Chapter 2, this is initially the result of the activation of the resident macrophages by bacterial components such as lipopolysaccharide (LPS) acting through Toll-like receptors on the macrophage. The cytokines and chemokines secreted by the activated macrophages, especially the cytokine tumor necrosis factor-α (TNF-α), induce numerous changes in the endothelial cells of nearby blood capillaries, a process known as endothelial cell activation. The cytokines cause the release of Weibel–Palade bodies from within the endothelial cells, which deliver P-selectin to the endothelial cell surface. Cytokines and chemokines also induce the synthesis and translation of RNA encoding E-selectin, which thus also appears on the endothelial cell surface.

These two selectins cause leukocytes to adhere to and roll on the endothelial surface in large numbers. Among these will be polymorphonuclear leukocytes, mainly neutrophils, and monocytes. The cytokines also induce the production of the adhesion molecule VCAM-1 on the endothelial cells, which binds to adhesion molecules on the leukocytes. This strengthens the interaction between leukocytes and endothelial cells, and aids the neutrophils and monocytes to enter the infected tissue in large numbers to form an inflammatory focus (see Fig. 2.36). As monocytes mature into tissue macrophages and become activated in their turn, more and more inflammatory cells are attracted into the infected tissue and the inflammatory response is maintained and reinforced. The inflammatory response can be thought of as putting up a flag on the endothelial cells to signal the presence of infection, but as yet, the response is entirely nonspecific for the pathogen antigens.

A second crucial effect of infection is the activation of potential professional antigen-presenting cells—the dendritic cells—that reside in most tissues. These take up antigen in the infected tissues and, as for macrophages, they are activated through innate immune receptors that respond to common pathogen constituents. For example, the combination of LPS and lipopolysaccharide-binding protein (LBP) binding to the cell-surface receptors CD14 and the Toll-like receptor TLR-4 induces the dendritic cells to mature into potent antigen-presenting cells. Activated dendritic cells increase their synthesis of MHC class II molecules and, most importantly, begin to express the co-stimulatory molecules CD80 and CD86 on their surface. As described in Chapter 8, these antigen-presenting cells are carried away from the infected tissue in lymph, along with their antigen cargo, to enter secondary lymphoid tissues, in which they can initiate the adaptive immune response. They arrive in large numbers at the draining lymph nodes, or other nearby lymphoid tissue, attracted by the chemokines ELC, MIP-3β, and SLC that are produced by lymph node stromal and high vascular endothelial cells.

Once dendritic cells arrive in the lymphoid tissues, they appear to have reached their final destination. They eventually die in these tissues, but before this their role is to activate antigen-specific naive T lymphocytes. Naive lymphocytes are continually passing through the lymph nodes, which they enter from the blood across the walls of high endothelial venules, as we will describe below. Those naive T cells that are able to recognize antigen on the surface of dendritic cells are activated and both divide and mature into effector cells that reenter the circulation. When there is a local infection, the changes induced by inflammation in the walls of nearby venules, as we will see later, induce these effector T cells to leave the blood vessel and migrate to the site of infection.

Thus the local release of cytokines and chemokines at the site of infection has far-reaching consequences. As well as recruiting neutrophils and macrophages, which are not specific for antigen, the changes induced in the blood vessel walls also enable newly activated effector T lymphocytes to enter infected tissue.

10-4 An adaptive immune response is initiated when circulating T cells encounter their corresponding antigen in draining lymphoid tissues and become activated.

The importance of the peripheral lymphoid organs in the initiation of adaptive immune responses was first shown by ingenious experiments in which a skin flap was isolated from the body wall so that it had a blood circulation but no lymphatic drainage. Antigen placed in the flap of skin did not elicit a T-cell response, showing that T cells do not become sensitized in the infected tissue itself. We now know that naive T lymphocytes are activated in the peripheral lymphoid organs by antigens brought there by dendritic cells. The immune response to pathogens that enter through the skin rather than across mucosal surfaces is generally believed to occur in the lymph nodes, which are sites of intersection of two pathways of circulation, those of the lymph and the blood (see Fig. 1.8).

As described in Chapter 8, immature dendritic cells in tissues take up antigens and are stimulated by infection to migrate to draining lymph nodes. Antigens introduced directly into the bloodstream are picked up by antigen-presenting cells in the spleen, and lymphocytes are activated in the splenic white pulp (see Fig. 1.9). The trapping of antigen by antigen-presenting cells that migrate to these lymphoid tissues, and the continuous recirculation of naive T cells through these tissues, ensure that rare antigen-specific T cells will encounter their specific antigen on an antigen-presenting cell surface. The unique architecture of the peripheral lymphoid organs virtually guarantees contact of foreign antigen with specific T-cell receptors in the lymph nodes and spleen or in mucosa-associated lymphoid tissues (MALT) (see Fig. 1.10).

Naive T cells enter the lymphoid organs in essentially the same way as described in Chapter 2 for the entry of phagocytes into sites of infection, except that selectin is expressed on the T cell rather than the endothelium. L-selectin on naive T cells binds to sulfated carbohydrates on proteins such as the vascular addressins GlyCAM-1 and CD34. CD34 is expressed on endothelial cells in many tissues but is properly glycosylated for L-selectin binding only on the high endothelial venule cells of lymph nodes (Fig. 10.7). Binding of L-selectin causes the lymphocyte to roll on the endothelial surface, and although the interaction is too weak to promote extravasation, it is critical for the lymphocyte to selectively home to the lymphoid organs. It is essential for the initiation of the stronger interactions that follow between the T cell and the high endothelium, which are mediated by molecules with a relatively broad tissue distribution.

Chemokines produced by the cells of the lymph node are also important for initiating strong adhesion. These chemokines bind to proteoglycan molecules in the extracellular matrix and high endothelial venule cell walls, and are recognized by receptors on the naive T cell (see Section 7-30). Stimulation by these locally-bound chemokines activates the adhesion molecule LFA-1 on the T cell, increasing its affinity for ICAM-2, which is expressed constitutively on all endothelial cells, and ICAM-1, which, in the absence of inflammation, is expressed only on the high endothelial venule cells of peripheral lymphoid tissues. The binding of LFA-1 to its ligands ICAM-1 and ICAM-2 has a major role in T-cell adhesion to and migration through the wall of the venule into the lymph node (see Fig. 10.7).

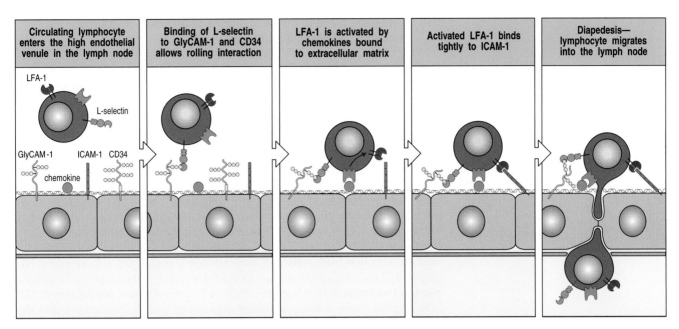

| Circulating lymphocyte enters the high endothelial venule in the lymph node | Binding of L-selectin to GlyCAM-1 and CD34 allows rolling interaction | LFA-1 is activated by chemokines bound to extracellular matrix | Activated LFA-1 binds tightly to ICAM-1 | Diapedesis— lymphocyte migrates into the lymph node |

Fig. 10.7 Lymphocytes in the blood enter lymphoid tissue by crossing the walls of high endothelial venules. The first step is the binding of L-selectin on the lymphocyte to sulfated carbohydrates of GlyCAM-1 and CD34 on the high endothelial cells. Local chemokines activate LFA-1 on the lymphocyte and cause it to bind tightly to ICAM-1 on the endothelial cell, allowing migration across the endothelium. For the lymphocyte to cross the high endothelial barrier successfully, migration has to lead to activation of matrix metalloproteinases, as with the migration of neutrophils out of the blood (see Fig. 2.36).

T-cells that have arrived in the T-cell zone via the high endothelial venules scan the surface of the antigen-presenting dendritic cells for specific peptide:MHC complexes. If they do not recognize antigen, they eventually leave the lymph node via an efferent lymphatic vessel. This returns them to the blood so that they can recirculate through other lymph nodes. Rarely, a naive T cell recognizes its specific peptide:MHC complex on the surface of a dendritic cell. This signals the activation of LFA-1, causing the T cell to adhere strongly to the dendritic cell and cease migrating. Binding to the peptide:MHC complexes and co-stimulatory molecules on the dendritic cell surface stimulates the naive T cell to proliferate and differentiate, resulting in the production of an expanded population of armed, antigen-specific effector T cells (see Fig. 8.4). The efficiency with which T cells screen each antigen-presenting cell in lymph nodes is very high, as can be seen by the rapid trapping of antigen-specific T cells in a single lymph node containing antigen: all of the antigen-specific T cells in a sheep were trapped in one lymph node within 48 hours of antigen deposition (Fig. 10.8).

Fig. 10.8 Trapping and activation of antigen-specific naive T cells in lymphoid tissue. Naive T cells entering the lymph node from the blood encounter many antigen-presenting dendritic cells in the lymph node cortex. T cells that do not recognize their specific antigen in the cortex leave via the efferent lymphatics and reenter the blood. T cells that do recognize their specific antigen bind stably to the dendritic cell and are activated through their T-cell receptors, resulting in the production of armed effector T cells. Lymphocyte recirculation and recognition is so effective that all of the specific naive T cells in circulation can be trapped by antigen in one node within 2 days. By 5 days after the arrival of antigen, activated effector T cells are leaving the lymph node in large numbers via the efferent lymphatics.

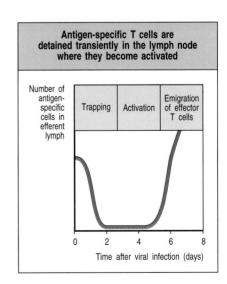

Antigen-specific T cells are detained transiently in the lymph node where they become activated

10-5 Cytokines made in the early phases of an infection influence the functional differentiation of CD4 T cells.

The differentiation of naive CD4 T cells into the two major classes of CD4 effector T cell occurs during the initial response of these cells to antigen in the peripheral lymphoid tissues. This step, at which a naive CD4 T cell becomes either an armed T_H1 cell or an armed T_H2 cell, has a critical impact on the outcome of an adaptive immune response, determining whether it will be dominated by macrophage activation or by antibody production.

The mechanisms that control this step in CD4 T-cell differentiation are not yet fully defined; however, it is clear that it can be profoundly influenced by cytokines present during the initial proliferative phase of T-cell activation. Experiments *in vitro* have shown that naive CD4 T cells initially stimulated in the presence of IL-12 and IFN-γ tend to develop into T_H1 cells (Fig. 10.9, left panels), in part because IFN-γ inhibits the proliferation of T_H2 cells. As IL-12, produced by dendritic cells and macrophages, and IFN-γ, produced by NK cells and CD8 T cells, predominate in the early phase of the response to viruses and to some intracellular bacteria, such as *Listeria* species (see Section 2-27), CD4 T-cell responses in these infections tend to be dominated by T_H1 cells. By contrast, CD4 T cells activated in the presence of IL-4, especially when IL-6 is also present, tend to differentiate into T_H2 cells. This is because IL-4 and IL-6 promote the differentiation of T_H2 cells, and IL-4 or IL-10, either alone or in combination, can also inhibit the generation of T_H1 cells.

One possible source of the IL-4 needed to generate T_H2 cells is a specialized subset of CD4 T cells that express the NK1.1 marker normally associated with NK cells; these cells are called **NK 1.1+ T cells**. They have a nearly invariant α:β T-cell receptor; in fact, essentially the same receptor seems to be used in the NK 1.1+ T cells of mice and their counterparts in humans. Unlike that of other CD4 T cells, the development of the NK 1.1+ T cells does not depend on the expression of MHC class II molecules. Instead, they recognize an MHC class IB molecule, CD1, which is not encoded within the MHC (see Section 5-18). In mice there are two CD1 genes (*CD1.1* and *CD1.2*), whereas in humans there are five (*CD1a–e*), of which only *CD1d* is homologous to the murine *CD1.1* and *CD1.2*. CD1 molecules are expressed by thymocytes, professional antigen-presenting cells, and intestinal epithelium.

Although the exact function of CD1 molecules is not well defined, CD1b is known to present a bacterial lipid, mycolic acid, to α:β T cells, whereas other CD1 molecules are recognized by γ:δ T cells. The activation of NK 1.1+ T cells is thought to depend on the expression of CD1 molecules induced in response to infection; whether all NK 1.1+ T cells recognize a specific antigen presented by these CD1 molecules is not known, but some at least are able to recognize glycolipid antigens presented by CD1d. Upon activation, these NK 1.1+ T cells secrete very large amounts of IL-4 and can therefore enhance the development of T_H2 cells (Fig. 10.9, right panels), which promotes the production of IgG1 (in mice) and IgE (in mice and humans) in subsequent humoral immune responses.

The differential capacity of pathogens to interact with dendritic cells, macrophages, NK cells, and NK 1.1+ T cells can therefore influence the overall balance of cytokines present early in the immune response, and thus determine whether T_H1 or T_H2 cells develop preferentially to bias the adaptive immune response toward a cellular or a humoral response. This can, in turn, determine whether the pathogen is eliminated or survives within the host; some pathogens may even have evolved to interact with the innate immune system so as to generate responses that are beneficial to them rather than to the host.

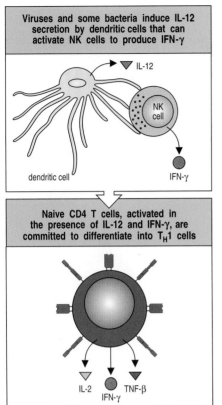

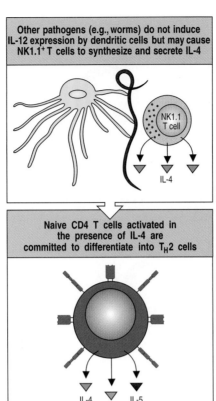

Fig. 10.9 The differentiation of naive CD4 T cells into different subclasses of armed effector T cells is influenced by cytokines elicited by the pathogen. Many pathogens, especially intracellular bacteria and viruses, activate dendritic cells and NK cells to produce IL-12 and IFN-γ, which cause proliferating CD4 T cells to differentiate into T$_H$1 cells. IL-4 can inhibit these responses. IL-4, produced by an NK 1.1$^+$ T cell in response to parasitic worms or other pathogens, acts on proliferating CD4 T cells to cause them to become T$_H$2 cells. The mechanisms by which these cytokines induce the selective differentiation of CD4 T cells is now the subject of intensive study. They may act either when the CD4 T cell is first activated by an antigen-presenting cell or during the subsequent proliferative phase.

10-6   Distinct subsets of T cells can regulate the growth and effector functions of other T-cell subsets.

The two subsets of CD4 T cells—T$_H$1 and T$_H$2—have very different functions: T$_H$2 cells are the most effective activators of B cells, especially in primary responses, whereas T$_H$1 cells are crucial for activating macrophages. It is also clear that the two CD4 T-cell subsets can regulate each other; once one subset becomes dominant, it is often hard to shift the response to the other subset. One reason for this is that cytokines from one type of CD4 T cell inhibit the activation of the other. Thus, IL-10, a product of T$_H$2 cells, can inhibit the development of T$_H$1 cells by acting on the antigen-presenting cell, whereas IFN-γ, a product of T$_H$1 cells, can prevent the activation of T$_H$2 cells (Fig. 10.10). If a particular CD4 T-cell subset is activated first or preferentially in a response, it can suppress the development of the other subset. The overall effect is that certain responses are dominated by either humoral (T$_H$2) or cell-mediated (T$_H$1) immunity. However, under many circumstances *in vivo*, there is a mixed T$_H$1 and T$_H$2 response.

This interplay of cytokines is important in human disease, but it has been explored at present mainly in certain mouse models, where such polarized responses are easier to study. For example, when BALB/c mice are experimentally infected with the protozoan parasite *Leishmania*, their CD4 T cells fail to differentiate into T$_H$1 effector cells; instead, the mice preferentially make T$_H$2 cells in response to this pathogen. These T$_H$2 cells are unable to activate macrophages to inhibit leishmanial growth, resulting in susceptibility to disease. By contrast, C57BL/6 mice respond by producing T$_H$1 cells that protect the host by activating infected macrophages to kill the *Leishmania*. The preferential activation of T$_H$2 rather than T$_H$1 cells in BALB/c mice can be reversed if IL-4 is blocked in the first days of infection by injecting anti-IL-4 antibody, but this treatment is ineffective after a week or so of infection.

Fig. 10.10 The two subsets of CD4 T cells each produce cytokines that can negatively regulate the other subset. T$_H$2 cells make IL-10, which acts on macrophages to inhibit T$_H$1 activation, perhaps by blocking macrophage IL-12 synthesis, and TGF-β, which acts directly on the T$_H$1 cells to inhibit their growth (left panels). T$_H$1 cells make IFN-γ, which blocks the growth of T$_H$2 cells (right panels). These effects allow either subset to dominate a response by suppressing outgrowth of cells of the other subset. A similar dichotomy in cytokine profile is seen in CD8 T cells (not shown), leading to the nomenclature T$_C$1 and T$_C$2 cells.

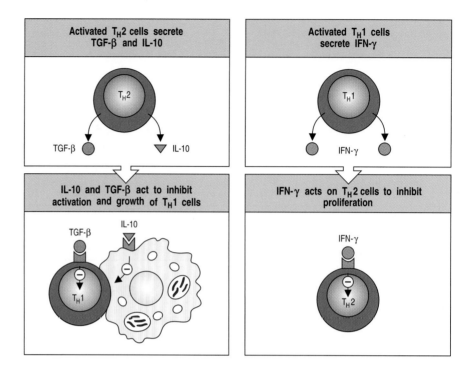

Because cytokines seem to regulate the balance between T$_H$1 and T$_H$2 cells, one might expect that it would be possible to shift this balance by administering appropriate cytokines. IL-2 and IFN-γ have been used to stimulate cell-mediated immunity in diseases such as lepromatous leprosy, and can cause both a local resolution of the lesion and a systemic change in T-cell responses. IL-12, which is a potent inducer of T$_H$1 cells, might be an even more attractive potential therapy.

CD8 T cells are also able to regulate the immune response by producing cytokines. It has become clear recently that effector CD8 T cells can, in addition to their familiar cytolytic function, also respond to antigen by secreting cytokines typical of either T$_H$1 or T$_H$2 cells. Such CD8 T cells, called T$_C$1 or T$_C$2 by analogy to the T$_H$ subsets, seem to be responsible for the development of leprosy in its lepromatous rather than its tuberculoid form, which we discuss in detail in Chapter 11. Patients with the less destructive tuberculoid leprosy make T$_C$1 cells, whose cytokines induce T$_H$1 cells, which can activate macrophages to rid the body of its burden of leprosy bacilli. Patients with lepromatous leprosy have CD8 T cells that suppress the T$_H$1 response by making IL-10 and TGF-β. Thus, the suppression of CD4 T cells by CD8 T cells that has been observed in various situations can be explained by their expression of different sets of cytokines.

10-7   The nature and amount of antigenic peptide can also affect the differentiation of CD4 T cells.

Another factor that influences the differentiation of CD4 T cells into distinct effector subsets is the amount and exact sequence of the antigenic peptide that initiates the response. Large amounts of peptide that achieve a high density on the surface of antigen-presenting cells tend to stimulate T$_H$1 cell responses, whereas low-density presentation tends to elicit T$_H$2 cell responses. Moreover, peptides that interact strongly with the T-cell receptor tend to stimulate T$_H$1-like responses, whereas peptides that bind weakly tend to stimulate T$_H$2-like responses (Fig. 10.11).

Fig. 10.11 The nature and amount of ligand presented to a CD4 T cell during primary stimulation can determine its functional phenotype. CD4 T cells presented with low levels of a ligand that binds the T-cell receptor poorly differentiate preferentially into T_H2 cells making IL-4 and IL-5. Such T cells are most active in stimulating naive B cells to differentiate into plasma cells and make antibody. T cells presented with a high density of a ligand that binds the T-cell receptor strongly differentiate into T_H1 cells that secrete IL-2, TNF-β, and IFN-γ, and are most effective in activating macrophages.

This difference could be very important in several circumstances. For instance, allergy is caused by the production of IgE antibody, which, as we learned in Chapter 9, requires high levels of IL-4 but does not occur in the presence of IFN-γ, a potent inhibitor of IL-4-driven class switching to IgE. Antigens that elicit IgE-mediated allergy are generally delivered in minute doses, and they elicit T_H2 cells that make IL-4 and no IFN-γ. It is also relevant that allergens do not elicit any of the known innate immune responses, which produce cytokines that tend to bias CD4 T-cell differentiation toward T_H1 cells. Finally, allergens are delivered to humans in minute doses across a thin mucosa, such as that of the lung. Something about this route of sensitization allows even potent generators of T_H1 responses like *Leishmania major* to induce T_H2 responses.

Most protein antigens that elicit CD4 T-cell responses stimulate the production of both T_H1 and T_H2 cells. This reflects the presence in most proteins of several different peptide sequences that can bind to MHC class II molecules and be presented to CD4 T cells. Some of these peptides are likely to bind to MHC class II molecules with high affinity, and consequently will be present at high density on the antigen-presenting cell, whereas others are likely to bind MHC class II molecules with low affinity and be present only at low density. Naive T cells specific for peptide antigens that have high affinity for MHC molecules are therefore likely to encounter a high density of their ligand, whereas others might only encounter a low density, and these differences could affect the subsequent response of the T cell. Indeed, it can be shown experimentally that some peptides in a protein tend to elicit the production of T_H2 cells, whereas other peptides tend to elicit T_H1 cells.

10-8 Armed effector T cells are guided to sites of infection by chemokines and newly expressed adhesion molecules.

The full activation of naive T cells takes 4–5 days and is accompanied by marked changes in the homing behavior of these cells. Armed effector cytotoxic CD8 T cells must travel from the lymph node, or other peripheral lymphoid tissue in which they have been activated, to attack and destroy infected cells. Armed effector CD4 T_H1 cells must also leave the lymphoid tissues to activate macrophages at the site of infection. Most of the antigen-specific armed effector T cells cease production of L-selectin, which mediates homing to the lymph nodes, while the expression of other adhesion molecules is increased (Fig. 10.12). One important change is a marked increase in the expression of the integrin $\alpha_4:\beta_1$, also known as VLA-4. This binds to the VCAM-1 molecule that is induced on activated endothelial cell surfaces and initiates the extravasation of the effector T cells. Thus if the innate immune response has already activated the endothelium at the site of infection, as described in Section 10-3, effector T cells will rapidly be recruited. At the early stage of the immune response, only a few of the effector T cells that enter the infected tissues will be expected to be specific for pathogen, as any effector T cell specific for any antigen will also be able to enter. However, specificity of the reaction is maintained, as only those effector T cells that recognize pathogen antigens will carry out their function, destroying infected cells or

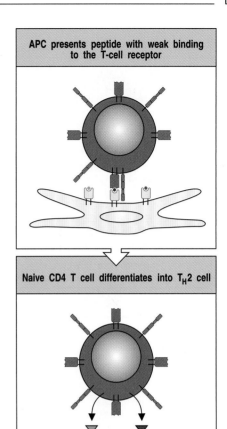

APC presents peptide with weak binding to the T-cell receptor

Naive CD4 T cell differentiates into T_H2 cell

IL-4   IL-5

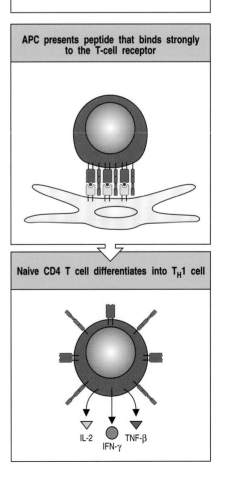

APC presents peptide that binds strongly to the T-cell receptor

Naive CD4 T cell differentiates into T_H1 cell

IL-2   TNF-β
IFN-γ

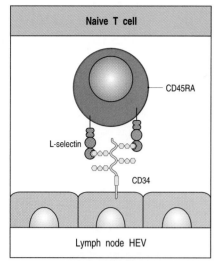

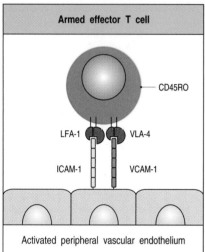

Fig. 10.12 Armed effector T cells change their surface molecules, allowing them to home to sites of infection. Naive T cells home to lymph nodes through the binding of L-selectin to sulfated carbohydrates displayed by various proteins, such as CD34 and GlyCAM-1 on the high endothelial venule (HEV, top panel). If they encounter antigen and differentiate into armed effector T cells, many lose expression of L-selectin, leave the lymph node about 4–5 days later, and now express VLA-4 and increased levels of LFA-1. These bind to VCAM-1 and ICAM-1, respectively, on peripheral vascular endothelium at sites of inflammation (bottom panel). On differentiating into effector cells, T cells also alter their splicing of the mRNA encoding the cell-surface molecule CD45. The CD45RO isoform expressed by effector T cells lacks one or more exons that encode extracellular domains present in the CD45RA isoform expressed by naive T cells, and makes effector T cells more sensitive to stimulation by specific antigen.

specifically activating pathogen-loaded macrophages. By the peak of an adaptive immune response, most of the recruited T cells will be specific for the infecting pathogen, as after several days of clonal expansion and differentiation these cells predominate in numbers.

Differential expression of adhesion molecules can direct different subsets of armed effector T cells to specific sites. Some, for example, migrate to the lamina propria of the gut, which involves the binding of both L-selectin and the $\alpha_4{:}\beta_7$ integrin expressed on the T cell to separate sites on MAdCAM-1. T cells that home to the epithelium of the gut express a novel integrin called $\alpha_e{:}\beta_7$ and bind to the E-cadherin expressed on epithelial cells. Cells that home to the skin, in contrast, express the cutaneous lymphocyte antigen (CLA), a glycosylated isoform of P-selectin glycolipid-1, and bind to E-selectin. As we will discuss later in this chapter, the peripheral immune system is compartmentalized such that different populations of lymphocytes migrate through different lymphoid compartments and—after activation—through the different tissues they serve. The selective expression of different homing receptors that bind to tissue-specific 'addressins' is the mechanism by which this is achieved.

Not all infections trigger innate immune responses that activate local endothelial cells, and it is not so clear how armed effector T cells are guided to the sites of infection in these cases. However, activated T cells seem to enter all tissues in very small numbers, perhaps via adhesive interactions such as the binding of P-selectin to P-selectin glycolipid-1, and could thus encounter their antigens even in the absence of a previous inflammatory response.

Effector T cells that recognize pathogen antigens in the tissues produce cytokines such as TNF-α, which activates endothelial cells to express E-selectin, VCAM-1, and ICAM-1, and chemokines such as RANTES (see Fig. 2.33), which can then act on effector T cells to activate their adhesion molecules. The increased levels of VCAM-1 and ICAM-1 on endothelial cells bind VLA-4 and LFA-1, respectively, on armed effector T cells, recruiting more of these cells into tissues that contain the antigen. At the same time, monocytes and polymorphonuclear leukocytes are recruited to these sites by adhesion to E-selectin. TNF-α and IFN-γ released by the activated T cells also act synergistically to change the shape of endothelial cells, allowing increased blood flow, increased vascular permeability, and increased emigration of leukocytes, fluid, and protein into a site of infection.

Thus one or a few specific effector T cells encountering antigen in a tissue can initiate a potent local inflammatory response that recruits both a greater number of specific armed effector cells and many more nonspecific inflammatory cells to that site. By contrast, effector T cells that enter the tissues but do not recognize their antigen are rapidly lost. They either enter afferent lymph in the tissues and return to the bloodstream, or undergo apoptosis. Most of the T cells in the afferent lymph that drains tissues are memory or effector T cells, which characteristically express the CD45RO isoform of the cell-surface molecule CD45 and lack L-selectin. Effector T cells and memory T cells have a similar phenotype, as we will discuss later, and both seem to be committed to migration through potential sites of infection. As well as allowing effector T cells to clear all sites of infection, this pattern of migration allows them to contribute, along with memory cells, to protecting the host against reinfection with the same pathogen (see Sections 10-11 and 10-12).

10-9 Antibody responses develop in lymphoid tissues under the direction of armed helper T cells.

Migration out of lymphoid tissues is clearly important for the effector actions of armed CD8 cytotoxic T cells and armed T_H1 cells. However, the most important functions of helper CD4 T cells, predominantly T_H2 cells, depend on their interactions with B cells, and these interactions occur in the lymphoid tissues themselves. B cells specific for a protein antigen cannot be activated to proliferate, form germinal centers, or differentiate into plasma cells until they encounter a helper T cell that is specific for one of the peptides derived from that antigen. Humoral immune responses to protein antigens thus cannot occur until after antigen-specific helper T cells have been generated.

One of the most interesting questions in immunology is how two antigen-specific lymphocytes—the naive antigen-binding B cell and the armed helper T cell—find one another to initiate a T-cell dependent antibody response. As we learned in Chapter 9, the likely answer lies in the migratory path of B cells through the lymphoid tissues and the presence of armed helper T cells on that path (Fig. 10.13).

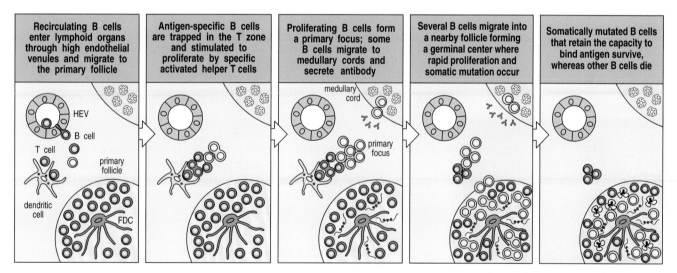

Fig. 10.13 The specialized regions of lymphoid tissue provide an environment where antigen-specific naive B cells can interact with armed helper T cells specific for the same antigen. The initial encounter of antigen-specific naive B cells with the appropriate helper T cells occurs in the T-cell areas in lymphoid tissue and stimulates the proliferation of B cells in contact with the helper T cells to form a primary focus, as shown in the first three panels. This also results in some isotype switching in the B cells. Some of the activated B-cell blasts then migrate to medullary cords, where they divide, differentiate into plasma cells, and secrete antibody for a few days (third panel). Other B-cell blasts migrate into primary lymphoid follicles, where they proliferate rapidly to form a germinal center under the influence of antigen and of helper T cells (fourth panel). The germinal center is the site of somatic hypermutation and selection of high-affinity B cells that are able to bind antigen better than lower-affinity cells and thus to survive either because they are protected from apoptotic signals delivered by T cells or/and because they are more capable of presenting antigen to T cells and thereby receiving positive signals such as IL-4 and CD40L (fifth panel). FDC, follicular dendritic cells.

If B cells binding their specific antigen in the T-cell zone of peripheral lymphoid organs receive specific signals from armed helper T cells, they proliferate in the T-cell areas (see Fig. 10.13, second panel). In the absence of T-cell signals, these antigen-specific B cells die within 24 hours of arriving in the T-cell zone.

About 5 days after primary immunization, primary foci of proliferating B cells appear in the T-cell areas, which correlates with the time needed for helper T cells to differentiate. Some of the B cells activated in the primary focus may migrate to the medullary cords of the lymph node, or to those parts of the red pulp that are next to the T-cell zones of the spleen, where they become plasma cells and secrete specific antibody for a few days (see Fig. 10.13, third panel). Others migrate to the follicle (see Fig. 10.13, fourth panel), where they proliferate further, forming a germinal center in which they undergo somatic hypermutation (see Sections 4-9 and 9-7). The antibodies secreted by B cells differentiating early in the response not only provide early protection; they may also be important in trapping antigen in the form of antigen:antibody complexes on the surface of the local follicular dendritic cells. The antigen:antibody complexes, which become coated with fragments of C3, are held by complement fragment receptors (CR1, CR2, and CR3) as well as by a nonphagocytic Fc receptor on the follicular dendritic cells. Antigen can be retained in lymphoid follicles in this form for very long periods. The function of this antigen is unclear, but it is likely that it regulates the long-term antibody response.

The proliferation, somatic hypermutation, and selection that occur in the germinal centers during a primary antibody response have been described in Chapter 9. The adhesion and chemokine molecules that govern the migratory behavior of B cells are likely to be very important to this process but, as yet, little is known of their nature or of the ligands to which they bind. The chemokine/receptor pair BLC/CXCR5, which controls B-cell migration into the follicle, may be important, particularly for B cells homing to the germinal center. Another chemokine receptor, CCR7, which is strongly expressed on T cells and weakly expressed on B cells, may play a role in temporarily directing B cells to the interface with the T-cell zone. The ligands for CCR7—MIP-3β and SLC—are highly expressed in the T zone (see Section 7-30) and could attract B cells that have regulated their CCR7 receptor upward.

10-10 Antibody responses are sustained in medullary cords and bone marrow.

The B cells activated in primary foci migrate either to adjacent follicles or to local extrafollicular sites of proliferation. B cells grow exponentially in these sites for 2–3 days and undergo six to seven cell divisions before the progeny come out of the cell cycle and form antibody-producing plasma cells *in situ* (Fig. 10.14, upper panel). Most of these plasma cells have a life-span of 2–3

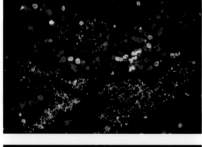

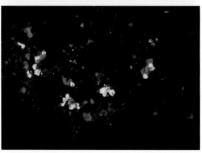

Fig. 10.14 Plasma cells are dispersed in medullary cords and bone marrow. In these sites they secrete antibody at high rates directly into the blood for distribution to the rest of the body. In the upper micrograph, plasma cells in lymph node medullary cords are stained green (with fluorescein anti-IgA) if they are secreting IgA, and red (with rhodamine anti-IgG) if they are secreting IgG. The plasma cells in these local extrafollicular sites are short lived (2–4 days). The lymphatic sinuses are outlined by green granular staining selective for IgA. In the lower micrograph, longer-lived plasma cells (3 weeks to 3 months or more) in the bone marrow are revealed with antibodies specific for light chains (fluorescein anti-λ and rhodamine anti-κ stain). Plasma cells secreting immunoglobulins containing λ light chains shown, on this micrograph, as yellow. Those secreting immunoglobulins containing κ light chains stain red. Photographs courtesy of P. Brandtzaeg.

days, after which they undergo apoptosis. About 10% of plasma cells in these extrafollicular sites live longer; their origin and ultimate fate are unknown. The B cells that migrate to the primary follicles to form germinal centers undergo isotype switching and affinity maturation before either becoming memory cells or leaving the germinal center to become relatively long-lived antibody-producing cells (see Sections 9-6 to 9-8).

These B cells leave germinal centers as plasmablasts (pre-plasma cells). Plasmablasts originating in the follicles of Peyer's patches and mesenteric lymph nodes migrate via lymph to the blood and then enter the lamina propria of the gut and other epithelial surfaces. Those originating in peripheral lymph node or splenic follicles migrate to the bone marrow (Fig. 10.14, lower panel). In these distant sites of antibody production, the plasmablasts differentiate into plasma cells that mostly have a life-span of months to years. These are thought to provide the antibody that can last in the blood for years after an initial immune response. Whether this supply of plasma cells is replenished by the continual but occasional differentiation of memory cells is not yet known. Studies of responses to nonreplicating antigens show that germinal centers are present for only 3–4 weeks after initial antigen exposure. Small numbers of B cells, however, continue to proliferate in the follicles for months. These may be the precursors of antigen-specific plasma cells in the mucosa and bone marrow throughout the subsequent months and years.

10-11 The effector mechanisms used to clear an infection depend on the infectious agent.

A primary adaptive immune response usually serves to clear the primary infection from the body and in most cases provides protection against reinfection with the same pathogen. However, as we will discuss further in Chapter 11, some pathogens evade complete clearance and persist for the life of the host, for example, *Leishmania*, toxoplasma, and herpes viruses. Figure 10.15 summarizes the different types of infection in humans and the ways in which they can be eliminated or held in check by a primary adaptive immune response. It also indicates the mechanisms involved in immunity to reinfection, or protective immunity, against these pathogens. Inducing protective immunity is the goal of vaccine development and to achieve this it is necessary to induce an adaptive immune response that has both the antigen specificity and the appropriate functional elements to combat the particular pathogen concerned. Protective immunity consists of two components—immune reactants, such as antibody or effector T cells generated in the initial infection or by vaccination, and long-lived immunological memory (Fig. 10.16), which we will consider in the last part of this chapter. The type of antibody or effector T cell that offers protection depends on the infectious strategy and lifestyle of the pathogen. Effective immunity against polio virus, for example, requires preexisting antibody, because the virus rapidly infects motor neurons and destroys them unless it is immediately neutralized by antibody and prevented from spreading within the body. Specific IgA on epithelial surfaces can also neutralize the virus before it enters the body. Thus, protective immunity can involve effector mechanisms (IgA in this case) that do not operate in the elimination of the primary infection (see Fig. 10.15).

Preformed reactants can also allow the immune system to respond more rapidly and efficiently to a second exposure to a pathogen. Thus, when opsonizing antibodies such as IgG1 are present (see Section 9-12), opsonization and phagocytosis of pathogens will be more efficient. If specific IgE is present, then pathogens will also be able to activate mast cells, rapidly initiating an inflammatory response through the release of histamine and leukotrienes.

Infectious agent	Disease	Humoral immunity				Cell-mediated Immunity	
		IgM	IgG	IgE	IgA	CD4 T cells (macrophages)	CD8 killer T cells
Viruses							
Variola	Smallpox						■
Varicella zoster	Chickenpox	◨	◨				◨
Epstein–Barr virus	Mononucleosis		◨				◨
Influenza virus	Influenza		■		■		◨
Mumps virus	Mumps		■				◨
Measles virus	Measles		■				◨
Polio virus	Poliomyelitis		◨				◨
Human immunodeficiency virus	AIDS		◨				◨
Bacteria							
Staphylococcus aureus	Boils	■	■				
Streptococcus pyogenes	Tonsilitis	■	■				
Streptococcus pneumoniae	Pneumonia	■	■				
Neisseria gonorrhoeae	Gonorrhea		◨		◨		
Neisseria meningitidis	Meningitis		■				
Corynebacterium diphtheriae	Diphtheria		◨				
Clostridium tetani	Tetanus		◨				
Treponema pallidum	Syphilis		◨	Transient			
Borrelia burgdorferi	Lyme disease		◨	Transient			
Salmonella typhi	Typhoid		■				
Vibrio cholerae	Cholera			◨			
Legionella pneumophila	Legionnaire's disease		◨			■	
Rickettsia prowazeki	Typhus					■	◨
Chlamydia trachomatis	Trachoma					◨	◨
Mycobacteria	Tuberculosis, leprosy					■	◨
Fungi							
Candida albicans	Candidiasis		◨			◨	
Protozoa							
Plasmodium spp.	Malaria		◨			◨	
Toxoplasma gondii	Toxoplasmosis		◨			◨	
Trypanosoma spp.	Trypanosomiasis		■				
Leishmania spp.	Leishmaniasis					■	
Worms	Schistosome	Schistosomiasis					◨	

Fig. 10.15 Different effector mechanisms are used to clear primary infections with different pathogens and to protect against subsequent reinfection. The pathogens are listed in order of increasing complexity, and the defense mechanisms used to clear a primary infection are identified by the red shading of the boxes where these are known. Yellow shading indicates a role in protective immunity. Paler shades indicate less well established mechanisms. Much has to be learned about such host–pathogen interactions. It is clear that classes of pathogens elicit similar protective immune responses, reflecting similarities in their lifestyles.

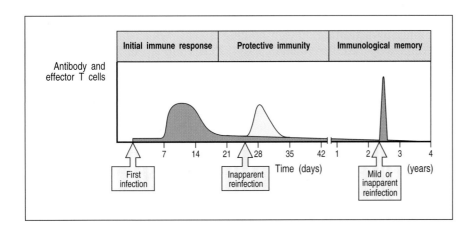

Fig. 10.16 Protective immunity consists of preformed immune reactants and immunological memory. Antibody levels and effector T-cell activity gradually decline after an infection is cleared. An early reinfection is rapidly cleared by these immune reactants. There are few symptoms but levels of immune reactants increase. Reinfection at later times leads to rapid increases in antibody and effector T cells owing to immunological memory, and infection can be mild or even inapparent.

10-12 Resolution of an infection is accompanied by the death of most of the effector cells and the generation of memory cells.

When an infection is effectively repelled by the adaptive immune system, two things occur. The first is the removal of most of the effector cells, as part of the restoration of tissue integrity. The immune system has well-developed mechanisms for getting rid of cells that have outlasted their usefulness. Most unwanted effector cells die by apoptosis, a process used by all multicellular eukaryotic organisms to remove unwanted cells.

The actions of effector cells remove the specific stimulus that originally recruited them. In the absence of this stimulus, they then undergo 'death by neglect,' removing themselves by apoptosis. The dying cells are rapidly cleared by macrophages, which recognize the membrane lipid phosphatidylserine. This lipid is normally found only on the inner surface of the plasma membrane, but in apoptotic cells it rapidly redistributes to the outer surface, where it can be recognized by specific receptors on phagocytes. Thus, not only does the ending of infection lead to the removal of the pathogen, it also leads to the loss of most of the pathogen-specific effector cells.

However, some of the effector cells are retained, and these provide the raw material for memory T-cell and B-cell responses. These are crucially important to the operation of the adaptive immune system, as we will argue in Chapter 15. The memory T cells, which we will consider at the end of this chapter, are retained virtually forever. However, the mechanisms underlying the decision to induce apoptosis in the majority of effector cells and retain only a few are not known. It seems likely that the answer will lie in the cytokines produced by the environment or by the T cells themselves.

Summary.

The adaptive immune response is required for effective protection of the host against pathogenic microorganisms. The response of the innate immune system to pathogens helps initiate the adaptive immune response, as interactions with these pathogens lead to the production of cytokines and the activation of dendritic cells to activated antigen-presenting cell status. The antigens of the pathogen are transported to local lymphoid organs by these migrating antigen-presenting cells and presented to antigen-specific naive T cells that continuously recirculate through the lymphoid organs. T-cell priming and the differentiation of armed effector T cells occur here on the surface of antigen-loaded dendritic cells, and the armed effector T cells either leave the lymphoid organ to effect cell-mediated immunity in sites of infection in the tissues, or remain in the lymphoid organ to participate in humoral immunity by activating antigen-binding B cells. Which response occurs is determined by the differentiation of CD4 T cells into T_H1 or T_H2 cells, which is in turn determined by the cytokines produced in the early nonadaptive phase. CD4 T-cell differentiation is also affected by ill-defined characteristics of the activating antigen and by its overall abundance. CD8 T cells play an important role in protective immunity, especially in protecting the host against infection by viruses and intracellular infections by *Listeria* and other microbial pathogens that have special means for entering the host cell's cytoplasm. Ideally, the adaptive immune response eliminates the infectious agent and provides the host with a state of protective immunity against reinfection with the same pathogen.

The mucosal immune system.

The immune system may be viewed as an organ that is distributed throughout the body to provide host defense against pathogens wherever these may enter or spread. Within the immune system, a series of anatomically distinct compartments can be distinguished, each of which is specially adapted to generate a response to pathogens present in a particular set of body tissues. The previous part of the chapter illustrated the general principles underlying the initiation of an adaptive immune response in the compartment comprising the peripheral lymph nodes and spleen. This is the compartment that responds to antigens that have entered the tissues or spread into the blood. A second compartment of the adaptive immune system of equal size to this, and located near the surfaces where most pathogens invade, is the mucosal immune system (commonly described by the acronym MALT). Two further distinct compartments are those of the body cavities (peritoneum and pleura) and the skin. Two key features define these compartments. The first is that immune responses induced within one compartment are largely confined in expression to that particular compartment. The second is that lymphocytes are restricted to particular compartments by expression of homing receptors that are bound by ligands, known as addressins, that are specifically expressed within the tissues of the compartment. We will illustrate the concept of compartmentalization of the immune system by considering the mucosal immune system. The mucosal surfaces of the body are particularly vulnerable to infection. They are thin and permeable barriers to the interior of the body because of their physiological activities in gas exchange (the lungs), food absorption (the gut), sensory activities (eyes, nose, mouth, and throat), and reproduction (uterus and vagina). The necessity for permeability of the surface lining these sites creates obvious vulnerability to infection and it is not surprising that the vast majority of infectious agents invade the human body through these routes.

A second important point to bear in mind when considering the immuno-biology of mucosal surfaces is that the gut acts as a portal of entry to a vast array of foreign antigens in the form of food. The immune system has evolved mechanisms to avoid a vigorous immune response to food antigens on the one hand and, on the other, to detect and kill pathogenic organisms gaining entry through the gut. To complicate matters further, most of the gut is heavily colonized by approximately 10^{14} commensal microorganisms, which live in symbiosis with their host. These bacteria are beneficial to their host in many ways. They provide protection against pathogenic bacteria by occupying the ecological niches for bacteria in the gut. They also serve a nutritional role in their host by synthesizing vitamin K and some of the components of the vitamin B complex. However, in certain circumstances they can also cause disease, as we will see later.

10-13 Mucosa-associated lymphoid tissue is located in anatomically defined microcompartments throughout the gut.

The mucosa-associated lymphoid tissues lining the gut are known as gut-associated lymphoid tissue or **GALT**. The tonsils and adenoids form a ring, known as Waldeyer's ring, at the back of the mouth at the entrance of the gut and airways. They represent large aggregates of mucosal lymphoid tissue, which often become extremely enlarged in childhood because of recurrent infections, and which in the past were victims of a vogue for surgical removal. A reduced IgA response to oral polio vaccination has been seen in individuals who have had their tonsils and adenoids removed, which illustrates the importance of this subcompartment of the mucosal immune system.

The other principal sites within the gut mucosal immune system for the induction of immune responses are the Peyer's patches of the small intestine, the appendix (which is another frequent victim of the surgeon's knife), and solitary lymphoid follicles of the large intestine and rectum. Peyer's patches are an extremely important site for the induction of immune responses in the small intestine and have a distinctive structure, forming domelike structures extending into the lumen of the intestine (see Fig. 1.10). The overlying layer of follicle-associated epithelium of the Peyer's patches contains specialized epithelial cells. These have microfolds on their luminal surface, instead of the microvilli present on the absorptive epithelial cells of the intestine, and are known as **microfold cells** or **M cells**. They are much less prominent than the absorptive gut epithelial cells, known as enterocytes, and form a membrane overlying the lymphoid tissue within the Peyer's patch. M cells lack a thick surface glycocalyx and do not secrete mucus. Hence they are adapted to interact directly with molecules and particles within the lumen of the gut.

M cells take up molecules and particles from the gut lumen by endocytosis or phagocytosis (Fig. 10.17). This material is then transported through the interior of the cell in vesicles to the basal cell membrane, where it is released into the extracellular space. This process is known as **transcytosis**. At their basal surface, the cell membrane of M cells is extensively folded around underlying lymphocytes and antigen-presenting cells, which take up the transported material released from the M cells and process it for antigen presentation.

Because M cells are much more accessible than enterocytes to particles within the gut, a number of pathogens target M cells to gain access to the subepithelial space, even though such pathogens then find themselves in the heart of the adaptive immune system of the intestine, the Peyer's patches. We will consider one of these pathogens in Section 10-19.

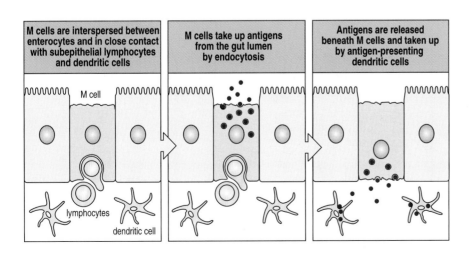

| M cells are interspersed between enterocytes and in close contact with subepithelial lymphocytes and dendritic cells | M cells take up antigens from the gut lumen by endocytosis | Antigens are released beneath M cells and taken up by antigen-presenting dendritic cells |

Fig. 10.17 M cells take up antigens from the lumen of the gut by endocytosis. The cell membrane at the base of these cells is folded around lymphocytes and dendritic cells within the Peyer's patches. Antigens are transported through M cells by the process of transcytosis and delivered directly to antigen-presenting cells and lymphocytes of the mucosal immune system.

10-14 The mucosal immune system contains a distinctive repertoire of lymphocytes.

In addition to the organized lymphoid tissue in which induction of immune responses occurs within the mucosal immune system, small foci of lymphocytes and plasma cells are scattered widely throughout the lamina propria of the gut wall. These represent the effector cells of the gut mucosal immune system. The life history of these cells is as follows. As naive lymphocytes, they emerge from the primary lymphoid organs of bone marrow and thymus to enter the inductive lymphoid tissue of the mucosal immune system via the bloodstream. They may encounter foreign antigens presented within the organized lymphoid tissue of the mucosal immune system and become activated to effector status. From these sites, the activated lymphocytes traffic via the lymphatics draining the intestines, pass through mesenteric lymph nodes, and eventually wind up in the thoracic duct, from where they circulate in the blood throughout the entire body (Fig. 10.18). They reenter the mucosal tissues from the small blood vessels lining the gut wall and other sites of MALT, such as the respiratory or reproductive mucosa, and the lactating breast; these vessels express the mucosal adressin MAdCAM-1. In this way, an immune response that may be started by foreign antigens presented in a limited number of Peyer's patches is disseminated throughout the mucosa of the body. This pathway of lymphocyte trafficking is distinct from and parallel to that of lymphocytes in the rest of the peripheral lymphoid system (see Fig. 1.11).

The distinctiveness of the mucosal immune system from the rest of the peripheral lymphoid system is further underlined by the different lymphocyte repertoires in the different compartments. The T cells of the gut can be divided into two types. One type bears the conventional α:β T-cell receptors in conjunction with either CD4 or CD8, and participates in conventional T-cell responses to foreign antigens as discussed in earlier chapters. The second class is made up of T cells with unusual surface phenotypes such as TCRγ:δ and CD8α:α TCRα:β. The receptors of these T cells do not bind to the normal MHC:peptide ligands but instead bind to a number of different

Fig. 10.18 Anatomy of mucosal immune responses. The left panel shows the afferent immune response. Antigen from pathogenic microorganisms is presented beneath mucosal surfaces to naive lymphocytes within organized mucosal lymphoid tissue, for example Peyer's patches. Activated lymphocytes leave this tissue via draining lymph nodes and reenter the circulation through the thoracic duct. The right panel shows the efferent immune response in which primed lymphocytes reenter mucosal tissues throughout the body from the circulation, thereby disseminating a mucosal immune response.

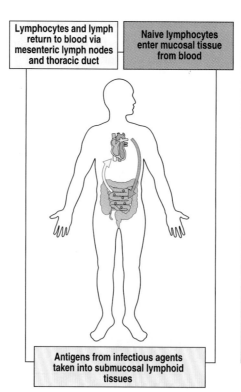

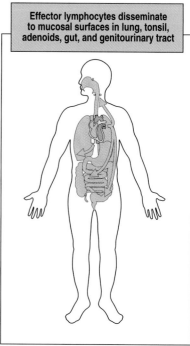

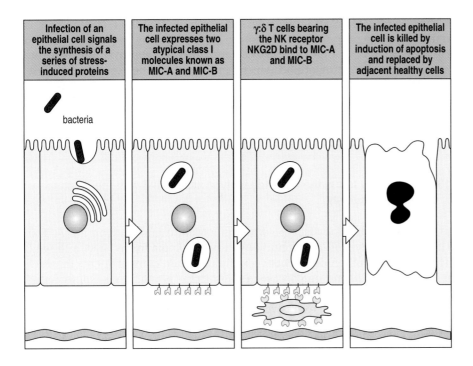

| Infection of an epithelial cell signals the synthesis of a series of stress-induced proteins | The infected epithelial cell expresses two atypical class I molecules known as MIC-A and MIC-B | γ:δ T cells bearing the NK receptor NKG2D bind to MIC-A and MIC-B | The infected epithelial cell is killed by induction of apoptosis and replaced by adjacent healthy cells |

Fig. 10.19 T cells of the mucosal immune system bearing γ:δ T-cell receptors and an activating NK receptor recognize and kill injured enterocytes. Infection or other injury to enterocytes, the epithelial cells lining the lumen of the gut, stimulates a stress response, which causes expression on the cell surface of two atypical MHC class IB molecules, known as MIC-A and MIC-B. Intraepithelial T cells carrying the NK receptor NKG2D bind MIC-A and MIC-B and induce apoptosis in the injured enterocytes. The dying enterocyte is removed from the epithelium and the local tissue injury is repaired.

ligands, including MHC class IB molecules. These highly specialized T cells are abundant in the epithelium of the gut and have a restricted repertoire of T-cell receptor specificities. Unlike conventional T cells, many of these cells do not undergo positive and negative selection in the thymus (see Chapter 7), and express receptors with sequences that have undergone no or minimal divergence from their germline-encoded sequences. These cells may be classified in phylogenetic terms as being at the interface between innate and adaptive immunity.

T cells bearing a γ:δ receptor are especially abundant in the gut mucosa compared with other lymphoid tissues. One subset of these γ:δ T cells in humans, which expresses a T-cell receptor that uses the $V_\delta 1$ gene segment, carries an activating C-type lectin NK receptor, NKG2D. This latter receptor binds to two MHC-like molecules—MIC-A and MIC-B—that are expressed on intestinal epithelial cells in response to cellular injury and stress. The injured cells may then be recognized and killed by this subset of γ:δ T cells (Fig. 10.19). This illustrates one of the key roles of T cells, which is to patrol and survey the body, destroying cells that express an abnormal phenotype as a result of stress or infection.

The $V_\delta 1$-containing receptor on these T cells may also play a part in allowing them to survey tissues for injured cells. Some human T cells expressing this receptor bind to CD1c, one of the isotypes of the CD1 family of MHC class I-like molecules that we encountered in Section 10-5. This protein, which shows increased expression on activated monocytes and dendritic cells, presents endogenous lipid and glycolipid antigens to some types of T cell. In response to antigen presentation by CD1c, these T cells secrete IFN-γ, which may have an important role in polarizing the response of conventional T cells bearing α:β receptors toward a T_H1 response. This is closely analogous, although opposite in effect, to the polarization toward T_H2 cells induced by secretion of IL-4 by NK 1.1⁺ T cells responding to CD1d discussed in Section 10-5.

A second group of specialized mucosal T cells, so far only characterized in mice, express α:β T-cell receptors together with a CD8 α:α homodimer, instead of the normal CD8 α:β heterodimer that characterizes MHC class I-restricted cytotoxic T cells. These cells can be found in the gut of mice lacking conventional MHC class I molecules, which shows that their development is not dependent on positive selection in the thymus by peptides bound to classical MHC class I molecules. They are, however, absent in mice lacking expression of β_2-microglobulin, which is necessary for the expression of MHC class IB molecules. The ligand recognized by these T cells in mice has been identified as the nonpolymorphic MHC class IB molecule known as Qa-2. These cells are likely to represent a further class of T cells that have a major role in maintaining the integrity of the gut mucosa by recognizing and destroying injured mucosal cells.

10-15 Secretory IgA is the antibody isotype associated with the mucosal immune system.

The dominant antibody isotype of the mucosal immune system is IgA. This class of antibody is found in humans in two isotypic forms, IgA1 and IgA2. The expression of IgA differs between the two main compartments in which it is found—blood and mucosal secretions. In the blood, IgA is mainly found as a monomer and the ratio of IgA1 to IgA2 is approximately 4:1. In mucosal secretions, IgA is almost exclusively produced as a dimer and the ratio of IgA1 to IgA2 is approximately 3:2. A number of common intestinal pathogens possess proteolytic enzymes that can digest IgA1, whereas IgA2 is much more resistant to digestion. The higher proportion of plasma cells secreting IgA2 in the gut lamina propria may therefore be the consequence of selective pressure by pathogens against individuals with low IgA2 levels in the gut. The mechanism of isotype switching to IgA is discussed in Section 9-14.

There are special mechanisms for the secretion of polymeric IgA and IgM antibody into the gut lumen (see Section 9-13). Polymeric IgA and IgM are synthesized throughout the gut by plasma cells located in the lamina propria and are transported into the gut by immature epithelial cells located at the base of the intestinal crypts. These express the polymeric immunoglobulin receptor on their basolateral surfaces. This receptor binds polymeric IgA or IgM and transports the antibody by transcytosis to the luminal surface of the gut. Upon reaching the luminal surface of the enterocyte, the antibody is released into the secretions by proteolytic cleavage of the extracellular domain of the polymeric IgA receptor. Secreted IgA and IgM bind to the mucus layer overlying the gut epithelium where they can bind to and neutralize gut pathogens and their toxic products (Fig. 10.20).

Fig. 10.20 The major antibody isotype present in the lumen of the gut is secretory polymeric IgA. This is synthesized by plasma cells in the lamina propria and transported into the lumen of the gut through epithelial cells at the base of the crypts. Polymeric IgA binds to the mucus layer overlying the gut epithelium and acts as an antigen-specific barrier to pathogens and toxins in the gut lumen.

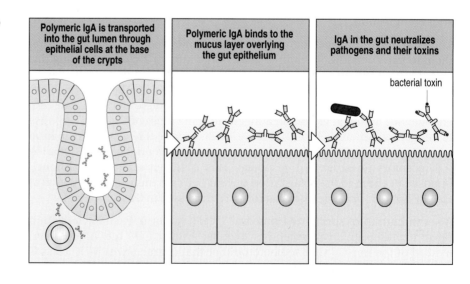

10-16 Most antigens presented to the mucosal immune system induce tolerance.

We are continuously exposed to a huge array of foreign antigens in the form of foods, but these do not normally induce an adaptive immune response. For example, IgA antibodies with high affinity to food antigens do not normally develop. This lack of response occurs despite the fact that the repertoire of lymphocyte antigen receptors has not been negatively selected to remove those specific for food antigens. This is because, like any other foreign antigen, food antigens do not play a part in the central mechanisms of lymphocyte tolerance to self, which are established in the thymus and bone marrow (see Chapter 7).

The feeding of foreign antigens leads typically to a state of specific and active unresponsiveness, a phenomenon known as **oral tolerance**. Thus, no antibody response follows the feeding of a foreign protein such as ovalbumin, although a strong antibody response to this protein can be induced by injecting it subcutaneously, especially if an adjuvant is given as well. However, the feeding of ovalbumin is followed by a prolonged period during which the administration of ovalbumin by injection, even in the presence of adjuvant, elicits no antibody response. This suppression is antigen-specific, because antibody responses to other injected antigens are not affected. These experiments show that there are antigen-specific mechanisms for suppressing peripheral immune responses to antigens delivered by mouth. The mechanisms of oral tolerance are partly understood but, before considering them, we will first discuss the contrasting immune responses that are seen in response to bacterial infections of the gut.

10-17 The mucosal immune system can mount an immune response to the normal bacterial flora of the gut.

We each harbor more than 400 species of commensal bacteria, which are present in the largest numbers in the colon and ileum. Despite the fact that these bacteria collectively weigh approximately 1 kg and outnumber us by approximately 10^{14} to 1, for most of the time we cohabit with our intestinal bacterial flora in a happy symbiotic relationship.

One protective activity of our normal gut flora is that of competition against pathogenic bacteria for space and nutrients, preventing their colonization of the gut (see Fig. 2.4). This activity is dramatically illustrated by one of the adverse effects of antibiotics. Taking an antibiotic kills large numbers of commensal gut bacteria and thereby offers an ecological niche to bacteria that would not otherwise be able to compete successfully with the normal flora and grow in the gut. One example of a bacterium that grows in the antibiotic-treated gut and can cause a severe infection is *Clostridium difficile*; this produces two toxins, which can cause severe bloody diarrhea associated with mucosal injury (Fig. 10.21).

There are some circumstances in which the normal bacterial inhabitants of the gut cause disease, for example, following breakdown of the integrity of the mucosa lining the gut. This may occur following poor blood flow in the gut, or following endotoxemia (see Chapter 2). In these circumstances, normally innocuous gut bacteria, such as nonpathogenic *Escherichia coli*, can cross the mucosa, invade the bloodstream, and cause fatal systemic infection. This illustrates the vital importance of the barrier to infection provided by the mucosal surfaces of the body. The normal gut flora also becomes an important cause of systemic infection in patients with immunodeficiency. This illustrates the role of the adaptive immune system in host defense against the flora of the gut, but also shows that this response does not result in the elimination of bacteria from the lumen of the gut, but rather a state resembling symbiosis.

Fig. 10.21 Treatment with antibiotics causes massive death of the commensal bacteria that normally colonize the colon. This allows pathogenic bacteria to proliferate and occupy an ecological niche that is normally occupied by harmless commensal bacteria. *Clostridium difficile* is an example of such a pathogen that produces toxins that may cause severe bloody diarrhea in patients treated with antibiotics.

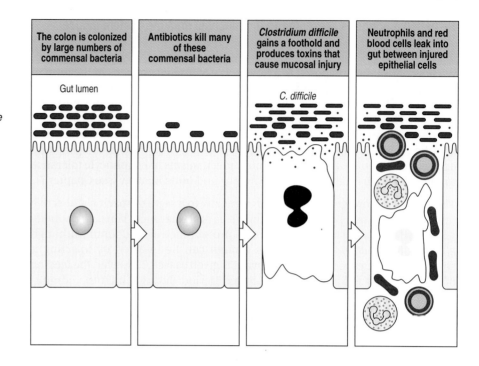

The colon is colonized by large numbers of commensal bacteria	Antibiotics kill many of these commensal bacteria	*Clostridium difficile* gains a foothold and produces toxins that cause mucosal injury	Neutrophils and red blood cells leak into gut between injured epithelial cells

The scale of the normal immune response to gut bacteria is illustrated by the study of animals delivered by Caesarian section into a sterile environment in which there is no colonization of the gut by microorganisms. These are known as **germ-free** or **gnotobiotic** animals. These animals have marked reductions in the size of all secondary lymphoid organs and reduced levels of antibodies of all isotypes.

10-18 Enteric pathogens cause a local inflammatory response and the development of protective immunity.

In spite of the array of innate immune mechanisms in the gut and stiff competition from the indigenous flora, the gut is a frequent site of infection by pathogenic microorganisms. These include many species of viruses, enteric pathogenic bacteria including *Salmonella, Yersinia, Shigella,* and *Listeria,* and protozoa such as *Entamoeba histolytica* and *Cryptosporidium.* These organisms cause disease in many different ways, but there are certain common features of infection that are crucial to understanding how these pathogens stimulate an immune response by the host, in contrast to the immunological tolerance shown to the foreign antigens ingested in food.

The most important consequence of infection in the gut, as elsewhere in the body, is the development of an inflammatory response. The release of cytokines and chemokines in this response is key to the induction of an adaptive immune response. The inflammatory mediators stimulate the maturation of dendritic cells and other antigen-presenting cells, so that they express the co-stimulatory molecules that provide the additional signals for activation and expansion of naive lymphocytes.

Some intestinal pathogens infect enterocytes, the absorptive cells that line much of the intestine. Enterocytes do not act as passive victims of infection but signal infection by releasing cytokines and chemokines. These include the chemokine IL-8, which is a potent neutrophil chemoattractant, and CC chemokines such as MCP-1, MIP-1α and β, and RANTES, which are chemoattractants for monocytes, eosinophils, and T lymphocytes (see Fig. 2.33). In this way, the onset of infection triggers an influx of inflammatory cells and lymphocytes, leading to the induction of an immune response to the antigens of the infectious agent. Injury and stress to the enterocytes lining the gut may also stimulate the expression of nonclassical MHC molecules, such as MIC-A and MIC-B (see Section 10-14). These act as ligands to the receptors on γ:δ T cells at the base of the crypts, which kill the infected mucosal cell, thereby promoting repair and recovery of the injured mucosa.

A number of pathogens directly exploit the M cell as a means of invasion. Some viruses are transported through the M cell by transcytosis and from the subepithelial space are able to establish systemic infection. For example, from this site polio and retroviruses enter intestinal neuronal cells and spread to the central nervous system. HIV, which we discuss in detail in Chapter 11, may use a similar route into the lymphoid tissue of the rectal mucosa, where it first encounters and infects macrophages.

Many of the most important enteric bacteria that cause infections in humans gain entry to the body through M cells. Invasion by this route delivers bacteria straight to the lymphoid system of the host. Depending on the pathogenicity of the organism and the strength of the host adaptive immune response, infections that breach the gut mucosa may be cleared with little tissue injury, cause a local inflammatory response, or invade the bloodstream or lymphatics and result in a systemic infection. Bacteria that specifically target M cells include *Salmonella typhi*, the causative agent of typhoid, and *S. typhimurium*, a major cause of bacterial food-poisoning. These bacteria cause a brisk local and systemic inflammatory response associated with the induction of T_H1-type T-cell responses and antibody responses of the IgG and IgA classes.

10-19 Infection by *Helicobacter pylori* causes a chronic inflammatory response, which may cause peptic ulcers, carcinoma of the stomach, and unusual lymphoid tumors.

There is one exceptional bacterial infection of the stomach, in which the inflammatory and immune response to the organism causes the disease instead of clearing the infection. *Helicobacter pylori*, which infects many millions of people around the world, adheres to the mucosa of the stomach and causes a local inflammatory response, with the release of IL-8 and the influx of leukocytes. In the majority of those infected there is no overt disease, but up to 5% of infected people make either one of two very different responses to the infection. In some, there is an excessive release of the hormone gastrin, which stimulates acid production and the development of peptic ulcers. In others, chronic inflammation has the opposite effect, leading to atrophy of the stomach, which is associated with reduced acid production and an increased risk of carcinoma of the stomach (Fig. 10.22). Rarely, lymphomas known as

Fig. 10.22 *Helicobacter pylori* infects the stomachs of many millions of people. In some, it stimulates the G cells of the stomach to secrete gastrin, which stimulates excess acid production by the stomach, causing peptic ulceration. In others, it causes chronic inflammation of the mucosa, leading to atrophy and loss of acid production. The chronic inflammation of the stomach wall in these individuals may lead to the development of gastric carcinoma.

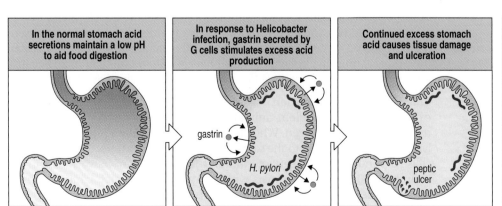
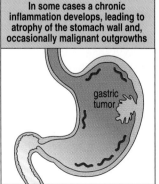

| In the normal stomach acid secretions maintain a low pH to aid food digestion | In response to Helicobacter infection, gastrin secreted by G cells stimulates excess acid production | Continued excess stomach acid causes tissue damage and ulceration | In some cases a chronic inflammation develops, leading to atrophy of the stomach wall and, occasionally malignant outgrowths |

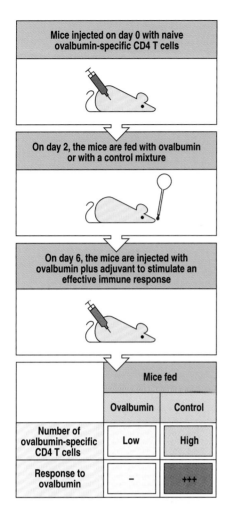

Fig. 10.23 T-cell anergy following feeding of antigen. Mice were injected with CD4 T cells bearing a transgenic receptor specific for an ovalbumin peptide. Two days later they were fed ovalbumin or a control protein. Four days later mice were injected with the relevant ovalbumin peptide with adjuvant. Eight days later the draining lymph node was harvested and the number of ovalbumin-specific transgenic T cells was measured and their proliferative response was assessed to stimulation by ovalbumin peptide *in vitro*. Mice fed with ovalbumin demonstrated a small reduction, compared with control-fed mice, in the number of transgenic T cells recovered, showing some deletion of T cells by orally fed antigen. However, many transgenic T cells remained in the ovalbumin-fed mice and these were refractory to stimulation by antigen *in vitro*, compared with the control-fed mice, which showed vigorous proliferative responses. These transgenic T cells had become anergic following oral intake of antigen.

MALT lymphomas, because of their origin from the mucosa-associated lymphoid tissue, arise from the B lymphocytes that accumulate in the chronic inflammatory lymphoid infiltrates of the stomach. These are very extraordinary tumors, because some of them, despite being monoclonal proliferations with a transformed phenotype, are still dependent on antigenic or other inflammatory stimulation by *H. pylori*. These tumors may regress if the *H. pylori* infection is effectively eliminated by antibiotics alone.

10-20 In the absence of inflammatory stimuli, the normal response of the mucosal immune system to foreign antigens is tolerance.

We have seen that there are two possible and opposite outcomes of exposure to foreign antigens entering through the mucosa of the gut. These are tolerance in the case of food antigens, which is contrasted with a vigorous antibody and T-cell response after exposure to pathogens. The essential difference between antigenic challenge by food compared with that by pathogens is that the pathogens cause inflammation, whereas food does not. Both the antigens within food and the antigens within pathogens are presented by antigen-presenting cells to T lymphocytes, but the contexts in which these two sources of antigen are presented are quite different.

Three different responses of T cells to the presentation of peptides derived from foods and other antigens delivered via the mucosa may account for the phenomenon of oral tolerance. The first is the deletion of antigen-specific T cells by the induction of apoptosis, which has been found to occur in experimental animals in response to oral intake of very large, and probably non-physiological, doses of antigen. This is probably not the most important mechanism of oral tolerance, although it may contribute.

The second response is anergy, in which T cells presented with peptide in the absence of co-stimulatory signals become refractory to further stimulation with antigen (see Section 8-11). The development of anergy in response to a food antigen was demonstrated by feeding ovalbumin to mice that had large numbers of T cells carrying a transgenically expressed T-cell receptor for an ovalbumin peptide epitope (see Appendix I, Section A-46). Following feeding of ovalbumin, T cells bearing the transgenic T-cell receptor could still be detected, but these were totally refractory to further stimulation by ovalbumin *in vitro* and *in vivo*, even when ovalbumin was injected systemically together with an adjuvant (Fig. 10.23).

The third response involves the development of regulatory T cells, which can actively suppress antigen-specific responses following rechallenge with antigen. One subset of T cells has been described that produces IL-4, IL-10, and TGF-β on stimulation with antigen. These cells have been called T$_H$3 cells. A similar population secretes TGF-β in an IL-10-dependent manner and has been named T$_R$1 (T regulatory 1 cells). This pattern of cytokine secretion in response to antigen-specific stimulation inhibits the development of T$_H$1 responses and is associated with low levels of antibody and virtually absent inflammatory T-cell responses. The γ:δ T cells that are abundant throughout the mucosal immune system may also have a role in oral tolerance, because tolerance appears to be reduced in mice lacking this subset of T lymphocytes.

Antigen-specific suppression is a form of oral tolerance that can be transferred experimentally to recipient animals by lymphocytes derived from animals that have been fed antigen. When an animal that has been injected with such lymphocytes is exposed to the same antigen for the first time, these regulatory cells respond to the antigen and inhibit the responses of naive T cells in the recipient animal. This contrasts with anergic T cells, which cannot transfer oral tolerance following transfer to a naive animal.

In order to understand the phenomenon of oral tolerance, it is essential to understand how orally delivered antigens are presented to T cells. Two routes of antigen presentation of soluble food antigens have been characterized that may induce T-cell responses favoring tolerance rather than immune activation. The first is presentation of soluble food antigens by the antigen-presenting cells of the gut and other peripheral lymphoid organs. In the absence of inflammatory stimuli, antigen presentation by dendritic cells favors the induction of tolerance rather than T-cell activation. Dendritic cells in Peyer's patches have been shown to express IL-10 and IL-4, in contrast to similar cells in peripheral lymph nodes which express IFN-γ and IL-12. However, this heterogeneity of cytokine responses does not fully explain tolerance to food antigens. These may be detected in the bloodstream after feeding and there is evidence that the induction of tolerance to food antigens takes place in lymph nodes and spleen as well as in the mucosal lymphoid system. The second possible route of presentation of food antigens is by the enterocytes of the gut, which express MHC class I and MHC class II molecules in the absence of co-stimulatory molecules and thus may induce anergy on presenting antigens to intraepithelial lymphocytes.

We will discuss each of these mechanisms of tolerance further in Chapter 13, where we consider how the loss of tolerance to self tissues may contribute to the development of autoimmune disease. As we will see in Chapter 14, one of the strategies for treating allergy and autoimmune disease is to attempt to manipulate the nature of the antigen-specific response to stimulate T cells with the properties of such regulatory T cells.

Summary.

The immune system can be divided into a series of functional anatomical compartments, of which the two most important are the peripheral lymphoid system made up of the conventionally studied spleen and lymph nodes, and the mucosal lymphoid system. Specific homing mechanisms for lymphocytes to each of these compartments serve to maintain a separate population of lymphocytes in each. The mucosal surfaces of the body are highly vulnerable to infection and possess a complex array of innate and adaptive mechanisms of immunity. The adaptive immune system of the mucosa-associated lymphoid tissues differs from that of the rest of the peripheral lymphoid system in several respects. The types and distribution of T cells differ, with significantly greater numbers of γ:δ T cells in the gut mucosa compared with peripheral lymph nodes and blood. The major antibody type secreted across the epithelial cells lining mucosal surfaces is secretory polymeric IgA. The mucosal lymphoid system is exposed to a vast array of foreign antigens from foods, from the commensal bacteria of the gut, and from pathogenic microorganisms and parasites. No immune response can normally be detected to food antigens. Indeed, soluble antigens taken by mouth may induce antigen-specific tolerance or antigen-specific suppression. In contrast, pathogenic microorganisms induce strong protective T_H1 responses. It is an important challenge to understand these contrasting specific immune responses. The key distinction between tolerance and the development of powerful protective adaptive immune responses is the context in which peptide antigen is presented to T lymphocytes in the mucosal immune system. In the absence of inflammation, presentation of peptide to T cells by MHC molecules on antigen-presenting cells occurs in the absence of co-stimulation. By contrast, pathogenic microorganisms induce inflammatory responses in the tissues, which stimulate the maturation and expression of co-stimulatory molecules on antigen-presenting cells. This form of antigen presentation to T cells favors development of a protective T_H1 response.

Immunological memory.

Having considered how an appropriate primary immune response is mounted to pathogens in both the peripheral lymphoid system and the mucosa-associated lymphoid tissues, we now turn to immunological memory, which is a feature of both compartments. Perhaps the most important consequence of an adaptive immune response is the establishment of a state of immunological memory. Immunological memory is the ability of the immune system to respond more rapidly and effectively to pathogens that have been encountered previously, and reflects the preexistence of a clonally expanded population of antigen-specific lymphocytes. Memory responses, which are called secondary, tertiary, and so on, depending on the number of exposures to antigen, also differ qualitatively from primary responses. This is particularly clear in the case of the antibody response, where the characteristics of antibodies produced in secondary and subsequent responses are distinct from those produced in the primary response to the same antigen. Memory T-cell responses have been harder to study, but can also be distinguished from the responses of naive or effector T cells. The principal focus of this section will be the altered character of memory responses, although we will also discuss emerging explanations of how immunological memory persists after exposure to antigen. A long-standing debate about whether specific memory is maintained by distinct populations of long-lived memory cells that can persist without residual antigen, or by lymphocytes that are under perpetual stimulation by residual antigen, appears to have been settled in favor of the former hypothesis.

10-21 Immunological memory is long-lived after infection or vaccination.

Most children in the United States are now vaccinated against measles virus; before vaccination was widespread, most were naturally exposed to this virus and suffered from an acute, unpleasant, and potentially dangerous viral illness. Whether through vaccination or infection, children exposed to the virus acquire long-term protection from measles. The same is true of many other acute infectious diseases: this state of protection is a consequence of immunological memory.

The basis of immunological memory has been hard to explore experimentally. Although the phenomenon was first recorded by the ancient Greeks and has been exploited routinely in vaccination programs for over 200 years, it is just now becoming clear that memory reflects a persistent population of specialized memory cells that is independent of the continued persistence of the original antigen that induced them. This mechanism of maintaining memory is consistent with the finding that only individuals who were themselves previously exposed to a given infectious agent are immune, and that memory is not dependent on repeated exposure to infection as a result of contacts with other infected individuals. This was established by observations made on remote island populations, where a virus such as measles can cause an epidemic, infecting all people living on the island at that time, after which the virus disappears for many years. On reintroduction from outside the island, the virus does not affect the original population but causes disease in those people born since the first epidemic. This means that immunological memory need not be maintained by repeated exposure to infectious virus.

Instead, it is most likely that memory is sustained by long-lived antigen-specific lymphocytes that were induced by the original exposure and that persist until a second encounter with the pathogen. It was thought that retained antigen, bound in immune complexes on follicular dendritic cells, might be crucial in maintaining these cells, but recent experiments suggest otherwise. While most of the memory cells are in a resting state, careful studies have shown that a small percentage are dividing at any one time. What stimulates this infrequent cell division is unclear. However, cytokines such as those produced either constitutively or during the course of antigen-specific immune responses directed at noncross-reactive antigens could be responsible. One such cytokine, IL-15, has been implicated in maintaining CD8 memory T cells. Regardless of cell division, the number of memory cells for a given antigen is highly regulated, remaining practically constant during the memory phase.

Immunological memory can be measured experimentally in various ways. Adoptive transfer assays (see Appendix I, Section A-42) of lymphocytes from animals immunized with simple, nonliving antigens have been favored for such studies, as the antigen cannot proliferate. When an animal is first immunized with a protein antigen, helper T-cell memory against that antigen appears abruptly and is at its maximal level after 5 days or so. Antigen-specific memory B cells appear some days later, because B-cell activation cannot begin until armed helper T cells are available, and B cells must then enter a phase of proliferation and selection in lymphoid tissue. By one month after immunization, memory B cells are present at their maximal levels. These levels are then maintained with little alteration for the lifetime of the animal. In these experiments, the existence of memory cells is measured purely in terms of the transfer of specific responsiveness from an immunized, or 'primed,' animal to an irradiated, immunoincompetent and nonimmunized recipient. In the following sections, we will look in more detail at the changes that occur in lymphocytes after antigen priming, and discuss the mechanisms that might account for these changes.

10-22 Both clonal expansion and clonal differentiation contribute to immunological memory in B cells.

Immunological memory in B cells can be examined by isolating B cells from immunized mice and restimulating them with antigen in the presence of armed helper T cells specific for the same antigen. The response of these primed B cells can be compared with the primary B-cell response seen on isolating B cells from unimmunized mice and stimulating them with antigen in the same way (Fig. 10.24). By these means, it is possible to show that antigen-specific memory B cells differ both quantitatively and qualitatively from naive B cells. B cells that can respond to antigen increase in frequency after priming by about 10- to 100-fold (see Fig. 10.24) and produce antibody of higher average affinity than unprimed B lymphocytes; the affinity of that

Fig. 10.24 The generation of secondary antibody responses from memory B cells is distinct from the generation of the primary antibody response. These responses can be studied and compared by isolating B cells from immunized and unimmunized donor mice, and stimulating them in culture in the presence of armed antigen-specific effector T cells. The primary response usually consists of antibody molecules made by plasma cells derived from a relatively large number of different precursor B cells. The antibodies are of relatively low affinity, with few somatic mutations. The secondary response derives from far fewer high-affinity precursor B cells, which have undergone significant clonal expansion. Their receptors and antibodies are of high affinity for the antigen and show extensive somatic mutation. Thus, there is usually only a 10- to 100-fold increase in the frequency of activatable B cells after priming; however, the quality of the antibody response is altered radically, such that these precursors induce a far more intense and effective response.

	Source of B cells	
	Unimmunized donor Primary response	**Immunized donor Secondary response**
Frequency of specific B cells	$1:10^4 - 1:10^5$	$1:10^3$
Isotype of antibody produced	IgM > IgG	IgG, IgA
Affinity of antibody	Low	High
Somatic hypermutation	Low	High

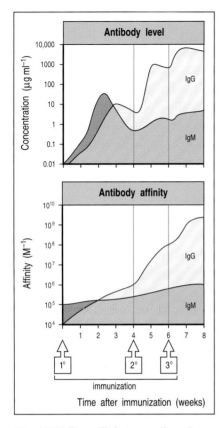

Fig. 10.25 The affinity as well as the amount of antibody increases with repeated immunization. The upper panel shows the increase in the level of antibody with time after primary, followed by secondary and tertiary, immunization; the lower panel shows the increase in the affinity of the antibodies. The increase in affinity (affinity maturation) is seen largely in IgG antibody (as well as in IgA and IgE, which are not shown) coming from mature B cells that have undergone isotype switching and somatic hypermutation to yield higher-affinity antibodies. Although some affinity maturation occurs in the primary antibody response, most arises in later responses to repeated antigen injections. Note that these graphs are on a logarithmic scale.

antibody continues to increase during the ongoing secondary and subsequent antibody responses (Fig. 10.25). The secondary antibody response is characterized in its first few days by the production of small amounts of IgM antibody and larger amounts of IgG antibody, with some IgA and IgE. These antibodies are produced by memory B cells that have already switched from IgM to these more mature isotypes and express IgG, IgA, or IgE on their surface, as well as a somewhat higher level of MHC class II molecules than is characteristic of naive B cells. Increased affinity for antigen and increased levels of MHC class II facilitate antigen uptake and presentation, and allow memory B cells to initiate their critical interactions with armed helper T cells at lower doses of antigen. Unlike memory T cells, which can traffic to tissues owing to changes in cell-surface molecules that affect migration and homing, it is thought that memory B cells continue to recirculate through the same secondary lymphoid compartments that contain naive B cells, principally the follicles of spleen, lymph node, and Peyer's patch. Some memory B cells can also be found in marginal zones, though it is not clear whether these represent a distinct subset of memory B cells.

The distinction between primary and secondary antibody responses is most clearly seen in those cases where the primary response is dominated by antibodies that are closely related and show few if any somatic hypermutations. This occurs in inbred mouse strains in response to certain haptens that are recognized by a limited set of naive B cells. The antibodies produced are encoded by the same V_H and V_L genes in all animals of the strain, suggesting that these variable regions have been selected during evolution for recognition of determinants on pathogens that happen to cross-react with some haptens. As a result of the uniformity of these primary responses, changes in the antibody molecules produced in secondary responses to the same antigens are easy to observe. These differences include not only numerous somatic hypermutations in antibodies containing the dominant variable regions but also the addition of antibodies containing V_H and V_L gene segments not detected in the primary response. These are thought to derive from B cells that were activated at low frequency during the primary response, and thus not detected, but that differentiated into memory B cells.

10-23 Repeated immunizations lead to increasing affinity of antibody owing to somatic hypermutation and selection by antigen in germinal centers.

Upon reexposure to the same antigen, a secondary immune response will ensue. In some ways, this resembles the primary immune response, with the initial proliferation of B cells and T cells at the interface between the T- and B-cell zones. The secondary response is characterized by early and vigorous generation of plasma cells, thus accounting for early profuse IgG production. Some B cells that have not yet undergone terminal differentiation can migrate into the follicle and become **germinal center B cells**. There, these B cells enter a second proliferative phase, during which the DNA encoding their immunoglobulin V domains again undergoes somatic hypermutation before the B cells differentiate into antibody-secreting plasma cells (see Section 9-7).

The antibodies produced by plasma cells in the primary and early secondary response have an important role in driving affinity maturation in the secondary response. In secondary and subsequent immune responses, any persisting antibodies produced by the B cells that differentiated in the primary response are immediately available to bind the newly introduced antigen. Some of these antibodies divert antigen to phagocytes for degradation and disposal (see Section 9-20). If there is sufficient preexisting antibody to clear or inactivate the pathogen, it is possible that no immune response will ensue. However, if there is a trace excess of antigen, B cells whose receptors bind the antigen

Fig. 10.26 The mechanism of affinity maturation in an antibody response. At the beginning of a primary response, B cells with receptors of a wide variety of affinities (K_A), most of which will bind antigen with low affinity, take up antigen, present it to helper T cells, and become activated to produce antibody of varying and relatively low affinity (top panel). These antibodies then bind and clear antigen, so that only those B cells with receptors of the highest affinity can continue to capture antigen and interact effectively with helper T cells. Such B cells will therefore be selected to undergo further expansion and clonal differentiation and the antibodies they produce will dominate a secondary response, (middle panel). These higher affinity antibodies will in turn compete for antigen and select for the activation of B cells bearing receptors of still higher affinity in the tertiary response (bottom panel).

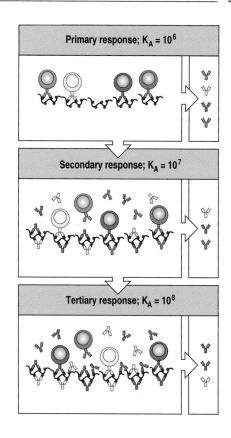

with sufficient avidity to compete with the preexisting antibody will take up the uncomplexed antigen, process it into peptide fragments, and present these peptides, bound to MHC class II molecules, to armed helper T cells surrounding and infiltrating the germinal centers (see Section 9-8). Contact between the B cells presenting antigenic peptides and armed helper T cells specific for the same peptide leads to an exchange of activating signals and the rapid proliferation of both activated antigen-specific B cells and helper T cells. Thus, only the higher-affinity memory B cells are efficiently stimulated in the secondary immune response. In this way, the affinity of the antibody produced rises progressively, as only B cells with high-affinity antigen receptors can bind antigen efficiently and be driven to proliferate by antigen-specific helper T cells (Fig. 10.26).

10-24 Memory T cells are increased in frequency and have distinct activation requirements and cell-surface proteins that distinguish them from armed effector T cells.

Because the T-cell receptor does not undergo isotype switching or affinity maturation, memory T cells have been far more difficult to characterize than memory B cells. Furthermore, it has proved hard to distinguish between effector T cells and memory T cells on the basis of their phenotype. After immunization, the number of T cells reactive to a given antigen increases markedly as effector T cells are produced, and then falls back to persist at a level significantly (100- to 1000-fold) above the initial frequency for the rest of the animal's or person's life (Fig. 10.27). These cells carry cell-surface proteins more characteristic of armed effector cells than of naive T cells. However, they are long-lived cells with distinct properties in terms of surface molecule expression, response to stimuli, and expression of genes that control cell survival. Therefore, they should be specifically designated memory T cells. In the case of B cells, the distinction between effector and memory cells is more obvious and has been recognized for some time because effector B cells, as we saw in Chapter 9, are terminally differentiated plasma cells that have already been activated to secrete antibody until they die.

A major problem in experiments aimed at establishing the existence of memory T cells is that most assays for T-cell effector function take several days, during which the putative memory T cells are reinduced to armed effector

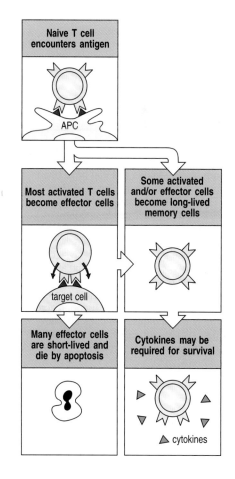

Fig. 10.27 Encounter with antigen generates effector T cells and long-lived memory T cells. On priming with antigen, a naive T cell divides and differentiates. Most of the progeny are relatively short-lived effector cells. However, some become long-lived memory T cells, which may be sustained by cytokines.

cell status. Thus, these assays do not distinguish preexisting effector cells from memory T cells. This problem does not apply to cytotoxic T cells, however, as cytotoxic effector T cells can program a target cell for lysis in 5 minutes. Memory CD8 T cells need to be reactivated to become cytotoxic, but they can do so without undergoing DNA synthesis, as shown by studies carried out in the presence of mitotic inhibitors. Recently, it has become possible to track particular clones of antigen-specific CD8 T cells by staining them with tetrameric MHC:peptide complexes (see Appendix I, Section A-28). It has been found that the number of antigen-specific CD8 T cells increases dramatically during an infection, and then drops by up to 100-fold; nevertheless, this final level is distinctly higher than before priming. These cells continue to express some markers characteristic of activated cells, like CD44, but stop expressing other activation markers, like CD69. In addition, they express more Bcl-2, a protein that promotes cell survival and may be responsible for the long half-life of memory CD8 cells. These cells are more sensitive to restimulation by antigen than are naive cells, and more quickly and more vigorously produce cytokines such as IFN-γ in response to such stimulation.

The issue is more difficult to address for CD4 T-cell responses, and the identification of memory CD4 T cells rests largely, but not entirely, on the existence of a long-lived population of cells that have surface characteristics of activated armed effector T cells (Fig. 10.28) but that are distinct from them in that they require additional restimulation before acting on target cells. Changes in three cell-surface proteins—L-selectin, CD44, and CD45—are particularly significant after exposure to antigen. L-selectin is lost on most memory CD4 T cells, whereas CD44 levels are increased on all memory T cells; these changes contribute to directing the migration of memory T cells from the blood into the tissues rather than directly into lymphoid tissues. The isoform of CD45 changes because of alternative splicing of exons that encode the extracellular domain of CD45 (Fig. 10.29), leading to isoforms that associate

Fig. 10.28 Many cell-surface molecules have altered expression on memory T cells. This is seen most clearly in the case of CD45, where there is a change in the isoforms expressed (see Fig. 10.29). Many of these changes are also seen on cells that have been activated to become armed effector T cells. The changes increase the adhesion of the T cell to antigen-presenting cells and to endothelial cells, particularly at sites of inflammation. They also increase the sensitivity of the memory T cell to antigen stimulation. The loss of CD62L on most or all memory T cells means that they are no longer able to enter lymph nodes across the high endothelial venules. Instead they migrate though the tissues and enter lymphoid compartments via the afferent lymph vessels.

Molecule	Other names	Relative expression on cells of indicated subset		Comments
		Naive	Memory	
LFA-3	CD58	1	>8	Ligand for CD2, involved in adhesion and signaling
CD2	T11	1	3	Mediates T-cell adhesion and activation
LFA-1	CD11a/CD18	1	3	Mediates leukocyte adhesion and signaling
α_4 integrin	VLA-4	1	4	Involved in T-cell homing to tissues
CD44	Ly24 Pgp-1	1	2	Lymphocyte homing to tissues
CD45RO		1	30	Lowest molecular weight isoform of CD45
CD45RA		10	1	High molecular weight isoform of CD45
L-selectin		High	Most low, some high	Lymph node homing receptor
CD3		1	1	Part of antigen-specific receptor complex

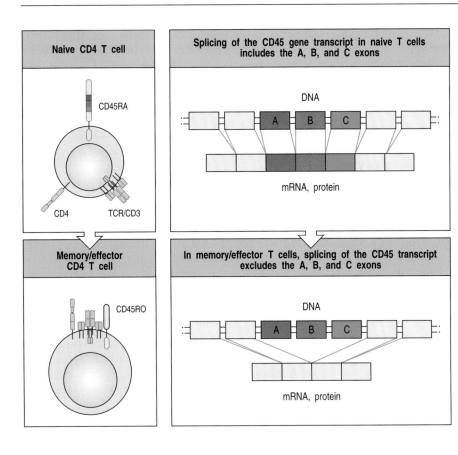

Fig. 10.29 Memory CD4 T cells express altered CD45 isoforms that regulate the interaction of the T-cell receptor with its co-receptors. CD45 is a transmembrane tyrosine phosphatase with three variable exons (A, B, and C) that encode part of its external domain. In naive T cells, high molecular weight isoforms (CD45RA) are found that do not associate with either the T-cell receptor (TCR/CD3) or co-receptors (CD4). In memory T cells, the variable exons are removed by alternative splicing of CD45 RNA, and this isoform, known as CD45RO, associates with both the T-cell receptor and the co-receptor. This receptor complex seems to transduce signals more effectively than the receptor on naive T cells.

Within the figure:

- Naive CD4 T cell — CD45RA, CD4, TCR/CD3
- Splicing of the CD45 gene transcript in naive T cells includes the A, B, and C exons — DNA; mRNA, protein
- Memory/effector CD4 T cell — CD45RO
- In memory/effector T cells, splicing of the CD45 transcript excludes the A, B, and C exons — DNA; mRNA, protein

with the T-cell receptor and facilitate antigen recognition. These changes are characteristic of cells that have been activated to become armed effector T cells (see Section 8-12), yet some of the cells on which these changes have occurred have many characteristics of resting CD4 T cells, suggesting that they represent memory CD4 T cells. Only after reexposure to antigen on an antigen-presenting cell do they achieve armed effector T-cell status, and acquire all the characteristics of armed T_H2 or T_H1 cells, secreting IL-4 and IL-5, or IFN-γ and TNF-β, respectively.

It thus seems reasonable to designate these cells as memory CD4 T cells, and to surmise that naive CD4 T cells can differentiate into armed effector T cells or into memory T cells that can later be activated to effector status. Recent experiments show that CD4 cells can differentiate into two types of memory cell, with distinct activation characteristics. One type are called effector memory cells because they can rapidly mature into effector CD4 T cells and secrete large amounts of IFN-γ, IL-4, and IL-5 early after restimulation. These cells lack the chemokine receptor CCR7 but express high levels of β-1 and β-2 integrins, as well as receptors for inflammatory chemokines. This profile suggests these effector memory cells are specialized for quickly entering inflamed tissues. The other type are called central memory cells. They express CCR7 and thus would be expected to recirculate more easily to T zones of secondary lymphoid tissues, as do naive T cells. These central memory cells are very sensitive to T-cell receptor cross-linking and quickly upregulate CD40L in response to it; however, they take longer to differentiate into effector cells and thus do not secrete as much cytokine as do effector memory cells early after restimulation. Interestingly, CD8 T cells can also be divided into analogous central and effector memory subsets.

As with memory CD8 T cells, the field will soon be revolutionized by direct staining of CD4 T cells with peptide:MHC class II oligomers (see Appendix I, Section A-28). This technique allows one not only to identify antigen-specific

CD4 T cells but also, using intracellular cytokine staining (see Appendix I, Section A-27), to determine whether they are T_H1 or T_H2 cells. These improvements in the identification and phenotyping of CD4 T cells will rapidly increase our knowledge of these hitherto mysterious cells, and could contribute valuable information on naive, memory, and effector CD4 T cells.

10-25 In immune individuals, secondary and subsequent responses are mediated solely by memory lymphocytes and not by naive lymphocytes.

In the normal course of an infection, a pathogen first proliferates to a level sufficient to elicit an adaptive immune response and then stimulates the production of antibodies and effector T cells that eliminate the pathogen from the body. Most of the armed effector T cells then die, and antibody levels gradually decline after the pathogen is eliminated, because the antigens that elicited the response are no longer present at the level needed to sustain it. We can think of this as feedback inhibition of the response. Memory T and B cells remain, however, and maintain a heightened ability to mount a response to a recurrence of infection with the same pathogen.

The antibody and memory T cells remaining in an immunized individual also prevent the activation of naive B and T cells by the same antigen. Such a response would be wasteful, given the presence of memory cells that can respond much more quickly. The suppression of naive lymphocyte activation can be shown by passively transferring antibody or memory T cells to naive recipients; when the recipient is then immunized, naive lymphocytes do not respond to the original antigen, but responses to other antigens are unaffected. This has been put to practical use to prevent the response of Rh⁻ mothers to their Rh⁺ children; if anti-Rh antibody is given to the mother before she reacts to her child's red blood cells, her response will be inhibited. The mechanism of this suppression is likely to involve the antibody-mediated clearance and destruction of the child's red blood cells, thus preventing naive B cells and T cells from mounting an immune response. Memory B-cell responses are not inhibited by antibody against the antigen, so the Rh⁻ mothers at risk must be identified and treated before a response has occurred. This is because memory B cells are much more sensitive, because of their high affinity and alterations in their B-cell receptor signaling requirements, to smaller amounts of antigen that cannot be efficiently cleared by the passive anti-Rh antibody. The ability of memory B cells to be activated to produce antibody even when exposed to preexisting antibody also allows secondary antibody responses to occur in individuals who are already immune.

Adoptive transfer of immune T cells (see Appendix I, Section A-42) to naive syngeneic mice also prevents the activation of naive T cells by antigen. This has been shown most clearly for cytotoxic T cells. It is possible that, once reactivated, the memory CD8 T cells regain cytotoxic activity sufficiently rapidly to kill the antigen-presenting cells that are required to activate naive CD8 T cells, thereby inhibiting the latter's activation.

These mechanisms might also explain the phenomenon known as **original antigenic sin**. This term was coined to describe the tendency of people to make antibodies only to epitopes expressed on the first influenza virus variant to which they are exposed, even in subsequent infections with variants that bear additional, highly immunogenic, epitopes (Fig. 10.30). Antibodies against the original virus will tend to suppress responses of naive B cells specific for the new epitopes. This might benefit the host by using only those B cells that can respond most rapidly and effectively to the virus. This pattern is broken only if the person is exposed to an influenza virus that lacks all epitopes seen in the original infection, as now no preexisting antibodies bind the virus and naive B cells are able to respond.

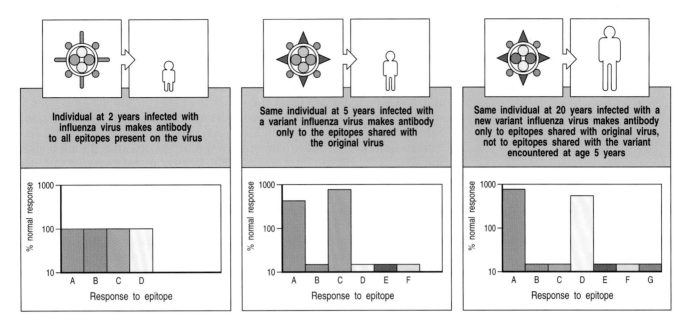

Individual at 2 years infected with
influenza virus makes antibody
to all epitopes present on the virus

Same individual at 5 years infected with
a variant influenza virus makes antibody
only to the epitopes shared with
the original virus

Same individual at 20 years infected with a
new variant influenza virus makes antibody
only to epitopes shared with original virus,
not to epitopes shared with the variant
encountered at age 5 years

Summary.

Protective immunity against reinfection is one of the most important consequences of adaptive immunity operating through the clonal selection of lymphocytes. Protective immunity depends not only on preformed antibody and armed effector T cells, but most importantly on the establishment of a population of lymphocytes that mediate long-lived immunological memory. The capacity of these cells to respond rapidly to restimulation with the same antigen can be transferred to naive recipients by primed B and T cells. The precise changes that distinguish naive, effector, and memory lymphocytes are now being characterized and, with the advent of receptor-specific reagents, the relative contributions of clonal expansion and differentiation to the memory phenotype are rapidly being clarified. Memory B cells can also be distinguished by changes in their immunoglobulin genes because of isotype switching and somatic hypermutation, and secondary and subsequent immune responses are characterized by antibodies with increasing affinity for the antigen.

Fig. 10.30 When individuals who have already been infected with one variant of influenza virus are infected with a second variant they make antibodies only to epitopes that were present on the initial virus. A child infected for the first time with an influenza virus at 2 years of age makes a response to all epitopes (left panel). At age 5 years, the same child exposed to a variant influenza virus responds preferentially to those epitopes shared with the original virus, and makes a smaller than normal response to new epitopes on the virus (middle panel). Even at age 20 years, this commitment to respond to epitopes shared with the original virus, and the subnormal response to new epitopes, is retained (right panel). This phenomenon is called 'original antigenic sin.'

Summary to Chapter 10.

Vertebrates resist infection by pathogenic microorganisms in several ways. The innate defenses can act immediately and may succeed in repelling the infection, but if not they are followed by a series of induced early responses which help to contain the infection as adaptive immunity develops. These first two phases of the immune response rely on recognizing the presence of infection using the nonclonotypic receptors of the innate immune system. They are covered in detail in Chapter 2, but are summarized in Fig. 10.31. Specialized subsets of T cells which may be viewed as intermediates between innate and adaptive immunity, since they bear a characteristic receptor rearrangement, are also important in determining the functional character of developing antigen-specific effector T cells. These include NK 1.1+ T cells which bias the CD4 T-cell response towards a T_H2 phenotype thereby promoting humoral immunity, and a subset of $\gamma.\delta$ T cells, particularly common in the gut, which secrete interferon-γ and therefore promote a T_H1 response. Both these subsets recognize CD1 molecules rather than MHC antigens. The mucosal immune system functions as a separate compartment of the peripheral immune system and has a distinctive anatomy and repertoire of lymphocytes. The gut-associated lymphoid tissue shows a remarkable ability to discriminate between different types of foreign antigens, responding differently to food antigens, innocuous gut flora, and pathogens, with only the latter provoking inflammation and a protective adaptive immune response. An adaptive immune response is mounted in the specialized lymphoid tissue that serves the particular site of infection and takes several days to develop, as T and B lymphocytes must encounter their specific antigen, proliferate, and differentiate into effector cells. T-cell dependent B-cell responses cannot be initiated until antigen-specific T cells have had a chance to proliferate and differentiate. Once an adaptive immune response has occurred, the antibodies and effector T cells are dispersed via the circulation and recruited into the infected tissues; the infection is usually controlled and the pathogen is contained or eliminated. The final effector mechanisms used to clear an infection depend on the type of infectious agent, and in most cases they are the same as those employed in the early phases of immune defense; only the recognition mechanism changes and is more selective (see Fig. 10.31). An effective adaptive immune response leads to a state of protective immunity. This state consists of the presence of effector cells and molecules produced in the initial response, and immunological memory. Immunological memory is manifest as a heightened ability to respond to pathogens that have been encountered previously and successfully eliminated. It is a property of memory T and B lymphocytes, which can transfer immune memory to naive recipients. The precise mechanism of immunological memory, which is arguably the most crucial feature of adaptive immunity, remains an active area of experimental science, and is now finally yielding its secrets. The artificial induction of protective immunity by vaccination, which includes immunological memory, is the most outstanding accomplishment of immunology in the field of medicine. Understanding of how this is accomplished is now catching up with its practical success. However, as we will see in Chapter 11, many pathogens do not induce protective immunity that completely eliminates the pathogen, so we will need to learn what prevents this before we can prepare effective vaccines against these pathogens.

Fig. 10.31 The components of the three phases of the immune response against different classes of microorganisms. The mechanisms of innate immunity that operate in the first two phases of the immune response have been covered in Chapter 2, while thymus-independent B-cell responses are covered in Chapter 9. The early phases contribute to the initiation of adaptive immunity, and influence the functional character of the antigen-specific effector T cells and antibodies that appear on the scene in the late phase of the response. There are striking similarities in the effector mechanisms at each phase of the response; the main change is in the recognition structures used.

	Phases of the immune response		
	Immediate (0–4 hours)	**Early (4–96 hours)**	**Late (after 96 hours)**
	Nonspecific Innate No memory No specific T cells	Nonspecific + specific Inducible No memory No specific T cells	Specific Inducible Memory Specific T cells
Barrier functions	Skin, epithelia	Local inflammation (C5a) Local TNF-α	IgA antibody in luminal spaces IgE antibody on mast cells Local inflammation
Response to extracellular pathogens	Phagocytes Alternative and MBL complement pathway	Mannan-binding lectin C-reactive protein T-independent B-cell antibody Complement	IgG antibody and Fc receptor-bearing cells IgG, IgM antibody + classical complement pathway
Response to intracellular bacteria	Macrophages	Activated NK-dependent macrophage activation IL-1, IL-6, TNF-α, IL-12	T-cell activation of macrophages by IFN-γ
Response to virus-infected cells	Natural killer (NK) cells	Interferon-α and -β IL-12-activated NK cells	Cytotoxic T cells IFN-γ

General references.

Brandtzaeg, P., Farstad, I.N., Johansen, F.E., Morton, H.C., Norderhaug, I.N., and Yamanaka, T.: **The B-cell system of human mucosae and exocrine glands.** *Immunol. Rev.* 1999, **171**:45-87.

Campbell, N., Yio, X.Y., So, L.P., Li, Y., and Mayer, L.: **The intestinal epithelial cell: processing and presentation of antigen to the mucosal immune system.** *Immunol. Rev.* 1999, **172**:315-324.

Fujihashi, K., Kweon, M.N., Kiyono, H., VanCott, J.L., van Ginkel, F.W., Yamamoto, M., and McGhee, J.R.: **A T cell/B cell/epithelial cell internet for mucosal inflammation and immunity.** *Springer Semin. Immunopathol.* 1997, **18**:477-494.

Section references.

10-1   The course of an infection can be divided into several distinct phases.

&

10-2   Infectious diseases are caused by diverse living agents that replicate in their hosts.

Mandell, G., Bennett, J., and Dolin, R., (eds): *Prinicipals and Practice of Infectious Diseases*, 5th edn. New York, Churchill Livingstone, 2000.

10-3   The nonspecific responses of innate immunity are necessary for an adaptive immune response to be initiated.

Fearon, D.T., and Carroll, M.C.: **Regulation of B lymphocyte responses to foreign and self-antigens by the CD19/CD21 complex.** *Annu. Rev. Immunol.* 2000, **18**:393-422.

Fearon, D.T., and Locksley, R.M.: **The instructive role of innate immunity in the acquired immune response.** *Science* 1996, **272**:50-53.

Janeway, C.A., Jr.: **The immune system evolved to discriminate infectious nonself from noninfectious self.** *Immunol. Today* 1992, **13**:11-16.

Medzhitov, R., and Janeway. C., Jr.: **Innate immunity.** *N. Engl. J. Med.* 2000, **343**:338-344.

10-4   An adaptive immune response is initiated when circulating T cells encounter their corresponding antigen in draining lymphoid tissues and become activated.

Finger, E.B., Purl, K.D., Alon, R., Lawrence, M.B., von Andrian, U.H., and Springer, T.A.: **Adhesion through L-selectin requires a threshold hydrodynamic shear.** *Nature* 1996, **379**:266-269.

Roake, J.A., Rao, A.S., Morris, P.J., Larson, C.P., Hankins, D.F., and Austyn, J.M.: **Dendritic cell loss from nonlymphoid tissues after systemic administration of lipopolysaccharide, tumor necrosis factor, and interleukin-1.** *J. Exp. Med.* 1995, **181**:2237-2247.

Shaw, S., Ebnet, K., Kaldjian, E.P., and Anderson, A.O.: **Orchestrated information transfer underlying leukocyte:endothelial interactions.** *Annu. Rev. Immunol.* 1996, **14**:155-177.

10-5   Cytokines made in the early phases of an infection influence the functional differentiation of CD4 T cells.

Bendelac, A., Rivera, M.N., Park, S.H., and Roark, J.H.: **Mouse CD1-specific NK1 T cells: development, specificity, and function.** *Annu. Rev. Immunol.* 1997, **15**:535-562.

Finkelman, F.D., Shea-Donohue, T., Goldhill, J., Sullivan, C.A., Morris, S.C., Madden, K.B., Gauser, W.C., and Urban, J.F., Jr.: **Cytokine regulation of host defense against parasitic intestinal nemotodes.** *Annu. Rev. Immunol.* 1997, **15**:505-533.

Hsieh, C.S., Macatonia, S.E., Tripp, C.S., Wolf, S.F., O'Garra, A., and Murphy, K.M.: **Development of T$_H$1 CD4$^+$ T cells through IL-12 produced by Listeria-induced macrophages.** *Science* 1993, **260**:547-549.

Moser, M., and Murphy, K.M.: **Dendritic cell regulation of T$_H$1-T$_H$2 development.** *Nature Immunol.* 2000, **1**:199-205.

10-6   Distinct subsets of T cells can regulate the growth and effector functions of other T-cell subsets.

Croft, M., Carter, L., Swain, S.L., and Dutton, R.W.: **Generation of polarized antigen-specific CD8 effector populations: reciprocal action of interleukin-4 and IL-12 in promoting type 2 versus type 1 cytokine profiles.** *J. Exp. Med.* 1994, **180**:1715-1728.

Nakamura, T., Kamogawa, Y., Bottomly, K., and Flavell, R.A.: **Polarization of IL-4- and IFN-gamma-producing CD4$^+$ T cells following activation of naive CD4$^+$ T cells.** *J. Immunol.* 1997, **158**:1085-1094.

Seder, R.A., and Paul, W.E.: **Acquistion of lymphokine producing phenotype by CD4$^+$ T cells.** *Annu. Rev. Immunol.* 1994, **12**:635-673.

10-7   The nature and amount of antigenic peptide can also affect the differentiation of CD4 T cells.

Constant, S.L., and Bottomly, K.: **Induction of Th1 and Th2 CD4$^+$ T cell responses: the alternative approaches.** *Annu. Rev. Immunol.* 1997, **15**:297-322.

Wang, L.F., Lin J.Y., Hsieh, K.H., and Lin, R.H.: **Epicutaneous exposure of protein antigen induces a predominant T$_H$2-like response with high IgE production in mice.** *J. Immunol.* 1996, **156**:4079-4082.

Wang, R., Abrams, S.I., Loh, D.Y., Hsieh, C.S., Murphy, K.M., and Russell, J.H.: **Separation of CD4$^+$ functional responses by peptide dose in Th1 and Th2 subsets expressing the same transgenic antigen receptor.** *Cell Immunol.* 1993, **148**:357-370.

10-8   Armed effector T cells are guided to sites of infection by chemokines and newly expressed adhesion molecules.

MacKay, C.R., Marston, W., and Dudler, L.: **Altered patterns of T-cell migration through lymph nodes and skin following antigen challenge.** *Eur. J. Immunol.* 1992, **22**:2205-2210.

Romanic, A.M., Graesser, D., Baron, J.L., Visintin, I., Janeway, C.A., Jr., and Madri, J.A.: **T cell adhesion to endothelial cells and extracellular matrix is modulated upon transendothelial cell migration.** *Lab. Invest.* 1997, **76**:11-23.

Sallusto, F., Kremmer, E., Palermo, B., Hoy, A., Ponath, P., Qin, S., Forster, R., Lipp, M., and Lanzavecchia, A.: **Switch in chemokine receptor expression upon TCR stimulation reveals novel homing potential for recently activated T cells.** *Eur. J. Immunol.* 1999, **29**:2037-2045.

10-9   Antibody responses develop in lymphoid tissues under the direction of armed helper T cells.

Jacob, J., Kassir, R., and Kelsoe, G.: ***In situ* studies of the primary immune response to (4-hydroxy-3-nitrophenyl)acetyl. I. the architecture and dynamics of responding cell population.** *J. Exp. Med.* 1991, **173**:1165-1175.

Kelsoe, G., and Zheng, B.: **Sites of B-cell activation *in vivo*.** *Curr. Opin. Immunol.* 1993, **5**:418-422.

Liu, Y.J., Zhang, J., Lane, P.J., Chan, E.Y., and MacLennan, I.C.: **Sites of specific B cell activation in primary and secondary responses to T cell-dependent and T cell-independent antigens.** *Eur. J. Immunol.* 1991, **21**:2951-2962.

MacLennan, I.C.M.: **Germinal centres.** *Annu. Rev. Immunol.* 1994, **12**:117-139.

10-10   Antibody responses are sustained in medullary cords and bone marrow.

Benner, R., Hijmans, W., and Haaijman, J.J.: **The bone marrow: the major source of serum immunoglobulins, but still a neglected site of antibody formation.** *Clin. Exp. Immunol.* 1981, **46**:1-8.

Manz, R.A., Thiel, A., and Radbruch, A.: **Lifetime of plasma cells in the bone marrow.** *Nature* 1997, **388**:133-134.

Slifka, M.K., Antia, R., Whitmire, J.K., and Ahmed, R.: **Humoral immunity due to long-lived plasma cells.** *Immunity* 1998, **8**:363-372.

Takahashi, Y., Dutta, P.R., Cerasoli, D.M., and Kelsoe, G.: *In situ* **studies of the primary immune response to (4-hydroxy-3- nitrophenyl)acetyl. V. Affinity maturation develops in two stages of clonal selection.** *J. Exp. Med.* 1998, **187**:885-895.

10-11 The effector mechanisms used to clear an infection depend on the infectious agent.

Mims, C.A.: *The Pathogenesis of Infectious Disease*, 3rd edn. London, Academic Press, 1987.

10-12 Resolution of an infection is accompanied by the death of most of the effector cells and the generation of memory cells.

Murali-Krishna, K., Altman, J.D., Suresh, M., Sourdive, D.J., Zajac, A.J., Miller, J.D., Slansky, J., and Ahmed, R.: **Counting antigen-specific CD8 T cells: a reeval-uation of bystander activation during viral infection.** *Immunity* 1998, **8**:177-187.

Webb, S., Hutchinson, J., Hayden, K., and Sprent, J.: **Expansion/deletion of mature T cells exposed to endogenous superantigens** *in vivo. J. Immunol.* 1994, **152**:586-597.

10-13 Mucosa-associated lymphoid tissue is located in anatomically defined microcompartments throughout the gut.

Brandtzaeg, P., Farstad, I.N., and Haraldsen, G.: **Regional specialization in the mucosal immune system: primed cells do not always home along the same track.** *Immunol. Today* 1999, **20**:267-277.

Hein, W.R.: **Organization of mucosal lymphoid tissue.** *Curr. Top. Microbiol. Immunol.* 1999, **236**:1-15.

Mowat, A.M., and Viney, J.L.: **The anatomical basis of intestinal immunity.** *Immunol. Rev.* 1997, **156**:145-166.

Neutra, M.R., Frey, A., and Kraehenbuhl, J.P.: **Epithelial M cells: gateways for mucosal infection and immunization.** *Cell* 1996, **86**:345-348.

10-14 The mucosal immune system contains a distinctive repertoire of lymphocytes.

Blumberg, R.S.: **Current concepts in mucosal immunity. II. One size fits all: nonclassical MHC molecules fulfill multiple roles in epithelial cell function.** *Am. J. Physiol.* 1998, **274**:G227-G231.

Hayday, A.C.: **[gamma][delta] cells: a right time and a right place for a conserved third way of protection.** *Annu. Rev. Immunol.* 2000, **18**:975-1026.

Park, S.H., Guy-Grand, D., Lemonnier, F.A., Wang, C.R., Bendelac, A., and Jabri, B.: **Selection and expansion of CD8alpha/alpha(1) T cell receptor alpha/beta(1) intestinal intraepithelial lymphocytes in the absence of both classical major histocompatibility complex class I and nonclassical CD1 molecules.** *J. Exp. Med.* 1999, **190**:885-890.

Porcelli, S.A., and Modlin, R.L.: **The CD1 system: antigen-presenting molecules for T cell recognition of lipids and glycolipids.** *Annu. Rev. Immunol.* 1999, **17**:297-329.

Roberts, S.J., Smith, A.L., West, A.B., Wen, L., Findly, R.C., Owen, M.J., and Hayday, A.C.: **T-cell alpha beta+ and gamma delta+ deficient mice display abnormal but distinct phenotypes toward a natural, widespread infection of the intestinal epithelium.** *Proc. Natl. Acad. Sci. USA* 1996, **93**:11774-11779.

10-15 Secretory IgA is the antibody isotype associated with the mucosal immune system.

Brandtzaeg, P.: **Molecular and cellular aspects of the secretory immunoglobulin system.** *APMIS* 1995, **103**:1-19.

Corthesy, B., and Kraehenbuhl, J.P.: **Antibody-mediated protection of mucosal surfaces.** *Curr. Top. Microbiol. Immunol.* 1999, **236**:93-111.

Lamm, M.E.: **Current concepts in mucosal immunity. IV. How epithelial transport of IgA antibodies relates to host defense.** *Am. J. Physiol.* 1998, **274**:G614-G617.

Mostov, K.E.: **Transepithelial transport of immunoglobulins.** *Annu. Rev. Immunol.* 1994, **12**:63-84.

10-16 Most antigens presented to the mucosal immune system induce tolerance.

Strobel, S., and Mowat, A.M.: **Immune responses to dietary antigens: oral tolerance.** *Immunol. Today* 1998, **19**:173-181.

10-17 The mucosal immune system can mount an immune response to the normal bacterial flora of the gut.

Berg, R.D.: **Bacterial translocation from the gastrointestinal tract.** *Adv. Exp. Med. Biol.* 1999, **473**:11-30.

Berg, R.D.: **The indigenous gastrointestinal microflora.** *Trends Microbiol.* 1996, **4**:430-435.

Bos, N.A., Kimura, H., Meeuwsen, C.G., De Visser, H., Hazenberg, M.P., Wostmann, B.S., Pleasants, J.R., Benner, R., and Marcus, D.M.: **Serum immunoglobulin levels and naturally occurring antibodies against carbohydrate antigens in germ-free BALB/c mice fed chemically defined ultrafiltered diet.** *Eur. J. Immunol.* 1989, **19**:2335-2339.

Kelly, C.P., and LaMont, J.T.: *Clostridium difficile* **infection.** *Annu. Rev. Med.* 1998, **49**:375-390.

10-18 Enteric pathogens cause a local inflammatory response and the development of protective immunity.

Holmes, K.V., Tresnan, D.B., and Zelus, B.D.: **Virus-receptor interactions in the enteric tract. Virus-receptor interactions.** *Adv. Exp. Med. Biol.* 1997, **412**:125-133.

Jepson, M.A., and Clark, M.A.: **Studying M cells and their role in infection.** *Trends Microbiol.* 1998, **6**:359-365.

Sansonetti, P.J., and Phalipon, A.: **M cells as ports of entry for enteroinvasive pathogens: mechanisms of interaction, consequences for the disease process.** *Semin. Immunol.* 1999, **11**:193-203.

Siebers, A., and Finlay, B.B.: **M cells and the pathogenesis of mucosal and systemic infections.** *Trends Microbiol.* 1996, **4**:22-29.

Vazquez-Torres, A., and Fang, F.C.: **Cellular routes of invasion by enteropathogens.** *Curr. Opin. Microbiol.* 2000, **3**:54-59.

10-19 Infection by *Helicobacter pylori* **causes a chronic inflammatory response, which may cause peptic ulcers, carcinoma of the stomach, and unusual lymphoid tumors.**

Blaser, M.J.: *Helicobacter pylori* **and gastric diseases.** *BMJ* 1998, **316**:1507-1510.

Isaacson, P.G.: **Gastric MALT lymphoma: from concept to cure.** *Ann. Oncol.* 1999, **10**:637-645.

Telford, J.L., Covacci, A., Rappuoli, R., and Chiara, P.: **Immunobiology of** *Helicobacter pylori* **infection.** *Curr. Opin. Immunol.* 1997, **9**:498-503.

Wotherspoon, A.C.: *Helicobacter pylori* **infection and gastric lymphoma.** *Br. Med. Bull.* 1998, **54**:79-85.

10-20 In the absence of inflammatory stimuli, the normal response of the mucosal immune system to foreign antigens is tolerance.

Chen, Y.H., and Weiner, H.L.: **Dose-dependent activation and deletion of antigen-specific T cells following oral tolerance.** *Ann. N.Y. Acad. Sci.* 1996, **778**:111-121.

Czerkinsky, C., Anjuere, F., McGhee, J.R., George-Chandy, A., Holmgren, J., Kieny, M.P., Fujiyashi, K., Mestecky, J.F., Pierrefite-Carle, V., Rask, C., and Sun, J.B.: **Mucosal immunity and tolerance: relevance to vaccine development.** *Immunol. Rev.* 1999, **170**:197-222.

Friedman, A.: **Induction of anergy in Th1 lymphocytes by oral tolerance. Importance of antigen dosage and frequency of feeding.** *Ann. N.Y. Acad. Sci.* 1996, **778**:103-110.

Levings, M.K., and Roncarolo, M.G.: **T-regulatory 1 cells: a novel subset of CD4 T cells with immunoregulatory properties.** *J. Allergy Clin. Immunol.* 2000, **106**:S109-S112.

MacDonald, T.T.: **Effector and regulatory lymphoid cells and cytokines in mucosal sites.** *Curr. Top. Microbiol. Immunol.* 1999, **236**:113-135.

Mayer, L.: **Current concepts in mucosal immunity. I. Antigen presentation in the intestine: new rules and regulations.** *Am. J. Physiol.* 1998, **274**:G7-G9.

Van Houten, N., and Blake, S.F.: **Direct measurement of anergy of antigen-specific T cells following oral tolerance induction.** *J. Immunol.* 1996, **157**:1337-1341.

10-21   Immunological memory is long-lived after infection or vaccination.

Black, F.L., and Rosen, L.: **Patterns of measles antibodies in residents of Tahiti and their stability in the absence of re-exposure.** *J. Immunol.* 1962, **88**:725-731.

Ku, C.C., Murakami, M., Sakamoto, A., Kappler, J., and Marrack, P.: **Control of homeostasis of CD8**[+] **memory T cells by opposing cytokines.** *Science* 2000, **288**:675-678.

Murali-Krishna, K., Lau, L.L., Sambhara, S., Lemonnier, F., Altman, J., and Ahmed, R.: **Persistence of memory CD8 T cells in MHC class I-deficient mice.** *Science* 1999, **286**:1377-1381.

Sprent, J., Tough, D.F., and Sun, S.: **Factors controlling the turnover of T memory cells.** *Immunol. Rev.* 1997, **156**:79-85.

Swain, S.L., Hu, H., and Huston, G.: **Class II-independent generation of CD4 memory T cells from effectors.** *Science* 1999, **286**:1381-1383.

10-22   Both clonal expansion and clonal differentiation contribute to immunological memory in B cells.

Berek, C., and Milstein, C.: **Mutation drift and repertoire shift in the maturation of the immune response.** *Immunol. Rev.* 1987, **96**:23-41.

Cumano, A., and Rajewsky, K.: **Clonal recruitment and somatic mutation in the generation of immunological memory to the hapten NP.** *EMBO J.* 1986, **5**:2459-2468.

McHeyzer-Williams, M.G., and Ahmed, R.: **B cell memory and the long-lived plasma cell.** *Curr. Opin. Immunol.* 1999, **11**:172-179.

Tarlinton, D.: **Germinal centers: form and function.** *Curr. Opin. Immunol.* 1998, **10**:245-251.

10-23   Repeated immunizations lead to increasing affinity of antibody owing to somatic hypermutation and selection by antigen in germinal centers.

Berek, C., Jarvis, J.M., and Milstein, C.: **Activation of memory and virgin B cell clones in hyperimmune animals.** *Eur. J. Immunol.* 1987, **17**:1121-1129.

Siskind, G.W., Dunn, P., and Walker, J.G.: **Studies on the control of antibody synthesis: II. Effect of antigen dose and of suppression by passive antibody on the affinity of antibody synthesized.** *J. Exp. Med.* 1968, **127**:55-66.

Szakal, A.K., Kosco, M.H., and Tew, J.G.: **Microanatomy of lymphoid tissue during humoral immune responses: structure function relationships.** *Annu. Rev. Immunol.* 1989, **7**:91-109.

10-24   Memory T cells are increased in frequency and have distinct activation requirements and cell-surface proteins that distinguish them from armed effector T cells.

Bradley, L.M., Atkins, G.G., and Swain, S.L.: **Long-term CD4**[+] **memory T cells from the spleen lack MEL-14, the lymph node homing receptor.** *J. Immunol.* 1992, **148**:324.

Michie, C.A., McLean, A., Alcock, C., and Beverly, P.C.L.: **Lifespan of human lymphocyte subsets defined by CD45 isoforms.** *Nature* 1992, **360**:264-265.

Novak, T.J., Farber, D., Leitenberg, D., Hong, S., Johnson, P., and Bottomly, K.: **Isoforms of the transmembrane tyrosine phosphatase CD45 differentially affect T-cell recognition.** *Immunity* 1994, **1**:81-92.

Rogers, P.R., Dubey, C., and Swain, S.L.: **Qualitative changes accompany memory T cell generation: faster, more effective responses at lower doses of antigen.** *J. Immunol.* 2000, **164**:2338-2346.

Sallusto, F., Lenig, D., Forster, R., Lipp, M., and Lanzavecchia, A.: **Two subsets of memory T lymphocytes with distinct homing potentials and effector functions.** *Nature* 1999, **401**:708-712.

10-25   In immune individuals, secondary and subsequent responses are mediated solely by memory lymphocytes and not by naive lymphocytes.

Fazekas de St Groth, B., and Webster, R.G.: **Disquisitions on original antigenic sin. I. Evidence in man.** *J. Exp. Med.* 1966, **140**:2893-2898.

Fridman, W.H.: **Regulation of B cell activation and antigen presentation by Fc receptors.** *Curr. Opin. Immunol.* 1993, **5**:355-360.

Pollack, W., Gorman, J.G., Freda, V.J., Ascari, W.Q., Allen, A.E., and Baker, W.J.: **Results of clinical trials of RhoGAm in women.** *Transfusion* 1968, **8**:151-153.

PART V

THE IMMUNE SYSTEM IN HEALTH AND DISEASE

Failures of Host Defense Mechanisms

In the normal course of an infection, the infectious agent triggers an innate immune response that causes symptoms, followed by an adaptive immune response that clears the infection and establishes a state of protective immunity. This does not always happen, however, and in this chapter we will examine three circumstances in which there are failures of host defense against infection: avoidance or subversion of a normal immune response by the pathogen; inherited failures of defense because of gene defects; and the acquired immune deficiency syndrome (AIDS), a generalized susceptibility to infection that is itself due to the failure of the host to control and eliminate the human immunodeficiency virus (HIV).

The propagation of a pathogen depends on its ability to replicate in a host and to spread to new hosts. Common pathogens must therefore grow without activating too vigorous an immune response and, conversely, must not kill the host too quickly. The most successful pathogens persist either because they do not elicit an immune response, or because they evade the response once it has occurred. Over millions of years of coevolution with their hosts, pathogens have developed various strategies for avoiding destruction by the immune system, and we have encountered some of them in earlier chapters. In the first part of this chapter we will examine these in more detail, and discuss some that have not yet been mentioned.

In the second part of the chapter we will turn to the **immunodeficiency diseases**, in which host defense fails. In most of these diseases, a defective gene results in the elimination of one or more components of the immune system, leading to heightened susceptibility to infection with particular classes of pathogen. Immunodeficiency diseases caused by defects in T- or B-lymphocyte development, phagocyte function, and components of the complement system have all been discovered. Finally, we will consider how the persistent infection of immune system cells by the human immunodeficiency virus, HIV, leads to the acquired immune deficiency syndrome, AIDS. The analysis of all these diseases has already made an important contribution to our understanding of host defense mechanisms and, in the longer term, might help to provide new methods of controlling or preventing infectious diseases, including AIDS.

Pathogens have evolved various means of evading or subverting normal host defenses.

Just as vertebrates have developed many different defenses against pathogens, so pathogens have evolved elaborate strategies to evade these defenses. Many pathogens use one or more of these strategies to evade the immune system. At the end of this chapter we will see how HIV succeeds in defeating the immune response by using several of them in combination.

11-1   Antigenic variation allows pathogens to escape from immunity.

One way in which an infectious agent can evade immune surveillance is by altering its antigens; this is particularly important for extracellular pathogens, against which the principal defense is the production of antibody against their surface structures. There are three ways in which **antigenic variation** can occur. First, many infectious agents exist in a wide variety of antigenic types. There are, for example, 84 known types of *Streptococcus pneumoniae*, an important cause of bacterial pneumonia. Each type differs from the others in the structure of its polysaccharide capsule. The different types are distinguished by serological tests and so are often known as **serotypes**. Infection with one serotype of such an organism can lead to type-specific immunity, which protects against reinfection with that type but not with a different serotype. Thus, from the point of view of the adaptive immune system, each serotype of *S. pneumoniae* represents a distinct organism. The result is that essentially the same pathogen can cause disease many times in the same individual (Fig. 11.1).

Fig. 11.1 Host defense against *Streptococcus pneumoniae* is type specific. The different strains of *S. pneumoniae* have antigenically distinct capsular polysaccharides. The capsule prevents effective phagocytosis until the bacterium is opsonized by specific antibody and complement, allowing phagocytes to destroy it. Antibody to one type of *S. pneumoniae* does not cross-react with the other types, so an individual immune to one type has no protective immunity to a subsequent infection with a different type. An individual must generate a new adaptive immune response each time he or she is infected with a different type of *S. pneumoniae*.

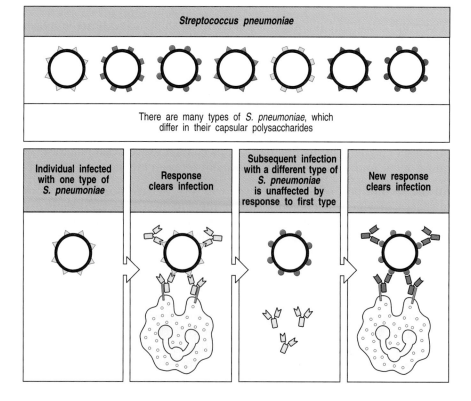

A second, more dynamic mechanism of antigenic variation is seen in the influenza virus. At any one time, a single virus type is responsible for most infections throughout the world. The human population gradually develops protective immunity to this virus type, chiefly by directing neutralizing antibody against the major surface protein of the influenza virus, its hemagglutinin. Because the virus is rapidly cleared from individual hosts, its survival depends on having a large pool of unprotected individuals among whom it spreads very readily. The virus might therefore be in danger of running out of potential hosts if it had not evolved two distinct ways of changing its antigenic type (Fig. 11.2).

The first of these, **antigenic drift**, is caused by point mutations in the genes encoding hemagglutinin and a second surface protein, neuraminidase. Every 2–3 years, a variant arises with mutations that allow the virus to evade neutralization by antibodies in the population; other mutations affect epitopes that are recognized by T cells and, in particular, CD8 T cells, so that cells infected with the mutant virus also escape destruction. Individuals who were previously infected with, and hence are immune to, the old variant are thus susceptible to the new variant. This causes an epidemic that is relatively mild because there is still some cross-reaction with antibodies and T cells produced against the previous variant of the virus, and therefore most of the population have some level of immunity (see Section 10-25).

Major influenza pandemics resulting in widespread and often fatal disease occur as the result of the second process, which is termed **antigenic shift**. This happens when there is reassortment of the segmented RNA genome of the influenza virus and related animal influenza viruses in an animal host, leading to major changes in the hemagglutinin protein on the viral surface. The resulting virus is recognized poorly, if at all, by antibodies and by T cells directed against the previous variant, so that most people are highly susceptible to the new virus, and severe infection results.

The third mechanism of antigenic variation involves programmed rearrangements in the DNA of the pathogen. The most striking example occurs in African trypanosomes, where changes in the major surface antigen occur repeatedly within a single infected host. Trypanosomes are insect-borne protozoa that replicate in the extracellular tissue spaces of the body and cause sleeping sickness in humans. The trypanosome is coated with a single type of glycoprotein, the variant-specific glycoprotein (VSG), which elicits a potent protective antibody response that rapidly clears most of the parasites. The trypanosome genome, however, contains about 1000 VSG genes, each encoding a protein with distinct antigenic properties. Only one of these is

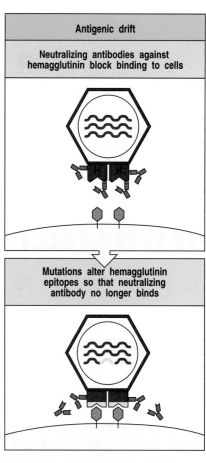

Antigenic drift

Neutralizing antibodies against hemagglutinin block binding to cells

Mutations alter hemagglutinin epitopes so that neutralizing antibody no longer binds

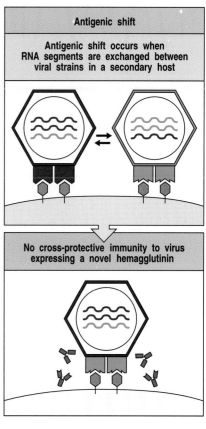

Antigenic shift

Antigenic shift occurs when RNA segments are exchanged between viral strains in a secondary host

No cross-protective immunity to virus expressing a novel hemagglutinin

Fig. 11.2 Two types of variation allow repeated infection with type A influenza virus. Neutralizing antibody that mediates protective immunity is directed at the viral surface protein hemagglutinin (H), which is responsible for viral binding to and entry into cells. Antigenic drift (top panels) involves the emergence of point mutants that alter the binding sites for protective antibodies on the hemagglutinin. When this happens, the new virus can grow in a host that is immune to the previous strain of virus. However, as T cells and some antibodies can still recognize epitopes that have not been altered, the new variants cause only mild disease in previously infected individuals. Antigenic shift (lower panels) is a rare event involving reassortment of the segmented RNA viral genomes of two different influenza viruses, probably in a bird. These antigen-shifted viruses have large changes in their hemagglutinin molecule and therefore T cells and antibodies produced in earlier infections are not protective. These shifted strains cause severe infection that spreads widely, causing the influenza pandemics that occur every 10–50 years. (There are eight RNA molecules in each viral genome but for simplicity only three are shown.)

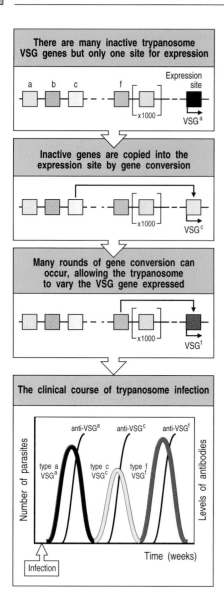

There are many inactive trypanosome VSG genes but only one site for expression

Inactive genes are copied into the expression site by gene conversion

Many rounds of gene conversion can occur, allowing the trypanosome to vary the VSG gene expressed

The clinical course of trypanosome infection

Fig. 11.3 Antigenic variation in trypanosomes allows them to escape immune surveillance. The surface of a trypanosome is covered with a variant-specific glycoprotein (VSG). Each trypanosome has about 1000 genes encoding different VSGs, but only the gene in a specific expression site within the telomere at one end of the chromosome is active. Although several genetic mechanisms have been observed for changing the VSG gene expressed, the usual mechanism is gene conversion. An inactive gene, which is not at the telomere, is copied and transposed into the telomeric expression site, where it becomes active. When an individual is first infected, antibodies are raised against the VSG initially expressed by the trypanosome population. A small number of trypanosomes spontaneously switch their VSG gene to a new type, and while the host antibody eliminates the initial variant, the new variant is unaffected. As the new variant grows, the whole sequence of events is repeated.

expressed at any one time by being placed into an active 'expression site' in the genome. The VSG gene expressed can be changed by gene rearrangement that places a new VSG gene into the expression site (Fig. 11.3). So, by having their own system of gene rearrangement that can change the VSG protein produced, trypanosomes keep one step ahead of an immune system capable of generating many distinct antibodies by gene rearrangement. A few trypanosomes with such changed surface glycoproteins thus evade the antibodies made by the host, and these soon grow and cause a recurrence of disease (see Fig. 11.3, bottom panel). Antibodies are then made against the new VSG, and the whole cycle repeats. This chronic cycle of antigen clearance leads to immune-complex damage and inflammation, and eventually to neurological damage, finally resulting in coma. This gives African trypanosomiasis its common name of sleeping sickness. These cycles of evasive action make trypanosome infections very difficult for the immune system to defeat, and they are a major health problem in Africa. Malaria is another major disease caused by a protozoan parasite that varies its antigens to evade elimination by the immune system.

Antigenic variation also occurs in bacteria: DNA rearrangements help to account for the success of two important bacterial pathogens—*Salmonella typhimurium*, a common cause of salmonella food poisoning, and *Neisseria gonorrhoeae*, which causes gonorrhea, a major sexually transmitted disease and an increasing public health problem in the United States. *S. typhimurium* regularly alternates its surface flagellin protein by inverting a segment of its DNA containing the promoter for one flagellin gene. This turns off expression of the gene and allows the expression of a second flagellin gene, which encodes an antigenically distinct protein. *N. gonorrhoeae* has several variable antigens, the most striking of which is the pilin protein, which, like the variable surface glycoproteins of the African trypanosome, is encoded by several variant genes, only one of which is active at any given time. Silent versions of the gene from time to time replace the active version downstream of the pilin promoter. All of these mechanisms help the pathogen to evade an otherwise specific and effective immune response.

11-2   Some viruses persist *in vivo* by ceasing to replicate until immunity wanes.

Viruses usually betray their presence to the immune system once they have entered cells by directing the synthesis of viral proteins, fragments of which are displayed on the surface MHC molecules of the infected cell, where they are detected by T lymphocytes. To replicate, a virus must make viral proteins, and rapidly replicating viruses that produce acute viral illnesses are therefore readily detected by T cells, which normally control them. Some viruses, however, can enter a state known as **latency** in which the virus is not being replicated. In the latent state, the virus does not cause disease but, because there are no viral peptides to flag its presence, the virus cannot be eliminated. Such latent infections can be reactivated and this results in recurrent illness.

Herpes viruses often enter latency. Herpes simplex virus, the cause of cold sores, infects epithelia and spreads to sensory neurons serving the area of infection. After an effective immune response controls the epithelial infection, the virus persists in a latent state in the sensory neurons. Factors such as sunlight, bacterial infection, or hormonal changes reactivate the virus, which then travels down the axons of the sensory neuron and reinfects the epithelial tissues (Fig. 11.4). At this point, the immune response again becomes active and controls the local infection by killing the epithelial cells, producing a new sore. This cycle can be repeated many times. There are two reasons why the sensory neuron remains infected: first, the virus is quiescent in the nerve and therefore few viral proteins are produced, generating few virus-derived peptides to present on MHC class I; second, neurons carry very low levels of MHC class I molecules, which makes it harder for CD8 T cells to recognize infected neurons and attack them. This low level of MHC class I expression might be beneficial, as it reduces the risk that neurons, which regenerate very slowly if at all, will be attacked inappropriately by CD8 T cells. It also makes neurons unusually vulnerable to persistent infections. Another example of this is provided by herpes zoster (or varicella zoster), the virus that causes chickenpox. This virus remains latent in one or a few dorsal root ganglia after the acute illness is over and can be reactivated by stress or immunosuppression to spread down the nerve and reinfect the skin. The reinfection causes the reappearance of the classic rash of varicella in the area of skin served by the infected dorsal root, a disease commonly called shingles. Herpes simplex reactivation is frequent, but herpes zoster usually reactivates only once in a lifetime in an immunocompetent host.

The Epstein–Barr virus (EBV), yet another herpes virus, enters latency in B cells after a primary infection that often passes without being diagnosed. In a minority of infected individuals, the initial acute infection of B cells is more severe, causing a disease known as infectious mononucleosis or glandular fever. EBV infects B cells by binding to CR2 (CD21), a component of the B-cell co-receptor complex. The infection causes most of the infected cells to proliferate and produce virus, leading in turn to the proliferation of antigen-specific T cells and the excess of mononuclear white cells in the blood that gives the disease its name. The infection is controlled eventually by specific CD8 T cells, which kill the infected proliferating B cells. A fraction of B lymphocytes become latently infected, however, and EBV remains quiescent in these cells. Latently infected cells express a viral protein, EBNA-1, which is needed to maintain the viral genome, but EBNA-1 interacts with the proteasome (see Section 5-3) to prevent its own degradation into peptides that would elicit a T-cell response.

Latently infected B cells can be isolated by taking B cells from individuals who have apparently cleared their EBV infection and placing them in tissue culture: in the absence of T cells, the latently infected cells that have retained the EBV genome transform infected B cells sometimes undergo malignant transformation, giving rise to a B-cell lymphoma called Burkitt's lymphoma (see Section 7-33). This is a rare event, and it seems likely that a crucial part of this process is a failure of T-cell surveillance. Further support for this hypothesis comes from the increased risk of EBV-associated B-cell lymphomas developing in patients with acquired and inherited immunodeficiencies of T-cell function (see Sections 11-15 and 11-26).

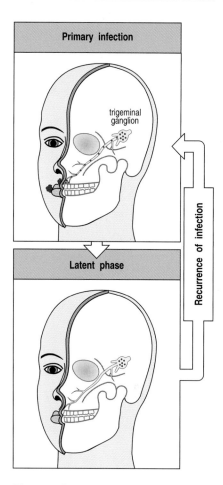

Fig. 11.4 Persistence and reactivation of herpes simplex virus infection. The initial infection in the skin is cleared by an effective immune response, but residual infection persists in sensory neurons such as those of the trigeminal ganglion, whose axons innervate the lips. When the virus is reactivated, usually by some environmental stress and/or alteration in immune status, the skin in the area served by the nerve is reinfected from virus in the ganglion and a new cold sore results. This process can be repeated many times.

Acute Infectious Mononucleosis

11-3 Some pathogens resist destruction by host defense mechanisms or exploit them for their own purposes.

Some pathogens induce a normal immune response but have evolved specialized mechanisms for resisting its effects. For instance, some bacteria that are engulfed in the normal way by macrophages have evolved means of avoiding destruction by these phagocytes; indeed, they use macrophages as their primary host. *Mycobacterium tuberculosis*, for example, is taken up by macrophages but prevents the fusion of the phagosome with the lysosome, protecting itself from the bactericidal actions of the lysosomal contents.

Other microorganisms, such as *Listeria monocytogenes*, escape from the phagosome into the cytoplasm of the macrophage, where they can multiply readily. They then spread to adjacent cells in tissues without emerging from the cell into the extracellular environment. They do this by hijacking the host cytoskeletal protein actin, which assembles into filaments at the rear of the bacterium. The actin filaments drive the bacteria forward into vacuolar projections to adjacent cells; these vacuoles are then lysed by the *Listeria*, releasing the bacteria directly into the cytoplasm of the adjacent cell. In this way they avoid attack by antibodies. Cells infected with *L. monocytogenes* are, however, susceptible to killing by cytotoxic T cells. The protozoan parasite *Toxoplasma gondii* can apparently generate its own vesicle, which isolates it from the rest of the cell because it does not fuse with any cellular vesicle. This might actually enable *T. gondii* to avoid making peptides derived from its proteins accessible for loading onto MHC molecules, and thus remain invisible to the immune system.

Two prominent spirochetal infections, **Lyme disease** and syphilis, avoid elimination by antibodies through less well understood mechanisms and establish a persistent and extremely damaging infection in tissues. Lyme disease is caused by the spirochete bacterium *Borrelia burgdorferi*, whereas syphilis, the more widespread and much the better understood of the two diseases, is caused by *Treponema pallidum*. *T. pallidum* is believed to avoid recognition by antibodies by coating its surface with host molecules until it has invaded tissues such as the central nervous system, where it is less easily reached by antibodies.

Finally, many viruses have evolved mechanisms to subvert various arms of the immune system. These range from capturing cellular genes for cytokines or cytokine receptors, to synthesizing complement-regulatory molecules, inhibiting MHC class I synthesis or assembly, or producing decoy proteins that mimic so-called TIR domains that we learned about in Section 6-15. This area is one of the most rapidly expanding areas in the field of host–pathogen relationships. Examples of how members of the herpes and poxvirus families subvert host responses are shown in Fig. 11.5.

11-4 Immunosuppression or inappropriate immune responses can contribute to persistent disease.

Toxic Shock
Syndrome

Many pathogens suppress immune responses in general. For example, staphylococci produce toxins, such as the **staphylococcal enterotoxins** and **toxic shock syndrome toxin-1**, that act as superantigens. Superantigens are proteins that bind the antigen receptors of very large numbers of T cells (see Section 7-26), stimulating them to produce cytokines that cause significant suppression of all immune responses. The details of this suppression are not understood. The stimulated T cells proliferate and then rapidly undergo apoptosis, leaving a generalized immunosuppression together with the deletion of many peripheral T cells.

Viral strategy	Specific mechanism	Result	Virus examples
Inhibition of humoral immunity	Virally encoded Fc receptor	Blocks effector functions of antibodies bound to infected cells	Herpes simplex Cytomegalovirus
	Virally encoded complement receptor	Blocks complement-mediated effector pathways	Herpes simplex
	Virally encoded complement control protein	Inhibits complement activation of infected cell	Vaccinia
Inhibition of inflammatory response	Virally encoded cytokine homologue, e.g., β-chemokine receptor	Sensitizes infected cells to effects of β-chemokine; advantage to virus unknown	Cytomegalovirus
	Virally encoded soluble cytokine receptor, e.g., IL-1 receptor homologue, TNF receptor homologue, interferon-γ receptor homologue	Blocks effects of cytokines by inhibiting their interaction with host receptors	Vaccinia Rabbit myxoma virus
	Viral inhibition of adhesion molecule expression, e.g., LFA-3 ICAM-1	Blocks adhesion of lymphocytes to infected cells	Epstein–Barr virus
	Protection from NFκB activation by short sequences that mimic TLRs	Blocks inflammatory responses elicited by IL-1 or bacterial pathogens	Vaccinia
Blocking of antigen processing and presentation	Inhibition of MHC class I expression	Impairs recognition of infected cells by cytotoxic T cells	Herpes simplex Cytomegalovirus
	Inhibition of peptide transport by TAP	Blocks peptide association with MHC class I	Herpes simplex
Immunosuppression of host	Virally encoded cytokine homologue of IL-10	Inhibits T_H1 lymphocytes Reduces interferon-γ production	Epstein–Barr virus

Fig. 11.5 Mechanisms of subversion of the host immune system by viruses of the herpes and pox families.

Many other pathogens cause mild or transient immunosuppression during acute infection. These forms of suppressed immunity are poorly understood but important, as they often make the host susceptible to secondary infections by common environmental microorganisms. A crucially important example of immune suppression follows trauma, burns, or even major surgery. The burned patient has a clearly diminished capability to respond to infection, and generalized infection is a common cause of death in these patients. The reasons for this are not fully understood.

Measles virus infection, in spite of the widespread availability of an effective vaccine, still accounts for 10% of the global mortality of children under 5 years old and is the eighth leading cause of death worldwide. Malnourished children are the main victims and the cause of death is usually secondary bacterial infection, particularly pneumonia caused by measles-induced immunosuppression. The immunosuppression that follows measles infection can last for several months and is associated with reduced T- and B-cell function. There is reduced or absent delayed-type hypersensitivity and, during this period of acquired immunodeficiency, children have markedly increased susceptibility to mycobacterial infection, reflecting the important role of macrophage activation by T_H1 cells in host defense against mycobacteria. An important mechanism for measles-induced immunosuppression is the infection of dendritic cells by measles virus. Infected dendritic cells cause unresponsiveness of T lymphocytes by mechanisms that are not yet understood, and it seems likely that this is the proximate cause of the immunosuppression induced by measles virus.

The most extreme case of immune suppression caused by a pathogen is the acquired immune deficiency syndrome caused by infection with HIV. The ultimate cause of death in AIDS is usually infection with an **opportunistic pathogen**, a term used to describe a microorganism that is present in the environment but does not usually cause disease because it is well controlled by normal host defenses. HIV infection leads to a gradual loss of immune competence, allowing infection with organisms that are not normally pathogenic.

Leprosy, which we discussed in Section 8-13, is a more complex case, in which the causal bacterium, *Mycobacterium leprae*, is associated either with the suppression of cell-mediated immunity or with a strong cell-mediated antibacterial response. This leads to two major forms of the disease—lepromatous and tuberculoid leprosy. In **lepromatous leprosy**, cell-mediated immunity is profoundly depressed, *M. leprae* are present in great profusion, and cellular immune responses to many antigens are suppressed. This leads to a phenotypic state in such patients called **anergy**, here meaning the absence of delayed-type hypersensitivity to a wide range of antigens unrelated to *M. leprae*. In **tuberculoid leprosy**, by contrast, there is potent cell-mediated immunity with macrophage activation, which controls but does not eradicate infection. Few viable microorganisms are found in tissues, the patients usually survive, and most of the symptoms and pathology are caused by the inflammatory response to these persistent microorganisms (Fig. 11.6). The difference between the two forms of disease might lie in a difference in the ratio of T_H1 to T_H2 cells, and this is thought to be caused by cytokines produced by CD8 T cells, as we learned in Section 10-6.

Lepromatous
Leprosy

11-5   Immune responses can contribute directly to pathogenesis.

Tuberculoid leprosy is just one example of an infection in which the pathology is caused largely by the immune response. This is true to some degree in most infections; for example, the fever that accompanies a bacterial infection is caused by the release of cytokines by macrophages. One medically important example of immunopathology is the wheezy broncheolitis caused by **respiratory syncytial virus** (**RSV**). Broncheolitis caused by RSV is the major cause of admission of young children to hospital in the Western world, with as many as 90,000 admissions and 4500 deaths each year in the United States alone. The first indication that the immune response to the virus might have a role in the pathogenesis of this disease came from the observation that young infants vaccinated with an alum-precipitated killed virus preparation suffered a worse disease than unvaccinated children. This occurred because the vaccine failed to induce neutralizing antibodies but succeeded in producing T_H2 cells. On infection, the T_H2 cells released interleukin (IL)-3, IL-4, and IL-5, which induced bronchospasm, increased mucus secretion, and tissue eosinophilia. Mice can be infected with RSV and develop a disease similar to that seen in humans.

Another example of a pathogenic immune response is the response to the eggs of the schistosome. Schistosomes are parasitic worms that lay eggs in the hepatic portal vein. Some of the eggs reach the intestine and are shed in the feces, spreading the infection; others lodge in the portal circulation of the liver, where they elicit a potent immune response leading to chronic inflammation,

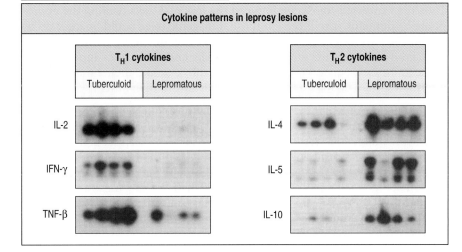

Infection with *Mycobacterium leprae* can result in different clinical forms of leprosy

There are two polar forms, tuberculoid and lepromatous leprosy, but several intermediate forms also exist

Tuberculoid leprosy	Lepromatous leprosy
Organisms present at low to undetectable levels	Organisms show florid growth in macrophages
Low infectivity	High infectivity
Granulomas and local inflammation. Peripheral nerve damage	Disseminated infection. Bone, cartilage, and diffuse nerve damage
Normal serum immunoglobulin levels	Hypergammaglobulinemia
Normal T-cell responsiveness. Specific response to *M. leprae* antigens	Low or absent T-cell responsiveness. No response to *M. leprae* antigens

Cytokine patterns in leprosy lesions

	T$_H$1 cytokines			T$_H$2 cytokines	
	Tuberculoid	Lepromatous		Tuberculoid	Lepromatous
IL-2			IL-4		
IFN-γ			IL-5		
TNF-β			IL-10		

Fig. 11.6 T-cell and macrophage responses to *Mycobacterium leprae* are sharply different in the two polar forms of leprosy. Infection with *M. leprae*, which stain as small dark red dots in the photographs, can lead to two very different forms of disease. In tuberculoid leprosy (left), growth of the organism is well controlled by T$_H$1-like cells that activate infected macrophages. The tuberculoid lesion contains granulomas and is inflamed, but the inflammation is local and causes only local effects, such as peripheral nerve damage. In lepromatous leprosy (right), infection is widely disseminated and the bacilli grow uncontrolled in macrophages; in the late stages of disease there is major damage to connective tissues and to the peripheral nervous system. There are several intermediate stages between these two polar forms. The cytokine patterns in the two polar forms of the disease are sharply different, as shown by the analysis of RNA isolated from lesions of four patients with lepromatous leprosy and four patients with tuberculoid leprosy (Northern blot, lower panel). Cytokines typically produced by T$_H$2 cells (IL-4, IL-5, and IL-10) dominate in the lepromatous form, whereas cytokines produced by T$_H$1 cells (IL-2, IFN-γ, and TNF-β) dominate in the tuberculoid form. It therefore seems that T$_H$1-like cells predominate in tuberculoid leprosy, and T$_H$2-like cells in lepromatous leprosy. IFN-γ would be expected to activate macrophages, enhancing killing of *M. leprae*, whereas IL-4 can actually inhibit the induction of bactericidal activity in macrophages. High levels of IL-4 would also explain the hypergamma-globulinemia observed in lepromatous leprosy. The determining factors in the initial induction of T$_H$1- or T$_H$2-like cells are suspected to be so-called T$_C$1 or T$_C$2 cells, by analogy to T$_H$1 and T$_H$2 cells. The mechanism for the anergy or generalized loss of effective cell-mediated immunity in lepromatous leprosy is not understood. Photographs courtesy of G. Kaplan; cytokine patterns courtesy of R.L. Modlin.

hepatic fibrosis, and eventually liver failure. This process reflects the excessive activation of T$_H$1 cells, and can be modulated by T$_H$2 cells, IL-4, or CD8 T cells, which can also act by producing IL-4.

In the case of the **mouse mammary tumor virus** (**MMTV**), a retrovirus that causes mammary tumors in mice, the immune response is required for the infective cycle of the pathogen (Fig. 11.7). MMTV is transferred from the

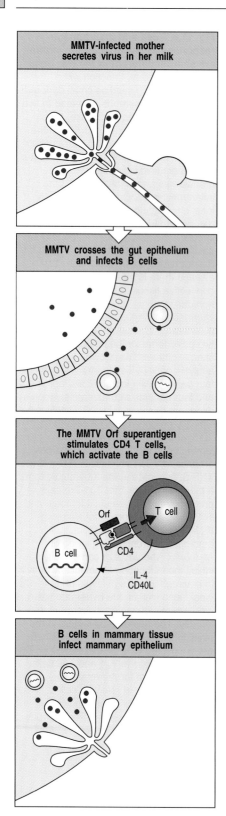

MMTV-infected mother secretes virus in her milk

MMTV crosses the gut epithelium and infects B cells

The MMTV Orf superantigen stimulates CD4 T cells, which activate the B cells

Orf

T cell

B cell

CD4

IL-4
CD40L

B cells in mammary tissue infect mammary epithelium

Fig. 11.7 Activation of T cells by the MMTV superantigen in mice is crucial for the virus life cycle. MMTV is transferred from mother to pup in milk, and crosses the gut epithelium to reach the lymphoid tissue of its new host and thus infect B lymphocytes. The superantigen encoded by MMTV, called the Orf or open reading frame, is expressed on the surface of the B cell and binds to appropriate T-cell receptor V_β domains on CD4 T cells. The superantigen also has binding sites for MHC class II molecules, so that a complex between superantigen, MHC molecule, T-cell receptor, and CD4 is formed, activating the T cell. The activated T cell produces the cytokine IL-4 and the cell-surface molecule CD40 ligand, and in turn activates the B cell to divide. This allows the virus to replicate within the B cell and subsequently to infect the mammary epithelium.

mother's mammary gland to her pups in milk. The virus then enters the B lymphocytes of the new host, where it must replicate to be transported to the mammary epithelium to continue its life cycle. As it is a retrovirus, however, MMTV can replicate only in dividing cells. The virus ensures that infected B cells will proliferate by causing them to express on their surface a super-antigen encoded within the MMTV genome. This superantigen enables the B cells to bypass the requirement for specific antigen and stimulate large numbers of CD4 T cells with the appropriate T-cell receptor V_β domain (see Section 5-15), causing them to produce cytokines and express CD40 ligand, which in turn stimulates the B cells to divide. The virus can then replicate in the B cells and infect the host's mammary epithelial cells.

One way to block this cycle of transmission is by deleting the particular subset of T cells carrying the V_β domain recognized by the viral superantigen. This has been done experimentally by taking mice that are normally susceptible to a particular MMTV virus, and using the superantigen gene from this virus to construct transgenic mice. As we learned in Section 7-26, superantigens that are expressed in the thymus induce the clonal deletion of developing T cells. Thus the expressed transgene induced the loss of T cells bearing the appropriate V_β domains. The B cells in these transgenic mice could be infected by the MMTV virus but could not activate any of the remaining T cells. Thus the infected B cells were not stimulated to divide, and could not support MMTV replication. Consequently, the transgenic mice, unlike their nontransgenic littermates, were unable to transmit the relevant strain of MMTV.

This mode of protection against MMTV might explain the finding that most mouse strains have MMTV genomes stably integrated into their DNA. These defective endogenous retroviruses have lost certain essential genes and are unable to produce virions, but they have retained the genes encoding their superantigens, which are expressed on the cells of the host. Although a section of the T-cell repertoire is lost as a result of carrying these endogenous retroviruses, the mice are protected against infection with nondefective MMTV encoding the same superantigen. There are several different strains of MMTV whose superantigens bind to different V_β domains, and these are matched by different endogenous MMTV strains. Mice containing different endogenous MMTV genomes delete different parts of their T-cell receptor repertoire, reducing the risk that whole mouse populations will be susceptible to a given MMTV strain. No human diseases dependent on such mechanisms have yet been described.

Summary.

Infectious agents can cause recurrent or persistent disease by avoiding normal host defense mechanisms or by subverting them to promote their own replication. There are many different ways of evading or subverting the immune response. Antigenic variation, latency, resistance to immune effector mechanisms, and suppression of the immune response all contribute to persistent and medically important infections. In some cases, the immune response is part of the problem; some pathogens use immune activation to spread infection, others would not cause disease if it were not for the immune response. Each of these mechanisms teaches us something about the nature of the immune response and its weaknesses, and each requires a different medical approach to prevent or to treat infection.

Inherited immunodeficiency diseases.

Immunodeficiencies occur when one or more components of the immune system is defective. The commonest cause of immune deficiency worldwide is malnutrition; however, in developed countries, most immunodeficiency diseases are inherited, and these are usually seen in the clinic as recurrent or overwhelming infections in very young children. Less commonly, acquired immunodeficiencies with causes other than malnutrition can manifest later in life. Although the pathogenesis of many of these acquired disorders has remained obscure, some are caused by known agents, such as drugs or irradiation that damage lymphocytes, or infection with measles or HIV. By examining which infections accompany a particular inherited or acquired immunodeficiency, we can see which components of the immune system are important in the response to particular infectious agents. The inherited immunodeficiency diseases also reveal how interactions between different cell types contribute to the immune response and to the development of T and B lymphocytes. Finally, these inherited diseases can lead us to the defective gene, often revealing new information about the molecular basis of immune processes and providing the necessary information for diagnosis, for genetic counseling, and eventually for gene therapy.

11-6   A history of repeated infections suggests a diagnosis of immunodeficiency.

Patients with immune deficiency are usually detected clinically by a history of recurrent infection. The type of infection is a guide to which part of the immune system is deficient. Recurrent infection by pyogenic bacteria suggests a defect in antibody, complement, or phagocyte function, reflecting the role of these parts of the immune system in host defense against such infections. By contrast, a history of recurrent viral infections is more suggestive of a defect in host defense mediated by T lymphocytes.

To determine the competence of the immune system in patients with possible immunodeficiency, a battery of tests is usually conducted (Fig. 11.8); these focus with increasing precision as the nature of the defect is narrowed down to a single element. The presence of the various cell types in blood is determined by routine hematology, often followed by FACS analysis (see Appendix I, Section A-22) of lymphocyte subsets, and the measurement of serum immunoglobulins. The phagocytic competence of freshly isolated polymorphonuclear leukocytes and monocytes is tested, and the efficiency of the complement system is determined by testing the dilution of serum required for lysis of 50% of antibody-coated red blood cells (this is denoted the CH_{50}).

In general, if such tests reveal a defect in one of these broad compartments of immune function, more specialized testing is then needed to determine the precise nature of the defect. Tests of lymphocyte function are often valuable,

Fig. 11.8 Evaluation of immune competence.

Evaluation of the cellular components of the human immune system			
	B cells	**T cells**	**Phagocytes**
Normal numbers (x10^9 per liter of blood)	Approximately 0.3	Total 1.0–2.5 CD4 0.5–1.6 CD8 0.3–0.9	Monocytes 0.15–0.6 Polymorphonuclear leukocytes Neutrophils 3.00–5.5 Eosinophils 0.05–0.25 Basophils 0.02
Measurement of function *in vivo*	Serum Ig levels Specific antibody levels	Skin test	—
Measurement of function *in vitro*	Induced antibody production in response to pokeweed mitogen	T-cell proliferation in response to phytohemagglutinin or to tetanus toxoid	Phagocytosis Nitro blue tetrazolium uptake Intracellular killing of bacteria
Specific defects	See Fig. 11.9	See Fig. 11.9	See Fig. 11.9

Evaluation of the humoral components of the human immune system				
	Immunoglobulins			**Complement**
Component	IgG	IgM	IgA	IgE	
Normal levels in adults	600–1400 mg dl^{-1}	40–345 mg dl^{-1}	60–380 mg dl^{-1}	0–200 IU ml^{-1}	CH$_{50}$ of 125–300 IU ml^{-1}

starting with the ability of polyclonal mitogens to induce T-cell proliferation and B-cell secretion of immunoglobulin in tissue culture (see Appendix I, Section A-31). These tests can eventually pinpoint the cellular defect in immunodeficiency.

11-7   Inherited immunodeficiency diseases are caused by recessive gene defects.

Before the advent of highly effective antibiotic therapy, it is likely that most individuals with inherited immune defects died in infancy or early childhood because of their susceptibility to particular classes of pathogen (Fig. 11.9). Such cases would not have been easy to identify, as many normal infants also died of infection. Thus, although many inherited immunodeficiency diseases have now been identified, the first immunodeficiency disease was not described until 1952. Most of the gene defects that cause these inherited immunodeficiencies are recessive and, for this reason, many of the known immunodeficiencies are caused by mutations in genes on the X chromosome. Recessive defects cause disease only when both chromosomes are defective. However, as males have only one X chromosome, all males who inherit an X chromosome carrying a defective gene will manifest disease, whereas female carriers, having two X chromosomes, are perfectly healthy. Immunodeficiency diseases that affect various steps in B- and T-lymphocyte development have been described, as have defects in surface molecules that are important for T- or B-cell function. Defects in phagocytic cells, in complement, in cytokines, in cytokine receptors, and in molecules that mediate effector responses also occur (see Fig. 11.9). Thus, immunodeficiency can be caused by defects in either the adaptive or the innate immune system. Individual examples of these diseases will be described in later sections.

More recently, the use of gene knockout techniques in mice has allowed the creation of many immunodeficient states that are adding rapidly to our knowledge of the contribution of individual molecules to normal immune function. Nevertheless, human immunodeficiency disease is still the best source of insight into normal pathways of host defense against infectious

Name of deficiency syndrome	Specific abnormality	Immune defect	Susceptibility
Severe combined immune deficiency	ADA deficiency	No T or B cells	General
	PNP deficiency	No T or B cells	General
	X-linked *scid*, γ_c chain deficiency	No T cells	General
	Autosomal *scid* DNA repair defect	No T or B cells	General
DiGeorge's syndrome	Thymic aplasia	Variable numbers of T and B cells	General
MHC class I deficiency	TAP mutations	No CD8 T cells	Chronic lung and skin inflammation
MHC class II deficiency	Lack of expression of MHC class II	No CD4 T cells	General
Wiskott–Aldrich syndrome	X-linked; defective WASP gene	Defective anti-polysaccharide antibody and impaired T cell activation responses	Encapsulated extracellular bacteria
X-linked agamma-globulinemia	Loss of Btk tyrosine kinase	No B cells	Extracellular bacteria, viruses
X-linked hyper-IgM syndrome	Defective CD40 ligand	No isotype switching	Extracellular bacteria *Pneumocystis carinii* *Cryptosporidium parvum*
Common variable immunodeficiency	Unknown; MHC-linked	Defective IgA and IgG production	Extracellular bacteria
Selective IgA	Unknown; MHC-linked	No IgA synthesis	Respiratory infections
Phagocyte deficiencies	Many different	Loss of phagocyte function	Extracellular bacteria and fungi
Complement deficiencies	Many different	Loss of specific complement components	Extracellular bacteria especially *Neisseria* spp.
Natural killer (NK) cell defect	Unknown	Loss of NK function	Herpes viruses
X-linked lympho-proliferative syndrome	SH2D1A mutant	Inability to control B cell growth	EBV-driven B cell tumors
Ataxia telangiectasia	Gene with PI 3-kinase homology	T cells reduced	Respiratory infections
Bloom's syndrome	Defective DNA helicase	T cells reduced Reduced antibody levels	Respiratory infections

Fig. 11.9 Human immunodeficiency syndromes. The specific gene defect, the consequence for the immune system, and the resulting disease susceptibilities are listed for some common and some rare human immunodeficiency syndromes. ADA, adenosine deaminase; PNP, purine nucleotide phosphorylase; TAP, transporters associated with antigen processing; WASP, Wiskott–Aldrich syndrome protein; EBV, Epstein–Barr virus; NK, natural killer.

diseases in humans. For example, a deficiency of antibody, of complement, or of phagocytic function each increases the risk of infection by certain pyogenic bacteria. This shows that the normal pathway of host defense against such bacteria is the binding of antibody followed by fixation of complement, which allows the uptake of opsonized bacteria by phagocytic cells. Breaking any one of the links in this chain of events leading to bacterial killing causes a similar immunodeficient state.

The study of immunodeficiency also teaches us about the redundancy of mechanisms of host defense against infectious disease. The first two humans to be discovered with a hereditary deficiency of complement were healthy immunologists. This teaches us two lessons. The first is that there are multiple protective immune mechanisms against infection; for example, although there is abundant evidence that complement deficiency increases susceptibility to pyogenic infection, not every human with complement deficiency suffers from recurrent infections. The second lesson concerns the phenomenon of **ascertainment artifact**. When an unusual observation is made in a patient with disease, there is a temptation to seek a causal link. However, no one would suggest that complement deficiency causes a genetic predisposition to becoming an immunologist. Complement deficiency was discovered in immunologists because they used their own blood in their experiments. If a particular measurement is made only in a highly selected group of patients with a particular disease, it is inevitable that the only abnormal results will be discovered in patients with that disease. This is an ascertainment artifact and emphasizes the importance of studying appropriate controls.

11-8   The main effect of low levels of antibody is an inability to clear extracellular bacteria.

Pyogenic, or pus-forming, bacteria have polysaccharide capsules that are not directly recognized by the receptors on macrophages and neutrophils that stimulate phagocytosis. They therefore escape immediate elimination by the innate immune response and are successful extracellular pathogens. Normal individuals can clear infections by such bacteria because antibody and complement opsonize the bacteria, making it possible for phagocytes to ingest and destroy them. The principal effect of deficiencies in antibody production is therefore a failure to control this class of bacterial infection. In addition, susceptibility to some viral infections, most notably those caused by enteroviruses, is also increased, because of the importance of antibodies in neutralizing infectious viruses that enter the body through the gut (see Chapter 10).

X-linked
Agammaglobulinemia

The first description of an immunodeficiency disease was Ogden C. Bruton's account, in 1952, of the failure of a male child to produce antibody. As this defect is inherited in an X-linked fashion and is characterized by the absence of immunoglobulin in the serum, it was called **Bruton's X-linked agammaglobulinemia** (**XLA**). The absence of antibody can be detected using immunoelectrophoresis (Fig. 11.10). Since then, many more diseases of antibody production have been described, most of them the consequence of failures in the development or activation of B lymphocytes. Infants with these diseases are usually identified as a result of recurrent infections with pyogenic bacteria such as *Streptococcus pneumoniae*.

The defective gene in XLA is now known to encode a protein tyrosine kinase called Btk (Bruton's tyrosine kinase) which is a member of the recently described family called Tec kinases (see Section 6-10). This protein is expressed in neutrophils as well as in B cells, although only B cells are defective in these patients, in whom B-cell maturation halts at the pre-B-cell stage.

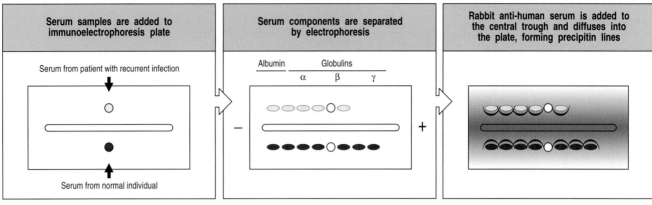

Fig. 11.10 Immunoelectrophoresis reveals the absence of several distinct immunoglobulin isotypes in serum from a patient with X-linked agammaglobulinemia (XLA). Serum samples from a normal control and from a patient with recurrent bacterial infection caused by the absence of antibody production, as reflected in an absence of gamma globulins, are separated by electrophoresis on an agar-coated slide. Antiserum raised against whole normal human serum and containing antibodies against many of its different proteins is put in a trough down the middle; each antibody forms an arc of precipitation with the protein it recognizes. The position of each arc is determined by the electrophoretic mobility of the serum protein; immunoglobulins migrate to the gamma globulin region of the gel. The absence of immunoglobulins in a patient who has X-linked agammaglobulinemia is shown in the photograph at the bottom, where several arcs are missing from the patient's serum (upper set). These are IgA, IgM, and several subclasses of IgG, each recognized in normal serum (lower set) by antibodies in the antiserum against human serum proteins. Photograph from the collection of the late C.A. Janeway Snr.

Thus it is likely that Btk is required to couple the pre-B-cell receptor (which consists of heavy chains, surrogate light chains, and Igα and Igβ) to nuclear events that lead to pre-B-cell growth and differentiation (see Section 7-9). In patients with Btk deficiencies, some B cells mature despite the defect in the signaling kinase, suggesting that signals transmitted by these kinases are not absolutely required.

As the gene responsible for XLA is found on the X chromosome, it is possible to identify female carriers by analyzing X-chromosome inactivation in their B cells. During development, female cells randomly inactivate one of their two X chromosomes. Because the product of a normal *btk* gene is required for normal B-lymphocyte development, only cells in which the normal allele of *btk* is active can develop into mature B cells. Thus, in female carriers of mutant *btk* genes, all B cells have the normal X chromosome as the active X. By contrast, the active X chromosomes in the T cells and macrophages of carriers are an equal mixture of the normal and *btk* mutant X chromosomes. This fact allowed female carriers of XLA to be identified even before the nature of *btk* was known. Nonrandom X inactivation only in B cells also demonstrates conclusively that the *btk* gene is required for normal B-cell development but not for the development of other cell types, and that Btk must act within B cells rather than on stromal cells or other cells required for B-cell development (Fig. 11.11).

The commonest humoral immune defect is the transient deficiency in immunoglobulin production that occurs in the first 6–12 months of life. The newborn infant has initial antibody levels comparable to those of the mother, because of the transplacental transport of maternal IgG (see Chapter 9). As the transferred IgG is catabolized, antibody levels gradually decrease until the infant begins to produce useful amounts of its own IgG at about 6 months

Fig. 11.11 The product of the *btk* gene is important for B-cell development. In X-linked agamma-globulinemia (XLA), a protein tyrosine kinase of the Tec family called Btk, encoded on the X chromosome, is defective. In normal individuals, B-cell development proceeds through a stage in which the pre-B-cell receptor consisting of μ:λ5:Vpre-B transduces a signal via Btk, triggering further B-cell development. In males with XLA, no signal can be transduced and, although the pre-B-cell receptor is expressed, the B cells develop no further. In female mammals, including humans, one of the two X chromosomes in each cell is permanently inactivated early in development. Because the choice of which chromosome to inactivate is random, half of the pre-B cells in a carrier female will have inactivated the chromosome with the wild-type *btk*. This means they can express only the defective *btk* gene, and cannot develop further. Therefore, in the carrier, mature B cells always have the nondefective X chromosome active. This is in sharp contrast to all other cell types, which have the nondefective X chromosome active in only half of the B cells. Nonrandom X chromosome inactivation in a particular cell lineage is a clear indication that the product of the X-linked gene is required for the development of cells of that lineage. It is also sometimes possible to identify the stage at which the gene product is required, by detecting the point in development at which X-chromosome inactivation develops bias. Using this kind of analysis, one can identify carriers of X-linked traits such as XLA without needing to know the nature of the mutant gene.

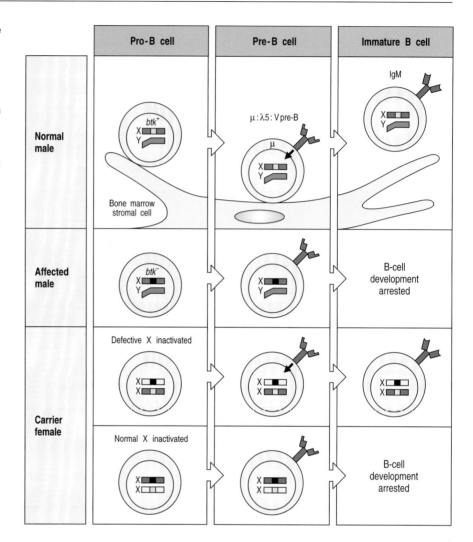

of age (Fig. 11.12). Thus, IgG levels are quite low between the ages of 3 months and 1 year and active IgG antibody responses are poor. In some infants this can lead to a period of heightened susceptibility to infection. This is especially true for premature babies, who begin with lower levels of maternal IgG and also reach immune competence later after birth.

The most common inherited form of immunoglobulin deficiency is selective IgA deficiency, which is seen in about 1 person in 800. Although no obvious disease susceptibility is associated with selective IgA defects, they are commoner in people with chronic lung disease than in the general population. Lack of IgA might thus result in a predisposition to lung infections with various pathogens and is consistent with the role of IgA in defense at the body's surfaces. The genetic basis of this defect is unknown but some data suggest that a gene of unidentified function mapping in the class III region of the MHC could be involved. A related syndrome called **common variable immunodeficiency**, in which there is usually a deficiency in both IgG and IgA, also maps to the MHC region.

People with pure B-cell defects resist many pathogens successfully. However, effective host defense against a subset of extracellular pyogenic bacteria, including staphylococci and streptococci, requires opsonization of these bacteria with specific antibody. These infections can be suppressed with

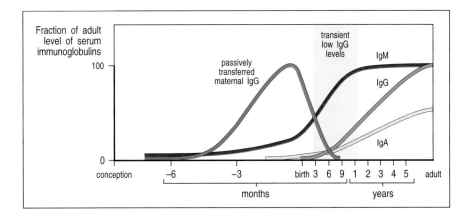

Fraction of adult
level of serum
immunoglobulins

100

0

conception   −6   −3   birth 3 6 9   1 2 3 4 5   adult

months   years

passively
transferred
maternal IgG

transient
low IgG
levels

IgM

IgG

IgA

Fig. 11.12 Immunoglobulin levels in newborn infants fall to low levels around 6 months of age. Newborn babies have high levels of IgG, transported across the placenta from the mother during gestation. After birth, the production of IgM starts almost immediately; the production of IgG, however, does not begin for about 6 months, during which time the total level of IgG falls as the maternally acquired IgG is catabolized. Thus, IgG levels are low from about the age of 3 months to 1 year, which can lead to susceptibility to disease.

antibiotics and periodic infusions of human immunoglobulin collected from a large pool of donors. As there are antibodies against many pathogens in this pooled immunoglobulin, it serves as a fairly successful shield against infection.

11-9   T-cell defects can result in low antibody levels.

Patients with **X-linked hyper IgM syndrome** have normal B- and T-cell development and high serum levels of IgM but make very limited IgM antibody responses against T-cell dependent antigens and produce immunoglobulin isotypes other than IgM and IgD only in trace amounts. This makes them highly susceptible to infection with extracellular pathogens. The molecular defect in this disease is in the CD40 ligand expressed on activated T cells, which therefore cannot engage the CD40 molecule on B cells; the B cells themselves are normal. We learned in Chapter 9 that CD40 ligand is critical in the T-cell dependent activation of B-cell proliferation and these patients show that CD40 ligand is also essential for the induction of the isotype switch and formation of germinal centers (Fig. 11.13). There are also defects in cell-mediated immunity in these individuals. For example, they are susceptible to infection with the opportunistic lung pathogen *Pneumocystis carinii*, which is normally killed by activated macrophages. The susceptibility is thought to be due, at least in part, to the inability of the T cells to deliver an activating signal to infected macrophages by engaging the CD40 expressed on these cells (see Section 8-29). A defect in T-cell activation could also contribute to the profound immunodeficiency suffered by these patients, as studies on mice that lack CD40 ligand have revealed a failure of antigen-specific T cells to expand in response to primary immunization with antigen.

In XLA, the hunt for the cause of the disease led to the discovery of a previously unidentified gene product. In X-linked hyper IgM syndrome, the gene for CD40 ligand was cloned independently and only then identified as the defective gene in this disorder. Thus, inherited immunodeficiencies can either lead us to new genes or help us to determine the roles of known genes in normal immune system function.

11-10   Defects in complement components cause defective humoral immune function.

Not surprisingly, the spectrum of infections associated with complement deficiencies overlaps substantially with that seen in patients with deficiencies in antibody production. Defects in the activation of C3, and in C3 itself, are associated with a wide range of pyogenic infections, emphasizing the important

Hyper IgM
Immunodeficiency

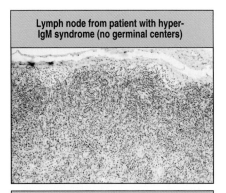

Lymph node from patient with hyper-IgM syndrome (no germinal centers)

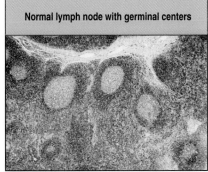

Normal lymph node with germinal centers

Fig. 11.13 Patients with X-linked hyper IgM syndrome are unable to activate their B cells fully. Lymphoid tissues in patients with hyper IgM syndrome are devoid of germinal centers (top panel), unlike a normal lymph node (bottom panel). B-cell activation by T cells is required both for isotype switching and for the formation of germinal centers, where extensive B-cell proliferation takes place. Photographs courtesy of R. Geha and A. Perez-Atayde.

role of C3 as an opsonin, promoting the phagocytosis of bacteria (Fig. 11.14). In contrast, defects in the membrane-attack components of complement (C5–C9) have more limited effects and result exclusively in susceptibility to *Neisseria* species. This indicates that host defense against these bacteria, which are capable of intracellular survival, is mediated by extracellular lysis by the membrane-attack complex of complement. Accurate data from large population studies in Japan, where endemic *N. meningitidis* infection is rare, show that the risk each year to a normal person of infection with this organism is approximately 1/2,000,000. This compares with a risk of 1/200 in the same population to a person with inherited deficiency of one of the membrane-attack complex proteins—a 10,000-fold increase in risk compared to a person with normal complement activity. The early components of the classical complement pathway are particularly important for the elimination of immune complexes and apoptotic cells, which can cause significant pathology in autoimmune diseases such as systemic lupus erythematosus. This aspect of inherited complement deficiency is discussed in Chapter 13.

Another set of diseases are caused by defects in complement control proteins (see Section 2-14). People lacking decay-accelerating factor (DAF) and CD59, which protect a person's own cell surfaces from complement activation, destroy their own red blood cells. This results in the disease paroxysmal nocturnal hemoglobinuria, as we learned in Chapter 2. A more striking consequence of the loss of a regulatory protein is seen in patients with C1-inhibitor defects. C1-inhibitor irreversibly inhibits the activity of several serine proteinase enzymes. These include two enzymes that participate in the contact activation system, factor XIIa (activated Hageman factor) and kallikrein, in addition to the two enzymes that together initiate the classical pathway of complement, C1r and C1s. Deficiency of C1-inhibitor leads to failure to regulate these two pathways. Their unregulated activity results in the excessive production of vasoactive mediators that cause fluid accumulation in the tissues and epiglottal swelling that can lead to suffocation. These mediators are bradykinin, produced by the cleavage of high molecular weight kininogen by kallikrein and the C2 kinin, produced by the activity of C1s on C2a. This syndrome is called **hereditary angioneurotic edema**.

Hereditary
Angioneurotic
Edema

Fig. 11.14 Defects in complement components are associated with susceptibility to certain infections and accumulation of immune complexes. Defects in the early components of the alternative pathway and in C3 lead to susceptibility to extracellular pathogens, particularly pyogenic bacteria. Defects in the early components of the classical pathway predominantly affect the processing of immune complexes and clearance of apoptotic cells, leading to immune-complex disease. Deficiency of mannose-binding lectin (MBL), the recognition molecule of the mannose-binding lectin pathway, is associated with bacterial infections, mainly in early childhood. Finally, defects in the membrane-attack components are associated only with susceptibility to strains of *Neisseria* species, the causative agents of meningitis and gonorrhea, implying that the effector pathway is important chiefly in defense against these organisms.

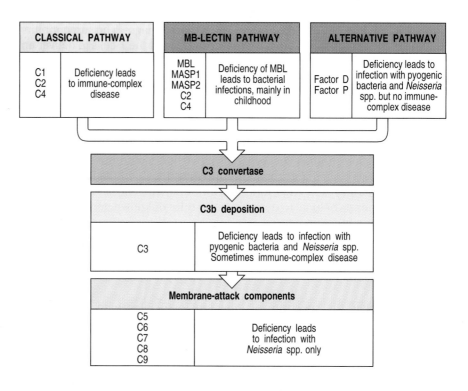

11-11 Defects in phagocytic cells permit widespread bacterial infections.

Defects in the recruitment of phagocytic cells to extravascular sites of infection can cause serious immunodeficiency. Leukocytes reach such sites by emigrating from blood vessels in a tightly regulated process consisting of three stages. The first is the rolling adherence of leukocytes to endothelial cells, through the binding of a fucosylated tetrasaccharide ligand known as sialyl-Lewis[x] to E-selectin and P-selectin. Sialyl-Lewis[x] is expressed on monocytes and neutrophils, whereas E-selectin and P-selectin are expressed on endothelium activated by mediators from the site of inflammation. The second stage is the tight adherence of the leukocytes to the endothelium through the binding of leukocyte β_2 integrins such as CD11b:CD18 (Mac-1:CR3) to counterreceptors on endothelial cells. The third and final stage is the transmigration of leukocytes through the endothelium along gradients of chemotactic molecules originating from the site of tissue injury. Neutrophil recruitment is illustrated in Fig. 2.36 and described in more detail in Section 2-22.

Deficiencies in the molecules involved in each of these stages can prevent neutrophils and macrophages from reaching sites of infection to ingest and destroy bacteria. Reduced rolling adhesion has been described in patients with a lack of sialyl-Lewis[x] caused by a deficiency in the fucosylation pathway responsible for its biosynthesis. Similarly, deficiencies in the leukocyte integrin common β_2 subunit CD18 have been identified and these prevent leukocyte migration to sites of infection because of a lack of tight leukocyte adhesion, causing the **leukocyte adhesion deficiency** syndrome. All these deficiencies lead to infections that are resistant to antibiotic treatment and that persist despite an apparently effective cellular and humoral adaptive immune response. A deficiency of neutrophils (neutropenia) associated with chemotherapy, malignancy, or aplastic anemia is also associated with a similar spectrum of severe pyogenic bacterial infections.

Most of the other known defects in phagocytic cells affect their ability to kill intracellular and/or ingested extracellular bacteria (Fig. 11.15). In **chronic granulomatous disease**, phagocytes cannot produce the superoxide radical and their antibacterial activity is thereby seriously impaired. Several different genetic defects, affecting any one of the four constituent proteins of the NADPH oxidase system, can cause this. Patients with this disease have chronic bacterial infections, which in some cases lead to the formation of granulomas. Deficiencies in the enzymes glucose-6-phosphate dehydrogenase and myeloperoxidase also impair intracellular killing and lead to a similar, although less severe, phenotype. Finally, in **Chediak–Higashi syndrome**, a

Leukocyte Adhesion Deficiency

Chronic Granulomatous Disease

Fig. 11.15 Defects in phagocytic cells are associated with persistence of bacterial infection. Defects in the leukocyte integrins with a common β_2 subunit (CD18) or defects in the selectin ligand, sialyl-Lewis[x], prevent phagocytic cell adhesion and migration to sites of infection (leukocyte adhesion deficiency). The respiratory burst is defective in chronic granulomatous disease, glucose-6-phosphate dehydrogenase (G6PD) deficiency, and myeloperoxidase deficiency. In chronic granulomatous disease, infections persist because macrophage activation is defective, leading to chronic stimulation of CD4 T cells and hence to granulomas. Vesicle fusion in phagocytes is defective in Chediak–Higashi syndrome. These diseases illustrate the critical role of phagocytes in removing and killing pathogenic bacteria.

Type of defect / name of syndrome	Associated infectious or other diseases
Leukocyte adhesion deficiency	Widespread pyogenic bacterial infections
Chronic granulomatous disease	Intracellular and extracellular infection, granulomas
G6PD deficiency	Defective respiratory burst, chronic infection
Myeloperoxidase deficiency	Defective intracellular killing, chronic infection
Chediak–Higashi syndrome	Intracellular and extracellular infection, granulomas

complex syndrome characterized by partial albinism, abnormal platelet function, and severe immunodeficiency, a defect in a gene encoding a protein involved in intracellular vesicle formation causes a failure to fuse lysosomes properly with phagosomes; the phagocytes in these patients have enlarged granules and impaired intracellular killing.

11-12 Defects in T-cell function result in severe combined immunodeficiencies.

Although patients with B-cell defects can deal with many pathogens adequately, patients with defects in T-cell development are highly susceptible to a broad range of infectious agents. This demonstrates the central role of T cells in adaptive immune responses to virtually all antigens. As such patients make neither specific T-cell dependent antibody responses nor cell-mediated immune responses, and thus cannot develop immunological memory, they are said to suffer from **severe combined immunodeficiency (SCID)**.

X-linked
SCID

Several different defects can lead to the SCID phenotype. In X-linked SCID, which is the commonest form of SCID, T cells fail to develop because of a mutation in the common γ chain of several cytokine receptors, including those for the interleukins IL-2, IL-4, IL-7, IL-9, and IL-15. We will examine this defect further in Section 11-13. The commonest forms of autosomally inherited SCID are due to **adenosine deaminase (ADA) deficiency** and **purine nucleotide phosphorylase (PNP) deficiency**. These enzyme defects affect purine degradation, and both result in an accumulation of nucleotide metabolites that are particularly toxic to developing T cells. B cells are also somewhat compromised in these patients.

Adenosine
Deaminase
Deficiency

One class of SCID individuals lack expression of all MHC class II gene products on their cells. This condition is also referred to as the **bare lymphocyte syndrome** as MHC class II molecules are not expressed on lymphocytes or thymic epithelial cells. As the thymus in such individuals lacks MHC class II molecules, CD4 T cells cannot be positively selected and therefore few develop. The antigen-presenting cells in these individuals also lack MHC class II molecules and so the few CD4 T cells that do develop cannot be stimulated by antigen. In these individuals, MHC class I expression is normal and CD8 T cells develop normally. However, such people suffer from severe combined immunodeficiency, illustrating the central importance of CD4 T cells in adaptive immunity to most pathogens. The syndrome is caused not by mutations in the MHC genes themselves, but by mutations in one of several different genes encoding gene-regulatory proteins that are required for the transcriptional activation of MHC class II promoters. Four complementing gene defects (known as Groups A, B, C, and D) have been defined in patients who fail to express MHC class II molecules, which implies that at least four different genes are required for normal MHC class II gene expression. One of these, named the **MHC class II transactivator**, or **CIITA**, is the gene mutated in Group A. The genes mutated in Groups B, C, and D are named RFXANK, RFX5, and RFXAP. These genes encode three proteins that are components of a multimeric transcriptional complex, RFX, which binds a sequence named an X box, present in the promoter of all MHC class II genes.

MHC Class II
Deficiency

In contrast, a more limited immunodeficiency, associated with chronic respiratory bacterial infections and skin ulceration with vasculitis, has been observed in a small number of patients showing almost complete absence of cell-surface MHC class I molecules. This condition has been labeled **bare lymphocyte syndrome (MHC class I)**. Affected individuals have normal levels of mRNA encoding MHC class I molecules and normal production of MHC class I proteins, but these proteins reach the cell surface in severely reduced numbers. The defect was shown to be similar to that in the *TAP* mutant cells mentioned in Section 5-2 and, indeed, affected patients have been found to have mutations in the *TAP1* or *TAP2* genes that encode the two subunits of the peptide transporter. The absence of MHC class I molecules at the cell surface leads to a lack of CD8 T cells expressing the α:β T cell receptor, but these patients do have CD8 T cells that bear the γ:δ receptor. It is surprising that they are not abnormally susceptible to viral infections, given the key role of MHC class I presentation and cytotoxic CD8 α:β T cells in the control of viral infections. However, there is evidence for TAP-independent pathways of antigen presentation by MHC class I molecules of certain peptides. The clinical phenotype of TAP1- and TAP2-deficient patients illustrates that these pathways may be sufficient to allow the control of viral infections.

Another set of defects leading to SCID are those that cause failures of DNA rearrangement in developing lymphocytes. For example, defects in either the *RAG-1* or *RAG-2* genes result in the arrest of lymphocyte development because of a failure to rearrange the antigen receptor genes. Thus there is a complete lack of T and B cells in mice with genetically engineered defects in the *RAG* genes, and in patients with autosomally inherited forms of SCID who lack a functional RAG protein. There are other patients with mutations in either the *RAG-1* or *RAG-2* genes who can nonetheless make a small amount of functional RAG protein, allowing a small amount of V(D)J recombination activity. They suffer from a distinctive and severe disease called Omenn's syndrome, in which, in addition to increased susceptibility to multiple opportunistic infections, there are also clinical features very similar to graft-versus-host disease (see Section 13-21) with rashes, eosinophilia, diarrhea, and enlargement of the lymph nodes. Normal or elevated numbers of T cells, all of which are activated, are found in these unfortunate children. A possible explanation for this phenotype is that very low levels of *RAG* activity allow some limited T-cell receptor gene recombination. However, no B cells are found and it may be that B cells have more stringent requirements for *RAG* activity. The T cells that are produced in these patients show an abnormal and highly restricted receptor repertoire, both in the thymus and in the periphery, where they have undergone clonal expansion and activation. The clinical features strongly suggest that these peripheral T cells are autoreactive and responsible for the graft versus host phenotype.

Omenn Syndrome

Another group of patients with autosomal SCID have a phenotype very similar to that of a mutant mouse strain called *scid*; *scid* mice suffer from an abnormal sensitivity to ionizing radiation as well as from severe combined immunodeficiency. They produce very few mature B and T cells, as there is a failure of DNA rearrangement in their developing lymphocytes; only rare VJ or VDJ joints are seen and most of these have abnormal features. The underlying defect has now been shown to be in the enzyme DNA-dependent protein kinase (DNA-PK), which binds to the end of the double-stranded breaks that occur during the process of antigen receptor gene rearrangement. These ends are found as DNA hairpin structures in the immature thymocytes of *scid* mice. Thus, it seems likely that DNA-PK is involved in resolving the hairpin structure (see Section 4-5).

Other defects in DNA repair and metabolizing enzymes are associated with a combination of immunodeficiency, increased sensitivity to the damaging effects of ionizing radiation, and cancer development. One example is **Bloom's syndrome**, a disease caused by mutations in a DNA helicase enzyme, which unwinds DNA. Another is **ataxia telangiectasia** (**AT**), in which the underlying defect is in a protein called ATM, which contains a kinase domain thought to be involved in intracellular signaling in response to DNA damage. Because repair of double-stranded DNA breaks and lymphocyte division are central to the function of the adaptive immune system, it is not surprising that defects such as these are associated with the development of immunodeficiency.

Finally, in patients with **DiGeorge's syndrome** the thymic epithelium fails to develop normally. Without the proper inductive environment T cells cannot mature, and both T-cell dependent antibody production and cell-mediated immunity are absent. Such patients have some serum immuno-globulin and variable numbers of B and T cells. As with all the severe combined immuo-deficiency diseases, it is the defect in T cells that is crucial. These diseases abundantly illustrate the central role of T cells in virtually all adaptive immune responses. In many cases B-cell development is normal, yet the response to nearly all pathogens is profoundly impaired.

11-13 Defective T-cell signaling, cytokine production, or cytokine action can cause immunodeficiency.

As we learned in Chapter 8, virtually all adaptive immune responses require the activation of antigen-specific T lymphocytes and their differentiation into cells producing cytokines that act on specific cytokine receptors. Several gene defects have been described that interfere with these processes. Thus, patients who lack CD3γ chains have low levels of surface T-cell receptors and defective T-cell responses. Patients making low levels of mutant CD3ε chains are also deficient in T-cell activation. Patients who make a defective form of the cytosolic protein tyrosine kinase ZAP-70, which transmits signals from the T-cell receptor (see Section 6-9) have recently been described. Their CD4 T cells emerge from the thymus in normal numbers, whereas CD8 T cells are absent. However, the CD4 T cells that mature fail to respond to stimuli that normally activate via the T-cell receptor and the patients are thus very immunodeficient.

Another group of patients show a lack of IL-2 production upon receptor ligation, and these patients have a severe immunodeficiency; however, T-cell development is normal in these individuals, as it is in mice in which mutations have been made in their IL-2 genes by gene knockout (see Appendix I, Section A-47). These IL-2-negative patients have heterogeneous defects; some of them fail to activate the transcription factor NFAT (see Section 6-11), which induces the transcription of several cytokine genes in addition to the IL-2 gene. This might explain why their immunodeficiency is more profound than that of mice whose IL-2 gene has been disrupted. IL-2-deficient mice can mount adaptive immune responses through an IL-2-independent pathway, possibly involving the cytokine IL-15, which shares many activities with IL-2; nevertheless, they are susceptible to a variety of infectious agents.

X-linked
SCID

In contrast to the normal development of T cells in patients deficient in IL-2, there is a failure of T-cell development in patients with **X-linked severe combined immunodeficiency** (**X-linked SCID**), which is caused by a defect in the γ chain of the IL-2 receptor. Thus, this disease showed that the common γ chain (γ_c) must be important in T-cell development for reasons unrelated to IL-2 binding or IL-2 responses. The demonstration that the IL-2 receptor γ_c chain is also part of other cytokine receptors, including the IL-7 receptor, helps to explain its role in early T-cell development. The γ chain seems to function in transducing the signal from this group of receptors and interacts with a kinase, JAK3 kinase, which is known to be defective in patients with an autosomally inherited immunodeficiency similar in pheno-type to X-linked SCID.

As in all serious T-cell deficiencies, X-linked SCID patients do not make effective antibody responses to most antigens, although their B cells seem normal. However, as the gene defect is on the X chromosome, one can determine whether the lack of B-cell function is solely a consequence of the lack of T-cell help by examining X-chromosome inactivation (see Section 11-7) in B cells of unaffected carriers. The majority of naive IgM-positive B cells from female carriers of X-linked SCID have inactivated the defective X chromosome rather than the normal one, showing that B-cell development is affected by, but not wholly dependent on, the common γ chain. However, mature memory B cells that have switched to isotypes other than IgM have inactivated the defective X chromosome almost without exception. This might reflect the fact that the IL-2 receptor γ chain is also part of the IL-4 receptor. Thus, B cells that lack this chain will have defective IL-4 receptors and will not proliferate in T-cell-dependent antibody responses. X-linked SCID is so severe that children who inherit it can survive only in a completely pathogen-free environment, unless given antibodies and successfully treated by bone marrow transplantation. A famous case in Houston became known as the 'bubble baby' because of the plastic bubble in which he was enclosed to protect him from infection.

Wiskott–Aldrich syndrome (WAS) is a disease that has shed new light on the molecular basis of T-cell signaling and its importance for immune function. The disease affects platelets and was first described as a blood-clotting disorder, but it is also associated with immunodeficiency due to impaired T-cell function, reduced T-cell numbers, and a failure of antibody responses to encapsulated bacteria. WAS is caused by a defective gene on the X chromosome, encoding a protein called WAS protein (WASP). This protein has been shown to bind Cdc42, a small GTP-binding protein that is known to regulate the organization of the actin cytoskeleton and to be important for the effective collaboration of T and B cells. WASP might have a role in regulating changes in the actin cytoskeleton in response to external stimuli. It has the ability to bind SH3 domains, which, as we saw in Chapter 6, have an affinity for amino acid sequences rich in proline that are found on some proteins of intracellular signaling pathways. In WAS patients, and in mice whose WASP gene has been knocked out, T cells fail to respond normally to mitogens or to the cross-linking of surface receptors. Cytotoxic T-cell responses are also impaired, and T-cell help for B-cell responses to polysaccharide antigens is lacking. WASP is expressed in all hematopoietic cell lineages and is likely to be a key regulator of lymphocyte and platelet development and function.

Wiskott–Aldrich Syndrome

11-14 The normal pathways for host defense against intracellular bacteria are illustrated by genetic deficiencies of IFN-γ and IL-12 and their receptors.

A small number of families have been identified containing several individuals who suffer from persistent and eventually fatal attacks by intracellular pathogens, especially mycobacteria and salmonellae. Typically these patients suffer from the ubiquitous, environmental nontuberculous strains of mycobacteria, such as *Mycobacterium avium*. They may also develop disseminated infection after vaccination with *Mycobacterium bovis* bacillus Calmette-Guérin, the strain of *M. bovis* that is used as a live vaccine against *M. tuberculosis*. The molecular bases of the susceptibility to these infections are null mutations in one of the following genes: IL-12, the IL-12 receptor β1 chain, or either of the two protein subunits, R1 and R2, of the receptor for IFN-γ. Similar susceptibility to intracellular bacterial infection is seen in mice with induced mutations in these same genes and also in mice lacking TNF-α or the TNF p55 receptor gene. All these genes must therefore play a critical part in the normal mechanisms of host defense against infection by these intracellular bacteria.

Interferon-γ Receptor Deficiency

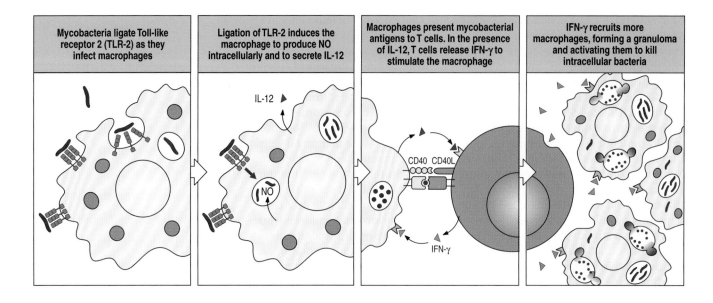

| Mycobacteria ligate Toll-like receptor 2 (TLR-2) as they infect macrophages | Ligation of TLR-2 induces the macrophage to produce NO intracellularly and to secrete IL-12 | Macrophages present mycobacterial antigens to T cells. In the presence of IL-12, T cells release IFN-γ to stimulate the macrophage | IFN-γ recruits more macrophages, forming a granuloma and activating them to kill intracellular bacteria |

Fig. 11.16 The expression of Toll-like receptor 2 (TLR-2) allows macrophages to respond effectively to mycobacteria. TLR-2 is activated on binding the polysaccharide coat of mycobacteria, stimulating internalization of the bound bacteria, and the expression of nitric oxide (NO) and cytokines such as IL-12 by the macrophage. Interaction of macrophages and T cells in the presence of IL-12 leads to T-cell secretion of IFN-γ, which activates the macrophage, leading to the death of the intracellular mycobacteria.

Interferon-γ
Receptor
Deficiency

Mycobacteria and salmonellae enter dendritic cells and macrophages, where they can reproduce and multiply. At the same time they provoke an immune response that involves several stages and eventually controls the infection with the help of CD4 T cells (Fig. 11.16). First, lipoproteins from the surface of the bacteria ligate receptors on macrophages and dendritic cells as they enter the cells. These receptors include the Toll-like receptors (see Section 2-16), particularly TLR-2, and their ligation stimulates nitric oxide (NO) production within the cells, which is toxic to the bacteria. Signaling by these Toll-like receptors also stimulates the release of IL-12 which, in turn, stimulates CD4 T cells to release IFN-γ and TNF-α. These cytokines activate and recruit more mononuclear phagocytic cells to the site of infection, resulting in the formation of granulomas. The key role of IFN-γ in activating macrophages to kill intracellular bacteria is illustrated dramatically by the failure to control infection in patients who are genetically deficient in either of the two subunits of this receptor. In the total absence of IFN-γ receptor expression, granuloma formation is much reduced, showing a role for this receptor in the development of granulomas. In contrast, if the underlying mutation is associated with the presence of low levels of functional receptor, granulomas form, but the macrophages within the granulomas are not sufficiently activated to be able to control the division and spread of the mycobacteria. It is important to appreciate that this cascade of cytokine reactions is occurring in the context of cognate interactions between the macrophages and dendritic cells harboring the intracellular bacteria and antigen-specific CD4 T cells. T-cell receptor ligation and co-stimulation of the phagocyte by, for example, CD40–CD40 ligand interaction are important components that augment the capacity of T cells to effectively activate the infected phagocytes to kill the intracellular bacteria (see Sections 8-26 and 8-28).

11-15 X-linked lymphoproliferative syndrome is associated with fatal infection by Epstein–Barr virus and with the development of lymphomas.

Epstein–Barr virus is a herpes virus that infects the majority of the human race and remains latent in B cells throughout life after primary infection. EBV infection can transform B lymphocytes and is used as a technique for immortalizing clones of B cells in the laboratory. This does not normally happen *in vivo* in humans because EBV infection is actively controlled and maintained in a latent state by cytotoxic T cells with specificity for B cells expressing EBV antigens (see Section 11-2). In the presence of T-cell immunodeficiency, this control mechanism can break down and a potentially lethal B-cell lymphoma may develop. One of the situations in which this occurs is the rare immuodeficiency, **X-linked lymphoproliferative syndrome**, which results from mutations in a gene named SH2-domain containing gene 1A (*SH2D1A*). Boys with this deficiency typically develop overwhelming EBV infection during childhood, and sometimes lymphomas. EBV infection in this condition is usually fatal and is associated with necrosis of the liver. Thus SH2D1A must play a vital, nonredundant role in the normal control of EBV infection.

The function of SH2D1A is partly understood. The SH2 domain of the protein interacts with the cytoplasmic tails of two transmembrane receptors, SLAM and 2B4, which are structurally homologous to each other, and to the T-cell adhesion molecule CD2 (see Section 8-4). SLAM (signaling lymphocyte activation molecule) is expressed on activated T cells, whereas 2B4 is found on T cells, B cells, and NK cells. Activation of these receptors initiates a signaling pathway by the recruitment of the tyrosine phosphatase, SHP-2 (see Section 6-14). It appears that the function of SH2D1A is to inhibit the recruitment of SHP-2 and thereby to inhibit cellular activation by SLAM and 2B4. There are two hypotheses to explain the pathogenesis of the fatal EBV infection seen in children with defects in SH2D1A. The first is that failure of T cells to kill B cells expressing antigens from multiplying EBV allows uncontrolled infection. The second is that B cells presenting EBV peptides uncontrollably activate T cells and that cytotoxic T cells cause tissue necrosis and death. Some cases of lymphoma in young boys have now been found associated with mutations in the *SH2D1A* gene in the absence of any evidence of EBV infection. This raises the possibility that *SH2D1A* may be a tumor suppressor gene in its own right, in addition to controlling a virus that can contribute to tumor formation.

11-16 Bone marrow transplantation or gene therapy can be useful to correct genetic defects.

It is frequently possible to correct the defects in lymphocyte development that lead to the SCID phenotype by replacing the defective component, generally by bone marrow transplantation. The major difficulties in these therapies result from MHC polymorphism. To be useful, the graft must share some MHC alleles with the host. As we learned in Section 7-20, the MHC alleles expressed by the thymic epithelium determine which T cells can be positively selected. When bone marrow cells are used to restore immune function to individuals with a normal thymic stroma, both the T cells and the antigen-presenting cells are derived from the graft. Therefore, unless the graft shares at least some MHC alleles with the recipient, the T cells that are selected on host thymic epithelium cannot be activated by graft-derived antigen-presenting cells (Fig. 11.17). There is also a danger that mature, post-thymic T cells in donor bone marrow might recognize the host as foreign and attack it, causing **graft-versus-host disease** (GVHD) (Fig. 11.18, top panel). This can be overcome by depleting the donor bone marrow of mature T cells. Bone marrow recipients are usually treated with irradiation that kills their own lymphocytes, thus making space for the grafted bone marrow cells and minimizing the threat of **host-versus-graft disease** (HVGD) (Fig. 11.18, third panel). In patients with the SCID phenotype, however, there is little problem with the host response to the transplanted bone marrow, as the patient is immunodeficient.

Now that specific gene defects are being identified, a different approach to correcting these inherited immune deficiencies can be attempted. The strategy involves extracting a sample of the patient's own bone marrow cells, inserting a normal copy of the defective gene into them, and returning them

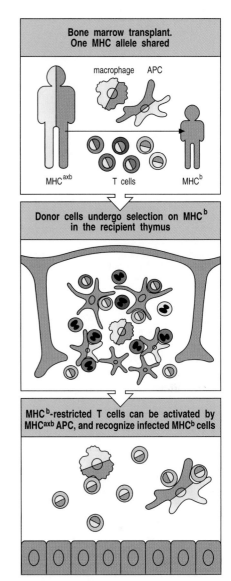

Fig. 11.17 Bone marrow donor and recipient must share at least some MHC molecules to restore immune function. In an allogeneic bone marrow transplant, the donor marrow cells share some MHC molecules with the recipient. The shared MHC type is designated b and illustrated in blue, the MHC type of the donor marrow that is not shared is designated a and shown in yellow. Donor lymphocytes are positively selected on MHC^b on thymic epithelial cells and negatively selected by the recipient stromal epithelial cells and at the cortico-medullary junction by encounter with dendritic cells derived from both the donor bone marrow and residual recipient dendritic cells. The negatively selected cells are shown as apoptotic cells. The antigen-presenting cells in the periphery can activate T cells that recognize MHC^b molecules; the activated T cells can then recognize infected MHC^b-bearing cells.

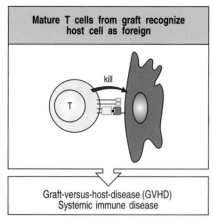

Graft-versus-host-disease (GVHD)
Systemic immune disease

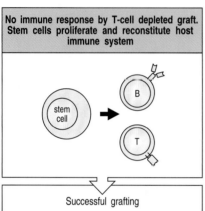

Successful grafting

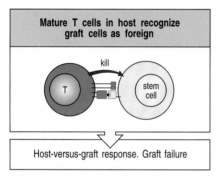

Host-versus-graft response. Graft failure

Fig. 11.18 Bone marrow grafting can be used to correct immuno-deficiencies caused by defects in lymphocyte maturation but two problems can arise. First, if there are mature T cells in the bone marrow, they can attack cells of the host by recognizing their MHC antigens, causing graft-versus-host disease (top panel). This can be prevented by T-cell depletion of the donor bone marrow (center panel). Second, if the recipient has competent T cells, these can attack the bone marrow stem cells (bottom panel). This causes failure of the graft by the usual mechanism of transplant rejection (see Chapter 13).

to the patient by transfusion. This approach, called **somatic gene therapy**, should correct the gene defect. Moreover, in immunodeficient patients, it might be possible to reinfuse the bone marrow into the patient without the usual irradiation used to suppress the recipient's bone marrow function. There is no risk of graft-versus-host disease in this case, although the host might respond to the replaced gene product and reject the engineered cells. Although this kind of approach is theoretically attractive, efficient transfer of genes into bone marrow stem cells is technically difficult and has been achieved only in mouse models. The first trials of gene therapy for correcting immunodeficiency, such as the treatment of a child with ADA deficiency at the National Institute of Health (NIH) in 1990, used the patient's lymphocytes as the vehicle for gene introduction. However, because most lymphocytes divide regularly, thus diluting out the new gene, the treatment had to be repeated regularly. In another study, bone marrow stem cells, obtained from cord blood from three patients with ADA deficiency were transduced with the ADA gene and reinfused. At the age of 4 years, these children expressed up to 10% of normal ADA levels only in T cells and not in other bone marrow-derived cells, and they remained immunodeficient in the absence of treatment with ADA enzyme replacement. More recently, however, there has been a successful attempt at correcting the phenotype of two X-linked SCID patients using a Moloney retrovirus-derived construct containing the γ_c chain to infect bone marrow stem cells.

Summary.

Genetic defects can occur in almost any molecule involved in the immune response. These defects give rise to characteristic deficiency diseases, which, although rare, provide a great deal of information about the development and functioning of the immune system in normal humans. Inherited immuo-deficiencies illustrate the vital role of the adaptive immune response and T cells in particular, without which both cell-mediated and humoral immunity fail. They have provided information about the separate roles of B lymphocytes in humoral immunity and of T lymphocytes in cell-mediated immunity, the importance of phagocytes and complement in humoral and innate immunity, and the specific functions of several cell-surface or signaling molecules in the adaptive immune response. There are also some inherited immune disorders whose causes we still do not understand. The study of these diseases will undoubtedly teach us more about the normal immune response and its control.

Acquired immune deficiency syndrome.

The first cases of the **acquired immune deficiency syndrome (AIDS)** were reported in 1981 but it is now clear that cases of the disease had been occurring unrecognized for at least 4 years before its identification. The disease is characterized by a susceptibility to infection with opportunistic pathogens or by the occurrence of an aggressive form of Kaposi's sarcoma or B-cell lymphoma, accompanied by a profound decrease in the number of CD4 T cells. As it seemed to be spread by contact with body fluids, it was early suspected to be caused by a new virus, and by 1983 the agent now known to be responsible for AIDS, called the **human immunodeficiency virus (HIV)**, was isolated and identified. It is now clear there are at least two types of HIV—HIV-1 and HIV-2—which are closely related to each other. HIV-2 is endemic in West Africa and is now spreading in India. Most AIDS worldwide, however, is

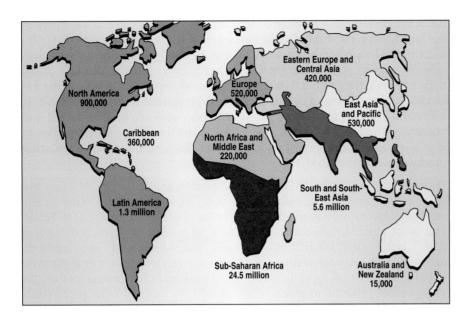

Fig. 11.19 HIV infection is spreading on all continents. The number of HIV-infected individuals is large (data are numbers of adults and children living with HIV/AIDS at the end of 1999, as estimated by the World Health Organization) and is increasing rapidly, especially in developing countries. It is estimated that 5.6 million individuals were newly infected with HIV during 1999.

caused by the more virulent HIV-1. Both viruses appear to have spread to humans from other primate species and the best evidence from sequence relationships suggests that HIV-1 has passed to humans on at least three independent occasions from the chimpanzee, *Pan troglodytes*, and HIV-2 from the sooty mangabey, *Cercocebus atys*.

HIV infection does not immediately cause AIDS, and the issues of how it does, and whether all HIV-infected patients will progress to overt disease, remain controversial. Nevertheless, accumulating evidence clearly implicates the growth of the virus in CD4 T cells, and the immune response to it, as the central keys to the puzzle of AIDS. HIV is a worldwide pandemic and, although great strides are being made in understanding the pathogenesis and epidemiology of the disease, the number of infected people around the world continues to grow at an alarming rate, presaging the death of many people from AIDS for many years to come. Estimates from the World Health Organization are that 16.3 million people have died from AIDS since the beginning of the epidemic and that there are currently around 34.3 million people alive with HIV infection (Fig. 11.19), of whom the majority are living in sub-Saharan Africa, where approximately 7% of young adults are infected. In some countries within this region, such as Zimbabwe and Botswana, over 25% of adults are infected.

AIDS in
Mother
and Child

11-17 Most individuals infected with HIV progress over time to AIDS.

Many viruses cause an acute but limited infection inducing lasting protective immunity. Others, such as herpes viruses, set up a latent infection that is not eliminated but is controlled adequately by an adaptive immune response. However, infection with HIV seems rarely, if ever, to lead to an immune response that can prevent ongoing replication of the virus. Although the initial acute infection does seem to be controlled by the immune system, HIV continues to replicate and infect new cells.

The initial infection with HIV generally occurs after transfer of body fluids from an infected person to an uninfected one. The virus is carried in infected CD4 T cells, dendritic cells, and macrophages, and as a free virus in blood, semen, vaginal fluid, or milk. It is most commonly spread by sexual intercourse, contaminated needles used for intravenous drug delivery, and the therapeutic

AIDS in
Mother
and Child

use of infected blood or blood products, although this last route of transmission has largely been eliminated in the developed world where blood products are screened routinely for the presence of HIV. An important route of virus transmission is from an infected mother to her baby at birth or through breast milk. In Africa, the perinatal transmission rate is approximately 25%, but this can largely be prevented by treating infected pregnant women with the drug zidovudine (AZT) (see Section 11-23). Mothers who are newly infected and breastfeed their infants transmit HIV 40% of the time, showing that HIV can also be transmitted in breast milk, but this is less common after the mother produces antibodies to HIV.

Primary infection with HIV is probably asymptomatic in 50% of cases but often causes an influenza-like illness with an abundance of virus in the peripheral blood and a marked drop in the numbers of circulating CD4 T cells. This acute viremia is associated in virtually all patients with the activation of CD8 T cells, which kill HIV-infected cells, and subsequently with antibody production, or **seroconversion**. The cytotoxic T-cell response is thought to be important in controlling virus levels, which peak and then decline, as the CD4 T-cell counts rebound to around 800 cells μl^{-1} (the normal value is 1200 cells μl^{-1}). At present, the best indicator of future disease is the level of virus that persists in the blood plasma once the symptoms of acute viremia have passed.

Most patients who are infected with HIV will eventually develop AIDS, after a period of apparent quiescence of the disease known as clinical latency or the asymptomatic period (Fig. 11.20). This period is not silent, however, for there is persistent replication of the virus, and a gradual decline in the function and numbers of CD4 T cells until eventually patients have few CD4 T cells left. At this point, which can occur anywhere between 2 and 15 years or more after the primary infection, the period of clinical latency ends and opportunistic infections begin to appear.

The typical course of an infection with HIV is illustrated in Fig. 11.21. However, it has become increasingly clear that the course of the disease can vary widely. Thus, although most people infected with HIV go on to develop AIDS and ultimately to die of opportunistic infection or cancer, this is not true of all individuals. A small percentage of people seroconvert, making antibodies against many HIV proteins, but do not seem to have progressive disease, in that their CD4 T-cell counts and other measures of immune competence are maintained. These long-term nonprogressors have unusually low levels of circulating virus and are being studied intensively to determine how they are

Fig. 11.20 Most HIV-infected individuals progress to AIDS over a period of years. The incidence of AIDS increases progressively with time after infection. Homosexuals and hemophiliacs are two of the groups at highest risk in the West—homosexuals from sexually transmitted virus and hemophiliacs from infected human blood used to replace clotting factor VIII. In Africa, spread is mainly by heterosexual intercourse. Hemophiliacs are now protected by the screening of blood products and the use of recombinant factor VIII. Neither homosexuals nor hemophiliacs who have not been infected with HIV show any evidence of AIDS. Most hemophiliacs in Western Europe and North America were exposed to HIV infection by inadvertent administration of contaminated blood products at the start of the HIV epidemic, with the peak of infection occurring in 1982–1983. From this infected population, there are robust cohort data on their progression to the development of AIDS, which are shown here. The age of the individual seems to play a significant role in the rate of progression of the development of HIV. More than 80% of those aged more than 40 at the time of infection progress to AIDS over 13 years, in comparison with approximately 50% of those aged less than 40 over a comparable time. There are a few individuals who, while infected with HIV, seem not to progress to develop AIDS. One protective mechanism is an inherited defect in the major HIV co-receptor, CCR5.

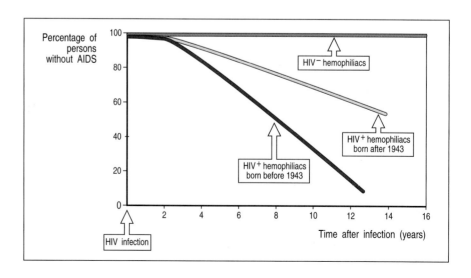

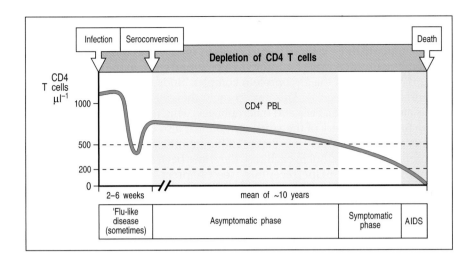

Fig. 11.21 The typical course of untreated infection with HIV. The first few weeks are typified by an acute influenza-like viral illness, sometimes called seroconversion disease, with high titers of virus in the blood. An adaptive immune response follows, which controls the acute illness and largely restores levels of CD4 T cells (CD4$^+$ PBL) but does not eradicate the virus. Opportunistic infections and other symptoms become more frequent as the CD4 T-cell count falls, starting at around 500 cells μl^{-1}. The disease then enters the symptomatic phase. When CD4 T-cell counts fall below 200 cells μl^{-1} the patient is said to have AIDS. Note that CD4 T-cell counts are measured for clinical purposes in cells per microliter (cells μl^{-1}), rather than cells per milliliter (cells ml^{-1}), the unit used elsewhere in this book.

able to control their HIV infection. A second group consists of seronegative people who have been highly exposed to HIV yet remain disease-free and virus-negative. Some of these people have specific cytotoxic lymphocytes and T$_H$1 lymphocytes directed against infected cells, which confirms that they have been exposed to HIV or possibly noninfectious HIV antigens. It is not clear whether this immune response accounts for clearing the infection, but it is a focus of considerable interest for the development and design of vaccines, which we will discuss later. There is a small group of people who are resistant to HIV infection because they carry mutations in a cell-surface receptor that is used as a co-receptor for viral entry, as we will see below.

We will return to discuss in more detail the interactions of HIV with the immune system and the prospects for manipulating them later in this chapter, but before doing so we must describe the viral life cycle and the genes and proteins on which it depends. Some of these proteins are the targets of the most successful drugs in use at present for the treatment of AIDS.

11-18 HIV is a retrovirus that infects CD4 T cells, dendritic cells, and macrophages.

HIV is an enveloped retrovirus whose structure is shown in Fig. 11.22. Each virus particle, or virion, contains two copies of an RNA genome, which are transcribed into DNA in the infected cell and integrated into the host cell chromosome. The RNA transcripts produced from the integrated viral DNA serve both as mRNA to direct the synthesis of the viral proteins and later as the RNA genomes of new viral particles, which escape from the cell by budding from the plasma membrane, each in a membrane envelope. HIV belongs to a group of retroviruses called the **lentiviruses**, from the Latin *lentus*, meaning slow, because of the gradual course of the diseases that they cause. These viruses persist and continue to replicate for many years before causing overt signs of disease.

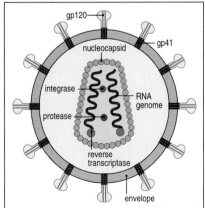

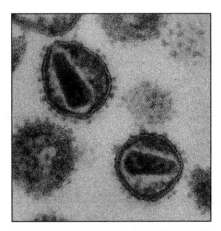

Fig. 11.22 The virion of human immunodeficiency virus (HIV). The virus illustrated is HIV-1, the leading cause of AIDS. The reverse transcriptase, integrase, and viral protease enzymes are packaged in the virion and are shown schematically in the viral capsid. In reality, many molecules of these enzymes are contained in each virion. Some structural proteins of the virus have been omitted for simplicity. Photograph courtesy of H. Gelderblom.

The ability of HIV to enter particular types of cell, known as the cellular tropism of the virus, is determined by the expression of specific receptors for the virus on the surface of those cells. HIV enters cells by means of a complex of two non-covalently associated viral glycoproteins, gp120 and gp41, in the viral envelope. The gp120 portion of the glycoprotein complex binds with high affinity to the cell-surface molecule CD4. This glycoprotein thereby draws the virus to CD4 T cells and to dendritic cells and macrophages, which also express some CD4. Before fusion and entry of the virus, gp120 must also bind to a co-receptor in the membrane of the host cell. Several different molecules may serve as a co-receptor for HIV entry, but in each case they have been identified as chemokine receptors. The chemokine receptors (see Chapters 2 and 10) are a closely related family of G protein-coupled receptors with seven transmembrane-spanning domains. Two chemokine receptors, known as CCR5, which is predominantly expressed on dendritic cells, macrophages, and CD4 T cells, and CXCR4, expressed on activated T cells, are the major co-receptors for HIV. After binding of gp120 to the receptor and co-receptor, the gp41 then causes fusion of the viral envelope and the plasma membrane of the cell, allowing the viral genome and associated viral proteins to enter the cytoplasm.

There are different variants of HIV, and the cell types that they infect are determined to a large degree by which chemokine receptor they bind as co-receptor. The variants of HIV that are associated with primary infections use CCR5, which binds the CC chemokines RANTES, MIP-1α, and MIP-1β (see Chapter 2), as a co-receptor, and require only a low level of CD4 on the cells they infect. These variants of HIV infect dendritic cells, macrophages, and T cells *in vivo*. However, they are often described simply as 'macrophage-tropic' because they infect macrophage but not T-cell lines *in vitro* and the cell tropism of different HIV variants was originally defined by their ability to grow in different cell lines.

In contrast, 'lymphocyte-tropic' variants of HIV infect only CD4 T cells *in vivo* and use CXCR4, which binds the CXC chemokine stromal-derived factor-1 (SDF-1), as a co-receptor. The lymphocyte-tropic variants of HIV can grow *in vitro* in T-cell lines, and require high levels of CD4 on the cells that they infect.

It appears that macrophage-tropic isolates of HIV are preferentially transmitted by sexual contact as they are the dominant viral phenotype found in newly infected individuals. Virus is disseminated from an initial reservoir of infected dendritic cells and macrophages and there is evidence for an important role for mucosal lymphoid tissue in this process. Mucosal epithelia, which are constantly exposed to foreign antigens, provide a milieu of immune system activity in which HIV replication occurs readily. Infection of CD4 T cells via CCR5 occurs early in the course of infection and continues to occur, with activated CD4 T cells accounting for the major production of HIV through-out infection. Late in infection, in approximately 50% of cases, the viral phenotype switches to a T-lymphocyte-tropic type that utilizes CXCR4 co-receptors, and this is followed by a rapid decline in CD4 T-cell count and progression to AIDS.

11-19 Genetic deficiency of the macrophage chemokine co-receptor for HIV confers resistance to HIV infection *in vivo*.

Further evidence for the importance of chemokine receptors in HIV infection has come from studies in a small group of individuals with high-risk exposure to HIV-1 but who remain seronegative. Cultures of lymphocytes and macrophages from these people were relatively resistant to macrophage-tropic HIV infection and were found to secrete high levels of RANTES, MIP-1α and MIP-1β in response to inoculation with HIV. In other experiments, the addition of these same chemokines to lymphocytes sensitive to HIV blocked their infection because of competition between these CC chemokines and the virus for the cell-surface receptor CCR5.

The resistance of these rare individuals to HIV infection has now been explained by the discovery that they are homozygous for an allelic, non-functional variant of CCR5 caused by a 32-base-pair deletion from the coding region that leads to a frameshift and truncation of the translated protein. The gene frequency of this mutant allele in Caucasoid populations is quite high at 0.09 (meaning that about 10% of the Caucasoid population are heterozygous carriers of the allele and about 1% are homozygous). The mutant allele has not been found in Japanese or black Africans from Western or Central Africa. Heterozygous deficiency of CCR5 might provide some protection against sexual transmission of HIV infection and a modest reduction in the rate of progression of the disease. In addition to the structural polymorphism of the gene, variation of the promoter region of the CCR5 gene has been found in both Caucasian and African Americans. Different promoter variants were associated with different rates of progression of disease.

These results provide a dramatic confirmation of experimental work suggesting that CCR5 is the major macrophage and T-lymphocyte co-receptor used by HIV to establish primary infection *in vivo*, and offers the possibility that primary infection might be blocked by therapeutic antagonists of the CCR5 receptor. Indeed, there is preliminary evidence that low molecular weight inhibitors of this receptor can block infection of macrophages by HIV *in vitro*. Such low molecular weight inhibitors might be the precursors of useful drugs that could be taken by mouth. Such drugs are very unlikely to provide complete protection against infection, as a very small number of individuals who are homozygous for the nonfunctional variant of CCR5 are infected with HIV. These individuals seem to have suffered from primary infection by CXCR4-using strains of the virus.

11-20 HIV RNA is transcribed by viral reverse transcriptase into DNA that integrates into the host cell genome.

One of the proteins that enters the cell with the viral genome is the viral reverse transcriptase, which transcribes the viral RNA into a complementary DNA (cDNA) copy. The viral cDNA is then integrated into the host cell genome by the viral integrase, which also enters the cell with the viral RNA. The integrated cDNA copy is known as the **provirus**. The infectious cycle up to the integration of the provirus is shown in Fig. 11.23. In activated CD4 T cells, virus replication is initiated by transcription of the provirus, as we will see in the next section. However, HIV can, like other retroviruses, establish a latent infection in which the provirus remains quiescent. This seems to occur in memory CD4 T cells and in dormant macrophages, and these cells are thought to be an important reservoir of infection.

The entire HIV genome consists of nine genes flanked by long terminal repeat sequences (LTRs), which are required for the integration of the provirus into the host cell DNA and contain binding sites for gene regulatory proteins that control the expression of the viral genes. Like other retroviruses, HIV has three major genes—*gag*, *pol*, and *env*. The *gag* gene encodes the structural proteins of the viral core, *pol* encodes the enzymes involved in viral replication and integration, and *env* encodes the viral envelope glycoproteins. The *gag* and *pol* mRNAs are translated to give polyproteins—long polypeptide chains that are then cleaved by the viral protease (also encoded by *pol*) into individual functional proteins. The product of the *env* gene, gp160, has to be cleaved by a host cell protease into gp120 and gp41, which are then assembled as trimers into the **viral envelope**.

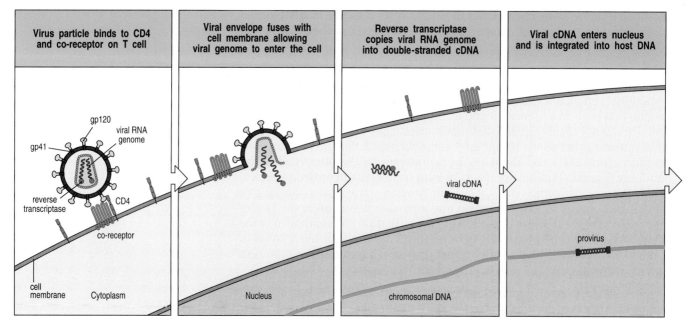

Fig. 11.23 The infection of CD4 T cells by HIV. The virus binds to CD4 using gp120, which is altered by CD4 binding so that it now also binds a specific seven-span chemokine receptor that acts as a co-receptor for viral entry. This binding releases gp41, which then causes fusion of the viral envelope with the cell membrane, and the release of the viral core into the cytoplasm. Once in the cytoplasm, the viral core releases the RNA genome, which is then reverse transcribed into double-stranded cDNA. The double-stranded cDNA migrates to the nucleus in association with the viral integrase and the Vpr protein, where it is integrated into the cell genome, becoming a provirus.

As shown in Fig. 11.24, HIV has six other, smaller, genes encoding proteins that affect viral replication and infectivity in various ways. We will discuss the function of two of these—Tat and Rev—in the following section.

Fig. 11.24 The genes and proteins of HIV-1. Like all retroviruses, HIV-1 has an RNA genome flanked by long terminal repeats (LTR) involved in viral integration and in regulation of the viral genome. The genome can be read in three frames and several of the viral genes overlap in different reading frames. This allows the virus to encode many proteins in a small genome. The three main protein products—Gag, Pol, and Env—are synthesized by all infectious retroviruses. The known functions of the different genes and their products are listed. The products of *gag*, *pol*, and *env* are known to be present in the mature viral particle, together with the viral RNA. The mRNAs for Tat, Rev, and Nef proteins are produced by splicing of viral transcripts, so their genes are split in the viral genome. In the case of Nef, only one exon, shown in yellow, is translated. The other gene products affect the infectivity of the virus in various ways that are not fully understood.

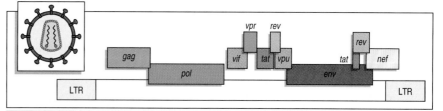

Gene		Gene product/function
gag	Group-specific antigen	Core proteins and matrix proteins
pol	Polymerase	Reverse transcriptase, protease, and integrase enzymes
env	Envelope	Transmembrane glycoproteins. gp120 binds CD4 and CCR5; gp41 is required for virus fusion and internalization
tat	Transactivator	Positive regulator of transcription
rev	Regulator of viral expression	Allows export of unspliced and partially spliced transcripts from nucleus
vif	Viral infectivity	Affects particle infectivity
vpr	Viral protein R	Transport of DNA to nucleus. Augments virion production. Cell cycle arrest
vpu	Viral protein U	Promotes intracellular degradation of CD4 and enhances release of virus from cell membrane
nef	Negative-regulation factor	Augments viral replication *in vivo* and *in vitro*. Downregulates CD4 and MHC class II

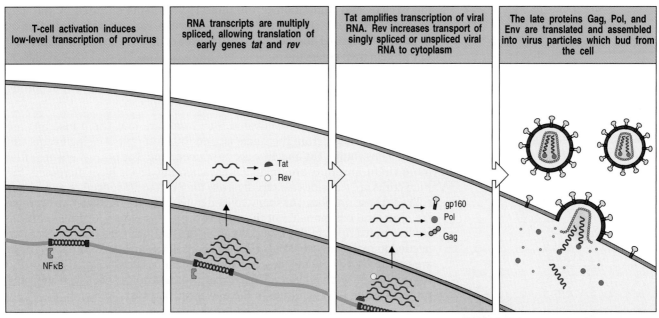

| T-cell activation induces low-level transcription of provirus | RNA transcripts are multiply spliced, allowing translation of early genes *tat* and *rev* | Tat amplifies transcription of viral RNA. Rev increases transport of singly spliced or unspliced viral RNA to cytoplasm | The late proteins Gag, Pol, and Env are translated and assembled into virus particles which bud from the cell |

11-21 Transcription of the HIV provirus depends on host cell transcription factors induced upon the activation of infected T cells.

The production of infectious virus particles from an integrated HIV provirus is stimulated by a cellular transcription factor that is present in all activated T cells. Activation of CD4 T cells induces the transcription factor NFκB, which binds to promoters not only in the cellular DNA but also in the viral LTR, thereby initiating the transcription of viral RNA by the cellular RNA polymerase. This transcript is spliced in various ways to produce mRNAs for the viral proteins. The Gag and Gag–Pol proteins are translated from unspliced mRNA; Vif, Vpr, Vpu, and Env are translated from singly spliced viral mRNA; Tat, Rev, and Nef are translated from multiply spliced mRNA. At least two of the viral genes, *tat* and *rev*, encode proteins, **Tat** and **Rev** respectively, that promote viral replication in activated T cells. Tat is a potent transcriptional regulator, which functions as an elongation factor that enables the transcription of viral RNA by the RNA polymerase II complex. Tat contains two binding sites, contained in one domain, named the transactivation domain. The first of these allows Tat to bind to a host cellular protein, cyclin T1. This binding reaction promotes the binding of the Tat protein through the second binding site in its transactivation domain to an RNA sequence in the LTR of the virus known as the transcriptional activation region (TAR). The consequence of this interaction is to greatly enhance the rate of viral genome transcription, by causing the removal of negative elongation factors that block the transcriptional activity of RNA polymerase II. The expression of cyclin T1 is greatly increased in activated compared with quiescent T lymphocytes. This, in conjunction with the increased expression of NFκB in activated T cells, may explain the ability of HIV to lie dormant in resting T cells and replicate in activated T cells (Fig. 11.25).

Eukaryotic cells have mechanisms to prevent the export from the cell nucleus of incompletely spliced mRNA transcripts. This could pose a problem for a retrovirus that is dependent on the export of unspliced, singly spliced, and multiply spliced mRNA species in order to translate the full complement of viral proteins. The Rev protein is the viral solution to this problem. Export from

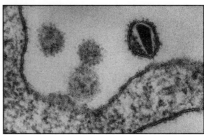

Fig. 11.25 Cells infected with HIV must be activated for the virus to replicate. Activation of CD4 T cells induces the expression of the transcription factor NFκB, which binds to the proviral LTR and initiates the transcription of the HIV genome into RNA. The first viral transcripts are processed extensively, producing spliced mRNAs encoding several regulatory proteins, including Tat and Rev. Tat both enhances transcription from the provirus and binds to the RNA transcripts, stabilizing them in a form that can be translated. The protein Rev binds the RNA transcripts and transports them to the cytosol. As levels of Rev increase, less extensively spliced and unspliced viral transcripts are transported out of the nucleus. The singly spliced and unspliced transcripts encode the structural proteins of the virus and the unspliced transcripts, which are the new viral genomes, are packaged with these to form many new virus particles. Photograph courtesy of H. Gelderblom.

the nucleus and translation of the three HIV proteins encoded by the fully spliced mRNA transcripts, Tat, Nef, and Rev, occurs early after viral infection by means of the normal host cellular mechanisms of mRNA export. The expressed Rev protein then enters the nucleus and binds to a specific viral RNA sequence, the Rev response element (RRE). Rev also binds to a host nucleocytoplasmic transport protein named Crm1, which engages a host pathway for exporting mRNA species through nuclear pores into the cytoplasm.

When the provirus is first activated, Rev levels are low, the transcripts are translocated slowly from the nucleus, and thus multiple splicing events can occur. Thus, more Tat and Rev are produced, and Tat in turn ensures that more viral transcripts are made. Later, when Rev levels have increased, the transcripts are translocated rapidly from the nucleus unspliced or only singly spliced. These unspliced or singly spliced transcripts are translated to produce the structural components of the viral core and envelope, together with the reverse transcriptase, the integrase, and the viral protease, all of which are needed to make new viral particles. The complete, unspliced transcripts that are exported from the nucleus late in the infectious cycle are required for the translation of *gag* and *pol* and are also destined to be packaged with the proteins as the RNA genomes of the new virus particles.

11-22 Drugs that block HIV replication lead to a rapid decrease in titer of infectious virus and an increase in CD4 T cells.

Studies with powerful drugs that completely block the cycle of HIV replication indicate that the virus is replicating rapidly at all phases of infection, including the asymptomatic phase. Two viral proteins in particular have been the target of drugs aimed at arresting viral replication. These are the viral reverse transcriptase, which is required for synthesis of the provirus, and the viral protease, which cleaves the viral polyproteins to produce the virion proteins and viral enzymes. Inhibitors of these enzymes prevent the establishment of further infection in uninfected cells. Cells that are already infected can continue to produce virions because, once the provirus is established, reverse transcriptase is not needed to make new virus particles, while the viral protease acts at a very late maturation step of the virus, and inhibition of the protease does not prevent virus from being released. However, in both cases, the released virions are not infectious and further cycles of infection and replication are prevented.

Because of the great efficacy of the protease inhibitors, it is possible to learn much about the kinetics of HIV replication *in vivo* by measuring the decline in viremia after the initiation of protease inhibitor therapy. For the first 2 weeks after starting treatment there is an exponential fall in plasma virus levels with a half-life of viral decay of about 2 days (Fig. 11.26). This phase reflects the decay in virus production from cells that were actively infected at the start of drug treatment, and indicates that the half-life of productively infected cells is similarly about 2 days. The results also show that free virus is cleared from the circulation very rapidly, with a half-life of about 6 hours. After 2 weeks, levels of virus in plasma have dropped by more than 95%, representing an almost total loss of productively infected CD4 lymphocytes. After this time, the rate of decline of plasma virus levels is much slower, reflecting the very slow decay of virus production from cells that provide a longer-lived reservoir of infection, such as dendritic cells and tissue macrophages, and from latently infected memory CD4 T cells that have been activated. Very long-term sources of infection might be CD4 memory T cells that continue to carry integrated provirus, and virus stored as immune complexes on follicular dendritic cells. These very long-lasting reservoirs of infection might prove to be resistant to drug therapy for HIV.

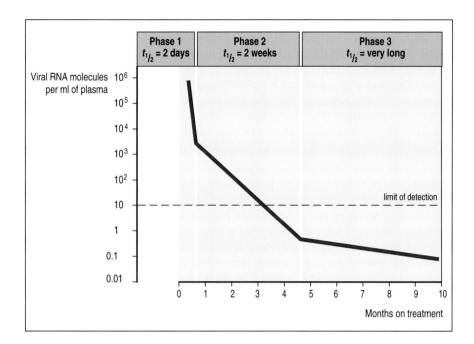

| Phase 1 $t_{1/2}$ = 2 days | Phase 2 $t_{1/2}$ = 2 weeks | Phase 3 $t_{1/2}$ = very long |

Viral RNA molecules per ml of plasma: 10^6, 10^5, 10^4, 10^3, 10^2, 10, 1, 0.1, 0.01

limit of detection

Months on treatment (0–10)

Fig. 11.26 Viral decay on drug treatment. The production of new HIV virus particles can be arrested for prolonged periods by combinations of protease inhibitors and viral reverse transcriptase inhibitors. After the initiation of such treatment, the virus produced by previously infected cells is no longer infectious, and virus production is curtailed as these cells die and no new cells are infected. The half-life of virus decay occurs in three phases. The first phase has a half-life of approximately 2 days and lasts for approximately 2 weeks, during which time viral production declines as the lymphocytes that were productively infected at the onset of treatment die. Released virus is rapidly cleared from the circulation, where it has a half-life ($t_{1/2}$) of 6 hours, and there is a decrease in virus levels in plasma of more than 95% during this first phase. The second phase lasts for about 6 months and has a half-life of about 2 weeks. During this phase, virus is released from infected macrophages and from resting, latently infected CD4 T cells stimulated to divide and develop productive infection. It is thought that there is then a third phase of unknown length that results from the reactivation of integrated provirus in memory T cells and other long-lived reservoirs of infection. This reservoir of latently infected cells might remain present for many years. Measurement of this phase of viral decay is impossible at present as viral levels in plasma are below detectable levels.

These studies show that most of the HIV present in the circulation of an infected individual is the product of rounds of replication in newly infected cells, and that virus from these productively infected cells is released into, and rapidly cleared from, the circulation at the rate of 10^9 to 10^{10} virions every day. This raises the question of what is happening to these virus particles: how are they removed so rapidly from the circulation? It seems most likely that HIV particles are opsonized by specific antibody and complement and removed by phagocytic cells of the mononuclear phagocyte system. Opsonized HIV particles can also be trapped on the surface of follicular dendritic cells, which are known to capture antigen:antibody complexes and retain them for prolonged periods (see Chapters 9 and 10).

The other issue raised by these studies is the effect of HIV replication on the population dynamics of CD4 T cells. The decline in plasma viremia is accompanied by a steady increase in CD4 T lymphocyte counts in peripheral blood: what is the source of the new CD4 T cells that appear once treatment is started? It seems highly unlikely that they are the recent progeny of stem cells that have developed in the thymus, because CD4 T cells are not normally produced in large numbers from the thymus even at its maximum rate of production in adolescents. Some investigators believe that these cells are emerging from sites of sequestration and add little to the total numbers of CD4 T cells in the body, whereas others advocate their origin from mature CD4 T cells that replicate, and argue that the production of such cells is an ongoing process that compensates for the continual loss of productively infected CD4 T cells.

11-23 HIV accumulates many mutations in the course of infection in a single individual and drug treatment is soon followed by the outgrowth of drug-resistant variants of the virus.

The rapid replication of HIV, with the generation of 10^9 to 10^{10} virions every day, coupled with a mutation rate of approximately 3×10^{-5} per nucleotide base per cycle of replication, leads to the generation of many variants of HIV in a single infected patient in the course of one day. Replication of a retroviral genome depends on two error-prone steps. Reverse transcriptase lacks the

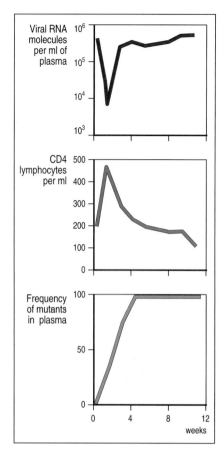

Fig. 11.27 Resistance of HIV to protease inhibitors. After the administration of a single protease inhibitor to a patient with HIV there is a precipitous fall in viral RNA levels in plasma with a half-life of approximately 2 days (top panel). This is accompanied by an initial rise in the number of CD4 T cells in peripheral blood (center panel). Within days of starting the drug, mutant drug-resistant variants can be detected in plasma (bottom panel) and in peripheral blood lymphocytes. After only 4 weeks of treatment, viral RNA levels and CD4 lymphocyte levels have returned to baseline levels, and 100% of plasma HIV is present as the drug-resistant mutant.

proofreading mechanisms associated with cellular DNA polymerases, and the RNA genomes of retroviruses are therefore copied into DNA with relatively low fidelity; the transcription of the proviral DNA into RNA copies by the cellular RNA polymerase is similarly a low-fidelity process. A rapidly replicating persistent virus that is going through these two steps repeatedly in the course of an infection can thereby accumulate many mutations, and numerous variants of HIV, sometimes called quasi-species, are found within a single infected individual. This very high variability was first recognized in HIV and has since proved to be common to the other lentiviruses.

As a consequence of its high variability, HIV rapidly develops resistance to antiviral drugs. When antiviral drugs are administered, variants of the virus that carry mutations conferring resistance to their effects emerge and expand until former levels of plasma virus are regained. Resistance to some of the protease inhibitors appears after only a few days (Fig. 11.27). Resistance to some of the nucleoside analogues that are potent inhibitors of reverse transcriptase develops in a similarly short time. By contrast, resistance to the nucleoside zidovudine (AZT), the first drug to be widely used for treating AIDS, takes months to develop. This is not because AZT is a more powerful inhibitor, but because resistance to zidovudine requires three or four mutations in the viral reverse transcriptase, whereas a single mutation can confer resistance to the protease inhibitors and other reverse-transcriptase inhibitors. As a result of the relatively rapid appearance of resistance to all known anti-HIV drugs, successful drug treatment might depend on the development of a range of antiviral drugs that can be used in combination. It might also be important to treat early in the course of an infection, thereby reducing the chances that a variant virus has accumulated all the necessary mutations to resist the entire cocktail. Current treatments follow this strategy and use combinations of viral protease inhibitors together with nucleoside analogues (see Fig. 11.26).

11-24 Lymphoid tissue is the major reservoir of HIV infection.

Although viral load and turnover are usually measured by detecting the viral RNA present in viral particles in the blood, the major reservoir of HIV infection is in lymphoid tissue, in which infected CD4 T cells, monocytes, macrophages, and dendritic cells are found. In addition, HIV is trapped in the form of immune complexes on the surface of follicular dendritic cells. These cells are not themselves infected but may act as a store of infective virions.

HIV infection takes different forms within different cells. As we have seen, more than 95% of the virus that can be detected in the plasma is derived from productively infected cells, which have a very short half-life of about 2 days. Productively infected CD4 lymphocytes are found in the T-cell areas of lymphoid tissue, and these are thought to succumb to infection in the course of being activated in an immune response. Latently infected memory CD4 cells that are activated in response to antigen presentation also produce virus. Such cells have a longer half-life of 2 to 3 weeks from the time that they are infected. Once activated, HIV can spread from these cells by rounds of replication in other activated CD4 T cells. In addition to the cells that are infected productively or latently, there is a further large population of cells infected by defective proviruses; such cells are not a source of infectious virus.

Macrophages and dendritic cells seem to be able to harbor replicating virus without necessarily being killed by it, and are therefore believed to be an important reservoir of infection, as well as a means of spreading virus to other tissues such as the brain. Although the function of macrophages as antigen-presenting cells does not seem to be compromised by HIV infection, it is thought that the virus causes abnormal patterns of cytokine secretion that could account for the wasting that commonly occurs in AIDS patients late in their disease.

11-25 An immune response controls but does not eliminate HIV.

Infection with HIV generates an adaptive immune response that contains the virus but only very rarely, if ever, eliminates it. The time course of various elements in the adaptive immune response to HIV is shown, together with the levels of infectious virus in plasma, in Fig. 11.28.

Seroconversion is the clearest evidence for an adaptive immune response to infection with HIV, but the generation of T lymphocytes responding to infected cells is thought by most workers in the field to be central in controlling the infection. Both CD8 cytotoxic T cells and T_H1 cells specifically responsive to infected cells are associated with the decline in detectable virus after the initial infection. These T-cell responses are unable to clear the infection completely and can cause some pathology. Nevertheless, there is evidence that the virus itself is cytopathic, and T-cell responses that reduce viral spread should therefore, on balance, reduce the pathology of the disease.

The ability of cytotoxic T lymphocytes to destroy HIV-infected cells is demonstrated by studies of peripheral blood cells from infected individuals, in which cytotoxic T cells specific for viral peptides can be shown to kill infected cells *in vitro*. *In vivo*, cytotoxic T cells can be seen to invade sites of HIV replication and they could, in theory, be responsible for killing many productively infected cells before any infectious virus can be released, thereby containing viral load at the quasi-stable levels that are characteristic of the asymptomatic period. The best evidence for the clinical importance of the control of HIV-infected cells by CD8 cytotoxic T cells comes from studies relating the numbers and activity of CD8 T cells to viral load. An inverse correlation was found between the number of CD8 T cells carrying a receptor specific for an HLA-A2-restricted HIV peptide and plasma RNA viral load. Similarly, patients with high levels of HIV-specific CD8 T cells showed slower progression of disease than those with low levels. There is also direct evidence from experiments in macaques infected with simian immuno-deficiency virus (SIV) that CD8 cytotoxic T cells control retrovirally-infected cells *in vivo*. Treatment of infected animals with depleting anti-CD8 monoclonal antibodies was followed by a large increase in viral load.

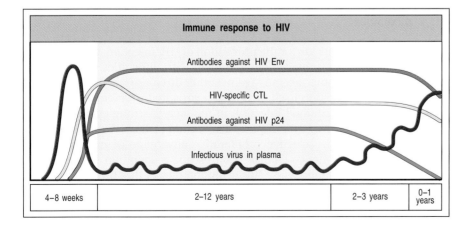

Fig. 11.28 The immune response to HIV. Infectious virus is present at relatively low levels in the peripheral blood of infected individuals during a prolonged asymptomatic phase, during which the virus is replicated persistently in lymphoid tissues. During this period, CD4 T-cell counts gradually decline, although antibodies and CD8 cytotoxic T cells directed against the virus remain at high levels. Two different antibody responses are shown in the figure, one to the envelope protein (Env) of HIV, and one to the core protein p24. Eventually, the levels of antibody and HIV-specific cytotoxic T lymphocytes (CTLs) also decline, and there is a progressive increase of infectious HIV in the peripheral blood.

Mutations that occur as HIV replicates can allow variants of the virus to escape recognition by antibody or cytotoxic T cells and can contribute to the failure of the immune system to contain the infection in the long term. Direct escape of virus-infected cells from killing by cytotoxic T lymphocytes has been shown by the occurrence of mutations of immunodominant viral peptides presented by MHC class I molecules. In other cases, variant peptides produced by the virus have been found to act as antagonists (see Section 6-12) for T cells responsive to the wild-type epitope, thus allowing both mutant and wild-type viruses to survive. Mutant peptides acting as antagonists have also been reported in hepatitis B virus infections, and similar mutant peptides might contribute to the persistence of some viral infections, especially when, as often happens, the immune response of an individual is dominated by T cells specific for a particular epitope.

11-26 HIV infection leads to low levels of CD4 T cells, increased susceptibility to opportunistic infection, and eventually to death.

There are three dominant mechanisms for the loss of CD4 T cells in HIV infection. First, there is evidence for direct viral killing of infected cells; second, there is increased susceptibility to the induction of apoptosis in infected cells; and third, there is killing of infected CD4 T cells by CD8 cytotoxic lymphocytes that recognize viral peptides.

When CD4 T-cell numbers decline below a critical level, cell-mediated immunity is lost, and infections with a variety of opportunistic microbes appear (Fig. 11.29). Typically, resistance is lost early to oral *Candida* species and to *Mycobacterium tuberculosis*, which shows as an increased prevalence of thrush (oral candidiasis) and tuberculosis. Later, patients suffer from shingles, caused by the activation of latent herpes zoster, from EBV-induced B-cell lymphomas, and from Kaposi's sarcoma, a tumor of endothelial cells that probably represents a response both to cytokines produced in the infection and to a novel herpes virus called HHV-8 that was identified in these lesions. Pneumonia caused by the fungus *Pneumocystis carinii* is common and often fatal. In the final stages of AIDS, infection with cytomegalovirus or *Mycobacterium avium* complex is more prominent. It is important to note that not all patients with AIDS get all these infections or tumors, and there are other tumors and infections that are less prominent but still significant. Rather, this is a list of the commonest opportunistic infections and tumors, most of which are normally controlled by robust CD4 T cell-mediated immunity that wanes as the CD4 T-cell counts drop toward zero (see Fig. 11.21).

11-27 Vaccination against HIV is an attractive solution but poses many difficulties.

A safe and effective vaccine for the prevention of HIV infection and AIDS is an attractive goal, but its achievement is fraught with difficulties that have not been faced in developing vaccines against other diseases. The first problem is the nature of the infection itself, featuring a virus that proliferates extremely rapidly and causes sustained infection in the face of strong cytotoxic T-cell and antibody responses. As we discussed in Section 11-25, HIV evolves in individual patients by the selective proliferative advantage of mutant virions encoding peptide sequence changes that escape recognition by antibodies and by cytotoxic T lymphocytes. This evolution means that the development of therapeutic vaccination strategies to block the development of AIDS in HIV-infected patients will be extremely difficult. Even after the viremia has been largely cleared by drug therapy, immune responses to HIV fail to prevent drug-resistant virus from rebounding and replicating at pretreatment levels.

Infections	
Parasites	*Toxoplasma* spp. *Cryptosporidium* spp. *Leishmania* spp. *Microsporidium* spp.
Intracellular bacteria	*Mycobacterium tuberculosis* *Mycobacterium avium intracellulare* *Salmonella* spp.
Fungi	*Pneumocystis carinii* *Cryptococcus neoformans* *Candida* spp. *Histoplasma capsulatum* *Coccidioides immitis*
Viruses	Herpes simplex Cytomegalovirus Varicella zoster

Malignancies
Kaposi's sarcoma Non-Hodgkin's lymphoma, including EBV-positive Burkitt's lymphoma Primary lymphoma of the brain

Fig. 11.29 A variety of opportunistic pathogens and cancers can kill AIDS patients. Infections are the major cause of death in AIDS, with respiratory infection with *Pneumocystis carinii* and mycobacteria being the most prominent. Most of these pathogens require effective macrophage activation by CD4 T cells or effective cytotoxic T cells for host defense. Opportunistic pathogens are present in the normal environment but cause severe disease primarily in immunocompromised hosts, such as AIDS patients and cancer patients. AIDS patients are also susceptible to several rare cancers, such as Kaposi's sarcoma and various lymphomas, suggesting that immune surveillance of their causative herpes viruses by T cells can normally prevent such tumors (see Chapter 14).

The second problem is our uncertainty over what form protective immunity to HIV might take. It is not known whether antibodies, cytotoxic T lymphocyte responses, or both are necessary to achieve protective immunity, and which epitopes might provide the targets of protective immunity. Third, if strong cytotoxic responses are necessary to provide protection against HIV, these might be difficult to develop and sustain through vaccination. Other effective viral vaccines rely on the use of live, attenuated viruses and there are concerns over the safety of pursuing this approach for HIV. Another possible approach is the use of DNA vaccination, a technique that we discuss in Section 14-25. Both of these approaches are being tested in animal models.

The fourth problem is the ability of the virus to persist in latent form as a transcriptionally silent provirus, which is invisible to the immune system. This might prevent the immune system from clearing the infection once it has been established. In summary, the ability of the immune system to clear infectious virus remains uncertain.

However, against this pessimistic background, there are grounds for hope that successful vaccines can be developed. Of particular interest are rare groups of people who have been exposed often enough to HIV to make it virtually certain that they should have become infected but who have not developed the disease. In some cases this is due to an inherited deficiency in the chemokine receptor used as co-receptor for HIV entry, as we explained in Section 11-19. However, this mutant chemokine receptor does not occur in Africa, where one such group has been identified. A small group of Gambian and Kenyan prostitutes who are estimated to have been exposed to many HIV-infected male partners each month for up to 5 years were found to lack antibody responses but to have cytotoxic T lymphocyte responses to a variety of peptide epitopes from HIV. These women seem to have been naturally immunized against HIV.

Although there is no perfect animal model for the development of HIV vaccines, one model system is based on simian immunodeficiency virus (SIV), which is closely related to HIV and infects macaques. SIV causes a similar disease to AIDS in Asian macaques such as the cynomolgus monkey, but does not cause disease in African cercopithecus monkeys such as the African green monkey, with which SIV has probably coexisted for up to a million years. Live attenuated SIV vaccines lacking the *nef* gene, and hybrid HIV–SIV viruses have been developed to test the principles of vaccination in primates, and both have proved successful in protecting primates against subsequent infection by fully virulent viruses. However, there are substantial difficulties to be overcome in the development of live attenuated HIV vaccines for use in at-risk populations, not least the worry of recombination between vaccine strains and wild-type viruses leading to reversion to a virulent phenotype. The alternative approach of DNA vaccination is being piloted in primate experiments, with some early signs of success.

Subunit vaccines, which induce immunity to only some proteins in the virus, have also been made. One such vaccine has been made from the envelope protein gp120 and has been tested on chimpanzees. This vaccine proved to be specific to the precise strain of virus used to make it, and was therefore useless in protection against natural infection. Subunit vaccines are also less efficient at inducing prolonged cytotoxic T-cell responses.

Finally, there are difficult ethical issues in the development of a vaccine. It would be unethical to conduct a vaccine trial without trying at the same time to minimize the exposure of a vaccinated population to the virus itself. However, the effectiveness of a vaccine can only be assessed in a population in which the exposure rate to the virus is high enough to assess whether vaccination is protective against infection. This means that initial vaccine trials might have to be conducted in countries where the incidence of infection is very high and public health measures have not yet succeeded in reducing the spread of HIV.

11-28 Prevention and education are one way in which the spread of HIV and AIDS can be controlled.

The one way in which we know we can protect against infection with HIV is by avoiding contact with body fluids, such as semen, blood, blood products, or milk from people who are infected. Indeed, it has been demonstrated repeatedly that this precaution, simple enough in the developed world, is sufficient to prevent infection, as health-care workers can take care of AIDS patients for long periods without seroconversion or signs of infection.

For this strategy to work, however, one must be able to test people at risk of infection with HIV periodically, so that they can take the steps necessary to avoid passing the virus to others. This, in turn, requires strict confidentiality and mutual trust. A barrier to the control of HIV is the reluctance of individuals to find out whether they are infected, especially as one of the consequences of a positive HIV test is stigmatization by society. As a result, infected individuals can unwittingly infect many others. Balanced against this is the success of therapy with combinations of the new protease inhibitors and reverse transcriptase inhibitors, which provides an incentive for potentially infected people to identify the presence of infection and gain the benefits of treatment. Responsibility is at the heart of AIDS prevention, and a law guaranteeing the rights of people infected with HIV might go a long way to encouraging responsible behavior. The rights of HIV-infected people are protected in the Netherlands and Sweden. The problem in the less-developed nations, where elementary health precautions are extremely difficult to establish, is more profound.

Summary.

Infection with the human immunodeficiency virus (HIV) is the cause of acquired immune deficiency syndrome (AIDS). This worldwide epidemic is now spreading at an alarming rate, especially through heterosexual contact in less-developed countries. HIV is an enveloped retrovirus that replicates in cells of the immune system. Viral entry requires the presence of CD4 and a particular chemokine receptor, and the viral cycle is dependent on transcription factors found in activated T cells. Infection with HIV causes a loss of CD4 T cells and an acute viremia that rapidly subsides as cytotoxic T-cell responses develop, but HIV infection is not eliminated by this immune response. HIV establishes a state of persistent infection in which the virus is continually replicating in newly infected cells. The current treatment consists of combinations of viral protease inhibitors together with nucleoside analogues and causes a rapid decrease in virus levels and a slower increase in CD4 T-cell counts. The main effect of HIV infection is the destruction of CD4 T cells, which occurs through the direct cytopathic effects of HIV infection and through killing by CD8 cytotoxic T cells. As the CD4 T-cell counts wane, the body becomes progressively more susceptible to opportunistic infection with intracellular microbes. Eventually, most HIV-infected individuals develop AIDS and die; however a small minority (3–7%), remain healthy for many years, with no apparent ill effects of infection. We hope to be able to learn from these individuals how infection with HIV can be controlled. The existence of such people and other people who have been naturally immunized against infection gives hope that it will be possible to develop effective vaccines against HIV.

Summary to Chapter 11.

Whereas most infections elicit protective immunity, most successful pathogens have developed some means of evading a fully effective immune response, and some result in serious, persistent disease. In addition, some individuals have inherited deficiencies in different components of the immune system, making them highly susceptible to certain classes of infectious agent. Persistent infection and immunodeficiency illustrate the importance of innate and adaptive immunity in effective host defense against infection and present huge challenges for future immunological research. The human immunodeficiency virus (HIV) combines the characteristics of a persistent infectious agent with the ability to create immunodeficiency in its human host, a combination that is usually slowly lethal to the patient. The key to fighting new pathogens like HIV is to develop our understanding of the basic properties of the immune system and its role in combating infection more fully.

General references.

Cohen, O.J., Kinter, A., and Fauci, A.S.: **Host factors in the pathogenesis of HIV disease.** *Immunol. Rev.* 1997, **159**:31-48.

Fischer, A., Cavazzana-Calvo, M., De-Saint-Basile, G., DeVillartay, J.P., Di-Santo, J.P., Hivroz, C., Rieux-Laucat, F., and Le-Deist, F.: **Naturally occurring primary deficiencies of the immune system.** *Annu. Rev. Immunol.* 1997, **15**:93-124.

Gao, F., Bailes, E., Robertson, D.L., Chen, Y., Rodenburg, C.M., Michael, S.F., Cummins, L.B., Arthur, L.O., Peeters, M., Shaw, G.M., Sharp, P.M., and Hahn, B.H.: **Origin of HIV-1 in the chimpanzee** *Pan troglodytes troglodytes. Nature* 1999, **397**:436-441.

Hill, A.V.: **The immunogenetics of human infectious diseases.** *Annu. Rev. Immunol.* 1998, **16**:593-617.

Kotwal, G.J.: **Microorganisms and their interaction with the immune system.** *J. Leukoc. Biol.* 1997, **62**:415-429.

Primary immunodeficiency diseases. Report of an IUIS Scientific Committee. International Union of Immunological Societies. *Clin. Exp. Immunol.* 1999, **118 Suppl 1**:1-28.

Royce, R.A., Sena, A., Cates, W., Jr., and Cohen, M.S.: **Sexual transmission of HIV.** *N. Engl. J. Med.* 1997, **336**:1072-1078.

Tortorella, D., Gewurz, B.E., Furman, M.H., Schust, D.J., and Ploegh, H.L.: **Viral subversion of the immune system.** *Annu. Rev. Immunol.* 2000, **18**:861-926.

Section references.

11-1   Antigenic variation allows pathogens to escape from immunity.

Clegg, S., Hancox, L.S., and Yeh, K.S.: *Salmonella typhimurium* **fimbrial phase variation and FimA expression.** *J. Bacteriol.* 1996, **178**:542-545.

Cossart, P.: **Host/pathogen interactions. Subversion of the mammalian cell cytoskeleton by invasive bacteria.** *J. Clin. Invest.* 1997, **99**:2307-2311.

Donelson, J.E., Hill, K.L., and El-Sayed, N.M.: **Multiple mechanisms of immune evasion by African trypanosomes.** *Mol. Biochem. Parasitol.* 1998, **91**:51-66.

Ito, T., Couceiro, J.N., Kelm, S., Baum, L.G., Krauss, S., Castrucci, M.R., Donatelli, I., Kida, H., Paulson, J.C., Webster, R.G., and Kawaoka, Y.: **Molecular basis for the generation in pigs of influenza A viruses with pandemic potential.** *J. Virol.* 1998, **72**:7367-7373.

Rudenko, G., Cross, M., and Borst, P.: **Changing the end: antigenic variation orchestrated at the telomeres of African trypanosomes.** *Trends Microbiol.* 1998, **6**:113-116.

Seifert, H.S., Wright, C.J., Jerse, A.E., Cohen, M.S., and Cannon, J.G.: **Multiple gonococcal pilin antigenic variants are produced during experimental human infections.** *J. Clin. Invest.* 1994, **93**:2744-2749.

Shu, L.L., Bean, W.J., and Webster, R.G.: **Analysis of the evolution and variation of the human influenza A virus nucleoprotein gene from 1933 to 1990.** *J. Virol.* 1993, **67**:2723-2729.

Webster, R.G., Bean, W.J., Gorman, O.T., Chambers, T.M., and Kawaoka, Y.: **Evolution and ecology of influenza A viruses.** *Microbiol. Rev.* 1992, **56**:152-179.

11-2   Some viruses persist *in vivo* by ceasing to replicate until immunity wanes.

Cohen, J.I.: **Epstein–Barr virus infection.** *N. Engl. J. Med.* 2000, **343**:481-492.

Ehrlich, R.: **Selective mechanisms utilized by persistent and oncogenic viruses to interfere with antigen processing and presentation.** *Immunol. Res.* 1995, **14**:77-97.

Garcia Blanco, M.A., and Cullen, B.R.: **Molecular basis of latency in pathogenic human viruses.** *Science* 1991, **254**:815-820.

Ho, D.Y.: **Herpes simplex virus latency: molecular aspects.** *Prog. Med. Virol.* 1992, **39**:76-115.

Longnecker, R., and Miller, C.L.: **Regulation of Epstein–Barr virus latency by latent membrane protein 2.** *Trends Microbiol.* 1996, **4**:38-42.

Nash, A.A.: **T cells and the regulation of herpes simplex virus latency and reactivation.** *J. Exp. Med.* 2000, **191**:1455-1458.

Steiner, I., and Kennedy, P.G.: **Molecular biology of herpes simplex virus type 1 latency in the nervous system.** *Mol. Neurobiol.* 1993, **7**:137-159.

Wensing, B., and Farrell, P.J.: **Regulation of cell growth and death by Epstein–Barr virus.** *Microbes. Infect.* 2000, **2**:77-84.

11-3 Some pathogens resist destruction by host defense mechanisms or exploit them for their own purposes.

Abendroth, A., and Arvin, A.: **Varicella-zoster virus immune evasion.** *Immunol. Rev.* 1999, **168**:143-156.

Alcami, A., and Koszinowski, U.H.: **Viral mechanisms of immune evasion.** *Immunol. Today* 2000, **21**:447-455.

Hengel, H., and Koszinowski, U.H.: **Interference with antigen processing by viruses.** *Curr. Opin. Immunol.* 1997, **9**:470-476.

Radolf, J.D.: **Role of outer membrane architecture in immune evasion by** *Treponema pallidum* **and** *Borrelia burgdorferi.* *Trends Microbiol.* 1994, **2**:307-311.

Sinai, A.P., and Joiner, K.A.: **Safe haven: the cell biology of nonfusogenic pathogen vacuoles.** *Annu. Rev. Microbiol.* 1997, **51**:415-462.

Smith, G.L.: **Virus proteins that bind cytokines, chemokines or interferons.** *Curr. Opin. Immunol.* 1996, **8**:467-471.

Smith, G.L., Symons, J.A., Khanna, A., Vanderplasschen, A., and Alcami, A.: **Vaccinia virus immune evasion.** *Immunol. Rev.* 1997, **159**:137-154.

11-4 Immunosuppression or inappropriate immune responses can contribute to persistent disease.

Bhardwaj, N.: **Interactions of viruses with dendritic cells: a double-edged sword.** *J. Exp. Med.* 1997, **186**:795-799.

Bloom, B.R., Modlin, R.L., and Salgame, P.: **Stigma variations: observations on suppressor T cells and leprosy.** *Annu. Rev. Immunol.* 1992, **10**:453-488.

Fleischer, B.: **Superantigens.** *APMIS* 1994, **102**:3-12.

Salgame, P., Abrams, J.S., Clayberger, C., Goldstein, H., Convit, J., Modlin, R.L., and Bloom, B.R.: **Differing lymphokine profiles of functional subsets of human CD4 and CD8 T cell clones.** *Science* 1991, **254**:279-282.

11-5 Immune responses can contribute directly to pathogenesis.

Cheever, A.W., and Yap, G.S.: **Immunologic basis of disease and disease regulation in schistosomiasis.** *Chem. Immunol.* 1997, **66**:159-176.

Doherty, P.C., Topham, D.J., Tripp, R.A., Cardin, R.D., Brooks, J.W., and Stevenson, P.G.: **Effector CD4[+] and CD8[+] T-cell mechanisms in the control of respiratory virus infections.** *Immunol. Rev.* 1997, **159**:105-117.

Openshaw, P.J.: **Immunopathological mechanisms in respiratory syncytial virus disease.** *Springer Semin. Immunopathol.* 1995, **17**:187-201.

Ross, R.: **Mouse mammary tumor virus and its interaction with the immune system.** *Immunol. Res.* 1998; **17**:209-216.

11-6 A history of repeated infections suggests a diagnosis of immuno-deficiency.

Rosen, F.S., Cooper, M.D., and Wedgwood, R.J.: **The primary immunodeficiencies.** *N. Engl. J. Med.* 1995, **333**:431-440.

11-7 Inherited immunodeficiency diseases are caused by recessive gene defects.

Fischer, A.: **Inherited disorders of lymphocyte development and function.** *Curr. Opin. Immunol.* 1996, **8**:445-447.

Kokron, C.M., Bonilla, F.A., Oettgen, H.C., Ramesh, N., Geha, R.S., and Pandolfi, F.: **Searching for genes involved in the pathogenesis of primary immunodeficiency diseases: lessons from mouse knockouts.** *J. Clin. Immunol.* 1997, **17**:109-126.

Smart, B.A., and Ochs, H.D.: **The molecular basis and treatment of primary immunodeficiency disorders.** *Curr. Opin. Pediatr.* 1997, **9**:570-576.

Smith, C.I., and Notarangelo, L.D.: **Molecular basis for X-linked immunodeficiencies.** *Adv. Genet.* 1997, **35**:57-115.

11-8 The main effect of low levels of antibody is an inability to clear extracellular bacteria.

Bruton, O.C.: **Agammaglobulinemia.** *Pediatrics* 1952, **9**:722-728.

Burrows, P.D., and Cooper, M.D.: **IgA deficiency.** *Adv. Immunol.* 1997, **65**:245-276.

Desiderio, S.: **Role of Btk in B cell development and signaling.** *Curr. Opin. Immunol.* 1997, **9**:534-540.

Fuleihan, R., Ramesh, N., and Geha, R.S.: **X-linked agammaglobulinemia and immunoglobulin deficiency with normal or elevated IgM: immunodeficiencies of B cell development and differentiation.** *Adv. Immunol.* 1995, **60**:37-56.

Lee, M.L., Gale, R.P., and Yap, P.L.: **Use of intravenous immunoglobulin to prevent or treat infections in persons with immune deficiency.** *Annu. Rev. Med.* 1997, **48**:93-102.

Notarangelo, L.D.: **Immunodeficiencies caused by genetic defects in protein kinases.** *Curr. Opin. Immunol.* 1996, **8**:448-453.

Ochs, H.D., and Wedgwood, R.J.: **IgG subclass deficiencies.** *Annu. Rev. Med.* 1987, **38**:325-340.

Preud'homme, J.L., and Hanson, L.A.: **IgG subclass deficiency.** *Immunodefic. Rev.* 1990, **2**:129-149.

11-9 T-cell defects can result in low antibody levels.

Ramesh, N., Seki, M., Notarangelo, L.D., and Geha, R.S.: **The hyper-IgM (HIM) syndrome.** *Springer Semin. Immunopathol.* 1998, **19**:383-389.

11-10 Defects in complement components cause defective humoral immune function.

Botto, M., Dell'Agnola, C., Bygrave, A.E., Thompson, E.M., Cook, H.T., Petry, F., Loos, M., Colten, H.R., and Rosen, F.S.: **Complement deficiencies.** *Annu. Rev. Immunol.* 1992, **10**:809-834.

Morgan, B.P., and Walport, M.J.: **Complement deficiency and disease.** *Immunol. Today* 1991, **12**:301-306.

Pandolfi, P.P., and Walport, M.J.: **Homozygous C1q deficiency causes glomerulonephritis associated with multiple apoptotic bodies.** *Nat. Genet.* 1998, **19**:56-59.

11-11 Defects in phagocytic cells permit widespread bacterial infections.

Fischer, A., Lisowska Grospierre, B., Anderson, D.C., and Springer, T.A.: **Leukocyte adhesion deficiency: molecular basis and functional consequences.** *Immunodefic. Rev.* 1988, **1**:39-54.

Jackson, S.H., Gallin, J.I., and Holland, S.M.: **The p47phox mouse knock-out model of chronic granulomatous disease.** *J. Exp. Med.* 1995, **182**:751-758.

Karsan, A., Cornejo, C.J., Winn, R.K., Schwartz, B.R., Way, W., Lannir, N., Gershoni-Baruch, R., Etzioni, A., Ochs, H.D., and Harlan, J.M.: **Leukocyte Adhesion Deficiency Type II is a generalized defect of** *de novo* **GDP-fucose biosynthesis. Endothelial cell fucosylation is not required for neutrophil rolling on human nonlymphoid endothelium.** *J. Clin. Invest.* 1998, **101**:2438-2445.

Malech, H.L., and Nauseef, W.M.: **Primary inherited defects in neutrophil function: etiology and treatment.** *Semin. Hematol.* 1997, **34**:279-290.

Rotrosen, D., and Gallin, J.I.: **Disorders of phagocyte function.** *Annu. Rev. Immunol.* 1987, **5**:127-150.

Spritz, R.A.: **Genetic defects in Chediak-Higashi syndrome and the beige mouse.** *J. Clin. Immunol.* 1998, **18**:97-105.

11-12 Defects in T-cell function result in severe combined immuno-deficiencies.

Bosma, M.J., and Carroll, A.M.: **The SCID mouse mutant: definition, characterization, and potential uses.** *Annu. Rev. Immunol.* 1991, **9**:323-350.

Gadola, S.D., Moins-Teisserenc, H.T., Trowsdale, J., Gross, W.L., and Cerundolo, V.: **TAP deficiency syndrome.** *Clin. Exp. Immunol.* 2000, **121**:173-178.

Gennery, A.R., Cant, A.J., and Jeggo, P.A.: **Immunodeficiency associated with DNA repair defects.** *Clin. Exp. Immunol.* 2000, **121**:1-7.

Grusby, M.J., and Glimcher, L.H.: **Immune responses in MHC class II-deficient mice.** *Annu. Rev. Immunol.* 1995, **13**:417-435.

Hirschhorn, R.: **Adenosine deaminase deficiency: molecular basis and recent developments.** *Clin. Immunol. Immunopathol.* 1995, **76**:S219-S227.

Lavin, M.F., and Shiloh, Y.: **The genetic defect in ataxia-telangiectasia.** *Annu. Rev. Immunol.* 1997, **15**:177-202.

Masternak, K., Barras, E., Zufferey, M., Conrad, B., Corthals, G., Aebersold, R., Sanchez, J.C., Hochstrasser, D.F., Mach, B., and Reith, W.: **A gene encoding a novel RFX-associated transactivator is mutated in the majority of MHC class II deficiency patients.** *Nat. Genet.* 1998, **20**:273-277.

Schwarz, K., and Bartram, C.R.: **V(D)J recombination pathology.** *Adv. Immunol.* 1996, **61**:285-326.

Steimle, V., Reith, W., and Mach, B.: **Major histocompatibility complex class II deficiency: a disease of gene regulation.** *Adv. Immunol.* 1996, **61**:327-340.

11-13 Defective T-cell signaling, cytokine production, or cytokine action can cause immunodeficiency.

Arnaiz Villena, A., Timon, M., Corell, A., Perez Aciego, P., Martin Villa, J.M., and Regueiro, J.R.: **Brief report: primary immunodeficiency caused by mutations in the gene encoding the CD3-gamma subunit of the T-lymphocyte receptor.** *N. Engl. J. Med.* 1992, **327**:529-533.

Castigli, E., Pahwa, R., Good, R.A., Geha, R.S., and Chatila, T.A.: **Molecular basis of a multiple lymphokine deficiency in a patient with severe combined immunodeficiency.** *Proc. Natl. Acad. Sci. USA* 1993, **90**:4728-4732.

DiSanto, J.P., Keever, C.A., Small, T.N., Nicols, G.L., O'Reilly, R.J., and Flomenberg, N.: **Absence of interleukin 2 production in a severe combined immunodeficiency disease syndrome with T cells.** *J. Exp. Med.* 1990, **171**:1697-1704.

DiSanto, J.P., Rieux Laucat, F., Dautry Varsat, A., Fischer, A., and de Saint Basile, G.: **Defective human interleukin 2 receptor gamma chain in an atypical X chromosome-linked severe combined immunodeficiency with peripheral T cells.** *Proc. Natl. Acad. Sci. USA* 1994, **91**:9466-9470.

Leonard, W.J.: **The molecular basis of X linked severe combined immunodeficiency.** *Annu. Rev. Med.* 1996, **47**:229-239.

Ochs, H.D.: **The Wiskott–Aldrich syndrome.** *Springer Semin. Immunopathol.* 1998, **9**:435-458.

Snapper, S.B., and Rosen, F.S.: **The Wiskott-Aldrich syndrome protein (WASP): roles in signaling and cytoskeletal organization.** *Annu. Rev. Immunol.* 1999, **17**:905-929.

11-14 The normal pathways for host defense against intracellular bacteria are illustrated by genetic deficiencies of IFN-γ and IL-12 and their receptors.

Altare, F., Durandy, A., Lammas, D., Emile, J.F., Lamhamedi, S., Le Deist, F., Drysdale, P., Jouanguy, E., Doffinger, R., Bernaudin, F., Jeppsson, O., Gollob, J.A., Meinl, E., Segal, A.W., Fischer, A., Kumararatne, D., and Casanova, J.L.: **Impairment of mycobacterial immunity in human interleukin-12 receptor deficiency.** *Science* 1998, **280**:1432-1435.

Altare, F., Lammas, D., Revy, P., Jouanguy, E., Doffinger, R., Lamhamedi, S., Drysdale, P., Scheel-Toellner, D., Girdlestone, J., Darbyshire, P., Wadhwa, M., Dockrell, H., Salmon, M., Fischer, A., Durandy, A., Casanova, J.L., and Kumararatne, D.S.: **Inherited interleukin 12 deficiency in a child with bacille Calmette-Guerin and *Salmonella enteritidis* disseminated infection.** *J. Clin. Invest.* 1998, **102**:2035-2040.

Gately, M.K., Renzetti, L.M., Magram, J., Stern, A.S., Adorini, L., Gubler, U., and Presky, D.H.: **The interleukin-12/interleukin-12-receptor system: role in normal and pathologic immune responses.** *Annu. Rev. Immunol.* 1998, **16**:495-521.

Jouanguy, E., Altare, F., Lamhamedi, S., Revy, P., Emile, J.F., Newport, M., Levin, M., Blanche, S., Seboun, E., Fischer, A., and Casanova, J.L.: **Interferon-gamma-receptor deficiency in an infant with fatal bacille Calmette-Guerin infection.** *N. Engl. J. Med.* 1996, **335**:1956-1961.

Jouanguy, E., Doffinger, R., Dupuis, S., Pallier, A., Altare, F., and Casanova, J.L.: **IL-12 and IFN-gamma in host defense against mycobacteria and salmonella in mice and men.** *Curr. Opin. Immunol.* 1999, **11**:346-351.

Newport, M.J., Huxley, C.M., Huston, S., Hawrylowicz, C.M., Oostra, B.A., Williamson, R., and Levin, M.: **A mutation in the interferon-gamma-receptor gene and susceptibility to mycobacterial infection.** *N. Engl. J. Med.* 1996, **335**:1941-1949.

11-15 X-linked lymphoproliferative syndrome is associated with fatal infection by Epstein–Barr virus and with the development of lymphomas.

Brandau, O., Schuster, V., Weiss, M., Hellebrand, H., Fink, F.M., Kreczy, A., Friedrich, W., Strahm, B., Niemeyer, C., Belohradsky, B.H., and Meindl, A.: **Epstein-Barr virus-negative boys with non-Hodgkin lymphoma are mutated in the SH2D1A gene, as are patients with X-linked lymphoproliferative disease (XLP).** *Hum. Mol. Genet.* 1999, **8**:2407-2413.

Coffey, A.J., Brooksbank, R.A., Brandau, O., Oohashi, T., Howell, G.R., Bye, J.M., Cahn, A.P., Durham, J., Heath, P., Wray, P., Pavitt, R., Wilkinson, J., Leversha, M., Huckle, E., Shaw-Smith, C.J., Dunham, A., Rhodes, S., Schuster, V., Porta, G., Yin, L., Serafini, P., Sylla, B., Zollo, M., Franco, B., and Bentley, D.R.: **Host response to EBV infection in X-linked lymphoproliferative disease results from mutations in an SH2-domain encoding gene.** *Nat. Genet.* 1998, **20**:129-135.

Howie, D., Sayos, J., Terhorst, C., and Morra, M.: **The gene defective in X-linked lymphoproliferative disease controls T cell dependent immune surveillance against Epstein-Barr virus.** *Curr. Opin. Immunol.* 2000, **12**:474-478.

Nichols, K.E., Harkin, D.P., Levitz, S., Krainer, M., Kolquist, K.A., Genovese, C., Bernard, A., Ferguson, M., Zuo, L., Snyder, E., Buckler, A.J., Wise, C., Ashley, J., Lovett, M., Valentine, M.B., Look, A.T., Gerald, W., Housman, D.E., and Haber, D.A.: **Inactivating mutations in an SH2 domain-encoding gene in X-linked lymphoproliferative syndrome.** *Proc. Natl. Acad. Sci. USA* 1998, **95**:13765-13770.

Sayos, J., Wu, C., Morra, M., Wang, N., Zhang, X., Allen, D., van Schaik, S., Notarangelo, L., Geha, R., Roncarolo, M.G., Oettgen, H., De Vries, J.E., Aversa, G., and Terhorst, C.: **The X-linked lymphoproliferative-disease gene product SAP regulates signals induced through the co-receptor SLAM.** *Nature* 1998, **395**:462-469.

11-16 Bone marrow transplantation or gene therapy can be useful to correct genetic defects.

Anderson, W.F.: **Human gene therapy.** *Nature* 1998, **392**:25-30.

Candotti, F., and Blaese, R.M.: **Gene therapy of primary immunodeficiencies.** *Springer Semin. Immunopathol.* 1998, **19**:493-508.

Cavazzana-Calvo, M., Hacein-Bey, S., De Saint, B.G., Gross, F., Yvon, E., Nusbaum, P., Selz, F., Hue, C., Certain, S., Casanova, J.L., Bousso, P., Deist, F.L., and Fischer, A.: **Gene therapy of human severe combined immunodeficiency (SCID)-X1 disease.** *Science* 2000, **288**:669-672.

Cournoyer, D., and Caskey, C.T.: **Gene therapy of the immune system.** *Annu. Rev. Immunol.* 1993, **11**:297-329.

Fischer, A., Haddad, E., Jabado, N., Casanova, J.L., Blanche, S., Le Deist, F., and Cavazzana-Calvo, M.: **Stem cell transplantation for immunodeficiency.** *Springer Semin. Immunopathol.* 1998, **19**:479-492.

Kohn, D.B., Hershfield, M.S., Carbonaro, D., Shigeoka, A., Brooks, J., Smogorzewska, E.M., Barsky, L.W., Chan, R., Burotto, F., Annett, G., Nolta, J.A., Crooks, G., Kapoor, N., Elder, M., Wara, D., Bowen, T., Madsen, E., Snyder, F.F., Bastian, J., Muul, L., Blaese, R.M., Weinberg, K., and Parkman, R.: **T lymphocytes with a normal ADA gene accumulate after transplantation of transduced autologous umbilical cord blood CD34+ cells in ADA-deficient SCID neonates.** *Nat. Med.* 1998, **4**:775-780.

Onodera, M., Ariga, T., Kawamura, N., Kobayashi, I., Ohtsu, M., Yamada, M., Tame, A., Furuta, H., Okano, M., Matsumoto, S., Kotani, H., McGarrity, G.J., Blaese, R.M., and Sakiyama, Y.: **Successful peripheral T-lymphocyte-directed gene transfer for a patient with severe combined immune deficiency caused by adenosine deaminase deficiency.** *Blood* 1998, **91**:30-36.

11-17 Most individuals infected with HIV progress over time to AIDS.

Baltimore, D.: **Lessons from people with nonprogressive HIV infection.** *N. Engl. J. Med.* 1995, **332**:259-260.

Barre-Sinoussi, F.: **HIV as the cause of AIDS.** *Lancet* 1996, **348**:31-35.

Kirchhoff, F., Greenough, T.C., Brettler, D.B., Sullivan, J.L., and Desrosiers, R.C.: **Brief report: absence of intact nef sequences in a long-term survivor with nonprogressive HIV-1 infection.** *N. Engl. J. Med.* 1995, **332**:228-232.

Pantaleo, G., Menzo, S., Vaccarezza, M., Graziosi, C., Cohen, O.J., Demarest, J.F., Montefiori, D., Orenstein, J.M., Fox, C., Schrager, L.K., Margolick, J.B., Buchbinder, S., Giorgi, J.V., and Fauci, A.S.: **Studies in subjects with long-term nonprogressive human immunodeficiency virus infection.** *N. Engl. J. Med.* 1995, **332**:209-216.

Peckham, C., and Gibb, D.: **Mother-to-child transmission of the human immunodeficiency virus.** *N. Engl. J. Med.* 1995, **333**:298-302.

Volberding, P.A.: **Age as a predictor of progression in HIV infection.** *Lancet* 1996, **347**:1569-70.

Wang, W.K., Essex, M., McLane, M.F., Mayer, K.H., Hsieh, C.C., Brumblay, H.G., Seage, G., and Lee, T.H.R.: **Pattern of gp120 sequence divergence linked to a lack of clinical progression in human immunodeficiency virus type 1 infection.** *Proc. Natl. Acad. Sci. USA* 1996, **93**:6693-6697.

11-18 HIV is a retrovirus that infects CD4 T cells, dendritic cells, and macrophages.

Chan, D.C., and Kim, P.S.: **HIV entry and its inhibition.** *Cell* 1998, **93**:681-684.

Connor, R.I., Sheridan, K.E., Ceradini, D., Choe, S., and Landau, N.R.: **Change in coreceptor use correlates with disease progression in HIV-1—infected individuals.** *J. Exp. Med.* 1997, **185**:621-628.

Grouard, G., and Clark, E.A.: **Role of dendritic and follicular dendritic cells in HIV infection and pathogenesis.** *Curr. Opin. Immunol.* 1997, **9**:563-567.

Moore, J.P., Trkola, A., and Dragic, T.: **Co-receptors for HIV-1 entry.** *Curr. Opin. Immunol.* 1997, **9**:551-562.

Unutmaz, D., and Littman, D.R.: **Expression pattern of HIV-1 coreceptors on T cells: implications for viral transmission and lymphocyte homing.** *Proc. Natl. Acad. Sci. USA* 1997, **94**:1615-1618.

Wyatt, R., and Sodroski, J.: **The HIV-1 envelope glycoproteins: fusogens, antigens, and immunogens.** *Science* 1998, **280**:1884-1888.

11-19 Genetic deficiency of the macrophage chemokine co-receptor for HIV confers resistance to HIV infection *in vivo.*

Berger, E.A., Murphy, P.M., and Farber, J.M.: **Chemokine receptors as HIV-1 coreceptors: roles in viral entry, tropism, and disease.** *Annu. Rev. Immunol.* 1999, **17**:657-700.

Littman, D.R.: **Chemokine receptors: keys to AIDS pathogenesis?** *Cell* 1998, **93**:677-680.

Liu, R., Paxton, W.A., Choe, S., Ceradini, D., Martin, S.R., Horuk, R., Macdonald, M.E., Stuhlmann, H., Koup, R.A., and Landau, N.R.: **Homozygous defect in HIV 1 coreceptor accounts for resistance of some multiply exposed individuals to HIV 1 infection.** *Cell* 1996, **86**:367-377.

Murakami, T., Nakajima, T., Koyanagi, Y., Tachibana, K., Fujii, N., Tamamura, H., Yoshida, N., Waki, M., Matsumoto, A., Yoshie, O., Kishimoto, T., Yamamoto, N., and Nagasawa, T.: **A small molecule CXCR4 inhibitor that blocks T cell line-tropic HIV-1 infection.** *J. Exp. Med.* 1997, **186**:1389-1393.

Nolan, G.P.: **Harnessing viral devices as pharmaceuticals: fighting HIV-1's fire with fire.** *Cell* 1997, **90**:821-824.

Samson, M., Libert, F., Doranz, B.J., Rucker, J., Liesnard, C., Farber, C.M., Saragosti, S., Lapoumeroulie, C., Cognaux, J., Forceille, C., Muyldermans, G., Verhofstede, C., Burtonboy, G., Georges, M., Imai, T., Rana, S., Yi, Y.J., Smyth, R.J., Collman, R.G., Doms, R.W., Vassart, G., and Parmentier, M.R.: **Resistance to HIV 1 infection in Caucasian individuals bearing mutant alleles of the CCR 5 chemokine receptor gene.** *Nature* 1996, **382**:722-725.

Yang, A.G., Bai, X., Huang, X.F., Yao, C., and Chen, S.: **Phenotypic knockout of HIV type 1 chemokine coreceptor CCR-5 by intrakines as potential therapeutic approach for HIV-1 infection.** *Proc. Natl. Acad. Sci. USA* 1997, **94**:11567-11572.

11-20 HIV RNA is transcribed by viral reverse transcriptase into DNA that integrates into the host cell genome.

Andrake, M.D., and Skalka, A.M.R.: **Retroviral integrase, putting the pieces together.** *J. Biol. Chem.* 1995, **271**:19633-19636.

Baltimore, D.: **The enigma of HIV infection.** *Cell* 1995, **82**:175-176.

McCune, J.M.: **Viral latency in HIV disease.** *Cell* 1995, **82**:183-188.

11-21 Transcription of the HIV provirus depends on host cell transcription factors induced upon the activation of infected T cells.

Cullen, B.R.: **Connections between the processing and nuclear export of mRNA: evidence for an export license?** *Proc. Natl. Acad. Sci. USA* 2000, **97**:4-6.

Cullen, B.R.: **HIV-1 auxiliary proteins: making connections in a dying cell.** *Cell* 1998, **93**:685-692.

Emerman, M., and Malim, M.H.: **HIV-1 regulatory/accessory genes: keys to unraveling viral and host cell biology.** *Science* 1998, **280**:1880-1884.

Fujinaga, K., Taube, R., Wimmer, J., Cujec, T.P., and Peterlin, B.M.: **Interactions between human cyclin T, Tat, and the transactivation response element (TAR) are disrupted by a cysteine to tyrosine substitution found in mouse cyclin T.** *Proc. Natl. Acad. Sci. USA* 1999, **96**:1285-1290.

Kinoshita, S., Su, L., Amano, M., Timmerman, L.A., Kaneshima, H., and Nolan, G.P.: **The T cell activation factor NF-ATc positively regulates HIV-1 replication and gene expression in T cells.** *Immunity* 1997, **6**:235-244.

Subbramanian, R.A., and Cohen, E.A.: **Molecular biology of the human immuno-deficiency virus accessory proteins.** *J. Virol.* 1994, **68**:6831-6835.

Trono, D.: **HIV accessory proteins: leading roles for the supporting cast.** *Cell* 1995, **82**:189-192.

11-22 Drugs that block HIV replication lead to a rapid decrease in titer of infectious virus and an increase in CD4 T cells.

Ho, D.D.: **Perspectives series: host/pathogen interactions. Dynamics of HIV-1 replication *in vivo*.** *J. Clin. Invest.* 1997, **99**:2565-2567.

Lipsky, J.J.: **Antiretroviral drugs for AIDS.** *Lancet* 1996, **348**:800-803.

Wei, X., Ghosh, S.K., Taylor, M.E., Johnson, V.A., Emini, E.A., Deutsch, P., Lifson, J.D., Bonhoeffer, S., Nowak, M.A., Hahn, B.H., Saag, M.S., and Shaw, G.M.: **Viral dynamics in human immunodeficiency virus type 1 infection.** *Nature* 1995, **373**:117-122.

11-23 HIV accumulates many mutations in the course of infection in a single individual and drug treatment is soon followed by the outgrowth of drug-resistant variants of the virus.

Bonhoeffer, S., May, R.M., Shaw, G.M., and Nowak, M.A.: **Virus dynamics and drug therapy.** *Proc. Natl. Acad. Sci. USA* 1997, **94**:6971-6976.

Condra, J.H., Schleif, W.A., Blahy, O.M., Gabryelski, L.J., Graham, D.J., Quintero, J.C., Rhodes, A., Robbins, H.L., Roth, E., Shivaprakash, M., Titus, D., Yang, T., Teppler, H., Squires, K.E., Deutsch, P.J., and Emini, E.A.: ***In vivo* emergence of HIV-1 variants resistant to multiple protease inhibitors.** *Nature* 1995, **374**:569-571.

Finzi, D., and Silliciano, R.F.: **Viral dynamics in HIV-1 infection.** *Cell* 1998, **93**:665-671.

Katzenstein, D.: **Combination therapies for HIV infection and genomic drug resistance.** *Lancet* 1997, **350**:970-971.

Moutouh, L., Corbeil, J., and Richman, D.D.: **Recombination leads to the rapid emergence of HIV 1 dually resistant mutants under selective drug pressure.** *Proc. Natl. Acad. Sci. USA* 1996, **93**:6106-6111.

11-24 Lymphoid tissue is the major reservoir of HIV infection.

Burton, G.F., Masuda, A., Heath, S.L., Smith, B.A., Tew, J.G., and Szakal, A.K.: **Follicular dendritic cells (FDC) in retroviral infection: host/pathogen perspectives.** *Immunol. Rev.* 1997, **156**:185-197.

Cameron, P., Pope, M., Granellipiperno, A., and Steinman, R.M.: **Dendritic cells and the replication of HIV 1.** *J. Leukoc. Biol.* 1996, **59**:158-171.

Chun, T.W., Carruth, L., Finzi, D., Shen, X., DiGiuseppe, J.A., Taylor, H., Hermankova, M., Chadwick, K., Margolick, J., Quinn, T.C., Kuo, Y.H., Brookmeyer, R., Zeiger, M.A., Barditch-Crovo, P., and Siliciano, R.F.: **Quantification of latent tissue reservoirs and total body viral load in HIV-1 infection.** *Nature* 1997, **387**:183-188.

Clark, E.A.: **HIV: dendritic cells as embers for the infectious fire.** *Curr. Biol.* 1996, **6**:655-657.

Dianzani, F., Antonelli, G., Riva, E., Uccini, S., and Visco, G.: **Plasma HIV viremia and viral load in lymph nodes.** *Nat. Med.* 1996, **2**:832-833.

Finzi, D., Blankson, J., Siliciano, J.D., Margolick, J.B., Chadwick, K., Pierson, T., Smith, K., Lisziewicz, J., Lori, F., Flexner, C., Quinn, T.C., Chaisson, R.E., Rosenberg, E., Walker, B., Gange, S., Gallant, J., and Siliciano, R.F.: **Latent infection of CD4+ T cells provides a mechanism for lifelong persistence of HIV-1, even in patients on effective combination therapy.** *Nat. Med.* 1999, **5**:512-517.

Haase, A.T.: **Population biology of HIV-1 infection: viral and CD4+ T cell demographics and dynamics in lymphatic tissues.** *Annu. Rev. Immunol.* 1999, **17**:625-656.

Knight, S.C., and Patterson, S.: **Bone marrow-derived dendritic cells, infection with human immunodeficiency virus, and immunopathology.** *Annu. Rev. Immunol.* 1997, **15**:593-615.

Orenstein, J.M., Fox, C., and Wahl, S.M.: **Macrophages as a source of HIV during opportunistic infections.** *Science* 1997, **276**:1857-1861.

Pierson, T., McArthur, J., and Siliciano, R.F.: **Reservoirs for HIV-1: mechanisms for viral persistence in the presence of antiviral immune responses and antiretroviral therapy.** *Annu. Rev. Immunol.* 2000, **18**:665-708.

Wong, J.K., Hezareh, M., Gunthard, H.F., Havlir, D.V., Ignacio, C.C., Spina, C.A., and Richman, D.D.: **Recovery of replication-competent HIV despite prolonged suppression of plasma viremia.** *Science* 1997, **278**:1291-1295.

11-25 An immune response controls but does not eliminate HIV.

Evans, D.T., O'Connor, D.H., Jing, P., Dzuris, J.L., Sidney, J., da Silva, J., Allen, T.M., Horton, H., Venham, J.E., Rudersdorf, R.A., Vogel, T., Pauza, C.D., Bontrop, R.E., DeMars, R., Sette, A., Hughes, A.L., and Watkins, D.I.: **Virus-specific cytotoxic T-lymphocyte responses select for amino-acid variation in simian immunodeficiency virus Env and Nef.** *Nat. Med.* 1999, **5**:1270-1276.

Goulder, P.J., Sewell, A.K., Lalloo, D.G., Price, D.A., Whelan, J.A., Evans, J., Taylor, G.P., Luzzi, G., Giangrande, P., Phillips, R.E., and McMichael, A.J.: **Patterns of immunodominance in HIV-1-specific cytotoxic T lymphocyte responses in two human histocompatibility leukocyte antigens (HLA)-identical siblings with HLA-A*0201 are influenced by epitope mutation.** *J. Exp. Med.* 1997, **185**:1423-33.

McMichael, A.: **T cell responses and viral escape.** *Cell* 1998, **93**:673-676.

Oldstone, M.B.: **HIV versus cytotoxic T lymphocytes—the war being lost.** *N. Engl. J. Med.* 1997, **337**:1306-1308.

Price, D.A., Goulder, P.J., Klenerman, P., Sewell, A.K., Easterbrook, P.J., Troop, M., Bangham, C.R., and Phillips, R.E.: **Positive selection of HIV-1 cytotoxic T lymphocyte escape variants during primary infection.** *Proc. Natl. Acad. Sci. USA* 1997, **94**:1890-1895.

Sattentau, Q.J.: **Neutralization of HIV 1 by antibody.** *Curr. Opin. Immunol.* 1996, **8**:540-545.

11-26 HIV infection leads to low levels of CD4 T cells, increased susceptibility to opportunistic infection, and eventually to death.

Badley, A.D., Dockrell, D., Simpson, M., Schut, R., Lynch, D.H., Leibson, P., and Paya, C.V.: **Macrophage-dependent apoptosis of CD4+ T lymphocytes from HIV-infected individuals is mediated by FasL and tumor necrosis factor.** *J. Exp. Med.* 1997, **185**:55-64.

Ho, D.D., Neumann, A.U., Perelson, A.S., Chen, W., Leonard, J.M., and Markowitz, M.: **Rapid turnover of plasma virions and CD4 lymphocytes in HIV-1 infection.** *Nature* 1995, **373**:123-126.

Katlama, C., and Dickinson, G.M.: **Update on opportunistic infections.** *AIDS* 1993, **7S1**:S185-S194.

Kedes, D.H., Operskalski, E., Busch, M., Kohn, R., Flood, J., and Ganem, D.R.: **The seroepidemiology of human herpesvirus 8 (Kaposi's sarcoma associated herpesvirus): distribution of infection in KS risk groups and evidence for sexual transmission.** *Nat. Med.* 1996, **2**:918-924.

Kolesnitchenko, V., Wahl, L.M., Tian, H., Sunila, I., Tani, Y., Hartmann, D.P., Cossman, J., Raffeld, M., Orenstein, J., Samelson, L.E., and Cohen, D.I.: **Human immunodeficiency virus 1 envelope-initiated G2-phase programmed cell death.** *Proc. Natl. Acad. Sci. USA* 1995, **92**:11889-11893.

Miller, R.: **HIV-associated respiratory diseases.** *Lancet* 1996, **348**:307-312.

Pantaleo, G. and Fauci, A.S.: **Apoptosis in HIV infection.** *Nat. Med.* 1995, **1**:118-120.

Zhong, W.D., Wang, H., Herndier, B., and Ganem, D.R.: **Restricted expression of Kaposi sarcoma associated herpesvirus (human herpesvirus 8) genes in Kaposi sarcoma.** *Proc. Natl. Acad. Sci. USA* 1996, **93**:6641-6646.

11-27 Vaccination against HIV is an attractive solution but poses many difficulties.

Bangham, C.R., and Phillips, R.E.: **What is required of an HIV vaccine?** *Lancet* 1997, **350**:1617-1621.

Burton, D.R.: **A vaccine for HIV type 1: the antibody perspective.** *Proc. Natl. Acad. Sci. USA* 1997, **94**:10018-10023.

Letvin, N.L.: **Progress in the development of an HIV-1 vaccine.** *Science* 1998, **280**:1875-1880.

MacQueen, K.M., Buchbinder, S., Douglas, J.M., Judson, F.N., McKirnan, D.J., and Bartholow, B.: **The decision to enroll in HIV vaccine efficacy trials: concerns elicited from gay men at increased risk for HIV infection.** *AIDS Res. Hum. Retroviruses* 1994, **10 Suppl 2**:S261-S264.

McMichael, A.J., and Hanke, T.: **Is an HIV vaccine possible?** *Nat. Med.* 1999, **5**:612-614.

Rowland Jones, S., Sutton, J., Ariyoshi, K., Dong, T., Gotch, F., McAdam, S., Whitby, D., Sabally, S., Gallimore, A., Corrah, T., Takiguchi, M., Schultz, T., McMichael, A., and Whittle, H.: **HIV-specific cytotoxic T-cells in HIV-exposed but uninfected Gambian women.** *Nat. Med.* 1995, **1**:59-64.

Rowland Jones, S., Tan, R., and McMichael, A.: **Role of cellular immunity in protection against HIV infection.** *Adv. Immunol.* 1997; **65**:277-346.

Salk, J., Bretscher, P.A., Salk, P.L., Clerici, M., and Shearer, G.M.: **A strategy for prophylactic vaccination against HIV.** *Science* 1993, **260**:1270-1272.

11-28 Prevention and education are one way in which the spread of HIV and AIDS can be controlled.

Coates, T.J., Aggleton, P., Gutzwiller, F., Des-Jarlais, D., Kihara, M., Kippax, S., Schechter, M., and van-den-Hoek, J.A.: **HIV prevention in developed countries.** *Lancet* 1996, **348**:1143-1148.

Decosas, J., Kane, F., Anarfi, J.K., Sodji, K.D., and Wagner, H.U.: **Migration and AIDS.** *Lancet* 1995, **346**:826-828.

Dowsett, G.W.: **Sustaining safe sex: sexual practices, HIV and social context.** *AIDS* 1993, **7 Suppl 1**:S257-S262.

Kimball, A.M., Berkley, S., Ngugi, E., and Gayle, H.: **International aspects of the AIDS/HIV epidemic.** *Annu. Rev. Public. Health* 1995, **16**:253-282.

Kirby, M.: **Human rights and the HIV paradox.** *Lancet* 1996, **348**:1217-1218.

Nelson, K.E., Celentano, D.D., Eiumtrakol, S., Hoover, D.R., Beyrer, C., Suprasert, S., Kuntolbutra, S., and Khamboonruang, C.: **Changes in sexual behavior and a decline in HIV infection among young men in Thailand.** *N. Engl. J. Med.* 1996, *335*:297-303.

Weniger, B.G., and Brown, T.: **The march of AIDS through Asia.** *N. Engl. J. Med.* 1996, *335*:343-345.

Allergy and Hypersensitivity

Allergic reactions occur when an individual who has produced IgE antibody in response to an innocuous antigen, or **allergen**, subsequently encounters the same allergen. The allergen triggers the activation of IgE-binding mast cells in the exposed tissue, leading to a series of responses that are characteristic of **allergy**. As we learned in Chapter 9, there are circumstances in which IgE is involved in protective immunity, especially in response to parasitic worms, which are prevalent in less developed countries. In the industrialized countries, however, IgE responses to innocuous antigens predominate and allergy is an important cause of disease (Fig. 12.1). Almost half the populations of North America and Europe have allergies to one or more common environmental antigens and, although rarely life-threatening, these cause much distress and lost time from school and work. Because of the medical importance of allergy in industrialized societies, much more is known about the pathophysiology of IgE-mediated responses than about the normal physiological role of IgE.

IgE-mediated allergic reactions			
Syndrome	Common allergens	Route of entry	Response
Systemic anaphylaxis	Drugs Serum Venoms Peanuts	Intravenous (either directly or following oral absorption into the blood)	Edema Increased vascular permeability Tracheal occlusion Circulatory collapse Death
Acute urticaria (wheal-and-flare)	Insect bites Allergy testing	Subcutaneous	Local increase in blood flow and vascular permeability
Allergic rhinitis (hay fever)	Pollens (ragweed, timothy, birch) Dust-mite feces	Inhalation	Edema of nasal mucosa Irritation of nasal mucosa
Asthma	Danders (cat) Pollens Dust-mite feces	Inhalation	Bronchial constriction Increased mucus production Airway inflammation
Food allergy	Tree nuts Peanuts Shellfish Milk Eggs Fish	Oral	Vomiting Diarrhea Pruritis (itching) Urticaria (hives) Anaphylaxis (rarely)

Fig. 12.1 IgE-mediated reactions to extrinsic antigens. All IgE-mediated responses involve mast-cell degranulation, but the symptoms experienced by the patient can be very different depending on whether the allergen is injected, inhaled, or eaten, and depending also on the dose of the allergen.

	Type I	Type II	Type III	Type IV		
Immune reactant	IgE	IgG	IgG	T_H1 cells	T_H2 cells	CTL
Antigen	Soluble antigen	Cell- or matrix-associated antigen	Soluble antigen	Soluble antigen	Soluble antigen	Cell-associated antigen
Effector mechanism	Mast-cell activation	FcR⁺ cells (phagocytes, NK cells)	FcR⁺ cells Complement	Macrophage activation	Eosinophil activation	Cytotoxicity
Example of hypersensitivity reaction	Allergic rhinitis, asthma, systemic anaphylaxis	Some drug allergies (e.g., penicillin)	Serum sickness, Arthus reaction	Contact dermatitis, tuberculin reaction	Chronic asthma, chronic allergic rhinitis	Contact dermatitis

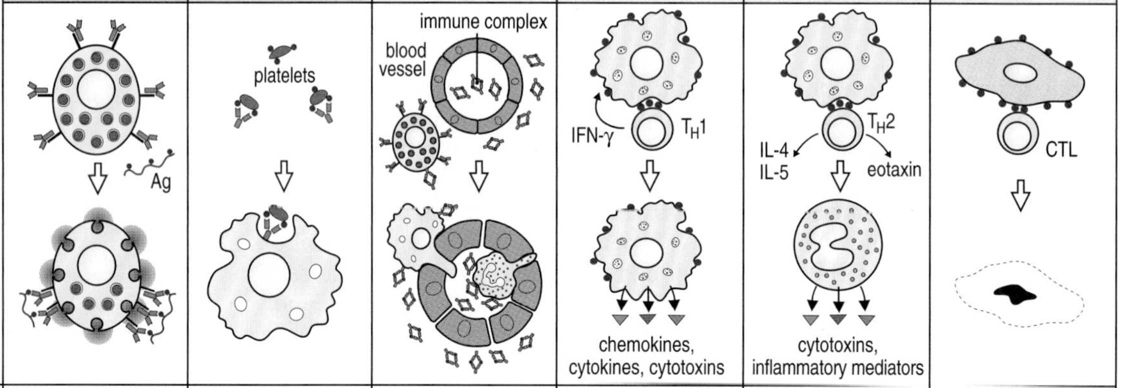

Fig. 12.2 There are four types of hypersensitivity reaction mediated by immunological mechanisms that cause tissue damage. Types I–III are antibody-mediated and are distinguished by the different types of antigens recognized and the different classes of antibody involved. Type I responses are mediated by IgE, which induces mast-cell activation, whereas types II and III are mediated by IgG, which can engage Fc-receptor and complement-mediated effector mechanisms to varying degrees, depending on the subclass of IgG and the nature of the antigen involved. Type II responses are directed against cell-surface or matrix antigens, whereas type III responses are directed against soluble antigens, and the tissue damage involved is caused by responses triggered by immune complexes. Type IV hypersensitivity reactions are T cell-mediated and can be subdivided into three groups. In the first group, tissue damage is caused by the activation of macrophages by T_H1 cells, which results in an inflammatory response. In the second, damage is caused by the activation by T_H2 cells of inflammatory responses in which eosinophils predominate; in the third, damage is caused directly by cytotoxic T cells (CTL).

The term allergy was originally defined by Clemens Von Pirquet as "an altered capacity of the body to react to a foreign substance," which was an extremely broad definition that included all immunological reactions. Allergy is now defined in a much more restricted manner as "disease following a response by the immune system to an otherwise innocuous antigen." Allergy is one of a class of immune system responses that are termed **hypersensitivity reactions**. These are harmful immune responses that produce tissue injury and may cause serious disease. Hypersensitivity reactions were classified into four types by Coombs and Gell (Fig. 12.2). Allergy is often equated with type I hypersensitivity (immediate-type hypersensitivity reactions mediated by IgE), and will be used in this sense here.

In this chapter we will first consider the mechanisms that favor the production of IgE. We then describe the pathophysiological consequences of the interaction between antigen and IgE that is bound by the high-affinity Fcε receptor (FcεRI) on mast cells. Finally, we will consider the causes and consequences of other types of immunological hypersensitivity reactions.

The production of IgE.

IgE is produced by plasma cells located in lymph nodes draining the site of antigen entry or locally, at the sites of allergic reactions, by plasma cells derived from germinal centers developing within the inflamed tissue. IgE differs from other antibody isotypes in being located predominantly in tissues, where it is tightly bound to the mast-cell surface through the high-affinity IgE receptor known as **FcεRI**. Binding of antigen to IgE cross-links these receptors and this causes the release of chemical mediators from the mast cells, which may lead to the development of a **type I hypersensitivity reaction**. Basophils and activated eosinophils also express FcεRI; they can therefore display surface-bound IgE and also take part in the production of type I hypersensitivity reactions. The factors that lead to an antibody response dominated by IgE are still being worked out. Here we will describe our current understanding of these processes before turning to the question of how IgE mediates allergic reactions.

12-1 Allergens are often delivered transmucosally at low dose, a route that favors IgE production.

There are certain antigens and routes of antigen presentation to the immune system that favor the production of IgE. CD4 T_H2 cells can switch the antibody isotype from IgM to IgE, or they can cause switching to IgG2 and IgG4 (human) or IgG1 and IgG3 (mouse) (see Section 9-4). Antigens that selectively evoke T_H2 cells that drive an IgE response are known as allergens.

Much human allergy is caused by a limited number of inhaled small-protein allergens that reproducibly elicit IgE production in susceptible individuals. We inhale many different proteins that do not induce IgE production; this raises the question of what is unusual about the proteins that are common allergens. Although we do not yet have a complete answer, some general principles have emerged (Fig. 12.3). Most allergens are relatively small, highly soluble proteins that are carried on desiccated particles such as pollen grains or mite feces. On contact with the mucosa of the airways, for example, the soluble allergen elutes from the particle and diffuses into the mucosa. Allergens are typically presented to the immune system at very low doses. It has been estimated that the maximum exposure of a person to the common pollen allergens in ragweed (*Artemisia artemisiifolia*) does not exceed 1 μg per year! Yet many people develop irritating and even life-threatening T_H2-driven IgE antibody responses to these minute doses of allergen. It is important to note that only some of the people who are exposed to these substances make IgE antibodies against them.

It seems likely that presenting an antigen transmucosally and at very low doses is a particularly efficient way of inducing T_H2-driven IgE responses. IgE antibody production requires T_H2 cells that produce interleukin-4 (IL-4) and IL-13 and it can be inhibited by T_H1 cells that produce interferon-γ (IFN-γ) (see Fig. 9.7). The presentation of low doses of antigen can favor the activation of T_H2 cells over T_H1 cells (see Section 10-7), and many common allergens are delivered to the respiratory mucosa by inhalation of a low dose. The dominant antigen-presenting cells in the respiratory mucosa are myeloid dendritic cells (see Section 7-29). These take up and process protein antigens very efficiently and become activated in the process. This in turn induces their migration to regional lymph nodes and differentiation into professional antigen-presenting cells with co-stimulatory activity that favors the differentiation of T_H2 cells.

Features of inhaled allergens that may promote the priming of T_H2 cells that drive IgE responses	
Protein	Only proteins induce T-cell responses
Enzymatically active	Allergens are often proteases
Low dose	Favors activation of IL-4-producing CD4 T cells
Low molecular weight	Allergen can diffuse out of particle into mucus
High solubility	Allergen can be readily eluted from particle
Stable	Allergen can survive in desiccated particle
Contains peptides that bind host MHC class II	Required for T-cell priming

Fig. 12.3 Properties of inhaled allergens. The typical characteristics of inhaled allergens are described in this table.

12-2   Enzymes are frequent triggers of allergy.

Several lines of evidence suggest that IgE is important in host defense against parasites (see Section 9-23). Many parasites invade their hosts by secreting proteolytic enzymes that break down connective tissue and allow the parasite access to host tissues, and it has been proposed that these enzymes are particularly active at promoting T_H2 responses. This idea receives some support from the many examples of allergens that are enzymes.

The major allergen in the feces of the house dust mite (*Dermatophagoides pteronyssimus*) (Fig. 12.4), which is responsible for allergy in approximately 20% of the North American population, is a cysteine protease homologous to papain, known as Der p 1. This enzyme has been found to cleave occludin, a protein component of intercellular tight junctions. This reveals one possible reason for the allergenicity of certain enzymes. By destroying the integrity of the tight junctions between epithelial cells, Der p 1 may gain abnormal access to subepithelial antigen-presenting cells, resident mast cells, and eosinophils (Fig. 12.5).

The allergenicity of Der p 1 may also be promoted by its proteolytic action on certain receptor proteins on B cells and T cells. It has been shown to cleave the α subunit of the IL-2 receptor, CD25, from T cells. Loss of IL-2 receptor activity might interfere with the maintenance of T_H1 cells, leading to a T_H2 bias (see Section 8-9).

The protease papain, derived from the papaya fruit, is used as a meat tenderizer and causes allergy in workers preparing the enzyme; such allergies are called occupational allergies. Another **occupational allergy** is the asthma caused by inhalation of the bacterial enzyme subtilisin, the 'biological' component of some laundry detergents. Injection of enzymatically active papain (but not of inactivated papain) into mice stimulates an IgE response. A closely related enzyme, chymopapain, is used medically to destroy intervertebral discs in patients with sciatica; the major (although rare) complication of this procedure is anaphylaxis, an acute systemic response to allergens (see Fig. 12.1).

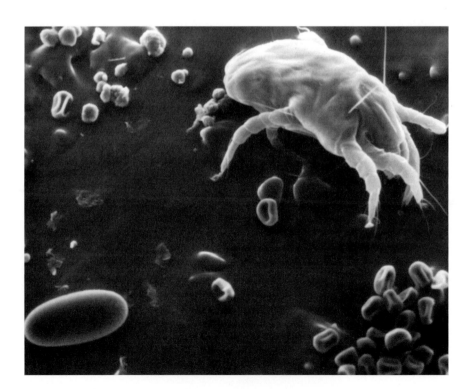

Fig. 12.4 Scanning electron micrograph of *D. pteronyssimus* with some of its fecal pellets. Photograph courtesy E.R. Tovey.

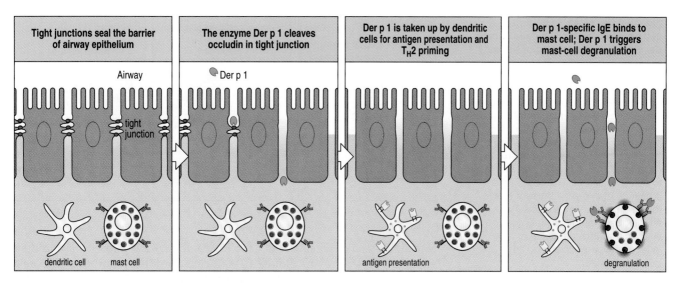

| Tight junctions seal the barrier of airway epithelium | The enzyme Der p 1 cleaves occludin in tight junction | Der p 1 is taken up by dendritic cells for antigen presentation and T$_H$2 priming | Der p 1-specific IgE binds to mast cell; Der p 1 triggers mast-cell degranulation |

Fig. 12.5 The enzymatic activity of some allergens enables penetration of epithelial barriers. The epithelial barrier of the airways is formed by tight junctions between the epithelial cells. Fecal pellets from the house dust mite, *D. pteronyssimus*, contain a proteolytic enzyme, Der p 1, that acts as an allergen. It cleaves occludin, a protein that helps maintain the tight junctions, and thus destroys the barrier function of the epithelium. Fecal antigens can then pass through and be taken up by dendritic cells in subepithelial tissue. Der p 1 is taken up by dendritic cells, which are activated and move to lymph nodes (not shown), where they act as antigen-presenting cells, inducing the production of T$_H$2 cells specific for Der p 1 and the production of Der p 1-specific IgE. Der p 1 may then bind directly to specific IgE on the resident mast cells, triggering mast-cell activation.

Not all allergens are enzymes, however; for example, two allergens identified from filarial worms are enzyme inhibitors. Many protein allergens derived from plants have been identified and sequenced, but their functions are currently obscure. Thus, there seems to be no systematic association between enzymatic activity and allergenicity.

12-3 Class switching to IgE in B lymphocytes is favored by specific signals.

There are two main components of the immune response leading to IgE production. The first consists of the signals that favor the differentiation of naive T$_H$0 cells to a T$_H$2 phenotype. The second comprises the action of cytokines and co-stimulatory signals from T$_H$2 cells that stimulate B cells to switch to producing IgE antibodies.

The fate of a naive CD4 T cell responding to a peptide presented by a dendritic cell is determined by the cytokines it is exposed to before and during this response, and by the intrinsic properties of the antigen, antigen dose, and route of presentation. Exposure to IL-4 favors the development of T$_H$2 cells and to IL-12 favors that of T$_H$1 cells. IgE antibodies are important in host defense against parasitic infections and this defense system is distributed anatomically mainly at the sites of entry of parasites—under the skin, under the epithelial surfaces of the airways (the mucosal-associated lymphoid tissues), and in the submucosa of the gut (the gut-associated lymphoid tissues). Cells of the innate and adaptive immune systems at these sites are specialized to secrete predominantly cytokines that drive T$_H$2 responses. The dendritic cells at these sites are of the myeloid phenotype (see Section 7-29); after taking up antigen they migrate to regional lymph nodes where their interaction with naive CD4 T cells drives the T cells to become T$_H$2 cells, which secrete IL-4 and IL-10. It is not known how myeloid dendritic cells induce this differentiation. One possibility is that they express a particular set of cytokines and co-stimulatory molecules yet to be characterized. Another is that they

Fig. 12.6 IgE class switching in B cells is initiated by T$_H$2 cells, which develop in the presence of an early burst of IL-4. In mice, IL-4 is secreted early in some immune responses by a small subset of CD4 T cells (NK1.1$^+$ CD4 T cells) that interact with antigen-presenting cells bearing the nonclassical MHC class I-like molecule CD1 (first panel). Naive T cells being primed by their first encounter with antigen carry receptors for IL-4 (IL-4R) and are driven to differentiate into T$_H$2 cells in the presence of this early burst of IL-4 (second panel). When these T$_H$2 effector cells interact with B cells specific for the same antigen, they induce isotype-switching so that IgE is produced.

activate a specialized subset of CD4 T cells, the NK1.1$^+$ subset, that produce abundant IL-4 that can induce CD4 T cells to differentiate into T$_H$2 cells following stimulation by antigen. These in turn induce B cells to produce IgE (Fig. 12.6).

Class switching of B cells to IgE production is induced by two separate signals, both of which can be provided by T$_H$2 cells (see Section 9-4). The first of these signals is provided by the cytokines IL-4 or IL-13, interacting with receptors on the B-cell surface. These transduce their signal by activation of the Janus family tyrosine kinases JAK1 and JAK3 (see Section 6-17) which ultimately lead to phosphorylation of the transcriptional regulator STAT6. Mice lacking functional IL-4, IL-13, or STAT6 all show impaired T$_H$2 responses and IgE switching, demonstrating the key importance of these signaling pathways. The second signal for IgE class switching is a co-stimulatory interaction between CD40 ligand on the T-cell surface with CD40 on the B-cell surface. This interaction is essential for all antibody class switching (see Section 9-3); patients with the X-linked hyper IgM syndrome have a deficiency of CD40 ligand and produce no IgG, IgA, or IgE.

The IgE response, once initiated, can be further amplified by basophils, mast cells, and eosinophils, which can also drive IgE production (Fig. 12.7). All three cell types express FcεRI, although eosinophils only express it when activated. When these specialized granulocytes are activated by antigen cross-linking of their FcεRI-bound IgE, they can express cell-surface CD40L and secrete IL-4; like T$_H$2 cells, therefore, they can drive class switching and IgE production by B cells (see Fig. 12.7). The interaction between these specialized granulocytes and B cells can occur at the site of the allergic reaction, as B cells are observed to form germinal centers at inflammatory foci. Blocking this amplification process is a goal of therapy, as allergic reactions can otherwise become self sustaining.

Fig. 12.7 Antigen binding to IgE on mast cells leads to amplification of IgE production. IgE secreted by plasma cells binds to the high-affinity IgE receptor on mast cells (illustrated here), basophils, and activated eosinophils. When the surface-bound IgE is cross-linked by antigen, these cells express CD40L and secrete IL-4, which in turn binds to IL-4 receptors (IL-4R) on the activated B cell, stimulating isotype switching by B cells and the production of more IgE. These interactions can occur *in vivo* at the site of allergen-triggered inflammation, for example in bronchial-associated lymphoid tissue.

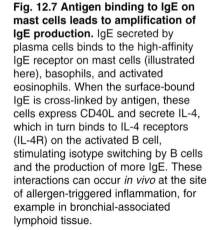

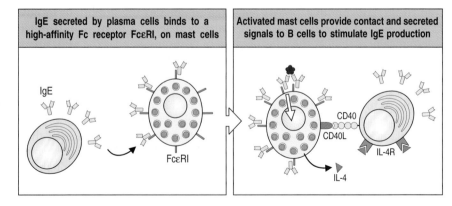

12-4   Genetic factors contribute to the development of IgE-mediated allergy, but environmental factors may also be important.

As many as 40% of people in Western populations show an exaggerated tendency to mount IgE responses to a wide variety of common environmental allergens. This state is called **atopy** and seems to be influenced by several genetic loci. **Atopic** individuals have higher total levels of IgE in the circulation and higher levels of eosinophils than their normal counterparts. They are more susceptible to allergic diseases such as hay fever and asthma. Studies of atopic families have identified regions on chromosomes 11q and 5q that appear to be important in determining atopy; candidate genes that could affect IgE responses are present in these regions. The candidate gene on chromosome 11 encodes the β subunit of the high-affinity IgE receptor, whereas on chromosome 5 there is a cluster of tightly linked genes that includes those for IL-3, IL-4, IL-5, IL-9, IL-12, IL-13, and granulocyte–macrophage colony-stimulating factor (GM-CSF). These cytokines are important in IgE isotype switching, eosinophil survival, and mast-cell proliferation. Of particular note, an inherited genetic variation in the promoter region of the IL-4 gene is associated with raised IgE levels in atopic individuals; the variant promoter will direct increased expression of a reporter gene in experimental systems. Atopy has also been associated with a gain-of-function mutation of the α subunit of the IL-4 receptor, which is associated with increased signaling following ligation of the receptor. It is too early to know how important these different polymorphisms are in the complex genetics of atopy.

Allergic Asthma

A second type of inherited variation in IgE responses is linked to the MHC class II region and affects responses to specific allergens. Many studies have shown that IgE production in response to particular allergens is associated with certain HLA class II alleles, implying that particular MHC:peptide combinations might favor a strong T_H2 response. For example, IgE responses to several ragweed pollen allergens are associated with haplotypes containing the MHC class II allele *DRB1*1501*. Many individuals are therefore generally predisposed to make T_H2 responses and specifically predisposed to respond to some allergens more than others. However, allergies to common drugs such as penicillin show no association with MHC class II or the presence or absence of atopy.

There is evidence that a state of atopy, and the associated susceptibility to asthma, rhinitis, and eczema, can be determined by different genes in different populations. Genetic associations found in one group of patients have frequently not been confirmed in patients of different ethnic origins. There are also likely to be genes that affect only particular aspects of allergic disease. For example, in asthma there is evidence for different genes affecting at least three aspects of the disease phenotype—IgE production, the inflammatory response, and clinical responses to particular types of treatment. Some of the best-characterized genetic polymorphisms of candidate genes associated with asthma are shown in Fig. 12.8, together with possible ways in which the genetic variation may affect the particular type of disease that develops and its response to drugs.

The prevalence of atopic allergy, and of asthma in particular, is increasing in economically advanced regions of the world, an observation that is best explained by environmental factors. The four main candidate environmental factors are changes in exposure to infectious diseases in early childhood, environmental pollution, allergen levels, and dietary changes. Alterations in exposure to microbial pathogens is the most plausible explanation at present for the increase in atopic allergy. Atopy is negatively associated with a history of infection with measles or hepatitis A virus, and with positive tuberculin

Fig. 12.8 Candidate susceptibility genes for asthma. May also affect response to bronchodilator therapy with β_2-adrenergic agonists. †Patients with alleles associated with reduced enzyme production failed to show a beneficial response to a drug inhibitor of 5-lipoxygenase. This is an example of a pharmacogenetic effect, in which genetic variation affects the response to medication.

Gene	Nature of polymorphism	Possible mechanism of association
IL-4	Promoter variant	Variation in expression of IL-4
IL-4 receptor α chain	Structural variant	Increased signaling in response to IL-4
High-affinity IgE receptor β chain	Structural variant	Variation in consequences of IgE ligation by antigen
MHC class II genes	Structural variants	Enhanced presentation of particular allergen-derived peptides
T-cell receptor α locus	Microsatellite markers	Enhanced T-cell recognition of certain allergen-derived peptides
β_2-Adrenergic receptor	Structural variants	Increased bronchial hyperreactivity*
5-Lipoxygenase	Promoter variant	Variation in leukotriene production†

skin tests (suggesting prior exposure and immune response to *Mycobacterium tuberculosis*). In contrast, there is evidence that children who have had attacks of bronchiolitis associated with respiratory syncytial virus (RSV) infection are more prone to the later development of asthma. Children hospitalized with this disease have a skewed ratio of cytokine production away from IFN-γ towards IL-4, the cytokine that induces T_H2 responses. It is possible that infection by an organism that evokes a T_H1 immune response early in life might reduce the likelihood of T_H2 responses later in life and vice versa. It might be expected that exposure to environmental pollution would worsen the expression of atopy and asthma. The best evidence shows the opposite effect, however. Children from the city of Halle in the former East Germany, which has severe air pollution, had a lower prevalence of atopy and asthma than an ethnically matched population from Munich, exposed to much cleaner air. This does not mean that polluted air is not bad for the lungs. The children from Halle had a higher overall prevalence of respiratory disease than their counterparts from Munich, but this was predominantly not allergic in origin.

While it is clear that allergy is related to allergen exposure, there is no evidence that the rising prevalence of allergy is due to any systematic change in allergen exposure. Nor is there any evidence that changes in diet can explain the increase in allergy in economically advanced populations.

Summary.

Allergic reactions are the result of the production of specific IgE antibody to common, innocuous antigens. Allergens are small antigens that commonly provoke an IgE antibody response. Such antigens normally enter the body at very low doses by diffusion across mucosal surfaces and therefore trigger a T_H2 response. The differentiation of naive allergen-specific T cells into T_H2 cells is also favored by the presence of an early burst of IL-4, which seems to be derived from a specialized subset of T cells. Allergen-specific T_H2 cells produce IL-4 and IL-13, which drive allergen-specific B cells to produce IgE. The specific IgE produced in response to the allergen binds to the high-affinity receptor for IgE on mast cells, basophils, and activated eosinophils. IgE production can be amplified by these cells because, upon activation, they produce IL-4 and CD40 ligand. The tendency to IgE over-production is influenced by genetic and environmental factors. Once IgE is produced in response to an allergen, reexposure to the allergen triggers an allergic response. We will describe the mechanism and pathology of allergic responses in the next part of the chapter.

Effector mechanisms in allergic reactions.

Allergic reactions are triggered when allergens cross-link preformed IgE bound to the high-affinity receptor FcεRI on mast cells. Mast cells line the body surfaces and serve to alert the immune system to local infection. Once activated, they induce inflammatory reactions by secreting chemical mediators stored in preformed granules, and by synthesizing leukotrienes and cytokines after activation occurs. In allergy, they provoke very unpleasant reactions to innocuous antigens that are not associated with invading pathogens that need to be expelled. The consequences of IgE-mediated mast-cell activation depend on the dose of antigen and its route of entry; symptoms range from the irritating sniffles of hay fever when pollen is inhaled, to the life-threatening circulatory collapse that occurs in systemic anaphylaxis (Fig. 12.9). The immediate allergic reaction caused by mast-cell degranulation is followed by a more sustained inflammation, known as the late-phase response. This late response involves the recruitment of other effector cells, notably T_H2 lymphocytes, eosinophils, and basophils, which contribute significantly to the immunopathology of an allergic response.

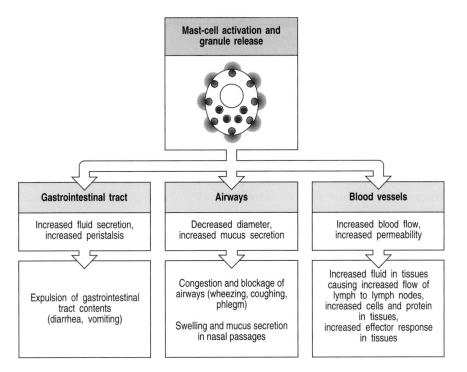

Fig. 12.9 Mast-cell activation has different effects on different tissues.

12-5 Most IgE is cell-bound and engages effector mechanisms of the immune system by different pathways from other antibody isotypes.

Most antibodies are found in body fluids and engage effector cells, through receptors specific for the Fc constant regions, only after binding specific antigen through the antibody variable regions. IgE, however, is an exception as it is captured by the high-affinity Fcε receptor in the absence of bound antigen. This means that IgE is mostly found fixed in the tissues on mast cells that bear this receptor, as well as on circulating basophils and activated eosinophils. The ligation of cell-bound IgE antibody by specific antigen triggers activation of these cells at the site of antigen entry into the tissues. The release of inflammatory lipid mediators, cytokines, and chemokines at sites of IgE-triggered reactions results in the recruitment of eosinophils and basophils to augment the type I response.

There are two types of IgE-binding Fc receptor. The first, FcεRI, is a high-affinity receptor of the immunoglobulin superfamily that binds IgE on mast cells, basophils, and activated eosinophils (see Section 9-22). When the cell-bound IgE antibody is cross-linked by a specific antigen, FcεRI transduces an activating signal. High levels of IgE, such as those that exist in subjects with allergic diseases or parasite infections, can result in a marked increase in FcεRI on the surface of mast cells, enhanced sensitivity of such cells to activation by low concentrations of specific antigen, and markedly increased IgE-dependent release of chemical mediators and cytokines.

The second IgE receptor, FcεRII, usually known as **CD23**, is a C-type lectin and is structurally unrelated to FcεRI; it binds IgE with low affinity. CD23 is present on many different cell types, including B cells, activated T cells, monocytes, eosinophils, platelets, follicular dendritic cells, and some thymic epithelial cells. This receptor was thought to be crucial for the regulation of IgE antibody levels; however, knockout mouse strains lacking the CD23 gene show no major abnormality in the development of polyclonal IgE responses. However the CD23 knockout mice have demonstrated a role for CD23 in enhancing the antibody response to a specific antigen in the presence of that same antigen complexed with IgE. This antigen-specific, IgE-mediated enhancement of antibody responses fails to occur in mice lacking the CD23 gene. This demonstrates a role for CD23 on antigen-presenting cells in the capture of antigen by specific IgE.

12-6 Mast cells reside in tissues and orchestrate allergic reactions.

Mast cells were described by Ehrlich in the mesentery of rabbits and named *Mastzellen* ('fattened cells'). Like basophils, mast cells contain granules rich in acidic proteoglycans that take up basic dyes. However, in spite of this resemblance, and the similar range of mediators stored in these basophilic granules, mast cells are derived from a different myeloid lineage than basophils and eosinophils. Mast cells are highly specialized cells, and are prominent residents of mucosal and epithelial tissues in the vicinity of small blood vessels and postcapillary venules, where they are well placed to guard against invading pathogens (see Sections 9-20 and 9-21). Mast cells are also found in subendothelial connective tissue. They home to tissues as agranular cells; their final differentiation, accompanied by granule formation, occurs after they have arrived in the tissues. The major growth factor for mast cells is stem-cell factor (SCF), which acts on the cell-surface receptor c-Kit (see Section 7-2). Mice with defective c-Kit lack differentiated mast cells and cannot make IgE-mediated inflammatory responses. This shows that such responses depend almost exclusively on mast cells.

Class of product	Examples	Biological effects
Enzyme	Tryptase, chymase, cathepsin G, carboxypeptidase	Remodel connective tissue matrix
Toxic mediator	Histamine, heparin	Toxic to parasites Increase vascular permeability Cause smooth muscle contraction
Cytokine	IL-4, IL-13	Stimulate and amplify T_H2 cell response
	IL-3, IL-5, GM-CSF	Promote eosinophil production and activation
	TNF-α (some stored preformed in granules)	Promotes inflammation, stimulates cytokine production by many cell types, activates endothelium
Chemokine	MIP-1α	Attracts monocytes, macrophages, and neutrophils
Lipid mediator	Leukotrienes C4, D4, E4	Cause smooth muscle contraction Increase vascular permeability Stimulate mucus secretion
	Platelet-activating factor	Attracts leukocytes Amplifies production of lipid mediators Activates neutrophils, eosinophils, and platelets

Fig. 12.10 Molecules released by mast cells on activation. Mast cells produce a wide variety of biologically active proteins and other chemical mediators. The enzymes and toxic mediators listed in the first two rows are released from the preformed granules. The cytokines, chemokines, and lipid mediators are synthesized after activation.

Mast cells express FcεRI constitutively on their surface and are activated when antigens cross-link IgE bound to these receptors (see Fig. 9.35). Degranulation occurs within seconds, releasing a variety of preformed inflammatory mediators (Fig. 12.10). Among these are **histamine**—a short-lived vasoactive amine that causes an immediate increase in local blood flow and vessel permeability—and enzymes such as mast-cell chymase, tryptase, and serine esterases. These enzymes can in turn activate matrix metalloproteinases, which break down tissue matrix proteins, causing tissue destruction. Large amounts of tumor necrosis factor (TNF)-α are also released by mast cells after activation. Some comes from stores in mast-cell granules; some is newly synthesized by the activated mast cells themselves. TNF-α activates endothelial cells, causing increased expression of adhesion molecules, which promotes the influx of inflammatory leukocytes and lymphocytes into tissues (see Section 2-22).

On activation, mast cells synthesize and release chemokines, lipid mediators such as leukotrienes and platelet-activating factor (PAF), and additional cytokines such as IL-4 and IL-13 which perpetuate the T_H2 response. These mediators contribute to both the acute and the chronic inflammatory responses. The lipid mediators, in particular, act rapidly to cause smooth muscle contraction, increased vascular permeability, and mucus secretion, and also induce the influx and activation of leukocytes, which contribute to the late-phase response. The lipid mediators derive from membrane phospholipids, which are cleaved to release the precursor molecule arachidonic acid. This molecule can be modified by two pathways to give rise to prostaglandins, thromboxanes, and leukotrienes. The leukotrienes, especially C4, D4, and E4, are important in sustaining inflammatory responses in the tissues. Many anti-inflammatory drugs are inhibitors of arachidonic acid metabolism. Aspirin, for example, is an inhibitor of the enzyme cyclooxygenase and blocks the production of prostaglandins.

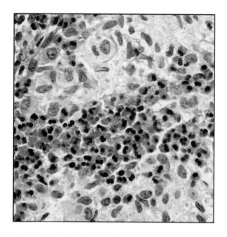

Fig. 12.11 Eosinophils can be detected easily in tissue sections by their bright refractile orange coloration. In this light micrograph, a large number of eosinophils are seen infiltrating a tumor of Langherhans' cells known as Langerhans' cell histiocytosis. The tissue section is stained with hematoxylin and eosin; it is the eosin that imparts the characteristic orange color to the eosinophils. Photograph courtesy of T. Krausz.

IgE-mediated activation of mast cells thus orchestrates an important inflammatory cascade that is amplified by the recruitment of eosinophils, basophils, and T$_H$2 lymphocytes. The physiological importance of this reaction is as a defense mechanism against certain types of infection (see Section 9-23). In allergy, however, the acute and chronic inflammatory reactions triggered by mast-cell activation have important pathophysiological consequences, as seen in the diseases associated with allergic responses to environmental antigens.

12-7   Eosinophils are normally under tight control to prevent inappropriate toxic responses.

Eosinophils are granulocytic leukocytes that originate in bone marrow. They are so called because their granules, which contain arginine-rich basic proteins, are colored bright orange by the acidic stain eosin (Fig. 12.11). Only very small numbers of these cells are normally present in the circulation; most eosinophils are found in tissues, especially in the connective tissue immediately underneath respiratory, gut, and urogenital epithelium, implying a likely role for these cells in defense against invading organisms. Eosinophils have two kinds of effector function. First, on activation they release highly toxic granule proteins and free radicals, which can kill microorganisms and parasites but can also cause significant tissue damage in allergic reactions. Second, activation induces the synthesis of chemical mediators such as prostaglandins, leukotrienes, and cytokines, which amplify the inflammatory response by activating epithelial cells, and recruiting and activating more eosinophils and leukocytes (Fig. 12.12).

Fig. 12.12 Eosinophils secrete a range of highly toxic granule proteins and other inflammatory mediators.

Class of product	Examples	Biological effects
Enzyme	Eosinophil peroxidase	Toxic to targets by catalyzing halogenation Triggers histamine release from mast cells
	Eosinophil collagenase	Remodels connective tissue matrix
Toxic protein	Major basic protein	Toxic to parasites and mammalian cells Triggers histamine release from mast cells
	Eosinophil cationic protein	Toxic to parasites Neurotoxin
	Eosinophil-derived neurotoxin	Neurotoxin
Cytokine	IL-3, IL-5, GM-CSF	Amplify eosinophil production by bone marrow Cause eosinophil activation
Chemokine	IL-8	Promotes influx of leukocytes
Lipid mediator	Leukotrienes C4, D4, E4	Cause smooth muscle contraction Increase vascular permeability Increase mucus secretion
	Platelet-activating factor	Attracts leukocytes Amplifies production of lipid mediators Activates neutrophils, eosinophils, and platelets

The activation and degranulation of eosinophils is strictly regulated, as their inappropriate activation would be very harmful to the host. The first level of control acts on the production of eosinophils by the bone marrow. Few eosinophils are produced in the absence of infection or other immune stimulation. But when T_H2 cells are activated, cytokines such as IL-5 are released that increase the production of eosinophils in the bone marrow and their release into the circulation. However, transgenic animals overexpressing IL-5 have increased numbers of eosinophils (**eosinophilia**) in the circulation but not in their tissues, indicating that migration of eosinophils from the circulation into tissues is regulated separately, by a second set of controls. The key molecules in this case are CC chemokines (see Section 2-20). Most of these cause chemotaxis of several types of leukocyte, but two are specific for eosinophils and have been named **eotaxin 1** and **eotaxin 2**.

The eotaxin receptor on eosinophils, CCR3, is a member of the chemokine family of receptors (see Section 6-16). This receptor also binds the CC chemokines MCP-3, MCP-4, and RANTES, which also induce eosinophil chemotaxis. The eotaxins and these other CC chemokines also activate eosinophils. Identical or similar chemokines also stimulate mast cells and basophils. For example, eotaxin attracts basophils and causes their degranulation, and MCP-1, which binds to CCR2, similarly activates mast cells in both the presence or absence of antigen. MCP-1 can also promote the differentiation of naive T_H0 cells to T_H2 cells; T_H2 cells also carry CCR3 and migrate toward eotaxin. These findings show that families of chemokines, as well as cytokines, can coordinate certain kinds of immune response.

A third set of controls regulates the state of eosinophil activation. In their nonactivated state, eosinophils do not express high-affinity IgE receptors and have a high threshold for release of their granule contents. After activation by cytokines and chemokines, this threshold drops, FcεRI is expressed, and the number of Fcγ receptors and complement receptors on the cell surface also increases. The eosinophil is now primed to carry out its effector activity, for example degranulation in response to antigen that cross-links specific IgE bound to FcεRI on the eosinophil surface.

The potential of eosinophils to cause tissue injury is illustrated by rare syndromes due to abnormally large numbers of eosinophils in the blood (**hypereosinophilia**). These syndromes are sometimes seen in association with T-cell lymphomas, in which unregulated IL-5 secretion drives a marked increase in the numbers of circulating eosinophils. The clinical manifestations of hypereosinophilia are damage to the endocardium (Fig. 12.13) and to nerves, leading to heart failure and neuropathy, both thought to be caused by the toxic effects of eosinophil granule proteins.

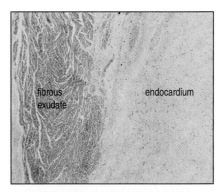

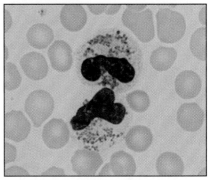

Fig. 12.13 Hypereosinophilia can cause injury to the endocardium. The top panel shows a section of the endocardium from a patient with hypereosinophilic syndrome. There is an organized fibrous exudate and the underlying endocardium is thickened by fibrous tissue. Although there are large numbers of circulating eosinophils, these cells are not seen in the injured endocardium, which is thought to be damaged by granules released from circulating eosinophils. The panel at the bottom shows two partly degranulated eosinophils (center) surrounded by erythrocytes in a peripheral blood film. Photographs courtesy of D. Swirsky and T. Krausz.

12-8 Eosinophils and basophils cause inflammation and tissue damage in allergic reactions.

In a local allergic reaction, mast-cell degranulation and T_H2 activation cause eosinophils to accumulate in large numbers and to become activated. Their continued presence is characteristic of chronic allergic inflammation and they are thought to be major contributors to tissue damage.

Basophils are also present at the site of an inflammatory reaction. Basophils share a common stem-cell precursor with eosinophils; growth factors for basophils are very similar to those for eosinophils and include IL-3, IL-5, and GM-CSF. There is evidence for reciprocal control of the maturation of the stem-cell population into basophils or eosinophils. For example, transforming growth factor (TGF)-β in the presence of IL-3 suppresses eosinophil differentiation and enhances that of basophils. Basophils are normally present in very low numbers in the circulation and seem to have a similar role to eosinophils in defense against pathogens. Like eosinophils, they are recruited to the sites of allergic reactions. Basophils express FcεRI on the cell surface and, on activation by cytokines or antigen, they release histamine and IL-4 from the basophilic granules after which they are named.

Eosinophils, mast cells, and basophils can interact with each other. Eosinophil degranulation releases **major basic protein**, which in turn causes degranulation of mast cells and basophils. This effect is augmented by any of the cytokines that affect eosinophil and basophil growth, differentiation, and activation, such as IL-3, IL-5, and GM-CSF.

12-9 An allergic reaction is divided into an immediate response and a late-phase response.

The inflammatory response after IgE-mediated mast-cell activation occurs as an immediate reaction, starting within seconds, and a late reaction, which takes up to 8–12 hours to develop. These reactions can be distinguished clinically (Fig. 12.14). The **immediate reaction** is due to the activity of histamine, prostaglandins, and other preformed or rapidly synthesized mediators that cause a rapid increase in vascular permeability and the contraction of smooth muscle. The **late-phase reaction** is caused by the induced synthesis and release of mediators including leukotrienes, chemokines, and cytokines from the activated mast cells (see Fig. 12.10). These recruit other leukocytes, including eosinophils and T$_H$2 lymphocytes, to the site of inflammation.

Fig. 12.14 Allergic reactions can be divided into an immediate response and a late-phase response. A wheal-and-flare allergic reaction develops within a minute or two of superficial injection of antigen into the epidermis and lasts for up to 30 minutes. The reaction to an intracutaneous injection of house dust mite antigen is shown in the upper left panel and is labeled 'HDM;' the area labeled 'saline' shows the absence of any response to a control injection of saline solution. A more widespread edematous response, as shown in the upper right panel, develops approximately 8 hours later and can persist for some hours. Similarly, the response to an inhaled antigen can be divided into early and late responses (bottom panel). An asthmatic response in the lungs with narrowing of the airways caused by the constriction of bronchial smooth muscle can be measured as a fall in the forced expired volume of air in one second (FEV$_1$). The immediate response peaks within minutes after antigen inhalation and then subsides. Approximately 8 hours after antigen challenge, there is a late-phase response that also results in a fall in the FEV$_1$. The immediate response is caused by the direct effects on blood vessels and smooth muscle of rapidly metabolized mediators such as histamine released by mast cells. The late-phase response is caused by the effects of an influx of inflammatory leukocytes attracted by chemokines and other mediators released by mast cells during and after the immediate response. Photographs courtesy of A.B. Kay.

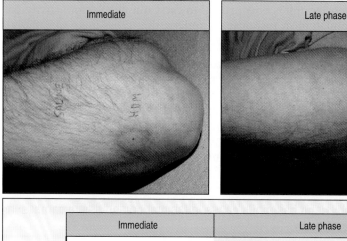

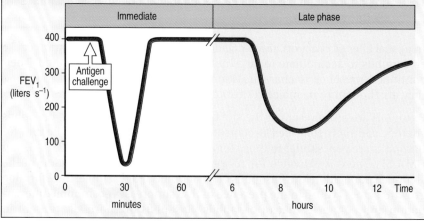

Although the late-phase reaction is clinically less marked than the immediate response, it is associated with a second phase of smooth muscle contraction, sustained edema, and the development of one of the cardinal features of allergic asthma: airway hyperreactivity to nonspecific bronchoconstrictor stimuli such as histamine and methacholine.

The late-phase reaction is an important cause of much serious long-term illness, as for example in chronic asthma. This is because the late reaction induces the recruitment of inflammatory leukocytes, especially eosinophils and T_H2 lymphocytes, to the site of the allergen-triggered mast-cell response. This late response can easily convert into a chronic inflammatory response if antigen persists and stimulates allergen-specific T_H2 cells, which in turn promote eosinophilia and further IgE production.

12-10 The clinical effects of allergic reactions vary according to the site of mast-cell activation.

When reexposure to allergen triggers an allergic reaction, the effects are focused on the site at which mast-cell degranulation occurs. In the immediate response, the preformed mediators released are short-lived, and their potent effects on blood vessels and smooth muscles are therefore confined to the vicinity of the activated mast cell. The more sustained effects of the late-phase response are also focused on the site of initial allergen-triggered activation, and the particular anatomy of this site may determine how readily the inflammation can be resolved. Thus, the clinical syndrome produced by an allergic reaction depends critically on three variables: the amount of allergen-specific IgE present; the route by which the allergen is introduced; and the dose of allergen (Fig. 12.15).

Acute Systemic Anaphylaxis

If an allergen is introduced directly into the bloodstream or is rapidly absorbed from the gut, the connective tissue mast cells associated with all blood vessels can become activated. This activation causes a very dangerous syndrome called **systemic anaphylaxis**. Disseminated mast-cell activation has a variety of potentially fatal effects: the widespread increase in vascular permeability leads to a catastrophic loss of blood pressure; airways constrict, causing difficulty in breathing; and swelling of the epiglottis can cause suffocation. This potentially fatal syndrome is called **anaphylactic shock**. It can occur if drugs are administered to people who have IgE specific for that drug, or after an insect bite in individuals allergic to insect venom. Some foods, for example peanuts or brazil nuts, can cause systemic anaphylaxis in susceptible individuals. This syndrome can be rapidly fatal but can usually be controlled by the immediate injection of epinephrine, which relaxes the smooth muscle and inhibits the cardiovascular effects of anaphylaxis.

The most frequent allergic reactions to drugs occur with penicillin and its relatives. In people with IgE antibodies against penicillin, administration of the drug by injection can cause anaphylaxis and even death. Great care should be taken to avoid giving a drug to patients with a past history of allergy to that drug or one that is closely related structurally. Penicillin acts as a hapten (see Section 9-2); it is a small molecule with a highly reactive β-lactam ring that is crucial for its antibacterial activity. This ring reacts with amino groups on host proteins to form covalent conjugates. When penicillin is ingested or injected, it forms conjugates with self proteins, and the penicillin-modified self peptides can provoke a T_H2 response in some individuals. These T_H2 cells then activate penicillin-binding B cells to produce IgE antibody to the penicillin hapten. Thus, penicillin acts both as the B-cell antigen and, by modifying self peptides, as the T-cell antigen. When penicillin is injected intravenously into an allergic individual, the penicillin-modified proteins can cross-link IgE molecules on the mast cells and cause anaphylaxis.

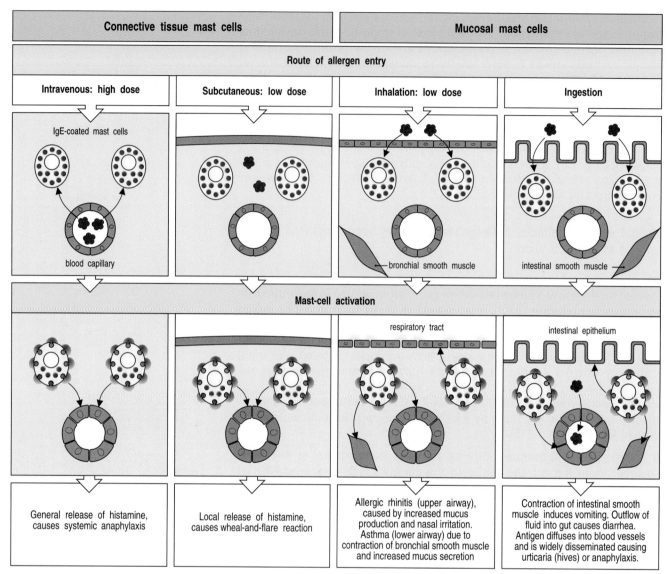

Connective tissue mast cells		Mucosal mast cells	
Route of allergen entry			
Intravenous: high dose	Subcutaneous: low dose	Inhalation: low dose	Ingestion

Mast-cell activation

| General release of histamine, causes systemic anaphylaxis | Local release of histamine, causes wheal-and-flare reaction | Allergic rhinitis (upper airway), caused by increased mucus production and nasal irritation. Asthma (lower airway) due to contraction of bronchial smooth muscle and increased mucus secretion | Contraction of intestinal smooth muscle induces vomiting. Outflow of fluid into gut causes diarrhea. Antigen diffuses into blood vessels and is widely disseminated causing urticaria (hives) or anaphylaxis. |

Fig. 12.15 The dose and route of allergen administration determine the type of IgE-mediated allergic reaction that results. There are two main anatomical distributions of mast cells: those associated with vascularized connective tissues, called connective tissue mast cells, and those found in submucosal layers of the gut and respiratory tract, called mucosal mast cells. In an allergic individual, all of these are loaded with IgE directed against specific allergens. The overall response to an allergen then depends on which mast cells are activated. Allergen in the bloodstream activates connective tissue mast cells throughout the body, resulting in the systemic release of histamine and other mediators. Subcutaneous administration of allergen activates only local connective tissue mast cells, leading to a local inflammatory reaction. Inhaled allergen, penetrating across epithelia, activates mainly mucosal mast cells, causing smooth muscle contraction in the lower airways; this leads to bronchoconstriction and difficulty in expelling inhaled air. Mucosal mast-cell activation also increases the local secretion of mucus by epithelial cells and causes irritation. Similarly, ingested allergen penetrates across gut epithelia, causing vomiting due to intestinal smooth muscle contraction and diarrhea due to outflow of fluid across the gut epithelium. Food allergens can also be disseminated in the bloodstream, causing urticaria (hives) when the food allergen reaches the skin.

12-11 Allergen inhalation is associated with the development of rhinitis and asthma.

Inhalation is the most common route of allergen entry. Many people have mild allergies to inhaled antigens, manifesting as sneezing and a runny nose. This is called **allergic rhinitis**, and results from the activation of mucosal mast cells beneath the nasal epithelium by allergens such as pollens that

Acute responses		Chronic response
Inflammatory mediators cause increased mucus secretion and smooth muscle contraction leading to airway obstruction	Recruitment of cells from the circulation	Chronic response caused by cytokines and eosinophil products

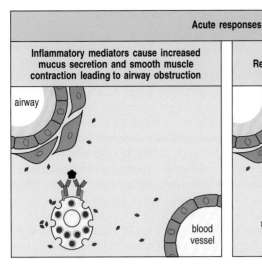
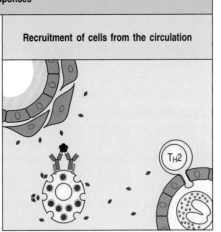
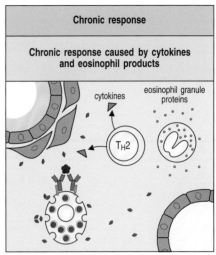

Fig. 12.16 The acute response in allergic asthma leads to T$_H$2-mediated chronic inflammation of the airways. In sensitized individuals, cross-linking of specific IgE on the surface of mast cells by an inhaled allergen triggers them to secrete inflammatory mediators, causing increased vascular permeability, contraction of bronchial smooth muscle, and increased mucus secretion. There is an influx of inflammatory cells, including eosinophils and T$_H$2 cells, from the blood. Activated mast cells and T$_H$2 cells secrete cytokines that augment eosinophil activation and degranulation, which causes further tissue injury and the entry of more inflammatory cells. The result is chronic inflammation, which can cause irreversible damage to the airways.

release their protein contents, which can then diffuse across the mucus membranes of the nasal passages. Allergic rhinitis is characterized by intense itching and sneezing, local edema leading to blocked nasal passages, a nasal discharge, which is typically rich in eosinophils, and irritation of the nose as a result of histamine release. A similar reaction to airborne allergens deposited on the conjunctiva of the eye is called **allergic conjunctivitis**. Allergic rhinitis and conjunctivitis are commonly caused by environmental allergens that are only present during certain seasons of the year. For example, hay fever is caused by a variety of allergens, including certain grass and tree pollens. Autumnal symptoms may be caused by weed pollen, such as that of ragweed. These reactions are annoying but cause little lasting damage.

A more serious syndrome is **allergic asthma**, which is triggered by allergen-induced activation of submucosal mast cells in the lower airways (Fig. 12.16). This leads within seconds to bronchial constriction and increased secretion of fluid and mucus, making breathing more difficult by trapping inhaled air in the lungs. Patients with allergic asthma often need treatment, and asthmatic attacks can be life-threatening. An important feature of asthma is chronic inflammation of the airways, which is characterized by the continued presence of increased numbers of T$_H$2 lymphocytes, eosinophils, neutrophils, and other leukocytes (Fig. 12.17).

Although allergic asthma is initially driven by a response to a specific allergen, the subsequent chronic inflammation seems to be perpetuated even in the apparent absence of further exposure to allergen. The airways become characteristically hyperreactive and factors other than reexposure to antigen can trigger asthma attacks. For example, the airways of asthmatics characteristically show hyperresponsiveness to environmental chemical irritants such as cigarette smoke and sulfur dioxide; viral or, to a lesser extent, bacterial respiratory tract infections can exacerbate the disease by inducing a T$_H$2-dominated local response.

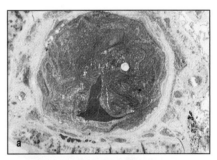
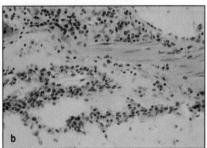

Fig. 12.17 Morphological evidence of chronic inflammation in the airways of an asthmatic patient. Panel a shows a section through a bronchus of a patient who died of asthma; there is almost total occlusion of the airway by a mucus plug. In panel b, a close-up view of the bronchial wall shows injury to the epithelium lining the bronchus, accompanied by a dense inflammatory infiltrate that includes eosinophils, neutrophils, and lymphocytes. Photographs courtesy of T. Krausz.

12-12 Skin allergy is manifest as urticaria or chronic eczema.

The same dichotomy between immediate and delayed responses is seen in cutaneous allergic responses. The skin forms an effective barrier to the entry of most allergens but it can be breached by local injection of small amounts of allergen, for example by a stinging insect. The entry of allergen into the epidermis or dermis causes a localized allergic reaction. Local mast-cell activation in the skin leads immediately to a local increase in vascular permeability, which causes extravasation of fluid and swelling. Mast-cell activation also stimulates the release of chemicals from local nerve endings by a nerve axon reflex, causing the vasodilation of surrounding cutaneous blood vessels, which causes redness of the surrounding skin. The resulting skin lesion is called a **wheal-and-flare reaction**. About 8 hours later, a more widespread and sustained edematous response appears in some individuals as a consequence of the late-phase response (see Fig. 12.14). A disseminated form of the wheal-and-flare reaction, known as **urticaria** or hives, sometimes appears when ingested allergens enter the bloodstream and reach the skin. Histamine released by mast cells activated by allergen in the skin causes large, itchy, red swellings of the skin.

Allergists take advantage of the immediate response to test for allergy by injecting minute amounts of potential allergens into the epidermal layer of the skin. Although the reaction after the administration of antigen by intraepidermal injection is usually very localized, there is a small risk of inducing systemic anaphylaxis. Another standard test for allergy is to measure levels of IgE antibody specific for a particular allergen in a sandwich ELISA (see Appendix I, Section A-6).

Although acute urticaria is commonly caused by allergens, the causes of chronic urticaria, in which the urticarial rash can recur over long periods, are less well understood. In up to a third of cases, it seems likely that chronic urticaria is an autoimmune disease caused by autoantibodies against the α chain of FcϵRI. This is an example of a type II hypersensitivity reaction in which an autoantibody against a cellular receptor triggers cellular activation, in this case causing mast-cell degranulation with resulting urticaria.

Atopic
Dermatitis

A more prolonged inflammatory response is sometimes seen in the skin, most often in atopic children. They develop a persistent skin rash called **eczema** or atopic dermatitis, due to a chronic inflammatory response similar to that seen in the bronchial walls of patients with asthma. The etiology of eczema is not well understood. T_H2 cells and IgE are involved, and it usually clears in adolescence, unlike rhinitis and asthma, which can persist throughout life.

12-13 Allergy to foods causes symptoms limited to the gut and systemic reactions.

When an allergen is eaten, two types of allergic response are seen. Activation of mucosal mast cells associated with the gastrointestinal tract leads to transepithelial fluid loss and smooth muscle contraction, causing diarrhea and vomiting. For reasons that are not understood, connective tissue mast cells in the dermis and subcutaneous tissues can also be activated after ingestion of allergen, presumably by allergen that has been absorbed into the bloodstream, and this results in urticaria. Urticaria is a common reaction when penicillin is given orally to a patient who already has penicillin-specific IgE antibodies. Ingestion of food allergens can also lead to the development of generalized anaphylaxis, accompanied by cardiovascular collapse and acute asthmatic symptoms. Certain foods, most importantly peanuts, tree nuts, and shellfish, are particularly associated with this type of life-threatening response.

12-14 Allergy can be treated by inhibiting either IgE production or the effector pathways activated by cross-linking of cell-surface IgE.

The approaches to the treatment and prevention of allergy are set out in Fig. 12.18. Two treatments are commonly used in clinical practice—one is desensitization and the other is blockade of the effector pathways. There are also several approaches still in the experimental stage. In **desensitization** the aim is to shift the antibody response away from one dominated by IgE toward one dominated by IgG; the latter can bind to the allergen and thus prevent it from activating IgE-mediated effector pathways. Patients are injected with escalating doses of allergen, starting with tiny amounts. This injection schedule gradually diverts the IgE-dominated response, driven by T_H2 cells, to one driven by T_H1 cells, with the consequent downregulation of IgE production. Recent evidence shows that desensitization is also associated with a reduction in the numbers of late-phase inflammatory cells at the site of the allergic reaction. A potential complication of the desensitization approach is the risk of inducing IgE-mediated allergic responses.

An alternative, and still experimental, approach to desensitization is vaccination with peptides derived from common allergens. This procedure induces T-cell anergy (see Section 8-11), which is associated with multiple changes in the T-cell phenotype, including downregulation of cytokine production and reduced expression of the CD3:T-cell receptor complex. IgE-mediated responses are not induced by the peptides because IgE, in contrast to T cells, can only recognize the intact antigen. A major difficulty with this approach is that an individual's responses to peptides are restricted by their MHC class II alleles (see Section 5-12); therefore, patients with different MHC class II molecules respond to different allergen-derived peptides. As the human population is outbred and expresses a wide variety of MHC class II alleles, the number of peptides required to treat all allergic individuals might be very large.

Another vaccination strategy that shows promise in experimental models of allergy is the use of oligodeoxynucleotides rich in unmethylated cytosine guanine dinucleotides (CpG) as adjuvants (see Section 14-19) for desensitization

Target step	Mechanism of treatment	Specific approach
T_H2 activation	Reverse T_H2/T_H1 balance	Injection of specific antigen or peptides. Administration of cytokines, e.g., IFN-γ, IL-10, IL-12, TGF-β. Use of adjuvants such as CpG oligodeoxynucleotides to stimulate T_H1 response
Activation of B cell to produce IgE	Block co-stimulation. Inhibit T_H2 cytokines	Inhibit CD40L. Inhibit IL-4 or IL-13
Mast-cell activation	Inhibit effects of IgE binding to mast cell	Blockade of IgE receptor
Mediator action	Inhibit effects of mediators on specific receptors. Inhibit synthesis of specific mediators	Antihistamine drugs. Lipooxygenase inhibitors
Eosinophil-dependent inflammation	Block cytokine and chemokine receptors that mediate eosinophil recruitment and activation	Inhibit IL-5. Block CCR3

Fig. 12.18 Approaches to the treatment of allergy. Possible methods of inhibiting allergic reactions are shown. Two approaches are in regular clinical use. The first is the injection of specific antigen in desensitization regimes, which are believed to divert the immune response to the allergen from a T_H2 to a T_H1 type, so that IgG is produced in place of IgE. The second clinically useful approach is the use of specific inhibitors to block the synthesis or effects of inflammatory mediators produced by mast cells.

regimes. These oligonucleotides mimic bacterial DNA sequences known as CpG motifs and strongly promote T_H1 responses. Their mechanism of action is discussed in Sections 14-19 and 8-6 and Appendix I, Section A-4.

The signaling pathways that enhance the IgE response in allergic disease are also potential targets for therapy. Inhibitors of IL-4, IL-5, and IL-13 would be predicted to reduce IgE responses, but redundancy between some of the activities of these cytokines might make this approach difficult to implement in practice. A second approach to manipulating the response is to give cytokines that promote T_H1-type responses. IFN-γ, IFN-α, IL-10, IL-12, and TGF-β have each been shown to reduce IL-4-stimulated IgE synthesis *in vitro*, and IFN-γ and IFN-α have been shown to reduce IgE synthesis *in vivo*.

Another target for therapeutic intervention might be the high-affinity IgE receptor. An effective competitor for IgE at this receptor could prevent the binding of IgE to the surfaces of mast cells, basophils, and eosinophils. Candidate competitors include humanized anti-IgE monoclonal antibodies, which bind to IgE and block its binding to the receptor, and modified IgE Fc constructs that bind to the receptor but lack variable regions and thus cannot bind antigen. Yet another approach would be to block the recruitment of eosinophils to sites of allergic inflammation. The eotaxin receptor CCR3 is a potential target for this type of therapy. The production of eosinophils in bone marrow and their exit into the circulation might also be reduced by a blockade of IL-5 action.

The mainstays of therapy at present, however, are drugs that treat the symptoms of allergic disease and limit the inflammatory response. Anaphylactic reactions are treated with epinephrine, which stimulates the reformation of endothelial tight junctions, promotes the relaxation of constricted bronchial smooth muscle, and also stimulates the heart. Inhaled bronchodilators that act on β-adrenergic receptors to relax constricted muscle are also used to relieve acute asthma attacks. Anti-histamines that block the histamine H_1 receptor reduce the urticaria that follows histamine release from mast cells and eosinophils. Relevant H_1 receptors include those on blood vessels that cause increased permeability of the vessel wall, and those on unmyelinated nerve fibers that are thought to mediate the itching sensation. In chronic allergic disease it is extremely important to treat and prevent the chronic inflammatory tissue injury. Topical or systemic corticosteroids (see Section 14-1) are used to suppress the chronic inflammatory changes seen in asthma, rhinitis, and eczema. However, what is really needed is a means of converting the T-cell response to the allergenic peptide antigen from predominantly T_H2 to predominantly T_H1. This topic is also discussed in Chapter 14.

Summary.

The allergic response to innocuous antigens reflects the pathophysiological aspects of a defensive immune response whose physiological role is to protect against helminthic parasites. It is triggered by antigen binding to IgE anti-bodies bound to the high-affinity IgE receptor FcϵRI on mast cells. Mast cells are strategically distributed beneath the mucosal surfaces of the body and in connective tissue. Antigen cross-linking the IgE on their surface causes them to release large amounts of inflammatory mediators. The resulting inflammation can be divided into early events, characterized by short-lived mediators such as histamine, and later events that involve leukotrienes, cytokines, and chemokines, which recruit and activate eosinophils and basophils. The late phase of this response can evolve into chronic inflammation, characterized by the presence of effector T cells and eosinophils, which is most clearly seen in chronic allergic asthma.

Hypersensitivity diseases.

Immunological responses involving IgG antibodies or specific T cells can also cause adverse hypersensitivity reactions. Although these effector arms of the immune response normally participate in protective immunity to infection, they occasionally react with noninfectious antigens to produce acute or chronic hypersensitivity reactions. We will describe common examples of such reactions in this part of the chapter.

12-15 Innocuous antigens can cause type II hypersensitivity reactions in susceptible individuals by binding to the surfaces of circulating blood cells.

Antibody-mediated destruction of red blood cells (hemolytic anemia) or platelets (thrombocytopenia) is an uncommon side-effect associated with the intake of certain drugs such as the antibiotic penicillin, the anti-cardiac arrhythmia drug quinidine, or the antihypertensive agent methyldopa. These are examples of **type II hypersensitivity reactions** in which the drug binds to the cell surface and serves as a target for anti-drug IgG antibodies that cause destruction of the cell (see Fig. 12.2). The anti-drug antibodies are made in only a minority of individuals and it is not clear why these individuals make them. The cell-bound antibody triggers clearance of the cell from the circulation, predominantly by tissue macrophages in the spleen, which bear Fcγ receptors.

12-16 Systemic disease caused by immune complex formation can follow the administration of large quantities of poorly catabolized antigens.

Type III hypersensitivity reactions can arise with soluble antigens. The pathology is caused by the deposition of antigen:antibody aggregates or **immune complexes** at certain tissue sites. Immune complexes are generated in all antibody responses but their pathogenic potential is determined, in part, by their size and the amount, affinity, and isotype of the responding antibody. Larger aggregates fix complement and are readily cleared from the circulation by the mononuclear phagocytic system. The small complexes that form at antigen excess, however, tend to deposit in blood vessel walls. There they can ligate Fc receptors on leukocytes, leading to leukocyte activation and tissue injury.

A local type III hypersensitivity reaction can be triggered in the skin of sensitized individuals who possess IgG antibodies against the sensitizing antigen. When antigen is injected into the skin, circulating IgG antibody that has diffused into the tissues forms immune complexes locally. The immune complexes bind Fc receptors on mast cells and other leukocytes, which creates a local inflammatory response with increased vascular permeability. The enhanced vascular permeability allows fluid and cells, especially polymorphonuclear leukocytes, to enter the site from the local vessels. This reaction is called an **Arthus reaction** (Fig. 12.19). The immune complexes also activate complement, releasing C5a, which contributes to the inflammatory reaction by ligating C5a receptors on leukocytes (see Sections 2-12 and 6-16). This causes their activation and chemotactic attraction to the site of inflammation. The Arthus reaction is absent in mice lacking the α or γ chain of the FcγRIII receptor (CD16) on mast cells, but remains largely unperturbed in complement-deficient mice, showing the primary importance of FcγRIII in triggering inflammatory responses via immune complexes.

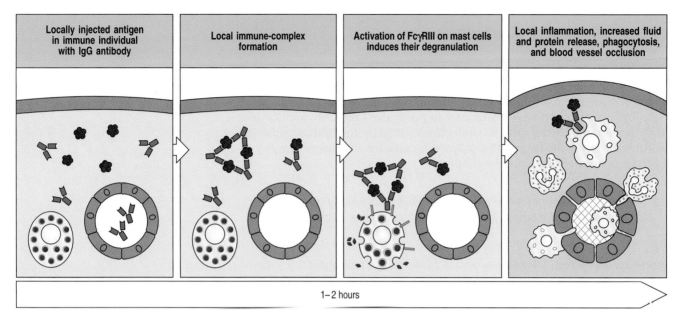

| Locally injected antigen in immune individual with IgG antibody | Local immune-complex formation | Activation of FcγRIII on mast cells induces their degranulation | Local inflammation, increased fluid and protein release, phagocytosis, and blood vessel occlusion |

1–2 hours

Fig. 12.19 The deposition of immune complexes in local tissues causes a local inflammatory response known as an Arthus reaction (type III hypersensitivity reaction). In individuals who have already made IgG antibody against an antigen, the same antigen injected into the skin forms immune complexes with IgG antibody that has diffused out of the capillaries. Because the dose of antigen is low, the immune complexes are only formed close to the site of injection, where they activate mast cells bearing Fcγ receptors (FcγRIII). As a result of mast-cell activation, inflammatory cells invade the site, and blood vessel permeability and blood flow are increased. Platelets also accumulate inside the vessel at the site, ultimately leading to vessel occlusion.

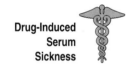

Drug-Induced Serum Sickness

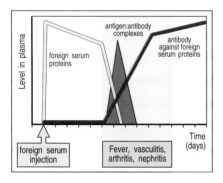

Fig. 12.20 Serum sickness is a classic example of a transient immune complex-mediated syndrome. An injection of a foreign protein or proteins leads to an antibody response. These antibodies form immune complexes with the circulating foreign proteins. The complexes are deposited in small vessels and activate complement and phagocytes, inducing fever and the symptoms of vasculitis, nephritis, and arthritis. All these effects are transient and resolve when the foreign protein is cleared.

A systemic type III hypersensitivity reaction, known as **serum sickness**, can result from the injection of large quantities of a poorly catabolized foreign antigen. This illness was so named because it frequently followed the administration of therapeutic horse antiserum. In the preantibiotic era, antiserum made by immunizing horses was often used to treat pneumococcal pneumonia; the specific anti-pneumococcal antibodies in the horse serum would help the patient to clear the infection. In much the same way, **antivenin** (serum from horses immunized with snake venoms) is still used today as a source of neutralizing antibodies to treat people suffering from the bites of poisonous snakes.

Serum sickness occurs 7–10 days after the injection of the horse serum, an interval that corresponds to the time required to mount a primary immune response that switches from IgM to IgG antibody against the foreign antigens in horse serum. The clinical features of serum sickness are chills, fever, rash, arthritis, and sometimes glomerulonephritis. Urticaria is a prominent feature of the rash, implying a role for histamine derived from mast-cell degranulation. In this case the mast-cell degranulation is triggered by the ligation of cell-surface FcγRIII by IgG-containing immune complexes.

The course of serum sickness is illustrated in Fig. 12.20. The onset of disease coincides with the development of antibodies against the abundant soluble proteins in the foreign serum; these antibodies form immune complexes with their antigens throughout the body. These immune complexes fix complement and can bind to and activate leukocytes bearing Fc and complement receptors; these in turn cause widespread tissue injury. The formation of immune complexes causes clearance of the foreign antigen and so serum sickness is usually a self-limiting disease. Serum sickness after a second dose of antigen follows the kinetics of a secondary antibody response and the onset of disease occurs typically within a day or two. Serum sickness is nowadays seen after the use of anti-lymphocyte globulin, employed as an immunosuppressive agent in transplant recipients, and also, rarely, after the administration of streptokinase, a bacterial enzyme that is used as a thrombolytic agent to treat patients with a myocardial infarction or heart attack.

A similar type of immunopathological response is seen in two other situations in which antigen persists. The first is when an adaptive antibody response fails to clear an infectious agent, for example in subacute bacterial endocarditis or chronic viral hepatitis. In this situation, the multiplying bacteria or viruses are continuously generating new antigen in the presence of a persistent antibody response that fails to eliminate the organism. Immune complex disease ensues, with injury to small blood vessels in many tissues and organs, including the skin, kidneys, and nerves. Immune complexes also form in autoimmune diseases such as systemic lupus erythematosus where, because the antigen persists, the deposition of immune complexes continues, and serious disease can result (see Section 13-7).

Some inhaled allergens provoke IgG rather than IgE antibody responses, perhaps because they are present at relatively high levels in inhaled air. When a person is reexposed to high doses of such inhaled antigens, immune complexes form in the alveolar wall of the lung. This leads to the accumulation of fluid, protein, and cells in the alveolar wall, slowing blood–gas interchange and compromising lung function. This type of reaction occurs in certain occupations such as farming, where there is repeated exposure to hay dust or mold spores. The disease that results is therefore called **farmer's lung**. If exposure to antigen is sustained, the alveolar membranes can become permanently damaged.

12-17 Delayed-type hypersensitivity reactions are mediated by T_H1 cells and CD8 cytotoxic T cells.

Unlike the immediate hypersensitivity reactions described so far, which are mediated by antibodies, **delayed-type hypersensitivity** or **type IV hypersensitivity reactions** are mediated by antigen-specific effector T cells. These function in essentially the same way as during a response to an infectious pathogen, as described in Chapter 8. The causes and consequences of some syndromes in which type IV hypersensitivity responses predominate are listed in Fig. 12.21. These responses can be transferred between experimental animals by purified T cells or cloned T-cell lines.

Type IV hypersensitivity reactions are mediated by antigen-specific effector T cells		
Syndrome	**Antigen**	**Consequence**
Delayed-type hypersensitivity	Proteins: Insect venom Mycobacterial proteins (tuberculin, lepromin)	Local skin swelling: Erythema Induration Cellular infiltrate Dermatitis
Contact hypersensitivity	Haptens: Pentadecacatechol (poison ivy) DNFB Small metal ions: Nickel Chromate	Local epidermal reaction: Erythema Cellular infiltrate Vesicles Intraepidermal abscesses
Gluten-sensitive enteropathy (celiac disease)	Gliadin	Villous atrophy in small bowel Malabsorption

Fig. 12.21 Type IV hypersensitivity responses. These reactions are mediated by T cells and all take some time to develop. They can be grouped into three syndromes, according to the route by which antigen passes into the body. In delayed-type hypersensitivity the antigen is injected into the skin; in contact hypersensitivity it is absorbed into the skin; and in gluten-sensitive enteropathy it is absorbed by the gut.

Fig. 12.22 The stages of a delayed-type hypersensitivity reaction. The first phase involves uptake, processing, and presentation of the antigen by local antigen-presenting cells. In the second phase, T_H1 cells that were primed by a previous exposure to the antigen migrate into the site of injection and become activated. Because these specific cells are rare, and because there is little inflammation to attract cells into the site, it can take several hours for a T cell of the correct specificity to arrive. These cells release mediators that activate local endothelial cells, recruiting an inflammatory cell infiltrate dominated by macrophages and causing the accumulation of fluid and protein. At this point, the lesion becomes apparent.

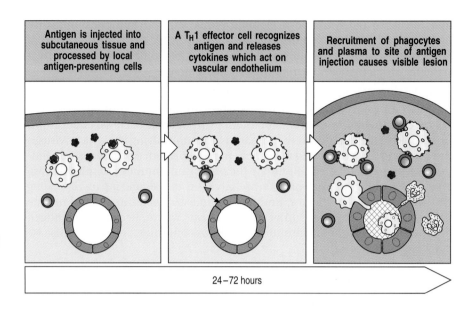

Antigen is injected into subcutaneous tissue and processed by local antigen-presenting cells

A T_H1 effector cell recognizes antigen and releases cytokines which act on vascular endothelium

Recruitment of phagocytes and plasma to site of antigen injection causes visible lesion

24–72 hours

The prototypic delayed-type hypersensitivity reaction is an artifact of modern medicine—the tuberculin test (see Appendix I, Section A-38). This is used to determine whether an individual has previously been infected with *Mycobacterium tuberculosis*. Small amounts of tuberculin—a complex mixture of peptides and carbohydrates derived from *M. tuberculosis*—are injected intradermally. In individuals who have previously been exposed to the bacterium, either by infection with the pathogen or by immunization with BCG, an attenuated form of *M. tuberculosis*, a local T cell-mediated inflammatory reaction evolves over 24–72 hours. The response is mediated by T_H1 cells, which enter the site of antigen injection, recognize complexes of peptide:MHC class II molecules on antigen-presenting cells, and release inflammatory cytokines, such as IFN-γ and TNF-β. The cytokines stimulate the expression of adhesion molecules on endothelium and increase local blood vessel permeability, allowing plasma and accessory cells to enter the site; this causes a visible swelling (Fig. 12.22). Each of these phases takes several hours and so the fully developed response appears only 24–48 hours after challenge. The cytokines produced by the activated T_H1 cells and their actions are shown in Fig. 12.23.

Very similar reactions are observed in several cutaneous hypersensitivity responses. These can be elicited by either CD4 or CD8 T cells, depending on the pathway by which the antigen is processed. Typical antigens that cause cutaneous hypersensitivity responses are highly reactive small molecules that can easily penetrate intact skin, especially if they cause itching that leads to scratching. These chemicals then react with self proteins, creating protein–hapten complexes that can be processed to hapten–peptide complexes, which can bind to MHC molecules that are recognized by T cells as foreign antigens. There are two phases to a cutaneous hypersensitivity response—sensitization and elicitation. During the sensitization phase, cutaneous Langerhans' cells take up and process antigen, and migrate to regional lymph nodes, where they activate T cells (see Fig. 8.15), with the consequent production of memory T cells, which end up in the dermis. In the elicitation phase, further exposure to the sensitizing chemical leads to antigen presentation to memory T cells in the dermis, with release of T-cell cytokines such as IFN-γ and IL-17. This stimulates the keratinocytes of the epidermis to release cytokines such as IL-1, IL-6, TNF-α and GM-CSF, and CXC chemokines including IL-8, interferon-inducible protein (IP)-9, IP-10, and MIG (monokine induced by IFN-γ). These cytokines and chemokines

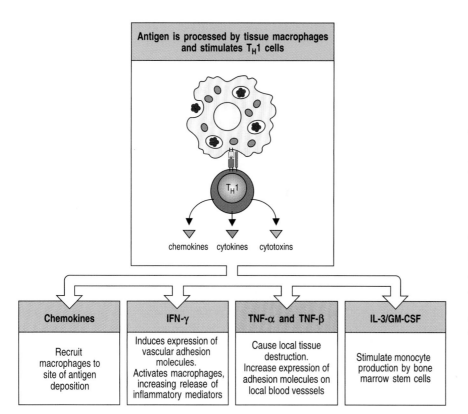

Antigen is processed by tissue macrophages and stimulates T$_H$1 cells

chemokines   cytokines   cytotoxins

Chemokines	IFN-γ	TNF-α and TNF-β	IL-3/GM-CSF
Recruit macrophages to site of antigen deposition	Induces expression of vascular adhesion molecules. Activates macrophages, increasing release of inflammatory mediators	Cause local tissue destruction. Increase expression of adhesion molecules on local blood vesssels	Stimulate monocyte production by bone marrow stem cells

Fig. 12.23 The delayed-type (type IV) hypersensitivity response is directed by chemokines and cytokines released by T$_H$1 cells stimulated by antigen. Antigen in the local tissues is processed by antigen-presenting cells and presented on MHC class II molecules. Antigen-specific T$_H$1 cells that recognize the antigen locally at the site of injection release chemokines and cytokines that recruit macrophages to the site of antigen deposition. Antigen presentation by the newly recruited macrophages then amplifies the response. T cells can also affect local blood vessels through the release of TNF-α and TNF-β, and stimulate the production of macrophages through the release of IL-3 and GM-CSF. Finally, T$_H$1 cells activate macrophages through the release of IFN-γ and TNF-α, and kill macrophages and other sensitive cells through the cell-surface expression of the Fas ligand.

enhance the inflammatory response by inducing the migration of monocytes into the lesion and their maturation into macrophages, and by attracting more T cells (Fig. 12.24).

The rash produced by contact with poison ivy (Fig. 12.25) is caused by a T-cell response to a chemical in the poison ivy leaf called pentadecacatechol. This compound is lipid-soluble and can therefore cross the cell membrane and

Contact Sensitivity to Poison Ivy

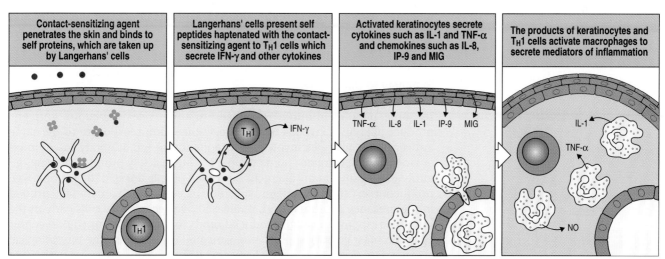

| Contact-sensitizing agent penetrates the skin and binds to self proteins, which are taken up by Langerhans' cells | Langerhans' cells present self peptides haptenated with the contact-sensitizing agent to T$_H$1 cells which secrete IFN-γ and other cytokines | Activated keratinocytes secrete cytokines such as IL-1 and TNF-α and chemokines such as IL-8, IP-9 and MIG | The products of keratinocytes and T$_H$1 cells activate macrophages to secrete mediators of inflammation |

IFN-γ

TNF-α IL-8   IL-1   IP-9   MIG

IL-1   TNF-α NO

Fig. 12.24 Elicitation of a delayed-type hypersensitivity response to a contact-sensitizing agent. The contact-sensitizing agent is a small highly reactive molecule that can easily penetrate intact skin. It binds covalently as a hapten to a variety of endogenous proteins, which are taken up and processed by Langerhans' cells, the major antigen-presenting cells of skin. These present haptenated peptides to effector T$_H$1 cells (which must have been previously primed in lymph nodes and then have traveled back to the skin). These then secrete cytokines such as IFN-γ that stimulate keratinocytes to secrete further cytokines and chemokines. These in turn attract monocytes and induce their maturation into activated tissue macrophages, which contribute to the inflammatory lesions depicted in Fig. 12.25.

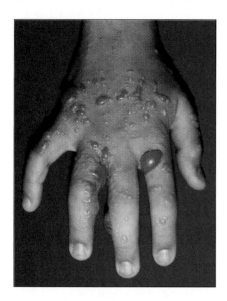

Fig. 12.25 Blistering skin lesions on hand of patient with poison ivy contact dermatitis. Photograph courtesy of R. Geha.

modify intracellular proteins. These modified proteins generate modified peptides within the cytosol, which are translocated into the endoplasmic reticulum and are delivered to the cell surface by MHC class I molecules. These are recognized by CD8 T cells, which can cause damage either by killing the eliciting cell or by secreting cytokines such as IFN-γ. The well-studied chemical picryl chloride produces a CD4 T-cell hypersensitivity reaction. It modifies extracellular self proteins, which are then processed by the exogenous pathway (see Section 5-5) into modified self peptides that bind to self MHC class II molecules and are recognized by T$_H$1 cells. When sensitized T$_H$1 cells recognize these complexes they can produce extensive inflammation by activating macrophages (see Fig. 12.24). As the chemicals in these examples are delivered by contact with the skin, the rash that follows is called a **contact hypersensitivity reaction**.

Some insect proteins also elicit delayed-type hypersensitivity response. However, the early phases of the host reaction to an insect bite are often IgE-mediated or the result of the direct effects of insect venoms. Important delayed-type hypersensitivity responses to divalent cations such as nickel have also been observed. These divalent cations can alter the conformation or the peptide binding of MHC class II molecules, and thus provoke a T-cell response. Finally, although this section has focused on the role of T cells in inducing delayed-type hypersensitivity reactions, there is evidence that antibody and complement may also play a part. Mice deficient in B cells, antibody, or complement show impaired contact hypersensitivity reactions. These requirements for B cells, antibody, and complement may reflect their role in the early steps of the elicitation of these reactions.

Summary.

Hypersensitivity diseases reflect normal immune mechanisms directed against innocuous antigens. They can be mediated by IgG antibodies bound to modified cell surfaces, or by complexes of antibodies bound to poorly catabolized antigens, as occurs in serum sickness. Hypersensitivity reactions mediated by T cells can be activated by modified self proteins, or by injected proteins such as those in the mycobacterial extract tuberculin. These T cell-mediated responses require the induced synthesis of effector molecules and develop more slowly, which is why they are termed delayed-type hypersensitivity.

Summary to Chapter 12.

In some people, immune responses to otherwise innocuous antigens produce allergic or hypersensitivity reactions upon reexposure to the same antigen. Most allergies involve the production of IgE antibody to common environmental allergens. Some people are intrinsically prone to making IgE antibodies against many allergens, and such people are said to be atopic. IgE production is driven by antigen-specific T$_H$2 cells, which are initially primed in the presence of a burst of IL-4 released by specialized T cells early in the immune response. The IgE produced binds to the high-affinity IgE receptor FcεRI on mast cells, basophils, and activated eosinophils. The physiological role of this system is to provide front-line defense against parasite pathogens but, in economically developed societies in which parasitic infections are uncommon, it is almost always involved in allergic reactions. Eosinophils and specific effector T cells have an extremely important role in chronic allergic inflammation, which is the major cause of the chronic morbidity of asthma. Antibodies of other isotypes and antigen-specific effector T cells contribute to hypersensitivity to other antigens.

General references.

Bernstein, D.I.: **Allergic reactions to workplace allergens.** *JAMA* 1997, **278**:1907-1913.

Costa, J.J., Weller, P.F., and Galli, S.J.: **The cells of the allergic response: mast cells, basophils, and eosinophils.** *JAMA* 1997, **278**:1815-1822.

Galli, S.J.: **Allergy.** *Curr. Biol.* 2000, **10**:R93-R95.

Galli, S.J.: **Complexity and redundancy in the pathogenesis of asthma: reassessing the roles of mast cells and T cells.** *J. Exp. Med.* 1997, **186**:343-347.

Kay, A.B.: *Allergy and Allergic Diseases.* Oxford, Blackwell Science, 1997.

Kay, A.B.: **Role of T cells in asthma.** *Chem. Immunol.* 1998, **71**:178-191.

Middleton, E., Jr., Reed, C.E., Ellis, E.F., Adkinson, N.F., Yunginger, J.W., and Busse, W.W.: *Allergy: Principles and Practice,* 4th edn. St Louis, Mosby, 1993.

Oettgen, H.C., and Geha, R.S.: **IgE in asthma and atopy: cellular and molecular connections.** *J. Clin. Invest.* 1999, **104**:829-835.

Ray, A., and Cohn, L.: **Th2 cells and GATA-3 in asthma: new insights into the regulation of airway inflammation.** *J. Clin. Invest.* 1999, **104**:985-993.

Romagnani, S.: **Atopic allergy and other hypersensitivities interactions between genetic susceptibility, innocuous and/or microbial antigens and the immune system.** *Curr. Opin. Immunol.* 1997, **9**:773-775.

Romagnani, S.: **The role of lymphocytes in allergic disease.** *J. Allergy Clin. Immunol.* 2000, **105**:399-408.

Rosen, F.S.: **Urticaria, angioedema, and anaphylaxis.** *Pediatr. Rev.* 1992, **13**:387-390.

Section references.

12-1 Allergens are often delivered transmucosally at low dose, a route that favors IgE production.

Lambrecht, B.N., De Veerman, M., Coyle, A.J., Gutierrez-Ramos, J.C., Thielemans, K., and Pauwels, R.A.: **Myeloid dendritic cells induce Th2 responses to inhaled antigen, leading to eosinophilic airway inflammation.** *J. Clin. Invest.* 2000, **106**:551-559.

O'Hehir, R.E., Garman, R.D., Greenstein, J.L., and Lamb, J.R.: **The specificity and regulation of T-cell responsiveness to allergens.** *Annu. Rev. Immunol.* 1991, **9**:67-95.

Parronchi, P., Macchia, D., Piccinni, M.P., Biswas, P., Simonelli, C., Maggi, E., Ricci, M., Ansari, A.A., and Romagnani, S.: **Allergen and bacterial antigen-specific T-cell clones established from atopic donors show a different profile of cytokine production.** *Proc. Natl. Acad. Sci. USA* 1991, **88**:4538-4542.

Sertl, K., Takemura, T., Tschachler, E., Ferrans, V.J., Kaliner, M.A., and Shevach, E.M.: **Dendritic cells with antigen-presenting capability reside in airway epithelium, lung parenchyma, and visceral pleura.** *J. Exp. Med.* 1986, **163**:436-451.

12-2 Enzymes are frequent triggers of allergy.

Garraud, O., Nkenfou, C., Bradley, J.E., Perler, F.B., and Nutman, T.B.: **Identification of recombinant filarial proteins capable of inducing polyclonal and antigen-specific IgE and IgG4 antibodies.** *J. Immunol.* 1995, **155**:1316-1325.

Grammer, L.C., and Patterson, R.: **Proteins: chymopapain and insulin.** *J. Allergy. Clin. Immunol.* 1984, **74**:635-640.

Kauffman, H.F., Tomee, J.F., van de Riet, M.A., Timmerman, A.J., and Borger, P.: **Protease-dependent activation of epithelial cells by fungal allergens leads to morphologic changes and cytokine production.** *J. Allergy Clin. Immunol.* 2000, **105**:1185-1193.

Shakib, F., Schulz, O., and Sewell, H.: **A mite subversive: cleavage of CD23 and CD25 by Der p 1 enhances allergenicity.** *Immunol. Today* 1998, **19**:313-316.

Thomas, W.R., Smith, W., and Hales, B.J.: **House dust mite allergen characterisation: implications for T-cell responses and immunotherapy.** *Int. Arch. Allergy Immunol.* 1998, **115**:9-14.

Wan, H., Winton, H.L., Soeller, C., Tovey, E.R., Gruenert, D.C., Thompson, P.J., Stewart, G.A., Taylor, G.W., Garrod, D.R., Cannell, M.B., and Robinson, C.: **Der p 1 facilitates transepithelial allergen delivery by disruption of tight junctions.** *J. Clin. Invest.* 1999, **104**:123-133.

12-3 Class switching to IgE in B lymphocytes is favored by specific signals.

Bacharier, L.B., and Geha, R.S.: **Molecular mechanisms of IgE regulation.** *J. Allergy Clin. Immunol.* 2000, **105**:S547-S558.

Bendelac, A., Rivera, M.N., Park, S.H., and Roark, J.H.: **Mouse CD1-specific NK1 T cells: development, specificity, and function.** *Annu. Rev. Immunol.* 1997, **15**:535-562.

Burd, P.R., Thompson, W.C., Max, E.E., and Mills, F.C.: **Activated mast cells produce interleukin 13.** *J. Exp. Med.* 1995, **181**:1373-1380.

Chen, H., and Paul, W.E.: **Cultured NK1.1⁺ CD4⁺ T cells produce large amounts of IL-4 and IFN-gamma upon activation by anti-CD3 or CD1.** *J. Immunol.* 1997, **159**:2240-2249.

Corry, D.B., and Kheradmand, F.: **Induction and regulation of the IgE response.** *Nature* 1999, **402**:B18-B23.

Gauchat, J.F., Henchoz, S., Fattah, D., Mazzei, G., Aubry, J.P., Jomotte, T., Dash, L., Page, K., Solari, R., Aldebert, D., et al.: **CD40 ligand is functionally expressed on human eosinophils.** *Eur. J. Immunol.* 1995, **25**:863-865.

Gauchat, J.F., Henchoz, S., Mazzei, G., Aubry, J.P., Brunner, T., Blasey, H., Life, P., Talabot, D., Flores Romo, L., Thompson, J., et al.: **Induction of human IgE synthesis in B cells by mast cells and basophils.** *Nature* 1993, **365**:340-343.

Hoey, T., and Grusby, M.J.: **STATs as mediators of cytokine-induced responses.** *Adv. Immunol.* 1999, **71**:145-162.

Kaplan, M.H., and Grusby, M.J.: **Regulation of T helper cell differentiation by STAT molecules.** *J. Leukoc. Biol.* 1998, **64**:2-5.

Romagnani, S., Parronchi, P., D'Elios, M.M., Romagnani, P., Annunziato, F., Piccinni, M.P., Manetti, R., Sampognaro, S., Mavilia, C., De-Carli, M., Maggi, E., and Del-Prete, G.F.: **An update on human Th1 and Th2 cells.** *Int. Arch. Allergy Immunol.* 1997, **113**:153-156.

Sallusto, F., Lenig, D., Mackay, C.R., and Lanzavecchia, A.: **Flexible programs of chemokine receptor expression on human polarized T helper 1 and 2 lymphocytes.** *J. Exp. Med.* 1998, **187**:875-883.

Sallusto, F., Mackay, C.R., and Lanzavecchia, A.: **Selective expression of the eotaxin receptor CCR3 by human T helper 2 cells.** *Science* 1997, **277**:2005-2007.

12-4 Genetic factors contribute to the development of IgE-mediated allergy, but environmental factors may also be important.

Cookson, W.: **The alliance of genes and environment in asthma and allergy.** *Nature* 1999, **402**:B5-B11.

Gern, J.E., Lemanske, R.F., Jr., and Busse, W.W.: **Early life origins of asthma.** *J. Clin. Invest.* 1999, **104**:837-843.

Hershey, G.K., Friedrich, M.F., Esswein, L.A., Thomas, M.L., and Chatila, T.A.: **The association of atopy with a gain-of-function mutation in the alpha subunit of the interleukin-4 receptor.** *N. Engl. J. Med.* 1997, **337**:1720-1725.

Matricardi, P.M., Rosmini, F., Ferrigno, L., Nisini, R., Rapicetta, M., Chionne, P., Stroffolini, T., Pasquini, P., and D'Amelio, R.: **Cross sectional retrospective study of prevalence of atopy among Italian military students with antibodies against hepatitis A virus.** *BMJ* 1997, **314**:999-1003.

Mitsuyasu, H., Yanagihara, Y., Mao, X.Q., Gao, P.S., Arinobu, Y., Ihara, K., Takabayashi, A., Hara, T., Enomoto, T., Sasaki, S., Kawai, M., Hamasaki, N., Shirakawa, T., Hopkin, J.M., and Izuhara, K.: **Cutting edge: dominant effect of Ile50Val variant of the human IL-4 receptor alpha-chain in IgE synthesis.** *J. Immunol.* 1999, **162**:1227-1231.

von Mutius, E., Martinez, F.D., Fritzsch, C., Nicolai, T., Roell, G., and Thiemann, H.H.: **Prevalence of asthma and atopy in two areas of West and East Germany.**

Am. J. Respir. Crit Care Med. 1994, **149**:358-364.

Shaheen, S.O., Aaby, P., Hall, A.J., Barker, D.J., Heyes, C.B., Shiell, A.W., and Goudiaby, A.: **Measles and atopy in Guinea-Bissau.** *Lancet* 1996, **347**:1792-1796.

Shirakawa, T., Enomoto, T., Shimazu, S., and Hopkin, J.M.: **The inverse association between tuberculin responses and atopic disorder.** *Science* 1997, **275**:77-79.

12-5   Most IgE is cell-bound and engages effector mechanisms of the immune system by different pathways from other antibody isotypes.

Heyman, B.: **Regulation of antibody responses via antibodies, complement, and Fc receptors.** *Annu. Rev. Immunol.* 2000, **18**:709-737.

Kinet, J.P.: **The high-affinity IgE receptor (FcεRI): from physiology to pathology.** *Annu. Rev. Immunol.* 1999, **17**:931-972.

Metzger, H.: **The receptor with high affinity for IgE.** *Immunol. Rev.* 1992, **125**:37-48.

Payet, M., and Conrad, D.H.: **IgE regulation in CD23 knockout and transgenic mice.** *Allergy* 1999, **54**:1125-1129.

Turner, H., and Kinet, J.P.: **Signalling through the high-affinity IgE receptor FcεRI.** *Nature* 1999, **402**:B24-B30.

12-6   Mast cells reside in tissues and orchestrate allergic reactions.

Austen, K.F.: **The Paul Kallos Memorial Lecture. From slow-reacting substance of anaphylaxis to leukotriene C4 synthase.** *Int. Arch. Allergy Immunol.* 1995, **107**:19-24.

Bingham, C.O., III, and Austen, K.F.: **Mast-cell responses in the development of asthma.** *J. Allergy Clin. Immunol.* 2000, **105**:S527-S534.

Charlesworth, E.N.: **The role of basophils and mast cells in acute and late reactions in the skin.** *Allergy* 1997, **52**:31-43.

Galli, S.J.: **Mast cells and basophils.** *Curr. Opin. Hematol.* 2000, **7**:32-39.

Galli, S.J.: **The Paul Kallos Memorial Lecture. The mast cell: a versatile effector cell for a challenging world.** *Int. Arch. Allergy Immunol.* 1997, **113**:14-22.

Mekori, Y.A., and Metcalfe, D.D.: **Mast cell-T cell interactions.** *J. Allergy Clin. Immunol.* 1999, **104**:517-523.

Metcalfe, D.D., Baram, D., and Mekori, Y.A.: **Mast cells.** *Physiol. Rev.* 1997, **77**:1033-1079.

Rodewald, H.R., Dessing, M., Dvorak, A.M., and Galli, S.J.: **Identification of a committed precursor for the mast cell lineage.** *Science* 1996, **271**:818-822.

Williams, C.M., and Galli, S.J.: **The diverse potential effector and immuno regulatory roles of mast cells in allergic disease.** *J. Allergy Clin. Immunol.* 2000, **105**:847-859.

12-7   Eosinophils are normally under tight control to prevent inappropriate toxic responses.

Assa'ad, A.H., Spicer, R.L., Nelson, D.P., Zimmermann, N., and Rothenberg, M.E.: **Hypereosinophilic syndromes.** *Chem. Immunol.* 2000, **76**:208-229.

Capron, M., and Desreumaux, P.: **Immunobiology of eosinophils in allergy and inflammation.** *Res. Immunol.* 1997, **148**:29-33.

Collins, P.D., Marleau, S., Griffiths Johnson, D.A., Jose, P.J., and Williams, T.J.: **Cooperation between interleukin-5 and the chemokine eotaxin to induce eosinophil accumulation *in vivo*.** *J. Exp. Med.* 1995, **182**:1169-1174.

Kay, A.B., Barata, L., Meng, Q., Durham, S.R., and Ying, S.: **Eosinophils and eosinophil-associated cytokines in allergic inflammation.** *Int. Arch. Allergy Immunol.* 1997, **113**:196-199.

Kita, H., and Gleich, G.J.: **Eosinophils and IgE receptors: a continuing controversy.** *Blood* 1997, **89**:3497-3501.

Matthews, A.N., Friend, D.S., Zimmermann, N., Sarafi, M.N., Luster, A.D., Pearlman, E., Wert, S.E., and Rothenberg, M.E.: **Eotaxin is required for the baseline level of tissue eosinophils.** *Proc. Natl. Acad. Sci. USA* 1998, **95**:6273-6278.

Palframan, R.T., Collins, P.D., Williams, T.J., and Rankin, S.M.: **Eotaxin induces a rapid release of eosinophils and their progenitors from the bone marrow.** *Blood* 1998, **91**:2240-2248.

Parker, C.W.: **Lipid mediators produced through the lipoxygenase pathway.** *Annu. Rev. Immunol.* 1987, **5**:65-84.

Rothenberg, M.E.: **Eosinophilia.** *N. Engl. J. Med.* 1998, **338**:1592-1600.

Rothenberg, M.E., MacLean, J.A., Pearlman, E., Luster, A.D., and Leder, P.: **Targeted disruption of the chemokine eotaxin partially reduces antigen-induced tissue eosinophilia.** *J. Exp. Med.* 1997, **185**:785-790.

12-8   Eosinophils and basophils cause inflammation and tissue damage in allergic reactions.

Dvorak, A.M.: **Cell biology of the basophil.** *Int. Rev. Cytol.* 1998, **180**:87-236.

Plager, D.A., Stuart, S., and Gleich, G.J.: **Human eosinophil granule major basic protein and its novel homolog.** *Allergy* 1998, **53**:33-40.

Schroeder, J.T., and MacGlashan, D.W., Jr.: **New concepts: the basophil.** *J. Allergy Clin. Immunol.* 1997, **99**:429-433.

Thomas, L.L.: **Basophil and eosinophil interactions in health and disease.** *Chem. Immunol.* 1995, **61**:186-207.

12-9   An allergic reaction is divided into an immediate response and a late-phase response.

Bentley, A.M., Kay, A.B., and Durham, S.R.: **Human late asthmatic reactions.** *Clin. Exp. Allergy* 1997, **27 Suppl 1**:71-86.

Cieslewicz, G., Tomkinson, A., Adler, A., Duez, C., Schwarze, J., Takeda, K., Larson, K.A., Lee, J.J., Irvin, C.G., and Gelfand, E.W.: **The late, but not early, asthmatic response is dependent on IL-5 and correlates with eosinophil infiltration.** *J. Clin. Invest.* 1999, **104**:301-308.

Liu, M.C., Hubbard, W.C., Proud, D., Stealey, B.A., Galli, S.J., Kagey Sobotka, A., Bleecker, E.R., and Lichtenstein, L.M.: **Immediate and late inflammatory responses to ragweed antigen challenge of the peripheral airways in allergic asthmatics. Cellular, mediator, and permeability changes.** *Am. Rev. Respir. Dis.* 1991, **144**:51-58.

Macfarlane, A.J., Kon, O.M., Smith, S.J., Zeibecoglou, K., Khan, L.N., Barata, L.T., McEuen, A.R., Buckley, M.G., Walls, A.F., Meng, Q., Humbert, M., Barnes, N.C., Robinson, D.S., Ying, S., and Kay, A.B.: **Basophils, eosinophils, and mast cells in atopic and nonatopic asthma and in late-phase allergic reactions in the lung and skin.** *J. Allergy Clin. Immunol.* 2000, **105**:99-107.

Pearlman, D.S.: **Pathophysiology of the inflammatory response.** *J. Allergy Clin. Immunol.* 1999, **104**:S132-S137.

Varney, V.A., Hamid, Q.A., Gaga, M., Ying, S., Jacobson, M., Frew, A.J., Kay, A.B., and Durham, S.R.: **Influence of grass pollen immunotherapy on cellular infiltration and cytokine mRNA expression during allergen-induced late-phase cutaneous responses.** *J. Clin. Invest.* 1993, **92**:644-651.

12-10   The clinical effects of allergic reactions vary according to the site of mast-cell activation.

Bochner, B.S., and Lichtenstein, L.M.: **Anaphylaxis.** *N. Engl. J. Med.* 1991, **324**:1785-1790.

deShazo, R.D., and Kemp, S.F.: **Allergic reactions to drugs and biologic agents.** *JAMA* 1997, **278**:1895-1906.

Dombrowicz, D., Flamand, V., Brigman, K.K., Koller, B.H., and Kinet, J.P.: **Abolition of anaphylaxis by targeted disruption of the high affinity immunoglobulin E receptor alpha chain gene.** *Cell* 1993, **75**:969-976.

Fernandez, M., Warbrick, E.V., Blanca, M., and Coleman, J.W.: **Activation and hapten inhibition of mast cells sensitized with monoclonal IgE anti-penicillin antibodies: evidence for two-site recognition of the penicillin derived determinant.** *Eur. J. Immunol.* 1995, **25**:2486-2491.

Kemp, S.F., Lockey, R.F., Wolf, B.L., and Lieberman, P.: **Anaphylaxis. A review of 266 cases.** *Arch. Intern. Med.* 1995, **155**:1749-1754.

Martin, T.R., Galli, S.J., Katona, I.M., and Drazen, J.M.: **Role of mast cells in anaphylaxis. Evidence for the importance of mast cells in the cardiopulmonary alterations and death induced by anti-IgE in mice.** *J. Clin. Invest.* 1989, **83**:1375-1383.

Oettgen, H.C., Martin, T.R., Wynshaw Boris, A., Deng, C., Drazen, J.M., and Leder, P.: **Active anaphylaxis in IgE-deficient mice.** *Nature* 1994, **370**:367-370.

Padovan, E.: **T-cell response in penicillin allergy.** *Clin. Exp. Allergy* 1998, **28 Suppl 4**:33-36.

Reisman, R.E.: **Insect stings.** *N. Engl. J. Med.* 1994, **331**:523-527.

Weltzien, H.U., and Padovan, E.: **Molecular features of penicillin allergy.** *J. Invest. Dermatol.* 1998, **110**:203-206.

12-11 Allergen inhalation is associated with the development of rhinitis and asthma.

Busse, W.W., Gern, J.E., and Dick, E.C.: **The role of respiratory viruses in asthma.** *Ciba Found. Symp.* 1997, **206**:208-213.

Cohn, L., Homer, R.J., Niu, N., and Bottomly, K.: **T helper 1 cells and interferon gamma regulate allergic airway inflammation and mucus production.** *J. Exp. Med.* 1999, **190**:1309-1318.

Drazen, J.M., Arm, J.P., and Austen, K.F.: **Sorting out the cytokines of asthma.** *J. Exp. Med.* 1996, **183**:1-5.

Elias, J.A., Zhu, Z., Chupp, G., and Homer, R.J.: **Airway remodeling in asthma.** *J. Clin. Invest.* 1999, **104**:1001-1006.

Galli, S.J.: **Complexity and redundancy in the pathogenesis of asthma: reassessing the roles of mast cells and T cells.** *J. Exp. Med.* 1997, **186**:343-347.

Haselden, B.M., Kay, A.B., and Larche, M.: **Immunoglobulin E-independent major histocompatibility complex-restricted T cell peptide epitope-induced late asthmatic reactions.** *J. Exp. Med.* 1999, **189**:1885-1894.

Holgate, S.T.: **Asthma: a dynamic disease of inflammation and repair.** *Ciba Found. Symp.* 1997, **206**:5-28; discussion 28-34, 106-110.

Naclerio. R., and Solomon, W.: **Rhinitis and inhalant allergens.** *JAMA* 1997, **278**:1842-1848.

Platts-Mills, T.A.: **The role of allergens in allergic airway disease.** *J. Allergy Clin. Immunol.* 1998, **101**:S364-S366.

12-12 Skin allergy is manifest as urticaria or chronic eczema.

Fiebiger, E., Hammerschmid, F., Stingl, G., and Maurer, D.: **Anti-FcεRIα auto-antibodies in autoimmune-mediated disorders. Identification of a structure–function relationship.** *J. Clin. Invest.* 1998, **101**:243-251.

Fiebiger, E., Stingl, G., and Maurer, D.: **Anti-IgE and anti-FcεRI auto-antibodies in clinical allergy.** *Curr. Opin. Immunol.* 1996, **8**:784-789.

Leung, D.Y.: **Immune mechanisms in atopic dermatitis and relevance to treatment.** *Allergy Proc.* 1991, **12**:339-346.

Ring, J., Bieber, T., Vieluf, D., Kunz, B., and Przybilla, B.: **Atopic eczema, Langerhans cells and allergy.** *Int. Arch. Allergy Appl. Immunol.* 1991, **94**:194-201.

Sabroe, R.A., and Greaves, M.W.: **The pathogenesis of chronic idiopathic urticaria.** *Arch. Dermatol.* 1997, **133**:1003-1008.

12-13 Allergy to foods causes symptoms limited to the gut and systemic reactions.

Bindslev-Jensen, C.: **Food allergy.** *BMJ* 1998, **316**:1299-1302.

Ewan, P.W.: **Clinical study of peanut and nut allergy in 62 consecutive patients: new features and associations.** *BMJ* 1996, **312**:1074-1078.

Nordlee, J.A., Taylor, S.L., Townsend, J.A., Thomas, L.A., and Bush, R.K.: **Identification of a Brazil-nut allergen in transgenic soybeans.** *N. Engl. J. Med.* 1996, **334**:688-692.

Rumsaeng, V., and Metcalfe, D.D.: **Food allergy.** *Semin. Gastrointest. Dis.* 1996, **7**:134-143.

Sampson, H.A.: **Food allergy.** *JAMA* 1997, **278**:1888-1894.

12-14 Allergy can be treated by inhibiting either IgE production or the effector pathways activated by cross-linking of cell-surface IgE.

Adkinson, N.F., Jr., Eggleston, P.A., Eney, D., Goldstein, E.O., Schuberth, K.C., Bacon, J.R., Hamilton, R.G., Weiss, M.E., Arshad, H., Meinert, C.L., Tonascia, J.,

and Wheeler, B.: **A controlled trial of immunotherapy for asthma in allergic children.** *N. Engl. J. Med.* 1997, **336**:324-331.

Askenase, P.W.: **Gee whiz: CpG DNA allergy therapy!** *J. Allergy Clin. Immunol.* 2000, **106**:37-40.

Barnes, P.J.: **Therapeutic strategies for allergic diseases.** *Nature* 1999, **402**:B31-B38.

Bertrand, C., and Geppetti, P.: **Tachykinin and kinin receptor antagonists: therapeutic perspectives in allergic airway disease.** *Trends Pharmacol. Sci.* 1996, **17**:255-259.

Creticos, P.S., Reed, C.E., Norman, P.S., Khoury, J., Adkinson, N.F., Jr., Buncher, C.R., Busse, W.W., Bush, R.K., Gadde, J., Li, J.T., et al.: **Ragweed immunotherapy in adult asthma.** *N. Engl. J. Med.* 1996, **334**:501-506.

Douglass, J.A., Thien, F.C., and O'Hehir, R.E.: **Immunotherapy in asthma.** *Thorax* 1997, **52 Suppl 3**:S22-S29.

Drazen, J.: **Clinical pharmacology of leukotriene receptor antagonists and 5-lipoxygenase inhibitors.** *Am. J. Respir. Crit. Care. Med.* 1998, **157**:S233-S237; discussion S247-S248.

Durham, S.R., Walker, S.M., Varga, E.M., Jacobson, M.R., O'Brien, F., Noble, W., Till, S.J., Hamid, Q.A., and Nouri-Aria, K.T.: **Long-term clinical efficacy of grass-pollen immunotherapy.** *N. Engl. J. Med.* 1999, **341**:468-475.

Kline, J.N.: **Effects of CpG DNA on Th1/Th2 balance in asthma.** *Curr. Top. Microbiol. Immunol.* 2000, **247**:211-225.

Milgrom, H., Fick, R.B., Jr., Su, J.Q., Reimann, J.D., Bush, R.K., Watrous, M.L., and Metzger, W.J.: **Treatment of allergic asthma with monoclonal anti-IgE antibody. rhuMAb-E25 Study Group.** *N. Engl. J. Med.* 1999, **341**:1966-1973.

van Neerven, R.J., Ebner, C., Yssel, H., Kapsenberg, M.L., and Lamb, J.R.: **T-cell responses to allergens: epitope-specificity and clinical relevance.** *Immunol. Today* 1996, **17**:526-532.

Sabroe, I., Peck, M.J., Van Keulen, B.J., Jorritsma, A., Simmons, G., Clapham, P.R., Williams, T.J., and Pease, J.E.: **A small molecule antagonist of chemokine receptors CCR1 and CCR3. Potent inhibition of eosinophil function and CCR3-mediated HIV-1 entry.** *J. Biol. Chem.* 2000, **275**:25985-25992.

12-15 Innocuous antigens can cause type II hypersensitivity reactions in susceptible individuals by binding to the surfaces of circulating blood cells.

Greinacher, A., Potzsch, B., Amiral, J., Dummel, V., Eichner, A., and Mueller Eckhardt, C.: **Heparin-associated thrombocytopenia: isolation of the antibody and characterization of a multimolecular PF4-heparin complex as the major antigen.** *Thromb. Haemost.* 1994, **71**:247-251.

Murphy, W.G., and Kelton, J.G.: **Immune haemolytic anaemia and thrombocytopenia: drugs and autoantibodies.** *Biochem. Soc. Trans.* 1991, **19**:183-186.

Petz, L.D.: **Drug-induced autoimmune hemolytic anemia.** *Transfus. Med. Rev.* 1993, **7**:242-254.

Salama, A., Santoso, S., and Mueller Eckhardt, C.: **Antigenic determinants responsible for the reactions of drug-dependent antibodies with blood cells.** *Br. J. Haematol.* 1991, **78**:535-539.

12-16 Systemic disease caused by immune complex formation can follow the administration of large quantities of poorly catabolized antigens.

Bielory, L., Gascon, P., Lawley, T.J., Young, N.S., and Frank, M.M.: **Human serum sickness: a prospective analysis of 35 patients treated with equine anti-thymocyte globulin for bone marrow failure.** *Medicine (Baltimore)* 1988, **67**:40-57.

Cochrane, C.G., and Koffler, D.: **Immune complex disease in experimental animals and man.** *Adv. Immunol.* 1973, **16**:185-264.

Davies, K.A., Mathieson, P., Winearls, C.G., Rees, A.J., and Walport, M.J.: **Serum sickness and acute renal failure after streptokinase therapy for myocardial infarction.** *Clin. Exp. Immunol.* 1990, **80**:83-88.

Lawley, T.J., Bielory, L., Gascon, P., Yancey, K.B., Young, N.S., and Frank, M.M.: **A prospective clinical and immunologic analysis of patients with serum sickness.** *N. Engl. J. Med.* 1984, **311**:1407-1413.

Ravetch, J.V., and Clynes, R.: **Divergent roles for Fc receptors and complement** *in vivo.* Annu. Rev. Immunol. 1998, **16**:421-432.

Schifferli, J.A., Ng, Y.C., and Peters, D.K.: **The role of complement and its receptor in the elimination of immune complexes.** N. Engl. J. Med. 1986, **315**:488-495.

Theofilopoulos, A.N., and Dixon, F.J.: **Immune complexes in human diseases: a review.** Am. J. Pathol. 1980, **100**:529-594.

12-17   Delayed-type hypersensitivity reactions are mediated by T$_H$1 cells and CD8 cytotoxic T cells.

Bernhagen, J., Bacher, M., Calandra, T., Metz, C.N., Doty, S.B., Donnelly, T., and Bucala, R.: **An essential role for macrophage migration inhibitory factor in the tuberculin delayed-type hypersensitivity reaction.** J. Exp. Med. 1996, **183**:277-282.

Grabbe, S., and Schwarz, T.: **Immunoregulatory mechanisms involved in elicitation of allergic contact hypersensitivity.** Immunol. Today 1998, **19**:37-44.

Kalish, R.S., Wood, J.A., and LaPorte, A.: **Processing of urushiol (poison ivy) hapten by both endogenous and exogenous pathways for presentation to T cells *in vitro.*** J. Clin. Invest. 1994, **93**:2039-2047.

Larsen, C.G., Thomsen, M.K., Gesser, B., Thomsen, P.D., Deleuran, B.W., Nowak, J., Skodt, V., Thomsen, H.K., Deleuran, M., Thestrup Pedersen, K., et al.: **The delayed-type hypersensitivity reaction is dependent on IL-8. Inhibition of a tuberculin skin reaction by an anti-IL-8 monoclonal antibody.** J. Immunol. 1995, **155**:2151-2157.

Muller, G., Saloga, J., Germann, T., Schuler, G., Knop, J., and Enk, A.H.: **IL-12 as mediator and adjuvant for the induction of contact sensitivity *in vivo.*** J. Immunol. 1995, **155**:4661-4668.

Tsuji, R.F., Geba, G.P., Wang, Y., Kawamoto, K., Matis, L.A., and Askenase, P.W.: **Required early complement activation in contact sensitivity with generation of local C5-dependent chemotactic activity, and late T cell interferon gamma: a possible initiating role of B cells.** J. Exp. Med. 1997, **186**:1015-1026.

Vollmer, J., Weltzien, H.U., and Moulon, C.: **TCR reactivity in human nickel allergy indicates contacts with complementarity-determining region 3 but excludes superantigen-like recognition.** J. Immunol. 1999, **163**:2723-2731.

Autoimmunity and Transplantation

We have learned in preceding chapters that the adaptive immune response is a critical component of host defense against infection and therefore essential for normal health. Unfortunately, adaptive immune responses are also sometimes elicited by antigens not associated with infectious agents, and this may cause serious disease. These responses are essentially identical to adaptive immune responses to infectious agents; only the antigens differ. In Chapter 12, we saw how responses to certain environmental antigens cause allergic diseases and other hypersensitivity reactions. In this chapter we will examine responses to two particularly important categories of antigen: responses to self tissue antigens, called **autoimmunity**, which can lead to **autoimmune diseases** characterized by tissue damage; and responses to transplanted organs that lead to **graft rejection**. We will examine these disease processes and the mechanisms that lead to the undesirable adaptive immune responses that are their root cause.

Autoimmune responses are directed against self antigens.

Autoimmune disease occurs when a specific adaptive immune response is mounted against self antigens. The normal consequence of an adaptive immune response against a foreign antigen is the clearance of the antigen from the body. Virus-infected cells, for example, are destroyed by cytotoxic T cells, whereas soluble antigens are cleared by formation of immune complexes of antibody and antigen, which are taken up by cells of the mononuclear phagocytic system such as macrophages. When an adaptive immune response develops against self antigens, however, it is usually impossible for immune effector mechanisms to eliminate the antigen completely, and so a sustained response occurs. The consequence is that the effector pathways of immunity cause chronic inflammatory injury to tissues, which may prove lethal. The mechanisms of tissue damage in autoimmune diseases are essentially the same as those that operate in protective immunity and in hypersensitivity diseases. Some common autoimmune diseases are listed in Fig. 13.1.

Adaptive immune responses are initiated by the activation of antigen-specific T cells, and it is believed that autoimmunity is initiated in the same way. T-cell responses to self antigens can inflict tissue damage either directly or indirectly. Cytotoxic T-cell responses and inappropriate activation of macrophages by T_H1 cells can cause extensive tissue damage, whereas inappropriate T-cell help to self-reactive B cells can initiate harmful autoantibody responses.

Fig. 13.1 Autoimmune diseases classified by the mechanism of tissue damage. Autoimmune diseases can be grouped in the same way as hypersensitivity reactions, according to the type of immune response and the mechanism by which it damages tissues. The immunopathological mechanisms are as illustrated for the hypersensitivity reactions in Fig. 12.2, with the exception of the type I IgE-mediated responses, which are not a known cause of autoimmune disease. Some additional autoimmune diseases in which the antigen is a cell-surface receptor, and the pathology is due to altered signaling, are listed later in Fig. 13.11. Several immunopathogenic mechanisms operate in parallel to cause many autoimmune diseases. This is illustrated in the case of rheumatoid arthritis, which appears in more than one category of immunopathological mechanism.

Some common autoimmune diseases classified by immunopathogenic mechanism		
Syndrome	Autoantigen	Consequence
Type II antibody to cell-surface or matrix antigens		
Autoimmune hemolytic anemia	Rh blood group antigens, I antigen	Destruction of red blood cells by complement and FcR⁺ phagocytes, anemia
Autoimmune thrombocytopenic purpura	Platelet integrin GpIIb:IIIa	Abnormal bleeding
Goodpasture's syndrome	Noncollagenous domain of basement membrane collagen type IV	Glomerulonephritis, pulmonary hemorrhage
Pemphigus vulgaris	Epidermal cadherin	Blistering of skin
Acute rheumatic fever	Streptococcal cell-wall antigens. Antibodies cross-react with cardiac muscle	Arthritis, myocarditis, late scarring of heart valves
Type III immune-complex disease		
Mixed essential cryoglobulinemia	Rheumatoid factor IgG complexes (with or without hepatitis C antigens)	Systemic vasculitis
Systemic lupus erythematosus	DNA, histones, ribosomes, snRNP, scRNP	Glomerulonephritis, vasculitis, rash
Rheumatoid arthritis	Rheumatoid factor IgG complexes	Arthritis
Type IV T cell-mediated disease		
Insulin-dependent diabetes mellitus	Pancreatic β-cell antigen	β-Cell destruction
Rheumatoid arthritis	Unknown synovial joint antigen	Joint inflammation and destruction
Experimental autoimmune encephalomyelitis (EAE), multiple sclerosis	Myelin basic protein, proteolipid protein, myelin oligodendrocyte glycoprotein	Brain invasion by CD4 T cells, weakness

Autoimmune responses are a natural consequence of the open repertoires of both B-cell and T-cell receptors, which allow them to recognize any pathogen. Although these repertoires are purged of most receptors that bind with high affinity to self antigens encountered during development, they still include receptors of lower affinity reactive to some self antigens. It is not known what triggers autoimmunity, but both environmental and genetic factors, especially MHC genotype, are clearly important. Transient autoimmune responses are common, but it is only when they are sustained and cause lasting tissue damage that they attract medical attention. In this section, we will examine the nature of autoimmune responses and how autoimmunity leads to tissue damage. In the last section of this chapter, we will examine the mechanisms by which self-tolerance is lost and autoimmune responses are initiated.

13-1   Specific adaptive immune responses to self antigens can cause autoimmune disease.

Early in the study of immunity it was realized that the powerful effector mechanisms used in host defense could, if turned against the host, cause severe tissue damage; **Ehrlich** termed this *horror autotoxicus*. Healthy individuals do not mount sustained adaptive immune responses to their own antigens and, although transient responses to damaged self tissues occur, these rarely cause additional tissue damage. But although self-tolerance is the general rule, sustained immune responses to self tissues occur in some individuals, and these autoimmune responses cause the severe tissue damage that Ehrlich predicted.

In certain genetically susceptible strains of experimental animals, autoimmune disease can be induced artificially by injection of 'self' tissues from a genetically identical animal mixed with strong adjuvants containing bacteria (see Appendix I, Section A-4). This shows that autoimmunity can be provoked by inducing a specific, adaptive immune response to self antigens and forms the basis for our understanding of how autoimmune disease arises. In humans, autoimmunity usually arises spontaneously; that is, we do not know what events initiate the immune response to self that leads to the autoimmune disease. There is evidence, as we will learn in the last part of this chapter, that some autoimmune disorders, such as rheumatic fever, may be triggered by infectious agents. There is, however, also evidence, particularly from animal models of autoimmunity, that many autoimmune disorders occur through internal dysregulation of the immune system without the participation of infectious agents.

13-2   Autoimmune diseases can be classified into clusters that are typically either organ-specific or systemic.

The classification of disease is an uncertain science, especially in the absence of a precise understanding of causative mechanisms. This is well illustrated by the difficulty in classifying the autoimmune diseases. It is useful to distinguish two major patterns of autoimmune disease, the diseases in which the expression of autoimmunity is restricted to specific organs of the body, known as 'organ-specific' autoimmune diseases, and those in which many tissues of the body are affected, the 'systemic' autoimmune diseases. Examples of organ-specific autoimmune diseases are Hashimoto's thyroiditis and Graves' disease, each predominantly affecting the thyroid gland, and type I insulin-dependent diabetes mellitus (IDDM), which affects the pancreatic islets. Examples of systemic autoimmune disease are systemic lupus erythematosus (SLE) and primary Sjögren's syndrome, in which tissues as diverse as the skin, kidneys, and brain may all be affected.

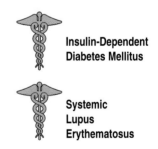

Insulin-Dependent Diabetes Mellitus

Systemic Lupus Erythematosus

The autoantigens recognized in these two categories of disease are themselves respectively organ-specific and systemic. Thus, Graves' disease is characterized by the production of antibodies to the thyroid-stimulating hormone (TSH) receptor in the thyroid gland; Hashimoto's thyroiditis by antibodies to thyroid peroxidase; and type I diabetes by anti-insulin antibodies. By contrast, SLE is characterized by the presence of antibodies to antigens that are ubiquitous and abundant in every cell of the body, such as anti-chromatin antibodies and antibodies to proteins of the pre-mRNA splicing machinery—the spliceosome complex—within the cell.

It is likely that the organ-specific and systemic autoimmune diseases have somewhat different etiologies, which provides a biological basis for their division into two broad categories. Evidence for the validity of this classification also comes from observations that different autoimmune diseases cluster

Organ-specific autoimmune diseases
Type I diabetes mellitus
Goodpasture's syndrome
Multiple sclerosis
Graves' disease Hashimoto's thyroiditis Autoimmune pernicious anemia Autoimmune Addison's disease Vitiligo Myasthenia gravis

Systemic autoimmune diseases
Rheumatoid arthritis
Scleroderma
Systemic lupus erythematosus Primary Sjögren's syndrome Polymyositis

Fig. 13.2 Some common autoimmune diseases classified according to their 'organ-specific' or 'systemic' nature. Diseases that tend to occur in clusters are grouped in single boxes. Clustering is defined as more than one disease affecting a single patient or different members of a family. Not all autoimmune diseases can be classified according to this scheme. For example, autoimmune hemolytic anemia may occur in isolation or in association with systemic lupus erythematosus (SLE).

within individuals and within families. The organ-specific autoimmune diseases frequently occur together in many combinations; for example, autoimmune thyroid disease and the autoimmune depigmenting disease vitiligo are often found in the same person. Similarly, SLE and primary Sjögren's syndrome can coexist within a single individual or among different members of a family.

These clusters of autoimmune diseases provide the most useful classification into different subtypes, each of which may turn out to have a distinct mechanism. A working classification of autoimmune diseases based on clustering is given in Fig. 13.2. It can be seen that the strict separation of diseases into 'organ-specific' and 'systemic' categories breaks down to some extent. Not all autoimmune diseases can be usefully classified in this manner. Autoimmune hemolytic anemia, for example, sometimes occurs as a solitary entity and could be classified as an organ-specific disease. In other circumstances it may occur in conjunction with SLE as part of a systemic autoimmune disease.

Although anyone can, in principle, develop an autoimmune disease, it seems that some individuals are more at risk than others of developing particular diseases. We will first consider those factors that contribute to susceptibility.

13-3 Susceptibility to autoimmune disease is controlled by environmental and genetic factors, especially MHC genes.

The best evidence in humans for susceptibility genes for autoimmunity comes from family studies, especially studies of twins. A semiquantitative technique for measuring what proportion of the susceptibility to a particular disease arises from genetic factors is to compare the incidence of disease in monozygotic and dizygotic twins. If a disease shows a high concordance in all twins, it could be caused by shared genetic or environmental factors. This is because both monozygotic and dizygotic twins tend to be brought up in shared environmental conditions. If the high concordance is restricted to monozygotic rather than dizygotic twins, however, then genetic factors are likely to be more important than environmental factors.

Studies with twins have been undertaken for several human diseases in which autoimmunity is important, including type I IDDM, rheumatoid arthritis, multiple sclerosis, and SLE. In each case, around 20% of pairs of monozygotic twins show disease concordance, compared with fewer than 5% of dizygotic twins. A similar technique is to compare the frequency of a disease such as diabetes in the siblings of patients who have diabetes with the frequency of that disease in the general population. The ratio of these two frequencies gives a measure of the heritability of the disease, although shared environmental factors within families could also be at least partly responsible for an increased frequency.

Results from both twin and family studies show an important role for both inherited and environmental factors in the induction of autoimmune disease. In addition to this evidence from humans, certain inbred mouse strains have

Associations of HLA serotype with susceptibility to autoimmune disease			
Disease	HLA allele	Relative risk	Sex ratio (♀:♂)
Ankylosing spondylitis	B27	87.4	0.3
Acute anterior uveitis	B27	10	<0.5
Goodpasture's syndrome	DR2	15.9	~1
Multiple sclerosis	DR2	4.8	10
Graves' disease	DR3	3.7	4–5
Myasthenia gravis	DR3	2.5	~1
Systemic lupus erythematosus	DR3	5.8	10–20
Type I insulin-dependent diabetes mellitus	DR3/DR4 heterozygote	~25	~1
Rheumatoid arthritis	DR4	4.2	3
Pemphigus vulgaris	DR4	14.4	~1
Hashimoto's thyroiditis	DR5	3.2	4–5

Fig. 13.3 Associations of HLA serotype and sex with susceptibility to autoimmune disease. The 'relative risk' for an HLA allele in an autoimmune disease is calculated by comparing the observed number of patients carrying the HLA allele with the number that would be expected, given the prevalence of the HLA allele in the general population. For type I insulin-dependent diabetes mellitus (IDDM), the association is in fact with the HLA-DQ gene, which is tightly linked to the DR genes but is not detectable by serotyping. Some diseases show a significant bias in the sex ratio; this is taken to imply that sex hormones are involved in pathogenesis. Consistent with this, the difference in the sex ratio in these diseases is greatest between the menarche and the menopause, when levels of such hormones are highest.

an almost uniform susceptibility to particular spontaneous or experimentally induced autoimmune diseases, whereas other strains do not. These findings have led to an extensive search for genes that determine susceptibility to autoimmune disease.

So far, susceptibility to autoimmune disease has been most consistently associated with MHC genotype. Human autoimmune diseases that show associations with HLA type are shown in Fig. 13.3. For most of these diseases, susceptibility is linked most strongly with MHC class II alleles, but in some cases there are strong associations with particular MHC class I alleles.

The association of MHC genotype with disease is assessed initially by comparing the frequency of different alleles in patients with their frequency in the normal population. For IDDM, this approach originally demonstrated an association with HLA-DR3 and HLA-DR4 alleles identified by serotyping (Fig. 13.4). Such studies also showed that the MHC class II allele HLA-DR2 has a dominant protective effect; individuals carrying HLA-DR2, even in

Fig. 13.4 Population studies show association of susceptibility to IDDM with HLA genotype. The HLA genotypes (determined by serotyping) of diabetic patients (bottom panel) are not representative of those found in the population (top panel). Almost all diabetic patients express HLA-DR3 and/or HLA-DR4, and HLA-DR3/DR4 heterozygosity is greatly over-represented in diabetics compared with controls. These alleles are linked tightly to HLA-DQ alleles that confer susceptibility to IDDM. By contrast, HLA-DR2 protects against the development of IDDM and is found only extremely rarely in diabetic patients. The small letter x represents any allele other than DR2, DR3, or DR4.

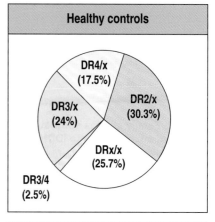

Healthy controls

DR4/x (17.5%)
DR2/x (30.3%)
DR3/x (24%)
DRx/x (25.7%)
DR3/4 (2.5%)

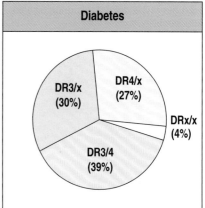

Diabetes

DR3/x (30%)
DR4/x (27%)
DRx/x (4%)
DR3/4 (39%)

association with one of the susceptibility alleles, rarely develop diabetes. Another way of determining whether MHC genes are important in auto-immune disease is to study the families of affected patients; it has been shown that two siblings affected with the same autoimmune disease are far more likely than expected to share the same MHC haplotypes (Fig. 13.5).

As HLA genotyping has become more exact through the sequencing of HLA alleles, disease associations that were originally discovered through HLA serotyping using antibodies have been defined more precisely. For example, the association between IDDM and the DR3 and DR4 alleles is now known to be due to their tight genetic linkage to DQβ alleles that confer susceptibility to disease. Indeed, disease susceptibility is most closely associated with polymorphisms at a particular position in the DQβ amino acid sequence. The most abundant DQβ amino acid sequence has an aspartic acid at position 57 that is able to form a salt bridge across the end of the peptide-binding cleft of the DQ molecule. By contrast, the diabetic patients in Caucasoid populations mostly have valine, serine, or alanine at that position and thus make DQ molecules that lack this salt bridge (Fig. 13.6). The nonobese diabetic (NOD) strain of mice, which develops spontaneous diabetes, also has a serine at that position in the homologous MHC class II molecule, known as I-A^{g7}.

The association of MHC genotype with autoimmune disease is not surprising, because autoimmune responses involve T cells, and the ability of T cells to respond to a particular antigen depends on MHC genotype. Thus the associations can be explained by a simple model in which susceptibility to an autoimmune disease is determined by differences in the ability of different allelic variants of MHC molecules to present autoantigenic peptides to autoreactive T cells. This would be consistent with what we know of T-cell involvement in particular diseases. In diabetes, for example, there are associations with both MHC class I and MHC class II alleles and this is consistent with the finding that both CD8 and CD4 T cells, which respond to antigens presented by MHC class I and MHC class II molecules, respectively, mediate the autoimmune response.

An alternative hypothesis for the association between MHC genotype and susceptibility to autoimmune diseases emphasizes the role of MHC alleles in shaping the T-cell receptor repertoire (see Chapter 7). This hypothesis proposes

Fig. 13.5 Family studies show strong linkage of susceptibility to IDDM with HLA genotype. In families in which two or more siblings have IDDM, it is possible to compare the HLA genotypes of affected siblings. Affected siblings share two HLA haplotypes much more frequently than would be expected if the HLA genotype did not influence disease susceptibility.

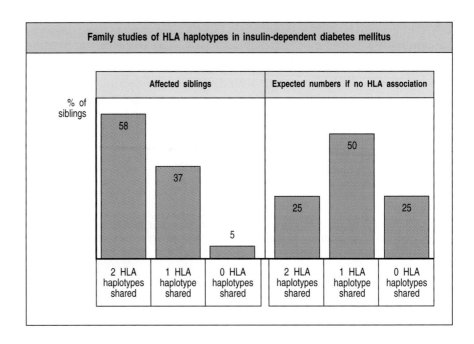

Fig. 13.6 Amino acid changes in the sequence of an MHC class II protein correlate with susceptibility to and protection from diabetes. The HLA-DQβ₁ chain contains an aspartic acid (Asp) at position 57 in most people; in Caucasoid populations, patients with IDDM more often have valine, serine, or alanine at this position instead, as well as other differences. Asp 57, shown in red on the backbone structure of the DQβ chain, forms a salt bridge (shown in green in the center panel) to an arginine residue (shown in pink) in the adjacent α chain (gray). The change to an uncharged residue (for example, alanine, shown in yellow in the bottom panel) disrupts this salt bridge, altering the stability of the DQ molecule. The nonobese diabetic (NOD) strain of mice, which develops spontaneous diabetes, shows a similar replacement of serine for aspartic acid at position 57 of the homologous I-Aβ chain, and NOD mice transgenic for β chains with Asp 57 have a marked reduction in diabetes incidence. Photographs courtesy of C. Thorpe.

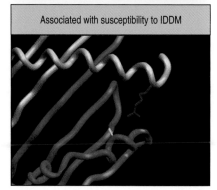

Position 57 of the DQβ chain affects susceptibility to insulin-dependent diabetes mellitus (IDDM)

α chain

Position 57

β chain

Associated with resistance to IDDM

Associated with susceptibility to IDDM

that self peptides associated with certain MHC molecules may drive the positive selection of developing thymocytes that are specific for particular auto-antigens. Such autoantigenic peptides might be expressed at too low a level or bind too poorly to self MHC molecules to drive negative selection in the thymus, but be present at a sufficient level or bind strongly enough to drive positive selection. This hypothesis is supported by observations that I-A^{g7}, the disease-associated MHC class II molecule in the diabetes-prone NOD mice, binds many peptides very poorly and may therefore be less effective in driving intrathymic negative selection of T cells that bind self peptides.

However, MHC genotype alone does not determine genetic susceptibility to disease. Identical twins, sharing all of their genes, are far more likely to develop the same autoimmune disease than MHC-identical siblings, demonstrating that genetic factors other than the MHC also affect whether an individual develops disease. Recent studies of the genetics of autoimmune diabetes in humans and mice have shown that there are several independently segregating disease susceptibility loci in addition to the MHC.

There is also evidence that variation in the level of a potential autoantigen within the thymus can influence disease development. In the case of human insulin, which can act as an autoantigen in type I IDDM, the level of transcription of the insulin gene shows genetic variation between individuals; this is associated with a polymorphic minisatellite sequence located upstream of the gene. Gene variants that are transcribed at a high level in the thymus tend to protect against the development of diabetes, whereas variants transcribed at a lower level are associated with disease susceptibility. This is because the expression of high levels of insulin in the thymus may cause the deletion of T cells specific for the insulin peptides (see Section 7-24).

13-4 The genes that have been associated with the development of systemic lupus erythematosus provide important clues to the etiology of the disease.

The major serological abnormality in SLE is the presence of autoantibodies to ubiquitous and abundant intracellular antigens, such as chromatin. How is tolerance broken to such all-pervasive self antigens? A number of genes have been implicated in the etiology of SLE in humans and mice (Fig. 13.7). These can be classified into three categories on the basis of their physiological function. The first comprises genes whose products are active in the body's mechanisms for disposing of dead and dying cells, which could provide a source of autoantigens. Genetic knockout in mice (see Appendix I, Section A-47)

Fig. 13.7 Genes encoding three functional classes of proteins are implicated in the etiology of SLE. The first class comprises genes encoding proteins that may have a role in the clearance of cellular debris and masking it from the immune system. The second comprises genes for products that modify the thresholds for lymphocyte activation and cell deletion. The third class includes genes encoding products that may affect the expression of inflammation in individual organs.

Some proteins implicated in the etiology of SLE		
Candidate activity	**Role**	**Deficiency in SLE**
Antigen clearance	Binding and clearance of autoantigens and immune complexes	Complement proteins: C1q, C1r, and C1s, C4>>C2 Serum IgM
	Masking or digestion of DNA and chromatin	Serum amyloid P component DNase 1
Tolerance induction	Threshold for lymphocyte activation	Lyn SHP-1 CD22 FcγRIIB
	Deletion of autoreactive lymphocytes	Fas and Fas ligand Cell-cycle inhibitor p21
Organ-specific manifestations of autoimmunity	Renal disease	FcγRIIB polymorphism FcγRIII polymorphism

of four genes in this category has produced animal models of SLE. One of these genes codes for the complement protein C1q, which, together with other complement proteins, is involved in the effective clearance of immune complexes and apoptotic cells. A second gene in this category encodes serum amyloid P component, which binds chromatin and may mask it from the immune system. Its deletion results in the development of antibodies against chromatin and development of glomerulonephritis caused by deposition of immune complexes of these antibodies in the kidney. Third, deletion of DNase I, an enzyme that digests extracellular chromatin, results in the development of anti-chromatin antibodies and glomerulonephritis. Fourth, a similar phenotype has been seen in mice in which the secretory portion of the immunoglobulin μ chain is deleted, and which thus lack secreted IgM, which may have an important role in the clearance of effete cells. However, the majority of cases of spontaneous SLE are likely to be influenced by far more complex genetic factors than these single-gene defects.

The second category of disease susceptibility genes for SLE includes those encoding proteins that regulate the thresholds for tolerance and activation of T and B lymphocytes, such as Fas, Fas ligand, the signaling molecule SHP-1, the B-cell inhibitory receptor CD22, FcγRIIB, and the cell-cycle inhibitor p21. The third category of genes encode proteins that could modify the expression of SLE in individual organs by their involvement in immune complex-mediated inflammation. Examples are the polymorphic genes for FcγRIIa and FcγRIII, where the variant proteins are thought to differ in their ability to bind immune complexes and are associated with the presence of nephritis in SLE.

A further very important factor in disease susceptibility to SLE is the hormonal status of the patient. Indeed, many autoimmune diseases show a strong sex bias (see Fig. 13.3). Where a bias towards disease in one sex is observed in experimental animals, castration or the administration of estrogen to males usually normalizes disease incidence between the two sexes. Furthermore, many autoimmune diseases that are more common in females show peak incidence in the years of active child bearing, when production of the female sex hormones estrogen and progesterone is at its greatest. A thorough understanding of how these genetic and hormonal factors contribute to disease susceptibility might allow us to prevent the autoimmune response.

13-5   Antibody and T cells can cause tissue damage in autoimmune disease.

Tissue injury in autoimmune disease results because the self antigen is an intrinsic component of the body and, consequently, the effector mechanisms of the immune system are directed at the body's own tissues. Also, because the adaptive immune response is incapable of removing the offending autoantigen from the body, the immune response persists, and there is a constant supply of new autoantigen, which amplifies the response. An important exception to this rule is type I IDDM, in which the autoimmune response destroys the target organ completely. This leads to a failure to produce insulin—one of the major autoantigens in this disease. Lack of insulin is in turn responsible for the phenotype of diabetes mellitus.

The mechanisms of tissue injury in autoimmunity can be classified according to the scheme adopted for hypersensitivity reactions (see Figs 13.1 and 12.2). As with the hypersensitivity reactions, tissue damage can be mediated by the effector actions of both T cells and B cells. The antigen, or group of antigens, against which the autoimmune response is directed, and the mechanism by which the antigen-bearing tissue is damaged, together determine the pathology and clinical expression of the disease (see Fig. 13.1).

Autoimmune diseases differ from hypersensitivity responses in that type I IgE-mediated responses do not seem to have a major role. IgE autoantibodies have, however, been found in autoimmune disease, and although there is no proof that they mediate any autoimmune disease, there are diseases where this may be so. For example, asthma and eosinophilia (see Chapter 12) are found in a rare autoimmune vasculitis, an inflammatory disease of blood vessels that is known as Churg–Strauss vasculitis.

By contrast, autoimmunity that damages tissues by mechanisms analogous to type II hypersensitivity reactions is quite common. In this form of auto-immunity, IgG or IgM responses to autoantigens located on cell surfaces or extracellular matrix cause the injury. In other cases of autoimmunity, tissue damage can be due to type III responses, which involve immune complexes containing autoantibodies to soluble autoantigens; these autoimmune diseases are systemic and are characterized by autoimmune vasculitis. Finally, in a number of organ-specific autoimmune diseases, T-cell responses are directly involved in causing the tissue damage. In most autoimmune diseases, several mechanisms of immunopathogenesis operate. Examples include SLE, type I IDDM, and rheumatoid arthritis, in which there is evidence that both T-cell and antibody-mediated pathways cause tissue injury. We will examine how autoantibodies cause tissue damage, before ending with a consideration of self-reactive T-cell responses and their role in autoimmune disease.

13-6   Autoantibodies against blood cells promote their destruction.

IgG or IgM responses to antigens located on the surface of blood cells lead to the rapid destruction of these cells. An example of this is **autoimmune hemolytic anemia**, where antibodies against self antigens on red blood cells trigger destruction of the cells, leading to anemia. This can occur in two ways (Fig. 13.8). Red cells with bound IgG or IgM antibody are rapidly cleared from the circulation by interaction with Fc or complement receptors, respectively, on cells of the fixed mononuclear phagocytic system; this occurs particularly in the spleen. Alternatively, the autoantibody-sensitized red cells are lysed by formation of the membrane-attack complex of complement. In **autoimmune thrombocytopenic purpura**, autoantibodies against the GpIIb:IIIa fibrinogen receptor on platelets can cause thrombocytopenia (a depletion of platelets), which can in turn cause hemorrhage.

Autoimmune
Hemolytic
Anemia

Fig. 13.8 Antibodies specific for cell-surface antigens can destroy cells. In autoimmune hemolytic anemias, red cells coated with IgG autoantibodies against a cell-surface antigen are rapidly cleared from the circulation by uptake by Fc receptor-bearing macrophages in the fixed mononuclear phagocytic system (left panel). Red cells coated with IgM autoantibodies fix C3 and are cleared by CR1- and CR3-bearing macrophages in the fixed mononuclear phagocytic system (not shown). Uptake and clearance by these mechanisms occurs mainly in the spleen. The binding of certain rare autoantibodies that fix complement extremely efficiently causes the formation of the membrane-attack complex on the red cells, leading to intravascular hemolysis (right panel).

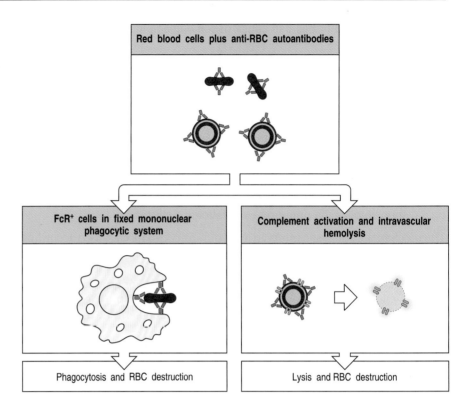

Red blood cells plus anti-RBC autoantibodies

FcR+ cells in fixed mononuclear phagocytic system

Complement activation and intravascular hemolysis

Phagocytosis and RBC destruction

Lysis and RBC destruction

Lysis of nucleated cells by complement is less common because these cells are better defended by complement regulatory proteins. These proteins protect cells against immune attack by interfering with the activation of complement components and their assembly into a membrane-attack complex (see Section 2-14). Although the activation of complement by the bound autoantibody can proceed to a limited degree, nucleated cells are able to resist lysis by exocytosis or endocytosis of parts of the cell membrane bearing the membrane-attack complex. Nevertheless, nucleated cells targeted by autoantibodies are still destroyed by cells of the mononuclear phagocytic system. Autoantibodies against neutrophils, for example, cause neutropenia, which increases susceptibility to infection with pyogenic bacteria. In all of these cases, accelerated clearance of autoantibody-sensitized cells is the cause of their depletion in the blood. One therapeutic approach to this type of autoimmunity is removal of the spleen, the organ in which the main clearance of red cells, platelets, and leukocytes occurs.

13-7 The fixation of sublytic doses of complement to cells in tissues stimulates a powerful inflammatory response.

The binding of IgG and IgM antibodies to cells in tissues causes inflammatory injury by a variety of mechanisms. One of these is fixation of complement. Although nucleated cells are relatively resistant to lysis by complement, the assembly of sublytic amounts of the membrane-attack complex on their surface provides a powerful activating stimulus. Depending on the type of cell, the interaction of sublytic doses of the membrane-attack complex with the cell membrane can cause cytokine release, generation of a respiratory burst, or the mobilization of membrane phospholipids to generate arachidonic acid—the precursor of prostaglandins and leukotrienes (lipid mediators of inflammation).

Most cells in tissues are fixed in place and cells of the inflammatory system are attracted to them by chemoattractant molecules. One such is the complement fragment C5a, which is released as a result of complement activation triggered by autoantibody binding. Other chemoattractants, such as leukotriene B4, can be released by the autoantibody-targeted cells. Inflammatory leukocytes are further activated by binding to autoantibody Fc regions and fixed complement C3 fragments on the tissue cells. Tissue injury can then result from the products of the activated leukocytes and by antibody-dependent cellular cytotoxicity mediated by natural killer (NK) cells (see Section 9-21).

A probable example of this type of autoimmunity is **Hashimoto's thyroiditis**, in which autoantibodies against tissue-specific antigens such as thyroid peroxidase and thyroglobulin are found at extremely high levels for prolonged periods. Direct T cell-mediated cytotoxicity, which we will discuss later, is probably also important in this disease.

13-8   Autoantibodies against receptors cause disease by stimulating or blocking receptor function.

A special class of type II hypersensitivity reaction occurs when the autoantibody binds to a cell-surface receptor. Antibody binding to a receptor can either stimulate the receptor or block its stimulation by its natural ligand. In **Graves' disease**, autoantibody against the thyroid-stimulating hormone receptor on thyroid cells stimulates the excessive production of thyroid hormone. The production of thyroid hormone is normally controlled by feedback regulation; high levels of thyroid hormone inhibit release of thyroid-stimulating hormone (TSH) by the pituitary. In Graves' disease, feedback inhibition fails because the autoantibody continues to stimulate the TSH receptor in the absence of TSH, and the patients become hyperthyroid (Fig. 13.9).

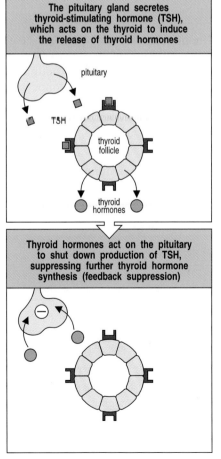 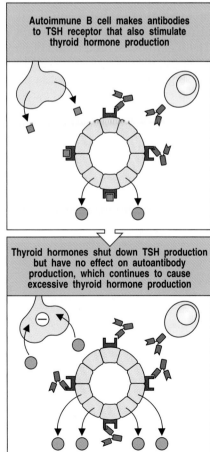

Fig. 13.9 Feedback regulation of thyroid hormone production is disrupted in Graves' disease. Graves' disease is caused by autoantibodies specific for the receptor for thyroid-stimulating hormone (TSH). Normally, thyroid hormones are produced in response to TSH and limit their own production by inhibiting the production of TSH by the pituitary (left panels). In Graves' disease, the autoantibodies are agonists for the TSH receptor and therefore stimulate production of thyroid hormones (right panels). The thyroid hormones inhibit TSH production in the normal way but do not affect production of the autoantibody; the excessive thyroid hormone production induced in this way causes hyperthyroidism.

Fig. 13.10 Autoantibodies inhibit receptor function in myasthenia gravis. In normal circumstances, acetylcholine released from stimulated motor neurons at the neuromuscular junction binds to acetylcholine receptors on skeletal muscle cells, triggering muscle contraction (left panel). Myasthenia gravis is caused by autoantibodies against the α subunit of the receptor for acetylcholine. These autoantibodies bind to the receptor without activating it and also cause receptor internalization and degradation (right panel). As the number of receptors on the muscle is decreased, the muscle becomes less responsive to acetylcholine.

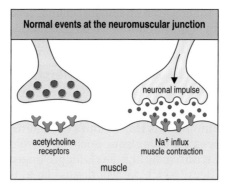

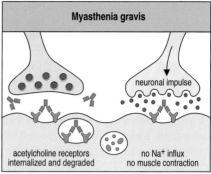

Myasthenia
Gravis

In **myasthenia gravis**, autoantibodies against the α chain of the nicotinic acetylcholine receptor, which is present on skeletal muscle cells at neuromuscular junctions, can block neuromuscular transmission. The antibodies are believed to drive the internalization and intracellular degradation of acetylcholine receptors (Fig. 13.10). Patients with myasthenia gravis develop potentially fatal progressive weakness as a result of their autoimmune disease. Diseases caused by autoantibodies that act as agonists or antagonists for cell-surface receptors are listed in Fig. 13.11.

13-9   Autoantibodies against extracellular antigens cause inflammatory injury by mechanisms akin to type II and type III hypersensitivity reactions.

Antibody responses to extracellular matrix molecules are infrequent, but can be very damaging when they occur. In **Goodpasture's syndrome**, an example of a type II hypersensitivity reaction (see Fig. 12.2), antibodies are formed against the α3 chain of basement membrane collagen (type IV collagen). These antibodies bind to the basement membranes of renal glomeruli (Fig. 13.12a) and, in some cases, to the basement membranes of pulmonary alveoli, causing a rapidly fatal disease if untreated. The autoantibodies bound to basement membrane ligate Fcγ receptors, leading to activation of monocytes, neutrophils, and tissue basophils and mast cells. These release chemokines that attract a further influx of neutrophils into the glomeruli, causing severe tissue injury (Fig. 13.12b). The autoantibodies also cause local activation of complement, which may amplify the tissue injury.

Immune complexes are produced whenever there is an antibody response to a soluble antigen (see Appendix I, Section A-8). Normally, they are cleared efficiently by red blood cells bearing complement receptors and by phagocytes

Fig. 13.11 Autoimmune diseases caused by autoantibodies against cell-surface receptors. These antibodies produce different effects depending on whether they are agonists (which stimulate) or antagonists (which inhibit) the receptor. Note that different autoantibodies against the insulin receptor can either stimulate or inhibit signaling.

Diseases mediated by autoantibodies against cell-surface receptors		
Syndrome	Antigen	Consequence
Graves' disease	Thyroid-stimulating hormone receptor	Hyperthyroidism
Myasthenia gravis	Acetylcholine receptor	Progressive weakness
Insulin-resistant diabetes	Insulin receptor (antagonist)	Hyperglycemia, ketoacidosis
Hypoglycemia	Insulin receptor (agonist)	Hypoglycemia

Fig. 13.12 Autoantibodies reacting with glomerular basement membrane cause the inflammatory glomerular disease known as Goodpasture's syndrome. The panels show sections of renal glomeruli in serial biopsies taken from patients with Goodpasture's syndrome. Panel a, glomerulus stained for IgG deposition by immuno-fluoresence. Anti-glomerular basement membrane antibody (stained green) is deposited in a linear fashion along the glomerular basement membrane.

The autoantibody causes local activation of cells bearing Fc receptors, complement activation, and influx of neutrophils. Panel b, hematoxylin and eosin staining of a section through a renal glomerulus shows that the glomerulus is compressed by formation of a crescent (C) of prolifer-ating mononuclear cells within the Bowman's capsule (B) and there is influx of neutrophils (N) into the glomerular tuft. Photographs courtesy of M. Thompson and D. Evans.

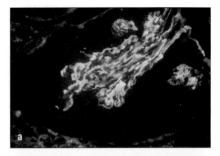
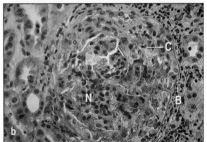

of the mononuclear phagocytic system that have both complement and Fc receptors, and such complexes cause little tissue damage. This clearance system can, however, fail in three circumstances. The first follows the injection of large amounts of antigen, leading to the formation of large amounts of immune complexes that overwhelm the normal clearance mechanisms. An example of this is serum sickness (see Section 12-16), which is caused by injection of large amounts of serum proteins. This is a transient disease, lasting only until the immune complexes have been cleared. The second circumstance is seen in chronic infections such as bacterial endocarditis, where the immune response to bacteria lodged on a cardiac valve is incapable of clearing infection. The persistent release of bacterial antigens from the valve infection in the presence of a strong antibacterial antibody response causes widespread immune-complex injury to small blood vessels in organs such as the kidney and the skin.

The third type of failure to clear immune complexes is seen in SLE. This is an immune complex-mediated disease characterized by chronic IgG antibody production directed at ubiquitous self antigens present in all nucleated cells. In SLE, a wide range of autoantibodies are produced to common cellular con-stituents. The main antigens are three intracellular nucleoprotein particles—the nucleosome, the spliceosome, and a small cytoplasmic ribonucleoprotein complex containing two proteins known as Ro and La (named after the first two letters of the surnames of the two patients in which autoantibodies against these proteins were discovered). In order for these autoantigens to participate in immune-complex formation, they must become extracellular. The autoantigens of SLE are exposed on dead and dying cells and released from injured tissues. In SLE, large quantities of antigen are available, so large numbers of small immune complexes are produced continuously and are deposited in the walls of small blood vessels in the renal glomerulus, in glomerular basement membrane (Fig. 13.13), in joints, and in other organs. This leads to activation of phagocytic cells through their Fc receptors.

Fig. 13.13 Deposition of immune complexes in the renal glomerulus causes renal failure in systemic lupus erythematosus (SLE). Panel a, a section through a renal glomerulus from a patient with SLE, shows that the deposition of immune complexes has caused thickening of the glomerular basement membrane, seen as the clear 'canals' running through the glomerulus. Panel b, a similar section stained with fluorescent anti-immunoglobulin, reveals immunoglobulin deposits in the basement membrane. Panel c, by electron microscopy the immune complexes are seen as dense protein deposits between the glomerular basement membrane and the renal epithelial cells. Polymorphonuclear neutrophilic leukocytes are also present, attracted by the deposited immune complexes. Photographs courtesy of M. Kashgarian.

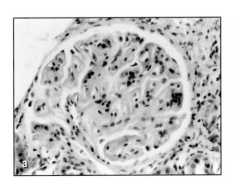
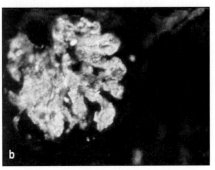
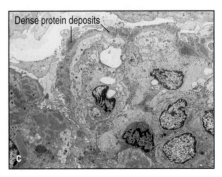

The consequent tissue damage releases more nucleoprotein complexes, which in turn form more immune complexes. Eventually, the inflammation induced in small blood vessel walls, especially in the kidney, can cause sufficient damage to kill the patient. Mice that lack FcγRIII illustrate the importance of Fc receptors in causing the inflammatory response to immune complexes. Such mice do not develop glomerulonephritis, despite deposition of immune complexes and C3, demonstrating the dominant role of Fc receptors in the autoimmune effector mechanisms of SLE.

13-10 Environmental cofactors can influence the expression of autoimmune disease.

The presence of an autoantibody by itself is not sufficient to cause autoimmune disease. For disease to occur, the autoantigen must be available for binding by the autoantibody. Two examples illustrate how the availability of auto-antigens and the resulting expression of disease can be modulated by environmental cofactors. In untreated Goodpasture's disease, as described in Section 13-9, autoantibodies against type IV collagen typically cause a fatal glomerulonephritis. Type IV collagen is distributed widely in basement mem-branes throughout the body, including those of the alveoli of the lung, the renal glomeruli, and the cochlea of the inner ear. All patients with Goodpasture's disease develop glomerulonephritis, about 40% develop pulmonary hemorrhage, but none become deaf.

This pattern of disease expression was explained when it was discovered that pulmonary hemorrhage was found almost exclusively in those patients who smoked cigarettes. What differs between basement membrane in glomeruli, alveoli, and the cochlea is the availability of the antigen to antibodies. The major function of glomerular basement membrane is the filtration of plasma, and the endothelium lining glomerular capillaries is fenestrated to allow access of plasma to the basement membrane. Glomerular basement mem-brane is therefore immediately accessible to circulating autoantibodies. In the alveoli, in contrast, the basement membrane separates the alveolar epithelium from the capillary endothelium, whose cells are joined together by tight junctions. Injury to the endothelial lining of pulmonary capillaries is therefore necessary before antibodies can gain access to the basement mem-brane. Cigarette smoke stimulates an inflammatory response in the lungs, which damages alveolar capillaries and exposes the autoantigen to antibody. Finally, in the inner ear, the cochlear basement membrane seems to remain inaccessible to autoantibodies at all times.

A second example of the importance of environmental influences on the expression of autoimmunity is the effect of infection on the vasculitis associ-ated with **Wegener's granulomatosis**. This disease, which is characterized by a severe necrotizing vasculitis, is strongly associated with the presence of autoantibodies to a granule proteinase of neutrophils (Fig. 13.14); the anti-bodies are known as anti-neutrophil cytoplasm antibodies (commonly abbreviated as ANCA). The autoantigen is proteinase-3, an abundant serine

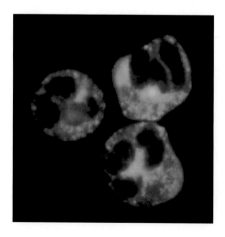

Fig. 13.14 Serum from patients with Wegener's granulomatosis contains autoantibodies reactive with neutrophil cytoplasmic granules. Normal neutrophils with permeabilized cell membranes have been incubated with serum from a patient with Wegener's granulomatosis. IgG antibodies in the serum reactive with cytoplasmic granules are detected by addition of fluorescein-conjugated antibodies against IgG. Photograph courtesy of C. Pusey.

proteinase of neutrophil granules. Although there is a general correlation between the levels of ANCA and the expression of disease, it is quite common to find patients with high levels of ANCA who remain asymptomatic. If such an individual develops an infection, however, this frequently induces a severe flare-up of the vasculitis.

It is thought that the reason for this is that resting neutrophils do not express proteinase-3 on the cell surface, and so in the absence of infection the antigen is inaccessible to anti-proteinase-3 autoantibodies. After infection, a variety of cytokines activate neutrophils, with translocation of proteinase-3 to the cell surface. Anti-proteinase-3 antibodies can now bind neutrophils and stimulate degranulation and release of free radicals. In parallel, activation of vascular endothelial cells by the infection causes the expression of vascular adhesion molecules, such as E-selectin, which promote the binding of activated neutrophils to vessel walls with resultant injury. In this way, a variety of nonspecific infections can exacerbate an autoimmune disease.

13-11 The pattern of inflammatory injury in autoimmunity can be modified by anatomical constraints.

We have seen that the distribution of organ injury in Goodpasture's syndrome can be explained by the accessibility of basement membrane collagen to autoantibodies and that environmental factors can influence the availability of antigen in different organs. Another example of how the expression of autoimmune inflammation can be modified by anatomical factors is seen in **membranous glomerulonephritis**. In this disease, patients develop heavy proteinuria (the excretion of protein in the urine), which can cause life-threatening depletion of plasma protein levels. Biopsy of an affected kidney reveals evidence of deposition of antibody and complement beneath the basement membrane of the glomerulus but, in contrast to Goodpasture's syndrome, there is no significant influx of inflammatory cells. The autoantigen in this disease has not been characterized. However, an excellent rodent model of membranous glomerulonephritis is **Heymann's nephritis**, in which autoantibodies against a glycoprotein on the surface of tubular epithelial cells of the kidney are induced by injection of tubular epithelial tissue. The proteinuria can be abolished by depletion of any of the proteins of the membrane-attack complex of complement but is unaltered by depletion of neutrophils. This shows that the antibodies deposited beneath the glomerular basement membrane in this disease cause tissue injury by activation of complement, but the glomerular basement membrane acts as a complete barrier to inflammatory leukocytes.

In other autoimmune diseases, high levels of autoantibodies against intracellular antigens can be found in the absence of any evidence of antibody-induced inflammation. One such example is a rare myositis (inflammation of muscle) associated with pulmonary fibrosis. Most patients with this disease have high levels of autoantibodies reactive with aminoacyl-tRNA synthetases, the intracellular enzymes responsible for loading tRNAs with amino acids. Addition of these autoantibodies to cell-free extracts *in vitro* stops translation and protein synthesis completely. There is, however, no evidence that these antibodies cause any injury *in vivo*, where it is unlikely that they can enter living cells. In this disease, the autoantibody is thought to be a marker of a particular pattern of tissue injury, and does not contribute to the immunopathology of the myositis. Other examples of autoantibodies that are useful diagnostic markers of the presence of disease, but that might play no part in causing organ injury, are antibodies against mitochondrial antigens associated with primary biliary cirrhosis and antibodies against smooth muscle antigens in chronic active hepatitis.

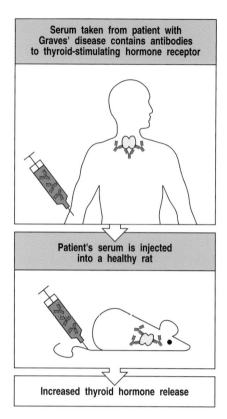

Serum taken from patient with Graves' disease contains antibodies to thyroid-stimulating hormone receptor

Patient's serum is injected into a healthy rat

Increased thyroid hormone release

Fig. 13.15 Serum from some patients with autoimmune disease can transfer the same disease to experimental animals. When the autoantigen is very similar in humans and mice or rats, the transfer of antibody from an affected human can cause the same symptoms in an experimental animal. For example, antibody from patients with Graves' disease frequently produces thyroid activation in rats.

13-12 The mechanism of autoimmune tissue damage can often be determined by adoptive transfer.

To classify a disease as autoimmune, one must show that an adaptive immune response to a self antigen causes the observed pathology. Initially, the demonstration that antibodies against the affected tissue could be detected in the serum of patients suffering from various diseases was taken as evidence that the diseases had an autoimmune basis. However, such auto-antibodies are also found when tissue damage is caused by trauma or infection, though these are typically of much lower affinity than those associated with autoimmune disease. This suggests that autoantibodies can result from, rather than be the cause of, tissue damage. Thus, one must show that the observed autoantibodies are pathogenic before classifying a disease as autoimmune.

It is often possible to transfer disease to experimental animals through the transfer of autoantibodies, causing pathology similar to that seen in the patient from whom the antibodies were obtained (Fig. 13.15). This does not always work, however, presumably because of species differences in autoantigen structure. Some autoimmune diseases can also be transferred from mother to fetus (Fig. 13.16) and are observed in the newborn babies of diseased mothers. When babies are exposed to IgG autoantibodies transferred across the placenta, they will often manifest pathology similar to the mother's (Fig. 13.17). This natural experiment is one of the best proofs that particular autoantibodies exert pathogenic effects. The symptoms of the disease in the newborn typically disappear rapidly as the maternal antibody is catabolized, although they may cause chronic organ injury, such as damage

Fig. 13.16 Some autoimmune diseases that can be transferred across the placenta by pathogenic IgG autoantibodies. These diseases are caused mostly by autoantibodies to cell-surface or tissue-matrix molecules. This suggests that an important factor determining whether an autoantibody that crosses the placenta causes disease in the fetus or newborn baby is the accessibility of the antigen to the auto-antibody. Autoimmune congenital heart block is caused by fibrosis of the developing cardiac conducting tissue, leading to slowing of the heart rate (bradycardia), and there is evidence that this expresses abundant Ro antigen (see Section 13-9). Ro protein is a constituent of an intracellular small cytoplasmic ribonucleoprotein. It is not yet known whether it is expressed at the cell surface of cardiac conducting tissue to act as a target for autoimmune tissue injury.

Autoimmune diseases transferred across the placenta to the fetus and newborn infant		
Disease	**Autoantibody**	**Symptom**
Myasthenia gravis	Anti-acetylcholine receptor	Muscle weakness
Graves' disease	Anti-thyroid-stimulating hormone (TSH) receptor	Hyperthyroidism
Thrombocytopenic purpura	Anti-platelet antibodies	Bruising and hemorrhage
Neonatal lupus rash and/or congenital heart block	Anti-Ro antibodies Anti-La antibodies	Photosensitive rash and/or bradycardia
Pemphigus vulgaris	Anti-desmoglein-3	Blistering rash

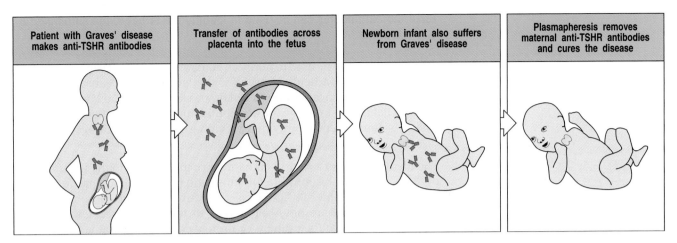

| Patient with Graves' disease makes anti-TSHR antibodies | Transfer of antibodies across placenta into the fetus | Newborn infant also suffers from Graves' disease | Plasmapheresis removes maternal anti-TSHR antibodies and cures the disease |

Fig. 13.17 Antibody-mediated autoimmune diseases can appear in the infants of affected mothers as a consequence of transplacental antibody transfer. In pregnant women, IgG antibodies cross the placenta and accumulate in the fetus before birth (see Fig. 9.22). Babies born to mothers with IgG-mediated autoimmune disease therefore frequently show symptoms similar to those of the mother in the first few weeks of life. Fortunately, there is little lasting damage as the symptoms disappear along with the maternal antibody. In Graves' disease, the symptoms are caused by antibodies against the thyroid-stimulating hormone receptor (TSHR). Children of mothers making thyroid-stimulating antibody are born with hyperthyroidism, but this can be corrected by replacing the plasma with normal plasma (plasmapheresis), thus removing the maternal antibody.

to the heart in babies of mothers with SLE or Sjögren's syndrome. The clearance can be speeded up by a complete exchange of the infant's blood or plasma (plasmapheresis), though this is of no clinical use after permanent injury has occurred, as in congenital heart block.

13-13 T cells specific for self antigens can cause direct tissue injury and have a role in sustained autoantibody responses.

Activated effector T cells specific for self peptide:self MHC complexes can cause local inflammation by activating macrophages or can damage tissue cells directly. Diseases in which these actions of T cells are likely to be important include type I IDDM, rheumatoid arthritis, and multiple sclerosis. Affected tissues in patients with these diseases are heavily infiltrated with T lymphocytes and activated macrophages. These autoimmune diseases are mediated by T cells specific for the autoantigen presented by self MHC. T cells are, of course, also required to sustain all autoantibody responses.

It is much more difficult to demonstrate the existence of autoreactive T cells than it is to demonstrate the presence of autoantibodies. First, autoreactive human T cells cannot be used to transfer disease to experimental animals because T-cell recognition is MHC-restricted and animals and humans have different MHC alleles. Second, it is difficult to identify the antigen recognized by a T cell; for example, autoantibodies can be used to stain self tissues to reveal the distribution of the autoantigen, whereas T cells cannot. Nevertheless, there is strong evidence for the involvement of autoreactive T cells in several autoimmune diseases. In type I IDDM, the insulin-producing β cells of the pancreatic islets are selectively destroyed by specific T cells. When such diabetic patients are transplanted with half a pancreas from an identical twin donor, the β cells in the grafted tissue are rapidly and selectively destroyed by CD8 T cells. Recurrence of disease can be prevented by the immunosuppressive drug cyclosporin A (see Chapter 14), which inhibits T-cell activation. Progress towards identifying the targets of such autoreactive T cells and proving that these cells cause disease will be discussed in Section 13-15.

13-14 Autoantibodies can be used to identify the target of the autoimmune process.

Autoantibodies can be used to purify an autoantigen so that it can be identified. This approach is particularly useful if the autoantibody causes disease in animals, from which large amounts of tissue can be obtained. Autoantibodies can also be used to examine the distribution of the target antigen in cells and tissues by immunohistology, often providing clues to the pathogenesis of the disease.

The identification of a critical autoantigen can also lead to the identification of the CD4 T cells responsible for stimulating autoantibody production. As we learned in Chapter 8, CD4 T cells selectively activate those B cells that bind epitopes that are physically linked to the peptide recognized by the T cell. It follows that the proteins or protein complexes isolated by means of autoantibodies should contain the peptide recognized by the autoreactive CD4 T cell. For example, in myasthenia gravis the autoantibodies that cause disease bind mainly to the α chain of the acetylcholine receptor and can be used to isolate the receptor from lysates of skeletal muscle cells. CD4 T cells that recognize peptide fragments of this receptor subunit can also be found in patients with myasthenia gravis (Fig. 13.18). Thus, both autoreactive B cells and autoreactive T cells are required for this disease.

The same phenomenon is seen in SLE. Tissue damage in this disease is caused by immune complexes of autoantibodies directed against a variety of nucleoprotein antigens (see Section 13-9). These autoantibodies show a high degree of somatic hypermutation, which has all the hallmarks of being antigen-driven (see Section 4-9), and the B cells that produce them can be shown to have undergone extensive clonal expansion. Thus, the autoantibodies have the characteristic properties of antibodies formed in response to chronic stimulation of B cells by antigen and specific CD4 T cells, strongly suggesting that they are produced in response to autoantigens containing peptides recognized by specific autoreactive CD4 T cells. Further evidence for this

Fig. 13.18 Autoimmune disease caused by antibodies also requires autoreactive T cells. Autoantibodies from the serum of myasthenia gravis patients immunoprecipitate the acetylcholine receptor from lysates of skeletal muscle cells (top panels). To be able to produce antibodies, the same patients should also have CD4 T cells that respond to a peptide derived from the acetylcholine receptor. To detect them, T cells from myasthenia gravis patients are isolated and grown in the presence of the acetylcholine receptor plus antigen-presenting cells of the correct MHC type (bottom panels). T cells specific for epitopes of the acetylcholine receptor are stimulated to proliferate and can thus be detected.

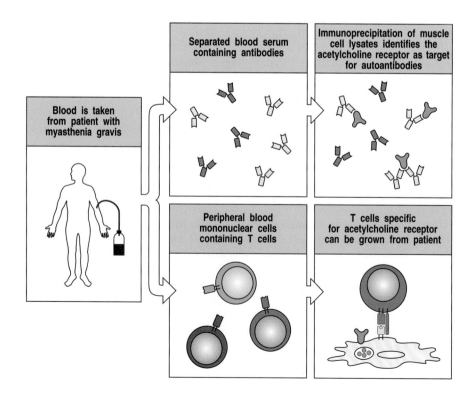

comes from the collective autoantibody specificities observed in individual patients. The autoantibodies in any one individual tend to bind all constituents of a particular nucleoprotein particle; this strongly suggests that there must be CD4 T cells present that are specific for a peptide constituent of this particle. A B cell whose receptor binds a component of this particle will internalize and process the particle, present the peptide to these autoreactive T cells, and receive help from them (Fig. 13.19). Such B-cell–T-cell interactions initiate the antibody response and promote clonal expansion and somatic hypermutation, thus accounting for the observed characteristics of the autoantibody response as well as the clustering of autoantibody specificities in individual patients. This allows the spreading of the autoimmune response to different components of multimolecular complexes, known as **antigen spreading** or **determinant spreading**.

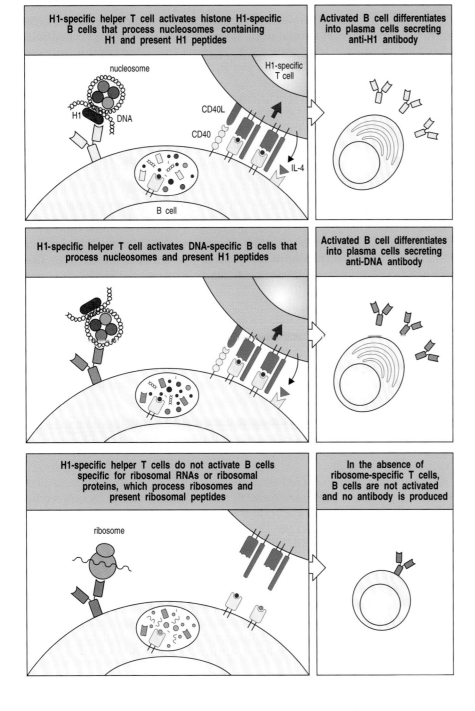

Fig. 13.19 Autoreactive helper T cells of one specificity can drive the production of autoantibodies with several different specificities, in a phenomenon known as antigen spreading. In an SLE patient, a B cell specific for the H1 histone protein in nucleosomes, for example, will bind and internalize the whole nucleosome, and present peptides derived from H1 histone as well as other peptides. This B cell can receive help from a T cell specific for one of the peptides derived from H1 (top panels). A B cell that recognizes the DNA in the nucleosome can also internalize the nucleosome, process it, and present the H1 peptide to that T cell and be activated by it (center panels). Thus, a single autoreactive helper T cell can stimulate a diverse antibody response, but the antibodies will be restricted to those specific for the constituents of a single type of particle. B cells able to bind ribosomes, for example, do not present the H1 peptide and so will not be activated to produce anti-ribosomal antibodies in this patient (bottom panels).

13-15 The target of T cell-mediated autoimmunity is difficult to identify owing to the nature of T-cell ligands.

Although there is good evidence that T cells are involved in many autoimmune diseases, the T cells that cause particular diseases are hard to isolate, and their targets are difficult to identify. Also, the cells are hyporesponsive and thus difficult to assay. It is also difficult to assay the T cells for their ability to cause disease, because any assay requires target cells of the same MHC genotype as the patient. This problem becomes more tractable in animal models. As many autoimmune diseases in animals are induced by immunization with self tissue, the nature of the autoantigen can be determined by fractionating an extract of the tissue and testing the fractions for their ability to induce disease. It is also possible to clone T-cell lines that will transfer the disease from an affected animal to another animal with the same MHC genotype.

This has made it possible to identify the autoantigens recognized by T cells in many experimental autoimmune diseases; they are individual peptides that bind to specific MHC molecules. In some cases, the peptide antigen, when made immunogenic, is able to elicit disease symptoms in animals of the appropriate MHC genotype. An example of such an experimental autoimmune disease is **experimental allergic encephalomyelitis (EAE)**, which can be induced in certain susceptible strains of mice and rats by injection of central nervous system tissue together with Freund's complete adjuvant. This disease resembles human **multiple sclerosis**, in which characteristic plaques of tissue injury are disseminated throughout the central nervous system. Plaques of active disease show infiltration of nervous tissue by lymphocytes, plasma cells, and macrophages, which cause destruction of the myelin sheaths that surround nerve cell axons in the brain and spinal cord.

Multiple Sclerosis

Further analysis of EAE showed that injection with various purified components of the myelin sheath, notably myelin basic protein (MBP), proteolipid protein (PLP), and myelin oligodendroglial protein (MOG), can induce EAE. The disease can be transferred to syngeneic animals by using cloned T-cell lines derived from animals with EAE (Fig. 13.20). Many of these cloned T-cell lines are stimulated by peptides of MBP. EAE can be caused by injection of these MBP peptides into animals that possess MHC alleles capable of presenting such peptides to T cells. Activated T cells specific for myelin proteins have also been identified in patients with multiple sclerosis. Although it has not yet been proved that these cells cause the demyelination in multiple sclerosis, this finding suggests that animal models might provide clues to the identity of autoantigenic proteins in human disease.

A variety of inflammatory autoimmune diseases can be mediated by T_H1 cells responding to self antigens. EAE, for example, can be caused by T_H1 cells specific for MBP, as shown by the ability of specific clones of T_H1, but not T_H2, cells to cause disease on adoptive transfer. Although MBP is an intracellular protein, it is processed for presentation by the vesicular pathway and thus its peptides are presented by MHC class II molecules and recognized by CD4 T cells. Another inflammatory autoimmune disease, **rheumatoid arthritis**, may be caused by T_H1 cells specific for an as yet unidentified antigen present in joints. Engagement with this antigen triggers the T cells to release lymphokines that initiate local inflammation within the joint. This causes swelling, accumulation of polymorphonuclear leukocytes and macrophages, and damage to cartilage, leading to the destruction of the joint. Rheumatoid arthritis is a complex disease and also involves antibodies, often including an IgM anti-IgG autoantibody called **rheumatoid factor**. Like the SLE autoantibodies described in Section 13-9, the rheumatoid factors isolated from the joints of patients with rheumatoid arthritis show evidence of a T-cell

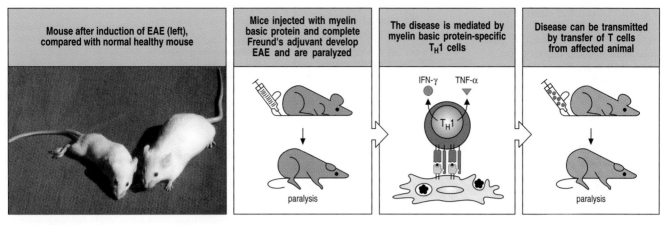

Fig. 13.20 T cells specific for myelin basic protein mediate inflammation of the brain in experimental autoimmune encephalomyelitis (EAE). This disease is produced in experimental animals by injecting them with isolated spinal cord homogenized in complete Freunds' adjuvant. EAE is due to an inflammatory reaction in the brain that causes a progressive paralysis affecting first the tail and hind limbs (as shown in the mouse on the left of the photograph, compared with a healthy mouse on the right) before progressing to forelimb paralysis and eventual death. One of the autoantigens identified in the spinal cord homogenate is myelin basic protein (MBP). Immunization with MBP alone in complete Freund's adjuvant can also cause these disease symptoms. Inflammation of the brain and paralysis are mediated by T_H1 cells specific for MBP. Cloned MBP-specific T_H1 cells can transfer symptoms of EAE to naive recipients provided that the recipients carry the correct MHC allele. In this system it has therefore proved possible to identify the peptide:MHC complex recognized by the T_H1 clones that transfer disease. Other purified components of the myelin sheath can also induce the symptoms of EAE, so there is more than one autoantigen in this disease.

dependent, antigen-driven B-cell response against the Fc portion of IgG. Some of the tissue damage in this disease is caused by the resultant IgM:IgG immune complexes.

Autoantigens recognized by CD4 T cells can be identified by adding cell extracts to cultures of blood mononuclear cells and testing for recognition by CD4 cells derived from an autoimmune patient. If the autoantigen is present in the cell extract, it should be effectively presented, as phagocytes in the blood cultures can take up extracellular protein, degrade it in intracellular vesicles, and present the resulting peptides bound to MHC class II molecules. Identification of autoantigenic peptides is, however, particularly difficult in autoimmune diseases caused by CD8 T cells, as autoantigens recognized by CD8 T cells are not effectively presented in such cultures. Peptides presented by MHC class I molecules must usually be made by the target cells themselves (see Chapter 5); intact cells of target tissue from the patient must therefore be used to study autoreactive CD8 T cells that cause tissue damage. Conversely, the pathogenesis of the disease can itself give clues to the identity of the antigen in some CD8 T cell-mediated diseases. For example, in type I IDDM, the insulin-producing β cells of the pancreatic islets of Langerhans seem to be specifically targeted and destroyed by CD8 T cells (Fig. 13.21). This suggests that a protein unique to β cells is the source of the peptide recognized by the pathogenic CD8 T cells. Studies in the NOD mouse model of type I diabetes have shown that peptides from insulin itself are recognized by pathogenic CD8 cells, confirming the role of insulin as one of the principal autoantigens in type I diabetes.

CD4 T cells also seem to be involved in type I IDDM, consistent with the linkage of disease susceptibility to particular MHC class II alleles (see Fig. 13.4). Identifying the autoantigen recognized by CD4 T cells in these diseases is an important goal. Not only might it help us to understand disease pathogenesis but it might also result in several innovative approaches to treatment (see Chapter 14).

Fig. 13.21 Selective destruction of pancreatic β cells in insulin-dependent diabetes mellitus (IDDM) indicates that the autoantigen is produced in β cells and recognized on their surface. In IDDM, there is highly specific destruction of insulin-producing β cells in the pancreatic islets of Langerhans, sparing other islet cell types (α and δ). This is shown schematically in the upper panels. In the lower panels, islets from normal (left) and diabetic (right) mice are stained for insulin (brown), which shows the β cells, and glucagon (black), which shows the α cells. Note the lymphocytes infiltrating the islet in the diabetic mouse (right) and the selective loss of the β cells (brown) whereas the α cells (black) are spared. The characteristic morphology of the islet is also disrupted with the loss of the β cells. Photographs courtesy of I. Visintin.

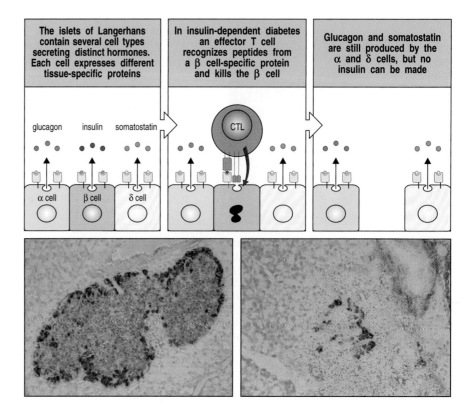

| The islets of Langerhans contain several cell types secreting distinct hormones. Each cell expresses different tissue-specific proteins | In insulin-dependent diabetes an effector T cell recognizes peptides from a β cell-specific protein and kills the β cell | Glucagon and somatostatin are still produced by the α and δ cells, but no insulin can be made |

glucagon   insulin   somatostatin

α cell   β cell   δ cell

CTL

Summary.

For a disease to be defined as autoimmune, the tissue damage must be shown to be caused by an adaptive immune response to self antigens. Autoimmune diseases can be mediated by autoantibodies and/or by auto-reactive T cells, and tissue damage can result from direct attack on the cells bearing the antigen, from immune-complex formation, or from local inflammation. Autoimmune diseases caused by antibodies that bind to cellular receptors, causing either excess activity or inhibition of receptor function, fall into a special class. T cells can be involved directly in inflammation or cellular destruction, and they are also required to sustain autoantibody responses. Similarly, B cells may be important antigen-presenting cells for sustaining autoantigen-specific T-cell responses. The most convincing proof that the immune response is causal in autoimmunity is transfer of disease by transferring the active component of the immune response to an appropriate recipient. The immediate challenge is to identify the autoantigens recognized by T cells in autoimmunity, and to use this information to control the activity of these T cells, or to prevent their activation in the first place. The deeper, more important question is how the autoimmune response is induced. Much has been learned about the induction of immune responses to tissue antigens by examining the response to nonself tissues in transplantation experiments. We will therefore examine the immune response to grafted tissues in the next part of the chapter before turning to the problem of how tolerance is normally maintained, and why immune responses to self antigens occur to cause autoimmune disease.

Responses to alloantigens and transplant rejection.

The transplantation of tissues to replace diseased organs is now an important medical therapy. In most cases, adaptive immune responses to the grafted tissues are the major impediment to successful transplantation. Rejection is caused by immune responses to alloantigens on the graft, which are proteins that vary from individual to individual within a species, and are thus perceived as foreign by the recipient. In blood transfusion, which was the earliest and is still the most common tissue transplant, blood must be matched for ABO and Rh blood group antigens to avoid the rapid destruction of mismatched red blood cells by antibodies (see Appendix I, Section A-11). Because there are only four major ABO types and two Rh blood types, this is relatively easy. When tissues containing nucleated cells are transplanted, however, T-cell responses to the highly polymorphic MHC molecules almost always trigger a response against the grafted organ. Matching the MHC type of donor and recipient increases the success rate of grafts, but perfect matching is possible only when donor and recipient are related and, in these cases, genetic differences at other loci still trigger rejection. In this section, we will examine the immune response to tissue grafts, and ask why such responses do not reject the one foreign tissue graft that is tolerated routinely—the mammalian fetus.

Fig. 13.22 Skin graft rejection is the result of a T cell-mediated anti-graft response. Grafts that are syngeneic are permanently accepted (first panels) but grafts differing at the MHC are rejected around 10–13 days after grafting (first-set rejection, second panels). When a mouse is grafted for a second time with skin from the same donor, it rejects the second graft faster (third panels). This is called a second-set rejection and the accelerated response is MHC-specific; skin from a second donor of the same MHC type is rejected equally fast, whereas skin from an MHC-different donor is rejected in a first-set pattern (not shown). Naive mice that are given T cells from a sensitized donor behave as if they had already been grafted (final panels).

13-16 Graft rejection is an immunological response mediated primarily by T cells.

The basic rules of tissue grafting were first elucidated by skin transplantation between inbred strains of mice. Skin can be grafted with 100% success between different sites on the same animal or person (an **autograft**), or between genetically identical animals or people (a **syngeneic graft**). However, when skin is grafted between unrelated or **allogeneic** individuals (an **allograft**), the graft is initially accepted but is then rejected about 10–13 days after grafting (Fig. 13.22). This response is called a **first-set rejection** and

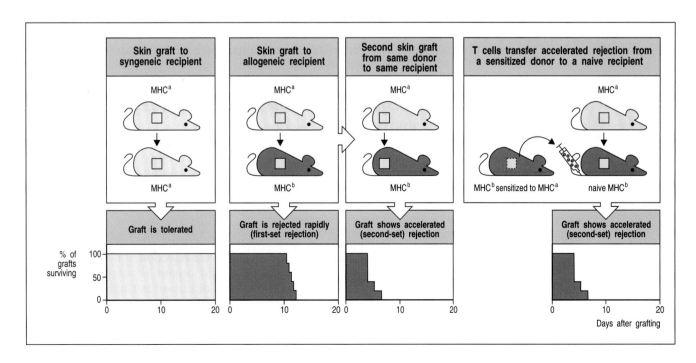

is quite consistent. It depends on a T-cell response in the recipient, because skin grafted onto *nude* mice, which lack T cells, is not rejected. The ability to reject skin can be restored to *nude* mice by the adoptive transfer of normal T cells.

When a recipient that has previously rejected a graft is regrafted with skin from the same donor, the second graft is rejected more rapidly (6–8 days) in a **second-set rejection** (see Fig. 13.22). Skin from a third-party donor grafted onto the same recipient at the same time does not show this faster response but follows a first-set rejection course. The rapid course of second-set rejection can be transferred to normal or irradiated recipients by transferring T cells from the initial recipient, showing that graft rejection is caused by a specific immunological reaction.

Immune responses are a major barrier to effective tissue transplantation, destroying grafted tissue by an adaptive immune response to its foreign proteins. These responses can be mediated by CD8 T cells, by CD4 T cells, or by both. Antibodies can also contribute to second-set rejection of tissue grafts.

13-17 Matching donor and recipient at the MHC improves the outcome of transplantation.

When donor and recipient differ at the MHC, the immune response, which is known as an alloreactive response as it is directed against antigens (alloantigens) that differ between members of the same species, is directed at the nonself allogeneic MHC molecule or molecules present on the graft. In most tissues, these will be predominantly MHC class I antigens. Once a recipient has rejected a graft of a particular MHC type, any further graft bearing the same nonself MHC molecule will be rapidly rejected in a second-set response. As we learned in Chapter 5, the frequency of T cells specific for any nonself MHC molecule is relatively high, making differences at MHC loci the most potent trigger of the rejection of initial grafts; indeed, the major histocompatibility complex was originally so named because of its central role in graft rejection.

Insulin-Dependent Diabetes Mellitus

Once it became clear that recognition of nonself MHC molecules is a major determinant of graft rejection, a considerable amount of effort was put into MHC matching between recipient and donor. Although HLA matching significantly improves the success rate of clinical organ transplantation, it does not in itself prevent rejection reactions. There are two main reasons for this. First, HLA typing is imprecise, owing to the polymorphism and complexity of the human MHC; unrelated individuals who type as HLA-identical with antibodies against MHC proteins rarely have identical MHC genotypes. This should not be a problem with HLA-identical siblings: because siblings inherit their MHC genes as a haplotype, one sibling in four should be truly HLA-identical. Nevertheless, grafts between HLA-identical siblings are invariably rejected, albeit more slowly, unless donor and recipient are identical twins. This rejection is the result of differences between minor histocompatibility antigens, which is the second reason for the failure of HLA matching to prevent rejection reactions. These minor histocompatibility antigens will be discussed in the next section.

Thus, unless donor and recipient are identical twins, all graft recipients must be given immunosuppressive drugs to prevent rejection. Indeed, the current success of clinical transplantation of solid organs is more the result of advances in immunosuppressive therapy, discussed in Chapter 14, than of improved tissue matching. The limited supply of cadaveric organs, coupled with the urgency of identifying a recipient once a donor organ becomes available, means that accurate matching of tissue types is achieved only rarely.

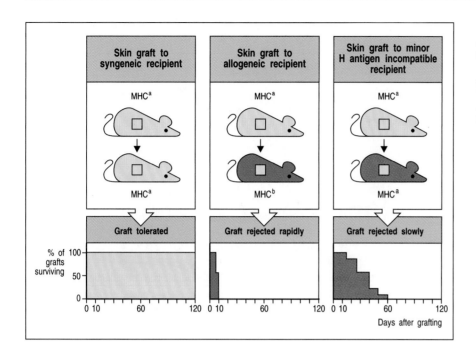

Fig. 13.23 Even complete matching at the MHC does not ensure graft survival. Although syngeneic grafts are not rejected (left panels), MHC-identical grafts from donors that differ at other loci (minor H antigen loci) are rejected (right panels), albeit more slowly than MHC-disparate grafts (center panels).

13-18 In MHC-identical grafts, rejection is caused by peptides from other alloantigens bound to graft MHC molecules.

When donor and recipient are identical at the MHC but differ at other genetic loci, graft rejection is not so rapid (Fig. 13.23). The polymorphic antigens responsible for the rejection of MHC-identical grafts are therefore termed **minor histocompatibility antigens** or **minor H antigens**. Responses to single minor H antigens are much less potent than responses to MHC differences because the frequency of responding T cells is much lower. However, most inbred mouse strains that are identical at the MHC differ at multiple minor H antigen loci, so grafts between them are still uniformly and relatively rapidly rejected. The cells that respond to minor H antigens are generally CD8 T cells, implying that most minor H antigens are complexes of donor peptides and MHC class I molecules. However, peptides bound to MHC class II molecules can also participate in the response to MHC-identical grafts.

Minor H antigens are now known to be peptides derived from polymorphic proteins that are presented by the MHC molecules on the graft (Fig. 13.24). MHC class I molecules bind and present a selection of peptides derived from proteins made in the cell, and if polymorphisms in these proteins mean that different peptides are produced in different members of a species, these can be recognized as minor H antigens. One set of proteins that induce minor H responses is encoded on the male-specific Y chromosome. Responses induced by these proteins are known collectively as H-Y. As these Y chromosome-specific genes are not expressed in females, female anti-male minor H responses occur; however, male anti-female responses are not seen, because both males and females express X-chromosome genes. One H-Y antigen has been identified in mice and humans as peptides from a protein encoded by the Y-chromosome gene *Smcy*. An X-chromosome homologue of *Smcy*, called *Smcx*, does not contain these peptide sequences, which are therefore expressed uniquely in males. The nature of the majority of minor H antigens, encoded by autosomal genes, is unknown, but one, HA-2, has been identified as a peptide derived from myosin.

Fig. 13.24 Minor H antigens are peptides derived from polymorphic cellular proteins bound to MHC class I molecules. Self proteins are routinely digested by proteasomes within the cell's cytosol, and peptides derived from them are delivered to the endoplasmic reticulum, where they can bind to MHC class I molecules and be delivered to the cell surface. If a polymorphic protein differs between the graft donor (shown in red on the left) and the recipient (shown in blue on the right), it can give rise to an antigenic peptide (red on the donor cell) that can be recognized by the recipient's T cells as nonself and elicit an immune response. Such antigens are the minor H antigens.

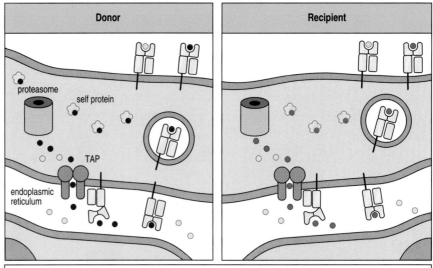

Polymorphic self proteins that differ in amino acid sequence between individuals give rise to minor H antigen differences between donor and recipient

The response to minor H antigens is in most ways analogous to the response to viral infection. However, an antiviral response eliminates only infected cells, whereas all cells in the graft express minor H antigens, and thus the entire graft is destroyed in the response against these antigens. Thus, even though MHC genotype might be matched exactly, polymorphism in any other protein could elicit potent T-cell responses that would destroy the entire graft. It is no wonder that successful transplantation requires the use of powerful immunosuppressive drugs.

13-19 There are two ways of presenting alloantigens on the transplant to the recipient's T lymphocytes.

We saw in Section 13-17 that alloreactive effector T cells that bind directly to allogeneic MHC class I molecules in mismatched organ grafts are an important cause of graft rejection; this is called direct allorecognition. Before they can cause rejection, naive alloreactive T cells must be activated by antigen-presenting cells that both bear the allogeneic MHC molecules and have co-stimulatory activity. Organ grafts carry with them antigen-presenting cells of donor origin, known as passenger leukocytes, and these are an important stimulus to alloreactivity (Fig. 13.25). This route for sensitization of the recipient to a graft seems to involve donor antigen-presenting cells leaving the graft and migrating via the lymph to regional lymph nodes. Here they can activate those host T cells that bear the corresponding T-cell receptors. The activated alloreactive effector T cells are then carried back to the graft, which they attack directly (Fig. 13.26). Indeed, if the grafted tissue is depleted of antigen-presenting cells by treatment with antibodies or by prolonged incubation, rejection occurs only after a much longer time. Also, if the site of grafting lacks lymphatic drainage, no response to the graft results.

A second mechanism of allograft recognition leading to graft rejection is the uptake of allogeneic proteins by the recipient's own antigen-presenting cells and their presentation to T cells by self MHC molecules (see Fig. 13.25). The recognition of allogeneic proteins presented in this way is known as indirect allorecognition, in contrast to the direct recognition by T cells of allogeneic MHC class I and class II molecules expressed on the graft itself. Among the graft-derived peptides presented by the recipient's antigen-presenting cells are the minor H antigens and also peptides from the foreign MHC molecules themselves, which are a major source of the polymorphic peptides recognized by the recipient's T cells.

The relative contributions of direct and indirect allorecognition in graft rejection are not known. Direct allorecognition is thought to be largely responsible for acute rejection, especially when MHC mismatches mean that the frequency of directly alloreactive recipient T cells is high. Furthermore, a direct cytotoxic T-cell attack on graft cells can be carried out only by T cells that recognize the graft MHC molecules directly. Nonetheless, T cells with indirect allospecificity can contribute to graft rejection by activating macrophages, which cause tissue injury and fibrosis, and are likely to be important in the development of an alloantibody response to a graft.

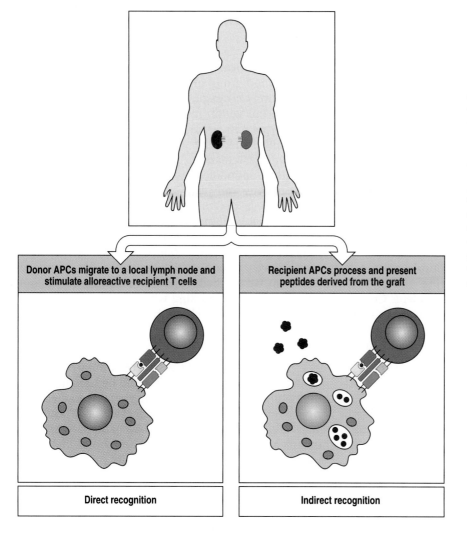

Fig. 13.25 Alloantigens in grafted organs are recognized in two different ways. Direct recognition of a grafted organ (red in upper panel) is mediated by T cells whose receptors have specificity for the allogeneic MHC class I or class II molecule in combination with peptide. These alloreactive T cells are stimulated by donor antigen-presenting cells (APC), which express both the allogeneic MHC molecule and co-stimulatory activity (bottom left panel). Indirect recognition of the graft (bottom right panel) is mediated by T cells whose receptors are specific for allogeneic peptides that are derived from the grafted organ. Proteins from the graft are processed by the recipient's antigen-presenting cells and are therefore presented by self (recipient) MHC class I or class II molecules.

Donor APCs migrate to a local lymph node and stimulate alloreactive recipient T cells

Recipient APCs process and present peptides derived from the graft

Direct recognition

Indirect recognition

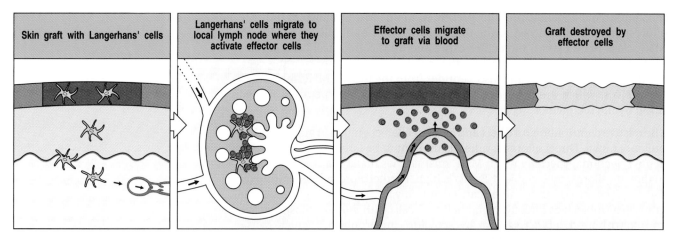

| Skin graft with Langerhans' cells | Langerhans' cells migrate to local lymph node where they activate effector cells | Effector cells migrate to graft via blood | Graft destroyed by effector cells |

Fig. 13.26 The initiation of graft rejection normally involves migration of donor antigen-presenting cells from the graft to local lymph nodes. The example of a skin graft is illustrated here, in which Langerhans' cells are the antigen-presenting cells. They display peptides from the graft on their surface. After traveling to a lymph node, these antigen-presenting cells encounter recirculating naive T cells specific for graft antigens, and stimulate these T cells to divide. The resulting activated effector T cells migrate via the thoracic duct to the blood and home to the grafted tissue, which they rapidly destroy. Destruction is highly specific for donor-derived cells, suggesting that it is mediated by direct cytotoxicity and not by nonspecific inflammatory processes.

13-20 Antibodies reacting with endothelium cause hyperacute graft rejection.

Antibody responses are also an important potential cause of graft rejection. Preexisting alloantibodies to blood group antigens and polymorphic MHC antigens can cause rapid rejection of transplanted organs in a complement-dependent reaction that can occur within minutes of transplantation. This type of reaction is known as **hyperacute graft rejection**. Most grafts that are transplanted routinely in clinical medicine are vascularized organ grafts linked directly to the recipient's circulation. In some cases, the recipient might already have circulating antibodies against donor graft antigens, produced in response to a previous transplant or a blood transfusion. Such antibodies can cause very rapid rejection of vascularized grafts because they react with antigens on the vascular endothelial cells of the graft and initiate the complement and blood clotting cascades. The vessels of the graft become blocked, causing its immediate death. Such grafts become engorged and purple-colored from hemorrhaged blood, which becomes deoxygenated (Fig. 13.27). This problem can be avoided by **cross-matching** donor and recipient. Cross-matching involves determining whether the recipient has antibodies that react with the white blood cells of the donor. If antibodies of this type are found, they are a serious contraindication to transplantation, as they lead to near-certain hyperacute rejection.

A very similar problem prevents the routine use of animal organs—**xenografts**—in transplantation. If xenogeneic grafts could be used, it would circumvent the major limitation in organ replacement therapy, namely the severe shortage of donor organs. Pigs have been suggested as a potential source of organs for xenografting as they are a similar size to humans and are easily farmed. Most humans and other primates have antibodies that react with endothelial cell antigens of other mammalian species, including pigs. When pig xenografts are placed in humans, these antibodies trigger hyperacute rejection by binding to the endothelial cells of the graft and initiating the complement and clotting cascades. The problem of hyperacute rejection

Fig. 13.27 Preexisting antibody against donor graft antigens can cause hyperacute graft rejection. In some cases, recipients already have antibodies to donor antigens, which are often blood group antigens. When the donor organ is grafted into such recipients, these antibodies bind to vascular endothelium in the graft, initiating the complement and clotting cascades. Blood vessels in the graft become obstructed by clots and leak, causing hemorrhage of blood into the graft. This becomes engorged and turns purple from the presence of deoxygenated blood.

is exacerbated in xenografts because complement-regulatory proteins such as CD59, DAF (CD55), and MCP (CD46) (see Section 2-14) work less efficiently across a species barrier; the complement-regulatory proteins of the xenogeneic endothelial cells cannot protect them from attack by human complement. A recent step toward xenotransplantation has been the development of transgenic pigs expressing human DAF. Preliminary experiments have shown prolonged survival of organs transplanted from these pigs into recipient cynomolgus monkeys, under cover of heavy immunosuppression. However, hyperacute rejection is only the first barrier faced by a xenotransplanted organ. The T lymphocyte-mediated graft rejection mechanisms might be extremely difficult to overcome with present immunosuppressive regimes.

13-21 The converse of graft rejection is graft-versus-host disease.

Allogeneic bone marrow transplantation is a successful therapy for some tumors derived from bone marrow precursors, such as certain leukemias and lymphomas. It may also be successful in the treatment of some primary immunodeficiency diseases (see Chapter 11) and inherited bone marrow diseases, such as the severe forms of thalassemia. In leukemia therapy, the recipient's bone marrow, the source of the leukemia, must first be destroyed by aggressive cytotoxic chemotherapy. One of the major complications of allogeneic bone marrow transplantation is graft-versus-host disease (GVHD), in which mature donor T cells that contaminate the allogeneic bone marrow recognize the tissues of the recipient as foreign, causing a severe inflammatory disease characterized by rashes, diarrhea, and pneumonitis. Graft-versus-host disease occurs not only when there is a mismatch of a major MHC class I or class II antigen but also in the context of disparities between minor H antigens. Graft-versus-host disease is a common complication in recipients of bone marrow transplants from HLA-identical siblings, who typically differ from each other in many polymorphic proteins encoded by genes unlinked to the MHC.

The presence of alloreactive T cells can easily be demonstrated experimentally by the mixed lymphocyte reaction (MLR), in which lymphocytes from a potential donor are mixed with irradiated lymphocytes from the potential recipient. If the donor lymphocytes contain alloreactive T cells, these will respond by cell division (Fig. 13.28). The MLR is sometimes used in the selection of donors for bone marrow transplants, when the lowest possible alloreactive response is essential. However, the limitation of the MLR in selection of bone marrow donors is that the test does not accurately quantitate alloreactive T cells. A more accurate test is a version of the limiting-dilution assay (see Appendix I, Section A-25), which precisely counts the frequency of alloreactive T cells.

Although graft-versus-host disease is usually harmful to the recipient of a bone marrow transplant, there can be some beneficial effects. Some of the therapeutic effect of bone marrow transplantation for leukemia can be due to a graft-versus-leukemia effect, in which the allogeneic bone marrow recognizes

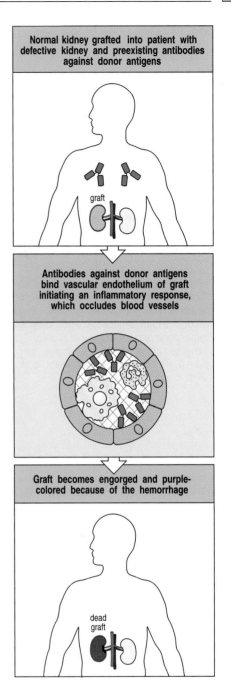

Normal kidney grafted into patient with defective kidney and preexisting antibodies against donor antigens

graft

Antibodies against donor antigens bind vascular endothelium of graft initiating an inflammatory response, which occludes blood vessels

Graft becomes engorged and purple-colored because of the hemorrhage

dead graft

Graft-Versus-Host Disease

Fig. 13.28 The mixed lymphocyte reaction (MLR) can be used to detect histoincompatibility. Lymphocytes from the two individuals who are to be tested for compatibility are isolated from peripheral blood. The cells from one person (yellow), which will also contain antigen-presenting cells, are either irradiated or treated with mitomycin C; they will act as stimulator cells but cannot now respond by DNA synthesis and cell division to antigenic stimulation by the other person's cells. The cells from the two individuals are then mixed (top panel). If the unirradiated lymphocytes (the responders, blue) contain alloreactive T cells, these will be stimulated to proliferate and differentiate to effector cells. Between 3 and 7 days after mixing, the culture is assessed for T-cell proliferation (bottom left panel), which is mainly the result of CD4 T cells recognizing differences in MHC class II molecules, and for the generation of activated cytotoxic T cells (bottom right panel), which respond to differences in MHC class I molecules.

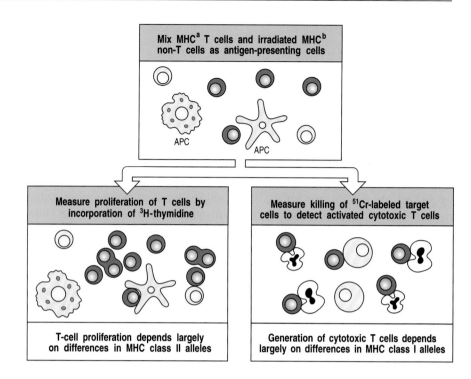

minor H antigens or tumor-specific antigens expressed by the leukemic cells, leading the donor cells to kill the leukemic cells. One such minor H antigen, HB-1, is a B-cell lineage marker that is expressed by acute lymphoblastic leukemia cells, which are B-lineage cells, and by B lymphocytes transformed with Epstein–Barr virus (EBV). One of the treatment options for suppressing the development of graft-versus-host disease is the elimination of mature T cells from the donor bone marrow *in vitro* before transplantation, thereby removing alloreactive T cells. Those T cells that subsequently mature from the donor marrow *in vivo* in the recipient are tolerant to the recipient's antigens. Although the elimination of graft-versus-host disease has benefits for the patient, there is an increase in the risk of leukemic relapse, which provides strong evidence in support of the graft-versus-leukemia effect.

13-22 Chronic organ rejection is caused by inflammatory vascular injury to the graft.

The success of modern immunosuppression means that approximately 85% of cadaveric kidney grafts are still functioning a year after transplantation. However, there has been no improvement in rates of long-term graft survival: the half-life for functional survival of renal allografts remains about 8 years. The major cause of late organ failure is chronic rejection, characterized by concentric arteriosclerosis of graft blood vessels, accompanied by glomerular and tubular fibrosis and atrophy.

Mechanisms that contribute to chronic rejection can be divided into those due to alloreactivity and those due to other pathways, and into early and late events after transplantation. Alloreactivity may occur days and weeks after transplantation and cause acute graft rejection. Alloreactive responses may also occur months to years after transplantation, and be associated with clinically hard-to-detect gradual loss of graft function. Other important causes of chronic graft rejection include ischemia–reperfusion injury, which occurs at the time of grafting but may have late adverse effects on the grafted organ, and later-developing adverse factors such as chronic cyclosporin toxicity or cytomegalovirus infection.

Infiltration of the graft vessels and tissues by macrophages, followed by scarring, are prominent histological features of late graft rejection. A model of injury has been developed in which alloreactive T cells infiltrating the graft secrete cytokines that stimulate the expression of endothelial adhesion molecules and also secrete chemokines such as RANTES (see Fig. 2.33), which attracts monocytes that mature into macrophages in the graft. A second phase of chronic inflammation then supervenes, dominated by macrophage products including interleukin (IL)-1, TNF-α and the chemokine MCP, which leads to further macrophage recruitment. These mediators conspire to cause chronic inflammation and scarring, which eventually leads to irreversible organ failure. Animal models of chronic rejection also show that alloreactive IgG antibodies may induce accelerated atherosclerosis in transplanted solid organs.

13-23 A variety of organs are transplanted routinely in clinical medicine.

Although the immune response makes organ transplantation difficult, there are few alternative therapies for organ failure. Three major advances have made it possible to use organ transplantation routinely in the clinic. First, the technical skill to carry out organ replacement surgery has been mastered by many people. Second, networks of transplantation centers have been organized to ensure that the few healthy organs that are available are HLA-typed and so matched with the most suitable recipient. Third, the use of powerful immunosuppressive drugs, especially cyclosporin A and FK-506, known as tacrolimus, to inhibit T-cell activation (see Chapter 14), has increased graft survival rates dramatically. The different organs that are transplanted in the clinic are listed in Fig. 13.29. Some of these operations are performed routinely with a very high success rate. By far the most frequently transplanted solid organ is the kidney, the organ first successfully transplanted between identical twins in the 1950s. Transplantation of the cornea is even more frequent; this tissue is a special case, as it is not vascularized, and corneal grafts between unrelated people are usually successful even without immunosuppression.

There are, however, many problems other than graft rejection associated with organ transplantation. First, donor organs are difficult to obtain; this is especially a problem when the organ involved is a vital one, such as the heart or liver. Second, the disease that destroyed the patient's organ might also destroy the graft. Third, the immunosuppression required to prevent graft rejection increases the risk of cancer and infection. Finally, the procedure is very costly. All of these problems need to be addressed before clinical transplantation can become commonplace. The problems most amenable to scientific solution are the development of more effective means of immunosuppression, the induction of graft-specific tolerance, and the development of xenografts as a practical solution to organ availability.

13-24 The fetus is an allograft that is tolerated repeatedly.

All of the transplants discussed so far are artefacts of modern medical technology. However, one tissue that is repeatedly grafted and repeatedly tolerated is the mammalian fetus. The fetus carries paternal MHC and minor H antigens that differ from those of the mother (Fig. 13.30), and yet a mother can successfully bear many children expressing the same nonself MHC proteins derived from the father. The mysterious lack of rejection of the fetus has puzzled generations of reproductive immunologists and no comprehensive explanation has yet emerged. One problem is that acceptance of the fetal allograft is so much the norm that it is difficult to study the mechanism that prevents rejection; if the mechanism for rejecting the fetus is rarely activated, how can one analyze the mechanisms that control it?

Tissue transplanted	5-year graft survival*	No. of grafts in USA (1999)
Kidney	80–90%	13,429 (12,483)
Liver	40–50%	4698
Heart	70%	2234 (2185)
Lung	30–40%	934 (885)
Cornea	~70%	~40,000†
Bone marrow	80%	23,500‡

Fig. 13.29 Tissues commonly transplanted in clinical medicine. All grafts except corneal and some bone marrow grafts require long-term immunosuppression. The number of organ grafts performed in the United States in 1999 is shown. Figures in brackets are for the organ alone, while the total figure includes combination transplants, e.g., heart and lungs. *The 5-year survival values are an average; closer matching between donor and recipient generally gives better survival. Data courtesy of United Network for Organ Sharing. † Data for 2000 courtesy of National Eye Institute. ‡ Data for 1998 courtesy of International Bone Marrow Transplant Registry.

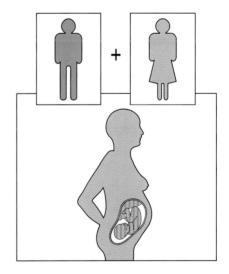

Fig. 13.30 The fetus is an allograft that is not rejected. Although the fetus carries MHC molecules derived from the father, and other foreign antigens, it is not rejected. Even when the mother bears several children to the same father, no sign of immunological rejection is seen.

Various hypotheses have been advanced to account for the tolerance shown to the fetus. It has been proposed that the fetus is simply not recognized as foreign. There is evidence against this hypothesis, as women who have borne several children usually make antibodies directed against the father's MHC proteins; indeed, this is the best source of antibodies for human MHC typing. However, the placenta, which is a fetus-derived tissue, seems to sequester the fetus away from the mother's T cells. The outer layer of the placenta, the interface between fetal and maternal tissues, is the trophoblast. This does not express classical MHC class I and class II proteins, making it resistant to recognition and attack by maternal T cells. Tissues lacking class I expression are, however, vulnerable to attack by NK cells (see Chapter 10). The trophoblast might be protected from attack by NK cells by expression of a nonclassical and minimally polymorphic HLA class I molecule—HLA-G. This protein has been shown to bind to the two major inhibitory NK receptors, KIR1 and KIR2, and to inhibit NK killing.

The placenta may also sequester the fetus from the mother's T cells by an active mechanism of nutrient depletion. The enzyme indoleamine 2,3-dioxygenase (IDO) is expressed at a high level by cells at the maternal–fetal interface. This enzyme catabolizes, and thereby depletes, the essential amino acid tryptophan at this site. T cells starved of tryptophan show reduced responsiveness. Inhibition of IDO in pregnant mice, using the inhibitor 1-methyltryptophan, causes rapid rejection of allogeneic but not syngeneic fetuses. This supports the hypothesis that maternal T cells, alloreactive to paternal MHC proteins, may be held in check in the placenta by tryptophan depletion.

It is likely that fetal tolerance is a multifactorial process. The trophoblast does not act as an absolute barrier between mother and fetus, and fetal blood cells can cross the placenta and be detected in the maternal circulation, albeit in very low numbers. There is direct evidence from experiments in mice for specific T-cell tolerance against paternal MHC alloantigens. Pregnant female mice whose T cells bear a transgenic receptor specific for a paternal alloantigen showed reduced expression of this T-cell receptor during pregnancy. These same mice lost the ability to control the growth of an experimental tumor bearing the same paternal MHC alloantigen. After pregnancy, tumor growth was controlled and the level of the T-cell receptor increased. This experiment demonstrates that the maternal immune system must have been exposed to paternal MHC alloantigens, and that the immune response to these antigens was temporarily suppressed.

Yet another factor that might contribute to maternal tolerance of the fetus is the secretion of cytokines at the maternal–fetal interface. Both uterine epithelium and trophoblast secrete cytokines, including transforming growth factor (TGF)-β, IL-4, and IL-10. This cytokine pattern tends to suppress T_H1 responses (see Chapter 10). Induction or injection of cytokines such as interferon (IFN)-γ and IL-12, which promote T_H1 responses in experimental animals, promote fetal resorption, the equivalent of spontaneous abortion in humans.

The fetus is thus tolerated for two main reasons: it occupies a site protected by a nonimmunogenic tissue barrier, and it promotes a local immunosuppressive response in the mother. We will see later that several sites in the body have these characteristics and allow prolonged acceptance of foreign tissue grafts. They are usually called immunologically privileged sites.

Summary.

Clinical transplantation is now an everyday reality, its success built on MHC matching, immunosuppressive drugs, and technical skill. However, even accurate MHC matching does not prevent graft rejection; other genetic differences between host and donor can result in allogeneic proteins whose peptides are presented as minor H antigens by MHC molecules on the grafted tissue, and responses to these can lead to rejection. As we lack the ability to specifically suppress the response to the graft without compromising host defense, most transplants require generalized immunosuppression of the recipient. This can be significantly toxic and increases the risk of cancer and infection. The fetus is a natural allograft that must be accepted—it almost always is—or the species will not survive. Tolerance to the fetus might hold the key to inducing specific tolerance to grafted tissues, or it might be a special case not applicable to organ replacement therapy.

Self-tolerance and its loss.

Tolerance to self is acquired by clonal deletion or inactivation of developing lymphocytes. Tolerance to antigens expressed by grafted tissues can be induced artificially, but it is very difficult to establish once a full repertoire of functional B and T lymphocytes has been produced, which occurs during fetal life in humans and around the time of birth in mice. We have already discussed the two important mechanisms of self-tolerance—clonal deletion by ubiquitous self antigens and clonal inactivation by tissue-specific antigens presented in the absence of co-stimulatory signals (see Chapters 7–8). These processes were first discovered by studying tolerance to nonself, where the absence of tolerance could be studied in the form of graft rejection. In this section, we will consider tolerance to self and tolerance to nonself as two aspects of the same basic mechanisms. These mechanisms consist of direct induction of tolerance in the periphery, either by deletion or by anergy. There is also a state referred to as **immunological ignorance**, in which T cells or B cells coexist with antigen without being affected by it. Finally, there are mechanisms of tolerance that involve T-cell–T-cell interactions, known variously as immune deviation or immune suppression. In an attempt to understand the related phenomena of autoimmunity and graft rejection, we also examine instances where tolerance to self is lost.

13-25 Many autoantigens are not so abundantly expressed that they induce clonal deletion or anergy but are not so rare as to escape recognition entirely.

We saw in Chapter 7 that clonal deletion removes immature T cells that recognize ubiquitous self antigens and in Chapter 8 that antigens expressed abundantly in the periphery induce anergy or clonal deletion in lymphocytes that encounter them on tissue cells. Most self proteins are expressed at levels that are too low to serve as targets for T-cell recognition and thus cannot serve as autoantigens. It is likely that very few self proteins contain peptides that are presented by a given MHC molecule at a level sufficiently high to be recognized by effector T cells but too low to induce tolerance. T cells able to recognize these rare antigens will be present in the individual but will not normally be activated. This is because their receptors only bind self peptides with very low affinity, or because they are exposed to levels of self peptide that are too low to deliver any signal to the T cell. Such T cells are said to be in a state of immunological ignorance. This state has been demonstrated experimentally using transgenic animals in which ovalbumin was expressed at high or very low concentrations in the pancreas. Lymphocytes reactive to ovalbumin were transferred to these animals. The lymphocytes transferred to animals expressing high levels of ovalbumin proliferated in response to oval-bumin presented by antigen-presenting cells and then died. In contrast, the

Fig. 13.31 T cells are ignorant of very low levels of autoantigen. Transgenic mice were developed that expressed ovalbumin in the pancreas at high or very low levels. CD8 lymphocytes specific for ovalbumin were injected into these mice. After 3 days, the regional lymph nodes draining the pancreas were isolated and the amount of proliferation of the ovalbumin-specific CD8 T cells was measured. In the mice expressing high levels of ovalbumin, these T cells were proliferating, in contrast to mice expressing low levels of ovalbumin, in which no proliferation was observed. After 4 weeks, the spleens were obtained, the ovalbumin-specific T cells were enumerated and their spontaneous proliferation and proliferation in response to ovalbumin *in vitro* was assessed. At this time no ovalbumin-specific T cells were recovered from the spleens of the mice expressing high levels of ovalbumin, and there was no proliferative response to ovalbumin *in vitro*. Thus in these mice the ovalbumin-specific T cells divided early, after encounter with ovalbumin in the periphery, and then died, illustrating the phenomenon of peripheral tolerance by deletion. In contrast, in the mice expressing very low levels of ovalbumin, the ovalbumin-specific T cells persisted without proliferating in the periphery, were recovered from the spleen at 4 weeks, and responded normally to ovalbumin presented *in vitro*. This silent persistence in the presence of low levels of autoantigen illustrates the phenomenon of T-cell ignorance.

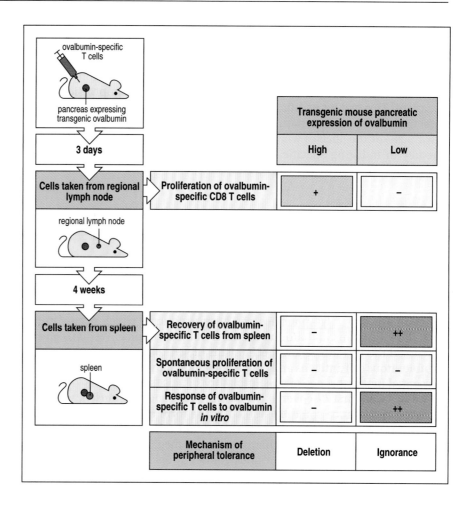

	Transgenic mouse pancreatic expression of ovalbumin	
	High	Low
Proliferation of ovalbumin-specific CD8 T cells	+	–
Recovery of ovalbumin-specific T cells from spleen	–	++
Spontaneous proliferation of ovalbumin-specific T cells	–	–
Response of ovalbumin-specific T cells to ovalbumin *in vitro*	–	++
Mechanism of peripheral tolerance	Deletion	Ignorance

lymphocytes transferred to animals expressing very low levels of pancreatic ovalbumin did not divide but persisted and could be stimulated normally when exposed to high levels of ovalbumin *in vitro* (Fig. 13.31).

In the organ-specific autoimmune diseases such as type I IDDM and Hashimoto's thyroiditis, autoimmunity is unlikely to reflect a general failure of the main mechanisms of tolerance—clonal deletion and clonal inactivation. For example, clonal deletion of developing lymphocytes mediates tolerance to self MHC molecules. If such tolerance were not induced, the reactions to self tissues would be similar to those seen in graft-versus-host disease (see Section 13-21). To estimate the impact of clonal deletion on the developing T-cell repertoire, we should remember that the frequency of T cells able to respond to any set of nonself MHC molecules can be as high as 10% (see Section 5-14), yet responses to self MHC antigens are not detected in naturally self-tolerant individuals. Moreover, mice given bone marrow cells from a foreign donor at birth, before significant numbers of mature T cells have appeared, can be rendered fully and permanently tolerant to the bone marrow donor's tissues, provided that the bone marrow donor's cells continue to be produced so as to induce tolerance in each new cohort of developing T cells (Fig. 13.32). This experiment, performed by **Medawar**, validated **Burnet's** prediction that developing lymphocytes collectively carrying a complete repertoire of receptors must be purged of self-reactive cells before they achieve functional maturity; it won them a Nobel Prize.

Clonal deletion reliably removes all T cells that can mount aggressive responses against self MHC molecules; the organ-specific autoimmune diseases, which involve rare T-cell responses to a particular self peptide

Fig. 13.32 Tolerance to allogeneic skin can be established in bone marrow chimeric mice. If mice are injected with allogeneic bone marrow at birth (top panel) before they achieve immune competence, they become chimeric, with T cells and antigen-presenting cells (APCs) deriving from both host and donor bone marrow stem cells. T cells developing in these mice are negatively selected on APCs of both host and donor (center panel), so that mature T cells are tolerant to the MHC molecules of the bone marrow donor. This allows the chimeric mouse to accept skin derived from the bone marrow donor. Such acquired tolerance is specific, because skin from an unrelated or 'third party' donor is rejected normally (bottom panel).

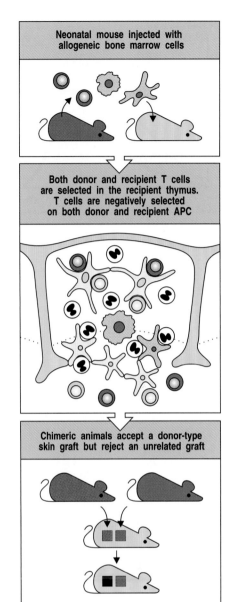

Neonatal mouse injected with allogeneic bone marrow cells

Both donor and recipient T cells are selected in the recipient thymus. T cells are negatively selected on both donor and recipient APC

Chimeric animals accept a donor-type skin graft but reject an unrelated graft

bound to a self MHC molecule, are therefore unlikely to reflect a general failure in clonal deletion; nor are they likely to be caused by a random failure in the mechanisms responsible for anergy. Rather, the lymphocytes that mediate autoimmune responses seem not to be subject to clonal deletion or inactivation. Such autoreactive cells are present in all of us, but they do not normally cause autoimmunity because they are activated only under special circumstances.

A striking demonstration that autoreactive T cells can be present in healthy individuals comes from a strain of mice carrying transgenes encoding an autoreactive T-cell receptor specific for a peptide of myelin basic protein bound to self MHC class II molecules. The autoreactive receptor is present on every T cell, yet the mice are healthy unless their T cells are activated. As the level of the specific peptide:MHC class II complex is low except in the central nervous system, a site not visited by naive T cells, the autoreactive T cells remain in a state of immunological ignorance. When these T cells are activated, for example, by deliberate immunization with myelin basic protein, as in EAE, they migrate into all tissues, including the central nervous system, where they recognize their myelin basic protein:MHC class II ligand. Recognition triggers cytokine production by the activated T cells, causing inflammation in the brain and the destruction of myelin and neurons that ultimately causes the paralysis in EAE.

It is likely that only a small fraction of proteins will be able to serve as autoantigens. An autoantigen must be presented by an MHC molecule at a level sufficient for the antigen to be recognized by effector T cells, but must not be presented to naive T cells at a level sufficient to induce tolerance. Many self proteins are expressed at levels too low to be detected even by effector T cells. It has been estimated that we can make approximately 10^5 proteins of average length 300 amino acids, capable of generating about 3×10^7 distinct self peptides. As there are rarely more than 10^5 MHC molecules per cell, and as the MHC molecules on a cell must bind 10–100 identical peptides for T-cell recognition to occur, fewer than 10,000 self peptides (<1 in 3000) can be presented by any given antigen-presenting cell at levels detectable by T cells. Thus, most peptides will be presented at levels that are insufficient to engage effector T cells, whereas many of the peptides that can be detected by T cells will be presented at a high enough level to induce clonal deletion or anergy. However, as shown in Fig. 13.33, a few peptides may fail to induce tolerance yet be present at high enough levels to be recognized by effector T cells. Autoreactivity probably arises most frequently when the antigen is expressed selectively in a tissue, as is the case of insulin in the pancreas, rather than ubiquitously, because tissue-specific antigens are less likely to induce clonal deletion of developing T cells in the thymus. The nature of such peptides will vary depending on the MHC genotype of the individual, because MHC polymorphism profoundly affects peptide binding (see Section 5-13). This argument leaves aside the crucial issue of how T cells specific for such autoantigens are activated to become effector T cells, which we will consider later.

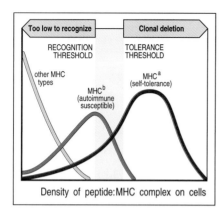

Fig. 13.33 An autoantigenic peptide will be presented at different levels on different MHC molecules. Peptides bind to different MHC molecules with varying affinities; in this example, a self peptide binds well to MHC^a, less well to MHC^b, and poorly to all other MHC types. The graph shows the number of cells displaying a given density of peptide:MHC complex for different MHC genotypes. It illustrates how the self peptide is displayed at levels that are too low for T-cell recognition in most MHC haplotypes (yellow), and at levels that would induce tolerance in the MHC^a haplotype (red). Only the MHC^b molecule (blue) presents the peptide at an intermediate level so that it can be recognized by T cells without inducing tolerance.

If it is true that only a few peptides can act as autoantigens, then it is not surprising that there are relatively few distinct autoimmune syndromes, and that all individuals with a particular autoimmune disease tend to recognize the same antigens. If all antigens could give rise to autoimmunity, one would expect that different individuals with the same disease might recognize different antigens on the target tissue, which does not seem to be the case. Finally, because the level of autoantigenic peptide presented is determined by polymorphic residues in MHC molecules that govern the affinity of peptide binding, this idea could also explain the association of autoimmune diseases with particular MHC genotypes (see Fig. 13.3).

13-26 The induction of a tissue-specific response requires the presentation of antigen by antigen-presenting cells with co-stimulatory activity.

As we learned in Chapter 8, only antigen-presenting cells that express co-stimulatory activity can initiate clonal expansion of T cells—an essential step in all adaptive immune responses, including graft rejection and, presumably, autoimmunity. In tissue grafts, it is the donor antigen-presenting cells in the graft that initially stimulate host T cells, starting the direct allorecognition response that leads to graft rejection. Antigen-presenting cells bearing both graft antigens and co-stimulatory activity travel to regional lymph nodes. Here they are examined by large numbers of naive host T cells and can activate those that bear specific receptors (see Fig. 13.26). Grafts depleted of antigen-presenting cells are tolerated for long periods, but are eventually rejected. This rejection is due to the recipient's T cells responding to graft antigens, both MHC and minor H antigens, after they have been processed and presented by recipient antigen-presenting cells (see Fig. 13.25).

The ability of the recipient's antigen-presenting cells to pick up antigens in tissues and initiate graft rejection may be relevant to the initiation of autoimmune tissue damage as well. Transplantation experiments show that host antigen-presenting cells can stimulate both cytotoxic T-cell responses and inflammatory T$_H$1 responses against the transplanted tissue; thus, tissue antigens can be taken up and presented in conjunction with both MHC class I and class II molecules by antigen-presenting cells. In autoimmunity, tissues may be similarly attacked by MHC class I-restricted cytotoxic T cells or injured by inflammatory damage mediated by T$_H$1 cells, as a consequence of the uptake and presentation of tissue antigens by such antigen-presenting cells.

To induce a response to tissue antigens, the antigen-presenting cell must express co-stimulatory activity. As we saw in Chapters 2 and 8, the expression of co-stimulatory molecules in antigen-presenting cells is regulated to occur in response to infection. Transient autoimmune responses are seen in the context of such events, and it is thought that one trigger for autoimmunity is infection. This is discussed further in Section 13-31.

13-27 In the absence of co-stimulation, tolerance is induced.

Activation of naive T cells requires interaction with cells expressing both the appropriate peptide:MHC complex and co-stimulatory molecules; in the absence of full co-stimulation, specific antigen recognition leads to partial T-cell activation, leading to T-cell anergy or deletion (see Section 8-11). Tissue cells are not known to express B7 or other co-stimulatory molecules, and can therefore induce tolerance. Experiments with transgenes show that the expression of foreign antigens in peripheral tissues can in some cases induce tolerance, whereas in other cases the foreign antigen seems not to be presented to naive T cells at a sufficient level and is ignored (see Fig. 13.31). Autoimmunity can be induced by coexpression of a foreign antigen and B7 in the same target tissue, but as B7 expression on peripheral tissue cells is not by itself a sufficient stimulus for autoimmunity, it is clear that the loss of tolerance to self tissues requires the coexpression of both a suitable target antigen and co-stimulatory molecules. As discussed in Section 13-25, antigens that are unable to induce clonal anergy or deletion, but that can nonetheless act as targets for effector T cells, can serve as autoantigens; these antigens are likely to be tissue-specific and relatively few.

By analogy with graft rejection, it seems likely that autoimmunity is initiated when a professional antigen-presenting cell picks up a tissue-specific autoantigen and migrates to the regional lymph node, where it is induced to express co-stimulatory activity. Once an autoantigen is expressed on a cell with co-stimulatory potential, naive ignorant T cells specific for the autoantigen can become activated and can home to the tissues, where they interact with their target antigens. At this stage, the absence of co-stimulatory molecules on tissue cells that present the autoantigen can again limit the response. Armed effector T cells kill only a limited number of antigen-expressing tissue cells if these lack co-stimulatory activity; after killing a few targets, the effector cell dies. Thus, not only can responses not be initiated in the absence of co-stimulatory activity, they also cannot be sustained. Therefore, in addition to the question of how autoimmunity is avoided, we must ask: Why does it ever occur? How are responses to self initiated, and how they are sustained?

13-28 Dominant immune suppression can be demonstrated in models of tolerance and can affect the course of autoimmune disease.

In some models of tolerance, it can be shown that specific T cells actively suppress the actions of other T cells that can cause tissue damage. Tolerance in these cases is dominant in that it can be transferred by T cells, which are usually called **suppressor T cells** or **regulatory T cells**. Furthermore, depletion of the suppressor T cells leads to aggravated responses to self or graft antigens. Although it is clear that immune suppression exists, the mechanisms responsible are highly controversial. Here, we will examine the phenomenon in several animal models.

Neonatal rats can be rendered tolerant to allogeneic skin grafts by prior injection of allogeneic bone marrow. This tolerance is highly specific and can be transferred to normal adult recipient rats. This shows that tolerance in this model is dominant and active, as the transferred cells prevent the lymphocytes of the recipient from mediating graft rejection. In order to transfer this tolerance, cells of both the allogeneic graft donor and the neonatal tolerized host must be transferred. Removal of either cell type abolishes the transfer of tolerance.

This finding is reminiscent of **Medawar's** studies of tolerance in neonatal bone marrow chimeric mice discussed in Section 13-25. In both cases, even injection of massive numbers of normal syngeneic lymphocytes, which would react vigorously against the foreign cells in the normal environment of the cell donor, did not break tolerance. Tolerance could be broken only by alloreactive cells from an animal that had been immunized with cells from the allogeneic donor before transfer; such cells probably break tolerance by killing the allogeneic donor cells. Thus, an active host response prevents graft rejection in this model. The tolerance is specific for cells of the original donor, and so the suppression must also be specific.

When mice transgenic for a T-cell receptor specific for myelin basic protein are crossed with RAG$^{-/-}$ mice, they spontaneously develop EAE. TCR transgenic mice that have functional *RAG* genes are able to rearrange their endogenous TCRα chain genes. Since TCRα chain expression is not allelically excluded (see Section 7-16), many T cells in such TCR transgenic mice nevertheless express receptors containing endogenous TCRα chains and have a diverse repertoire. In the RAG$^{-/-}$ mice, no such rearrangements can occur, and the only T-cell receptor expressed is encoded by the transgenes. The observation that, when the background population of diverse T cells is lost, the mice develop an autoimmune disease suggests that this population contains cells normally capable of suppressing the autoimmune disease. Such cells have been shown in an increasing number of autoimmune diseases, and their isolation as cloned T-cell lines is a major goal for people who study the induction of autoimmunity. The reason for this renewed interest in T cell-mediated regulation is that the intentional induction of such cells could be a major advance in the prevention of autoimmune disease.

The mechanisms of tolerance in these animal models are not fully understood. There is evidence for the existence of CD4-positive regulatory cells that can inhibit autoimmune disease. These cells have an activated phenotype and can be identified by the expression of CD25, the IL-2 receptor α chain. Depletion of these cells from the peripheral immune system of normal mice leads to the development of insulin-dependent diabetes, thyroiditis, and gastritis. Similarly, if thymocytes depleted of CD25$^+$ CD4 cells are adoptively transferred into athymic recipients, autoimmune disease results. It is not known how these cells mediate their effects, but one possibility is by secreting cytokines that inhibit other lymphocytes.

In the NOD mouse model for type I IDDM, transfer of a particular insulin-specific T-cell clone can prevent the destruction of pancreatic β cells by autoreactive T cells. This suggests that the insulin-specific T cells can suppress the activity of other autoaggressive T cells in an antigen-dependent manner. They do this by homing to the islet, where they react with insulin peptides presented on the macrophages or dendritic cells. This stimulates the secretion of cytokines, prominent among which is TGF-β, a known immunosuppressive cytokine. There are interesting hints that such cells naturally affect the course of the autoimmune response that causes human diabetes; β-cell destruction in humans occurs over a period of several years before diabetes is manifest, yet when new islets are transplanted from an identical twin into his or her diabetic sibling, they are destroyed within weeks. This suggests that, in the normal course of the disease, specific T cells protect the β-cells from attack by effector T cells and the disease therefore progresses slowly. It might be that after the host islets have been destroyed, these protective mechanisms decline in activity but the effector cells responsible for β-cell destruction do not.

If specific suppression of autoimmune responses could be elicited at will, autoimmunity would not be a problem. Feeding with specific antigen is known to elicit a local immune response in the intestinal mucosa, and responses to the same antigen given subsequently by a systemic route are suppressed (see Section 10-20). This response has been exploited in experimental autoimmune diseases by feeding proteins from target tissues to mice. Mice fed with insulin are protected from diabetes; mice fed with myelin basic protein are resistant to EAE (Fig. 13.34).

EAE is usually caused by T_H1 cells that produce IFN-γ in response to myelin basic protein. In mice fed this protein, CD4 T cells found in the brain instead produce cytokines such as TGF-β and IL-4 (see Fig. 13.34). TGF-β, in particular, suppresses the function of inflammatory T_H1 lymphocytes. In these cases, the protection seems to be tissue-specific rather than antigen-specific. Thus, feeding with myelin basic protein will protect against EAE elicited by immunization with other brain antigens. Feeding with antigen might induce the production of T cells producing TGF-β and IL-4 because these cytokines are also required for IgA production against antigens ingested in food. If feeding with antigen works as a clinical therapy, it would have the advantage over treatments with immunosuppressive drugs that it does not alter the general immune competence of the host. Unfortunately, early studies of this approach to treatment in humans with multiple sclerosis or rheumatoid arthritis have shown minimal, if any, benefit. These strategies may prove more successful in preventing the onset of disease than reversing established disease. However, this approach requires the identification of patients at the very onset of disease or those who are at extremely high risk of developing disease and adds impetus to studies to identify the disease susceptibility genes for the development of autoimmune diseases.

Another strategy for controlling immune responses is to manipulate the cytokine balance that determines whether a CD4 T-cell response is predominantly T_H1 or T_H2. It is possible experimentally to switch T_H1 to T_H2 responses by administration of IL-4, and vice versa using IFN-γ. This is known as **immune deviation** and is discussed further in Chapter 14.

Unlike human diabetes, which follows a chronic progressive course in humans, multiple sclerosis is a chronic relapsing disease with acute episodes followed by periods of quiescence. This suggests a balance between auto-immune and protective T cells that can alter at different stages of the disease. However, it remains to be proven whether the specific suppressive cells discussed in this section exist naturally and contribute to self-tolerance, or

Fig. 13.34 Antigen given orally can lead to protection against autoimmune disease. Experimental allergic encephalomyelitis (EAE) is induced in mice by immunization with spinal cord homogenate in complete Freund's adjuvant (upper panels); the disease is mediated by T_H1 cells specific for myelin antigens. These cells produce the cytokine IFN-γ (top left photograph, where the brown staining reveals the presence of IFN-γ), but not of TGF-β. These T cells are presumably responsible for the damage that results in paralysis. When mice are first fed with myelin basic protein (MBP), later immunization with spinal cord or MBP fails to induce the disease (lower panels). In these orally tolerized mice, IFN-γ-producing cells are absent (lower left photograph), whereas TGF-β-producing T cells (lower right photograph, brown staining) are found in the brain in place of the autoaggressive T_H1 cells and presumably protect the brain from autoimmune attack. Photographs courtesy of S. Khoury, W. Hancock, and H. Weiner.

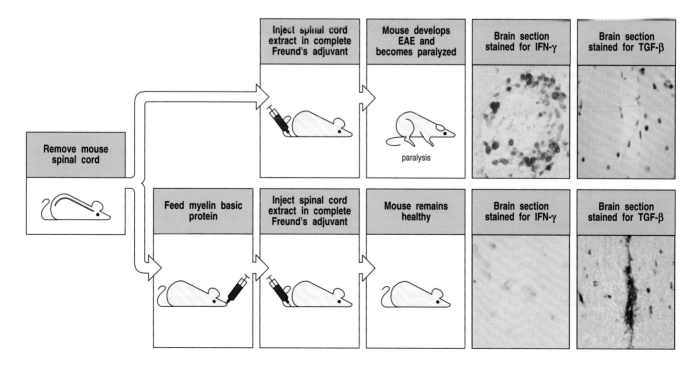

whether they arise only upon artificial stimulation or in response to autoimmune attack. Nevertheless, because they can play an active, dominant part in self-tolerance, they are particularly attractive potential mediators for immunotherapy of autoimmune disease.

13-29 Antigens in immunologically privileged sites do not induce immune attack but can serve as targets.

Tissue grafts placed in some sites in the body do not elicit immune responses. For instance, the brain and the anterior chamber of the eye are sites in which tissues can be grafted without inducing rejection. Such locations are termed **immunologically privileged sites** (Fig. 13.35). It was originally believed that immunological privilege arose from the failure of antigens to leave privileged sites and induce immune responses. Subsequent studies have shown, however, that antigens do leave immunologically privileged sites, and that these antigens do interact with T cells; but instead of eliciting a destructive immune response, they induce tolerance or a response that is not destructive to the tissue. Immunologically privileged sites seem to be unusual in three ways. First, the communication between the privileged site and the body is atypical in that extracellular fluid in these sites does not pass through conventional lymphatics, although proteins placed in these sites do leave them and can have immunological effects. Naive lymphocytes, similarly, may be excluded by the tissue barriers of privileged sites, such as the blood–brain barrier. Second, humoral factors, presumably cytokines, that affect the course of an immune response are produced in privileged sites and leave them together with antigens. The anti-inflammatory cytokine TGF-β seems to be particularly important in this regard: antigens mixed with TGF-β seem to induce mainly T-cell responses that do not damage tissues, such as T_H2 rather than T_H1 responses. Third, the expression of Fas ligand by the tissues of immunologically privileged sites may provide a further level of protection by inducing apoptosis of Fas-bearing lymphocytes that enter these sites. This mechanism of protection is not fully understood, as it appears that under some circumstances the expression of Fas ligand by tissues may trigger an inflammatory response by neutrophils.

Paradoxically, the antigens sequestered in immunologically privileged sites are often the targets of autoimmune attack; for example, brain autoantigens such as myelin basic protein are targeted in multiple sclerosis. It is therefore clear that this antigen does not induce tolerance due to clonal deletion of the self-reactive T cells. Mice only become diseased when they are deliberately immunized with myelin basic protein, in which case they become acutely sick, show severe infiltration of the brain with specific T_H1 cells, and often die.

Thus, at least some antigens expressed in immunologically privileged sites induce neither tolerance nor activation, but if activation is induced elsewhere they can become targets for autoimmune attack (see Section 13-25). It seems plausible that T cells specific for antigens sequestered in immunologically privileged sites are more likely to remain in a state of immunological ignorance. Further evidence for this comes from the eye disease **sympathetic ophthalmia** (Fig. 13.36). If one eye is ruptured by a blow or other trauma, an autoimmune response to eye proteins can occur, although this happens only rarely. Once the response is induced, it often attacks both eyes. Immunosuppression and removal of the damaged eye, the source of antigen, is frequently required to preserve vision in the undamaged eye.

It is not surprising that effector T cells can enter immunologically privileged sites: such sites can become infected, and effector cells must be able to enter these sites during infection. As we learned in Chapter 10, effector T cells enter

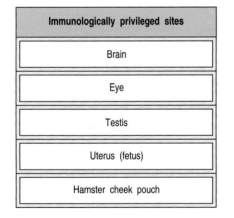

Immunologically privileged sites
Brain
Eye
Testis
Uterus (fetus)
Hamster cheek pouch

Fig. 13.35 Some body sites are immunologically privileged. Tissue grafts placed in these sites often last indefinitely, and antigens placed in these sites do not elicit destructive immune responses.

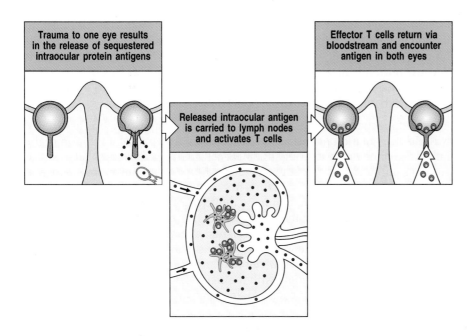

Fig. 13.36 Damage to an immuno-logically privileged site can induce an autoimmune response. In the disease sympathetic ophthalmia, trauma to one eye releases the sequestered eye antigens into the surrounding tissues, making them accessible to T cells. The effector cells that are elicited attack the traumatized eye, and also infiltrate and attack the healthy eye. Thus, although the sequestered antigens do not induce a response by themselves, if a response is induced elsewhere they can serve as targets for attack.

most or all tissues after activation, but accumulations of cells are seen only when antigen is recognized in the site, triggering the production of cytokines that alter tissue barriers.

13-30 B cells with receptors specific for peripheral autoantigens are held in check by a variety of mechanisms.

During B-cell development in the bone marrow, B-cell antigen receptors specific for self molecules are produced as a consequence of the random generation of the repertoire. If a self molecule is expressed in the bone marrow in an appropriate form, clonal deletion and receptor editing can remove all of these self-reactive B cells while they are still immature (see Sections 7-9 and 7-10). There are, however, many self molecules available only in the periphery whose expression is restricted to particular organs. An example is thyroglobulin (the precursor of thyroxine), which is expressed only in the thyroid and at extremely low levels in plasma. Back-up mechanisms exist to ensure that B cells reactive to these self molecules do not cause autoimmune disease. When a mature B cell in the periphery encounters self molecules that bind its receptor, four proposed mechanisms could bring about the observed nonreactivity. Failure of any one of these mechanisms could lead to autoimmunity.

First, B cells that recognize a self antigen arrest their migration in the T-cell zone of peripheral lymphoid tissues (Fig. 13.37), just like B cells that bind a

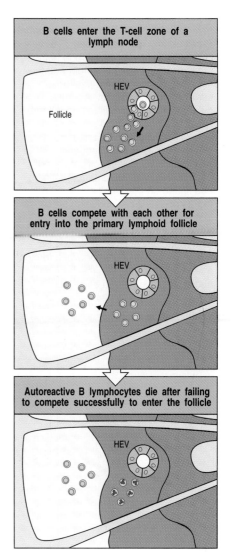

Fig. 13.37 Autoreactive B cells do not compete effectively in peripheral lymphoid tissue to enter primary lymphoid follicles. In the top panel, B cells are seen entering the T-cell zone of a lymph node through high endothelial venules (HEVs). Those with reactivity for foreign antigens are shown in yellow; autoreactive cells are gray. The autoreactive cells fail to compete with B cells specific for foreign antigens for exit from the T-cell zone and entry into primary follicles (center panel). The autoreactive B cells that fail to receive survival signals most likely undergo apoptosis *in situ* in the T-cell zone (bottom panel).

foreign antigen (see Chapter 9). However, in contrast to the response to foreign antigens, in which activated effector CD4 T cells are present, B cells binding self antigens will not be able to interact with helper CD4 T cells because normally no such cells exist for self antigens. This lack of interaction prevents the B cells from migrating out of the T-cell zones into the follicles; instead, the trapped self-reactive B cells undergo apoptosis.

A second mechanism is the induction of B-cell anergy, which is associated with downregulation of surface IgM expression and partial inhibition of the linked signaling pathways (Fig. 13.38). B-cell anergy can be induced by exposure to soluble circulating antigen (see Section 7-17); if mice are inoculated intravenously with protein solutions from which all protein aggregates have been rigorously removed to eliminate multivalent complexes, the peripheral B cells that bind these proteins can be inactivated. The lifespan of anergic B cells is short as they are usually eliminated after failing to enter the primary lymphoid follicles or after interacting with antigen-specific T cells, as described below. This form of B-cell tolerance can therefore be viewed as a form of delayed B-cell deletion, with the significant difference that there may be autoimmune diseases such as SLE in which anergic cells can be rescued.

The third mechanism depends on the presence of T cells that are specific for the self antigen and express Fas ligand. In the rare instances when such an autoreactive T cell matures and is activated, it is able to kill autoreactive anergic B cells in a Fas-dependent manner (see Fig. 13.38). In the absence of the normal pathways of co-stimulation, anergized B lymphocytes that have been chronically exposed to self antigen show enhanced sensitivity to apoptosis after ligation of Fas by Fas ligand. They are therefore not subject to the stimulatory signals that oppose apoptosis in B cells whose surface receptors have just been ligated by foreign antigen. The importance of this mechanism is nicely illustrated by the consequences of mutation in the genes for Fas or Fas ligand. Mice deficient in Fas or Fas ligand develop severe autoimmune disease, similar to SLE, associated with the production of a similar spectrum of autoantibodies.

Finally, there is evidence for a distinct mechanism for dealing with B cells that develop self-reactive specificities as a result of somatic hypermutation during a response to a foreign antigen (Fig. 13.39). At a crucial phase at the height of the germinal center reaction, an encounter with a large dose of soluble antigen causes a wave of apoptosis in germinal center B cells within a few hours. Thus B cells that become reactive for abundant soluble self antigens could be removed.

All these mechanisms reemphasize the fact that the mere existence in the body of some B lymphocytes with receptor specificities directed against self

Fig. 13.38 Peripheral B-cell anergy. An autoreactive B cell encounters its soluble autoantigen in the periphery (first panel), which leads to the development of B-cell anergy (second panel). This is characterized by a reduction in both the expression of surface IgM and the signaling pathways after ligation of surface immunoglobulin. A further mechanism for maintaining peripheral B-cell tolerance is shown in the two panels on the right. If an anergized, self-reactive B cell subsequently encounters a T cell specific for a self peptide from the relevant autoantigen, Fas ligand on the T-cell surface binds to Fas on the B cell, inducing apoptosis in the B cell. If an anergized B cell takes up a foreign antigen as a result of immunological cross-reactivity or formation of complexes between foreign and self molecules, this mechanism aborts activation of the self-reactive B cell by T cells that recognize the foreign moiety.

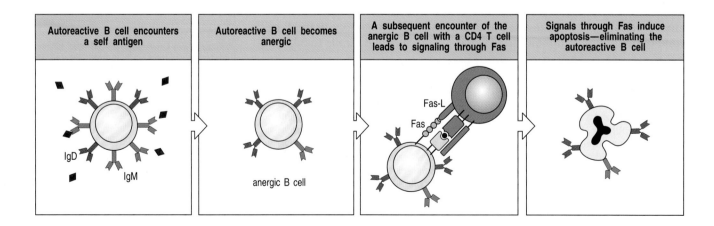

Fig. 13.39 Elimination of autoreactive B cells in germinal centers. During somatic hypermutation in germinal centers (depicted in the top panel), B cells with autoreactive B-cell receptors can arise. Ligation of these receptors by soluble autoantigen induces apoptosis of the autoreactive B cell by signaling through the B-cell antigen receptor in the absence of helper T cells.

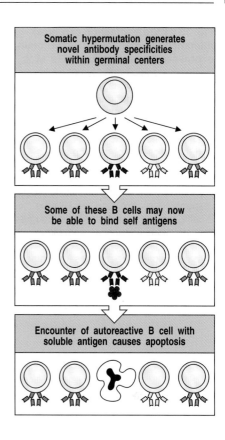

is not in itself harmful. Before an immune response can be initiated they need to receive effective help, the B-cell receptors must be ligated, and their intracellular signaling machinery must be set to respond normally.

13-31 Autoimmunity may sometimes be triggered by infection.

Human autoimmune diseases often appear gradually, making it difficult to find out how the process is initiated. Nevertheless, there is a strong suspicion that infections can trigger autoimmune disease in genetically susceptible individuals. Indeed, many experimental autoimmune diseases are induced by mixing tissue cells with adjuvants that contain bacteria. For example, to induce EAE, the spinal cord or myelin basic protein used for immunization must be emulsified in complete Freund's adjuvant, which includes killed *Mycobacterium tuberculosis* (see Appendix I, Section A-4). When the mycobacteria are omitted from the adjuvant, not only is no disease elicited, but the animals become refractory to any subsequent attempt to induce the disease by antigen in complete Freund's adjuvant, and this resistance can be transferred to syngeneic recipients by T cells (Fig. 13.40). Infection is important in the induction of disease in several other known cases. For example,

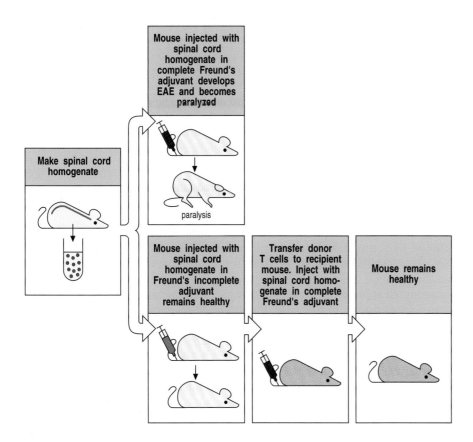

Fig. 13.40 Bacterial adjuvants are required for induction of experimental autoimmune disease. Mice immunized with spinal cord homogenate in complete Freund's adjuvant, which contains large numbers of killed *Mycobacterium tuberculosis*, get experimental allergic encephalomyelitis (EAE). Mice immunized with the same antigen in incomplete Freund's adjuvant, which lacks the *M. tuberculosis*, not only do not become diseased but are actually protected from subsequent disease induction. Moreover, T cells from these mice can transfer protection from disease to naive, syngeneic recipients.

Fig. 13.41 Virus infection can break tolerance to a transgenic viral protein expressed in pancreatic β cells. Mice made transgenic for the lymphocytic choriomeningitis virus (LCMV) nucleoprotein under the control of the rat insulin promoter express the nucleoprotein in their pancreatic β cells, but do not respond to this protein and therefore do not develop an autoimmune diabetes. However, if the transgenic mice are infected with LCMV, a potent antiviral cytotoxic T-cell response is elicited, and this kills the β cells, leading to diabetes. It is thought that infectious agents can sometimes elicit T-cell responses that cross-react with self peptides (a process known as molecular mimicry) and that this could cause autoimmune disease in a similar way.

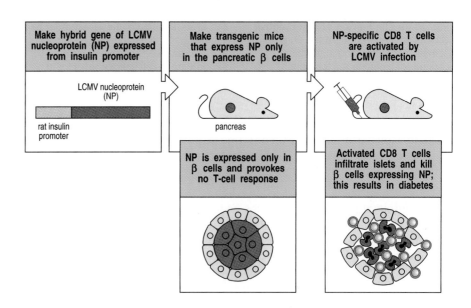

transgenic mice that express a T-cell receptor specific for myelin basic protein (see Section 13-25) often develop spontaneous autoimmune encephalo-myelitis if they become infected. One possible mechanism for this loss of tolerance is that the infectious agents induce co-stimulatory activity on antigen-presenting cells expressing low levels of peptides from myelin basic protein, thus activating the autoreactive T cells.

It has also been suggested that autoimmunity can be initiated by a mechanism known as **molecular mimicry**, in which antibodies or T cells generated in response to an infectious agent cross-react with self antigens. To show that infectious agents can trigger responses that can destroy tissues, mice were made transgenic for a viral nuclear protein driven by the insulin promoter, so that the protein was expressed only in pancreatic β cells. As the amount of viral protein expressed was low, the T cells that recognized this protein remained immunologically ignorant. That is, they were neither tolerant to the viral protein nor activated by it, and the animals showed no sign of disease. If they were infected with the live virus, however, they responded by making cytotoxic CD8 T cells specific for the viral protein, and these armed CD8 cytotoxic T cells could then destroy the β cells, causing diabetes (Fig. 13.41).

It has long been known that molecular mimicry can operate in antibody-mediated autoimmunity; microbial antigens can elicit antibody responses that react not only with the antigens on the pathogen but also with host anti-gens of similar structure. This type of response occurs after infection with some *Streptococcus* species. These elicit antibodies that cross-react with kidney, joint, and heart antigens to produce **rheumatic fever**. Such responses are usually transient and do not lead to sustained autoantibody production, as the helper T cells involved are specific for the microbe and not for self proteins. Host proteins that form a complex with bacteria can induce a similar transient response; in this case, the antibody response is not cross-reactive but the bacterium is acting as a carrier, allowing B cells that express an autoreactive receptor to receive inappropriate T-cell help. These and some other mechanisms that could allow an infectious agent to break tolerance are summarized in Fig. 13.42. All of these mechanisms can be shown to act in experimental systems, and there is some evidence for their importance in human autoimmune disease as well.

The argument that some autoimmune diseases might be initiated by infection is strengthened by the fact that there are several human autoimmune diseases in which a prior infection with a specific agent or class of agents leads to a particular disease (Fig. 13.43). Disease susceptibility in these cases is determined largely by MHC genotype.

In most autoimmune diseases, however, there is still no firm evidence that a particular infectious agent is associated with disease onset. Furthermore, a number of animal models of autoimmunity show that infection is not necessary for certain diseases to develop. Indeed, in some cases, infection may prevent or delay disease onset. Murine models of SLE and diabetes show that these diseases develop in genetically predisposed animals housed in germ-free conditions. The prevalence of diabetes in nonobese diabetic (NOD) mice is higher in mice raised in pathogen-free environments than in colonies housed in environments where infectious diseases are not rigorously excluded.

In humans, SLE is a very rare disease in African populations living in West Africa. By contrast, SLE has a prevalence of as high as one in 500 African-American women living in the West Coast of the United States. It has been proposed that the high prevalence of infection, particularly by parasites, in West Africa may protect against the development of SLE. In support of this idea, infection of mice genetically predisposed to the development of SLE with murine strains of malaria delays the onset of SLE disease. The mechanism of this protective effect is not known.

Mechanism	Disruption of cell or tissue barrier	Infection of antigen-presenting cell	Binding of pathogen to self protein	Molecular mimicry	Superantigen
Effect	Release of sequestered self antigen; activation of nontolerized cells	Induction of co-stimulatory activity on antigen-presenting cells	Pathogen acts as carrier to allow anti-self response	Production of cross-reactive antibodies or T cells	Polyclonal activation of autoreactive T cells
Example	Sympathetic ophthalmia	Effect of adjuvants in induction of EAE	? Interstitial nephritis	Rheumatic fever ? Diabetes ? Multiple sclerosis	? Rheumatoid arthritis

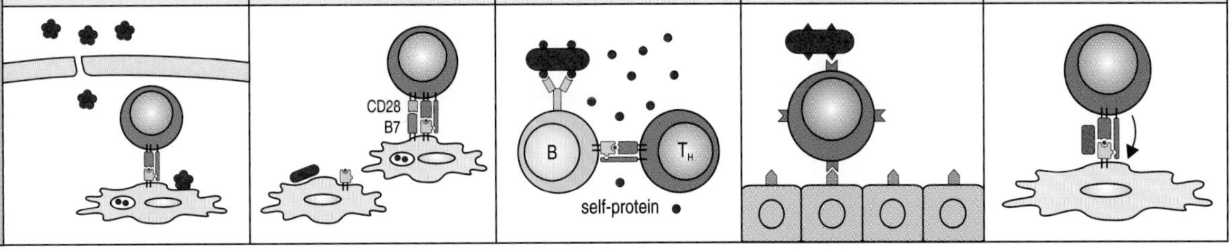

Fig. 13.42 There are several ways in which infectious agents could break self-tolerance. Because some antigens are sequestered from the circulation, either behind a tissue barrier or within the cell, an infection that breaks cell and tissue barriers might expose hidden antigens (first panel). A second possibility is that infectious agents might trigger expression of co-stimulators on antigen-presenting cells that have taken up tissue antigens, thereby inducing an autoimmune response (second panel). In some cases, infectious agents might bind to self proteins. Because the infectious agent induces a helper T-cell response, any B cell that recognizes the self protein will also receive help (third panel). Such responses should be self-limiting once the infectious agent is eliminated, because at this point the T-cell help will no longer be provided. Molecular mimicry might result in infectious agents inducing either T- or B-cell responses that can cross-react with self antigens (fourth panel). Polyclonal T-cell activation by a bacterial superantigen could overcome clonal anergy, allowing an autoimmune process to begin (last panel).

Fig. 13.43 Association of infection with autoimmune diseases. Several autoimmune diseases occur after specific infections and are presumably triggered by the infection. The case of post-streptococcal disease is best known but is now rare because effective antibiotic therapy of group A streptococcal infection usually prevents postinfection complications. Most of these postinfection autoimmune diseases also show susceptibility linkage to the MHC.

Associations of infection with immune-mediated tissue damage		
Infection	HLA association	Consequence
Group A *Streptococcus*	?	Rheumatic fever (carditis, polyarthritis)
Chlamydia trachomatis	HLA-B27	Reiter's syndrome (arthritis)
Shigella flexneri, Salmonella typhimurium, Salmonella enteritidis, Yersinia enterocolitica, Campylobacter jejuni	HLA-B27	Reactive arthritis
Borrelia burgdorferi	HLA-DR2, DR4	Chronic arthritis in Lyme disease

Summary.

Tolerance to self is a normal state that is maintained chiefly by clonal deletion of developing T and B cells and clonal deletion or inactivation of mature peripheral T and B cells. In addition, some antigens are ignored by the immune system, either because they are present at too low a level, or because they are present in immunologically privileged sites. When the state of self-tolerance is disrupted, autoimmunity can result. The process of clonal deletion influences the kinds of autoimmune disease that can occur. One group of autoantigens are those that do not trigger clonal deletion in the thymus, either because they are not abundant enough or because they are tissue-specific and not expressed in the thymus. Autoimmunity to these antigens, such as insulin, causes organ-specific autoimmune diseases such as type I diabetes mellitus. In the second category of autoimmune diseases, the systemic autoimmune diseases such as SLE, tolerance is broken to ubiquitous self antigens. Predisposition to these diseases may be due to inherited abnormalities in the regulation of immune responses and in the waste disposal mechanisms for removing dying cells at sites of inflammation. A third mechanism of self-tolerance—dominant suppression—has been noted in several experimental systems of autoimmunity and graft rejection; if this mechanism could be understood, it might be possible to use it to prevent both graft rejection and autoimmunity, which are closely related problems.

Summary to Chapter 13.

The response to noninfectious antigens causes three types of medical problem: allergy (the subject of Chapter 12), and autoimmunity and graft rejection, the subjects of this chapter. These responses have many features in common because all use the normal mechanisms of the adaptive immune response to produce symptoms and pathology. What is unique to these syndromes is their initiation and the nature of the antigens recognized, not the underlying nature of the response itself. For each of these undesirable categories of response, the question is how to control them without adversely affecting protective immunity to infection. The answer might lie in a more complete understanding of the regulation of the immune response, especially the suppressive mechanisms that seem to be important in tolerance. The deliberate control of the immune response is examined further in the next chapter.

General references.

Charlton, B., Auchincloss, H., Jr., and Fathman, C.G.: **Mechanisms of transplantation tolerance.** *Annu. Rev. Immunol.* 1994, **12**:707-734.

Goodnow, C.C.: **Balancing immunity and tolerance: deleting and tuning lymphocyte repertoires.** *Proc. Natl. Acad. Sci. USA* 1996, **93**:2264-2271.

Goodnow, C.C.: **Glimpses into the balance between immunity and self-tolerance.** *Ciba Found. Symp.* 1997, **204**:190-202.

Green, E.A., and Flavell, R.A.: **The initiation of autoimmune diabetes.** *Curr. Opin. Immunol.* 1999, **11**:663-669.

Mellor, A.L., and Munn, D.H.: **Immunology at the maternal-fetal interface: lessons for T cell tolerance and suppression.** *Annu. Rev. Immunol.* 2000, **18**:367-391.

Moller, G. (ed): **Chronic autoimmune diseases.** *Immunol. Rev.* 1995, **144**: 1-314.

O'Garra, A., Steinman, L., and Gijbels, K.: **CD4+ T-cell subsets in autoimmunity.** *Curr. Opin. Immunol.* 1997, **9**:872-883.

Steinman, L.: **A few autoreactive cells in an autoimmune infiltrate control a vast population of nonspecific cells: a tale of smart bombs and the infantry.** *Proc. Natl. Acad. Sci. USA* 1996, **93**:2253-2256.

Tan, E.M.: **Autoantibodies in pathology and cell biology.** *Cell* 1991, **67**:841-842.

Wong, F.S., and Janeway, C.A., Jr.: **Insulin-dependent diabetes mellitus and its animal models.** *Curr. Opin. Immunol.* 1999, **11**:643-647.

Section references.

13-1   Specific adaptive immune responses to self antigens can cause autoimmune disease.

Naparstek, Y., and Plotz, P.H.: **The role of autoantibodies in autoimmune disease.** *Annu. Rev. Immunol.* 1993, **11**:79-104.

Steinman, L.: **Multiple sclerosis: a coordinated immunological attack against myelin in the central nervous system.** *Cell* 1996, **85**:299-302.

13-2   Autoimmune diseases can be classified into clusters that are typically either organ-specific or systemic.

Aichele, P., Bachmann, M.F., Hengartner, H., and Zinkernagel, R.M.: **Immunopathology or organ-specific autoimmunity as a consequence of virus infection.** *Immunol. Rev.* 1996, **152**:21-45.

Bach, J.F.: **Organ-specific autoimmunity.** *Immunol. Today* 1995, **16**:353-355.

King, C., and Sarvetnick, N.: **Organ-specific autoimmunity.** *Curr. Opin. Immunol.* 1997, **9**:863-871.

13-3   Susceptibility to autoimmune disease is controlled by environmental and genetic factors, especially MHC genes.

Bennett, S.T., and Todd, J.A.: **Human type 1 diabetes and the insulin gene: principles of mapping polygenes.** *Annu. Rev. Genet.* 1996, **30**:343-370.

Gonzalez, A., Katz, J.D., Mattei, M.G., Kikutani, H., Benoist, C., and Mathis D.: **Genetic control of diabetes progression.** *Immunity* 1997, **7**:873-883.

Haines, J.L., Ter Minassian, M., Bazyk, A., Gusella, J.F., Kim, D.J., Terwedow, H., et al.: **A complete genomic screen for multiple sclerosis underscores a role for the major histocompatibility complex. The Multiple Sclerosis Genetics Group.** *Nat. Genet.* 1996, **13**:469-471.

McDevitt, H.O.: **Discovering the role of the major histocompatibility complex in the immune response.** *Annu. Rev. Immunol.* 2000, **18**:1-17.

Schmidt, D., Verdaguer, J., Averill, N., and Santamaria, P.: **A mechanism for** the major histocompatibility complex-linked resistance to autoimmunity. *J. Exp. Med.* 1997, **186**:1059-1075.

Todd, J.A., and Steinman, L.: **The environment strikes back.** *Curr. Opin. Immunol.* 1993, **5**:863-865.

Vyse, T.J., and Todd, J.A.: **Genetic analysis of autoimmune disease.** *Cell* 1996, **85**:311-318.

Waksman, B.H.: **Multiple sclerosis. More genes versus environment.** *Nature* 1995, **377**:105-106.

Wicker, L.S.: **Major histocompatibility complex-linked control of autoimmunity.** *J. Exp. Med.* 1997, **186**:973-975.

13-4   The genes that have been associated with the development of systemic lupus erythematosus provide important clues to the etiology of the disease.

Bickerstaff, M.C., Botto, M., Hutchinson, W.L., Herbert, J., Tennent, G.A., Bybee, A., Mitchell, D.A., Cook, H.T., Butler, P.J., Walport, M.J., and Pepys, M.B.: **Serum amyloid P component controls chromatin degradation and prevents antinuclear autoimmunity.** *Nat. Med.* 1999, **5**:694-697.

Boes, M., Schmidt, T., Linkemann, K., Beaudette, B.C., Marshak-Rothstein, A., and Chen, J.: **Accelerated development of IgG autoantibodies and autoimmune disease in the absence of secreted IgM.** *Proc. Natl. Acad. Sci. USA* 2000, **97**:1184-1189.

Cornall, R.J., Cyster, J.G., Hibbs, M.L., Dunn, A.R., Otipoby, K.L., Clark, E.A., and Goodnow, C.C.: **Polygenic autoimmune traits: Lyn, CD22, and SHP-1 are limiting elements of a biochemical pathway regulating BCR signaling and selection.** *Immunity* 1998, **8**:497-508.

Ehrenstein, M.R., Cook, H.T., and Neuberger, M.S.: **Deficiency in serum immunoglobulin (Ig)M predisposes to development of IgG autoantibodies.** *J. Exp. Med.* 2000, **191**:1253-1258.

Napirei, M., Karsunky, H., Zevnik, B., Stephan, H., Mannherz, H.G., and Moroy, T.: **Features of systemic lupus erythematosus in Dnase1-deficient mice.** *Nat. Genet.* 2000, **25**:177-181.

Wakeland, E.K., Wandstrat, A.E., Liu, K., and Morel, L.: **Genetic dissection of systemic lupus erythematosus.** *Curr. Opin. Immunol.* 1999, **11**:701-707.

Walport, M.J.: **Lupus, DNase and defective disposal of cellular debris.** *Nat. Genet.* 2000, **25**:135-136.

Whitacre, C.C., Reingold, S.C., and O'Looney, P.A.: **A gender gap in autoimmunity.** *Science* 1999, **283**:1277-1278.

13-5   Antibody and T cells can cause tissue damage in autoimmune disease.

Couser, W.G.: **Pathogenesis of glomerulonephritis.** *Kidney Int. Suppl.* 1993, **42**:S19-S26.

Jennette, J.C., and Falk, R.J.: **The pathology of vasculitis involving the kidney.** *Am. J. Kidney Dis.* 1994, **24**:130-141.

McFarland, H.F.: **Significance of autoreactive T cells in diseases such as multiple sclerosis using an innovative primate model.** *J. Clin. Invest.* 1994, **94**:921-922.

Rapoport, B.: **Pathophysiology of Hashimoto's thyroiditis and hypothyroidism.** *Annu. Rev. Med.* 1991, **42**:91-96.

13-6   Autoantibodies against blood cells promote their destruction.

Beardsley, D.S., and Ertem, M.: **Platelet autoantibodies in immune thrombocytopenic purpura.** *Transfus. Sci.* 1998, **19**:237-244.

Clynes, R., and Ravetch, J.V.: **Cytotoxic antibodies trigger inflammation through Fc receptors.** *Immunity* 1995, **3**:21-26.

Domen, R.E.: **An overview of immune hemolytic anemias.** *Cleve. Clin. J. Med.* 1998, **65**:89-99.

Silberstein, L.E.: **Natural and pathologic human autoimmune responses to carbohydrate antigens on red blood cells.** *Springer Semin. Immunopathol.* 1993, **15**:139-153.

13-7   The fixation of sublytic doses of complement to cells in tissues stimulates a powerful inflammatory response.

Brandt, J., Pippin, J., Schulze, M., Hansch, G.M., Alpers, C.E., Johnson, R.J., Gordon, K., and Couser, W.G.: **Role of the complement membrane attack complex (C5b-9) in mediating experimental mesangioproliferative glomeru-lonephritis.** *Kidney Int.* 1996, **49**:335-343.

Hansch, G.M.: **The complement attack phase: control of lysis and non-lethal effects of C5b-9.** *Immunopharmacol.* 1992, **24**:107-117.

Shin, M.L., and Carney, D.F.: **Cytotoxic action and other metabolic conse-quences of terminal complement proteins.** *Prog. Allergy* 1988, **40**:44-81.

13-8   Autoantibodies against receptors cause disease by stimulating or blocking receptor function.

Bahn, R.S., and Heufelder, A.E.: **Pathogenesis of Graves' ophthalmopathy.** *N. Engl. J. Med.* 1993, **329**:1468-1475.

Feldmann, M., Dayan, C., Grubeck Loebenstein, B., Rapoport, B., and Londei, M.: **Mechanism of Graves thyroiditis: implications for concepts and therapy of autoimmunity.** *Int. Rev. Immunol.* 1992, **9**:91-106.

McIntosh, R., Watson, P., and Weetman, A.: **Somatic hypermutation in autoimmune thyroid disease.** *Immunol. Rev.* 1998, **162**:219-231.

Vincent, A., Lily, O., and Palace, J.: **Pathogenic autoantibodies to neuronal proteins in neurological disorders.** *J. Neuroimmunol.* 1999, **100**:169-180.

13-9   Autoantibodies against extracellular antigens cause inflammatory injury by mechanisms akin to type II and type III hypersensitivity reactions.

Bayer, A.S., and Theofilopoulos, A.N.: **Immunopathogenetic aspects of infec-tive endocarditis.** *Chest* 1990, **97**:204-212.

Clynes, R., Dumitru, C., and Ravetch, J.V.: **Uncoupling of immune complex formation and kidney damage in autoimmune glomerulonephritis.** *Science* 1998, **279**:1052-1054.

Kotzin, B.L.: **Systemic lupus erythematosus.** *Cell* 1996, **85**:303-306.

Lawley, T.J., Bielory, L., Gascon, P., Yancey, K.B., Young, N.S., and Frank, M.M.: **A prospective clinical and immunologic analysis of patients with serum sickness.** *N. Engl. J. Med.* 1984, **311**:1407-1413.

Mamula, M.J.: **Lupus autoimmunity: from peptides to particles.** *Immunol. Rev.* 1995, **144**:301-314.

Tan, E.M.: **Antinuclear antibodies: diagnostic markers for autoimmune diseases and probes for cell biology.** *Adv. Immunol.* 1989, **44**:93-151.

13-10   Environmental cofactors can influence the expression of autoimmune disease.

Donaghy, M., and Rees, A.J.: **Cigarette smoking and lung haemorrhage in glomerulonephritis caused by autoantibodies to glomerular basement membrane.** *Lancet* 1983, **2**:1390-1393.

Kallenberg, C.G., Brouwer, E., Weening, J.J., and Tervaert, J.W.: **Antineutrophil cytoplasmic antibodies: current diagnostic and pathophysio-logical potential.** *Kidney Int.* 1994, **46**:1-15.

Pinching, A.J., Rees, A.J., Russell, B.A., Lockwood, C.M., Mitchison, R.S., and Peters, D.K.: **Relapses in Wegener's granulomatosis: the role of infection.** *BMJ* 1980, **281**:836-838.

Rees, A.J., Lockwood, C.M., and Peters, D.K.: **Enhanced allergic tissue injury in Goodpasture's syndrome by intercurrent bacterial infection.** *BMJ* 1977, **2**:723-726.

Turner, A.N., and Rees, A.J.: **Goodpasture's disease and Alport's syndromes.** *Annu. Rev. Med.* 1996, **47**:377-386.

13-11   The pattern of inflammatory injury in autoimmunity can be modified by anatomical constraints.

Cavallo, T.: **Membranous nephropathy. Insights from Heymann nephritis.**
Am. J. Pathol. 1994, **144**:651-658.

Couser, W.G., and Abrass, C.K.: **Pathogenesis of membranous nephropathy.** *Annu. Rev. Med.* 1988, **39**:517-530.

Nagaraju, K., Raben, N., Loeffler, L., Parker, T., Rochon, P.J., Lee, E., Danning, C., Wada, R., Thompson, C., Bahtiyar, G., Craft, J., Hooft, V.H., and Plotz, P.: **Conditional up-regulation of MHC class I in skeletal muscle leads to self-sustaining autoimmune myositis and myositis-specific autoantibodies.** *Proc. Natl. Acad. Sci. USA* 2000, **97**:9209-9214.

Plotz, P.H., Rider, L.G., Targoff, I.N., Raben, N., O'Hanlon, T.P., and Miller, F.W.: **NIH conference. Myositis: immunologic contributions to understanding cause, pathogenesis, and therapy.** *Ann. Intl. Med.* 1995, **122**:715-724.

Tan, E.M.: **Do autoantibodies inhibit function of their cognate antigens *in vivo*?** *Arthritis Rheum.* 1989, **32**:924-925.

13-12   The mechanism of autoimmune tissue damage can often be determined by adoptive transfer.

Bottazzo, G.F., and Doniach, D.: **Autoimmune thyroid disease.** *Ann. Rev. Med.* 1986, **37**:353-359.

Vernet der Garabedian, B., Lacokova, M., Eymard, B., Morel, E., Faltin, M., Zajac, J., Sadovsky, O., Dommergues, M., Tripon, P., and Bach, J.F.: **Association of neonatal myasthenia gravis with antibodies against the fetal acetylcholine receptor.** *J. Clin. Invest.* 1994, **94**:555-559.

Vincent, A., Newsom Davis, J., Wray, D., Shillito, P., Harrison, J., Betty, M., Beeson, D., Mills, K., Palace, J., Molenaar, P., et al.: **Clinical and experimental observations in patients with congenital myasthenic syndromes.** *Ann. N.Y. Acad. Sci.* 1993, **681**:451-460.

Willcox, N.: **Myasthenia gravis.** *Curr. Opin. Immunol.* 1993, **5**:910-917.

13-13   T cells specific for self antigens can cause direct tissue injury and have a role in sustained autoantibody responses.

Haskins, K., and Wegmann, D.: **Diabetogenic T-cell clones.** *Diabetes* 1996, **45**:1299-1305.

Nepom, G.T.: **Glutamic acid decarboxylase and other autoantigens in IDDM.** *Curr. Opin. Immunol.* 1995, **7**:825-830.

Tisch, R., and McDevitt, H.: **Insulin-dependent diabetes mellitus.** *Cell* 1996, **85**:291-297.

Zamvil, S., Nelson, P., Trotter, J., Mitchell, D., Knobler, R., Fritz, R., and Steinman, L.: **T-cell clones specific for myelin basic protein induce chronic relapsing paralysis and demyelination.** *Nature* 1985, **317**:355-358.

Zekzer, D., Wong, F.S., Ayalon, O., Altieri, M., Shintani, S., Solimena, M., and Sherwin, R.S.: **GAD-reactive CD4+ Th1 cells induce diabetes in NOD/SCID mice.** *J. Clin. Invest.* 1998, **101**:68-73.

13-14   Autoantibodies can be used to identify the target of the autoimmune process.

James, J.A., Gross, T., Scofield, R.H., and Harley, J.B.: **Immunoglobulin epitope spreading and autoimmune disease after peptide immunization: Sm B/B'-derived PPPGMRPP and PPPGIRGP induce spliceosome autoimmunity.** *J. Exp. Med.* 1995, **181**:453-461.

Lu, L., Kaliyaperumal, A., Boumpas, D.T., and Datta, S.K.: **Major peptide autoepitopes for nucleosome-specific T cells of human lupus.** *J. Clin. Invest.* 1999, **104**:345-355.

McCluskey, J., Farris, A.D., Keech, C.L., Purcell, A.W., Rischmueller, M., Kinoshita, G., Reynolds, P., and Gordon, T.P.: **Determinant spreading: lessons from animal models and human disease.** *Immunol. Rev.* 1998, **164**:209-229.

Protti, M.P., Manfredi, A.A., Horton, R.M., Bellone, M., and Conti Tronconi, B.M.: **Myasthenia gravis: recognition of a human autoantigen at the molecular level.** *Immunol. Today* 1993, **14**:363-368.

Roth, R., Gee, R.J., and Mamula, M.J.: **B lymphocytes as autoantigen-presenting cells in the amplification of autoimmunity.** *Ann. N.Y. Acad. Sci.* 1997, **815**:88-104.

13-15 The target of T cell-mediated autoimmunity is difficult to identify owing to the nature of T-cell ligands.

Bell, R.B., and Steinman, L.: **Trimolecular interactions in experimental autoimmune demyelinating disease and prospects for immunotherapy.** *Semin. Immunol.* 1991, **3**:237-245.

Feldman, M., Brennan, F.M., and Maini, R.N.: **Rheumatoid arthritis.** *Cell* 1996, **85**:307-310.

Hafler, D.A., and Weiner, H.L.: **Immunologic mechanisms and therapy in multiple sclerosis.** *Immunol. Rev.* 1995, **144**:75-107.

Hafler, D.A., Saadeh, M.G., Kuchroo, V.K., Kilford, E., and Steinman, L.: **TCR usage in human and experimental demyelinating disease.** *Immunol. Today* 1996, **17**:152-159.

Steinman, L.: **Multiple sclerosis: a coordinated immunological attack against myelin in the central nervous system.** *Cell* 1996, **85**:299-302.

Tisch, R., and McDevitt, H.O.: **Antigen-specific immunotherapy: is it a real possibility to combat T-cell-mediated autoimmunity?** *Proc. Natl. Acad. Sci. USA* 1994, **91**:437-438.

Wong, F.S., Karttunen, J., Dumont, C., Wen, L., Visintin, I., Pilip, I.M., Shastri, N., Pamer, E.G., and Janeway, C.A., Jr.: **Identification of an MHC class I-restricted autoantigen in type 1 diabetes by screening an organ-specific cDNA library.** *Nat. Med.* 1999, **5**:1026-1031.

13-16 Graft rejection is an immunological response mediated primarily by T cells.

Lafferty, K.J.: **A contemporary view of transplantation tolerance: an immunologist's perspective.** *Clin. Transplant.* 1994, **8**:181-187.

Lechler, R., Gallagher, R.B., and Auchincloss, H.: **Hard graft? Future challenges in transplantation.** *Immunol. Today* 1991, **12**:214-216.

Lee, R.S., and Auchincloss, H., Jr.: **Mechanisms of tolerance to allografts.** *Chem. Immunol.* 1994, **58**:236-258.

Rosenberg, A.S., and Singer, A.: **Cellular basis of skin allograft rejection: an** *in vivo* **model of immune-mediated tissue destruction.** *Annu. Rev. Immunol.* 1992, **10**:333-358.

Shi, C., Lee, W.S., He, Q., Zhang, D., Fletcher, D.L., Jr., Newell, J.B., and Haber, E.: **Immunologic basis of transplant-associated arteriosclerosis.** *Proc. Natl. Acad. Sci. USA* 1996, **93**:4051-4056.

13-17 Matching donor and recipient at the MHC improves the outcome of transplantation.

Benichou, G., Takizawa, P.A., Olson, C.A., McMillan, M., and Sercarz, E.E.: **Donor major histocompatibility complex (MHC) peptides are presented by recipient MHC molecules during graft rejection.** *J. Exp. Med.* 1992, **175**:305-308.

Mickelson, E.M., Petersdorf, E., Anasetti, C., Martin, P., Woolfrey, A., and Hansen, J.A.: **HLA matching in hematopoietic cell transplantation.** *Hum. Immunol.* 2000, **61**:92-100.

Opelz, G.: **Factors influencing long-term graft loss. The Collaborative Transplant Study.** *Transplant. Proc.* 2000, **32**:647-649.

Opelz, G., and Wujciak, T.: **The influence of HLA compatibility on graft survival after heart transplantation. The Collaborative Transplant Study.** *N. Engl. J. Med.* 1994, **330**:816-819.

Opelz, G., Wujciak, T., and Dohler, B.: **Is HLA matching worth the effort? Collaborative Transplant Study.** *Transplant. Proc.* 1999, **31**:717-720.

Petersdorf, E.W., Gooley, T.A., Anasetti, C., Martin, P.J., Smith, A.G., Mickelson, E.M., Woolfrey, A.E., and Hansen, J.A.: **Optimizing outcome after unrelated marrow transplantation by comprehensive matching of HLA class I and II alleles in the donor and recipient.** *Blood* 1998, **92**:3515-3520.

13-18 In MHC-identical grafts, rejection is caused by peptides from other alloantigens bound to graft MHC molecules.

Goulmy, E.: **Human minor histocompatibility antigens: new concepts for marrow transplantation and adoptive immunotherapy.** *Immunol. Rev.* 1997, **157**:125-140.

den Haan, J.M., Meadows, L.M., Wang, W., Pool, J., Blokland, E., Bishop, T.L., Reinhardus, C., Shabanowitz, J., Offringa, R., Hunt, D.F., Engelhard, V.H., and Goulmy, E.: **The minor histocompatibility antigen HA-1: a diallelic gene with a single amino acid polymorphism.** *Science* 1998, **279**:1054-1057.

Mutis, T., Gillespie, G., Schrama, E., Falkenburg, J.H., Moss, P., and Goulmy, E.: **Tetrameric HLA class I-minor histocompatibility antigen peptide complexes demonstrate minor histocompatibility antigen-specific cytotoxic T lymphocytes in patients with graft-versus-host disease.** *Nat. Med.* 1999, **5**:839-842.

Scott, D.M., Ehrmann, I.E., Ellis, P.S., Chandler, P.R., and Simpson, E.: **Why do some females reject males? The molecular basis for male-specific graft rejection.** *J. Mol. Med.* 1997, **75**:103-114.

Warrens, A.N., Lombardi, G., and Lechler, R.I.: **Presentation and recognition of major and minor histocompatibility antigens.** *Transpl. Immunol.* 1994, **2**:103-107.

13-19 There are two ways of presenting alloantigens on the transplant to the recipient's T lymphocytes.

Auchincloss, H., Jr., and Sultan, H.: **Antigen processing and presentation in transplantation.** *Curr. Opin. Immunol.* 1996, **8**:681-687.

Gould, D.S., and Auchincloss, H., Jr.: **Direct and indirect recognition: the role of MHC antigens in graft rejection.** *Immunol. Today* 1999, **20**:77-82.

13-20 Antibodies reacting with endothelium cause hyperacute graft rejection.

Auchincloss, H., Jr., and Sachs, D.H.: **Xenogeneic transplantation.** *Annu. Rev. Immunol.* 1998, **16**:433-470.

Dorling, A., Riesbeck, K., Warrens, A., and Lechler, R.: **Clinical xenotransplantation of solid organs.** *Lancet* 1997, **349**:867-871.

Joziasse, D.H., and Oriol, R.: **Xenotransplantation: the importance of the Galα1,3Gal epitope in hyperacute vascular rejection.** *Biochim. Biophys. Acta* 1999, **1455**:403-418.

Kissmeyer Nielsen, F., Olsen, S., Petersen, V.P., and Fjeldborg, O.: **Hyperacute rejection of kidney allografts, associated with pre-existing humoral antibodies against donor cells.** *Lancet* 1966, **2**:662-665.

Robson, S.C., Schulte am, E.J., and Bach, F.H.: **Factors in xenograft rejection.** *Ann. N.Y. Acad. Sci.* 1999, **875**:261-276.

Sharma, A., Okabe, J., Birch, P., McClellan, S.B., Martin, M.J., Platt, J.L., and Logan, J.S.: **Reduction in the level of Gal(α1,3)Gal in transgenic mice and pigs by the expression of an α(1,2)fucosyltransferase.** *Proc. Natl. Acad. Sci. USA* 1996, **93**:7190-7195.

Williams, G.M., Hume, D.M., Hudson, R.P., Jr., Morris, P.J., Kano, K., and Milgrom, F.: **"Hyperacute" renal-homograft rejection in man.** *N. Engl. J. Med.* 1968, **279**:611-618.

13-21 The converse of graft rejection is graft-versus-host disease.

Barrett, A.J., and van Rhee, F.: **Graft-versus-leukaemia.** *Baillieres Clin. Haematol.* 1997, **10**:337-355.

Dazzi, F., and Goldman, J.: **Donor lymphocyte infusions.** *Curr. Opin. Hematol.* 1999, **6**:394-399.

Flowers, M.E., Kansu, E., and Sullivan, K.M.: **Pathophysiology and treatment of graft-versus-host disease.** *Hematol. Oncol. Clin. North Am.* 1999, **13**:1091-1112.

Goulmy, E., Schipper, R., Pool, J., Blokland, E., Flakenburg, J.H., Vossen, J., Grathwohl, A., Vogelsang, G.B., van Houwelingen, H.C., and van Rood, J.J.: **Mismatches of minor histocompatibility antigens between HLA-identical donors and recipients and the development of graft-versus-host disease after bone marrow transplantation.** *N. Engl. J. Med.* 1996, **334**:281-285.

Murphy, W.J., and Blazar, B.R.: **New strategies for preventing graft-versus-host disease.** *Curr. Opin. Immunol.* 1999, **11**:509-515.

Porter, D.L., and Antin, J.H.: **The graft-versus-leukemia effects of allogeneic cell therapy.** *Annu. Rev. Med.* 1999, **50**:369-386.

13-22 Chronic organ rejection is caused by inflammatory vascular injury to the graft.

Hancock, W.W., Buelow, R., Sayegh, M.H., and Turka, L.A.: **Antibody-induced transplant arteriosclerosis is prevented by graft expression of anti-oxidant and anti-apoptotic genes.** *Nat. Med.* 1998, **4**:1392-1396.

Lautenschlager, I., Soots, A., Krogeus, L., Kauppinen, H., Saarinen, O., Bruggeman, C., and Ahonen, J.: **CMV increases inflammation and accelerates chronic rejection in rat kidney allografts.** *Transplant. Proc.* 1997, **29**:802-803.

Orosz, C.G., and Peletier, R.P.: **Chronic remodeling pathology in grafts.** *Curr. Opin. Immunol.* 1997, **9**:676-680.

Paul, L.C.: **Current knowledge of the pathogenesis of chronic allograft dysfunction.** *Transplant. Proc.* 1999, **31**:1793-1795.

Takada, M., Nadeau, K.C., Shaw, G.D., Marquette, K.A., and Tilney, N.L.: **The cytosine-adhesion molecule cascade in ischemia/reperfusion injury of the rat kidney. Inhibition by a soluble P-selectin ligand.** *J. Clin. Invest.* 1997, **99**:2682-2690.

Womer, K.L., Vella, J.P., and Sayegh, M.H.: **Chronic allograft dysfunction: mechanisms and new approaches to therapy.** *Semin. Nephrol.* 2000, **20**:126-147.

13-23 A variety of organs are transplanted routinely in clinical medicine.

Murray, J.E.: **Human organ transplantation: background and consequences.** *Science* 1992, **256**:1411-1416.

13-24 The fetus is an allograft that is tolerated repeatedly.

Carosella, E.D., Rouas-Freiss, N., Paul, P., and Dausset, J.: **HLA-G: a tolerance molecule from the major histocompatibility complex.** *Immunol. Today* 1999, **20**:60-62.

Flanagan, J.R., Murata, M., Burke, P.A., Shirayoshi, Y., Appella, E., Sharp, P.A., and Ozato, K.: **Negative regulation of the major histocompatibility complex class I promoter in embryonal carcinoma cells.** *Proc. Natl. Acad. Sci. USA* 1991, **88**:3145-3149.

Mellor, A.L., and Munn, D.H.: **Immunology at the maternal-fetal interface: lessons for T cell tolerance and suppression.** *Annu. Rev. Immunol.* 2000, **18**:367-391.

Munn, D.H., Zhou, M., Attwood, J.T., Bondarev, I., Conway, S.J., Marshall, B., Brown, C., and Mellor, A.L.: **Prevention of allogeneic fetal rejection by tryptophan catabolism.** *Science* 1998, **281**:1191-1193.

Parham, P.: **Immunology: keeping mother at bay.** *Curr. Biol.* 1996, **6**:638-641.

Pazmany, L., Mandelboim, O., Vales Gomez, M., Davis, D.M., Reyburn, H.T., and Strominger, J.L.: **Protection from natural killer cell-mediated lysis by HLA-G expression on target cells.** *Science* 1996, **274**:792-795.

Schust, D.J., Tortorella, D., and Ploegh, H.L.: **HLA-G and HLA-C at the feto-maternal interface: lessons learned from pathogenic viruses.** *Semin. Cancer Biol.* 1999, **9**:37-46.

13-25 Many autoantigens are not so abundantly expressed that they induce clonal deletion or anergy but are not so rare as to escape recognition entirely.

Billingham, R.E., Brent, L., and Medawar, P.B.: **Actively acquired tolerance of foreign cells.** *Nature* 1953, **172**:603-606.

Goverman, J., Woods, A., Larson, L., Weiner, L.P., Hood, L., and Zaller, D.M.: **Transgenic mice that express a myelin basic protein-specific T cell receptor develop spontaneous autoimmunity.** *Cell* 1993, **72**:551-560.

Heath, W.R., Kurts, C., Miller, J.F., and Carbone, F.R.: **Cross-tolerance: a pathway for inducing tolerance to peripheral tissue antigens.** *J. Exp. Med.* 1998, **187**:1549-1553.

Katz, J.D., Wang, B., Haskins, K., Benoist, C., and Mathis, D.: **Following a diabetogenic T cell from genesis through pathogenesis.** *Cell* 1993, **74**:1089-1100.

Kurts, C., Sutherland, R.M., Davey, G., Li, M., Lew, A.M., Blanas, E., Carbone, F.R., Miller, J.F., and Heath, W.R.: **CD8 T cell ignorance or tolerance to islet antigens depends on antigen dose.** *Proc. Natl. Acad. Sci. USA* 1999, **96**:12703-12707.

Margulies, D.H.: **Interactions of TCRs with MHC-peptide complexes: a quantitative basis for mechanistic models.** *Curr. Opin. Immunol.* 1997 **9**:390-395.

Miller, J.F., and Heath, W.R.: **Self-ignorance in the peripheral T-cell pool.** *Immunol. Rev.* 1993, **133**:131-150.

Wang, R., Nelson, A., Kimachi, K., Grey, H.M., and Farr, A.G.: **The role of peptides in thymic positive selection of class II major histocompatibility complex-restricted T cells.** *Proc. Natl. Acad. Sci. USA* 1998, **95**:3804-3809.

13-26 The induction of a tissue-specific response requires the presentation of antigen by antigen-presenting cells with co-stimulatory activity.

Anderson, D.E., Sharpe, A.H., and Hafler, D.A.: **The B7-CD28/CTLA-4 costimulatory pathways in autoimmune disease of the central nervous system.** *Curr. Opin. Immunol.* 1999, **11**:677-683.

Bluestone, J.A.: **Costimulation and its role in organ transplantation.** *Clin. Transplant.* 1996, **10**:104-109.

Guerder, S., Picarella, D.E., Linsley, P.S., and Flavell, R.A.: **Costimulator B7-1 confers antigen-presenting-cell function to parenchymal tissue and in conjunction with tumor necrosis factor alpha leads to autoimmunity in transgenic mice.** *Proc. Natl. Acad. Sci. USA* 1994, **91**:5138-5142.

Lafferty, K.J., Prowse, S.J., Simeonovic, C.J., and Warren, H.S.: **Immunobiology of tissue transplantation: a return to the passenger leukocyte concept.** *Annu. Rev. Immunol.* 1983, **1**:143-173.

13-27 In the absence of co-stimulation, tolerance is induced.

Chai, J.G., Vendetti, S., Amofah, E., Dyson, J., and Lechler, R.: **CD152 ligation by CD80 on T cells is required for the induction of unresponsiveness by costimulation-deficient antigen presentation.** *J. Immunol.* 2000, **165**:3037-3042.

Hammerling, G.J., Schonrich, G., Ferber, I., and Arnold, B.: **Peripheral tolerance as a multi-step mechanism.** *Immunol. Rev.* 1993, **133**:93-104.

Lenschow, D.J., and Bluestone, J.A.: **T cell co-stimulation and *in vivo* tolerance.** *Curr. Opin. Immunol.* 1993, **5**:747-752.

Marelli-Berg, F.M., and Lechler, R.I.: **Antigen presentation by parenchymal cells: a route to peripheral tolerance?** *Immunol. Rev.* 1999, **172**:297-314.

Miller, J.F., and Basten, A.: **Mechanisms of tolerance to self.** *Curr. Opin. Immunol.* 1996, **8**:815-821.

13-28 Dominant immune suppression can be demonstrated in models of tolerance and can affect the course of autoimmune disease.

Olsson, T.: **Critical influences of the cytokine orchestration on the outcome of myelin antigen-specific T-cell autoimmunity in experimental autoimmune encephalomyelitis and multiple sclerosis.** *Immunol. Rev.* 1995, **144**:245-268.

Qin, S., Cobbold, S.P., Pope, H., Elliott, J., Kioussis, D., Davies, J., and Waldmann, H.: **"Infectious" transplantation tolerance.** *Science* 1993, **259**:974-977.

Rocken, M., and Shevach E.M.: **Immune deviation—the third dimension of nondeletional T cell tolerance.** *Immunol. Rev.* 1996, **149**:175-194.

Sakaguchi, S.: **Regulatory T cells: key controllers of immunologic self-tolerance.** *Cell* 2000, **101**:455-458.

Shevach, E.M.: **Regulatory T cells in autoimmunity.** *Annu. Rev. Immunol.* 2000, **18**:423-449.

Tian, J., Atkinson, M.A., Clare Salzler, M., Herschenfeld, A., Forsthuber, T., Lehmann, P.V., and Laufman, D.L.: **Nasal administration of glutamate decarboxylase (GAD65) peptides induces TH2 responses and prevents murine insulin-dependent diabetes.** *J. Exp. Med.* 1996, **183**:1561-1567.

Weiner, H.L.: **Oral tolerance for the treatment of autoimmune diseases.** *Annu. Rev. Med.* 1997, **48**:341-351.

13-29 Antigens in immunologically privileged sites do not induce immune attack but can serve as targets.

Alison, J., Georgiou, H.M., Strasser, A., and Vaux, D.L.: **Transgenic expression of CD95 ligand on islet beta cells induces a granulocytic infiltration but does not confer immune privilege upon islet allografts.** *Proc. Natl. Acad. Sci. USA* 1997, **94**: 3943-3947.

Ferguson, T.A., and Griffith, T.S.: **A vision of cell death: insights into immune privilege.** *Immunol. Rev.* 1997, **156**:167-184.

Green, D.R., and Ware, C.F.: **Fas-ligand: privilege and peril.** *Proc. Natl. Acad. Sci. USA* 1997, **94**:5986-5990.

Streilein, J.W., Ksander, B.R., and Taylor, A.W.: **Immune deviation in relation to ocular immune privilege.** *J. Immunol.* 1997, **158**:3557-3560.

13-30 B cells with receptors specific for peripheral autoantigens are held in check by a variety of mechanisms.

Akkaraju, S., Canaan, K., and Goodnow, C.C.: **Self-reactive B cells are not eliminated or inactivated by autoantigen expressed on thyroid epithelial cells.** *J. Exp. Med.* 1997, **186**:2005-2012.

Chen, C., Prak, E.L., and Weigert, M.: **Editing disease-associated autoantibodies.** *Immunity* 1997, **6**:97-105.

Fulcher, D.A., and Basten, A.: **B-cell activation versus tolerance—the central role of immunoglobulin receptor engagement and T-cell help.** *Int. Rev. Immunol.* 1997, **15**:33-52.

Goodnow, C.C., Cyster, J.G., Hartley, S.B., Bell, S.E., Cooke, M.P., Healy, J.I., Akkaraju, S., Rathmell, J.C., Pogue, S.L., and Shokat, K.P.: **Self-tolerance checkpoints in B lymphocyte development.** *Adv. Immunol.* 1995, **59**:279-368.

Rathmell, J.C., Cooke, M.P., Ho, W.Y., Grein, J., Townsend, S.E., Davis, M.M., and Goodnow, C.C.: **CD95 (Fas)-dependent elimination of self-reactive B cells upon interaction with CD4⁺ T cells.** *Nature* 1995, **376**:181-184.

Shokat, K.M., and Goodnow, C.C.: **Antigen-induced B-cell death and elimination during germinal-centre immune responses.** *Nature* 1995, **375**:334-338.

13-31 Autoimmunity may sometimes be triggered by infection.

Aichele, P., Bachmann, M.F., Hengarter, H., and Zinkernagel, R.M.: **Immunopathology or organ-specific autoimmunity as a consequence of virus infection.** *Immunol. Rev.* 1996, **152**:21-45.

Barnaba, V., and Sinigaglia, F.: **Molecular mimicry and T cell-mediated autoimmune disease.** *J. Exp. Med.* 1997, **185**:1529-1531.

Horwitz, M.S., and Sarvetnick, N.: **Viruses, host responses, and autoimmunity.** *Immunol. Rev.* 1999, **169**:241-253.

Moens, U., Seternes, O.M., Hey, A.W., Silsand, Y., Traavik, T., Johansen, B., and Rekvig, O.P.: *In vivo* **expression of a single viral DNA-binding protein generates systemic lupus erythematosus-related autoimmunity to double-stranded DNA and histones.** *Proc. Natl. Acad. Sci. USA* 1995, **92**:12393-12397.

Rocken, M., Urban, J.F., and Shevach, E.M.: **Infection breaks T-cell tolerance.** *Nature* 1992, **359**:79-82.

Steinhoff, U., Burkhart, C., Arnheiter, H., Hengartner, H., and Zinkernagel, R.: **Virus or a hapten-carrier complex can activate autoreactive B cells by providing linked T help.** *Eur. J. Immunol.* 1994, **24**:773-776.

Steinman, L., and Conlon, P.: **Viral damage and the breakdown of self-tolerance.** *Nat. Med.* 1997, **3**:1085-1087.

Steinman, L., and Oldstone, M.B.: **More mayhem from molecular mimics.** *Nat. Med.* 1997, **3**:1321-1322.

Von Herrath, M.G., and Oldstone, M.B.: **Virus-induced autoimmune disease.** *Curr. Opin. Immunol.* 1996, **8**:878-895.

Von Herrath, M.G., Evans, C.F., Horwitz, M.S., and Oldstone, M.B.: **Using transgenic mouse models to dissect the pathogenesis of virus-induced autoimmune disorders of the islets of Langerhans and the central nervous system.** *Immunol. Rev.* 1996, **152**:111-143.

Wucherpfennig, K.W., and Strominger, J.L.: **Molecular mimicry in T cell-mediated autoimmunity: viral peptides activate human T cell clones specific for myelin basic protein.** *Cell* 1995, **80**:695-705.

Manipulation of the Immune Response

Most of this book has been concerned with the mechanisms whereby the immune system successfully protects us from disease. In the preceding three chapters, however, we have seen examples of the failure of immunity to some important infections, and conversely, with allergy and autoimmunity, how inappropriate immune responses can themselves cause disease. We have also discussed the problems arising from immune responses to grafted tissues.

In this chapter we will consider the ways in which the immune system can be manipulated or controlled, both to suppress unwanted immune responses in autoimmunity, allergy, and graft rejection, and to stimulate protective immune responses to some of the diseases that, at present, largely elude the immune system. It has long been felt that it should be possible to deploy the powerful and specific mechanisms of adaptive immunity to destroy tumors, and we will discuss the present state of progress toward that goal. In the final section of the chapter we will discuss present vaccination strategies and how a more rational approach to the design and development of vaccines promises to increase their efficacy and widen their usefulness and application.

Extrinsic regulation of unwanted immune responses.

The unwanted immune responses that occur in autoimmune disease, transplant rejection, and allergy present slightly different problems, and the approach to developing effective treatment is correspondingly different for each. We have already discussed the treatment of allergy in Chapter 12: the problems in this case are due to the production of IgE, and the goals are, accordingly, to treat the adverse consequences of an IgE response, or to induce the production of IgG instead of IgE against the allergenic antigens. In autoimmune disease and graft rejection the problem is an immune response to tissue antigens, and the goal is to downregulate the response to avoid damage to the tissues or disruption of their function. From the point of view of management, the single most important difference between allograft rejection and autoimmunity is that allografts are a deliberate surgical intervention and the immune response to them can be foreseen, whereas autoimmune responses are not detected until they are already established. Effective treatment of an established immune response is much harder to achieve than prevention of a response before it has had a chance to develop, and autoimmune diseases are generally harder to control than a *de novo* immune response to an allograft. The relative difficulty of suppressing established immune responses is seen in animal models of autoimmunity, in which methods able to prevent the induction of autoimmune disease generally fail to halt established disease.

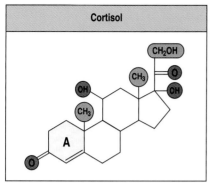

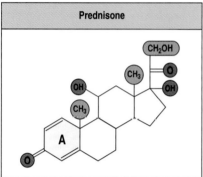

Fig. 14.1 The structure of the anti-inflammatory corticosteroid drug prednisone. Prednisone is a synthetic analogue of the natural adrenocorticosteroid cortisol. Introduction of the 1,2 double bond into the A ring increases anti-inflammatory potency approximately fourfold compared with cortisol, without modifying the sodium-retaining activity of the compound.

Current treatments for immunological disorders are nearly all empirical in origin, using immunosuppressive drugs identified by screening large numbers of natural and synthetic compounds. The drugs currently used to suppress the immune system can be divided into three categories: first, powerful anti-inflammatory drugs of the corticosteroid family such as prednisone; second, cytotoxic drugs such as azathioprine and cyclophosphamide; and third, fungal and bacterial derivatives, such as cyclosporin A, FK506 (tacrolimus), and rapamycin (sirolimus), which inhibit signaling events within T lymphocytes. These drugs are all very broad in their actions and inhibit protective functions of the immune system as well as harmful ones. Opportunistic infection is therefore a common complication of immunosuppressive drug therapy. The ideal immunosuppressive agent would be one that targets the specific part of the adaptive immune response that causes tissue injury. Paradoxically, antibodies themselves, by virtue of their exquisite specificity, might offer the best possibility for the therapeutic inhibition of specific immune responses. We will also consider experimental approaches to controlling specific immune responses by manipulating the local cytokine environment or by manipulating antigen so as to divert the response from a pathogenic pathway to an innocuous one. We have discussed in Chapters 12 and 13 how the pathological responses that cause allergy, autoimmunity, or graft rejection can be prevented by innocuous, nonpathological T-cell responses.

14-1 Corticosteroids are powerful anti-inflammatory drugs that alter the transcription of many genes.

Corticosteroid drugs are powerful anti-inflammatory agents that are used widely to suppress the harmful effects of immune responses of autoimmune or allergic origin, as well as those induced by graft rejection. **Corticosteroids** are pharmacological derivatives of members of the glucocorticoid family of steroid hormones; one of the most widely used is **prednisone**, which is a synthetic analogue of cortisol (Fig. 14.1). Cortisol acts through intracellular receptors that are expressed in almost every cell of the body. On binding hormone, these receptors regulate the transcription of specific genes, as illustrated in Fig. 14.2.

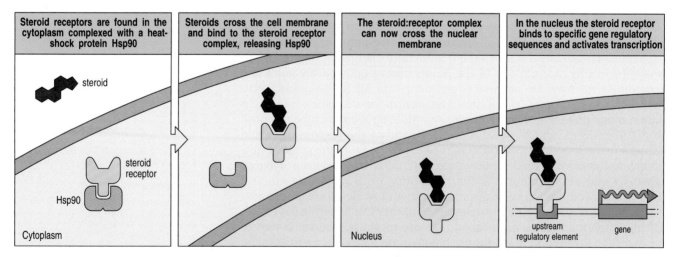

Fig. 14.2 Mechanism of steroid action. Corticosteroids are lipid-soluble molecules that enter cells by diffusing across the plasma membrane and bind to their receptors in the cytosol. Binding of corticosteroid to the receptor displaces a dimer of a heat-shock protein named Hsp90, exposing the DNA-binding region of the receptor. The steroid:receptor complex then enters the nucleus and binds to specific DNA sequences in the promoter regions of steroid-responsive genes. Corticosteroids exert their numerous effects by modulating the transcription of a wide variety of genes.

The expression of as many as 1% of genes in the genome may be regulated by glucocorticoids, which can either induce or, less commonly, suppress the transcription of responsive genes. The pharmacological effects of corticosteroid drugs result from exposure of the glucocorticoid receptors to supraphysiological concentrations of ligand. The abnormally high level of ligation of glucocorticoid receptors causes exaggerated glucocorticoid-mediated responses, which have both beneficial and toxic effects.

Given the large number of genes regulated by corticosteroids and that different genes are regulated in different tissues, it is hardly surprising that the effects of steroid therapy are very complex. The beneficial effects are anti-inflammatory and are summarized in Fig. 14.3; however, there are also many adverse effects, including fluid retention, weight gain, diabetes, bone mineral loss, and thinning of the skin. The use of corticosteroids to control disease requires a careful balance between helping the patient by reducing the inflammatory manifestations of disease and avoiding harm from the toxic side-effects of the drug. For this reason, corticosteroids used in transplant recipients and to treat inflammatory autoimmune and allergic disease are often administered in combination with other drugs in an effort to keep the dose and toxic effects to a minimum. In autoimmunity and allograft rejection, corticosteroids are commonly combined with cytotoxic immunosuppressive drugs.

14-2   Cytotoxic drugs cause immunosuppression by killing dividing cells and have serious side-effects.

The two cytotoxic drugs most commonly used as immunosuppressants are **azathioprine** and **cyclophosphamide** (Fig. 14.4). Both interfere with DNA synthesis and have their major pharmacological action on dividing tissues. They were developed originally to treat cancer and, after observations that they were cytotoxic to dividing lymphocytes, were found to be immunosuppressive as well. The use of these compounds is limited by a range of toxic effects on tissues that have in common the property of continuous cell division. These effects include decreased immune function, as well as anemia,

Corticosteroid therapy	
Effect on	**Physiological effects**
↓ IL-1, TNF-α, GM-CSF ↓ IL-3, IL-4, IL-5, IL-8	↓ Inflammation ↓ caused by cytokines
↓ NOS	↓ NO
↓ Phospholipase A$_2$ ↓ Cyclooxygenase type 2 ↑ Lipocortin-1	↓ Prostaglandins ↓ Leukotrienes
↓ Adhesion molecules	Reduced emigration of leukocytes from vessels
↑ Endonucleases	Induction of apoptosis in lymphocytes and eosinophils

Fig. 14.3 Anti-inflammatory effects of corticosteroid therapy. Corticosteroids regulate the expression of many genes, with a net anti-inflammatory effect. First, they reduce the production of inflammatory mediators, including cytokines, prostaglandins, and nitric oxide. Second, they inhibit inflammatory cell migration to sites of inflammation by inhibiting the expression of adhesion molecules. Third, corticosteroids promote the death by apoptosis of leukocytes and lymphocytes.

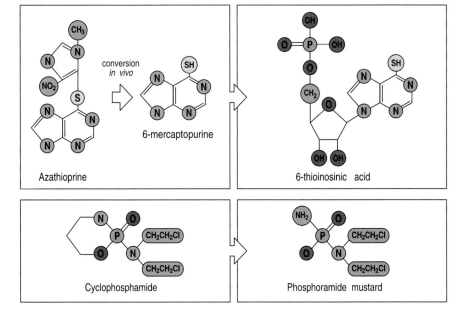

Fig. 14.4 The structure and metabolism of the cytotoxic immunosuppressive drugs azathioprine and cyclophosphamide. Azathioprine was developed as a modification of the anti-cancer drug 6-mercaptopurine; by blocking the reactive thiol group, the metabolism of this drug is slowed down. It is slowly converted *in vivo* to 6-mercaptopurine, which is then metabolized to 6-thio-inosinic acid, which blocks the pathway of purine bio-synthesis. Cyclophosphamide was similarly developed as a stable pro-drug, which is activated enzymatically in the body to phosphoramide mustard, a powerful and unstable DNA-alkylating agent.

leukopenia, thrombocytopenia, damage to intestinal epithelium, hair loss, and fetal death or injury. As a result of their toxicity, these drugs are used at high doses only when the aim is to eliminate all dividing lymphocytes, and in these cases treated patients require subsequent bone marrow transplantation to restore their hematopoietic function. They are used at lower doses, and in combination with other drugs such as corticosteroids, to treat unwanted immune responses.

Azathioprine is converted *in vivo* to a purine antagonist that interferes with the synthesis of nucleic acids and is toxic to dividing cells. It is metabolized to 6-thioinosinic acid, which competes with inosine monophosphate, thereby blocking the synthesis of adenosine monophosphate and guanosine monophosphate and thus inhibiting DNA synthesis. It is less toxic than cyclophosphamide, which is metabolized to phosphoramide mustard, which alkylates DNA. Cyclophosphamide is a member of the nitrogen mustard family of compounds, which were originally developed as chemical weapons. With this pedigree goes a range of highly toxic effects including inflammation of and hemorrhage from the bladder, known as hemorrhagic cystitis, and induction of bladder neoplasia.

14-3   Cyclosporin A, FK506 (tacrolimus), and rapamycin (sirolimus) are powerful immunosuppressive agents that interfere with T-cell signaling.

There are now relatively nontoxic alternatives to the cytotoxic class of drugs that can be used for immunosuppression in transplant patients. The systematic study of products from bacteria and fungi has led to the development of a large number of important medicines including the two immuno-suppressive drugs **cyclosporin A** and **FK506** or **tacrolimus**, which are now widely used to treat transplant recipients. Cyclosporin A is a cyclic decapeptide derived from a soil fungus from Norway, *Tolypocladium inflatum*. FK506, now known as tacrolimus, is a macrolide compound from the filamentous bacterium *Streptomyces tsukabaensis* found in Japan; macrolides are compounds that contain a many-membered lactone ring to which is attached one or more deoxy sugars. Another *Streptomyces* macrolide, called **rapamycin** or **sirolimus**, is being evaluated in clinical studies and is also likely to become important in the prevention of transplant rejection; rapamycin is derived from *Streptomyces hygroscopicus*, found on Easter Island ('Rapa ui' in Polynesian—hence the name of the drug). All three compounds exert their pharmacological effects by binding to members of a family of intracellular proteins known as the immunophilins, forming complexes that interfere with signaling pathways important for the clonal expansion of lymphocytes (see Chapter 6).

Cyclosporin A and tacrolimus block T-cell proliferation by inhibiting the phosphatase activity of a Ca^{2+}-activated enzyme called **calcineurin** at nanomolar concentrations. Their mechanism of action, which we will discuss further in the next section, revealed a role for calcineurin in transmitting signals from the T-cell receptor to the nucleus. Both drugs reduce the expression of several cytokine genes that are normally induced on T-cell activation (Fig. 14.5). These include interleukin (IL)-2, whose synthesis by T lymphocytes is an important growth signal for T cells. Cyclosporin A and tacrolimus inhibit T-cell proliferation in response to either specific antigens or allogeneic cells and are used extensively in medical practice to prevent the rejection of allogeneic organ grafts. Although the major immunosuppressive effects of both drugs are probably the result of inhibition of T-cell proliferation, they also act on other cells and have a large variety of other immunological effects (see Fig. 14.5), some of which might turn out to be important pharmacologically.

Immunological effects of cyclosporin A and tacrolimus	
Cell type	Effects
T lymphocyte	Reduced expression of IL-2, IL-3, IL-4, GM-CSF, TNF-α Reduced proliferation following decreased IL-2 production Reduced Ca²⁺-dependent exocytosis of granule-associated serine esterases Inhibition of antigen-driven apoptosis
B lymphocyte	Inhibition of proliferation secondary to reduced cytokine production by T lymphocytes Inhibition of proliferation following ligation of surface immunoglobulin Induction of apoptosis following B-cell activation
Granulocyte	Reduced Ca²⁺-dependent exocytosis of granule-associated serine esterases

Fig. 14.5 Cyclosporin A and tacrolimus inhibit lymphocyte and some granulocyte responses.

Cyclosporin A and tacrolimus are effective, but they are not problem-free. First, as with the cytotoxic agents, they affect all immune responses indiscriminately. The only way of controlling their immunosuppressive action is by varying the dose; at the time of grafting, high doses are required but, once a graft is established, the dose can be decreased to allow useful protective immune responses while maintaining adequate suppression of the residual response to the grafted tissue. This is a difficult balance that is not always achieved. Furthermore, although T cells are particularly sensitive to the actions of these drugs, their molecular targets are found in other cell types and therefore these drugs have effects on many other tissues. Cyclosporin A and tacrolimus are both toxic to kidneys and other organs. Finally, treatment with these drugs is expensive because they are complex natural products that must be taken for prolonged periods. Thus there is room for improvement in these compounds, and better and less expensive analogues are being sought. Nevertheless, at present, they are the drugs of choice in clinical transplantation, and they are also being tested in a variety of autoimmune diseases, especially those that, like graft rejection, are mediated by T cells.

14-4 Immunosuppressive drugs are valuable probes of intracellular signaling pathways in lymphocytes.

The mechanism of action of cyclosporin A and tacrolimus is now fairly well understood. Each binds to a different group of immunophilins: cyclosporin A to the cyclophilins, and tacrolimus to the FK-binding proteins (FKBP). These immunophilins are peptidyl-prolyl cis–trans isomerases but their isomerase activity does not seem to be relevant to the immunosuppressive activity of the drugs that bind them. Rather, the immunophilin:drug complexes bind and inhibit the Ca²⁺-activated serine/threonine phosphatase calcineurin. Calcineurin is activated in T cells when intracellular calcium ion levels rise after T-cell receptor binding; on activation it dephosphorylates the NFATc family of transcription factors in the cytoplasm, allowing them to migrate to the nucleus, where they form complexes with nuclear partners including the transcription factor AP-1, and induce transcription of genes including those for IL-2, CD40 ligand, and Fas ligand (Fig. 14.6, and see Sections 6-11 and 8-10). This pathway is inhibited by cyclosporin A and tacrolimus, which thus inhibit the clonal expansion of activated T cells. Calcineurin is found in other cells besides T cells but at higher levels; T cells are therefore particularly susceptible to the inhibitory effects of these drugs.

Fig. 14.6 Cyclosporin A and tacrolimus inhibit T-cell activation by interfering with the serine/threonine-specific phosphatase calcineurin. Signaling via T-cell receptor-associated tyrosine kinases leads to the activation and increased synthesis of the transcription factor AP-1 and other partner proteins, as well as increasing the concentration of Ca^{2+} in the cytoplasm (left panels). The Ca^{2+} binds to calcineurin and thereby activates it to dephosphorylate the cytoplasmic form of members of the family of nuclear factors of activated T cells (NFATc). Once dephosphorylated, the active NFATc family members migrate to the nucleus to form a complex with AP-1 and other partner proteins; the NFATc:AP-1 complexes can then induce the transcription of genes required for T-cell activation, including the IL-2 gene. When cyclosporin A (CsA) or tacrolimus are present, they form complexes with their immunophilin targets, cyclophilin (CyP) and FK-binding protein (FKBP), respectively (right panels). The complex of cyclophilin with cyclosporin A can bind to calcineurin and block its ability to activate NFATc family members. The complex of tacrolimus with FKBP binds to calcineurin at the same site, also blocking its activity.

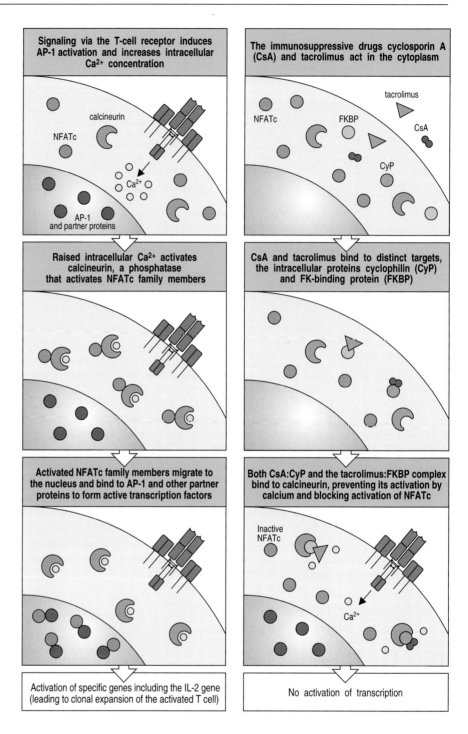

Signaling via the T-cell receptor induces AP-1 activation and increases intracellular Ca^{2+} concentration

The immunosuppressive drugs cyclosporin A (CsA) and tacrolimus act in the cytoplasm

Raised intracellular Ca^{2+} activates calcineurin, a phosphatase that activates NFATc family members

CsA and tacrolimus bind to distinct targets, the intracellular proteins cyclophilin (CyP) and FK-binding protein (FKBP)

Activated NFATc family members migrate to the nucleus and bind to AP-1 and other partner proteins to form active transcription factors

Both CsA:CyP and the tacrolimus:FKBP complex bind to calcineurin, preventing its activation by calcium and blocking activation of NFATc

Activation of specific genes including the IL-2 gene (leading to clonal expansion of the activated T cell)

No activation of transcription

Rapamycin has a different mode of action from either cyclosporin A or tacrolimus. Like tacrolimus, it binds to the FKBP family of immunophilins. However, the rapamycin:immunophilin complex has no effect on calcineurin activity but, instead, blocks the signal transduction pathway triggered by ligation of the IL-2 receptor. Rapamycin also inhibits lymphocyte proliferation driven by IL-4 and IL-6, implying a common postreceptor pathway of signaling by these cytokines. The rapamycin:immunophilin complex acts by binding to a protein kinase named mTOR (mammalian target of rapamycin; also known as FRAP, RAFT1, and RAPT1). This kinase phosphorylates two downstream intracellular targets. The first is another kinase, p70 S6 kinase, which in turn regulates the translation of many proteins. The second is PHAS-1, a repressor of protein translation, which is inhibited by phosphorylation mediated by mTOR. Both PHAS-1 and p70 S6 kinase appear to mediate the effects of rapamycin in lymphocytes. Because rapamycin has different pharmacological activities from cyclosporin A and tacrolimus, trials are being undertaken to see if combination therapy involving rapamycin given together with either cyclosporin A or tacrolimus might provide more effective and safer treatment than the use of just one of these drugs. The rationale for such studies is that it may be possible to use lower amounts of each drug when used in combination, compared with the amounts required for treatment with a single agent. This might be a means of reducing unwanted side-effects.

14-5   Antibodies against cell-surface molecules have been used to remove specific lymphocyte subsets or to inhibit cell function.

Cytotoxic drugs kill all proliferating cells and therefore indiscriminately affect all types of activated lymphocyte and any other cell that is dividing. Cyclosporin A, tacrolimus, and rapamycin are more selective, but still inhibit most adaptive immune responses. In contrast, antibodies can interfere with immune responses in a nontoxic and much more specific manner. The potential of antibodies for removal of unwanted lymphocytes is demonstrated by **anti-lymphocyte globulin**, a preparation of immunoglobulin from horses immunized with human lymphocytes, which has been used for many years to treat acute graft rejection episodes. Anti-lymphocyte globulin does not, however, discriminate between useful lymphocytes and those responsible for unwanted responses. Moreover, horse immunoglobulin is highly antigenic in humans and the large doses used in therapy are often followed by the development of serum sickness, caused by the formation of immune complexes of horse immunoglobulin and human anti-horse immunoglobulin antibodies (see Chapter 12). Nevertheless, anti-lymphocyte globulins are still in use to treat acute rejection and have stimulated the quest for monoclonal antibodies to achieve more specifically targeted effects.

Immunosuppressive monoclonal antibodies act by one of two general mechanisms. Some monoclonal antibodies trigger the destruction of lymphocytes *in vivo*, and are referred to as depleting antibodies, whereas others are nondepleting and act by blocking the function of their target protein without killing the cell that bears it. IgG monoclonal antibodies that cause lymphocyte depletion target these cells to macrophages and NK cells, which bear Fc receptors and which respectively kill the lymphocytes by phagocytosis and antibody-dependent cytotoxicity. Many antibodies are being tested for their ability to inhibit allograft rejection and to modify the expression of autoimmune disease. We will discuss some of these examples after looking at the measures being taken to prepare monoclonal antibodies for therapy in humans.

14-6   Antibodies can be engineered to reduce their immunogenicity in humans.

The major impediment to therapy with monoclonal antibodies in humans is that these antibodies are most readily made by using mouse cells, and humans rapidly develop antibody responses to mouse antibodies. This not only blocks the actions of the mouse antibodies but leads to allergic reactions, and if treatment is continued can result in anaphylaxis (see Section 12-10). Once this has happened, future treatment with any mouse monoclonal antibody is ruled out. This problem can, in principle, be avoided by making antibodies that are not recognized as foreign by the human immune system, and three strategies are being explored for their construction. One approach is to clone human V regions into a phage display library and select for binding to human cells, as described in Appendix I (see Section A-13). In this way, monoclonal antibodies that are entirely human in origin can be obtained. Second, mice lacking endogenous immunoglobulin genes can be made transgenic (see Appendix I, Section A-46) for human immunoglobulin heavy- and light-chain loci by using yeast artificial chromosomes. B cells in these mice have receptors encoded by human immunoglobulin genes but are not tolerant to most human proteins. In these mice, it is possible to induce human monoclonal antibodies against epitopes on human cells or proteins.

Finally, one can graft the complementarity-determining regions (CDRs) of a mouse monoclonal antibody, which form the antigen-binding loops, onto the framework of a human immunoglobulin molecule, a process known as **humanization**. Because antigen-binding specificity is determined by the structure of the CDRs (see Chapter 3), and because the overall frameworks of mouse and human antibodies are so similar, this approach produces a monoclonal antibody that is antigenically identical to human immunoglobulin but binds the same antigen as the mouse monoclonal antibody from which the CDR sequences were derived. These recombinant antibodies are far less immunogenic in humans than the parent mouse monoclonal antibodies, and thus they can be used for the treatment of humans with far less risk of anaphylaxis.

14-7   Monoclonal antibodies can be used to inhibit allograft rejection.

Antibodies specific for various physiological targets have been used in attempts to prevent the development of allograft rejection by inhibiting the development of harmful inflammatory and cytotoxic responses. One approach is to perfuse the organ before transplantation with antibodies that react with antigen-presenting cells and thus target them for destruction within the mononuclear phagocytic system. Depletion of antigen-presenting cells in the graft by this method is effective at preventing allograft rejection in animal models, although there is no convincing evidence that it is successful in humans. Antibodies have, however, been used to treat episodes of graft rejection in humans. Anti-CD3 antibodies are moderately effective as an adjunct to immunosuppressive drugs in the treatment of episodes of transplanted kidney rejection.

A further approach to inhibiting allograft rejection is the blockade of the co-stimulatory signals needed to activate T cells that recognize donor antigens. In animal studies of graft rejection, a fusion protein made from CTLA-4 and the Fc portion of human immunoglobulin, which binds to both B7.1 and B7.2 (see Section 8-5), has allowed the long-term survival of certain grafted tissues. Even more effective in a primate model of renal allograft rejection was the use of a humanized monoclonal antibody against the CD40 ligand (CD154), present on T cells (see Section 8-17). CD40 ligand binds to CD40, expressed on dendritic and endothelial cells, stimulating these cells to secrete cytokines such as IL-6, IL-8, and IL-12. The mechanism of the immuno-suppressive effect of anti-CD40 ligand antibody is not known, but it is most likely to be a consequence of blocking the activation of dendritic cells by T helper cells recognizing donor antigens.

Monoclonal antibodies against other targets have also had some success in preventing graft rejection in animals. Of particular interest are certain nondepleting anti-CD4 antibodies: when given for a short time during primary exposure to grafted tissue, these lead to a state of tolerance in the recipient (Fig. 14.7). This tolerant state is an example of the dominant immune suppression discussed in Section 13-27. It is long-lived and can be transferred

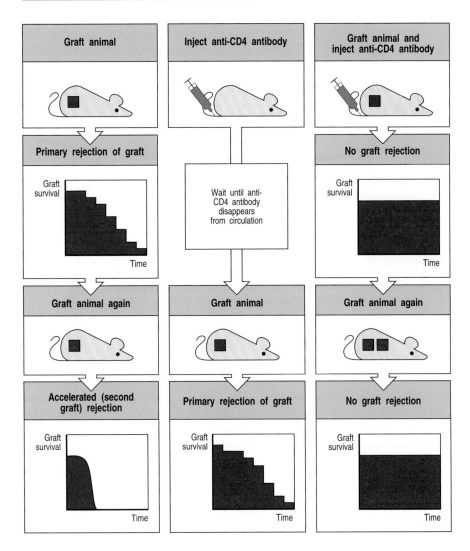

| Graft animal | Inject anti-CD4 antibody | Graft animal and inject anti-CD4 antibody |

Fig. 14.7 A tissue graft given together with anti-CD4 antibody can induce specific tolerance. Mice grafted with tissue from a genetically different mouse reject that graft. Having been primed to respond to the antigens in the graft, they then reject a subsequent graft of identical tissue more rapidly (left panels). Mice injected with anti-CD4 antibody alone can recover immune competence when the antibody disappears from the circulation, as shown by a normal primary rejection of graft tissue (center panels). However, when tissue is grafted and anti-CD4 antibody is administered at the same time, the primary rejection response is markedly inhibited (right panels). An identical graft made later in the absence of anti-CD4 antibody is not rejected, showing that the animal has become tolerant to the graft antigen. This tolerance can be transferred with T cells to naive recipients (not shown).

to naive recipients by CD4 T cells producing cytokines typical of T_H2 cells, although T cells producing other patterns of cytokines might also be involved (see Section 14-9). The presence of anti-CD4 antibody at the time of transplantation might favor the development of a nondamaging T_H2 response, rather than an inflammatory T_H1 response, because of a reduced strength of interaction between the graft cell antigens and responding naive T cells, as discussed in Section 10-7.

In human bone marrow transplantation, depleting antibodies directed at mature T lymphocytes have proved particularly useful. Elimination of mature T lymphocytes from donor bone marrow before infusion into a recipient is very effective at reducing the incidence of graft-versus-host disease (see Section 13-21). In this disease, the T lymphocytes in the donor bone marrow recognize the recipient as foreign and mount a damaging alloreaction against the recipient, causing rashes, diarrhea, and pneumonia, which is often fatal.

14-8   Antibodies can be used to alleviate and suppress autoimmune disease.

Autoimmune disease is detected only once the autoimmune response has caused tissue damage or has disturbed specific physiological functions. There are three main approaches to treatment. First, anti-inflammatory therapy

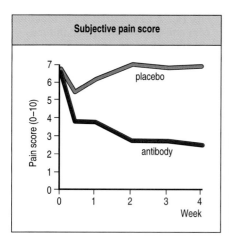

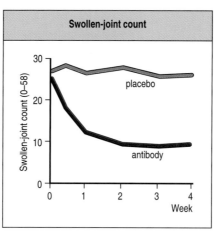

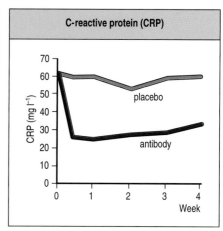

Fig. 14.8 Anti-inflammatory effects of anti-TNF-α therapy in rheumatoid arthritis. The clinical course of 24 patients was followed for 4 weeks after treatment with either a placebo or a monoclonal antibody against TNF-α at a dose of 10 mg kg⁻¹. The antibody therapy was associated with a reduction in both subjective and objective parameters of disease activity (as measured by pain score and swollen-joint count, respectively) and in the systemic inflammatory acute-phase response, measured as a fall in the concentration of the acute-phase reactant C-reactive protein. Data courtesy of R.N. Maini.

can reduce tissue injury caused by an inflammatory autoimmune response; second, immunosuppressive therapy can be aimed at reducing the auto-immune response; and third, treatment can be directed specifically at compensating for the result of the damage. For example, diabetes, which is induced by autoimmune attack on pancreatic β cells, is treated by insulin replacement therapy. Anti-inflammatory therapy for autoimmune disease includes the use of anti-cytokine antibodies; anti-TNF-α antibodies induce striking temporary remissions in rheumatoid arthritis (Fig. 14.8). Antibodies can also be used to block cell migration to sites of inflammation; for example, anti-CD18 antibodies prevent leukocytes adhering tightly to vascular endothelium and reduce inflammation in animal models of disease.

The ultimate goal of immunotherapy for autoimmune disease is specific intervention to restore tolerance to the relevant autoantigens. Two experimental approaches are under investigation. The first aims at blocking the specific response to autoantigen. One way to attempt this is to identify the clonally restricted T-cell receptors or immunoglobulin carried by the lymphocytes that cause disease, and to target these with antibodies directed against idiotypic determinants on the relevant antigen receptor. Another way is to identify particular MHC class I or class II molecules responsible for presenting peptides from autoantigens and to inhibit their antigen-presenting function selectively with antibodies or blocking peptides. This approach has been successful in some animal models of autoimmunity, for example experimental autoimmune encephalomyelitis (EAE) (Fig. 14.9), in which it seems that a limited number of clones of T cells, responding to a single peptide, might cause disease. However, autoimmune disease in humans and most animal models is driven by a polyclonal response to autoantigens by T and B lymphocytes. For this reason, immunotherapy based on the identification of specific receptors carried by pathogenic lymphocytes is unlikely to succeed. Immunotherapy based on the identification of the particular MHC molecules that drive an autoimmune response is more likely to be effective, but such therapy would also inhibit some protective immune responses.

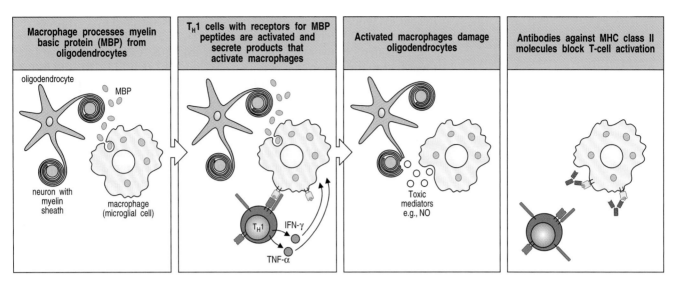

| Macrophage processes myelin basic protein (MBP) from oligodendrocytes | T_H1 cells with receptors for MBP peptides are activated and secrete products that activate macrophages | Activated macrophages damage oligodendrocytes | Antibodies against MHC class II molecules block T-cell activation |

Fig. 14.9 Anti-MHC class II antibody can inhibit the development of experimental autoimmune encephalomyelitis. In mice with experimental autoimmune encephalomyelitis (EAE), macrophages process myelin basic protein (MBP) and present MBP peptides to T_H1 lymphocytes in conjunction with co-stimulatory signals. Activated T_H1 cells secrete cytokines, which activate macrophages. The activated macrophages can, in turn, injure the oligodendrocytes. Antibodies against MHC class II molecules block this process by blocking the interaction between T_H1 cells and antigen-presenting macrophages.

14-9 Modulation of the pattern of cytokine expression by T lymphocytes can inhibit autoimmune disease.

The second approach to immunotherapy for autoimmune disease is to try to turn a pathological autoimmune response into an innocuous one. This approach is being pursued experimentally because, as we learned in Chapter 13, tolerance to tissue antigens does not always depend on the absence of a T-cell response; instead, it can be actively maintained by T cells secreting cytokines that suppress the development of a harmful, inflammatory T-cell response. As the pattern of cytokines expressed by T lymphocytes is critical in determining the perpetuation and expression of autoimmune disease, the manipulation of cytokine expression offers a way of controlling it. There are various techniques, collectively known as **immune modulation**, that can affect cytokine expression by T lymphocytes. These involve manipulating the cytokine environment in which T-cell activation takes place, or manipulating the way antigen is presented, as these factors have been observed to influence the differentiation and cytokine-secreting function of the activated T cells (see Sections 8-13, 10-5, and 10-6).

As discussed in earlier chapters, CD4 T lymphocytes can be subdivided into two major subsets, the T_H1 cells, which secrete interferon (IFN)-γ, and the T_H2 cells, which secrete IL-4, IL-5, IL-10, and transforming growth factor (TGF)-β. In many cases, autoimmune disease is associated with the activation of T_H1 cells, which activate macrophages and drive an inflammatory immune response. In animal models of experimentally induced autoimmune disease, such as EAE, the relative activation of the T_H1 and T_H2 subsets of T lymphocytes can be manipulated to give either a T_H1 response and disease, or a T_H2 response that confers protection against disease. The preferential activation of T_H1 or T_H2 cells can be achieved by direct manipulation of the cytokine environment or by administering antigen by particular routes, for example by feeding (see Section 14-10).

Recent evidence shows that patterns of cytokines secreted by T lymphocytes are very complicated and that the T_H1 and T_H2 subdivision of T lymphocytes is a considerable oversimplification. For example, CD4 T cells have been identified that develop in a cytokine environment rich in IL-10, and in turn secrete high levels of IL-10 and little IL-2 and IL-4. This pattern of cytokine secretion has bystander effects on other T cells and suppresses antigen-induced activation of other CD4 T lymphocytes. These cells have been provisionally designated T_r1 cells (T regulatory cells 1).

Another subset of T cells with immunosuppressive bystander effects secretes TGF-β as the dominant cytokine and has been designated T_H3. Such cells might be predominantly of mucosal origin and activated by the mucosal presentation of antigen (see Section 14-10).

A further subset of T cells also seems to be implicated in immunoregulation. These are the NK1.1$^+$ CD4 T cells, so named because they bear the receptor NK1.1, which is usually found on NK cells. NK1.1$^+$ T cells, which we discussed in Section 10-5, recognize antigens, including lipid antigens, presented by the class I-like molecule CD1 (see Section 5-18) and respond by secreting IL-4. Thus, when stimulated, the NK1.1$^+$ T cells can act to promote T_H2 responses. Although there is no direct evidence that NK1.1$^+$ T cells are involved in immunomodulation in humans, in mice that suffer spontaneous autoimmune disease this population of cells is either missing or decreased. Furthermore, transfer of NK1.1$^+$ T cells into such mice prevents the onset of the autoimmune disease.

Immune modulation aims to alter the balance between different subsets of responding T cells such that helpful responses are promoted and damaging responses are suppressed. As a therapy for autoimmunity it has the advantage that one might not need to know the precise nature of the autoantigen stimulating the autoimmune disease. This is because the administration of cytokines or antigen to modulate the immune response causes changes in the pattern of cytokine expression that have bystander effects on lymphocytes with the presumed autoreactive receptors. However, the drawback of this approach is the unpredictability of the results. In murine models of diseases such as diabetes and EAE, most of the results suggest that a T_H2 response can protect against T_H1-mediated disease, but there is evidence that T_H2 lymphocytes can also contribute to the pathology of these diseases.

An additional problem is the difficulty of modulating established immune responses. Experiments in animals have shown that anti-cytokine antibodies (or recombinant cytokines) present at the time of immunization with an autoantigen can sometimes divert a pathogenic immune response. In contrast, the modification of an ongoing immune response is much harder to achieve with this approach, although there have been some examples of experimental success, as we will see later.

14-10 Controlled administration of antigen can be used to manipulate the nature of an antigen-specific response.

When the target antigen of an unwanted response is identified, it is possible to manipulate the response by using antigen directly rather than by using antibodies or relying on the bystander effects discussed in the previous section. This is because the way in which antigen is presented to the immune system affects the nature of the response, and the induction of one type of response to an antigen can inhibit a pathogenic response to the same antigen.

As mentioned in Chapter 12, this principle has been applied with some success to the treatment of allergies caused by an IgE response to very low doses of antigen. Repeated treatment of allergic individuals with higher doses of allergen seems to divert the allergic response to one dominated by T cells that favor the production of IgG and IgA antibodies. These antibodies are thought to desensitize the patient by binding the small amounts of allergen normally encountered and preventing it from binding to IgE.

With T cell-mediated autoimmune disease, there has been considerable interest in using peptide antigens to suppress pathogenic responses. The type of CD4 T-cell response induced by a peptide depends on the way in which it is presented to the immune system. For instance, peptides given orally tend to prime T_H2 T cells that make IL-4 or T cells that make predominantly TGF-β without activating T_H1 cells or inducing a great deal of systemic antibody. These mucosal immune responses have relatively little pathogenic potential. Indeed, experiments in animal models indicate that they can protect against induced autoimmune disease. Experimental autoimmune encephalomyelitis is induced by injection of myelin basic protein in complete Freund's adjuvant and resembles multiple sclerosis, whereas collagen arthritis is similarly induced by injection of collagen type II and has features in common with rheumatoid arthritis. Oral administration of myelin basic protein or type II collagen inhibits the development of disease in animals (see Fig. 13.34), and has some beneficial effects in reducing the activity of preestablished disease. Trials using this approach in humans with multiple sclerosis or rheumatoid arthritis have found marginal therapeutic effects. Intravenous delivery of peptides can also inhibit inflammatory responses stimulated by the same peptide presented in a different context. When a soluble peptide is given intravenously, it binds preferentially to MHC class II molecules on resting B cells and tends to induce anergy in T_H1 cells. Thus, a careful choice of the dose or structure of antigen, or its route of administration, can allow us to control the type of response that results. Whether such approaches can be effective in manipulating the established immune responses driving human autoimmune diseases remains to be seen.

Summary.

Existing treatments for unwanted immune responses, such as allergic reactions, autoimmunity, and graft rejection, depend largely on three types of drug. Anti-inflammatory drugs, of which the most potent are the corticosteroids, are used for all three types of response. These have a broad spectrum of actions, however, and a correspondingly wide range of toxic side-effects; their dose must be controlled carefully. They are therefore normally used in combination with either cytotoxic or immunosuppressive drugs. The cytotoxic drugs kill all dividing cells and thereby prevent lymphocyte proliferation, but they suppress all immune responses indiscriminately and also kill other types of dividing cells. The immunosuppressive drugs act by intervening in the intracellular signaling pathways of T cells. They are less generally toxic than the cytotoxic drugs, but they also suppress all immune responses indiscriminately. They are also much more expensive than cytotoxic drugs.

Immunosuppressive drugs are now the drugs of choice in the treatment of transplant patients, where they can be used to suppress the immune response to the graft before it has become established. Autoimmune responses are already well established at the time of diagnosis and, in consequence, are much more difficult to suppress. They are therefore less responsive to the immunosuppressive drugs and, for that reason, they are usually controlled with a combination of corticosteroids and cytotoxic drugs. In animal experiments, attempts have been made to target immunosuppression more specifically, by blocking the response to autoantigen with the use of antibodies or antigenic peptides, or by diverting the immune response into a non-pathogenic pathway by manipulating the cytokine environment, or by administering antigen through the oral route where a nonpathogenic immune response is likely to be invoked. None of these treatments is yet proven in humans, and most require that the relevant antigen be known. For that reason, and because they are relatively ineffective against established immune responses, the promise of these approaches in animal models might be difficult to realize in a clinical context.

Using the immune response to attack tumors.

Cancer is one of the three leading causes of death in industrialized nations. As treatments for infectious diseases and the prevention of cardiovascular disease continue to improve, and the average life expectancy increases, cancer is likely to become the most common fatal disease in these countries. Cancers are caused by the progressive growth of the progeny of a single transformed cell. Therefore, curing cancer requires that all the malignant cells be removed or destroyed without killing the patient. An attractive way to achieve this would be to induce an immune response against the tumor that would discriminate between the cells of the tumor and their normal cell counterparts. Immunological approaches to the treatment of cancer have been attempted for over a century, with tantalizing but unsustainable results. Experiments in animals have, however, provided evidence for immune responses to tumors and have shown that T cells are a critical mediator of tumor immunity. More recently, advances in our understanding of antigen presentation and the molecules involved in T-cell activation have provided new immunotherapeutic strategies based on a better molecular understanding of the immune response. These are showing some success in animal models and are now being tested in human patients.

14-11 The development of transplantable tumors in mice led to the discovery that mice could mount a protective immune response against tumors.

The finding that tumors could be induced in mice after treatment with chemical carcinogens or irradiation, coupled with the development of inbred strains of mice, made it possible to undertake the key experiments that led to the discovery of immune responses to tumors. These tumors could be transplanted between mice, and the experimental study of tumor rejection has generally been based on the use of such tumors. If these bear MHC molecules foreign to the mice into which they are transplanted, the tumor cells are readily recognized and destroyed by the immune system, a fact that was exploited to develop the first MHC-congenic strains of mice. Specific immunity to tumors must therefore be studied within inbred strains, so that host and tumor can be matched for their MHC type.

Transplantable tumors in mice exhibit a variable pattern of growth when injected into syngeneic recipients. Most tumors grow progressively and eventually kill the host. However, if mice are injected with irradiated tumor cells that cannot grow, they are frequently protected against subsequent injection with a normally lethal dose of viable cells of the same tumor. There seems to be a spectrum of immunogenicity among transplantable tumors: injections of irradiated tumor cells seem to induce varying degrees of protective immunity against a challenge injection of viable tumor cells at a distant site. These protective effects are not seen in T-cell deficient mice but can be conferred by adoptive transfer of T cells from immune mice, showing the need for T cells to mediate all these effects.

Fig. 14.10 Tumor rejection antigens are specific to individual tumors. Mice immunized with an irradiated tumor and challenged with viable cells of the same tumor can, in some cases, reject a lethal dose of that tumor (left panels). This is the result of an immune response to tumor rejection antigens. If the immunized mice are challenged with viable cells of a different tumor, there is no protection and the mice die (right panels).

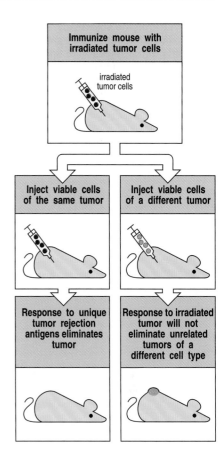

These observations indicate that the tumors express antigenic peptides that can become targets of a tumor-specific T-cell response. The antigens expressed by experimentally induced murine tumors, often termed **tumor-specific transplantation antigens (TSTAs)**, or **tumor rejection antigens (TRAs)**, are usually specific for an individual tumor. Thus immunization with irradiated tumor cells from tumor X protects a syngeneic mouse from challenge with live cells from tumor X but not from challenge with a different syngeneic tumor Y, and vice versa (Fig. 14.10).

14-12 T lymphocytes can recognize specific antigens on human tumors.

Tumor rejection antigens are peptides of tumor-cell proteins that are presented to T cells by MHC molecules. These peptides can become the targets of a tumor-specific T-cell response because they are not displayed on the surface of normal cells, at least not at levels sufficient to be recognized by T cells. Six different categories of tumor rejection antigens can be distinguished and examples of each of these are given in Fig. 14.11. The first category consists of antigens that are strictly tumor specific. These antigens are the result of point mutations or gene rearrangements, which often arise as part of the process of oncogenesis. Point mutations may evoke a T-cell response either by allowing *de novo* binding of a peptide to MHC class I molecules or by creating a new epitope for T cells by modification of a peptide that already binds class I molecules (Fig. 14.12). A special class of tumor-specific antigen in the case of B- and T-cell tumors, which are derived from single clones of lymphocytes, are the idiotypic sequences unique to the antigen receptor expressed by the clone.

The second category comprises proteins encoded by genes that are normally expressed only in male germ cells, which do not express MHC molecules and therefore cannot present peptides from these molecules to T lymphocytes. Tumor cells show widespread abnormalities of gene expression, including the activation of these genes and thus the presentation of these proteins to T cells; hence, these proteins are effectively tumor specific in their expression as antigens (Fig. 14.13).

The third category of tumor rejection antigen is comprised of differentiation antigens encoded by genes that are only expressed in particular types of tissue. The best examples of these are the differentiation antigens expressed in melanocytes and melanoma cells; a number of these antigens are proteins involved in the pathways of production of the black pigment, melanin. The fourth category is comprised of antigens that are strongly overexpressed in tumor cells compared with their normal counterparts (see Fig. 14.13). An example is HER-2/neu (also known as c-Erb-2), which is a receptor tyrosine kinase homologous to the epidermal growth factor receptor. This receptor is overexpressed in many adenocarcinomas, including breast and ovarian cancers, where it is linked with a poor prognosis. MHC class I-restricted, CD8-positive cytotoxic T lymphocytes have been found infiltrating solid tumors overexpressing HER-2/neu but are not capable of destroying such tumors *in vivo*. The fifth category of tumor rejection antigens is comprised of molecules that display abnormal posttranslational modifications. An example is underglycosylated mucin, MUC-1, which is expressed by a number of tumors, including breast and pancreatic cancers.

Fig. 14.11 Proteins selectively expressed in human tumors are candidate tumor rejection antigens. The molecules listed here have all been shown to be recognized by cytotoxic T lymphocytes raised from patients with the tumor type listed.

Potential tumor rejection antigens have a variety of origins			
Class of antigen	**Antigen**	**Nature of antigen**	**Tumor type**
Tumor-specific mutated oncogene or tumor-suppressor	Cyclin-dependent kinase 4	Cell-cycle regulator	Melanoma
	β-Catenin	Relay in signal transduction pathway	Melanoma
	Caspase-8	Regulator of apoptosis	Squamous cell carcinoma
Germ cell	MAGE-1 MAGE-3	Normal testicular proteins	Melanoma Breast Glioma
Differentiation	Tyrosinase	Enzyme in pathway of melanin synthesis	Melanoma
	Surface Ig	Specific antibody after gene rearrangements in B-cell clone	Lymphoma
Abnormal gene expression	HER-2/neu	Receptor tyrosine kinase	Breast Ovary
Abnormal post-translational modification	MUC-1	Underglycosylated mucin	Breast Pancreas
Oncoviral protein	HPV type 16, E6 and E7 proteins	Viral transforming gene products	Cervical carcinoma

Proteins encoded by viral oncogenes comprise the sixth category of tumor rejection antigen. These oncoviral proteins are viral proteins that may play a critical role in the oncogenic process and, because they are foreign, they can evoke a T-cell response. Examples of such proteins are the human papilloma type 16 virus proteins, E6 and E7, which are expressed in cervical carcinoma.

Although each of these categories of tumor rejection antigen may evoke an anti-tumor response *in vitro* and *in vivo*, it is exceptional for such a response to be able to spontaneously eliminate an established tumor. It is the goal of tumor immunotherapy to harness and augment such responses to treat cancer more effectively. In this respect, the spontaneous remission occasionally observed in cases of malignant melanoma and renal cell carcinoma, even when disease is quite advanced, offers hope that this goal is achievable.

In melanoma, tumor-specific antigens were discovered by culturing irradiated tumor cells with autologous lymphocytes, a reaction known as the mixed lymphocyte–tumor cell culture. From such cultures, cytotoxic T lymphocytes could be identified that would kill, in an MHC-restricted fashion, tumor cells bearing the relevant tumor-specific antigen. Melanomas have been studied in detail using this approach. Cytotoxic T cells reactive against melanoma peptides have been cloned and used to characterize melanomas by the array of tumor-specific antigens displayed. These studies have yielded three

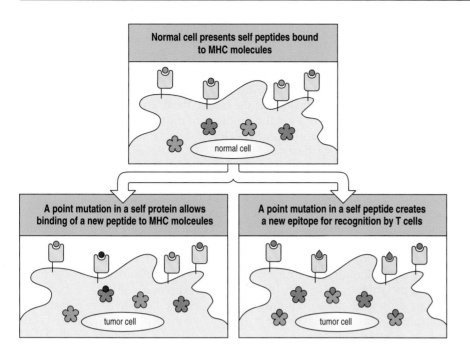

Fig. 14.12 Tumor rejection antigens may arise by point mutations in self proteins, which occur during the process of oncogenesis. In some cases a point mutation in a self protein may allow a new peptide to associate with MHC class I molecules (lower left panel). In other cases, a point mutation occurring within a self peptide that can bind self MHC proteins causes the expression of a new epitope for T-cell binding (lower right panel). In both cases, these mutated peptides will not have induced tolerance by the clonal deletion of developing T cells and can be recognized by mature T cells.

important findings. The first is that melanomas carry at least five different antigens that can be recognized by cytotoxic T lymphocytes. The second is that cytotoxic T lymphocytes reactive against melanoma antigens are not expanded *in vivo*, suggesting that these antigens are not immunogenic *in vivo*. The third is that the expression of these antigens can be selected against *in vitro* and possibly also *in vivo* by the presence of specific cytotoxic T cells. These discoveries offer hope for tumor immunotherapy, an indication that these antigens are not naturally strongly immunogenic, and also a caution about the possibility of selecting, *in vivo*, tumor cells that can escape recognition and killing by cytotoxic T cells.

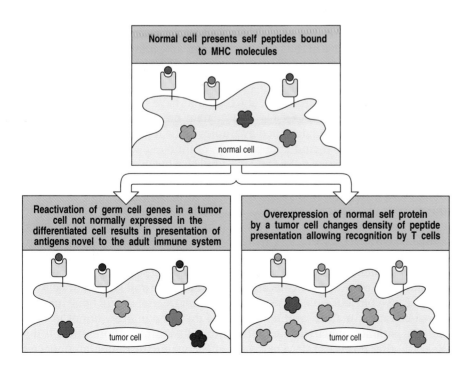

Fig. 14.13 Tumor rejection antigens are peptides of cell proteins presented by self MHC class I molecules. This figure shows two ways in which tumor rejection antigens may arise from unmutated proteins. In some cases, proteins that are normally expressed only in male germ cell tissues are reexpressed by the tumor cells (lower left panel). As these proteins are normally expressed only during germ cell development, and in cells lacking MHC antigens, T cells are not tolerant of these self antigens and can respond to them as though they were foreign proteins. In other tumors, over-expression of a self protein increases the density of presentation of a normal self peptide on tumor cells (lower right panel). Such peptides are then presented at high enough levels to be recognized by T cells. It is often the case that the same germ cell or self proteins are overexpressed in many tumors of a given tissue origin, giving rise to shared tumor rejection antigens.

Consistent with these findings, functional melanoma-specific T cells can be propagated from peripheral blood lymphocytes, from tumor-infiltrating lymphocytes, or by draining the lymph nodes of patients in whom the melanoma is growing. Interestingly, none of the peptides recognized by these T cells derives from the mutant proto-oncogenes or tumor suppressor genes that are likely to be responsible for the initial transformation of the cell into a cancer cell, although a few are the products of mutant genes. The rest derive from normal genes but are displayed at levels detectable by T cells for the first time on tumor cells, as illustrated in Fig. 14.13. Antigens of the MAGE family are not expressed in any normal adult tissues, with the exception of the testis, which is an immunologically privileged site. They probably represent early developmental antigens reexpressed in the process of tumorigenesis. Only a minority of melanoma patients have T cells reactive against the MAGE antigens, indicating that these antigens are either not expressed or are not immunogenic in most cases. The most common melanoma antigens are peptides from the enzyme tyrosinase or from three other proteins—gp100, MART1, and gp75. These are differentiation antigens specific to the melanocyte lineage from which melanomas arise. It is likely that overexpression of these antigens in tumor cells leads to an abnormally high density of specific peptide:MHC complexes and this makes them immunogenic. Although in most cases tumor rejection antigens are presented as peptides complexed with MHC class I molecules, tyrosinase has been shown to stimulate CD4 T-cell responses in some melanoma patients by being ingested and presented by cells expressing MHC class II.

Tumor rejection antigens shared between most examples of a tumor, and against which tolerance can be broken, represent candidate antigens for tumor vaccines. The MAGE antigens are candidates because of their limited tissue distribution and their shared expression by many melanomas. It might seem dangerous to use tumor vaccines based on antigens that are not truly tumor-specific because of the risk of inducing autoimmunity. Often, however, the tissues from which tumors arise are dispensable; the prostate is perhaps the best example of this. With melanoma, however, some melanocyte-specific tumor rejection antigens are also expressed in certain retinal cells, in the inner ear, in the brain, and in the skin. Despite this, melanoma patients receiving immunotherapy with whole tumor cells or tumor-cell extracts, although occasionally developing vitiligo—a destruction of pigmented cells in the skin that correlates well with a good response to the tumor—do not develop abnormalities in the visual, vestibular, and central nervous systems, perhaps because of the low level of expression of MHC class I molecules in these sites.

In addition to the human tumor antigens that have been shown to induce cytotoxic T-cell responses (see Fig. 14.11), many other candidate tumor rejection antigens have been identified by studies of the molecular basis of cancer development. These include the products of mutated cellular oncogenes or tumor suppressors, such as Ras and p53, and also fusion proteins, such as the Bcr–Abl tyrosine kinase that results from the chromosomal translocation (t9;22) found in chronic myeloid leukemia. It is intriguing that, in each of these cases, no specific cytotoxic T-cell response has been identified in cultures of autologous lymphocytes with tumor cells bearing these mutated antigens. However, cytotoxic T lymphocytes specific for these antigens can be developed in vitro by using peptide sequences derived either from the mutated sequence or from the fusion sequence of these common oncogenic proteins; these cytotoxic T cells are able to recognize and kill tumor cells. In chronic myeloid leukemia, it is known that, after treatment and bone marrow transplantation, mature lymphocytes from the bone marrow donor infused into the patient can help to eliminate any residual tumor. At present, it is not clear whether this is a graft-versus-host effect, where the donor lymphocytes are responding to **alloantigens** expressed on the leukemia cells, or whether there is a specific anti-leukemic response. The ability to prime the donor cells against leukemia-specific peptides offers the prospect of enhancing the anti-leukemic effect while minimizing the risk of graft-versus-host disease. It is a challenge for immunologists to understand why these mutated proteins do not prime cytotoxic T cells in the patients in which the tumors arise. They are excellent targets for therapy, as they are unique to the tumor and have a causal role in oncogenesis.

14-13 Tumors can escape rejection in many ways.

Burnet called the ability of the immune system to detect tumor cells and destroy them '**immune surveillance**.' However, it is difficult to show that tumors are subject to surveillance by the immune system; after all, cancer is a common disease, and most tumors show little evidence of immunological control. The incidence of the common tumors in mice that lack lymphocytes is little different from their incidence in mice with normal immune systems; the same is true for humans deficient in T cells. The major tumor types that occur with increased frequency in immunodeficient mice or humans are virus-associated tumors; immune surveillance thus seems to be critical for control of virus-associated tumors, but the immune system does not normally respond to the novel antigens deriving from the multiple genetic alterations in spontaneous tumors. The goal in the development of anti-cancer vaccines is to break the tolerance of the immune system for antigens expressed mainly or exclusively by the tumor.

It is not surprising that spontaneously arising tumors are rarely rejected by T cells, as in general they probably lack either distinctive antigenic peptides or the adhesion or co-stimulatory molecules needed to elicit a primary T-cell response. Moreover, there are other mechanisms whereby tumors can avoid immune attack or evade it when it occurs (Fig. 14.14). Tumors tend to be genetically unstable and can lose their antigens by mutation; in the event of an immune response, this instability might generate mutants that can escape the immune response. Some tumors, such as colon and cervical cancers, lose the expression of a particular MHC class I molecule (Fig. 14.15), perhaps through immunoselection by T cells specific for a peptide presented by that

Mechanisms by which tumors escape immune recognition		
Low immunogenicity	**Antigenic modulation**	**Tumor-induced immune suppression**
No peptide:MHC ligand No adhesion molecules No co-stimulatory molecules	Antibody against tumor cell-surface antigens can induce endocytosis and degradation of the antigen. Immune selection of antigen-loss variants	Factors (e.g., TGF-β) secreted by tumor cells inhibit T cells directly

Fig. 14.14 Tumors can escape immune surveillance in a variety of ways. First, tumors can have low immunogenicity (left panel). Some tumors do not have peptides of novel proteins that can be presented by MHC molecules, and therefore appear normal to the immune system. Others have lost one or more MHC molecules, and most do not express co-stimulatory proteins, which are required to activate naive T cells. Second, tumors can initially express antigens to which the immune system responds but lose them by antibody-induced internalization or antigenic variation. When tumors are attacked by cells responding to a particular antigen, any tumor that does not express that antigen will have a selective advantage (center panel). Third, tumors often produce substances, such as TGF-β, that suppress immune responses directly (right panel).

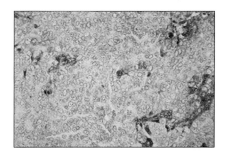

Fig. 14.15 Loss of MHC class I expression in a prostatic carcinoma. Some tumors can evade immune surveillance by loss of expression of MHC class I molecules, preventing their recognition by CD8 T cells. A section of a human prostate cancer that has been stained with a peroxidase-conjugated antibody to HLA class I is shown. The brown stain correlating with HLA class I expression is restricted to infiltrating lymphocytes and tissue stromal cells. The tumor cells that occupy most of the section show no staining. Photograph courtesy of G. Stamp.

MHC class I molecule. In experimental studies, when a tumor loses expression of all MHC class I molecules, it can no longer be recognized by cytotoxic T cells, although it might become susceptible to NK cells (Fig. 14.16). However, tumors that lose only one MHC class I molecule might be able to avoid recognition by specific CD8 cytotoxic T cells while remaining resistant to NK cells, conferring a selective advantage *in vivo*.

Fig. 14.16 Tumors that lose expression of all MHC class I molecules as a mechanism of escape from immune surveillance are more susceptible to NK cell killing. Regression of transplanted tumors is largely due to the actions of cytotoxic T cells (CTLs), which recognize novel peptides bound to MHC class I antigens on the surface of the cell (left panels). NK cells have inhibitory receptors that bind MHC class I molecules, so variants of the tumor that have low levels of MHC class I, although less sensitive to CD8 cytotoxic T cells, become susceptible to NK cells (center panels). Although *nude* mice lack T cells, they have higher than normal levels of NK cells, and so tumors that are sensitive to NK cells grow less well in *nude* mice than in normal mice. Transfection with MHC class I genes can restore both resistance to NK cells and susceptibility to CD8 cytotoxic T cells (right panels). However, tumors that lose only one MHC class I molecule can escape a specific cytotoxic CD8 T-cell response while remaining NK resistant. The bottom panels show scanning electron micrographs of NK cells attacking leukemia cells. Left panel: shortly after binding to the target cell, the NK cell has put out numerous microvillous extensions and established a broad zone of contact with the leukemia cell. The NK cell is the smaller cell on the left in both photographs. Right panel: 60 minutes after mixing, long microvillous processes can be seen extending from the NK cell (bottom left) to the leukemia cell and there is extensive damage to the leukemia cell membrane; the plasma membrane of the leukemia cell has rolled up and fragmented under the NK cell attack. Photographs courtesy of J.C. Hiserodt.

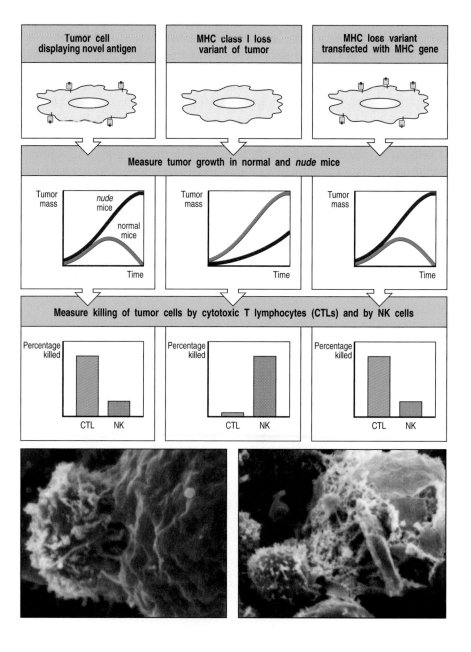

Yet another way in which tumors might evade rejection is by making immunosuppressive cytokines. Many tumors make these, although in most cases little is known of their precise nature. TGF-β was first identified in the culture supernatant of a tumor (hence its name, transforming growth factor-β) and, as we have seen, tends to suppress inflammatory T-cell responses and cell-mediated immunity, which are needed to control tumor growth. A number of tumors of different tissue origins, such as melanoma, ovarian carcinoma, and B-cell lymphoma, have been shown to produce the immunosuppressive cytokine IL-10, which can reduce dendritic cell development and activity. Thus, there are many different ways in which tumors avoid recognition and destruction by the immune system.

14-14 Monoclonal antibodies against tumor antigens, alone or linked to toxins, can control tumor growth.

The advent of monoclonal antibodies suggested the possibility of targeting and destroying tumors by making antibodies against tumor-specific antigens (Fig. 14.17). This depends on finding a tumor-specific antigen that is a cell-surface molecule. Some of the cell-surface molecules that have been targeted in experimental clinical trials are shown in Fig. 14.18. So far there has been limited success with this approach, although, as an adjunct to other therapies, it holds promise. Some striking initial results have been reported in the treatment of breast cancer with a humanized monoclonal antibody, known as Herceptin, which targets a growth factor receptor, HER-2/neu, that is over-expressed in about a quarter of breast cancer patients. As we discussed in Section 14-12, this overexpression accounts for HER-2/neu evoking an antitumor T-cell response, although HER-2/neu is also associated with a poorer prognosis. It is thought that Herceptin acts by blocking interaction between the receptor and its natural ligand and by downregulating the level of expression of the receptor. The effects of this antibody can be potentiated when it is combined with conventional chemotherapy. A second monoclonal antibody that has promise for the treatment of non-Hodgkin's B-cell lymphoma binds to CD20 and is known as Rituximab. Ligation and clustering of CD20

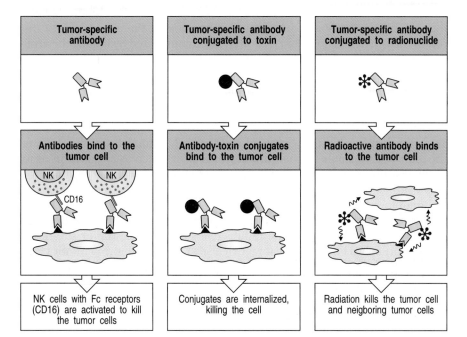

Fig. 14.17 Monoclonal antibodies that recognize tumor-specific antigens might be used in a variety of ways to help eliminate tumors. Tumor-specific antibodies of the correct isotypes might be able to direct the lysis of the tumor cells by NK cells, activating the NK cells via their Fc receptors (left panels). A more useful strategy might be to couple the antibody to a powerful toxin (center panels). When the antibody binds to the tumor cell and is endo-cytozed, the toxin is released from the antibody and can kill the tumor cell. If the antibody is coupled to a radionuclide (right panels), binding of the antibody to a tumor cell will deliver a dose of radiation sufficient to kill the tumor cell. In addition, nearby tumor cells could also receive a lethal radiation dose, even though they did not bind the antibody.

Tumor-specific antibody | Tumor-specific antibody conjugated to toxin | Tumor-specific antibody conjugated to radionuclide

Antibodies bind to the tumor cell | Antibody-toxin conjugates bind to the tumor cell | Radioactive antibody binds to the tumor cell

NK   NK

CD16

NK cells with Fc receptors (CD16) are activated to kill the tumor cells | Conjugates are internalized, killing the cell | Radiation kills the tumor cell and neigboring tumor cells

Fig. 14.18 Examples of tumor antigens that have been targeted by monoclonal antibodies in therapeutic trials. CEA, carcinoembryonic antigen.

Tumor tissue origin	Type of antigen	Antigen	Tumor type
Lymphoma/ leukemia	Differentiation antigen	CD5 Idiotype CAMPATH-1 (CDw52)	T-cell lymphoma B-cell lymphoma T- and B-cell lymphoma
	B-cell signaling receptor	CD20	Non-Hodgkin's B-cell lymphoma
Solid tumors	Cell-surface antigens Glycoprotein Carbohydrate	CEA, mucin-1 Lewis^y CA-125	Epithelial tumors (breast, colon, lung) Epithelial tumors Ovarian carcinoma
	Growth factor receptor	Epidermal growth factor receptor p185^{HER2} IL-2 receptor	Lung, breast, head, and neck tumors Breast, ovarian tumors T- and B-cell tumors
	Stromal extracellular antigen	FAP-α Tenascin Metalloproteinases	Epithelial tumors Glioblastoma multiforme Epithelial tumors

transduces a signal that causes lymphocyte apoptosis. Monoclonal antibodies coupled to γ-emitting radioisotopes have also been used to image tumors, for the purpose of diagnosis and monitoring tumor spread (Fig. 14.19).

The first reported successful treatment of a tumor with monoclonal antibodies used anti-idiotypic antibodies to target B-cell lymphomas whose surface immunoglobulin expressed the corresponding idiotype. The initial course of treatment usually leads to a remission, but the tumor always reappears in a mutant form that no longer binds to the antibody used for the initial treatment. This case represents a clear example of genetic instability enabling a tumor to evade treatment.

Other problems with tumor-specific or tumor-selective monoclonal antibodies as therapeutic agents include inefficient killing of cells after binding of the monoclonal antibody and inefficient penetration of the antibody into the tumor mass. The first problem can often be circumvented by linking the antibody to a toxin, producing a reagent called an **immunotoxin**; two favored toxins are ricin A chain and *Pseudomonas* toxin. Both approaches require the antibody to be internalized to allow the cleavage of the toxin from the antibody in the endocytic compartment, allowing the toxin chain to penetrate and kill the cell.

Two other approaches using monoclonal antibody conjugates involve linking the antibody molecule to chemotherapeutic drugs such as adriamycin or to radioisotopes. In the first case, the specificity of the monoclonal antibody for a cell-surface antigen on the tumor concentrates the drug to the site of the tumor. After internalization, the drug is released in the endosomes and exerts its cytostatic or cytotoxic effect. Monoclonal antibodies linked to radionuclides (see Fig. 14.17) concentrate the radioactive source in the tumor site. Both these approaches have the advantage of also killing neighboring tumor cells, because the released drug or radioactive emissions can affect cells adjacent to those that actually bind the antibody. Ultimately, combinations of toxin-, drug-, or radionuclide-linked monoclonal antibodies, together with vaccination strategies aimed at inducing T cell-mediated immunity, might provide the most effective cancer immunotherapy.

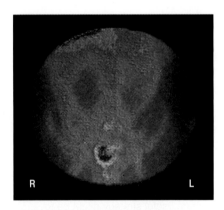

Fig. 14.19 Recurrent colorectal cancer can be detected with a radiolabeled monoclonal antibody against carcinoembryonic antigen. A patient with a possible recurrence of a colorectal cancer was injected intravenously with an indium-111-labeled monoclonal antibody to carcinoembryonic antigen. The recurrent tumor is seen as two red spots located in the pelvic region. The blood vessels are faintly outlined by circulating antibody that has not bound to the tumor. Photograph courtesy of A.M. Peters.

14-15 Enhancing the immunogenicity of tumors holds promise for cancer therapy.

Although vaccines based on tumor antigens are, in principle, the ideal approach to T cell-mediated cancer immunotherapy, it may be many decades before the dominant tumor antigens for common cancers are identified. Even then, it is not clear how widely the relevant epitopes will be shared between tumors, and peptides of tumor rejection antigens will be presented only by particular MHC alleles. To be effective, a tumor vaccine may therefore need to include a range of tumor antigens. MAGE-1 antigens, for example, are recognized only by T cells in melanoma patients expressing the HLA-A1 haplotype. However, the range of MAGE-type proteins that has now been characterized encompasses peptide epitopes presented by many HLA class I and II molecules.

Until recently, most cancer vaccines have used the individual patient's tumor removed at surgery as a source of vaccine antigens. These cell-based vaccines are prepared by mixing either irradiated tumor cells or tumor extracts with bacterial adjuvants such as BCG or *Corynebacterium parvum*, which enhance their immunogenicity (see Appendix I, Section A-4). Such vaccines have generated modest therapeutic results in melanomas but have, in general, been disappointing.

Where candidate tumor rejection antigens have been identified, for example in melanoma, experimental vaccination strategies include the use of whole proteins, peptide vaccines based on sequences recognized by cytotoxic T lymphocytes (either administered alone or presented by the patient's own dendritic cells), and recombinant viruses encoding these peptide epitopes. A novel experimental approach to tumor vaccination is the use of heat-shock proteins isolated from tumor cells. The underlying principle of this therapy is that one of the physiological activities of heat-shock proteins is to act as intra-cellular chaperones of antigenic peptides. There is evidence for receptors on the surface of professional antigen-presenting cells that take up certain heat-shock proteins together with any bound peptides. Uptake of heat-shock proteins via these receptors delivers the accompanying peptide into the antigen-processing pathways leading to peptide presentation by MHC class I molecules. This experimental technique for tumor vaccination has the advantage that it does not depend on any prior knowledge of the nature of the relevant tumor rejection antigens, but the disadvantage that the heat-shock proteins purified from the cell carry very many peptides, so that any tumor rejection antigen might constitute only a tiny fraction of the peptides bound to the heat-shock protein.

A further experimental approach to tumor vaccination in mice is to increase the immunogenicity of tumor cells by introducing genes that encode co-stimulatory molecules or cytokines. This is intended to make the tumor itself more immunogenic. The basic scheme of such experiments is shown in Fig. 14.20. A tumor cell transfected with the gene encoding the co-stimulatory molecule B7 (see Section 8-5) is implanted in a syngeneic animal. These B7-positive cells can activate tumor-specific naive T cells to become armed effector T cells able to reject the tumor cells. They are also able to stimulate further proliferation of the armed effector cells that reach the site of implantation. These T cells can then target the tumor cells whether they express B7 or not; this can be shown by reimplanting nontransfected tumor cells, which are also rejected.

The second strategy, that of introducing cytokine genes into tumors so that they secrete the relevant cytokine, is aimed at attracting antigen-presenting cells to the tumor and takes advantage of the paracrine nature of cytokines. In mice, the most effective tumor vaccines so far are tumor cells that secrete

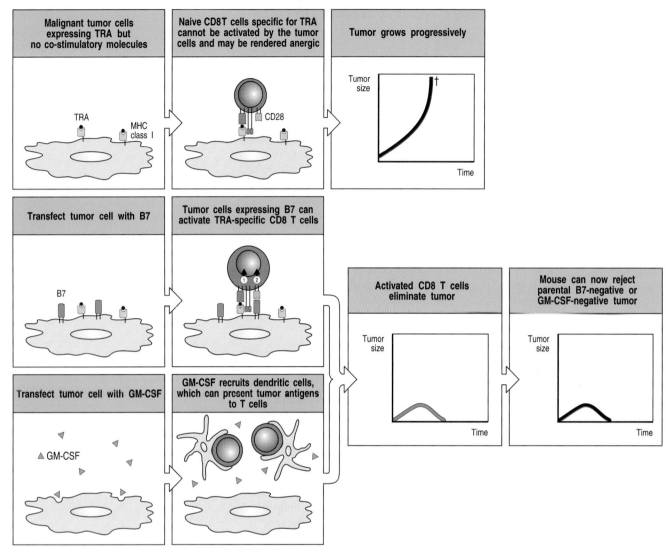

Fig. 14.20 Transfection of tumors with the gene for B7 or for GM-CSF enhances tumor immunogenicity. A tumor that does not express co-stimulatory molecules will not induce an immune response, even though it might express tumor rejection antigens (TRAs), because naive CD8 T cells specific for the TRA cannot be activated by the tumor. The tumor therefore grows progressively in normal mice and eventually kills the host (top panels). If such tumor cells are transfected with a co-stimulatory molecule, such as B7, TRA-specific CD8 T cells now receive both signal 1 and signal 2 from the same cell (see Section 8-5) and can therefore be activated (center panels). The same effect can be obtained by transfecting the tumor with the gene encoding GM-CSF, which attracts and stimulates the differentiation of dendritic cell precursors (bottom panels). Both these strategies have been tested in mice and shown to elicit memory T cells, although results with GM-CSF are more impressive. Because TRA-specific CD8 cells have now been activated, even the original B7-negative or GM-CSF negative tumor cells can be rejected.

granulocyte-macrophage colony-stimulating factor (GM-CSF), which induces the differentiation of hematopoietic precursors into dendritic cells and attracts these to the site. GM-CSF is also thought to function as an adjuvant, activating the dendritic cells. It is believed that these cells process the tumor antigens and migrate to the local lymph nodes, where they induce potent anti-tumor responses. The B7-transfected cells seem less potent in inducing anti-tumor responses, perhaps because the bone marrow-derived dendritic cells express more of the molecules required to activate naive T cells than do B7-transfected tumor cells. In addition, the tumor cells do not share the dendritic cells' special ability to migrate into the T-cell areas of the lymph nodes, where they are optimally placed to interact with passing naive T cells (see Section 8-6).

The potency of dendritic cells in activating T-cell responses provides the rationale for yet another strategy for vaccinating against tumors. The use of antigen-pulsed autologous dendritic cells to stimulate therapeutically useful cytotoxic T-cell responses to tumors has been developed in experimental models, and there have been initial trials in humans with cancer.

Clinical trials are in progress to determine the safety and efficacy of many of these approaches in human patients. What is uncertain is whether people with established cancers can generate sufficient T-cell responses to eliminate all their tumor cells under circumstances in which any tumor-specific naive T cells might have been rendered tolerant to the tumor. Moreover, there is always the risk that immunogenic transfectants will elicit an autoimmune response against the normal tissue from which the tumor derived.

Summary.

Tumors represent outgrowths of a single abnormal cell, and animal studies have shown that some tumors elicit specific immune responses that suppress their growth. These seem to be directed at MHC-bound peptides derived from antigens that might be mutated, inappropriately expressed, or overexpressed in the tumor cells. T-cell deficient individuals, however, do not develop more tumors than normal individuals. This is probably chiefly because most tumors do not make distinctive antigenic proteins or do not express the co-stimulatory molecules necessary to initiate an adaptive immune response. Tumors also have other ways of avoiding or suppressing immune responses, such as ceasing to express MHC class I molecules, or making immunosuppressive cytokines. Monoclonal antibodies have been developed for tumor immunotherapy by conjugation to toxins or to cytotoxic drugs or radionuclides, which are thereby delivered at high dose specifically to the tumor cells. More recently, attempts have been made to develop vaccines based on tumor cells taken from patients and made immunogenic by the addition of adjuvants, or by pulsing autologous dendritic cells with tumor-cell extracts or tumor antigens. This approach has been extended in animal experiments to transfection of tumor cells with genes encoding co-stimulatory molecules or cytokines that attract and activate dendritic cells.

Manipulating the immune response to fight infection.

Infection is the leading cause of death in the human population. The two most important contributions to public health in the past 100 years have been sanitation and vaccination, which together have dramatically reduced deaths from infectious disease. Modern immunology grew from the success of Jenner's and Pasteur's vaccines against smallpox and chicken cholera, respectively, and its greatest triumph has been the global eradication of smallpox, announced by the World Health Organization in 1980. A global campaign to eradicate polio is now under way.

Adaptive immunity to a specific infectious agent can be achieved in several ways. One early strategy was to deliberately cause a mild infection with the unmodified pathogen. This was the principle of variolation, in which the inoculation of a small amount of dried material from a smallpox pustule would cause a mild infection followed by long-lasting protection against reinfection. However, infection following variolation was not always mild: fatal smallpox ensued in about 3% of cases, which would not meet modern criteria for safety. Jenner's achievement was the realization that infection with a bovine analogue of smallpox, vaccinia (from *vacca*—a cow), which caused cowpox, would provide protective immunity against smallpox in humans without the risk of significant disease. He named the process **vaccination**, and Pasteur, in his honour, extended the term to the stimulation of protection to other infectious agents. Humans are not a natural host of vaccinia, which establishes only a brief and limited subcutaneous infection but contains antigens that stimulate an immune response that is cross-reactive with smallpox antigens and thereby confers protection from the human disease.

This established the general principles of safe and effective vaccination, and vaccine development in the early part of the 20th century followed two empirical pathways. The first was the search for attenuated organisms with reduced pathogenicity that would stimulate protective immunity; the second was the development of vaccines based on killed organisms and, subsequently, purified components of organisms that would be as effective as live whole organisms. Killed vaccines were desirable because any live vaccine, including vaccinia, can cause lethal systemic infection in the immunosuppressed.

Immunization is now considered so safe and so important that most states in the United States require all children to be immunized against measles, mumps, and polio viruses with live attenuated vaccines, as well as against tetanus (caused by *Clostridium tetani*), diphtheria (caused by *Corynebacterium diphtheriae*), and whooping cough (caused by *Bordetella pertussis*), with inactivated toxins or toxoids prepared from these bacteria (see Fig. 1.33). More recently, a vaccine has become available against *Haemophilus* B, one of the causative agents of meningitis. Current vaccination schedules for children in the United States are shown in Fig. 14.21. Impressive as these accomplishments are, there are still many diseases for which we lack effective vaccines, as shown in Fig. 14.22. Even where a vaccine such as measles or polio can be used effectively in developed countries, technical and economic problems can prevent its widespread use in developing countries, where mortality from

Current immunization schedule for children (USA)										
Vaccine given	1 month	2 months	4 months	6 months	12 months	15 months	18 months	4–6 years	11–12 years	14–16 years
Diphtheria–tetanus–pertussis (DTP/DTaP)		■	■	■		■	■	■	■ *	■
Inactivated polio vaccine		■	■	■	■	■	■	■		
Measles/mumps/rubella (MMR)					■	■		■		
Pneumococcal conjugate		■	■	■	■	■				
Haemophilus B conjugate (HiBC)		■	■	■	■	■				
Hepatitis B	■	■		■	■	■	■			
Varicella					■	■	■			

Fig. 14.21 Recommended childhood vaccination schedules (in red) in the United States. Each red bar denotes a time range during which a vaccine dose should be given. Bars spanning multiple months indicate a range of times during which the vaccine may be given. * Tetanus and diptheria toxoids only.

Some diseases for which effective vaccines are not yet available		
Disease	**Estimated annual mortality**	**Estimated annual incidence**
Malaria*	1,086,000	300–500 million
Schistosomiasis	14,000	no numbers available
Worm infestation	16,000	no numbers available
Tuberculosis	1,498,000	~8 million
Diarrheal disease	2,213,000	~4,100 million
Respiratory disease	4,039,000	~362 million
HIV/AIDS	2,673,000	~2 million
Measles†	875,000	~44 million

Fig. 14.22 Diseases for which effective vaccines are still needed. *The number of people infected is estimated at ~200 million, of which 20 million have severe disease. †Current measles vaccines are effective but heat-sensitive, which makes their use difficult in tropical countries. Estimated mortality data for 1999 from *World Health Report 2000* (World Health Organization).

these diseases is still high. The development of vaccines therefore remains an important goal of immunology and the latter half of the 20th century saw a shift to a more rational approach, based on a detailed molecular understanding of microbial pathogenicity, analysis of the protective host response to pathogenic organisms, and the understanding of the regulation of the immune system to generate effective T- and B-lymphocyte responses.

14-16 There are several requirements for an effective vaccine.

The particular requirements for successful vaccination vary according to the nature of the infecting organism. For extracellular organisms, antibody provides the most important adaptive mechanism of host defense, whereas for control of intracellular organisms, an effective CD8 T-lymphocyte response is also essential. The ideal vaccination provides host defense at the point of entry of the infectious agent; stimulation of mucosal immunity is therefore an important goal of vaccination against those many organisms that enter through mucosal surfaces.

Effective protective immunity against some microorganisms requires the presence of preexisting antibody at the time of exposure to the infection. For example, the clinical manifestations of tetanus and diphtheria are entirely due to the effects of extremely powerful exotoxins (see Fig. 9.23). Preexisting antibody against the bacterial exotoxin is necessary to provide a defense against these diseases. Preexisting antibodies are also required to protect against some intracellular pathogens, such as the poliomyelitis virus, which infect critical host cells within a short period after entering the body and are not easily controlled by T lymphocytes once intracellular infection is established.

Immune responses to infectious agents usually involve antibodies directed at multiple epitopes and only some of these antibodies confer protection. The particular T-cell epitopes recognized can also affect the nature of the

Features of effective vaccines	
Safe	Vaccine must not itself cause illness or death
Protective	Vaccine must protect against illness resulting from exposure to live pathogen
Gives sustained protection	Protection against illness must last for several years
Induces neutralizing antibody	Some pathogens (such as poliovirus) infect cells that cannot be replaced (e.g., neurons). Neutralizing antibody is essential to prevent infection of such cells
Induces protective T cells	Some pathogens, particularly intracellular, are more effectively dealt with by cell-mediated responses
Practical considerations	Low cost-per-dose Biological stability Ease of administration Few side-effects

Fig. 14.23 There are several criteria for an effective vaccine.

response. For example, as we saw in Chapter 11, the predominant epitope recognized by T cells after vaccination with respiratory syncytial virus induces a vigorous inflammatory response but fails to elicit neutralizing antibodies and thus causes pathology without protection. Thus, an effective vaccine must lead to the generation of antibodies and T cells directed at the correct epitopes of the infectious agent. For some of the modern vaccine techniques, in which only one or a few epitopes are used, this consideration is particularly important.

A number of very important additional constraints need to be satisified by a successful vaccine (Fig. 14.23). First, it must be safe. Vaccines must be given to huge numbers of people, relatively few of whom are likely to die of, or sometimes even catch, the disease that the vaccine is designed to prevent. This means that even a low level of toxicity is unacceptable. Second, the vaccine must be able to produce protective immunity in a very high proportion of the people to whom it is given. Third, because it is impracticable to give large or dispersed rural populations regular 'booster' vaccinations, a successful vaccine must generate long-lived immunological memory. This means that both B and T lymphocytes must be primed by the vaccine. Fourth, vaccines must be very cheap if they are to be administered to large populations. Vaccines are one of the most cost-effective measures in health care, but this benefit is eroded as the cost-per-dose rises.

An effective vaccination program provides herd immunity—by lowering the number of susceptible members of a population, the natural reservoir of infected individuals in that population falls, reducing the probability of transmission of infection. Thus, even nonvaccinated members of a population can be protected from infection if the majority are vaccinated.

14-17 The history of vaccination against *Bordetella pertussis* illustrates the importance of developing an effective vaccine that is perceived to be safe.

The history of vaccination against the bacterium that causes whooping cough, *Bordetella pertussis*, provides a good example of the challenges of developing and disseminating an effective vaccine. At the turn of the 20th century, whooping cough killed approximately 0.5% of American children under the age of 5 years. In the early 1930s, a trial of a killed, whole bacterial cell vaccine on the Faroe Islands provided evidence of a protective effect. In the United States, systematic use of a whole-cell vaccine in combination with diphtheria and tetanus toxoids (the DPT vaccine) since the 1940s resulted in a decline in the annual infection rate from 200 to less than 2 cases per 100,000 of the population. First vaccination with DPT was typically given at the age of 3 months.

Whole-cell pertussis vaccine causes side-effects, typically redness, pain, and swelling at the site of the injection; less commonly, vaccination is followed by high temperature and persistent crying. Very rarely, fits and a short-lived sleepiness or a floppy unresponsive state ensue. During the 1970s, widespread concern developed after several anecdotal observations that encephalitis leading to irreversible brain damage might very rarely follow pertussis vaccination. In Japan, in 1972, approximately 85% of children were given the pertussis vaccine, and fewer than 300 cases of whooping cough and no deaths were reported. As a result of two deaths after vaccination in Japan in 1975, DPT was temporarily suspended and then reintroduced with the first vaccination at 2 years of age rather than 3 months. In 1979 there were approximately 13,000 cases of whooping cough and 41 deaths. The possibility that pertussis vaccine very rarely causes severe brain damage has been studied extensively and expert consensus is that pertussis vaccine is not a primary cause of brain injury. There is no doubt that there is greater morbidity from whooping cough than from the vaccine.

The public and medical perception that whole-cell pertussis vaccination might be unsafe provided a powerful incentive to develop safer pertussis vaccines. Study of the natural immune response to *B. pertussis* showed that infection induced antibodies against four components of the bacterium— pertussis toxin, filamentous hemagglutinin, pertactin, and fimbrial antigens. Immunization of mice with these antigens in purified form protected them against challenge with pertussis. This has led to the development of acellular pertussis vaccines, all of which contain purified pertussis toxoid, that is, toxin inactivated by chemical treatment, for example with hydrogen peroxide or formaldehyde, or more recently by genetic engineering of the toxin. Some also contain one or more of the filamentous hemagglutinin, pertactin, and fimbrial antigens. Current evidence shows that these are probably as effective as whole-cell pertussis vaccine and are free of the common minor side-effects of the whole-cell vaccine.

The main messages of the history of pertussis vaccination are, first, that vaccines must be extremely safe and free of side-effects; second, that the public and the medical profession must perceive the vaccine to be safe; and third, that careful study of the nature of the protective immune response can lead to acellular vaccines that are safer than and as effective as whole-cell vaccines.

14-18 Conjugate vaccines have been developed as a result of understanding how T and B cells collaborate in an immune response.

Although acellular vaccines are inevitably safer than vaccines based on whole organisms, a fully effective vaccine cannot normally be made from a single isolated constituent of a microorganism, and it is now clear that this is because of the need to activate more than one cell type to initiate an immune response. One consequence of this insight has been the development of conjugate vaccines. We have already described briefly one of the most important of these in Section 9-2.

Many bacteria, including *Neisseria meningitidis* (meningococcus), *Streptococcus pneumoniae* (pneumococcus), and *Haemophilus* species, have an outer capsule composed of polysaccharides that are species- and type-specific for particular strains of the bacterium. The most effective defense against these microorganisms is opsonization of the polysaccharide coat with antibody. The aim of vaccination is therefore to elicit antibodies against the polysaccharide capsules of the bacteria.

Capsular polysaccharides can be harvested from bacterial growth medium and, because they are T-cell independent antigens, they can be used on their own as vaccines. However, young children under the age of 2 years cannot make good T-cell independent antibody responses and cannot be vaccinated effectively with polysaccharide vaccines. An efficient way of overcoming this problem (see Fig. 9.4) is to chemically conjugate bacterial polysaccharides to protein carriers, which provide peptides that can be recognized by antigen-specific T cells, thus converting a T-cell independent response into a T-cell dependent anti-polysaccharide antibody response. By using this approach, various conjugate vaccines have been developed against *Haemophilus influenzae*, an important cause of serious childhood chest infections and meningitis, and these are now widely applied.

14-19 The use of adjuvants is another important approach to enhancing the immunogenicity of vaccines.

Purified antigens are not usually strongly immunogenic on their own and most acellular vaccines require the addition of **adjuvants**, which are defined as substances that enhance the immunogenicity of antigens (see Appendix I, Section A-4). For example, tetanus toxoid is not immunogenic in the absence of adjuvants, and tetanus toxoid vaccines often contain aluminum salts, which bind polyvalently to the toxoid by ionic interactions and selectively stimulate antibody responses. Pertussis toxin, produced by *B. pertussis*, has adjuvant properties in its own right and, when given mixed as a toxoid with tetanus and diphtheria toxoids, not only vaccinates against whooping cough but also acts as an adjuvant for the other two toxoids. This mixture makes up the DPT triple vaccine given to infants in the first year of life.

Many important adjuvants are sterile constituents of bacteria, particularly of their cell walls. For example, Freund's complete adjuvant, widely used in experimental animals to augment antibody responses, is an oil and water emulsion containing killed mycobacteria. A complex glycolipid, muramyl dipeptide, which can be extracted from mycobacterial cell walls or synthesized, contains much of the adjuvant activity of whole killed mycobacteria. Other bacterial adjuvants include killed *B. pertussis*, bacterial polysaccharides, bacterial heat-shock proteins, and bacterial DNA. Many of these adjuvants cause quite marked inflammation and are not suitable for use in vaccines for humans.

It is thought that most, if not all, adjuvants act on antigen-presenting cells, especially on dendritic cells, and reflect the importance of these cells in initiating immune responses. As we learned in Section 8-6, dendritic cells are widely distributed throughout the body, where they act as sentinels to detect potential pathogens at their portals of entry. These tissue dendritic cells take up antigens from their environment by phagocytosis and macropinocytosis, and they are tuned to respond to the presence of infection by migrating into lymphoid tissue and presenting these antigens to T cells. They appear to detect the presence of pathogens in two main ways. The first of these is direct, and follows the ligation and activation of receptors for invading micro-organisms. These include receptors of the complement system, Toll-like receptors (TLRs), and other pattern recognition receptors of the innate immune system. There is much that we still have to learn about the direct mechanisms of detection of infectious agents. For example, bacterial DNA containing unmethylated CpG dinucleotide motifs, bacterial heat-shock proteins, and muramyl dipeptides each have powerful activating effects on antigen-presenting cells, and, while there is indirect evidence that many adjuvants use various TLRs, it is not known how they are detected. When dendritic cells are activated through direct interactions with the products of infectious agents, they respond by secreting cytokines and expressing co-stimulatory molecules, which in turn stimulate the activation and differentiation of antigen-specific T cells.

The second mechanism of stimulation of dendritic cells by invading organisms is indirect and involves their activation by cytokine signals derived from the inflammatory response triggered by infection (see Chapter 2). Cytokines such as GM-CSF are particularly effective in activating dendritic cells to express co-stimulatory signals and, in the context of viral infection, dendritic cells also express interferon (IFN)-α and IL-12.

Adjuvants trick the immune system into responding as though there were an active infection, and just as different classes of infectious agent stimulate different types of immune response (see Chapter 10), different adjuvants may promote different types of response, for example, an inflammatory T_H1 response or an antibody-dominated response. Some adjuvants, for example, pertussis toxin, stimulate mucosal immune responses, which are particularly important in defense against organisms entering through the digestive or respiratory tracts. These adjuvants have been discussed earlier when we described mucosal immunity and will be further discussed in Section 14-26.

Following our increased understanding of the mechanisms of action of adjuvants, rational approaches to improving the activity of vaccines in clinical settings are being implemented. One approach is to coadminister cytokines. For example, IL-12 is a cytokine produced by macrophages, dendritic cells, and B cells that stimulates T lymphocytes and NK cells to release IFN-γ and promotes a T_H1 response. It has been used as an adjuvant to promote protective immunity against the protozoan parasite *Leishmania major*. Certain strains of mice are susceptible to severe cutaneous and systemic infection by *L. major*; these mice mount an immune response that is predominantly T_H2 in type and is ineffective in eliminating the organism (see Section 10-6). The coadministration of IL-12 with a vaccine containing leishmania antigens generated a T_H1 response and protected the mice against challenge with *L. major*. The use of IL-12 to promote a T_H1 response has also proved valuable in reducing the pathogenic consequences of experimental parasitic infection by *Schistosoma mansoni* and will be considered in Section 14-27. These are important examples of how an understanding of the regulation of immune responses can enable rational intervention to enhance the effectiveness of vaccines.

14-20 Live-attenuated viral vaccines are usually more potent than 'killed' vaccines and can be made safer by using recombinant DNA technology.

Most antiviral vaccines currently in use consist of inactivated or live attenuated viruses. Inactivated, or 'killed,' viral vaccines consist of viruses treated so that they are unable to replicate. Live-attenuated viral vaccines are generally far more potent, perhaps because they elicit a greater number of relevant effector mechanisms, including cytotoxic CD8 T cells: inactivated viruses cannot produce proteins in the cytosol, so peptides from the viral antigens cannot be presented by MHC class I molecules and thus cytotoxic CD8 T cells are not generated by these vaccines. Attenuated viral vaccines are now in use for polio, measles, mumps, rubella, and varicella.

Traditionally, attenuation is achieved by growing the virus in cultured cells. Viruses are usually selected for preferential growth in nonhuman cells and, in the course of selection, become less able to grow in human cells (Fig. 14.24). Because these attenuated strains replicate poorly in human hosts, they induce immunity but not disease when given to people. Although attenuated virus strains contain multiple mutations in genes encoding several of their proteins, it might be possible for a pathogenic virus strain to reemerge by a further series of mutations. For example, the type 3 Sabin polio vaccine strain differs at only 10 of 7429 nucleotides from a wild-type progenitor strain. On extremely rare occasions, reversion of the vaccine to a neurovirulent strain can occur, causing paralytic disease in the unfortunate recipient.

Attenuated viral vaccines can also pose particular risks to immunodeficient recipients in whom they often behave as virulent opportunistic infections. Immunodeficient infants who are vaccinated with live-attenuated polio before their inherited immunoglobulin deficiencies have been diagnosed are at risk because they cannot clear the virus from their gut, and there is therefore an increased chance that mutation of the virus will lead to fatal paralytic

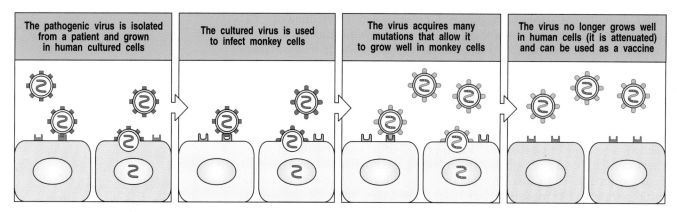

The pathogenic virus is isolated from a patient and grown in human cultured cells

The cultured virus is used to infect monkey cells

The virus acquires many mutations that allow it to grow well in monkey cells

The virus no longer grows well in human cells (it is attenuated) and can be used as a vaccine

Fig. 14.24 Viruses are traditionally attenuated by selecting for growth in nonhuman cells. To produce an attenuated virus, the virus must first be isolated by growing it in cultured human cells. The adaptation to growth in cultured human cells can cause some attenuation in itself; the rubella vaccine, for example, was made in this way. In general, however, the virus is then adapted to growth in cells of a different species, until it grows only poorly in human cells. The adaptation is a result of mutation, usually a combination of several point mutations. It is usually hard to tell which of the mutations in the genome of an attenuated viral stock are critical to attenuation. An attenuated virus will grow poorly in the human host, and will therefore produce immunity but not disease.

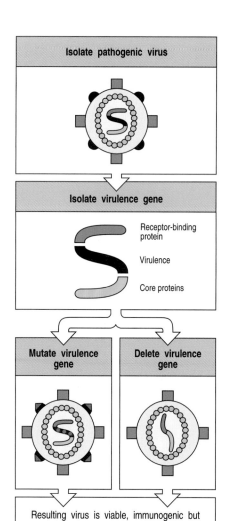

Isolate pathogenic virus

Isolate virulence gene

Receptor-binding protein

Virulence

Core proteins

Mutate virulence gene

Delete virulence gene

Resulting virus is viable, immunogenic but avirulent. It can be used as a vaccine

disease. For the same reason, patients with immunoglobulin deficiencies show an abnormal susceptibility to chronic infection by opportunistic enteroviruses, and can develop chronic, and ultimately lethal, echovirus encephalitis if mutation of the virus leads to neurovirulence.

An empirical approach to attenuation is still in use but might be superseded by two new approaches that use recombinant DNA technology. One is the isolation and *in vitro* mutagenesis of specific viral genes. The mutated genes are used to replace the wild-type gene in a reconstituted virus genome, and this deliberately attenuated virus can then be used as a vaccine (Fig. 14.25). The advantage of this approach is that mutations can be engineered so that reversion to wild type is virtually impossible.

Such an approach might be useful in developing live influenza vaccines. As we learned in Chapter 11, the influenza virus can reinfect the same host several times, because it undergoes antigenic shift and thus escapes the original immune response. The current approach to vaccination against influenza is to use a killed virus vaccine that is reformulated annually on the basis of the prevalent strains of virus. The vaccine is moderately effective, reducing mortality in elderly populations and morbidity in healthy adults. The ideal influenza vaccine would be an attenuated live organism that matched the prevalent virus strain. This could be created by first introducing a series of attenuating mutants into the gene encoding a viral polymerase protein, PB2. The mutated gene segment from the attenuated virus could then be substituted for the wild-type gene in a virus carrying the relevant hemagglutinin and neuraminidase antigenic variants of the current epidemic or pandemic strain. This last procedure could be repeated as necessary to keep pace with the antigenic shift of the virus.

Fig. 14.25 Attenuation can be achieved more rapidly and reliably with recombinant DNA techniques. If a gene in the virus that is required for virulence but not for growth or immunogenicity can be identified, this gene can be either multiply mutated (left lower panel) or deleted from the genome (right lower panel) by using recombinant DNA techniques. This procedure creates an avirulent (nonpathogenic) virus that can be used as a vaccine. The mutations in the virulence gene are usually large, so that it is very difficult for the virus to revert to the wild type.

14-21 Live-attenuated bacterial vaccines can be developed by selecting nonpathogenic or disabled mutants.

Similar approaches are being used for bacterial vaccine development. *Salmonella typhi*, the causative agent of typhoid, has been manipulated to develop a live vaccine. A strain of wild-type bacteria was mutated using nitrosoguanidine; a new strain was selected to be defective in the enzyme UDP-galactose epimerase, thus blocking the pathway for synthesis of lipopolysaccharide, an important determinant of bacterial pathogenesis. Recent approaches to the rational design of attenuated *Salmonella* vaccines have involved the specific targeting of genes encoding enzymes in the biosynthetic pathways of amino acids containing aromatic rings, such as tyrosine and phenylalanine. Mutating these genes makes auxotrophic organisms, which are dependent for growth on an external supply of an essential nutrient that wild-type bacteria would be capable of biosynthesizing. These bacteria grow poorly in the gut but should survive long enough as a vaccine to induce an effective immune response.

It is not only vaccination of humans against *Salmonella* that is important. Modern methods of mass production of chickens for food has led to extensive infection of poultry with *Salmonella* strains that are pathogenic to humans and an increasingly important cause of food poisoning. Thus, in parts of the world where typhoid is prevalent, vaccinating humans has a high priority. In other parts, where food poisoning caused by *Salmonella typhimurium* and *S. enteritidis* infection is common, vaccination of chickens would contribute to public health.

14-22 Attenuated microorganisms can serve as vectors for vaccination against many pathogens.

An effective live-attenuated typhoid vaccine would not only be valuable in its own right but could also serve as a vector for presenting antigens from other organisms. Attenuated strains of *Salmonella* have been used as carriers of heterologous genes encoding tetanus toxoid and antigens from organisms as diverse as *Listeria monocytogenes*, *Bacillus anthracis*, *Leishmania major*, *Yersinia pestis*, and *Schistosoma mansoni*. Each of these has been used as an oral vaccine to protect mice against experimental challenge with the respective pathogen.

Viral vectors can similarly be engineered to carry heterologous peptides or proteins from other microorganisms. Although vaccinia is no longer needed to protect against the development of smallpox, it remains a candidate as an avirulent carrier of heterologous antigens. Genes encoding protective antigens from several different organisms could be placed in a single vaccine strain. This approach makes it possible to immunize individuals against several different pathogens at once, but such a vaccine could not be used twice because the vaccinia vector itself generates long-lasting immunity that would neutralize its effectiveness on a second administration; this is an example of the phenomenon called 'original antigenic sin' (see Fig. 10.30). The development of successful heterologous vaccines requires the identification of protective antigens; it therefore depends on the analytical power of recombinant DNA methods, as well as their use to manipulate gene structure.

Plant viruses, which are nonpathogenic to humans, have been used as a source of novel vaccine vectors. These viruses can be engineered to incorporate heterologous peptide antigens into chimeric coat proteins. The success of this approach relies on the successful identification of protective peptide antigens as well as the immunogenicity of the vaccine. Using this strategy, mice have been protected against lethal challenge with rabies virus by prior feeding with spinach leaves infected by recombinant alfalfa mosaic virus incorporating a rabies virus peptide. Popeye may need rejuvenation as a role model to encourage children to eat spinach.

14-23 Synthetic peptides of protective antigens can elicit protective immunity.

One route to vaccine development is the identification of the T-cell peptide epitopes that stimulate protective immunity. This can be approached in two ways. One possibility is to synthesize systematically overlapping peptides from immunogenic proteins and to test each in turn for its ability to stimulate protective immunity. An alternative, but no less arduous approach—'reverse' immunogenetics—has been used in developing a vaccine against malaria (Fig. 14.26).

The immunogenicity of T-cell peptide epitopes depends on their specific associations with particular polymorphic variants of MHC molecules. The starting point for the studies on malaria was an association between the human MHC class I molecule HLA-B53 and resistance to cerebral malaria— a relatively infrequent complication of infection but one that is usually fatal. The hypothesis is that these MHC molecules are protective because they present peptides that are particularly good at evoking cytotoxic T lymphocytes. A direct route to identifying the relevant peptides is to elute them from MHC molecules of cells infected with the pathogen. In HLA-B53, a high proportion of the peptides eluted had proline in the second of nine positions; this information was used to identify candidate protective peptides from four proteins of *Plasmodium falciparum* expressed in the early phase of hepatocyte infection, an important phase of infection to target in an effective immune response. One of the candidate peptides, from liver stage antigen-1, is recognized by cytotoxic T cells when bound to HLA-B53.

This approach is being extended to other MHC class I and class II molecules associated with protective immune responses against infection. Recently, a protective peptide epitope was eluted from MHC class II molecules in *Leishmania*-infected macrophages and used as a guide to isolate the gene from *Leishmania*. The gene was then used to make a protein-based vaccine that primed mice from susceptible strains for responses to *Leishmania*.

These results show considerable promise, but they also illustrate one of the major drawbacks to the approach. A malaria peptide that is restricted by HLA-B53 might not be immunogenic in an individual lacking HLA-B53: indeed, this presumably accounts for the higher susceptibility of these individuals to natural infections. Because of the very high polymorphism of MHC molecules in humans it will be necessary to identify panels of protective T-cell epitopes and construct vaccines containing arrays of these to develop vaccines that will protect the majority of a susceptible population.

Particular HLA molecule found to have most affinity for nonapeptides with proline as second residue

Candidate nonapeptides with proline as second residue are identified

proline

Assembly of HLA protein in the presence of each of the candidate peptides is assayed

Proliferation assay conducted with lymphocytes from infected patients

Peptide identified as having potential for vaccine development

Fig. 14.26 'Reverse' immunogenetics can be used to identify protective T-cell epitopes against infectious diseases. Population studies show that the MHC class I variant HLA-B53 is associated with resistance to cerebral malaria. Self nonapeptides were eluted from HLA-B53 and found to have a strong preference for proline at the second position. Candidate nonapeptide sequences containing proline at position 2 were then identified in several malarial protein sequences and synthesized. These synthetic nonapeptides were then tested to see whether they fitted well into the peptide groove of HLA-B53 by assaying whether HLA-B53 would assemble to form a stable cell-surface heterodimer in the presence of peptide. Peptide sequences identified by this approach were then tested to see whether they would induce the proliferation of T cells from patients infected by malaria. Such sequences are good candidates for incorporation into vaccines.

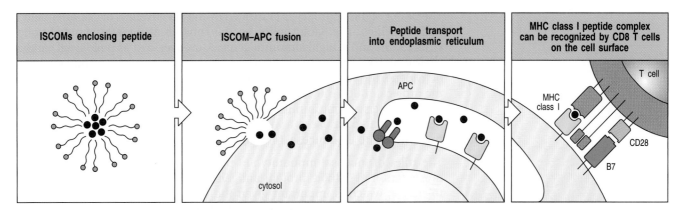

| ISCOMs enclosing peptide | ISCOM–APC fusion | Peptide transport into endoplasmic reticulum | MHC class I peptide complex can be recognized by CD8 T cells on the cell surface |

Fig. 14.27 ISCOMs can be used to deliver peptides to the MHC class I processing pathway. ISCOMs (immune stimulatory complexes) are lipid micelles that will fuse with cell membranes. Peptides trapped in ISCOMs can be delivered to the cytosol of an antigen-presenting cell (APC), allowing the peptide to be transported into the endoplasmic reticulum, where it can be bound by newly synthesized MHC class I molecules and hence transported to the cell surface as peptide:MHC class I complexes. This is a possible means for delivering vaccine peptides to activate CD8 cytotoxic T cells. ISCOMs can also be used to deliver proteins to the cytosol of other types of cell, where they can be processed and presented as though they were a protein produced by the cell.

There are other problems with peptide vaccines. Peptides are not strongly immunogenic and it is particularly difficult to generate MHC class I-specific responses by *in vivo* immunization with peptides. One approach to this problem is to integrate peptides by genetic engineering into carrier proteins within a viral vector, such as hepatitis B core antigen, which are then processed *in vivo* through natural antigen-processing pathways. A second possible technique is the use of **ISCOMs** (immune stimulatory complexes). These are lipid carriers that act as adjuvants but have minimal toxicity. They seem to load peptides and proteins into the cell cytoplasm, allowing MHC class I-restricted T-cell responses to peptides to develop (Fig. 14.27). These carriers are being developed for use in human immunization. Another approach to delivering protective peptides, which we discussed in the previous section, is the genetic engineering of infectious microorganisms to create vaccines that stimulate immunity without causing disease.

14-24 The route of vaccination is an important determinant of success.

Most vaccines are given by injection. This route has two disadvantages, the first practical, the second immunological. Injections are painful and expensive, requiring needles, syringes, and a trained injector. They are unpopular with the recipient, reducing vaccine uptake, and mass vaccination by this approach is laborious. The immunological drawback is that injection may not be the most effective way of stimulating an appropriate immune response as it does not mimic the usual route of entry of the majority of pathogens against which vaccination is directed.

Many important pathogens infect mucosal surfaces or enter the body through mucosal surfaces. Examples include respiratory microorganisms such as *B. pertussis*, rhinoviruses and influenza viruses, and enteric microorganisms such as *Vibrio cholerae*, *Salmonella typhi*, enteropathogenic *Escherichia coli*, and *Shigella*. The enteric microorganisms are particularly important pathogens in underdeveloped countries. It is therefore important to understand how these organisms stimulate mucosal immunity and to develop vaccines that behave similarly. To this end, there are efforts to develop vaccines that can be administered to the mucosa orally or by nasal inhalation.

The power of this approach is illustrated by the effectiveness of live-attenuated polio vaccines. The Sabin polio vaccine consists of three attenuated polio virus strains and is highly immunogenic. Moreover, just as polio itself can be transmitted by fecal contamination of public swimming pools and other failures of hygiene, the vaccine can be transmitted from one individual to another by the orofecal route. Infection with *Salmonella* likewise stimulates a powerful mucosal and systemic immune response and, as we saw in Section 14-21, has been attenuated for use as a vaccine and carrier of heterologous antigens for presentation to the mucosal immune system.

The rules of mucosal immunity are poorly understood. On the one hand, presentation of soluble protein antigens by the oral route often results in tolerance, which is important given the enormous load of foodborne and airborne antigens presented to the gut and respiratory tract. As discussed in Sections 14-10 and 13-28, the ability to induce tolerance by oral or nasal administration of antigens is being explored as a therapeutic mechanism for reducing unwanted immune responses. On the other hand, the mucosal immune system can respond to and eliminate mucosal infections such as pertussis, cholera, and polio. The proteins from these microorganisms that stimulate immune responses are therefore of special interest. One group of powerfully immunogenic proteins at mucosal surfaces is a group of bacterial toxins that have the property of binding to eukaryotic cells and are protease-resistant. A recent finding of potential practical importance is that certain of these molecules, such as the *E. coli* heat-labile toxin and pertussis toxin, have adjuvant properties that are retained even when the parent molecule has been engineered to eliminate its toxic properties. These molecules can be used as adjuvants for oral or nasal vaccines. In mice, nasal insufflation of either of these mutant toxins together with tetanus toxoid resulted in the development of protection against lethal challenge with tetanus toxin.

14-25 Protective immunity can be induced by injecting DNA encoding microbial antigens and human cytokines into muscle.

The latest development in vaccination has come as a surprise even to the scientists who first developed the method. The story begins with attempts to use nonreplicating bacterial plasmids encoding proteins for gene therapy: proteins expressed *in vivo* from these plasmids were found to stimulate an immune response. When DNA encoding a viral immunogen is injected

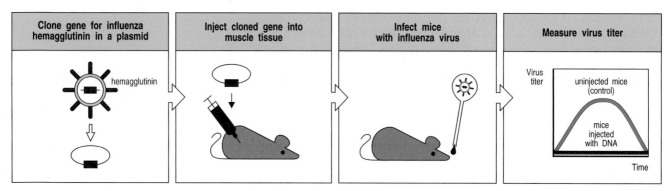

Fig. 14.28 DNA vaccination by injection of DNA encoding a protective antigen and cytokines directly into muscle. Influenza hemagglutinin contains both B- and T-cell epitopes. When a DNA plasmid containing the gene for hemagglutinin is injected directly into muscle, an influenza-specific immune response consisting of both antibody and cytotoxic CD8 T cells results. The response can be enhanced by including a plasmid encoding GM-CSF in the injection. The plasmid DNAs are presumably expressed by some of the cells in the muscle tissue into which they are injected, provoking an immune response that involves both antibody and cytotoxic T cells. The details of this process are not yet understood.

intramuscularly, it leads to the development of antibody responses and cytotoxic T cells that allow the mice to reject a later challenge with whole virus (Fig. 14.28). This response does not appear to damage the muscle tissue, is safe and effective, and, because it uses only a single microbial gene, does not carry the risk of active infection. This procedure has been termed '**DNA vaccination**.' DNA coated onto minute metal projectiles can be administered by 'biolistic' (biological ballistic) gun, so that several metal particles penetrate the skin and enter the muscle beneath. This technique has been shown to be effective in animals and might be suitable for mass immunization, although it has yet to be tested in humans. Mixing in plasmids that encode cytokines such as GM-CSF makes immunization with genes encoding protective antigens much more effective, as was seen earlier for tumor immunity.

14-26 The effectiveness of a vaccine can be enhanced by targeting it to sites of antigen presentation.

An important way of enhancing the effectiveness of a vaccine is to target it efficiently to antigen-presenting cells. This is an important mechanism of action of vaccine adjuvants. There are three complementary approaches. The first is to prevent proteolysis of the antigen on its way to antigen-presenting cells. Preserving antigen structure is an important reason why so many vaccines are given by injection rather than by the oral route, which exposes the vaccine to digestion in the gut. The second and third approaches are to target the vaccine selectively, once in the body, to antigen-presenting cells and to devise methods of engineering the selective uptake of the vaccine into antigen-processing pathways within the cell.

Techniques to enhance the uptake of antigens by antigen-presenting cells include coating the antigen with mannose to enhance uptake by mannose receptors on antigen-presenting cells, and presenting the antigen as an immune complex to take advantage of antibody and complement binding by Fc and complement receptors. The effects of DNA vaccination have been enhanced experimentally by injecting DNA encoding antigen coupled to CTLA-4, which enables the selective binding of the expressed protein to antigen-presenting cells carrying B7, the receptor for CTLA-4 (see Section 8-5).

A more complicated set of strategies involves targeting vaccine antigens selectively into antigen-presenting pathways within the cell. For example, human papillomavirus E7 antigen has been coupled to the signal peptide that targets a lysosomal-associated membrane protein to lysosomes and endosomes. This directs the E7 antigen directly to the intracellular compartments in which antigens are cleaved to peptides before binding to MHC class II molecules (see Section 5-5). A vaccinia virus incorporating this chimeric antigen induced a greater response in mice to E7 antigen than did vaccinia incorporating wild-type E7 antigen alone. A second approach is the use of ISCOMs, which seem to encourage the entry of peptides into the cytoplasm, thus enhancing the loading of peptides onto MHC class I molecules (see Section 14-23).

An improved understanding of the mechanisms of mucosal immunity (see Chapter 10) has led to the development of techniques to target antigens to M cells overlying Peyer's patches (see Fig. 1.10). These specialized epithelial cells lack the mucin barrier and digestive properties of other mucosal epithelial cells. Instead, they can bind and endocytose macromolecules and microorganisms, which are transcytosed intact and delivered to the underlying lymphoid tissue. In view of these properties, it is not surprising that some pathogens target M cells to gain entry to the body. The counterattack by immunologists is to gain a detailed molecular understanding of this mechanism of bacterial pathogenesis and subvert it as a delivery system for vaccines. For example, the outer membrane fimbrial proteins of *Salmonella typhimurium* have a key role in the binding of these bacteria to M cells. It might be possible to use these fimbrial proteins or, ultimately, just their binding motifs, as targeting agents for vaccines. A related strategy to encourage the uptake of mucosal vaccines by M cells is to encapsulate antigens in particulate carriers that are taken up selectively by M cells.

14-27 An important question is whether vaccination can be used therapeutically to control existing chronic infections.

There are many chronic diseases in which infection persists because of a failure of the immune system to eliminate disease. These can be divided into two groups, those infections in which there is an obvious immune response that fails to eliminate the organism, and those in which the infection seems to be invisible to the immune system and evokes a barely detectable immune response.

In the first category, the immune response is often partly responsible for the pathogenic effects. Infection by the helminth *Schistosoma mansoni* is associated with a powerful T_H2-type response, characterized by high IgE levels, circulating and tissue eosinophilia, and a harmful fibrotic response to schistosome ova, leading to hepatic fibrosis. Other common parasites, such as *Plasmodium* and *Leishmania* species, cause damage because they are not eliminated effectively by the immune response in many patients. Mycobacteria causing tuberculosis and leprosy cause persistent intracellular infection; a T_H1 response helps to contain these infections but also causes granuloma formation and tissue necrosis (see Fig. 8.43). Among viruses, hepatitis B and hepatitis C infections are commonly followed by persistent viral carriage and hepatic injury, resulting in death from hepatitis or from hepatoma. HIV infection, as we have seen in Chapter 11, persists despite an ongoing immune response.

There is a second category of chronic infection, predominantly viral, in which the immune response fails to clear infection because of the relative invisibility of the infectious agent to the immune system. A good example is herpes simplex type 2, which is transmitted venereally, becomes latent in nerve tissue, and causes genital herpes, which is frequently recurrent. This invisibility seems to be caused by a viral protein, ICP-47, which binds to the TAP complex and inhibits peptide transport into the endoplasmic reticulum in infected cells (see Chapter 4). Thus viral peptides are not presented to the immune system by MHC class I molecules. Another example in this category of chronic infection is genital warts, caused by certain papilloma viruses to which very little immune response is evoked.

There are two main immunological approaches to the treatment of chronic infection. One is to try to boost or change the pattern of the host immune response by using cytokine therapy. The second is to attempt therapeutic vaccination to see whether the host immune response can be supercharged by immunization with antigens from the infectious agent in combination with adjuvant. There has been substantial pharmaceutical investment in therapeutic vaccination but it is too early to know whether the approach will be successful.

Some promise for the cytokine therapy approach comes from the experimental treatment of leprosy: one can clear certain leprosy lesions by the injection of cytokines directly into the lesion, which may cause reversal of the type of leprosy seen. Another example in which cytokine therapy has been shown to be effective in treating an established infection depends on combining a cytokine with an anti-parasitic drug. In a proportion of mice infected with

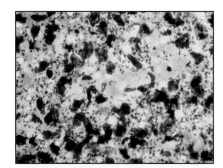

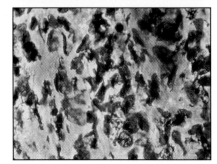

Fig. 14.29 Treatment with anti-IL-4 antibody at the time of infection with *Leishmania major* allows normally susceptible mice to clear the infection. The top panel shows a hematoxylin–eosin-stained section through the footpad of a mouse of the BALB/c strain infected with *Leishmania major* (small red dots). Large numbers of parasites are present in tissue macrophages. The bottom panel shows a similar preparation from a mouse infected in the same experiment but simultaneously treated with a single injection of anti-IL-4 monoclonal antibody. Very few parasites are present. Photographs courtesy of R.M. Locksley.

Leishmania and subsequently treated with a combination of drug therapy and IL-12, the immune response deviated from a T_H2 to a T_H1 pattern and the infection was cleared. In most of the animal studies, however, it seems that the anti-cytokine antibody or the cytokine needs to be present at the first encounter with the antigen to modulate the response effectively. For example, in experimental leishmaniasis in mice, susceptible BALB/c mice injected with anti-IL-4 antibody at the time of infection clear their infection (Fig. 14.29). However, if administration of anti-IL-4 antibody is delayed by just one week, there is progressive growth of the parasite and a dominant T_H2 response.

14-28 Modulation of the immune system might be used to inhibit immunopathological responses to infectious agents.

We have mentioned several times the possibility of modulating immunity by cytokine manipulation of the immune response. This approach is being explored as a means of inhibiting harmful immune responses to a number of important infections. As we have seen in the preceding section, the liver fibrosis in schistosomiasis results from the powerful T_H2-type response. The coadministration of *S. mansoni* ova together with IL-12 does not protect mice against subsequent infection with *S. mansoni* cercariae but has a striking effect in reducing hepatic granuloma formation and fibrosis in response to ova. IgE levels are reduced, with reduced tissue eosinophilia, and the cytokine response indicates the activation of T_H1 rather than T_H2 cells. Although these results indicate that it might be possible to use a combination of antigen and cytokines to vaccinate against the pathology of diseases for which a fully protective vaccine is unavailable, they do not solve the difficulty of applying this approach in patients whose infection is already established.

Summary.

The greatest triumphs of modern immunology have come from vaccination, which has eradicated or virtually eliminated several human diseases. It is the single most successful manipulation of the immune system so far, because it takes advantage of the immune system's natural specificity and inducibility. Nevertheless, there are many important infectious diseases for which there is still no effective vaccine. The most effective vaccines are based on attenuated live microorganisms, but these carry some risk and are potentially lethal to immunosuppressed or immunodeficient individuals. Better techniques for developing live-attenuated vaccines, or vaccines that incorporate both immunogenic components and protective antigens of pathogens, are therefore being sought. Most current viral vaccines are based on live attenuated virus, but many bacterial vaccines are based on components of the microorganism, including components of the toxins that it produces. Protective response to carbohydrate antigens can be enhanced by conjugation to a protein. Vaccines based on peptide epitopes are still at an experimental stage and have the problem that the peptide is likely to be specific for particular variants of the MHC molecules to which they must bind, as well as being only very weakly immunogenic. A vaccine's immunogenicity often depends on adjuvants that can help, directly or indirectly, to activate antigen-presenting cells that are necessary for the initiation of immune responses. The development of oral vaccines is particularly important for stimulating immunity to the many pathogens that enter through the mucosa. Cytokines have been used experimentally as adjuvants to boost the immunogenicity of vaccines or to bias the immune response along a specific path.

Summary to Chapter 14.

One of the great future challenges of immunology is the control of the immune response, so that unwanted immune responses can be suppressed and desirable responses elicited. Current methods of suppressing unwanted responses rely, to a great extent, on drugs that suppress adaptive immunity indiscriminately and are thus inherently flawed. We have seen in this book that the immune system can suppress its own responses in an antigen-specific manner and, by studying these endogenous regulatory events, it might be possible to devise strategies to manipulate specific responses while sparing general immune competence. This should allow the development of new treatments that selectively suppress the responses that lead to allergy, autoimmunity, or the rejection of grafted organs. Similarly, as we understand more about tumors and infectious agents, better strategies to mobilize the immune system against cancer and infection should become possible. To achieve all this, we need to learn more about the induction of immunity and the biology of the immune system, and to apply what we have learned to human disease.

General references.

Gilboa, E.: **The makings of a tumor rejection antigen.** *Immunity* 1999, **11**:263-270.

Miller, J.F., Heath, W.R., Allison, J., Morahan, G., Hoffmann, M., Kurts, C., and Kosaka, H.: **T cell tolerance and autoimmunity.** *Ciba Found. Symp.* 1997, **204**:159-171.

Nossal, G.J.: **Host immunobiology and vaccine development.** *Lancet* 1997, **350**:1316-1319.

Pardoll, D.M.: **Therapeutic vaccination for cancer.** *Clin. Immunol.* 2000, **95**:S44-S62.

Plotkin, S.A., and Mortimer, E.A.: *Vaccines,* 2nd edn. Philadelphia, W.B. Saunders Co., 1994.

Rappuoli, R.: **Rational design of vaccines.** *Nat. Med.* 1997, **3**:374-376.

Rocken, M., and Shevach., E.M.: **Immune deviation—the third dimension of nondeletional T cell tolerance.** *Immunol. Rev.* 1996, **149**:175-194.

Steinman, L.: **Multiple approaches to multiple sclerosis.** *Nat. Med.* 2000, **6**:15-16.

Waldmann, T.A.: **Immune receptors: targets for therapy of leukemia/lymphoma, autoimmune diseases and for the prevention of allograft rejection.** *Annu. Rev. Immunol.* 1992, **10**:675-704.

Section references.

14-1 Corticosteroids are powerful anti-inflammatory drugs that alter the transcription of many genes.

Barnes, P.J.: **Anti-inflammatory actions of glucocorticoids: molecular mechanisms.** *Clin. Sci. (Colch.)* 1998, **94**:557-572.

Boumpas, D.T., Chrousos, G.P., Wilder, R.L., Cupps, T.R., and Balow, J.E.: **Glucocorticoid therapy for immune-mediated diseases: basic and clinical correlates.** *Ann. Intern. Med.* 1993, **119**:1198-1208.

Cronstein, B.N., Kimmel, S.C., Levin, R.I., Martiniuk, F., and Weissmann, G.: **Corticosteroids are transcriptional regulators of acute inflammation.** *Trans. Assoc. Am. Physicians.* 1992, **105**:25-35.

Cupps, T.R., and Fauci, A.S.: **Corticosteroid-mediated immunoregulation in man.** *Immunol. Rev.* 1982, **65**:133-155.

14-2 Cytotoxic drugs cause immunosuppression by killing dividing cells and have serious side-effects.

Aarbakke, J., Janka-Schaub, G., and Elion, G.B.: **Thiopurine biology and pharmacology.** *Trends Pharmacol. Sci.* 1997, **18**:3-7.

Zhu, L.P., Cupps, T.R., Whalen, G., and Fauci, A.S.: **Selective effects of cyclophosphamide therapy on activation, proliferation, and differentiation of human B cells.** *J. Clin. Invest.* 1987, **79**:1082-1090.

14-3 Cyclosporin A, FK506 (tacrolimus), and rapamycin (sirolimus) are powerful immunosuppressive agents that interfere with T-cell signaling.

Brazelton, T.R., and Morris, R.E.: **Molecular mechanisms of action of new xenobiotic immunosuppressive drugs: tacrolimus (FK506), sirolimus (rapamycin), mycophenolate mofetil and leflunomide.** *Curr. Opin. Immunol.* 1996, **8**:710-720.

Crabtree, G.R.: **Generic signals and specific outcomes: signaling through Ca²⁺, calcineurin, and NF-AT.** *Cell* 1999, **96**:611-614.

14-4 Immunosuppressive drugs are valuable probes of intracellular signaling pathways in lymphocytes.

Abraham, R.T.: **Mammalian target of rapamycin: immunosuppressive drugs uncover a novel pathway of cytokine receptor signaling.** *Curr. Opin. Immunol.* 1998, **10**:330-336.

Brown, E.J., and Schreiber, S.L.: **A signaling pathway to translational control.** *Cell* 1996, **86**:517-520.

Cutler, N.S., Heitman, J., and Cardenas, M.E.: **TOR kinase homologs function in a signal transduction pathway that is conserved from yeast to mammals.** *Mol. Cell Endocrinol.* 1999, **155**:135-142.

14-5 Antibodies against cell-surface molecules have been used to remove specific lymphocyte subsets or to inhibit cell function.

Cobbold, S.P., Qin, S., Leong, L.Y., Martin, G., and Waldmann, H.: **Reprogramming the immune system for peripheral tolerance with CD4 and CD8 monoclonal antibodies.** *Immunol. Rev.* 1992, **129**:165-201.

Honey, K., Cobbold, S.P., and Waldmann, H.: **Dominant regulation: a common mechanism of monoclonal antibody induced tolerance?** *Immunol. Res.* 1999, **20**:1-14.

14-6 Antibodies can be engineered to reduce their immunogenicity in humans.

Hayden, M.S., Gilliland, L.K., and Ledbetter, J.A.: **Antibody engineering.** *Curr. Opin. Immunol.* 1997, **9**:201-212.

Little, M., Kipriyanov, S.M., Le Gall, F., and Moldenhauer, G.: **Of mice and men: hybridoma and recombinant antibodies.** *Immunol. Today* 2000, **21**:364-370.

Winter, G., Griffiths, A.D., Hawkins, R.E., and Hoogenboom, H.R.: **Making antibodies by phage display technology.** *Annu. Rev. Immunol.* 1994, **12**:433-455.

14-7 Monoclonal antibodies can be used to inhibit allograft rejection.

Charlton, B., Auchincloss, H., Jr., and Fathman, C.G.: **Mechanisms of transplantation tolerance.** *Annu. Rev. Immunol.* 1994, **12**:707-734.

Chatenoud, L.: **Tolerogenic antibodies and fusion proteins to prevent graft rejection and treat autoimmunity.** *Mol. Med. Today* 1998, **4**:25-30.

Kirk, A.D., Burkly, L.C., Batty, D.S., Baumgartner, R.E., Berning, J.D., Buchanan, K., Fechner, J.H., Jr., Germond, R.L., Kampen, R.L., Patterson, N.B., Swanson, S.J., Tadaki, D.K., TenHoor, C.N., White, L., Knechtle, S.J., and Harlan, D.M.: **Treatment with humanized monoclonal antibody against CD154 prevents acute renal allograft rejection in nonhuman primates.** *Nat. Med.* 1999, **5**:686-693.

Lechler, R., and Bluestone, J.A.: **Transplantation tolerance—putting the pieces together.** *Curr. Opin. Immunol.* 1997, **9**:631-633.

Waldmann, H., and Cobbold, S.: **How do monoclonal antibodies induce tolerance? A role for infectious tolerance?** *Annu. Rev. Immunol.* 1998, **16**:619-644.

14-8 Antibodies can be used to alleviate and suppress autoimmune disease.

Bach, J.F.: **Tolerance induction in transplantation and autoimmune diseases.** *Mol. Med. Today* 1995, **1**:302-303.

Feldmann, M., Elliott, M.J., Woody, J.N., and Maini, R.N.: **Anti-tumor necrosis factor-alpha therapy of rheumatoid arthritis.** *Adv. Immunol.* 1997, **64**:283-350.

Karin, N., Mitchell, D.J., Brocke, S., Ling, N., and Steinman, L.: **Reversal of experimental autoimmune encephalomyelitis by a soluble peptide variant of a myelin basic protein epitope: T cell receptor antagonism and reduction of interferon gamma and tumor necrosis factor alpha production.** *J. Exp. Med.* 1994, **180**:2227-2237.

Moreau, T., Coles, A., Wing, M., Thorpe, J., Miller, D., Moseley, I., Issacs, J., Hale, G., Clayton, D., Scolding, N., Waldmann, H., and Compston, A.: **CAMPATH-IH in multiple sclerosis.** *Mult. Scler.* 1996, **1**:357-365.

Riethmuller, G., Rieber, E.P., Kiefersauer, S., Prinz, J., van der Lubbe, P., Meiser, B., Breedveld, F., Eisenburg, J., Kruger, K., Deusch, K. et al.: **From anti-lymphocyte serum to therapeutic monoclonal antibodies: first experiences with a chimeric CD4 antibody in the treatment of autoimmune disease.** *Immunol. Rev.* 1992, **129**:81-104.

14-9   Modulation of the pattern of cytokine expression by T lymphocytes can inhibit autoimmune disease.

Adorini, L., and Sinigaglia, F.: **Pathogenesis and immunotherapy of auto-immune diseases.** *Immunol. Today* 1997, **18**:209-211.

Bach, J.F.: **Cytokine-based immunomodulation of autoimmune diseases: an overview.** *Transplant. Proc.* 1996, **28**:3023-3025.

Coffman, R.L., Mocci, S., and O'Garra, A.: **The stability and reversibility of Th1 and Th2 populations.** *Curr. Top. Microbiol. Immunol.* 1999, **238**:1-12.

Fukaura, H., Kent, S.C., Pietrusewicz, M.J., Khoury, S.J., Weiner, H.L., and Hafler, D.A.: **Induction of circulating myelin basic protein and proteolipid protein-specific transforming growth factor-beta1-secreting Th3 T cells by oral administration of myelin in multiple sclerosis patients.** *J. Clin. Invest.* 1996, **98**:70-77.

Hafler, D.A., Kent, S.C., Pietrusewicz, M.J., Khoury, S.J., Weiner, H.L., and Fukaura, H.: **Oral administration of myelin induces antigen-specific TGF-beta 1 secreting T cells in patients with multiple sclerosis.** *Ann. N.Y. Acad. Sci.* 1997, **835**:120-131.

Levings, M.K., and Roncarolo, M.G.: **T-regulatory 1 cells: a novel subset of CD4 T cells with immunoregulatory properties.** *J. Allergy Clin. Immunol.* 2000, **106**:S109-S112.

O'Garra, A.: **Cytokines induce the development of functionally heterogeneous T helper cell subsets.** *Immunity* 1998, **8**:275-283.

Smith, J.A., and Bluestone, J.A.: **T cell inactivation and cytokine deviation promoted by anti-CD3 mAbs.** *Curr. Opin. Immunol.* 1997, **9**:648-654.

14-10   Controlled administration of antigen can be used to manipulate the nature of an antigen-specific response.

Fairchild, P.J.: **Altered peptide ligands: prospects for immune intervention in autoimmune disease.** *Eur. J. Immunogenet.* 1997, **24**:155-167.

Liblau, R., Tisch, R., Bercovici, N., and McDevitt, H.O.: **Systemic antigen in the treatment of T-cell-mediated autoimmune diseases.** *Immunol. Today* 1997, **18**:599-604.

Magee, C.C., and Sayegh, M.H.: **Peptide-mediated immunosuppression.** *Curr. Opin. Immunol.* 1997, **9**:669-675.

Weiner, H.L.: **Oral tolerance for the treatment of autoimmune diseases.** *Annu. Rev. Med.* 1997, **48**:341-351.

Wraith, D.C.: **Antigen-specific immunotherapy of autoimmune disease: a commentary.** *Clin. Exp. Immunol.* 1996, **103**:349-352.

14-11   The development of transplantable tumors in mice led to the discovery that mice could mount a protective immune response against tumors.

Jaffee, E.M., and Pardoll, D.M.: **Murine tumor antigens: is it worth the search?** *Curr. Opin. Immunol.* 1996, **8**:622-627.

14-12   T lymphocytes can recognize specific antigens on human tumors.

Boon, T., Coulie, P.G., and Van den Eynde, B.: **Tumor antigens recognized by T cells.** *Immunol. Today* 1997, **18**:267-268.

Chaux, P., Vantomme, V., Stroobant, V., Thielemans, K., Corthals, J., Luiten, R., Eggermont, A.M., Boon, T., and van der Bruggen, P.: **Identification of MAGE-3 epitopes presented by HLA-DR molecules to CD4(+) T lymphocytes.** *J. Exp. Med.* 1999, **189**:767-778.

de Smet, C., Lurquin, C., Lethe, B., Martelange, V., and Boon, T.: **DNA methylation is the primary silencing mechanism for a set of germ line- and tumor-specific genes with a CpG-rich promoter.** *Mol. Cell Biol.* 1999, **19**:7327-7335.

Disis, M.L., and Cheever, M.A.: **HER-2/neu oncogenic protein: issues in vaccine development.** *Crit. Rev. Immunol.* 1998, **18**:37-45.

Disis, M.L., and Cheever, M.A.: **Oncogenic proteins as tumor antigens.** *Curr. Opin. Immunol.* 1996, **8**:637-642.

Robbins, P.F., and Kawakami, Y.: **Human tumor antigens recognized by T cells.** *Curr. Opin. Immunol.* 1996, **8**:628-636.

14-13   Tumors can escape rejection in many ways.

Bodmer, W.F., Browning, M.J., Krausa, P., Rowan, A., Bicknell, D.C., and Bodmer, J.G.: **Tumor escape from immune response by variation in HLA expression and other mechanisms.** *Ann N.Y. Acad. Sci.* 1993, **690**:42-49.

Ikeda, H., Lethe, B., Lehmann, F., van Baren, N., Baurain, J.F., de-Smet, C., Chambost, H., Vitale, M., Moretta, A., Boon, T., and Coulie, P.G.: **Characterization of an antigen that is recognized on a melanoma showing partial HLA loss by CTL expressing an NK inhibitory receptor.** *Immunity* 1997, **6**:199-208.

Koopman, L.A., Corver, W.E., van der Slik, A.R., Giphart, M.J., and Fleuren, G.J.: **Multiple genetic alterations cause frequent and heterogeneous human histocompatibility leukocyte antigen class I loss in cervical cancer.** *J. Exp. Med.* 2000, **191**:961-976.

Tada, T., Ohzeki, S., Utsumi, K., Takiuchi, H., Muramatsu, M., Li, X.F., Shimizu, J., Fujiwara, H., and Hamaoka, T.: **Transforming growth factor-beta-induced inhibition of T cell function. Susceptibility difference in T cells of various phenotypes and functions and its relevance to immunosuppression in the tumor-bearing state.** *J. Immunol.* 1991, **146**:1077-1082.

Torre Amione, G., Beauchamp, R.D., Koeppen, H., Park, B.H., Schreiber, H., Moses, H.L., and Rowley, D.A.: **A highly immunogenic tumor transfected with a murine transforming growth factor type beta 1 cDNA escapes immune surveillance.** *Proc. Natl. Acad. Sci. USA* 1990, **87**:1486-1490.

14-14   Monoclonal antibodies against tumor antigens, alone or linked to toxins, can control tumor growth.

Bendandi, M., and Longo, D.L.: **Biologic therapy for lymphoma.** *Curr. Opin. Oncol.* 1999, **11**:343-350.

Cragg, M.S., French, R.R., and Glennie, M.J.: **Signaling antibodies in cancer therapy.** *Curr. Opin. Immunol.* 1999, **11**:541-547.

Fan, Z., and Mendelsohn, J.: **Therapeutic application of anti-growth factor receptor antibodies.** *Curr. Opin. Oncol.* 1998, **10**:67-73.

Houghton, A.N., and Scheinberg, D.A.: **Monoclonal antibody therapies—a 'constant' threat to cancer.** *Nat. Med.* 2000, **6**:373-374.

Kreitman, R.J.: **Immunotoxins in cancer therapy.** *Curr. Opin. Immunol.* 1999, **11**:570-578.

Weiner, L.M.: **Monoclonal antibody therapy of cancer.** *Semin. Oncol.* 1999, **26**:43-51.

14-15   Enhancing the immunogenicity of tumors holds promise for cancer therapy.

Bendandi, M., Gocke, C.D., Kobrin, C.B., Benko, F.A., Sternas, L.A., Pennington, R., Watson, T.M., Reynolds, C.W., Gause, B.L., Duffey, P.L., Jaffe, E.S., Creekmore, S.P., Longo, D.L., and Kwak, L.W.: **Complete molecular remissions induced by patient-specific vaccination plus granulocyte-monocyte colony-stimulating factor against lymphoma.** *Nat. Med.* 1999, **5**:1171-1177.

Fong, L., and Engleman, E.G.: **Dendritic cells in cancer immunotherapy.** *Annu. Rev. Immunol.* 2000, **18**:245-273.

Hellstrom, K.E., Gladstone, P., and Hellstrom, I.: **Cancer vaccines: challenges and potential solutions.** *Mol. Med. Today* 1997, **3**:286-290.

Kugler, A., Stuhler, G., Walden, P., Zoller, G., Zobywalski, A., Brossart, P., Trefzer, U., Ullrich, S., Muller, C.A., Becker, V., Gross, A.J., Hemmerlein, B., Kanz, L., Muller, G.A., and Ringert, R.H.: **Regression of human metastatic renal cell carcinoma after vaccination with tumor cell-dendritic cell hybrids.** *Nat. Med.* 2000, **6**:332-336.

Li, Y., Hellstrom, K.E., Newby, S.A., and Chen, L.: **Costimulation by CD48 and B7-1 induces immunity against poorly immunogenic tumors.** *J. Exp. Med.* 1996, **183**:639-644.

Melief, C.J., Offringa, R., Toes, R.E., and Kast, W.M.: **Peptide-based cancer vaccines.** *Curr. Opin. Immunol.* 1996, **8**:651-657.

Pardoll, D.M.: **Cancer vaccines.** *Nat. Med.* 1998, **4**:525-531.

Pardoll, D.M.: **Paracrine cytokine adjuvants in cancer immunotherapy.** *Annu. Rev. Immunol.* 1995, **13**:399-415.

Przepiorka, D., and Srivastava, P.K.: **Heat shock protein–peptide complexes as immunotherapy for human cancer.** *Mol. Med. Today* 1998, **4**:478-484.

Ragnhammar, P.: **Anti-tumoral effect of GM-CSF with or without cytokines and monoclonal antibodies in solid tumors.** *Med. Oncol.* 1996, **13**:167-176.

Schuler, G., and Steinman, R.M.: **Dendritic cells as adjuvants for immune-mediated resistance to tumors.** *J. Exp. Med.* 1997, **186**:1183-1187.

14-16   There are several requirements for an effective vaccine.

Ada, G.L.: **The immunological principles of vaccination.** *Lancet* 1990, **335**:523-526.

Anderson, R.M., Donnelly, C.A., and Gupta, S.: **Vaccine design, evaluation, and community-based use for antigenically variable infectious agents.** *Lancet* 1997, **350**:1466-1470.

Levine, M.M., and Levine, O.S.: **Influence of disease burden, public perception, and other factors on new vaccine development, implementation, and continued use.** *Lancet* 1997, **350**:1386-1392.

Nichol, K.L., Lind, A., Margolis, K.L., Murdoch, M., McFadden, R., Hauge, M., Magnan, S., and Drake, M.: **The effectiveness of vaccination against influenza in healthy, working adults.** *N. Engl. J. Med.* 1995, **333**:889-893.

Rabinovich, N.R., McInnes, P., Klein, D.L., and Hall, B.F.: **Vaccine technologies: view to the future.** *Science* 1994, **265**:1401-1404.

14-17   The history of vaccination against *Bordetella pertussis* illustrates the importance of developing an effective vaccine that is perceived to be safe.

Decker, M.D., and Edwards, K.M.: **Acellular pertussis vaccines.** *Pediatr. Clin. North Am.* 2000, **47**:309-335.

Mortimer, E.A.: **Pertussis vaccines**, in Plotkin, S.A., and Mortimer, E.A.: *Vaccines*, 2nd edn. Philadelphia, W.B. Saunders Co., 1994.

Poland, G.A.: **Acellular pertussis vaccines: new vaccines for an old disease.** *Lancet* 1996, **347**:209-210.

14-18   Conjugate vaccines have been developed as a result of understanding how T and B cells collaborate in an immune response.

van den Dobbelsteen, G.P., and van Rees, E.P.: **Mucosal immune responses to pneumococcal polysaccharides: implications for vaccination.** *Trends Microbiol.* 1995, **3**:155-159.

Kroll, J.S., and Booy, R.: ***Haemophilus influenzae*: capsule vaccine and capsulation genetics.** *Mol. Med. Today* 1996, **2**:160-165.

Peltola, H., Kilpi, T., and Anttila, M.: **Rapid disappearance of *Haemophilus influenzae* type b meningitis after routine childhood immunisation with conjugate vaccines.** *Lancet* 1992, **340**:592-594.

Rosenstein, N.E., and Perkins, B.A.: **Update on *Haemophilus influenzae* serotype b and meningococcal vaccines.** *Pediatr. Clin. North Am.* 2000, **47**:337-352.

14-19   The use of adjuvants is another important approach to enhancing the immunogenicity of vaccines.

Alving, C.R., Koulchin, V., Glenn, G.M., and Rao, M.: **Liposomes as carriers of peptide antigens: induction of antibodies and cytotoxic T lymphocytes to conjugated and unconjugated peptides.** *Immunol. Rev.* 1995, **145**:5-31.

Audibert, F.M., and Lise, L.D.: **Adjuvants: current status, clinical perspectives and future prospects.** *Immunol. Today* 1993, **14**:281-284.

Gupta, R.K., and Siber, G.R.: **Adjuvants for human vaccines—current status, problems and future prospects.** *Vaccine* 1995, **13**:1263-1276.

Hartmann, G., Weiner, G.J., and Krieg, A.M.: **CpG DNA: a potent signal for growth, activation, and maturation of human dendritic cells.** *Proc. Natl. Acad. Sci. USA* 1999, **96**:9305-9310.

Rhodes, J., Chen, H., Hall, S.R., Beesley, J.E., Jenkins, D.C., Collins, P., and Zheng, B.: **Therapeutic potentiation of the immune system by costimulatory Schiff-base-forming drugs.** *Nature* 1995, **376**:71-75.

Scott, P., and Trinchieri, G.: **IL-12 as an adjuvant for cell-mediated immunity.** *Semin. Immunol.* 1997, **9**:285-291.

Takahashi, H., Takeshita, T., Morein, B., Putney, S., Germain, R.N., and Berzofsky, J.A.: **Induction of CD8⁺ cytotoxic T cells by immunization with purified HIV-1 envelope protein in ISCOMs.** *Nature* 1990, **344**:873-875.

Vogel, F.R.: **Immunologic adjuvants for modern vaccine formulations.** *Ann. N.Y. Acad. Sci.* 1995, **754**:153-160.

14-20   Live-attenuated viral vaccines are usually more potent than 'killed' vaccines and can be made safer by using recombinant DNA technology.

Brochier, B., Kieny, M.P., Costy, F., Coppens, P., Bauduin, B., Lecocq, J.P., Languet, B., Chappuis, G., Desmettre, P., Afiademanyo, K., et al.: **Large-scale eradication of rabies using recombinant vaccinia-rabies vaccine.** *Nature* 1991, **354**:520-522.

Parkin, N.T., Chiu, P., and Coelingh, K.: **Genetically engineered live attenuated *influenza* A virus vaccine candidates.** *J. Virol.* 1997, **71**:2772-2778.

14-21   Live-attenuated bacterial vaccines can be developed by selecting nonpathogenic or disabled mutants.

Guleria, I., Teitelbaum, R., McAdam, R.A., Kalpana, G., Jacobs, W.R., Jr., and Bloom, B.R.: **Auxotrophic vaccines for tuberculosis.** *Nat. Med.* 1996, **2**:334-337.

Hassan, J.O., Curtiss, R., III.: **Effect of vaccination of hens with an avirulent strain of *Salmonella typhimurium* on immunity of progeny challenged with wild-type *Salmonella* strains.** *Infect. Immun.* 1996, **64**:938-944.

Levine, M.M., Galen, J., Barry, E., Noriega, F., Tacket, C., Sztein, M., Chatfield, S., Dougan, G., Losonsky, G., and Kotloff, K.: **Attenuated *Salmonella typhi* and *Shigella* as live oral vaccines and as live vectors.** *Behring Inst. Mitt.* 1997, **98**:120-123.

14-22   Attenuated microorganisms can serve as vectors for vaccination against many pathogens.

Chatfield, S.N., Roberts, M., Dougan, G., Hormaeche, C., and Khan, C.M.: **The development of oral vaccines against parasitic diseases utilizing live attenuated *Salmonella*.** *Parasitology* 1995, **110 Suppl**:S17-S24.

Cirillo, J.D., Stover, C.K., Bloom, B.R., Jacobs, W.R., Jr., and Barletta, R.G.: **Bacterial vaccine vectors and bacillus Calmette-Guerin.** *Clin. Infect. Dis.* 1995, **20**:1001-1009.

Moss, B.: **Genetically engineered poxviruses for recombinant gene expression, vaccination, and safety.** *Proc. Natl. Acad. Sci. USA* 1996, **93**:11341-11348.

Paoletti, E.: **Applications of pox virus vectors to vaccination: an update.** *Proc. Natl. Acad. Sci. USA* 1996, **93**:11349-11353.

14-23   Synthetic peptides of protective antigens can elicit protective immunity.

Berzofsky, J.A.: **Epitope selection and design of synthetic vaccines. Molecular approaches to enhancing immunogenicity and cross-reactivity of engineered vaccines.** *Ann. N.Y. Acad. Sci.* 1993, **690**:256-264.

Berzofsky, J.A.: **Mechanisms of T cell recognition with application to vaccine design.** *Mol. Immunol.* 1991, **28**:217-223.

Davenport, M.P., and Hill, A.V.: **Reverse immunogenetics: from HLA-disease associations to vaccine candidates.** *Mol. Med. Today* 1996, **2**:38-45.

Hoffman, S.L., Rogers, W.O., Carucci, D.J., and Venter, J.C.: **From genomics to vaccines: malaria as a model system.** *Nat. Med.* 1998, **4**:1351-1353.

Modelska, A., Dietzschold, B., Sleysh, N., Fu, Z.F., Steplewski, K., Hooper, D.C., Koprowski, H., and Yusibov, V.: **Immunization against rabies with plant-derived antigen.** *Proc. Natl. Acad. Sci. USA* 1998, **95**:2481-2485.

14-24   The route of vaccination is an important determinant of success.

Burnette, W.N.: **Bacterial ADP-ribosylating toxins: form, function, and** recombinant vaccine development. *Behring Inst. Mitt.* 1997, **98**:434-441.

Douce, G., Fontana, M., Pizza, M., Rappuoli, R., and Dougan, G.: **Intranasal immunogenicity and adjuvanticity of site-directed mutant derivatives of cholera toxin.** *Infect. Immun.* 1997, **65**:2821-2828.

Dougan, G.: **The molecular basis for the virulence of bacterial pathogens: implications for oral vaccine development.** *Microbiology* 1994, **140**:215-224.

Dougan, G., Ghaem-Maghami, M., Pickard, D., Frankel, G., Douce, G., Clare, S., Dunstan, S., and Simmons, C.: **The immune responses to bacterial antigens encountered *in vivo* at mucosal surfaces.** *Philos. Trans. R. Soc. Lond. B. Biol. Sci.* 2000, **355**:705-712.

Ivanoff, B., Levine, M.M., and Lambert, P.H.: **Vaccination against typhoid fever: present status.** *Bull. World Health Organ.* 1994, **72**:957-971.

Levine, M.M.: **Modern vaccines. Enteric infections.** *Lancet* 1990, **335**:958-961.

14-25 Protective immunity can be induced by injecting DNA encoding microbial antigens and human cytokines into muscle.

Donnelly, J.J., Ulmer, J.B., Shiver, J.W., and Liu, M.A.: **DNA vaccines.** *Annu. Rev. Immunol.* 1997, **15**:617-648.

Gurunathan, S., Klinman, D.M., and Seder, R.A.: **DNA vaccines: immunology, application, and optimization.** *Annu. Rev. Immunol.* 2000, **18**:927-974.

14-26 The effectiveness of a vaccine can be enhanced by targeting it to sites of antigen presentation.

Deliyannis, G., Boyle, J.S., Brady, J.L., Brown, L.E., and Lew, A.M.: **A fusion DNA vaccine that targets antigen-presenting cells increases protection from viral challenge.** *Proc. Natl. Acad. Sci. USA* 2000, **97**:6676-6680.

Hahn, H., Lane-Bell, P.M., Glasier, L.M., Nomellini, J.F., Bingle, W.H., Paranchych, W., and Smit, J.: **Pilin-based anti-Pseudomonas vaccines: latest developments and perspectives.** *Behring Inst. Mitt.* 1997, **98**:315-325.

Neutra, M.R.: **Current concepts in mucosal immunity. V. Role of M cells in transepithelial transport of antigens and pathogens to the mucosal immune system.** *Am. J. Physiol.* 1998, **274**:G785-G791.

Shen, Z., Reznikoff, G., Dranoff, G., and Rock, K.L.: **Cloned dendritic cells can present exogenous antigens on both MHC class I and class II molecules.** *J. Immunol.* 1997, **158**:2723-2730.

Tan, M.C., Mommaas, A.M., Drijfhout, J.W., Jordens, R., Onderwater, J.J., Verwoerd, D., Mulder, A.A., van-der-Heiden, A.N., Scheidegger, D., Oomen, L.C., Ottenhoff, T.H., Tulp, A., Neefjes, J.J., and Koning, F.: **Mannose receptor-mediated uptake of antigens strongly enhances HLA class II-restricted antigen presentation by cultured dendritic cells.** *Eur. J. Immunol.* 1997, **27**:2426-2435.

Thomson, S.A., Burrows, S.R., Misko, I.S., Moss, D.J., Coupar, B.E., and Khanna, R.: **Targeting a polyepitope protein incorporating multiple class II-restricted viral epitopes to the secretory/endocytic pathway facilitates immune recognition by CD4+ cytotoxic T lymphocytes: a novel approach to vaccine design.** *J. Virol.* 1998, **72**:2246-2252.

14-27 An important question is whether vaccination can be used therapeutically to control existing chronic infections.

Burke, R.L.: **Contemporary approaches to vaccination against herpes simplex virus.** *Curr. Top. Microbiol. Immunol.* 1992, **179**:137-158.

Grange, J.M., and Stanford, J.L.: **Therapeutic vaccines.** *J. Med. Microbiol.* 1996, **45**:81-83.

Hill, A., Jugovic, P., York, I., Russ, G., Bennink, J., Yewdell, J., Ploegh, H., and Johnson, D.: **Herpes simplex virus turns off the TAP to evade host immunity.** *Nature* 1995, **375**:411-415.

Modlin, R.L.: **Th1-Th2 paradigm: insights from leprosy.** *J. Invest. Dermatol.* 1994, **102**:828-832.

Reiner, S.L., and Locksley, R.M.: **The regulation of immunity to *Leishmania major*.** *Annu. Rev. Immunol.* 1995, **13**:151-177.

Stanford, J.L.: **The history and future of vaccination and immunotherapy for leprosy.** *Trop. Geogr. Med.* 1994, **46**:93-107.

14-28 Modulation of the immune system might be used to inhibit immunopathological responses to infectious agents.

Biron, C.A., and Gazzinelli, R.T.: **Effects of IL-12 on immune responses to microbial infections: a key mediator in regulating disease outcome.** *Curr. Opin. Immunol.* 1995, **7**:485-496.

Grau, G.E., and Modlin, R.L.: **Immune mechanisms in bacterial and parasitic diseases: protective immunity versus pathology.** *Curr. Opin. Immunol.* 1991, **3**:480-485.

Kaplan, G.: **Recent advances in cytokine therapy in leprosy.** *J. Infect. Dis.* 1993, **167 Suppl 1**:S18-S22.

Locksley, R.M.: **Interleukin 12 in host defense against microbial pathogens.** *Proc. Natl. Acad. Sci. USA* 1993, **90**:5879-5880.

Murray, H.W.: **Interferon-gamma and host antimicrobial defense: current and future clinical applications.** *Am. J. Med.* 1994, **97**:459-467.

Sher, A., Gazzinelli, R.T., Oswald, I.P., Clerici, M., Kullberg, M., Pearce, E.J., Berzofsky, J.A., Mosmann, T.R., James, S.L., and Morse, H.C.: **Role of T-cell derived cytokines in the downregulation of immune responses in parasitic and retroviral infection.** *Immunol. Rev.* 1992, **127**:183-204.

Sher, A., Jankovic, D., Cheever, A., and Wynn, T.: **An IL-12-based vaccine approach for preventing immunopathology in schistosomiasis.** *Ann. N.Y. Acad. Sci.* 1996, **795**:202-207.

Evolution of the Immune System: Past, Present, and Future,

by Charles A. Janeway, Jr.

We began this book with an overview of immunology and its fascination for scientists over the last century. In this chapter, we will reexamine the basic facts of immunology in terms of how they evolved over the eons. We will start, as we did in this book, with the evolution of the innate immune system, which is almost as old as the first multicellular organisms on the planet. Then we will look at the fascinating question of the evolution of the adaptive immune system in vertebrates, of which much has been learned over the last few years, especially by people who study gene rearrangement. The third part of the chapter will analyze how immunological memory has played a critical role in the evolution of the vertebrates. Immunological memory makes resistance to reinfection possible and so is obviously important for survival on a planet teeming with pathogens; and those organisms endowed with it survive better than those that lack it. Most vertebrates have it, and the theory is that the possession of immunological memory is an important factor in enabling the vertebrates to dominate the Earth. Whether or not this is a good thing is open to question, but that is outside the scope of this book.

Finally, the question is, or rather the questions are, what will the future of immunology look like? How will it be affected by the human genome project, for instance? How many genes will we find to be devoted to host defense? I am willing to bet that the genes that mediate defense against infection will make up a fair proportion of the human genome, somewhere between 1% and 10%, because host defense is such a fundamental property for life on Earth. Because the immune system has been studied so intensively over the last century, and especially the last twenty-five years, I would argue that it is the best-characterized system in biology. Yet there are vast areas of ignorance about how the immune system functions and what genes are responsible for the important choices that have to be made in immune responses. I think we have many problems to solve, especially in the clinical realm. Research in these areas will bring us much farther along than we are now, so young students and investigators need not worry that knowing the human genome, for instance, will solve all immunological problems. Indeed, the real point of immunology as a science is to figure out all the mysteries of host defense and its obverse, allograft rejection. Furthermore, although we have learned a lot about why and how cancer happens, we still have very crude ideas about how to treat it. If we could get the immune system to mount a specific response against the body's tumor cells, while sparing neighboring tissue cells, we would have basically cured cancer. What a wonderful goal, and what a praiseworthy (and indeed prizeworthy) achievement!

So I continue to be optimistic for the careers of those young scientists who, as I did once, despair that all the interesting problems have been solved when it is likely that only a small fraction of a percent have been. There are great achievements still to come, such as a vaccine for AIDS, a cure for autoimmune diseases like insulin-dependent diabetes, and control of many of the infectious diseases that have ravaged the Earth for centuries and continue to do so. It has been the dream of many that the achievements in immunology will add years to the lives of many people. One consequence of the elimination of disease would make the control of the world's population a primary goal for the future, and immunology may contribute to the control of population by vaccinating against fertility. All these things and many as yet unimagined benefits can be expected to excite and invigorate the students of the future.

Evolution of the innate immune system.

The innate immune system is well developed in the fruit fly *Drosophila melanogaster*, a favorite model organism for many aspects of biological research, and in many other invertebrates, including the nematode worm, *Caenorhabditis elegans*. What these organisms share in common with the vertebrates are the genes that encode intracellular signaling pathways leading from the cell surface to the activation of the transcription factor NFκB (see Chapter 6). Each organism has a cassette of genes that encode the proteins of this pathway. That makes us believe that the activation of NFκB is the original and central signaling pathway of activation in innate immunity, leading in turn to the activation of a set of genes that depend on NFκB for their transcription. This pathway is a universal pathway that leads to activation in all host defense systems, as we will learn in this part of the chapter.

Innate immunity has its origins in early eukaryotes such as the amoeba.

Many of us grew up marveling at the amoeba's abundance in pond water; if you look at amoebas under a high-power lens, you can see them wandering around on the slide, but you will also see that they are feeding on microorganisms, just like a culture of macrophages. It seems as if the amoeba is the earliest form of macrophage, and perhaps gave rise, by an unknown evolutionary pathway, to the modern macrophage. Innate immunity in eukaryotes can be thought of as arising from the need of a unicellular microorganism such as an amoeba to discriminate between food and other amoebas. If you think about it, any amoeba that could not make this distinction would be bound to consume itself and vanish from the face of the Earth. Therefore, we can infer a specific surface receptor on amoebas that acts to discriminate between food, which can eagerly be engulfed, from what is another amoeba, or even another part of the same amoeba. The nature of this presumed receptor is not yet known, but it must be highly specific and must discriminate self from nonself, which is one of the most basic functions of the immune system.

Like macrophages, amoebas move around under the microscope seemingly at random, unless exposed to a chemoattractant. Then, they all head in the same direction. In this respect also, amoebas behave like macrophages, and may well have occupied the coelomic cavity of early multicellular organisms as useful passengers. All vertebrates and many invertebrates have a population of phagocytic cells that patrol their blood vessels and tissues, as described in Chapter 2, and which have much in common with amoebas. It is possible that such phagocytic cells, the probable ancestors of macrophages, could derive from a population of cells within the multicellular organism that retained an ancestral, unicellular morphology—a form of evolutionary neoteny—the expression of primitive traits—in which the ontogeny of the macrophage would recapitulate its phylogeny.

One further mystery about macrophages is whether evolutionarily they are the source of dendritic cells and lymphocytes. The origin of lymphocytes, which we will deal with later, is a mystery in itself. But the origin of dendritic cells, which appear to have no other function than to present antigen to T cells, and therefore must have arisen after the evolution of the T lymphocyte, is also curious. Did they evolve simultaneously as antigen-presenting cells with their target, the thymus-derived T cell, and what is their function in the absence of T cells? These questions also seem important to me.

Sophisticated means of host defense were hard-wired in the genome by the time organisms diverged into plants and animals.

Genomic analysis of plants and animals provides evidence that a sophisticated mechanism of host defense was in existence by the time the ancestors of plants and animals diverged. This system, shared by plants and animals, is the Toll pathway of NFκB activation of gene function. This pathway has been demonstrated conclusively in fruit flies such as *Drosophila* and in vertebrates such as mice and humans, and is also believed to occur in plants, where the evidence for it is less direct. The necessary DNA sequences are, however, found in all three classes of organisms—invertebrates, vertebrates, and plants. More compellingly, there is evidence in all three groups that the products of these shared genes interact in similar pathways with a role in host defense. In the fruit fly, where this genetic module was discovered as the organizer of the dorsal–ventral axis during embryonic development, it was subsequently shown also to be essential for host defense. We will discuss *Drosophila* in the next section, as it provides the most complete story. For now, we will take it as the foundation of host defense in all organisms except prokaryotes.

When the molecules of the Toll pathway were looked for in the mouse, they were relatively easy to identify. On searching a library of mouse expressed sequence tags (ESTs), several fragments of a gene for mouse Toll were isolated, now known as mouse **Toll-like receptor 4** (**TLR-4**). It turns out that the mouse has ten Toll-like receptor genes, each seemingly involved in a variety of host defense functions.

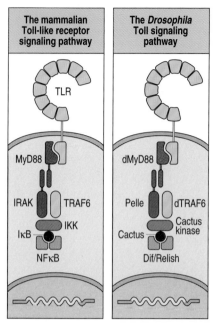

Fig. 1 A comparison of the *Drosophila* and mammalian Toll signaling pathways. The components of the mammalian Toll-like receptor signaling pathway that culminates in the activation of NFκB have direct parallels in the components of the signaling pathway from the Toll receptor of *Drosophila*. The intracellular domain of the Toll-like receptors interacts with a homologous domain in the adaptor protein MyD88. A similar interaction occurs between the intracellular domain of Toll and dMyD88. The next step in both signaling pathways occurs via the interaction of death domains, between MyD88 and IRAK in the mammalian cells and between dMyD88 and Pelle in *Drosophila*. Both IRAK and Pelle are serine kinases. At this point the mammalian signaling pathway passes through an adaptor, TRAF6, which is activated by IRAK and in turn activates IKK. IKK phosphorylates the inhibitor of NFκB, IκB, targeting it for degradation and releasing the active dimeric transcription factor, NFκB. In *Drosophila*, homologues of MyD88, TRAF6, and of a kinase that is homologous to IKK that phosphorylates the *Drosophila* IκB homologue Cactus are found. Moreover, the terminal parts of the pathway are also homologous between *Drosophila* and mammals; phosphorylation of Cactus initiates its degradation and the release of the Dif/Relish dimer, which is a transcription factor and homologue of NFκB.

The first mouse *Toll* gene isolated turned out to be defective in two mouse strains that cannot respond to bacterial lipopolysaccharide (LPS), one of the pathogen-associated molecular patterns, or PAMPs, recognized by innate immune system **pattern recognition receptors**. These mice lack TLR-4 function; in one strain the defect is due to a point mutation in the so-called TIR domain (Toll/IL-1 receptor domain, since it is found in both Toll and IL-1 receptors), while in the other strain it is due to a null mutation that abolishes expression of the gene altogether. These mice are exceptionally susceptible to infection with gram-negative bacteria, which carry LPS on their surface, and cannot mount an adaptive immune response against them. This was, in a sense, the first proof that the loss of innate immunity had a discernible effect on the adaptive immune response, and served as a proof in principle that the adaptive immune response depended on an effective innate immune response, at least in some cases. The question remaining is: How general is this basic principle?

Fruit flies illustrate the virtues of a nonclonal system of host defense.

The common fruit fly, *Drosophila melanogaster*, is a wonderful model for studying aspects of host defense that are obscured by the adaptive immune response in vertebrates. The study of insect immunology, which was pioneered by research groups in Sweden and France, clearly demonstrated the relative efficacy of a nonclonal system of host defense. One of the most obvious advantages was the absence of autoimmune diseases, which were therefore clearly shown to depend on adaptive immunity.

One of the most surprising results to emerge from an analysis of fly immunity to infection with various microorganisms was that there was a primitive form of specific recognition. For instance, flies bearing different *Toll* mutations were susceptible to infection with different types of pathogens, leading to the belief that the innate immune system had developed its own specificity sensors. This is still being investigated, but it looks as though a general specificity system based on variations in *Toll* and other pattern-recognition receptors exists in the fruit fly, and thus, may exist in humans as well. How extensive these variations are, and whether they are important in animals such as mice and humans, are as yet unknown.

Many genes that operate in fruit fly immunity also operate in humans and plants and appear to be universal components of host defense.

Many of the genes involved in immunity to disease in flies have homologues that also operate in humans, for example, *Toll* genes in the fly, *Drosophila* and *TLR* genes in man. Homologues of these genes have also been found in mice, sharks, nematodes, and plants. Furthermore, they are involved in host defense in all of the species in which they have been studied. Their identification in plants is most impressive, where the genes were identified by positional cloning of disease-resistance genes, a very demanding technique.

In fruit flies, there is a very strict order of Toll pathway gene products starting with Toll and going on to dMyD88, Pelle, Cactus, and Dif/Relish, all of which are cytoplasmic proteins involved in the transmission of the signal from Toll, a cell-surface receptor, to the nucleus to induce the activation of specific sets of genes. The same order is found in the homologues of the Toll pathway found in the innate immune system in vertebrates (Fig. 1). The plant genes do not seem to be arranged in the same order. However, if one examines the plant genome carefully, there are signs of all these signaling elements. There are clear-cut examples of some genes that are separate in the fly and fused in plants, and vice versa. This consistency of function, structure, order, and purpose over such a wide evolutionary range is a most impressive example of the evolution of a biological function, save for essential processes such as DNA and RNA replication and cell division. It makes one feel certain that the Toll receptors are performing the same defensive function throughout evolution. In mammals, as we learned in Chapter 2, their role is to induce co-stimulatory molecules on the surfaces of myeloid cells that have taken up pathogens, and thus pave the way for the induction of the adaptive immune response, which we will discuss in the next section.

Summary.

The innate immune system exists to provide early defense against pathogen attack, and to alert the adaptive immune system to the fact that pathogen invasion has begun. This dual function appears to operate through a very ancient signaling pathway, the Toll pathway, that long predates the adaptive immune system, and is present in the fruit fly, vertebrates, and, most probably, also in plants. Another component of innate immunity, the phagocytic cells such as macrophages that scavenge incoming pathogens, could have their origins in unicellular amoeba-like eukaryotes.

Evolution of the adaptive immune response.

The evolution of adaptive immunity is one of the greatest biological enigmas of all time. For a long time, the origin of adaptive immunity was shrouded in mystery, but the fog is surely rising and a clear picture is beginning to emerge. As we will see in this part of the chapter, the evolution of adaptive immunity appears to have been made possible by the invasion of a putative immunoglobulin-like gene by a transposable element, almost certainly a retroposon. This conferred on the ancestral gene the ability to undergo gene rearrangement, and thus generate diversity. There are, however, still many unanswered questions, and probably many others that have not yet been thought of. The first key question is what was the nature of the piece of DNA that was invaded? It must have resembled a member of the immunoglobulin gene superfamily, and may have already been functioning as some type of antigen receptor, for it to operate appropriately in its changed form; this narrows the field considerably. Second, what was the nature of the cell in which this receptor was expressed? The retroposon itself must have integrated into the host DNA within a germ cell, in order for the two genes *RAG-1* and *RAG-2* to be inherited together with their targets, called recognition signal sequences, or RSS for short. As *RAG-1* and *RAG-2* are inherited as a tightly linked pair of genes, while there are at least seven locations to which the ends of the retroposon dispersed (these are the T-cell receptor α, β, γ, and δ chain loci plus the immunoglobulin H, κ, and λ loci), there must be powerful positive selection for these to persist, and that suggests that there was a significant advantage to the organism in expressing a somatically rearranging receptor. What then was the function of this receptor and the function of the cell type in which it was expressed, that could then make good use of this new diversity in recognition? It must have been something like a lymphocyte, but was it more like a macrophage, a polymorphonuclear leukocyte, an NK cell, or some other cell unlike a lymphocyte that no longer exists in vertebrates? And finally, how did the signaling machinery that we learned about in Chapter 6 develop to support this new device, a receptor gene that could rearrange its gene segments? The answers to these questions will, I predict, fill thousands of papers in the years to come.

Fig. 2 The integration of a transposable element into a cell-surface receptor gene was the event that ultimately gave rise to the immunoglobulin and T-cell receptor genes and their somatic recombination. Transposable elements are closed circles of extra chromosomal DNA, or episomes, that can insert themselves into the genome and later excise themselves to recreate the closed circle. In this way they can move from site to site in the genome. The transposable element must contain two functional elements, the first being transposase enzymes that mediate the integration and excision of the DNA circle from the genome, the second being specific recognition sequences for those transposases; these are sometimes called integration/excision sequences (left panel). The transposases cleave the transposable element between the two integration/excision sequences. They also cleave genomic DNA at a random site, and then ligate the ends of the integration/excision sequences with the cut ends of the genomic DNA (center panel). The excision of the transposable element is the reverse of this process, the transposases cutting between the ends of the integration/excision sequences and the genomic DNA, and then joining together the two integration/excision sequences. In the evolution of the immunoglobulin and T-cell receptor genes, an initial integration event into the middle of a cell-surface receptor has been followed by rearrangements of DNA that has separated the transposase enzymes, that we now know as the *RAG-1* and *RAG-2* genes, from the integration/excision sequences, that we now term the recombination signal sequences, or RSS (right panel). However, the enzymes retain their ancestral function and the *RAG* gene products will catalyze the insertion of RSS containing fragments of DNA into the genome.

Adaptive immunity appears abruptly in the cartilaginous fish.

It has been known for at least 50 years that all jawed fish can mount an adaptive immune response. On the other hand, hagfish and lampreys, which are jawless vertebrates, lack all signs of an adaptive immune system: they do not have organized lymphoid tissue, they lack primary immune responses, and most importantly, they do not exhibit immunological memory. By contrast, even cartilaginous fish, the earliest jawed fish to survive to the present day, have organized lymphoid tissue, albeit primitive, T-cell receptors and immunoglobulins, and the ability to mount adaptive immune responses. What makes the two phylogenetically related groups so different? And why are they so different? That is the mystery of the evolution of adaptive immunity, and what a mystery it is!

It was only in 1998 that the answers to these questions began to become apparent. In jawed fish and all 'higher' vertebrates, adaptive immunity is possible because of what I like to think of as the immunological 'Big Bang,' which occurred in some ancestor of the jawed fish. A transposable element invaded a stretch of DNA, presumably a gene that was similar to an immunoglobulin gene or a T-cell receptor gene, and rapidly segregated the transposon sequences encoding the recombinase enzymes used for the invasion from the recognition sequences for these enzymes (Fig. 2). These remnants of the original transposon became the recombination signal sequences of immunoglobulin and T-cell receptor genes. Invasion by a retrotransposon had been speculated on for years as an explanation of the presence of the *RAG* genes, which encode recombination enzymes essential for the rearrangement of immunoglobulin and T-cell receptor genes (see Chapter 4). Like other retroposons, *RAG* genes lack introns. The action of RAG proteins at the recognition signal sequences was well known for many years. But it was only in 1998 that two laboratories, working independently, came to the startling conclusion that they had their hands on the key discovery to explain the origin of immunoglobulin and T-cell receptor rearrangement.

The discovery was that the present-day RAG proteins can indeed catalyze a transposition event. While conducting experiments to look at the enzyme mechanisms involved in gene rearrangement, researchers in two independent laboratories, using different assays, both noticed that the stretch of DNA containing the recombination signal sequences was being inserted into other DNA fragments, a process identical to that of transposition (see Fig. 2). This simultaneous discovery was no accident but, as often happens in science, the result of having two very smart people working on the same important

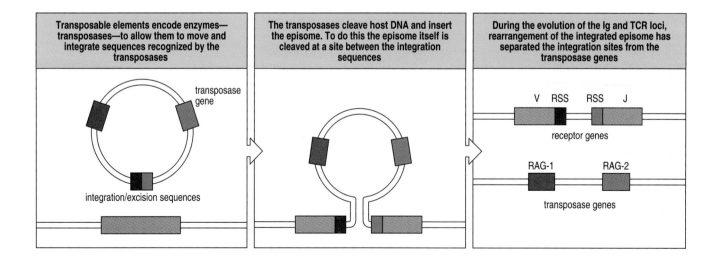

| Transposable elements encode enzymes—transposases—to allow them to move and integrate sequences recognized by the transposases | The transposases cleave host DNA and insert the episome. To do this the episome itself is cleaved at a site between the integration sequences | During the evolution of the Ig and TCR loci, rearrangement of the integrated episome has separated the integration sites from the transposase genes |

problem. The discoveries were published in the same month in two leading journals (and should, I believe, qualify the two scientists leading the two independent groups for a Nobel Prize).

Gene rearrangement is used to control gene expression.

One need only turn back to the earlier chapters on adaptive immunity (see Chapters 3–7), where we described the almost infinite variability of the antigen receptors and how it is achieved by gene rearrangement, to see how important gene rearrangement is. And why is it key? Without it, there would be no variable antigen receptors and no possibility of clonal selection—and thus no adaptive immunity.

The discovery of gene rearrangement as a universal generator of diversity solved one of the great early mysteries of immunology: how could the in formation in a human, mouse, or fish genome encode so many unique proteins—the immunoglobulins and T-cell receptors? There were heated discussions during annual workshops about whether the entire repertoire could be encoded in the germline, or whether some genetic trick that had only been seen in immunoglobulins (at that date) would turn out to be true for T-cell receptors as well. Eventually, the molecular genetic discoveries of Susumu Tonegawa and others put this debate to rest.

But the discovery of rearranging gene segments raised other issues about the control of gene expression. How could one go from a silent V gene segment to one that produced a protein just by rearranging it? The answer we give in this book, which is accepted by most, but not all, immunologists, is that gene rearrangement also controls gene expression, and so can do the whole job alone. That raises further problems, which were dealt with in Chapter 7, such as how can the same enzymes work on different genes in T cells and B cells, thus producing T-cell receptors in one and immunoglobulins in the other. This discrimination is now thought to be a function of the different transcription factors produced by early B cells and early T cells, which open up different parts of the chromosomes and make them accessible to the recombination enzymes. Those produced in early B-lineage cells operated on early B cells to produce surface immunoglobulin receptors, and not on T cells, and vice versa.

Animals generate antigen receptor diversity in many different ways.

Most animals that we are familiar with generate a large part of their antigen receptor diversity as humans do, by putting together gene segments in different combinations, as described in earlier chapters. However, we noted a few exceptions in passing (see Section 4-10), and it is useful to return to these now and see how far they violate the law that gene segments must be joined according to very strict rules. These rules are the 12/23 rule of gene segment joining, and the requirement for RAG proteins (see Chapter 7). The requirement for RAG proteins is essentially absolute. Mice and people who lack RAG proteins have a total absence of immunoglobulin and T-cell receptors, as manifest in Omenn's syndrome, a human disease that is caused by the mutations in one or the other RAG protein. The 12/23 rule is more subtle, and to my knowledge no patient has been discovered with a deficiency in this rule, but it would be expected to give a phenotype much like that in Omenn's syndrome.

These examples from patients tell us that our ideas about the importance of genes and proteins that we learned about in mice are generally true for humans as well. We must not, however, be too confident that every lesson learned from mice will immediately apply to humans or to other species, since there are many examples where that is not the case.

Some animals use gene rearrangement to always join together the same V and J gene segment initially, and then go on to diversify this recombined V region in various ways. In chickens and rabbits, the recombined V region is diversified by gene conversion (see Section 4-10) in the bursa of Fabricius (in chickens) or another intestinal lymphoid organ (in rabbits) (Fig. 3). Other animals generate their diverse repertoire mainly by somatic hypermutation of a fairly invariant recombined V region, as does the sheep in its ileal Peyer's patch. Some primitive fish have multiple copies of discrete V_L–J_L–C_L and V_H–D_H–J_H–C_H cassettes, and activate rearrangement in different copies, while carcharine sharks have multiple 'rearranged' V_L regions in the germline genome and apparently generate diversity by activating transcription of different copies. All these animals have survived in a hostile environment because they have the other benefit of adaptive immunity, namely, immunological memory. This is the single greatest advantage conferred on animals that have rearranging gene segments, and is the focus of the next part of this chapter.

Summary.

Gene rearrangement has been known since the early 1970s, and the 'immunological Big Bang' has been known for about a few years. Although these two events well explain the development of adaptive immunity, in the next section we will learn about the enormous value that immunological memory has for preserving the two genetic elements of adaptive immunity: recognition signal sequences and RAG proteins. Gene rearrangement is the result of a chance insertion of a retroposon into an unknown cell that must have been either a sperm or an ovum, as it is inserted in the germline. Somehow, still not explained adequately, this retroposon had the good luck to invade a member of the immunoglobulin gene family, and to carry its invasive ends that allow retroposons to move from cell to cell into the right place in the target primordial immunoglobulin gene. At the same time, the *RAG* genes were preserved, presumably from the same retroposon, but were carried by a different chromosome. These two events, which allowed adaptive immunity to occur, also made immunological memory possible, but they did not make it necessary. Therefore, we ask in the next section of this chapter how immunological memory evolved. What were the evolutionary forces acting on the immune system that not only guaranteed the survival of the

Fig. 3 The organization of immunoglobulin genes is different in different species, but all are capable of generating a diverse repertoire of receptors. The organization of the immunoglobulin heavy-chain genes in mammals, where there are separated clusters of repeated V, D, and J gene segments is not the only solution to the problem of generating a diverse repertoire of receptors. Other species have found alternative solutions. In 'primitive' species, such as the shark, the locus consists of multiple repeats of a basic unit composed of a V gene segment, one or two D gene segments, a J gene segment, and a C gene segment. A more extreme version of this organization is found in the λ light-chain locus of some elasmobranch species, where the repeated unit consists of already rearranged VJ–C genes, from which a random choice is made for expression. Finally, in chickens, there is a single rearranging heavy-chain locus, but multiple copies of V-segment pseudogenes. Diversity in this system is created by gene conversion, where sequences from the V pseudogenes are copied onto the single rearranged VH gene.

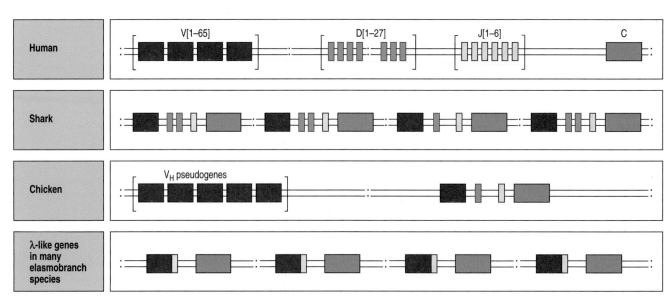

species that were lucky enough to inherit this trait, but expanded them to the multitudes that occupy the Earth's surface? And what contribution did the possession of immunological memory make to the ability of vertebrates to occupy most of the ecological niches currently present (although we should not forget that insects and many other invertebrates cohabit with us)?

The importance of immunological memory in fixing adaptive immunity in the genome.

The previously described events that allowed adaptive immunity to occur also made immunological memory possible. However, immunological memory is an invariable feature of all adaptive immune responses in all vertebrates that have evolved beyond the hagfish and the lamprey. The reason is that immunological memory confers a tremendous survival advantage, as it confers the ability to respond more rapidly and more effectively to a second and subsequent challenge by the same pathogen. Moreover, the ability to mount an antibody response, and then to maintain it, protects against many infectious agents. In this part of the chapter, we ask why immunological memory is so important to survival of the organisms that have it, and why it has been preserved throughout the time of the vertebrate radiation, of which we humans consider ourselves the model of the finished product.

Immunological memory is the hallmark of adaptive immunity.

Immunological memory is the property of remembering specific adaptive immune responses, and making a greater and more rapid response in the future to the same pathogen (see Fig. 1.20). This property had been noted by Thucidydes in his account of the Peloponnesian war and had been taken advantage of medically by, for example, infecting individuals with dried material taken from smallpox lesions, a process known as variolation, after the smallpox virus *Variola major*. Today we would interpret this as immunization with an attenuated or killed virus, but at the time the process was known, in most cases, only to produce a mild form of the disease that would protect the sufferer from a subsequent infection with a more virulent form. This was used by the Chinese for many centuries, by the Turks to protect their daughter's beauty, and was communicated to England in the diaries of the Lady Mary Wortly Montague. Jenner subsequently introduced vaccination against smallpox by injecting material from cowpox lesions into the skin. He was counting on the oral reports of milkmaids, who said that their excellent complexions, free of the ravages of smallpox, were due to exposure to cowpox. Of course, he knew nothing of the mechanisms of the protection, but the same basic procedure was used up until the late 1970s, when, thanks to the efforts of a large number of field workers, the last case of smallpox was eradicated. This was a field trial for adaptive immunity and immune memory, and it was a spectacular success.

When we ask why cowpox should protect against smallpox, and what this tells us about the immune system, we learn that cowpox is closely related to the smallpox virus, sharing some of its antigens, and sets up a state of protective immunity to both viruses. The initial infection with cowpox is mild, and can be contained by the primary response of the immune system, but it sets up the conditions for a more potent secondary response that is now able to control the more virulent smallpox infection. The same lesson can be learnt from vaccination against polio with either the Salk vaccine, which is a formalin-killed vaccine, or the later-developed Sabin vaccine, which is a live attenuated poliovirus that establishes a superficial infection that is eradicated. A worldwide effort is underway to eradicate polio by vaccination. It is in its early stages, and the campaign will undoubtedly take longer to accomplish than for smallpox (see Chapter 1), because several different strains of poliovirus have to be eradicated rather than the single strain of smallpox. There are also questions about the best way to carry out vaccination against polio, and about the safety and efficacy of current vaccine stocks. Nonetheless, the World Health Organization (WHO) is pressing forward with its efforts to eradicate polio from the face of the Earth, and we all hope that it succeeds.

Why does vaccination provide long-term immunity to reinfection? Immunological memory provides the answer. What allows the adaptive immune system to make useful responses to attenuated organisms? It is the development of immunological memory that again makes it possible to think of these effects.

Why then does the response to an attenuated pathogen or to a mild infection protect the individual from a fully virulent infection? The answer lies in the nature of immunological memory itself. The adaptive immune response may be thought of in three phases or developmental stages (not to be confused with the three phases of innate immunity, innate induced responses, and the adaptive immune response described in Chapter 1). The first is the naive lymphocyte phase, which accounts for many of the cells in the immune system, and in particular all the newly formed lymphocytes, which have not yet encountered their specific antigens (Fig. 4). The second phase is the phase of the primary immune response, during which the selected lymphocytes expand in numbers very remarkably and differentiate into effector cells. In the response to lymphocytic choriomeningitis virus (LCMV), the numbers of virus-specific CD8 T cells increase more than 10^5-fold! That is a remarkable expansion, and it happens in a relatively short time, because of rapid cell division. The adaptive immune response is thus a powerful way of markedly increasing by clonal selection, the right combination of gene segments to deal with the particular pathogen, and then to rapidly expand the cell population containing them to mount a primary response that it is hoped has, and usually have, two effects. One is the elimination of the infectious agent, and the other is the generation of memory cells that can rapidly and specifically respond to any reinfection. Memory cells form the third phase of the immune response, and when such LCMV-specific memory T cells are boosted with LCMV, they again undergo rapid proliferation, so that an effective secondary immune response is developed even more rapidly—within a few days. The ability to 'remember' a previous response at the cellular level is quite unusual in biology, and represents a remarkable feat of genetic and biological engineering. That is the virtue of a clonal system of host defense. Vertebrates, which have both innate and adaptive immunity, have the combined benefits of nonclonal and clonal immunity, and can therefore survive over a long lifetime in a pathogen-filled environment.

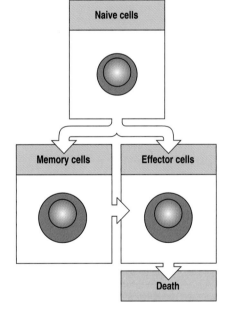

Fig. 4 The three phases of the adaptive immune response, naive, memory, and effector cells.
Lymphocytes of the immune system can exist in one of three phases. All cells initially are naive lymphocytes, until antigenic stimulation changes their fate. Some become effector cells, which die once they have completed their effector functions. Others become memory cells, which can also mature into effector cells on reexposure to antigen.

Immunological memory allows survival in a world filled with pathogens.

As a consequence of the more rapid and intensified secondary immune response, second and subsequent infections with potentially pathogenic microorganisms are often asymptomatic or are mild and of limited duration. The advantage of immunological memory, therefore, is that it allows us to survive without recurring debilitating disease even in a world teeming with pathogens. Viruses usually encountered in childhood give rise to protective immunity, which is relatively easy to mimic in a vaccine with either live attenuated viruses or killed virus particles. A complete list of vaccines currently recommended by physicians in the United States is given in Fig. 14.21. The same vaccines are recommended by most physicians in the first and second world, where they have proved highly successful in reducing childhood illness and mortality. It is in the third world that vaccines are crying out to be developed, as shown by the list of diseases that kill mostly third-world children, and where no effective vaccine exists (see Fig. 14.22).

At this time, immunologists need to focus their attention on two tasks. The first is the preparation and distribution of vaccines to third-world countries, so that their children can have a disease-free childhood and a much longer life expectancy. This is the key to population control in the long term, as parents must first have confidence in the survival of their offspring before population control can become acceptable. The second is the study of the natural course of diseases in these countries, in order to discover ways of preventing them from occurring in the first place. Vaccines are at least part of the solution. I am optimistic about this, as mankind is capable of great feats, as shown by the eradication of smallpox by vaccination. But we need leadership and dedication and understanding, which are human qualities that are in short supply.

Immunological memory for self proteins leads to autoimmune disease.

Immunological memory also has its down side, which is a propensity to cause autoimmune disease. We have discussed autoimmunity in Chapter 13, but the contribution of memory cells has been given little emphasis. It is our opinion that without immune memory, there would be no autoimmune disease. Many types of autoimmune disease are found especially in older people. There is, however, at least one exception, and that is autoimmune diabetes, which occurs most frequently in teenagers, and can occur in children as young as 3–4 years old. Its occurrence in childhood has led to attempts to prevent diabetes by immunological means, almost all of which have failed. We believe that this failure is much like the failure to prevent hemolytic disease of the newborn in mothers who have already had a Rh$^+$ child; it is too late to close the stable door, because the colt has already bolted, or more scientifically, an antibody or T-cell response has already been induced.

I and my colleagues would like to try primary prevention measures on children from families that have had one or more diabetic child already, and we think we have the correct antigen for this trial in insulin itself. Early on in work on auto-immune diabetes, certain insulin peptides were found that triggered the CD4 T cells that apparently caused diabetes in mice. Subsequently, it was found that the same peptide, in a shortened form, could be recognized by CD8 T cells, which had the ability to attack the pancreatic β cells themselves. Thus, one of the key proteins recognized in insulin-dependent diabetes mellitus is the insulin molecule itself. An initial trial of insulin injections to create a state of tolerance to these insulin peptides is underway. The jury is still out on whether they will work, but already there is a glimmer of hope in the results.

Summary.

There is a lot of evidence that immunological memory is one of the cardinal features conferred on the individual by the existence of adaptive immunity. It is seen in all vertebrate groups that have undergone the immunological Big Bang, and it is found in all vertebrate species that have been tested for it. It also has a down side, in that it contributes to autoimmune disease, which, as we learned in Chapter 13, occurs only in species that have adaptive immunity and is a consequence of their adaptive immune responses.

Future directions of research in immunobiology.

The future of immunobiology appears to be assured. The early successes of immunization by Jenner and Pasteur laid the way for the incredible flowering of immunobiology in the 20th century. Now it seems proper to ask what questions are going to be answered by immunobiology in the 21st century, and what remains to be done.

One of the big questions is whether immunologists can figure out a way to really understand innate immunity, given that it is a system in which there is really no specific product to measure. Have we discovered this product in the ability of Toll to activate co-stimulatory molecules? We don't think so. Immunologists will need to spend years, decades, or even centuries to figure out all the ins and outs of innate immunity. Another big question is how adaptive immunity really works. It seems that apoptosis plays a big part in controlling the adaptive immune system, but does anyone really know how to measure its contribution *in vivo*? Certainly we do not.

Future studies in cancer immunology promise untold benefits, but we do not see them even remotely on the horizon. Imagine how it would feel to come up with a general cure for cancer, one of mankind's oldest and deadliest enemies. Such advances are, of course, a double-edged sword, since decreasing deaths from cancer and infectious disease may well lead to an increasing population that exceeds the capacity of the available natural and social resources. Nevertheless, we should think positively about the benefits of immunological research, because we should all be interested in the benefits that can flow from such investigations, and not the possible negative outcomes. So this last part of the book will be written from an optimist's viewpoint that all research is aimed at increasing the benefits to mankind.

Future studies should vastly expand our knowledge of innate immunity.

Immunologists traditionally like to use specificity controls in their experiments, which is one of the beauties of immunological experimental design. However, when one moves from experiments that test the functioning of the adaptive immune system to those that test the innate immune system, one is at a loss for proper controls. Therefore, the questions have to be posed in different ways. Fortunately, we live in an era when the DNA sequences of whole genomes can be produced in relatively short times, and that will help us distinguish the wheat from the chaff.

The analysis of whole genomes is just getting underway, so we do not yet know what it will be able to do. The entire genomes of several pathogenic microorganisms have already been analyzed, with encouraging results and many surprises. Recently, the full genome of the important model organism *Drosophila melanogaster* was published, resulting in a series of articles, including one on the implications for *Drosophila* immunity from the knowledge of the entire genome of the fly. This organism has only innate immunity, and is therefore not expected to tell us much about adaptive immunity, but we will eventually be able to identify all the genes that make the fruit fly such a successful species with just innate immunity to protect it.

Among mammals, we can look forward to the complete sequences of a single human genome which was recently released, and the sequence of a mouse genome should soon be available. However, knowing the entire human genome will not explain immunology, and especially not how the immune system functions. That level of understanding of such a complex system requires studies at the functional, molecular, and genetic levels, so we are a long way from a complete understanding of immunobiology, although we may be getting closer to understanding immunogenetics.

Future studies should refine our knowledge of adaptive immunity.

We already have a reasonably complete understanding of adaptive immunity. We know that antigen is taken up by dendritic cells in the tissues and brought to lymph nodes in the form of peptide:MHC complexes. We know that the presence of infection is necessary both to process the antigen and to induce co-stimulatory molecules on the dendritic cell surface. But it was only in 1999 that we learned about DC-SIGN, an important cell-surface molecule that is made by activated dendritic cells and that recognizes naive T cells by binding to the cell-adhesion molecule ICAM-3 (see Fig. 8.8). Thus, we can expect many more surprises from basic immunobiology.

In fact, I would guess that there are more surprises to come than in the total amount of what we already think we know. I remember a colleague, who later went on to fame and glory by discovering the genes that transport peptide across the endoplasmic reticulum membrane, declaring confidently that immunology was "finished as an experimental science…" in 1989, reflecting Niels Jerne's famous statement that he was "waiting for the end" in 1967. On the contrary, I would say that immunology is still struggling to explain major phenomena such as discrimination of self from nonself, and self tolerance. These are not easy questions and they will not be answered by simple experiments. Nor will the observation that the innate immune system can induce the expression of co-stimulatory molecules on the antigen-presenting cell surface answer all the questions about the impact of innate immunity on the adaptive immune response. There is more to it than the little we now know, or even that we think we know.

Future studies of tumor immunity hold great promise for an immunological cure for cancer.

The analysis of immunity to tumors is still in its infancy. This is one of the applied questions about which my colleague spoke so disparagingly in 1989. Yet imagine if we could harness the adaptive immune system to the detection and rejection of cancer. Would that not be astonishingly beneficial? It would allow the body's own system of defense to guard the body against one of nature's deadliest killers. An immune response induced to cancer that would allow the killer T cells to discriminate between the cancerous cells that are threatening to overwhelm the body from the normal cells that are needed for important bodily functions would truly be a wonderful thing.

The reasons my colleague gave for his scorn for this exciting idea are hard to fathom. He stated that the practical applications of immunology were of no interest to him, only of interest to those of a practical bent. But what is immunology for but to solve practical problems? Quite apart from the fact that we have arguably learned more over the past ten years than over the previous hundred, and that we have all worked harder to provide results that would stand the test of time than we did earlier, the rewards in terms of applied immunology have been harvested many times over. Let us not scorn the benefits of applied immunology, although I, too, pay most attention to the impractical world of mouse immunology. I am hopeful that major advances in tumor biology and tumor immunobiology will lead to wonderful new treatments for cancer—these cannot arrive too soon.

Future vaccine development should greatly increase our ability to prevent infectious disease.

The development of new vaccines is another goal for the future. What an accomplishment it will be to prepare vaccines against pathogens that are highly variable, such as the human immunodeficiency virus (HIV) that causes AIDS. HIV is highly variable in its genome, and hence in its protein antigens. This turns out to be typical behavior for a retrovirus that can multiply itself about a thousandfold every hour, giving off variant clones that rapidly become dominant in the infection. Using an attenuated retrovirus as a vaccine is not an option. Nor is the use of a killed virus, as all tests of such vaccines have been totally ineffective. So we must invent a new kind of vaccine altogether. What kind of vaccine will it be? We do not know, or we would already have it. But we can speculate on the types of vaccine that might work.

One possibility is a vaccine virus that lacks one or more genes essential for causing disease. This strategy has been tried with the *nef* gene, which was naturally deleted in certain rare HIV strains isolated from long-term non-progressors, in which they appear to produce no disease. This gave hope that one could learn from long-term nonprogressors the secret of vaccinating the public with an equally benign virus. So far there are no reports of people who resist HIV infection for their entire lifetime, except for those who lack the chemokine receptor CCR5 that functions as a co-receptor for the virus along with CD4 (see Chapter 11). So the challenge is to improve on nature in the 21st century. How will these advances come? We cannot guess, but that there will be advances is sure. Just look at any recent issue of any immunology journal, and there are advances galore!

Future studies of autoimmunity and graft rejection should allow control of immune responses to one's own body or to a piece borrowed from someone else.

Graft rejection and autoimmunity are two sides to the same coin; in both cases, an adaptive immune response is made to tissue, either one's own or tissue received from a donor. These two adaptive immune responses tell us much about the nature of the T-cell receptor repertoire. The existence of alloreactivity (see Chapter 5) tells us that this interaction with nonself MHC molecules is a general property of T cells and is not simply based on their selection on self peptide:self MHC molecules. Similarly, autoimmunity tells us that autoreactive cells are present in all of us, but that they are normally suppressed by another set of T cells, commonly called T regulatory cells although they might as well be called T suppressor cells. These cells are CD4 T cells that secrete cytokines, although the precise cytokine profile is not yet firmly established. One cloned line that has suppressor properties in the murine model of diabetes secretes T_H1-type cytokines along with TGF-β, which is, as we learned in Chapter 14, a highly immunosuppressive cytokine. Other cytokines, including IL-10, also suppress the immune response. In general, we have only touched on the nature of cells that suppress immune responses, in part because their existence became highly controversial in the 1970s and 1980s. They have been largely avoided, until someone thought to make them fashionable again by calling them Treg cells. That has opened the way to recommencing the study of T cells that regulate immune responses by recognizing antigen in a conventional way, and which can regulate the immune response by releasing cytokines in a directed manner, like all effector T cells. The only thing that distinguishes a Treg cell from other T cells is the set of cytokines it secretes. If we could learn how to control Treg as we have learned how to control other sets of T cells, then we could dream of controlling autoimmune diseases and allograft rejection by treating patients with Treg, or, better still, inducing Treg in the patient her or himself.

Summary.

This section of the chapter differs from the rest in trying to foresee the future of immunology research. In trying to do so, it is bound to be wrong. So many discoveries are made in other fields that use immunological techniques and methods, and the scope of what is yet to be discovered is so great, that one takes a great risk of being proved wrong even before the book is published. Nevertheless, we think it is worth trying, as research is charging ahead at an unprecedented rate. We hope that our speculations help some student with an unconventional idea to at least try it out; who knows, it might work!

Summary of the Afterword.

The first part of this chapter dealt with the evolution of the innate immune response, while the second part dealt with the far more complicated adaptive immune response. But both immune responses are highly complicated in their details, and we still do not know much about the extent of innate immunity. We are also being surprised all the time by the adaptive immune response, which gets more and more complex as we learn more about it. Rather than an increase in knowledge simplifying, as happens often in physics, the immune system seems to grow ever more complex. It was a big surprise to most that the 12/23 rule and the RAG recombinases appeared abruptly in jawed fish, but missed the agnathans. In studying the adaptive immune response and its most important consequence—which is immunological memory—we will learn a great deal. As we look to clinical immunology, the study and control of immune responses in humans, we could learn to control AIDS, cancer, autoimmunity, and graft rejection, which would guarantee health and long life for everyone. And what does the future contain? More surprises for sure.

Immunologists' Toolbox

Immunization.

Natural adaptive immune responses are normally directed at antigens borne by pathogenic microorganisms. The immune system can, however, also be induced to respond to simple nonliving antigens, and experimental immunologists have focused on the responses to these simple antigens in developing our understanding of the immune response. The deliberate induction of an immune response is known as **immunization**. Experimental immunizations are routinely carried out by injecting the test antigen into the animal or human subject. The route, dose, and form in which antigen is administered can profoundly affect whether a response occurs and the type of response that is produced, and are considered in Sections A-1–A-4. The induction of protective immune responses against common microbial pathogens in humans is often called **vaccination**, although this term is correctly only applied to the induction of immune responses against smallpox by immunizing with the cross-reactive cowpox virus, vaccinia (see Chapter 14).

To determine whether an immune response has occurred and to follow its course, the immunized individual is monitored for the appearance of immune reactants directed at the specific antigen. Immune responses to most antigens elicit the production of both specific antibodies and specific effector T cells. Monitoring the antibody response usually involves the analysis of relatively crude preparations of **antiserum** (plural: **antisera**). The **serum** is the fluid phase of clotted blood, which, if taken from an immunized individual, is called antiserum because it contains specific antibodies against the immunizing antigen as well as other soluble serum proteins. To study immune responses mediated by T cells, blood lymphocytes or cells from lymphoid organs such as the spleen are tested; T-cell responses are more commonly studied in experimental animals than in humans.

Any substance that can elicit an immune response is said to be **immunogenic** and is called an **immunogen**. There is a clear operational distinction between an immunogen and an antigen. An **antigen** is defined as any sustance that can bind to a specific antibody. All antigens therefore have the potential to elicit specific antibodies, but some need to be attached to an immunogen in order to do so. This means that although all immunogens are antigens, not all antigens are immunogenic. The antigens used most frequently in experimental immunology are proteins, and antibodies to proteins are of enormous utility in experimental biology and medicine. Purified proteins are, however, not always highly immunogenic and to provoke an immune response have to be administered with an adjuvant (see Section A-4). Carbohydrates, nucleic acids, and other types of molecule are all potential antigens, but will often only induce an immune response if attached to a protein carrier. Thus, the immunogenicity of protein antigens determines the outcome of virtually every immune response.

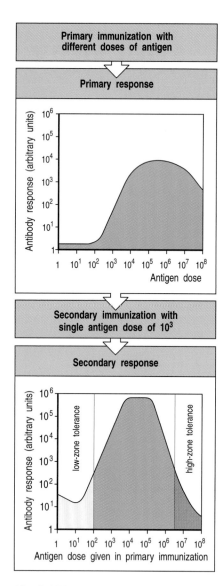

Primary immunization with
different doses of antigen

Primary response

Secondary immunization with
single antigen dose of 10^3

Secondary response

Fig. A.1 The dose of antigen used in an initial immunization affects the primary and secondary antibody response. The typical antigen dose–response curve shown here illustrates the influence of dose on both a primary antibody response (amounts of antibody produced expressed in arbitrary units) and the effect of the dose used for priming on a secondary antibody response elicited by a dose of antigen of 10^3 arbitrary mass units. Very low doses of antigen do not cause an immune response at all. Slightly higher doses appear to inhibit specific antibody production, an effect known as low-zone tolerance. Above these doses there is a steady increase in the response with antigen dose to reach a broad optimum. Very high doses of antigen also inhibit immune responsiveness to a subsequent challenge, a phenomenon known as high-zone tolerance.

Antisera generated by immunization with even the simplest antigen will contain many different antibody molecules that bind to the immunogen in slightly different ways. Some of the antibodies in an antiserum are cross-reactive. A **cross-reaction** is defined as the binding of an antibody to an antigen other than the immunogen; most antibodies cross-react with closely related antigens but, on occasion, some bind antigens having no clear relationship to the immunogen. These cross-reacting antibodies can create problems when the antiserum is used to detect a specific antigen. They can be removed from an antiserum by **absorption** with the cross-reactive antigen, leaving behind the antibodies that bind only to the immunogen. Absorption can be performed by affinity chromatography using immobilized antigen, a technique that is also used for purification of antibodies or antigens (see Section A-5). Most problems of cross-reactivity can be avoided, however, by making monoclonal antibodies (see Section A-12).

Although almost any structure can be recognized by antibody as an antigen, usually only proteins elicit fully developed adaptive immune responses. This is because proteins have the ability to engage T cells, which contribute to inducing most antibody responses and are required for immunological memory. Proteins engage T cells because the T cells recognize antigens as peptide fragments of proteins bound to major histocompatibility complex (MHC) molecules (see Section 3-11). An adaptive immune response that includes immunological memory can be induced by nonpeptide antigens only when they are attached to a protein carrier that can engage the necessary T cells (see Section 9-2 and Fig. 9.4).

Immunological memory is produced as a result of the initial or **primary immunization**, which evokes the **primary immune response**. This is also known as **priming**, as the animal or person is now 'primed' like a pump to mount a more potent response to subsequent challenges with the same antigen. The response to each immunization is increasingly intense, so that **secondary**, **tertiary**, and subsequent responses are of increasing magnitude (Fig. A.1). Repetitive challenge with antigen to achieve a heightened state of immunity is known as **hyperimmunization**.

Certain properties of a protein that favor the priming of an adaptive immune response have been defined by studying antibody responses to simple natural proteins like hen egg-white lysozyme and to synthetic polypeptide antigens (Fig. A.2). The larger and more complex a protein, and the more distant its relationship to self proteins, the more likely it is to elicit a response. This is because such responses depend on the proteins being degraded into peptides that can bind to MHC molecules, and on the subsequent recognition of these peptide:MHC complexes by T cells. The larger and more distinct the protein antigen, the more likely it is to contain such peptides. Particulate or aggregated antigens are more immunogenic because they are taken up more efficiently by the specialized antigen-presenting cells responsible for initiating a response. Indeed small soluble proteins are unable to induce a response unless they are made to aggregate in some way. Many vaccines, for example, use aggregated protein antigens to potentiate the immune response.

A-1 Haptens.

Small organic molecules of simple structure, such as phenyl arsonates and nitrophenyls, do not provoke antibodies when injected by themselves. However, antibodies can be raised against them if the molecule is attached covalently, by simple chemical reactions, to a protein carrier. Such small molecules were termed **haptens** (from the Greek *haptein*, to fasten) by the immunologist Karl Landsteiner, who first studied them in the early 1900s. He found that animals immunized with a hapten–carrier conjugate produced

Factors that influence the immunogenicity of proteins		
Parameter	Increased immunogenicity	Decreased immunogenicity
Size	Large	Small (MW<2500)
Dose	Intermediate	High or low
Route	Subcutaneous > intraperitoneal > intravenous or intragastric	
Composition	Complex	Simple
Form	Particulate	Soluble
	Denatured	Native
Similarity to self protein	Multiple differences	Few differences
Adjuvants	Slow release	Rapid release
	Bacteria	No bacteria
Interaction with host MHC	Effective	Ineffective

Fig. A.2 Intrinsic properties and extrinsic factors that affect the immunogenicity of proteins.

three distinct sets of antibodies (Fig. A.3). One set comprised hapten-specific antibodies that reacted with the same hapten on any carrier, as well as with free hapten. The second set of antibodies was specific for the carrier protein, as shown by their ability to bind both the hapten-modified and unmodified carrier protein. Finally, some antibodies reacted only with the specific conjugate of hapten and carrier used for immunization. Landsteiner studied mainly the antibody response to the hapten, as these small molecules could be synthesized in many closely related forms. He observed that antibodies raised against a particular hapten bind that hapten but, in general, fail to bind even very closely related chemical structures. The binding of haptens by anti-hapten antibodies has played an important part in defining the precision of antigen binding by antibody molecules. Anti-hapten antibodies are also important medically as they mediate allergic reactions to penicillin and other compounds that elicit antibody responses when they attach to self proteins (see Section 12-10).

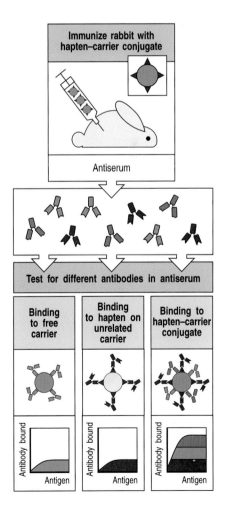

Fig. A.3 Antibodies can be elicited by small chemical groups called haptens only when the hapten is linked to an immunogenic protein carrier. Three types of antibodies are produced. One set (blue) binds the carrier protein alone and is called carrier-specific. One set (red) binds to the hapten on any carrier or to free hapten in solution and is called hapten-specific. One set (purple) only binds the specific conjugate of hapten and carrier used for immunization, apparently binding to sites at which the hapten joins the carrier, and is called conjugate-specific. The amount of antibody of each type in this serum is shown schematically in the graphs at the bottom; note that the original antigen binds more antibody than the sum of anti-hapten and anti-carrier antibodies owing to the additional binding of conjugate-specific antibody.

A-2   Routes of immunization.

The route by which antigen is administered affects both the magnitude and the type of response obtained. The most common routes by which antigen is introduced experimentally or as a vaccine into the body are injection into tissue by **subcutaneous** (**s.c.**) injection between the epidermis and dermal layers, or by **intradermal** (**i.d.**) injection, or **intramuscular** (**i.m.**) injection; by **intravenous** (**i.v.**) injection or transfusion directly into the bloodstream; into the gastrointestinal tract by **oral** administration; into the respiratory tract by **intranasal** (**i.n.**) administration or inhalation.

Antigens injected subcutaneously generally elicit the strongest responses, most probably because the antigen is taken up by Langerhans' cells and efficiently presented in local lymph nodes, and so this is the method most commonly used when the object of the experiment is to elicit specific antibodies or T cells against a given antigen. Antigens injected or transfused directly into the bloodstream tend to induce immune unresponsiveness or tolerance unless they bind to host cells or are in the form of aggregates that are readily taken up by antigen-presenting cells.

Antigen administration via the gastrointestinal tract is used mostly in the study of allergy. It has distinctive effects, frequently eliciting a local antibody response in the intestinal lamina propria, while producing a systemic state of tolerance that manifests as a diminished response to the same antigen if subsequently administered in immunogenic form elsewhere in the body. This 'split tolerance' may be important in avoiding allergy to antigens in food, as the local response prevents food antigens from entering the body, while the inhibition of systemic immunity helps to prevent the formation of IgE antibodies, which are the cause of such allergies (see Chapter 12).

Introduction of antigen into the respiratory tract is also used mainly in the study of allergy. Protein antigens that enter the body through the respiratory epithelium tend to elicit allergic responses, for reasons that are not clear.

A-3   Effects of antigen dose.

The magnitude of the immune response depends on the dose of immunogen administered. Below a certain threshold dose, most proteins do not elicit any immune response. Above the threshold dose, there is a gradual increase in the response as the dose of antigen is increased, until a broad plateau level is reached, followed by a decline at very high antigen doses (see Fig. A.1). As most infectious agents enter the body in small numbers, immune responses are generally elicited only by pathogens that multiply to a level sufficient to exceed the antigen dose threshold. The broad response optimum allows the system to respond to infectious agents across a wide range of doses. At very high antigen doses the immune response is inhibited, which may be important in maintaining tolerance to abundant self proteins such as plasma proteins. In general, secondary and subsequent immune responses occur at lower antigen doses and achieve higher plateau values, which is a sign of immunological memory. However, under some conditions, very low or very high doses of antigen may induce specific unresponsive states, known respectively as acquired **low-zone** or **high-zone tolerance**.

A-4   Adjuvants.

Most proteins are poorly immunogenic or nonimmunogenic when administered by themselves. Strong adaptive immune responses to protein antigens almost always require that the antigen be injected in a mixture known as an

adjuvant. An adjuvant is any substance that enhances the immunogenicity of substances mixed with it. Adjuvants differ from protein carriers in that they do not form stable linkages with the immunogen. Furthermore, adjuvants are needed primarily for initial immunizations, whereas carriers are required to elicit not only primary but also subsequent responses to haptens. Commonly used adjuvants are listed in Fig. A.4.

Adjuvants can enhance immunogenicity in two different ways. First, adjuvants convert soluble protein antigens into particulate material, which is more readily ingested by antigen-presenting cells such as macrophages. For example, the antigen can be adsorbed on particles of the adjuvant (such as alum), made particulate by emulsification in mineral oils, or incorporated into the colloidal particles of ISCOMs. This enhances immunogenicity somewhat, but such adjuvants are relatively weak unless they also contain bacteria or bacterial products. Such microbial constituents are the second means by which adjuvants enhance immunogenicity, and although their exact contribution to enhancing immunogenicity is unknown, they are clearly the more important component of an adjuvant. Microbial products may signal macrophages or dendritic cells to become more effective antigen-presenting cells (see Chapter 2). One of their effects is to induce the production of inflammatory cytokines and potent local inflammatory responses; this effect is probably intrinsic to their activity in enhancing responses, but precludes their use in humans.

Nevertheless, some human vaccines contain microbial antigens that can also act as effective adjuvants. For example, purified constituents of the bacterium *Bordetella pertussis*, which is the causal agent of whooping cough, are used as both antigen and adjuvant in the triplex DPT (diphtheria, pertussis, tetanus) vaccine against these diseases.

Adjuvants that enhance immune responses		
Adjuvant name	Composition	Mechanism of action
Incomplete Freund's adjuvant	Oil-in-water emulsion	Delayed release of antigen; enhanced uptake by macrophages
Complete Freund's adjuvant	Oil-in-water emulsion with dead mycobacteria	Delayed release of antigen; enhanced uptake by macrophages; induction of co-stimulators in macrophages
Freund's adjuvant with MDP	Oil-in-water emulsion with muramyldipeptide (MDP), a constituent of mycobacteria	Similar to complete Freund's adjuvant
Alum (aluminum hydroxide)	Aluminum hydroxide gel	Delayed release of antigen; enhanced macrophage uptake
Alum plus *Bordetella pertussis*	Aluminum hydroxide gel with killed *B. pertussis*	Delayed release of antigen; enhanced uptake by macrophages; induction of co-stimulators
Immune stimulatory complexes (ISCOMs)	Matrix of Quil A containing viral proteins	Delivers antigen to cytosol; allows induction of cytotoxic T cells

Fig. A.4 Common adjuvants and their use. Adjuvants are mixed with the antigen and usually render it particulate, which helps to retain the antigen in the body and promotes uptake by macrophages. Most adjuvants include bacteria or bacterial components that stimulate macrophages, aiding in the induction of the immune response. ISCOMs (immune stimulatory complexes) are small micelles of the detergent Quil A; when viral proteins are placed in these micelles, they apparently fuse with the antigen-presenting cell, allowing the antigen to enter the cytosol. Thus, the antigen-presenting cell can stimulate a response to the viral protein, much as a virus infecting these cells would stimulate an anti-viral response.

The detection, measurement, and characterization of antibodies and their use as research and diagnostic tools.

B cells contribute to adaptive immunity by secreting antibodies, and the response of B cells to an injected immunogen is usually measured by analyzing the specific antibody produced in a **humoral immune response**. This is most conveniently achieved by assaying the antibody that accumulates in the fluid phase of the blood or **plasma**; such antibodies are known as circulating antibodies. Circulating antibody is usually measured by collecting blood, allowing it to clot, and then isolating the serum from the clotted blood. The amount and characteristics of the antibody in the resulting antiserum are then determined using the assays we will describe in Sections A-5–A-11.

The most important characteristics of an antibody response are the specificity, amount, isotype or class, and affinity of the antibodies produced. The **specificity** determines the ability of the antibody to distinguish the immunogen from other antigens. The amount of antibody can be determined in many different ways and is a function of the number of responding B cells, their rate of antibody synthesis, and the persistence of the antibody after production. The persistence of an antibody in the plasma and extracellular fluid bathing the tissues is determined mainly by its **isotype** (see Sections 4-15 and 9-12); each isotype has a different half-life *in vivo*. The isotypic composition of an antibody response also determines the biological functions these antibodies can perform and the sites in which antibody will be found. Finally, the strength of binding of the antibody to its antigen in terms of a single antigen-binding site binding to a monovalent antigen is termed its **affinity** (the total binding strength of a molecule with more than one binding site is called its **avidity**). Binding strength is important, since the higher the affinity of the antibody for its antigen, the less antibody is required to eliminate the antigen, as antibodies with higher affinity will bind at lower antigen concentrations. All these parameters of the humoral immune response help to determine the capacity of that response to protect the host from infection.

Antibody molecules are highly specific for their corresponding antigen, being able to detect one molecule of a protein antigen out of more than 10^8 similar molecules. This makes antibodies both easy to isolate and study, and invaluable as probes of biological processes. Whereas standard chemistry would have great difficulty in distinguishing between two such closely related proteins as human and pig insulin, or two such closely related structures as *ortho*- and *para*-nitrophenyl, antibodies can be made that discriminate between these two structures absolutely. The value of antibodies as molecular probes has stimulated the development of many sensitive and highly specific techniques to measure their presence, to determine their specificity and affinity for a range of antigens, and to ascertain their functional capabilities. Many standard techniques used throughout biology exploit the specificity and stability of antigen binding by antibodies. Comprehensive guides to the conduct of these antibody assays are available in many books on immunological methodology; we will illustrate here only the most important techniques, especially those used in studying the immune response itself.

Some assays for antibody measure the direct binding of the antibody to its antigen. Such assays are based on **primary interactions**. Others determine the amount of antibody present by the changes it induces in the physical state

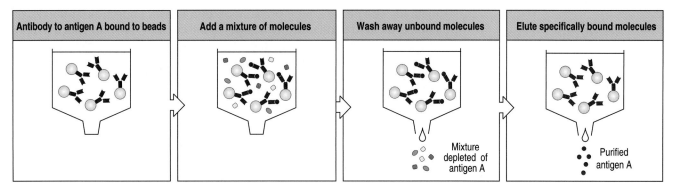

| Antibody to antigen A bound to beads | Add a mixture of molecules | Wash away unbound molecules | Elute specifically bound molecules |

Mixture depleted of antigen A

Purified antigen A

Fig. A.5 Affinity chromatography uses antigen–antibody binding to purify antigens or antibodies. To purify a specific antigen from a complex mixture of molecules, a monoclonal antibody is attached to an insoluble matrix, such as chromatography beads, and the mixture of molecules is passed over the matrix. The specific antibody binds the antigen of interest; other molecules are washed away. Specific antigen is then eluted by altering the pH, which can usually disrupt antibody–antigen bonds. Antibodies can be purified in the same way on beads coupled to antigen (not shown).

of the antigen, such as the precipitation of soluble antigen or the clumping of antigenic particles; these are called secondary interactions. Both types of assay can be used to measure the amount and specificity of the antibodies produced after immunization, and both can be applied to a wide range of other biological questions.

As assays for antibody were originally conducted with antisera from immune individuals, they are commonly referred to as **serological assays**, and the use of antibodies is often called **serology**. The amount of antibody is usually determined by antigen-binding assays after titration of the antiserum by serial dilution, and the point at which binding falls to 50% of the maximum is usually referred to as the **titer** of an antiserum.

A-5   Affinity chromatography.

Specific antibody can be isolated from an antiserum by **affinity chromatography**, which exploits the specific binding of antibody to antigen held on a solid matrix (Fig. A.5). Antigen is bound covalently to small, chemically reactive beads, which are loaded into a column, and the antiserum is allowed to pass over the beads. The specific antibodies bind, while all the other proteins in the serum, including antibodies to other substances, can be washed away. The specific antibodies are then eluted, typically by lowering the pH to 2.5 or raising it to greater than 11. Antibodies bind stably under physiological conditions of salt concentration, temperature, and pH, but the binding is reversible as the bonds are noncovalent. Affinity chromatography can also be used to purify antigens from complex mixtures by using beads coated with specific antibody. The technique is known as affinity chromatography because it separates molecules on the basis of their affinity for one another.

A-6   Radioimmunoassay (RIA), enzyme-linked immunosorbent assay (ELISA), and competitive inhibition assay.

Radioimmunoassay (RIA) and **enzyme-linked immunosorbent assay (ELISA)** are direct binding assays for antibody (or antigen) and both work on the same principle, but the means of detecting specific binding is different. Radioimmunoassays are commonly used to measure the levels of hormones in blood and tissue fluids, while ELISA assays are frequently used in viral diagnostics, for example in detecting cases of HIV infection. For both these

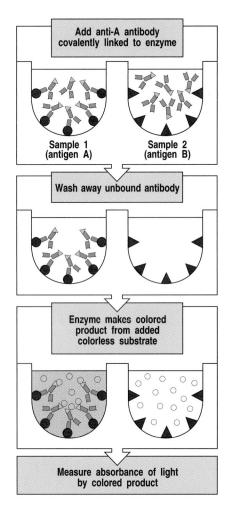

Fig. A.6 The principle of the enzyme-linked immunosorbent assay (ELISA). To detect antigen A, purified antibody specific for antigen A is linked chemically to an enzyme. The samples to be tested are coated onto the surface of plastic wells to which they bind nonspecifically; residual sticky sites on the plastic are blocked by adding irrelevant proteins (not shown). The labeled antibody is then added to the wells under conditions where nonspecific binding is prevented, so that only binding to antigen A causes the labeled antibody to be retained on the surface. Unbound labeled antibody is removed from all wells by washing, and bound antibody is detected by an enzyme-dependent color-change reaction. This assay allows arrays of wells known as microtiter plates to be read in fiberoptic multichannel spectrometers, greatly speeding the assay. Modifications of this basic assay allow antibody or antigen in unknown samples to be measured as shown in Figs A.7 and A.29 (see also Section A-10).

methods one needs a pure preparation of a known antigen or antibody, or both, in order to standardize the assay. We will describe the assay with a sample of pure antibody, which is the more usual case, but the principle is similar if pure antigen is used instead. In RIA for an antigen, pure antibody against that antigen is radioactively labeled, usually with ^{125}I; for the ELISA, an enzyme is linked chemically to the antibody. The unlabeled component, which in this case would be antigen, is attached to a solid support, such as the wells of a plastic multiwell plate, which will adsorb a certain amount of any protein.

The labeled antibody is allowed to bind to the unlabeled antigen, under conditions where nonspecific adsorption is blocked, and any unbound antibody and other proteins are washed away. Antibody binding in RIA is measured directly in terms of the amount of radioactivity retained by the coated wells, whereas in ELISA, binding is detected by a reaction that converts a colorless substrate into a colored reaction product (Fig. A.6). The color change can be read directly in the reaction tray, making data collection very easy, and ELISA also avoids the hazards of radioactivity. This makes ELISA the preferred method for most direct-binding assays. Labeled anti-immunoglobulin antibodies (see Section A-10) can also be used in RIA or ELISA to detect binding of unlabeled antibody to unlabeled antigen-coated plates. In this case, the labeled anti-immunoglobulin antibody is used in what is termed a 'second layer.' The use of such a second layer also amplifies the signal, as at least two molecules of the labeled anti-immunoglobulin antibody are able to bind to each unlabeled antibody. RIA and ELISA can also be carried out with unlabeled antibody stuck to the plates and labeled antigen added.

A modification of ELISA known as a **capture** or **sandwich ELISA** (or more generally as an **antigen-capture assay**) can be used to detect secreted products such as cytokines. Rather than the antigen being directly attached to a plastic plate, antigen-specific antibodies are bound to the plate. These are able to bind antigen with high affinity, and thus concentrate it on the surface of the plate, even with antigens that are present in very low concentrations in the initial mixture. A separate labeled antibody that recognizes a different epitope to the immobilized first antibody is then used to detect the bound antigen.

These assays illustrate two crucial aspects of all serological assays. First, at least one of the reagents must be available in a pure, detectable form in order to obtain quantitative information. Second, there must be a means of separating the bound fraction of the labeled reagent from the unbound, free fraction so that the percentage of specific binding can be determined. Normally, this separation is achieved by having the unlabeled partner trapped on a solid support. Labeled molecules that do not bind can then be washed away, leaving just the labeled partner that has bound. In Fig. A.6, the unlabeled antigen is attached to the well and the labeled antibody is trapped by binding to it. The separation of bound from free is an essential step in every assay that uses antibodies.

RIA and ELISA do not allow one to measure directly the amount of antigen or antibody in a sample of unknown composition, as both depend on the binding of a pure labeled antigen or antibody. There are various ways around this problem, one of which is to use a **competitive inhibition assay**, as shown in Fig. A.7. In this type of assay, the presence and amount of a particular antigen in an unknown sample is determined by its ability to compete with a labeled reference antigen for binding to an antibody attached to a plastic well. A standard curve is first constructed by adding varying amounts of a known, unlabeled standard preparation; the assay can then measure the amount of antigen in unknown samples by comparison with the standard. The competitive binding assay can also be used for measuring antibody in a sample of unknown composition by attaching the appropriate antigen to the plate and measuring the ability of the test sample to inhibit the binding of a labeled specific antibody.

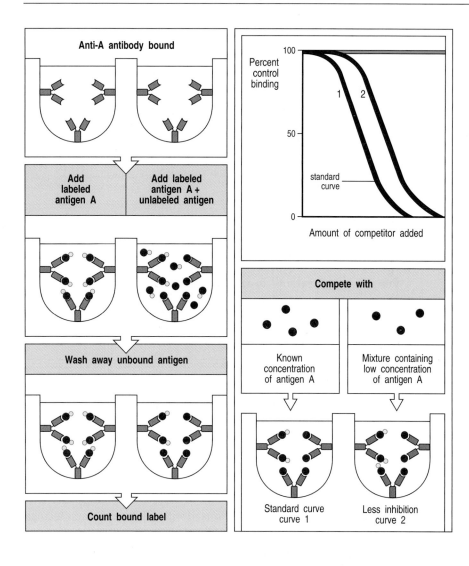

Anti-A antibody bound

Add labeled antigen A | Add labeled antigen A + unlabeled antigen

Wash away unbound antigen

Count bound label

Percent control binding

Amount of competitor added

standard curve

Compete with

Known concentration of antigen A | Mixture containing low concentration of antigen A

Standard curve curve 1 | Less inhibition curve 2

Fig. A.7 Competitive inhibition assay for antigen in unknown samples. A fixed amount of unlabeled antibody is attached to a set of wells, and a standard reference preparation of a labeled antigen is bound to it. Unlabeled standard or test samples are then added in various amounts and the displacement of labeled antigen is measured, generating characteristic inhibition curves. A standard curve is obtained by using known amounts of unlabeled antigen identical to that used as the labeled species, and comparison with this curve allows the amount of antigen in unknown samples to be calculated. The green line on the graph represents a sample lacking any substance that reacts with anti-A antibodies.

A-7 Hemagglutination and blood typing.

The direct measurement of antibody binding to antigen is used in most quantitative serological assays. However, some important assays are based on the ability of antibody binding to alter the physical state of the antigen it binds to. These secondary interactions can be detected in a variety of ways. For instance, when the antigen is displayed on the surface of a large particle such as a bacterium, antibodies can cause the bacteria to clump or **agglutinate**. The same principle applies to the reactions used in blood typing, only here the target antigens are on the surface of red blood cells and the clumping reaction caused by antibodies against them is called **hemagglutination** (from the Greek *haima*, blood).

Hemagglutination is used to determine the ABO blood group of blood donors and transfusion recipients. Clumping or agglutination is induced by antibodies or agglutinins called anti-A or anti-B that bind to the A or B blood-group substances, respectively (Fig. A.8). These blood-group antigens are arrayed in many copies on the surface of the red blood cell, allowing the cells to agglutinate when cross-linked by antibodies. Because hemagglutination involves the cross-linking of blood cells by the simultaneous binding of antibody molecules to identical antigens on different cells, this reaction demonstrates that each antibody molecule has at least two identical antigen-binding sites.

Fig. A.8 Hemagglutination is used to type blood groups and match compatible donors and recipients for blood transfusion. Common gut bacteria bear antigens that are similar or identical to blood-group antigens, and these stimulate the formation of antibodies to these antigens in individuals who do not bear the corresponding antigen on their own red blood cells (left column); thus, type O individuals, who lack A and B, have both anti-A and anti-B antibodies, while type AB individuals have neither. The pattern of agglutination of the red blood cells of a transfusion donor or recipient with anti-A and anti-B antibodies reveals the individual's ABO blood group. Before transfusion, the serum of the recipient is also tested for antibodies that agglutinate the red blood cells of the donor, and vice versa, a procedure called a cross match, which may detect potentially harmful antibodies to other blood groups that are not part of the ABO system.

Serum from individuals of type	Red blood cells from individuals of type			
	O	**A**	**B**	**AB**
	Express the carbohydrate structures			
	R–GlcNAc–Gal Fuc	R–GlcNAc–Gal–GalNAc Fuc	R–GlcNAc–Gal–Gal Fuc	R–GlcNAc–Gal–GalNAc Fuc + R–GlcNAc–Gal–Gal Fuc
O Anti-A and anti-B antibodies	no agglutination	agglutination	agglutination	agglutination
A Anti-B antibodies	no agglutination	no agglutination	agglutination	agglutination
B Anti-A antibodies	no agglutination	agglutination	no agglutination	agglutination
AB No antibodies to A or B	no agglutination	no agglutination	no agglutination	no agglutination

A-8 Precipitin reaction.

When sufficient quantities of antibody are mixed with soluble macromolecular antigens, a visible precipitate consisting of large aggregates of antigen cross-linked by antibody molecules can form. The amount of precipitate depends on the amounts of antigen and antibody, and on the ratio between them (Fig. A.9). This **precipitin reaction** provided the first quantitative assay for antibody, but is now seldom used in immunology. However, it is important to understand the interaction of antigen with antibody that leads to this reaction, as the production of **antigen:antibody complexes**, also known as **immune complexes**, *in vivo* occurs in almost all immune responses and occasionally can cause significant pathology (see Chapters 12 and 13).

In the precipitin reaction, various amounts of soluble antigen are added to a fixed amount of serum containing antibody. As the amount of antigen added increases, the amount of precipitate generated also increases up to a maximum and then declines (see Fig. A.9). When small amounts of antigen are added,

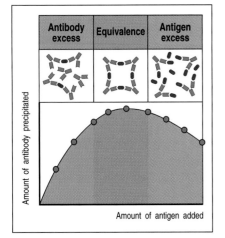

Fig. A.9 Antibody can precipitate soluble antigen. Analysis of the precipitate can generate a precipitin curve. Different amounts of antigen are added to a fixed amount of antibody, and precipitates form by antibody cross-linking of antigen molecules. The precipitate is recovered and the amount of precipitated antibody measured; the supernatant is tested for residual antigen or antibody. This defines zones of antibody excess, equivalence, and antigen excess. At equivalence, the largest antigen:antibody complexes form. In the zone of antigen excess, some of the immune complexes are too small to precipitate. These soluble immune complexes can cause pathological damage to small blood vessels when they form *in vivo* (see Chapter 13).

Fig. A.10 Different antibodies bind to distinct epitopes on an antigen molecule. The surface of an antigen possesses many potential antigenic determinants or epitopes, distinct sites to which an antibody can bind. The number of antibody molecules that can bind to a molecule of antigen at one time defines the antigen's valence. Steric considerations can limit the number of different antibodies that bind to the surface of an antigen at any one time (center and bottom panels) so that the number of epitopes on an antigen is always greater than or equal to its valence.

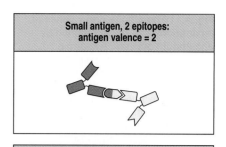

Small antigen, 2 epitopes: antigen valence = 2

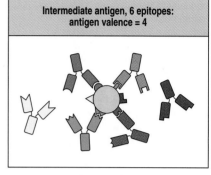

Intermediate antigen, 6 epitopes: antigen valence = 4

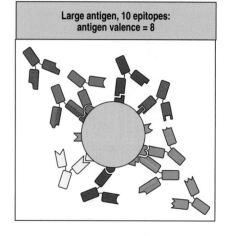

Large antigen, 10 epitopes: antigen valence = 8

antigen:antibody complexes are formed under conditions of antibody excess so that each molecule of antigen is bound by antibody and cross-linked to other molecules of antigen. When large amounts of antigen are added, only small antigen:antibody complexes can form and these are often soluble in this zone of antigen excess. Between these two zones, all of the antigen and antibody is found in the precipitate, generating a zone of equivalence. At equivalence, very large lattices of antigen and antibody are formed by cross-linking. While all antigen:antibody complexes can potentially produce disease, the small, soluble immune complexes formed in the zone of antigen excess may persist and cause pathology *in vivo*.

The precipitin reaction is affected by the number of binding sites that each antibody has for antigen, and by the maximum number of antibodies that can be bound by an antigen molecule or particle at any one time. These quantities are defined as the **valence** of the antibody and the valence of the antigen: the valence of both the antibodies and the antigen must be two or greater before any precipitation can occur. The valence of an antibody depends on its structural class (see Section 4-19).

Antigen will be precipitated only if it has several antibody-binding sites. This condition is usually satisfied in macromolecular antigens, which have a complex surface with binding sites for several different antibodies. The site on an antigen to which each distinct antibody molecule binds is called an **antigenic determinant** or an **epitope**. Steric considerations limit the number of distinct antibody molecules that can bind to a single antigen molecule at any one time however, because antibody molecules binding to epitopes that partially overlap will compete for binding. For this reason, the valence of an antigen is almost always less than the number of epitopes on the antigen (Fig. A.10).

A-9 Equilibrium dialysis: measurement of antibody affinity and avidity.

The affinity of an antibody is the strength of binding of a monovalent ligand to a single antigen-binding site on the antibody. The affinity of an antibody that binds small antigens, such as haptens, that can diffuse freely across a dialysis membrane can be determined directly by the technique of **equilibrium dialysis**. A known amount of antibody, whose molecules are too large to cross a dialysis membrane, is placed in a dialysis bag and offered various amounts of antigen. Molecules of antigen that bind to the antibody are no longer free to diffuse across the dialysis membrane, so only the unbound molecules of antigen equilibrate across it. By measuring the concentration of antigen inside the bag and in the surrounding fluid, one can determine the amount of the antigen that is bound as well as the amount that is free when equilibrium has been achieved. Given that the amount of antibody present is known, the affinity of the antibody and the number of specific binding sites for the antigen per molecule of antibody can be determined from this information. The data are usually analyzed using **Scatchard analysis** (Fig. A.11); such analyses were used to demonstrate that a molecule of IgG antibody has two identical antigen-binding sites.

Fig. A.11 The affinity and valence of an antibody can be determined by equilibrium dialysis. A known amount of antibody is placed in the bottom half of a dialysis chamber and exposed to different amounts of a diffusible monovalent antigen, such as a hapten. At equilibrium, the concentration of free antigen will be the same on each side of the membrane, so that at each concentration of antigen added, the fraction of the antigen bound is determined from the difference in concentration of total antigen in the top and bottom chambers. This information can be transformed into a Scatchard plot as shown here. In Scatchard analysis, the ratio r/c (where r = moles of antigen bound per mole of antibody and c = molar concentration of free antigen) is plotted against r. The number of binding sites per antibody molecule can be determined from the value of r at infinite free-antigen concentration, where r/free = 0, in other words at the x-axis intercept. The analysis of a monoclonal IgG antibody molecule, in which there are two identical antigen-binding sites per molecule, is shown in the left panel. The slope of the line is determined by the affinity of the antibody molecule for its antigen; if all the antibody molecules in a preparation are identical, as for this monoclonal antibody, then a straight line is obtained whose slope is equal to $-K_a$, where K_a is the association (or affinity) constant and the dissociation constant $K_d = 1/K_a$. However, antiserum raised even against a simple antigenic determinant such as a hapten contains a heterogeneous population of antibody molecules (see Section A-1). Each antibody molecule would, if isolated, make up part of the total and give a straight line whose x-axis intercept is less than two, as this antibody molecule contains only a fraction of the total binding sites in the population (middle panel). As a mixture, they give curved lines with an x-axis intercept of two for which an average affinity ($\overline{K}_a$) can be determined from the slope of this line at a concentration of antigen where 50% of the sites are bound, or at x = 1 (right panel). The association constant determines the equilibrium state of the reaction Ag + Ab = Ag:Ab, where antigen = Ag and antibody = Ab, and K_a = [Ag:Ab]/[Ag][Ab]. This constant reflects the 'on' and 'off' rates for antigen binding to the antibody; with small antigens such as haptens, binding is usually as rapid as diffusion allows, whereas differences in 'off' rates determine the affinity constant. However, with larger antigens the 'on' rate may also vary as the interaction becomes more complex.

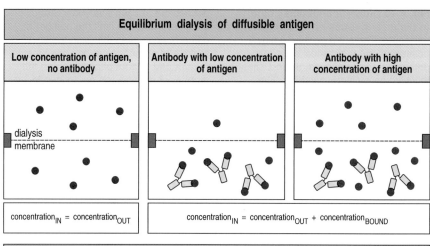

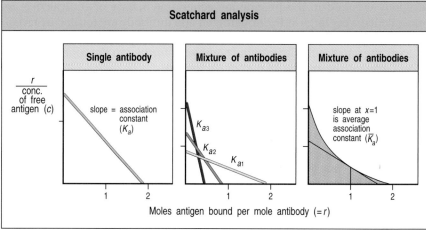

Whereas affinity measures the strength of binding of an antigenic determinant to a single antigen-binding site, an antibody reacting with an antigen that has multiple identical epitopes or with the surface of a pathogen will often bind the same molecule or particle with both of its antigen-binding sites. This increases the apparent strength of binding, since both binding sites must release at the same time in order for the two molecules to dissociate. This is often referred to as **cooperativity** in binding, but it should not be confused with the cooperative binding found in a protein such as hemoglobin, in which binding of ligand at one site enhances the affinity of a second binding site for its ligand. The overall strength of binding of an antibody molecule to an antigen or particle is called its avidity (Fig. A.12). For IgG antibodies, bivalent binding can significantly increase avidity; in IgM antibodies, which have ten identical antigen-binding sites, the affinity of each site for a monovalent antigen is usually quite low, but the avidity of binding of the whole antibody to a surface such as a bacterium that displays multiple identical epitopes can be very high.

A-10   Anti-immunoglobulin antibodies.

A general approach to the detection of bound antibody that avoids the need to label each preparation of antibody molecules is to detect bound, unlabeled antibody with a labeled antibody specific for immunoglobulins themselves. Immunoglobulins, like other proteins, are immunogenic when used to immunize individuals of another species. The majority of **anti-immunoglobulin antibodies** raised in this way recognize conserved features shared by all

Fig. A.12 The avidity of an antibody is its strength of binding to intact antigen. When an IgG antibody binds a ligand with multiple identical epitopes, both binding sites can bind the same molecule or particle. The overall strength of binding, called avidity, is greater than the affinity, the strength of binding of a single site, since both binding sites must dissociate at the same time for the antibody to release the antigen. This property is very important in the binding of antibody to bacteria, which usually have multiple identical epitopes on their surfaces.

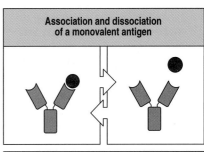

Association and dissociation of a monovalent antigen

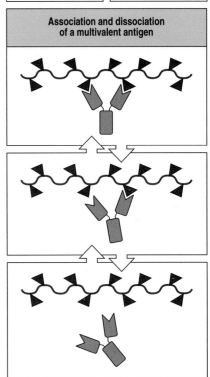

Association and dissociation of a multivalent antigen

immunoglobulin molecules of the immunizing species. Others can be specific for immunoglobulin chains, heavy or light chains, for example, or for individual isotypes. Antibodies raised by immunization of goats with mouse IgG are commonly used in experimental immunology. Such goat anti-mouse IgG antibodies can be purified using affinity chromatography, then labeled and used as a general probe for bound IgG antibodies. Anti-immunoglobulin antisera have found many uses in clinical medicine and biological research since their introduction. Fluorescently labeled anti-immunoglobulin antibodies are now widely used both in immunology and other areas of biology as secondary reagents for detecting specific antibodies bound, for example, to cell structures (see Sections A-14 and A-16). Labeled anti-immunoglobulin antibodies can also be used in radioimmunoassay or ELISA (see Section A-6) to detect binding of unlabeled antibody to antigen-coated plates.

Antibodies specific for individual immunoglobulin isotypes can be produced by immunizing an animal of a different species with a pure preparation of one isotype and then removing those antibodies that cross-react with immunoglobulins of other isotypes by using affinity chromatography (see Section A-5). **Anti-isotype antibodies** can be used to measure how much antibody of a particular isotype in an antiserum reacts with a given antigen. This reaction is particularly important for detecting small amounts of specific antibodies of the IgE isotype, which are responsible for most allergies. The presence in an individual's serum of IgE binding to an antigen correlates with allergic reactions to that antigen.

An alternative approach to detecting bound antibodies exploits bacterial proteins that bind to immunoglobulins with high affinity and specificity. One of these, **Protein A** from the bacterium *Staphylococcus aureus,* has been exploited widely in immunology for the affinity purification of immunoglobulin and for the detection of bound antibody. The use of standard second reagents such as labeled anti-immunoglobulin antibodies or Protein A to detect antibody bound specifically to its antigen allows great savings in reagent labeling costs, and also provides a standard detection system so that results in different assays can be compared directly.

A-11   Coombs tests and the detection of Rhesus incompatibility.

These tests use anti-immunoglobulin antibodies (see Section A-10) to detect the antibodies that cause **hemolytic disease of the newborn,** or **erythroblastosis fetalis**. Anti-immunoglobulin antibodies were first developed by Robin Coombs and the test for this disease is still called the Coombs test. Hemolytic disease of the newborn occurs when a mother makes IgG antibodies specific for the **Rhesus** or **Rh blood group antigen** expressed on the red blood cells of her fetus. Rh-negative mothers make these antibodies when they are exposed to Rh-positive fetal red blood cells bearing the paternally inherited Rh antigen. Maternal IgG antibodies are normally transported across the placenta to the fetus, where they protect the newborn infant against infection. However, IgG anti-Rh antibodies coat the fetal red blood cells, which are then destroyed by phagocytic cells in the liver, causing a hemolytic anemia in the fetus and newborn infant.

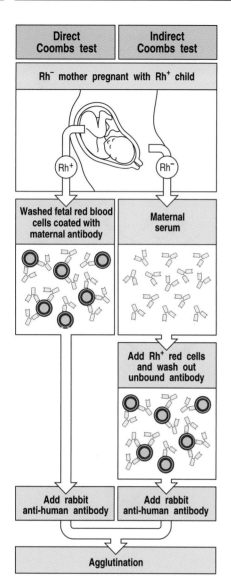

Direct Coombs test	Indirect Coombs test
Rh⁻ mother pregnant with Rh⁺ child	
Washed fetal red blood cells coated with maternal antibody	Maternal serum
	Add Rh⁺ red cells and wash out unbound antibody
Add rabbit anti-human antibody	Add rabbit anti-human antibody
Agglutination	

Fig. A.13 The Coombs direct and indirect anti-globulin tests for antibody to red blood cell antigens. A Rh⁻ mother of a Rh⁺ fetus can become immunized to fetal red blood cells that enter the maternal circulation at the time of delivery. In a subsequent pregnancy with a Rh⁺ fetus, IgG anti-Rh antibodies can cross the placenta and damage the fetal red blood cells. In contrast to anti-Rh antibodies, maternal anti-ABO antibodies are of the IgM isotype and cannot cross the placenta, and so do not cause harm. Anti-Rh antibodies do not agglutinate red blood cells but their presence on the fetal red blood cell surface can be shown by washing away unbound immunoglobulin and then adding antibody to human immunoglobulin, which agglutinates the antibody-coated cells. Anti-Rh antibodies can be detected in the mother's serum in an indirect Coombs test; the serum is incubated with Rh⁺ red blood cells, and once the antibody binds, the red blood cells are treated as in the direct Coombs test.

Since the Rh antigens are widely spaced on the red blood cell surface, the IgG anti-Rh antibodies cannot fix complement and cause lysis of red blood cells *in vitro*. Furthermore, for reasons that are not fully understood, antibodies to Rh blood group antigens do not agglutinate red blood cells as do antibodies to the ABO blood group antigens. Thus, detecting these antibodies was difficult until anti-human immunoglobulin antibodies were developed. With these, maternal IgG antibodies bound to the fetal red blood cells can be detected after washing the cells to remove unbound immunoglobulin that is present in the fetal serum. Adding anti-human immunoglobulin antibodies to the washed fetal red blood cells agglutinates any cells to which maternal antibodies are bound. This is the **direct Coombs test** (Fig. A.13), so called because it directly detects antibody bound to the surface of the fetal red blood cells. An **indirect Coombs test** is used to detect nonagglutinating anti-Rh antibody in maternal serum; the serum is first incubated with Rh-positive red blood cells, which bind the anti-Rh antibody, after which the antibody-coated cells are washed to remove unbound immunoglobulin and are then agglutinated with anti-immunoglobulin antibody (see Fig. A.13). The indirect Coombs test allows Rh incompatibilities that might lead to hemolytic disease of the newborn to be detected, and this knowledge allows the disease to be prevented (see Section 10-25). The Coombs test is also commonly used to detect antibodies to drugs that bind to red blood cells and cause hemolytic anemia.

A-12 Monoclonal antibodies.

The antibodies generated in a natural immune response or after immunization in the laboratory are a mixture of molecules of different specificities and affinities. Some of this heterogeneity results from the production of antibodies that bind to different epitopes on the immunizing antigen, but even antibodies directed at a single antigenic determinant such as a hapten can be markedly heterogeneous, as shown by **isoelectric focusing**. In this technique, proteins are separated on the basis of their isoelectric point, the pH at which their net charge is zero. By electrophoresing proteins in a pH gradient for long enough, each molecule migrates along the pH gradient until it reaches the pH at which it is neutral and is thus concentrated (focused) at that point. When antiserum containing anti-hapten antibodies is treated in this way and then transferred to a solid support such as nitrocellulose paper, the anti-hapten antibodies can be detected by their ability to bind labeled hapten. The binding of antibodies of various isoelectric points to the hapten shows that even antibodies that bind the same antigenic determinant can be heterogeneous.

Antisera are valuable for many biological purposes but they have certain inherent disadvantages that relate to the heterogeneity of the antibodies they contain. First, each antiserum is different from all other antisera, even if raised in a genetically identical animal by using the identical preparation of antigen and the same immunization protocol. Second, antisera can be produced in only limited volumes, and thus it is impossible to use the identical serological reagent in a long or complex series of experiments or clinical tests. Finally, even antibodies purified by affinity chromatography (see Section A-5) may include minor populations of antibodies that give unexpected cross-reactions, which confound the analysis of experiments. To avoid these problems, and to harness the full potential of antibodies, it was necessary to develop a way of making an unlimited supply of antibody molecules of homogeneous structure and known specificity. This has been achieved through the production of monoclonal antibodies from cultures of hybrid antibody-forming cells or, more recently, by genetic engineering.

Biochemists in search of a homogeneous preparation of antibody that they could subject to detailed chemical analysis turned early to proteins produced by patients with multiple myeloma, a common tumor of plasma cells. It was known that antibodies are normally produced by plasma cells and since this disease is associated with the presence of large amounts of a homogeneous gamma globulin called a **myeloma protein** in the patient's serum, it seemed likely that myeloma proteins would serve as models for normal antibody molecules. Thus, much of the early knowledge of antibody structure came from studies on myeloma proteins. These studies showed that monoclonal antibodies could be obtained from immortalized plasma cells. However, the antigen specificity of most myeloma proteins was unknown, which limited their usefulness as objects of study or as immunological tools.

This problem was solved by Georges Köhler and Cesar Milstein, who devised a technique for producing a homogeneous population of antibodies of known antigenic specificity. They did this by fusing spleen cells from an immunized mouse to cells of a mouse myeloma to produce hybrid cells that both proliferated indefinitely and secreted antibody specific for the antigen used to immunize the spleen cell donor. The spleen cell provides the ability to make specific antibody, while the myeloma cell provides the ability to grow indefinitely in culture and secrete immunoglobulin continuously. By using a myeloma cell partner that produces no antibody proteins itself, the antibody produced by the hybrid cells comes only from the immune spleen cell partner. After fusion, the hybrid cells are selected using drugs that kill the myeloma parental cell, while the unfused parental spleen cells have a limited life-span and soon die, so that only hybrid myeloma cell lines or **hybridomas** survive. Those hybridomas producing antibody of the desired specificity are then identified and cloned by regrowing the cultures from single cells (Fig. A.14). Since each hybridoma is a **clone** derived from fusion with a single B cell, all the antibody molecules it produces are identical in structure, including their antigen-binding site and isotype. Such antibodies are called **monoclonal antibodies**. This technology has revolutionized the use of antibodies by providing a limitless supply of antibody of a single and known specificity. Monoclonal antibodies are now used in most serological assays, as diagnostic probes, and as therapeutic agents. So far, however, only mouse monoclonals are routinely produced and efforts to use this same approach to make human monoclonal antibodies have met with very limited success.

A-13 Phage display libraries for antibody V-region production.

This is a technique for producing antibody-like molecules. Gene segments encoding the antigen-binding variable or V domains of antibodies are fused to genes encoding the coat protein of a bacteriophage. Bacteriophage containing such gene fusions are used to infect bacteria, and the resulting phage particles have coats that express the antibody-like fusion protein, with

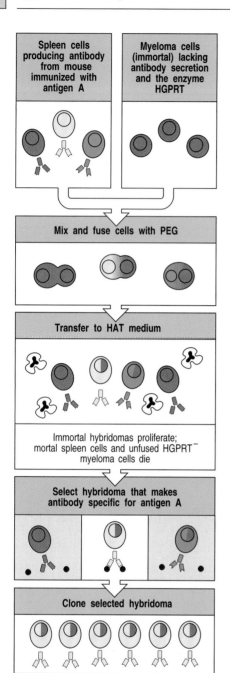

| Spleen cells producing antibody from mouse immunized with antigen A | Myeloma cells (immortal) lacking antibody secretion and the enzyme HGPRT |

Mix and fuse cells with PEG

Transfer to HAT medium

Immortal hybridomas proliferate; mortal spleen cells and unfused HGPRT⁻ myeloma cells die

Select hybridoma that makes antibody specific for antigen A

Clone selected hybridoma

Fig. A.14 The production of monoclonal antibodies. Mice are immunized with antigen A and given an intravenous booster immunization three days before they are killed, in order to produce a large population of spleen cells secreting specific antibody. Spleen cells die after a few days in culture. In order to produce a continuous source of antibody they are fused with immortal myeloma cells by using polyethylene glycol (PEG) to produce a hybrid cell line called a hybridoma. The myeloma cells are selected beforehand to ensure that they are not secreting antibody themselves and that they are sensitive to the hypoxanthine-aminopterin-thymidine (HAT) medium that is used to select hybrid cells because they lack the enzyme hypoxanthine:guanine phosphoribosyl transferase (HGPRT). The HGPRT gene contributed by the spleen cell allows hybrid cells to survive in the HAT medium, and only hybrid cells can grow continuously in culture because of the malignant potential contributed by the myeloma cells. Therefore, unfused myeloma cells and unfused spleen cells die in the HAT medium, as shown here by cells with dark, irregular nuclei. Individual hybridomas are then screened for antibody production, and cells that make antibody of the desired specificity are cloned by growing them up from a single antibody-producing cell. The cloned hybridoma cells are grown in bulk culture to produce large amounts of antibody. As each hybridoma is descended from a single cell, all the cells of a hybridoma cell line make the same antibody molecule, which is thus called a monoclonal antibody.

the antigen-binding domain displayed on the outside of the bacteriophage. A collection of recombinant phage, each displaying a different antigen-binding domain on its surface, is known as a **phage display library**. In much the same way that antibodies specific for a particular antigen can be isolated from a complex mixture by affinity chromatography (see Section A-5), phage expressing antigen-binding domains specific for a particular antigen can be isolated by selecting the phage in the library for binding to that antigen. The phage particles that bind are recovered and used to infect fresh bacteria. Each phage isolated in this way will produce a monoclonal antigen-binding particle analogous to a monoclonal antibody (Fig. A.15). The genes encoding the antigen-binding site, which are unique to each phage, can then be recovered from the phage DNA and used to construct genes for a complete antibody molecule by joining them to parts of immunoglobulin genes that encode the invariant parts of an antibody. When these reconstructed antibody genes are introduced into a suitable host cell line, such as the nonantibody-producing myeloma cells used for hybridomas, the transfected cells can secrete antibodies with all the desirable characteristics of monoclonal antibodies produced from hybridomas.

In much the same way that a collection of phage can display a wide variety of potential antigen-binding sites, the phage can also be engineered to display a wide variety of antigens; such a library is known as an **antigen display library**. In such cases, the antigens displayed are often short peptides encoded by chemically synthesized DNA sequences that have mixtures of all four nucleotides in some positions, so that all possible amino acids are incorporated. It is not usual for every position in a peptide to be allowed to vary in this way, since the number of different peptide sequences increases dramatically with the number of variable positions; there are over 2×10^{10} possible sequences of eight amino acids.

A-14 Immunofluorescence microscopy.

Since antibodies bind stably and specifically to their corresponding antigen, they are invaluable as probes for identifying a particular molecule in cells, tissues, or biological fluids. Antibody molecules can be used to locate their target molecules accurately in single cells or in tissue sections by a variety of

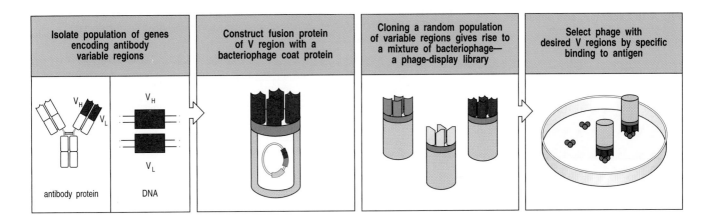

| Isolate population of genes encoding antibody variable regions | Construct fusion protein of V region with a bacteriophage coat protein | Cloning a random population of variable regions gives rise to a mixture of bacteriophage— a phage-display library | Select phage with desired V regions by specific binding to antigen |

different labeling techniques. When the antibody itself, or the anti-immunoglobulin antibody used to detect it, is labeled with a fluorescent dye the technique is known as **immunofluorescence microscopy**. As in all sero-logical techniques, the antibody binds stably to its antigen, allowing unbound antibody to be removed by thorough washing. As antibodies to proteins recognize the surface features of the native, folded protein, the native structure of the protein being sought usually needs to be preserved, either by using only the most gentle chemical fixation techniques or by using frozen tissue sections that are fixed only after the antibody reaction has been performed. Some anti-bodies, however, bind proteins even if they are denatured, and such antibodies will bind specifically even to protein in fixed tissue sections.

The fluorescent dye can be covalently attached directly to the specific anti-body, but more commonly, the bound antibody is detected by fluorescent anti-immunoglobulin, a technique known as **indirect immunofluorescence**. The dyes chosen for immunofluorescence are excited by light of one wave-length, usually blue or green, and emit light of a different wavelength in the visible spectrum. The most common fluorescent dyes are fluorescein, which emits green light, Texas Red and Peridinin chlorophyll protein (PerCP), which emit red light, and rhodamine and phycoerythrin (PE) which emit orange/red light (Fig. A.16). By using selective filters, only the light coming from the dye or fluorochrome used is detected in the fluorescence microscope (Fig. A.17). Although Albert Coons first devised this technique to identify the plasma cell as the source of antibody, it can be used to detect the distribution of any protein. By attaching different dyes to different antibodies, the distri-bution of two or more molecules can be determined in the same cell or tissue section (see Fig. A.17).

The recent development of the **confocal fluorescent microscope**, which uses computer-aided techniques to produce an ultrathin optical section of a cell or tissue, gives very high resolution immunofluorescence microscopy without the need for elaborate sample preparation. The resolution of the confocal microscope can be further increased using low-intensity illumination so that two photons are required to excite the fluorochrome. A pulsed laser beam is used, and only when it is focused into the focal plane of the microscope is the intensity sufficient to excite fluorescence. In this way the fluorescence emission itself can be restricted to the optical section.

One important development in the area of microscopy has been the use of **time-lapse video microscopy**, in which sensitive digital video cameras record the movement of fluorescently labeled molecules in cell membranes and their redistribution when cells come into contact with each other. Cell-surface molecules can be fluorescently labeled in two main ways. One is by the binding of fluorochrome-labeled Fab fragments of antibodies specific for the

Fig. A.15 The production of antibodies by genetic engineering. Short primers to consensus sequences in heavy- and light-chain variable (V) regions of immunoglobulin genes are used to generate a library of heavy- and light-chain V-region DNAs by the polymerase chain reaction, with spleen DNA as the starting material. These heavy- and light-chain V-region genes are cloned randomly into a filamentous phage such that each phage expresses one heavy-chain and one light-chain V region as a surface fusion protein with antibody-like properties. The resulting phage display library is multiplied in bacteria, and the phage are then bound to a surface coated with antigen. The unbound phage are washed away; the bound phage are recovered, multiplied in bacteria, and again bound to antigen. After a few cycles, only specific high-affinity antigen-binding phage are left. These can be used like antibody molecules, or their V genes can be recovered and engineered into anti-body genes to produce genetically engineered antibody molecules (not shown). This technology may replace the hybridoma technology for producing monoclonal antibodies and has the advantage that humans can be used as the source of DNA.

Fig. A.16 Excitation and emission wavelengths for common fluorochromes.

Excitation and emission wavelengths of some commonly used fluorochromes		
Probe	Excitation (nm)	Emission (nm)
R-phycoerythrin (PE)	480; 565	578
Fluorescein	495	519
PerCP	490	675
Texas Red	589	615
Rhodamine	550	573

Fig. A.17 Immunofluorescence microscopy. Antibodies labeled with a fluorescent dye such as fluorescein (green triangle) are used to reveal the presence of their corresponding antigens in cells or tissues. The stained cells are examined in a microscope that exposes them to blue or green light to excite the fluorescent dye. The excited dye emits light at a characteristic wavelength, which is captured by viewing the sample through a selective filter. This technique is applied widely in biology to determine the location of molecules in cells and tissues. Different antigens can be detected in tissue sections by labeling antibodies with dyes of distinctive color. Here, antibodies to the protein glutamic acid decarboxylase (GAD) coupled to a green dye are shown to stain the β cells of pancreatic islets of Langerhans. The α cells do not make this enzyme and are labeled with antibodies to the hormone glucagon coupled with an orange fluorescent dye. GAD is an important antigen in diabetes, a disease in which the insulin-secreting β cells of the islets of Langerhans are destroyed by an immune attack on self tissues (see Chapter 13). Photograph courtesy of M. Solimena and P. De Camilli.

protein of interest; the other is by generating a fusion protein, in which the protein of interest has been attached to one of a family of fluorescent proteins obtained from jellyfish. The first of these fluorescent proteins to come into wide use was green fluorescent protein (GFP), isolated from the jellyfish, *Aequorea victoria*. Other variants of this protein with different properties are now available, and the list of available fluorescent labels now includes red, blue, cyan, or yellow fluorescent proteins. Using cells transfected with the genes encoding such fusion proteins, it is possible to show the redistribution of T-cell receptors, co-receptors, adhesion molecules, and other signaling molecules, such as CD45, that takes place when a T cell makes contact with a target cell. It is clear from these observations that the interface between the T cell and its target does not represent a simple apposition of two cell membranes, but is an actively organized and dynamic structure, often now referred to as the 'immunological synapse.'

A-15   Immunoelectron microscopy.

Antibodies can be used to detect the intracellular location of structures or particular proteins at high resolution by electron microscopy, a technique known as **immunoelectron microscopy**. Antibodies against the required antigen are labeled with gold particles and then applied to ultrathin sections, which are then examined in the transmission electron microscope. Antibodies labeled with gold particles of different diameters enable two or more proteins to be studied simultaneously (see Fig. 5.8). The difficulty with this technique is in staining the ultrathin section adequately, as few molecules of antigen are present in each section.

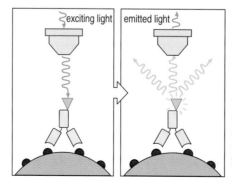

A-16 Immunohistochemistry.

An alternative to immunofluorescence (see Section A-14) for detecting a protein in tissue sections is **immunohistochemistry**, in which the specific antibody is chemically coupled to an enzyme that converts a colorless substrate into a colored reaction product *in situ*. The localized deposition of the colored product where antibody has bound can be directly observed under a light microscope. The antibody binds stably to its antigen, allowing unbound antibody to be removed by thorough washing. This method of detecting bound antibody is analogous to ELISA (see Section A-6) and frequently uses the same coupled enzymes, the difference in detection being primarily that in immunohistochemistry the colored products are insoluble and precipitate at the site where they are formed. Horseradish peroxidase and alkaline phosphatase are the two enzymes most commonly used in these applications. Horseradish peroxidase oxidises the substrate diaminobenzidine to produce a brown precipitate, while alkaline phosphatase can produce red or blue dyes depending on the substrates used; a common substrate is 5-bromo-4-chloro-3-indolyl phosphate plus nitroblue tetrazolium (BCIP/NBT), which gives rise to a dark blue or purple stain. As with immunofluorescence, the native structure of the protein being sought usually needs to be preserved, so that it will be recognized by the antibody. Tissues are fixed by the most gentle chemical fixation techniques or frozen tissue sections are used that are fixed only after the antibody reaction has been performed.

A-17 Immunoprecipitation and co-immunoprecipitation.

In order to raise antibodies to membrane proteins and other cellular structures that are difficult to purify, mice are often immunized with whole cells or crude cell extracts. Antibodies to the individual molecules are then obtained by using these immunized mice to produce hybridomas making monoclonal antibodies (see Section A-12) that bind to the cell type used for immunization. To characterize the molecules identified by the antibodies, cells of the same type are labeled with radioisotopes and dissolved in nonionic detergents that disrupt cell membranes but do not interfere with antigen–antibody interactions. This allows the labeled protein to be isolated by binding to the antibody in a reaction known as **immunoprecipitation**. The antibody is usually attached to a solid support, such as the beads that are used in affinity chromatography (see Section A-5), or to Protein A. Cells can be labeled in two main ways for immunoprecipitation analysis. All the proteins in a cell can be labeled metabolically by growing the cell in radioactive amino acids that are incorporated into cellular protein (Fig. A.18). Alternatively, one can label only

Fig. A.18 Cellular proteins reacting with an antibody can be characterized by immunoprecipitation of labeled cell lysates. All actively synthesized cellular proteins can be labeled metabolically by incubating cells with radioactive amino acids (shown here for methionine) or one can label just the cell-surface proteins by using radioactive iodine in a form that cannot cross the cell membrane or by a reaction with the small molecule biotin, detected by its reaction with labeled avidin (not shown). Cells are lysed with detergent and individual labeled cell-associated proteins can be precipitated with a monoclonal antibody attached to beads. After unbound proteins have been washed away, the bound protein is eluted in the detergent sodium dodecyl sulfate (SDS), which dissociates it from the antibody and also coats the protein with a strong negative charge, allowing it to migrate according to its size in polyacrylamide gel electrophoresis (PAGE). The positions of the labeled proteins are determined by auto-radiography using X-ray film. This technique of SDS-PAGE can be used to determine the molecular weight and subunit composition of a protein. Patterns of protein bands observed with metabolic labeling are usually more complex than those revealed by radioiodination, owing to the presence of precursor forms of the protein (right panel). The mature form of a surface protein can be identified as being the same size as that detected by surface iodination or biotinylation (not shown).

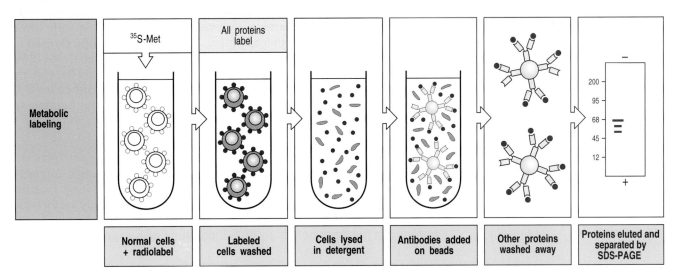

| Metabolic labeling | Normal cells + radiolabel | Labeled cells washed | Cells lysed in detergent | Antibodies added on beads | Other proteins washed away | Proteins eluted and separated by SDS-PAGE |

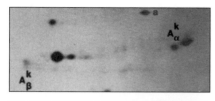
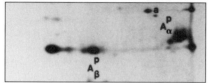

Fig. A.19 Two-dimensional gel electrophoresis of MHC class II molecules. Proteins in mouse spleen cells have been labeled metabolically (see Fig. A.18), precipitated with a monoclonal antibody against the mouse MHC class II molecule H2-A, and separated by isoelectric focusing in one direction and SDS-PAGE in a second direction at right angles to the first; hence the term two-dimensional gel electrophoresis. This allows one to distinguish molecules of the same molecular weight on the basis of their charge. The separated proteins are detected by autoradiography. The MHC class II molecules are composed of two chains, α and β, and in the different MHC class II molecules these have different isoelectric points (compare upper and lower panels). The MHC genotype of mice is indicated by lower case superscripts (k,p). Actin, a common contaminant, is marked a. Photographs courtesy of J.F. Babich.

the cell-surface proteins by radioiodination under conditions that prevent iodine from crossing the plasma membrane and labeling proteins inside the cell, or by a reaction that labels only membrane proteins with biotin, a small molecule that is detected readily by labeled avidin, a protein found in egg whites that binds biotin with very high affinity.

Once the labeled proteins have been isolated by the antibody, they can be characterized in several ways. The most common is polyacrylamide gel electrophoresis (PAGE) of the proteins after dissociating them from antibody in the strong ionic detergent sodium dodecyl sulfate (SDS), a technique generally abbreviated as **SDS-PAGE**. SDS binds relatively homogeneously to proteins, conferring a charge that allows the electrophoretic field to drive protein migration through the gel. The rate of migration is controlled mainly by protein size (see Fig. A.18). Proteins of differing charges can be separated using isoelectric focusing (see Section A-12). This technique can be combined with SDS-PAGE in a procedure known as **two-dimensional gel electrophoresis**. For this, the immunoprecipitated protein is eluted in urea, a nonionic solubilizing agent, and run on an isoelectric focusing gel in a narrow tube of polyacrylamide. This first-dimensional isoelectric focusing gel is then placed across the top of an SDS-PAGE slab gel, which is then run vertically to separate the proteins by molecular weight (Fig. A.19). Two-dimensional gel electrophoresis is a powerful technique that allows many hundreds of proteins in a complex mixture to be distinguished from one another.

Immunoprecipitation and the related technique of immunoblotting (see Section A-18) are useful for determining the molecular weight and isoelectric point of a protein as well as its abundance, distribution, and whether, for example, it undergoes changes in molecular weight and isoelectric point as a result of processing within the cell.

A-18   Immunoblotting (Western blotting).

Like immunoprecipitation (see Section A-17), **immunoblotting** is used for identifying the presence of a given protein in a cell lysate, but it avoids the problem of having to label large quantities of cells with radioisotopes. Unlabeled cells are placed in detergent to solubilize all cell proteins and the lysate is run on SDS-PAGE to separate the proteins (see Section A-17). The size-separated proteins are then transferred from the gel to a stable support such as a nitrocellulose membrane. Specific proteins are detected by treatment with antibodies able to react with SDS-solubilized proteins (mainly those that react with denatured sequences); the bound antibodies are revealed by anti-immunoglobulin antibodies labeled with radioisotopes or an enzyme. The term **Western blotting** as a synonym for immunoblotting arose because the comparable technique for detecting specific DNA sequences is known as Southern blotting, after Ed Southern who devised it, which in turn provoked the name Northern for blots of size-separated RNA, and Western for blots of size-separated proteins. Western blots have many applications in basic research and clinical diagnosis. They are often used to test sera for the presence of antibodies to specific proteins, for example to detect antibodies to different constituents of the human immunodeficiency virus, HIV (Fig. A.20).

Co-immunoprecipitation is an extension of the immunoprecipitation technique which is used to determine whether a given protein interacts physically with another given protein. Cell extracts containing the presumed interaction complex are first immunoprecipitated with antibody against one of the proteins. The material identified by this means is then tested for the presence of the other protein by immunoblotting with a specific antibody.

Fig. A.20 Western blotting is used to identify antibodies to the human immunodeficiency virus (HIV) in serum from infected individuals. The virus is dissociated into its constituent proteins by treatment with the detergent SDS, and its proteins are separated using SDS-PAGE. The separated proteins are transferred to a nitrocellulose sheet and reacted with the test serum. Anti-HIV antibodies in the serum bind to the various HIV proteins and are detected with enzyme-linked anti-human immunoglobulin, which deposits colored material from a colorless substrate. This general methodology will detect any combination of antibody and antigen and is used widely, although the denaturing effect of SDS means that the technique works most reliably with antibodies that recognize the antigen when it is denatured.

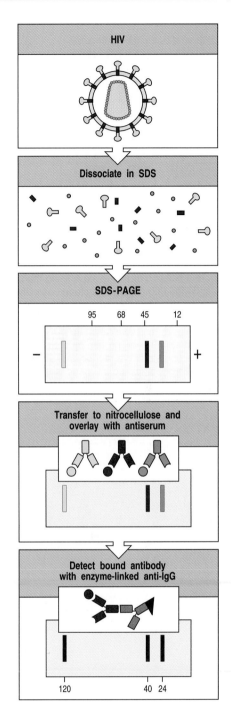

A-19 Use of antibodies in the isolation and identification of genes and their products.

As a first step in isolating the gene that codes for a particular protein, antibodies specific for the protein can be used to isolate the purified protein from cells using affinity chromatography (see Section A-5). Small amounts of amino acid sequence can then be obtained from the protein's amino-terminal end or from peptide fragments generated by proteolysis. The information in these amino acid sequences is used to construct a set of synthetic oligonucleotides corresponding to the possible DNA sequences, which are then used as probes to isolate the gene encoding the protein from either a library of DNA sequences complementary to mRNA (a cDNA library) or a genomic DNA library (a library of chromosomal DNA fragments).

An alternative approach to gene identification uses antibodies to identify the protein product of a gene that has been introduced into a cell that does not normally express it. This technique is most commonly applied to the identification of genes encoding cell-surface proteins. A set of cDNA-containing expression vectors is first made from a cDNA library prepared from the total mRNA from a cell type that does express the protein of interest. The vectors are used to transfect a cell type that does not normally express the protein of interest, and the vector drives expression of the cDNA it contains without integrating into the host cell DNA. Cells expressing the required protein after transfection are then isolated by binding to specific antibodies that detect the presence of the protein on the cell surface. The vector containing the required gene can then be recovered from these cells (Fig. A.21).

The recovered vectors are then introduced into bacterial cells where they replicate rapidly, and these amplified vectors are used in a second round of transfection in mammalian cells. After several cycles of transfection, isolation, and amplification in bacteria, single colonies of bacteria are picked and the vectors prepared from cultures of each colony are used in a final transfection to identify a cloned vector carrying the cDNA of interest, which is then isolated and characterized. This methodology has been used to isolate many genes encoding cell-surface molecules.

The full amino acid sequence of the protein can be deduced from the nucleotide sequence of its cDNA, and this often gives clues to the nature of the protein and its biological properties. The nucleotide sequence of the gene and its regulatory regions can be determined from genomic DNA clones. The gene can be manipulated and introduced into cells by transfection for larger-scale production and functional studies. This approach has been used to characterize many immunologically important proteins, such as the MHC glycoproteins.

The converse approach is taken to identify the unknown protein product of a cloned gene. The gene sequence is used to construct synthetic peptides of 10–20 amino acids that are identical to part of the deduced protein sequence,

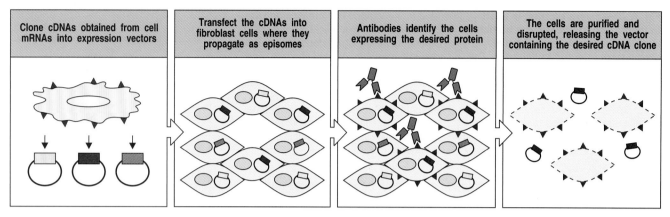

| Clone cDNAs obtained from cell mRNAs into expression vectors | Transfect the cDNAs into fibroblast cells where they propagate as episomes | Antibodies identify the cells expressing the desired protein | The cells are purified and disrupted, releasing the vector containing the desired cDNA clone |

Fig. A.21 The gene encoding a cell-surface molecule can be isolated by expressing it in fibroblasts and detecting its protein product with monoclonal antibodies. Total mRNA from a cell line or tissue expressing the protein is isolated, converted into cDNA, and cloned as cDNAs in a vector designed to direct expression of the cDNA in fibroblasts. The entire cDNA library is used to transfect cultured fibroblasts. Fibroblasts that have taken up cDNA encoding a cell-surface protein express the protein on their surface; they can be isolated by binding a monoclonal antibody against that protein. The vector containing the gene is isolated from the cells that express the antigen and used for more rounds of transfection and reisolation until uniform positive expression is obtained, ensuring that the correct gene has been isolated. The cDNA insert can then be sequenced to determine the sequence of the protein it encodes and can also be used as the source of material for large-scale expression of the protein for analysis of its structure and function. The method illustrated is limited to cloning genes for single-chain proteins (that is, those encoded by only one gene) that can be expressed in fibroblasts. It has been used to clone many genes of immunological interest such as that for CD4.

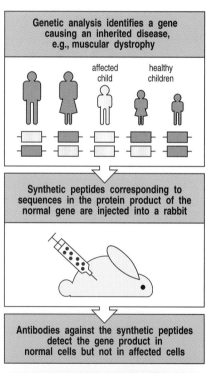

| Genetic analysis identifies a gene causing an inherited disease, e.g., muscular dystrophy |

affected child   healthy children

| Synthetic peptides corresponding to sequences in the protein product of the normal gene are injected into a rabbit |

| Antibodies against the synthetic peptides detect the gene product in normal cells but not in affected cells |

and antibodies are then raised against these peptides by coupling them to carrier proteins; the peptides behave as haptens. These anti-peptide antibodies often bind the native protein and so can be used to identify its distribution in cells and tissues and to try to ascertain its function (Fig. A.22). This approach to identifying the function of a gene is often called 'reverse genetics' as it works from gene to phenotype rather than from phenotype to gene, which is the classical genetic approach. The great advantage of reverse genetics over the classical approach is that it does not require a detectable phenotypic genetic trait in order to identify a gene.

Antibodies can also be used in the determination of the function of gene products. Some antibodies are able to act as agonists, when the binding of the antibody to the molecule mimics the binding of the natural ligand and activates the function of the gene product. For example, antibodies to the CD3 molecule have been used to stimulate T cells, replacing the interaction of the T-cell receptor with MHC:peptide antigens in cases where the specific peptide antigen is not known. Conversely, antibodies can function as antagonists, inhibiting the binding of the natural ligand and thus blocking the function.

Fig. A.22 The use of antibodies to detect the unknown protein product of a known gene is called reverse genetics. When the gene responsible for a genetic disorder such as Duchenne muscular dystrophy is isolated, the amino acid sequence of the unknown protein product of the gene can be deduced from the nucleotide sequence of the gene, and synthetic peptides representing parts of this sequence can be made. Antibodies are raised against these peptides and purified from the antiserum by affinity chromatography on a peptide column (see Fig. A.5). Labeled antibody is used to stain tissue from individuals with the disease and from unaffected individuals to determine differences in the presence, amount, and distribution of the normal gene product. The product of the dystrophin gene is present in normal mouse skeletal muscle cells, as shown in the bottom panel (red fluorescence), but is missing from the cells of mice bearing the mutation *mdx*, the mouse equivalent of Duchenne muscular dystrophy (not shown). Photograph ($\times$ 15) courtesy of H.G.W. Lidov and L. Kunkel.

Isolation of lymphocytes.

A-20   Isolation of peripheral blood lymphocytes by Ficoll-Hypaque™ gradient.

The first step in studying lymphocytes is to isolate them so that their behavior can be analyzed *in vitro*. Human lymphocytes can be isolated most readily from peripheral blood by density centrifugation over a step gradient consisting of a mixture of the carbohydrate polymer Ficoll™ and the dense iodine-containing compound metrizamide. This yields a population of mononuclear cells at the interface that has been depleted of red blood cells and most polymorphonuclear leukocytes or granulocytes (Fig. A.23). The resulting population, called **peripheral blood mononuclear cells**, consists mainly of lymphocytes and monocytes. Although this population is readily accessible, it is not necessarily representative of the lymphoid system, as only recirculating lymphocytes can be isolated from blood.

A particular cell population can be isolated from a sample or culture by binding to antibody-coated plastic surfaces, a technique known as **panning**, or by removing unwanted cells by treatment with specific antibody and complement to kill them. Cells can also be passed over columns of antibody-coated, nylon-coated steel wool and different populations differentially eluted. This technique extends affinity chromatography to cells, and is now a very popular way to separate cells. All these techniques can also be used as a pre-purification step prior to sorting out highly purified populations by FACS (see Section A-22).

A-21   Isolation of lymphocytes from tissues other than blood.

In experimental animals, and occasionally in humans, lymphocytes are isolated from lymphoid organs, such as spleen, thymus, bone marrow, lymph nodes, or mucosal-associated lymphoid tissues, most commonly the palatine tonsils in humans (see Fig. 1.7). A specialized population of lymphocytes resides in surface epithelia; these cells are isolated by fractionating the epithelial layer after its detachment from the basement membrane. Finally, in situations where local immune responses are prominent, lymphocytes can be isolated from the site of the response itself. For example, in order to study the autoimmune reaction that is thought to be responsible for rheumatoid arthritis, an inflammatory response in joints, lymphocytes are isolated from the fluid aspirated from the inflamed joint space.

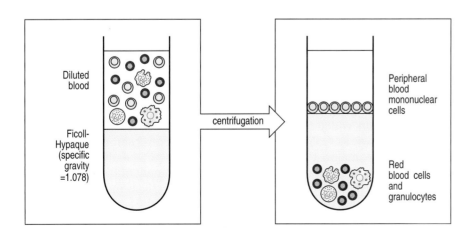

Fig. A.23 Peripheral blood mononuclear cells can be isolated from whole blood by Ficoll-Hypaque™ centrifugation. Diluted anticoagulated blood (left panel) is layered over Ficoll-Hypaque™ and centrifuged. Red blood cells and polymorphonuclear leukocytes or granulocytes are more dense and centrifuge through the Ficoll-Hypaque™, while mononuclear cells consisting of lymphocytes together with some monocytes band over it and can be recovered at the interface (right panel).

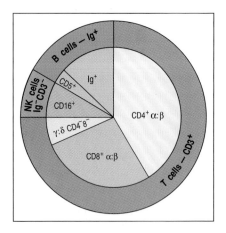

Fig. A.24 The distribution of lymphocyte subpopulations in human peripheral blood. As shown on the outside of the circle, lymphocytes can be divided into T cells bearing T-cell receptors (detected with anti-CD3 antibodies), B cells bearing immuno-globulin receptors (detected with anti-immunoglobulin antibodies), and null cells including natural killer (NK) cells, that label with neither. Further divisions of the T-cell and B-cell populations are shown inside. Using anti-CD4 and anti-CD8 antibodies, α:β T cells can be subdivided into two populations, whereas γ:δ T cells are identified with antibodies against the γ:δ T-cell receptor and mainly lack CD4 and CD8. A minority population of B cells express CD5 on their surface (see Section 7-28).

A-22 Flow cytometry and FACS analysis.

Resting lymphocytes present a deceptively uniform appearance, all being small round cells with a dense nucleus and little cytoplasm (see Fig. 1.5). However, these cells comprise many functional subpopulations, which are usually identified and distinguished from each other on the basis of their differential expression of cell-surface proteins, which can be detected using specific antibodies (Fig. A.24). B and T lymphocytes, for example, are identified unambiguously and separated from each other by antibodies to the constant regions of the B- and T-cell antigen receptors. T cells are further subdivided on the basis of expression of the co-receptor proteins CD4 and CD8.

An immensely powerful tool for defining and enumerating lymphocytes is the flow cytometer, which detects and counts individual cells passing in a stream through a laser beam. A flow cytometer equipped to separate the identified cells is called a **fluorescence-activated cell sorter** (**FACS**). These instruments are used to study the properties of cell subsets identified using monoclonal antibodies to cell-surface proteins. Individual cells within a mixed population are first tagged by treatment with specific monoclonal antibodies labeled with fluorescent dyes, or by specific antibodies followed by labeled anti-immunoglobulin antibodies. The mixture of labeled cells is then forced with a much larger volume of saline through a nozzle, creating a fine stream of liquid containing cells spaced singly at intervals. As each cell passes through a laser beam it scatters the laser light, and any dye molecules bound to the cell will be excited and will fluoresce. Sensitive photomultiplier tubes detect both the scattered light, which gives information on the size and granularity of the cell, and the fluorescence emissions, which give information on the binding of the labeled monoclonal antibodies and hence on the expression of cell-surface proteins by each cell (Fig. A.25).

In the cell sorter, the signals passed back to the computer are used to generate an electric charge, which is passed from the nozzle through the liquid stream at the precise time that the stream breaks up into droplets, each containing no more than a single cell; droplets containing a charge can then be deflected from the main stream of droplets as they pass between plates of opposite charge, so that positively charged droplets are attracted to a negatively charged plate, and vice versa. In this way, specific subpopulations of cells, distinguished by the binding of the labeled antibody, can be purified from a mixed population of cells. Alternatively, to deplete a population of cells, the same fluorochrome can be used to label different antibodies directed at marker proteins expressed by the various undesired cell types. The cell sorter can be used to direct labeled cells to a waste channel, retaining only the unlabeled cells.

When cells are labeled with a single fluorescent antibody, the data from a flow cytometer are usually displayed in the form of a histogram of fluorescence intensity versus cell numbers. If two or more antibodies are used, each coupled to different fluorescent dyes, then the data are more usually displayed in the form of a two-dimensional scatter diagram or as a contour diagram, where the fluorescence of one dye-labeled antibody is plotted against that of a second, with the result that a population of cells labeling with one antibody can be further subdivided by its labeling with the second antibody (see Fig. A.25). By examining large numbers of cells, flow cytometry can give quantitative data on the percentage of cells bearing different molecules, such as surface immunoglobulin, which characterizes B cells, the T-cell receptor-associated molecules known as CD3, and the CD4 and CD8 co-receptor proteins that distinguish the major T-cell subsets. Likewise, FACS analysis has been instrumental in defining stages in the early development of B and T cells. As the power of the FACS technology has grown, progressively more anti-bodies labeled with distinct fluorescent dyes can be used at the same time. Three-, four-, and even five-color analyses can now be handled by very powerful machines. FACS analysis has been applied to a broad range of problems in immunology; indeed, it played a vital role in the early identification of AIDS as a disease in which T cells bearing CD4 are depleted selectively (see Chapter 11).

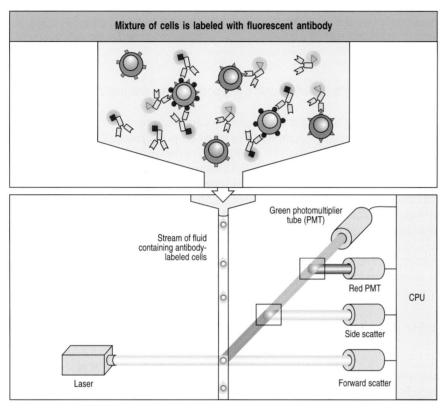

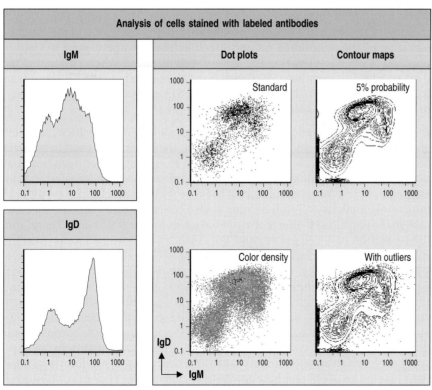

Fig. A.25 The FACS™ allows individual cells to be identified by their cell-surface antigens and to be sorted. Cells to be analyzed by flow cytometry are first labeled with fluorescent dyes (top panel). Direct labeling uses dye-coupled antibodies specific for cell-surface antigens (as shown here), while indirect labeling uses a dye-coupled immunoglobulin to detect unlabeled cell-bound antibody. The cells are forced through a nozzle in a single-cell stream that passes through a laser beam (second panel). Photomultiplier tubes (PMTs) detect the scattering of light, which is a sign of cell size and granularity, and emissions from the different fluorescent dyes. This information is analyzed by computer (CPU). By examining many cells in this way, the number of cells with a specific set of characteristics can be counted and levels of expression of various molecules on these cells can be measured. The lower part of the figure shows how this data can be represented, using the expression of two surface immunoglobulins, IgM and IgD, on a sample of B cells from a mouse spleen. The two immunoglobulins have been labeled with different colored dyes. When the expression of just one type of molecule is to be analyzed (IgM or IgD), the data is usually displayed as a histogram, as in the left-hand panels. Histograms display the distribution of cells expressing a single measured parameter (for example, size, granularity, fluorescence intensity). When two or more parameters are measured for each cell (IgM and IgD), various types of two-color plots can be used to display the data, as shown in the right-hand panel. All four plots represent the same data. The horizontal axis represents intensity of IgM fluorescence and the vertical axis the intensity of IgD fluorescence. Two-color plots provide more information than histograms; they allow recognition, for example, of cells that are 'bright' for both colors, 'dull' for one and bright for the other, dull for both, negative for both, and so on. For example, the cluster of dots in the extreme lower left portions of the plots represents cells that do not express either immunoglobulin, and are mostly T cells. The standard dot plot (upper left) places a single dot for each cell whose fluorescence is measured. It is good for picking up cells that lie outside the main groups but tends to saturate in areas containing a large number of cells of the same type. A second means of presenting these data is the color dot plot (lower left), which uses color density to indicate high-density areas. A contour plot (upper right) draws 5% 'probability' contours, with 5% of the cells lying between each contour providing the best monochrome visualization of regions of high and low density. The lower right plot is a 5% probability contour map which also shows outlying cells as dots.

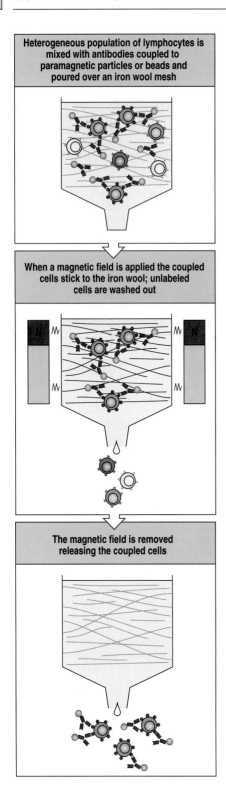

Heterogeneous population of lymphocytes is mixed with antibodies coupled to paramagnetic particles or beads and poured over an iron wool mesh

When a magnetic field is applied the coupled cells stick to the iron wool; unlabeled cells are washed out

The magnetic field is removed releasing the coupled cells

Fig. A.26 Lymphocyte subpopulations can be separated physically by using antibodies coupled to paramagnetic particles or beads. A mouse monoclonal antibody specific for a particular cell-surface molecule is coupled to paramagnetic particles or beads. It is mixed with a heterogeneous population of lymphocytes and poured over an iron wool mesh in a column. A magnetic field is applied so that the antibody-bound cells stick to the iron wool while cells which have not bound antibody are washed out; these cells are said to be negatively selected for lack of the molecule in question. The bound cells are released by removing the magnetic field; they are said to be positively selected for presence of the antigen recognized by the antibody.

A-23   Lymphocyte isolation using antibody-coated magnetic beads.

Although the FACS is superb for isolating small numbers of cells in pure form, when large numbers of lymphocytes must be prepared quickly, mechanical means of separating cells are preferable. A powerful and efficient way of isolating lymphocyte populations is to couple paramagnetic beads to monoclonal antibodies that recognize distinguishing cell-surface molecules. These antibody-coated beads are mixed with the cells to be separated, and run through a column containing material that attracts the paramagnetic beads when the column is placed in a strong magnetic field. Cells binding the magnetically labeled antibodies are retained; cells lacking the appropriate surface molecule can be washed away (Fig. A.26). The bound cells are positively selected for expression of the particular cell-surface molecule, while the unbound cells are negatively selected for its absence.

A-24   Isolation of homogeneous T-cell lines.

The analysis of specificity and effector function in T cells depends heavily on the study of monoclonal populations of T lymphocytes. These can be obtained in four distinct ways. First, as for B-cell hybridomas (see Section A-12), normal T cells proliferating in response to specific antigen can be fused to malignant T-cell lymphoma lines to generate **T-cell hybrids**. The hybrids express the receptor of the normal T cell, but proliferate indefinitely owing to the cancerous state of the lymphoma parent. T-cell hybrids can be cloned to yield a population of cells all having the same T-cell receptor. When stimulated by their specific antigen these cells release cytokines such as the T-cell growth factor interleukin-2 (IL-2), and the production of cytokines is used as an assay to assess the antigen specificity of the T-cell hybrid.

T-cell hybrids are excellent tools for the analysis of T-cell specificity, as they grow readily in suspension culture. However, they cannot be used to analyze the regulation of specific T-cell proliferation in response to antigen because they are continually dividing. T-cell hybrids also cannot be transferred into an animal to test for function *in vivo* because they would give rise to tumors. Functional analysis of T-cell hybrids is also confounded by the fact that the malignant partner cell affects their behavior in functional assays. Therefore, the regulation of T-cell growth and the effector functions of T cells must be studied using **T-cell clones**. These are clonal cell lines of a single T-cell type and antigen specificity, which are derived from cultures of heterogeneous T cells, called **T-cell lines**, whose growth is dependent on periodic restimulation with specific antigen and, frequently, on the addition of T-cell growth factors (Fig. A.27). T-cell clones also require periodic restimulation with antigen and are more tedious to grow than T-cell hybrids but, because their growth depends on specific antigen recognition, they maintain antigen specificity, which is often lost in T-cell hybrids. Cloned T-cell lines can be used for studies of effector function both *in vitro* and *in vivo*. In addition, the proliferation of T cells, a critical aspect of clonal selection, can be characterized only in cloned T-cell lines, where such growth is dependent on antigen recognition. Thus, both types of monoclonal T-cell line have valuable applications in experimental studies.

Studies of human T cells have relied largely on T-cell clones because a suitable fusion partner for making T-cell hybrids has not been identified. However, a human T-cell lymphoma line, called Jurkat, has been characterized extensively because it secretes IL-2 when its antigen receptor is cross-linked with anti-receptor monoclonal antibodies. This simple assay system has yielded much information about signal transduction in T cells. One of the Jurkat cell line's most interesting features, shared with T-cell hybrids, is that it stops growing when its antigen receptor is cross-linked. This has allowed mutants lacking the receptor or having defects in signal transduction pathways to be selected simply by culturing the cells with anti-receptor antibody and selecting those that continue to grow. Thus, T-cell tumors, T-cell hybrids, and cloned T-cell lines all have valuable applications in experimental immunology.

Finally, primary T cells from any source can be isolated as single, antigen-specific cells by limiting dilution rather than by first establishing a mixed population of T cells in culture as a T-cell line and then deriving clonal subpopulations. During the growth of T-cell lines, particular T-cell clones can come to dominate the cultures and give a false picture of the number and specificities in the original sample. Direct cloning of primary T cells avoids this artifact.

Characterization of lymphocyte specificity, frequency, and function.

B cells are relatively easy to characterize as they have only one function—antibody production. T cells are more difficult to characterize as there are several different classes with different functions. It is also technically more difficult to study the membrane-bound T-cell receptors than the antibodies secreted in large amounts by B cells. All the methods in this part of the appendix can be used for T cells. Some are also used to detect and count B cells.

On many occasions it is important to know the frequency of antigen-specific lymphocytes, especially T cells, in order to measure the efficiency with which an individual responds to a particular antigen, for example, or the degree to which specific immunological memory has been established. There are a number of methods for doing this, either by detecting the cells directly by the specificity of their receptor, or by detecting activation of the cells to provide some particular function, such as cytokine secretion or cytotoxicity.

The first technique of this type to be established was the limiting-dilution culture (see Section A-25), in which the frequency of specific T or B cells responding to a particular antigen could be estimated by plating the cells into 96-well plates at increasing dilutions and measuring the number of wells in which there was no response. However, in this type of assay it became laborious to ask detailed questions about the phenotype of the responding cells, and to compare responses from different cell subpopulations.

A simpler assay for measuring the responses of T-cell populations has been developed from a variant of the antigen-capture ELISA method (see Section A-6), called the ELISPOT assay (see Section A-26). It assays T cells on the basis of cytokine production. In the ELISPOT assay, cytokine secreted by individual

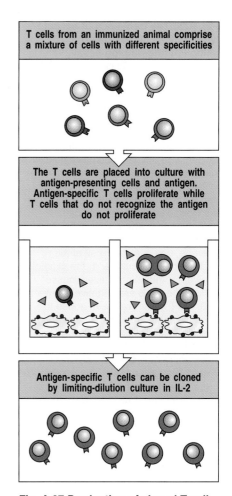

Fig. A.27 Production of cloned T-cell lines. T cells from an immune donor, comprising a mixture of cells with different specificities, are activated with antigen and antigen-presenting cells. Single responding cells are cultured by limiting dilution in the T-cell growth factor IL-2, which selectively stimulates the responding cells to proliferate. From these single cells, cloned lines specific for antigen are identified and can be propagated by culture with antigen, antigen-presenting cells, and IL-2.

Fig. A.28 The frequency of specific lymphocytes can be determined using limiting-dilution assay. Varying numbers of lymphoid cells from normal or immunized mice are added to individual culture wells and stimulated with antigen and antigen-presenting cells (APCs) or polyclonal mitogen and added growth factors. After several days, the wells are tested for a specific response to antigen, such as cytotoxic killing of target cells. Each well that initially contained a specific T cell will make a response to its target, and from the Poisson distribution one can determine that when 37% of the wells are negative, each well contained, on average, one specific T cell at the beginning of the culture. In the example shown, for the unimmunized mouse 37% of the wells are negative when 160,000 T cells have been added to each well; thus the frequency of antigen-specific T cells is 1 in 160,000. When the mouse is immunized, 37% of the wells are negative when only 1100 T cells have been added; hence the frequency of specific T cells after immunization is 1 in 1100, an increase in responsive cells of 150-fold.

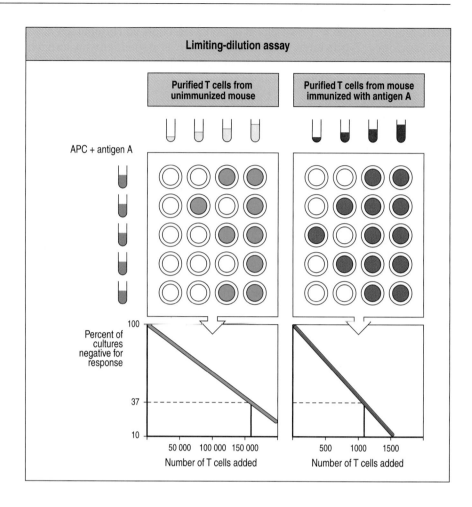

activated T cells is immobilized as discrete spots on a plastic plate, which are counted to give the number of activated T cells. The ELISPOT assay suffers from many of the same problems as the limiting-dilution assay in giving information about the nature of the activated cells, and it can be difficult to determine whether individual cells are capable of secreting mixtures of cytokines. It was therefore important to develop assays that could make these measurements on single cells. Measurements based on flow cytometry (see Section A-22) proved the answer, with the development of methods for detecting fluorescently labeled cytokines within activated T cells. The drawback of intracellular cytokine staining (see Section A-27) was that the T cells have to be killed and permeabilized by detergents to enable the cytokines to be detected. This led to the more sophisticated technique of capturing secreted labeled cytokines on the surfaces of the living T cells (see Section A-27).

Finally, methods for directly detecting T cells on the basis of the specificity of their receptor, using fluorochrome-tagged tetramers of specific MHC:peptide complexes (see Section A-28), have revolutionized the study of T-cell responses in a similar way to the use of monoclonal antibodies.

A-25 Limiting-dilution culture.

The response of a lymphocyte population is a measure of the overall response, but the frequency of lymphocytes able to respond to a given antigen can be determined only by limiting-dilution culture. This assay makes use of the Poisson distribution, a statistical function that describes how objects are distributed at random. For instance, when a sample of heterogeneous T cells is distributed equally into a series of culture wells, some wells will receive no T cells specific for a given antigen, some will receive one specific T cell, some two, and so on. The T cells in the wells are activated with specific antigen, antigen-presenting cells, and growth factors. After allowing several days for their growth and differentiation, the cells in each well are tested for a response to antigen, such as cytokine release or the ability to kill specific target cells (Fig. A.28). The assay is replicated with different numbers of T cells in the samples. The logarithm of the proportion of wells in which there is no response is plotted against the number of cells initially added to each well. If cells of one type, typically antigen-specific T cells because of their rarity, are the only limiting factor for obtaining a response, then a straight line is obtained. From the Poisson distribution, it is known that there is, on average, one antigen-specific cell per well when the proportion of negative wells is 37%. Thus, the frequency of antigen-specific cells in the population equals the reciprocal of the number of cells added to each well when 37% of the wells are negative. After priming, the frequency of specific cells goes up substantially, reflecting the antigen-driven proliferation of antigen-specific cells. The limiting-dilution assay can also be used to measure the frequency of B cells that can make antibody to a given antigen.

A-26   ELISPOT assays.

A modification of the ELISA antigen-capture assay (see Section A-6), called the **ELISPOT assay**, has provided a powerful tool for measuring the frequency of T-cell responses. Populations of T cells are stimulated with the antigen of interest, and are then allowed to settle onto a plastic plate coated with antibodies to the cytokine that is to be assayed (Fig. A.29). If an activated T cell is secreting that cytokine, it is captured by the antibody on the plastic plate. After a period the cells are removed, and a second antibody to the cytokine is added to the plate to reveal a circle of bound cytokine surrounding the position of each activated T cell; counting each spot, and knowing the number of T cells originally added to the plate allows a simple calculation of the frequency of T cells secreting that particular cytokine, giving the ELISPOT assay its name. ELISPOT can also be used to detect specific antibody secretion by B cells, in this case by using antigen-coated surfaces to trap specific antibody and labeled anti-immunoglobulin to detect the bound antibody.

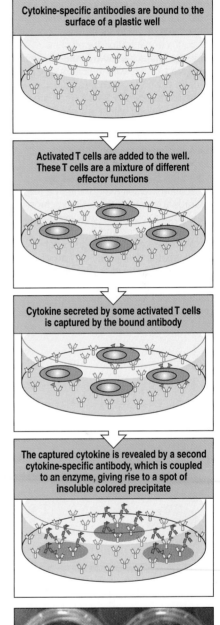

Fig. A.29 The frequency of cytokine-secreting T cells can be determined by the ELISPOT assay. The ELISPOT assay is a variant of the ELISA assay in which antibodies bound to a plastic surface are used to capture cytokines secreted by individual T cells. Usually, cytokine specific antibodies are bound to the surface of a plastic tissue-culture well and the unbound antibodies are removed (top panel). Activated T cells are then added to the well and settle onto the antibody-coated surface (second panel). If a T cell is secreting the appropriate cytokine, this will then be captured by the antibody molecules on the plate surrounding the T cell (third panel). After a period of time the T cells are removed, and the presence of the specific cytokine is detected using an enzyme-labeled second antibody specific for the same cytokine. Where this binds, a colored reaction product can be formed (fourth panel). Each T cell that originally secreted cytokine gives rise to a single spot of color, hence the name of the assay. The results of such an ELISPOT assay for T cells secreting IFN-γ in response to different stimuli are shown in the last panel. In this example, T cells from a patient with melanoma were stimulated with the mitogen PHA (top left), a peptide from cytomegalovirus (top right), a peptide from a melanoma tumor associated antigen (lower left), and with no peptide (lower right). You can see the greater response to the cytomegalovirus peptide compared to the melanA peptide by the greater number of spots. Photograph courtesy of C. Smith and R. Dunbar.

Fig. A.30 Cytokine-secreting cells can be identified by intracellular cytokine staining. The cytokines secreted by activated T cells can be determined by using fluorochrome-labeled antibodies to detect cytokine molecules that have been allowed to accumulate inside the cell. The accumulation of cytokine molecules, to allow them to reach a high enough concentration for efficient detection, is achieved by treating the activated T cells with inhibitors of protein export. In such treated cells, proteins destined to be secreted are instead retained within the endoplasmic reticulum (first panel). These treated cells are then fixed, to cross-link the proteins inside the cell and in the cell membranes, so that they are not lost when the cell is permeabilized by dissolving the cell membrane in a mild detergent (center panel). Fluorochrome-labeled antibodies can now enter the permeabilized cell and bind to the cytokines inside the cell (last panel). Cells labeled in this way can also be labeled with antibodies that bind to cell-surface proteins to determine which subsets of T cells are secreting particular cytokines.

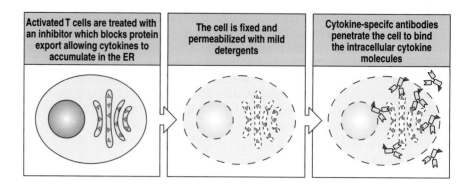

A-27   Identification of functional subsets of T cells by staining for cytokines.

One problem with the detection of cytokine production on a single-cell level is that the cytokines are secreted by the T cells into the surrounding medium, and any association with the originating cell is lost. Two methods have been devised that allow the cytokine profile produced by individual cells to be determined. The first, that of **intracellular cytokine staining** (Fig. A.30), relies on the use of metabolic poisons that inhibit protein export from the cell. The cytokine thus accumulates within the endoplasmic reticulum and vesicular network of the cell. If the cells are subsequently fixed and rendered permeable by the use of mild detergents, antibodies can gain access to these intracellular compartments and detect the cytokine. The T cells can be stained for other markers simultaneously, and thus the frequency, for example, of IL-10-producing CD25$^+$ CD4 T cells can be easily obtained.

A second method, which has the advantage that the cells being analyzed are not killed in the process, is called **cytokine capture**. This technique uses hybrid antibodies, in which the two separate heavy- and light-chain pairs from different antibodies are combined to give a mixed antibody molecule in which the two antigen-binding sites recognize different ligands (Fig. A.31).

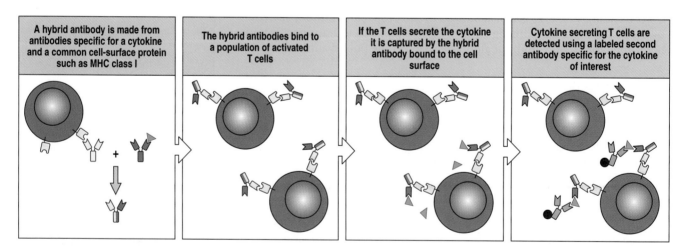

Fig. A.31 Hybrid antibodies containing cell-specific and cytokine-specific binding sites can be used to assay cytokine secretion by living cells and to purify cells secreting particular cytokines. Hybrid antibodies can be made by mixing together heavy- and light-chain pairs from antibodies of different specificities, for example, an antibody to an MHC class I molecule and an antibody specific for a cytokine such as IL-4 (first panel). The hybrid antibodies are then added to a population of activated T cells, and bind to each cell via the MHC class I binding arm (second panel). If some of the cells in the population are secreting the appropriate cytokine, IL-4, this is captured by the cytokine specific arm of the hybrid antibody (third panel). The presence of the cytokine can then be revealed, for example using a fluorochrome-labeled second antibody specific for the same cytokine, but binding to a different site to the one used for the hybrid antibody (last panel). Such labeled cells may be analyzed by flow cytometry, or can be isolated using a fluorescence activated cell sorter. Alternatively, the second cytokine specific antibody may be coupled to magnetic beads, and the cytokine producing cells isolated magnetically.

In the bispecific antibodies used to detect cytokine production, one of the antigen-binding sites is specific for a T-cell surface marker, while the other is specific for the cytokine in question. The bispecific antibody binds to the T cells through the binding site for the cell-surface marker, leaving the cytokine-binding site free. If that T cell is secreting the particular cytokine, it is captured by the bound antibody before it diffuses away from the surface of the cell. It can then be detected by adding a fluorochrome-labeled second antibody specific for the cytokine to the cells.

A-28 Identification of T-cell receptor specificity using MHC:peptide tetramers.

For many years, the ability to identify antigen-specific T cells directly through their receptor specificity eluded immunologists. Foreign antigen could not be used directly to identify T cells, since, unlike B cells, they do not recognize antigen alone but rather the complexes of peptide fragments of antigen bound to self MHC molecules. Moreover, the affinity of interaction between the T-cell receptor and the MHC:peptide complex was in practice so low that attempts to label T cells with their specific MHC:peptide complexes routinely failed. The breakthrough in labeling antigen-specific T cells came with the idea of making multimers of the MHC:peptide complex, so as to increase the avidity of the interaction.

Peptides can be biotinylated using the bacterial enzyme BirA, which recognizes a specific amino acid sequence. Recombinant MHC molecules containing this target sequence are used to make MHC:peptide complexes which are then biotinylated. Avidin, or the bacterial counterpart streptavidin, contains four sites that bind biotin with extremely high affinity. Mixing the biotinylated MHC:peptide complex with avidin or streptavidin results in the formation of an **MHC:peptide tetramer**—four specific MHC:peptide complexes bound to a single molecule of streptavidin (Fig. A.32). Routinely, the streptavidin moiety is labeled with a fluorochrome to allow detection of those T cells capable of binding the MHC:peptide tetramer.

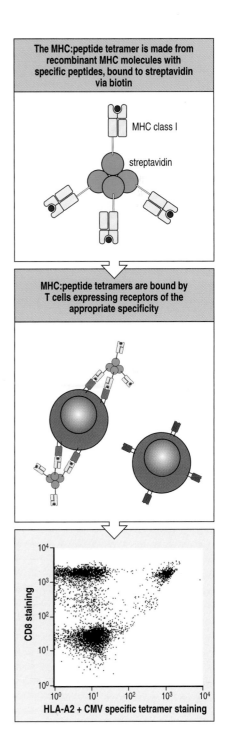

The MHC:peptide tetramer is made from recombinant MHC molecules with specific peptides, bound to streptavidin via biotin

MHC class I

streptavidin

MHC:peptide tetramers are bound by T cells expressing receptors of the appropriate specificity

CD8 staining

HLA-A2 + CMV specific tetramer staining

Fig. A.32 MHC:peptide complexes coupled to streptavidin to form tetramers are able to stain antigen-specific T cells. MHC:peptide tetramers are formed from recombinant refolded MHC:peptide complexes containing a single defined peptide epitope. The MHC molecules can be chemically derivatized to contain biotin, but more usually, the recombinant MHC heavy chain is linked to a bacterial biotinylation sequence, a target for the *E. coli* enzyme BirA, which is used to add a single biotin group to the MHC molecule. Streptavidin is a tetramer, each subunit having a single binding site for biotin, hence the streptavidin/MHC:peptide complex creates a tetramer of MHC:peptide complexes (top panel). While the affinity between the T-cell receptor and its MHC:peptide ligand is too low for a single complex to bind stably to a T cell, the tetramer, by being able to make a more avid interaction with multiple MHC:peptide complexes binding simultaneously, is able to bind to T cells whose receptors are specific for the particular MHC:peptide complex (center panel). Routinely, the streptavidin molecules are coupled to a fluorochrome, so that the binding to T cells can be monitored by flow cytometry. In the example shown in the bottom panel, T cells have been stained simultaneously with antibodies specific for CD3 and CD8, and with a tetramer of HLA-A2 molecules containing a cytomegalovirus peptide. Only the CD3+ cells are shown, with the staining of CD8 displayed on the vertical axis and the tetramer staining displayed along the horizontal axis. The CD8− cells (mostly CD4+) on the bottom left of the figure show no specific tetramer staining, while the bulk of the CD8+ cells, on the top left, likewise show no tetramer staining. However, a discrete population of tetramer positive CD8+ cells, at the top right of the panel, comprising some 5% of the total CD8+ cells, can clearly be demonstrated. Data courtesy of Ms G. Aubert

MHC:peptide tetramers have been used to identify populations of antigen-specific T cells in, for example, patients with acute Epstein–Barr virus infections (infectious mononucleosis), showing that up to 80% of the peripheral T cells in infected individuals can be specific for a single MHC:peptide complex. They have also been used to follow responses over longer timescales in individuals with HIV or, in the example we show, cytomegalovirus infections. These reagents have also been important in identifying the cells responding, for example, to nonclassical class I molecules such as HLA-E or HLA-G, in both cases showing that these nonclassical molecules are recognized by subsets of NK receptors.

A-29 Assessing the diversity of the T-cell repertoire by 'spectratyping.'

The extent of the diversity of the T-cell repertoire, either generally or during specific immune responses, is often of interest. In particular, as T cells do not undergo somatic hypermutation and affinity maturation in the same way that B cells do, the relationship between the repertoire of T cells making a primary response to antigen and the repertoire of T cells involved in secondary and subsequent responses to antigen has been difficult to determine. This information has usually been obtained through the laborious process of cloning the T cells involved in specific responses (see Section A-24), and the cloning and sequencing of their T-cell receptors.

It is possible, however, to estimate the diversity of T-cell responses by making use of the junctional diversity generated when T-cell receptors are created by somatic recombination, a technique known as **spectratyping**. Variability in the length of the CDR3 segments is created during the recombination process, both by variation in the exact positions at which the junctions between gene segments occur, and by variation in the number of N-nucleotides added. Both these processes result in the length of the V_β CDR3 varying by up to nine amino acids. The problem in detecting this variability is that there are 24 families of V_β gene segments in humans and it is not possible to design a single oligonucleotide primer that will anneal to all of these families. Specific oligonucleotide primers can, however, be designed for each family, and these can be used in the polymerase chain reaction (PCR), together with a primer specific for the C_β region, to amplify, for each individual family, a segment of the mRNA for the T-cell receptor β chain that spans the CDR3 region. A population of TCR V_β genes will therefore show a distribution, or 'spectrum,' of CD3 lengths, and will give rise to PCR products of different lengths that can be resolved, usually by polyacrylamide gel electrophoresis (Fig. A.33). The deletion and addition of nucleotides during the generation of T-cell receptors by rearrangement is random, and so in a normal individual the CDR3 lengths follow a Gaussian distribution. Deviations from this Gaussian distribution, such as an excess of one particular CDR3 length, indicate the presence of clonal expansions of T cells, such as occurs during a T-cell response.

A-30 Biosensor assays for measuring the rates of association and disassociation of antigen receptors for their ligands.

Two of the important questions that are always asked of any receptor–ligand interaction is: what is the strength of binding, or affinity, of the interaction, and what are the rates of association and disassociation? Traditionally, measurements of affinity have been made by equilibrium binding measurements (see Section A-9), and measurements of rates of binding have been difficult to obtain. Equilibrium binding assays also cannot be performed on T-cell receptors, which have large macromolecular ligands and which cannot be isolated and purified in large quantity.

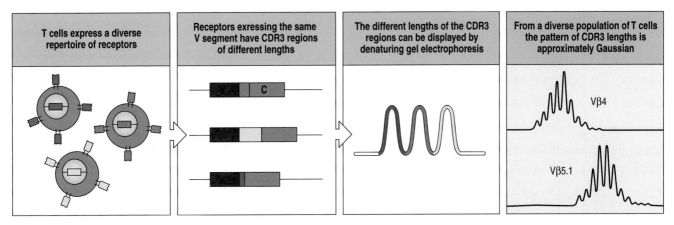

| T cells express a diverse repertoire of receptors | Receptors exressing the same V segment have CDR3 regions of different lengths | The different lengths of the CDR3 regions can be displayed by denaturing gel electrophoresis | From a diverse population of T cells the pattern of CDR3 lengths is approximately Gaussian |

Fig. A.33 The diversity of the T-cell receptor repertoire can be displayed by spectratyping, a PCR-based technique that separates different receptors on the basis of their CDR3 length. The process of generation of T-cell receptors is stochastic, giving rise to a population of mature T cells whose receptors are clonally distributed (first panel). In each of the cells expressing a particular V_β gene segment, all of the differences between the unique receptors are restricted to the CDR3 region, where there will be differences in length as well as sequence as a consequence of the imprecision of the rearrangement process (second panel). Using sets of primers for the PCR reaction that are specific for individual V_β gene segments at one end and for a conserved part of the C region at the other, it is possible to generate a set of DNA fragments that span the CDR3 region. If these are separated by denaturing acrylamide gel electrophoresis then a series of bands are formed or, since these fragments can be labeled with fluorochromes and analyzed by automated gel readers, a series of peaks corresponding to the different length fragments (third panel). The pattern of peaks obtained in this way is known as a spectratype. From a diverse population of cells, the distribution of fragment lengths is Gaussian, as shown in the last panel, where the spectratypes of two different V_β regions from the same individual are shown. In this case, both of the patterns are approximately Gaussian; deviations from a Gaussian distribution may indicate expansion of particular clones of T cells, perhaps in response to antigenic challenge. Data courtesy of Dr. L. McGreavey.

It is now possible to measure binding rates directly, by following the binding of ligands to receptors immobilized on gold-plated glass slides, using a phenomenon known as surface plasmon resonance to detect the binding (Fig. A.34). A full explanation of surface plasmon resonance is beyond the scope of this textbook, as it is based on advanced physical and quantum mechanical principles. Briefly, it relies on the total internal reflection of a beam of light from the surface of a gold-coated glass slide. As the light is reflected, some of its energy excites electrons in the gold coating and these excited electrons are in turn affected by the electric field of any molecules binding to the surface of the glass coating. The more molecules that bind to the surface, the greater the effect on the excited electrons, and this in turn affects the reflected light beam. The reflected light thus becomes a sensitive measure of the number of atoms bound to the gold surface of the slide.

If a purified receptor is immobilized on the surface of the gold-coated glass slide, to make a biosensor 'chip,' and a solution containing the ligand is flowed over that surface, the binding of ligand to the receptor can be followed until it reaches equilibrium (see Fig. A.34). If the ligand is then washed out, dissociation of ligand from the receptor can easily be followed and the dissociation rate calculated. A new solution of the ligand at a different concentration can then be flowed over the chip and the binding once again measured. The affinity of binding can be calculated in a number of ways in this type of assay. Most simply, the ratio of the rates of association and dissociation will give an estimate of the affinity, but more accurate estimates can be obtained from the measurements of the binding at different concentrations of ligand. From measurements of binding at equilibrium, a Scatchard plot (see Fig. A.11) will give a measurement of the affinity of the receptor–ligand interaction.

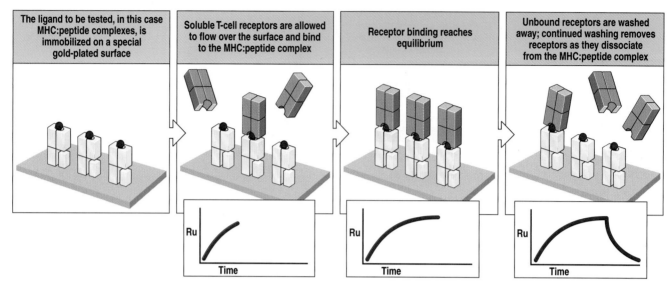

| The ligand to be tested, in this case MHC:peptide complexes, is immobilized on a special gold-plated surface | Soluble T-cell receptors are allowed to flow over the surface and bind to the MHC:peptide complex | Receptor binding reaches equilibrium | Unbound receptors are washed away; continued washing removes receptors as they dissociate from the MHC:peptide complex |

Fig. A.34 Measurement of receptor–ligand interactions can be made in real time using a biosensor. Biosensors are able to measure the binding of molecules on the surface of gold-plated glass chips through the indirect effects of the binding on the total internal reflection of a beam of polarized light at the surface of the chip. Changes in the angle and intensity of the reflected beam are measured in 'resonance units' (Ru) and plotted against time in what is termed a 'sensorgram.' Depending on the exact nature of the receptor–ligand pair to be analyzed, either the receptor or the ligand can be immobilized on the surface of the chip. In the example shown, MHC:peptide complexes are immobilized on such a surface (first panel).

T-cell receptors in solution are now allowed to flow over the surface, and to bind to the immobilized MHC:peptide complexes (second panel). As the T-cell receptors bind, the sensorgram (inset panel below the main panel) reflects the increasing amount of protein bound. As the binding reaches either saturation or equilibrium (third panel), the sensorgram shows a plateau, as no more protein binds. At this point, unbound receptors can be washed away. With continued washing, bound receptors now start to dissociate and are removed in the flow of the washing solution (last panel). The sensorgram now shows a declining curve, reflecting the rate at which the receptor and ligand dissociation occurs.

A-31 Stimulation of lymphocyte proliferation by treatment with polyclonal mitogens or specific antigen.

To function in adaptive immunity, rare antigen-specific lymphocytes must proliferate extensively before they differentiate into functional effector cells in order to generate sufficient numbers of effector cells of a particular specificity. Thus, the analysis of induced lymphocyte proliferation is a central issue in their study. It is, however, difficult to detect the proliferation of normal lymphocytes in response to specific antigen because only a minute proportion of cells will be stimulated to divide. Great impetus was given to the field of lymphocyte culture by the finding that certain substances induce many or all lymphocytes of a given type to proliferate. These substances are referred to collectively as **polyclonal mitogens** because they induce mitosis in lymphocytes of many different specificities or clonal origins. T and B lymphocytes are stimulated by different polyclonal mitogens (Fig. A.35). Polyclonal mitogens seem to trigger essentially the same growth response mechanisms as antigen. Lymphocytes normally exist as resting cells in the G_0 phase of the cell cycle. When stimulated with polyclonal mitogens, they rapidly enter the G_1 phase and progress through the cell cycle. In most studies, lymphocyte proliferation is most simply measured by the incorporation of ^{3}H-thymidine into DNA. This assay is used clinically for assessing the ability of lymphocytes from patients with suspected immunodeficiencies to proliferate in response to a nonspecific stimulus.

Once lymphocyte culture had been optimized using the proliferative response to polyclonal mitogens as an assay, it became possible to detect antigen-specific T-cell proliferation in culture by measuring ^{3}H-thymidine uptake in response to an antigen to which the T-cell donor had been previously immunized (Fig. A.36). This is the assay most commonly used for

Mitogen	Responding cells
Phytohemagglutinin (PHA) (red kidney bean)	T cells
Concanavalin (ConA) (Jack bean)	T cells
Pokeweed mitogen (PWM) (Pokeweed)	T and B cells
Lipopolysaccharide (LPS) (*Escherichia coli*)	B cells (mouse)

Fig. A.35 Polyclonal mitogens, many of plant origin, stimulate lymphocyte proliferation in tissue culture. Many of these mitogens are used to test the ability of lymphocytes in human peripheral blood to proliferate.

assessing T-cell responses after immunization, but it reveals little about the functional capabilities of the responding T cells. These must be ascertained by functional assays, as outlined in Sections A-33 and A-34.

A-32 Measurements of apoptosis by the TUNEL assay.

Apoptotic cells can be detected by a procedure known as TUNEL staining. In this technique, the 3′ ends of the DNA fragments generated in apoptotic cells are labeled with biotin-coupled uridine by using the enzyme terminal deoxynucleotidyl transferase (TdT). The biotin label is then detected with enzyme tagged streptavidin, which binds to biotin. When the colorless substrate of the enzyme is added to a tissue section or cell culture, it is reacted upon to produce a colored precipitate only in cells that have undergone apoptosis (Fig. A.37). This technique has revolutionized the detection of apoptotic cells.

A-33 Assays for cytotoxic T cells.

Activated CD8 T cells generally kill any cells that display the specific peptide:MHC class I complex they recognize. Thus CD8 T-cell function can be determined using the simplest and most rapid T-cell bioassay—the killing of a target cell by a cytotoxic T cell. This is usually detected in a ^{51}Cr-release assay. Live cells will take up, but do not spontaneously release, radioactively labeled sodium chromate, Na$_2$^{51}CrO$_4$. When these labeled cells are killed, the radioactive chromate is released and its presence in the supernatant of mixtures of target cells and cytotoxic T cells can be measured (Fig. A.38). In a similar assay, proliferating target cells such as tumor cells can be labeled with ^{3}H-thymidine, which is incorporated into the replicating DNA. On attack by a cytotoxic T cell, the DNA of the target cells is rapidly fragmented and retained in the filtrate, while large, unfragmented DNA is collected on a filter, and one can measure either the release of these fragments or the retention of ^{3}H-thymidine in chromosomal DNA. These assays provide a rapid, sensitive, and specific measure of the activity of cytotoxic T cells.

A-34 Assays for CD4 T cells.

CD4 T-cell functions usually involve the activation rather than the killing of cells bearing specific antigen. The activating effects of CD4 T cells on B cells or macrophages are mediated in large part by nonspecific mediator proteins called cytokines, which are released by the T cell when it recognizes antigen. Thus, CD4 T-cell function is usually studied by measuring the type and amount of these released proteins. As different effector T cells release different amounts and types of cytokines, one can learn about the effector potential of that T cell by measuring the proteins it produces.

Cytokines can be detected by their activity in biological assays of cell growth, where they serve either as growth factors or growth inhibitors. A more specific assay is a modification of ELISA known as a capture or sandwich ELISA (see Section A-6). In this assay, the cytokine is characterized by its ability to bridge between two monoclonal antibodies reacting with different epitopes on the cytokine molecule. Cytokine-secreting cells can also be detected by ELISPOT (see Section A-26).

Sandwich ELISA and ELISPOT avoid a major problem of cytokine bioassays, the ability of different cytokines to stimulate the same response in a bioassay. Bioassays must always be confirmed by inhibition of the response with neutralizing monoclonal antibodies specific for the cytokine. Another way of

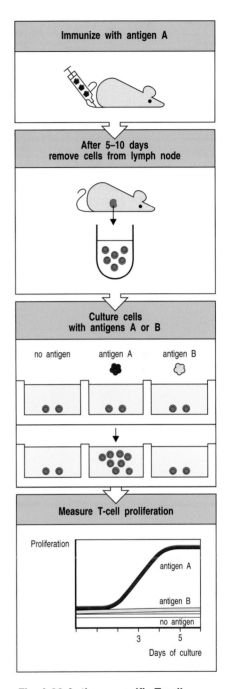

Fig. A.36 Antigen-specific T-cell proliferation is used frequently as an assay for T-cell responses. T cells from mice or humans that have been immunized with an antigen (A) proliferate when they are exposed to antigen A and antigen-presenting cells but not when cultured with unrelated antigens to which they have not been immunized (antigen B). Proliferation can be measured by incorporation of ^{3}H-thymidine into the DNA of actively dividing cells. Antigen-specific proliferation is a hallmark of specific CD4 T-cell immunity.

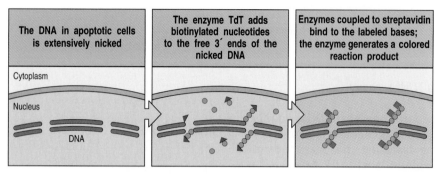

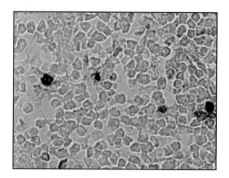

| The DNA in apoptotic cells is extensively nicked | The enzyme TdT adds biotinylated nucleotides to the free 3′ ends of the nicked DNA | Enzymes coupled to streptavidin bind to the labeled bases; the enzyme generates a colored reaction product |

Fig. A.37 Fragmented DNA can be labeled by terminal deoxynucleotidyl transferase (TdT) to reveal apoptotic cells. When cells undergo programmed cell death, or apoptosis, their DNA becomes fragmented (left panel). The enzyme TdT is able to add nucleotides to the ends of DNA fragments; most commonly in this assay, biotin-labeled nucleotides (usually dUTP) are added (second panel). The biotinylated DNA can be detected by using streptavidin, which binds to biotin, coupled to enzymes that convert a colorless substrate into a colored insoluble product (third panel). Cells stained in this way can be detected by light microscopy, as shown in the photograph of apoptotic cells (stained red) in the thymic cortex. Photograph courtesy of R. Budd and J. Russell.

identifying cells actively producing a given cytokine is to stain them with a fluorescently tagged anti-cytokine monoclonal antibody and identify and count them by FACS (see Section A-22).

A quite different approach to detecting cytokine production is to determine the presence and amount of the relevant cytokine mRNA in stimulated T cells. This can be done for single cells by *in situ* hybridization and for cell populations by **reverse transcriptase–polymerase chain reaction (RT–PCR)**. Reverse transcriptase is an enzyme used by certain RNA viruses, such as the human immunodeficiency virus (HIV-1) that causes AIDS, to convert an RNA genome into a DNA copy, or cDNA. In RT–PCR, mRNA is isolated from cells and cDNA copies made using reverse transcriptase. The desired cDNA is then selectively amplified by PCR using sequence-specific primers. When the products of the reaction are subjected to electrophoresis on an agarose gel, the amplified DNA can be visualized as a band of a specific size. The amount of amplified cDNA sequence will be proportional to its representation in the mRNA; stimulated T cells actively producing a particular cytokine will produce large amounts of that particular mRNA and thus give correspondingly large amounts of the selected cDNA on RT–PCR. The level of cytokine mRNA in the original tissue is usually determined by comparison with the outcome of RT–PCR on the mRNA produced by a so-called 'housekeeping gene' expressed by all cells.

A-35   DNA microarrays.

Any cell expresses, at any one time, many hundreds or even thousands of genes. Some of the products are expressed at high levels, the actin that forms the cytoskeleton of the cell is one example, while others may only be expressed in a few copies per cell. Different cell types, or cells at different stages of

Fig. A.38 Cytotoxic T-cell activity is often assessed by chromium release from labeled target cells. Target cells are labeled with radioactive chromium as $Na_2^{51}CrO_4$, washed to remove excess radioactivity and exposed to cytotoxic T cells. Cell destruction is measured by the release of radioactive chromium into the medium, detectable within 4 hours of mixing target cells with T cells.

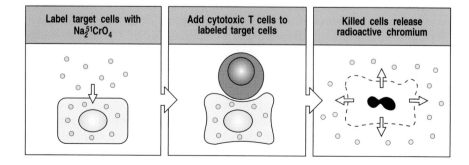

| Label target cells with $Na_2^{51}CrO_4$ | Add cytotoxic T cells to labeled target cells | Killed cells release radioactive chromium |

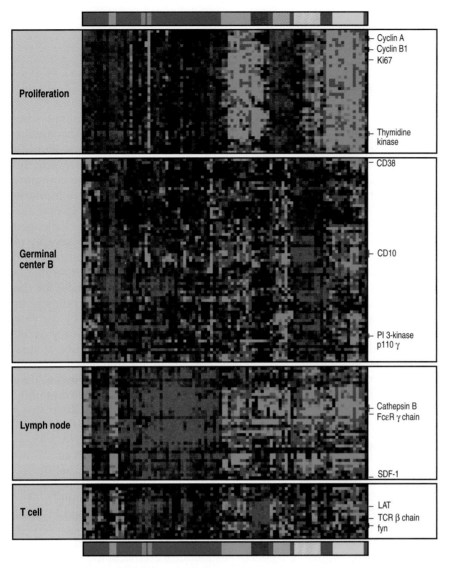

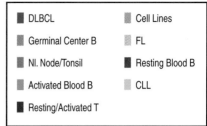

Proliferation
— Cyclin A
— Cyclin B1
— Ki67
— Thymidine kinase

Germinal center B
— CD38
— CD10
— PI 3-kinase p110 γ

Lymph node
— Cathepsin B
— FcεR γ chain
— SDF-1

T cell
— LAT
— TCR β chain
— fyn

■ DLBCL   ▨ Cell Lines

▨ Germinal Center B   ▨ FL

▨ NI. Node/Tonsil   ■ Resting Blood B

▨ Activated Blood B   □ CLL

■ Resting/Activated T

Fig. A.39 DNA microarrays allow a rapid, simultaneous screening of a great many genes for changes in expression between different cells. In the experiment whose results are shown here, nearly 18,000 cDNA clones made from lymphoid cells and lymphoid tumors were arrayed horizontally across the chip. Some of these cDNAs represent the products of known genes, and examples of these are indicated on the right hand edge of the diagram. To these cDNAs mRNA was hybridized from 96 normal cells, cell lines and lymphoid tumor cells, with the mRNA from each individual cell arrayed vertically. The types of cells used are indicated by the colored bars at the top, with the colors explained in the key. The malignant cells used were from diffuse large B-cell lymphoma (DLBCL), follicular lymphoma (FL) or mantle cell lymphoma, and chronic lymphocytic leukemia (CLL). A number of established lymphoid cell lines were included. Normal cells are represented by resting B cells from peripheral blood (Resting Blood B), B cells activated by cross-linking cell-surface IgM with or without additional cytokines and co-stimulation (Activated blood B) germinal center B cells from tonsil (Germinal Center B), and normal, noninflamed tonsil and lymph node (NI. Node/Tonsil) were used as representatives of different stages in B-cell maturation. Normal T cells, CD4 T cells, either resting or stimulated with PMA and ionomycin (Resting/Activated T) were also used. Each point on the array represents, therefore, the hybridization of the mRNA from one of these cell lines to one of the cDNAs, and is displayed in color to represent the level of expression of the mRNA in question, green being those expressed at lower levels than in a control cell, while red represents those expressed at a higher level. The data shown has been clustered by patterns of expression of the various genes, to give clusters of genes upregulated in proliferating cells, germinal center B cells, lymph node B cells, and in T cells. Courtesy of Dr. L. M. Staudt.

maturation, or even tumor cells compared to their normal counterparts, will express different sets of genes, and trying to identify these differences is an important field of research, in immunology as well as in other areas of biology. One important new technique in analyzing these differences makes use of arrays of hundreds of DNA sequences attached to a glass surface—a so-called **DNA microarray** or 'DNA chip.' The array contains a range of DNA sequences from known genes, arranged in a fixed pattern, and the differential expression of those genes in a particular cell type or tissue is tested by exposing the array to labeled mRNA (or cDNA made from it) from the tissue. Hybridization of labeled mRNAs to their corresponding DNA sequences in the array is detected by standard techniques and the whole technique is readily automated. Many different samples can be examined in parallel, which makes this a powerful analytical technique, as can be seen from the example we illustrate in Fig. A.39. Here the DNA microarray has been constructed with almost 18,000 cDNA clones known to be expressed in either B or T cells and B-cell tumors. This array was then probed with fluorochrome-labeled cDNAs from 96 normal and malignant cells, and the level of expression of approximately 18,000 genes in each of the cell lines was measured simultaneously. In this particular case the patterns of expression of the different genes revealed that the malignant B cells formed discrete subtypes, which were then found to have different clinical prognoses.

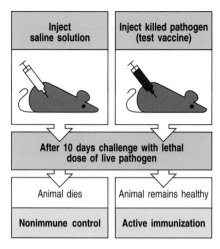

Fig. A.40 *In vivo* assay for the presence of protective immunity after vaccination in animals. Mice are injected with the test vaccine or a control such as saline solution. Different groups are then challenged with lethal or pathogenic doses of the test pathogen or with an unrelated pathogen as a specificity control (not shown). Unimmunized animals die or become severely infected. Successful vaccination is seen as specific protection of immunized mice against infection with the test pathogen. This is called active immunity and the process is called active immunization.

Detection of immunity *in vivo*.

A-36   Assessment of protective immunity.

An adaptive immune response against a pathogen often confers long-lasting immunity against infection with that pathogen; successful vaccination achieves the same end. The very first experiment in immunology, Jenner's successful vaccination against smallpox, is still the model for assessing the presence of such protective immunity. The assessment of protective immunity conferred by vaccination has three essential steps. First, an immune response is elicited by immunization with a candidate vaccine. Second, the immunized individuals, along with unimmunized controls, are challenged with the infectious agent (Fig. A.40). Finally, the prevalence and severity of infection in the immunized individual is compared with the course of the disease in the unimmunized controls. For obvious reasons, such experiments are usually carried out first in animals, if a suitable animal model for the infection exists. However, eventually a trial must be carried out in humans. In this case, the infectious challenge is usually provided naturally by carrying out the trial in a region where the disease is prevalent. The efficacy of the vaccine is determined by assessing the prevalence and severity of new infections in the immunized and control populations. Such studies necessarily give less precise results than a direct experiment but, for most diseases, they are the only way of assessing a vaccine's ability to induce protective immunity in humans.

A-37   Transfer of protective immunity.

The tests described in Section A-36 show that protective immunity has been established, but cannot show whether it involves humoral immunity, cell-mediated immunity, or both. When these studies are carried out in inbred mice, the nature of protective immunity can be determined by transferring serum or lymphoid cells from an immunized donor animal to an unimmunized syngeneic recipient (that is, a genetically identical animal of the same inbred strain) (Fig. A.41). If protection against infection can be conferred by the transfer of serum, the immunity is provided by circulating antibodies and is called **humoral immunity**. Transfer of immunity by antiserum or purified antibodies provides immediate protection against many pathogens and against toxins such as those of tetanus and snake venom. However, although protection is immediate, it is temporary, lasting only so long as the transferred antibodies remain active in the recipient's body. This type of transfer is therefore called **passive immunization**. Only **active immunization** with antigen can provide lasting immunity. Moreover, the recipient may become immunized to the antiserum used to transfer immunity. Horse or sheep sera are the usual sources of anti-snake venoms used in humans, and repeated administration can lead either to serum sickness (see Section 12-16) or, if the recipient becomes allergic to the foreign serum, to anaphylaxis (see Section 12-10).

Protection against many diseases cannot be transferred with serum but can be transferred by lymphoid cells from immunized donors. The transfer of lymphoid cells from an immune donor to a normal syngeneic recipient is called **adoptive transfer** or **adoptive immunization**, and the immunity transferred is called **adoptive immunity**. Immunity that can be transferred only with lymphoid cells is called **cell-mediated immunity**. Such cell transfers must be between genetically identical donors and recipients, such as members of the same inbred strain of mouse, so that the donor lymphocytes are not rejected by the recipient and do not attack the recipient's tissues. Adoptive transfer of immunity is used clinically in humans in experimental approaches to cancer therapy or as an adjunct to bone marrow transplantation; in these cases, the patient's own T cells, or the T cells of the bone marrow donor, are given.

A-38 The tuberculin test.

Local responses to antigen can indicate the presence of active immunity. Active immunity is often studied *in vivo*, especially in humans, by injecting antigens locally in the skin. If a reaction appears, this indicates the presence of antibodies or immune lymphocytes that are specific for that antigen; the **tuberculin test** is an example of this. When people have had tuberculosis they develop cell-mediated immunity that can be detected as a local response when their skin is injected with a small amount of tuberculin, an extract of *Mycobacterium tuberculosis*, the pathogen that causes tuberculosis. The response typically appears a day or two after the injection and consists of a raised, red, and hard (or indurated) area in the skin, which then disappears as the antigen is degraded.

A-39 Testing for allergic responses.

Local **intracutaneous** injections of minute doses of the antigens that cause allergies are used to determine which antigen triggers a patient's allergic reactions. Local responses that happen in the first few minutes after antigen injection in immune recipients are called **immediate hypersensitivity reactions**, and they can be of several forms, one of which is the wheal-and-flare response (see Fig. 12.14). Immediate hypersensitivity reactions are mediated by specific antibodies of the IgE class formed as a result of earlier exposures to the antigen. Responses that take hours to days to develop, such as the tuberculin test, are referred to as **delayed-type hypersensitivity** responses and are caused by preexisting immune T cells. This latter type of response was observed by Jenner when he tested vaccinated individuals with a local injection of vaccinia virus.

These tests work because the local deposit of antigen remains concentrated in the initial site of injection, eliciting responses in local tissues. They do not cause generalized reactions if sufficiently small doses of antigen are used. However, local tests carry a risk of systemic allergic reactions, and they should be used with caution in people with a history of hypersensitivity.

A-40 Assessment of immune responses and immunological competence in humans.

The methods used for testing immune function in humans are necessarily more limited than those used in experimental animals, but many different tests are available. They fall into several groups depending on the reason the patient is being studied.

Assessment of protective immunity in humans generally relies on tests conducted *in vitro*. To assess humoral immunity, specific antibody levels in the patient's serum are assayed by RIA or, more commonly, ELISA (see Section A-6), using the test microorganism or a purified microbial product as antigen. To test for humoral immunity against viruses, antibody production is often measured by the ability of serum to neutralize the infectivity of live virus for tissue culture cells. In addition to providing information about protective

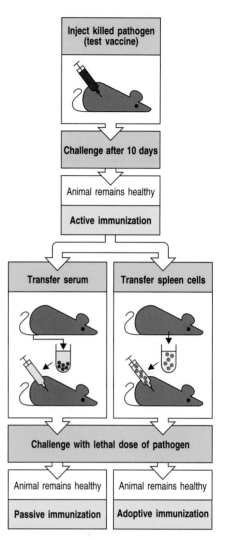

Fig. A.41 Immunity can be transferred by antibodies or by lymphocytes. Successful vaccination leads to a long-lived state of protection against the specific immunizing pathogen. If this immune protection can be transferred to a normal syngeneic recipient with serum from an immune donor, then immunity is mediated by antibodies; such immunity is called humoral immunity and the process is called passive immunization. If immunity can only be transferred by infusing lymphoid cells from the immune donor into a normal syngeneic recipient, then the immunity is called cell-mediated immunity and the transfer process is called adoptive transfer or adoptive immunization. Passive immunity is short-lived, as antibody is eventually catabolized, but adoptively transferred immunity is mediated by immune cells, which can survive and provide longer-lasting immunity.

immunity, the presence of antibody to a particular pathogen indicates that the patient has been exposed to it, making such tests of crucial importance in epidemiology. At present, testing for antibody to HIV is the main screening test for infection with this virus, critical both for the patient and in blood banking, where blood from infected donors must be excluded from the supply. Essentially similar tests are used in investigating allergy, where allergens are used as the antigens in tests for specific IgE antibody by ELISA or RIA (see Section A-6), which may be used to confirm the results of skin tests.

Cell-mediated immunity, that is immunity mediated by T cells, is technically more difficult to measure than humoral immunity. This is principally because T cells do not make a secreted antigen-binding product, so there is no simple binding assay for their antigen-specific responses. T-cell activity can be divided into an induction phase, in which T cells are activated to divide and differentiate, and an effector phase, in which their function is expressed. Both phases require that the T cell interacts with another cell and that it recognizes specific antigen displayed in the form of peptide:MHC complexes on the surface of this interacting cell. In the induction phase, the interaction must be with an antigen-presenting cell able to deliver co-stimulatory signals, whereas, in the effector phase, the nature of the target cell depends on the type of effector T cell that has been activated. Most commonly, the presence of T cells that have responded to a specific antigen is detected by their subsequent *in vitro* proliferation when reexposed to the same antigen (see Section A-31).

T-cell proliferation indicates only that cells able to recognize that antigen have been activated previously; it does not reveal what effector function they mediate. The effector function of a T cell is assayed by its effect on an appropriate target cell. Assays for cytotoxic CD8 T cells (see Section A-33) and for cytokine production by CD4 T cells (see Sections A-26, A-27, and A-34) are used to characterize the immune response. Cell-mediated immunity to infectious agents can also be tested by skin test with extracts of the pathogen, as in the tuberculin test (see Section A-36). These tests provide information about the exposure of the patient to the disease and also about their ability to mount an adaptive immune response to it.

Patients with immune deficiency (see Chapter 11) are usually detected clinically by a history of recurrent infection. To determine the competence of the immune system in such patients, a battery of tests are usually conducted (see Appendix V); these focus with increasing precision as the nature of the defect is narrowed down to a single element. The presence of the various cell types in blood is determined by routine hematology, often followed by FACS analysis of lymphocyte subsets (see Section A-22), and the measurement of serum immunoglobulins. The phagocytic competence of freshly isolated polymorphonuclear leukocytes and monocytes is tested, and the efficiency of the complement system (see Chapters 2 and 9) is determined by testing the dilution of serum required for lysis of 50% of antibody-coated red blood cells (this is denoted the CH_{50}).

In general, if such tests reveal a defect in one of the broad compartments of immune function, more specialized testing is then needed to determine the precise nature of the defect. Tests of lymphocyte function are often valuable, starting with the ability of polyclonal mitogens to induce T-cell proliferation and B-cell secretion of immunoglobulin in tissue culture (see Section A-31). These tests can eventually pinpoint the cellular defect in immunodeficiency.

In patients with autoimmune diseases (see Chapter 13), the same parameters are usually analyzed to determine whether there is a gross abnormality in the immune system. However, most patients with such diseases show few abnormalities in general immune function. To determine whether a patient is producing antibody against their own cellular antigens, the most informative test is to react their serum with tissue sections, which are then examined for bound antibody by indirect immunofluorescence using anti-human immunoglobulin labeled with fluorescent dye (see Section A-14). Most autoimmune diseases are associated with the production of broadly characteristic patterns of autoantibodies directed at self tissues. These patterns aid in the diagnosis of the disease and help to distinguish autoimmunity from tissue inflammation due to infectious causes.

It is also possible to investigate allergies by administration of possible allergens by routes other than intracutaneous administration. Allergen may be given by inhalation to test for asthmatic allergic responses (see Fig. 12.14); this is mainly done for experimental purposes in studies of the mechanisms and treatment of asthma. Similarly, food allergens may be given by mouth. The administration of allergens is potentially very dangerous because of the risk of causing anaphylaxis, and must only be carried out by trained and experienced investigators in an environment in which full resuscitation facilities are available.

A-41   The Arthus reaction.

This is an experimental method using only animal models for studying the formation of immune complexes in tissues and how immune complexes cause inflammation (see Section 12-16). The original reaction described by Maurice Arthus was induced by the repeated injection of horse serum into rabbits. Initial injections of horse serum into the skin induced no reaction, but later injections, following the production of antibodies to the proteins in horse serum, induced an inflammatory reaction at the site of injection after several hours, characterized by the presence of edema, hemorrhage, and neutrophil infiltration, which frequently progressed to tissue necrosis. Most investigators now use passive models of the Arthus reaction in which either antibody is infused systemically and antigen given locally (passive Arthus reaction) or antigen is infused systemically and antibody injected locally (reverse passive Arthus reaction).

Manipulation of the immune system.

A-42   Adoptive transfer of lymphocytes.

Ionizing radiation from X-ray or γ-ray sources kills lymphoid cells at doses that spare the other tissues of the body. This makes it possible to eliminate immune function in a recipient animal before attempting to restore immune function by adoptive transfer, and allows the effect of the adoptively transferred cells to be studied in the absence of other lymphoid cells. James Gowans originally used this technique to prove the role of the lymphocyte in immune responses. He showed that all active immune responses could be transferred to irradiated recipients by small lymphocytes from immunized donors. This technique can be refined by transferring only certain lymphocyte subpopulations, such as B cells, CD4 T cells, and so on. Even cloned T-cell lines have been tested for their ability to transfer immune function, and have been shown to confer adoptive immunity to their specific antigen. Such adoptive transfer studies are a cornerstone in the study of the intact immune system, as they can be carried out rapidly, simply, and in any strain of mouse.

A-43   Hematopoietic stem-cell transfers.

All cells of hematopoietic origin can be eliminated by treatment with high doses of X rays, allowing replacement of the entire hematopoietic system, including lymphocytes, by transfusion of donor bone marrow or purified hematopoietic stem cells from another animal. The resulting animals are called **radiation bone marrow chimeras** from the Greek word *chimera*, a mythical animal that had the head of a lion, the tail of a serpent, and the body of a goat. This technique is used experimentally to examine the development of lymphocytes, as opposed to their effector functions, and has been particularly important in studying T-cell development. Essentially the same technique is used in humans to replace the hematopoietic system when it fails, as in aplastic anemia or after nuclear accidents, or to eradicate the bone marrow and replace it with normal marrow in the treatment of certain cancers. In man, bone marrow is the main source of hematopoietic stem cells, but increasingly they are being obtained from peripheral blood after the donor has been treated with hematopoietic growth factors such as GM-CSF, or from umbilical cord blood, which is rich in such stem cells.

A-44   *In vivo* depletion of T cells.

The importance of T-cell function *in vivo* can be ascertained in mice with no T cells of their own. Under these conditions, the effect of a lack of T cells can be studied, and T-cell subpopulations can be restored selectively to analyze their specialized functions. T lymphocytes originate in the thymus, and neonatal **thymectomy**, the surgical removal of the thymus of a mouse at birth, prevents T-cell development from occurring because the export of most functionally mature T cells only occurs after birth in the mouse. Alternatively, adult mice can be thymectomized and then irradiated and reconstituted with bone marrow; such mice will develop all hematopoietic cell types except mature T cells.

The recessive *nude* mutation in mice is caused by a mutation in the gene for the transcription factor Wnt and in homozygous form causes hairlessness and absence of the thymus. Consequently, these animals fail to develop T cells from bone marrow progenitors. Grafting thymectomized or *nude/nude* mice with thymic epithelial elements depleted of lymphocytes allows the graft recipients to develop normal mature T cells. This procedure allows the role of the nonlymphoid thymic stroma to be examined; it has been crucial in determining the role of thymic stromal cells in T-cell development (see Chapter 7).

A-45   *In vivo* depletion of B cells.

There is no single site of B-cell development in mice, so techniques such as thymectomy cannot be applied to the study of B-cell function and development in rodents. However, **bursectomy**, the surgical removal of the Bursa of Fabricius in birds, can inhibit the development of B cells in these species. In fact, it was the effect of thymectomy versus bursectomy that led to the naming

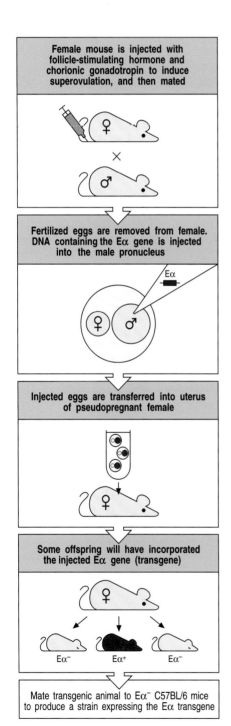

| Female mouse is injected with follicle-stimulating hormone and chorionic gonadotropin to induce superovulation, and then mated |

| Fertilized eggs are removed from female. DNA containing the Eα gene is injected into the male pronucleus |

Eα

| Injected eggs are transferred into uterus of pseudopregnant female |

| Some offspring will have incorporated the injected Eα gene (transgene) |

Eα⁻   Eα⁺   Eα⁻

| Mate transgenic animal to Eα⁻ C57BL/6 mice to produce a strain expressing the Eα transgene |

Fig. A.42 The function and expression of genes can be studied *in vivo* by using transgenic mice. DNA encoding a protein of interest, here the mouse MHC class II protein Eα, is purified and microinjected into the male pronuclei of fertilized eggs. The eggs are then implanted into pseudopregnant female mice. The resulting offspring are screened for the presence of the transgene in their cells, and positive mice are used as founders that transmit the transgene to their offspring, establishing a line of transgenic mice that carry one or more extra genes. The function of the Eα gene used here is tested by breeding the transgene into C57BL/6 mice that carry an inactivating mutation in their endogenous Eα gene.

of T cells for thymus-derived lymphocytes and B cells for bursal-derived lymphocytes. There are no known spontaneous mutations (analogous to the *nude* mutation) in mice that produce animals with T cells but no B cells. However, such mutations exist in humans, leading to a failure to mount humoral immune responses or make antibody. The diseases produced by such mutations are called agammaglobulinemias because they were originally detected as the absence of gamma globulins. The genetic basis for one form of this disease in humans has now been established (see Chapter 11), and some features of the disease can be reproduced in mice by targeted disruption of the corresponding gene (see Section A-47). Several different mutations in crucial regions of immunoglobulin genes have already been produced by gene targeting and have provided mice lacking B cells.

A-46 Transgenic mice.

The function of genes has traditionally been studied by observing the effects of spontaneous mutations in whole organisms and, more recently, by analyzing the effects of targeted mutations in cultured cells. The advent of gene cloning and *in vitro* mutagenesis now make it possible to produce specific mutations in whole animals. Mice with extra copies or altered copies of a gene in their genome can be generated by **transgenesis**, which is now a well established procedure. To produce **transgenic mice**, a cloned gene is introduced into the mouse genome by microinjection into the male pronucleus of a fertilized egg, which is then implanted into the uterus of a pseudopregnant female mouse. In some of the eggs, the injected DNA becomes integrated randomly into the genome, giving rise to a mouse that has an extra genetic element of known structure, the transgene (Fig. A.42).

The transgene, to be studied in detail, needs to be introduced onto a stable, well-characterized genetic background. However, it is difficult to prepare transgenic embryos successfully in inbred strains of mice, and transgenic mice are routinely prepared in F_2 embryos (that is, the embryo formed after the mating of two F_1 animals). The transgene must then be bred onto a well-characterized genetic background; this requires 10 generations of back-crossing with an inbred strain to assure that the integrated transgene is largely (>99%) free of heterogeneous genes from the founder mouse of the transgenic mouse line (Fig. A.43).

This technique allows one to study the impact of a newly discovered gene on development, to identify the regulatory regions of a gene required for its normal tissue-specific expression, to determine the effects of its over-expression or expression in inappropriate tissues, and to find out the impact of mutations on gene function. Transgenic mice have been particularly useful in studying the role of T-cell and B-cell receptors in lymphocyte development, as described in Chapter 7.

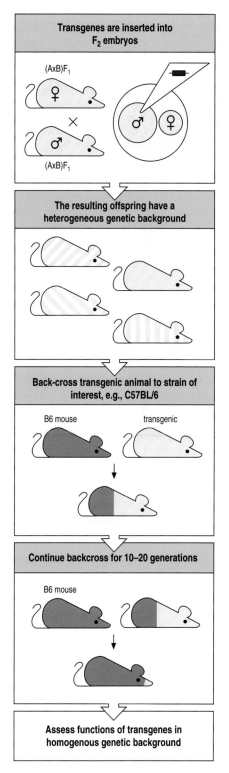

Fig. A.43 The breeding of transgenic co-isogeneic or congenic mouse strains. Transgenic mouse strains are routinely made in F_2 mice. To produce mice on an inbred background, the transgene is introgressively back-crossed onto a standard strain, usually C57BL/6 (B6). The presence of the transgene is tracked by carrying out PCR on genomic DNA extracted from the tail of young mice. After 10 generations of back-crossing, mice are >99% genetically identical, so that any differences observed between the mice are likely to be due to the transgene itself. The same technique can be used to breed a gene knockout into a standard strain of mice, as most gene knockouts are made in the 129 strain of mice (see Fig. A.45). The mice are then intercrossed and homozygous knockout mice detected by an absence of an intact copy of the gene of interest (as determined by PCR).

Fig. A.44 The deletion of specific genes can be accomplished by homologous recombination. When pieces of DNA are introduced into cells, they can integrate into cellular DNA in two different ways. If they randomly insert into sites of DNA breaks, the whole piece is usually integrated, often in several copies. However, extrachromosomal DNA can also undergo homologous recombination with the cellular copy of the gene, in which case only the central, homologous region is incorporated into cellular DNA. Inserting a selectable marker gene such as resistance to neomycin (*neo^r*) into the coding region of a gene does not prevent homologous recombination, and it achieves two goals. First, any cell that has integrated the injected DNA is protected from the neomycin-like antibiotic G418. Second, when the gene recombines with homologous cellular DNA, the *neo^r* gene disrupts the coding sequence of the modified cellular gene. Homologous recombinants can be discriminated from random insertions if the gene for herpes simplex virus thymidine kinase (HSV-tk) is placed at one or both ends of the DNA construct, which is often known as a 'targeting construct' because it targets the cellular gene. In random DNA integrations, HSV-tk is retained. HSV-tk renders the cell sensitive to the antiviral agent ganciclovir. However, as HSV-tk is not homologous to the target DNA, it is lost from homologous recombinants. Thus, cells that have undergone homologous recombination are uniquely both G418 and ganciclovir resistant, and survive in a mixture of the two antibiotics. The presence of the disrupted gene has to be confirmed by Southern blotting or by PCR using primers in the *neo^r* gene and in cellular DNA lying outside the region used in the targeting construct. By using two different resistance genes one can disrupt the two cellular copies of a gene, making a deletion mutant (not shown).

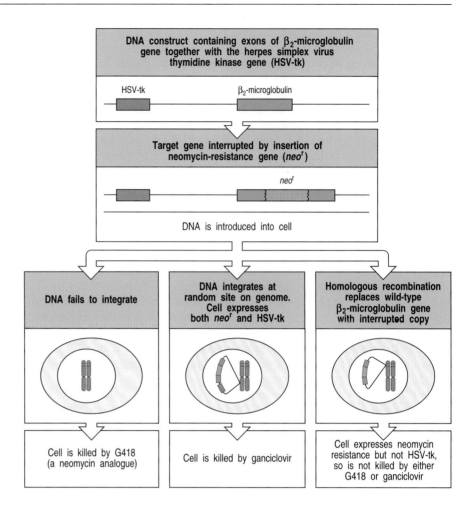

A-47 Gene knockout by targeted disruption.

In many cases, the functions of a particular gene can be fully understood only if a mutant animal that does not express the gene can be obtained. Whereas genes used to be discovered through identification of mutant phenotypes, it is now far more common to discover and isolate the normal gene and then determine its function by replacing it *in vivo* with a defective copy. This procedure is known as **gene knockout**, and it has been made possible by two fairly recent developments: a powerful strategy to select for targeted mutation by homologous recombination, and the development of continuously growing lines of **embryonic stem cells** (**ES cells**). These are embryonic cells which, on implantation into a blastocyst, can give rise to all cell lineages in a chimeric mouse.

The technique of **gene targeting** takes advantage of the phenomenon known as **homologous recombination** (Fig. A.44). Cloned copies of the target gene are altered to make them nonfunctional and are then introduced into the ES cell where they recombine with the homologous gene in the cell's genome, replacing the normal gene with a nonfunctional copy. Homologous recombination is a rare event in mammalian cells, and thus a powerful selection strategy is required to detect those cells in which it has occurred. Most commonly, the introduced gene construct has its sequence disrupted by an inserted antibiotic-resistance gene such as that for neomycin resistance. If this construct undergoes homologous recombination with the endogenous copy of the gene, the endogenous gene is disrupted but the antibiotic-resistance gene remains functional, allowing cells that have incorporated the gene to be selected in culture for resistance to the neomycin-like drug G418. However, antibiotic resistance on its own shows only that the cells have taken up and integrated the neomycin-resistance gene. To be able to select for those cells in which homologous recombination has occurred, the ends of the construct usually carry the thymidine kinase gene from the herpes simplex virus (HSV-tk). Cells that incorporate DNA randomly usually retain the entire DNA construct including HSV-tk, whereas homologous recombination between the construct and cellular DNA, the desired result, involves the exchange of homologous DNA sequences so that the nonhomologous HSV-tk genes at the ends of the construct are eliminated. Cells carrying HSV-tk are killed by the antiviral drug ganciclovir, and so cells with homologous recombinations have the unique feature of being resistant to both neomycin and ganciclovir, allowing them to be selected efficiently when these drugs are added to the cultures (see Fig. A.44).

This technique can be used to produce homozygous mutant cells in which the effects of knocking-out a specific gene can be analyzed. Diploid cells in which both copies of a gene have been mutated by homologous recombination can be selected after transfection with a mixture of constructs in which the gene to be targeted has been disrupted by one or other of two different antibiotic-resistance genes. Having obtained a mutant cell with a functional defect, the defect can be ascribed definitively to the mutated gene if the mutant phenotype can be reverted with a copy of the normal gene transfected into the mutant cell. Restoration of function means that the defect in the mutant gene has been complemented by the normal gene's function. This technique is very powerful as it allows the gene that is being transferred to be mutated in precise ways to determine which parts of the protein are required for function.

To knock out a gene *in vivo*, it is only necessary to disrupt one copy of the cellular gene in an ES cell. ES cells carrying the mutant gene are produced by targeted mutation (see Fig. A.44), and injected into a blastocyst which is reimplanted into the uterus. The cells carrying the disrupted gene become incorporated into the developing embryo and contribute to all tissues of the resulting chimeric offspring, including those of the germline. The mutated gene can therefore be transmitted to some of the offspring of the original chimera, and further breeding of the mutant gene to homozygosity produces mice that completely lack the expression of that particular gene product (Fig. A.45). The effects of the absence of the gene's function can then be studied. In addition, the parts of the gene that are essential for its function can be identified by determining whether function can be restored by introducing different mutated copies of the gene back into the genome by transgenesis.

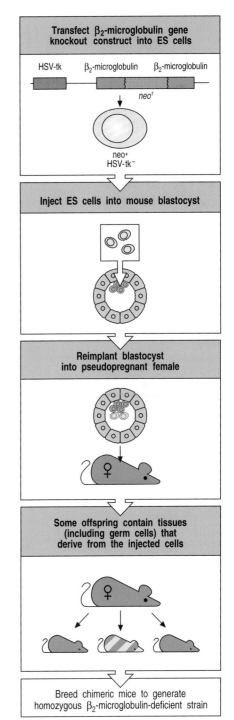

Fig. A.45 Gene knockout in embryonic stem cells enables mutant mice to be produced. Specific genes can be inactivated by homologous recombination in cultures of embryonic stem cells (ES cells). Homologous recombination is carried out as described in Fig. A.44. In this example, the gene for β2-microglobulin in ES cells is disrupted by homologous recombination with a targeting construct. Only a single copy of the gene needs to be disrupted. ES cells in which homologous recombination has taken place are injected into mouse blastocysts. If the mutant ES cells give rise to germ cells in the resulting chimeric mice (striped in the figure), then the mutant gene can be transferred to their offspring. By breeding the mutant gene to homozygosity, a mutant phenotype is generated. These mutant mice are usually of the 129 strain as gene knockout is generally conducted in ES cells derived from the 129 strain of mice. In this case, the homozygous mutant mice lack MHC class I molecules on their cells, as MHC class I molecules have to pair with β2-microglobulin for surface expression. The β2-microglobulin-deficient mice can then be bred with mice transgenic for subtler mutants of the deleted gene, allowing the effect of such mutants to be tested *in vivo*.

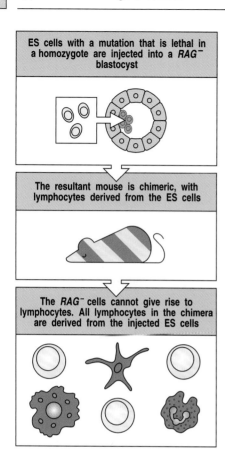

ES cells with a mutation that is lethal in a homozygote are injected into a *RAG⁻* blastocyst

The resultant mouse is chimeric, with lymphocytes derived from the ES cells

The *RAG⁻* cells cannot give rise to lymphocytes. All lymphocytes in the chimera are derived from the injected ES cells

Fig. A.46 The role of recessive lethal genes in lymphocyte function can be studied using *RAG*-deficient chimeric mice. ES cells homozygous for the lethal mutation are injected into a *RAG*-deficient blastocyst (top panel). The *RAG*-deficient cells can give rise to all the tissues of a normal mouse except lymphocytes, and so can compensate for any deficiency in the developmental potential of the mutant ES cells (middle panel). If the mutant ES cells are capable of differentiating into hematopoietic stem cells, that is, if the gene function that has been deleted is not essential for this developmental pathway, then all the lymphocytes in the chimeric mouse will be derived from the ES cells (bottom panel), as *RAG*-deficient mice cannot make lymphocytes of their own.

The manipulation of the mouse genome by gene knockout and transgenesis is revolutionizing our understanding of the role of individual genes in lymphocyte development and function.

Because the most commonly used ES cells are derived from a poorly characterized strain of mice known as strain 129, the analysis of the function of a gene knockout often requires extensive back-crossing to another strain, just as in transgenic mice (see Fig. A.43). One can track the presence of the mutant copy of the gene by the presence of the *neo*ʳ gene. After sufficient back-crossing, the mice are intercrossed to produce mutants on a stable genetic background.

A problem with gene knockouts arises when the function of the gene is essential for the survival of the animal; in such cases the gene is termed a **recessive lethal gene** and homozygous animals cannot be produced. However, by making chimeras with mice that are deficient in B and T cells, it is possible to analyze the function of recessive lethal genes in lymphoid cells. To do this, ES cells with homozygous lethal loss-of-function mutations are injected into blastocysts of mice lacking the ability to rearrange their antigen-receptor genes because of a mutation in their recombinase-activating genes (*RAG* knockout mice). As these chimeric embryos develop, the *RAG*-deficient cells can compensate for any developmental failure resulting from the gene knockout in the ES cells in all except the lymphoid lineage. So long as the mutated ES cells can develop into hematopoietic progenitors in the bone marrow, the embryos will survive and all of the lymphocytes in the resulting chimeric mouse will be derived from the mutant ES cells (Fig. A.46).

A second powerful technique achieves tissue-specific or developmentally regulated gene deletion by employing the DNA sequences and enzymes used by bacteriophage P1 to excise itself from a host cell's genome. Integrated bacteriophage P1 DNA is flanked by recombination signal sequences called *loxP* sites. A recombinase, Cre, recognizes these sites, cuts the DNA and joins the two ends, thus excising the intervening DNA in the form of a circle. This mechanism can be adapted to allow the deletion of specific genes in a transgenic animal only in certain tissues or at certain times in development. First, *loxP* sites flanking a gene, or perhaps just a single exon, are introduced by homologous recombination (Fig. A.47). Usually, the introduction of these sequences into flanking or intronic DNA does not disrupt the normal function of the gene. Mice containing such *loxP* mutant genes are then mated with mice made transgenic for the Cre recombinase, under the control of a tissue-specific or inducible promoter. When the Cre recombinase is active, either in the appropriate tissue or when induced, it excises the DNA between the inserted *loxP* sites, thus inactivating the gene or exon. Thus, for example, using a T-cell specific promoter to drive expression of the Cre recombinase, a gene can be deleted only in T cells, while remaining functional in all other cells of the animal. This is an extremely powerful genetic technique that while still in its infancy, was used to demonstrate the importance of B-cell receptors in B-cell survival. It is certain to yield exciting results in the future.

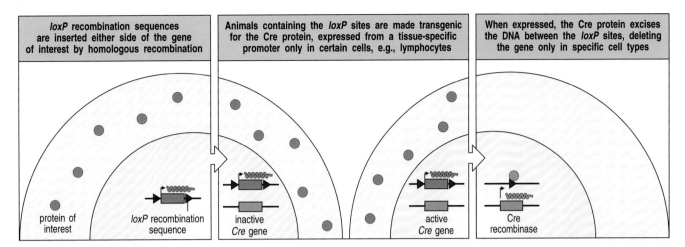

Fig. A.47 The P1 bacteriophage recombination system can be used to eliminate genes in particular cell lineages. The P1 bacteriophage protein Cre excises DNA that is bounded by recombination signal sequences called *loxP* sequences. These sequences can be introduced at either end of a gene by homologous recombination (left panel). Animals carrying genes flanked by *loxP* can also be made transgenic for the gene for the Cre protein, which is placed under the control of a tissue-specific promoter so that it is expressed only in certain cells or only at certain times during development (middle panel). In the cells in which the Cre protein is expressed, it recognizes the *loxP* sequences and excises the DNA lying between them (right panel). Thus, individual genes can be deleted only in certain cell types or only at certain times. In this way, genes that are essential for the normal development of a mouse can be analyzed for their function in the developed animal and/or in specific cell types. Genes are shown as boxes, RNA as squiggles, and proteins as colored balls.

APPENDICES
II – V

Appendix II. CD antigens.					
CD antigen	Cellular expression	Molecular weight (kDa)	Functions	Other names	Family relationships
CD1a,b,c,d	Cortical thymocytes, Langerhans' cells, dendritic cells, B cells (CD1c), intestinal epithelium, smooth muscle, blood vessels (CD1d)	43–49	MHC class I-like molecule, associated with β_2-microglobulin. Has specialized role in presentation of lipid antigens		Immunoglobulin
CD2	T cells, thymocytes, NK cells	45–58	Adhesion molecule, binding CD58 (LFA-3). Binds Lck intracellularly and activates T cells	T11, LFA-2	Immunoglobulin
CD3	Thymocytes, T cells	γ: 25–28 δ: 20 ε: 20	Associated with the T-cell antigen receptor (TCR). Required for cell-surface expression of and signal transduction by the TCR	T3	Immunoglobulin
CD4	Thymocyte subsets, T_H1 and T_H2 T cells (about two thirds of peripheral T cells), monocytes, macrophages	55	Co-receptor for MHC class II molecules. Binds Lck on cytoplasmic face of membrane. Receptor for HIV-1 and HIV-2 gp120	T4, L3T4	Immunoglobulin
CD5	Thymocytes, T cells, subset of B cells	67		T1, Ly1	Scavenger receptor
CD6	Thymocytes, T cells, B cells in chronic lymphatic leukemia	100–130	Binds CD166	T12	Scavenger receptor
CD7	Pluripotential hematopoietic cells, thymocytes, T cells	40	Unknown, cytoplasmic domain binds PI 3-kinase on cross-linking. Marker for T cell acute lymphatic leukemia and pluripotential stem cell leukemias		Immunoglobulin
CD8	Thymocytes subsets, cytotoxic T cells (about one third of peripheral T cells)	α: 32–34 β: 32–34	Co-receptor for MHC class I molecules. Binds Lck on cytoplasmic face of membrane	T8, Lyt2,3	Immunoglobulin
CD9	Pre-B cells, monocytes, eosinophils, basophils, platelets, activated T cells, brain and peripheral nerves, vascular smooth muscle	24	Mediates platelet aggregation and activation via FcγRIIa, may play a role in cell migration		Tetraspanning membrane protein, also called transmembrane 4 (TM4)
CD10	B- and T-cell precursors, bone marrow stromal cells	100	Zinc metalloproteinase, marker for pre-B acute lymphatic leukemia (ALL)	Neutral endopeptidase, common acute lymphocytic leukemia antigen (CALLA)	
CD11a	Lymphocytes, granulocytes, monocytes and macrophages	180	αL subunit of integrin LFA-1 (associated with CD18); binds to CD54 (ICAM-1), CD102 (ICAM-2), and CD50 (ICAM-3)	LFA-1	Integrin α

CD antigen	Cellular expression	Molecular weight (kDa)	Functions	Other names	Family relationships
CD11b	Myeloid and NK cells	170	αM subunit of integrin CR3 (associated with CD18): binds CD54, complement component iC3b, and extracellular matrix proteins	Mac-1	Integrin α
CD11c	Myeloid cells	150	αX subunit of integrin CR4 (associated with CD18); binds fibrinogen	CR4, p150, 95	Integrin α
CD11d	Leukocytes	125	αD subunits of integrin; associated with CD18; binds to CD50		Integrin α
CDw12	Monocytes, granulocytes, platelets	90–120	Unknown		
CD13	Myelomonocytic cells	150–170	Zinc metalloproteinase	Aminopeptidase N	
CD14	Myelomonocytic cells	53–55	Receptor for complex of lipopoly-saccharide and lipopolysaccharide binding protein (LBP)		
CD15	Neutrophils, eosinophils, monocytes		Terminal trisaccharide expressed on glycolipids and many cell-surface glycoproteins	Lewis^x (Le^x)	
CD15s	Leukocytes, endothelium		Ligand for CD62E, P	Sialyl-Lewis^x (sLe^x)	poly-N-acetyl-lactosamine
CD15u			Sulphated CD15		Carbohydrate structures
CD16	Neutrophils, NK cells, macrophages	50–80	Component of low affinity Fc receptor, FcγRIII, mediates phagocytosis and antibody-dependent cell-mediated cytotoxicity	FcγRIII	Immunoglobulin
CDw17	Neutrophils, monocytes, platelets		Lactosyl ceramide, a cell-surface glycosphingolipid		
CD18	Leukocytes	95	Intergrin β2 subunit, associates with CD11a, b, c, and d		Integrin β
CD19	B cells	95	Forms complex with CD21 (CR2) and CD81 (TAPA-1); co-receptor for B cells—cytoplasmic domain binds cytoplasmic tyrosine kinases and PI 3-kinase		Immunoglobulin
CD20	B cells	33–37	Oligomers of CD20 may form a Ca^{2+} channel; possible role in regulating B-cell activation		Contains 4 transmembrane segments
CD21	Mature B cells, follicular dendritic cells	145	Receptor for complement component C3d, Epstein–Barr virus. With CD19 and CD81, CD21 forms co-receptor for B cells	CR2	Complement control protein (CCP)
CD22	Mature B cells	α: 130 β: 140	Binds sialoconjugates	BL-CAM	Immunoglobulin
CD23	Mature B cells, activated macrophages, eosinophils, follicular dendritic cells, platelets	45	Low-affinity receptor for IgE, regulates IgE synthesis; ligand for CD19:CD21:CD81 co-receptor	FcεRII	C-type lectin
CD24	B cells, granulocytes	35–45	Unknown	Possible human homologue of mouse heat stable antigen (HSA)	
CD25	Activated T cells, B cells, and monocytes	55	IL-2 receptor α chain	Tac	CCP
CD26	Activated B and T cells, macrophages	110	Exopeptidase, cleaves N terminal X-Pro or X-Ala dipeptides from polypeptides	Dipeptidyl peptidase IV	Type II transmembrane glycoprotein

CD antigen	Cellular expression	Molecular weight (kDa)	Functions	Other names	Family relationships
CD27	Medullary thymocytes, T cells, NK cells, some B cells	55	Binds CD70; can function as a co-stimulator for T and B cells		TNF receptor
CD28	T-cell subsets, activated B cells	44	Activation of naive T cells, receptor for co-stimulatory signal (signal 2) binds CD80 (B7.1) and CD86 (B7.2)	Tp44	Immunoglobulin and CD86 (B7.2)
CD29	Leukocytes	130	Integrin β1 subunit, associates with CD49a in VLA-1 integrin		Integrin β
CD30	Activated T, B, and NK cells, monocytes	120	Binds CD30L (CD153); cross-linking CD30 enhances proliferation of B and T cells	Ki-1	TNF receptor
CD31	Monocytes, platelets, granulocytes, T-cell subsets, endothelial cells	130–140	Adhesion molecule, mediating both leukocyte–endothelial and endothelial–endothelial interactions	PECAM-1	Immunoglobulin
CD32	Monocytes, granulocytes, B cells, eosinophils	40	Low affinity Fc receptor for aggregated immunoglobulin:immune complexes	FcγRII	Immunoglobulin
CD33	Myeloid progenitor cells, monocytes	67	Binds sialoconjugates		Immunoglobulin
CD34	Hematopoietic precursors, capillary endothelium	105–120	Ligand for CD62L (L-selectin)		Mucin
CD35	Erythrocytes, B cells, monocytes, neutrophils, eosinophils, follicular dendritic cells	250	Complement receptor 1, binds C3b and C4b, mediates phagocytosis	CR1	CCP
CD36	Platelets, monocytes, endothelial cells	88	Platelet adhesion molecule; involved in recognition and phagocytosis of apoptosed cells	Platelet GPIV, GPIIIb	
CD37	Mature B cells, mature T cells, myeloid cells	40–52	Unknown, may be involved in signal transduction; forms complexes with CD53, CD81, CD82, and MHC class II		Transmembrane 4
CD38	Early B and T cells, activated T cells, germinal center B cells, plasma cells	45	NAD glycohydrolase, augments B cell proliferation	T10	
CD39	Activated B cells, activated NK cells, macrophages, dendritic cells	78	Unknown, may mediate adhesion of B cells		
CD40	B cells, macrophages, dendritic cells, basal epithelial cells	48	Binds CD154 (CD40L); receptor for co-stimulatory signal for B cells, promotes growth, differentiation, and isotype switching of B cells, and cytokine production by macrophages and dendritic cells		TNF receptor
CD41	Platelets, megakaryocytes	Dimer: GPIIba: 125 GPIIbb: 22	αIIb integrin, associates with CD61 to form GPIIb, binds fibrinogen, fibronectin, von Willebrand factor, and thrombospondin	GPIIb	Integrin α
CD42a,b,c,d	Platelets, megakaryocytes	a: 23 b: 135, 23 c: 22 d: 85	Binds von Willebrand factor, thrombin; essential for platelet adhesion at sites of injury	a: GPIX b: GPIbα c: GPIbβ d: GPV	Leucine-rich repeat
CD43	Leukocytes, except resting B cells	115–135 (neutrophils) 95–115 (T cells)	Has extended structure, approx. 45 nm long and may be anti-adhesive	Leukosialin, sialophorin	Mucin
CD44	Leukocytes, erythrocytes	80–95	Binds hyaluronic acid, mediates adhesion of leukocytes	Hermes antigen, Pgp-1	Link protein

CD antigen	Cellular expression	Molecular weight (kDa)	Functions	Other names	Family relationships
CD45	All hematopoietic cells	180–240 (multiple isoforms)	Tyrosine phosphatase, augments signaling through antigen receptor of B and T cells, multiple isoforms result from alternative splicing (see below)	Leukocyte common antigen (LCA), T200, B220	Fibronectin type III
CD45RO	T-cell subsets, B-cell subsets, monocytes, macrophages	180	Isoform of CD45 containing none of the A, B, and C exons		Fibronectin type II
CD45RA	B cells, T-cell subsets (naive T cells), monocytes	205–220	Isoforms of CD45 containing the A exon		Fibronectin type II
CD45RB	T-cell subsets, B cells, monocytes, macrophages, granulocytes	190–220	Isoforms of CD45 containing the B exon	T200	Fibronectin type II
CD46	Hematopoietic and non-hematopoietic nucleated cells	56/66 (splice variants)	Membrane co-factor protein, binds to C3b and C4b to permit their degradation by Factor I	MCP	CCP
CD47	All cells	47–52	Adhesion molecule, thrombospondin receptor	IAP, MER6, OA3	Immunoglobulin superfamily
CD48	Leukocytes	40–47	Putative ligand for CD244	Blast-1	Immunoglobulin
CD49a	Activated T cells, monocytes, neuronal cells, smooth muscle	200	$\alpha 1$ integrin, associates with CD29, binds collagen, laminin-1	VLA-1	Integrin α
CD49b	B cells, monocytes, platelets, megakaryocytes, neuronal, epithelial and endothelial cells, osteoclasts	160	$\alpha 2$ integrin, associates with CD29, binds collagen, laminin	VLA-2, platelet GPIa	Integrin α
CD49c	B cells, many adherent cells	125, 30	$\alpha 3$ integrin, associates with CD29, binds laminin-5, fibronectin, collagen, entactin, invasin	VLA-3	Integrin α
CD49d	Broad distribution includes B cells, thymocytes, monocytes, granulocytes, dendritic cells	150	$\alpha 4$ integrin, associates with CD29, binds fibronectin, MAdCAM-1, VCAM-1	VLA-4	Integrin α
CD49e	Broad distribution includes memory T cells, monocytes, platelets	135, 25	$\alpha 5$ integrin, associates with CD29, binds fibronectin, invasin	VLA-5	Integrin α
CD49f	T lymphocytes, monocytes, platelets, megakaryocytes, trophoblasts	125, 25	$\alpha 6$ integrin, associates with CD29, binds laminin, invasin, merosine	VLA-6	Integrin α
CD50	Thymocytes, T cells, B cells, monocytes, granulocytes	130	Binds integrin CD11a/CD18	ICAM-3	Immunoglobulin
CD51	Platelets, megakaryocytes	125, 24	αV integrin, associates with CD61, binds vitronectin, von Willebrand factor, fibrinogen, and thrombospondin; may be receptor for apoptotic cells	Vitronectin receptor	Integrin α
CD52	Thymocytes, T cells, B cells (not plasma cells), monocytes, granulocytes, spermatozoa	25	Unknown, target for antibodies used therapeutically to deplete T cells from bone marrow	CAMPATH-1, HE5	
CD53	Leukocytes	35–42	Unknown	MRC OX44	Transmembrane 4
CD54	Hematopoietic and non-hematopoietic cells	75–115	Intercellular adhesion molecule (ICAM)-1 binds CD11a/CD18 integrin (LFA-1) and CD11b/CD18 integrin (Mac-1), receptor for rhinovirus	ICAM-1	Immunoglobulin
CD55	Hematopoietic and non-hematopoietic cells	60–70	Decay accelerating factor (DAF), binds C3b, disassembles C3/C5 convertase	DAF	CCP
CD56	NK cells	135–220	Isoform of neural cell-adhesion molecule (NCAM), adhesion molecule	NKH-1	Immunoglobulin
CD57	NK cells, subsets of T cells, B cells, and monocytes		Oligosaccharide, found on many cell-surface glycoproteins	HNK-1, Leu-7	

CD antigen	Cellular expression	Molecular weight (kDa)	Functions	Other names	Family relationships
CD58	Hematopoietic and non-hematopoietic cells	55–70	Leukocyte function-associated antigen-3 (LFA-3), binds CD2, adhesion molecule	LFA-3	Immunoglobulin
CD59	Hematopoietic and non-hematopoietic cells	19	Binds complement components C8 and C9, blocks assembly of membrane-attack complex	Protectin, Mac inhibitor	Ly-6
CD60a			Disialyl ganglioside D3 (GD3)		Carbohydrate structures
CD60b			9-*O*-acetyl-GD3		Carbohydrate structures
CD60c			7-*O*-acetyl-GD3		Carbohydrate structures
CD61	Platelets, megakaryocytes, macrophages	110	Intergrin β3 subunit, associates with CD41 (GPIIb/IIIa) or CD51 (vitronectin receptor)		Integrin β
CD62E	Endothelium	140	Endothelium leukocyte adhesion molecule (ELAM), binds sialyl-Lewis^x, mediates rolling interaction of neutrophils on endothelium	ELAM-1, E-selectin	C-type lectin, EGF, and CCP
CD62L	B cells, T cells, monocytes, NK cells	150	Leukocyte adhesion molecule (LAM), binds CD34, GlyCAM, mediates rolling interactions with endothelium	LAM-1, L-selectin, LECAM-1	C-type lectin, EGF, and CCP
CD62P	Platelets, megakaryocytes, endothelium	140	Adhesion molecule, binds CD162 (PSGL-1), mediates interaction of platelets with endothelial cells, monocytes and rolling leukocytes on endothelium	P-selectin, PADGEM	C-type lectin, EGF, and CCP
CD63	Activated platelets, monocytes, macrophages	53	Unknown, is lysosomal membrane protein translocated to cell surface after activation	Platelet activation antigen	Transmembrane 4
CD64	Monocytes, macrophages	72	High-affinity receptor for IgG, binds IgG3>IgG1>IgG4>>>IgG2, mediates phagocytosis, antigen capture, ADCC	FcγRI	Immunoglobulin
CD65	Myeloid cells		Oligosaccharide component of a ceramide dodecasaccharide		
CD66a	Neutrophils	160–180	Unknown, member of carcino-embryonic antigen (CEA) family (see below)	Biliary glyco-protein-1 (BGP-1)	Immunoglobulin
CD66b	Granulocytes	95–100	Unknown, member of carcino-embryonic antigen (CEA) family	Previously CD67	Immunoglobulin
CD66c	Neutrophils, colon carcinoma	90	Unknown, member of carcino-embryonic antigen (CEA) family	Nonspecific cross-reacting antigen (NCA)	Immunoglobulin
CD66d	Neutrophils	30	Unknown, member of carcino-embryonic antigen (CEA) family		Immunoglobulin
CD66e	Adult colon epithelium, colon carcinoma	180–200	Unknown, member of carcino-embryonic antigen (CEA) family	Carcino-embryonic antigen (CEA)	Immunoglobulin
CD66f	Unknown		Unknown, member of carcino-embryonic antigen (CEA) family	Pregnancy specific glycoprotein	Immunoglobulin
CD68	Monocytes, macrophages, neutrophils, basophils, large lymphocytes	110	Unknown	Macrosialin	Mucin

CD antigen	Cellular expression	Molecular weight (kDa)	Functions	Other names	Family relationships
CD69	Activated T and B cells, activated macrophages and NK cells	28, 32 homodimer	Unknown, early activation antigen	Activation inducer molecule (AIM)	C-type lectin
CD70	Activated T and B cells, and macrophages	75, 95, 170	Ligand for CD27, may function in co-stimulation of B and T cells	Ki-24	TNF
CD71	All proliferating cells, hence activated leukocytes	95 homodimer	Transferrin receptor	T9	
CD72	B cells (not plasma cells)	42 homodimer	Unknown	Lyb-2	C-type lectin
CD73	B-cell subsets, T-cell subsets	69	Ecto-5′-nucleotidase, dephosphorylates nucleotides to allow nucleoside uptake		
CD74	B cells, macrophages, monocytes, MHC class II positive cells	33, 35, 41, 43 (alternative initiation and splicing)	MHC class II-associated invariant chain	Ii, Iγ	
CD75	Mature B cells, T-cells subsets		Lactosamines, ligand for CD22, mediates B-cell–B-cell adhesion		
CD75s			α-2,6-sialylated lactosamines		Carbohydrate structures
CD77	Germinal center B cells		Neutral glycosphingolipid (Galα1→4Galβ1→4Galcβ1 →ceramide), binds Shiga toxin, cross-linking induces apoptosis	Globotriaocyl-ceramide (Gb3) Pk blood group	
CD79α,β	B cells	α: 40–45 β: 37	Components of B-cell antigen receptor analogous to CD3, required for cell-surface expression and signal transduction	Igα, Igβ	Immunoglobulin
CD80	B-cell subset	60	Co-stimulator, ligand for CD28 and CTLA-4	B7 (now B7.1), BB1	Immunoglobulin
CD81	Lymphocytes	26	Associates with CD19, CD21 to form B cell co-receptor	Target of anti-proliferative antibody (TAPA-1)	Transmembrane 4
CD82	Leukocytes	50–53	Unknown	R2	Transmembrane 4
CD83	Dendritic cells, B cells, Langerhans' cells	43	Unknown	HB15	Immunoglobulin
CDw84	Monocytes, platelets, circulating B cells	73	Unknown	GR6	Immunoglobulin
CD85	Dendritic cells		ILT/LIR family	GR4	Immunoglobulin superfamily
CD86	Monocytes, activated B cells, dendritic cells	80	Ligand for CD28 and CTLA4	B7.2	Immunoglobulin
CD87	Granulocytes, monocytes, macrophages, T cells, NK cells, wide variety of nonhematopoietic cell types	35–59	Receptor for urokinase plasminogen activator	uPAR	Ly-6
CD88	Polymorphonuclear leukocytes, macrophages, mast cells	43	Receptor for complement component C5a	C5aR	G protein-coupled receptor
CD89	Monocytes, macrophages, granulocytes, neutrophils, B-cell subsets, T-cell subsets	50–70	IgA receptor	FcαR	Immunoglobulin
CD90	CD34+ prothymocytes (human), thymocytes, T cells (mouse)	18	Unknown	Thy-1	Immunoglobulin
CD91	Monocytes, many non-hematopoietic cells	515, 85	α2-macroglobulin receptor		EGF, LDL receptor

CD antigen	Cellular expression	Molecular weight (kDa)	Functions	Other names	Family relationships
CD92	Neutrophils, monocytes, platelets, endothelium	70	Unknown	GR9	
CD93	Neutrophils, monocytes, endothelium	120	Unknown	GR11	
CD94	T-cell subsets, NK cells	43	Unknown	KP43	C-type lectin
CD95	Wide variety of cell lines, *in vivo* distribution uncertain	45	Binds TNF-like Fas ligand, induces apoptosis	Apo-1, Fas	TNF receptor
CD96	Activated T cells, NK cells	160	Unknown	T-cell activation increased late expression (TACTILE)	Immunoglobulin
CD97	Activated B and T cells, monocytes, granulocytes	75–85	Binds CD55	GR1	EGF, G protein-coupled receptor
CD98	T cells, B cells, natural killer cells, granulocytes, all human cell lines	80, 45 heterodimer	May be amino acid transporter	4F2, FRP-1	
CD99	Peripheral blood lymphocytes, thymocytes	32	Unknown	MIC2, E2	
CD100	Hematopoietic cells	150 homodimer	Unknown	GR3	Semaphorin
CD101	Monocytes, granulocytes, dendritic cells, activated T cells	120 homodimer	Unknown	BPC#4	Immunoglobulin
CD102	Resting lymphocytes, monocytes, vascular endothelium cells (strongest)	55–65	Binds CD11a/CD18 (LFA-1) but not CD11b/CD18 (Mac-1)	ICAM-2	Immunoglobulin
CD103	Intraepithelial lymphocytes, 2–6% peripheral blood lymphocytes	150, 25	αE integrin	HML-1, α6, αE integrin	Integrin α
CD104	CD4$^-$ CD8$^-$ thymocytes, neuronal, epithelial, and some endothelial cells, Schwann cells, trophoblasts	220	Integrin β4 associates with CD49f, binds laminins	β4 integrin	Integrin β
CD105	Endothelial cells, activated monocytes and macrophages, bone marrow cell subsets	90 homodimer	Binds TGF-β	Endoglin	
CD106	Endothelial cells	100–110	Adhesion molecule, ligand for VLA-4 ($\alpha_4\beta_1$ integrin)	VCAM-1	Immunoglobulin
CD107a	Activated platelets, activated T cells, activated neutrophils, activated endothelium	110	Unknown, is lysosomal membrane protein translocated to the cell surface after activation	Lysosomal associated membrane protein-1 (LAMP-1)	
CD107b	Activated platelets, activated T cells, activated neutrophils, activated endothelium	120	Unknown, is lysosomal membrane protein translocated to the cell surface after activation	LAMP-2	
CD108	Erythrocytes, circulating lymphocytes, lymphoblasts	80	Unknown	GR2, John Milton-Hagen blood group antigen	
CD109	Activated T cells, activated platelets, vascular endothelium	170	Unknown	Platelet activation factor, GR56	
CD110	Platelets		MPL, TPO R		
CD111	Myeloid cells		PPR1/Nectin1		
CD112	Myeloid cells		PRR2		
CD114	Granulocytes, monocytes	150	Granulocytes colony-stimulating factor (G-CSF) receptor		Immunoglobulin, fibronectin type III

CD antigen	Cellular expression	Molecular weight (kDa)	Functions	Other names	Family relationships
CD115	Monocytes, macrophages	150	Macrophage colony-stimulating factor (M-CSF) receptor	M-CSFR, c-fms	Immunoglobulin, tyrosine kinase
CD116	Monocytes, neutrophils, eosinophils, endothelium	70–85	Granulocyte-macrophage colony-stimulating factor (GM-CSF) receptor α chain	GM-CSFRα	Cytokine receptor, fibronectin type III
CD117	Hematopoietic progenitors	145	Stem-cell factor (SCF) receptor	c-Kit	Immunoglobulin, tyrosine kinase
CD118	Broad cellular expression		Interferon-α, β receptor	IFN-α, βR	
CD119	Macrophages, monocytes, B cells, endothelium	90–100	Interferon-γ receptor	IFN-γR	Fibronectin type III
CD120a	Hematopoietic and nonhematopoietic cells, highest on epithelial cells	50–60	TNF receptor, binds both TNF-α and TNF-β	TNFR-I	TNF receptor
CD120b	Hematopoietic and nonhematopoietic cells, highest on myeloid cells	75–85	TNF receptor, binds both TNF-α and TNF-β	TNFR-II	TNF receptor
CD121a	Thymocytes, T cells	80	Type I interleukin-1 receptor, binds IL-1α and IL-1β	IL-1R type I	Immunoglobulin
CDw121b	B cells, macrophages, monocytes	60–70	Type II interleukin-1 receptor, binds IL-1α and IL-1β	IL-1R type II	Immunoglobulin
CD122	NK cells, resting T-cell subsets, some B-cell lines	75	IL-2 receptor β chain	IL-2Rβ	Cytokine receptor, fibronectin type III
CD123	Bone marrow stem cells, granulocytes, monocytes, megakaryocytes	70	IL-3 receptor α chain	IL-3Rα	Cytokine receptor, fibronectin type III
CD124	Mature B and T cells, hematopoietic precursor cells	130–150	IL-4 receptor	IL-4R	Cytokine receptor, fibronectin type III
CD125	Eosinophils, basophils, activated B cells	55–60	IL-5 receptor	IL-5R	Cytokine receptor, fibronectin type III
CD126	Activated B cells and plasma cells (strong), most leukocytes (weak)	80	IL-6 receptor α subunit	IL-6Rα	Immunoglobulin, cytokine receptor, fibronectin type III
CD127	Bone marrow lymphoid precursors, pro-B cells, mature T cells, monocytes	68–79, possibly forms homodimers	IL-7 receptor	IL-7R	Fibronectin type III
CDw128	Neutrophils, basophils, T-cell subsets	58–67	IL-8 receptor	IL-8R	G protein-coupled receptor
CD129	Not yet assigned				
CD130	Most cell types, strong on activated B cells and plasma cells	130	Common subunit of IL-6, IL-11, oncostatin-M (OSM) and leukemia inhibitory factor (LIF) receptors	IL-6Rβ, IL-11Rβ, OSMRβ, LIFRβ, IFRβ	Immunoglobulin, cytokine receptor, fibronectin type III
CDw131	Myeloid progenitors, granulocytes	140	Common β subunit of IL-3, IL-5, and GM-CSF receptors	IL-3Rβ, IL-5Rβ, GM-CSFRβ	Cytokine receptor, fibronectin type III
CD132	B cells, T cells, NK cells, mast cells, neutrophils	64	IL-2 receptor γ chain, common subunit of IL-2, IL-4, IL-7, IL-9, and IL-15 receptors		Cytokine receptor
CD133	Stem/progenitor cells		AC133		
CD134	Activated T cells	50	May act as adhesion molecule costimulator	OX40	TNF receptor
CD135	Multipotential precursors, myelomonocytic and B-cell progenitors	130, 155	Growth factor receptor	FLK2, STK-1	Immunoglobulin, tyrosine kinase
CDw136	Monocytes, epithelial cells, central and peripheral nervous system	180	Chemotaxis, phagocytosis, cell growth, and differentiation	MSP-R, RON	Tyrosine kinase

CD antigen	Cellular expression	Molecular weight (kDa)	Functions	Other names	Family relationships
CDw137	T and B lymphocytes, monocytes, some epithelial cells		Co-stimulator of T-cell proliferation	ILA (induced by lymphocyte activation), 4-1BB	TNF receptor
CD138	B cells		Heparan sulphate proteoglycan binds collagen type I	Syndecan-1	
CD139	B cells	209, 228	Unknown		
CD140a,b	Stromal cells, some endothelial cells	a: 180 b: 180	Platelet derived growth factor (PDGF) receptor α and β chains		
CD141	Vascular endothelial cells	105	Anticoagulant, binds thrombin, the complex then activates protein C	Thrombomodulin fetomodulin	C-type lectin, EGF
CD142	Epidermal keratinocytes, various epithelial cells, astrocytes, Schwann cells. Absent from cells in direct contact with plasma unless induced by inflammatory mediators	45–47	Major initiating factor of clotting. Binds Factor VIIa; this complex activates Factors VII, IX, and X	Tissue factor, thromboplastin	Fibronectin type III
CD143	Endothelial cells, except large blood vessels and kidney, epithelial cells of brush borders of kidney and small intestine, neuronal cells, activated macrophages and some T cells. Soluble form in plasma	170–180	Zn^{2+} metallopeptidase dipeptidyl peptidase, cleaves angiotensin I and bradykinin from precursor forms	Angiotensin converting enzyme (ACE)	
CD144	Endothelial cells	130	Organizes adherens junction in endothelial cells	Cadherin-5, VE-cadherin	Cadherin
CD145	Endothelial cells, some stromal cells	25, 90, 110	Unknown		
CD146	Endothelium	130	Potential adhesion molecule, localized at cell–cell junctions	MCAM, MUC18, S-ENDO	Immunoglobulin
CD147	Leukocytes, red blood cells, platelets, endothelial cells	55–65	Potential adhesion molecule	M6, neurothelin, EMMPRIN, basigin, OX-47	Immunoglobulin
CD148	Granulocytes, monocytes, dendritic cells, T cells, fibroblasts, nerve cells	240–260	Contact inhibition of cell growth	HPTPη	Fibronectin type III, protein tyro-sine phosphatase
CD150	Thymocytes, activated lymphocytes	75–95	Unknown	SLAM	Immunoglobulin
CD151	Platelets, megakaryocytes, epithelial cells, endothelial cells	32	Associates with $\beta1$ integrins	PETA-3, SFA-1	Transmembrane 4
CD152	Activated T cells	33	Receptor for B7.1 (CD80), B7.2 (CD86); negative regulator of T-cell activation	CTLA-4	Immunoglobulin
CD153	Activated T cells, activated macrophages, neutrophils, B cells	38–40	Ligand for CD30, may co-stimulate T cells	CD30L	TNF
CD154	Activated CD4 T cells	30 trimer	Ligand for CD40, inducer of B cell proliferation and activation	CD40L, TRAP, T-BAM, gp39	TNF receptor
CD155	Monocytes, macrophages, thymocytes, CNS neurons	80–90	Normal function unknown; receptor for poliovirus	Poliovirus receptor	Immunoglobulin
CD156a	Neutrophils, monocytes	69	Unknown, may be involved in integrin leukocyte extravasation	MS2, ADAM 8 (A disintegrin and metallo-protease)	
CD156b			TACE/ADAM17. Adhesion structures		
CD157	Granulocytes, monocytes, bone marrow stromal cells, vascular endothelial cells, follicular dendritic cells	42–45 (50 on monocytes)	ADP-ribosyl cyclase, cyclic ADP-ribose hydrolase	BST-1	
CD158	NK cells		KIR family		

CD antigen	Cellular expression	Molecular weight (kDa)	Functions	Other names	Family relationships
CD158a	NK-cell subsets	50 or 58	Inhibits NK cell cytotoxicity on binding MHC class I molecules	p50.1, p58.1	Immunoglobulin
CD158b	NK-cell subsets	50 or 58	Inhibits NK cell cytotoxicity on binding HLA-Cw3 and related alleles	p50.2, p58.2	Immunoglobulin
CD159a	NK cells		Binds CD94 to form NK receptor; inhibits NK cell cytotoxicity on binding MHC class I moolecules	NKG2A	
CD160	T cells			BY55	
CD161	NK cells, T cells	44	Regulates NK cytotoxicity	NKRP1	C-type lectin
CD162	Neutrophils, lymphocytes, monocytes	120 homodimer	Ligand for CD62P	PSGL-1	Mucin
CD162R	NK cells			PEN5	
CD163	Monocytes, macrophages	130	Unknown	M130	
CD164	Epithelial cells, monocytes, bone marrow stromal cells	80	Unknown	MUC-24 (multi-glycosylated protein 24)	Mucin
CD165	Thymocytes, thymic epithelial cells, CNS neurons, pancreatic islets, Bowman's capsule	37	Adhesion between thymocytes and thymic epithelium	Gp37, AD2	
CD166	Activated T cells, thymic epithelium, fibroblasts, neurons	100–105	Ligand for CD6, involved integrin neurite extension	ALCAM, BEN, DM-GRASP, SC-1	Immunoglobulin
CD167a	Normal and transformed epithelial cells	63, 64 dimer	Binds collagen	DDR1, trkE, cak, eddr1	Receptor tyrosine kinase, discoidin-related
CD168	Breast cancer cells	Five isoforms: 58, 60, 64, 70, 84	Adhesion molecule. Receptor for hyaluronic acid-mediated motility—mediated cell migration	RHAMM	
CD169	Subsets of macrophages	185	Adhesion molecule. Binds sialylated carbohydrates. May mediate macrophage binding to granulocytes and lymphocytes	Sialoadhesin	Immunoglobulin superfamily, sialoadhesin family
CD170	Neutrophils	67 homodimer	Adhesion molecule. Sialic acid-binding Ig-like lectin (Siglec). Cytoplasmic tail contains ITIM motifs	Siglec-5, OBBP2, CD33L2	Immunoglobulin superfamily, sialoadhesin family
CD171	Neurons, Schwann cells, lymphoid and myelomonocytic cells, B cells, CD4 T cells (not CD8 T cells)	200–220, exact MW varies with cell type	Adhesion molecule, binds CD9, CD24, CD56, also homophilic binding	L1, NCAM-L1	Immunoglobulin superfamily
CD172a		115–120	Adhesion molecule; the transmembrane protein is a substrate of activated receptor tyrosine kinases (RTKs) and binds to SH2 domains	SIRP, SHPS1, MYD-1, SIRP-α-1, protein tyrosine phosphatase, non-receptor type substrate 1 (PTPNS1)	Immunoglobulin superfamily
CD173	All cells		Blood group H type 2. Carbohydrate moiety		
CD174	All cells		Lewis y Blood group. Carbohydrate moiety		
CD175	All cells		Tn Blood group. Carbohydrate moiety		
CD175s	All cells		Sialyl-Tn Blood group. Carbohydrate moiety		
CD176	All cells		TF Blood group. Carbohydrate moiety		

CD antigen	Cellular expression	Molecular weight (kDa)	Functions	Other names	Family relationships
CD177	Myeloid cells	56–64	NB1 is a GPI-linked neutrophil-specific antigen, found on only a subpopulation of neutrophils present in NB1-positive adults (97% of healthy donors) NB1 is first expressed at the myelocyte stage of myeloid differentiation	NB1	
CD178	Activated T cells	38–42	Fas ligand; binds to Fas to induce apoptosis	FasL	TNF superfamily
CD179a	Early B cells	16–18	Immunoglobulin iota chain associates noncovalently with CD179b to form a surrogate light chain which is a component of the pre-B-cell receptor that plays a critical role in early B-cell differentiation	VpreB, IGVPB, IGι	Immunoglobulin superfamily
CD179b	B cells	22	Immunoglobulin λ-like polypeptide 1 associates noncovalently with CD179a to form a surrogate light chain that is selectively expressed at the early stages of B-cell development. Mutations in the CD179b gene have been shown to result in impairment of B-cell development and agammaglobulinemia in humans	IGLL1, λ5 (IGL5), IGVPB, 14.	Immunoglobulin superfamily
CD180	B cells	95–105	Type 1 membrane protein consisting of extracellular leucine-rich repeats (LRR). Is associated with a molecule called MD-1 and forms the cell-surface receptor complex, RP105/MD-1, which by working in concert with TLR4, controls B-cell recognition and signaling of lipopolysaccharide (LPS)	LY64, RP105	Toll-like receptors (TLR)
CD183	Particularly on malignant B cells from chronic lymphoproliferative disorders	46–52	CXC chemokine receptor involved in chemotaxis of malignant B lymphocytes. Binds INP10 and MIG[3]	CXCR3, G protein-coupled receptor 9 (GPR 9)	Chemokine receptors, G protein-coupled receptor superfamily
CD184	Preferentially expressed on the more immature CD34+ haematopoietic stem cells	46–52	Binding to SDF-1 (LESTR/fusin); acts as a cofactor for fusion and entry of T-cell line; trophic strains of HIV-1	CXCR4, NPY3R, LESTR, fusin, HM89	Chemokine receptors, G protein-coupled receptor superfamily
CD195	Promyelocytic cells	40	Receptor for a CC type chemokine. Binds to MIP-1α, MIP-1β and RANTES. May play a role in the control of granulocytic lineage proliferation or differentiation. Acts as co-receptor with CD4 for primary macrophage-tropic isolates of HIV-1	CMKBR5, CCR5, CKR-5, CC-CKR-5, CKR5	Chemokine receptors, G protein-coupled receptor superfamily
CDw197	Activated B and T lymphocytes, strongly upregulated in B cells infected with EBV and T cells infected with HHV6 or 7	46–52	Receptor for the MIP-3β chemokine; probable mediator of EBV effects on B lymphocytes or of normal lymphocyte functions	CCR7. EBI1 (Epstein–Barr virus induced gene 1), CMKBR7, BLR2	Chemokine receptors, G protein-coupled receptor superfamily
CD200	Normal brain and B-cell lines	41 (rat thymocytes) 47 (rat brain)	Antigen identified by MoAb MRC OX-2. Nonlineage molecules. Function unknown	MOX-2, MOX-1	Immunoglobulin superfamily
CD201	Endothelial cells	49	Endothelial cell-surface receptor (EPCR) that is capable of high-affinity binding of protein C and activated protein C. It is downregulated by exposure of endothelium to tumor necrosis factor	EPCR	CD1 major histocompatibility complex family
CD202b	Endothelial cells	140	Receptor tyrosine kinase, binds angiopoietin-1; important in angiogenesis, particularly for vascular network formation in endothelial cells. Defects in TEK are associated with inherited venous malformations; the TEK signaling pathway appears to be critical for endothelial cell–smooth muscle cell communication in venous morphogenesis	VMCM, TEK (tyrosine kinase, endothelial), TIE2 (tyrosine kinase with Ig and EGF homology domains), VMCM1	Immunoglobulin superfamily, tyrosine kinase

CD antigen	Cellular expression	Molecular weight (kDa)	Functions	Other names	Family relationships
CD203c	Myeloid cells (uterus, basophils, and mast cells)	101	Belongs to a series of ectoenzymes that are involved in hydrolysis of extracellular nucleotides. They catalyze the cleavage of phosphodiester and phosphosulfate bonds of a variety of molecules, including deoxynucleotides, NAD, and nucleotide sugars	NPP3, B10, PDNP3, PD-Iβ, gp130RB13-6	Type II transmembrane proteins, Ecto-nucleotide pyrophosphatase/phosphodiesterase (E-NPP) family
CD204	Myeloid cells	220	Mediate the binding, internalization, and processing of a wide range of negatively charged macromolecules. Implicated in the pathologic deposition of cholesterol in arterial walls during atherogenesis	Macrophage scavenger R (MSR1)	Scavenger receptor family, collagen-like
CD205	Dendritic cells	205	Lymphocyte antigen 75; putative antigen-uptake receptor on dendritic cells	LY75, DEC-205, GP200-MR6	Type I transmembrane protein
CD206	Macrophages, endothelial cells	175–190	Type I membrane glycoprotein; only known example of a C-type lectin that contains multiple C-type CRDs (carbohydrate-recognition domains); it binds high-mannose structures on the surface of potentially pathogenic viruses, bacteria, and fungi	Macrophage mannose receptor (MMR), MRC1	C-type lectin superfamily
CD207	Langerhans' cells	40	Type II transmembrane protein; Langerhans' cell specific C-type lectin; potent inducer of membrane superimposition and zippering leading to BG (Birbeck granules) formation	Langerin	C-type lectin superfamily
CD208	Interdigitating dendritic cells in lymphoid organs	70–90	Homologous to CD68, DC-LAMP is a lysosomal protein involved in remodeling of specialized antigen-processing compartments and in MHC class II-restricted antigen presentation. Upregulated in mature DCs induced by CD40L, TNF-α and LPS.	D lysosome-associated membrane protein, DC-LAMP	Major histocompatibility complex family
CD209	Dendritic cells	44	C-type lectin; binds ICAM3 and HIV-1 envelope glycoprotein gp120 enables T-cell receptor engagement by stabilization of the DC/T-cell contact zone, promotes efficient infection in *trans* cells that express CD4 and chemokine receptors; type II transmembrane protein	DC-SIGN (dendritic cell-specific ICAM3-grabbing non-integrin)	C-type lectin superfamily
CDw210	B cells, T helper cells, and cells of the monocyte/macrophage lineage	90–110	Interleukin 10 receptor α and β	IL-10Rα, IL-10RA, HIL-10R, IL-10Rβ, IL-10RB, CRF2-4, CRFB4	Class II cytokine receptor family
CD212	Activated CD4, CD8, and NK cells	130	IL-12 receptor β chain; a type I transmembrane protein involved in IL-12 signal transduction.	IL-12R IL-12RB	Haemopoietin cytokine receptor superfamily
CD213a1	B cells, monocytes, fibroblasts, endothelial cells	60–70	Receptor which binds Il-13 with a low affinity; together with IL-4Rα can form a functional receptor for IL-13, also serves as an alternate accessory protein to the common cytokine receptor gamma chain for IL-4 signaling	IL-13Rα 1, NR4, IL-13Ra	Haemopoietic cytokine receptor superfamily
CD213a2	B cells, monocytes, fibroblasts, endothelial cells		IL-13 receptor which binds as a monomer with high affinity to interleukin-13 (IL-13), but not to IL-4; human cells expressing IL-13RA2 show specific IL-13 binding with high affinity	IL-13Rα 2, IL-13BP	Haemopoietic cytokine receptor superfamily
CDw217	Activated memory T cells	120	Interleukin 17 receptor homodimer	IL-17R, CTLA-8	Chemokine/cytokine receptors

CD antigen	Cellular expression	Molecular weight (kDa)	Functions	Other names	Family relationships
CD220	Nonlineage molecules	α: 130 β: 95	Insulin receptor; integral transmembrane glycoprotein comprised of two α and two β subunits; this receptor binds insulin and has a tyrosine-protein kinase activity—autophosphorylation activates the kinase activity	Insulin receptor	Insulin receptor family of tyrosine-protein kinases, EGFR family
CD221	Nonlineage molecules	α: 135 β: 90	Insulin-like growth factor I receptor binds insulin-like growth factor with a high affinity. It has tyrosine kinase activity and plays a critical role in transformation events. Cleavage of the precursor generates α and β subunits.	IGF1R, JTK13	Insulin receptor family of tyrosine-protein kinases, EGFR family
CD222	Nonlineage molecules	250	Ubiquitously expressed multifunctional type I transmembrane protein. Its main functions include internalization of IGF-II, internalization or sorting of lysosomal enzymes and other M6P-containing proteins	IGF2R, CIMPR, CI-MPR, IGF2R, M6P-R (Mannose-6-phosphate receptor)	Mammalian lectins
CD223	Activated T and NK cells	70	Involved in lymphocyte activation; binds to HLA class-II antigens; role in down-regulating antigen specific response; close relationship of LAG3 to CD4	Lymphocyte-activation gene 3 LAG-3	Immunoglobulin superfamily
CD224	Nonlineage molecules	62 (unprocessed precursor)	Predominantly a membrane-bound enzyme; plays a key role in the γ-glutamyl cycle, a pathway for the synthesis and degradation of glutathione. This enzyme consists of two polypeptide chains, which are synthesized in precursor form from a single polypeptide	γ-glutamyl transferase, GGT1, D22S672 D22S732	γ-glutamyl transferase protein family
CD225	Leukocytes and endothelial cells	16–17	Interferon-induced transmembrane protein 1 is implicated in the control of cell growth. It is a component of a multimeric complex involved in the transduction of antiproliferative and homotypic adhesion signals	Leu 13, IFITM1, IFI17	IFN-induced transmembrane proteins
CD226	NK cells, platelets, monocytes, and a subset of T cells	65	Adhesion glycoprotein; mediates cellular adhesion to other cells bearing an unidentified ligand and cross-linking CD226 with antibodies causes cellular activation	DNAM-1 (PTA1), DNAX, TLiSA1	Immunoglobulin superfamily
CD227	Human epithelial tumors, such as breast cancer	122 (nonglycosylated)	Epithelial mucin containing a variable number of repeats with a length of twenty amino acids, resulting in many different alleles. Direct or indirect interaction with actin cytoskeleton.	PUM (peanut-reactive urinary mucin), MUC.1, mucin 1	Mucin
CD228	Predominantly in human melanomas	97	Tumor-associated antigen (melanoma) identified by monoclonal antibodies 133.2 and 96.5, involved in cellular iron uptake.	Melanotransferrin, P97	Transferrin superfamily
CD229	Lymphocytes	90–120	May participate in adhesion reactions between T lymphocytes and accessory cells by homophilic interaction	Ly9	Immunoglobulin superfamily (CD2 subfamily)
CD230	Expressed both in normal and infected cells	27–30	The function of PRP is not known. It is encoded in the host genome found in high quantity in the brain of humans and animals infected with neurodegenerative diseases known as transmissible spongiform encephalopathies or prion diseases (Creutzfeld–Jakob disease, Gerstmann–Strausler–Scheinker syndrome, fatal familial insomnia)	CJD, PRIP, prion protein (p27-30)	Prion family

CD antigen	Cellular expression	Molecular weight (kDa)	Functions	Other names	Family relationships
CD231	T-cell acute lymphoblastic leukemia, neuroblastoma cells and normal brain neuron	150	The function of CD231 is currently unknown. It is cell-surface glycoprotein which is a specific marker for T-cell acute lymphoblastic leukemia. Also found on neuroblastomas	TALLA-1, TM4SF2, A15, MXS1, CCG-B7	Transmembrane 4 superfamily (TM4SF also known as tetraspanins)
CD232	Nonlineage molecules	200	Receptor for an immunologically active semaphorin (virus-encoded semaphorin protein receptor)	VESPR, PLXN, PLXN-C1	Plexin family
CD233	Erythroid cells	93	Band 3 is the major integral glycoprotein of the erythrocyte membrane. It has two functional domains. Its integral domain mediates a 1:1 exchange of inorganic anions across the membrane, whereas its cytoplasmic domain provides binding sites for cytoskeletal proteins, glycolytic enzymes, and hemoglobin. Multifunctional transport protein	SLC4A1, Diego blood group, D1, AE1, EPB3	Anion exchanger family
CD234	Erythroid cells and nonerythroid cells	35	Fy-glycoprotein; Duffy blood group antigen; nonspecific receptor for many chemokines such as IL-8, GRO, RANTES, MCP-1 and TARC. It is also the receptor for the human malaria parasites *Plasmodium vivax* and *Plasmodium knowlesi* plays a role in inflammation and in malaria infection	GPD, CCBP1, DARC (duffy antigen/receptor for chemokines)	Family 1 of G protein-coupled receptors, chemokine receptors superfamily
CD235a	Erythroid cells	31	Major carbohydrate rich sialoglycoprotein of human erythrocyte membrane which bears the antigenic determinants for the MN and Ss blood groups. The N-terminal glycosylated segment, which lies outside the erythrocyte membrane, has MN blood group receptors and also binds influenza virus	Glycophorin A, GPA, MNS	Glycophorin A family
CD235b	Erythroid cells	GYPD is smaller than GYPC (24 kD vs 32 kD)	This protein is a minor sialoglycoprotein in human erythrocyte membranes. Along with GYPA, GYPB is responsible for the MNS blood group system. The Ss blood group antigens are located on glycophorin B	Glycophorin B, MNS, GPB	Glycophorin A family
CD236	Erythroid cells	24	Glycophorin C (GPC) and glycophorin D (GPD) are closely related sialoglycoproteins in the human red blood cell (RBC) membrane. GPD is a ubiquitous shortened isoform of GPC, produced by alternative splicing of the same gene. The Webb and Duch antigens, also known as glycophorin D, result from single point mutations of the glycophorin C gene	Glycophorin D, GPD, GYPD	Type III membrane proteins
CD236R	Erythroid cells	32	Glycophorin C (GPC) is associated with the Gerbich (Ge) blood group deficiency. It is a minor red cell-membrane component, representing about 4% of the membrane sialoglycoproteins, but shows very little homology with the major red cell-membrane glycophorins A and B. It plays an important role in regulating the mechanical stability of red cells and is a putative receptor for the merozoites of *Plasmodium falciparum*	Glycophorin C, GYPC, GPC	Type III membrane proteins
CD238	Erythroid cells	93	KELL blood group antigen; homology to a family of zinc metalloglycoproteins with neutral endopeptidase activity, type II transmembrane glycoprotein	KELL	Belongs to peptidase family m13 (zinc metalloprotease); also known as the neprilysin subfamily

CD antigen	Cellular expression	Molecular weight (kDa)	Functions	Other names	Family relationships
CD239	Erythroid cells	78	A type I membrane protein. The human F8/G253 antigen, B-CAM, is a cell surface glycoprotein that is expressed with restricted distribution pattern in normal fetal and adult tissues, and is upregulated following malignant transformation in some cell types. Its overall structure is similar to that of the human tumor marker MUC 18 and the chicken neural adhesion molecule SC1	B-CAM (B-cell adhesion molecule), LU, Lutheran blood group	Immunoglobulin superfamily
CD240CE	Erythroid cells	45.5	Rhesus blood group, CcEe antigens. May be part of an oligomeric complex which is likely to have a transport or channel function in the erythrocyte membrane. It is highly hydrophobic and deeply buried within the phospholipid bilayer	RHCE, RH30A, RHPI, Rh4	Rh family
CD240D	Erythroid cells	45.5 (product—30)	Rhesus blood group, D antigen. May be part of an oligomeric complex which is likely to have a transport or channel function in the erythrocyte membrane. Absent in the Caucasian RHD-negative phenotype	RhD, Rh4, RhPI, RhII, Rh30D	Rh family
CD241	Erythroid cells	50	Rhesus blood group-associated glycoprotein RH50, component of the RH antigen multisubunit complex; required for transport and assembly of the Rh membrane complex to the red blood cell surface. highly homologous to RH, 30kD components. Defects in RhAg are a cause of a form of chronic hemolytic anemia associated with stomatocytosis, and spherocytosis, reduced osmotic fragility, and increased cation permeability	RhAg, RH50A	Rh family
CD242	Erythroid cells	42	Intercellular adhesion molecule 4, Landsteiner-Wiener blood group. LW molecules may contribute to the vaso-occlusive events associated with episodes of acute pain in sickle cell disease	ICAM-4, LW	Immunoglobulin superfamily, intercellular adhesion molecules (ICAMs)
CD243	Stem/progenitor cells	170	Multidrug resistance protein 1 (P-glycoprotein). P-gp has been shown to utilise ATP to pump hydrophobic drugs out of cells, thus increasing their intracellular concentration and hence their toxicity. The MDR 1 gene is amplified in multidrug resistant cell lines	MDR-1, p-170	ABC superfamily of ATP-binding transport proteins
CD244	NK cells	66	2B4 is a cell-surface glycoprotein related to CD2 and implicated in the regulation of natural killer and T-lymphocyte function. It appears that the primary function of 2B4 is to modulate other receptor–ligand interactions to enhance leukocyte activation	2B4, NK cell activation inducing ligand (NAIL)	Immunoglobulin superfamily
CD245	T cells	220–240	Cyclin E/Cdk2 interacting protein p220. NPAT is involved in a key S phase event and links cyclical cyclin E/Cdk2 kinase activity to replication-dependent histone gene transcription. NPAT gene may be essential for cell maintenance and may be a member of the housekeeping genes	NPAT	

CD antigen	Cellular expression	Molecular weight (kDa)	Functions	Other names	Family relationships
CD246	Expressed in the small intestine, testis, and brain but not in normal lymphoid cells	177 kDa; after glycosylation, produces a 200 kDa mature glyco-protein	Anaplastic (CD30+ large cell) lymphoma kinase; plays an important role in brain development, involved in anaplastic nodal non Hodgkin lymphoma or Hodgkin's disease with translocation t(2;5)(p23;q35) or inv2(23;q35). Oncogenesis via the kinase function is activated by oligomerization of NPM1-ALK mediated by the NPM1 part	ALK	Insulin receptor family of tyrosine-protein kinases
CD247	T cells, NK cells	16	T-cell receptor ζ; has a probable role in assembly and expression of the TCR complex as well as signal transduction upon antigen triggering. TCRζ together with TCRα:β and γ.δ heterodimers and CD3-γ, -δ, and -ε, forms the TCR-CD3 complex. The ζ chain plays an important role in coupling antigen recognition to several intracellular signal-transduction pathways. Low expression of the anti-gen results in impaired immune response	ζ chain, CD3Z	Immunoglobulin superfamily

Compiled by Laura Herbert, Royal Free Hospital, London. Data based on CD designations made at the 7th Workshop on Human Leucocyte Differentiation Antigens, provided by Protein Reviews on the Web (www.ncbi.nlm.nih.gov/prow/).

Appendix III. Cytokines and their receptors.						
Family	**Cytokine (alternative names)**	**Size (no. of amino acids) and form**	**Receptors (c denotes common subunit)**	**Producer cells**	**Actions**	**Effect of cytokine or receptor knock-out (where known)**
Hematopoietins (four-helix bundles)	Epo (erythropoietin)	165, monomer*	EpoR	Kidney cells hepatocytes	Stimulates erythroid progenitors	Epo or EpoR: embryonic lethal
	IL-2 (T-cell growth factor)	133, monomer	CD25 (α), CD122 (β), CD132 (γc)	T cells	T-cell proliferation	IL-2: deregulated T-cell proliferation, colitis IL-2Rα: incomplete T-cell development IL-2Rβ: increased T-cell autoimmunity IL-2γ_c: severe combined immunodeficiency
	IL-3 (multicolony CSF)	133, monomer	CD123, βc	T cells, thymic epithelial cells	Synergistic action in early hematopoiesis	IL-3: impaired eosinophil development. Bone marrow unresponsive to IL-5, GM-CSF
	IL-4 (BCGF-1, BSF-1)	129, monomer	CD124, CD132 (γc)	T cells, mast cells	B-cell activation, IgE switch supresses T_H1 cells	IL-4: decreased IgE synthesis
	IL-5 (BCGF-2)	115, homodimer	CD125, βc	T cells, mast cells	Eosinophil growth, differentiation	IL-5: decreased IgE, IgG1 synthesis (in mice); decreased levels of IL-9, IL-10, and eosinophils
	IL-6 (IFN-β_2, BSF-2, BCDF)	184, monomer	CD126, CD130	T cells, macrophages, endothelial cells	T- and B-cell growth and differentiation, acute phase protein production, fever	IL-6: decreased acute phase reaction, reduced IgA production
	IL-7	152, monomer*	CD127, CD132 (γc)	Non-T cells	Growth of pre-B cells and pre-T cells	IL-7: early thymic and lymphocyte expansion severely impaired
	IL-9	125, monomer	IL-9R, CD132 (γc)	T cells	Mast-cell enhancing activity stimulates T_H2	
	IL-11	178, monomer	IL-11R, CD130	Stromal fibroblasts	Synergistic action with IL-3 and IL-4 in hematopoiesis	
	IL-13 (P600)	132, monomer	IL-13R, CD132 (γc) (may also include CD24)	T cells	B-cell growth and differentiation, inhibits macrophage inflammatory cytokine production and T_H1 cells	IL-13: defective regulation of isotype-specific responses
	G-CSF	?, monomer*	G-CSFR	Fibroblasts and monocytes	Stimulates neutrophil development and differentiation	Defective myelopoiesis, neutropenia
	IL-15 (T-cell growth factor)	114, monomer	IL-15R, CD122 (IL-Rβ) CD132 (γc)	Many non-T cells	IL-2-like, stimulates growth of intestinal epithelium, T cells, and NK cells	
	GM-CSF (granulocyte-macrophage colony-stimulating factor)	127, monomer*	CD116, βc	Macrophages, T cells	Stimulates growth and differentiation of myelomonocytic lineage cells, particularly dendritic cells	GM-CSF, GM-CSFR: pulmonary alveolar proteinosis
	OSM (OM, oncostatin M)	196, monomer	OSMR or LIFR, CD130	T cells, macrophages	Stimulates Kaposi's sarcoma cells, inhibits melanoma growth	

*May function as dimers

Family	Cytokine (alternative names)	Size (no. of amino acids) and form	Receptors (c denotes common subunit)	Producer cells	Actions	Effect of cytokine or receptor knock-out (where known)
	LIF (leukemia inhibitory factor)	179, monomer	LIFR, CD130	Bone marrow stroma, fibroblasts	Maintains embryonic stem cells, like IL-6, IL-11, OSM	LIFR: die at or soon after birth; decreased hematopoietic stem cells
Interferons	IFN-γ	143, homodimer	CD119, IFNGR2	T cells, natural killer cells	Macrophage activation, increased expression of MHC molecules and antigen processing components, Ig class switching, supresses T_H2	IFN-γ, IFN-γR: decreased resistance to bacterial infection, especially mycobacteria and certain viruses, impaired T_H1 responses
	IFN-α	166, monomer	CD118, IFNAR2	Leukocytes	Antiviral, increased MHC class I expression	IFN-α: impaired antiviral defences
	IFN-β	166, monomer	CD118, IFNAR2	Fibroblasts	Antiviral, increased MHC class I expression	
Immunoglobulin superfamily	B7.1 (CD80)	262, dimer	CD28, CTLA-4	Antigen-presenting cells	Co-stimulation of T-cell responses	CD28: decreased T-cell responses
	B7.2 (B70, CD86)		CD28, CTLA-4	Antigen-presenting cells	Co-stimulation of T-cell responses	B7.2: decreased co-stimulator response to alloantigen. CTLA-4: massive lymphoproliferation, early death
TNF family	TNF-α (cachectin)	157, trimers	p55, p75 CD120a, CD120b	Macrophages, NK cells, T cells	Local inflammation, endothelial activation	TNF-αR: resistance to septic shock, susceptibility to *Listeria* STNFαR: periodic febrile attacks
	TNF-β (lymphotoxin, LT, LT-α)	171, trimers	p55, p75 CD120a, CD120b	T cells, B cells	Killing, endothelial activation	TNF-β: absent lymph nodes, decreased antibody, increased IgM
	LT-β	Transmembrane, trimerizes with TNF-β, (LT-α)	LTβR or HVEM	T cells, B cells	Lymph node development	Defective development of peripheral lymph nodes, Peyer's patches, and spleen
	CD40 ligand (CD40L)	Trimers	CD40	T cells, mast cells	B-cell activation, class switching	CD40L: poor antibody response, no class switching, diminished T-cell priming (hyper IgM syndrome)
	Fas ligand (FasL)	Trimers	CD95 (Fas)	T cells, stroma?	Apoptosis, Ca^{2+}-independent cytotoxicity	Fas, FasL: mutant forms lead to lymphoproliferation, and autoimmunity
	CD27 ligand (CD27L)	Trimers (?)	CD27	T cells	Stimulates T-cell proliferation	
	CD30 ligand (CD30L)	Trimers (?)	CD30	T cells	Stimulates T- and B-cell proliferation	CD30: increased thymic size, alloreactivity
	4-1BBL	Trimers (?)	4-1BB	T cells	Co-stimulates T and B cells	
	Trail	281, aa trimers	DR4, DR5 DCR1, DCR2 and OPG	T cells, monocytes	Apoptosis of activated T cells and tumor cells	

Family	Cytokine (alternative names)	Size (no. of amino acids) and form	Receptors (c denotes common subunit)	Producer cells	Actions	Effect of cytokine or receptor knock-out (where known)
	OPG-L (RANK-L)	316 aa trimers	RANK/OPG	Osteoblasts, T cells	Stimulates osteoclasts and bone resorption	OPG-L: osteopetrotic runted, toothless OPG: osteoporosis
Unassigned	TGF-β	112, homo- and heterotrimers	TGF-βR	Chondrocytes, monocytes, T cells	Inhibits cell growth, anti-inflammatory induces IgA secretion	TGFβ: lethal inflammation
	IL-1α	159, monomer	CD121a (IL-1RI) and CD121b (IL-1RII)	Macrophages, epithelial cells	Fever, T-cell activation, macrophage activation	IL-1RI: decreased IL-6 production
	IL-1β	153, monomer	CD121a (IL-1RI) and CD121b (IL-1RII)	Macrophages, epithelial cells	Fever, T-cell activation, macrophage activation	IL-1β: impaired acute phase response
	IL-1 RA	?, monomer	CD121a	Monocytes, macrophages, neutrophils, hepatocytes	Binds to but doesn't trigger IL-1 receptor, acts as a natural antagonist of IL-1 function	IL-1RA: reduced body mass, increased sensitivity to endotoxins (septic shock)
	IL-10 (cytokine synthesis inhibitor F)	160, homodimer	IL-10Rα, CRF2-4 (IL-10Rβ)	T cells, macrophages, EBV-transformed B cells	Potent suppressant of macrophage functions	Il-10 or CRF2-4: reduced growth, anemia, chronic enterocolitis
	IL-12 (NK cell stimulatory factor)	197 and 306, heterodimer	IL-12Rβ1 IL-12Rβ2	B cells, macrophages	Activates NK cells, induces CD4 T-cell differentiation to T_H1-like cells	IL-12: impaired in IFN-γ production and in T_H1 responses
	MIF	115, monomer		T cells, pituitary cells	Inhibits macrophage migration, stimulates macrophage activation, induces steroid resistance	MIF: resistance to septic shock
	IL-16	130, homotetramer	CD4	T cells, mast cells, eosinophils	Chemoattractant for CD4 T cells, monocytes, and eosinophils, antiapoptotic for IL-2-stimulated T cells	
	IL-17 (mCTLA-8)	150, monomer		CD4 memory cells	Induce cytokine production by epithelia, endothelia, and fibroblasts	
	IL-18 (IGIF, interferon-γ inducing factor)	157, monomer	Il-1Rrp (IL-1R related protein)	Activated macrophages and Kupffer cells	Induces IFN-γ production by T cells and NK cells, favors T_H1 induction and later T_H2 responses	Defective NK activity and T_H1 responses

Chemokine	Systematic name	Chromosome	Target cell	Specific receptor
†ELR⁺ CXC	**CXCL**			
IL-8	8	4	Neutrophil, basophil, T cell	CXCR1, 2
GROα	1	4	Neutrophil	CXCR2 >> 1
GROβ	2	4	Neutrophil	CXCR2
GROγ	3	4	Neutrophil	CXCR2
ENA-78	5	4	Neutrophil	CXCR2
LDGF-PBP	7	4	Fibroblast, neutrophil	CXCR2
GCP-2	6	4	Neutrophil	CXCR2
†ELR⁻ CXC				
PF4	4	4	Fibroblast	Unknown
Mig	9	4	Activated T cell	CXCR3
IP-10	10	4	Activated T cell ($T_H1 > T_H2$)	CXCR3
SDF-1α/β	12	10	CD34⁺ bone marrow cell, T cell, dendritic cell, B cell, naive B cell, activated CD4 T cell	CXCR4
BUNZO/STRC33	16	17	T cell, NK T cell	CXCR6
I-TAC	11	4	Activated T cell	CXCR3
BLC/BCA-1	13	4	Naive B cells, activated CD4 T cells	CXCR5
CC	**CCL**			
MIP-1α	3	17	Monocyte/macrophage, T cell ($T_H1 > T_H2$), NK cell, basophil, immature dendritic cell, bone marrow cell	CCR1, 5
MIP-1β	4	17	Monocyte/macrophage, T cell ($T_H1 > T_H2$), NK cell, basophil, immature dendritic cell, bone marrow cell	CCR1, 5
MDC	22	16	Immature dendritic cell, IA NK cell, T cell ($T_H2 > T_H1$), thymocyte	CCR4
TECK	25	19	Macrophage, thymocytes, dendritic cell	CCR9
TARC	17	16	T-cell ($T_H2 > T_H1$), immature dendritic cell, IA NK cell, T cell ($T_H2 > T_H1$), thymocyte	CCR4
RANTES	5	17	Monocyte/macrophage, T cell (memory T cell > T cell; $T_H1 > T_H2$), NK cell, basophil, eosinophil, dendritic cell	CCR1, 3, 5
HCC-1	14	17	Monocyte	CCR1
HCC-4	16	17	Monocyte	CCR1
DC-CK1	18	17	Naive T cell > T cell	Unknown
MIP-3α	20	2	T cell (memory T cell > T cell), peripheral blood mononuclear cell, bone marrow cell–dendritic cell	CCR6
MIP-3β	19	9	Naive T cell, mature dendritic cell, B cell	CCR7
MCP-1	2	17	T cell, monocyte, basophil	CCR2
MCP-2	8	17	T cell, monocyte, eosinophil, basophil	CCR2
MCP-3	7	17	T cell, monocyte, eosinophil, basophil, dendritic cell	CCR2
MCP-4	13	17	T cell, monocyte, eosinophil, basophil, dendritic cell	CCR2, 3
None	12	(11)	Eosinophil, monocyte, T cell	CCR2
Eotaxin	11	17	Eosinophil	CCR3
Eotaxin-2/MPIF-2	24	?	T cell (?), eosinophil, basophil	CCR3
I-309	1	17	Neutrophil (TCA-3 only), T cell	CCR8
MIP-5/HCC-2	15	17	T cell, monocyte, neutrophil (?), dendritic cell	CCR1, 3
MPIF-1	23	17 (?)	Monocyte, T cell (resting), neutrophil (?)	Unknown
6Ckine	21	9	Naive T cell, B cell, mesangial cells (?)	CCR7
CTACK	27	9	T cell	CCR10
MEC	28	5	T cell, eosinophil	CCR10, 3
C and CX3C				
Lymphotactin	XCL 1	1 (1)	T cell, NK cell	XCR1
Fractalkine	CX3CL 1	16	T cell, monocyte, neutrophil (?)	CX3CR1

† ELR refers to the three amino acids that precede the first cysteine residue of the CXC motif. If these amino acids are Glu-Leu-Arg (ie ELR⁺), then the chemokine is chemotactic for neutrophils while if they are not (ELR⁻) then the chemokine is chemotactic for lymphocytes.

Appendix V. Immunological Constants.

Evaluation of the cellular components of the human immune system			
	B cells	**T cells**	**Phagocytes**
Normal numbers ($\times 10^9$ per liter of blood)	Approximately 0.3	Total 1.0–2.5 CD4 0.5–1.6 CD8 0.3–0.9	Monocytes 0.15–0.6 Polymorphonuclear leukocytes Neutrophils 3.00–5.5 Eosinophils 0.05–0.25 Basophils 0.02
Measurement of function *in vivo*	Serum Ig levels Specific antibody levels	Skin test	—
Measurement of function *in vitro*	Induced antibody production in response to pokeweed mitogen	T-cell proliferation in response to phytohemagglutinin or to tetanus toxoid	Phagocytosis Nitro blue tetrazolium uptake Intracellular killing of bacteria
Specific defects	See Fig. 11.8	See Fig. 11.8	See Fig. 11.8

Evaluation of the humoral components of the human immune system				
	Immunoglobulins			**Complement**
Component	IgG	IgM	IgA	IgE	
Normal levels	600–1400 mg dl^{-1}	40–345 mg dl^{-1}	60–380 mg dl^{-1}	0–200 IU ml^{-1}	CH$_{50}$ of 125–300 IU ml^{-1}

BIOGRAPHIES

Emil von Behring (1854–1917) discovered antitoxin antibodies with Shibasaburo Kitasato.

Baruj Benacerraf (1920–) discovered immune response genes and collaborated in the first demonstration of MHC restriction.

Jules Bordet (1870–1961) discovered complement as a heat-labile component in normal serum that would enhance the anti-microbial potency of specific antibodies.

Frank MacFarlane Burnet (1899–1985) proposed the first generally accepted clonal selection hypothesis of adaptive immunity.

Jean Dausset (1916–) was an early pioneer in the study of the human major histocompatibility complex or HLA.

Peter Doherty (1940–) and **Rolf Zinkernagal** (1944–) showed that antigen recognition by T cells is MHC-restricted, thereby establishing the biological role of the proteins encoded by the major histocompatibility complex and leading to an understanding of antigen processing and its importance in the recognition of antigen by T cells.

Gerald Edelman (1929–) made crucial discoveries about the structure of immunoglobulins, including the first complete sequence of an antibody molecule.

Paul Ehrlich (1854–1915) was an early champion of humoral theories of immunity, and proposed a famous side-chain theory of antibody formation that bears a striking resemblance to current thinking about surface receptors.

James Gowans (1924–) discovered that adaptive immunity is mediated by lymphocytes, focusing the attention of immunologists on these small cells.

Michael Heidelberger (1888–1991) developed the quantitative precipitin assay, ushering in the era of quantitative immunochemistry.

Edward Jenner (1749–1823) described the successful protection of humans against smallpox infection by vaccination with cowpox or vaccinia virus. This founded the field of immunology.

Niels Jerne (1911–1994) developed the hemolytic plaque assay and several important immunological theories, including an early version of clonal selection, a prediction that lymphocyte receptors would be inherently biased to MHC recognition, and the idiotype network.

Shibasaburo Kitasato (1892–1931) discovered antibodies in collaboration with Emil von Behring.

Robert Koch (1843–1910) defined the criteria needed to characterize an infectious disease, known as Koch's postulates.

Georges Köhler (1946–1995) pioneered monoclonal antibody production from hybrid antibody-forming cells with Cesar Milstein.

Karl Landsteiner (1868–1943) discovered the ABO blood group antigens. He also carried out detailed studies of the specificity of antibody binding using haptens as model antigens.

Peter Medawar (1915–1987) used skin grafts to show that tolerance is an acquired characteristic of lymphoid cells, a key feature of clonal selection theory.

Elie Metchnikoff (1845–1916) was the first champion of cellular immunology, focusing his studies on the central role of phagocytes in host defense.

Cesar Milstein (1927–) pioneered monoclonal antibody production with Georges Köhler.

Louis Pasteur (1822–1895) was a French microbiologist and immunologist who validated the concept of immunization first studied by Jenner. He prepared vaccines against chicken cholera and rabies.

Rodney Porter (1920–1985) worked out the polypeptide structure of the antibody molecule, laying the groundwork for its analysis by protein sequencing.

George Snell (1903–1996) worked out the genetics of the murine major histocompatibility complex and generated the congenic strains needed for its biological analysis, laying the groundwork for our current understanding of the role of the MHC in T-cell biology.

Susumu Tonegawa (1939–) discovered the somatic recombination of immunological receptor genes that underlies the generation of diversity in human and murine antibodies and T-cell receptors.

GLOSSARY

In the context of immunoglobulins, α is the type of heavy chain in IgA.

α:β T cell: see **T cell**.

α:β T-cell receptor: see **T-cell receptor**.

The **ABO blood group system** antigens are expressed on red blood cells. They are used for typing human blood for transfusion. Individuals who do not express A or B antigens on their red blood cells naturally form antibodies that interact with them.

The removal of antibodies specific for one antigen from an antiserum to render it specific for another antigen or antigens is called **absorption**.

Accessory effector cells in adaptive immunity are cells that aid in the response but do not directly mediate specific antigen recognition. They include phagocytes, mast cells, and NK cells.

The **acquired immune deficiency syndrome (AIDS)** is a disease caused by infection with the human immunodeficiency virus (HIV-1). AIDS occurs when an infected patient has lost most of his or her CD4 T cells, so that infections with opportunistic pathogens occur.

Acquired immune response: see **adaptive immune response**.

Immunization with antigen is called **active immunization** to distinguish it from the transfer of antibody to an unimmunized individual, which is called passive immunization.

Acute lymphoblastic leukemia is a highly aggressive, undifferentiated form of lymphoid malignancy derived from a progenitor cell that is thought to be able to give rise to both T and B lineages of lymphoid cells. Most of these leukemias show partial differentiation toward the B-cell lineage (so called B-ALL) whereas a minority show features of T cells (T-ALL).

Acute-phase proteins are found in the blood shortly after the onset of an infection. These proteins participate in early phases of host defense against infection. An example is the mannan-binding lectin.

The **acute-phase response** is a change in the blood that occurs during early phases of an infection. It includes the production of acute-phase proteins and also of cellular elements.

The **adaptive immune response** or **adaptive immunity** is the response of antigen-specific lymphocytes to antigen, including the development of immunological memory. Adaptive immune responses are generated by clonal selection of lymphocytes. Adaptive immune responses are distinct from innate and nonadaptive phases of immunity, which are not mediated by clonal selection of antigen-specific lymphocytes. Adaptive immune responses are also known as **acquired immune responses**.

The **adaptor proteins** are key linkers between receptors and down-stream members of signaling pathways. These proteins are functionally heterogeneous, but all use a similar domain, known as an SH2 domain, as the means of interacting with the phoshotyrosine residues generated by the receptor-associated tyrosine kinases. The protein known as Vav is an adaptor protein of this type.

The **adenoids** are mucosal-associated lymphoid tissues located in the nasal cavity.

The enzyme defect **adenosine deaminase deficiency** leads to the accumulation of toxic purine nucleosides and nucleotides, resulting in the death of most developing lymphocytes within the thymus. It is a cause of severe combined immunodeficiency.

Adhesion molecules mediate the binding of one cell to other cells or to extracellular matrix proteins. Integrins, selectins, members of the immunoglobulin gene superfamily, and CD44 and related proteins are all adhesion molecules important in the operation of the immune system.

An **adjuvant** is any substance that enhances the immune response to an antigen with which it is mixed.

Adoptive immunity is immunity conferred on a naive or irradiated recipient by transfer of lymphoid cells from an actively immunized donor. This is called **adoptive transfer** or **adoptive immunization**.

Afferent lymphatic vessels drain fluid from the tissues and carry macrophages and dendritic cells from sites of infection in most parts of the body to the lymph nodes.

Affinity is the strength of binding of one molecule to another at a single site, such as the binding of a monovalent Fab fragment of antibody to a monovalent antigen. See also **avidity**.

Affinity chromatography is the purification of a substance by means of its affinity for another substance immobilized on a solid support. For example, an antigen can be purified by affinity chromatography on a column of beads to which specific antibody molecules are covalently linked.

Affinity maturation refers to the increase in the affinity for the specific antigen of the antibodies produced during the course of a humoral immune response. It is particularly prominent in secondary and subsequent immunizations.

Agammaglobulinemia: see **X-linked agammaglobulinemia (XLA)**.

Agglutination is the clumping together of particles, usually by antibody molecules binding to antigens on the surfaces of adjacent particles. Such particles are said to **agglutinate**. When the particles are red blood cells, the phenomenon is called hemagglutination.

Agonist peptides are peptide antigens that activate their specific T cells, inducing them to make cytokines and to proliferate. They are thought to differ from antagonist peptides by their ability to induce T-cell receptor dimerization.

AIDS: see **acquired immune deficiency syndrome**.

Alleles are variants of a single genetic locus.

Allelic exclusion refers to the fact that in a heterozygous individual, only one of the alternative C-region alleles of the heavy or light chain is expressed in a single B cell and in an immunoglobulin molecule. The term has come to be used more generally to describe the expression of a single receptor specificity in cells with the potential to express two or more receptors.

Allergens are antigens that elicit hypersensitivity or allergic reactions.

Allergic asthma is constriction of the bronchial tree due to an allergic reaction to inhaled antigen.

The lining of the eye, called the conjunctiva, manifests **allergic conjunctivitis** in sensitized individuals exposed to allergens.

An **allergic reaction** is a response to innocuous environmental antigens, or allergens, due to preexisting antibody or primed T cells. There are various immune mechanisms of allergic reactions, but the most common is the binding of allergen to IgE antibody on mast cells, which causes asthma, hay fever, and other common allergic reactions.

Allergic rhinitis is an allergic reaction in the nasal mucosa, also known as hay fever, that causes runny nose, sneezing, and tears.

Allergy is a symptomatic reaction to a normally innocuous environmental antigen. It results from the interaction between the antigen and antibody or primed T cells produced by earlier exposure to the same antigen.

Alloantigens are classically defined as polymorphisms at the MHC locus; they stimulate intense reactions to allografted tissues, hence their naming.

Two individuals or two mouse strains that differ at the MHC are said to be **allogeneic**. The term can also be used for allelic differences at other loci. Rejection of grafted tissues from unrelated donors usually results from T-cell responses to allogeneic MHC molecules expressed by the grafted tissues. See also **syngeneic**; **xenogeneic**.

An **allograft** is a graft of tissue from an allogeneic or nonself donor of the same species; such grafts are invariably rejected unless the recipient is immunosuppressed.

Alloreactivity describes the stimulation of T cells by MHC molecules other than self; it marks the recognition of allogeneic MHC molecules. Such responses are also called **alloreactions**.

Allotypes are allelic polymorphisms that can be detected by antibodies specific for the polymorphic gene products; in immunology, **allotypic** differences in immunoglobulin molecules were important in deciphering the genetics of antibodies.

An **altered peptide ligand**, or partial agonist, is a peptide, usually closely related to an agonist peptide in amino acid sequence, that induces only a partial response from T cells specific for the agonist peptide.

The **alternative pathway** of complement activation is not triggered by antibody, as is the classical pathway of complement activation, but by the binding of complement protein C3b to the surface of a pathogen; it is therefore a feature of innate immunity. The alternative pathway also amplifies the classical pathway of complement activation.

Anaphylactic shock or systemic anaphylaxis is an allergic reaction to systemically administered antigen that causes circulatory collapse and suffocation due to tracheal swelling. It results from binding of antigen to IgE antibody on connective tissue mast cells throughout the body, leading to the disseminated release of inflammatory mediators.

Anaphylatoxins are small fragments of complement proteins, released by cleavage during complement activation. These small fragments are recognized by specific receptors, and they recruit fluid and inflammatory cells to sites of their release. The fragments C5a, C3a, and C4a are all anaphylatoxins, listed in order of decreasing potency *in vivo*.

Peptide fragments of antigens are bound to specific MHC class I molecules by **anchor residues**. These are residues of the peptide that have amino acid side chains that bind into pockets lining the peptide-binding groove of the MHC class I molecule. Each MHC class I molecule binds different patterns of anchor residues, called anchor motifs, giving some specificity to peptide binding. Anchor residues exist but are less obvious for peptides that bind to MHC class II molecules.

Anergy is a state of nonresponsiveness to antigen. People are said to be **anergic** when they cannot mount delayed-type hypersensitivity reactions to challenge antigens, whereas T cells and B cells are said to be **anergic** when they cannot respond to their specific antigen under optimal conditions of stimulation.

Antagonist peptides are peptides, usually closely related in sequence to an agonist peptide, that inhibit the response of a cloned T-cell line specific for the agonist peptide.

An **antibody** is a protein that binds specifically to a particular substance—its antigen. Each antibody molecule has a unique structure that enables it to bind specifically to its corresponding antigen, but all antibodies have the same overall structure and are known collectively as immunoglobulins or Igs. Antibodies are produced by plasma cells in response to infection or immunization, and bind to and neutralize pathogens or prepare them for uptake and destruction by phagocytes.

Antibody combining site: see **antigen-binding site**.

The **antibody repertoire** or immunoglobulin repertoire describes the total variety of antibodies in the body of an individual.

Antibody-dependent cell-mediated cytotoxicity (**ADCC**) is the killing of antibody-coated target cells by cells with Fc receptors that recognize the constant region of the bound antibody. Most ADCC is mediated by NK cells that have the Fc receptor FcγRIII or CD16 on their surface.

An **antigen** is any molecule that can bind specifically to an antibody. Their name arises from their ability to **gen**erate **anti**bodies. However, some antigens do not, by themselves, elicit antibody production; those antigens that can induce antibody production are called immunogens.

Both libraries of cDNA clones in expression vectors and bacteriophage libraries encoding random peptide sequences have been used to identify the targets of specific antibodies and, in some cases, of T cells. Such libraries are termed **antigen display libraries**.

Antigen:antibody complexes are noncovalently associated groups of antigen and antibody molecules that can vary in size from small soluble complexes to large insoluble complexes that precipitate out of solution; they are also known as immune complexes.

The **antigen-binding site** of an antibody, or **antibody combining site**, is found at the surface of the antibody molecule that makes physical contact with the antigen. Antigen-binding sites are made up of six hypervariable loops, three from the light-chain V region and three from the heavy-chain V region.

In an **antigen-capture assay**, the antigen binds to a specific antibody, and its presence is detected by a second antibody that must be labeled and directed at a different epitope.

An **antigenic determinant** is the portion of an antigenic molecule that is bound by a given antibody or antigen receptor; it is also known as an epitope.

Influenza virus varies from year to year by a process of **antigenic drift** in which point mutations of viral genes cause small differences in the structure of viral surface antigens. Periodically, influenza viruses undergo an **antigenic shift** through reassortment of their segmented genome with another influenza virus, changing their surface antigens radically. Such antigenic shift variants are not recognized by individuals immune to influenza, so when antigenic shift variants arise, there is widespread and serious disease.

Many pathogens evade the adaptive immune response by **antigenic variation** in which new antigens are displayed that are not recognized by antibodies or T cells elicited in earlier infections.

Antigen presentation describes the display of antigen as peptide fragments bound to MHC molecules on the surface of a cell; T cells recognize antigen when it is presented in this way.

Antigen-presenting cells (**APCs**) are highly specialized cells that can process antigens and display their peptide fragments on the cell surface together with molecules required for T-cell activation. The main antigen-presenting cells for T cells are dendritic cells, macrophages, and B cells.

Antigen processing is the degradation of proteins into peptides that can bind to MHC molecules for presentation to T cells. All antigens except peptides must be processed into peptides before they can be presented by MHC molecules.

T and B lymphocytes collectively bear on their surface highly diverse **antigen receptors** capable of recognizing a wide diversity of antigens. Each individual lymphocyte bears receptors of a single antigen specificity.

Antigen spreading: see **epitope spreading**.

Anti-immunoglobulin antibodies are antibodies against immunoglobulin constant domains, useful for detecting bound antibody molecules in immunoassays and other applications. These can be divided into **anti-isotype antibodies** made in a different species, **anti-allotype antibodies** made in the same species against allotypic variants, and **anti-idiotype antibodies**, made against unique determinants to a single antibody.

Anti-lymphocyte globulin is antibody raised in another species against human T cells.

An **antiserum** (plural: **antisera**) is the fluid component of clotted blood from an immune individual that contains antibodies against the molecule used for immunization. Antisera contain heterogeneous collections of antibodies, which bind the antigen used for immunization, but each has its own structure, its own epitope on the antigen, and its own set of cross-reactions. This heterogeneity makes each antiserum unique.

Bites by poisonous snakes can be treated by identification of the snake and injection of an **antivenin** specific for that snake's venom.

Aplastic anemia is a failure of bone marrow stem cells so that formation of all cellular elements of the blood ceases; it can be treated by bone marrow transplantation.

Apoptosis, or programmed cell death, is a form of cell death in which the cell activates an internal death program. It is characterized by nuclear DNA degradation, nuclear degeneration and condensation, and the phagocytosis of cell residua. Proliferating cells frequently undergo apoptosis, which is a natural process in development, and proliferating lymphocytes undergo high rates of apoptosis in development and during immune responses. Apoptosis contrasts with necrosis, death from without, which occurs in situations such as poisoning and anoxia.

The **appendix** is a gut-associated lymphoid tissue located at the beginning of the colon.

In this book, we have termed primed effector T cells **armed effector T cells**, because these cells can be triggered to perform their effector functions immediately on contact with cells bearing the peptide:MHC complex for which they are specific. They contrast with memory T cells, which need to be activated by antigen-presenting cells to differentiate into effector T cells before they can mediate effector responses.

The **Arthus reaction** is a skin reaction in which antigen is injected into the dermis and reacts with IgG antibodies in the extracellular spaces, activating complement and phagocytic cells to produce a local inflammatory response.

Ascertainment artifact refers to data that seem to demonstrate some finding, but fail to do so because they are collected from a population that is selected in a biased fashion.

Ataxia telangiectasia is a disease characterized by staggering, multiple disorganized blood vessels, and an immunodeficiency in a protein called ATM, which contains a kinase thought to be important in signaling of double-stranded DNA breaks.

Atopic allergy, or **atopy**, is the increased tendency seen in some people to produce immediate hypersensitivity reactions (usually mediated by IgE antibodies) against innocuous substances.

Pathogens are said to be **attenuated** when they can grow in their host and induce immunity without producing serious clinical disease.

Antibodies specific for self antigens are called **autoantibodies**.

A graft of tissue from one site to another on the same individual is called an **autograft**.

Diseases in which the pathology is caused by adaptive immune responses to self antigens are called **autoimmune diseases**.

Autoimmune hemolytic anemia is a pathological condition with low levels of red blood cells (anemia), which is caused by autoantibodies that bind red blood cell surface antigens and target the red blood cell for destruction.

An adaptive immune response directed at self antigens is called an **autoimmune response**; likewise, adaptive immunity specific for self antigens is called **autoimmunity**.

In the disease **autoimmune thrombocytopenic purpura**, antibodies against a patient's platelets are made. Antibody binding to platelets causes them to be taken up by cells with Fc receptors and complement receptors, causing a fall in platelet counts that leads to purpura (bleeding).

Autoreactivity describes immune responses directed at self antigens.

Avidity is the sum total of the strength of binding of two molecules or cells to one another at multiple sites. It is distinct from affinity, which is the strength of binding of one site on a molecule to its ligand.

The **avidity hypothesis** of T-cell selection in the thymus states that T cells must have a measurable affinity for self MHC molecules in order to mature, but not so great an affinity as to cause activation of the cell when it matures, as this would require that the cell be deleted to maintain self tolerance.

Azathioprine is a potent immunosuppressive drug that is converted to its active form *in vivo* and then kills rapidly proliferating cells, including lymphocytes responding to grafted tissues.

4-1BB is a member of the TNF receptor family that specifically binds to 4-1BB ligand.

4-1BB ligand (**4-1BBL**) is a member of the TNF family that binds to 4-1BB.

β **barrel**: see β **sheet**.

A **B cell**, or **B lymphocyte**, is one of the two major types of lymphocyte. The antigen receptor on B lymphocytes, usually called the B-cell receptor, is a cell-surface immunoglobulin. On activation by antigen, B cells differentiate into cells producing antibody molecules of the same antigen specificity as this receptor. B cells are divided into two classes. **B-1 cells**, also known as CD5 B cells, are a class of atypical, self-renewing B cells found mainly in the peritoneal and pleural cavities in adults. They have a far less diverse repertoire of receptors than do **B-2 cells**, also known as conventional B cells, which are generated in the bone marrow throughout life, emerging to populate the blood and lymphoid tissues.

A β **sheet** is one of the fundamental structural building blocks of proteins, consisting of adjacent, extended strands of amino acids (β **strands**) that are bonded together by interactions between backbone amide and carbonyl groups. β Sheets can be parallel, in which case the adjacent strands run in the same direction, or antiparallel, where adjacent strands run in opposite directions. All immunoglobulin domains are made up of antiparallel β-sheet structures. A β **barrel** or a β **sandwich** is another way of describing the structure of the immunoglobulin domain.

The light chain of the MHC class I proteins is called β_2-**microglobulin.** It binds noncovalently to the heavy or α chain.

The major T-cell co-stimulatory molecules are the **B7 molecules**, **B7.1** (CD80) and **B7.2** (CD86). They are closely related members of the immunoglobulin gene superfamily and both bind to the CD28 molecule on T cells. They are expressed differentially on various antigen-presenting cell types. We use the term **B7 molecules** to refer to both B7.1 and B7.2.

Many infectious diseases are caused by **bacteria**, which are prokaryotic microorganisms that exist as many different species and strains. Bacteria can live on body surfaces, in extracellular spaces, in cellular vesicles, or in the cytosol, and different bacterial species cause distinctive infectious diseases.

BALT: see **bronchial-associated lymphoid tissue**.

Bare lymphocyte syndrome is an immunodeficiency disease in which MHC class II molecules are not expressed on cells as a result of one of several different regulatory gene defects. Patients with bare lymphocyte syndrome are severely immunodeficient and have few CD4 T cells. In MHC class I deficiency, or **bare lymphocyte syndrome (MHC class I)**, the most common defect is mutation in either TAP1 or TAP2.

Basophils are white blood cells containing granules that stain with basic dyes, and which are thought to have a function similar to mast cells.

Bb is the large active fragment of complement component factor B. It is produced when factor B is captured by bound C3b and cleaved by factor D. Bb remains associated with C3b and is the serine protease component of the alternative pathway C3 convertase.

The **B-cell antigen receptor**, or **B-cell receptor (BCR)**, is the cell-surface receptor of B cells for specific antigen. It is composed of a transmembrane immunoglobulin molecule associated with the invariant Igα and Igβ chains in a noncovalent complex.

A complex of CD19, TAPA-1, and CR2 makes up the **B-cell co-receptor**; co-ligation of this complex with the B-cell antigen receptor increases responsiveness to antigen by about 100-fold.

The **B-cell corona** in the spleen is the zone of the white pulp primarily made up of B cells.

B-cell mitogens are substances that cause B cells to proliferate.

The protein known as Bcl-2 protects cells from apoptosis by binding to the mitochondrial membrane. It is encoded by the *bcl-2* gene, which was discovered at the breakpoint of an oncogenic chromosomal translocation in B-cell leukemia.

Blk: see **tyrosine kinase**.

The adaptor protein **BLNK** operates in B cells in the same way as LAT does in T cells; it has multiple tyrosine residues that are targets for phosphorylation, and recruits signaling molecules to membrane lipid rafts.

Blood group antigens are surface molecules on red blood cells that are detectable with antibodies from other individuals. The major blood group antigens are called ABO and Rh (Rhesus), and are used in routine blood banking to type blood. There are many other blood group antigens that can be detected in cross-matching.

In transfusion medicine, **blood typing** is used to determine whether donor and recipient have the same ABO and Rh blood group antigens. A cross-match, in which serum from the donor is tested on the cells of the recipient, and vice versa, is used to rule out other incompatibilities. Transfusion of incompatible blood causes a transfusion reaction, in which red blood cells are destroyed and the released hemoglobin causes toxicity.

Bloom's syndrome is a disease characterized by low T-cell numbers, reduced antibody levels, and an increased susceptibility to respiratory infections, cancer, and radiation damage. It is caused by mutations in a DNA helicase.

B lymphocyte: see **B cell**.

B-lymphocyte chemokine (BLC) is a CXC chemokine that attracts B cells and activated T cells into the follicles of peripheral lymphoid tissues by binding to the CXCR5 receptor.

BLyS is a secreted member of the TNF family of cytokines. It is secreted by T cells and plays critical roles in germinal center and plasma-cell formation, and possibly dendritic-cell maturation.

The **bone marrow** is the site of hematopoiesis, the generation of the cellular elements of blood, including red blood cells, monocytes, poly-morphonuclear leukocytes, and platelets. The bone marrow is also the site of B-cell development in mammals and the source of stem cells that give rise to T cells upon migration to the thymus. Thus, bone marrow transplantation can restore all the cellular elements of the blood, including the cells required for adaptive immunity.

A **bone marrow chimera** is formed by transferring bone marrow from one mouse to an irradiated recipient mouse, so that all of the lymphocytes and blood cells are of donor genetic origin. Bone marrow chimeras have been crucial in elucidating the development of lymphocytes and other blood cells.

A **booster immunization** is commonly given after a primary immunization, to increase the titer of antibodies.

Bradykinin is a vasoactive peptide that is produced as a result of tissue damage and acts as an inflammatory mediator.

The lymphoid cells and organized lymphoid tissues in the respiratory tract have been termed the **bronchial-associated lymphoid tissues (BALT)**. These tissues are very important in the induction of immune responses to inhaled antigens and to respiratory infection.

Bruton's X-linked agammaglobulinemia: see **X-linked agamma-globulinemia**.

Burkitt's lymphoma is caused by Epstein–Barr virus (EBV) and occurs mainly in sub-Saharan Africa.

The **bursa of Fabricius** is an outpouching of the cloaca found in birds. It is an aggregate of epithelial tissue and lymphoid cells and is the site of intense early B-cell proliferation. The bursa of Fabricius is required for B-cell development in birds, as its removal or **bursectomy** early in life causes an absence of B cells in adult birds. An equivalent structure has not been detected in mammals, where B-cell development follows a different pathway.

The constant regions of the polypeptide chains of immunoglobulin molecules are made up of one or more constant domains or **C domains** of similar structure; each immunoglobulin chain also has a single variable or V domain.

C region: see **constant region**.

The **C1 complex** of complement components comprises one molecule of **C1q** bound to two molecules each of the zymogens **C1r** and **C1s**. C1q initiates the classical pathway of complement activation by binding to a pathogen surface or to bound antibody. This binding activates the associated C1r, which in turn cleaves and activates C1s. The active form of C1s then cleaves the next two components in the pathway, **C4** and **C2**.

C1 inhibitor (C1INH) is a protein that inhibits the activity of activated complement component C1 by binding to and inactivating its C1r:C1s enzymatic activity. It also inhibits other serine proteases including kallikrein. Deficiency in C1INH is the cause of the disease hereditary angioneurotic edema, in which the production of vasoactive peptides, kinins, leads to subcutaneous and laryngeal swelling.

The complement fragment **C3b** is the major product of the C3 convertase, and the principal effector molecule of the complement system. It has a highly reactive thioester bond which allows it to bind covalently to the surface on which it is generated. Once bound, it acts as an opsonin to promote the destruction of pathogens by phagocytes and removal of immune complexes; C3b is bound by the complement receptor CR1, while its proteolytic derivative, iC3b, is bound by the complement receptors CR1, CR2, and CR3.

The generation of the enzyme **C3 convertase** on the surface of a pathogen or cell is a crucial step in complement activation. The classical pathway C3 convertase is formed from membrane-bound C4b complexed with the protease C2b. The alternative pathway of complement activation uses a different but homologous C3 convertase, formed from membrane-bound C3b complexed with the protease Bb. These C3 convertases have the same activity, catalyzing the deposition of large numbers of C3b molecules that bind covalently to the pathogen surface, leading to opsonization and the activation of the effector cascade that causes membrane lesions.

C3dg is a breakdown product of C3b that remains attached to the microbial surface, where it can bind to CD21, the complement receptor CR2.

C4b-binding protein can inactivate the classical pathway C3 convertase if it forms on host cells, by displacing C2b from the C4b:C2b complex. It binds to C4b attached to host cells, but cannot bind C4b attached to pathogens. This is because it has a second binding site specific for sialic acid, a terminal sugar on vertebrate cell surfaces, but not on pathogens.

C5 is an inactive complement component that is cleaved by the **C5 convertase** to release the potent inflammatory peptide **C5a** and a larger fragment, **C5b**, that initiates the formation of a membrane-attack complex from the terminal components of complement.

The receptor for the C5a fragment of complement, the **C5a receptor**, is a seven-transmembrane spanning receptor that couples to a heterotrimeric G protein. Similar receptors bind to **C3a** and **C4a**.

The complement components **C6**, **C7**, and **C8** form a complex with the active complement fragment **C5b** in the late events of complement activation. This complex inserts into the membrane and induces polymerization of **C9** to form a pore known as the membrane-attack complex.

The cytosolic serine/threonine phosphatase **calcineurin** has a crucial role in signaling via the T-cell receptor. The immunosuppressive drugs cyclosporin A and tacrolimus (also known as FK506) form complexes with cellular proteins called immunophilins that bind and inactivate calcineurin, suppressing T-cell responses.

The protein **calnexin** is an 88 kDa protein found in the endoplasmic reticulum. It binds to partly folded members of the immunoglobulin superfamily of proteins and retains them in the endoplasmic reticulum until folding is completed.

Calreticulin is the molecular chaperone that binds initially to MHC class I, MHC class II, and other proteins that contain immunoglobulin-like domains, such as the T-cell and B-cell antigen receptors.

Antibodies or antigens can be measured in various **capture** assays. In these assays, antigens are captured by antibodies bound to plastic (or vice versa). Antibody binding to a plate-bound antigen can be measured using labeled antigen or anti-immunoglobulin. Antigen binding to plate-bound antibody can be measured by using an antibody that binds to a different epitope on the antigen.

Carriers are foreign proteins to which small nonimmunogenic antigens, or haptens, can be coupled to render the hapten immunogenic. *In vivo*, self proteins can also serve as carriers if they are correctly modified by the hapten; this is important in allergy to drugs.

Caseation necrosis is a form of necrosis seen in the center of large granulomas, such as the granulomas in tuberculosis. The term comes from the white cheesy appearance of the central necrotic area.

Caspases are a family of closely related cysteine proteases that cleave proteins at aspartic acid residues. They have important roles in apoptosis.

CD: see **clusters of differentiation** and Appendix II.

The **CD3 complex** is the complex of α:β or γ:δ T-cell receptor chains with the invariant subunits CD3γ, δ, and ε, and the dimeric ζ chains.

The cell-surface protein **CD4** is important for recognition by the T-cell receptor of antigenic peptides bound to MHC class II molecules. It acts as a co-receptor by binding to the lateral face of the MHC class II molecules.

CD4 T cells are T cells that carry the co-receptor protein CD4. They recognize peptides derived from intravesicular sources, which are bound to MHC class II molecules, and differentiate into CD4 T_H1 and CD4 T_H2 effector cells that activate macrophages and B-cell responses to antigen.

CD5 B cells are a class of atypical, self-renewing B cells found mainly in the peritoneal and pleural cavities in adults. They have a far less diverse receptor repertoire than conventional B cells, and since they are the first B cells to be produced they are also known as B-1 cells.

The cell-surface protein **CD8** is important for recognition by the T-cell receptor of antigenic peptides bound to MHC class I molecules. It acts as a co-receptor by binding to the lateral face of MHC class I molecules.

CD8 T cells are T cells that carry the co-receptor CD8. They recognize antigens, for example viral antigens, that are synthesized in the cytoplasm of a cell. Peptides derived from these antigens are transported by TAP, assembled with MHC class I molecules in the endoplasmic reticulum, and displayed as peptide:MHC class I complexes on the cell surface. CD8 T cells differentiate into cytotoxic CD8 T cells.

CD19:CR2:TAPA-1 complex: see **B-cell co-receptor**.

CD23: see Appendix II.

CD28: see Appendix II.

CD30 and CD30L: see Appendix II.

CD34: see Appendix II.

B-cell growth is triggered in part by the binding of **CD40 ligand**, also known as **CD154**, expressed on activated helper T cells, to **CD40** on the B-cell surface.

CD45, or the leukocyte common antigen, is a transmembrane tyrosine phosphatase found on all leukocytes. It is expressed in different isoforms on different cell types, including the different subtypes of T cells. These isoforms are commonly denoted by the designation of CD45R followed by the exon whose presence gives rise to distinctive antibody-binding patterns.

CD59: see **protectin**.

CDR: see **complementarity determining region**.

Cell-adhesion molecules (**CAMs**) are cell-surface proteins that are involved in binding cells together in tissues and also in less permanent cell–cell interactions.

Cell-mediated immunity, or a **cell-mediated immune response**, describes any adaptive immune response in which antigen-specific T cells have the main role. It is defined operationally as all adaptive immunity that cannot be transferred to a naive recipient with serum antibody. Cf. **humoral immunity**.

Cell-surface immunoglobulin is the B-cell receptor for antigen. See also **B-cell antigen receptor**.

Cellular immunology is the study of the cellular basis of immunity.

Central lymphoid organs are sites of lymphocyte development. In humans, B lymphocytes develop in bone marrow, whereas T lymphocytes develop within the thymus from bone marrow-derived progenitors. They are also sometimes known as the primary lymphoid organs.

Central tolerance is tolerance that is established in lymphocytes developing in central lymphoid organs. Cf. **peripheral tolerance**.

Centroblasts are large, rapidly dividing cells found in germinal centers, and are the cells in which somatic hypermutation is believed to occur. Antibody-secreting and memory B cells derive from these cells.

Centrocytes are the small B cells in germinal centers that derive from centroblasts. They may mature into antibody-secreting plasma cells or memory B cells, or may undergo apoptosis, depending on their receptor's interaction with antigen.

Chediak–Higashi syndrome is caused by a defect in a protein involved in intracellular vesicle fusion. Phagocytic cell function is affected as lysosomes fail to fuse properly with phagosomes and there is impaired killing of ingested bacteria.

Chemokines are small chemoattractant proteins that stimulate the migration and activation of cells, especially phagocytic cells and lymphocytes. They have a central role in inflammatory responses. Chemokines and their receptors are listed in Appendix IV.

Most lymphoid tumors, and many other tumors, bear **chromosomal translocations** that mark points of breakage and rejoining of different chromosomes. These chromosomal breaks are particularly frequent in lymphomas and leukemias.

Chronic granulomatous disease is an immunodeficiency disease in which multiple granulomas form as a result of defective elimination of bacteria by phagocytic cells. It is caused by defects in the NADPH oxidase system of enzymes that generate the superoxide radical involved in bacterial killing.

Chronic lymphocytic leukemias (CLLs) are B-cell tumors that are found in the blood. The great majority express CD5 and unmutated V genes and are therefore thought to arise from B-1 cells.

The cell-surface receptor called **c-Kit** is found on many immature hematopoietic cells. It is a receptor for CSF.

The **class II-associated invariant chain peptide (CLIP)** is a peptide of variable length cleaved from the class II invariant chain by proteases. It remains associated with the MHC class II molecule in an unstable form until it is removed by the HLA-DM protein.

Class II transactivator (CIITA): see **MHC class II transactivator**.

Class switching: see **isotype switching**.

Classes: see **isotypes**.

The **classical pathway** of complement activation is the pathway activated by C1 binding either directly to bacterial surfaces or to antibody, that serves as a means of flagging the bacteria as foreign. See also **alternative pathway**; **mannan-binding lectin**.

Clonal deletion is the elimination of immature lymphocytes on binding to self antigens to produce tolerance to self, as required by the clonal selection theory. Clonal deletion is the main mechanism of central tolerance and can also occur in peripheral tolerance.

Clonal expansion is the proliferation of antigen-specific lymphocytes in response to antigenic stimulation and precedes their differentiation into effector cells. It is an essential step in adaptive immunity, allowing rare antigen-specific cells to increase in number so that they can effectively combat the pathogen that elicited the response.

The **clonal selection theory** is a central paradigm of adaptive immunity. It states that adaptive immune responses derive from individual antigen-specific lymphocytes that are self-tolerant. These specific lymphocytes proliferate in response to antigen and differentiate into antigen-specific effector cells that eliminate the eliciting pathogen, and memory cells to sustain immunity. The theory was formulated by Sir Macfarlane Burnet and in earlier forms by Niels Jerne and David Talmage.

A **clone** is a population of cells all derived from a single progenitor cell.

A **cloned T-cell line** is a continuously growing line of T cells derived from a single progenitor cell. Cloned T-cell lines must be stimulated with antigen periodically to maintain growth. They are useful for studying T-cell specificity, growth, and effector functions.

A feature unique to individual cells or members of a clone is said to be **clonotypic**. Thus, a monoclonal antibody that reacts with the receptor on a cloned T-cell line is said to be a clonotypic antibody and to recognize its clonotype or the clonotypic receptor of that cell. See also **idiotype**; **idiotypic**.

Clusters of differentiation (CD) are groups of monoclonal antibodies that identify the same cell-surface molecule. The cell-surface molecule is designated CD followed by a number (e.g. CD1, CD2, etc.). For a current listing of CDs see Appendix II.

The **coagulation system** is a proteolytic cascade of plasma enzymes that triggers blood clotting when blood vessels are damaged.

The expression of a gene is said to be **codominant** when both alleles at one locus are expressed in roughly equal amounts in heterozygotes. Most genes show this property, including the highly polymorphic MHC genes.

A **coding joint** is formed by the imprecise joining of a V gene segment to a (D)J gene segment in immunoglobulin or T-cell receptor genes.

Co-immunoprecipitation: see **immunoprecipitation analysis**.

Co-isogenic: see **congenic**.

Collectins are a structurally related family of calcium-dependent sugar-binding proteins or lectins containing collagen-like sequences. An example is mannan-binding lectin.

Antigen receptors manifest two distinct types of **combinatorial diversity** generated by the combination of separate units of genetic information. Receptor gene segments are joined in many different combinations to generate diverse receptor chains, and then two different receptor chains (heavy and light in immunoglobulins; α and β or γ and δ in T-cell receptors) are combined to make the antigen-recognition site.

The **common γ chain** (γ_c) is a transmembrane polypeptide chain (CD132) that is common to a subgroup of class I cytokine receptors. It plays a key role in the intracellular signaling mediated by these receptors as shown by gene knockout.

Common lymphoid progenitors are stem cells that give rise to all lymphocytes. They are derived from pluripotent hematopoietic stem cells.

Common variable immunodeficiency is a relatively common deficiency in antibody production whose pathogenesis is not yet understood. There is a strong association with genes mapping within the MHC.

Competitive binding assays are serological assays in which unknowns are detected and quantitated by their ability to inhibit the binding of a labeled known ligand to its specific antibody. This is also referred to as a competitive inhibition assay.

When known sources of antibody or antigen are used as competitive inhibitors of antigen–antibody interactions, this assay is referred to as a **competitive inhibition assay**.

The **complement** system is a set of plasma proteins that act together to attack extracellular forms of pathogens. **Complement activation** can occur spontaneously on certain pathogens or by antibody binding to the pathogen. The pathogen becomes coated with complement proteins that facilitate pathogen removal by phagocytes and can also kill certain pathogens directly.

Complement receptors (CRs) are cell-surface proteins on various cells that recognize and bind complement proteins that have bound an antigen such as a pathogen. Complement receptors on phagocytes allow them to identify pathogens coated with complement proteins for uptake and destruction. Complement receptors include CR1, CR2, CR3, CR4, and the receptor for C1q.

The **complementarity determining regions (CDRs)** of immunoglobulins and T-cell receptors are the parts of these molecules that determine their specificity and make contact with specific ligand. The CDRs are the most variable part of the molecule, and contribute to the diversity of these molecules. There are three such regions (CDR1, CDR2, and CDR3) in each V domain.

In **confocal fluorescent microscopy**, it is possible to use optics to produce images at very high resolution by having two origins of fluorescent light that come together only at one plane of a thicker section.

Conformational epitopes, or discontinuous epitopes, on a protein antigen are formed from several separate regions in the primary sequence of a protein brought together by protein folding. Antibodies that bind conformational epitopes bind only native folded proteins.

Congenic strains of mice are genetically identical at all loci except one. Each strain is generated by the repetitive back-crossing of mice carrying the desired trait onto a strain that provides the genetic background for the set of congenic strains. The most important congenic strains in immunology are the congenic strains, developed by George Snell, that differ from each other at the MHC.

Conjugate vaccines are vaccines made from capsular polysaccharides bound to proteins of known immunogenicity, such as tetanus toxoid.

The **constant region (C region)** of an immunoglobulin or T-cell receptor is that part of the molecule that is relatively constant in amino acid sequence between different molecules. In an antibody molecule the constant regions of each chain are composed of one or more C domains. The constant region of an antibody determines its particular effector function. Cf. **variable region.**

A **contact hypersensitivity reaction** is a form of delayed-type hypersensitivity in which T cells respond to antigens that are introduced by contact with the skin. Poison ivy hypersensitivity is a contact hypersensitivity reaction due to T-cell responses to the chemical antigen pentadecacatechol in poison ivy leaves.

Continuous epitopes, or linear epitopes, are antigenic determinants on proteins that are contiguous in the amino acid sequence and therefore do not require the protein to be folded into its native conformation for antibody to bind. The epitopes detected by T cells are continuous.

A **convertase** is an enzymatic activity that converts a complement protein into its reactive form by cleaving it. Generation of the C3 convertase is the pivotal event in complement activation.

The **Coombs test** is a test for antibody binding to red blood cells. Red blood cells that are coated with antibody are agglutinated if they are exposed to an anti-immunoglobulin antibody. The Coombs test is important in detecting the nonagglutinating antibodies against red blood cells produced by Rh incompatibility in pregnancy.

Two binding sites are said to demonstrate **cooperativity** in binding to their ligand when the binding of ligand to one site enhances the binding of ligand to the second site.

A **co-receptor** is a cell-surface protein that increases the sensitivity of the antigen receptor to antigen by binding to associated ligands and participating in signaling for activation. CD4 and CD8 are MHC-binding co-receptors on T cells, whereas CD19 is part of a complex that makes up a co-receptor on B cells.

Corona: see **B-cell corona.**

Corticosteroids are a family of drugs related to steroids that are naturally produced in the adrenal cortex, such as cortisone. Corticosteroids can kill lymphocytes, especially developing thymocytes, inducing apoptotic cell death. They are useful anti-inflammatory, anti-lymphoid tumor, and immunosuppressive agents.

The proliferation of lymphocytes requires both antigen binding and the receipt of a **co-stimulatory signal**. Co-stimulatory signals are delivered to T cells by the **co-stimulatory molecules**, B7.1 and B7.2, related molecules that are expressed on the surface of the cell presenting antigen, and which bind the T-cell surface molecule CD28. B cells may receive co-stimulatory signals from common pathogen components such as LPS, from complement fragments, or from CD40 ligand expressed on the surface of an activated antigen-specific helper T cell.

Cowpox is the common name of the disease produced by vaccinia virus, used by Edward Jenner in the successful vaccination against smallpox, which is caused by the related variola virus.

CR: see **complement receptors.**

CR1 (CD35) is one of several receptors on cells for various components of complement. It is used to remove immune complexes from the plasma.

CR2 (CD21) is part of the B-cell co-receptor complex along with CD19 and CD81. It binds to antigens that have various breakdown products of C3, especially C3d, bound to them and, by cross-linking to the B-cell receptor, enhances sensitivity to antigen by at least a hundredfold. It is also used by the Epstein–Barr virus to invade B cells and produce the symptoms of infectious mononucleosis.

CR3 (CD11b:CD18) is a β_2 integrin that functions both as an adhesion molecule and as a complement receptor. It binds iC3b, and stimulates phagocytosis.

CR4 (CD11c:CD18) is a β_2 integrin that binds iC3b and stimulates phagocytosis.

C-reactive protein is an acute-phase protein that binds to phosphorylcholine, which is a constituent of the C-polysaccharide of the bacterium *Streptococcus pneumoniae*, hence its name. Many other bacteria also have surface phosphorylcholine that is accessible to C-reactive protein, so the protein can bind many different bacteria and opsonize them for uptake by phagocytes. C-reactive protein does not bind to mammalian tissues.

Cross-matching is used in blood typing and histocompatibility testing to determine whether donor and recipient have antibodies against each other's cells that might interfere with successful transfusion or grafting.

A **cross-reaction** is the binding of antibody to an antigen not used to elicit that antibody. Thus, if antibody raised against antigen A also binds antigen B, it is said to cross-react with antigen B. The term is used generically to describe the reactivity of antibodies or T cells with antigens other than the eliciting antigen.

The protein known as **Csk** or **C-terminal Src kinase** is constitutively active in lymphocytes and has the function of phosphorylating the C-terminal tyrosine of Src-family tyrosine kinases, thus inactivating them.

CTLA-4 is the high-affinity receptor for B7 molecules on T cells.

Cutaneous lymphocyte antigen (CLA) is a cell-surface molecule that is involved in lymphocyte homing to the skin in humans.

Cutaneous T-cell lymphoma is a malignant growth of T cells that home to the skin.

Cyclophosphamide is a DNA alkylating agent that is used as an immunosuppressive drug. It acts by killing rapidly dividing cells, including lymphocytes proliferating in response to antigen.

Cyclosporin A is a powerful immunosuppressive drug that inhibits signaling from the T-cell receptor, preventing T-cell activation and effector function. It binds to cyclophilin, and this complex binds to and inactivates the serine/threonine phosphatase calcineurin.

Cytokine capture: see **antigen capture.**

Cytokine receptors are cellular receptors for cytokines. Binding of the cytokine to the cytokine receptor induces new activities in the cell, such as growth, differentiation, or death. Cytokine receptors are listed in Appendix III.

Cytokines are proteins made by cells that affect the behavior of other cells. Cytokines made by lymphocytes are often called lymphokines or interleukins (abbreviated IL), but the generic term cytokine is used in this book and most of the literature. Cytokines act on specific cytokine receptors on the cells that they affect. Cytokines and their receptors are listed in Appendix III. See also **chemokines.**

T cells that can kill other cells are called **cytotoxic T cells**. Most cytotoxic T cells are MHC class I-restricted CD8 T cells, but CD4 T cells can also kill in some cases. Cytotoxic T cells are important in host defense against cytosolic pathogens.

Cytotoxins are proteins made by cytotoxic T cells that participate in the destruction of target cells. Perforins and granzymes are the major defined cytotoxins.

In the context of immunoglobulins, δ is the type of heavy chain in IgD.

D gene segments, or **diversity gene segments**, are short DNA sequences that join the V and J gene segments in rearranged immunoglobulin heavy-chain genes and in T-cell receptor β and δ chain genes. See **gene segments**.

Dark zone: see **germinal centers.**

Death domains were originally defined in proteins encoded by genes involved in programmed cell death or apoptosis, and are now known to be involved in protein–protein interactions.

The **decay-accelerating factor (DAF** or **CD55)** is a cell-surface molecule that protects cells from lysis by complement. Its absence causes the disease paroxysmal nocturnal hemoglobinuria.

Defective endogenous retroviruses are partial retroviral genomes integrated into host cell DNA and carried as host genes. There are a great many defective endogenous retroviruses in the mouse genome.

Delayed-type hypersensitivity, or type IV hypersensitivity, is a form of cell-mediated immunity elicited by antigen in the skin and is mediated by CD4 T_H1 cells. It is called delayed-type hypersensitivity because the reaction appears hours to days after antigen is injected. Cf. **immediate hypersensitivity**.

Dendritic cells, also known as interdigitating reticular cells, are found in T-cell areas of lymphoid tissues. They have a branched or dendritic morphology and are the most potent stimulators of T-cell responses. Nonlymphoid tissues also contain dendritic cells, but these are not able to stimulate T-cell responses until they are activated and migrate to lymphoid tissues. The dendritic cell derives from bone marrow precursors. It is distinct from the follicular dendritic cell that presents antigen to B cells.

Dendritic epidermal T cells (dETCs), are a specialized class of $\gamma{:}\delta$ T cells found in the skin of mice and some other species, but not humans. All dETCs have the same $\gamma{:}\delta$ T-cell receptor; their function is unknown.

Desensitization is a procedure in which an allergic individual is exposed to increasing doses of allergen in hopes of inhibiting their allergic reactions. It probably involves shifting the balance between CD4 T_H1 and T_H2 cells and thus changing the antibody produced from IgE to IgG.

Determinant spreading: see **epitope spreading**.

Diacylglycerol (DAG) is most commonly released from inositol phospholipids by the action of phospholipase C-γ. Diacylglycerol production is stimulated by the ligation of many receptors and it acts as an intracellular signaling molecule, activating cytosolic protein kinase C, which further propagates the signal.

Diapedesis is the movement of blood cells, particularly leukocytes, from the blood across blood vessel walls into tissues.

The **differential signaling hypothesis** proposes that qualitatively different antigens might mediate the positive and negative selection of T cells in the thymus. Cf. **avidity hypothesis**.

Differentiation antigens are proteins detected on some cells by means of specific antibodies. Many differentiation antigens have important functional roles characteristic of the differentiated phenotypes of the cell on which they are expressed, such as cell-surface immunoglobulin on B cells.

DiGeorge's syndrome is a recessive genetic immunodeficiency disease in which there is a failure to develop thymic epithelium, and is associated with absent parathyroid glands and large vessel anomalies. It seems to be due to a developmental defect in neural crest cells.

The **direct Coombs test** uses anti-immunoglobulin to agglutinate red blood cells as a way of detecting whether they are coated with antibody *in vivo* due to autoimmunity or maternal anti-fetal immune responses (see **Coombs test**; **indirect Coombs test**).

Discontinuous epitopes: see **conformational epitopes**.

DNA microarrays are created by placing a different DNA on a small part of a microchip, and using them to assess RNA expression in normal or malignant cells.

When vaccinating with plasmid DNA, it was seen that an adaptive immune response to the encoded protein occurred, leading to the term **DNA vaccination**. This lead to the realization that bacterial DNA, which is loaded with unmethylated CpG dinucleotides, was adjuvant for this type of vaccination.

The genetic defect in *scid* mice, which cannot rearrange their T- or B-cell receptor genes and have a severe combined immunodeficiency

phenotype, is in the enzyme **DNA-dependent kinase**. This enzyme is part of a complex of proteins that bind to the hairpin ends of double-stranded breaks in DNA, and its catalytic subunit is critical for VDJ recombination.

In tissue grafting experiments, the grafted tissues come from a **donor** and are placed in a recipient or host.

Double-negative thymocytes are immature T cells within the thymus that lack expression of the two co-receptors, CD4 and CD8. In a normal thymus, these represent about 5% of thymocytes.

Double-positive thymocytes are an intermediate stage in T-cell development in the thymus and are characterized by expression of both the CD4 and the CD8 co-receptor proteins. They represent the majority (~80%) of thymocytes.

The term **draining lymph node** is used for any lymph node that is downstream of a site of infection and thus receives antigens and microbes from the site via the lymphatic system. Draining lymph nodes often enlarge enormously during an immune response and can be palpated; they were originally called swollen glands.

In the context of immunoglobulins, ε (epsilon) is the heavy chain of IgE.

The **early induced responses** or early nonadaptive responses are a series of host defense responses that are triggered by infectious agents early in infection. They are distinct from innate immunity because there is an inductive phase, and from adaptive immunity in that they do not operate by clonal selection of rare antigen-specific lymphocytes.

Early pro-B cell: see **pro-B cell**.

The common skin disease **eczema** is seen mainly in children; its etiology is poorly understood.

In immunology, **edema** is the swelling caused by the entry of fluid and cells from the blood into the tissues, which is one of the cardinal features of the process of inflammation.

Effector lymphocytes can mediate the removal of pathogens from the body without the need for further differentiation, as distinct from naive lymphocytes, which must proliferate and differentiate before they can mediate effector functions, and memory cells, which must differentiate and often proliferate before they become effector cells. They are also called armed effector cells in this book, to indicate that they can be triggered to effector function by antigen binding alone.

Effector mechanisms are those processes by which pathogens are destroyed and cleared from the body. Innate and adaptive immune responses use most of the same effector mechanisms to eliminate pathogens.

Lymphocytes leave a lymph node through the **efferent lymphatic vessel**.

Electrophoresis is the movement of molecules in a charged field. In immunology, many forms of electrophoresis are used to separate molecules, especially protein molecules, to determine their charge, size, and subunit composition.

ELISA: see **enzyme-linked immunosorbent assay**.

ELISPOT assay is an adaptation of ELISA in which cells are placed over antibodies or antigens attached to a plastic surface. The antigen or antibody traps the cells' secreted products, which can then be detected using an enzyme-coupled antibody that cleaves a colorless substrate to make a localized colored spot.

Embryonic stem (ES) cells are mouse embryonic cells that will grow continuously in culture and that retain the ability to contribute to all cell lineages. ES cells can be genetically manipulated in tissue culture and then inserted into mouse blastocysts to generate mutant lines of mice; most often, genes are deleted in ES cells by homologous recombination and the mutant ES cells are then used to generate gene knockout mice. They have also been used to clone sheep, and could soon be used to replace body parts in humans.

Encapsulated bacteria have thick carbohydrate coats that protect them from phagocytosis. Encapsulated bacteria can cause extracellular infections and are effectively engulfed and destroyed by phagocytes only if they are first coated with antibody and complement produced in an adaptive immune response.

Cytokines that can induce a rise in body temperature are called **endogenous pyrogens**, as distinct from exogenous substances such as endotoxin from gram-negative bacteria that induce fever by triggering endogenous pyrogen synthesis and release.

Antigen taken up by phagocytosis generally enters the **endosomes**, the acidified vesicles present in cells. Protein antigens entering by this route are presented by MHC class II molecules.

Endotoxins are bacterial toxins that are released only when the bacterial cell is damaged, as opposed to exotoxins, which are secreted bacterial toxins. The most important endotoxin is the lipopolysaccharide of gram-negative bacteria, which is a potent inducer of cytokine synthesis and the proximal cause of endotoxic shock.

Within genomic DNA, there are specific sequences that act as cell-specific **enhancers** of RNA transcription.

The **enzyme-linked immunosorbent assay (ELISA)** is a serological assay in which bound antigen or antibody is detected by a linked enzyme that converts a colorless substrate into a colored product. The ELISA assay is widely used in biology and medicine as well as in immunology.

Eosinophils are white blood cells thought to be important chiefly in defense against parasitic infections; they are activated by the lymphocytes of the adaptive immune response. The level of eosinophils in the blood is normally quite low. It can increase markedly in several situations, such as atopy, resulting in **eosinophilia**, an abnormally large number of eosinophils in the blood.

Eotaxin-1 and **eotaxin-2** are CC chemokines that act specifically on eosinophils.

An **epitope** is a site on an antigen recognized by an antibody or an antigen receptor; epitopes are also called antigenic determinants. A T-cell epitope is a short peptide derived from a protein antigen. It binds to an MHC molecule and is recognized by a particular T cell. B-cell epitopes are antigenic determinants recognized by B cells and are typically discontinuous in the primary structure.

Epitope spreading describes the fact that responses to autoantigens tend to become more diverse as the response persists. This is also called determinant spreading or antigen spreading.

The **Epstein–Barr virus (EBV)** is a herpesvirus that selectively infects human B cells by binding to complement receptor 2 (CR2, also known as CD21). It causes infectious mononucleosis and establishes a lifelong latent infection in B cells that is controlled by T cells. Some B cells latently infected with EBV will proliferate *in vitro* to form lymphoblastoid cell lines.

The affinity of an antibody for its antigen can be determined by **equilibrium dialysis**, a technique in which antibody in a dialysis bag is exposed to varying amounts of a small antigen able to diffuse across the dialysis membrane. The amount of antigen inside and outside the bag at the equilibrium diffusion state is determined by the amount and affinity of the antibody in the bag.

E-rosettes are human T cells that will bind to treated red blood cells from sheep; the many red blood cells bound to each T cell give it the appearance of a rosette and increase its buoyant density so that the T cells can be isolated by gradient centrifugation. E-rosetting is often used for isolating human T cells.

Erp57 is a chaperone molecule involved in loading peptide onto MHC class I molecules in the endoplasmic reticulum.

Erythroblastosis fetalis is a severe form of Rh hemolytic disease in which maternal anti-Rh antibody enters the fetus and produces a hemolytic anemia so severe that the fetus has mainly immature erythroblasts in the peripheral blood.

E-selectin: see **selectins**.

Experimental allergic encephalomyelitis (EAE) is an inflammatory disease of the central nervous system that develops after mice are immunized with neural antigens in a strong adjuvant.

The movement of cells or fluid from within blood vessels to the surrounding tissues is called **extravasation**.

IgG antibody molecules can be cleaved into three fragments by the enzyme papain. Two of these are identical **Fab fragments**, so called because they are the **F**ragment with specific **a**ntigen **b**inding. The Fab fragment consists of the light chain and the N-terminal half of the heavy chain held together by an interchain disulfide bond. Another protease, pepsin, cuts in the same general region of the antibody molecule as papain but on the carboxy-terminal side of the disulfide bonds. This produces the **F(ab')₂ fragment**, in which the two arms of the antibody molecule remain linked. See also **Fc fragment**.

FACS®: see **fluorescence-activated cell sorter**.

Factor B, factor D, factor H, factor I, and **factor P** are all components of the alternative pathway of complement activation. **Factor B** plays a role very similar to that of C2b in the classical pathway. **Factor D** is a serine protease that cleaves factor B. **Factor H** is an inhibitory protein with a role similar to decay-accelerating factor. **Factor I** is a protease that breaks down various components of the alternative pathway. **Factor P**, or properdin, is a positive regulatory component of the alternative pathway. It stabilizes the C3 convertase of the alternative pathway on the surface of bacterial cells.

Farmer's lung is a hypersensitivity disease caused by the interaction of IgG antibodies with large amounts of an inhaled allergen in the alveolar wall of the lung, causing alveolar wall inflammation and compromising gas exchange.

Fas is a member of the TNF receptor family; it is expressed on certain cells and makes them susceptible to killing by cells expressing **Fas ligand**, a cell-surface member of the TNF family of proteins. Binding of Fas ligand to Fas triggers apoptosis in the Fas-bearing cell.

IgG antibody molecules can be cleaved into three fragments by the enzyme papain. One of these is the **Fc fragment**, so-called for **F**ragment **c**rystallizable. The Fc fragment consists of the C-terminal halves of the two heavy chains disulfide-bonded to each other by the residual hinge region. See also **Fab fragments**.

Fc receptors are receptors for the Fc portion of immunoglobulin isotypes. They include the Fcγ and Fcε receptors.

The high-affinity **Fcε receptor (FcεRI)** on the surface of mast cells and basophils binds free IgE. When antigen binds this IgE and cross-links FcεRI, it causes mast-cell activation.

Fcγ receptors, including FcγRI, RII, and RIII, are cell-surface receptors that bind the Fc portion of IgG molecules. Most Fcγ receptors bind only aggregated IgG, allowing them to discriminate bound antibody from free IgG. They are expressed on phagocytes, B lymphocytes, NK cells, and follicular dendritic cells. They have a key role in humoral immunity, linking antibody binding to effector cell functions.

When tissue or organ grafts are placed in an unmatched recipient, they are rejected by a **first set rejection**, which is an immune response by the host against foreign antigens in the graft. Cf. **second set rejection**.

FK506: see **tacrolimus**.

Individual cells can be characterized and separated in a machine called a **fluorescence-activated cell sorter** (**FACS®**) that measures cell size, granularity, and fluorescence due to bound fluorescent antibodies as single cells pass in a stream past photodetectors. The analysis of single cells in this way is called flow cytometry and the instruments that carry out the measurements and/or sort cells are called flow cytometers or cell sorters.

Peripheral lymphoid tissues, such as lymph nodes and Peyer's patches, contain large areas of B cells called **follicles**, which are organized around follicular dendritic cells.

A **follicular center cell lymphoma** is a type of B-cell lymphoma that tends to grow in the follicles of lymphoid tissues.

The **follicular dendritic cells** of lymphoid follicles are cells of uncertain origin. They are characterized by long branching processes that make intimate contact with many different B cells. They have Fc receptors that are not internalized by receptor-mediated endocytosis and thus hold antigen:antibody complexes on the surface for long periods. These cells are crucial in selecting antigen-binding B cells during antibody responses.

The V domains of immunoglobulins and T-cell receptors contain relatively invariant **framework regions** that provide a protein scaffold for the hypervariable regions that make contact with antigen.

Fungi are single-celled and multicellular eukaryotic organisms, including the yeasts and molds, that can cause a variety of diseases. Immunity to fungi is complex and involves both humoral and cell-mediated responses.

Fv: see **single-chain Fv**.

Fyn: see **tyrosine kinase**.

In the context of immunoglobulins, γ is the heavy chain of IgG.

G proteins are intracellular proteins that bind GTP and convert it to GDP in the process of cell signal transduction. There are two kinds of G protein, the heterotrimeric (α, β, γ) receptor-associated G proteins, and the small G proteins, such as Ras and Raf, that act downstream of many transmembrane signaling events.

GALT: see **gut-associated lymphoid tissues**.

Most T lymphocytes have α:β heterodimeric T-cell receptors, but some bear a distinct γ:δ **T-cell receptor** composed of different antigen-recognition chains, γ and δ, assembled in a γ:δ heterodimer. Cells bearing these receptors are called γ:δ **T cells** and their specificity and function are not yet clear.

Plasma proteins can be separated on the basis of electrophoretic mobility into albumin and the α, β, and γ globulins. Most antibodies migrate in electrophoresis as γ **globulins** (or **gamma globulins**), and patients who lack antibodies are said to have agammaglobulinemia.

GEF: see **guanine-nucleotide exchange factor**.

In birds and rabbits, immunoglobulin receptor diversity is generated mainly by **gene conversion**, in which homologous inactive V gene segments exchange short sequences with an active, rearranged V-region gene.

Gene knockout is jargon for gene disruption by homologous recombination.

The V domains of the polypeptide chains of antigen receptors are encoded in sets of **gene segments** that must first undergo somatic recombination to form a complete V-domain exon. There are three types of gene segment: V gene segments that encode the first 95 amino acids, D gene segments that encode about 5 amino acids, and J gene segments that form the last 10–15 amino acids of the V region. There are multiple copies of each type of gene segment in the germline DNA, but only one is expressed for each type of receptor chain in a receptor-bearing lymphocyte.

A gene can be specifically disrupted by a technique known as **gene targeting** or gene knockout. Usually this involves homologous recombination in embryonic stem cells followed by the preparation of chimeric mice by injection of these cells into the blastocyst.

Gene therapy is the correction of a genetic defect by the introduction of a normal gene into bone marrow or other cell types. It is also known as somatic gene therapy because it does not affect the germline genes of the individual.

Genetic immunization is a novel technique for inducing adaptive immune responses. Plasmid DNA encoding a protein of interest is injected into muscle; for unknown reasons, it is expressed and elicits antibody and T-cell responses to the protein encoded by the DNA.

Mice that are raised in the complete absence of intestinal and other flora are called **germ-free** or **gnotobiotic** mice. Such mice have very depleted immune systems, but they can respond virtually normally to any specific antigen, provided it is mixed with a strong adjuvant.

Germinal centers in secondary lymphoid tissues are sites of intense B-cell proliferation, selection, maturation, and death during antibody responses. Germinal centers form around follicular dendritic cell networks when activated B cells migrate into lymphoid follicles. They can be divided by morphology into the dark zone, which is rich in proliferating B lymphocytes, and a light zone, which contains FDCs and centrocytes.

Immunoglobulin and T-cell receptor genes are said to be in the **germline configuration** in the DNA of germ cells and in all somatic cells in which somatic recombination has not occurred.

The **germline diversity** of antigen receptors is due to the inheritance of multiple gene segments that encode V domains; such diversity is distinguished from the diversity that is generated during gene rearrangement or after receptor gene expression, which is somatically generated.

One theory of antibody diversity, the **germline theory**, proposed that each antibody was encoded in a separate germline gene. This is now known not to happen in people, mice, and most other organisms, but appears to happen in Elasmobranchs, which have rearranged genes in the germline.

GlyCAM-1 is a mucinlike molecule found on the high endothelial venules of lymphoid tissues. It is an important ligand for the L-selectin molecule expressed on naive lymphocytes, directing these cells to leave the blood and enter the lymphoid tissues.

Gnotobiotic: see **germ-free**.

Goodpasture's syndrome is an autoimmune disease in which autoantibodies against basement membrane or type IV collagen are produced and cause extensive vasculitis. It is rapidly fatal.

Tissue and organ grafts between genetically distinct individuals almost always elicit an adaptive immune response that causes **graft rejection**, the destruction of the grafted tissue by attacking lymphocytes.

When mature T lymphocytes are injected into a nonidentical immunoincompetent recipient, they can attack the recipient, causing a **graft-versus-host** (GVH) reaction; in human patients, mature T cells in allogeneic bone marrow grafts can cause **graft-versus-host disease** (GVHD).

Granulocyte: see **polymorphonuclear leukocyte**.

Granulocyte-macrophage colony-stimulating factor (GM-CSF) is a cytokine involved in the growth and differentiation of myeloid and monocytic lineage cells, including dendritic cells, monocytes and tissue macrophages, and cells of the granulocyte lineage.

A **granuloma** is a site of chronic inflammation usually triggered by persistent infectious agents such as mycobacteria or by a non-degradable foreign body. Granulomas have a central area of macrophages, often fused into multinucleate giant cells, surrounded by T lymphocytes.

Granzymes are serine proteases produced by cytotoxic T cells and are involved in inducing apoptosis in the target cell.

Graves' disease is an autoimmune disease in which antibodies against the thyroid-stimulating hormone receptor cause overproduction of thyroid hormone and thus hyperthyroidism.

The **guanine-nucleotide exchange factors** (GEFs) are proteins that can remove the bound GDP from small G proteins; this allows GTP to bind and activate the G protein.

The **gut-associated lymphoid tissues** (GALT) are lymphoid tissues closely associated with the gastrointestinal tract, including the palatine tonsils, Peyer's patches, and intraepithelial lymphocytes. The GALT has a distinctive biology related to its exposure to antigens from food and normal intestinal microbial flora.

H antigens or **histocompatibility antigens**, are known as major histocompatibility antigens when they encode molecules that present foreign peptides to T cells and as minor H antigens when they present polymorphic self peptides to T cells. See also **histocompatibility**.

The major histocompatibility complex of the mouse is called **H-2** (for **histocompatibility-2**). Haplotypes are designated by a lower-case superscript, as in H-2^b.

A **haplotype** is a linked set of genes associated with one haploid genome. The term is used mainly in connection with the linked genes of the major histocompatibility complex, which are usually inherited as one haplotype from each parent. Some MHC haplotypes are over-represented in the population, a phenomenon known as linkage disequilibrium.

Haptens are molecules that can bind antibody but cannot by themselves elicit an adaptive immune response. Haptens must be chemically linked to protein carriers to elicit antibody and T-cell responses.

Hashimoto's thyroiditis is an autoimmune disease characterized by persistent high levels of antibody against thyroid-specific antigens. These antibodies recruit NK cells to the tissue, leading to damage and inflammation.

All immunoglobulin molecules have two types of chain, a **heavy chain (H chain)** of 50–70 kDa and a light chain of 25 kDa. The basic immunoglobulin unit consists of two identical heavy chains and two identical light chains. Heavy chains come in a variety of heavy-chain classes or isotypes, each of which confers a distinctive functional activity on the antibody molecule.

Helper CD4 T cells are CD4 T cells that can help B cells make antibody in response to antigenic challenge. The most efficient helper T cells are also known as T$_H$2 cells, which make the cytokines IL-4 and IL-5. Some experts refer to all CD4 T cells, regardless of function, as helper T cells; we do not accept this usage because function can be determined only in cellular assays, and some CD4 T cells kill the cells they interact with.

A **hemagglutinin** is any substance that causes red blood cells to agglutinate, a process known as **hemagglutination**. The hemagglutinins in human blood are antibodies that recognize the ABO blood group antigens. Influenza and some other viruses have hemagglutinin molecules that bind to glycoproteins on host cells to initiate the infectious process.

Hematopoiesis is the generation of the cellular elements of blood, including the red blood cells, leukocytes, and platelets. These cells all originate from pluripotent **hematopoietic stem cells** whose differentiated progeny divide under the influence of **hematopoietic growth factors**.

A **hematopoietic lineage** is any developmental series of cells that derives from hematopoietic stem cells and results in the production of mature blood cells.

Hemolytic disease of the newborn: see **erythroblastosis fetalis**.

Recombination signal sequences (RSS) flanking gene segments consist of a seven-nucleotide **heptamer** and a nine-nucleotide nonamer of conserved sequence, separated by 12 or 23 nucleotides. RSSs form the target for the site-specific RAG-1:RAG-2 recombinase that joins the gene segments.

Hereditary angioneurotic edema is the clinical name for a genetic deficiency of the C1 inhibitor of the complement system. In the absence of C1 inhibitor, spontaneous activation of the complement system can cause diffuse fluid leakage from blood vessels, the most serious consequence of which is epiglottal swelling leading to suffocation.

Individuals **heterozygous** for a particular gene have two different alleles of that gene.

An excellent model for membranous glomerulonephritis is **Heymann's nephritis**, a disease induced by injecting animals with tubular epithelial tissue.

High endothelial venules (HEVs) are specialized venules found in lymphoid tissues. Lymphocytes migrate from blood into lymphoid tissues by attaching to and migrating across the **high endothelial cells** of these vessels.

Tolerance to injected protein antigens occurs at low or high doses of antigen. Tolerance induced by the injection of high doses of antigen is called **high-zone tolerance**, whereas tolerance produced with low doses of antigen is called low-zone tolerance.

The **hinge region** of antibody molecules is a flexible domain that joins the Fab arms to the Fc piece. The flexibility of the hinge region in IgG and IgA molecules allows the Fab arms to adopt a wide range of angles, permitting binding to epitopes spaced variable distances apart.

Histamine is a vasoactive amine stored in mast cell granules. Histamine released by antigen binding to IgE molecules on mast cells causes dilation of local blood vessels and smooth muscle contraction, producing some of the symptoms of immediate hypersensitivity reactions. Antihistamines are drugs that counter histamine action.

Histocompatibility is literally the ability of tissues (Greek: *histos*) to get along with each other. It is used in immunology to describe the genetic systems that determine the rejection of tissue and organ grafts resulting from immunological recognition of histocompatibility (H) antigens.

Histocompatibility-2: see **H-2**.

HIV: see **human immunodeficiency virus**.

HLA, the acronym for **H**uman **L**eukocyte **A**ntigen, is the genetic designation for the human MHC. Individual loci are designated by upper-case letters, as in HLA-A, and alleles are designated by numbers, as in HLA-A*0201.

The invariant **HLA-DM** molecule in humans is involved in loading peptides onto MHC class II molecules. It is encoded in the MHC within a set of genes resembling MHC class II genes. A homologous protein in mice is called H-2M.

Hodgkin's disease is an immune system tumor characterized by large cells called Reed-Sternberg cells, which derive from mutated B-lineage cells. It exists in at least two polar forms, Hodgkin's lymphoma and nodular sclerosis.

Homeostasis is a generic term describing the status of physiological normality. In the case of lymphocytes, homeostasis refers to an uninfected individual who has normal numbers of lymphocytes.

Cellular genes can be disrupted by **homologous recombination** with copies of the gene into which erroneous sequences have been inserted. When these exogenous DNA fragments are introduced into cells, they recombine selectively with the cellular gene through remaining regions of sequence homology, replacing the functional gene with a nonfunctional copy.

Host-versus-graft disease (HVGD) is another name for the allograft rejection reaction. The term is used mainly in relation to bone marrow transplantation.

The **human immunodeficiency virus (HIV)** is the causative agent of the acquired immune deficiency syndrome (AIDS). HIV is a retrovirus of the lentivirus family that selectively infects macrophages and CD4 T cells, leading to their slow depletion, which eventually results in immunodeficiency.

Human leukocyte antigen: see **HLA**

Humanization is a term used to describe the genetic engineering of mouse hypervariable loops of a desired specificity into otherwise human antibodies. The DNA encoding hypervariable loops of mouse monoclonal antibodies or V regions selected in phage display libraries is inserted into the framework regions of human immunoglobulin genes. This allows the production of antibodies of a desired specificity that do not cause an immune response in humans treated with them.

Humoral immunity is the antibody-mediated specific immunity made in a **humoral immune response**. Humoral immunity can be transferred to unimmunized recipients by using immune serum containing specific antibody.

Monoclonal antibodies are most commonly produced from **hybridomas**. These are hybrid cell lines formed by fusing a specific antibody-producing B lymphocyte with a myeloma cell that is selected for its ability to grow in tissue culture and for an absence of immunoglobulin chain synthesis.

Hyperacute graft rejection of an allogenic tissue graft is an immediate reaction caused by natural preformed antibodies that react against antigens on the graft. The antibodies bind to endothelium and trigger the blood clotting cascade, leading to an engorged, ischemic graft and rapid loss of the organ.

Hypereosinophilia is an abnormal state in which there are extremely large numbers of eosinophils in the blood.

Repetitive immunization to achieve a heightened state of immunity is called **hyperimmunization**.

Immune responses to innocuous antigens that lead to symptomatic reactions upon reexposure are called **hypersensitivity reactions**. These can cause **hypersensitivity diseases** if they occur repetitively. This state of heightened reactivity to antigen is called **hypersensitivity**. Hypersensitivity reactions are classified by mechanism: type I hypersensitivity reactions involve IgE antibody triggering of mast cells; type II hypersensitivity reactions involve IgG antibodies against cell-surface or matrix antigens; type III hypersensitivity reactions involve antigen:antibody complexes; and type IV hypersensitivity reactions are T cell-mediated.

The **hypervariable (HV) regions** of immunoglobulin and T-cell receptor V domains are small regions that make contact with the antigen and differ extensively from one receptor to the next. Cf. **framework regions**.

The inactive complement fragment **iC3b** is produced by cleavage of C3b and is the first step in C3b inactivation.

The **ICAMs** (intercellular adhesion molecules) are cell-surface ligands for the leukocyte integrins and are crucial in the binding of lymphocytes and other leukocytes to certain cells, including antigen-presenting cells and endothelial cells. They are members of the immunoglobulin superfamily. **ICAM-1** is the most prominent ligand for the integrin CD11a:CD18 or LFA-1. It is rapidly inducible on endothelial cells by infection, and plays a major role in local inflammatory responses. **ICAM-2** is constitutively expressed at relatively low levels by endothelium. **ICAM-3** is expressed only on leukocytes and is thought to play an important part in adhesion between T cells and antigen-presenting cells, particularly dendritic cells.

All antibody molecules belong to a family of plasma proteins called **immunoglobulins (Ig)**. Membrane-bound immunoglobulin serves as the specific antigen receptor on B lymphocytes.

Iccosomes are small fragments of membrane coated with immune complexes that fragment off the processes of follicular dendritic cells in lymphoid follicles early in a secondary or subsequent antibody response.

ICOS is a CD28-related protein that is induced on activated T cells and can enhance T-cell responses. It binds a ligand known as LICOS, which is distinct from the B7 molecules.

Each immunoglobulin molecule has the potential of binding a variety of antibodies directed at its unique features or **idiotype**. An idiotype is made up of a series of **idiotopes** from idiotype epitopes.

Lymphocyte antigen receptors can recognize one another through idiotope–anti-idiotope interactions, forming an **idiotypic network** of receptors that may be important for the generation and maintenance of the receptor repertoire. The proposed components of idiotype networks exist, but their functional significance is uncertain.

IFN: see **interferons**.

Ig: standard abbreviation for **immunoglobulin**. Different immunoglobulin isotypes are called IgM, IgD, IgG, IgA, and IgE.

Igα, Igβ: see **B-cell antigen receptor**.

IgA is the class of immunoglobulin characterized by α heavy chains. IgA antibodies are secreted mainly by mucosal lymphoid tissues.

IgD is the class of immunoglobulin characterized by δ heavy chains. It appears as surface immunoglobulin on mature naive B cells but its function is unknown.

IgE is the class of immunoglobulin characterized by ε heavy chains. It is involved in allergic reactions.

IgG is the class of immunoglobulin characterized by γ heavy chains. It is the most abundant class of immunoglobulin found in the plasma.

IgM is the class of immunoglobulin characterized by μ heavy chains. It is the first immunoglobulin to appear on the surface of B cells and the first to be secreted.

IL: see **interleukins**.

Immature B cells are B cells that have rearranged a heavy- and a light-chain V-region gene and express surface IgM, but have not yet matured sufficiently to express surface IgD as well.

Tissues throughout the body contain **immature dendritic cells**, which only leave the tissues in response to an inflammatory mediator or an infection. See also **dendritic cells**.

During allergic reactions, there are normally two phases: the first happens almost immediately and is called the **immediate reaction**. Hypersensitivity reactions that occur within minutes of exposure to antigen are called **immediate hypersensitivity reactions**; such reactions are antibody mediated. Cf. **delayed-type hypersensitivity**.

When large amounts of antigen are injected into the blood, they are initially removed slowly by normal catabolic processes that also degrade plasma proteins. However, if the antigen elicits an antibody response, then antigen is removed at an accelerated rate as antigen:antibody complexes, a process known as **immune clearance**.

The binding of antibody to a soluble antigen forms an **immune complex**. Large immune complexes form when sufficient antibody is available to cross-link the antigen; these are readily cleared by the reticuloendothelial system of cells bearing Fc and complement receptors. Small, soluble immune complexes that form when antigen is in excess can be deposited in and damage small blood vessels.

Immune deviation is a term used to describe the polarization of an immune response to one dominated by T_H1 or T_H2 by the injection of antigen.

Immune modulation is a general term encompassing various alterations in an immune response.

The **immune response** is the response made by the host to defend itself against a pathogen.

Immune response (Ir) genes are genetic polymorphisms that control the intensity of the immune response to a particular antigen. Virtually all Ir phenotypes are due to the differential binding of peptide fragments of antigen to MHC molecules, especially MHC class II molecules. The term is little used now. An **immune response (Ir) gene defect** is usually, but not always, due to failure to bind an immunogenic peptide, so that no T-cell response is observed.

It has been proposed that most tumors that arise are detected and eliminated by **immune surveillance** mediated by lymphocytes specific for tumor antigens. There is little evidence for the efficacy of this proposed process, but it remains an important concept in tumor immunology.

The **immune system** is the name used to describe the tissues, cells, and molecules involved in adaptive immunity, or sometimes the totality of host defense mechanisms.

Immunity is the ability to resist infection.

Immunization is the deliberate provocation of an adaptive immune response by introducing antigen into the body. See also **active immunization; passive immunization**.

Immunobiology is the study of the biological basis for host defense against infection.

Immunoblotting is a common technique in which proteins separated by gel electrophoresis are blotted onto a nitrocellulose membrane and revealed by the binding of specific labeled antibodies.

Immunodeficiency diseases are a group of inherited or acquired disorders in which some aspect or aspects of host defense are absent or functionally defective.

Immunodiffusion is the detection of antigen or antibody by the formation of an antigen:antibody precipitate in a clear agar gel.

Immunoelectrophoresis is a technique in which antigens are first separated by their electrophoretic mobility and are then detected and identified by immunodiffusion.

Immunofluorescence is a technique for detecting molecule using antibodies labeled with fluorescent dyes. The bound fluorescent antibody can be detected by microscopy, or by flow cytometry depending on the application being used. **Indirect immunofluorescence** uses anti-immunoglobulin antibodies labeled with fluorescent dyes to detect the binding of a specific unlabeled antibody.

There are three ways of detecting molecules in tissues: **immunofluorescent microscopy** that reveals the presence of any molecule against which you have a specific antibody; **immunohistochemistry**, in which one links an enzyme that produces a change in a molecule that is visible under the microscope; and **immunoelectronmicroscopy**, in which different sized gold particles are linked to antibodies and detected as bound gold particles.

Any molecule that can elicit an adaptive immune response on injection into a person or animal is called an **immunogen**. In practice, only proteins are fully **immunogenic** because only proteins can be recognized by T lymphocytes.

Immunogenetics was originally the analysis of genetic traits by means of antibodies against genetically polymorphic molecules such as blood group antigens or MHC proteins. Immunogenetics now refers to the genetic analysis, by any technique, of molecules important in immunology.

Immunoglobulin A: see **IgA**.

Immunoglobulin D: see **IgD**.

Many proteins are partly or entirely composed of protein domains known as **immunoglobulin domains** or **Ig domains** because they were first described in antibody molecules. Immunoglobulin domains are characteristic of proteins of the immunoglobulin superfamily, which includes antibodies, T-cell receptors, MHC molecules, and many other proteins described in this book. The immunoglobulin domain consists of a sandwich of two β sheets held together by a disulfide bond and called the **immunoglobulin fold**. There are two main types of immunoglobulin domain: C domains and V domains. Domains less closely related to the canonical Ig domains are sometimes also called **immunoglobulin-like domains**.

Immunoglobulin G: see **IgG**.

Immunoglobulin E: see **IgE**.

Immunoglobulin M: see **IgM**.

The **immunoglobulin repertoire**, also known as the antibody repertoire, is the total variety of immunoglobulin molecules in the body of an individual.

Many proteins involved in antigen recognition and cell–cell interaction in the immune system and other biological systems are members of a protein family called the **immunoglobulin superfamily**, or **Ig superfamily**, because their shared structural features were first defined in immunoglobulin molecules. All members of the immunoglobulin superfamily have at least one immunoglobulin or immunoglobulin-like domain.

The detection of antigens in tissues by means of visible products produced by the degradation of a colorless substrate by antibody-linked enzymes is called **immunohistochemistry**. This technique has the advantage that it can be combined with other stains to be viewed in the light microscope, whereas immunofluorescence microscopy requires a special dark-field or UV microscope.

Immunological ignorance describes a form of self tolerance in which reactive lymphocytes and their target antigen are both detectable within an individual, yet no autoimmune attack occurs. Most autoimmune diseases probably reflect the loss of other lymphocytes known as regulatory or suppressor T cells.

When an antigen is encountered more than once, the adaptive immune response to each subsequent encounter is speedier and more effective, a crucial feature of protective immunity known as **immunological memory**. Immunological memory is specific for a particular antigen and is long-lived.

Allogeneic tissue placed in certain sites in the body, such as the brain, does not elicit graft rejection. Such sites are called **immunologically privileged sites**. Immunological privilege results from the effects of both physical barriers to cell and antigen migration, and soluble immunosuppressive mediators such as certain cytokines.

Immunology is the study of all aspects of host defense against infection and of adverse consequences of immune responses.

Immunophilins are proteins with peptidyl-prolyl *cis–trans* isomerase activity that bind the immunosuppressive drugs cyclosporin A, tacrolimus, and rapamycin.

Soluble proteins, or membrane proteins solubilized in detergents, can be labeled and then detected by **immunoprecipitation analysis** using specific antibodies. The immunoprecipitated labeled protein is usually detected by SDS-PAGE followed by autoradiography. When proteins that do not react directly with the antibody used are nevertheless precipitated, they are said to co-immunoprecipitate.

The T and B cell antigen receptors are associated with transmembrane molecules with **immunoreceptor tyrosine-based activation motifs** (**ITAMs**) in their cytoplasmic domains. These tyrosine-containing motifs are sites of tyrosine phosphorylation and of association with tyrosine kinases and other phosphotyrosine-binding moieties involved in receptor signaling. Related motifs with opposing effects are **immunoreceptor tyrosine-based inhibitory motifs** (**ITIMs**), which recruit phosphatases to the receptor site that remove the phosphate groups added by tyrosine kinases.

The ability of the immune system to sense and regulate its own responses is called **immunoregulation**.

Compounds that inhibit adaptive immune responses are called **immunosuppressive drugs**. They are used mainly in the treatment of graft rejection and severe autoimmune disease.

Immunotoxins are antibodies that are chemically coupled to toxic proteins usually derived from plants or microbes. The antibody targets the toxin moiety to the required cells. Immunotoxins are being tested as anticancer agents and as immunosuppressive drugs.

The **indirect Coombs test** is a variation of the direct Coombs test in which an unknown serum is tested for antibodies against normal red blood cells by first mixing the two and then washing out the serum from the red blood cells and reacting them with anti-immunoglobulin antibody. If antibody in the unknown serum binds to the red blood cells, agglutination by anti-immunoglobulin occurs.

Indirect immunofluorescence: see **immunofluorescence**.

Macrophages and many other cells have an **inducible NO synthase**, or **iNOS**, that is induced by many different stimuli to activate NO synthesis. This is a major mechanism of host resistance to intracellular infection in mice, and probably in humans as well.

Infectious mononucleosis, or glandular fever, is the common form of infection with the Epstein–Barr virus. It consists of fever, malaise, and swollen lymph nodes.

Inflammation is a general term for the local accumulation of fluid, plasma proteins, and white blood cells that is initiated by physical injury, infection, or a local immune response. This is also known as an **inflammatory response**. Acute inflammation is the term used to describe early and often transient episodes, whereas chronic inflammation occurs when the infection persists or during autoimmune diseases. Many different forms of inflammation are seen in different diseases. The cells that invade tissues undergoing inflammatory responses are often called **inflammatory cells** or an **inflammatory infiltrate**.

Influenza hemagglutinin: see **hemagglutinin**.

The early phases of the host response to infection depend on **innate immunity** in which a variety of innate resistance mechanisms recognize and respond to the presence of a pathogen. Innate immunity is present in all individuals at all times, does not increase with repeated exposure to a given pathogen, and discriminates between a group of related pathogens.

When inositol phospholipid is cleaved by phospholipase C-γ, it yields **inositol trisphosphate (IP$_3$)** and diacylglycerol. Inositol trisphosphate releases calcium ions from intracellular stores in the endoplasmic reticulum.

In **insulin-dependent diabetes mellitus (IDDM)**, the β cells of the pancreatic islets of Langerhans are destroyed so that no insulin is produced. The disease is believed to result from an autoimmune attack on the β cells.

Integrins are heterodimeric cell-surface proteins involved in cell–cell and cell–matrix interactions. They are important in adhesive interactions between lymphocytes and antigen-presenting cells and in lymphocyte and leukocyte migration into tissues. The **β$_1$-integrins**, or very late antigens (VLA), are a family of integrins with shared β$_1$ chains and different α chains that mediate adhesion to other cells and to extracellular matrix proteins.

Intercellular adhesion molecules: see **ICAMs**.

Interdigitating reticular cells: see **dendritic cells**.

Interferons are cytokines that can induce cells to resist viral replication. **Interferon-α (IFN-α)** and **interferon-β (IFN-β)** are produced by leukocytes and fibroblasts, respectively, as well as by other cells, whereas **interferon-γ (IFN-γ)** is a product of CD4 T$_H$1 cells, CD8 T cells, and NK cells. IFN-γ has as its primary action the activation of macrophages.

Interleukin, abbreviated **IL**, is a generic term for cytokines produced by leukocytes. We use the more general term cytokine in this book, but the term interleukin is used in the naming of specific cytokines such as IL-2. The interleukins are listed in Appendix III.

Interleukin-2 (IL-2) is the cytokine that is most central to the development of an adaptive immune response.

Staining for cytokines in cells that produce them can be achieved by permiablizing the cell and reacting it with a labelled fluorescent anti-cytokine antibody. This procedure is called **intracellular cytokine staining**.

Injections can be administered by a number of routes: **intracutaneous** (intradermal)—entering the skin or dermis; **subcutaneous**—entering below the skin or dermis; **intramuscular**—entering the muscle; **intranasal**—by way of the nose; and **intravenous**—entering a vein.

Intrathymic dendritic cells: see **dendritic cells**.

The major histocompatibility complex (MHC) class II proteins are assembled in the endoplasmic reticulum with the **invariant chain (Ii)**, which is involved in shielding the MHC class II molecules from binding peptides and in delivering them to cellular vesicles. There Ii is degraded, leaving the MHC class II molecules able to bind peptide fragments of antigen.

ISCOMs are **immune stimulatory complexes** of antigen held within a lipid matrix that acts as an adjuvant and enables the antigen to be taken up into the cytoplasm after fusion of the lipid with the plasma membrane.

Isoelectric focusing is an electrophoretic technique in which proteins migrate in a pH gradient until they reach the place in the gradient at which their net charge is neutral—their isoelectric point. Uncharged proteins no longer migrate; thus each protein is focused at its isoelectric point.

The first antibodies produced in a humoral immune response are IgM, but activated B cells subsequently undergo **isotype switching** or class switching to secrete antibodies of different isotypes: IgG, IgA, and IgE. Isotype switching does not affect antibody specificity significantly, but alters the effector functions that an antibody can engage. Isotype switching occurs by recombination involving the deletion of DNA between the rearranged V region and the selected C-region exon at so-called S regions.

Immunoglobulins are made in several distinct **isotypes** or classes—IgM, IgG, IgD, IgA, and IgE—each of which has a distinct heavy-chain C region encoded by a distinct C-region gene. The isotype of an antibody determines the effector mechanisms that it can engage on binding antigen. The different heavy-chain C regions are encoded in exons 3′ to the V(D)J rearrangement site. This allows the same antibody heavy-chain V region to link up with different heavy-chain C-region isotypes as a result of somatic recombination.

Isotypic exclusion describes the use of one or other of the light-chain isotypes, κ or λ, by a given B cell or antibody.

The **J gene segments**, or **joining gene segments**, are found some distance 5′ to the C genes in immunoglobulin and T-cell receptor loci. A V and in the case of IgH chains, TCRβ chains and TCRδ chains, a D gene segment must also rearrange to a J gene segment to form a complete V-region exon.

Cytokine receptors signal via **Janus kinases (JAKs)**—tyrosine kinases that are activated by the aggregation of cytokine receptors. These kinases phosphorylate proteins known as STATs, for **S**ignal **T**ransducers and **A**ctivators of **T**ranscription. STATs are normally found in the cytosol, but move to the nucleus on phosphorylation and activate a variety of genes.

Junctional diversity is the diversity present in antigen-specific receptors that is created during the process of joining V, D, and J gene segments.

In the context of immunoglobulins, **κ** is one of the two classes of light-chain.

Killer activatory receptors (KARs) are cell-surface receptors on NK cells or cytotoxic T cells; they are receptors that can activate killing by these cells.

Killer T cell is a commonly used term for cytotoxic T cell.

The **kinin system** is an enzymatic cascade of plasma proteins that is triggered by tissue damage to produce several inflammatory mediators, including the vasoactive peptide bradykinin.

The cell-surface receptor **c-Kit**, present on developing B cells and other developing white blood cells, binds to the stem cell factor on bone marrow stromal cells. Kit has protein tyrosine kinase activity.

Kupffer cells are phagocytes lining the hepatic sinusoids; they remove debris and dying cells from the blood, but are not known to elicit immune responses.

In the context of immunoglobulins, **λ** is one of the two is one of the two classes of light-chain.

λ5: see **pre-B-cell receptor**.

L chain: see **light chain**.

Langerhans' cells are phagocytic dendritic cells found in the epidermis. They can migrate from the epidermis to regional lymph nodes via the afferent lymphatics. In the lymph node they differentiate into dendritic cells.

The **large pre-B cells** have a cell-surface **pre-B-cell receptor**, which is lost on the transition to small pre-B cells, in which light-chain gene rearrangement occurs.

LAT: see **linker of activation in T cells**.

The **late pro-B cell** is the stage in B-cell development in which V_H to DJ_H joining occurs.

Some viruses can enter a cell but not replicate, a state known as **latency**. Latency can be established in various ways; when the virus is reactivated and replicates, it can produce disease.

In type I immediate hypersensitivity reactions, the **late-phase reaction** persists and is resistant to treatment with antihistamine.

The tyrosine kinase **Lck** associates most strongly with the cytoplasmic tails of CD4 and CD8. It plays a central role in signal transduction and activation of the T-cell receptor.

LCMV: see **lymphocytic choriomeningitis virus**.

Lentiviruses are a group of retroviruses that include the human immunodeficiency virus, HIV-1. They cause disease after a long incubation period and can take years to become apparent.

Leprosy is caused by *Mycobacterium leprae* and occurs in a variety of forms. There are two polar forms: **lepromatous leprosy**, which is characterized by abundant replication of leprosy bacilli and abundant antibody production without cell-mediated immunity; and **tuberculoid leprosy**, in which few organisms are seen in the tissues, there is little or no antibody, but cell-mediated immunity is very active. The other forms of leprosy are intermediate between the polar forms.

Leukemia is the unrestrained proliferation of a malignant white blood cell characterized by very high numbers of the malignant cells in the blood. Leukemias can be lymphocytic, myelocytic, or monocytic.

Leukocyte is a general term for a white blood cell. Leukocytes include lymphocytes, polymorphonuclear leukocytes, and monocytes.

Leukocyte adhesion deficiency is an immunodeficiency disease in which the common β chain of the leukocyte integrins is not produced. This mainly affects the ability of leukocytes to enter sites of infection with extracellular pathogens, so that infections cannot be effectively eradicated.

Leukocyte common antigen: see **CD45**.

The **leukocyte functional antigens** or **LFAs**, are cell adhesion molecules initially defined with monoclonal antibodies: **LFA-1** is a β_2 integrin; **LFA-2** is a member of the immunoglobulin superfamily, as is **LFA-3,** now called CD58. LFA-1 is particularly important in T-cell adhesion to endothelial cells and antigen-presenting cells.

Leukocyte integrins: see **leukocyte functional antigens**.

Leukocytosis is the presence of increased numbers of leukocytes in the blood. It is commonly seen in acute infection.

Leukotrienes are lipid mediators of inflammation that are derived from arachidonic acid. They are produced by macrophages and other cells.

LFA-1, LFA-2, LFA-3: see **leukocyte functional antigens**.

LICOS is the ligand for ICOS, a CD28-related protein that is induced on activated T cells and can enhance T-cell responses. LICOS is produced on activated dendritic cells, monocytes, and B cells.

The immunoglobulin **light chain** (**L chain**) is the smaller of the two types of polypeptide chain that make up all immunoglobulins. It consists of one V and one C domain, and is disulfide-bonded to the heavy chain. There are two classes of light chain, known as κ and λ.

Light zone: see **germinal centers**.

Linear epitope: see **continuous epitopes**.

Alleles at linked loci within the major histocompatibility complex are said to be in **linkage disequilibrium** if they are inherited together more frequently than predicted from their individual frequencies.

Epitopes recognized by B cells and helper T cells must be physically linked for the helper T cell to activate the B cell. This is called **linked recognition**.

The adaptor protein known as **linker of activation in T cells** (**LAT**) is a cytoplasmic protein with several tyrosines that become phosphorylated by the tyrosine kinase ZAP-70. It becomes associated with membrane lipid rafts and coordinates downstream signaling events in T-cell activation.

Low-zone tolerance: see **high-zone tolerance**.

A molecule of bacterial lipopolysaccharide (LPS) has first to be bound by the **LPS-binding protein** (**LBP**) before it can interact with CD14, an LPS:LBP-binding protein on cells such as macrophages.

L-selectin is an adhesion molecule of the selectin family found on lymphocytes. L-selectin binds to CD34 and GlyCAM-1 on high endothelial venules to initiate the migration of naive lymphocytes into lymphoid tissue. Also called CD62L.

Lyme disease is a chronic infection with *Borrelia burgdorferi*, a spirochete that can evade the immune response.

Lymph is the extracellular fluid that accumulates in tissues and is carried by lymphatic vessels back through the lymphatic system to the thoracic duct and into the blood.

Lymph nodes are a type of peripheral or secondary lymphoid organ. They are found in many locations thoughout the body where lymphatic vessels converge, and are sites where adaptive immune responses are initiated. Antigen-presenting cells and antigen delivered by the lymphatic vessels from a site of infection are displayed to the many recirculating lymphocytes that migrate through the lymph nodes. Some of these lymphocytes will recognize the antigen and respond to it, triggering an adaptive immune response.

The **lymphatic system** is the system of lymphoid channels and tissues that drains extracellular fluid from the periphery via the thoracic duct to the blood. It includes the lymph nodes, Peyer's patches, and other organized lymphoid elements apart from the spleen, which communicates directly with the blood.

Lymphatic vessels, or **lymphatics**, are thin-walled vessels that carry lymph through the lymphatic system.

A **lymphoblast** is a lymphocyte that has enlarged and increased its rate of RNA and protein synthesis.

All adaptive immune responses are mediated by **lymphocytes**. Lymphocytes are a class of white blood cells that bear variable cell-surface receptors for antigen. These receptors are encoded in rearranging gene segments. There are two main classes of lymphocyte—B lymphocytes (B cells) and T lymphocytes (T cells)—which mediate humoral and cell-mediated immunity, respectively. Small lymphocytes have little cytoplasm and condensed nuclear chromatin; on antigen recognition, the cell enlarges to form a lymphoblast and then proliferates and differentiates into an antigen-specific effector cell.

Lymphocyte function-associated antigens: see **leukocyte functional antigens.**

The **lymphocyte receptor repertoire** is the totality of the highly variable antigen receptors carried by B and T lymphocytes.

Lymphocytic choriomeningitis virus (**LCMV**) is a virus that causes a nonbacterial meningitis in mice and occasionally in humans. It is used extensively in experimental studies.

Dendritic cells can arise from myeloid cells, in which case they are called myeloid dendritic cells, or from lymphoid tissues, in which case they are called **lymphoid dendritic cells**. Functional differences exist between these two lineages.

Lymphoid organs are organized tissues characterized by very large numbers of lymphocytes interacting with a nonlymphoid stroma. The central or primary lymphoid organs, where lymphocytes are generated, are the thymus and bone marrow. The main peripheral or secondary lymphoid organs, in which adaptive immune responses are initiated, are the lymph nodes, spleen, and mucosal-associated lymphoid tissues such as tonsils and Peyer's patches.

Lymphokines are cytokines produced by lymphocytes.

Lymphomas are tumors of lymphocytes that grow in lymphoid and other tissues but do not enter the blood in large numbers. There are many types of lymphoma, which represent the transformation of various developmental stages of B or T lymphocytes.

Lymphopoiesis is the differentiation of lymphoid cells from a common lymphoid progenitor.

Lymphotoxin (LT) is also known as tumor necrosis factor-β (TNF-β), a cytokine secreted by inflammatory CD4 T cells that is directly cytotoxic for some cells.

Lyn: see **tyrosine kinase**.

Lysosomes are acidified organelles that contain many degradative hydrolytic enzymes. Material taken up into endosomes is eventually delivered to lysosomes.

Lytic granules containing perforin and granzymes are a defining characteristic of armed effector cytotoxic cells.

In the context of immunoglobulins, **μ** is the heavy chain of IgM.

Antigens and pathogens enter the body from the intestines through cells called microfold or **M cells**, which are specialized for this function. They are found over the gut-associated lymphoid tissue, or GALT. They may provide a route of infection for HIV.

Mac-1 is another name for the leukocyte integrin CD11b:CD18 (or complement receptor 2 (CR2)).

Macroglobulin describes plasma proteins that are globulins of high molecular weight, including immunoglobulin M (IgM) and α_2-macroglobulin.

Resting macrophages will not destroy certain intracellular bacteria unless the macrophage is activated by a T cell. **Macrophage activation** is important in controlling infection and also causes damage to neighboring tissues.

The **macrophage mannose receptor** is highly specific for certain carbohydrates that occur on the surface of some pathogens but not on host cells.

Macrophages are large mononuclear phagocytic cells important in innate immunity, in early non-adaptive phases of host defense, as antigen-presenting cells, and as effector cells in humoral and cell-mediated immunity. They are migratory cells deriving from bone marrow precursors and are found in most tissues of the body. They have a crucial role in host defense.

Dendritic cells are unique in being able to carry out **macropinocytosis**, a process in which large amounts of extracellular fluid are taken up in single vesicles. This is one means of antigen uptake.

MadCAM-1 is the mucosal cell adhesion molecule-1 or mucosal addressin that is recognized by the lymphocyte surface proteins L-selectin and VLA-4, allowing specific homing of lymphocytes to mucosal tissues.

Eosinophils can be triggered to release their **major basic protein**, which can then act on mast cells to cause their degranulation.

The **major histocompatibility complex (MHC)** is a cluster of genes on human chromosome 6 or mouse chromosome 17. It encodes a set of membrane glycoproteins called the MHC molecules. The MHC class I molecules present peptides generated in the cytosol to CD8 T cells, and the MHC class II molecules present peptides degraded in intracellular vesicles to CD4 T cells. The MHC also encodes proteins involved in antigen processing and other aspects of host defense. The MHC is the most polymorphic gene cluster in the human

genome, having large numbers of alleles at several different loci. Because this polymorphism is usually detected by using antibodies or specific T cells, the MHC molecules are often called major histocompatibility antigens.

MALT: see **mucosal-associated lymphoid tissue**.

The **mannan-binding lectin (MBL)**, also called mannose-binding protein, is an acute-phase protein that binds to mannose residues. It can opsonize pathogens bearing mannose on their surfaces and can activate the complement system via the **mannan-binding lectin pathway (MB-lectin pathway)** an important part of innate immunity.

The follicular **mantle zone** is a rim of B lymphocytes that surrounds lymphoid follicles. The precise nature and role of mantle zone lymphocytes have not yet been determined.

MAP kinases: see **mitogen-activated protein kinases**.

The **marginal zone** of the lymphoid tissue of the spleen lies at the border of the white pulp. It contains a unique population of B cells, the **marginal zone B cells**, which do not circulate and are distinguished by a distinct set of surface proteins.

The components of the MB-lectin pathway of complement activation include two serine proteases, **MASP-1** and **MASP-2**, that bind to mannan-binding lectin and play the same role in cleaving C4 as do C1r and C1s in the classical pathway.

Mast cells are large cells found in connective tissues throughout the body, most abundantly in the submucosal tissues and the dermis. They contain large granules that store a variety of mediator molecules including the vasoactive amine histamine. Mast cells have high-affinity Fcε receptors (FcεRI) that allow them to bind IgE monomers. Antigen-binding to this IgE triggers mast-cell degranulation and mast-cell activation, producing a local or systemic immediate hypersensitivity reaction. Mast cells have a crucial role in allergic reactions. **Mastocytosis** indicates an overproduction of mast cells.

Mature B cells are B cells that have acquired surface IgM and IgD and have become able to respond to antigen.

MBL; MB-lectin pathway: see **mannan-binding lectin**.

The **medulla** is generally the central or collecting point of an organ. The thymic medulla is the central area of each thymic lobe, rich in bone marrow-derived antigen-presenting cells and the cells of a distinctive medullary epithelium. The medulla of the lymph node is a site of macrophage and plasma cell concentration through which the lymph flows on its way to the efferent lymphatics.

Membrane cofactor of proteolysis (MCP or CD46) is a host-cell membrane protein that acts in conjunction with factor I to cleave C3b to its inactive derivative iC3b and thus prevent convertase formation.

B cells carry on their surfaces many molecules of **membrane immunoglobulin (mIg)** of a single specificity, which acts as the receptor for antigen.

The **membrane-attack complex** is made up of the terminal complement components, which assemble to generate a membrane-spanning hydrophilic pore, damaging the membrane.

Membranous glomerulonephritis is a disease of the kidneys characterized by proteinuria and heavy deposits of antibody and complement.

Memory cells are the lymphocytes that mediate immunological memory. They are more sensitive to antigen than are naive lymphocytes and respond rapidly on reexposure to the antigen that originally induced them. Both **memory B cells** and **memory T cells** have been defined.

MHC; MHC class I; MHC class II: see **major histocompatibility complex**.

The **MHC class IB** molecules encoded within the MHC are not highly polymorphic like the MHC class I and MHC class II molecules, and present a restricted set of antigens.

The **MHC class II compartment (MIIC)** is a site in the cell where MHC class II molecules accumulate, encounter HLA-DM, and bind antigenic peptides, before migrating to the surface of the cell.

The protein that activates the transcription of MHC class II genes, the **MHC class II transactivator (CIITA)**, is one of several defective genes in the disease bare lymphocyte syndrome, in which MHC class II molecules are lacking on all cells.

Various specialized strains of mice are used to explore the role of MHC polymorphism *in vivo*. These are called **MHC congenic**, meaning that the mice differ only at the MHC complex, **MHC recombinant**, meaning that the mice have a crossover within the MHC, or **MHC mutant**, meaning that they are mutant at one or more loci.

MHC genes are inherited in most cases as an **MHC haplotype**, the set of genes in a haploid genome inherited from one parent. Thus, if the parents are designated as ab and cd, then the offspring are most likely to be ac, ad, bc, or bd.

MHC molecules is the general name given to the highly polymorphic glycoproteins encoded by MHC class I and MHC class II genes, which are involved in the presentation of peptide antigens to T cells. They are also known as histocompatibility antigens.

The development of **MHC:peptide tetramers** held together by fluorescent streptavidin, which has four binding sites for the biotin attached to the tail of the MHC molecule, has made it possible to stain specific T cells in any species.

MHC-restricted antigen recognition, or **MHC restriction**, refers to the fact that a given T cell will recognize a peptide antigen only when it is bound to a particular MHC molecule. Normally, as T cells are stimulated only in the presence of self MHC molecules, antigen is recognized only as peptides bound to self MHC molecules.

MIC molecules are MHC class I-like molecules that are expressed in the gut under conditions of stress and are encoded within the class I region of the human MHC. They are not found in mice.

Microfold cells: see **M cells**.

Microorganisms are microscopic organisms, unicellular except for some fungi, that include bacteria, yeasts and other fungi, and protozoa, all of which can cause human disease.

mIg: see **membrane immunoglobulin**.

Anti-carbohydrate antibodies can bind either the ends or the middles of polysaccharide chains; the latter antibodies are called **middle-binders**.

MIIC: see **MHC class II compartment**.

Minor histocompatibility antigens (minor H antigens) are peptides of polymorphic cellular proteins bound to MHC molecules that can lead to graft rejection when they are recognized by T cells.

Minor lymphocyte stimulatory (Mls) loci: see **Mls antigens**.

Mitogen-activated protein kinases (MAP kinases) are kinases that become phosphorylated and activated on cellular stimulation by a variety of ligands, and lead to new gene expression by phosphorylating key transcription factors.

When lymphocytes from two unrelated individuals are cultured together, the T cells proliferate in response to the allogeneic MHC molecules on the cells of the other donor. This **mixed lymphocyte reaction** is used in testing for histocompatibility.

Mls antigens are non-MHC antigens that provoke strong primary mixed lymphocyte responses. They are encoded by **minor lymphocyte stimulatory (Mls) loci,** which are endogenous mammary tumor viruses integrated in the mouse genome. Mls antigens are encoded in the 3' long terminal repeat of the integrated virus and act as superantigens. They stimulate a large number of T lymphocytes by binding to the V_β domain of all T-cell receptors bearing the V_β for which the superantigen is specific.

MMTV: see **mouse mammary tumor virus**.

It has been proposed that infectious agents could provoke autoimmunity by **molecular mimicry**, the induction of antibodies and T cells that react against the pathogen but also cross-react with self antigens.

Monoclonal antibodies are produced by a single clone of B lymphocytes. Monoclonal antibodies are usually produced by making hybrid antibody-forming cells from a fusion of nonsecreting myeloma cells with immune spleen cells.

Monocytes are white blood cells with a bean-shaped nucleus; they are precursors of macrophages.

Some antibodies recognize all allelic forms of a polymorphic molecule such as an MHC class I protein; these antibodies are thus said to recognize a **monomorphic** epitope.

An individual lymphocyte carries antigen receptors of a single antigen specificity and thus has the property of **monospecificity** in response to antigen.

Mouse mammary tumor virus (MMTV) is a retrovirus that encodes a viral superantigen; integrated copies of related viruses encode the endogenous superantigens known as Mls antigens.

Mucins are highly glycosylated cell-surface proteins. Mucinlike molecules are bound by L-selectin in lymphocyte homing.

All of the body's internal epithelial organs are lined with epithelium that is coated with mucus, and are therefore called **mucosal epithelia**. This system is the site of entry for virtually all antigens, and is protected by a unique set of lymphoid organs.

The **mucosal-associated lymphoid tissue (MALT)** comprises all lymphoid cells in epithelia and in the lamina propria lying below the body's mucosal surfaces. The main sites of mucosal-associated lymphoid tissues are the gut-associated lymphoid tissues (GALT), and the bronchial-associated lymphoid tissues (BALT).

Multiple myeloma is a tumor of plasma cells, almost always first detected as multiple foci in bone marrow. Myeloma cells produce a monoclonal immunoglobulin, called a myeloma protein, that is detectable in the patient's plasma.

Multiple sclerosis is a neurological disease characterized by focal demyelination in the central nervous system, lymphocytic infiltration in the brain, and a chronic progressive course. It is caused by an autoimmune response to various antigens found in the myelin sheath.

Myasthenia gravis is an autoimmune disease in which autoantibodies against the acetylcholine receptor on skeletal muscle cells cause a block in neuromuscular junctions, leading to progressive weakness and eventually death.

Dendritic cells can arise from myeloid cells, in which case they are called **myeloid dendritic cells**, or from lymphoid tissues, in which case they are called lymphoid dendritic cells. Functional differences exist between these two lineages.

Myeloid progenitors are cells in bone marrow that give rise to the granulocytes and macrophages of the immune system.

Myeloma proteins are immunoglobulins secreted by myeloma tumors and are found in the patient's plasma.

Myelopoiesis is the production of monocytes and polymorphonuclear leukocytes in bone marrow.

Naive lymphocytes are lymphocytes that have never encountered their specific antigen and thus have never responded to it, as distinct from memory or effector lymphocytes. All lymphocytes leaving the central lymphoid organs are naive lymphocytes, those from the thymus being **naive T cells** and those from bone marrow being **naive B cells**.

Natural killer cells (NK cells) are large granular, non-T, non-B lymphocytes, which kill certain tumor cells. NK cells are important in innate immunity to viruses and other intracellular pathogens, as well as in antibody-dependent cell-mediated cytotoxicity (ADCC).

Necrosis is the death of cells or tissues due to chemical or physical injury, as opposed to apoptosis, which is a biologically programmed form of cell death. Necrosis leaves extensive cellular debris that needs to be removed by phagocytes, whereas apoptosis does not.

During intrathymic development, thymocytes that recognize self are deleted from the repertoire, a process known as **negative selection**. Autoreactive B cells undergo a similar process in bone marrow.

Antibodies that can inhibit the infectivity of a virus or the toxicity of a toxin molecule are said to **neutralize** them. Such antibodies are known as **neutralizing antibodies** and the process of inactivation as **neutralization**.

Neutropenia describes the situation in which there are fewer neutrophils in the blood than normal.

Neutrophils, also known as **neutrophilic polymorphonuclear leukocytes**, are the major class of white blood cell in human peripheral blood. They have a multilobed nucleus and neutrophilic granules. Neutrophils are phagocytes and have an important role in engulfing and killing extracellular pathogens.

NFAT: see **nuclear factor of activated T cells**.

The transcription factor called **NFκB** is made up of two chains of 50 kDa and 65 kDa. It is found under normal circumstances in the cytosol, where it is bound to a third chain called IκB, which is an inhibitor of NFκB transcription.

NK1.1 CD4 T cells are a small subset of T cells that express the NK1.1 marker, a molecule normally found on NK cells. NK1.1 T cells also express α:β T-cell receptors of limited diversity and either the co-receptor molecule CD4 or no co-receptor. They are enriched in the liver, and produce cytokines shortly after stimulation.

NK cells: see **natural killer cells**.

Nodular sclerosis: see **Hodgkin's disease**.

N-nucleotides are inserted into the junctions between gene segments of T-cell receptor and immunoglobulin heavy-chain V-region genes during gene segment joining. These N-regions are not encoded in either gene segment, but are inserted by the enzyme terminal deoxynucleotidyl transferase (TdT). They markedly increase the diversity of these receptors.

Recombination signal sequences (RSS) flanking gene segments consist of a seven-nucleotide heptamer and a nine-nucleotide **nonamer** of conserved sequence, separated by 12 or 23 nucleotides. RSSs form the target for the site-specific recombinase that joins the gene segments in antigen receptor gene rearrangement.

When T- and B-cell receptor gene segments rearrange, they often form **nonproductive rearrangements** that cannot encode a protein because the coding sequences are in the wrong translational reading frame.

N-regions: see **N-nucleotides**.

The transcription factor called **nuclear factor of activated T cells** (**NFAT**) is a complex of a protein called NFATc, as it is held in the cytosol by serine/threonine phosphorylation, and the Fos/Jun dimer known as AP-1. It moves from the cytosol to the nucleus on cleavage of the phosphate residues by calcineurin, a serine/threonine protein phosphatase.

The *nude* mutation of mice produces hairlessness and defective formation of the thymic stroma, so that nude mice, which are homozygous for this mutation, have no mature T cells.

When a person's work induces allergy, this is called **occupational allergy**.

Oncogenes are genes involved in regulating cell growth. When these genes are defective in structure or expression, they can cause cells to grow continuously to form a tumor.

An **opportunistic pathogen** is a microorganism that causes disease only in individuals with compromised host defense mechanisms, as occurs in AIDS.

Opsonization is the alteration of the surface of a pathogen or other particle so that it can be ingested by phagocytes. Antibody and complement opsonize extracellular bacteria for destruction by neutrophils and macrophages.

The feeding of foreign antigens leads typically to a state of specific and active unresponsiveness, a phenomenon known as **oral tolerance**.

An autoimmune disease that targets a specific organ is said to be **organ-specific**.

Original antigenic sin describes the tendency of humans to make antibody responses to those epitopes shared between the original strain of a virus and subsequent related viruses, while ignoring other highly immunogenic epitopes on the second and subsequent viruses.

Lymphocyte subpopulations can be isolated by **panning** on petri dishes coated with monoclonal antibodies against cell-surface markers, to which the lymphocytes bind.

The **paracortical area**, or **paracortex**, is the T-cell area of lymph nodes, lying just below the follicular cortex, which is primarily composed of B cells.

Parasites are organisms that obtain sustenance from a live host. In medical practice, the term is restricted to worms and protozoa, the subject matter of parasitology.

Paroxysmal nocturnal hemoglobinuria (**PNH**) is a disease in which complement regulatory proteins are defective, so that complement activation leads to episodes of spontaneous hemolysis.

Partial agonist peptides: see **altered peptide ligands**.

Passive hemagglutination is a technique for detecting antibody in which red blood cells are coated with antigen and the antibody is detected by agglutination of the coated red blood cells.

The injection of antibody or immune serum into a naive recipient is called **passive immunization**. Cf. **active immunization**.

Pathogenic microorganisms, or **pathogens**, are microorganisms that can cause disease when they infect a host.

Pathology is the scientific study of disease. The term pathology is also used to describe detectable damage to tissues.

The term **pattern recognition receptors** (**PRRs**) is used to define receptors that bind to **pathogen-associated molecular patterns** (**PAMPs**).

Pattern-recognition molecules are receptors of the innate immune system that recognize common molecular patterns on pathogen surfaces.

The cell-adhesion molecule **PECAM** (**CD31**) is found both on lymphocytes and at endothelial cell junctions. It is believed that CD31–CD31 interactions enable leukocytes to leave blood vessels and enter tissues.

Pentadecacatechol is the chemical substance in the leaves of the poison ivy plant that causes the cell-mediated immunity associated with hypersensitivity to poison ivy.

Pentraxins are a family of acute-phase proteins formed of five identical subunits, to which C-reactive protein and serum amyloid protein belong.

Perforin is a protein that can polymerize to form the membrane pores that are an important part of the killing mechanism in cell-mediated cytotoxicity. Perforin is produced by cytotoxic T cells and NK cells and is stored in granules that are released by the cell when it contacts a specific target cell.

The **periarteriolar lymphoid sheath** (**PALS**) is part of the inner region of the white pulp of the spleen, and contains mainly T cells.

Peripheral blood mononuclear cells are lymphocytes and monocytes isolated from peripheral blood, usually by Ficoll-Hypaque™ density centrifugation.

Peripheral or **secondary lymphoid organs** are the lymph nodes, spleen, and mucosal-associated lymphoid tissues, in which immune responses are induced, as opposed to the central lymphoid organs, in which lymphocytes develop.

Peripheral tolerance is tolerance acquired by mature lymphocytes in the peripheral tissues, as opposed to central tolerance, which is acquired by immature lymphocytes during their development.

Peyer's patches are aggregates of lymphocytes along the small intestine, especially the ileum.

Antibody-like phage can be produced by cloning immunoglobulin V-region genes in filamentous phage, which thus express antigen-binding domains on their surfaces, forming a **phage display library**. Antigen-binding phage can be replicated in bacteria and used like antibodies. This technique can be used to develop novel antibodies of any specificity.

Phagocytosis is the internalization of particulate matter by cells. Usually, the **phagocytic cells** or **phagocytes** are macrophages or neutrophils, and the particles are bacteria that are taken up and destroyed. The ingested material is contained in a vesicle called a **phagosome**, which then fuses with one or more lysosomes to form a **phagolysosome**. The lysosomal enzymes are important in pathogen destruction and degradation to small molecules.

Phosphatidylinositol bisphosphate (**PIP$_2$**) is a membrane-associated phospholipid that is cleaved by phospholipase C-γ to give the signaling molecules diacylglycerol and inositol trisphosphate.

Phospholipase C-γ is a key enzyme in signal transduction. It is activated by protein tyrosine kinases that are themselves activated by receptor ligation, and activated phospholipase C-γ cleaves inositol phospholipid into inositol trisphosphate and diacylglycerol.

Plasma is the fluid component of blood containing water, electrolytes, and the plasma proteins.

Plasma cells are terminally differentiated B lymphocytes and are the main antibody-secreting cells of the body. They are found in the medulla of the lymph nodes, in splenic red pulp, and in bone marrow.

A **plasmablast** is a B cell in a lymph node that already shows some features of a plasma cell.

Platelet activating factor (**PAF**) is a lipid mediator that activates the blood clotting cascade and several other components of the innate immune system.

Platelets are small cell fragments found in the blood that are crucial for blood clotting. They are formed from megakaryocytes.

P-nucleotides are nucleotides found in junctions between gene segments of the rearranged V-region genes of antigen receptors. They are an inverse repeat of the sequence at the end of the adjacent gene segment, being generated from a hairpin intermediate during recombination, and hence are called palindromic or P-nucleotides.

Poison ivy is a plant whose leaves contain pentadecacatechol, a potent contact sensitizing agent and a frequent cause of contact hypersensitivity.

Antigen activates specific lymphocytes, whereas all mitogens, by definition, activate most or all lymphocytes, a process known as **polyclonal activation** because it involves multiple clones of diverse specificity. Such mitogens are known as **polyclonal mitogens**.

The major histocompatibility complex is both **polygenic**, containing several loci encoding proteins of identical function, and polymorphic, having multiple alleles at each locus.

The **poly-Ig receptor** binds polymeric immunoglobulins, especially IgA, at the basolateral membrane of epithelia and transports them across the cell, where they are released from the apical surface. This transcytosis transfers IgA from its site of synthesis to its site of action at epithelial surfaces.

The **polymerase chain reaction** or **PCR** uses high temperature and unique thermostable enzymes to replicate DNA. It has revolutionized molecular biology.

Polymorphism literally means existing in a variety of different shapes. Genetic polymorphism is variability at a gene locus in which the variants occur at a frequency of greater than 1%. The major histocompatibility complex is the most polymorphic gene cluster known in humans.

Polymorphonuclear leukocytes are white blood cells with multilobed nuclei and cytoplasmic granules. There are three types of polymor-pho-nuclear leukocyte: the neutrophils with granules that stain with neutral dyes, the eosinophils with granules that stain with eosin, and the basophils with granules that stain with basic dyes.

Some antibodies show **polyspecificity**, the ability to bind to many different antigens. This is also known as **polyreactivity**.

Only those developing T cells whose receptors can recognize antigens presented by self MHC molecules can mature in the thymus, a process known as **positive selection**. All other developing T cells die before reaching maturity.

Expression of the **pre-B-cell receptor**, or **pre-B-cell receptor complex**, is a critical event in B-cell development. Expression of this receptor, which is a complex of at least five proteins, causes the pre-B cell to enter the cell cycle, to turn off the *RAG* genes, to degrade the RAG proteins, and to expand by several cell divisions. Then the signal ceases, and the pre-B cell is ready to rearrange its light chains.

During B-cell development, **pre-B cells** are cells that have rearranged their heavy-chain genes but not their light-chain genes.

The **precipitin reaction** was the first quantitative technique for measuring antibody production. The amount of antibody is determined from the amount of precipitate obtained with a fixed amount of antigen. The precipitin reaction also can be used to define antigen valence and zones of antibody or antigen excess in mixtures of antigen and antibody.

Prednisone is a synthetic steroid with potent anti-inflammatory and immunosuppressive activity used in treating acute graft rejection, autoimmune disease, and lymphoid tumors.

In T-cell development, TCR β chains expressed by CD44^{low}CD25$^+$ thymocytes pair with a surrogate α chain called pTα (pre-T-cell α) to form a **pre-T-cell receptor** that exits the endoplasmic reticulum via the Golgi as a complex with the CD3 molecules.

During T-dependent antibody responses, a **primary focus** of B-cell activation forms in the vicinity of the margin between T and B cell areas of lymphoid tissue. Here, the T and B cells interact and B cells can differentiate directly into antibody-forming cells or migrate to lymphoid follicles for further proliferation and differentiation.

Lymphoid tissues contain lymphoid follicles made up of follicular dendritic cells and B lymphocytes. The **primary follicles** contain resting B lymphocytes and are the site at which germinal centers form when they are entered by activated B cells, forming secondary follicles.

The **primary immune response** is the adaptive immune response to an initial exposure to antigen. **Primary immunization**, also known as priming, generates both the primary immune response and immunological memory.

The binding of antibody molecules to antigen is called a **primary interaction**, as distinct from secondary interactions in which binding is detected by some associated change such as the precipitation of soluble antigen or agglutination of particulate antigen.

Primary lymphoid organ: see **central lymphoid organ**.

Priming of antigen-specific naive lymphocytes occurs when antigen is presented to them in an immunogenic form; the primed cells will differentiate either into armed effector cells or into memory cells that can respond in second and subsequent immune responses.

During B-cell development, **pro-B cells** are cells that have displayed

B-cell surface marker proteins but have not yet completed heavy-chain gene rearrangement. They are divided into early pro-B cells and late pro-B cells.

Any lymphocyte receptor chain can be rearranged in either of two ways, productive and nonproductive. **Productive rearrangements** are in the correct reading frame for the receptor chain in question.

Progenitors are the more differentiated progeny of stem cells that give rise to distinct subsets of mature blood cells and lack the capacity for self-renewal posssed by true stem cells.

Programmed cell death: see **apoptosis**.

Properdin: see **factor P**.

Prostaglandins, like leukotrienes, are lipid products of the metabolism of arachidonic acid that have a variety of effects on a variety of tissues, including activities as inflammatory mediators.

Cytosolic proteins are degraded by a large catalytic multisubunit protease called a **proteasome**. It is thought that peptides that are presented by MHC class I molecules are generated by the action of proteasomes, and two interferon-inducible subunits of some proteasomes are encoded in the MHC.

Protectin (CD59) is a cell-surface protein that protects host cells from being damaged by complement. It inhibits the formation of the membrane-attack complex by preventing the binding of C8 and C9 to the C5b,6,7 complex.

Protective immunity is the resistance to specific infection that follows infection or vaccination.

Protein A is a membrane component of *Staphylococcus aureus* that binds to the Fc region of IgG and is thought to protect the bacteria from IgG antibodies by inhibiting their interactions with complement and Fc receptors. It is useful for purifying IgG antibodies.

Protein kinase C (**PKC**) is a family of serine/threonine kinases that are activated by diacylglycerol and calcium as a result of signaling via many different receptors.

Protein kinases add phosphate groups to proteins, and **protein phosphatases** remove these phosphate groups. Enzymes that add phosphate groups to tyrosine residues are called **protein tyrosine kinases**. These enzymes have crucial roles in signal transduction and regulation of cell growth. Their activity is regulated by a second set of molecules called **protein tyrosine phosphatases** that remove the phosphate from the tyrosine residues.

Proto-oncogenes are cellular genes that regulate growth control. When mutated or aberrantly expressed, they can contribute to the malignant transformation of cells, leading to cancer. Cf. **oncogenes**.

A **provirus** is the DNA form of a retrovirus when it is integrated into the host cell genome, where it can remain transcriptionally inactive for long periods of time.

P-selectin: see **selectins**.

pTα: see **pre-T-cell receptor**.

The **pulmonary surfactant A** and **D** proteins are members of the pentraxin family that play various roles in the acute-phase response.

Purine nucleotide phosphorylase (PNP) deficiency is an enzyme defect that results in severe combined immunodeficiency. This enzyme is important in purine metabolism, and its deficiency causes the accumulation of purine nucleosides, which are toxic for developing T cells, causing the immune deficiency.

Pus is the mixture of cell debris and dead neutrophils that is present in wounds and abscesses infected with extracellular encapsulated bacteria.

Bacteria with large capsules are difficult for phagocytes to ingest. Such encapsulated bacteria often produce pus at the site of infection, and are thus called **pyogenic bacteria**. Pyogenic organisms used to

kill many young people. Now, pyogenic infections are largely limited to the elderly.

The **12/23 rule** states that gene segments of immunoglobulin or T-cell receptors can be joined only if one has a recognition signal sequence with a 12 base pair spacer, and the other has a 23 base pair spacer.

Rac: see **small G proteins**.

Radiation bone marrow chimeras are mice that have been heavily irradiated and then reconstituted with bone marrow cells of a different strain of mouse, so that the lymphocytes differ genetically from the environment in which they develop. Such chimeric mice have been important in studying lymphocyte development.

Antigen–antibody interaction can be studied by **radioimmunoassay** (**RIA**) in which antigen or antibody is labeled with radioactivity. An unlabeled antigen or antibody is attached to a solid support such as a plastic surface, and the fraction of the labeled antibody or antigen retained on the surface is determined in order to measure binding.

Raf: see **small G proteins**.

The recombination activating genes *RAG-1* and *RAG-2* encode the proteins **RAG-1** and **RAG-2,** which are critical to receptor gene rearrangement. Mice lacking either of these genes cannot form receptors and thus have no lymphocytes.

Rapamycin, or sirolimus, is an immunosuppressive drug that blocks cytokine action.

Ras: see **small G proteins**.

Antigen receptor expression requires gene segment **rearrangement** in developing lymphocytes. The expressed V-region sequences are composed of rearranged gene segments.

The replacement of a light chain of a self-reactive antigen receptor on immature B cells with a light chain that does not confer autoreactivity is known as **receptor editing**. This has also been shown with heavy chains.

The antigen receptors of lymphocytes are associated with **receptor-associated tyrosine kinases**, mainly of the Src family, which bind to receptor tails via their SH2 domains.

Receptor-mediated endocytosis is the internalization into endosomes of molecules bound to cell-surface receptors. Antigens bound to B-cell receptors are internalized by this process.

A **recessive lethal gene** encodes a protein that is needed for the human or animal to develop to adulthood; when both copies are defective, the human or animal dies *in utero* or early after birth.

In any situation in which cells or tissues are transplanted, they come from a donor and are placed in a **recipient** or host.

Recombination activating genes: see **RAG-1** and **RAG-2**.

Recombination signal sequences (**RSSs**) are short stretches of DNA that flank the gene segments that are rearranged to generate a V-region exon. They always consist of a conserved heptamer and nonamer separated by 12 or 23 base pairs. Gene segments are only joined if one is flanked by an RSS containing a 12 base pair spacer and the other is flanked by an RSS containing a 23 base pair spacer—the 12/23 rule of gene segment joining.

The nonlymphoid area of the spleen in which red blood cells are broken down is called the **red pulp**.

Reed-Sternberg cells are large malignant B cells that are found in Hodgkin's disease.

Regulatory or **suppressor T cells** can inhibit T-cell responses.

When neutrophils and macrophages take up opsonized particles, this triggers a metabolic change in the cell called the **respiratory burst**. It leads to the production of a number of mediators.

The virus known as **respiratory syncytial virus (RSV)** is a human pathogen that is a common cause of severe chest infection in young children, often associated with wheezing, as well as in immunocompromised patients and patients with AIDS.

The **Rev** protein is the product of the *rev* gene of the human immunodeficiency virus (HIV). The Rev protein promotes the passage of viral RNA from nucleus to cytoplasm during HIV replication.

The enzyme **reverse transcriptase** is an essential component of retroviruses, as it translates the RNA genome into DNA before integration into host cell DNA. Reverse transcriptase also enables RNA sequences to be converted into complementary DNA (cDNA), and so to be cloned, and thus is an essential reagent in molecular biology.

The **reverse transcriptase–polymerase chain reaction** (RT-PCR) is used to amplify RNA sequences. The enzyme reverse transcriptase is used to convert an RNA sequence into a cDNA sequence, which is then amplified by PCR.

The **Rhesus (Rh) blood group antigen** is a red cell membrane antigen that is also detectable on the red blood cells of rhesus monkeys. Anti-Rh antibodies do not agglutinate human red blood cells, so antibody to Rh antigen must be detected by using a Coombs test.

Rheumatic fever is caused by antibodies elicited by infection with some *Streptococcus* species. These antibodies cross-react with kidney, joint, and heart antigens.

Rheumatoid arthritis is a common inflammatory joint disease that is probably due to an autoimmune response. The disease is accompanied by the production of **rheumatoid factor**, an IgM anti-IgG antibody that can also be produced in normal immune responses.

The technique of **sandwich ELISA** uses antibody bound to a surface to trap a protein by binding to one of its epitopes. The trapped protein is then detected by an enzyme-linked antibody specific for a different epitope on the protein's surface. This gives the assay a high degree of specificity.

Scatchard analysis is a mathematical analysis of equilibrium binding that allows the affinity and valence of a receptor–ligand interaction to be determined.

Scavenger receptors on macrophages and other cells bind to numerous ligands and remove them from the blood. The Kupffer cells in the liver are particularly rich in scavenger receptors.

SCID, *scid*: see **severe combined immunodeficiency**.

SDS-PAGE is the common abbreviation for polyacrylamide gel electrophoresis (PAGE) of proteins dissolved in the detergent sodium dodecyl sulfate (SDS). This technique is widely used to characterize proteins, especially after labeling and immunoprecipitation.

A **secondary antibody response** is the antibody response induced by a second or booster injection of antigen—a **secondary immunization**. The secondary response starts sooner after antigen injection, reaches higher levels, is of higher affinity than the primary response, and is dominated by IgG antibodies. Therefore, the response to each immunization is increasingly intense, so the secondary, tertiary, and subsequent responses are of increasing magnitude.

When the recipient of a first tissue or organ graft has rejected that graft, a second graft from the same donor is rejected more rapidly and vigorously in what is called a **second set rejection**.

The co-stimulatory signal required for lymphocyte activation is often called a **second signal**, with the first signal coming from the binding of antigen by the antigen receptor. Both signals are required for the activation of most lymphocytes.

Secondary interactions: see **primary interactions**.

The **secretory component** attached to IgA antibodies in body secretions is a fragment of the poly-Ig receptor left attached to the IgA after transport across epithelial cells.

A cell is said to be **selected** by antigen when its receptors bind that antigen. If the cell starts to proliferate as a result, this is called clonal selection, and the cell founds a clone; if the cell is killed as a result of binding antigen, this is called negative selection or clonal deletion.

Selectins are a family of cell-surface adhesion molecules of leukocytes and endothelial cells that bind to sugar moieties on specific glycoproteins with mucinlike features.

Antigens in the body of an individual are by convention called **self antigens**. Lymphocytes are screened during their immature stages for reactivity with self antigens, and those that do respond undergo apoptosis.

Tolerance is the failure to respond to an antigen; when that antigen is borne by self tissues, tolerance is called **self tolerance**. See also **tolerance**.

Allergic reactions require prior immunization, called **sensitization**, by the allergen that elicits the acute response. Allergic reactions occur only in **sensitized** individuals.

Sepsis is infection of the bloodstream. This is a very serious and frequently fatal condition. Infection of the blood with gram-negative bacteria triggers **septic shock** through the release of the cytokine TNF-α.

A **sequence motif** is a pattern of nucleotides or amino acids shared by different genes or proteins that often have related functions. Sequence motifs observed in peptides that bind a particular MHC glycoprotein are based on the requirements for particular amino acids to achieve binding to that MHC molecule.

Seroconversion is the phase of an infection when antibodies against the infecting agent are first detectable in the blood.

Serology is the use of antibodies to detect and measure antigens using **serological assays**, so called because these assays were originally carried out with serum, the fluid component of clotted blood, from immunized individuals.

Mast cells in mice store the small soluble mediator **serotonin** in their granules.

Different strains of bacteria and other pathogens can sometimes be distinguished by their **serotype**, the ability of an immune serum to agglutinate or lyse some strains of bacteria and not others.

Serpins are a large family of protease inhibitors.

Serum is the fluid component of clotted blood.

Serum sickness occurs when foreign serum or serum proteins are injected into a person. It is caused by the formation of immune complexes between the injected protein and the antibodies formed against it. It is characterized by fever, arthralgias, and nephritis.

Severe combined immune deficiency (SCID) is an immune deficiency disease in which neither antibody nor T-cell responses are made. It is usually the result of T-cell deficiencies. The *scid* mutation causes severe combined immune deficiency in mice.

SH2 domain: see **Src-family tyrosine kinases**.

A **signal joint** is formed by the precise joining of recognition signal sequences in the process of somatic recombination that generates T-cell receptor and immunoglobulin genes.

Signal transducers and activators of transcription (STATs): see **Janus kinases**.

Signal transduction describes the general process by which cells perceive changes in their environment. In lymphocytes, the most important changes are those occurring during infection that generate antigens that stimulate the cells of the immune system to bring about an adaptive immune response.

A **single-chain Fv** fragment, comprising a V region of a heavy chain linked by a stretch of synthetic peptide to a V region of a light chain, can be made by genetic engineering.

During T-cell maturation in the thymus, mature T cells are detected by the expression of either the CD4 or the CD8 co-receptor and are therefore called **single-positive thymocytes**.

Sirolimus is the drug name that has been adopted for the chemical rapamycin; the two terms are used interchangeably in the literature.

The chemokine known as **SLC** is produced by lymphatic vessels and attracts dendritic cells.

Small G proteins are monomeric G proteins such as Ras, that act as intracellular signaling molecules downstream of many transmembrane signaling events. They bind GTP in their active form, and hydrolyze it to GDP to become inactive.

Small pre-B cells: see **large pre-B cells**.

Smallpox is an infectious disease, caused by the virus variola, that once killed at least 10% of infected people. It has now been eradicated by vaccination.

When immunologists discovered that antibodies were variable, they entertained various theories, including the **somatic diversification theory** that postulated that the genes for immunoglobulin were constant, and that they diversified in somatic cells. This turned out to be partly true, as somatic hypermutation is now well established. However, other theories were needed to explain other features of antibody diversity, including somatic gene rearrangement and isotype switching.

Somatic gene therapy: see **gene therapy**.

During B-cell responses to antigen, the V-region DNA sequence undergoes **somatic hypermutation,** resulting in the generation of variant immunoglobulins, some of which bind antigen with a higher affinity. This allows the affinity of the antibody response to increase. These mutations affect only somatic cells, and are not inherited through germline transmission.

During lymphocyte development, receptor gene segments undergo **somatic recombination** to generate intact V-region exons that encode the V region of each immunoglobulin and T-cell receptor chain. These events occur only in somatic cells; the changes are not inherited.

Spacer: see **12-23 rule**.

The **specificity** of an antibody determines its ability to distinguish the immunogen from other antigens.

One uses **spectratyping** to define certain types of DNA gene segments that give a repetitive spacing of three nucleotides, or one codon.

The **spleen** is an organ in the upper left side of the peritoneal cavity containing a red pulp, involved in removing senescent blood cells, and a white pulp of lymphoid cells that respond to antigens delivered to the spleen by the blood.

The **Src-family tyrosine kinases** are receptor-associated protein tyrosine kinases. They have several domains, called Src-homology 1, 2, and 3. The SH1 domain contains the active site of the kinase, the SH2 domain can bind to phosphotyrosine residues, and the SH3 domain is involved in interactions with proline-rich regions in other proteins.

Staphylococcal enterotoxins (SEs) cause food poisoning and also stimulate many T cells by binding to MHC class II molecules and the V_β domain of certain T-cell receptors; the staphylococcal enterotoxins are thus superantigens.

STATs: see **Janus kinases**.

Stem-cell factor (SCF) is a transmembrane protein on bone marrow stromal cells that binds to c-Kit, a signaling receptor carried on developing B cells and other developing white blood cells.

The development of B lymphocytes and T lymphocytes occurs in association with **stromal cells**, which provide various soluble and cell-bound signals to the developing lymphocyte.

Antigens can be injected into the **subcutaneous** layer to induce an adaptive immune response. See also **intracutaneous**.

Superantigens are molecules that stimulate a subset of T cells by binding to MHC class II molecules and V_β domains of T-cell receptors, stimulating the activation of T cells expressing particular V_β gene segments. The staphylococcal enterotoxins are one of the sources of superantigens.

Suppressor T cells: see **regulatory T cells**.

The membrane-bound immunoglobulin that acts as the antigen receptor on B cells is often known as **surface immunoglobulin (sIg)**.

The **surrogate light chain** is made up of two molecules called VpreB and λ5. Together, this chain can pair with an in-frame heavy chain, move to the cell surface, and signal for pre-B-cell growth.

When isotype switching occurs, the active heavy-chain V-region exon undergoes somatic recombination with a 3′ constant-region gene at a **switch region** of DNA. These DNA joints do not need to occur at precise sites, because they occur in intronic DNA. Thus, all switch recombinations are productive.

Syk: see **tyrosine kinase**.

When one eye is damaged, there is often an autoimmune response that damages the other eye, a syndrome known as **sympathetic ophthalmia**.

A **syngeneic graft** is a graft between two genetically identical individuals. It is accepted as self.

Systemic anaphylaxis is the most dangerous form of immediate hypersensitivity reaction. It involves antigen in the bloodstream triggering mast cells all over the body. The activation of these mast cells causes widespread vasodilation, tissue fluid accumulation, epiglottal swelling, and often death.

Systemic lupus erythematosus (SLE) is an autoimmune disease in which autoantibodies against DNA, RNA, and proteins associated with nucleic acids form immune complexes that damage small blood vessels, especially of the kidney.

T cells, or **T lymphocytes**, are a subset of lymphocytes defined by their development in the thymus and by heterodimeric receptors associated with the proteins of the CD3 complex. Most T cells have α:β heterodimeric receptors but γ:δ T cells have a γ:δ heterodimeric receptor.

TACI (transmembrane activator and CAML-interactor) is a member of the TNF-receptor family and is one of two major receptors for BLyS. It is found on B cells, dendritic cells, and T cells and is probably the important receptor for receiving signals from BLyS.

Tacrolimus, or FK506, is an immunosuppressive polypeptide drug that inactivates T cells by inhibiting signal transduction from the T-cell receptor. Tacrolimus and cyclosporin A are the most commonly used immunosuppressive drugs in organ transplantation.

TAP1 and **TAP2** (transporters associated with antigen processing) are ATP-binding cassette proteins involved in transporting short peptides from the cytosol into the lumen of the endoplasmic reticulum, where they associate with MHC class I molecules.

Tapasin, or the **TAP-associated protein**, is a key molecule in the assembly of MHC class I molecules; a cell deficient in this protein has only unstable MHC class I molecules on the cell surface.

The functions of effector T cells are always assayed by the changes that they produce in antigen-bearing **target cells**. These cells can be B cells, which are activated to produce antibody; macrophages, which are activated to kill bacteria or tumor cells; or labeled cells that are killed by cytotoxic T cells.

The **Tat** protein is a product of the *tat* gene of HIV. It is produced when latently infected cells are activated, and it binds to a transcriptional enhancer in the long terminal repeat of the provirus, increasing the transcription of the proviral genome.

T-cell antigen receptor: see **T-cell receptor**.

A **T-cell clone** is derived from a single progenitor T cell. See also **cloned T-cell line**.

T-cell hybrids are formed by fusing an antigen-specific, activated T cell with a T-cell lymphoma. The hybrid cells bear the receptor of the specific T-cell parent and grow in culture like the lymphoma.

T-cell lines are cultures of T cells grown by repeated cycles of stimulation, usually with antigen and antigen-presenting cells. When single T cells from these lines are propagated, they can give rise to T-cell clones or cloned T-cell lines.

The **T-cell receptor (TCR)** consists of a disulfide-linked heterodimer of the highly variable α and β chains expressed at the cell membrane as a complex with the invariant CD3 chains. T cells carrying this type of receptor are often called α:β T cells. An alternative receptor made up of variable γ and δ chains is expressed with CD3 on a subset of T cells. Both of these receptors are expressed with a disulfide-linked homodimer of ζ chains.

The **T-cell zones** in lymphoid tissues are enriched in T cells and are distinct from the B-cell zones and the stromal elements.

Activation of the lymphocyte antigen receptors is linked to activation of PLC-γ through members of the **Tec kinase** family of src-like tyrosine kinases. Other Tec kinases are Btk in B cells, which is mutated in the human immunodeficiency disease X-linked agammaglobulinemia (XLA), and Itk in T lymphocytes.

The complement system can be activated directly or by antibody, but both pathways converge with the activation of the **terminal complement components,** which may assemble to form the membrane-attack complex.

The enzyme **terminal deoxynucleotidyl transferase (TdT)** inserts nontemplated or N-nucleotides into the junctions between gene segments in T-cell receptor and immunoglobulin V-region genes. The N-nucleotides contribute greatly to junctional diversity in V regions.

When the same antigen is injected a third time, the response elicited is called a **tertiary response** and the injection a **tertiary immunization**.

T$_H$1 cells are a subset of CD4 T cells that are characterized by the cytokines they produce. They are mainly involved in activating macrophages, and are sometimes called inflammatory CD4 T cells.

T$_H$2 cells are a subset of CD4 T cells that are characterized by the cytokines they produce. They are mainly involved in stimulating B cells to produce antibody, and are often called helper CD4 T cells.

The term **T$_H$3 cell** has been used to describe unique cells that produce mainly transforming growth factor-β in response to antigen; they develop predominantly in the mucosal immune response to antigens that are presented orally.

The lymph from most of the body, except the head, neck, and right arm, is gathered in a large lymphatic vessel, the **thoracic duct,** which runs parallel to the aorta through the thorax and drains into the left subclavian vein. The thoracic duct thus returns the lymphatic fluid and lymphocytes back into the peripheral blood circulation.

Surgical removal of the thymus is called **thymectomy**.

The **thymic anlage** is the tissue from which the thymic stroma develops during embryogenesis.

The **thymic cortex** is the outer region of each thymic lobule in which thymic progenitor cells proliferate, rearrange their T-cell receptor genes, and undergo thymic selection, especially positive selection on **thymic cortical epithelial cells**.

The **thymic stroma** consists of epithelial cells and connective tissue that form the essential microenvironment for T-cell development.

Thymocytes are lymphoid cells found in the thymus. They consist mainly of developing T cells, although a few thymocytes have achieved functional maturity.

The **thymus,** the site of T-cell development, is a lymphoepithelial organ in the upper part of the middle of the chest, just behind the breastbone.

Some antigens elicit responses only in individuals that have T cells; they are called **thymus-dependent antigens** or **TD antigens**. Other antigens can elicit antibody production in the absence of T cells and are called **thymus-independent antigens** or **TI antigens**. There are two types of TI antigen: the **TI-1 antigens,** which have intrinsic B-cell activating activity, and the **TI-2 antigens,** which seem to activate B cells by having multiple identical epitopes that cross-link the B-cell receptor.

T cell and T lymphocyte are short designations for **thymus-dependent lymphocyte,** the lymphocyte population that fails to develop in the absence of a functioning thymus.

One uses **time-lapse video microscopy** to examine processes in biology, that are anywhere from cell migration (fast) to a flower blossoming (slow).

During the process of germinal center formation, cells called **tingible body macrophages** appear. These are phagocytic cells engulfing apoptotic B cells, which are produced in large numbers during the height of the germinal center response.

All dendritic cells arise from hematopoietic progenitors that migrate to various locations all over the body. Here, they are referred to as **tissue dendritic cells**

Transplantation of organs or **tissue grafts** such as skin grafts is used medically to repair organ or tissue deficits.

Some autoimmune diseases attack particular tissues, such as the β cell in the islets of Langerhans in autoimmune diabetes mellitus; such diseases are called **tissue-specific autoimmune disease**.

The **titer** of an antiserum is a measure of its concentration of specific antibodies based on serial dilution to an end point, such as a certain level of color change in an ELISA assay.

TLR1-10: see **Toll-like receptors**.

T lymphocytes: see **T cells**.

TNF: see **lymphotoxin; tumor necrosis factor-α**.

There are several members of the **TNF receptor (TNFR)** family. Some lead to apoptosis of the cell on which they are expressed (TNFR-I, II, Fas), while others lead to activation (CD40, 4-1BB). All of them signal as trimeric proteins.

Tolerance is the failure to respond to an antigen; the immune system is said to be **tolerant** to self antigens. Tolerance to self antigens is an essential feature of the immune system; when tolerance is lost, the immune system can destroy self tissues, as happens in autoimmune disease.

The **Toll pathway** is an ancient signaling pathway that activates the transcription factor NFκB by degrading its inhibitor IκB.

All members of the Toll family of receptors so far discovered have been homologous to the original Toll in *Drosophila*. They have been named **Toll-like receptors (TLR)** followed by a number, as in **Toll-like receptor 4** or **TLR-4**.

The palatine **tonsils** that lie on either side of the pharynx are large aggregates of lymphoid cells organized as part of the mucosal- or gut-associated immune system.

Toxic shock syndrome is a systemic toxic reaction caused by the massive production of cytokines by CD4 T cells activated by the bacterial superantigen **toxic shock syndrome toxin-1 (TSST-1)**, which is secreted by *Staphylococcus aureus*.

Inactivated toxins called **toxoids** are no longer toxic but retain their immunogenicity so that they can be used for immunization.

T$_R$1: see **regulatory T cells**.

The family of proteins known as TNF receptor-associated factors, or

TRAFs, consists of at least six members that bind to various TNF family receptors or TNFRs. They share a domain known as a TRAF domain, and have a crucial role as signal transducers between upstream members of the TNFR family and downstream transcription factors.

The active transport of molecules across epithelial cells is called **transcytosis**. Transcytosis of IgA molecules involves transport across intestinal epithelial cells in vesicles that originate on the basolateral surface and fuse with the apical surface in contact with the intestinal lumen.

The insertion of small pieces of DNA into cells is called **transfection**. If the DNA is expressed without integrating into host cell DNA, this is called a transient transfection; if the DNA integrates into host cell DNA, then it replicates whenever host cell DNA is replicated, producing a stable transfection.

Foreign genes can be placed in the mouse genome by **transgenesis**. This generates **transgenic mice** that are used to study the function of the inserted gene, or transgene, and the regulation of its expression.

Some cancers have chromosomal **translocations**, in which a piece of one chromosome is abnormally linked to another chromosome.

The grafting of organs or tissues from one individual to another is called **transplantation**. The **transplanted organs** or grafts can be rejected by the immune system unless the host is tolerant to the graft antigens or immunosuppressive drugs are used to prevent rejection.

Transporters associated with antigen processing: see **TAP1** and **TAP2**.

The **tuberculin test** is a clinical test in which a purified protein derivative (PPD) of *Mycobacterium tuberculosis*, the causative agent of tuberculosis, is injected subcutaneously. PPD elicits a delayed-type hypersensitivity reaction in individuals who have had tuberculosis or have been immunized against it.

Tuberculoid leprosy: see **leprosy**.

Tumor immunology is the study of host defenses against tumors, usually studied by tumor transplantation.

Tumor necrosis factor-α (TNF-α) is a cytokine produced by macrophages and T cells that has multiple functions in the immune response. It is the defining member of the TNF family of cytokines. These cytokines function as cell-associated or secreted proteins that interact with receptors of the **tumor necrosis factor receptor (TNFR)** family, which in turn communicates with the interior of the cell via components known as TRAFs (tumor necrosis factor receptor-associated factors).

Tumor necrosis factor-β (TNF-β): see **lymphotoxin**.

Tumors transplanted into syngeneic recipients can grow progressively or can be rejected through T-cell recognition of **tumor-specific transplantation antigens (TSTA)** or **tumor rejection antigens**. TSTA are peptides of mutant or overexpressed cellular proteins bound to MHC class I molecules on the tumor cell surface.

The **TUNEL assay** (**T**dT-dependent d**U**TP–biotin **n**ick **e**nd **l**abeling) identifies apoptotic cells *in situ* by the characteristic fragmentation of their DNA. Biotin-tagged dUTP added to the free 3′ ends of the DNA fragments by the enzyme TdT can be detected by immunohistochemical staining with enzyme-linked streptavidin.

In **two-dimensional gel electrophoresis,** proteins are separated by isoelectric focusing in one dimension, followed by SDS-PAGE on a slab gel at right-angles to the first dimension. This can separate and identify large numbers of distinct proteins.

Hypersensitivity reactions are classified by mechanism: **type I hypersensitivity reactions** involve IgE antibody triggering of mast cells; **type II hypersensitivity reactions** involve IgG antibodies against cell surface or matrix antigens; **type III hypersensitivity reactions** involve antigen:antibody complexes; and **type IV hypersensitivity reactions** are T cell-mediated.

A **tyrosine kinase** is an enzyme that specifically phosphorylates tyrosine residues in proteins. They are critical in T- and B-cell activation. The kinases that are critical for B-cell activation are Blk, Fyn, Lyn, and Syk. The tyrosine kinases that are critical for T-cell activation are called Lck, Fyn, and ZAP-70.

Urticaria is the technical term for hives, which are red, itchy skin welts usually brought on by an allergic reaction.

The variable region of the polypeptide chains of an immunoglobulin or T-cell receptor is composed of a single N-terminal **V domain**. Paired V domains form the antigen-binding sites of immunoglobulins and T-cell receptors.

The first 95 amino acids or so of immunoglobulin or T-cell receptor V domains are encoded in inherited **V gene segments**. There are multiple different V gene segments in the germline genome. To produce a complete exon encoding a V domain, one V gene segment must be rearranged to join up with a J or a rearranged DJ gene segment.

Vaccination is the deliberate induction of adaptive immunity to a pathogen by injecting a **vaccine**, a dead or attenuated (nonpathogenic) form of the pathogen.

The first effective vaccine was **vaccinia**, a cowpox virus that causes a limited infection in humans that leads to immunity to the human smallpox virus.

The **valence** of an antibody or antigen is the number of different molecules that it can combine with at one time.

The **variability** of a protein is a measure of the difference between the amino acid sequences of different variants of that protein. The most variable proteins known are antibodies and T-cell receptors.

Variability plot: see **Wu and Kabat plot**.

Variable gene segments: see **V gene segments**.

The **variable region (V region)** of an immunoglobulin or T-cell receptor is formed of the amino-terminal domains of its component polypeptide chains. These are called the variable domains (V domains) and are the most variable parts of the molecule. They contain the antigen-binding sites.

Vascular addressins are molecules on endothelial cells to which leukocyte adhesion molecules bind. They have a key role in selective homing of leukocytes to particular sites in the body.

Vav: see **adaptor proteins**.

The enzyme that joins the gene segments of B-cell and T-cell receptor genes is called the **V(D)J recombinase**. It is made up of several enzymes, but most important are the products of the recombinase-activating genes *RAG-1* and *RAG-2* whose protein products RAG-1 and RAG-2 are expressed in developing lymphocytes and make up the only known lymphoid-specific components of the V(D)J recombinase.

The process of **V(D)J recombination** is found exclusively in lymphocytes in vertebrates, and allows the recombination of different gene segments into sequences encoding complete protein chains of immunoglobulins and T-cell receptors.

The **very late antigens (VLA)** are members of the β_1 family of integrins involved in cell–cell and cell–matrix interactions. Some VLAs are important in leukocyte and lymphocyte migration.

Vesicles are small membrane-bounded compartments within the cytosol.

Many viruses that are produced by mammalian cells are enclosed in a **viral envelope** of host cell membrane lipid and proteins bound to the viral core by viral envelope proteins.

Virions are complete virus particles, the form in which viruses spread from cell to cell or from one individual to another.

Viruses are pathogens composed of a nucleic acid genome enclosed in a protein coat. They can replicate only in a living cell, as they do not possess the metabolic machinery for independent life.

VpreB: see **pre-B-cell receptor**.

Antibodies against the neutrophil granule proteinase-3 are formed in **Wegener's granulomatosis**, an autoimmune disease in which there is severe necrotizing vasculitis. The presence of anti-neutrophil cytoplasmic antigen, or ANCA, helps in the diagnosis of this disease.

Weibel–Palade bodies are granules within endothelial cells that contain P-selectin. Activation of the endothelial cell by mediators such as histamine and C5a leads to rapid translocation of P-selectin to the cell surface.

In **Western blotting**, a mixture of proteins is separated, usually by gel electrophoresis and transferred by blotting to a nitrocellulose membrane; labeled antibodies are then used as probes to detect specific proteins.

When small amounts of allergen are injected into the dermis of an allergic individual, a **wheal-and-flare reaction** is observed. This consists of a raised area of skin containing fluid and a spreading, red, itchy circular reaction.

The discrete areas of lymphoid tissue in the spleen are known as the **white pulp**.

The **Wiskott–Aldrich syndrome** is characterized by defects in the cytoskeleton of cells due to a mutation in the protein WASP. Patients with this disease are highly susceptible to pyogenic bacteria.

A **Wu and Kabat plot**, or variability plot, is generated from the amino acid sequences of related proteins by plotting the variability of the sequence against amino acid residue number. Variability is the number of different amino acids observed at a position divided by the frequency of the most common amino acid.

Animals of different species are **xenogeneic**.

The use of **xenografts**, organs from a different species, is being explored as a solution to the severe shortage of human organs for transplantation. The main problem with xenografting is the presence of natural antibodies against xenograft antigens; attempts are being made to modify these reactions by creating transgenic animals.

Mice with mutations in the *btk* gene have a deficiency in making antibodies, especially in primary responses. These mice are called **xid**, for X-linked immunodeficiency, the mouse equivalent of X-linked agammaglobulinemia, the human form of this disease.

X-linked agammaglobulinemia (XLA) is a genetic disorder in which B-cell development is arrested at the pre-B-cell stage and no mature B cells or antibodies are formed. The disease is due to a defect in the gene encoding the protein tyrosine kinase *btk*.

X-linked hyper IgM syndrome is a disease in which little or no IgG, IgE, or IgA antibody is produced and even IgM responses are deficient, but serum IgM levels are normal to high. It is due to a defect in the gene encoding the CD40 ligand or CD154.

X-linked lymphoproliferative syndrome is a rare immunodeficiency that results from mutations in a gene named SH2-domain containing gene 1A (SH2D1A). Boys with this deficiency typically develop overwhelming EBV infection during childhood, and sometimes lymphomas.

X-linked severe combined immunodeficiency (X-linked SCID) is a disease in which T-cell development fails at an early intrathymic stage and no production of mature T cells or T-cell dependent antibody occurs. It is due to a defect in a gene that encodes the γ_c chain that is a component of the receptors for several different cytokines.

The protein called **ZAP-70** is found in T cells and is a relative of Syk in B cells. It contains two SH2 domains that, when bound to the phosphorylated ζ chain, leads to activation of kinase activity. The main cellular substrate of ZAP-70 is a large adaptor protein called LAT.

INDEX

Figures occurring in the Afterword are labeled in the form **Fig. 1**, and may be found between pages 600–606; those in Appendix I are labeled in the form **Fig. A.1**, and those in Chapters 1 to 14 are labeled in the form **Fig. 1.1**, **Fig. 2.1** etc.